High-Performance Liquid Chromatography of Peptides and Proteins: Separation, Analysis, and Conformation

Editors

Colin T. Mant
Research Associate
Department of Biochemistry
University of Alberta
Edmonton, Alberta, Canada

Robert S. Hodges
Professor
Department of Biochemistry
University of Alberta
Edmonton, Alberta, Canada

CRC Press
Taylor & Francis Group
Boca Raton London New York

CRC Press is an imprint of the
Taylor & Francis Group, an **informa** business

CRC Press
Taylor & Francis Group
6000 Broken Sound Parkway NW, Suite 300
Boca Raton, FL 33487-2742

First issued in paperback 2019

© 1991 by Taylor & Francis Group, LLC
CRC Press is an imprint of Taylor & Francis Group, an Informa business

No claim to original U.S. Government works

ISBN-13: 978-0-8493-6549-2 (hbk)
ISBN-13: 978-0-367-40296-9 (pbk)
Library of Congress Card Number 91-17549

This book contains information obtained from authentic and highly regarded sources. Reasonable efforts have been made to publish reliable data and information, but the author and publisher cannot assume responsibility for the validity of all materials or the consequences of their use. The authors and publishers have attempted to trace the copyright holders of all material reproduced in this publication and apologize to copyright holders if permission to publish in this form has not been obtained. If any copyright material has not been acknowledged please write and let us know so we may rectify in any future reprint.

Except as permitted under U.S. Copyright Law, no part of this book may be reprinted, reproduced, transmitted, or utilized in any form by any electronic, mechanical, or other means, now known or hereafter invented, including photocopying, microfilming, and recording, or in any information storage or retrieval system, without written permission from the publishers.

For permission to photocopy or use material electronically from this work, please access www.copyright.com (http://www.copyright.com/) or contact the Copyright Clearance Center, Inc. (CCC), 222 Rosewood Drive, Danvers, MA 01923, 978-750-8400. CCC is a not-for-profit organization that provides licenses and registration for a variety of users. For organizations that have been granted a photocopy license by the CCC, a separate system of payment has been arranged.

Trademark Notice: Product or corporate names may be trademarks or registered trademarks, and are used only for identification and explanation without intent to infringe.

Library of Congress Cataloging-in-Publication Data

High-performance liquid chromatography of peptides and proteins:
separation, analysis, and conformation / editors, Colin T. Mant,
  Robert S. Hodges.
      p. cm.
  Includes bibliographical references and index.
  ISBN 0-8493-6549-X
  1. High performance liquid chromatography. 2. Proteins—Analysis.
  3. Peptides—Analysis. I. Mant, Colin T. II. Hodges, Robert S.
QP519.9.H53H5355 1991
574.19'245—dc20

91-17549
CIP

Visit the Taylor & Francis Web site at
http://www.taylorandfrancis.com

and the CRC Press Web site at
http://www.crcpress.com

PREFACE

On the face of it, preparing what the editors wished to be a comprehensive and practical publication encompassing high-performance liquid chromatography (HPLC) of peptides and proteins appeared to be a somewhat daunting ambition. There is certainly no scarcity of material on the topic at hand (from sources including published papers, books, industrial communications, and even word-of-mouth), a reflection of the explosive growth in the application of HPLC to peptide/protein separations during the past decade. However, to paraphrase a comment in one of the enclosed articles ("the approach to their separation must be tailored to the separation goals"), we felt that our approach to the book had to be tailored to our perceived goal of a practical teaching manual for both the novice HPLC user as well as more experienced chromatographers. Thus, in addition to being a hands-on teaching manual and textbook, we also wanted the book to represent the perfect reference guide and training tool for personnel in well-established HPLC laboratories.

With these aims in mind, we requested our contributors, scientists of international renown, to help us prepare a series of precise, easy-to-read articles, emphasizing the practical approach to HPLC with minimal theory, although with clear expression of the underlying principles to peptide and protein separations. We felt that by having a large number of contributors (117) producing a substantial number (82) of short articles (containing clear and simplistic figures and only key references), with the concomitant advantage of expressing different viewpoints and approaches, we could produce a publication representing the flavor of the HPLC field as a whole, instead of just a confined vision. Although coordinating this considerable number of contributors and articles necessarily took time and put some stress on us as editors, it was undoubtedly worth it. All of the major modes of microbore, ultrafast, and analytical HPLC (size-exclusion, ion-exchange, reversed-phase, hydrophobic interaction, affinity, and immunoaffinity chromatography) are discussed, together with a presentation of preparative HPLC, including displacement techniques. Problem-solving approaches to the separation of various classes of biologically active peptides and proteins (e.g., peptide hormones, viral proteins, hydrophobic integral membrane proteins, receptor proteins and enzymes) are thoroughly explored. The importance of peptide standards for monitoring column and instrument performance and for optimizing separation conditions is heavily emphasized. In order to maximize the practical potential of HPLC, several articles focus on the choice of the correct detection method (electrochemical, UV, fluorescence) as well as the need for a proper knowledge of approaches to column and instrument maintenance and trouble-shooting. A section of predictive approaches, an area of ever-growing interest, deals with computer simulation of both peptide separations and peptide structure. Also included is the value of HPLC for protein conformation/folding studies and protein structure determination and verification by peptide mapping. Aids to HPLC users include descriptions of trace enrichment techniques, batch extraction of peptides, and sample preparation methods. Finally, various complementary techniques to HPLC are described, including amino acid analysis, preparation of proteins for microsequence analysis, sequencing techniques, and capillary zone electrophoresis.

Good intentions are all very well, but to produce the kind of publication we envisaged required the support and cooperation of all our contributors. One of the prime rewards of this whole endeavor has been our making the acquaintance of a host of scientists we may not otherwise have met, considering our diverse interests. Our sincere thanks to one and all.

Colin Mant
Bob Hodges
University of Alberta
Edmonton, Canada

ACKNOWLEDGMENTS

We are grateful for the continuing support of the Medical Research Council of Canada and the Alberta Heritage Foundation for Medical Research for providing the funding and equipment necessary for us to maintain international competitiveness in the HPLC field. We also thank the co-directors (Dr. Kay and Dr. Smillie) of the MRC Group for their support and understanding of the time taken to complete this worthwhile project.

Our thanks to secretaries Vicki Luxton and Dawn Lockwood for putting up with an enormous typing burden.

Finally, our thanks to wives Janice Mant and Phyllis Hodges for their encouragement and patience with the whole venture. Like us, they believe it was worth it. For this reason, and many others, we dedicate this book to them.

THE EDITORS

Dr. Colin T. Mant, Ph.D., is a Research Associate in the laboratory of Dr. Robert S. Hodges at the University of Alberta, Edmonton, Alberta, Canada.

Dr. Mant attended the University of Manchester Institute of Science and Technology (UMIST) in Manchester, England and received a B.Sc. (Hons.) degree in biochemistry from the Victoria University of Manchester in 1975. He carried out his postgraduate work at the University of Kent at Canterbury, England, under the supervision of Dr. Ronald B. Cain. This work, involving research into the microbial catabolism of herbicide-related synthetic pyrimidines and carried out in collaboration with I.C.I. (Plant Protection) Ltd. in Berkshire, England, led to Dr. Mant receiving his Ph.D. in Biochemistry in 1982.

Dr. Mant joined the laboratory of Dr. Hodges in 1982 as a Medical Research Council of Canada postdoctoral fellow (1982 to 1987) and Research Associate (1987 to present) in the renowned Medical Research Council of Canada Group in Protein Structure and Function. He has more than 45 publications in the area of HPLC methodology.

Apart from the day-to-day management of the HPLC projects in Dr. Hodges' group, now an important part of the total research carried out in the editors' laboratory, Dr. Mant also acts as key collaborator with Dr. Hodges on many of the other research interests of the group, particularly on projects involving protein design, computer simulation, and muscle regulation. He is also a local organizer for the 13th American Peptide Symposium, to be held in Edmonton in 1993.

Dr. Robert S. Hodges, Ph.D., is Professor of Biochemistry and a member of the Medical Research Council of Canada Group in Protein Structure and Function at the University of Alberta, Edmonton, Alberta, Canada.

Dr. Hodges graduated in 1965 from the University of Saskatchewan, Saskatoon, Saskatchewan, Canada with a B.Sc. degree in honors biochemistry. After graduation, he worked for 2 years as Research Scientific Service Officer, Department of National Defence, Defence Research Board, Government of Canada. In 1971, he obtained his Ph.D. degree in biochemistry from the University of Alberta. This work was supervised by Dr. L. B. Smillie and involved the amino acid sequence determination of the first two-stranded α-helical coiled-coil (tropomyosin). This research led to the hypothesis that a hydrophobic 3—4 repeat was responsible for the formation and stabilization of this unique structure. After graduation, he worked as a Medical Research Council of Canada postdoctoral fellow (1971 to 1973) and Research Associate (1973 to 1974) in the laboratory of Professor Bruce Merrifield, Nobel Prize winner in Chemistry in 1984, at the Rockefeller University, New York. The research involved structure-function studies of the enzyme ribonuclease A through the chemical synthesis of protein fragments using the Merrifield solid-phase method.

In 1974, Dr. Hodges was appointed Assistant Professor of Biochemistry and became one of the five founding members of the Medical Research Council of Canada Group In Protein Structure and Function. The deficiencies of traditional chromatographic methods for purification of synthetic peptides, proteins, and protein fragments were evident to the group and Dr. Hodges purchased their first HPLC in 1979 to maintain international competitiveness. He became an Associate Professor in 1977 and Professor in 1984.

The demand on Dr. Hodges' research laboratory for synthetic peptides and peptide-conjugates by the Canadian research community resulted in the formation, in 1985, of a cost-recovery service facility, the Alberta Peptide Institute, of which he remains Director.

In 1986, Dr. Hodges formed S.P.I. Synthetic Peptides Incorporated, a University of Alberta spin-off research and development company. The objectives of S.P.I. under his supervision as President, are to commercialize peptide-based products (peptide pharmaceuticals, peptide diagnostics, and synthetic vaccines). The company also markets HPLC peptide standards, HPLC software (ProDigest LC) and peptide/protein structure prediction software. S.P.I. offers a custom synthesis and HPLC methodology development service for the pharmaceutical and biotechnology industry.

In 1990, the Government of Canada began a new program to promote excellence in Canadian science, The Networks of Centres of Excellence (involving universities and industrial partners across Canada). Dr. Hodges is a participant in two of these programs. First, the Protein Engineering Network: 3D-Structure, Function and Design. The second, the Bacterial Diseases Network: Molecular Strategies for the Study and Control of Bacterial Pathogens of Humans, Animals, Fish and Plants.

Dr. Hodges in 1986 received special recognition for contributions to biotechnology in "A Tribute to Biotechnology" sponsored by the Government of Alberta. In 1988, he joined the Editorial Board of the journal *Peptide Research*. In 1990, he became a member of the program committee for the 12th American Peptide Symposium and will chair the 13th American Peptide Symposium in Edmonton, in 1993. In 1991, he was elected a Fellow of the Royal Society of Canada.

It is noteworthy that HPLC methodology has played and will continue to play a critical role in the success of biotechnology. Dr. Hodges has published more than 50 HPLC methodological papers from over 170 research papers. His current major research interests include the *de novo* design of peptides and proteins with novel structural and biological properties, immunogenicity/ antigenicity of peptides and proteins, development of synthetic vaccines to viral and bacterial pathogens and understanding muscle regulation. Peptide synthesis and HPLC are the major methodologies utilized to achieve his goals and the importance of HPLC is reflected by the dozen HPLC instruments utilized by his group.

CONTRIBUTORS

Andrew J. Alpert
President
PolyLC, Inc.
Columbia, Maryland

Firoz D. Antia
Research Assistant
Department of Chemical Engineering
Yale University
New Haven, Connecticut

Vivian Arrizon-Lopez
Senior Application Chemist
Research and Applications Department
Beckman Instruments, Inc.
Palo Alto, California

John E. Battersby
Senior Research Associate
Department of Analytical Chemistry
Genentech, Inc.
San Francisco, California

Wayne J. Becktel
Assistant Professor
Department of Biochemistry
The Ohio State University
Columbus, Ohio

Suzanne Benjannet
Research Assistant
Department of Molecular
    Neuroendocrinology
Clinical Research Institute of
    Montreal
Montreal, Quebec, Canada

Hugh P. J. Bennett
Associate Professor
Department of Medicine
McGill University
and
Department of Endocrine Laboratory
Royal Victoria Hospital
Montreal, Quebec, Canada

Ron Biehler
Senior Applications Chemist
Department of Applications
Beckman Instruments, Inc.
Palo Alto, California

James A. Black
Facility Manager
Alberta Peptide Institute
University of Alberta
Edmonton, Alberta, Canada

Michael M. Brasseur
Manager
Department of Peptide Chemistry
Magainin Sciences, Inc.
Plymouth Meeting, Pennsylvania

T. W. Lorne Burke
Technologist
Department of Biochemistry
University of Alberta
Edmonton, Alberta, Canada

Klaus Büttner
Post-Doctoral Fellow
Torrey Pines Institute for Molecular
    Studies
San Diego, California

Paul J. Cachia
Senior Research Scientist and Manager
Peptide Synthesis and Services Division
Synthetic Peptides, Inc.
and
Research Associate
Department of Biochemistry
University of Alberta
Edmonton, Alberta, Canada

Michael R. Carpenter
Technologist
Department of Biochemistry
University of Alberta
Edmonton, Alberta, Canada

Rosanne C. Chloupek
Senior Research Associate
Department of Analytical Chemistry
Genentech, Inc.
San Francisco, California

Michel Chrétien
Scientific Director and Chief Executive
    Officer
Laboratory of Molecular
    Neuroendocrinology
Clinical Research Institute of Montreal
Montreal, Quebec, Canada

Joel C. Colburn
Product Manager
Department of Research and Development
Applied Biosystems, Inc.
San Jose, California

J. Myron Crawford
Associate in Research
Department of Molecular Biophysics and
    Biochemistry
Yale University
New Haven, Connecticut

Judy Cummings
Applications Specialist
Research and Applications Department
Beckman Instruments, Inc.
Palo Alto, California

Raymond D. DeAngelis
Associate in Research II
Department of Molecular Biophysics and
    Biochemistry
Yale University
New Haven, Connecticut

C. Dewaele
Chief HPLC Assistant
Laboratory of Organic Chemistry
University of Gent
Gent, Belgium

John W. Dolan
President
LC Resources Inc.
Lafayette, California
and
Editor
LC Troubleshooting
LC/GC Magazine
Eugene, Oregon

Jing Dong
Research Scholar
Hormone Receptor Laboratory
University of Louisville
Louisville, Kentucky

Stefán Einarsson
Research Associate
Icelandic Fisheries
    Laboratories
Reykjavik, Iceland

Heinz Engelhardt
Professor
Department of Angewandte Physikalische
    Chemie
Universität des Saarlandes
Saarbrücken, Federal Republic of
    Germany

Ulrich Esser
Hartmann & Braun
Frankfurt, Germany

Petr Folk
Research Associate
Department of General
    Physiology
Charles University
Prague, Czechoslovakia

Helene Hagestam Freiser
Senior Research Scientist
SynChrom, Inc.
Lafayette, Indiana

Robert Galyean
Research Assistant
The Clayton Foundation Laboratories for
    Peptide Biology
The Salk Institute
La Jolla, California

Heinz Goetz
Application Chemist
Hewlett-Packard
Waldbronn, West Germany

Karen M. Gooding
Analytical Director and President
SynChrom, Inc.
Lafayette, Indiana

Paul D. Grossman
Scientist
Department of Research and Development
Applied Biosystems, Inc.
Foster City, California

Joachim R. Grün
Chemist
Max-Planck Institute for Molecular
Genetics
Berlin, Germany

William S. Hancock
Staff Scientist
Department of Analytical
    Chemistry
Genentech
San Francisco, California

Göran Hansson
Graduate Student
Department of Analytical and Marine
    Chemistry
University of Göteborg
Göteborg, Sweden

Jon Harbaugh
Product Planning Manager
Department of Product
    Planning
Beckman Instruments, Inc.
Palo Alto, California

Milton T. W. Hearn
Professor of Biochemistry
Director, Centre for Bioprocess
    Technology
Monash University
Clayton, Victoria, Australia

Richard A. Henry
President
Keystone Scientific, Inc.
Bellefonte, Pennsylvania

Robert S. Hodges
Professor
Department of Biochemistry
University of Alberta
Edmonton, Alberta, Canada

Carl A. Hoeger
Senior Research Associate
The Clayton Foundation Laboratories for
    Peptide Biology
The Salk Institute
La Jolla, California

Anita L. Hong
Department of Custom Peptide Synthesis
Applied Biosystems, Inc.
Foster City, California

Csaba Horváth
Professor and Chairman
Department of Chemical Engineering
Yale University
New Haven, Connecticut

Richard A. Houghten
Director
Torrey Pines Institute for Molecular
    Studies
San Diego, California

Victor J. Hruby
Regents Professor
Department of Chemistry
and
Department of Biochemistry
University of Arizona
Tucson, Arizona

Salman M. Hyder
Research Scientist
Department of Pharmacology
University of Texas Medical School
Houston, Texas

Niggi Iberg
Research Associate
Department of Surgical Research,
Children's Hospital
Harvard Medical School
Boston, Massachusetts

Richard H. Ingraham
Senior Scientist
Department of Medicinal Chemistry
Boehringer Ingelheim Pharmaceutical, Inc.
Ridgefield, Connecticut

Toshiaki Isobe
Instructor in Biochemistry
Department of Chemistry, Faculty of
   Science
Tokyo Metropolitan University
Setagaya-ku, Tokyo, Japan
and
Instructor
Department of Physiology
Saitama Medical School
Iruma-gun, Saitama, Japan

Djuro Josić
Research Scientist
Institut feur Molekularbiologie
Freie Universitaet Berlin
Berlin, Germany

Krishna K. Kalghatgi
Research Scientist
Department of Chemical Engineering
Yale University
New Haven, Connecticut

Barry L. Karger
Professor and Director
Chemistry Department and Barnett
   Institute
Northeastern University
Boston, Massachusetts

Andrew Kawasaki
Research Associate
Department of Chemistry
University of Arizona
Tucson, Arizona

Cyril M. Kay
Professor
Department of Biochemistry
University of Alberta
Edmonton, Alberta, Canada

Djohan Kesuma
Chemist
Department of Custom Peptide Synthesis
Applied Biosystems
Foster City, California

Michael Klagsbrun
Associate Professor
Department of Biological Chemistry and
   Surgery
Harvard Medical School
Boston, Massachussets

I. Koornneef
Technician
Hubrecht Laboratory
Netherlands Institute for Developmental
   Biology
Utrecht, The Netherlands
Present status
Department of Biochemistry
University of Nÿmegen
Nÿmegen, The Netherlands

Herman Krapf
Product Group Manager
Separation Products Division
Pharmacia/LKB
Biotechnology AB
Uppsala, Sweden

Michael Kunitani
Senior Scientist and Director
Department of Analytical
   Chemistry
Cetus Corp., Inc.
Emeryville, California

Henk H. Lauer
Lauer Labs
Emmen, The Netherlands

Claude Lazure
Associate Director
Laboratory of Molecular
    Neuroendocrinology
Clinical Research Institute of Montreal
Montreal, Quebec, Canada

Kok K. Lee
Postdoctoral Fellow
Synthetic Peptides, Inc.
Edmonton, Alberta, Canada

Maryann Lee
Staff Researcher II
Department of Pharmaceutical Analysis I
Syntex Research
Palo Alto, California

Albert Light
Professor
Department of Chemistry
Purdue University
West Lafayette, Indiana

Susanne Linde
Staff Scientist
Department for Protein Chemistry
Hagedorn Research Laboratory
Gentofte, Denmark

Margaret A. Lindorfer
Senior Research Associate
Institute of Molecular Biology
University of Oregon
Eugene, Oregon

Karen Lockhart
Supervisor
Department of Pharmaceutical Analysis 1
Syntex Research
Palo Alto, California

D. C. Lommen
LC Resources, Inc.
Lafayette, California

Mary B. LoPresti
Associate in Research
Department of Molecular Biophysics and
    Biochemistry
Yale University
New Haven, Connecticut

Colin T. Mant
Research Associate
Department of Biochemistry
University of Alberta
Edmonton, Alberta, Canada

Heidrun Matern
Ph.D.
Department of Internal Medicine III
Aachen University of Technology
Aachen, West Germany

Siegfried Matern
Professor of Medicine, Director
Department of Internal Medicine III
Aachen University of Technology
Aachen, West Germany

Richard A. McClintock
Research Assistant
The Clayton Foundation Laboratories for
    Peptide Biology
The Salk Institute
La Jolla, California

Stephen Moring
Research and Development Chemist
Department of Research and
    Development
Applied Biosystems, Inc.
Foster City, California

Robert L. Moritz
Senior Research Officer
Joint Protein Structure Laboratory
Ludwig Institute for Cancer Research
(Melbourne Branch)
and
The Walter and Eliza Hall Institute for
    Medical Research
P. O. Royal Melbourne Hospital
Parkville, Victoria, Australia

Michael Nattriss
Instrument Technologist
Department of Biochemistry
MRC Group in Protein Structure and
    Function
Edmonton, Alberta, Canada

Sai M. Ngai
Graduate Student
Department of Biochemistry
University of Alberta
Edmonton, Alberta, Canada

Cecilia Nguyen
Chemist II
Department of Pharmaceutical
    Analysis I
Syntex Research
Palo Alto, California

R. Nieuwland
Graduate Student
Hubrecht Laboratory
Netherlands Institute for Developmental
    Biology
Utrecht, The Netherlands
Present status
Research Associate
Department of Haematology
Academic Hospital Utrecht
Utrecht, The Netherlands

Mark P. Nowlan
Senior Research Scientist
SynChrom, Inc.
Lafayette, Indiana

Kerry D. Nugent
President
Michrom Bioresources
Pleasanton, California

John M. Ostresh
Director of Research
Multiple Peptide Systems
San Diego, California

J. M. Robert Parker
Research Associate
Department of Biochemistry
University of Alberta
Edmonton, Alberta, Canada

Matthew V. Piserchio
Keystone Scientific, Inc.
Bellefonte, Pennsylvania

Terry M. Phillips
Director
Department of Immunochemistry
George Washington University
    Medical Center
Washington, D.C.

Frank W. Putnam
Distinguished Professor Emeritus
Department of Biology
Indiana University
Bloomington, Indiana

Richard Reinhardt
Senior Scientist
Max-Planck-Institute for Molecular
    Genetics
Berlin, Germany

Werner Reutter
Professor
Institut feur Molekularbiologie
Freie Universitaet Berlin
Berlin, Germany

Jean E. Rivier
Professor
The Clayton Foundation Laboratories
    for Peptide Biology
The Salk Institute
La Jolla, California

James A. Rochemont
Research Associate
Department of Biochemical
    Neuroendocrinology
Clinical Research Institute of Montreal
Montreal, Quebec, Canada

Timothy D. Schlabach
Product Manager
Spectra-Physics
San Jose, California

Nabil G. Seidah
Director
Department of Biochemical
    Neuroendocrinology
Clinical Research Institute of Montreal
Montreal, Quebec, Canada

Paul Semchuk
Technologist I
Department of Biochemistry
University of Alberta
Edmonton, Alberta, Canada

William Shalongo
Department of Biochemistry
University of Iowa
Iowa City, Iowa

Richard J. Simpson
Head
Joint Protein Structure Laboratory
Ludwig Institute for Cancer Research
(Melbourne Branch)
and
The Walter and Eliza Hall Institute for
    Medical Research
P. O. Royal Melbourne Hospital
Parkville, Victoria, Australia

Lawrence B. Smillie
Professor
Department of Biochemistry
University of Alberta
Edmonton, Alberta, Canada

Lloyd R. Snyder
Vice President
LC Resources, Inc.
Orinda, California

Earle Stellwagen
Professor
Department of Biochemistry
University of Iowa
Iowa City, Iowa

Kathryn L. Stone
Lab Manager
Department of Molecular Biophysics and
    Biochemistry
Yale University
New Haven, Connecticut

Nobuhiro Takahashi
Chief Scientist
Department of Life Science
Toa Nenryo Kogyo K. K.
Iruma-gun, Saitama, Japan

Geza Toth
Research Associate
Department of Chemistry
University of Arizona
Tucson, Arizona

Klaus K. Unger
Professor
Institut für Anorganische and Analytische
    Chemie
Johannes Gutenberg-Universität
Mainz, Federal Republic of Germany

Jennifer E. Van Eyk
Ph.D. candidate
Department of Biochemistry
University of Alberta
Edmonton, Alberta, Canada

A. J. M. van den Eijnden-van Raaij
Doctor
Hubrecht Laboratory
Netherlands Institute for Developmental
    Biology
Utrecht, The Netherlands

E. J. J. van Zoelen
Doctor
Hubrecht Laboratory
Netherlands Institute for Developmental
    Biology
Utrecht, The Netherlands
Present status
Professor
Department of Cell Biology
University of Nÿmegen
Nÿmegen, The Netherlands

M. Verzele
Professor
Laboratory of Organic Chemistry
University of Gent
Gent, Belgium

C. Timothy Wehr
Manager, Chemistry R & D
High Performance Electrophoresis
Bio-Rad Laboratories
Richmond, California

Benny S. Welinder
Head
Department for Protein Chemistry
Hagedorn Research Laboratory
Gentofte, Denmark

Gjalt W. Welling
Senior Research Scientist
Department of Medical Microbiology
Rijksuniversiteit Groningen
Groningen, The Netherlands

Sytske Welling-Wester
Senior Research Scientist
Department of Medical Microbiology
Rijksuniversiteit Groningen
Groningen, The Netherlands

Ronald D. Wiehle
Research Associate
Hormone Receptor Laboratory
University of Louisville
Louisville, Kentucky

Kenneth R. Williams
Professor (Adj.) of Research
Department of Molecular Biophysics and
    Biochemistry
Yale University
New Haven, Connecticut

K. J. Wilson
Vice President, General Manager
Applied Biosystems
Foster City, California

James L. Wittliff
Professor
Hormone Receptor Laboratory
University of Louisville
Louisville, Kentucky

Shiaw-Lin Wu
Scientist
Department of Medicinal and Analytical
    Chemistry
Genentech, Inc.
San Francisco, California

Paul Zavitsanos
Application Engineer
Hewlett-Packard
Montreal, Canada

Nian E. Zhou
Postdoctoral Fellow
Department of Biochemistry
University of Alberta
Edmonton, Alberta, Canada

Lynne R. Zieske
Scientist
Department of Research and Development
Applied Biosystems, Inc.
Foster City, California

G. J. T. Zwartkruis
Research Associate
Hubrecht Laboratory
Netherlands Institute for Developmental
    Biology
Utrecht, The Netherlands

TABLE OF CONTENTS

Section I
Practical Aspects of High-Performance
Liquid Chromatography

PROPERTIES OF PEPTIDES/PROTEINS AND PRACTICAL IMPLICATIONS

Robert S. Hodges and Colin T. Mant

I. INTRODUCTION

This article is designed to provide an introduction to the characteristics of peptides and proteins which determine their retention behavior in HPLC. Reversed-phase chromatography (RPC) is the favored mode of HPLC for peptide separations and the uniqueness of peptide and protein molecules is highlighted by comparing their retention behavior in RPC with that of a series of alkylphenones. In addition, the differences in retention behavior of these two classes of molecules during RPC, a result of the mechanism by which they interact with the hydrophobic stationary phase, serve to illustrate the difference in approach required to separate peptides or proteins compared to that required for small organic molecules.

II. CHARACTERISTICS OF PEPTIDES AND PROTEINS

A. PRIMARY STRUCTURE OF PEPTIDES AND PROTEINS

Peptides and proteins are an unique class of molecules consisting of amino acids as the fundamental units, or building blocks. The 20 naturally occurring amino acids found in peptides/proteins vary dramatically in the properties of their side-chains or R groups. Thus, the properties of peptides and proteins (solubility, structure and function) are characteristic of their amino acid composition and sequence (primary peptide/protein structure). The side-chains are generally classified according to their polarity (nonpolar or hydrophobic vs. polar or hydrophilic). The polar side-chains are divided into three main groups: uncharged polar, potentially positively charged or basic side-chains, and potentially negatively charged or acidic R groups. Within any single group, there are considerable variations in the size, shape and properties of the side-chains. Peptides/proteins generally contain both ionizable acidic and basic side-chains. Thus, they have a characteristic isoelectric point and their overall net charge and polarity in aqueous solution will vary with pH. The hydrophobicity/hydrophilicity as well as the number of charged groups present become important factors in the separation of peptides and proteins.

B. HYDROPHOBICITY/HYDROPHILICITY OF AMINO ACID SIDE-CHAINS

A hydrophobicity/hydrophilicity scale for the 20 amino acid residues found in proteins was derived by Guo et al.[1] at pH 2 and pH 7 from reversed-phase peptide retention data of model synthetic peptides: Ac-Gly-X-X-(Leu)$_3$-(Lys)$_2$-amide, where X is substituted by the 20 amino acids found in proteins. These researchers argued that an accurate measure of the hydrophobicity/hydrophilicity of an amino acid side-chain could be obtained from the interaction of a peptide with the hydrophobic stationary phase in an aqueous environment. The relative hydrophobicity/hydrophilicity values of the retention coefficients determined at pH 2 and pH 7 were very logical and did not show the large discrepancies for various amino acid side-chains observed in previous studies.[2] For instance, the retention data (Table 1) show that the aromatic and large bulky aliphatic side-chains are the most hydrophobic among the 20 amino acid side-chains. In addition, the neutral amino acids have very similar hydrophilicity/hydrophobicity

TABLE 1

**Relative Hydrophobicity/Hydrophilicity Scale for Amino Acid
Side-Chains Obtained from Observed Retention Behavior of
Peptides in RPC**

Hydrophobicity/ hydrophilicity	Amino acid residue	Retention coefficient (min)[a]	
		pH 2	pH 7
Most hydrophobic	Trp	8.8	9.5
	Phe	8.1	9.0
	Leu	8.1	9.0
	Ile	7.4	8.3
	Met	5.5	6.0
	Val	5.0	5.7
	Tyr	4.5	4.6
	Cys	2.6	2.6
	Pro	2.0	2.2
	Ala	2.0	2.2
	Glu*	1.1	−1.3
	Thr	0.6	0.3
	Asp*	0.2	−2.6
	Gln	0.0	0.0
	Ser	−0.2	−0.5
	Gly	−0.2	−0.2
	Arg*	−0.6	+0.9
	Asn	−0.6	−0.8
	His*	−2.1	+2.2
Most hydrophilic	Lys*	−2.1	−0.2

Note: The stars denote residues exhibiting major shifts in relative hydrophobicity/hydrophilicity values on changing the pH from pH 2 to pH 7 (or vice versa).

[a] Synthetic model peptides (sequence shown in text) were run on analytical C_8 and C_{18} columns at two pH values: pH 2 conditions, linear AB gradient (1% B/min) at a flow-rate of 1 ml/min, where eluent A is 0.1% aq. TFA and eluent B is 0.1% TFA in acetonitrile; pH 7 conditions, linear AB gradient (1.67% B/min, equivalent to 1% acetonitrile/min) at a flow-rate of 1 ml/min, where eluent A consisted of aq. 10 mM $(NH_4)_2HPO_4$ and eluent B consisted of 60% aq. acetonitrile, both eluents containing 0.1 M $NaClO_4$.

From Guo, D. et al., *J. Chromatogr.*, 359, 499, 1986. With permission.

values at pH 2.0 and pH 7.0, as expected, even though they were determined using drastically different chromatographic conditions. At pH 2.0, amino acids with basic side-chains (positively charged) have a negative contribution to retention, while protonated, acidic side-chains (uncharged) make little or a slightly positive contribution.

The relative elution order of the peptides was as expected. For example, the -(Glu)$_2$-peptide was more hydrophobic than the -(Asp)$_2$-peptide. Similarly, -(Ala)$_2$->-(Gly)$_2$, -(Thr)$_2$- (Ser)$_2$- (Ile)$_2$->-(Val)$_2$-, and -(Gln)$_2$->-(Asn)$_2$-. All of these peptides differed by two methylene groups. Interestingly, the differences in the values of the retention coefficients of alanine compared to glycine, and isoleucine compared to valine were 2.2 and 2.4 min, respectively. In both cases, an additional methylene group has been added to the side-chain of glycine or valine, and this group was accessible to interact with the hydrophobic stationary phase. In contrast, the retention coefficients for glutamic acid and aspartic acid, glutamine and asparagine, and threonine and serine differed by only 0.9, 0.6, and 0.8 min, respectively, on adding the extra methylene group. This can be explained by the fact that the extra methylene group was not as accessible when added to the amino acid side-chain between the peptide backbone and the hydrophilic functional groups. The -(Leu)$_2$- peptide was more hydrophobic than the -(Ile)$_2$-peptide, although these peptides contained the same number of carbon atoms. Since isoleucine is β-branched, the β-carbon is close to the peptide backbone and not as available to interact with the hydrophobic stationary phase compared to the conformation of the leucine side-chain.

The most striking changes in the retention coefficients in raising the pH from 2 to 7 were seen in the values for Glu, Asp, His, Arg and Lys. At pH 7.0, the side chains of the acidic residues (Glu, Asp) are completely ionized, making their relatively large shift in retention reasonable. The largest shift was seen for histidine, which loses its positive charge above pH 6 to 6.5. The only unexpected result was the decrease in hydrophilicity of Lys and Arg at pH 7 relative to pH 2. This may be due to some ionic interactions, at pH 7, with negatively charged silanols which were not fully suppressed by the conditions used for chromatography.

C. SIZE AND CONFORMATION OF PEPTIDES AND PROTEINS

The distinction between a peptide, polypeptide, or protein is somewhat arbitrary with peptides usually defined as molecules containing 50 amino acids or less. Molecules of greater than 50 amino acids usually have a stable three-dimensional structure in aqueous solution and are referred to as proteins. Conformation can be an important factor in peptides as well as proteins and should always be a consideration when choosing the conditions for chromatography. Though secondary structure (e.g., α-helix or β-sheet) is generally absent even in benign aqueous conditions for small peptides (up to approximately 15 residues), the potential for a defined secondary, tertiary or quaternary structure increases with increasing polypeptide chain length and, for peptides greater than 20 to 35 residues, folding to internalize hydrophobic residues to stabilize the folded structure in aqueous solution is likely to become a significant conformational feature. Similarly, the presence of disulfide bridge(s) would be expected to effect conformation of even small peptides and thus a peptide's retention behavior in HPLC would be altered from its retention time in the fully reduced state. This has been well documented by the research of Lee et al.[3]

D. DETECTION OF PEPTIDES AND PROTEINS

Peptide bonds absorb light strongly in the far ultraviolet (<220 nm) providing a convenient means of detection (usually 210 to 220 nm). In addition, the aromatic side chains of tyrosine, phenylalanine, and tryptophan absorb light in the 250 to 290 nm ultraviolet range. Most polypeptides and proteins contain tyrosine residues, which provides a rapid means of estimating their concentrations in solution. UV detection is the most widely used and convenient method for detection of peptides/proteins in HPLC. This subject is dealt with in greater detail elsewhere in this volume (Section IX; "Methods of Detection").

III. REVERSED-PHASE CHROMATOGRAPHIC BEHAVIOR OF PEPTIDES/PROTEINS VS. SMALL ORGANIC MOLECULES

The excellent resolving power and versatility of RPC has led to its becoming the dominant mode of HPLC for both analytical and preparative separations of peptides in recent years. Figure 1 demonstrates that the interaction behavior of peptides with hydrophobic stationary phases is quite different to that of small organic molecules such as alkylphenones. This is illustrated graphically in Figure 1, where the logarithm of the capacity factor (k') has been plotted against the % acetonitrile required for isocratic elution of the component of interest from an analytical C$_8$ reversed-phase column. Mixtures of peptides (P1 to P5) and alkylphenones (A1 to A4) were chromatographed isocratically at a flow-rate of 1 ml/min in 0.05% aq. trifluoroacetic acid (TFA) containing various percentage concentrations (v/v) of acetonitrile. The capacity factor (k') was then calculated using the equation: k' = (t$_r$-t$_o$)/t$_o$, where t$_r$ is the retention time of the component

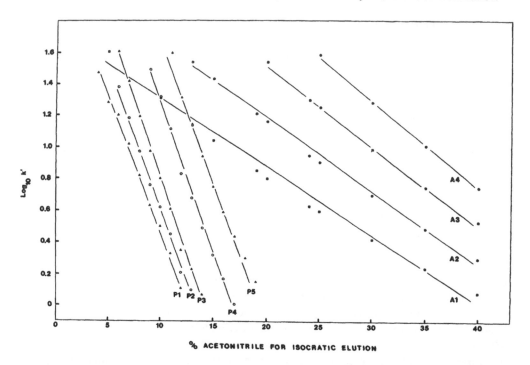

FIGURE 1. Effect of acetonitrile concentration on the $\log_{10} k'$ (capacity factor) values of peptides and alkylphenones during isocratic elution. Column: Aquapore RP 300 C_8 (220 × 4.6 mm I.D., 7-μm particle size, 300-Å pore size; Brownlee Labs., Santa Clara, CA). HPLC instrument: the instrumentation consisted of a Varian Vista Series 5000 liquid chromatograph (Varian, Walnut Creek, CA) coupled to a Hewlett-Packard (Avondale, PA) HP1040A detection system, HP85B computer, HP9121 disc drive, HP2225A Thinkjet printer and HP7470 plotter. Conditions: isocratic elution with 0.05% aq. TFA containing various percentage concentrations (v/v) of acetonitrile; flow-rate, 1 ml/min; temperature, 26°C; absorbance at 210 nm. P1 to P5 denote decapeptides 1 to 5, respectively (sequence variations are described in the text; these peptides were obtained from Synthetic Peptides Incorporated, University of Alberta, Edmonton, Alberta, Canada). A1 to A4 denote propiophenone, butyrophenone, valerophenone and hexanophenone, respectively. The capacity factor (k′) was calculated using the equation: $k' = (t_r - t_o)/t_o$, where t_r is the observed retention time of the peptide or alkylphenone and t_o is the elution time of unretained compounds (uracil, in this case). (From Mant, C. T., Burke, T. W. L., and Hodges, R. S., *Chromatographia*, 24, 565, 1987. With permission).

of interest and t_o is the time taken to elute unretained compounds. Retention time values become less accurate as they approach t_o and values which were $<2t_o$ were not included in the plots shown in Figure 1. The sequences of the five peptides, P1-P5, varied in composition as follows: peptides P2 and P3 differ by only a methyl group (-Gly³-Gly⁴- and -Ala³-Gly⁴-, respectively); peptides P3 and P4 differ by two carbon atoms (-Ala³-Gly⁴- and -Val³-Gly⁴-, respectively); peptides P4 and P5 differ by 3 carbon atoms (-Val³-Gly⁴- and -Val³-Val⁴-, respectively). All peptides contained a Nα-acetylated N-terminal and a C-terminal amide, except peptide P1, which was identical to peptide P3 but had a free α-amino group. The alkylphenones, A1- to A4-denoting propiophenone, butyrophenone, valerophenone and hexanophenone, respectively, differ by a methylene group from each other. Under conditions where solutes are partitioning with the hydrophobic stationary phase, the logarithm of their capacity factor (k′) values should show a linear dependence with concentration of organic modifier in the mobile phase.[4-8] The plots for the five peptides in Figure 1 formed a series of essentially parallel lines, as did the plots for the alkylphenones. However, the slopes for the peptides were markedly steeper than for those of the alkylphenones. Thus, a 5% (v/v) increase in the acetonitrile concentration caused a seven-fold decrease in the k′ values for the peptides; in contrast, the k′ values of the alkylphenones changed only by a factor of 1.7. The partitioning rate of the alkylphenones with the hydrophobic stationary phase is much greater than the partitioning rate of peptides. Peptides partition only

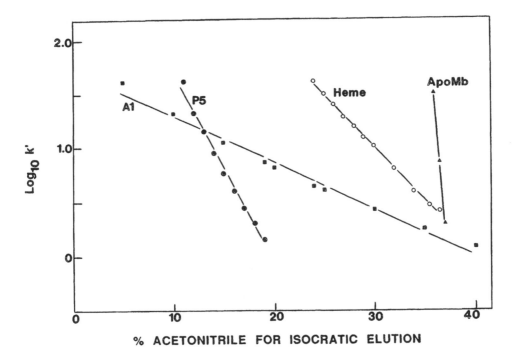

FIGURE 2. Effect of acetonitrile concentration on the $\log_{10} k'$ (capacity factor) values of peptides, alkylphenones and proteins during isocratic elution. HPLC instrument, column and conditions: same as Figure 1. Details of capacity factor calculations are given in Figure 1. P5 denotes decapeptide 5 (see text); A1 denotes propiophenone; apoMb denotes apomyoglobin; heme is the heme group from myoglobin (this group was detected at a wavelength of 400 nm).

to a very limited extent[1,7] and approach an adsorption/desorption mechanism of interaction, due to multi-site binding.

Figure 2 compares the log k' vs.% acetonitrile plot of the decapeptide P5 and propiophenone (A1) (Figure 1) with that of the protein myoglobin (specifically, apomyoglobin) and the heme group from myoglobin. It is immediately obvious that the increase in polypeptide chain length from 10 residues for peptide P5 to 153 residues for myoglobin results in an increase in the steepness of the slope close to vertical. The increase in size of the heme group (~550 daltons) compared to propiophenone has also increased the slope term. As the slopes approach the vertical, the interaction mechanism approaches 100% adsorption/desorption.

The results shown in Figures 1 and 2 clearly demonstrate that the acetonitrile concentration range available for practical isocratic separations of peptides/proteins is very narrow compared to smaller organic molecules. For example, the alkylphenones can be separated isocratically over a wide range of acetonitrile concentrations (30 to 50%; in contrast, the peptides can be separated isocratically within a reasonable time period (within 60 min) only over a very narrow range of acetonitrile concentrations (10 to 13%). In other words, the partitioning range, or partitioning window, available for efficient peptide/protein separations is very narrow compared to that of small organic molecules. We have defined the partitioning window of a particular sample component as the range of isocratic organic solvent concentration available to separate the component of interest (baseline resolution) from its nearest impurities. Since this window is very narrow for peptides and proteins, optimizing isocratic conditions for practical and reproducible peptide/protein separations can prove difficult. In addition, although the peak width of small organic molecules increases as the isocratic concentrations are decreased, the peaks remain symmetrical. In contrast, the peptide peaks not only broaden with decreasing acetonitrile concentrations but also tail severely.[7] Multi-site interactions of peptides/proteins with the hydrophobic sorbent can be used to explain the narrow partitioning window. For the above reasons, optimization of peptide/protein separations in analytical and preparative reversed-phase chromatography generally involves the manipulation of gradient rate rather than isocratic elution conditions. A more detailed discussion of this topic can be found in Reference 7.

Many examples of peptide purification require that the desired peptide(s) be resolved from small organic molecule contaminants. This is particularly common following solid-phase peptide synthesis. It is well known that flow-rate and, particularly, gradient-rate variations can greatly enhance peptide resolution during analytical and preparative RPC separations.[9-12] This generally takes the form of decreasing the gradient-rate (keeping flow-rate constant) to improve the separation of adjacent peptide peaks. However, due to the difference in the mechanism of interaction of peptides/proteins with hydrophobic stationary phases compared to small organic molecules, a different approach than merely shallowing the gradient-rate may be required. This is demonstrated later in this volume,[11] during preparative purification of a synthetic peptide from a crude peptide product. Although low gradient-rates served to separate the desired peptide from synthetic peptide impurities, a considerably higher gradient-rate was required to resolve the peptide from contaminating small organic molecules.

In summary, it is important for the researcher to understand the differences in the different mechanisms of interaction of peptides/proteins with HPLC stationary phases (specifically, in this article, hydrophobic stationary phases of RPC), compared to other types of compounds, as this may effect profoundly the correct approach to a particular separation problem.

ACKNOWLEDGMENTS

This work was supported by the Medical Research Council of Canada and by equipment grants from the Alberta Heritage Foundation for Medical Research.

REFERENCES

1. **Guo, D., Mant, C.T., Taneja, A.K., Parker, J.M.R., and Hodges, R.S.,** Prediction of peptide retention times in reversed-phase high-performance liquid chromatography. I. Determination of retention coefficients of amino acid residues using model synthetic peptides, *J. Chromatogr.,* 359, 499, 1986.
2. **Mant, C.T. and Hodges, R.S.,** Optimization and prediction of peptide retention behavior in reversed-phase chromatography, in *HPLC of Proteins, Peptides and Polynucleotides,* Hearn, M.T.W., Ed., VCH Publishers, USA (in press).
3. **Lee, K.K., Black, J.A., and Hodges, R.S.,** Separation of intrachain disulfide bridged peptides from their reduced forms by reversed-phase chromatography, *this publication.*
4. **Abrahamsson, M. and Gröningsson, K.,** High-performance liquid chromatography of the tetradecapeptide somatostatin, *J. Liq. Chromatogr.,* 3, 495, 1980.
5. **Vigh, Gy., Varga-Puchony, Z., Hlavay, J., and Papp-Hites, E.,** Factors influencing the retention of insulins in reversed-phase high-performance liquid chromatography systems, *J. Chromatogr.,* 236, 51, 1982.
6. **Bij, K.E., Horváth, Cs., Melander, W.R., and Nahum, A.,** Surface silanols in silica-bonded hydrocarbonaceous stationary phases. II. Irregular retention behavior and effect of silanol masking, *J. Chromatogr.,* 203, 65, 1981.
7. **Mant, C.T., Burke, T.W.L., and Hodges, R.S.,** Optimization of peptide separations in reversed-phase HPLC: isocratic versus gradient elution, *Chromatographia,* 24, 565, 1987.
8. **Hearn, M.T.W.,** High-performance liquid chromatography of peptides and proteins: general principles and basic theory, this publication.

9. **Hodges, R.S. and Mant, C.T.,** Standard chromatographic conditions for size-exclusion, ion-exchange, reversed-phase and hydrophobic interaction chromatography, this publication.
10. **Burke, T.W.L., Mant, C.T., and Hodges, R.S.,** The effect of varying flow-rate, gradient-rate and detection wavelength on peptide elution profiles in reversed-phase chromatography, this publication.
11. **Hodges, R.S., Burke, T.W.L., Mant, C.T., and Ngai, S.M.,** Preparative reversed-phase gradient elution chromatography on analytical columns, this publication.
12. **Burke, T.W.L., Black, J.A., Mant, C.T., and Hodges, R.S.,** Preparative reversed-phase shallow gradient approach to the purification of closely-related peptide analogs on analytical instrumentation, this publication.

STANDARD CHROMATOGRAPHIC CONDITIONS FOR SIZE EXCLUSION, ION-EXCHANGE, REVERSED-PHASE AND HYDROPHOBIC INTERACTION CHROMATOGRAPHY

Robert S. Hodges and Colin T. Mant

I. INTRODUCTION

The purpose of this article is to introduce to the researcher standard operating conditions that should be first attempted in the separation of any peptide or protein mixture by size-exclusion (SEC), ion-exchange (IEC), reversed-phase (RPC) or hydrophobic interaction (HIC) chromatography. The columns selected have all been used successfully in the authors' laboratory and are representative of the many columns that are available from alternate manufacturers' that may perform equally well. It is extremely important to evaluate the performance of each new column when it enters the laboratory. In addition, it is essential to use a set of standards that have been designed specifically to test column performance and interact with various stationary phases in a manner similar to the molecules of interest. The peptide standards utilized in this article have been described in detail in this volume as part of each chapter describing a different mode of HPLC. If the researcher using these standards and standard operating conditions cannot generate an equivalent elution profile on a new column as shown here, then the new column would be suspect. The standards are used routinely in the authors' laboratory to monitor column performance throughout the lifetime of the column. The standard elution profiles are maintained in a log book for each column to provide an excellent record of changes in column performance with time. This procedure enables easy comparison of the same column (batch to batch) from the same manufacturer or of columns from different manufacturers. Since the stationary phases for a given HPLC mode differ widely from one manufacturer to another in terms of the functional ligand used, ligand density, particle size and pore size, as well as column dimensions, the peptide standards become a useful diagnostic of these differences. Columns that do not work successfully with the standards are much more easily returned to the manufacturer than reporting to the manufacturer the unsuccessful separation of a particular complex and less well-defined sample mixture of interest.

II. SIZE-EXCLUSION CHROMATOGRAPHY

High-performance size-exclusion chromatography has seen widespread application to the separation of proteins and polypeptides in the past few years. However, until quite recently, relatively little attention has been paid to its potential for resolving peptides in the 200 to 5000 dalton range (2 to 50 residues). Much of the reluctance to employ SEC for peptide applications appears to stem from the existence of more powerful peptide resolving modes of HPLC (IEC and, particularly, RPC). In addition, HPLC columns with the correct fractionation range for peptide separations will probably require pore-diameters less than those currently available (designed mainly for protein separations). Peptides tend to be eluted at the lower end of the fractionation range of these columns and, in our experience, any peptide resolution obtained tends to deteriorate quite rapidly with time. However, SEC may be extremely useful as one step, usually the first, in a multistep (or multidimensional) approach to the resolution of a complex peptide mixture.[1,2]

Separation of peptides or proteins by a mechanism based solely on molecular size (ideal SEC) occurs only when there is no interaction between the solutes and the column matrix. Most modern SEC columns are weakly anionic (negatively charged) and slightly hydrophobic, resulting in solute/packing interactions and, hence, giving rise to deviations from ideal size-exclusion behavior, i.e., non-ideal SEC.[3,4] Although these non-ideal, mixed-mode, effects may provide an advantage for specific peptide or protein separations, they must be suppressed if predictable solute elution behavior is required. Thus, electrostatic effects (and hydrophobic effects)[3,4] between solutes and the column matrix may be minimized by the addition of salts to the mobile phase.

Figure 1 compares the elution profiles of three synthetic peptide standards (1, 2, and 5; 10, 20, and 50 residues, respectively) obtained with two different silica-based size-exclusion columns (panels A and B) and one column containing a non-silica, agarose-based packing (panel C). The mobile phase buffer employed is typical of standard run conditions applicable to both peptide and protein separations, and is a good place to start.

STANDARD RUN CONDITIONS FOR SEC

- Isocratic elution — 50 mM aq. KH_2PO_4, pH 6.5, containing 0.1 M potassium chloride
- Temperature — room temperature
- Flow-rate — 0.5 ml/min

All three columns in Figure 1 exhibited similar peptide elution profiles under ideal size-exclusion conditions, i.e., conditions under which a linear log MW vs. elution time relationship is obtained. The longer retention times and somewhat broader peptide peaks obtained with the agarose-based packing (panel C) are probably due to its significantly larger column volume compared to the two silica-based columns. No size-exclusion column which has passed through our hands has improved on the peptide separations shown in Figure 1, highlighting the need for columns designed specifically for peptide applications.

It should be noted that the above suggested standard run conditions are non-denaturing. Many proteins and large peptides may deviate from ideal size-exclusion behavior due to conformational effects. In addition, the tendency of peptides or protein fragments to maintain or reform a particular conformation as opposed to a random coil configuration in non-denaturing media will complicate retention time prediction. Under these circumstances, and where predictable elution behavior is required, the addition of a denaturing agent (e.g., 8 M urea) to the mobile phase buffer is recommended.[3,4]

III. CATION-EXCHANGE CHROMATOGRAPHY

Strong cation-exchange chromatography is probably the most useful mode of high-performance ion-exchange chromatography for peptide/protein separations.[2,5,6] The main advantage of strong cation-exchangers, generally containing sulfonate functionalities, lies in their ability to retain their negatively charged character in the acidic to neutral pH range. Lowering the pH to 3.0 results in almost complete protonation of the side chain carboxyl groups of the acidic residues (glutamic acid, E; aspartic acid, D; pKa values ~4.0), emphasizing any basic character of the solute. In contrast, at neutral pH, these acidic residues are completely ionized. Thus, by

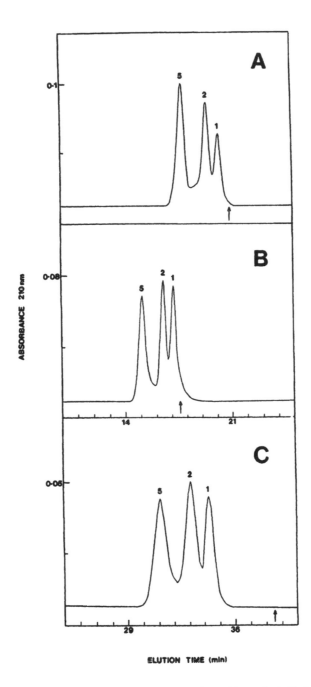

FIGURE 1. Comparison of ideal SEC profiles of a mixture of synthetic peptide standards obtained with different size-exclusion columns. Columns: (A) Altex Spherogel TSK G2000 SW (300×7.5 mm I.D.; 10-μm particle size, 130-Å pore size; Beckman Instruments, Berkeley, CA), (B) SynChropak GPC 60 (300 × 7.8 mm I.D., 10 μm, 60 Å; SynChrom, Lafayette, IN, U.S.) and (C) Pharmacia Superose 12 (300 × 10 mm I.D., 10 μm; Pharmacia, Dorval, Canada). HPLC instrument: the HPLC instrument consisted of a Varian Vista Series 5000 liquid chromatograph (Varian, Walnut Creek, CA) coupled to a Hewlett-Packard (Avondale, PA) HP1040A detection system, HP85B computer, HP9121 disc drive, HP2225A Thinkjet printer and HP7470A plotter. Mobile phase: 50 mM KH$_2$PO$_4$, pH 6.5, containing 0.1 M KCl; flow-rate, 0.5 ml/min; temperature, 26°C. Peptide standards 1, 2, and 5 contain 10, 20, and 50 residues, respectively (sequences shown elsewhere in this volume[3]). The standards were obtained from Synthetic Peptides Incorporated, Department of Biochemistry, University of Alberta, Edmonton, Alberta, Canada. (From Mant, C. T. and Hodges, R. S., *J. Chromatogr.*, 397, 99, 1987. With permission.)

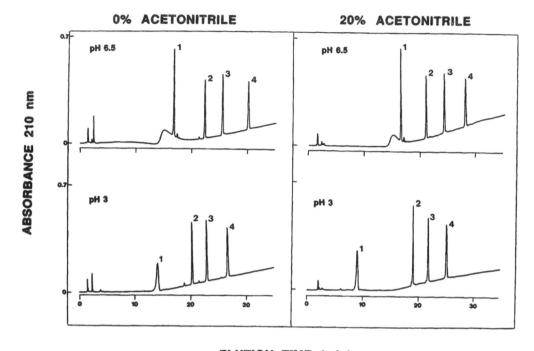

FIGURE 2. Strong cation-exchange chromatography of a mixture of synthetic peptide standards at pH 6.5 (top panels) and pH 3.0 (bottom panels). Column: PolySULFOETHYL A (200 × 4.6 mm I.D., 5-μm particle size, 200-Å pore size, PolyLC, Columbia, MD). HPLC instrument: same as Figure 1, except for an HP9000 Series 300 computer. Conditions: linear AB gradient (20 mM KCl/min, following 10-min isocratic elution with buffer A), where buffer A is 5 mM KH$_2$PO$_4$, pH 3.0 or 6.5, and buffer B is buffer A containing 0.5 M KCl; for the runs with acetonitrile (right hand profiles), both buffers contained 20% (v/v) acetonitrile; flow-rate, 1 ml/min; temperature, 26°C. Peptide standards 1 to 4 contain +1 to +4 net charge, respectively (sequences shown elsewhere in this volume[5]). These 11-residue standards were obtained from Synthetic Peptides Incorporated.

manipulating the pH of the mobile phase, the net charge of a peptide/protein may be varied, which can dramatically affect the relative retention behavior of peptides or proteins in a complex mixture.

Although the major separation mode of ion-exchange chromatography is electrostatic in nature, ion-exchange packings may often exhibit significant hydrophobic characteristics, giving rise to mixed mode contributions to solute separations. Though mixed-mode effects can enhance peptide or protein separations, it has been shown that removal of non-specific hydrophobic interactions may be necessary just to elute peptides from the ion-exchange matrix.[5,6] A non-polar organic solvent such as acetonitrile can be added to the mobile phase buffers to suppress hydrophobic interactions between the solute and the ion-exchange packing.

Peptides and proteins may be removed from the ion-exchange sorbents by either gradient or isocratic elution. Linear AB gradient elution is generally the elution mode of choice when attempting to separate mixtures of peptides or proteins with a wide range of net charges. Gradient elution of peptides/proteins is usually performed with salt gradients of either sodium or potassium chloride in phosphate, tris or citrate mobile phase buffers.

Figure 2 shows the separation of four synthetic cation-exchange (1 to 4) peptide standards on a silica-based strong cation-exchange column at pH 3.0 and 6.5 in the presence and absence of 20% acetonitrile. The peptide standards were specifically designed to contain only basic residues with no acidic residues present and, thus, their retention behavior on a strong cation-exchanger should be unaffected by pH variation. Any variation in retention behavior is indicative of pH effects on the cation-exchange matrix.[5-7] Such effects are a result of the manufacturing procedures used to prepare the packing material. Many silica-based ion-exchangers have a bonded coating containing the sulfonic acid functionality. However, such coatings may also contain other functional groups which have not been totally derivatized. If they are pH sensitive, they may introduce unwanted positively charged groups on the matrix which can block the sulfonate functionalities by ion-pair formation and, thus, affect retention behavior of the components of interest.

STANDARD RUN CONDITIONS FOR CEC

- Linear AB gradient
  - Buffer A = 5 mM aq. KH_2PO_4, pH 6.5 or 3.0
  - Buffer B = 5 mM aq. KH_2PO_4 containing 0.5 M KCl, pH 6.5 or 3.0

If acetonitrile is required, add 20% acetonitrile to both buffers during preparation so as not to dilute the salt concentrations.

- Gradient rate — 10-min isocratic hold of buffer A, followed by a linear increasing salt gradient of 20 mM KCl/min. For example,

Time (min)	Mobile phase composition
0	100% A
10	100% A
35	100% B

- Flow rate — 1 ml/min for standard analytical columns (4- to 4.6-mm I.D.)
- Temperature — room temperature

From Figure 2, all four peptide standards (1, 2, 3, and 4; +1, +2, +3, and +4 net charge, respectively) were removed by the gradient at pH 6.5. At pH 3.0, there was a slight decrease in peptide retention times and peptide 1 (+1 net charge) was now eluted during the initial 10-min isocratic hold of buffer A, instead of by the gradient (as seen at pH 6.5). The presence of 20% acetonitrile also served to reduce peptide retention times at both pH values, albeit not significantly, indicating that some small degree of hydrophobic peptide/packing interaction was present in the absence of the solvent. The sharpness of the peptide peaks (i.e., narrow band widths) eluted by the gradient indicates an efficient column. In fact, this is a very good strong cation-exchange column, with only minor pH sensitivity and hydrophobic characteristics. In addition, its extremely advantageous ability to retain a weakly basic (+1 net charge) peptide is not, in our experience, a widespread characteristic of commercially available cation-exchange columns.

Finally, it should be noted that the run conditions shown above for pH 6.5 are also suitable as initial conditions for anion-exchange separations of peptides and proteins.

IV. REVERSED-PHASE CHROMATOGRAPHY

The excellent resolving power of reversed-phase chromatography (RPC) has resulted in it becoming the predominant HPLC technique for peptide and protein (particularly the former) separations in recent years. It is usually superior to other modes of HPLC with respect to speed and efficiency. In addition, the availability of volatile mobile phase solvents makes it ideal for both analytical and preparative separations.

Figure 3 shows the elution profiles of a series of six synthetic peptide reversed-phase standards on conventional analytical (top panel), narrowbore (middle panel) and microbore (bottom panel) columns containing the same silica-based C_8 packing. The most widely-used mobile phase system for peptide and protein separations is the aqueous acetonitrile system containing the hydrophobic ion-pairing reagent, trifluoroacetic acid (TFA). Apart from the powerful resolving capability of this system, the low pH (pH 2.0) helps to suppress any undesirable ionic interactions between the solute and packing due to the presence of non-derivatized silanols (negatively charged above ~ pH 4.0 to 4.5).[8,9] Silica-based packings, containing octyl (C_8) or octadecyl (C_{18}) ligands are still the most widely employed, with a 300- Å pore size matrix and a 5- to 10-μm range of particle size having good general utility for analytical peptide and protein separations in RPC.

The elution profiles shown in Figure 3 were obtained with the above mobile phase, using a linear AB gradient (1% B/min), the usual elution method for peptides and proteins in RPC, at flow-rates of 1 ml/min, 0.2 ml/min and 0.1 ml/min for the analytical, narrowbore and microbore columns, respectively. These represent typical flow-rates for the respective column internal diameters,[10] and produced similar resolution and run times of the peptide standards. At the standard gradient-rate of 1% acetonitrile/min, the mixture was well resolved. The major advantage of the decrease in column diameter is to increase sensitivity, as indicated by the decrease in relative sample load required to maintain similar peak detection.[10]

A. STANDARD RUN CONDITIONS FOR RPC

- Linear AB gradient
  - Eluent A = 0.1% aqueous TFA
  - Eluent B = 0.1% TFA in acetonitrile
- Gradient rate — 1% B/min. For example,

Time (min)	Mobile phase composition
0	100% A
50	50% A, 50% B

- Flow-rate — 1 ml/min for standard analytical columns (4- to 4.6-mm I.D.)
- Temperature — room temperature

In general, room temperature is adequate for most peptide and protein separations. Heating blocks for columns or heating compartments are not generally necessary. The effect of temperature on peptide separations has been described by Guo et al.[8] and Mant and Hodges.[9] Although increasing column temperature can provide slightly improved peptide and protein resolution, which may be occasionally advantageous, this improvement must be balanced against possible solute degradation and acceleration of column aging at elevated temperatures. Manufacturers are developing new stationary phases with increased stability of the linkage joining the ligand to the support. With the trend to manufacture new supports with smaller particle sizes (1 to 3 μm) for increased resolution and non-porous supports for ultrafast performance, the requirement for high temperature runs may become more prominent in the future. Elevated temperatures do improve mass transfer of solute between the stationary and mobile phases, as well as reducing the viscosity of the mobile phase.

B. EFFECT OF VARYING GRADIENT-RATE AND FLOW-RATE ON PEPTIDE ELUTION PROFILES IN RPC

The status of RPC as the dominant mode of HPLC for peptide and protein separations is reflected by the large number of RPC applications detailed in this volume compared to the other

17

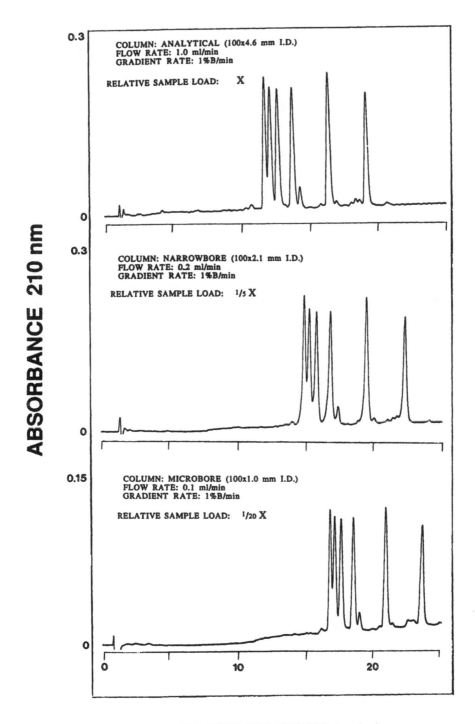

FIGURE 3. Elution profiles of synthetic peptide standards on reversed-phase C₈ columns of varying internal diameters. Columns: Aquapore RP300 C₈ [100 × 4.6 (top), 2.1 (middle) or 1.0 (bottom) mm I.D., 7 μm, 300 Å; Brownlee Labs., Santa Clara, CA]. HPLC instrument: same as Figure 2, except for an HP1090 liquid chromatograph with autosampler. Conditions: linear AB gradient (1% B/min), where eluent A is 0.1% aq. trifluoroacetic acid (TFA) and eluent B is 0.1% TFA in acetonitrile; flow-rate, 1.0 (top), 0.2 (middle) or 0.1 (bottom) ml/min; temperature, 26°C. The 10-residue peptide standards (sequences shown elsewhere in this volume[10,11]) were obtained from Synthetic Peptides Incorporated.

major HPLC modes. For this reason, it is worthwhile at this point to consider some of the approaches to optimizing conditions for RPC available to the researcher. Thus, the following presents a brief overview on the effects of varying gradient-rate and flow-rate on parameters such as peptide retention time, peak width, peak height, and peptide resolution.

It is extremely important to understand the effect of varying flow-rate and gradient rate on peptide separations. The selection of optimal conditions may be critical in establishing an analytical or preparative method. The results in Figure 4A show the effect of varying gradient-rate at constant flow-rate on the parameters of a single peptide component. A standard aq. TFA/acetonitrile mobile phase has been employed. It is immediately obvious that decreasing the gradient-rate from 8% B/min to 0.2% B/min decreases peak height and increases peak width in proportion (10-fold increase in peak width and a corresponding 10-fold decrease in peak height). A decreasing gradient-rate results in an increase in retention time of the peptide component of interest (Figures 4A and 5, panels A and B). The decrease in gradient-rate also improves resolution of a peptide mixture (Figure 5, panels A and B).

The effect of increasing the flow-rate at constant gradient-rate on the parameters of a single peptide component is shown in Figure 4B. As the flow-rate is increased from 0.1 to 2.0 ml/min, there is a decrease in peak width and a decrease in peak height. A 20-fold increase in flow-rate causes an approximate 8-fold decrease in peak height and an approximate 2-fold decrease in peak width. An increase in flow-rate also causes a decrease in retention time (Figures 4B and 5, panels C and D). Although there is a loss in detection sensitivity, there is a corresponding increase in resolution with increasing flow-rate (Figure 5, panels C and D).

The selection of optimal conditions in reversed-phase separation depends on the requirements of the researcher and which compromises can be made regarding run time (speed), resolution, sensitivity (detector response) and sample load. For example, if the best analytical separation is required, with sensitivity and/or run time not a serious issue, then, as observed in Figure 5, high flow-rates and low gradient-rates would be selected to achieve maximum resolution. However, it must be remembered that the alteration of run conditions to improve one aspect of a separation is usually at the expense of another.

All of these aspects of RPC are considered in greater detail elsewhere in this volume.[10,11]

V. HYDROPHOBIC INTERACTION CHROMATOGRAPHY

As the name suggests, the protein resolving capabilities of hydrophobic interaction chromatography (HIC) are based on the strength of hydrophobic interactions between the protein and a non-polar stationary phase. However, this technique differs from RPC, which also utilizes solute/stationary phase hydrophobic interactions as the basis of its separation mechanism, in the solvent conditions utilized and the relative hydrophobicity and density of the bonded ligand. Unlike RPC, where the column packings and mobile phases tend to denature (unfold) proteins, the stationary phase packings and run conditions of HIC are intended to maintain proteins in their native conformation. Thus, ligands of lower hydrophobicity, lower ligand densities (typically ~1/10 that of RPC columns) and non-denaturing aqueous mobile phases of high ionic strength are all designed to promote interactions between the stationary phase of a HIC column and the protein in its native, folded state, i.e., proteins are eluted in order of increasing surface hydrophobicity.

Protein separations on HIC columns are typically carried out by employing linear decreasing salt gradients at neutral pH. The presence of a high salt concentration, typically 1.5 to 2.0 M $(NH_4)_2SO_4$, in the starting buffer promotes interactions between the weakly hydrophobic stationary phase and the (generally) weakly hydrophobic protein surface. As the salt concentration decreases with time, the least strongly sorbed proteins are eluted first.

A: FLOW RATE: 1 ml/min

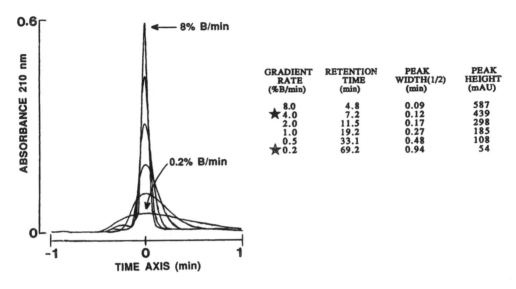

GRADIENT RATE (%B/min)	RETENTION TIME (min)	PEAK WIDTH(1/2) (min)	PEAK HEIGHT (mAU)
8.0	4.8	0.09	587
★ 4.0	7.2	0.12	439
2.0	11.5	0.17	298
1.0	19.2	0.27	185
0.5	33.1	0.48	108
★ 0.2	69.2	0.94	54

B: GRADIENT RATE: 1% B/min

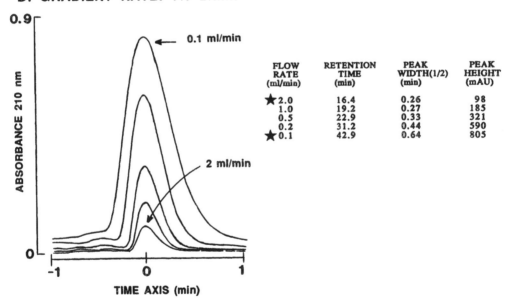

FLOW RATE (ml/min)	RETENTION TIME (min)	PEAK WIDTH(1/2) (min)	PEAK HEIGHT (mAU)
★ 2.0	16.4	0.26	98
1.0	19.2	0.27	185
0.5	22.9	0.33	321
0.2	31.2	0.44	590
★ 0.1	42.9	0.64	805

FIGURE 4. Effect of varying flow-rate or gradient-rate on the reversed-phase elution profile of a peptide peak. Column: SynChropak RP-P C_{18} (250 × 4.6 mm I.D., 6.5 μm, 300 Å). HPLC instrument: same as Figure 3. Conditions: linear AB gradients, where eluent A is 0.1% aq. TFA and eluent B is 0.1% TFA in acetonitrile; temperature, 26°C. Upper profiles: gradient-rates of 0.2 to 8.0% B/min at a constant flow-rate of 1 ml/min. Lower profiles: constant gradient-rate of 1% B/min at flow-rates of 0.1. to 2.0 ml/min. The data marked by stars correspond to the peak profiles denoted by arrows at varying gradient-rates (upper profiles) or flow-rates (lower profiles). All the data presented here is derived from the peptide indicated by a star in Figure 5.

STANDARD RUN CONDITIONS FOR HIC

- Linear AB gradient
  - Buffer A = 0.1 M NaH$_2$PO$_4$, pH 7.0, containing 1.7 M (NH$_4$)$_2$SO$_4$
  - Buffer B = 0.1 M NaH$_2$PO$_4$, pH 7.0

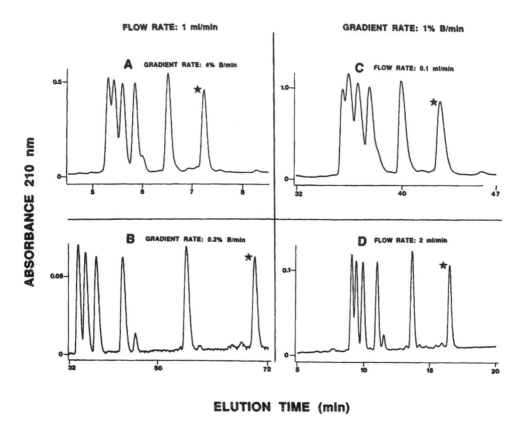

FIGURE 5. Effect of varying flow-rate or gradient-rate on the reversed-phase elution profile of a mixture of synthetic peptide standards. Column and HPLC instrument: same as Figure 4. Conditions: same as Figure 4, with representative profiles at gradient-rates of 4% (panel A) and 0.2% (panel B) B/min at constant flow-rate (1 ml/min), or a constant gradient-rate of 1% B/min at flow-rates of 0.1 (panel C) and 2.0 (panel D) ml/min. The data presented in Figure 4 was derived from the peptide indicated by the star. The 10-residue peptide standards (sequences shown elsewhere in this volume[10,11]) were obtained from Synthetic Peptides Incorporated.

- Gradient rate — 6.67% B/min. For example,

Time (min)	Mobile phase composition
0	100% A
15	100% B

- Flow-rate — 1 ml/min
- Temperature — room temperature

Figure 6 shows the elution profile obtained, using the above run conditions, of cytochrome c (C), myoglobin (M) and lysozyme (L) on a Bio-Gel TSK Phenyl-5PW column.[12] These three proteins (sold by Bio-Rad as a protein standard mixture for HIC) possess several features which make them attractive as HIC (and RPC) standards. The three proteins, for instance, are well retained and well separated by HIC columns. Cytochrome c and myoglobin both possess a heme group, enabling detection at 400 nm (as well as 210 and 280 nm). As detailed later in this volume,[13] the proteins are very useful probes of the denaturing effects of more hydrophobic stationary phases and/or mobile phases. The non-covalently bound heme group of myoglobin (the heme group of cytochrome c is covalently linked), for instance, is particularly sensitive to

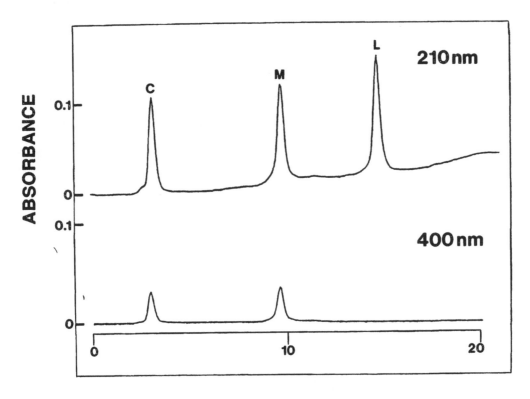

FIGURE 6. Hydrophobic interaction chromatography of a mixture of cytochrome c (C), myoglobin (M) and lysozyme (L). Column: Bio-Gel TSK Phenyl-5PW HIC column (75 × 7.5 mm I.D., 10-μm, 1000 Å; Bio-Rad Labs., Richmond, CA). HPLC instrument: same as Figure 2. Conditions: linear AB gradient (100% A to 100% B over 15 min; 6.67% B/min), where buffer A is 0.1 M NaH$_2$PO$_4$, pH 7.0, containing 1.7 M (NH$_4$)$_2$SO$_4$, and buffer A is 0.1 M NaH$_2$PO$_4$, pH 7.0; flow-rate, 1 ml/min; absorbance at 210 nm (upper profile) or 400 nm (lower profile); temperature, 26°C.

changes in the hydrophobicity of the stationary phase and/or mobile phase eluents, where its dissociation from the protein portion of the myoglobin molecule (apomyoglobin) may be easily followed by comparing absorbances at 210 (or 280 nm) and 400 nm.[13]

VI. CLOSING REMARKS

This article has described standard run conditions for SEC, IEC, RPC, and HIC which the authors recommend as a good starting point for most peptide and protein separations in HPLC, especially so if the researcher is not sure exactly what to expect for a particular sample mixture or from a particular column. These initial run conditions may well prove to be perfectly suited to the researcher's needs without any further optimization. However, when guidance to optimization of a particular separation is required, or where mobile phases other than those described here are necessary for specific separation problems, the reader only needs to refer to the relevant sections elsewhere in this book to tap into a wealth of information.

ACKNOWLEDGMENTS

This work was supported by the Medical Research Council of Canada and equipment grants from the Alberta Heritage Foundation for Medical Research.

REFERENCES

1. **Mant, C.T. and Hodges, R.S.,** General method for the separation of cyanogen bromide digests of proteins by high-performance liquid chromatography: rabbit skeletal troponin I, *J. Chromatogr.,* 326, 364, 1985.
2. **Mant, C.T. and Hodges, R.S.,** Optimization of peptide separations in high-performance liquid chromatography, *J. Liq. Chromatogr.,* 12, 139, 1989.
3. **Mant, C.T. and Hodges, R.S.,** Requirements for peptide standards to monitor ideal and non-ideal behavior in size-exclusion chromatography, this publication.
4. **Mant, C.T. and Hodges, R.S.,** Size-exclusion high-performance liquid chromatography of peptides. Requirement for peptide standards to monitor column performance and non-ideal behavior, *J. Chromatogr.,* 397, 99, 1987.
5. **Mant, C.T. and Hodges, R.S.,** The use of peptide standards for monitoring ideal and non-ideal behavior in cation-exchange chromatography, this publication.
6. **Burke, T.W.L., Mant, C.T., Black, J.A., and Hodges, R.S.,** Strong cation-exchange HPLC of peptides: (1) effect of non-specific hydrophobic interactions; (2) linearization of peptide retention behavior, *J. Chromatogr.,* 476, 377, 1989.
7. **Mant, C.T. and Hodges, R.S.,** Separation of peptides by strong cation-exchange high-performance liquid chromatography, *J. Chromatogr.,* 327, 147, 1985.
8. **Guo, D., Mant, C.T., Taneja, A.K., Parker, J.M.R., and Hodges, R.S.,** Prediction of peptide retention times in reversed-phase high-performance liquid chromatography. I. Determination of retention coefficients of amino acid residues of model synthetic peptides, *J. Chromatogr.,* 359, 499, 1986.
9. **Mant, C.T. and Hodges, R.S.,** Requirement for peptide standards to monitor column performance and the effect of column dimensions, organic modifiers and temperature in reversed-phase chromatography, this publication.
10. **Burke, T.W.L., Mant, C.T., and Hodges, R.S.,** Comparison of peptide resolution on conventional, narrow bore and microbore reversed-phase columns, this publication.
11. **Burke, T.W.L., Mant, C.T., and Hodges, R.S.,** The effect of varying flow-rate, gradient-rate and detection wavelength on peptide elution profiles in reversed-phase chromatography, this publication.
12. Technical bulletin (Bio-Radiations, No. 48, March 1984) from Bio-Rad Labs., Richmond, CA.
13. **Mant, C.T. and Hodges, R.S.,** Effects of HPLC solvents and hydrophobic matrices on denaturation of proteins, this publication.

PREVENTIVE MAINTENANCE AND TROUBLESHOOTING LC INSTRUMENTATION

John W. Dolan

I. INTRODUCTION

Preventive maintenance and troubleshooting of modern liquid chromatography (LC) instrumentation are closely related tasks. While LC downtime cannot be eliminated, one can minimize problems by practicing preventive maintenance in order to correct problems before they impair system operation. The constraints of this chapter do not allow for an in-depth discussion of maintenance and troubleshooting, but the key items to consider are reviewed here. For a comprehensive discussion of LC instrument operation and troubleshooting, see Reference 1. Topical discussions can be found in Reference 2 on a continuing basis. Additionally, a LC trouble-shooting expert system[3] that runs on IBM-compatible computers is available to help interactively solve LC problems. Finally, the operation and service manuals for the LC instrument are excellent sources of maintenance and troubleshooting information.

Presently, there is considerable controversy about the nature of unwanted sample interactions with LC system components. Some workers claim reduced protein denaturation and higher recoveries when LC systems contain "biocompatible" components, primarily titanium, Hastalloy C, plastics, glass, and some ceramics. A review of this topic can be found in Reference 4. Because this topic is poorly understood, no firm recommendations are appropriate in this chapter.

The present discussion is organized around the various LC system modules, in sequence from the mobile-phase reservoir through to the data system.

II. MOBILE-PHASE RESERVOIRS

The most common problem associated with mobile-phase reservoirs is contamination of the rest of the system by particulates from the mobile phase or as a result of microbial growth in the mobile phase. Particulates can block tubing, filters, and frits and can result in unreliable check-valve operation in the pump. Cleanliness is the key to prevention of these problems.

A. RESERVOIR CLEANING

The brown glass bottles in which HPLC-grade solvents are purchased are the most common reservoirs. These can be used directly without further cleaning, and should be discarded on a regular basis (e.g., once a month) in order to prevent a build-up of contaminants. When non-disposable reservoirs are used, such as laboratory glassware or customized reservoir containers, these should be thoroughly washed on a similar schedule. If reservoirs are washed, use normal lab washing procedures, being conscientious about rinsing thoroughly with HPLC-grade water. Also, be wary of inadvertently contaminating the reservoir with metal (e.g., from chromic acid) or organic (e.g., detergent) residues. Whenever particulate material is noticed in the reservoir, the reservoir should be cleaned or replaced.

B. MOBILE PHASE REAGENTS

Mobile phase reagents of the best quality that can be obtained should be used for the LC separation of biopolymers. This is especially important when low wavelength UV detection (e.g., <220 nm) is used, because interference and noise problems are accentuated under these conditions. Organic solvents should be of HPLC-grade. HPLC-grade water should be purchased or prepared in-house using a high-quality water purification system (e.g., Milli-Q). All other buffers, salts, and additives should be HPLC-grade, if available, or of the highest quality that can be obtained. The quality of trifluoroacetic acid (TFA) is especially important, because impurities originally in the TFA or resulting from aging can cause excessive baseline noise. Finally, all reagents should be checked for miscibility; some aqueous buffers can precipitate inside the LC system when they are mixed with certain organic solvents, even though the components are soluble in bulk solution.

C. MOBILE PHASE FILTRATION

Filter mobile phase through a 0.5 μm membrane filter before it is added to the reservoir; in some cases it will be necessary to use a 0.2 μm (biological) filter. The only case in which filtration is not necessary is when pure HPLC-grade solvents are used. This is because it is possible to contaminate these solvents inadvertently during filtration (they generally are filtered through a 0.2 μm filter before they are bottled). During filtration, be cautious about contaminating the mobile phase through use of dirty or dusty glassware, rubber stoppers, Tygon tubing and other common sources of contamination. Following filtration, the mobile phase should be thoroughly degassed (see Section III.A below).

D. MICROBIAL GROWTH

When proteins and peptides are being separated, microbial growth in the mobile phase reservoir is common, because the mobile phase is often a good microbial growth medium. Be especially careful when using acetate buffers (which provide a good carbon source) and/or low organic solvent content (high organic solvent content suppresses growth). A small amount of azide (e.g., 0.04% sodium azide) can be added to the mobile phase to suppress microbial growth. When using aqueous mobile phases, use fresh mobile phase each day. Buffers can be made up as concentrates and stored under refrigeration, then conveniently diluted each day before use.

E. INLET-LINE FILTERS

The filter on the end of the solvent inlet line leading to the pump serves two purposes. First, it helps to keep particulate matter from reaching the pump. A 10 μm porosity filter should be used (2 and 5 μm filters are available, but can restrict the flow enough to cause cavitation problems in the pump inlet line). This filter is not a substitute for mobile-phase filtration, but rather an added precaution to keep dust and other inadvertent contaminants from reaching the pump. The second function of the filter is to act as a weight to keep the inlet line at the bottom of the reservoir. This allows the reservoir to be pumped nearly dry without drawing air into the pump. Most commonly, porous stainless-steel filters are used on the inlet lines, but plastic filters are available for use when metals must be avoided. Plastic filters are generally available from companies that sell biocompatible LC equipment and supplies (e.g., Beckman/Altex). When plastic filters are used, the inlet line must be held below the surface using another technique, such as an undersized hole in the reservoir cap or a piece of Parafilm wrapped around the top of the reservoir.

III. PUMPS

The pump(s) of the LC system has more moving parts than any other module, with the possible exception of some autosamplers. Because of this mechanical complexity, the pump is the most error-prone component of most LC systems. However, with the appropriate preventive maintenance practices, pump downtime can be kept to a minimum. The pumping mechanism consists of a mechanically driven piston that moves back and forth in a pump head plus an inlet and outlet check valve to regulate flow. A pump seal allows the piston to move freely, yet prevents mobile phase from leaking out around the piston. The check valves and the pump seal are the pump parts most likely to give problems. The problems are caused by bubbles, dirt and normal wear, in order of importance. Each of these causes is discussed below.

A. AIR BUBBLES

Air bubbles in the pump head are the most common problem with LC operation today. The presence of air causes the check valve balls to float, so that they do not operate properly. This results in erratic or low delivery rates by the pump accompanied by pressure fluctuations. Microbubbles can cause erratic mobile-phase proportioning (with resultant retention time problems) without affecting check valve operation.[5]

The best way to avoid bubble problems is to degas the mobile phase thoroughly, preferably by helium sparging. Continued sparging (or use of a helium-pressurized reservoir) will prevent air from redissolving during LC operation. Be sure that all the fittings on the low-pressure side of the pump are tight so that air is not drawn in.

Bubbles can be removed from the pump head by using one or more of the following techniques. The easiest method is to open the solvent purge valve and increase the flow rate. In many cases, this will force the bubbles out of the pump. Sometimes it will be necessary to loosen an outlet check valve while the pump is operating in order to remove the air bubbles. When the air has been removed, lower the flow rate to the normal operating flow and then close the purge valve and/or tighten the outlet check valve or fitting to prevent air from re-entering the pump. If these bleeding techniques are unsuccessful, it will be necessary to change to helium degassed solvent (e.g. methanol for reversed-phase operation) to purge the air. This is an effective method for removing air trapped in the column or detector, as well. If a slight positive helium pressure is maintained on the mobile phase reservoirs, the pump will tend to be self-priming, thus eliminating problems associated with air bubbles.

B. CHECK VALVE PROBLEMS

The second most common pump problem is check valve malfunction. In addition to the bubble problems discussed above, dirt is a major check valve enemy. Particulate matter can enter the system as a result of using unfiltered mobile phase and/or not using inlet frits on the tubing connecting the reservoir and the pump. Preventing contamination of the mobile phase was discussed in the previous section. Particulate matter in the check valves prevents the ball from properly sealing with the valve seat, and the valve leaks. The symptoms are the same as for air bubbles - fluctuating or low pressures plus flow rate problems. To clean a check valve, remove it and flush it with clean solvent (e.g., methanol), then water. Now sonicate the valve for 15 to 20 min in a 50% dilution of nitric acid (be careful!). Rinse the valve with water, followed by methanol, and then reinstall it. In most cases, this procedure will remove the contaminant from the check valve. If it does not, replace the valve with a new one.

C. PUMP SEALS

The final common pump problem is worn pump seals. By design, the pump seal does not seal completely around the piston, so the piston is damp behind the seal. When buffer salts are used in the mobile phase, evaporation of the mobile phase, once the pump is turned off, will result in a buffer deposit on the piston. (This problem can also occur if the pump is allowed to run dry.) The crystalline buffer residue acts as an abrasive on the pump seal when the pump is restarted, damaging the seal and possibly scratching the piston. Eventually the seal leaks or sheds enough particulate matter to cause blockage problems downstream. To prevent abrasive damage resulting from buffer residues, flush behind the pump seal at the end of each day's LC operation. Many LC systems have flushing ports at the rear of the pump head for this purpose. Use a syringe to flush 5 to 10 ml of warm water into this space behind the head, catching the waste in a beaker. Next, follow this with 5 to 10 ml of methanol or isopropanol to remove any residual water. Workers who practice this simple flushing procedure find that pump seals last indefinitely.

IV. FITTINGS AND TUBING

The various LC modules are connected most commonly with stainless-steel tubing and compression fittings. When biocompatibility is a concern, Teflon, Tefzel, PEEK (polyether-etherketone), or other plastics are used instead of stainless steel. For convenience, many workers use fittings that can be finger tightened instead of conventional compression fittings. To minimize extra-column dispersion, all tubing through which the sample passes should be of 0.010 in. I.D. or smaller and in as short lengths as are convenient. Larger I.D. (e.g., 0.020 in.) tubing is suitable for other system connections (e.g., connecting the pump to the injector), but lengths should be kept short in order to minimize the gradient delay volume. The keys to minimizing fittings problems are (a) assemble the fittings so that the tubing contacts the bottom of the fitting body, and (b) tighten the fitting sufficiently to prevent leaks, but not so much as to distort the fitting body. References 6 to 8 discuss the proper use and interchangeability of LC fittings.

V. INJECTION VALVES AND AUTOSAMPLERS

Samples are introduced into the LC column via a manual injector or an autosampler. Manual injection valves require little maintenance and seldom cause problems. The key to preventive maintenance of these devices is to keep them clean. Be sure to flush any residual sample or buffer from the valve at the end of each day's work. Valve leakage can occur if the valve seal gets scratched or if excessive pressures (e.g., >5000 psi) are used. The most common source of seal damage results from particulate matter from a precolumn (saturator column) that is used ahead of the valve in some applications. Microscopic silica particles wash from these columns and can act as an abrasive on the relatively soft seal surface. Prevent this problem by using an in-line filter containing a 0.5 µm frit ahead of the injector. Sample valves can be rebuilt easily using a kit and instructions supplied by the valve vendor.

Dirty samples are another source of particulate matter that can scratch injection valves. To determine if a sample needs to be filtered, hold it up to the light — any opalescence or visible particles indicate that filtration is required. Using a syringe-mounted membrane filter (e.g., 2 µm porosity), samples can be filtered directly into autosampler vials.

Autosamplers are more mechanically complex than manual injectors, so they are more susceptible to problems. The autosampler injector valve can be maintained and fixed in the same manner as the manual valves discussed above. Sample needle blockage is the most common auto-sampler problem. Often this is due to a piece of the vial septum becoming lodged in the needle. Use of septa with a Teflon surface on both sides will minimize this problem. If the autosampler is thoroughly flushed at the end of each day's use and the fittings are kept snug, few problems should be encountered. Because of the variety of commercial units, troubleshooting autosamplers is beyond the scope of this chapter. Consult the operation and service manuals for complete details on the repair of autosamplers. A general discussion of autosampler troubleshooting can be found in References 9 and 10.

VI. COLUMNS

The maintenance and troubleshooting of LC columns is covered in a separate article ("Tools and techniques to extend LC column lifetime" by K.D. Nugent and J.W. Dolan). The keys to long column life are (a) inject only samples that are free from particulate matter, (b) use sample preparation techniques to remove as many interfering materials as is practical, (c) always use an in-line filter (0.5 μm porosity) just downstream from the injector, and (d) use a guard column whenever possible. These practices will keep contaminants from reaching the analytical column, thus prolonging its lifetime. Regular column flushing (and, in some cases backflushing) plus frit replacement when necessary will help to recover lost performance from analytical columns.

VII. DETECTORS

UV detectors are the most widely used LC detectors with protein and peptide samples. Because the samples often require low-wavelength operation (e.g., <220 nm), detector problems are more common than with assays that operate at 254 nm. Several designs and numerous models of UV detectors exist, including fixed-wavelength, variable-wavelength, scanning and diode-array detectors. In spite of these differences, many problems are common to all types of detectors.

A. EXCESSIVE NOISE

Excessive baseline noise is the most common complaint about detectors for biological applications. It has been said that mobile phases containing trifluoroacetic acid (TFA), acetonitrile and water are the most stringent test of mobile phase mixing and pump performance. Low wavelength detection seems to accentuate any problems associated with mobile phase mixing or baseline noise. Unfortunately, from a troubleshooting standpoint, the combination of TFA/acetonitrile mobile phases plus low wavelength are the norm for the present applications. As a result of this, a significantly higher level of short-term baseline noise can be expected than is observed with the same mobile phase at higher wavelengths. As was mentioned in Section II.B, the purity and age of the TFA can affect the baseline noise. It is also common to see the baseline rise during the gradient, as a result of different absorbances between the A and B solvents.

B. PREVENTING BASELINE PROBLEMS

There are a number of practices that can be used to minimize baseline problems, although these problems cannot be eliminated entirely.

1. Mixing

For protein and peptide separations, gradients in the 5 to 80% organic solvent range are common; it is at the extremes of this range that mixing is most difficult. Mixing of TFA/acetonitrile mobile phases can be enhanced by premixing the A and B mobile phases as much as possible. Premix the A solvent so that it contains 5% acetonitrile and the B solvent so that it contains 20% water. Partially premixed mobile phases mix more easily on line, and result in smaller baseline disturbances. In some cases, a supplemental mixer can be added to the system to improve mobile phase mixing.

2. Matching UV Absorbance

The baseline rise observed with TFA/acetonitrile gradients results from the difference in absorbance of acetonitrile and water. Most of this difference can be compensated for by adding 15% more TFA to the water reservoir than to the acetonitrile reservoir. For example, make solvent A 5% acetonitrile, 95% water with 0.115% TFA; solvent B should be 80% acetonitrile, 20% water with 0.1% TFA.

3. Wavelength Adjustment

Use the highest wavelength possible for detection, in order to minimize the background noise. Because many of the endogenous substances in sample matrices have poor UV absorbance, increasing the wavelength will reduce their response. Of course, one must balance the loss in sensitivity for the sample peaks with the reduced baseline and background noise so that the optimum signal-to-noise level is obtained.

4. Degassing

Finally, thoroughly degassing and continuously sparging (or maintaining a positive helium pressure in the reservoirs) will reduce the baseline noise by removing dissolved air. Continuous sparging should be done with care. Excessively vigorous sparging can result in a change in composition of the mobile phase through selective loss of one or more components. Generally, vigorous sparging for about 5 min followed by a gentle trickle of helium will be sufficient. A snugly fitting reservoir cap with a small (e.g., 1 mm) vent hole will help prevent air from redissolving. For a more detailed discussion of degassing under these conditions (see Reference 5).

5. Detector Lamps

As the deuterium lamps used in UV detectors age, the background noise increases. This is more pronounced at low wavelengths than at high wavelengths. As a result, it may be necessary to replace the lamp before the normal 500 to 1000 hr lifetime expected with higher-wavelength applications.

C. DETECTOR CELL PROBLEMS

The remaining detector problems are associated with the detector cell. Dirt and bubbles are the primary enemies of the detector cell. Sample components can gradually coat the windows of the detector cell, increasing the background absorbance, and thus decreasing the sensitivity and increasing the background noise. While these problems cannot be eliminated, they can be minimized by introducing only cleaned-up samples and flushing the system with strong solvent at the end of each day's operation. The sample cell can be cleaned by drawing aqueous nitric acid (e.g., 1:1 nitric acid:water) through it. Detector cell rebuilding is beyond the capabilities of most laboratories; it is better to send the cell to the manufacturer for a factory rebuild.

Air bubbles can get caught in the detector cell, causing sharp spikes in the baseline when they pass through the optical path. Innocuous microbubbles are more likely to collect and coalesce to form larger bubbles when the cell is dirty. For this reason, it is a good practice to solvent-flush the cell regularly to remove any contaminants and to always use degassed mobile phase. Bubble problems also can be minimized by using degassed mobile phase and mounting a backpressure restrictor on the detector cell outlet.

D. OTHER DETECTOR PROBLEMS

Most of the remaining detector problems are electronic in nature — these are beyond the scope of this chapter. First check to be sure that all the cables are properly attached and that the fuses have not blown. Reference 11 presents some simple techniques to isolate sources of extraneous electronic noise in the lab. Then consult the operation and service manuals or call the manufacturer for technical assistance.

VIII. DATA SYSTEMS

Troubleshooting recorders and data systems is beyond the scope of this chapter. First check the obvious causes of problems: loose cables, blown fuses, improperly set switches, insufficient paper, or improperly executed methods. If excessive baseline noise is observed, minimize mobile phase problems (see detector discussion) and optionally check for sources of electronic noise.[11] If these cures are not successful, consult the operation and maintenance guide for the system or call the manufacturer for technical assistance.

IX. SUMMARY

In conclusion, preventive maintenance and troubleshooting of LC systems is mostly common sense coupled with knowledge of a few basic procedures. First, make sure that everything that goes into the system is clean as possible. This means filtering the mobile phase and using proper sample preparation techniques. Inlet line frits, in-line filters and guard columns can improve system reliability. Second, make sure that nothing unwanted stays in the system. The LC system should be flushed daily with mobile phase free of buffers, salts, or other additives to remove additive residues. Also, flush the column regularly with strong solvent to remove retained materials, and wash behind the pump seals to remove any buffer residues. Third, keep air out of the LC system. Thoroughly degas the mobile phase, preferably with helium sparging. This will improve pump and detector reliability and will help to remove any bubbles that inadvertently get into the system. Finally, be aware of how the LC system works under normal conditions so that problems are spotted quickly when they do occur. Note the pressure each day in order to anticipate the need to replace a frit or in-line filter. Check the system for leaks — these can be eliminated by gently tightening the fittings. By anticipating problems, they can be fixed before they cause a loss of data or force system shutdown.

REFERENCES

1. **Dolan, J.W. and Snyder, L.R.,** *Troubleshooting LC Systems*, Human, Clifton, NJ, 1989.
2. **Dolan, J.W.,** LC Troubleshooting, *LC/GC*, a monthly feature.
3. **Jupille, T.H. and Buglio, B.,** *The HPLC Doctor*, LC Resources, Lafayette, CA, 1986.
4. **Hattangadi, S.,** The selling of biocompatibility: a case of need versus want?, *LC/GC*, 7, 108, 1989.
5. **Dolan, J.W.,** Mobile phase proportioning problems: a case study, *LC/GC*, 7, 18, 1989.
6. **Dolan, J.W. and Upchurch, P.,** Troubleshooting LC Fittings, Part I, *LC/GC*, 6, 788, 1986.
7. **Dolan, J.W. and Upchurch, P.,** Troubleshooting LC Fittings, Part II, *LC/GC*, 6, 886, 1986.
8. **Upchurch, P.,** *HPLC Fittings*, Upchurch Scientific, Oak Harbor, WA, 1988.
9. **Dolan, J.W.,** Troubleshooting autosamplers, Part I, *LC/GC*, 5, 92, 1987.
10. **Dolan, J.W.,** Troubleshooting autosamplers, Part II, *LC/GC*, 5, 224, 1987.
11. **Fleming, L.H., Milsap, J.P., and Reynolds, N.C., Jr.,** Recognizing and eliminating noise problems in liquid chromatography, *LC/GC*, 6, 978, 1988.

TOOLS AND TECHNIQUES TO EXTEND LC COLUMN LIFETIMES

Kerry D. Nugent and John W. Dolan

I. INTRODUCTION

There are currently several hundred columns commercially available for preparative and analytical liquid chromatography (LC) applications for proteins and peptides. It is impossible to list all of the potential problems associated with maintaining these columns, but there are a few general guidelines that will help extend the useful life of LC columns.

LC columns are available for all chromatographic modes used with proteins and peptides, including size-exclusion (gel filtration), ion-exchange (anion- and cation-exchange), reversed-phase, hydrophobic interaction, and affinity chromatography. Although their uses are different, most of these LC columns are similar in that they are packed with small (3- to 20-μm particle size) silica or polymeric porous (60- to 4000-Å pore size) particles coated with a bonded stationary phase.

This chapter presents some practices that will help improve the performance and lifetime of LC columns used for protein and peptide analysis. Topics include sample preparation, mobile phase composition, in-line filters, guard columns, and precolumns. Troubleshooting techniques cover how to deal with blocked frits, column voids, column contamination, and bonded phase loss.

II. VALIDATION

The most important skill in column maintenance is the ability to recognize if a problem exists. Typical signs of column degradation are changes in system pressure, band retention, selectivity, column plate number, peak tailing, and/or peak splitting.[1] Although each of these can be indicative of a deteriorating column, they can also be caused by non-column problems, including mobile phase contaminants, gradient reproducibility, extra column volumes, temperature changes, secondary retention effects and sample degradation.

The best way to assess the quality of an LC column is to validate its performance when new and continue to use the same validation protocol throughout the lifetime of the column (whenever a problem is suspected). Most commercial LC columns for proteins and peptides are tested by the manufacturer; a copy of the column performance results and test conditions is usually supplied with the column. Because it may be difficult to duplicate the manufacturer's test conditions, an alternative is to use an appropriate standard test mixture such as reported by Mant and co-workers.[2-4]

III. PREVENTIVE MAINTENANCE

The best way to maximize the performance and lifetime of an LC column is to practice regular preventive maintenance. The most important factors when working with proteins and peptides are the sample preparation technique and the mobile phase composition used. With proper attention to detail in these areas, plus the appropriate use of in-line devices such as filters, guard columns and precolumns, the useful life of an LC column can be extended greatly.

A. SAMPLE PREPARATION

The most powerful means of preventing column problems in the LC separation of biopolymers is to take the time to prepare properly the samples to be injected. The primary techniques for preparing biopolymer samples for modern LC are extensions of those used in the classical LC separations, but often require more rigor in order to minimize damage to expensive, small particle LC columns. These techniques, which have been reviewed recently by Wehr,[5] are usually designed to meet several objectives which help prevent column problems. These include: (a) simplification of the sample by removing unwanted materials that could harm the column; (b) putting the sample in the correct form for injection by proper solubilization or by altering its physicochemical state; (c) optimizing the sample concentration for proper column loading; and (d) removal of particulate materials (usually down to a particle diameter of 0.2 μm) to prevent blockage of columns. In properly preparing samples, *it is not sufficient to ensure that all the material to be injected is in solution. The sample solvent must also be compatible with the mobile phase being used, so that the sample or buffer does not precipitate in the pores of the column packing.* If biological activity must be maintained, care must be used in solubilizing the sample to prevent denaturation. If LC is being used for analytical purposes, however, much greater gains can be achieved by using denaturation as a technique to improve sample solubilization and thus extend LC column life.[6]

B. MOBILE PHASE COMPATIBILITY

Most commercial LC columns come with a set of instructions for proper use and care. If these guidelines and a few other general rules are followed, many of the problems with column degradation due to the mobile phase can be prevented. When working with silica-based LC columns, the most important parameter to consider is pH. Most manufacturers recommend using a mobile phase pH between 2 and 7.5, but even within these limits problems can occur. At pH >7, silica will slowly dissolve and greatly shorten column life. Although one can use a precolumn (a silica-based column upstream from the injector to presaturate the mobile phase with dissolved silica), this can also create problems.[1] It is better to use a polymer-based column or one of the newer stabilized silica columns designed for applications above pH 7.

In the reversed-phase separation of proteins and peptides, the universal solvent system is acetonitrile, water, and trifluoroacetic acid (TFA) at pH 2. Although these conditions are widely used with silica-based columns, they are also ideal for stationary phase removal. Glajch and co-workers have reported losses of up to 80% of the original stationary phase in 96 hr.[7] The extent of the loss varies with the column type. The user should be aware of this degradation process and its impact on both column life and column performance (especially retention time repeatability). Selecting the proper silica-based or polymeric reversed-phase column will help minimize this problem (see article by Nugent entitled "Commercially Available Columns and Packings for Reversed-Phase Chromatography of Peptides and Proteins").

When polymeric packings are used, one must be aware of the solvents that can be used with a particular column. Using an incompatible solvent can shrink or swell the packing particles, causing the packing bed to collapse. Many different types of polymers are used as base supports for biopolymer LC columns; carefully read and follow the manufacturer's recommendations for use.

For all types of LC columns, it is important that the mobile phase components are miscible with each other and with the injection solvent. This is especially true for gradients using organic solvents and buffers; mobile-phase strength effects can cause precipitation of sample or buffer inside the column, resulting in column blockage or a packing void. *LC columns should not be stored in a mobile phase that contains salts, acids or bases.* All columns should be flushed with 10 to 15 column volumes of either an organic/water mixture or pure water with 0.05% sodium azide. These storage conditions suppress microbial growth which can degrade columns even in storage.

C. IN-LINE FILTERS AND GUARD COLUMNS

Even after taking all of the sample pretreatment and mobile phase precautions mentioned above, it is strongly recommended that an in-line filter and/or guard column be placed between the injector and the main LC column. For size-exclusion chromatography and with other modes using fairly pure samples, an in-line filter (0.5 μm porosity) is recommended. This will help prevent blockage of the column inlet frit with particulates from the sample, mobile phase or from instrument wear (e.g., pump seals and injector rotors). When working with complex samples, a guard column is usually best, but several factors must be evaluated when working with biopolymers.

Guard columns usually are used to remove materials that are irreversibly bound to the column packing or are so large that they block either the column frit or the packing bed (often due to aggregation of improperly solubilized biopolymers). *The best guard column to use is one that is prepacked with the exact same material (same support, particle size, pore size, and bonded phase) as the separation column.* This column has the highest potential for removal of all materials that could harm the main column and prevents chemical differences between the guard and separating columns from changing the separation. This is especially critical with reversed-phase separation of proteins, because column length has little effect on the separation.[8,9] In this case, the best thing to do is to prepare the sample properly, use an in-line filter and only a very short separating column.

Several manufacturers sell bulk pellicular material to allow users to pack their own guard columns; these have been used extensively with small molecule LC separations. For biopolymer separations, these materials, although much less expensive, do not do an adequate job of protecting the main column, and any differences in packing chemistry could have a dramatic effect on the resulting separation. In general, if a matched guard column is not available for the analytical column, an in-line filter plus judicious sample preparation are the best defenses against separation problems.

IV. TROUBLESHOOTING

Even with rigorous attention to preventative maintenance, most LC columns used for biopolymers will deteriorate with use. The troubleshooting practices presented here can help extend the useful life of an aging column. The common causes of column degradation are: a blockage of the frit or packing bed at the head of the column, a void in the packing bed, a contaminant bound to the packing material, or a loss in the bonded phase of the packing. Each of these problems is discussed below.

A. COLUMN BLOCKAGE

A significant rise in system backpressure is the first indication of a blocked frit at the head of the column (if an in-line filter is used, check it first). Often the contamination can be removed by backflushing the column (attach the outlet to the tubing from the injector and leave the inlet free and flush the column with 20 column volumes of mobile phase). Some columns, however, are not stable to reversal (the manufacturer's data sheet should state this). In this case, or if column reversal is not successful at lowering the pressure, replace the inlet frit (follow the guidelines provided by the column manufacturer). A rise in backpressure can also result from a column void (a loss in resolution accompanies the pressure increase), or from chemical contamination of the column (generally accompanied by a change in retention). If either of these causes is suspected, consult the appropriate section below.

B. COLUMN VOIDS

Column voids are the result of packing bed collapse due to mechanical shocks, pressure surges, packing dissolution, or poor column packing technique. Voids are diagnosed by peak broadening or splitting into a shoulder or a second peak (confirm this by rerunning a sample of the column test mixture). If a void is suspected, carefully open the top of the column and look for a cavity in the packing bed. If a void exists, most workers prefer to discard the column. A void can be filled by adding packing material (of the same type as used in the column) or an inert material such as silanized glass beads. Fill the void with a slurry of packing material (if glass beads are used, first ensure that the packing bed is level), reassemble the column, and install it backwards to put the filled void at the column outlet. This will give the best chance of restoring column performance and extending column life.[10] If this fails to improve the peak shape, the column should be discarded or repacked.

C. COLUMN CONTAMINATION

Biochemical samples generally contain many components that can bind strongly to most column packings. Although such contamination can be associated with pressure problems, the usual result is a change in the retention characteristics of the column. If sample contamination is suspected, the column should be backflushed with an appropriate wash solvent to remove the interfering material.[1] Examples include *the use of nonpolar solvents to remove lipid contaminants, EDTA to remove metal ion contaminants, and detergents or chaotropes (urea, guanidine, etc.) to remove bound proteins.*

D. BONDED PHASE LOSS

The gradual loss of bonded phase is usually characterized by a decrease in the retention times for all bands in the chromatogram. It can also cause increased tailing of basic species due to increased silanol interactions on silica-based columns. As reported by Glajch and co-workers,[7] this loss may be slow and not affect column performance even when up to 50% of the stationary phase is removed. If bonded phase losses are suspected (validate this with the standard column test mixture), it is best to discard the column. Although Mant and co-workers have reported on the successful recoating of certain reversed-phase columns,[11] including follow-up rederivatization of the same column,[12] this procedure is not a long-term solution to bonded-phase loss, since gradual dissolution of the silica matrix will make further regeneration impractical at some point. For further details, see article entitled "On-Line Derivatization of Silica Supports for Regeneration of Reversed-Phase Columns" by Mant and Hodges.

V. CONCLUSIONS

LC columns for biopolymer analysis are expensive, so the necessary precautions should be taken in order to extend their life and minimize the actual cost per sample. The first step in preventing LC column problems is to understand as much as possible about the column characteristics and those of the sample. A proper validation protocol will help to isolate quickly column problems from other system- and sample-related problems. Proper preventative maintenance, including sample pretreatment, regular column flushing, and proper storage techniques, is critical for ensuring good performance and long column life. Finally, as problems do arise, use a systematic approach to troubleshooting and repair in order to minimize downtime and to maximize the potential benefits of each corrective procedure.

REFERENCES

1. **Wehr, C.T. and Majors, R.E.,** Regeneration of biopolymer columns, *LC/GC*, 5, 942, 1987.
2. **Mant, C.T. and Hodges, R.S.,** Requirement for peptide standards in reversed-phase HPLC, *LC/GC*, 4, 250, 1986.
3. **Burke, T.W.L., Mant, C.T., Black, J.A., and Hodges, R.S.,** Strong cation-exchange HPLC of peptides: (1) effect of non-specific hydrophobic interactions; (2) linearization of peptide retention behaviors, *J. Chromatogr.*, 476, 377, 1989.
4. **Mant, C.T., Parker, J.M.R., and Hodges, R.S.,** Size-exclusion HPLC of peptides: requirement for peptide standards to monitor column performance and non-ideal behavior, *J. Chromatogr.*, 397, 99, 1987.
5. **Wehr, C.T. and Majors, R.E.,** Sample preparation for biopolymer separations, *LC/GC*, 5, 548, 1987.
6. **Nugent, K.D., Burton, W.G., Slattery, T.K., Johnson, B.F., and Snyder, L.R.,** Separation of proteins by reversed-phase HPLC: II. Optimizing sample pretreatment and mobile phase conditions, *J. Chromatogr.*, 443, 381, 1988.
7. **Glajch, J.L., Kirkland, J.J., and Köhler, J.,** Effect of column degradation on the reversed-phase HPLC of peptides and proteins, *J. Chromatogr.*, 184, 81, 1987.
8. **Pearson, J.D.,** HPLC column length designed for submicrogram scale protein isolation, *Anal. Biochem.*, 152, 189, 1986.
9. **Burton, W.G., Nugent, K.D., Slattery, T.K., Summers, B.R., and Snyder, L.R.,** Separation of proteins by reversed-phase HPLC. I. Optimizing the column, *J. Chromatogr.*, 443, 363, 1988.
10. **Vendrell, J. and Aviles, F.X.,** Regeneration of reversed-phase high-performance liquid chromatographic columns by flow reversal, *J. Chromatogr.*, 356, 420, 1986.
11. **Mant, C.T., Parker, J.M.R., and Hodges, R.S.,** On-line derivatization to restore reversed-phase column performance in peptide separations, *LC/GC*, 4, 1004, 1986.
12. **Mant, C.T. and Hodges, R.S.,** On-line derivatization of silica supports for regeneration and preparation of reversed-phase columns, *J. Chromatogr.*, 409, 155, 1987.

MOBILE PHASE PREPARATION AND COLUMN MAINTENANCE

Colin T. Mant and Robert S. Hodges

I. INTRODUCTION

The rapid evolution of HPLC has necessitated the development of specialized instrumentation (pumps, detectors, etc.) and column packings, the expensive nature of which stresses the
need for careful maintenance. A critical part of maintaining efficient instrument and column
performance lies in the careful preparation of the mobile phase. In addition, column lifetime will
be extended through proper cleaning and storage procedures.

This article provides a brief overview of practical aspects of mobile phase preparation and
column maintenance, representing a compendium of general, well-recognized concerns combined with suggestions from our own experience as well as manufacturers' recommendations.

II. MOBILE PHASE PREPARATION

Several liters of a mobile phase may pass through a column in daily use and any impurities
in these mobile phases may have an adverse effect on column performance unless care is taken
to minimize their presence. Mobile phase purity is a particularly important consideration in the
successful application of HPLC for separation of biopolymers such as peptides and proteins,
where UV detection at wavelengths below 220 nm is common and elution profiles will be
especially sensitive to the presence of any UV-absorbing or fluorescing impurities.

Mobile phase purity requirements will vary depending on the application, e.g., the requirements for isocratic elution are not as stringent as for gradient elution. Thus, for applications
employing isocratic elution, the detector can often be balanced electronically or by the use of
a blank in the reference cell (as long as the linear dynamic range of the detector is not exceeded).
Since any UV-absorbing or -fluorescing impurities will usually reach a steady-state equilibrium
within the column, the detector baseline offset may remain relatively high, but constant.[1] In
contrast, the presence of UV- absorbing or fluorescing impurities in the mobile phase may
seriously affect gradient elution profiles of peptides and proteins. During gradient elution, the
chromatographic run generally begins with a weak eluent, followed by an increase in eluent
strength (i.e., increasing ionic strength in IEC, increasing non-polarity of mobile in RPC) over
a period of time (the exception to this being HIC, where mobile phase ionic strength is decreased
with time). If the weaker eluent contains UV-active or fluorescent impurities that are strongly
retained by the stationary phase, they are concentrated at the top of the column during the
equilibration period or during the initial stages of the gradient run.[1] As the eluent strength is
increased, these impurities are eluted from the column and detected, resulting in spurious peaks
and/or rapidly rising baselines. If the stronger eluent contains the impurities, similar results may
be obtained, although of reduced magnitude, (some baseline drift is typical), since less of it is
generally passed through the column and the impurities in the stronger eluent are more quickly
eluted as the solvent strength is increased.[1]

Prior to the first run performed on a previously stored column, a gradient run in the absence
of sample (i.e., a "blank" run) should be carried out. In fact, even before the blank run, we suggest
that the column should be subjected to a rapid gradient wash [e.g., 100% eluent A (the weak

eluent) to 100% eluent B (the strong eluent) in 15 min, followed by rapid reequilibration back to the desired starting conditions]. This will serve to remove any impurities from the column that may have accumulated during storage. The subsequent blank run should always be run out to 100% eluent B, in case any strongly adsorbed impurities are present in the mobile phase. A welcome bonus to these initial runs without sample is the fact that subsequent runs with the sample(s) will be more reproducible, since the column is then thoroughly conditioned.[2]

Potential sources of mobile phase impurities include the mobile phase solvents (water, organic modifier, etc.), additives (salts, ion-pairing reagents, etc.) and glassware, including the mobile phase reservoirs themselves, and these will now be discussed separately.

A. MOBILE PHASE SOLVENTS
1. Water

Almost without exception, mobile phases employed for peptide and protein separations in HPLC are aqueous-based, with a greater proportion (often 100%, depending on the HPLC mode and application) of water passing through a given column than any other solvent. Clean, pure water is an obvious requirement for HPLC applications, and HPLC-grade water can either be purchased or purified on-site, depending on demand. Due to the relatively high cost of buying HPLC-grade water, and where there is sufficient demand, there is an increasing tendency to make use of water purification systems such as a Milli-Q system (Millipore, Bedford, MA) or a HP 661A water purifier (Hewlett-Packard, Avondale, PA), both of which have been successfully employed in the authors' laboratory in a cost-effective manner. However, in situations where only relatively small volumes of water are consumed (e.g., where HPLC is a seldom-used technique or for micro-LC techniques), the researcher is probably better off purchasing HPLC-grade water (usually in 4 l bottles). Another alternative for obtaining water pure enough for employment in HPLC, and where water consumption is low to moderate, is to take advantage of the solvent purifying capabilities of a reversed-phase packing. During the early days of HPLC use in the authors' laboratory, for instance, laboratory-supplied double distilled and deionized water was further purified by passage through a semi-preparative C_{18} column (having first filtered the water through a 0.22 μm filter) retained specifically for this one purpose. Any undesirable contaminants were retained by the column and, at regular intervals, were removed from the column by elution with acetonitrile or isopropanol. This procedure worked well enough until demand for pure water exceeded supply, due to an expansion of the numbers of HPLC instruments in the laboratory combined with a growing requirement for semi-preparative scale peptide purification.

2. Other Solvents

The term "other solvents" almost invariably refers to non-polar solvents (e.g., acetonitrile, methanol, isopropanol) employed either as the organic modifier or occasionally, as a means of suppressing non-ideal hydrophobic interactions in SEC[2] or IEC.[3] All non-aqueous solvents commonly employed in HPLC are available in a highly pure form (spectroscopic or HPLC-grade). These solvents are not cheap and, for those fortunate enough to have easy access to the proper facilities (and whose large consumption of expensive solvents is a serious economic burden), purification of lesser-grade solvents may be an option.[1] However, for the average researcher, and particularly where solvent consumption is low to moderate, there is no reasonable substitute for simply purchasing the required pure solvents.

3. Glassware

If one has taken the trouble to purchase high-quality water or other solvents, or has purified the solvents on-site, it would obviously be unfortunate (as well as a waste of time and money) to then contaminate them through the use of dirty glassware. The potential for this occurrence is usually during the preparation of the mobile phase eluents (e.g., buffer preparation) and/or their storage in glass reservoirs. As pointed out earlier in this book by Dolan,[3] when non-disposable reservoirs are used (e.g., laboratory glassware or customized reservoir containers), they should be thoroughly washed using normal laboratory washing procedures, being particularly careful about rinsing well with HPLC-grade water.

Dolan[3] also discusses the importance of keeping the mobile phase free of particulate matter. He notes that filtration of purchased HPLC-grade solvents is not necessary, since they are generally filtered through a 0.2 μm filter prior to bottling. This is also true of water-purification systems such as the Milli-Q and the HP 661A systems. However, care must be taken not to contaminate a mobile phase essentially free of particulate matter, with glassware insufficiently rinsed during cleaning. It is almost impossible to remove all traces of dust from reservoirs and mobile phases, but in-line filters, both before and after the pump, and prior to the column, ensure that this problem remains negligible.[3]

B. MOBILE PHASE ADDITIVES

As a general guideline, any additives to a (pure) mobile phase solvent should be HPLC-grade, if available, or of the highest quality that can be obtained.[3] The suitability of a particular additive can be quickly determined by a UV scan of an aqueous solution of the additive at the highest concentration at which it is likely to be employed (e.g., 0.1% trifluoroacetic acid for RPC, 1 M KCl for IEC). If there are obvious UV-absorbing contaminants in the reagent, it can either be discarded in favour of a cleaner preparation or, where possible, it can be purified.

What follows in this section is a description of some common mobile phase additives and some of our experiences with their purification and/or use.

1. Ion-Pairing Reagents

Trifluoroacetic acid (TFA) is by far the most extensively-used mobile phase additive in RPC. The quality of this reagent [and the lesser-used heptafluorobutyric acid (HFBA)] is especially important, since impurities originally in the TFA or resulting from aging can cause excessive baseline noise.[3] Highly pure TFA and HFBA are readily available from many commercial sources (HPLC-grade, spectrophotometric-grade and sequenator-grade are all suitable for HPLC use). Simple distillation of less pure (analytical grade) TFA has also produced a reagent of perfectly acceptable quality for HPLC. Since the volumes of TFA consumed during HPLC are extremely small (0.1% concentrations are typical), the purification of just 1 liter, for example, will serve for a very large number of analytical runs. Even if larger volumes are required, the distillation process is quite straightforward.

In the case of phosphoric acid (H_3PO_4), occasionally used as a non-volatile, weak hydrophilic ion-pairing reagent in place of the more hydrophobic TFA,[4] we have generally used analytical-grade reagent without any problems. The low concentration (0.1%) of the acid typically employed for RPC is probably a factor in being able to achieve this.

2. Buffer Components

Triethylamine is a reagent frequently employed in both volatile and non-volatile buffer systems over a wide pH range.[5] In its capacity as a major buffer component, this reagent is frequently employed at concentrations of 0.20 to 0.25 M, making purity of this reagent an important consideration. Highly pure triethylamine can be readily purchased, although the authors' laboratory has regularly used lesser-grade reagent, following distillation over ninhydrin, with no discernible problems.

Without exception, we have always found that analytical-grade phosphate-based buffer salts, frequently employed in this laboratory in SEC, IEC, HIC, and affinity chromatography (and occasionally in RPC), require some form of purification prior to their use. This is particularly apparent when they are used for gradient elution in IEC or RPC. If no effort is made to clean them up prior to gradient elution, contaminants in the analytical reagents produce unwanted peaks and drifting baselines. This is easily avoided by a simple clean-up procedure involving the extraction of contaminants by a chelating resin. We routinely prepare a stock solution (e.g., usually 1 liter of 1 to 2 M aq. KH_2PO_4 for IEC), add the chelating resin (BioRad Chelex-100 in our case; BioRad Lab., Richmond, CA) (~10 g/l of solution) and stir for about 1 hr. The phosphate solution is then aliquoted, diluted as desired and filtered through a 0.22 μm filter. It is then ready for use. The remainder of the phosphate solution is stored at 4°C over the resin until further use. Note that the stock phosphate solution, or any phosphate-based buffers for that matter, should not be stored for long periods of time, since bacterial growth occurs even at this low temperature.

3. Salts

This laboratory frequently employs NaCl or KCl, either as the displacing salts for IEC or as a means to suppress non-ideal adsorptive behavior in SEC. We have never found any particular need to use other than analytical-grade salts for these purposes, following a quick spectral check of a solution of the reagents when they are first purchased.

Highly pure ammonium sulfate, frequently used at concentrations of 1.0 to 2.0 M for HIC applications, is available commercially. For instance, this laboratory regularly employs HPLC-grade $(NH_4)_2SO_4$ from BioRad.

For the occasional RPC applications carried out at neutral pH in the authors' laboratory, 0.1 M sodium perchlorate $(NaClO_4)$ is routinely added to the mobile phase buffer to suppress any non-ideal behavior (i.e., ionic interactions) with free silanols on silica-based columns. In our experience, chelexing of analytical-grade $NaClO_4$ has never successfully removed all UV-absorbing contaminants from this reagent. Instead, it has been necessary to pass a stock solution of the perchlorate through a preparative C_{18} column (following filtration through a 0.22 μm filter) to remove any impurities. The purified perchlorate solution is then diluted as required. This procedure, of course, can also be applied to the clean-up of any salts, including phosphate-based buffer salts. However, it is not as convenient as simple chelexing (see above).

4. Urea

For any studies requiring complete denaturation of proteins or peptides, (e.g., to ensure predictable retention behavior during SEC), urea concentrations in the 6 to 8 M range are quite typical. At these kinds of high concentrations, reagent purity is obviously vital. Although highly purified urea is commercially available, it tends to be substantially more expensive than the analytical-grade reagent, an important point considering the large amount of this reagent which can be consumed even during moderate use. However, analytical-grade urea can be purified to a level suitable for HPLC applications by a very straightforward procedure. Following preparation of the concentrated aqueous urea solution (usually requiring some heat, since dissolving urea in water is a highly endothermic process), the solution is stirred over a mixed-bed resin [BioRad AG 501-X8 (20 to 50 mesh) in our case] (~10 g/l of solution) for about 30 min to 1 hr to remove dissolved impurities. The resin is removed by filtering through a sintered glass funnel and the supernatant is subsequently filtered through a 0.22 μm filter. The urea solution is now ready for use.

III. STORAGE AND MAINTENANCE OF COLUMNS

This section deals with general hints on how best to maintain HPLC columns and is drawn from both our own experiences as well as manufacturers' guidelines. We are not concerned here with such aspects of column protection as in-line filters, guard columns or column frits which are expertly dealt with elsewhere in this book.[3,6,7]

A. PRETREATMENT OF SILICA-BASED RPC COLUMNS

Certain precautions should be observed when employing new reversed-phase columns for the first time. The experience of this laboratory suggests that new silica-based reversed-phase packings may contain hydrophobic ligands which are not covalently bonded to the silica support. Thus, gradient elution to high concentrations of organic solvent in the mobile phase may gradually elute these adsorbed ligands from the support, possibly resulting in their precipitation and blocking of lines or detector flow cells. The authors' experience with 10 mm internal diameter (ID) and 21.4 mm ID C_{18} columns also suggests that this problem is more serious with larger columns than with analytical sizes due to the significantly larger volume of packing material. This laboratory routinely washes new reversed-phase columns (not attached to the detector) with several column volumes of, sequentially, 100% isopropanol, 50:50 (v/v) isopropanol-water, and finally, 100% water. The column is then ready for use.

B. COLUMN CLEANING

One of the biggest problems encountered during HPLC of peptides and proteins (and, indeed, other biomolecules also) is the accumulation of strongly retained compounds on the column. Although the adsorbed compounds may not necessarily cause back pressure problems, they may change the retention characteristics of the column,[8] often resulting in deterioration of solute resolution. The most general technique for removing retained solutes from all types of HPLC packings is gradient elution from 0.1% aqueous TFA to 0.1% TFA in isopropanol,[9] with repetitive gradients sometimes necessary. If this general technique does not produce entirely satisfactory results (proteins are particularly hard to remove due to the multiplicity of sorption mechanisms[8]), cleaning procedures specific to a particular type or source of column may be required. In the case of proteins, best success is often achieved with protein-solubilizing agents such as surfactants or concentrated solutions of salts or chaotropic reagents.

For SEC columns, several column volumes of 20% DMSO in water may prove effective in removing adsorbed solutes. Alternatively, appropriate buffers containing 0.5 to 1.0 M NaCl or denaturants such as 6 M guanidine hydrochloride (GuHCl) or 8 M urea may be employed. Using detergents such as 0.1% SDS is another alternative. More vigorous conditions may be applied to polymer-based columns and flushing these supports with strong base (1.0 to 2 M NaOH) is possible.

IEC columns are usually flushed with 1.0 to 2.0 M salt solutions to remove strongly retained solutes. In addition, chelating reagents such as 0.1 M EDTA have also proved useful on occasion. Polymer-based columns may also be flushed with 2 M NaOH followed by high salt elution. One specific cleaning procedure suggested for Partisil IEC silica-based columns (Whatman) involves washing with concentrated buffer, followed by an acid wash (0.5 M H_3PO_4, 0.1 M H_2SO_4, or 0.1 M HNO_3), EDTA and, finally, methanol. Manufacturers instructions for cleaning a polyether-based strong cation-exchange column (Pharmacia Mono S HR 5/5) include washing the column with a few ml of 0.5% SDS in 0.5 M NaOH (removes precipitated or denatured samples); the SDS is then removed with 80 to 100% methanol. In difficult cases, the manufacturer suggested employing 2 to 4 500-µl injections of TFA for dissolution of precipitated samples.

The general procedure of aqueous TFA to TFA/isopropanol gradient elution described above is usually the most effective cleaning procedure for RPC (and HIC) columns. Chelating reagents such as 0.1 M EDTA are also occasionally useful. In a manner similar to that noted above for non-silica-based SEC and IEC columns, pH-stable polymer-based RPC columns may be flushed with 0.1 to 2 M NaOH, depending on the manufacturer's instructions.

An example of the excellent results that may be achieved through employment of a stringent washing procedure is illustrated in Figure 1 for a silica-based reversed-phase C_8 column. The column (an Aquapore RP-300 C_8 from Brownlee Labs., Santa Clara, CA) had been extensively used over a long period of time. Although the retention times of a series of reversed-phase peptide standards[10] had not decreased significantly from when the column was new, their peak shapes were showing some deterioration (Figure 1A). This peak skewing is more apparent for

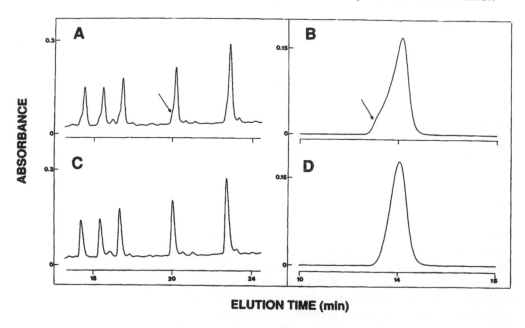

FIGURE 1.   Regeneration of reversed-phase column performance by a clean-up procedure. Column: analytical C_8 column (220×4.6 mm I.D., 7-μm particle size, 300-Å pore size). Conditions: Panels A and C, linear AB gradient (1% B/min) at a flow-rate of 1 ml/min, where eluent A is 0.1% aqueous TFA and eluent B is 0.1% TFA in acetonitrile; Panels B and D, isocratic elution with 45% aqueous acetonitrile, containing 0.1% TFA, at a flow-rate of 1 ml/min. Samples: Panels A and C, five synthetic decapeptide reversed-phase standards[10] (obtained from Synthetic Peptides Incorporated, Department of Biochemistry, University of Alberta, Edmonton, Alberta, Canada); Panels B and D, anthracene. Panels A and C show elution profiles obtained prior to column cleaning; Panels B and D show elution profiles obtained following a column clean-up procedure (described in the text). Arrows denote peak skewing.

a compound which partitions down the whole length of the column, such as anthracene (Figure 1B). In our experience, such peak skewing is usually a result of channels forming in the column bed (or occasionally a column void, as discussed below) the only solution for which is column repacking. However, this apparently was not the case for this particular column. Figures 1C and 1D show, respectively, the elution profiles of the peptide standards and anthracene following a wash procedure involving flushing the column with, sequentially, 50:50 (v/v) isopropanol/ toluene and 100% isopropanol. The improvement in the elution profiles of the peptide standards (Figure 1C) and, particularly, anthracene (Figure 1D) is immediately apparent, with the column performing almost as well as when new.

N.B. Two important points should be stressed concerning cleaning procedures:

1. Before employing any of the above procedures, the manufacturer's guidelines should be consulted for column compatibility with solvents and reagents
2. One should be sure that successive solvents are compatible, e.g., a solvent of high organic content should not be followed directly by a high-salt aqueous solvent.

C. REGENERATION OF SILICA-BASED REVERSED-PHASE COLUMNS BY ON-LINE DERIVATIZATION

Employment of silica-based reversed-phase columns containing straight-chain alkyl functional groups (e.g., C_4, C_8, C_{18}), particularly at low pH, results in a gradual loss of these functional groups, resulting in a decrease in solute retention times as well as peak broadening. As Nugent and Dolan[6] point out, the universal solvent system of acetonitrile, water and TFA at pH 2.0 is ideal for stationary phase removal. However, in many cases, the stationary phase can be regenerated by on-line rederivatization with alkylsilane reagents, and this is discussed in detail by Mant and Hodges[11] elsewhere in this book.

D. COLUMN VOIDS

Peak shape deterioration, such as asymmetrical peaks, broadened peaks or peak doublets, often indicates the presence of a column void. Voids usually form with silica-based columns when the silica matrix dissolves leaving a gap at the top of the column bed. This frequently happens when mobile phases close to neutral or alkaline mobile phases are used on a regular basis. The way to verify the presence of a void is to remove the top (inlet) column fitting carefully and check the column bed. Shallow voids (1 to 3 mm) are generally not a problem to repair. Wehr and Majors[8] suggest removing just enough packing to form a level surface, then adding a small amount of packing material (either the same or similar to that of the original) in a slurry. The added slurry should extend above the top of the column in a slight mound so that the slurry will be slightly compressed when the column end fitting is replaced and tightened.[8] We have had success with this procedure for analytical SEC, IEC, and particularly, RPC columns, using either microparticulate or guard column-type pellicular packings. When using pellicular materials, we have found that dry-packing (as opposed to slurry packing) generally suffices.

Wehr and Majors[8] point out that if the depth of the void is greater than a few millimetres, the above topping-off procedure probably will not work (the column then being discarded or requiring complete repacking). However, even if the void is fairly deep (up to ~10 mm), attempting to repair the column is still worthwhile, considering the expense of having to buy a new one. An excellent example of this is illustrated in Figure 2 for a semi-preparative C_{18} column. Figure 2A shows the analytical elution profile of five synthetic peptide standards[10] on this extensively-used column. Quite clearly, this column was exhibiting serious peak deterioration (the profile should have been similar to that illustrated in Figure 1C for the cleaned analytical C_8 column). On removing the top column fitting, a void of at least 10 mm in depth was observed. Similar column packing (C_{18} material) to that already in the column was slurried in isopropanol and added to the column bed, followed by replacement of the column fitting. The improvement in the peptide elution profile of this repaired column is dramatic (Figure 2B). Figure 2C shows the highly symmetrical anthracene peak obtained on this regenerated column, an indicator of a well-packed column bed. The efficiency of the column (~7300 plates), calculated from the anthracene peak, is now similar to that when the column was new.

E. COLUMN STORAGE

HPLC columns, whether containing silica- or polymer-based packings, should not be stored in a mobile phase that contains salts, acids or bases.

Recommended storage conditions for both silica- and polymer-based SEC columns are 10 to 25% aqueous alcohol (usually methanol) to prevent bacterial growth.

IEC columns are generally stored in 10 to 20% or 100% alcohol (usually methanol).

RPC columns should be stored in a high concentration of organic solvent. The columns are then cleaned as well as stored after each days use. The concentration of organic solvent should be significantly higher than the concentrations used in the mobile phase and >50% aqueous organic solvent is recommended for most purposes. For long term storage, 100% methanol is recommended by many manufacturers.

If columns are to be stored for long periods of time, they should be sealed to prevent drying out.

N.B. In a manner similar to that stated previously concerning column cleaning, manufacturers' guidelines should be consulted concerning column storage conditions.

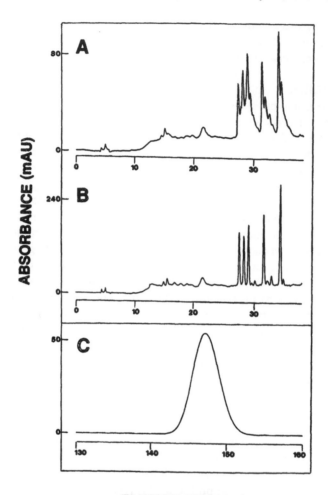

ELUTION TIME (min)

FIGURE 2. Regeneration of reversed-phase column performance by
repairing a column void. Column: semi-preparative C_{18} column (300
× 7.5 mm I.D.). Conditions: Panels A and B, linear AB gradient (1%
B/min) at a flow-rate of 2 ml/min, where eluent A is 0.05% aqueous
TFA and eluent B is 0.05% TFA in acetonitrile; Panel C, isocratic
elution with 45% aqueous acetonitrile, containing 0.05% TFA, at a
flow-rate of 2 ml/min. Samples: Panels A and B, five synthetic
decapeptide reversed-phase standards;[10] Panel C, anthracene. Panel A
shows elution profile of peptide standards prior to void repair; Panels
B and C show elution profiles obtained following slurry packing of C_{18}
material into the column void.

ACKNOWLEDGMENTS

This work was supported by research and equipment grants from the Medical Research
Council of Canada and the Alberta Heritage Foundation for Medical Research, respectively.

REFERENCES

1. Johnson, E. and Stevenson, B., *Basic Liquid Chromatography*, Varian Associates, Inc., Palo Alto, CA, 1978.
2. Chloupek, R.C., Battersby, J.E., and Hancock, W.S., Practical considerations for assessing product quality of biosynthetic proteins by HPLC, this publication.
3. Dolan, J.W., Preventive maintenance and troubleshooting LC instrumentation, this publication.
4. Mant, C.T. and Hodges, R.S., The effects of anionic ion-pairing reagents on peptide retention in reversed-phase chromatography, this publication.
5. Rivier, J.E., Use of trialkylammonium phosphate buffers in reversed-phase HPLC for high resolution and recovery of peptides and proteins, *J. Liq. Chromatogr.*, 1, 343, 1978.
6. Nugent, K.D. and Dolan, J.W., Tools and techniques to extend LC column lifetimes, this publication.
7. Henry, R., HPLC guard columns, this publication.
8. Wehr, C.T. and Majors, R.E., Regeneration of biopolymer columns, *LC.GC.*, 5, 942, 1987.
9. Regnier, F.E., HPLC of proteins, *Methods Enzymol.*, 91, 137, 1983.
10. Mant, C.T. and Hodges, R.S., Requirement for peptide standards to monitor column performance and the effect of column dimensions, organic modifiers and temperature in reversed-phase chromatography, this publication.
11. Mant, C.T. and Hodges, R.S., On-line derivatization of silica supports for regeneration of reversed-phase columns, this publication.

HPLC GUARD COLUMNS

Richard Henry

I. INTRODUCTION

Placing a small column in front of a more expensive analytical or preparative column to prolong its usefulness has become popular in HPLC. Columns receiving protection can be referred to as primary columns with the understanding that they can be either analytical or preparative in dimension. The term primary is used because they are first in value and in importance of creating the separation. While seldom, if ever, needed or employed in gas chromatography (GC), some estimates suggest that as many as 50% of HPLC users employ guard protection in some form. The location of a guard column is shown in Figure 1.

II. WHAT GUARD COLUMNS REMOVE

Guard devices can be thought of as more-or-less expendable columns designed to remove anything that will interfere with the separation or shorten the lifetime of the primary column. Some things that fall into this category include:

1. Particles that will clog the primary column and necessitate a frit change, which can be a delicate procedure.
2. Compounds, including ions, that will be strongly adsorbed by the packing and eventually cause baseline drift, spurious peaks, loss of resolution and change in selectivity.
3. Compounds, including ions, which can form a precipitate upon contact with the mobile phase or packing.
4. Compounds that are co-eluted with and interfere with the detection and quantitation of the compound(s) of interest.

III. OFF-LINE/ON-LINE DISTINCTION

Solid-Phase Extraction (SPE) can be viewed as off-line guard protection using low-pressure, disposable glass or plastic columns. Disposable, off-line columns, sometimes referred to as clean-up columns, can remove all of the categories listed above; however, they can add time to the analysis and cause sample loss and sample dilution. Disposable clean-up columns are filled with inexpensive, inefficient porous packings which generally do not separate closely-related compounds. This article will not deal with this important category of off-line sample clean-up.

On-line methods require high-pressure, stainless steel column designs. They are typically connected between the injector and primary column by small internal diameter (ID) capillary tubing, are very convenient and provide quantitative sample transfer. Several integral designs which do not require capillary connection are also available. On-line guard columns are too expensive to be disposable and are expected to last for numerous injections. They have an added advantage in that they can provide some protection from adverse mobile phase effects. Because the sample must pass through them on the way to the primary separating column, on-line guard columns must be filled with high-performance packings. They can be small-diameter, porous or larger-diameter, solid-core materials. Difficult samples may require the employment of both off-line and on-line guard devices.

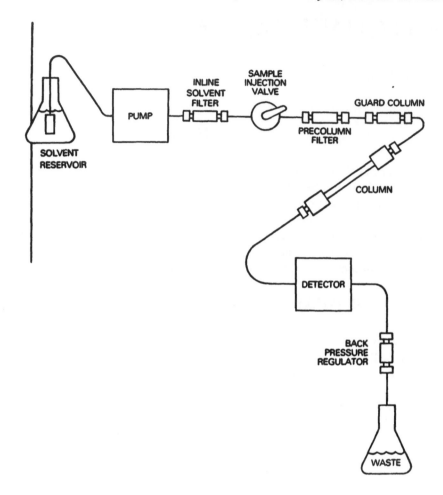

FIGURE 1. Location of guard devices in an HPLC system.

In most cases, an on-line guard column should have little or no effect on the separation, as shown in Figure 2. It is always important to verify this by injection of standards before and after guard installation. Notice in Figure 2 the increase in retention time due to the added length and volume of the guard device.

IV. TYPES OF ON-LINE GUARD DEVICES

Many convenient designs are available in both empty and prepacked versions. Table 1 lists several categories of guard columns which are commercially available, and Figure 3 shows corresponding examples of each type.

The dimension of the guard column and type of packing combine to determine overall performance. A guard column may be tested alone (without the primary column) to determine the amount of retention and resolution added to the system. A short guard column containing high surface area (high capacity), porous packing will add substantially to the separation as shown in Figure 4. A guard column containing low surface area (low capacity), pellicular packing will add virtually nothing to the separation, but will also provide less guard protection.

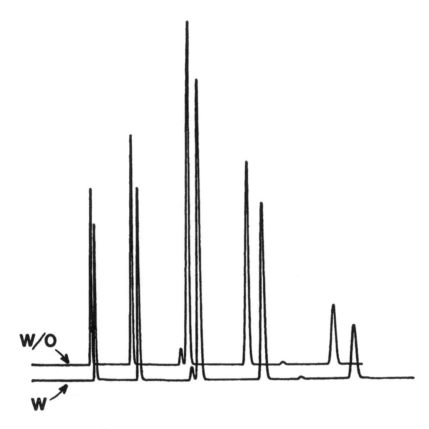

FIGURE 2. Effect of guard column on a separation. Primary column: 250 × 4.6 mm I.D. column containing ODS Hypersil. Guard column: 10 × 4 mm I.D. drop-in cartridge (an example of a column fitting with integral guard cartridge; see Figure 3); w/o and w denote without and with guard, respectively. Sample: theophylline, *p*-nitroaniline, methyl benzoate, phenetole, *o*-xylene. Eluent: 60% aq. acetonitrile. Flow-rate: 1.25 ml/min.

TABLE 1
Guard Column Categories

1. Tap-fill (dry-pack)
   Ferrule or threaded types
   Typical length, 2—5 cm
   Typical I.D., 2 mm and greater
   Pellicular or porous packings, large diameter (≥20 μm)
2. Prepacked conventional
   Ferrule or threaded types
   Typical length, 2—5 cm
   Typical I.D., 2 mm and greater
   Porous packings, small diameter (3—10 μm)
3. Prepacked cartridge
   Ferrule, threaded, or holder types
   Typical length, 1—2 cm
   Typical I.D., 2 mm and greater
   Porous packings, small diameter (3—10 μm)

CONVENTIONAL GUARD
(Categories 1 and 2)

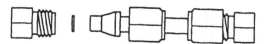

THREADED GUARD
(Categories 1 and 2)

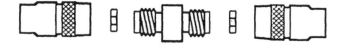

DIRECT-CONNECT GUARD
(Category 1)

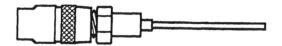

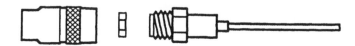

FIGURE 3. Some examples of guard column designs. Categories 1, 2, and 3 refer to guard column categories shown in Table 1. Examples A and D can be tap-filled by the user. A and B are also commonly available prepacked. E and F are examples of prepacked cartridges with two types of holders (in-line or integral with column fitting, respectively).

DIRECT-CONNECT GUARD WITH FINGER-TIGHTEN FITTING
(Category 3)

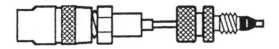

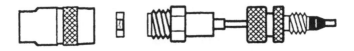

COLUMN GUARD AND HOLDER
(Category 3)

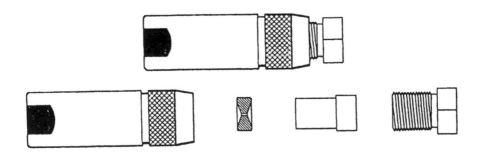

COLUMN FITTING WITH INTEGRAL GUARD CARTRIDGE
(Category 3)

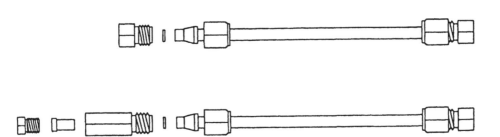

FIGURE 3 (continued)

COLUMN FITTING WITH
INTEGRAL FILTER

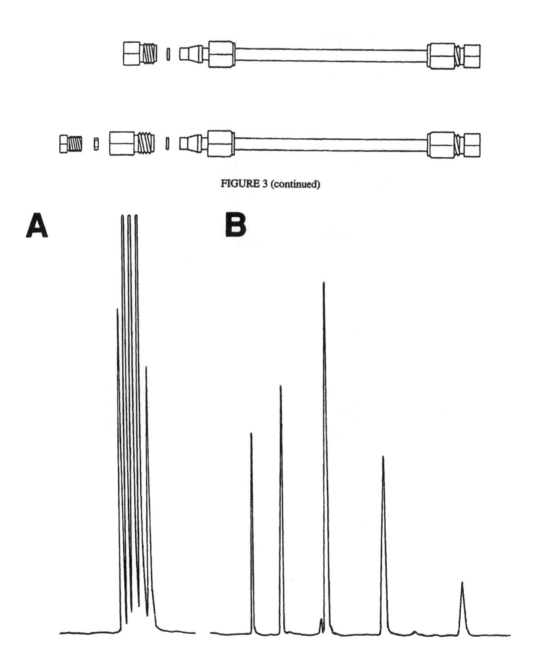

FIGURE 3 (continued)

A

B

FIGURE 4. Contribution of guard column to retention and resolution. Primary column: 250 × 4.6 mm I.D. column containing ODS Hypersil. Guard column: 20 × 4 mm threaded guard (see Figure 3). Test sample and run conditions: see Figure 2. The test sample was run with the guard column alone (A) or attached to the primary column (B).

TABLE 2
Types of Packing Materials

1. Small, porous (identical to primary column)
   High capacity; heavy protection
   Acts as extension; can increase system performance
   Usually prepacked; must be well-packed
2. Small, porous, wide-pore
   Lower capacity; some effect on separation
   Usually prepacked; must be well-packed
   Limited surface modifications available
3. Pellicular or solid-core
   Low capacity; no effect on separation
   Large particle (30 μ m); dry-pack method
   Limited surface modifications available
4. Large, porous
   High capacity; noticeable affect on separation (performance loss)
   Dry-pack method
   Very limited surface modifications available

V. CHOICE OF PACKING

When available, the same packing that is in the primary column is a good choice; however, low surface-area, pellicular (solid-core) or wide-pore packings can also be excellent choices, particularly when one desires a low capacity guard packing that does not introduce much of its own selectivity into the separation process. Table 2 shows some types of packings that have been used effectively in guard devices.

The graph in Figure 5 shows the increased retention contributed by different lengths of guard columns up to 5 cm packed with three different types of packings from Table 2. The most common choice is category 1, a packing similar or identical to the analytical column; however, a guard column with active sites can degrade a separation even if it is well-packed. If the same packing cannot be selected for guard use, a very inert material should be selected to avoid problems. Figure 6 shows how an unsuitable guard packing can degrade the separation of sensitive samples such as pharmaceuticals and biochemicals.

VI. WHEN TO CHANGE A GUARD COLUMN

Due to its location at the column system inlet, a guard device sees the impact of high pressure and high sample concentration so it must be changed regularly. Everyone understands this function, but the rules for when to change a guard device are not as clearly defined. Some common indicators include:

1. High pressure-drop; a packed guard column may not be needed. Consider a primary column that has a convenient frit replacement design or a precolumn guard filter.
2. Loss of performance measured by such things as resolution, efficiency, symmetry, etc.
3. Baseline drift, spurious peaks, split peaks, etc.

Some HPLC users prefer to change guard columns at regular intervals before significant loss of performance is noted. If particle contamination or sample precipitation is a problem, the use

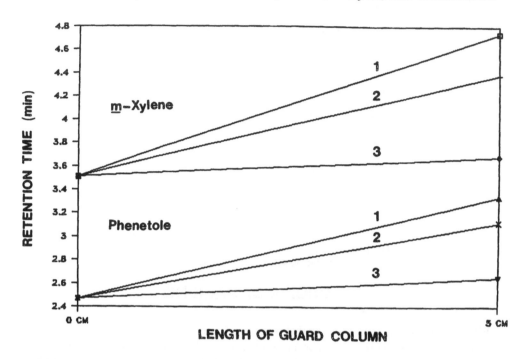

FIGURE 5. Effect of guard column length and packing on retention time. Primary column: 100×4.6 mm I.D. containing 5-μm particle size ODS Hypersil. Guard column packings: ODS Hypersil, 5-μm particle size, 200 m²/g surface area (1); DELTABOND ODS wide-pore, 5-μm, 100 m²/g (2); pellicular ODS, 40 μm, 10 m²/g (3); packed in 4 mm I.D. columns. Run conditions: see Figure 2.

of a precolumn guard filter as shown in Figure 7 is strongly encouraged, alone or in addition to a guard column. The addition of a guard filter alone will not detectably increase solute retention time, although this should be determined by testing before and after installation. In principle, on-line guard devices that prolong system performance and reduce cost are very attractive; however, the effect of installing any extra device must be tested and fully appreciated. The simplest device to install with very few adverse effects is a low-volume, well-swept line filter that is available from many sources. When a guard column is added, the user must appreciate that coupled columns (two or more columns in series) will be created, with the net effect being the sum of what the sample experiences in each column. While the common guard column choices have been summarized above, there may also be many situations where a unique approach can be taken to solve a specific problem. For example, an ion-exchange guard may be employed with a reversed-phase column to remove ionic contaminants, or a cyano guard may be added to a reversed-phase column to adjust selectivity and improve resolution as well as to provide guard protection. A very important principle to remember is that the net, combined result of coupled columns will always be the sum of the retention times observed by testing each individual column alone. Although this treatment has placed emphasis on guard devices for standard analytical HPLC columns, the need for guard filters and columns may be even greater when smaller-bore or larger-bore (preparative) columns are used. The principles of selection and testing remain the same regardless of column inner diameter and flow rate.

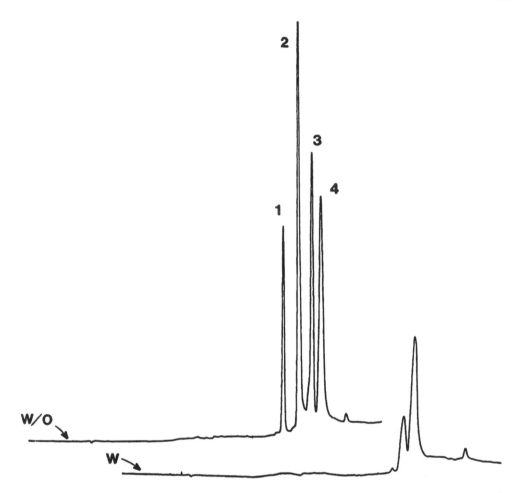

FIGURE 6. Degradation of performance by mismatched guard column with active sites. Primary column: DELTA-BOND ODS, 5-µm particle size (250×4.6 mm I.D.). Guard column: conventional ODS, 5 µm (50×4 mm I.D.). Sample: ribonuclease A, insulin, lysozyme and albumin (peaks 1 to 4, respectively). Conditions: linear AB gradient (6.25% B/min), where eluent A is 0.1% aq. trifluoroacetic acid (TFA) and eluent B is 0.1% TFA in 95% aq. acetonitrile. The sample was run with (w) and without (w/o) a guard attached to the primary column.

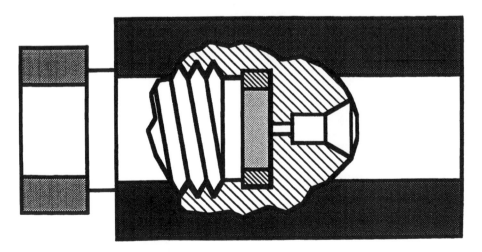

FIGURE 7. Diagram of a precolumn guard filter.

ON-LINE DERIVATIZATION OF SILICA SUPPORTS FOR REGENERATION OF REVERSED-PHASE COLUMNS

Colin T. Mant and Robert S. Hodges

I. INTRODUCTION

Extensive use of silica-based reversed-phase columns generally results in the gradual removal of hydrophobic ligands from the silica and, subsequently, in an increased concentration of underivatized silanols and an increased potential for ionic interaction with positively charged peptide residues. The results are a gradual deterioration in column performance and, eventually, an end of the column's useful life. The high cost of column replacement emphasizes the desirability of prolonging or regenerating the active life of a column.

Several researchers have demonstrated the viability of on-line silanization of deteriorated reversed-phase packings.[1-7] This procedure generally involves pumping a solution of the desired silane reagent (usually a chloroalkylsilane) through the column in an attempt to rederivatize exposed silanols and, hence, recover column performance.

This article describes a rapid, on-line procedure for the regeneration of silica-based C_8 and C_{18} reversed-phase columns and demonstrates the value of peptide standards for monitoring improvements in column performance.

II. EXPERIMENTAL

A. MATERIALS

Chlorodimethyloctylsilane was obtained from Aldrich (Milwaukee, WI). Reagent grade dichloromethane and pyridine were redistilled prior to use. HPLC-grade water, acetonitrile and isopropanol were obtained from Fisher Scientific (Fairlawn, NJ).

Molecular Sieves (Type 4A, Grade 514, 8 to 12 mesh beads, 4-Å pore size) were obtained from Fisher Scientific. Membrane filters (47 mm diameter discs, 0.2-μm pore size), resistant to organic solvents, were obtained from S.P.E. (Rexdale, Canada).

A mixture of five synthetic decapeptide standards was obtained from Synthetic Peptides Incorporated (Dept. of Biochemistry, University of Alberta, Edmonton, Canada).

B. REGENERATION PROCEDURE

The regeneration of deteriorated reversed-phase packings described in this article was carried out at room temperature (26°C) using a single-piston, Vista series 5000 LC pump (Varian, Walnut Creek, CA). The pump and column were disconnected from the detector. The columns were flushed successively at 2 ml/min with 30 ml of water, acetonitrile and dichloromethane, the dichloromethane having been previously dried over molecular sieves and filtered through a 0.2 μm filter. The silanization solution (20 ml of 0.4 M chlorodimethyloctylsilane and 0.8 M pyridine in dry dichloromethane) was then pumped through the columns at 0.2 ml/min. Following silanization, the columns were flushed successively with 30 ml (2 ml/min) of dichloromethane and isopropanol (if the backpressure on the columns is too high at a flow-rate of 2 ml/min [especially with the viscous isopropanol], the flow-rate can be reduced, as long as the columns are washed thoroughly with each solvent). Finally, the columns were equilibrated in 0.05% aq. trifluoroacetic acid (TFA) and flushed at 1 ml/min with a linear AB gradient (2% B/min), where Eluent A was 0.05% aq. TFA and Eluent B was 0.05% TFA in acetonitrile. The regenerated columns were then ready for chromatography. At least one, generally two, blank gradients, with the detector attached, were run before chromatographing the peptide standards, to ensure that all impurities have been removed from the column.

It should be noted that a white precipitate may appear at the column outlet during rederivatization. This precipitate is pyridine hydrochloride, formed by reaction of the pyridine with hydrogen chloride (released during reaction of the silane reagent with free silanols) and precipitating out of solution due to evaporation of the volatile dichloromethane. To prevent clogging of narrow stainless steel lines, it was found worthwhile to remove these lines from the column outlets.

III. RESULTS AND DISCUSSION

The incorporation of a base such as pyridine in the silanization reaction mixture is advantageous in neutralizing hydrogen chloride that is released during reaction of the chlorodi-methyloctylsilane with free silanols on the silica surface. Hydrogen chloride interacts with stainless steel, a situation not compatible with the use of stainless steel columns, lines and injectors. In addition, hydrogen chloride may also bind preferentially to unreacted sites, thereby impeding further reaction. Pyridine is often incorporated into reaction mixtures for laboratory-scale production of reversed-phase packings.[9-15] Apart from the acid-binding properties of the base, which shifts the equilibrium of the reaction to the product side, the base also favorably affects the kinetics of the silanization reaction.[9] The 2:1 molar ratio of base to silane used in the regeneration procedure (see Experimental) was designed to ensure the removal of all hydrogen chloride produced during regeneration. The high reactivity of silane reagents with water necessitated an initial flushing of deteriorated columns with acetonitrile, followed by equilibration of the columns in dichloromethane, a solvent suggested as a good medium for silanization reactions.[9] The silanization solution was applied through a 10-ml injection loop (2×10 ml injections) rather than directly through the LC pump.[4] Following rederivatization and removal of the silane reagent by a post-reaction dichloromethane wash, the columns were flushed with isopropanol to remove any adsorbed alkylsilane groups not covalently bonded to the silica matrix. The final linear gradient wash was designed to remove any reaction products or solvent impurities retained by the column during regeneration.

The effectiveness of the regeneration procedure was monitored by chromatographing a series of five synthetic peptide standards both before and after regeneration. The hydrophobicity of the standards (S1-S5) increases only slightly from S2 to S5, enabling a very precise determination of the resolving powers of reversed-phase columns[16] (for further details, see article by Mant and Hodges entitled "Requirement for Peptide Standards to Monitor Column Performance and the Effect of Column Dimensions, Organic Modifiers and Temperature in Reversed-Phase Chromatography").

A. REGENERATION OF ANALYTICAL REVERSED-PHASE COLUMNS

Figure 1, left panel, demonstrates chromatographic profiles of the peptide standards obtained before (top) and after (bottom) on-line rederivatization of an extensively used C_{18} column (250 $\times$ 4.1 mm I.D.). The peptides were chromatographed at pH 2.0 and the increase in retention times and the improvement in peak shape of the peptides following regeneration is very clear. In fact, the elution profile of the standards on the treated column resembled very closely the profile obtained with the column when it was new.

Rederivatization of the C_{18} packing with the octylsilane reagent produced a mixed C_8/C_{18}

59

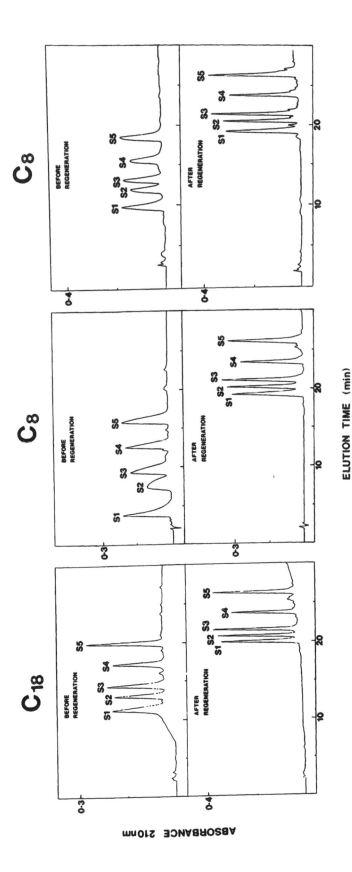

FIGURE 1. Monitoring of reversed-phase analytical column performance before and after regeneration. Columns: Left panel, SynChropak RP-P C$_{18}$ (SynChrom, Linden, IN), 250 × 4.1 mm I.D., 6.5-μm particle size, 300-Å pore size; middle and right panels, SynChropak RP-8 C$_8$, 250 × 4.1 mm I.D., 6.5 μm, 300 Å. Mobile phase: linear AB gradient (1% B/min), where Eluent A is 0.05% aq. TFA (left and middle panels) or 0.1% aq. TFA (right panel) and Eluent B is 0.05% TFA in acetonitrile (pH 2.0); flow-rate, 1 ml/min; 26°C. The HPLC runs were carried out on a Varian Vista Series 5000 liquid chromatograph (Varian, Walnut Creek, CA) coupled to a Hewlett-Packard (Avondale, PA) HP 1040A detection system, HP85B computer, HP9121 disc drive, HP2225A Thinkjet printer and HP7470A plotter. S1 to S5 are synthetic reversed-phase decapeptide standards (ca. 20 nmol of each).

bonded phase. Although chlorodimethyloctadecylsilane has been successfully used in the authors' laboratory to derivatize silica supports[7] (producing a C_{18} bonded phase), there are more problems associated with its use than the corresponding octylsilane. The octadecylsilane reagent is a solid which requires dissolving in dichloromethane and careful filtering before use. Once in solution, it was found to have a tendency to precipitate out of solution if not utilized quickly. In contrast, the liquid octylsilane reagent simply has to be made up to the required dilution and it is ready to use without the problem of possible reagent precipitation. This laboratory regularly regenerates C_{18} bonded phases with the octylsilane reagent and the resulting mixed bonded phases exhibit the same resolving power for peptide separations as the original C_{18} sorbent when new.

Figure 1, middle panel, shows chromatographic profiles (at pH 2.0) of the peptide standards obtained before (top) and after regeneration of a deteriorated C_8 column (250×4.1 mm I.D.). This column posed a more severe test of the regeneration procedure than the above C_{18} packing. The performance of the C_8 column had deteriorated so badly during extensive use that the peptide standards all exhibited broad peaks and much reduced retention times at pH 2.0 (top). The improvement in column performance following regeneration was dramatic (bottom) and the elution profile of the peptide standards again resembled that obtained when the column was new.

Figure 1, right panel (top) shows the chromatographic profiles (at pH 2.0) of the peptide standards on the same C_8 column as above (Figure 1, middle panel) after several months of use following its original regeneration. The performance of the column had deteriorated again, but its peptide resolving power was quickly restored by a second rederivatization treatment (Figure 1, right panel, bottom). Thus, the useful life of a column may well be extended indefinitely with further treatments when column performance has deteriorated once more. Gradual dissolution of the silica matrix over a long period of use, however, will probably make further regeneration impractical at some point.

B. REGENERATION OF SEMIPREPARATIVE REVERSED-PHASE COLUMNS

Semipreparative columns (8 to 12 mm I.D.) are substantially more costly than analytical columns (4.1 to 4.6 mm I.D.) of comparable length and cost savings would be even more appreciable if their useful lifetimes could be extended. Figure 2 demonstrates the effectiveness of the regeneration treatment on the elution profiles (at pH 2.0) of the peptide standards on a deteriorated semipreparative C_{18} column (250×10 mm I.D.). The elution profile of the peptides showed an improvement following one silanization treatment of the column (Figure 2, middle panel), producing a mixed C_8/C_{18} bonded phase, compared to the profile obtained prior to regeneration (top panel). However, a second treatment was required to produce the satisfactory elution profile shown in the bottom panel of Figure 2. The necessity for this second silanization treatment is not surprising when one considers that the volume of the semipreparative column (Figure 2) was greater by about six-fold than that of the analytical columns (Figure 1). Apparently, while the level of silane reagent in a single silanization treatment was sufficient to regenerate completely the performance of the analytical columns, it was not enough to rederivatize the significantly higher volume of silica support (and, presumably, greater number of underivatized silanols) present in the larger column.

C. CHARACTERISTICS OF REVERSED-PHASE COLUMNS BEFORE AND AFTER REGENERATION

A comparison of column performance characteristics (theoretical plates, plate height) before and after regeneration is shown in Table 1. The peptide elution profiles obtained on the longer analytical C_8 column (250 mm in length) are shown in Figure 1, middle and right panels, although the column characteristics for the regenerated column only refer to those obtained after a second rederivatization (right panel). The shorter analytical C_8 column (150 mm in length) contained identical column packing to the 250-mm column. The peptide elution profiles

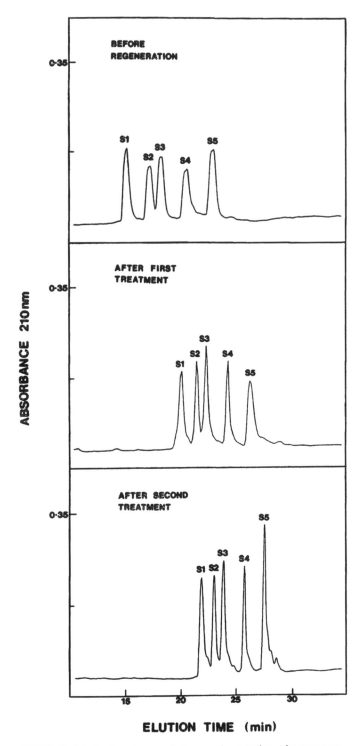

FIGURE 2. Monitoring of reversed-phase semipreparative column performance. Column, SynChropak RP-P C₁₈, 250 × 10 mm I.D., 6.5-μm particle size, 300-Å pore size. Mobile phase: linear AB gradient (1% B/min), where Eluent A is 0.05% aq. TFA and Eluent B is 0.05% TFA in acetonitrile; flow-rate, 2 ml/min; 26°C. Instrumentation as described in Figure 1. S1 to S5 are synthetic reversed-phase decapeptide standards (ca. 40 nmol of each). (From Mant, C. T. and Hodges, R. S., *J. Chromatog.*, 409, 155, 1987. With permission.)

TABLE 1
Characteristics of Silica-Based Reversed-Phase Columns Before and After Regeneration

Column[a]	t_R (min)	k'	N	H
C_8 (250 × 4.1 mm I.D.)				
(see Figure 1, right panel)				
New[b]	—	—	5965	0.042
Deteriorated column	9.9	1.9	1600	0.156
After regeneration	32.9	8.7	6500	0.038
C_8 (150 × 4.1 mm I.D.)				
New	—	—	4800	0.031
Deteriorated column	4.5	1.2	1100	0.136
After regeneration	12.4	5.0	4430	0.034
C_{18} (250 × 10 mm I.D.)				
(see Figure 2)				
New	—	—	6800	0.037
Deteriorated column	28.0	1.2	2100	0.119
After regeneration	56.2	3.5	5700	0.044

Note: The test solution containing uracil and anthracene was eluted with acetonitrile/water (45:55) at 1 ml/min and monitored at 254 nm. t_R denotes retention time of anthracene; the capacity factor (k') was calculated using the equation: $k' = (t_R-t_o/t_o)$, where t_R is the retention time of anthracene and t_o is the retention time of an unretained compound (uracil); N, the number of theoretical plates was calculated using the equation: $N = 5.5 \, (t_R/W_{1/2})^2$, where t_R is the retention time of anthracene and $W_{1/2}$ is the peak width at half height; the plate height (H) was calculated using the equation: H = L/N, where L is the length of the column (mm) and N is the number of theoretical plates.

[a] All columns were obtained from SynChrom (Linden, IN)
[b] Manufacturer's average values (N and H), obtained with a test mixture of CMP, UMP, GMP, and AMP using 0.1 M KH_2PO_4/0.05 M NaCl, pH 4.5, as the eluent.

obtained on the semipreparative C_{18} column are shown in Figure 2. The results shown in Table 1 underline the excellent improvement in the chromatographic profiles of the peptide standards following rederivatization of exposed silanols. The column performances approached or exceeded the values supplied by the manufacturer for the same columns when new.

D. STABILITY OF REGENERATED REVERSED-PHASE COLUMNS

Hydrolysis of the siloxane bond (-Si-O-Si-) formed between silanols on the silica surface and the alkylsilane reagent occurs preferentially in highly aqueous mobile phases, while the hydrophobic ligand tends to be eluted from the column at higher organic solvent concentrations.[17] Since their regeneration, all four columns described in this article (Figures 1 and 2, Table 1) have been used extensively, with only a gradual loss in their performance capabilities for separating peptides and proteins, indicating excellent stability of these rederivatized reversed-phase packings.

The results of a more controlled investigation of the stability of on-line derivatized C_8 and C_{18} reversed-phase packings are shown in Figure 3. The C_8 and C_{18} packings used in this investigation were prepared from underivatized silica in a manner identical to that described for rederivatization of deteriorated packings (see Experimental), using either chlorodimethyloctylsilane or chlorodimethyloctadecylsilane as the derivatizing reagent for the C_8 and C_{18} packings, respectively.[7] The columns were purged at 1 ml/min with 0.1% aq. TFA over a considerable length of time (100 hr). At various intervals, the columns were subjected to a linear gradient wash (0 to 100% B at 4% B/min; Eluent A is 0.1% aq. TFA and Eluent B is 0.1% TFA in acetonitrile) to remove any adsorbed, non-covalently bound ligands. Figure 3 demonstrates elution profiles of the peptide standards, S1 to S5, at pH 2.0 on the C_{18} and C_8 columns after 0

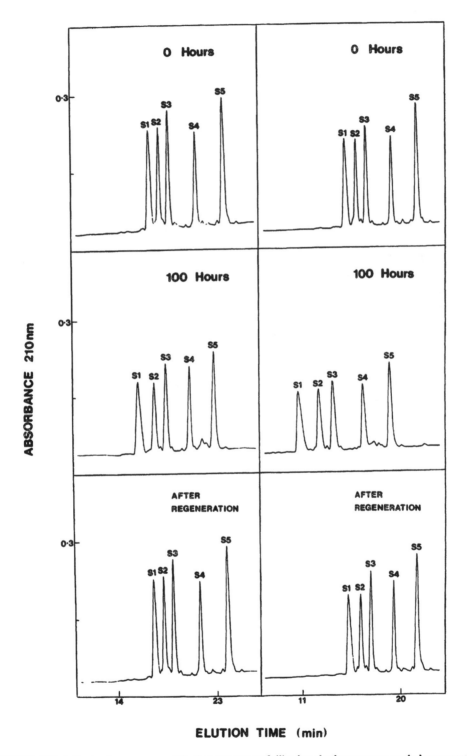

FIGURE 3. Effect of extensive aqueous mobile phase treatment of silica-based columns on reversed-phase separation of peptides at pH 2.0. Columns: C_{18} bonded phase (left) and C_8 bonded phase (right), 75×4.6 mm I.D., 6.5-μm particle size, 300-Å pore size. Elution profiles of a mixture of peptide standards (ca. 20 nmol of each) on freshly-prepared columns (top), the columns following 100 hr elution with 0.1% aq. TFA at 1 ml/min (middle), and the columns following regeneration by on-line silanization (bottom). Mobile phase: linear AB gradient (1% B/min), where Eluent A is 0.05% aq. TFA and Eluent B is 0.05% TFA in acetonitrile; flow-rate, 1 ml/min; 26°C. Instrumentation as described in Figure 1.

hr (top panels) and 100 hr (middle panels) treatment with 0.1% aq. TFA. The number of theoretical plates (N), measured as described in Table 1, showed a decrease from 2800 (C_{18}) and 1900 (C_8) at 0 h to 2300 (C_{18}) and 1400 (C_8) at 100 h. These moderate decreases in column efficiencies had little effect on the retention and resolution of the peptides and reflected the excellent stability of the packings. In addition, the slight loss in peptide retention (Figure 3, middle panels) was quickly restored following a further on-line silanization treatment (Figure 3, lower panels). Hydrolytic attack on siloxane bonds decreases with increasing ligand chain lengths, due to more efficient shielding by the longer alkyl chains.[5,15,17] Thus, the slightly more severe deterioration of the C_8 column compared to the C_{18} column was not unexpected.

IV. CONCLUSIONS

The on-line derivatization procedure described in this article is a rapid and cheap method for restoration of silica-based reversed-phase columns. The improved performance of the regenerated columns (Figures 1 and 2, Table 1) and the stability of both these columns and the packings prepared from underivatized silica (Figure 3) were particularly impressive considering the mild conditions used for the derivatization procedure. Application of the silanization procedure alone should generally suffice to restore efficient performance of deteriorated columns. In fact, the restoration of column performance by on-line derivatization confirmed that the deterioration in performance was due to chain loss, not packing problems. It should be noted, however, that packing irregularities in a column, resulting from extensive use, may occasionally have a derogatory effect on the performance of the rederivatized column. Under these circumstances, silanization coupled with removal and repacking of the rederivatized reversed-phase packing should completely restore column performance.

ACKNOWLEDGMENTS

This work was supported by research and equipment grants from the Medical Research Council of Canada and the Alberta Heritage Foundation for Medical Research, respectively.

REFERENCES

1. **Gilpin, R.K., Camilla, D.J., and Janicki, C.A.,** Preparation and use of *in situ* chemically bonded small-particle silica as packings in high-pressure liquid chromatography, *J. Chromatogr.*, 121, 13, 1976.
2. **Olieman, C., Sedlick, E., and Voskamp, D.,** *In situ* silylation of an octadecylsilyl-silica stationary phase applied to the analysis of peptides, such as secretin and glucagon, *J. Chromatogr.*, 207, 421, 1981.
3. **Wilson, T.D.,** Potential *in situ* regeneration of octadecyl-silica columns, *J. Chromatogr.*, 253, 260, 1982.
4. **Verzele, M., DeConinck, M., Dewaele, C., and Geeraert, E.,** On-column endcapping and derivatization in reversed-phase high performance liquid chromatography, *Chromatographia*, 19, 443, 1984.
5. **Abbott, S.R. and Simpson, R.A.,** Changes in column retention time in HPLC, *LC.GC.*, 4, 12, 1986.
6. **Mant, C.T., Parker, J.M.R., and Hodges, R.H.,** On-line derivatization to restore reversed-phase column performance in peptide separations, *LC.GC.*, 4, 1004, 1986.
7. **Mant, C.T. and Hodges, R.S.,** On-line derivatization of silica supports for regeneration and preparation of reversed-phase columns, *J. Chromatog.*, 409, 155, 1987.
8. **Unger, K.K.,** *Porous Silica*, Elsevier, Amsterdam, New York, 1979.
9. **Kinkel, J.N. and Unger, K.K.,** Role of solvent and base in the silanization reaction of silicas for reversed-phase high-performance liquid chromatography, *J. Chromatogr.*, 316, 193, 1984.

10. Hemetsberger, H., Maasfeld, W., and Ricken, H., The effect of chain length of bonded organic phases in reversed-phase high performance liquid chromatography, *Chromatographia*, 9, 303, 1976.
11. Hemetsberger, H., Kellermann, M., and Ricken, H., Behaviour of chemically bonded alkylmethyldichlorosilanes to silica gel in reversed-phase high-performance liquid chromatography, *Chromatographia*, 12, 726, 1977.
12. Hemetsberger, H., Behrensmeyer, P., Henning, J., and Ricken, H., Reversed-phase, high-performance liquid chromatography: effect of the structure of the chemically bonded hydrocarbon ligand on retention and selectivity, *Chromatographia*, 12, 71, 1979.
13. Spacek, P., Kubín, M., Vozka, S., and Porsch, B., Influence of the amount of bonded non-polar phase and the length of attached alkyl chains on retention characteristics of silica-based sorbents for reversed-phase high performance liquid chromatography, *J. Liq. Chromatogr.*, 3, 1465, 1980.
14. Marshall, D.B., Stutler, K.A., and Lochmüller, C.H., Synthesis of LC reversed phases of higher efficiency by initial partial deactivation of the silica surface. *J. Chromatogr. Sci.*, 22, 217, 1984.
15. Köhler, J., Chase, D.B., Farlee, R.D., Vega, A.J., and Kirkland, J.J., Comprehensive characterization of some silica-based stationary phases for high-performance liquid chromatography, *J. Chromatogr.*, 352, 275, 1986.
16. Mant, C.T. and Hodges, R.S., Requirement for peptide standards in reversed-phase high-performance liquid chromatography, *LC.GC.*, 2, 250, 1986.
17. Glajch, J.L., Kirkland, J.J., and Köhler, J., Effect of column degradation on the reversed-phase high-performance liquid chromatographic separation of peptides and protein, *J. Chromatogr.*, 384, 81, 1987.

Section II
Theoretical Aspects of High-Performance Liquid Chromatography

HPLC TERMINOLOGY: PRACTICAL AND THEORETICAL

Colin T. Mant and Robert S. Hodges

I. TYPES OF CHROMATOGRAPHY

This book deals with the separation technique known as liquid chromatography (LC) or, more specifically, high-performance liquid chromatography (HPLC) and Figure 1 shows the general outline of the various chromatographic techniques discussed in this book. Those areas within solid boxes are discussed in detail. Hydrophilic interaction chromatography is defined below and may represent a useful addition to our chromatographic repertoire for the separation of peptides and proteins.

What we have tried to do in this article is list the most common terms with which the novice chromatographer should be familiar when starting in HPLC. This list is not intended to be an in-depth listing and does not define all the theoretical terms which are discussed in detail in the remaining articles in this chapter. This list will also be restricted to those terms most often encountered in HPLC of peptides and proteins. For an excellent general discussion of chromatography, see Reference 1.

A. SIZE-EXCLUSION CHROMATOGRAPHY (SEC)

This technique is a major LC mode with the separation based upon the molecular size of the solute in solution. It is also known as gel-permeation, gel-filtration, steric-exclusion, or gel-chromatography. SEC is fundamentally the easiest mode of chromatography to understand and perform. The SEC packing is a gel with an inert porous surface. Soft gels such as Sephadex, a crosslinked dextran gel which has found wide spread use in open column chromatography, cannot withstand pressures exceeding 1 or 2 atmospheres. Only the semi-rigid (e.g., polystyrene) or rigid packings (silica) are useful in HPLC. SEC is unique among LC techniques in that it depends upon the physical restriction of the solute flowing into and out of pores of the column packing. Large molecules which cannot enter the pores pass through the column unretained and are eluted first. A solute which is capable of diffusing completely into all the pores is said to permeate the packing totally and is eluted last. Other solutes will selectively permeate the pores, depending on their relative size, and be eluted with a retention value between the two extremes.

B. ION-EXCHANGE CHROMATOGRAPHY (IEC)

In this mode of chromatography, the separation depends upon the exchange of ions between the mobile phase and the ionic sites of the packing (cationic or anionic). Figure 2 shows the ion-exchange process schematically for cation-exchange (CEX or CEC) and anion-exchange (AEX or AEC) chromatography.

The stationary phase matrix has a functional group with a fixed ionic charge covalently attached to it. There is an exchangeable counterion from the mobile phase buffer to preserve charge neutrality, since the mobile phase usually contains a large number of counterions opposite in charge to the surface ionic group. This counterion is in equilibrium with the matrix charged group in the form of an ion-pair. The presence of a sample ion of the same ionic charge as the counterion sets up an equilibrium. The sample ion can exchange with the counterion to be the partner of the covalently attached charge on the matrix. When the sample ion is paired with

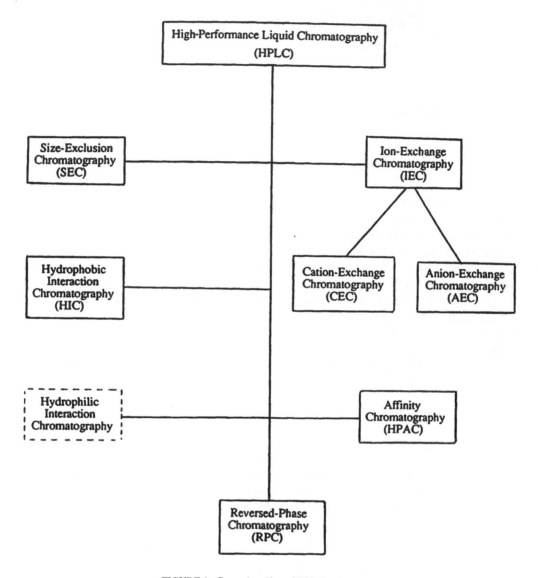

FIGURE 1. General outline of HPLC techniques.

the charged group on the matrix, the ion does not move down the column. The retention of the sample ion is based on the affinity of different ions for the site on the matrix and on a number of other solution parameters (pH, ionic strength, counterion type, etc.). For example, if the mobile phase buffer contains sodium chloride, the counterion is Na^+ in the case of the cation-exchange process and Cl^- in the anion-exchange process.

C. PARTITION CHROMATOGRAPHY

This technique depends upon the partitioning of the solute between two liquid phases. One of the liquid phases is held stationary on the solid-support (stationary phase) while the other is allowed to flow freely down the column (mobile phase). In the early stages of LC, the stationary phase (polar or non-polar) was coated onto an inert support and packed into a column and the mobile phase was passed through the column. This form of partition chromatography is called liquid-liquid chromatography.

To meet the need for more durable columns, packings with the stationary phase chemically

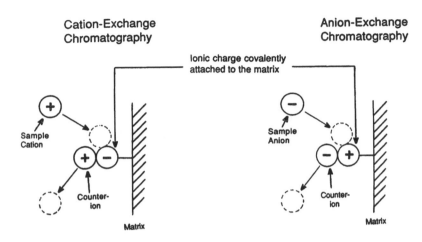

Cation-Exchange
Chromatography

Anion-Exchange
Chromatography

Ionic charge covalently
attached to the matrix

Sample
Cation

Counter-
ion

Matrix

Sample
Anion

Counter-
ion

Matrix

FIGURE 2. Schematic representation of ion-exchange chromatography.

bonded to the inert support were developed. This form of partition chromatography is called bonded-phase chromatography and is the most popular LC mode. Partition chromatography, whether liquid-liquid or bonded-phase chromatography, is termed "normal phase" if the stationary phase is more polar than the mobile phase and "reversed-phase" if the mobile phase is more polar than the stationary phase. The most popular support is microparticulate silica gel and the most popular bonded-phase is an organosilane, such as octadecyl, for reversed-phase chromatography. The majority of HPLC is carried out on chemically bonded phases. The three types of partition chromatography discussed below are reversed-phase chromatography (RPC), hydrophobic interaction chromatography (HIC), and hydrophilic interaction chromatography (HILIC).

1. Reversed-Phase Chromatography (RPC)

For the novice chromatographer, the term "reversed-phase" can cause confusion. To avoid this, simply remember that RPC is a form of partition chromatography where the chemically bonded-phase is hydrophobic or non-polar (e.g., an octadecyl group, Figure 3), and the starting mobile phase (e.g., water) must be more polar than the stationary phase. This method of chromatography is "reversed" from normal phase chromatography, where the stationary phase is hydrophilic or polar and the starting mobile phase is more non-polar than the stationary phase, hence the term "reversed-phase chromatography".

In this technique, the selectivity is often significantly different from that achieved in normal-phase separations. The elution order is generally related to the increasing hydrophobic nature of the solute. The more soluble a solute is in water or the more hydrophilic the solute, the faster it will be eluted. In contrast, the more hydrophobic the solute, the slower it will be eluted. In general, the elution order will be the "reverse" of a normal phase separation.

The most commonly used mobile phase eluent for peptide and protein separations is water and a water-miscible organic solvent such as methanol or acetonitrile. There are many variations of RPC in which various mobile phase additives are used to impart a different selectivity. For example, the addition of trifluoroacetic acid to the mobile phase allows ion-pairing with the cations of the sample. This hydrophobic anionic ion-pairing reagent can drastically change the separation compared to other anionic ion-pairing reagents like phosphate. Similarly, cationic ion-pairing reagents can affect the separation of anions by RPC (e.g., the addition of buffers containing tetrabutyl ammonium salts).

Gradients are run by increasing the concentration of organic modifier (e.g., methanol or acetonitrile) in the mobile phase to achieve elution of solute molecules in order of increasing hydrophobicity.

Support   Non-polar Ligand   Bonded-Phase

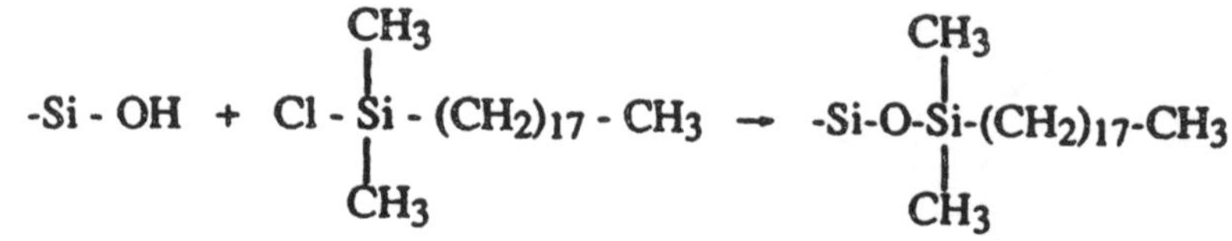

$$-Si-OH \;+\; Cl-\underset{\underset{CH_3}{|}}{\overset{\overset{CH_3}{|}}{Si}}-(CH_2)_{17}-CH_3 \;\rightarrow\; -Si-O-\underset{\underset{CH_3}{|}}{\overset{\overset{CH_3}{|}}{Si}}-(CH_2)_{17}-CH_3$$

Silica   Monochlorodimethyl-   C-18 or octadecyl
                    octadecyl silane   packing

FIGURE 3. Scheme outlining formation of silica-based C_{18} reversed-phase packing.

2. Hydrophobic Interaction Chromatography (HIC)

For the novice chromatographer, the terms "reversed-phase" and "hydrophobic interaction chromatography" can cause confusion. Both methods are partition chromatography involving a chemically bonded non-polar or hydrophobic phase with the mobile phase being more polar than the stationary phase. Both methods involve hydrophobic interactions between the solute and the stationary phase. What distinguishes these methods are:

1. The ligand density of the chemically bonded hydrophobic ligands on the stationary phase. In HIC the ligand density is approximately one-tenth of that used in a reversed-phase sorbent.
2. The mobile phase used in HIC is usually an aqueous solution of high salt concentration at neutral pH and the separation is effected by changing the salt concentration. Gradients are run by decreasing the salt concentration with time which increases the solubility of the solute in the mobile phase. This technique is somewhat analogous to "salting out" molecules from solution.
3. The method is generally used for the separation of proteins in their native state. The low hydrophobic ligand density and aqueous buffers at neutral pH tend to be non-denaturing, leaving proteins in their native states. Thus, the separation is based upon differences in the surface hydrophobicities of proteins in their native states.

By comparison, in RPC, the high hydrophobic ligand density and the use of low pH and organic modifiers tend to denature proteins and the separation is based on the overall hydrophobicity of the protein rather than surface hydrophobicity.

3. Hydrophilic Interaction Chromatography (HILIC)

This method of chromatography is a form of partition chromatography where the chemically bonded-phase is hydrophilic or polar and the mobile phase must be more non-polar than the stationary phase. This method of chromatography is identical to "normal-phase" chromatography but we prefer the term "hydrophilic interaction chromatography" which is more descriptive of the method and implies the opposite mobile and stationary phase to HIC and RPC. The chemically bonded hydrophilic sorbents are new and thus the use of these columns for the separation of peptides and proteins is just beginning[2] and will not be discussed further in this book.

In this technique, the elution order is generally related to the increasing hydrophilic nature of the solute, i.e., HILIC is complementary to RPC, where the elution order is related to the increasing hydrophobicity of the solute. The more soluble a solute is in water or the more hydrophilic the solute, the slower it will be eluted. In contrast, the more hydrophobic the solute, the faster it will be eluted. In general, the elution order will be the opposite to a reversed-phase or hydrophobic interaction separation.

The gradients are run by decreasing the concentration of organic modifier (e.g., acetonitrile) in the mobile phase to achieve elution of the solute molecules in order of increasing hydrophilicity. It should be noted that, as with RPC, HILIC may be employed with totally volatile mobile phases. The starting concentration of organic modifier is high (usually 70% or greater in water). High concentrations of a non-polar solvent will denature proteins and many proteins may be insoluble under such conditions, perhaps placing some limitations on the method for protein separations. However, the method may provide a useful alternative for the separation of molecules known to interact weakly or not at all with a reversed-phase sorbent. At present, it is too early to define the limitations of this method to the separation of peptides and proteins, although it does appear to be particularly useful for the separation of phosphorylated peptides.[2]

D. AFFINITY CHROMATOGRAPHY (HPAC)

High-performance affinity chromatography (HPAC) or high-performance immunoaffinity chromatography (HPIAC) are techniques where a biospecific adsorbent is prepared by coupling a specific ligand (such as an enzyme substrate, antigen or hormone) for the macromolecule of interest to a solid support. This immobilized ligand will interact only with molecules that can selectively bind to it. Molecules that will not bind are eluted unretained. The retained compound can later be released in a purified state.

II. LIST OF CHROMATOGRAPHIC TERMS

Alkyl chain length — Refers to the alkyl chain length of the chemically bonded-phase of a reversed-phase sorbent. For example, C_8 and C_{18} refer, respectively, to eight carbon (octyl) and eighteen carbon (octadecyl) n-alkyl chains covalently attached to the support.

Analysis time — Analysis time should include not only the time required to carry out separation of the desired components, but also the time required to elute longer retained components from a column plus the time required to reequilibrate the system prior to a subsequent run.

Analytical — Usually refers to analysis of a sample for identification of sample components by their chromatographic retention behavior and/or quantitation by integration of peak areas. The components generally are not recovered following their elution from an HPLC column (unlike **Preparative chromatography**). Analytical runs are almost universally carried out on **Microbore, Narrowbore, or Analytical columns**.

Analytical columns — Columns of 4 to 4.6 mm I.D. used in HPLC.

Asymmetry — Factor describing the shape of a chromatographic peak (Figure 4). Theory assumes a Gaussian shape and that peaks are symmetrical. The peak asymmetry factor is the ratio (at 10% of the peak height) of the distance between the peak apex and the back side of the chromatographic curve to the distance between the peak apex and the front side of the chromatographic curve. A value >1 is a tailing peak, while a value <1 is a fronting peak.

Band broadening — The dilution of the chromatographic band as it moves down the column. The peak is injected as a slug, and, if not for the process of band broadening, each separated component would be eluted as a narrow slug of pure compound. The measure of band broadening is band width, t_w, or more correctly, the number of theoretical plates in the column, N.

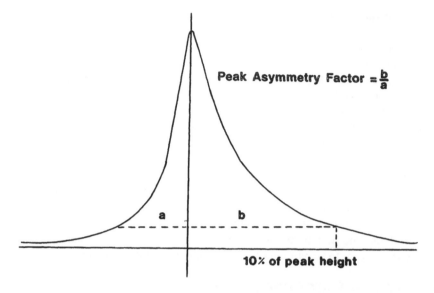

FIGURE 4. Determination of peak asymmetry.

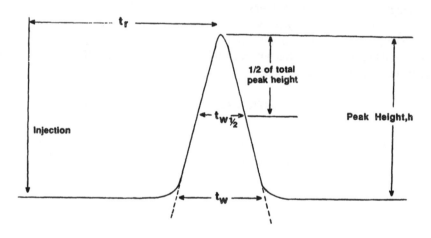

FIGURE 5. Various chromatographic peak parameters.

Bandspreading — See **Band broadening**.

Band width (t_w) — The width of the chromatographic band (in time units) during elution from a column (Figure 5). It is usually measured at the baseline by drawing tangents to the sides of the Gaussian curve representing the peak. Small band widths usually represent efficient separations. Also referred to as **Peak width**.

Band width at half height — The width of the chromatographic band (in time units), halfway between the apex of the peak and the baseline of the elution profile, during elution from a column (Figure 5).

Baseline drift — Baseline fluctuations, more prevalent in **Gradient elution** than **Isocratic elution** separations, due either to an imbalance in absorbance characteristics of the two (or more) eluents employed in **Gradient elution** or to UV-active or fluorescent impurities in the mobile phase.

Biocompatible — Term to indicate that the column or instrument component will not irreversibly or strongly adsorb or deactivate biomolecules such as proteins.

Blank run — A chromatographic run carried out in the absence of a sample.

Breakthrough volume — The volume at which a particular solute pumped continuously through a column will begin to be eluted. The breakthrough volume is useful in determining the total sample capacity of the column for a particular solute.

Capacity — See **Sample capacity**.

Capacity factor — A chromatographic parameter that measures the degree of retention. See k′ for calculation method.

Capillary tubing — Tubing for connecting various parts of the chromatograph. Most capillary tubing used in HPLC is <0.020 in. in internal diameter (I.D.). The smallest useful I.D. is about 0.004 in.

Cartridge column — A type of column that has no endfittings and is held in a cartridge holder. The column is a tube: the packing is contained by frits in each end of the tube. Cartridges are easy to change and are less expensive and more convenient than conventional columns with endfittings.

Chain length — The length of carbon chain in the hydrocarbon portion of a reversed-phase packing. It is expressed as the number of carbon atoms (e.g., C_8, C_{18}).

Channeling — Occurs when voids created in the packing material of a column may cause mobile phase and accompanying solutes to move more rapidly than the average flow velocity, resulting in band broadening. The voids are created by poor packing or by erosion of the packed bed.

Chromatogram — A plot of detector signal output vs. time or elution volume during the chromatographic process.

Column back pressure — The pressure above gravity at the head of the column. Expressed in pounds per square inch (psi), bar, atmospheres (atm), or megapascals (MPa; 1 Atm = 14.7 lb in^{-2} = 1.013 Bar = 0.1013 MPa.

Column chromatography — Any form of chromatography that uses a column or tube to hold the stationary phase.

Column performance — Refers to the efficiency of a column; measured as the number of theoretical plates (N) for a given test compound.

Contaminant — Any unwanted component in a sample mixture.

Counterion — In an ion-exchange process, the ion in solution used to displace the ion of interest from the ionic site. In ion-pairing, it is the ion of opposite charge added to the mobile phase to form a neutral ion pair in solution.

Coverage — Refers to the amount of bonded phase on a silica support in bonded-phase chromatography. Coverage is usually described in μmol/m^2 or in terms of %C.

Cross-contamination — Generally refers to the presence, in the solute peak(s) of interest, of contaminating solutes eluted adjacent to the desired compound(s).

Dead volume (V_d) — The volume outside of the column packing itself. The interstitial volume (intraparticle volume + interparticle volume) plus extracolumn volume (contributed by injector, detector, connecting tubing and endfittings) all combine to create the dead volume. This volume can be determined by injecting an inert compound (i.e., a compound that does not interact with the column packing). Also abbreviated V_o.

Degassing — The process of removing dissolved gas from the mobile phase before or during use. Dissolved gas may come out of solution in the detector cell and cause baseline spikes and noise. Dissolved air can affect electrochemical detectors (by reaction) or fluorescence detectors (by quenching). Degassing is carried out by heating the solvent or by vacuum (in a vacuum flask), or on-line using evacuation of a tube made from a gas-permeable substance such as PTFE, or by helium sparging (most widely used method).

Displacement chromatography — Chromatographic process in which the sample is placed onto the head of the column and is then displaced by a compound that is more strongly sorbed than the compounds of the original mixture. Sample molecules are displaced by each other and by the more strongly sorbed compound. The result is that the eluted sample solute zones may be sharpened; displacement techniques have been used mainly in preparative HPLC applications.

Distribution coefficient (D or K_D) — See **Partition coefficient**.

Eddy diffusion term — The contribution to plate height that is due to molecules travelling along different paths through the column; depends on the particle size and geometry of the packing.

Efficiency — See **N**.

Effluent — Same as **Eluate**.

Eluate — Combination of mobile phase and solute exiting column; also called effluent.

Eluent — Mobile phase used to carry out a separation.

Eluite — A solute which is being eluted from a column.

Elution — The process of passing mobile phase through the column to transport solutes.

Elution chromatography — The most commonly used chromatographic method. The sample is applied to the head of the column and individual molecules are separated and eluted at the end of the column.

Elution volume (V_r) — Refers to the volume of mobile phase required to elute a solute from the column. $V_r = F \times t_r$, where F is the flow-rate and t_r is the retention time for the solute.

Endcapping — A column is said to be endcapped when a small silylating agent (e.g., trimethyl-chlorosilane) is used to bond residual silanol groups on a packing surface. Most often used with reversed-phase packings. May cut down on undesirable adsorption of basic or ionic compounds.

Endfitting — The fitting at the end of the column that connects it to the injector or detector via capillary tubing. Most HPLC endfittings contain a frit to hold the packing and have a low dead volume for minimum band spreading. Usually made of stainless steel.

Exclusion limit — In SEC, the upper limit of molecular weight (or size), beyond which molecules will be eluted at the same retention volume, called the exclusion volume. Many SEC packings are referred to by their exclusion limit. For example, a 10^5 column of porous silica gel will exclude any compounds with a molecular weight higher than 100,000, based on a polystyrene calibration standard.

Exclusion volume (V_e) — The retention volume of a molecule on an SEC packing; all molecules larger than the size of the largest pore are totally excluded and are eluted at the interstitial volume of the column.

Extracolumn effects — The band-broadening effects of parts of the chromatographic system outside of the column itself. Extracolumn effects must be minimized in order to maintain the efficiency of the column. Areas of band broadening can include the injector, connecting tubing, endfittings, frits, detector cell volume and internal detector tubing. The variances of all of these contributions are additive.

Fast LC — The use in HPLC of short columns (3 to 7 cm in length) with conventional internal diameters (2 to 6 mm), packed with small particles (3- or 5-μm particle diameter). Separation times in minutes, sometimes seconds, are common.

Flow-rate (F) — The volumetric rate of flow of mobile phase through an LC column. For a conventional HPLC column of 4.6-mm internal diameter, typical flow rates are 1 to 2 ml/min.

Fractionation range — In SEC, refers to the operating range of a gel or packing. This is the range in which the packing can separate molecules based on their size. Molecules that are too large to diffuse into the pores are excluded. Molecules that can diffuse into all of the pores totally permeate the packing, eluting (unseparated) at the permeation volume.

Frit — The porous element at either end of a column that serves to contain the column packing. It is placed at the very ends of the column tube or, more commonly, in the endfittings. Frits are made from stainless steel or other inert metal or plastic, such as porous PTFE (polytetrafluoroethylene) or polypropylene.

Frontal analysis or frontal chromatography — Chromatographic technique that involves continuous addition of sample to the column with the result that only the least-sorbed compound,

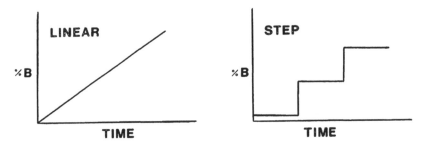

FIGURE 6. Gradient elution. Left: Linear (increasing) gradient. Right: stepwise (increasing) gradient.

which moves at the fastest rate, is obtained in a pure state. The second least-sorbed compound is eluted with the first-eluted compound, the third least-sorbed compound with the first and second compound, etc., until the original sample is eluted at the column exit. Frontal analysis is seldom used and is mainly a preparative technique.

Gaussian curve — A standard error curve, based on a mathematical function, that is a symmetrical, bell-shaped band or peak. Most chromatographic theory assumes a Gaussian peak.

Gradient delay time (t_d) — The time for the gradient to reach the top of the column from the proportioning valve via the pump, solvent mixer and injection loop. The time for the gradient to reach the detector (t_g)is the gradient delay time to the column top plus t_o, the time for an unretained peak to be eluted.

Gradient elution — Technique for decreasing separation time by increasing mobile phase strength (or decreasing mobile phase strength in HIC) over time during the chromatographic separation. Also known as solvent programming. Gradients can be continuous or stepwise (Figure 6). Binary, ternary, and quaternary solvent gradients have been used routinely in HPLC. Linear binary gradients are most common in peptide/protein separations.

Gradient-rate — Change in mobile phase composition with time, e.g., change in concentration of organic modifier with time (RPC), change in buffer ionic strength with time (IEC, HIC).

Guard column — A small column placed between the injector and the analytical column. Protects the analytical column against contamination by sample particulates and, perhaps, by strongly retained species. The guard column is usually packed with the same material as the analytical column and is often of the same I.D. It is much shorter, costs less, and is usually discarded when it becomes contaminated.

H — Same as **HETP**.

HETP — Height equivalent to a theoretical plate is the preferred measure of column efficiency because it allows a comparison between columns of different lengths.

$$H = HETP = \frac{L}{N}$$

where L is the length of the column, usually in mm, and N is the number of theoretical plates. For a typical HPLC column well-packed with 5-μm particles, HETP or H values are usually between 0.01 and 0.03 mm. The table below illustrates how length, H, and N values are related for an HPLC column packed with 13-μm particles.[1]

Column length (mm)	H (mm)	N
150	0.23	652
300	0.25	1200
500	0.24	2083

Hydrophilic — "Water-loving" refers both to stationary phases that are compatible with water and to water-soluble molecules in general. Most columns used to separate proteins are hydrophilic in nature (RPC and HIC packings are hydrophobic) and should not denature proteins in the aqueous environment.

Hydrophobic — "Water-hating" refers both to stationary phases that are not compatible with water and to molecules in general that have little affinity for water. Hydrophobic molecules have few polar functional groups; most are hydrocarbons or have high hydrocarbon content.

In-line filter — A device that prevents particulate matter from damaging the column. Modern in-line filters can be placed between the injector and the column without contributing to band broadening. A filter in this position is used to prevent sample particulates from entering the packed bed or the inlet frit.

Inlet — The initial part of the column, where the solvent and sample enter. There is usually an inlet frit that holds the packing in place and, in some cases, protects the packed bed.

Internal diameter — Columns are generally available only in a limited number of internal diameters (I.D.). 1 mm, 2 mm, and 4 mm I.D. columns are generally considered to be microbore, narrowbore and analytical columns, respectively. 10 mm I.D. columns are usually referred to as semi-preparative. Columns of 1″ I.D. or greater are referred to as preparative columns. Of course, all columns can be used in a preparative mode.

Interstitial volume (V_o) — The total volume of mobile phase within the length of the column. It is made up of the intraparticle volume (inside the packing itself) and interparticle volume (between the packing particles). Same as void volume. Also abbreviated V_i or V_m.

Ion-pair chromatography — Form of chromatography in which ions in solution can be "paired" or neutralized and separated as an ion pair on a reversed-phase column. Ion-pairing agents (or reagents) are usually ionic compounds that contain a hydrocarbon chain that imparts a certain hydrophobicity so that the ion pair can be retained on a reversed-phase column. Ion-pairing can also occur in normal-phase chromatography when one part of the pair is loaded onto a sorbent, but this technique is not as popular as the RPC technique.

Ion-pairing reagents — See **Ion-pair chromatography**.

Irregular packing — Refers to the shape of a silica gel-based packing. Irregular silicas are available in microparticulate sizes. The packings are made by grinding silica gel into small particles and then sizing them into narrow fractions using classification machinery. Spherical packings are now used more often than irregular packings in HPLC, but less-expensive irregular packings are still widely used in prep LC.

Isocratic — Use of a constant-composition mobile phase in liquid chromatography.

Isocratic elution — Elution of solutes from a column with a mobile phase of constant composition.

k′ — Capacity factor; permits normalization of retention behavior of a solute. Can be calculated from the equation where $k' = (t_r - t_o)/t_o$, where t_r is the retention time for the sample peak, and t_o is the elution time of an unretained component. See chapter by Hearn entitled "High-Performance Liquid Chromatography of Peptides and Proteins: General Principles and Basic Theory".[3]

Ligand — In ligand-exchange chromatography, refers to the molecule added to the mobile phase that acts as the chelating agent. In affinity chromatography, refers to the biospecific material (enzyme, antigen or hormone) coupled to the support (carrier) to form the affinity column.

Linear velocity (u) — The velocity of the mobile phase moving through the column. Expressed in cm/s. Related to flow rate by the cross-sectional area of the column. Sometimes expressed as v.

Loading — The amount of stationary phase coated or bonded onto a solid support. In liquid-liquid chromatography, the milligram amount of liquid phase per gram of packing. In RPC, the loading may be expressed in μmol/m² or in %C. See **Coverage**.

Macroporous resin — Cross-linked ion-exchange resins that have both micropores of molecular dimensions and macropores several hundred Å wide. These are highly porous resins with large internal surface areas accessible to large molecules.

Mass transfer — The process of solute movement into and out of the stationary phase or mobile phase. The faster the process of mass transfer, the better the efficiency of the column. In HPLC, mass transfer is the most important factor affecting column efficiency. It is increased by the use of small-particle packings, thin layers of stationary phase, low-viscosity mobile phases and high temperatures.

Mean pore diameter — The average pore diameter of the pore in a porous packing. The pore diameter is important in that it must allow free diffusion of solute molecules into and out of the pore so that the solute can interact with the stationary phase. In SEC, the packings have different pore diameters, and therefore molecules of different sizes can be separated. For a typical adsorbent such as silica gel, 60- and 100-Å pore diameters are most popular. For packings used for the separation of biomolecules by RPC, IEC, and HIC, pore diameters ≥300 Å are most often used.

Microbore columns — Refers to columns with smaller-than-usual internal diameters (<2 mm) used in HPLC.

Microparticulate — Refers to small particles used as HPLC stationary phases. Generally, packings with a particle diameter <10 μm that are totally porous are considered microparticles.

Microporous — Refers to porous resins containing pores which correspond to, or approach, molecular sizes of solutes. See Microreticular resin.

Microporous resin — Same as **Microreticular resin**.

Microreticular resin — Cross-linked synthetic ion-exchange resins with pore openings that correspond to molecular sizes. Diffusion into the narrow pores can be impaired, and low exchange rates, as well as poor performance, can occur, especially for large molecules.

Minimum plate height — The minimum of the curve that results from a plot of height equivalent to a theoretical plate (HETP or H) vs. linear velocity (u). This value represents the most theoretical plates that can be obtained for a certain column and mobile phase system. Usually occurs at very low flow rates.

Mobile phase — The solvent that moves the solute through the column.

Mobile phase strength — Refers to the strength of the mobile phase in terms of, e.g., organic modifier concentration in RPC, buffer ionic strength in IEC, HIC.

Mode of chromatography — There are many modes of chromatography. For example, size-exclusion, reversed-phase, hydrophobic interaction, hydrophilic interaction, ion-exchange, etc.

Modifier — Additive that changes the character of the mobile phase. For example, in reversed-phase chromatography, water is the weak solvent; acetonitrile, the strong solvent, is sometimes called the modifier.

Monomeric phase — Refers to a bonded phase in which single molecules are bonded to a support. For silica gel, monomeric phases are prepared by the reaction of an alkyl or aryl monochlorosilane. Polymeric phases are generally prepared from a di- or trichlorosilane reactant.

Multidimensional chromatography — The use of two or more columns or chromatographic techniques to effect a better separation. This may be useful for sample cleanup, increased resolution and increased throughput. Also called multicolumn chromatography.

Multimodal chromatography — Manifestation of more than one mode of separation on a particular class of sorbent. For instance, ion-exchange packings frequently exhibit non-specific hydrophobic interactions with solutes; size-exclusion packings tend to exhibit hydrophobic and anionic characteristics; silica-based reversed-phase columns may exhibit non-specific ionic interactions with solutes (particularly basic solutes) due to the presence of underivatized silanols which are negatively charged above pH values >4.0 to 4.5. These non-specific interactions can be suppressed through mobile phase manipulation to ensure "ideal" solute retention behavior,

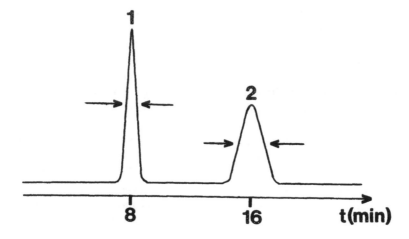

$$N_1 \; = \; 5.5\left(\dfrac{t_{r,1}}{t_{w\frac{1}{2},1}} \right)^2 \; = \; 5.5\left(\dfrac{8}{0.5} \right)^2 \; = \; 1400$$

$$N_2 \; = \; 5.5\left(\dfrac{t_{r,2}}{t_{w\frac{1}{2},2}} \right)^2 \; = \; 5.5\left(\dfrac{16}{1} \right)^2 \; = \; 1400$$

FIGURE 7. Effect of peak retention time on column efficiency calculation.

i.e., solute separation is effected through a single chromatography mode. However, the presence of non-specific interactions ("non-ideal" behavior) should not necessarily be dismissed as undesirable, since the manifestation of more than one separation mechanism by a particular packing may be advantageous for the resolution of specific sample mixtures.

N — The number of theoretical plates.

$$N = 16\left(\frac{t_r}{t_w} \right)^2$$

where t_r is retention time, and t_w is the peak width at baseline. A measure of the efficiency of a column. Sometimes measured as

$$N = 5.5\left(\frac{t_r}{t_{w1/2}} \right)^2$$

where $t_{w1/2}$ is the peak width at half height (Figure 7). The measurement of N can be in serious error if the peaks are very asymmetric. Normal practice is to treat chromatographic peaks as Gaussian in shape. Generally, the measurement involving peak width at half height has been found to be most useful since it can be carried out on peaks that are not completely resolved or display slight tailing. See chapter by Hearn entitled "High-Performance Liquid Chromatography of Peptides and Proteins; General Principles and Basic Theory".[3]

The value of N is a useful measure of the performance of the chromatographic column because to a first approximation, N is independent of retention time, i.e., the number of theoretical plates calculated for peak 1 emerging at x minutes will be the same as the value calculated for peak 2 emerging at y minutes ($N_1 \cong N_2$) for compounds of similar structure. This is the result of a proportional increase in peak width with increased retention time (Figure 7).

Narrow-bore columns — Columns of 2 to 3 mm I.D. used in HPLC.

Nonporous particle — Refers to a solid particle used as a support for a porous-coated or bonded phase.

Octadecylsilane (ODS) — The most popular reversed-phase packing in HPLC. Octadecylsilane phases are bonded to silica or polymeric packings. Both monomeric and polymeric phases are available.

Organic modifier — Water-miscible organic solvent added to an aqueous mobile phase to effect separations in reversed-phase HPLC.

Overload — In preparative chromatography, the overload condition is defined as the mass of sample injected onto the column at which efficiency and resolution begin to be affected if the sample size is further increased. See **Sample capacity**.

Packing — The adsorbent, gel, or solid support used in the HPLC column. Most HPLC packings are <10 μm in average diameter.

Particle size or particle diameter (d_p) — The average particle size of the packing in an LC column. A 5-μm column would be packed with particles having definite particle size distribution; packings are never monodisperse. See Particle-size distribution.

Particle-size distribution — A measure of the distribution of the particles used to pack the LC column. In HPLC, a narrow particle-size distribution is desirable. A particle-size distribution of $d_p \pm 10\%$ would mean that 90% of the particles fall between 9 and 11 μm for an average 10-μm d_p packing.

Partition chromatography — Separation process in which one of the liquid phases is held stationary on a solid support while the other is allowed to flow freely down the column. Solutes partition themselves between the two phases based on their individual partition coefficients. Liquid-liquid chromatography is an example.

Partition coefficient (K) — The amount of solute in the stationary phase relative to the amount of solute in the mobile phase. Can be the distribution coefficient, K_D.

Partitioning window — In RPC, the range of isocratic organic modifier concentration available to separate a desired component (baseline resolution) from its nearest impurities.

Peak shape — Describes the profile of a chromatographic peak. Theory assumes a Gaussian peak shape (perfectly symmetrical); peak asymmetry factor describes shape as a ratio. See **Asymmetry**.

Peak width (t_w) — Same as **Band width**.

Peak width at half height — Same as **Band width at half height**.

Pellicular — A pellicular particle is generally composed of an impervious, rigid central core with a thin, porous outer layer of porous sorbent. Pellicular packings have lower efficiencies and sample capacities than **Microporous packings**.

Plate height (H) — See **HETP**.

Plates — Refers to theoretical plates in a packed column. See **Theoretical plate**.

Polymeric packings — Packings based on polymeric materials, usually in the form of spherical beads. Common polymers used in LC are polystyrene-divinylbenzene, polyacrylamide, polymethylacrylate, polyethyleneoxide, polydextran, and polysaccharide.

Polystyrene-divinylbenzene resin (PDVB) — The most common polymer base for ion-exchange chromatography. Ionic groups are incorporated by various chemical reactions. Neutral PDVB beads are used in reversed-phase chromatography. Porosity and mechanical stability can be altered by varying the cross-linking through the variation of the divinylbenzene content.

Pore diameter — Same as **Mean pore diameter**.

Pore volume — The total volume of the pores in a porous packing; usually expressed in ml/g. Porosity. For a porous adsorbent, the ratio of the volume of the interstices to the volume of the solid particles. The pore volume is also used as a measure of porosity.

Post-column derivatization — Post-HPLC derivatization of amino acids or peptides for the purpose of detection.

Precolumn — A small column placed between the pump and the injector. It removes particulate matter that may be in the mobile phase, chemically sorbs substances that might interfere with the separation, or, as a saturator column, presaturates the mobile phase with stationary phase to prevent stripping of the column. Its volume has little effect on isocratic elution but contributes a delay to the gradient in gradient elution.

Precolumn derivatization — Pre-HPLC derivatization of amino acids or peptides for the purpose of detection and/or separation.

Preparative chromatography — The process of using liquid chromatography to isolate a sufficient amount of material for other experimental or functional purposes. For pharmaceutical or biotechnological purifications, columns several feet in diameter can be used for multiple grams of material. For isolating just a few micrograms of a valuable natural product, an analytical column can be used. Both are preparative chromatographic approaches.

Recovery — The amount of solute (sample) that is eluted from a column relative to the amount injected.

Reduced plate height (h) — Used to measure efficiencies of columns. A reversed-phase column with h value ≤ 6 is considered to be well packed. $h = H/dp$, where dp is the particle diameter and H is the height equivalent to a theoretical plate.

Regeneration — Returning the packing in the column to its initial state after gradient elution. Mobile phase is passed through the column stepwise or in a gradient. The stationary phase is solvated to its original condition. In ion-exchange chromatography, regeneration involves replacing ions taken up in the exchange process with the original ions that occupied the ion-exchange sites. Regeneration can also refer to bringing back any column to its original state (e.g., the removal of impurities with a strong solvent).

Reproducibility — Run-to-run consistency of an elution profile of an identical sample on the same column and under the same condition.

Residual silanols — The silanol (-Si-OH) groups that remain on the surface of a packing after a phase is chemically bonded onto its surface. These silanol groups may not be accessible to the reacting bulky organosilane (e.g., octadecyldimethylchlorosilane) but may be accessible to small polar compounds. They are often removed by endcapping with a small organosilane such as trimethylchlorosilane. See **Endcapping**.

Resin — A solid packing used in ion-exchange separations. The most popular resins are polystyrene-divinylbenzene copolymers of small particle size (<10 μm). Ionic functionalities are incorporated into the resin.

Resolution (R_s) — Ability of a column to separate chromatographic peaks. One attempts to achieve the best resolution possible. Resolution can be calculated in two ways:

$$(1) \qquad R_S = \frac{t_{r,2} - t_{r,1}}{\dfrac{t_{w,2} + t_{w,1}}{2}} = \frac{2\Delta t}{t_{w,2} + t_{w,1}}$$

where $t_{r,2}$ and $t_{r,1}$ are retention times of the retained components measured at the peak maximum and Δt is the difference between $t_{r,2}$ and $t_{r,1}$. The values of $t_{w,2}$ and $t_{w,1}$ are the peak widths in units of time measurement at the base as shown in Figure 8. If the peaks are eluted with similar retention times, their peak widths are usually nearly equal. In this case, it is possible to approximate R_s. Thus, if $t_{w,2} \cong t_{w,1}$ then

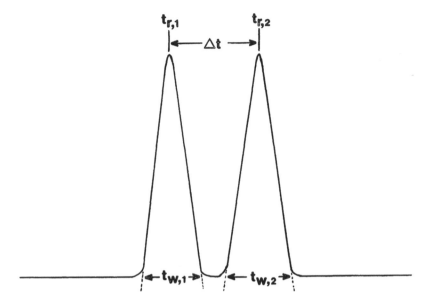

FIGURE 8. Determination of peak resolution.

$$R_s = \frac{\Delta t}{t_{w,2}} \text{ or } \frac{\Delta t}{t_{w,1}}$$

It is often more convenient to measure peak widths at half height. In this case,

$$R_s = \frac{1.176 \Delta t}{W_1 + W_2}$$

where Δt is the difference in retention time between two protein peaks (1 and 2) and W_1 and W_2 are their peak widths at half height. These equations are satisfied if the units of time and width are the same, such as minutes.

$$(2) \qquad R_s = \frac{1}{4}(N)^{1/2}(\alpha - 1)(1/1 + k')$$

See chapter by Hearn entitled "High-Performance Liquid Chromatography of Peptides and Proteins: General Principles and Basic Theory" for detailed discussion.[3] The value of resolution depends upon two factors: narrowness of the peak and the distance between peak maxima. Figure 9 illustrates how column efficiency and selectivity affect resolution. One way of improving resolution is to increase column efficiency which decreases the width of the peak. This is illustrated by comparing Figure 9A and 9B. Column efficiency is a function of such column parameters as mobile phase flow rate, particle size of the column packing and the viscosity of the solvent. Another way to improve resolution is to change column selectivity (compare Figure 9A and C). Column selectivity is a result of the interaction of the solute with the solvent and the column packing.

If the resolution is 0.6 or greater, the chromatographer should be able to identify clearly the number of components present. In general, a resolution of 1.0 or greater is required for good qualitative or quantitative work. This is especially true if the ratio of components deviates from 1:1 (Figure 10).

Retention time (t_r) — The time between injection and the appearance of the peak maximum.

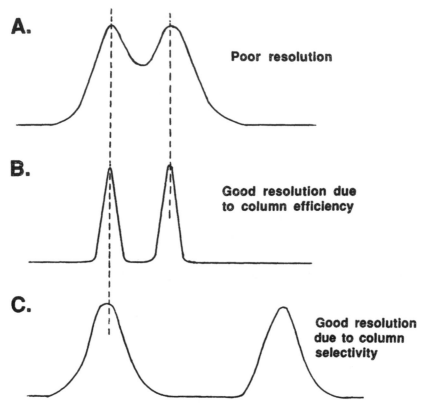

FIGURE 9. Comparison of column efficiency, selectivity and resolution.

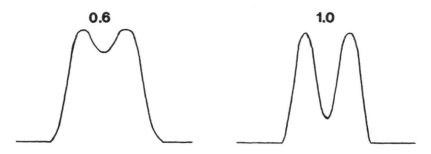

FIGURE 10. Illustration of relationship between observed peak separation and calculated resolution.

See chapter by Hearn entitled "High-Performance Liquid Chromatography of Peptides and Proteins: General Principles and Basic Theory".[3]

Retention volume (V_r) — The volume of mobile phase required to elute a substance from the column. $V_r = t_r$ (F) where F is the flow-rate. See chapter by Hearn entitled "High-Performance Liquid Chromatography of Peptides and Proteins: General Principles and Basic Theory".[3]

Sample capacity — Refers to the amount of sample that can be injected onto an LC column without overload. Often expressed as grams of sample per gram of packing. Overload is defined as the sample mass injected at which the column efficiency falls to 90% of its normal value.

Sample displacement chromatography — A preparative chromatographic process, employed to date only in RPC, related to **Displacement chromatography**, but without the requirement for a displacer compound unrelated to the solutes of interest. Sample molecules are displaced by each other and the major separation process takes place in water. See chapter by

Mant and Hodges entitled "Preparative Reversed-Phase Sample Displacement Chromatography of Synthetic Peptides"[4] for more details.

Sample recovery — Amount of desired (pure) product obtained from a sample mixture following preparative HPLC. Often expressed as a percentage of original sample load.

Saturator column — See **Precolumn.**

Selectivity (α) — The selectivity of a sorbent in all chromatographic modes can be defined as the relative separation achieved between adjacent solute peaks and will be reflected in the overall performance of the chromatographic system. Fixed by a certain stationary phase and mobile phase composition, the selectivity is given by the ratio of capacity factors for adjacent peaks. Thus,

$$\alpha = \frac{k'_1}{k'_2}$$

As the value of α approaches unity, selectivity declines. See chapter by Hearn entitled "High-Performance Liquid Chromatography of Peptides and Proteins: General Principles and Basic Theory" for a detailed discussion.[3]

Semipreparative chromatography — Refers to preparative liquid chromatography carried out on columns slightly larger than analytical size (4 to 5 mm I.D.), i.e., usually in the 6 to 10 mm I.D. range. Normal injection size is in the milligram to low-gram range. This is a somewhat subjective definition.

Sensitivity — The detector response for a given sample size.

Silanol — The Si-OH group found on the surface of silica gel. There are different strengths of silanols, depending on their location and relationship to each other. The strongest silanols are acidic and often lead to undesirable interactions with basic compounds during chromatography.

Silica gel — The most commonly used packing in liquid chromatography. It has an amorphous structure, is porous, and consists of siloxane and silanol groups. It is used as a bare packing for adsorption, as the support in liquid-liquid chromatography or for chemically bonded phases, and, with various pore sizes, as packing in size-exclusion chromatography. Microparticulate silicas of 5- and 10-μm average particle diameter are used in HPLC.

Siloxane — The Si-O-Si bond. A principal bond found in silica gel or for attachment of a silylated compound or bonded phase. Stable except at high pH values.

Silylation — The reaction of an organochloro- or organoalkoxysilane with a compound containing a reactive group. In liquid chromatography, it refers to the process of derivatizing the solute before chromatography in order to make it detectable or to prevent unwanted stationary phase interactions. It can also refer to the process of adding a chemically bonded phase to a solid support or to deactivating the packing to cut down on surface activity.

Slurry packing — The technique most often used to pack HPLC columns with microparticles. The packing is suspended in a slurry (10% wt/vol) and is rapidly pumped into the column. Special high-pressure pumps are used.

Solute — The dissolved component of a mixture that is to be separated in the chromatographic column.

Solvent strength — Refers to the ability of a solvent to elute a particular solute or compound from a column.

Sorbent — Refers to an adsorption packing used in liquid chromatography. A common sorbent is silica gel.

Spherical packing — Refers to spherical solid packing materials. Spherical packings are generally preferred over irregular particles.

Stationary phase — The immobile phase involved in the chromatographic process. The stationary phase in liquid chromatography can be a solid, a bonded or coated phase on a solid support, or a wall-coated phase. The stationary phase used often characterizes the LC mode. For example, silica gel is used in adsorption chromatography, an octadecylsilane bonded phase in reversed-phase chromatography, etc.

Step gradient — See **Gradient elution** and **Stepwise elution**.

Stepwise elution — Use of eluents of different compositions during the chromatographic run. These eluents are added in a stepwise manner with a pump, or by a selector valve. **Gradient elution** incorporates continuous changing of solvent composition. See Figure 6.

Support — Refers to solid particles. Support can be naked or coated or can have a chemically bonded phase in HPLC.

Surface area — In an adsorbent, refers to the total area of the solid surface. The surface area of a typical porous adsorbent such as silica gel can vary from 100 to 600 m^2/g. A standard 250 × 4.6 mm I.D. column requires approximately 3 g of packing.

Surface coverage — Usually refers to the mass of stationary phase per unit area of an LC support. Often expressed in μmol/m^2 of surface. Sometimes the %C is given as an indicator of surface coverage.

Swelling — Process in which resins and gels increase their volume because of their solvent environment. Solvent enters ion-exchange resin to dilute ions; in gels, solvent penetrates pores. If swelling occurs in packed columns, blockage or increased back pressure can occur. In addition, column efficiency can be affected.

Tailing — The phenomenon in which the normal Gaussian peak has an asymmetry factor >1. The peak will have skew in the trailing edge. Tailing is caused by sites on the packing that have a stronger-than-normal retention for the solute. A typical example of a tailing phenomenon is the strong adsorption of amines on the residual silanol groups of a low-coverage reversed-phase packing.

Theoretical plate — A concept described by Martin and Synge. Relates chromatographic separation to the theory of distillation. Measure of column efficiency (see **N**). Length of column relating to this concept is called height equivalent to a theoretical plate (HETP). See **HETP**.

t_0 — The elution time of an unretained component. See chapter by Hearn entitled "High-Performance Liquid Chromatography of Peptides and Proteins: General Principles and Basic Theory".[3]

Total permeation volume (V_p) — The retention volume on an SEC packing in which all molecules smaller than the smallest pore will be eluted. In other words, at V_p, all molecules totally permeate all of the pores and are eluted together.

Velocity (u) — Same as **Linear velocity**.

Void — The formation of a space, usually at the head of the column, caused by a settling or dissolution of the packing. A void in the column leads to decreased efficiency and loss of resolution. Even a small void can be disastrous for small microparticulate columns. The void can sometimes be removed by filling it with glass beads or porous packing.

Void time (t_m or t_0) — The time for elution of an unretained peak.

Void volume (V_I) — The total volume of mobile phase in the column; the remainder of the column is taken up by packing material. Can be determined by injecting an unretained substance that measures void volume plus extracolumn volume. Also referred to as interstitial volume. V_0 or V_m are sometimes used as symbols.

Wall effect — The consequence of the looser packing density near the walls of the rigid HPLC column. Mobile phase has a tendency to flow slightly faster near the wall because of the decreased permeability. The solute molecules that happen to be near the wall are carried along faster than the average of the solute band and, consequently, band spreading results.

The authors would like to acknowledge R.E. Majors as the source of much of this material.[5]

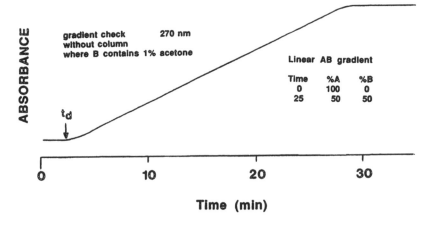

FIGURE 11. Monitoring of gradient linearity. Conditions: linear AB gradient (2% B/min), where eluent A is water and eluent B is 1% aq. acetone; flow-rate, 1 ml/min; detection at 270 nm. The term t_d denotes gradient delay time. There is no separating column present.

III. EVALUATION OF INSTRUMENT AND COLUMN PERFORMANCE

A. GRADIENT LINEARITY

Linear gradient elution is the elution mode of choice for the majority of IEC, RPC, and HIC applications in HPLC of peptides and proteins: IEC and RPC require linear increases in salt (ionic strength) or organic solvent concentration, respectively, while HIC requires a linear decrease in salt concentration. It is important to check periodically the gradient-making capabilities of the HPLC instrumentation, since consistent run-to-run reproducibility will only be maintained if the gradient remains linear. In addition, a potentially satisfactory peptide or protein separation may not be obtained if the supposed linear gradient is, in fact, unpredictably non-linear.

The technique for checking whether a particular HPLC pump is delivering linear gradients is quite straightforward. A linear AB gradient is run, where eluent B has some strong absorbance characteristic, e.g., acetone absorbs strongly at 270 nm. As the concentration of acetone increases with time, the absorbance of 270 nm should increase in a linear manner if eluents A and B are being mixed properly. The results of such a gradient check are shown in Figure 11, where eluent A is water and eluent B is 1% aq. acetone (2% B/min). Note that the presence of a column is not required during this gradient monitoring. It is clear that this particular HPLC instrument is delivering a gradient of very satisfactory linearity. The point at which the baseline starts to rise, t_d, is the gradient delay time of the HPLC instrument and is discussed below.

The gradient check may also be carried out in the presence of the mobile phase employed for a particular HPLC mode of interest. For instance, if RPC is to be employed, the gradient check can be carried out with 0.1% aq. TFA as eluent A and 1% acetone in 0.1% TFA/acetonitrile. This ensures that the gradient check is valid even in the presence of the mobile phase of the desired HPLC mode. If desired, the separating column may be present during this gradient monitoring; however, the linearity at the start of the gradient may be affected by adsorption of acetone to the stationary phase.

B. DETERMINATION OF t_o AND V_o

The term t_o denotes the time for unretained compounds to travel from the injector loop to the detector, via the separating column. To determine t_o the column is equilibrated with the starting eluent particular to the HPLC mode employed. When a stable baseline is observed, a TFA

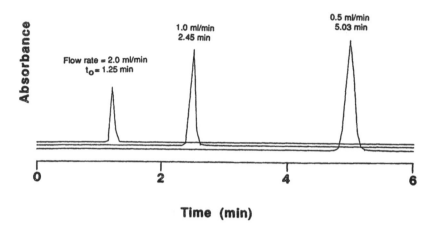

FIGURE 12. Determination of the elution time of an unretained peak, t_o Column: Conditions: isocratic elution with 0.1% aq. TFA at flow-rates of 0.5,1.0 or 2.0 ml/min. A 10-μl volume of an aqueous TFA solution (1 μl TFA/100 μl water) is injected when the baseline is steady and the resulting TFA peak is detected at 210 nm.

solution is injected (1 μl TFA/100 μl water; inject 10 μl). The apex of the observed TFA peak, detected at 210 nm, equals t_o. If desired, uracil provides a convenient substitute for TFA in this procedure in RPC.

As shown in Figure 12, t_o is dependent on flow-rate. The results shown in this figure were obtained on an analytical C_8 reversed-phase column, with 0.1% aq. TFA as the starting eluent. The interstitial column volume, V_o, is determined from t_o and the flow-rate (F) employed to determine t_o. Thus:

$$V_o = t_o \cdot F$$

C. DETERMINATION OF GRADIENT DELAY TIME, t_d

The gradient delay time is the time for the gradient to reach the top of the column from the proportioning valve via the pump, solvent mixer, solvent lines, injection loop, and any in-line filters. Knowledge of this value enables the researcher to calculate the percentage of eluent B in an AB gradient required to elute a particular sample component. It also enables the researcher to determine whether a solute is eluted prior to the start of the gradient, i.e., isocratic elution with the starting eluent.

The value of t_d may be determined from the point at which the gradient starts to rise in a gradient linearity check (in the absence of a column) (Figure 11).

Another, perhaps more accurate, method for determining t_d is illustrated in Figure 13. In the absence of any column, the HPLC system is equilibrated with aqueous TFA (0.05% aq. TFA in this case) (eluent A). When the baseline is steady, the mobile phase is switched from 100% eluent A to 50% eluent A and 50% eluent B (0.05% TFA in acetonitrile) at time = 0 min. With detection at 210 nm, there is a sudden increase in absorbance as acetonitrile, which absorbs at 210 nm, enters the detector. An S-curve is generated (rather than a vertical line) and t_d is determined from the mid-point of this curve. Figure 13 shows the S-curves obtained when carrying out this procedure on two different HPLC instruments. The instrument which produced the left-hand profile had a value for t_d of only 0.7 min; in contrast, the instrument which produced the shallow right-hand profile had a t_d value of 3 min. The shallow S-curve is a result of a large volume mixer in this instrument.

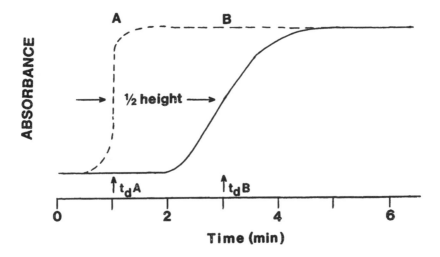

FIGURE 13. Determination of gradient delay time, t_d. Conditions: an initial isocratic elution at a flow-rate of 1 ml/min with 0.05% aq. TFA (eluent A); when the baseline is steady, the mobile phase is switched from 100% eluent A to 50% eluent A and 50% eluent B (0.05% TFA in acetonitrile); detection at 210 nm. The value for t_d is determined by the midpoint of the resulting S-curve. The two profiles are obtained from two different HPLC systems; $t_d = 0.7$ min and 3.0 min for the left- and right-hand profiles, respectively. There is no separating column present.

D. DETERMINATION OF COLUMN EFFICIENCY

The efficiency of a column is a measure of band broadening as a component travels down the column during elution. Efficiency is expressed by the equation:

$$N = 5.5\left(\frac{t_r}{W\frac{1}{2}}\right)^2$$

where the value for N is called the number of theoretical plates of the column, t_r is the retention time of the component and $W\frac{1}{2}$ is the width of the peak (minutes) at half height (Figure 7). The higher the value of N, the greater the efficiency of the column, and the narrower the component peak.

This laboratory regularly employs anthracene as a test solute to determine column efficiency.[6] The column is equilibrated with aqueous acetonitrile (usually 35 to 45% aq. acetonitrile, v/v) and the injected sample of anthracene is then eluted isocratically at a flow-rate of 1 or 2 ml/min (analytical or semi-preparative column, respectively). The anthracene peak is detected at 254 nm. For a given column internal diameter, the organic solvent strength (% acetonitrile) employed during isocratic elution depends on the retentive power and/or length of the column. For the most accurate determinations of column efficiency, it is important that the anthracene peak is not eluted too early or extremely late (k' values between 2 and 15 are appropriate). The capacity factor, $k' = t_r - t_o/t_o$ where t_o is the elution time of an unretained component.

An organic molecule such as anthracene also provides a useful probe of packing irregularities in a column. A skewed anthracene peak is a good indicator of a badly packed column.[7] Occasionally, such skewing may also be due to the column packing containing a high degree of adsorbed materials. This peak skewing may disappear following a rigorous column cleaning procedure.[7]

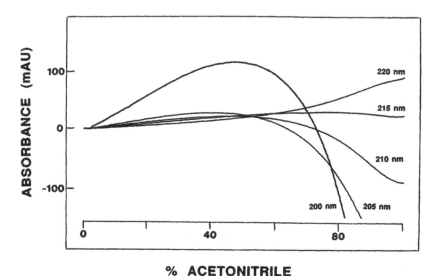

% ACETONITRILE

FIGURE 14. Baseline drift during linear AB gradient elution with RPC mobile phase eluents. HPLC instrument: Hewlett-Packard (Avondale, PA) HP1090 liquid chromatograph coupled to an HP1040A diode array detector, HP9000 Series 300 computer, HP9133 disc drive, HP2225A Thinkjet printer and HP7440A plotter. Conditions: linear AB gradient (1% B/min), where eluent A is 0.1% aq. TFA and eluent B is 0.1% TFA in acetonitrile; flow-rate, 1 ml/min; detection at 200, 205, 210, 215, and 220 nm. The baselines were recorded in the presence of an in-line filter, but in the absence of a separating column.

E. MAINTAINING FLAT BASELINES IN GRADIENT ELUTION RPC EMPLOYING TFA/ACETONITRILE MOBILE PHASES

The advantages of aqueous TFA/acetonitrile mobile phases for protein and, particularly, peptide separations are well documented.[8,9] Indeed, the powerful resolving capabilities of this mobile phase system have resulted in its being employed for the majority of peptide separations carried out by RPC. Since peptides and proteins generally show a high absorbance below 225 nm, due to peptide bond absorbance, whereas TFA does not, it might be expected that a high sensitivity of peptide and protein detection would be achieved when TFA is used in the mobile phase. This is, indeed, the case when employing TFA in isocratic systems. However, as Winkler has shown,[10-12] and as illustrated in Figure 14, the detector's sensitivity can be severely impaired by a strong baseline drift, the extent and type of drift depending on the wavelength employed for detection. The baseline profiles shown in Figure 14 were obtained by monitoring a linear AB gradient (1% B/min, where eluent A is 0.1% aq. TFA and eluent B is 0.1% TFA in acetonitrile) at various wavelengths in the absence of a separating column. In agreement with the results of Winkler,[10-12] baseline drift is minimized at a detection wavelength of 215 nm. At a higher wavelength, there is an ascending baseline; at shorter wavelengths, the baseline ascends and then changes direction and descends (this is particularly obvious at 200 nm).

Baseline drift can be a problem where only small amounts of sample are available and, hence, high detection sensitivity is required. Although detection at 215 nm produces the least baseline drift, this wavelength is less sensitive to peptide bond absorbance than lower wavelengths. As demonstrated by Burke et al. elsewhere in this book,[13] there is an approximate 4-fold decrease in detection sensitivity on each 10-nm increase in detection wavelength.

A common method of attempting to resolve the problem of baseline drift is to use inequivalent concentrations of TFA in eluent A (water) and eluent B (acetonitrile), i.e., lower the level of TFA in the higher absorbing organic solvent. Figure 15 illustrates the effect of manipulating the TFA concentration in eluent B on the baseline profiles obtained during linear AB gradient elution. The concentration of TFA was held constant (0.1%) in eluent A; eluent B contained 0.1, 0.08

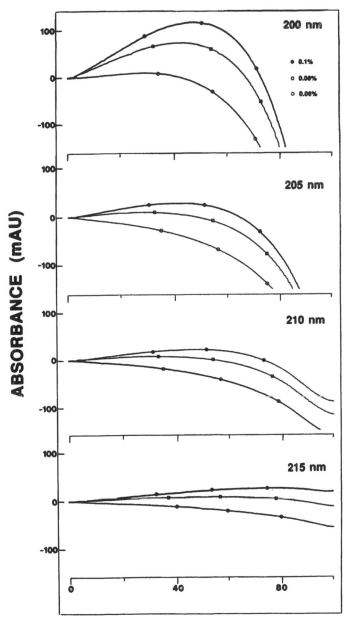

FIGURE 15. Effect on baseline profiles of dissimilar TFA concentrations in RPC mobile phase eluents. HPLC instrument: same as Figure 14. Conditions: linear AB gradient (1% B/min), where eluent A is 0.1% aq. TFA and eluent B is 0.06% (○), 0.08% (□) or 0.1% (●) TFA in acetonitrile; flow-rate, 1 ml/min; detection at 200, 205, 210, and 215 nm. The baselines were recorded in the presence of an in-line filter but in the absence of a separating column.

or 0.06% TFA. The optimum baseline profiles at different detection wavelengths (in terms of the flattest baseline over the greatest range of acetonitrile concentration in the gradient) were obtained at 0.06% TFA (200 nm) or 0.08% TFA (205, 210, and 215 nm) in eluent B. The baseline with least drift was clearly still obtained at a detection wavelength of 215 nm, and is almost flat

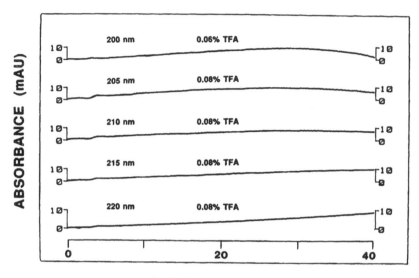

FIGURE 16. Maintenance of steady RPC baseline profiles at different wavelengths. HPLC instrument: same as Figure 14. Conditions: linear AB gradient (1% B/min) up to a concentration of 40% B, where eluent A is 0.1% aq. TFA and eluent B is 0.06% (200 nm) or 0.08% TFA (205, 210, 215, and 220 nm) in acetonitrile; flow-rate, 1 ml/min. The baselines were recorded in the presence of an in-line filter but in the absence of a separating column.

over the entire acetonitrile concentration range (0 to 100%). Thus, if the whole range of the gradient must be exploited, as for the elution of hydrophobic peptides or proteins, 215 nm represents a useful detection wavelength. However, a closer examination of the optimum baseline profiles shown in Figure 15 reveals that, even at detection wavelengths below 215 nm, a very respectable baseline may still be obtained over a wide acetonitrile concentration range, i.e., about 0 to 40% (200 nm), 0 to 50% (205 nm) or 0 to 60% (210 nm). This is more apparent in Figure 16 which shows the baselines obtained over a 40% acetonitrile concentration range for detection wavelengths of 200 to 220 nm. The baseline drift for all five detection wavelengths shown in Figure 16 never strayed beyond a range of 10 mAU. The majority of peptides are eluted from reversed-phase columns prior to a 50% acetonitrile concentration and, in addition, optimum peptide resolution is generally obtained between a 15 to 40% concentration of acetonitrile in the mobile phase.[14] Thus, good baselines at sensitive detection wavelengths can generally be obtained for most peptide applications if some care is taken during preparation of the mobile phase. This is well illustrated in Figure 17, which shows linear AB gradient (1% B/min) elution profiles of a mixture of five synthetic decapeptide reversed-phase standards using detection wavelengths of 200 or 215 nm. Eluent A was 0.1% aq. TFA and eluent B was 0.06% TFA (200 nm) or 0.08% TFA (215 nm) in acetonitrile. These peptides were all eluted prior to a 25% concentration of acetonitrile in the gradient, and baselines at both wavelengths were perfectly acceptable over this concentration range. However, the sensitivity of detection of the peptides at the lower wavelength is clearly much superior.

ACKNOWLEDGMENTS

This work was supported by the Medical Research Council of Canada and by equipment grants from the Alberta Heritage Foundation for Medical Research.

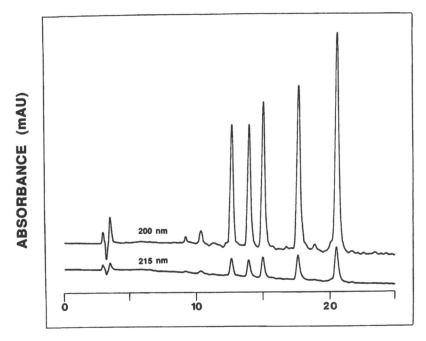

% ACETONITRILE

FIGURE 17. Effect of detection wavelength on sensitivity of peptide detection in RPC. HPLC instrument: same as Figure 14. Column: Aquapore RP300 C$_8$, 100 × 4.6 mm I.D., 7-μm particle size, 300-Å pore size (Brownlee Labs., CA). Conditions: linear AB gradient (1% B/min), where eluent A is 0.1% aq. TFA and eluent B is 0.06% (200 nm) or 0.08% TFA (215 nm) in acetonitrile; flow-rate, 1 ml/min. Sample: mixture of five synthetic decapeptide reversed-phase standards (obtained from Synthetic Peptides Incorporated, Department of Biochemistry, University of Alberta, Edmonton, Alberta, Canada).

REFERENCES

1. Johnson, E.L. and Stevenson, R., *Basic Liquid Chromatography*, Varian Associates, Walnut Creek, CA, 1978.
2. Alpert, A.J., Hydrophilic-interaction chromatography for the separation of peptides, nucleic acids and other polar compounds, *J. Chromatogr.*, 499, 177, 1990.
3. Hearn, M.T.W., High-performance liquid chromatography of peptides and proteins: general principles and basic theory, this publication.
4. Mant, C.T. and Hodges, R.S., Preparative reversed-phase sample displacement chromatography of synthetic peptides, this publication.
5. Majors, R.E., Glossary of liquid chromatography column terms, *LC.GC*, 6, 94, 1988.
6. Mant, C.T. and Hodges, R.S., On-line derivatization of silica supports for regeneration and preparation of reversed-phase columns, *J. Chromatogr.*, 409, 155, 1987.
7. Mant, C.T. and Hodges, R.S., Mobile phase preparation and column maintenance, this publication.
8. Regnier, F.E., High-performance liquid chromatography of proteins, *Methods Enzymol.*, 91, 137, 1983.
9. Mant, C.T. and Hodges, R.S., HPLC of peptides, in *High-Performance Liquid Chromatography of Biological Macromolecules: Methods and Applications*, Gooding, K. and Regnier, F., Eds., Marcel Dekker, New York, 1989, p. 301.
10. Winkler, G., Wolschann, P., Briza, P., Heinz, F.X., and Kunz, C., Spectral properties of trifluoroacetic acid-acetonitrile gradient systems for separation of picomole quantities of peptides by reversed-phase high-performance liquid chromatography, *J. Chromatogr.*, 347, 83, 1985.

11. **Winkler, G., Briza, P., and Kunz, C.,** Spectral properties of some ion-pairing reagents commonly used in reversed-phase high-performance liquid chromatography of proteins and peptides in acetonitrile gradient systems, *J. Chromatogr.*, 361, 191, 1986.
12. **Winkler, G.,** Increasing the sensitivity of UV detection in protein and peptide separations when using TFA-acetonitrile gradients, *LC.GC.*, 5, 1044, 1987.
13. **Burke, T.W.L., Mant, C.T., and Hodges, R.S.,** The effect of varying flow-rate, gradient-rate and detection wavelength on peptide elution profiles in reversed-phase chromatography, this publication.
14. **Hermodson, M. and Mahoney, W.C.,** Separation of peptides by reversed-phase high-performance liquid chromatography, *Methods Enzymol.*, 91, 352, 1983.

HIGH-PERFORMANCE LIQUID CHROMATOGRAPHY OF PEPTIDES AND PROTEINS: GENERAL PRINCIPLES AND BASIC THEORY

Milton T.W. Hearn

I. INTRODUCTION

The past decade has witnessed a phenomenal growth in the application of rapid, high resolution liquid chromatographic methods in peptide and protein chemistry. These techniques, which are now central to all areas of this expanding scientific field, are better known by their acronym HPLC or high-performance liquid chromatography. What sets HPLC apart from the early, more classical aspects of liquid chromatography in its various modes of separation selectivity has been the recent ability of the research or industrial practitioner to achieve *high resolution* separations of peptides and proteins with *short separation times* with medium to high sample throughputs. In common with classical liquid chromatography, the different modes of HPLC are classified according to the type of *stationary phase or sorbent* and the type of *mobile phase* or *eluent*. Knowledge of the properties of the stationary phase and mobile phase defines, in general terms, the nature of the retention mechanisms which control the separation. The different modes of HPLC separation encompass size-exclusion or gel-permeation (SEC, GPC) chromatography and the adsorption chromatographic modes which include hydrophobic interaction (HIC), reversed-phase (RPC), ion-exchange (IEC), metal chelate (MIC), biomimetic affinity (BMC), biospecific affinity (BAC) chromatography, and other group specific forms of adsorption behavior. Biomimetic affinity chromatography (BMC) can be distinguished from biospecific affinity chromatography (BAC) insofar that the former technique involves the serendipitous interaction of a chemical ligand which mimics a biological process, while the latter technique is based on biorecognition phenomena which occur *in vivo*. Sequential use of two or more of these different separation modes gives rise to the application areas known as *multidimensional HPLC*, while manifestation of more than one mode of separation with a particular class of sorbent is known as *multimodal HPLC*.

The development during the 1970s of reliable and rapid liquid chromatographic techniques and highly reproducible instrumentation, which can be used with direct or indirect high sensitivity detection methods, has consequently met a number of pressing expectations not previously attainable with the classical soft polysaccharide gels such as the crosslinked, derivatized dextrans and other mechanically compressible chromatographic media. This chapter is concerned specifically with one aspect of these developments, namely general considerations of the basic theory which underpin the successes of HPLC in peptide protein chemistry as evident today.

II. GENERAL CHROMATOGRAPHIC CONSIDERATIONS

The separation of a mixture of polar, ionizable substances such as peptides and proteins in a chromatographic column is the result of two events that occur within the column. The first event, which controls the average solute retention, is embodied in the concept of mass distribution along a chromatographic bed (or column) of length L, operated at a mobile phase flow-rate, F. As the solutes move down the column, individual components interact with the mobile phase and stationary phase to different extents. When the interaction of a specific solute with the stationary phase is very strong, that is when the *equilibrium distribution coefficient*, K, is large, then that solute will be retained to a greater extent than another component which interacts less strongly. Solute zones corresponding to each component will therefore migrate through the column at *different* velocities. This differential migration is thus a function of the equilibrium distribution *coefficient* established by the solutes between the stationary and mobile phase, the effective diffusivities of the solutes, their linear flow (or so-called superficial) velocities and related physicochemical properties.

In the case of size-exclusion HPLC (SEC), this differential migration arises as a consequence of the extent to which the solutes can permeate, by diffusion from the bulk mobile phase, to within the pore chambers of the stationary phase. Ideally, the stationary phase in SEC has been so prepared to not itself retard the transport of the solutes through chemical equilibrium interactions at its surface. In contrast, in the adsorption modes of HPLC, the surface of the stationary phase has been chemically modified to allow selective retardation of the solutes. Ideally, the sorbent surface in adsorption HPLC permits separation by only one retention process. In practice, this is rarely achieved with the consequence that most adsorption HPLC sorbents exhibit multimodal characteristics. These differences in retention which constitute the basis of selectivity of the chromatographic system will thus depend on the nature and magnitude of the intermolecular forces established between the solutes, the stationary phase, and the mobile phase. In addition, retention behavior will be affected by the hydrodynamic characteristics and fluid dynamic properties of the chromatographic system and the solute components.

The second event involved in solute migration is associated with the broadening of all solute peaks or zones. This dispersion occurs in both static and flowing liquid systems and is controlled, *inter alia*, by the respective diffusivities of the solutes. The concept of zone dispersion thus incorporates all the kinetic processes associated with the adsorption-desorption and the motion of the solutes through the pore and interstitial spaces within and around the stationary phase. As the individual solute zones move down the column, a number of dispersive effects come into effect as a consequence of the nonhomogeneity of the column bed, nonlinear flow characteristics, diffusion, and inappropriate kinetics of solute distribution between the mobile and the stationary phase. These dispersion effects collectively give rise to zone broadening which progressively increases throughout the period the solute spends traversing the chromatographic bed.

III. RETENTION RELATIONSHIPS

The time taken for a solute to completely pass through a chromatographic bed is called the *retention time*, t_r. As illustrated in Figure 1, the retention time of the solute, t_r, is measured as the time taken for the solute to move from one end of the column following injection, to emerge from the other end of the column and be immediately detected as an eluted zone with a peak maximum. Typically, the elution of the solute zone is revealed with spectrophotometers operating at an absorbance maximum. In order to enable comparisons to be made between columns of different dimensions or selectivities, the retention time, t_r, of a solute is usually compared with reference to the time taken for a different solute (or solvent) molecule to move through the column bed *without* any interaction, i.e., compared to the time taken for an inert component to be eluted through the void volume of the column, t_o. This comparison permits normalization of retention behavior of a solute in the form of a (unitless) *capacity factor*, k′ such that

$$k' = (t_r - t_o)/t_o \qquad (1)$$

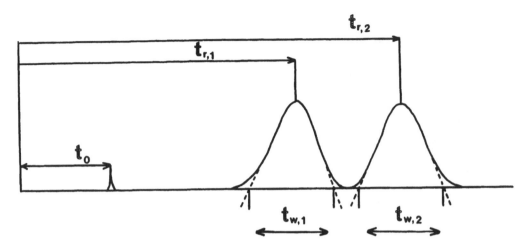

FIGURE 1. Representative chromatogram illustrating the relationship between the column void time, t_o, retention times for two components, $t_{r,1}$ and $t_{r,2}$ respectively, and the peak widths (in time units) at 4σ for each component, $t_{w,1}$ and $t_{w,2}$. From these data can be calculated the two fundamental chromatographic parameters: the capacity factor, k', and the number of theoretical plates (a measure of column efficiency), N, using Equations 1 and 15, respectively. The selectivity, α, and resolution, R_s, achieved between adjacent components can then be derived from k' and N using Equations 6 and 22, respectively.

or

$$t_r = t_o(1+k')$$ (2)

Alternatively, the capacity factor can be expressed in terms of elution volumes since the retention times, t_r and t_o, are related to the elution volumes and flow rate (F) of the chromatographic system through the relationships

$$V_r = t_r(F), \quad V_o = t_o(F)$$ (3)

hence

$$k' = (V_r - V_o)/V_o$$ (4)

or

$$V_r = V_o(1+k')$$ (5)

The significance of the capacity factor, k', in defining a particular chromatographic separation process is now extensively documented in the literature both in empirical as well as thermodynamic terms. For example, *selectivity*, α, in all chromatographic modes can be defined as the relative separation achieved between adjacent solute peaks and will be reflected in the overall performance of the chromatographic system. In particular, selectivity, α, is given by the ratio of capacity factors for adjacent peaks, i.e.

$$\alpha = k'_1/k'_2$$ (6)

With reference to Figure 1, it can be seen that selectivity of the two components in this example is ideal, i.e., there is base line separation between the two components. However, a significant period of separation time is lost in this separation insofar that the first component requires a substantial time to be eluted after the time void, t_0. The difference between $t_{r,1}$ and t_0, i.e., ($t_{r,1}$ − t_0), thus represents underutilized separation performance. One of the most important decisions an investigator can make with regard to a particular chromatographic system is the choice of the k′ range over which the separation will be complete. Clearly, if the data shown in Figure 1 were to represent a practical situation, then the first action of an investigator would be to examine chromatographic conditions which provide a smaller value of $t_{r,1}$ (or k′$_1$) without affecting selectivity. From practical considerations, the range 0<k′<10 is recommended for all chromatographic separations with peptides and proteins. As the value of α approaches unity (i.e., as α → 1), selectivity declines. An equally important task of selecting chromatographic conditions thus centers on the choice of α values which must be achieved for a particular set of k′ values.

The capacity factor, k′, can also be defined as the ratio n_s/n_m, where n_s is the total number of moles of the solute in the stationary phase and n_m the total number of moles of solute in the mobile phase, i.e.

$$k' = n_s / n_m \qquad (7)$$

or alternatively

$$k' = \frac{[X]_s V_s}{[X]_m V_m} \qquad (8)$$

where $[X]_s$, $[X]_m$ refer to the concentrations (in moles/liter) of the solute in the stationary and mobile phase, respectively, and V_s, V_m refer to the volumes of the stationary phase and mobile phase, respectively. The ratio $[X]_s / [X]_m$ is the *equilibrium distribution coefficient*, K, while the ratio V_s/V_m defines the phase ratio, Φ, of the chromatographic system. Hence, the capacity factor can also take the form:

$$k' = \Phi[X]_s / [X]_m \qquad (9)$$

or

$$k' = \Phi K \qquad (10)$$

Knowledge of the phase ratio, Φ, thus permits equilibrium distribution coefficients to be calculated directly from retention data. Conversely, for defined values of k′ and K, stationary phases of sorbents and columns can be selected with the desired characteristics to provide chromatographic systems with the calculated phase ratio. Research over the past decade has focused on the development of a very large number of different elution conditions and types of stationary phases for peptide and protein separations in attempts to maximize column selectivities.

Because of their atomic composition and molecular structures, peptides and proteins will invariably be retained by the HPLC sorbents through the participation of composite electrostatic, hydrogen bonding, van der Waals and hydrophobic interactions as well as the hydrodynamic behaviour of the solute in the distribution phenomenon. The nature of the predominant distribution mechanism in the various modes of HPLC selectivity will be dependent on the physical and chemical characteristics of the stationary phase as well as the nature of the molecular forces which hold the solute molecules within the mobile and stationary zones. High affinity of the peptide or protein solute for the stationary phase will be manifested as large values of K and hence large k′ values, whereas for an ideal size-exclusion mode with no electrostatic or hydrophobic interactions, K values will range between 0 and 1. In order to permit an operational capacity factor range of 0<k′<10 to be achieved with a particular sorbent, the mobile phase composition must be carefully selected. This can be achieved by either selecting a mobile phase of fixed composition (called an *isocratic mobile phase*) with appropriate elutropic strength to enable all components to be eluted, or by using a mobile phase of changing composition (called a *gradient mobile phase*).

Since the equilibrium distribution coefficient, K, is related through the Gibbs equation to the overall energy change, $\Delta G°$, for the separation event, then the capacity factor also takes on a fundamental thermodynamic complexion through the dependencies:

$$\Delta G° = -RT \ln K \tag{11}$$

$$k' = \Phi K \text{ from equation} \tag{10}$$

and hence,

$$\ln k' = \ln \Phi - \frac{\Delta G°}{RT} \tag{12}$$

where R is the gas constant and T the absolute temperature. Differences in the molecular characteristics and interactive behavior of peptide and protein solutes can thus be revealed in chromatographic separation, either through differences in unitary free energies of the solutes, or as changes in selectivity. Selectivity and free energy changes are linked through Equations 6 and 12 to give the relationship

$$\log \alpha = \frac{\Delta(\Delta G°)}{RT} \tag{13}$$

The term $\log \alpha$ (also known as the group coefficient, r) can be considered to represent a functional (or interactive) group contribution term, analogous to other structural or substituent terms such as the ΔR_m term used in thin layer chromatography (TLC), or the Hansch II term and the Hammett P electronic term used in structure-function dependencies. The dependency of $\log \alpha$ (or r) on $\Delta\Delta G°$ also forms the basis of several algorithmic methods now available to predict peptide and protein retention in interactive HPLC.

Full extension of the thermodynamic consequences of these theoretical dependencies involving the capacity factor are outside the scope of this chapter. It should, however, be noted here that these dependencies also form the basis of detailed, quantitative studies on solvent-peptide/protein interactions, conformational analysis of peptides and proteins in chromatographic systems and the design of novel chromatographic media currently underway in a variety of research laboratories. The interested reader can find further information on these aspects in References 1 to 11 as well as in the article by Hearn entitled "High Performance Liquid Chromatography of Peptides and Proteins: Quantitative Relationships for Isocratic and Gradient Elution of Peptides and Proteins".

IV. ZONE BROADENING RELATIONSHIP

The extent of zone broadening of a solute in a chromatographic system is reflected in the column efficiency, usually expressed in terms of the number of theoretical plates, N, or the height equivalent, HETP value or H. The theoretical plate number N of a column is defined as

$$N = \left(\frac{t_r^2}{\sigma_t^2}\right) \tag{14}$$

or

$$N = 16\left(\frac{t_r}{t_w}\right)^2 \tag{15}$$

where t_r is the elution time, and σ_t^2 the peak variance of the eluted zone in time units. For practical convenience, the peak variance, σ_t^2 is often replaced by the peak width, t_w, at baseline (e.g., the detector response line for zero sample concentration). For Gaussian peaks, t_w approximately corresponds to 4σ ($4 \times$ peak standard deviation). It should, however, be appreciated that by taking t_w as equal to 4σ, only 95% of the true peak area of a Gaussian peak will be integrated.

The theoretical plate number, N, can also be expressed in terms of k' such that

$$N = 16\left[\frac{(1+k')}{t_w}t_o\right]^2 \tag{16}$$

The value of N is dependent on a variety of chromatographic and solute parameters, including the column length, L, the particle diameter, d_p, the linear flow velocity, u (equivalent to L/t_r), and the solutes' diffusivities, (D_m and D_s) in the bulk mobile phase and within the stationary phase respectively. Often column efficiency is described in terms of H and the column length, L, whereby

$$N = L / H \tag{17}$$

A major task of experimental practice with HPLC techniques is to choose conditions which maximize the N-value and minimize the H-value. This task thus necessitates adequate control over the various processes which control zone broadening (or peak spreading). These processes are depicted in Figure 2 and include (1) eddy diffusion, (2) mobile phase mass transfer, (3) longitudinal molecular diffusion, (4) stagnant mobile phase mass transfer, and (5) stationary phase mass transfer. From a practical standpoint, these processes can partially be controlled with a column bed which has been well packed with particles of narrow particle size and pore size distributions. Because peptides and proteins have small effective diffusivities in HPLC systems (compared to low molecular weight organic analytes) the major problems associated with achieving high efficiencies, particularly with the adsorption modes of HPLC, invariably can be traced to inadequate control over stagnant mobile phase mass transfer, film diffusion, and stationary phase mass transfer kinetics. Film diffusion involves differential mass migration into stagnant mobile phase regions or into stationary phase films. Both effects will contribute to molecular zone spreading depending on the experimental conditions.

In order to permit comparison of column efficiencies with columns of identical bed dimensions packed with sorbent particles of different average diameter, the height equivalent H can be redefined in terms of the reduced plate height, h, while the linear flow velocity, u (= L/t_o), can also be redefined in terms of a reduced velocity such that

$$h = H / d_p \text{ and } v = ud_p / D_m \tag{18}$$

The contributions of the mass transport effects illustrated in Figure 2 which cause zone broadening have been formalized in terms of the dependency of h and v through the well known van Dempter/Knox equations,[12,13] i.e., in the form

$$h = Av^{1/3} + B/v + Cv \tag{19}$$

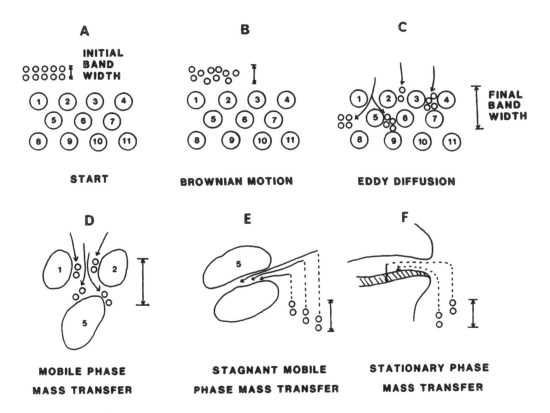

FIGURE 2. Various contributions to the molecular dispersion of a peak zone for a peptide or protein solute with a porous HPLC sorbent.

where the A-term encompasses eddy diffusion and mobile phase mass transfer effects and is a measure of the packing quality of the chromatographic bed; the B-term encompasses the longitudinal molecular diffusion effects, while the C-term incorporates mass transfer resistances within the stationary phase microenvironment. With well packed columns, operating at optimal flow rates and careful selected elution conditions, h-values approaching 2 to 5 times the particle diameter, d_p, should be realizable by most investigators. The major challenge today for high efficiency separations of peptide and proteins with available HPLC packing technology remains proper control of the C-term effects. Guidelines and solutions to these problems have been documented elsewhere in this volume.

Zone broadening of solutes in chromatographic systems arises from two causes, an intra-column cause as discussed above, and an extra-column cause due to the instrumentation characteristics. For this reason, it is not typically feasible to achieve the theoretical minimum h-value anticipated on the basis of considering the chromatographic bed (or column) characteristics. Rather, it is necessary to include a system (or instrumental) effect such that

$$\sigma_t^2 = \sigma^2_{column} + \sigma^2_{extra} \qquad (20)$$

where σ^2_{column} and σ^2_{extra} are the peak variances arising due to column broadening induced by column effects and by extra-column effects, respectively. Careful attention to extra-column influences (choice of tubing, type of injector, design of the flow through detector cell, etc.) can reduce the impact of σ^2_{extra} on the overall h-value. With microbore HPLC, in particular, it is essential that $\sigma^2_{extra} << \sigma^2_{column}$, otherwise high-speed, sensitive analysis becomes a meaningless exercise.

V. RESOLUTION

As is evident from the above discussion, solute retention expressed in terms of the capacity factor, k′, is governed by thermodynamic considerations, whereas zone broadening or peak dispersion, which is most conveniently expressed in terms of the reduced plate height h, arises from kinetic, time-dependent phenomena. *Resolution* between components depends on both k′ and h and can be expressed as

$$R_s = \frac{1}{4}\left(\frac{L}{hd_p}\right)^{1/2}(\alpha - 1)\left(\frac{1}{1+k'}\right)$$

(21)

or

$$R_s = \frac{1}{4}(N)^{1/2}(\alpha - 1)\left(\frac{1}{1+k'}\right)$$

(22)

The importance of Equations 21 and 22 resides in the linking of the three essential separation parameters which dictate the quality of a chromatographic experiment, namely relative retention, relative selectivity, and the extent of zone spreading. Larger values of R_s per unit time correspond to separations of higher performance, while lower values show poorer performance.

As is evident from Equations 21 and 22, R_s can be varied by changing N (and h) or k′ (and α). The power of instrumental HPLC permits the investigator to implement very effective and often simple modulation of k′ (or α) changes, while changes in N (or h) will often be (technically) more demanding. To provide efficient high resolution separations, R_s values ≥1 are required. Peak broadening of adjacent zones often represents the major analytical or preparative difficulty limiting resolution. Three strategies can be employed to enhance resolution: either (1) by increasing α, (2) by varying k′ over the range 1 to 10, or (3) by increasing N and thus decreasing h. The selection of experimental conditions which permit an increase in α-values, such as changes in the composition of the stationary phase or mobile phase or variation of temperature, have typically been based on empirical experience. Increasingly, intuitive computer-assisted methods are being applied to the optimization of resolution. Such methods form the basis of computer-assisted Expert systems in HPLC (see the section entitled "Prediction and Computer Simulation of Peptide and Protein Separations and Structure for Research and Training" for discussion of these important developments). An important consequence of increasing α is that retention times can be decreased while still increasing (or at least maintaining) R_s values. This tactic finds its most potent expression in the use of ion-pair techniques with RPC and in high-performance affinity chromatography (HPAC). Clearly, differences in α-values should be large and never unity (where *no* resolution is possible) within an operationally optimal range of k′ values. If this is not achieved, then the separation will be costly in terms of time, reagents, eluents and instrument usage.

With regard to the effect of k′ on R_s, it should be noted that resolution is proportional to the term k′/(1+k′). Consequently, little will be gained by choosing operational conditions with very large k′ values (i.e., values of k′>>20), where the penalty of excessive band broadening and long separation times will become excessive. Although gradient elution can remedy these consequences of large k′ values, as a guideline, resolution optima for complex peptide or protein mixtures should be achieved over the range 1≤k′≤10 by appropriate changes in the mobile phase elution strength, i.e., by increasing the ionic strength in IEC, the organic modifier content in RPC, or by decreasing the ionic strength in HIC, etc. For a particular column and flow-rate, each peptide or protein will have an optimal k′ value for maximum resolution. Resolution changes can also be achieved under the above conditions by varying column length or by changes in the elution conditions during the separation, i.e., by using gradient elution.

As is evident from Equation 21, R_s depends also on $(L/hd_p)^{1/2}$. Because the column is packed with sorbent particles of average diameter, d_p, when the mobile phase has a linear flow velocity, u, a pressure drop Δp will exist across the column. This pressure drop across the column is given by the relationship

$$\Delta p = \frac{u\eta L}{\varepsilon_p d_p^2} \tag{23}$$

where u is the linear velocity of the eluent, η the viscosity of the mobile phase and ε_p the specific permeability of the column. To a good approximation, ε_p can be estimated from $(d_p)^2/1000$. Attention must therefore be given to the choice of eluent composition (i.e., viscosity) and linear flow velocity for a particular column and stationary phase if the pressure drop is not to become excessive. For two well-packed columns of identical dimensions, packed with porous particles of nominal diameter 5 and 10 μm, and operated with the same mobile phase conditions, flow rate and temperature, the effective difference in pressure drop will be related to the ratio dp,1/dp,2, the differences in column permeability (εp,1 and εp,2), and the ratio of t_o,1/t_o,2. For the above example, the effective difference in pressure drop will be approximately 3- to 3.5-fold depending on the irregularity of the particle shape.

Although the resolution equations predict R_s will increase as the particle size, d_p, decreases, the trade off in the choice of stationary phases of small average d_p is the rapid increase in pressure drop. Thus, the high Δp with 3 to 5 μm particles suggests they should be packed into short, i.e., <25 cm, columns, although increasing interest with 1 to 2 μm d_p particles for rapid, high resolution analytical separations is emerging. To maintain a fixed pressure drop, Δp, across the bed as d_p of the packing is decreased, either the column length, L, or the eluent viscosity, η, must be decreased or the particle permeability, ε_p, increased proportionately. From practical considerations, adjustment in L and η are the preferred options, since significant increases in ε_p have other effects, e.g., reduced mechanical stability of the particle. It should be noted that changes in η will either mean a change in selectivity because a mobile phase of different composition is required or alternatively a change in the operational temperature of the mobile phase.

Flow rate, F, is a parameter which impacts on both N (or h) and Δp. As an operational variable, it has its greatest effect on resolution in size exclusion (SEC) separations. In SEC separations, variation of flow rate represents a key experimental tool to control resolution (the other being the stationary phase permeability and the connectivity of pores) for a particle of defined diameter packed into a column of specified dimensions. In most applications, the options to achieve optimal resolution with a particular separation selectivity and flow rate without allowing the column pressure drop to exceed practical experimental limits are to vary Δp while keeping L constant (i.e., by decreasing F and u, or increasing d_p), or by decreasing L and d_p with Δp constant. The scenario of increased column length, L, without changes in η requires improved engineering of the instrumentation, increased system cost and the use of particles of greater mechanical strength. For many practical reasons, this last scenario for improving resolution is not sustainable. The choice of these options obviously must hinge on the level of resolution and the separation performance per unit time demanded by the specific analytical or preparative problem. How these and other choices are decided will be discussed further in the article entitled "High-Performance Liquid Chromatography of Peptides and Proteins: Quantitative Relationships for Isocratic and Gradient Elution of Peptides and Proteins" by Hearn, which examines more advanced theoretical considerations for improving resolution and analyzing peptide and protein behavior in HPLC systems, and elsewhere in this volume.

VI. SUMMARY

This chapter has examined general considerations of the basic theory of HPLC as it applies to the separation of peptides and proteins. The concepts of the solute retention, t_r, and the capacity factor, k', as well as zone broadening and the theoretical plate number, N, or reduced plate height, h, have been introduced. From k' and N (or h) it is possible to determine quantitatively the selectivity, α, and the resolution, R_s, for a particular separation. The influences of specific chromatographic parameters, such as average particle diameter, d_p, flow rate, F, linear flow velocity, u, pressure drop Δp, column length, L, and extra column effects on zone broadening and resolution have been summarized. With this basic theory, investigators entering the realm of HPLC will have the ability to commence quantitative interpretation of their separation results, thus enabling them to improve more precisely the performance of their specific separation tasks.

REFERENCES

1. **Hearn, M.T.W. and Aguilar, M.I.,** Reversed-phase high performance liquid chromatography of peptides and proteins, in *Modern Physical Methods in Biochemistry,* Neuberger, A. and Van Deenen, L.L.M., Eds., Elsevier, Amsterdam, 1988, p. 107.
2. **Chicz, R.M. and Regnier, F.E.,** Surface-mediated retention effects of subtilisin site-specific variants in cation-exchange chromatography, *J. Chromatogr.,* 443, 193, 1988.
3. **Hodder, A.N., Aguilar, M.I., and Hearn, M.T.W.,** Identification and characterization of coulombic interactive regions on sperm whale myoglobin by high performance anion-exchange chromatography and computer graphics, *J. Chromatogr.,* 507, 33, 1990.
4. **Fausnaugh, J.L. and Regnier, F.E.,** Solute and mobile phase contributions to retention in hydrophobic interaction chromatography, *J. Chromatogr.,* 359, 131, 1986.
5. **Fausnaugh-Pollitt, J., Therenon, G., Janis, L., and Regnier, F.E.,** Chromatographic resolution of lysozyme variants, *J. Chromatogr.,* 443, 221, 1988.
6. **Wu, S.-L., Benedek, K., and Karger, B.L.,** Thermal behavior of proteins in high-performance hydrophobic-interaction chromatography. On-line spectroscopic and chromatographic characterization, *J. Chromatogr.,* 359, 3, 1986.
7. **Lu, X.M., Benedek, K., and Karger, B.L.,** Conformational effects in the high-performance liquid chromatography of proteins. Further studies of the reversed-phase chromatographic behavior of ribonuclease A, *J. Chromatogr.,* 359, 19, 1986.
8. **Unger, K.K.,** *Packings and Stationary Phases in Chromatographic Techniques,* Marcel Dekker, New York, 1989.
9. **Purcell, A.W., Aguilar, M.I., and Hearn, M.T.W.,** High-performance liquid chromatography of amino acids, peptides and proteins. XCI. The influence of temperature on the chromatographic behavior of peptides related to human growth hormone, *J. Chromatogr.,* 476, 125, 1989.
10. **Lork, K.D., Unger, K.K., Brückner, H., and Hearn, M.T.W.,** Retention behavior of paracelsin peptides on reversed-phase silicas with varying *n*-alkyl chain length and ligand density, *J. Chromatogr.,* 476, 135, 1989.
11. **Drake, A.F., Fung, M.A., and Simpson, C.F.,** Protein conformation changes as the result of binding to reversed-phase chromatography column materials, *J. Chromatogr.,* 476, 159, 1989.
12. **Guiochon, G.,** Preparation and operation of liquid chromatographic columns of very high efficiency, *J. Chromatogr.,* 185, 3, 1979.
13. **Stout, R.W., DeStefano, J.J., and Snyder, L.R.,** High-performance liquid chromatographic column efficiency as a function of particle composition and geometry and capacity factor, *J. Chromatogr.,* 282, 263, 1983.

HIGH-PERFORMANCE LIQUID CHROMATOGRAPHY OF PEPTIDES AND PROTEINS: QUANTITATIVE RELATIONSHIPS FOR ISOCRATIC AND GRADIENT ELUTION OF PEPTIDES AND PROTEINS

Milton T.W. Hearn

I. INTRODUCTION

In the article entitled "High-Performance Liquid Chromatography of Peptides and Proteins: General Principles and Basic Theory" by Hearn, basic considerations on the retention and zone broadening behavior of peptides and proteins in HPLC systems were discussed. This article examines the consequences of secondary chemical equilibria on both the thermodynamic and kinetic behavior of peptides and proteins in HPLC systems. In addition, this chapter provides additional theoretical framework for the assessment of regular and non-regular elution behavior of peptides and proteins particularly evident with the adsorption modes of HPLC.

As previously noted in the article entitled "High-Performance Liquid Chromatography of Peptides and Proteins: General Principles and Basic Theory" by Hearn, the aim of all chromatographic separations may be defined as the achievement of an optimal combination of speed of elution, sample size, and resolution of the solutes. Good resolution can only be obtained if there is adequate control over the differential migration rates of the group of peptide or protein solutes as they move down a column (e.g., control of column selectivity) and over the extent of zone dispersion for each of the solutes (e.g., control of column efficiency). Most current models of peptide and protein retention in HPLC are based on the concept of reversible, near equilibrium interaction between the solute and the sorbent. In the case of size-exclusion HPLC (SEC) differential "interaction" is taken to mean complete or partial permeation into the pores of the stationary phase which ideally exhibits no interactive characteristics involving chemical equilibria.

Under conditions involving the chromatographic interaction of a peptide or protein, P, with a stationary phase ligand, L, the equilibrium (or near equilibrium) process can be represented by the relationship

$$P(Solv)_a + nL(Solv)_b \rightleftharpoons P(Solv)_{(a-f)} \cdot nL + (nb + F)Solv \qquad (1)$$

where $P(Solv)_a$ represents a peptide or protein, involving a moles of solvent (water, etc.) or solvated ions; $nL(Solv)_b$ represents the solvated ligand, L, containing b moles of bound solvent (water, ions etc.); f is the number of solvent (water, etc.) molecules released from the solvated peptide or protein; and $P(Solv)_{(a-f)} \cdot nL$ represents the partially solvated peptide or protein adsorbed to ligands, L, immobilized at the surface of the sorbent. Hence, the dependency of peptide or protein retention on the number of ligands, n, and the eluent composition in an isocratic elution with a column of phase ratio, ϕ can be given by

$$k' = \Phi \bullet K_d [L(Solv)]^n [Solv]^{-(nb+f)} \tag{2}$$

where K_d is the equilibrium distribution constant, and the term $[Solv]^{-(nb+f)}$, reflects the effective valency $(nb+f)$ and the concentration of the displacing species which may be an organic solvent, salt ion, etc., required to desorb the peptide or protein from the immobilized ligand which may have hydrophobic, coulombic, or other chemical characteristics.[1,2] An alternative form of this nonmechanistic relationship can be written as the logarithmic expression

$$\log k' = A + B(\log 1/c) + C(\log 1/c)^2 + ... \tag{3}$$

where c is the concentration of the displacing species, and the terms A,B,C are system coefficients which incorporate the various constants of Equation 2. Under conditions of linear elution chromatography, i.e., where the retention behavior is totally dominated by a single type of selectivity mode over the capacity factor range $0 < k' < 10$ and k' is the concentration of the peptide or protein bound to the sorbent divided by the concentration of the peptide or protein in the mobile phase, e.g., k' is proportional to $[P]_s/[P]_m$ over the sample loading range, regular, stoichiometric chromatography is observed. Such chromatographic elution can then be interpreted in terms of isocratic or gradient elution relationships depending on the nature of the mobile phase composition.

II. ISOCRATIC RELATIONSHIPS

If the operational conditions are chosen such that the capacity factor is within the range $1 \leq k' \leq 20$, then the experimentally observed dependency of k' on the concentration of the displacing species can be approximated by the linear form

$$\log k' = A + B(\log 1/c) \tag{4}$$

for the different selectivity modes of HPLC, this general relationship can be expressed to reflect the following modalities: for reversed-phase separations, as

$$\log k' = \log k_{o,RP} - M\log[solvent] \tag{5}$$

for hydrophobic interaction separations, as

$$\log k' = \log k_{o,HIC} - H\log[salt] \tag{6}$$

for electrostatic (ion-exchange) separations, as

$$\log k' = \log k_{o,IEX} - Z\log[salt] \tag{7}$$

In each of the above empirical equations, the terms $\log k_{o,RP}$, $\log k_{o,HIC}$, $\log k_{o,IEX}$, are related to the distribution coefficients of the solutes when the concentration of the displacing substance falls to zero, i.e., when $c \to O$. In the case of reversed-phase HPLC (RPC), an empirical relationship between the capacity factor and the mole fraction of organic modifier, φ, can also be used to assess retention behavior. This dependency for peptide or protein separation in RPC takes the form

$$\log k' = \log k'_o - S\varphi \tag{8}$$

where $\log k_o'$ is the extrapolated capacity factor when $\varphi = 0$, and S is the slope of the plot of log k' vs. φ, over a defined k' range (usually $0 < k' < 20$).

These retention relationships (Equations 4 to 8) between log k' and the concentration or mole fraction of the displacing species form the basis for evaluating both optimization protocols and selectivity changes in a quantitative manner with peptides or proteins separated by adsorption HPLC. In particular, the values of the slope terms, S (derived for reversed-phase), Z (derived for ion-exchange) and H (derived for hydrophobic interaction) as well as the intercept terms log $k_{o,RP}$, log $k_{o,IEX}$ and log $k_{o,HIC}$ are prognostic indicators of the solute's behavior in a particular chromatographic system. Procedures to determine these essential parameters have been extensively documented in recent years in the scientific literature.[1-10] In brief, evaluation of these dependencies can be simply achieved through analysis of sets of data accumulated from multiple chromatographic experiments in which the elution strength of the mobile phase is systematically changed, i.e., the concentration of displacer is varied in such a manner to result in solute retention changes over the range $1 < k' < 20$. With automated HPLC equipment, the time taken for such studies is relatively short when the substantial benefits of the derived knowledge are considered, i.e., the required experimental time can take from 2 to 48 hr but save months of empirical 'recipe' experimentation. In particular, the derived data provide the basis to improve resolution as well as to reveal "non-ideal" or secondary retention behavior associated with, for example, conformational changes, aggregation or solute-specific buffer ion interactions.

The composite interplay between size-exclusion phenomena, hydrophobic, solvational, and coulombic interaction processes is a feature of all current HPLC chromatographic stationary phases. Depending on the magnitudes of these retention dependencies, the retention behavior of a peptide or protein in an interactive HPLC system can be formalized in terms which incorporate these multimodal contributions to the overall retention and selectivity process. A common relationship in which this formalism has found expression is given by

$$\ln k' = \ln\left[\zeta_{sec} k'_{sec} + \zeta_{vdw} k'_{vdw} + \zeta_{es} k'_{es} + \ldots\ldots\right] \tag{9}$$

In the case of RPC of peptides and proteins, this multimodal dependency can be represented by

$$\ln k' = \ln\left[\rho_{sec} k'_{sec} + \rho_{vdw} k'_o e^{-S\varphi} + \rho_p k'_p e^{-D(1-\varphi)} + \ldots\right] \tag{10}$$

where $\rho_{sec} k'_{sec}$ corresponds to the size-exclusion term, $\rho_{vdw} k'_{vdw}$ $(= S_{vdw} k_o e^{-S\varphi})$ to the hydrophobic (solvophobic) term, and $\rho_{es} k'_{es} = \rho_p k'_p e^{-D(1-\varphi)}$ to the polar, coulombic term for different mole fraction values, φ, of the organic solvent modifier. The coefficients S and D correspond to solute-specific parameters. The S value for a particular biopolymer is derived from the slope of the plots of the logarithmic capacity factor (log k') vs. the reciprocal logarithmic concentration of the organic solvent modifier or directly from the plots of log k' vs. the mole fraction, φ, of the organic solvent modifier in reversed-phase separations. The D parameter is derived from the same plot of log k' vs. reciprocal logarithmic concentration of organic solvent modifier over the φ-value range where the k' values increase as the φ-value increases, whilst k'_o and k'_p correspond to the solute capacity factors in neat water and pure solvent, respectively. Analogous methods can be employed to determine the Z- and H-value for a particular biopolymer from the log k' vs. reciprocal logarithmic concentration of displacing ion in ion-exchange and hydrophobic interaction separations.

Depending on the magnitude of the S, D, Z, H, k'_o and k'_p parameters, a variety of solute retention vs. mobile phase elutropic strength scenarios can be anticipated from Equations. 9 and 10 and various examples have been experimentally demonstrated. Figure 1 represents four limiting cases of such retention dependencies. Case b is typified by shallow log k' vs. ζ (or log $1/[c]$) dependencies with small log k' values at ζ (or[c]) = 0 and represents a commonly observed situation with small polar peptides separated under reversed-phase or ion-exchange HPLC

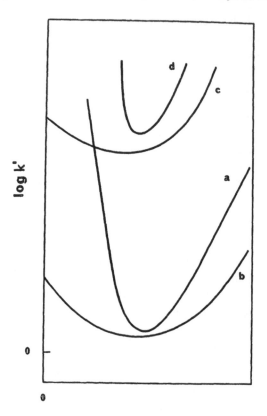

FIGURE 1. Schematic representation of the retention dependencies for peptides or proteins chromatographed on mixed-mode support media. The figure illustrates four case histories for the dependency of the logarithmic capacity factor (log k') on the mole fraction, ζ, of the displacing species. As the contact area associated with the solute-ligand interaction increases, the slopes of the log k' vs. ζ plots increase, resulting in a narrowing of the elution window over which the solute will desorb. Cases a and b are typically observed for RPC and IEC of polar peptides and small, polar globular proteins, while Cases c and d are more representative of the RPC and IEC behavior of highly hydrophobic polypeptides and non-polar globular proteins, including membrane proteins, respectively.

conditions. Peptides which exhibit this type of retention behavior can be chromatographed over a wide variety of isocratic elution conditions. Case c, which again exhibits shallow dependencies in terms of the log k' vs. ζ (or log l[c]) dependency but with large values of log k'_o, is representative of situations found with some low molecular weight, disulfide-rich proteins in ion-exchange HPLC (IEC) or very hydrophobic peptides under some reversed-phase conditions, and globular enzymes in affinity displacement ion exchange or in substrate analog displacement in affinity chromatography, where the substrate/analog or displacing species is again typically of low molecular weight. Some examples of polypeptide and protein displacement in hydrophobic interaction chromatography (HIC) also correspond to this Case c. Case a represents a typical scenario for many polypeptides and globular or fibrous proteins separated by reversed-phase and hydrophobic interaction techniques as well as with most polymer- and silica-based anion- and cation-exchange HPLC techniques. The hallmarks of Case a are: (1) a very narrow elution range, (2) very limited opportunity for isocratic elution, (3) substantial sensitivity to loading conditions, and (4) good recovery only provided changes in elution conditions are not made too rapidly. From practical considerations, these limiting chromatographic conditions are frequently chosen such that the minima of the plot of the logarithmic k' vs. ζ (or log l/[c]) corresponds to k' values equal to or less than unity and the flow rate is kept relatively low. Typically, this criterion is more easily achieved in ion-exchange than reversed-phase separations. Case d represents a worse-case scenario. In this case, the retention is large, the elution window narrow, and low recoveries can be anticipated. This case is typical of multimeric membrane proteins in RPC and IEC. A chromatographic sorbent of different selectivity is required if Case d is to ever approximate Case a. If a suitable sorbent is not available, then drastic changes in the mobile phase composition will be required, i.e., use of exotic detergent composites in IEC, betaines or zwitterionic pairing ions in RPC. With peptides or proteins with retention behavior characterized by Case a or d, gradient elution is also usually a mandatory requirement if elution is to occur.

A further important consequence of the multimodal behavior of peptides and proteins with HPLC stationary phases and sorbents typified by the Case histories a to d shown in Figure 1 is the opportunity these dependencies afford to affect elution by both *increasing* as well as *decreasing* the concentration of the displacing species over defined mole fraction ranges. This capability can be exploited in the use of gradient elution or step elution with increasing concentrations of displacer from a low value but also retro-gradient or inverse gradient elution with decreasing concentrations of displacer from a high value. Such methods are increasingly being employed in sample recovery in polypeptide microsequencing studies (see the sections entitled "Reversed-Phase Chromatography" and "Analysis of Peptides and Proteins").

The determination of S, D, Z, H, k'_o and k'_p values for a large variety of polypeptides and proteins, encompassing differences in amino acid composition, sequence, size, and hydrophobicity, can be derived from chromatographic data obtained with isocratic elution. However, these measurements are time consuming and can require high experimental precision. Furthermore, many polypeptides and proteins exhibit skewed peaks when eluted under isocratic conditions from modern HPLC sorbents. As a consequence, accurate determination of the average elution times and the peak variances requires calculation of the first and second moments of the peak zone. Although isocratic conditions can in some instances be employed in the separation of different polypeptides under well-defined isocratic elution conditions, complex mixtures of peptides, and proteins are routinely separated on n-alkylsilicas and other types of HPLC sorbents under gradient elution conditions.

Over the past decade, a variety of investigations have addressed the development of optimization models for the quantitative interpretation and prediction of isocratic and gradient elution data of low and high molecular weight solutes by HPLC. Of these different theoretical treatments, the Snyder linear solvent strength (LSS) gradient model[11,12] has, in particular, been shown to provide a very useful basis for evaluating and anticipating the retention behavior of polypeptides and proteins separated under regular gradient elution conditions. Gradient elution represents a logical alternative when isocratic elution methods are unsuitable, since it permits separation times to be reduced, peak volumes to be decreased and, because selectivity is being continuously varied, high resolution to be achieved. Gradient elution can also be used in an analogous manner to isocratic methods to characterize further the physicochemical basis of polypeptide retention in adsorption HPLC.

The same chromatographic variables relevant to retention, resolution, and bandwidth in isocratic elution also govern gradient elution. As shown by Snyder and others, the physicochemical parameters such as solute S, D, Z, and k'_o values can be determined through the application of the LSS theory. Under such conditions, the retention time, t_g, for a polypeptide chromatographed under gradient elution conditions with a binary gradient from solvent A to solvent B can be related[3-7,12] to the gradient steepness parameter, b, through the expression

$$t_g = \left(t_o / b\right)\left[\log 2.3 k_o' \, b\right] + t_o + t_d \tag{11}$$

where t_o is the column dead time and t_d is the gradient elapse time required for the change in solvent B to reach the column inlet. The determination of b values in linear solvent strength (LSS) systems can be easily achieved by using different gradient times or flow-rates. Thus, when solutes are also chromatographed over a range of gradient times in the same column with the same limits of mobile phase composition, the gradient steepness parameter can be derived by using the relationship

$$b_1 = \frac{t_o \log \beta}{t_{g1} - \left(t_{g2} / \beta\right) + \left(t_o\left[t_{G1} - t_{G2}\right] / t_{G2}\right)} \tag{12}$$

where t_{g1} and t_{g2} are the gradient retention times of the solute for two different gradient times t_{G1} and t_{G2}, respectively, and the coefficient β represents the ratio of the respective gradient times $(= t_{G2}/t_{G1})$. In the case of solutes chromatographed under linear solvent strength gradient conditions at two different flow rates, F_1 and F_2, the b value can be determined from the following expression

$$b_1 = \frac{\log\left[F_2 / F_1\right]}{X_1 - X_2\left(F_1 / F_2\right)} \tag{13}$$

where

$$X_1 = \frac{t_{g1} - t_{o1}}{t_{o1}} \text{ and } X_2 = \frac{t_{g2} - t_{o2}}{t_{o2}} \tag{14}$$

Evaluation of the b values from retention data determined over various gradient times allows the calculation of a range of median capacity factors, $\bar{k}$, (i.e., the capacity factor of the solute as it migrates past the midpoint of the column in gradient elution) can be derived from the expression

$$\bar{k} = 1 / 1.5b \tag{15}$$

In the case of RPC, the corresponding median organic mole fraction $\bar{\varphi}$ can be determined according to

$$\bar{\varphi} = \left[t_{g1} - t_o - \left(t_o / b\right)\log 2\right] / t^o G \tag{16}$$

For reversed-phase gradient elution separations, the relationship between the median capacity factor, $\bar{k}$ and the median mole fraction, $\bar{\varphi}$ can be related to the isocratic parameters through the empirical expressions

$$\log k = \log k_o' - S\varphi \qquad \text{isocratic elution} \tag{17}$$

$$\log \bar{k} = \log k_o' - S\bar{\varphi} \qquad \text{gradient elution} \tag{18}$$

with values of $\log k'_o$ and S derived by linear regression analysis.

Similarly, in the case of gradient elution in IEC, the concentration, $\bar{c}$, of the displacing salt when the sample band reaches the midpoint of the column can be given by[13,14]

$$\bar{c} = c_o + \left[t_g - t_o - t_e - 0.3\left(t_o / b \right) \right] \Delta c / t_G \qquad (19)$$

where $\Delta c\ (= c_f - c_o)$ is the difference between the final and initial salt concentration of the gradient. Thus, over a limited range of salt concentration, the relationship between log k′ and c can be expressed as

$$\log k' = \log K + Z \log (1 / c) \qquad \text{isocratic elution (20)}$$

$$\log \bar{k} = \log K + Z \log (1 / \bar{c}) \qquad \text{gradient elution (21)}$$

Consequently, values of Z and log K can be obtained by using linear regression analysis of the data form log $\bar{k}$ vs. log 1/c plots. The values of S, Z etc., thus obtained can then be utilized in optimization protocols and to characterize peptide/protein retention behavior in interactive HPLC (see the article entitled "Computer Simulation as a Tool for Optimizing Gradient Separations" by Snyder, Dolan, and Lommen for examples). In particular, when the data from isocratic and gradient elution experiments are completely superimposible, then the solute's retention behavior satisfies the criteria of 'ideal' linear elution chromatography.

Figures 2 and 3 illustrate experimental examples of data obtained by reversed-phase and anion-exchange HPLC in which the linear dependencies of log k (or log $\bar{k}$) on φ (or $\bar{\varphi}$) or log $\bar{k}$ (or log k′) on log (1/c̄) [or log 1/c], respectively, for several different peptides and proteins are graphically documented.[7,10] In these cases, the very high correlation of the data derived from isocratic and gradient elution protocols indicates that in these specific peptide and protein cases, regular elution behavior is followed where secondary equilibria processes either do not occur, or alternatively, occur with half lives substantially different to the mass transport half-lives. In such cases of regular retention behavior, peak widths also reach minimal values. It should be noted, however, that where secondary equilibrium processes become significant, e.g., where conformational effects dominate or inappropriate secondary chemical phenomena exist due to inappropriate pH, buffer concentration, choice of ion-pairing reagent, etc., non-linear dependencies of log k′ on φ (or log 1/c) are usually observed, as illustrated in Figure 1. Furthermore, significant divergence between isocratic elution behavior and gradient elution behavior will be observed when secondary equilibrium effects make important contributions to the overall retention process.

Variations in the chemical characteristics of the desorbing solvent or salt represent a very effective method to change the shape of the log k′ vs. φ (or log 1/c) dependency. Anticipation from the log k′ vs. φ dependencies of the appropriate choice of the displacing solvent of displacing salt can, thus, have very practical consequences. However, this choice can also have more profound physicochemical consequences in terms of optimizing the interaction of the solvent molecules or ions with the peptide or protein solute to ensure high recovery of mass or activity. These consequences can take the form of stabilization of the secondary and tertiary structure of the solutes through specific choice of the eluent composition or through variation of the equilibrium distribution co-efficient to optimize the peptide (or protein) interaction with the sorbent. An example of this latter effect is shown in Figure 4, where the linear dependency between the ion- exchange slope parameter, Z, and the logarithmic equilibrium distribution coefficient (log K_c) (see the article entitled "High-Performance Liquid Chromatography of Peptides and Proteins: General Principles and Basic Theory" by Hearn) is illustrated for two proteins — carbonic anhydrase and ovalbumin eluted from an anion-exchange HPLC column with different salts.[10] These and related data demonstrate that, depending on the salt employed, different selectivities and retention behavior can be achieved in a *predictable* manner. Guidelines based on consideration of the chaotropic and kosmotropic potential of different salts and solvents have been derived from such analyses and can be used to optimize both resolution and recovery.

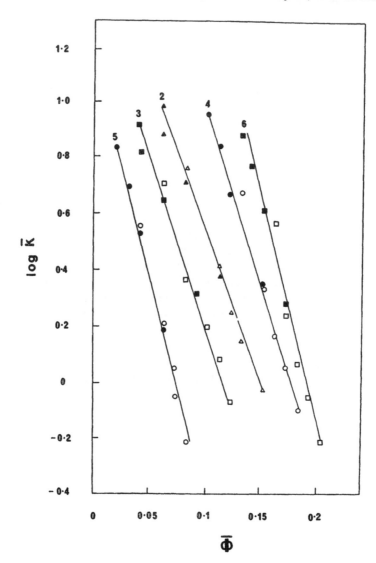

FIGURE 2. Plots of log k̄ vs. φ̄, based on gradient elution chromatography of LHRH-related peptides (2 to 6) [2, <EHW; 3, <EHWS; 4, <EHWSY; 5, GLRPG-amide; 6, < EHWSYGLRPG-amide]. The plots were derived from best-fit analysis of the data points shown, where t_G = 20, 30, 40, 60, and 120 min and F = 1 ml/min (open points), and F = 1, 2, 3, and 4 ml/min and t_G = 40 min (filled points). Other chromatographic conditions were: column, Bakerbond widepore C_4 (25 × 0.46 cm); d_p 5 μm; linear gradient of 0.1% aq. TFA (Solvent A) and water-acetonitrile (50:50 v/v) containing 0.1% TFA; temperature 20°C. (From Hearn, M. T. W. and Aguilar, M. I., *J. Chromatogr.*, 397, 47, 1987. With permission.)

IV. ZONE BROADENING BEHAVIOR

According to the van Dempter/Knox equations, the dispersion of a zone in a chromatographic system of reduced velocity, ν, is given by (see the article entitled "High-Performance Liquid Chromatography of Peptides and Proteins: General Principles and Basic Theory" by Hearn).

$$h = Av^{1/3} + B/v + Cv \qquad (22)$$

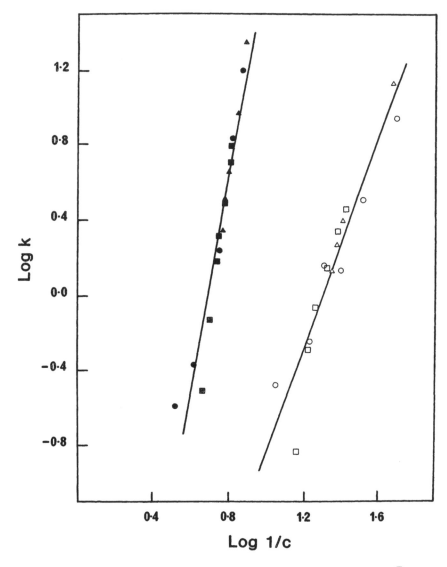

FIGURE 3. Retention plots for isocratic (log k′ vs. log 1/c) and gradient experiments (log $\bar{k}$ vs. log 1/$\bar{c}$) for ovalbumin and carbonic anhydrase, eluted with sodium bromide (isocratic and gradient data were derived from experiments with varied salt concentrations at pH 9.6 and a flow-rate of 1 ml/min). Other chromatographic conditions were: column, Mono Q (HR5/5); d_p 9 μm; mobile phase, 0.02 *M* piperazine, pH 9.6 (eluent A) and 0.02 *M* piperazine –0.3 *M* NaBr (eluent B); temperature, 20°C. (From Hodder, A. N., Aguilar, M. I., and Hearn, M. T. W., *J. Chromatogr.*, 476, 391, 1989. With permission.)

Under the flow rate conditions employed to separate peptides and proteins in modern HPLC columns and equipment, the reduced velocities are typically greater than 100. As a consequence, the $Av^{1/3}$ and B/v terms tend to become insignificant and the dependency of h on v is dominated by the Knox C-term, i.e., h ≈ Cv. When linear velocity conditions apply, the plate number, N, can then be approximated to

$$N = \frac{D_m t_o}{C d p^2} \tag{23}$$

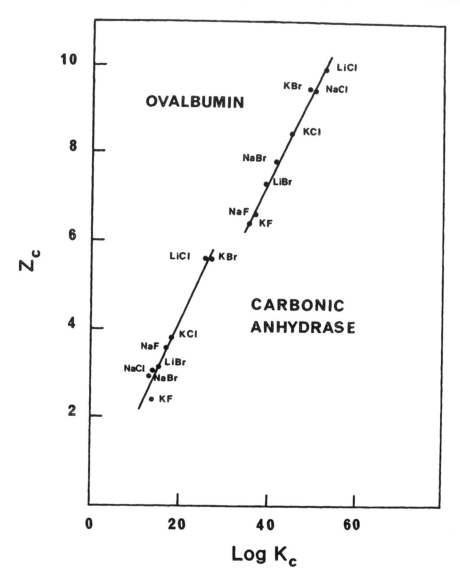

FIGURE 4. Plots of Z vs. log K_c, the ion exchange distribution coefficient as c → O, for ovalbumin and carbonic anhydrase, separated under conditions of varied gradient time with different displacing salts. (From Hodder, A. N., Aguilar, M. I., and Hearn, M. T. W., J. Chromatogr., 476, 391, 1989. With permission.)

while the resistance to mass transfer term, C, can be given by

$$C = \frac{[1 - x + k']/(1 + k')^2}{15\rho a' + 15\rho b' k' - 19.2\rho x} \qquad (24)$$

where x is the interstitial column volume fraction, typically about 0.67 for modern large pore stationary phases of average particle diameter $d_p \approx 5$ to 10 μm; a′ is the column packing parameter, typically about 1.1; b′ is the surface diffusion parameter; and ρ is the restricted diffusion parameter. With a knowledge of υ, k′, ρ, a′, b′ and D_m, it is thus possible to calculate the N (or h) value of a particular solute at a defined k′ with a specific stationary phase. The D_m value can be conveniently estimated from the Chang-Wilke equation such that

$$D_m = \frac{8.34 \times 10^{-10} T}{\eta M W^{0.33}} \qquad (25)$$

Procedures for the determination of a′, b′ and ρ for a particular chromatographic system and set of solutes can be found in References 12, 15, and 16.

In gradient elution, it is convenient to define zone broadening in terms of the *peak capacity* (PC) of the system with an average resolution, R_s equal to 1 for all pairs of solutes. Under such conditions

$$PC = t_G F / 4\sigma_v \qquad (26)$$

while the peak variance, σ_v can be related to k for linear solvent strength gradient conditions through the dependency

$$\sigma_v = [(k/2) + 1] G t_o F N^{-1/2} \qquad (27)$$

where F is the flow rate, N the number of theoretical plates, t_o the void time, and G the band compression factor, which arises from the increase in solvent strength across the solute zone as the gradient develops along the column, and is given by

$$G^2 = \left[1 + 2.3b + 1/3(2.3b)^2\right](1 + 2.3b) \qquad (28)$$

Peptides and proteins which show regular behavior, i.e., those that do not exhibit slow secondary equilibria, conformational interconversion, etc., and are thus adequately described by the near equilibrium, linear elution model, can be expected to exhibit zone broadening properties such that the calculated and experimental observed values of the peak variance ($\sigma_{v,calc}/\sigma_{v,exp}$) will be close to unity. In other cases, divergences can lead to σ_v,calc/$\sigma_{v,exp}$ ratios significantly less than unity. Such latter behavior, from the point of view of resolution, is far from optimal but nevertheless can provide essential information on the structural, interactive and conformational properties of peptides and proteins in HPLC systems. The interested reader is referred to several recent studies published elsewhere on this topic.[16-19]

V. APPLICATION OF CHROMATOGRAPHIC THEORETICAL CONSIDERATIONS

The value of the above quantitative relationships and associated analytical approaches can be found in the following extensions of chromatographic theory and experimentation as it applies to peptide and protein separations:

- Determination of conformational behavior during the separation
- Definition of interactive contact areas
- Monitoring the interplay of slow secondary chemical equilibria such as that occurring in ion-pair and dynamic liquid-liquid ion exchange with phases containing ionic additives
- Optimization of resolution for both isocratic and gradient elution
- Evaluation of retention mechanisms in terms of structure-retention dependencies
- Determination of data bases of functional group coefficients for retention prediction
- Design of novel stationary phase sorbents

Several of these aspects have been discussed elsewhere in this volume (see the sections entitled "Analysis of Peptide and Protein Conformation by HPLC", "Introduction to Narrow

Bore, Microbore and Rapid HPLC Analysis Techniques", and "Prediction and Computer Simulation of Peptide and Protein Separations and Structure for Research and Teaching"). In this section, two facts of these analytical approaches will be specifically examined, namely, the rationale for the use of ionic mobile phase additives, and approaches available to assess conformational behavior of peptides and proteins separated by adsorption HPLC.

A. USE OF ION-PAIR REAGENTS

As is evident from the earlier discussion, water-rich elements form the basis of all mobile phases used in the different HPLC modes. Selectivity modulation can then be implemented by the appropriate choice of mobile phase additives. These may be water-soluble organic solvents, e.g., acetonitrile, methanol, 2-propanol, etc., ionizable buffers, surfactants, and other ionic substances commonly called ion-pair reagents. Superimposed on these selectivity options are additional thermodynamic possibilities, e.g., temperature or pH modulation. In all chromatographic cases involving the addition of chemical additives to the mobile phase, the potential thus exists to manipulate selectivity and resolution through phenomena associated with secondary chemical equilibria. In particular, the nature and concentration of ionic additives can have dramatic effects on α and N. Variation of pH or water content of the mobile phase in interactive HPLC may not allow adequate control over column selectivity for peptides and proteins. Recourse must be made to the use of other approaches, particularly if the mobile phase has a low buffering capacity. In IEC and HIC modes, advantage can be taken of changes in ionic strength of the mobile phase using "salting in", "salting out" or salt desorption effects. However, in RPC, it is now much more common to control retention not through ionic strength per se, but through judicious choice of the appropriate co- and counter-ions. Similar methods are often extended to IEC in order to permit ion-specific control over selectivity. Ionic reagents which complex with peptides or proteins when added to the mobile phase are known as ion-pair, ion-interaction or hetaeric compounds and may be simple compounds such as trifluoroacetate or tetramethylammonium salts or, alternatively, surfactant or betaine compounds such as sodium heptylsulfonate or trimethylglycine. In physicochemical terms, the addition of ionic (and also non-ionic complexing reagents) will affect crucial mobile phase parameters such as surface tension, γ, dielectric constant, ε, and viscosity, η. As discussed in the article entitled "High-Performance Liquid Chromatography of Peptides and Proteins: General Principles and Basic Theory" by Hearn, all of these parameters will influence resolution through direct effects on selectivity and peak efficiencies. Generally, peptides and proteins will engage multiple ionic interactions with these complexing reagents with the consequence that power function dependencies will exist between retention and the concentration of hetaeric reagent. Whether ion-pair formation, i.e., stoichiometric condensation of the hetaeric ions with the peptide or protein in the bulk mobile phase, or dynamic ion-exchange distribution, i.e., stoichiometric ion-exchange with preadsorption of hetaeric ions onto the sorbent, occurs is still a matter of much research. However, the mathematical form of these dependencies can be explicitly expressed for monovalent -1 or $+1$ net charge interactions as:

$$k' = \frac{\left(k'_o + \beta[X]_m\right)}{\left(1 + K_2[X]_m\right)\left(1 + K_3[X]_m\right)} \tag{29}$$

where k'_o is the capacity factor in the absence of hetaeric ions, $[X]_m$ is the concentration of hataeric ion, and β, K_2, and K_3 correspond to the distribution coefficient terms for ion pair formation in the bulk eluent, distribution of the hetaeric ion onto the ligand, dynamic ion-exchange, and adsorption of the complexed ion pair onto the ligand. The derivation of Equation 29 can be found in References 20 and 21. An important prediction of this equation is that asymptotically limiting dependencies of k' on $[X]_m$ are anticipated. When the hetaeric ion, X,

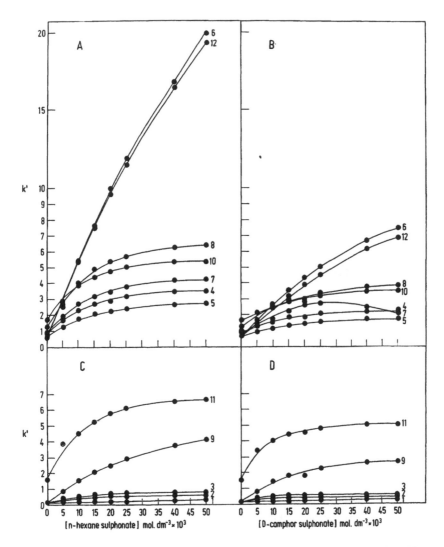

FIGURE 5. Dependence of the capacity factor of protonated peptides on the concentration of the pairing ion in the mobile phase, separated on a μBondapak C$_{18}$ column at a flow-rate of 2 ml/min. The mobile phases were (A) and (B) 25% methanol-water-50 mM NaH$_2$PO$_4$ — 15 mM H$_3$PO$_4$ titrated with 100 mM NaOH to pH 3.0, containing various concentrations of the ion pairing reagents n-hexanesulfonic acid and D-camphorsulfonic acid. Peptide key; 1 = Gly; 2 = Gly-Gly; 3 = Ala-Gly; 4 = Gly-Phe; 5 = Gly-Phe-amide; 6 = Arg-Phe; 7 = Val-Leu; 8 = Phe-Leu-amide; 9 = Ala-Lys; 10 = Gly-Leu-Tyr; 11 = Gly-Gly-Tyr-amide; 12 = Arg-Phe-Ala. (From Purcell, A. W., Aguilar, M. I., and Hearn, M. T. W., *J. Chromatogr.*, 476, 113, 1989. With permission.)

is of opposite charge to the charged peptide or protein, then increasing values of k′ will occur with increasing concentrations of X up to a limiting value (cf. Figure 5), while if the hetaeric ion, X, is of the same charge as the charged peptide or protein, then decreasing values of k′ will occur as [X] increases. Both scenarios have been observed experimentally. Operational concentrations of X are typically in the range 1 to 15 mM, i.e., up to ca. 0.1% w/v or 0.1% v/v. Table 1 lists a selection of common ion pairing or hetaeric ions now in wide usage. Many of these reagents will be familiar to peptide and protein chemists as the common additives employed in RPC. However, their capabilities go well beyond their attenuation or enhancement of retention with a single hetaeric substance. Selectivity, α, and efficiency, N, can be rationally manipulated as a function of concentration. With a set of hetaeric substances, their sequential use in columns

TABLE 1
Selection of Ionic Species (Hetaerons)
which Modify Retention Characteristics of
Peptides and Proteins in RPC, HIC, and
IEC

Cation	Anion
NH_4^+	ClO_4^-
Pyridinium	SO_4^{2-}
$C_3H_9NH_3^+$	BO_3^-
$C_{12}H_{25}NH_3^+$	HCO_2^-
$(HOCH_2CH_2)_3NH^+$	$CF_3CO_2^-$
$(CH_3CH_2)_3NH^+$	$C_3F_7CO_2^-$
N-Methylmorpholinium	$C_6F_{13}CO_2^-$
N-Methylpiperidinium	$C_4H_9SO_3^-$
Piperazinium	$C_6H_{13}SO_3^-$
$(CH_3)_4N^+$	$C_{12}H_{25}SO_3^-$
$(C_4H_9)_4N^+$	$C_{12}H_{25}SO_3^-$
Cetylpyridinium	D-Camphor-SO_3^-
Trimethylglycine	pTosylSO_3^-

Note: This selection does not include the chaotropic or kosmotropic ions. Generally, this selection is compatible
with UV detection below 235 nm or fluorometric detection. The concentration of these reagents in the mobile
phase is usually 1 to 5 mM, i.e., $\leq 0.01\%$ (v/v/ or w/v).

operating in tandem provides the multidimensionality needed for very high resolution analysis or purifications of low abundance peptide or protein components in complex samples.

Further readings on the role of hetaeric or ion pair reagents in peptide and protein chemistry can be found in References 21 and 22 and are also documented extensively elsewhere in this volume as specific applications.

B. ASSESSMENT OF CONFORMATION BEHAVIOR OF PEPTIDES AND PROTEINS BY HPLC TECHNIQUES

The dynamic folding/unfolding behavior of peptides and proteins in bulk solution has attracted enormous attention from protein chemists for over 40 years. In the process of changing their conformations in response to variation in the solution composition or temperature, all peptides and proteins make available two of their fundamental features. Conformational changes are associated with variation of the hydrodynamic volume of these solutes as well as changes in the relative accessibility of the amino acids (and other structural moieties) which form the secondary and tertiary hierarchial structure of the peptide or protein. Changes in hydrodynamic volume of a peptide and protein are most readily monitored by SEC. Typically, such changes are affected by variation of the concentration of a denaturant, e.g., urea or guanidine hydrochloride. From plots of the permeation coefficient, K_p, vs. [denaturant], transition values corresponding to denaturation, subunit dissociation, etc., can be readily obtained. Similarly, kinetic data can be extracted with appropriate experimental protocols. Adsorption HPLC can be used in an analogous manner (for application examples see the section in this volume entitled "Analysis of Peptide and Protein Conformation by HPLC"). When used in these ways, HPLC techniques provide a rapid analytical tool to monitor conformational behavior that occurs in the bulk solution (or mobile phase).

However, HPLC techniques have other attributes since they also permit the effect of the stationary phase or sorbent to be monitored. In particular, the influence of the sorbent surface

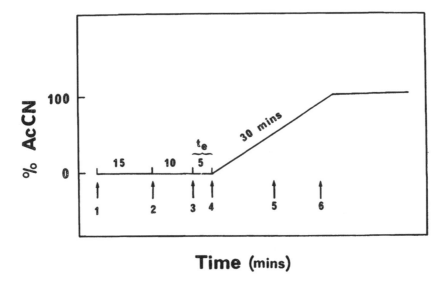

Time (mins)

FIGURE 6. Diagrammatic representation of loading conditions used to study the effect of preincubation, column residence time and solvent exposure on the conformational stability of proteins in gradient elution HPLC.

on the conformational status can be determined from an analysis of the retention dependencies and apparent rate constants for stationary phase induced effects evaluated. The experimental design for these investigations is relatively simple, particularly if a gradient elution system is used in conjunction with an adsorptive mode of HPLC. The solute is incubated for varying times in the mobile phase and, following adsorption onto the stationary phase surface, the respective k' (or $\bar{k}$) values are determined as shown in Section III, and plotted vs. log [l/c] (or log [l/$\bar{c}$], φ, $\bar{\varphi}$). Figure 6 illustrates the essential features of this protocol. If denaturation, subunit dissociation or other significant long term changes in tertiary folding, has occurred under the chromatographic conditions, then one or more of the following events will be evident: (1) more than one zone for the peptide may be observed, (2) k' and $\bar{k}$ will change with time of incubation, (3) significant changes in slopes of the log k' vs. log l/c plots at critical c-values, (4) distorted peak shapes which vary with time of incubation, and (5) dramatic changes in recovery often referred to as "irreversible binding". Figures 7 and 8 illustrate several of these features from data obtained from RPC and IEC studies with peptides and proteins.

Because the mechanism(s) of conformational changes of peptides and proteins which occur at the surface of HPLC packings is not known in detail, at this stage only apparent rate constants and apparent pathways can be deduced (see, for example, References 18, and 23 to 25). However, important criteria on the influence of a chromatographic parameter on recovery and activity of a particular polypeptide can be established from these experiments. This knowledge is particularly useful in scaling-up purification tasks with labile polypeptides or proteins. Several theoretical models have been reported recently for the evaluation of on-column effects which result in peak shape interconversions. At present, a two-stage unfolding model is most widely used for such evaluations. However, in the absence of comprehensive data on the precise molecular mechanisms involved, these treatments can result in 'lumped' kinetic terms due to the nature of the experimental methods and detection procedures. For readers wishing to be more acquainted with these recent developments, their attention is directed to References 15, 18, 19, and 23 to 26.

Temperature effects also provide a very useful method to study conformational behavior. If Equation 10 from the article entitled "High-Performance Liquid Chromatography of Peptides and Proteins: General Principles and Basic Theory" by Hearn is recalled (i.e., $k' = \Phi K$), then the dependency of k' on T can be given by

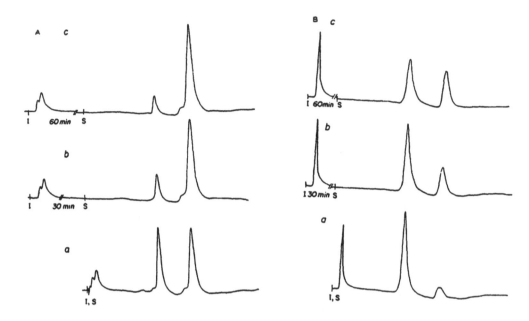

FIGURE 7. Chromatographic behavior of papain as a function of on-column incubation time at mobile phase compositions of (A) 10 mM H$_3$PO$_4$ (pH 2.2) and B 1-propanol-water (5.4:95.6, v/v) in which the total H$_3$PO$_4$ concentration is 10 mM. Conditions: Column = C$_4$ bonded phase on 10 μm, LiChrospher SI-500; mobile phase A = 10 mM H$_3$PO$_4$ (pH 2.2); mobile phase B = 1-propanol water (45:55, v/v) in which the total H$_3$PO$_4$ concentration is 10 mM; gradient rate = 3% propanol/min, 15-min linear gradient; flow-rate = 1 ml/min; sample = 20 mg/ml papain in mobile phase A, 6 μl injected; detection at 280 nm; column temperature = 5°C. 1 = Injection time; S = start of gradient. Incubation times: a = 0 min; b = 30 min; c = 60 min. (From Benedek, K., Dong, S., and Karger, B. L., *J. Chromatogr.*, 317, 227, 1984. With permission.)

$$\log k' = \frac{-\Delta H°_{assoc}}{RT} + \frac{-\Delta S°_{assoc}}{R} + \log \Phi \tag{30}$$

Hence, van't Hoff plots (i.e., log k' vs. 1/T) can be used to evaluate both $-\Delta H°_{assoc}$ and $\Delta S°_{assoc}$ for polypeptides and proteins in different conformational states. Such measurements are very useful prognosticators of the extent of flexibility or constraint associated with a particular peptide or protein conformation. In addition, van't Hoff plots enable enthalpy- or entropy-driven association of the solute with the ligand to be discriminated. In particular, the dependence of retention on temperature in RPC can be employed to evaluate the helical content of amphipathic and non-amphipathic peptides over a wide range of mobile phase conditions.[17]

VI. SUMMARY

This chapter has briefly examined more advanced aspects of the theory of HPLC as it applies to the separation of peptides and proteins. The concepts of isocratic and gradient elution have been linked and a useful theoretical approach based on the linear solvent strength concepts elaborated. From an evaluation of the log k' vs. log [displacer] dependencies, various system and physicochemical parameters can be determined and used to improve resolution. The effect of chromatographic parameters on zone broadening in gradient and isocratic elution has been further examined. Two examples of the participation of secondary chemical equilibria, namely ion pair interactions and conformational interconversions have also been introduced and their effect on retention and zone broadening behavior discussed. These applications demonstrate the value of quantitative methods in studies on the dynamics of peptides and proteins in HPLC

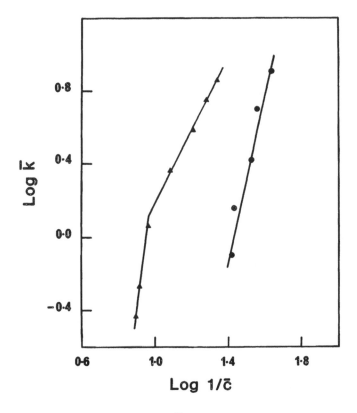

FIGURE 8. Retention plots of log k̄ vs. log l/c̄ for lysozyme eluted under gradient conditions with LiBr as the displacer salt. Gradient data were derived with varied gradient time (o) or varied flow-rate (Δ). The biphasic nature of the plots for varied flow rates, possibly attributable to solute re-orientation during these experiments, is illustrated. (From Hodder, A., Aguilar, M.I., and Hearn, M. T. W., *J. Chromatogr.*, 510, 1990, in press. With permission.)

systems. More comprehensive treatments of these, and related, aspects of the role of HPLC in elucidating the physicochemical behavior of peptides and proteins are found in References 17 to 27.

REFERENCES

1. **Geng, X. and Regnier, F.E.,** Retention model for proteins in reversed-phase liquid chromatography, *J. Chromatogr.*, 296, 15, 1984.
2. **Hearn, M.T.W.,** High resolution reversed-phase chromatography, in *Protein Purification: Principles, High Resolution Methods and Applications*, Janson, J.C. and Ryden, L., Eds, VCH Press, FL, 1989, p. 175.
3. **Stadalius, M.A., Gold, H.S., and Snyder, L.R.,** Optimization model for the gradient elution separation of peptide mixtures by reversed-phase high-performance liquid chromatography, *J. Chromatogr.*, 296, 31, 1984.
4. **Hearn, M.T.W. and Grego, B.,** High-performance liquid chromatography of amino acids, peptides and proteins. LV. Studies on the origin of band broadening of polypeptides and proteins separated by reversed-phase high-performance liquid chromatography, *J. Chromatogr.*, 296, 61, 1984.
5. **Stadalius, M.A., Quarry, M.A., and Snyder, L.R.,** Optimization model for the gradient elution separation of peptide mixtures by reversed-phase high-performance liquid chromatography. Application to method development and the choice of column configuration, *J. Chromatogr.*, 327, 93, 1985.

6. **Aguilar, M.I., Hodder, A.N., and Hearn, M.T.W.,** High-performance liquid chromatography of amino acids, peptides and proteins. LXV. Studies on the optimization of the reversed-phase gradient elution of polypeptides: evaluation of retention relationships with β-endorphin-related polypeptides, *J. Chromatogr.,* 327, 115, 1985.
7. **Hearn, M.T.W. and Aguilar, M.I.,** High-performance liquid chromatography of amino acids, peptides and proteins. LXVIII. Evaluation of retention and bandwidth relationships of peptides related to luteinising hormone-releasing factor, separated by gradient elution reversed-phase high-performance liquid chromatography, *J. Chromatogr.,* 359, 31, 1986.
8. **Fausnaugh, J.L. and Regnier, F.E.,** Solute and mobile phase contributions to retention in hydrophobic interaction chromatography of proteins, *J. Chromatogr.,* 359, 131, 1986.
9. **Drager, R.R. and Regnier, F.E.,** Application of the stoichiometric displacement model of retention to anion-exchange chromatography of nucleic acids, *J. Chromatogr.,* 359, 147, 1986.
10. **Hodder, A.N., Aguilar, M.I., and Hearn, M.T.W.,** High-performance liquid chromatography of amino acids, peptides and proteins. LXXXIX. The influence of different displacer salts on the retention properties of proteins separated by gradient anion-exchange chromatography, *J. Chromatogr.,* 476, 391, 1989.
11. **Stadalius, M.A. and Snyder, L.R.,** HPLC separations of large molecules: a general model, in *HPLC-Advances and Perspectives,* Vol. 4, Horváth, Cs., Ed., Academic Press, New York, 1986, p. 195.
12. **Snyder, L.R.,** Gradient elution, in *HPLC-Advances and Perspectives,* Vol. 1, Horváth, Cs., Ed., Academic Press, New York, 1980, p. 207.
13. **Stout, R.W., Sivakoff, S.I., Ricker, R.D., and Snyder, L.R.,** Separation of proteins by gradient elution from ion-exchange columns. Optimizing experimental conditions, *J. Chromatogr.,* 353, 439, 1986.
14. **Hearn, M.T.W., Hodder, A.N., and Aguilar, M.I.,** High-performance liquid chromatography of amino acids, peptides and proteins. LXXXVII. Comparison of retention and bandwidth properties of proteins eluted by gradient and isocratic anion-exchange chromatography, *J. Chromatogr.,* 458, 27, 1988.
15. **Hearn, M.T.W. and Aguilar, M.I.,** High-performance liquid chromatography of amino acids, peptides and proteins. LXXIII. Investigations on the relationships between molecular structure, retention and band-broadening properties of polypeptides separated by reversed-phase high-performance liquid chromatography, *J. Chromatogr.,* 397, 47, 1987.
16. **Hodder, A., Aguilar, M.I., and Hearn, M.T.W.,** Identification and characterization of coulombic interactive regions of hen lysozyme by high-performance anion-exchange chromatography and computer graphics, *J. Chromatogr.,* 517, 317, 1990.
17. **Purcell, A.W., Aguilar, M.I., and Hearn, M.T.W.,** High-performance liquid chromatography of amino acids, peptides and proteins. XC. Investigations into the relationship between structure and reversed-phase high-performance liquid chromatography retention behavior of peptides related to human growth hormone, *J. Chromatogr.,* 476, 113, 1989.
18. **Hearn, M.T.W., Hodder, A.N., and Aguilar, M.I.,** High-performance liquid chromatography of amino acids, peptides and proteins. LXVI. Investigations on the effects of chromatographic dwell in the reversed-phase high-performance liquid chromatographic separation of proteins, *J. Chromatogr.,* 327, 47, 1985.
19. **Hearn, M.T.W. and Aguilar, M.I.,** Reversed-phase high performance liquid chromatography of peptides and proteins, in *Modern Physical Methods in Biochemistry,* Neuberger, A. and van Deenen, L.L.M., Eds., Elsevier, Amsterdam, The Netherlands, 1988, p. 107.
20. **Horváth, Cs., Melander, W., and Molnar, I.,** Liquid chromatography of ionogenic substances with nonpolar stationary phases, *Anal. Chem.,* 49, 142, 1977.
21. **Hearn, M.T.W.,** Ion-pair chromatography of amino acids, peptides and proteins, in *Ion-Pair Chromatography,* Hearn, M.T.W., Ed., Marcel Dekker, New York, 1985, p. 207.
22. **Hearn, M.T.W.,** High performance liquid chromatogrraphy of peptides, in *HPLC-Advances and Perspectives,* Horváth, Cs., Ed., Academic Press, New York, 1983, p. 87.
23. **Lu, X.M., Benedek, K., and Karger, B.L.,** Conformational effects in the high-performance liquid chromatography of proteins. Further studies on the reversed-phase chromatographic behavior of ribonuclease A., *J. Chromatogr.,* 359, 19, 1986.
24. **Benedek, K., Dong, S., and Karger, B.L.,** Kinetics of unfolding of proteins on hydrophobic surfaces in reversed-phase liquid chromatography, *J. Chromatogr.,* 317, 227, 1984.
25. **Fridman, M., Aguilar, M.I., and Hearn, M.T.W.,** A comparative study of the equilibrium refolding of bovine, porcine and human growth hormone using size-exclusion chromatography, *J. Chromatogr.,* 512, 57, 1990.
26. **Hearn, M.T.W.,** Current status of high performance liquid chromatographic techniques for biopolymer analysis and purification, in *High-Performance Liquid Chromatography of Peptides, Proteins and Polynucleotides,* Hearn, M.T.W., Ed., VCH Press, FL, 1990, 1—35.
27. **Hearn, M.T.W.,** Chromatotopography: peptide dynamics at solid-liquid interfaces, in *Peptides: Chemistry, Structure and Biology,* Rivier, J. and Marshall, G., Eds., Escom Press, Leiden, 1990, 415—422.

Section III
Size-Exclusion Chromatography

REQUIREMENT FOR PEPTIDE STANDARDS TO MONITOR IDEAL AND NON-IDEAL BEHAVIOR IN SIZE-EXCLUSION CHROMATOGRAPHY

Colin T. Mant and Robert S. Hodges

I. INTRODUCTION

Despite the widespread application of high-performance size-exclusion chromatography in recent years, relatively little attention has been paid to its potential for resolving peptides in the 200 to 5000 dalton range (2 to 50 residues). It would be extremely useful to extend SEC to peptide separations, possibly as the first step in the resolution of a complex peptide mixture.[1,2]

Separation of peptides by a mechanism based solely on peptide size (ideal SEC) occurs only when there is no interaction between the solute and the column matrix. By definition, under ideal size-exclusion conditions, no molecule will be retained beyond the total permeation volume of the column (a combination of the void volume of the column, i.e., the volume outside the pores of the packing, and the internal pore volume of the size-exclusion matrix, i.e., the volume within the pores of the packing). However, although high-performance size-exclusion columns are designed to minimize non-specific interactions, most modern SEC columns are weakly anionic (negatively charged) and slightly hydrophobic, resulting in deviations from ideal size-exclusion behavior, i.e., non-ideal SEC (nSEC).[3-7] This non-ideal behavior can take the form of a solute being retained on the column longer than would be expected from its size, due either to hydrophobic interactions with the column packing or to an electrostatic attraction between the negatively charged column matrix and any positively charged character of the solute. Alternatively, a solute with negatively charged characteristics may be eluted sooner than expected due to charge repulsion.

The majority of reports in the literature concerning SEC of peptides and proteins tend to describe chromatographic conditions designed to ensure a pure size-exclusion process, a prerequisite if predictable solute elution behavior is required. On the other hand, it is often overlooked that the non-ideal properties of size-exclusion columns can be advantageous in the separation of peptides, adding another dimension to the peptide or protein resolving power of a SEC column.[6,8,9] Thus, what is occasionally viewed as a column limitation may become a useful analytical tool.

The need for standards to monitor column chromatography is well established, and protein standards are frequently used to demonstrate the resolving power of SEC columns or to calibrate columns for molecular weight determinations. However, the use of peptide standards to monitor the peptide resolving capability of SEC columns has only recently been addressed.[8] This article describes the design and application of a series of synthetic peptide size-exclusion standards and demonstrates their versatility in monitoring both ideal and non-ideal SEC column performance for peptide separations.

TABLE 1
Characteristics of Synthetic Size-Exclusion Peptide Standards

Peptide sequence[a]	No. of repeating units (n)	No. of residues	Net charge
Ac-(Gly-Leu-Gly-Ala-Lys-Gly-Ala-Gly-Val-Gly)$_n$-amide	1	10	+1
Ac-(Gly-Leu-Gly-Ala-Lys-Gly-Ala-Gly-Val-Gly)$_n$-amide	2	20	+2
Ac-(Gly-Leu-Gly-Ala-Lys-Gly-Ala-Gly-Val-Gly)$_n$-amide	3	30	+3
Ac-(Gly-Leu-Gly-Ala-Lys-Gly-Ala-Gly-Val-Gly)$_n$-amide	4	40	+4
Ac-(Gly-Leu-Gly-Ala-Lys-Gly-Ala-Gly-Val-Gly)$_n$-amide	5	50	+5

[a]   Ac = N^α-acetyl; amide = C^α-amide.

II. EXPERIMENTAL

A. PEPTIDE SYNTHESIS

The peptide standards described in this article were synthesized on a Beckman Model 990 peptide synthesizer (Beckman Instruments, Berkeley, CA) using the general procedure for solid-phase synthesis described by Parker and Hodges.[10] The size-exclusion standards are available from Synthetic Peptides Incorporated (Dept. of Biochemistry, University of Alberta, Edmonton, Alberta, Canada, T6G 2H7).

B. COLUMNS

Three size-exclusion columns are described in this article: (1) Altex Spherogel TSK G2000SW, 300 × 7.5 mm I.D., 10-μm particle size, 130-Å pore size (Beckman Instruments), (2) SynChropak GPC60, 300 × 7.8 mm I.D., 10-μm, 60-Å (SynChrom, Linden, IN), and (3) Pharmacia Superose 12, 300 × 10 mm I.D., 10-μm (Pharmacia, Dorval, Canada).

III. RESULTS AND DISCUSSION

A. DESIGN OF PEPTIDE STANDARDS

The synthetic peptide standards shown in Table 1 were designed specifically for assessing the peptide resolving power of size-exclusion columns and are capable of monitoring both ideal and non-ideal chromatographic behavior. The five peptide standards make up a polymer series, with a repeating ten-residue unit. The increasing size of the standards (10, 20, 30, 40, and 50 residues [peptides 1, 2, 3, 4, and 5 respectively]; 800 to 4000 dalton) enables the accurate molecular-weight calibration of a column during ideal SEC; the increasingly basic character of the standards (1 to 5 positively charged residues in peptides 1 to 5, respectively) makes them sensitive to the anionic character of a size-exclusion column matrix; the increasing hydrophobicity of the standards enables a determination of column hydrophobicity. In addition, the high glycine content of the standards minimizes or eliminates any tendency towards secondary structure; thus, the peptides remain in a random coil configuration in both denaturing and non-denaturing mobile phase eluents. The basic character of the peptide standards also ensures good solubility in aqueous solvents.

Figure 1 verifies that the synthetic polymer series can, indeed, exhibit ideal size-exclusion behavior, an important consideration prior to their use in comparing the separation properties of different SEC columns. The chromatographic profile of the five standards on a Spherogel TSK G2000SW silica-based column, coupled with the linear character of the $\log_{10}$MW vs. peptide retention time plot, clearly demonstrates the ability of the polymer series to monitor pure size-exclusion behavior on SEC columns.

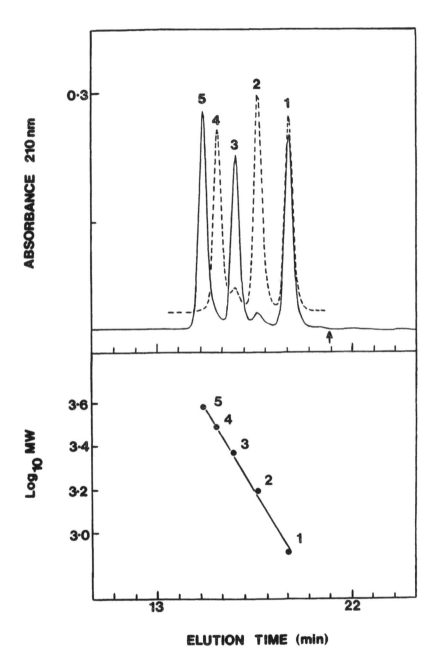

FIGURE 1. SEC of a mixture of synthetic peptide standards. The HPLC instrument consisted of a Varian Vista Series 5000 liquid chromatograph (Varian, Walnut Creek, CA) coupled to a Hewlett-Packard (Avondale, PA) HP1040A detection system, HP85B computer, HP9121 disc drive, HP2225A Thinkjet printer and HP7470A plotter. Top: elution profile of peptide standards obtained with an Altex Spherogel TSK G2000SW column (300 × 7.5 mm I.D.); mobile phase, 0.1% aq. trifluoroacetic acid (TFA), pH 2.0; flow-rate, 0.5 ml/min; temperature, 26°C. Bottom: plot of log molecular weight vs. retention time of the peptide standards. The peptide mixture consisted of five peptides with the sequence Ac-(G-L-G-A-K-G-A-G-V-G)$_n$-amide, where n denotes the number of decapeptides (1 to 5; see Table 1). The arrow denotes the elution time for the total permeation volume of the column (determined by the elution time of TFA). (From Mant, C. T., Parker, J. M. R., and Hodges, R. S., *J. Chromatogr.*, 397, 99, 1987.)

B. MONITORING OF NON-IDEAL SEC WITH PEPTIDE STANDARDS

The ideal size-exclusion behavior of the peptide standards shown in Figure 1 was obtained with a volatile 0.1% aq. trifluoroacetic acid (TFA) eluent. The choice of a volatile mobile phase is certainly advantageous in peptide separations, since it enables lyophilization of peptide fractions prior to immediate analysis or their direct application to ion-exchange or reversed-phase columns during multistep HPLC. The results shown in Figure 1 suggest that non-specific interactions of the standards with the TSK G2000SW packing were minimal in 0.1% aq. TFA (pH 2.0). Similar results were reported by Mant and Hodges[1] and Lau et al.[11] for a TSK G2000SW packing. However, it should be noted that hydrophobic and/or electrostatic characteristics may be more pronounced with one size-exclusion column than with another,[3,8] and the use of a simple, volatile mobile phase may be insufficient to suppress these interactions. Both the silica-based SynChropak GPC60 column and the agarose-based Superose 12 column, for instance, exhibited non-ideal behavior with 0.1% aq. TFA as eluent. However, the TSK G2000SW exhibited non-ideal behavior at pH 6.5 similar to that found with the GPC60 and Superose 12 columns.

Non-ideal behavior on the silica-based SynChropak GPC60 column is clearly illustrated in Figure 2. In the top panel of both the pH 2 (Figure 2, left) and pH 6.5 (Figure 2, right) chromatographic profiles, the separation of peptide standards 1, 2, and 5 (+1, +2, +5 net charge, respectively) appears to be based on an ion-exchange rather than a size-exclusion mechanism. The peptide elution order is reversed from that expected of a size-exclusion mechanism, with the smallest peptide being eluted first (n = 1; 10 residues) and the largest peptide being eluted last (n = 5; 50 residues). In addition, all three peptides were retained longer than the total permeation volume of the column (denoted by arrow). Electrostatic effects between solutes and the size-exclusion column matrix may be minimized by the addition of salts to the eluent. Aqueous buffers containing 0.1 to 0.4 *M* salt are commonly employed as the mobile phase for SEC. The eluting solvent which produced the chromatographic profiles at pH 2 (Figure 2, left) was 0.1% aq. TFA containing 10 m*M* (top panel), 50 m*M* (middle panel), or 200 m*M* (bottom panel) potassium chloride. As the ionic strength of the mobile phase increased, electrostatic effects were gradually overcome until an essentially ideal size-exclusion mechanism was apparent at a KCl concentration of 200 m*M*.

Different peptide mixtures may require eluents of markedly different pH values for optimal SEC separation. The effect of pH variations may not only affect the net charge of a particular peptide, but may also influence the non-specific solute/column matrix interactions. For instance, the non-specific interactions observed on the SynChropak GPC60 column at pH 2 (Figure 2, left) were even more pronounced at pH 6.5 (Figure 2, right), where a higher salt concentration was required at the higher pH to suppress the ion-exchange behavior of the column. This is particularly well demonstrated by comparing the 50 m*M* KCl profiles (Figure 2, middle panels). At pH 2, (left), a pure size-exclusion separation had almost been obtained, the three peptide standards having been eluted in the correct order and at retention times less than the total permeation volume of the column. In contrast, at pH 6.5 (right), all three peptides were eluted as a single peak and are still being retained slightly more than the total permeation volume. It is also apparent from Figure 2 that non-specific interactions were more pronounced at pH 6.5 in the presence of 25 m*M* KCl (right, top panel) than they were at pH 2 (left) in the presence of only 10 m*M* KCl (left, top panel). However, the presence of 200 m*M* KCl in the mobile phase at pH 6.5 (right, bottom panel) was again sufficient (cf. pH 2; left, bottom panel) to ensure ideal size-exclusion behavior of the peptide standards.

It should be noted that an increase in the KCl concentration to 1 *M* did not improve the chromatographic profile of the peptide standards at either pH 2 or 6.5. In fact, there was a slight deterioration in peptide resolution and an increase in peptide retention times, possibly due to promotion of hydrophobic interactions at this high salt level. These results supported the view that high-ionic-strength solvents (> 0.6 *M*) should generally be avoided in SEC.[7,8]

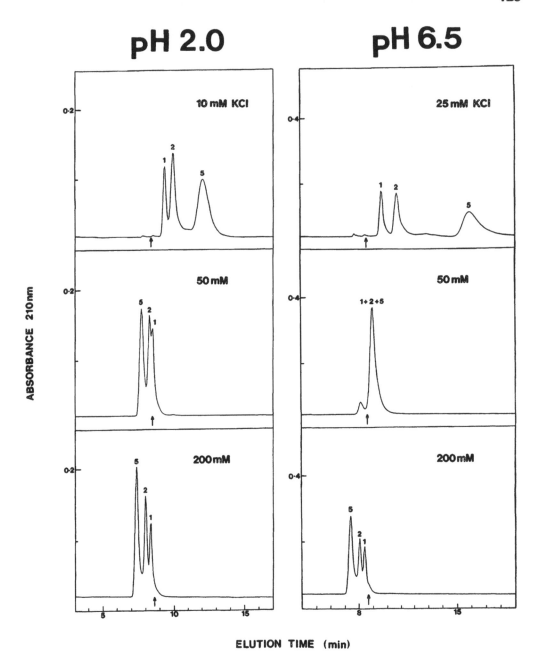

FIGURE 2. Effect of salt on non-specific interactions in SEC at pH 2 (left) and pH 6.5 (right). Column: SynChropak GPC60 (300×7.8 mm I.D.). Mobile phase: left, 0.1% aq. TFA (pH 2), containing 10, 50, or 200 mM potassium chloride; right, 5 mM KH$_2$PO$_4$ (pH 6.5), containing 25, 50, or 200 mM potassium chloride; flow-rate, 1 ml/min. Other details shown in Figure 1.

Several researchers have demonstrated the utility of adding volatile organic solvents, such as acetonitrile or trifluoroethanol, to 0.1% aq. TFA for effective separations of peptides and proteins.[11-15] A volatile aqueous triethylammonium formate buffer (0.25 M TEAF, pH 3) containing 30% acetonitrile has also proved useful for the separation of peptide and proteins.[16] Apart from their UV transparency, these organic modifiers decreased non-specific hydrophobic interactions of peptides with the size-exclusion matrix. Figure 3 demonstrates the effect of

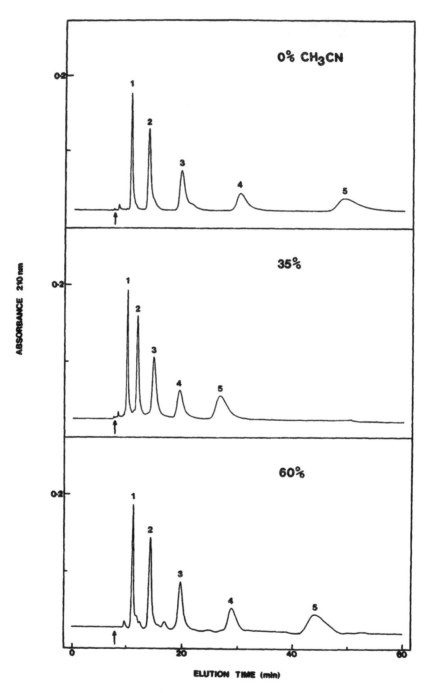

FIGURE 3. Effect of increasing acetonitrile concentrations on non-specific interactions in SEC. Column: SynChropak GPC60 (300 × 7.8 mm I.D.). Mobile phase: 0.1% aq. TFA, containing 0, 35, or 60% acetonitrile (CH₃CN); flow-rate, 1 ml/min. Other details shown in Figure 1. (From Mant, C. T., Parker, J. M. R., and Hodges, R. S., *J. Chromatogr.*, 397, 99, 1987. With permission.)

increasing acetonitrile (CH₃CN) concentrations on the elution profiles of the five peptide standards on the GPC60 column. As the concentration of the organic modifier increased, the retention times of all five peptides decreased to a minimum, presumably as hydrophobic interactions were overcome, and then proceeded to increase again. The peptide elution profiles were, in fact, very similar at the 0% (Figure 3, top panel) and 60% (Figure 3, bottom panel) levels of acetonitrile. The results suggested that, following reduction of peptide elution times to a minimum (with ca. 35% acetonitrile), further increases in acetonitrile concentration were possibly promoting ionic interactions of the standards with the column packing. Ideal size-exclusion behavior of the peptide standards was never obtained on this column with the volatile 0.1% aq. TFA/acetonitrile mobile phase; as mentioned above, the minimum elution times of the peptide standards were obtained in the presence of ca. 35% acetonitrile in the mobile phase (Figure 3, middle panel). In contrast, the addition of salt to the 0.1% aq. TFA mobile phase (Figure 2, left) did produce pure size-exclusion behavior of the peptide standards. These results indicated that not only was the GPC60 exhibiting both electrostatic and hydrophobic interactions with the peptide standards, but also that electrostatic interactions were dominant. The inclusion of 35% (v/v) acetonitrile in 0.1% aq. TFA containing 200 mM KCl did not improve the pure size-exclusion profile of peptide standards 1, 2, and 5, demonstrated in Figure 2, left (bottom panel). This suggested that the addition of salt was disrupting hydrophobic as well as electrostatic interactions.[8]

C. IDEAL SEC OF PEPTIDE STANDARDS

The ability to predict the position and/or elution order of peptides during SEC of a peptide mixture greatly simplifies preliminary isolation of peptides from a chemical or proteolytic protein digest.[2,8] Under conditions of ideal SEC, large peptide fragments, resulting from incomplete protein digestion, can be quickly identified and removed. The peptide standards enable an accurate comparison of the resolving power of different SEC columns under conditions which promote pure size-exclusion behavior. As stated previously, aqueous solvents and buffers containing 100 to 400 mM salt are commonly employed as the mobile phase for SEC. In fact, efficient separation of peptide mixtures in their absence is generally the exception rather than the rule. Figure 4 compares the elution profiles of peptide standards 1, 2, and 5 obtained with a Spherogel TSK G2000SW column (Panel A), a SynChropak GPC60 column (Panel B) and a Superose 12 column (Panel C), using a non-denaturing mobile-phase buffer (50 mM KH$_2$PO$_4$, pH 6.5, with 0.1 M KCl). Under these conditions, all three columns exhibited similar, and ideal, peptide elution profiles. The longer retention times and somewhat broader peptide peaks obtained with the Superose 12 are probably due to its significantly larger column volume compared to the two silica-based columns. The linearity of the log$_{10}$MW vs. peptide elution time plots, shown in Figure 5, underlines the pure size-exclusion behavior of the peptide standards on all three columns investigated.

It is interesting to note that the ideal size-exclusion profile of peptide standards 1, 2, and 5 on the TSK G2000SW column is somewhat improved when 0.1% aq. TFA is used as the eluting solvent (Figure 1, Panel A), compared to the profiles obtained in a non-denaturing phosphate buffer (Figure 4). The presence of salt in the eluent apparently has an as yet unexplained deleterious effect on the resolution of peptides at the lower end of the fractionating range of the TSK G2000SW column.

IV. CONCLUSIONS

The adoption of a common set of peptide standards by chromatographers utilizing SEC could greatly facilitate an accurate comparison of results between different researchers. Indeed, for anyone involved in peptide separations, peptide standards are essential for:

1. Identifying non-specific interactions between peptides and the size-exclusion packing
2. Monitoring column performance (efficiency and resolution)
3. Monitoring the effect of mobile phase composition and pH
4. Monitoring the effect of flow-rate

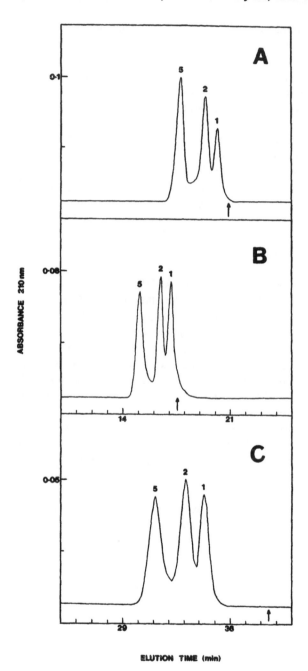

FIGURE 4. Comparison of ideal SEC profiles of a mixture of synthetic peptide standards obtained with different size-exclusion columns. Columns: (A) Altex Spherogel TSK G2000SW (300×7.5 mm I.D.), (b) SynChropak GPC60 (300 x 7.8 mm I.D.), and (C) Pharmacia Superose 12 (300 × 10 mm I.D.). Mobile phase, 50 mM KH$_2$PO$_4$ (pH 6.5) /100 mM potassium chloride; flow-rate, 0.5 ml/min; temperature 26°C. Other details shown in Figure 1. (From Mant, C. T., Parker, J. M. R., and Hodges, R. S., *J. Chromatogr.*, 397, 99, 1987.)

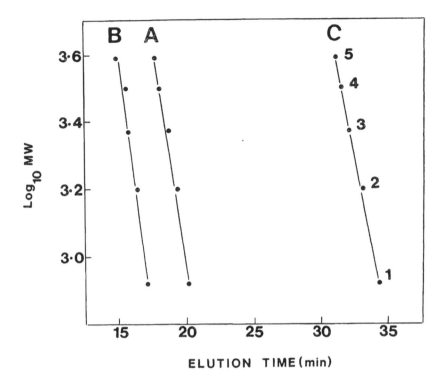

FIGURE 5. Plots of log MW vs. retention times of peptide standards obtained with different size-exclusion columns. Columns (A, B, C) and conditions as for Figure 4.

5. Monitoring the effect of changing column dimensions (length and diameter)
6. Monitoring the effect of particle size and pore size
7. Comparing packing materials (different packings or identical packings from different manufacturers)

The polymer series of synthetic standards described in this article has proved extremely beneficial in enabling rapid development of the optimal conditions for SEC of peptides and for monitoring the resolving power of various size-exclusion columns.

ACKNOWLEDGMENTS

This work was supported by the Medical Research Council of Canada and by equipment grants from the Alberta Heritage Foundation for Medical Research.

REFERENCES

1. **Mant, C.T. and Hodges, R.S.,** General method for the separation of cyanogen bromide digests of proteins by high-performance liquid chromatography: rabbit skeletal troponin I, *J. Chromatogr.*, 326, 349, 1985.
2. **Mant, C.T. and Hodges, R.S.,** Optimization of peptide separations by high-performance liquid chromatography, *J. Liq. Chromatogr.*, 12, 139, 1989.
3. **Pfannkoch, E., Lu, K.C., Regnier, F.E., and Barth, H.G.,** Characterization of some commercial high performance size-exclusion chromatography columns for water-soluble polymers, *J. Chromatogr. Sci.*, 18, 430, 1980.

4. **Engelhardt, H. and Mathes, D.,** High-performance liquid chromatography of proteins using chemically-modified silica supports, *Chromatographia,* 14, 325, 1981.

5. **Engelhardt, H., Ahr, G., and Hearn, M.T.W.,** Experimental studies with a bonded N-acetylaminopropylsilica stationary phase for the aqueous high-performance exclusion chromatography of polypeptides and proteins, *J. Liq. Chromatogr.,* 4, 1361, 1981.

6. **Kopaciewicz, W. and Regnier, F.E.,** Non-ideal size-exclusion chromatography of proteins: effects of pH at low ionic strength, *Anal. Biochem.,* 126, 8, 1982.

7. **Regnier, F.E.,** High-performance liquid chromatography of proteins, *Methods Enzymol.,* 91, 137, 1983.

8. **Mant, C.T., Parker, J.M.R., and Hodges, R.S.,** Size-exclusion high-performance liquid chromatography of peptides: requirement for peptide standards to monitor column performance and non-ideal behaviour, *J. Chromatogr.,* 397, 99, 1987.

9. **Engelhardt, H.,** Size-exclusion chromatography of proteins, this publication.

10. **Parker, J.M.R. and Hodges, R.S.,** I. Photoaffinity probes provide a general method to prepare synthetic peptide-conjugates, *J. Protein Chem.,* 3, 465, 1985.

11. **Lau, S.Y.M., Taneja, A.K., and Hodges, R.S.,** Effects of high-performance liquid chromatographic solvents and hydrophobic matrices on the secondary and quaternary structure of a model protein. Reversed-phase and size-exclusion high-performance liquid chromatography, *J. Chromatogr.,* 317, 129, 1984.

12. **Swergold, G.D., Rosen, O.M., and Rubin, C.S.,** Hormonal regulation of the phosphorylation of ATP citrate lyase in 3T3-L1 adipocytes, *J. Biol. Chem.,* 257, 4207, 1982.

13. **Swergold, G.D. and Rubin, C.S.,** High-performance gel-permeation chromatography of polypeptides in a volatile solvent: rapid resolution and molecular weight estimations of proteins and peptides on a column of TSK-G3000-PW, *Anal. Biochem.,* 131, 295, 1983.

14. **Taneja, A.K., Lau, S.Y.M., and Hodges, R.S.,** Separation of hydrophobic peptide polymers by size-exclusion and reversed-phase high-performance liquid chromatography, *J. Chromatogr.,* 317, 1, 1984.

15. **Sasaki, H., Matsuda, T., Ishikawa, O., Takamatsu, T., Tanaka, K., Kato, Y., and Hashimoto, T.,** New series of TSK-GEL PW type for high-performance gel filtration chromatography, *Scientific Report of Toyo Soda Mfg. Co. Ltd.,* 29, 37, 1985.

16. **Rivier, J.E.,** Evaluation of triethylammonium phosphate and formate-acetonitrile mixtures as eluents for high-performance gel permeation chromatography, *J. Chromatogr.,* 202, 211, 1980.

HIGH-PERFORMANCE SIZE-EXCLUSION CHROMATOGRAPHY OF PROTEINS

Karen M. Gooding and Helene Hagestam Freiser

I. INTRODUCTION

Size-exclusion chromatography (SEC) has been a primary tool used by investigators in the life sciences for purification of proteins. To a great extent, the popularity of this technique has been due to its speed, simplicity, and versatility. SEC is frequently used as a first step in a purification scheme to separate the molecules of interest from those which have radically different size characteristics. SEC can also be used to change the buffer environment of a sample and thus can be used for desalting. Finally, SEC can be implemented as a means of estimating molecular size and shape.

To date, the majority of SEC separations in the life sciences have been performed using carbohydrate gel columns which have been operated in cold rooms. In the past decade, however, there has been a trend to substitute high-performance size-exclusion chromatography (HPSEC) methods for those using soft carbohydrate gels because of the revolutionary improvements of HPSEC in speed of analysis, efficiency, and resolution.[1-3] Short separation times result from the isocratic separations, thus minimizing problems with protein stability, even at room temperature.

II. CONTRIBUTIONS OF COLUMN SUPPORT

A. GENERAL CHARACTERISTICS

The goal of high-performance size-exclusion chromatography (HPSEC) is to obtain maximum resolution of components of a mixture by separating them according to size. It is therefore imperative in SEC that elution be due strictly to size and all partition mechanisms be eliminated. This is generally accomplished by both designing a neutral column support and modifying the mobile phase appropriately. Two general categories of HPSEC supports have been developed for protein and peptide analysis. These are surface-modified silica-based supports and polymeric supports. Table 1 lists some commercial products from each of these categories. No one support in this list is ideal for all solutes, but they all were developed to have the maximum number of the following ideal characteristics:[4]

1. Rigidity and ability to withstand pressures of 100 atm or higher
2. Macroporosity, to allow penetration of proteins and/or peptides
3. Uniform pore diameter and geometry
4. High pore volume
5. Total hydrophilicity to prevent solvophobic partitioning
6. Elimination of ionic moieties to prevent ion-exchange partitioning
7. 5 to 10 micron particle size and spherical shape to minimize mass transfer limitations and maximize resolution
8. Good packability
9. Chemical stability over a broad pH range and with many different mobile phases

TABLE 1
Selected Manufacturers of High Performance Size
Exclusion Chromatography Supports

Product	Manufacturer
Silica-based	
SynChropak GPC	SynChrom, Inc.
TSK SW	Toya Soda Manufacturing Co.
Zorbax GF	DuPont Company
Polymeric	
Separon HEMA	Tessek Ltd.
Sepharose, Superose	Pharmacia-LKB Biotechnology
Shodex OH-pak	Showa Denko D. K.
TSK PW	Toya Soda Manufacturing Co.

10. Non-biodegradable composition
11. Inexpensive composition

None of the supports listed in Table 1 possesses all of these attributes; however, many of them encompass a majority. The major differences are found between the silica-based and polymeric categories of supports. Silica-based HPSEC supports meet most of the above criteria; however, their pH stability is generally limited to pH 3 to 8 and they have a slight ionic character. To be suitable for protein analysis, silica-based HPSEC supports must be well covered with a hydrophilic organic layer so that the native silica surface cannot interact with the biomolecules. Polymeric supports are less rigid than silica-based and they have a hydrophobic character which is chiefly introduced by the crosslinking process. Some of these also have some ionic properties. Because proteins and peptides are inherently both hydrophobic and ionic, properties of the supports which would adsorb solutes by these mechanisms must be minimized and/or suppressed by mobile phase modification.

B. RELEVANT PROPERTIES
I. Definition of Terms

Before discussing separation in HPSEC, it is necessary to define the terms required to calculate resolution. The totally excluded or void volume (V_o) of a column is the volume outside of the pores of the support, that is, between the support particles. The internal volume (V_i) is the volume within the pores of the support. The total volume (V_t) is the total accessible volume of the column or the sum of the void volume and the internal volume:

$$V_t = V_o + V_i$$

In ideal SEC, all solutes should be eluted at a volume (V_e) which is between that of a molecule which is excluded from the pores (V_o) and that of a molecule with complete access to all the pore volume (V_t). A distribution coefficient which describes the elution by SEC is called K_d and is defined in the following manner:

$$V_e = V_o + K_d V_i$$

It is obvious that K_d is a constant which shows the proportion of internal volume to which the solute has access. If a support functions in a SEC mode, K_d will be proportional to the logarithm of the molecular weight (MW) for values of K_d between 0.2 and 0.8.

Figure 1A illustrates a typical HPSEC separation of proteins. At pH values near neutrality, DNA functions well as a marker for void volume (V_o) and glycyl tyrosine is a good noninteractive marker for the included or total volume (V_t).[5] Figure 1B shows the linear relationship on this column between log MW and K_d for the proteins shown in Figure 1A, in addition to a few other standard proteins. It can be seen that the curve is not linear near the extremes in K_d; therefore, it is generally best to choose HPSEC columns on which the solutes of interest are eluted on the linear portion of the curve.

2. Pore Diameter

The pore diameter is the property of the support which actually determines the exclusion limits. Figure 2 shows several of the same proteins previously seen in Figure 1A when they were run on a 100-Å support instead of a 300-Å support. It can be seen that the K_d values are approximately 0.2 units lower on the 100-Å support. Generally, a given molecule will have a lower elution volume on a support with smaller pores. Homogeneity of pore diameter is reflected in the slope of the linear portion of the calibration curve — the smaller the absolute value of the slope, the better the pore homogeneity. Supports with lower slopes have a higher potential for resolution by HPSEC.[4,5]

3. Pore Volume

Another important physical property of an HPSEC support is the pore volume, which must be adequately large to have a high peak capacity. In HPSEC, it is desirable to have a minimum peak capacity of at least 7 to 10 peaks which are separated with a resolution of 1.0. Although a large pore volume is desirable for high resolution in HPSEC, an increase in pore volume results in an increase in fragility of the support matrix due to the concomitant decrease in the volume of solid. The porosity (ϕ) is a measure of the pore volume relative to the solid matrix and is calculated by:

$$\phi = V_i / V_o$$

An acceptable range for porosity for HPSEC supports is from 1.0 to 1.5.

4. Particle Size and Efficiency

The column efficiency or plate count is more important in HPSEC than in interactive modes of chromatography because the molecules are eluted isocratically instead of during a gradient in a bind/release mechanism. The narrower the peaks, the more will fit into the V_i and the better the resolution potential. Generally, supports for HPSEC have particle diameters of 5 to 10 μm to maximize resolution. When microparticulate HPSEC supports are packed well, the void volume is minimized and the porosity is maximized — resulting in the greatest potential for resolution.

It is important to realize that the plate count for small test molecules does not necessarily predict the plate count for proteins or polymers. Besides the differences caused by diffusion coefficients, which will be discussed later, there can be another decrease in plates for macromolecules which appears to be caused by pore geometry. Because no pores are ideal in structure, band spreading can be caused by restricted mass transfer for large molecules whose dimensions frequently approach either those of the pores or those of channels within the pores.

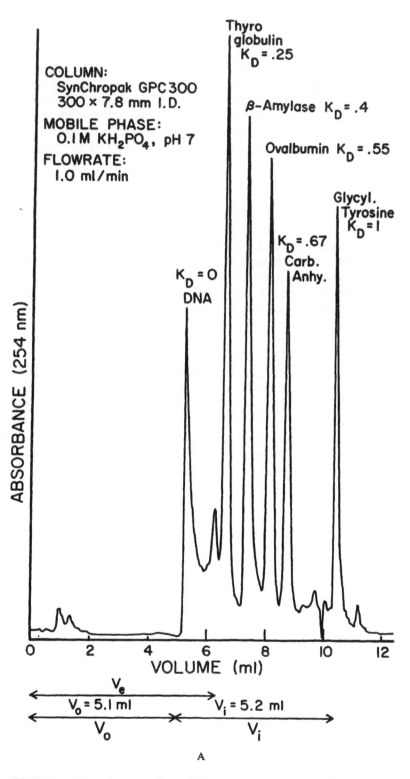

FIGURE 1. High-performance size-exclusion chromatography (HPSEC) of proteins.
Column: SynChropak GPC300 (SynChrom, Lafayette, IN), 300 × 7.8 mm I.D., 5-μm
particle size, 300-Å pore size. Mobile phase: 0.1 *M* potassium phosphate, pH 7; flow-rate,
1 ml/min. (A) typical profile of proteins obtained by HPSEC. (B) calibration curve for the
SynChropak GPC300.

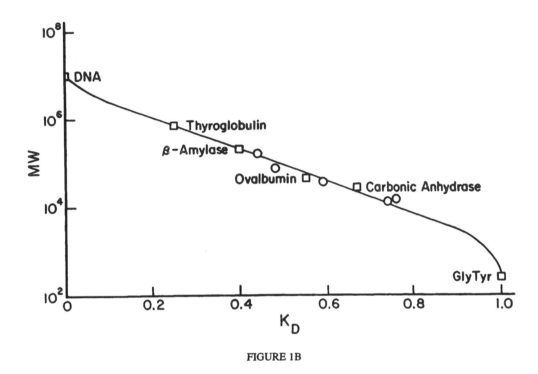

FIGURE 1B

III. CONTRIBUTIONS OF SOLUTE

A. HYDRODYNAMIC VOLUME

In actuality, the characteristic of molecules which is differentiated during SEC is the hydrodynamic volume and shape rather than solely the molecular weight. The importance of this distinction is illustrated in Figure 3 which depicts calibration curves for both linear sulfonated polystyrenes and globular proteins. The displacement of the calibration curve shows that a linear polymer has a much larger hydrodynamic volume than a globular protein of the same molecular weight. This effect emphasizes the fact that in order to achieve accuracy in molecular weight estimations, it is necessary to calibrate a SEC column with polymers which have the same shape characteristics as those of the solute. Soluble proteins and membrane proteins which are suitable for use as HPSEC standards can be found in references such as that by LeMaire.[6]

B. DIFFUSION COEFFICIENT

The low diffusion coefficients of proteins and other macromolecules result in their having significantly lower plate counts than small molecules such as the dipeptide, glycyl tyrosine. The peak widths for large molecules are also more adversely affected by increases in flow-rate or linear velocity than small molecules. Because the diffusion coefficient has an inverse relationship to the molecular weight of a molecule, the adverse effect of high linear velocity increases with the molecular size of the molecule, as can be seen in Figure 4. As a consequence, it is generally preferable to analyze proteins at low linear velocities (1 to 4 cm/min) for maximum resolution. This would translate to flow rates of about 0.1 to 0.5 ml/min for 4.6 mm I.D. columns or 0.4 to 1.5 ml/min for 7.8 mm I.D. columns.

C. SAMPLE SIZE

Of the various modes of chromatography used to analyze proteins, SEC has the lowest loading capacity in terms of both sample weight and volume. This limitation is due to the isocratic nature of SEC and to its volume dependence. Although the sample weight limitations

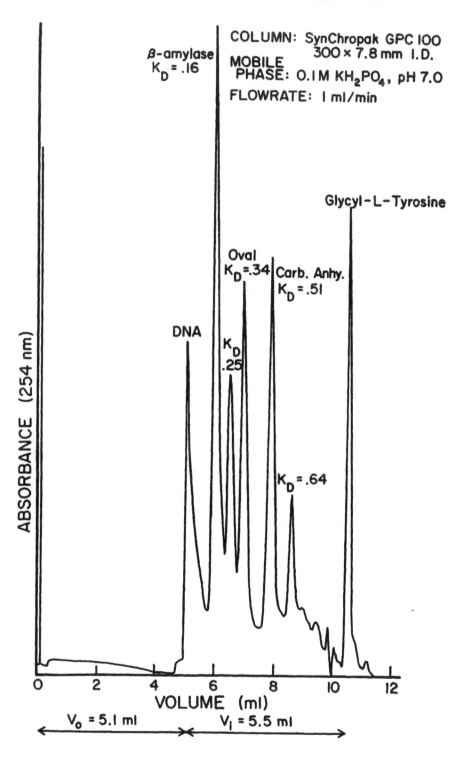

FIGURE 2.   HPSEC of standard proteins on a 100-Å pore diameter support. Column: SynChropak GPC100, 300 × 7.8 mm I.D., 5-μm particle size, 100-Å pore size. Mobile phase: same as for Figure 1A.

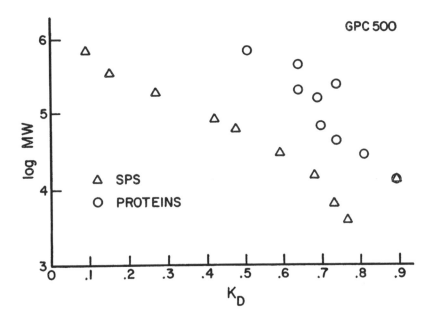

FIGURE 3. Effect of solute shape on calibration curves. Column: SynChropak GPC500, 250 × 4.6 mm I.D., 7-μm particle size, 500-Å pore size. Mobile phase: 0.1 M sodium sulfate for sulfonated polystyrenes (Δ) or 0.1 M potassium phosphate for standard proteins (o); flow-rate, 0.5 ml/min.

are partially based on the molecular weight of the solute, approximately 2 to 4 mg of protein can be loaded onto an HPSEC column which has dimensions of 300 × 7.8 mm I.D. without causing significant band broadening. Likewise, sample volumes of about 2% of the column volume can be injected before efficiency deteriorates.

IV. CONTRIBUTIONS OF MOBILE PHASE

The mobile phase is extremely important in HPSEC because it not only can cause a solute to have interaction with the support and be eluted in a nonideal manner (not truly by SEC), but it also can cause a solute to change its size or shape.

A. ELIMINATION OF NON-IDEAL INTERACTIONS

It is necessary to eliminate all interactions between a column support and a solute in order to achieve ideal SEC and truly separate solutes by size or hydrodynamic volume. Ionic interactions can be reduced by increasing the ionic strength of the mobile phase and/or reducing the pH. In Table 2 it can be seen that an ionic strength of at least 0.24 was needed to elute arginine with a K_d of 1.00 on any of the HPSEC columns in this 1980 study.[2] Because increasing the ionic strength likewise promotes hydrophobic interactions, this potential drawback must always be taken into consideration. Reduction of mobile phase pH is also generally effective for reducing ionic interactions on HPSEC supports because their residual charges tend to be negative — caused by either silanols on silica-based supports or carboxylic groups on some polymeric supports.[7] The range of pH adjustments is usually restricted, however, because many biological macromolecules, including proteins, are only stable at neutral pH.

Hydrophobic interactions can be decreased or eliminated by decreasing the ionic strength of the mobile phase or adding organics to the mobile phase. One must be alert, however, that these adjustments may increase ionic interactions concurrently. In Table 3, it can be seen that raising the ionic strength of the mobile phase increased the hydrophobic interactions of phenylethanol with all of the HPSEC supports investigated.[2]

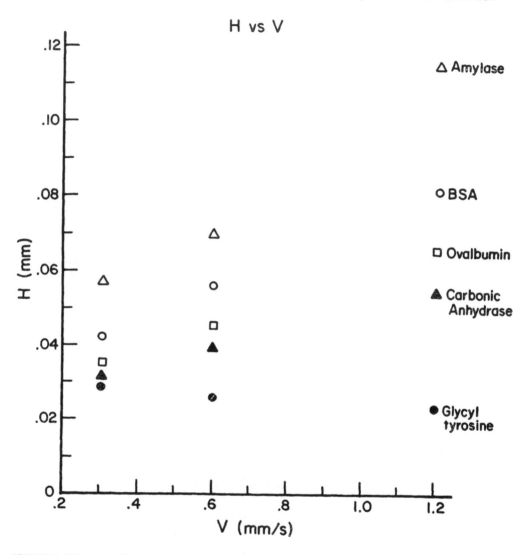

FIGURE 4. Effect of mobile phase linear velocity (V) on band spreading for solutes with different molecular weights. Column: SynChropak GPC500, 250 × 4.6 mm I.D., 7-μm particle size, 500-Å pore size. Mobile phase: 0.1 M potassium phosphate, pH 7. The plate height (H) was calculated using the equation: H = L/N, where L is the length of the column (mm) and N is the number of theoretical plates.

The inherent ionic and hydrophobic properties of biomolecules, as well as those of HPSEC supports, have resulted in the general use of mobile phases composed of buffers at neutral pH with ionic strengths from 0.05 to 0.1. After the appropriate mobile phase has been selected for an HPSEC analysis, calibration curves should be run in the same solvent system because changes in salt composition can also affect the hydrodynamic volume of polymers.[8]

Peptides have proven to be especially difficult to analyze by HPSEC. This has been due to many factors including their frequent hydrophobic or cationic properties, their nonstandard conformations which range from linear to globular, and the scarcity of commercial HPSEC columns which are suitable for biomolecules with molecular weights less than 10,000. Hodges et al. have developed peptide standards for HPSEC and have identified solvent systems which allow these standards to be analyzed without interactions with the support.[9]

B. ADDITION OF DETERGENTS

The addition of detergents to the mobile phase can cause multi-subunit proteins to divide into

TABLE 2
The K_d of Arginine on Various Columns as a Function of Mobile Phase Ionic Strength

Column	Mobile phase ionic strength[a]					
	0.026	0.12	0.24	0.60	1.20	2.40
TSK SW 3000	1.30	1.05	1.02	1.00	—	0.98
SynChropak GPC 100	1.35	1.06	1.01	—	—	0.98
LiChrosorb Diol	1.53	1.15	1.05	0.99	—	1.07
TSK SW 2000	1.57	1.06	1.02	0.99	—	0.98
Waters I-125	1.70	1.23	1.16	1.08	—	1.05
Waters Bondagel	1.75	1.11	1.06	1.02	—	1.00
Shodex OH pak B-804	2.06	1.16	1.07	1.02	1.02	—

[a] The mobile phase consisted of pH 7.05 phosphate buffer of the ionic strengths indicated above.

From Pfannkoch, E. et al., *J. Chromatogr. Sci.*, 18, 430, 1980 with the kind permission of Preston Publications.

TABLE 3
The K_d of Phenylethanol on Various Columns as a Function of Mobile Phase Ionic Strength

Column	Mobile phase ionic strength[a]					
	0.026	0.12	0.24	0.60	1.20	2.40
SynChropak GPC 100	1.44	1.49	1.53	1.63	1.81	2.33
TSK SW 3000	1.47	1.50	1.53	1.61	1.81	2.35
Waters I-125	1.83	1.88	1.88	2.03	2.29	3.03
TSK SW 2000	1.95	2.02	2.10	2.30	2.71	4.01
LiChrosorb Diol	2.49	2.56	2.64	2.93	3.52	5.31
Waters Bondagel	5.32	5.19	5.37	5.97	7.44	11.47
Shodex OH pak B-804	6.36	6.65	6.96	8.47	10.96	—

[a] The mobile phase consisted of pH 7.05 phosphate buffer of the ionic strengths indicated above.

From Pfannkoch, E. et al., *J. Chromatogr. Sci.*, 18, 430, 1980 with the kind permission of Preston Publications.

individual subunits, can change a globular molecule into one which is elongated, or can enlarge the size of a protein by adding a layer to the outside. Sodium dodecyl sulfate (SDS) is an example of a detergent which separates proteins into their subunits and also creates a layer on the outside surface of proteins, making them appear to have a higher molecular weight. The calibration curve for proteins using a mobile phase containing 0.1% SDS is offset from the standard curve in a non-denaturing mobile phase in a similar manner to the displacement that was seen in Figure 3 for linear polymers.

V. GUIDELINES TO ACHIEVING IDEAL HPSEC

In conclusion, HPSEC can be used effectively as a technique for protein analysis if the following general guidelines are met:

1.   Select a column support designed for HPSEC of proteins and/or peptides
2.   Select a pore diameter which results in a K_d from 0.2 to 0.8

3. Select a mobile phase in which there is no solute-support interaction
4. Calibrate the column in the mobile phase selected in Step 3 using molecules of the same shape as the solute

REFERENCES

1. **Regnier, F.E. and Noel, R.,** Glycerolpropylsilane bonded phases in the steric exclusion chromatography of biological macromolecules, *J. Chromatogr. Sci.*, 14, 316, 1976.
2. **Fukano, K., Komiya, K., Sasaki, H., and Hashimoto, T.,** Evaluation of new supports for high-pressure aqueous GPC: TSK-Gel SW type columns, *J. Chromatogr.*, 166, 47, 1978.
3. **Regnier, F.E. and Gooding, K.M.,** High performance liquid chromatography of proteins, *Anal. Biochem.*, 103, 1, 1980.
4. **Gooding, K.M. and Regnier, F.E.,** Size exclusion chromatography, in *HPLC of Biomolecules*, Marcel Dekker, New York, 1990.
5. **Pfannkoch, E., Lu, K.C., Regnier, F.E., and Barth, H.G.,** Characterization of some commercial high prerformance size-exclusion chromatography columns for water-soluble polymers, *J. Chromatogr. Sci.*, 18, 430, 1980.
6. **LeMaire, M., Aggerbeck, L.P., Monteilhet, C., Andersen, J.P., and Moller, J.V.,** The use of high-performance liquid chromatography for the determination of size and molecular weight of proteins: a caution and a list of membrane proteins suitable as standards, *Anal. Biochem.*, 154, 525, 1986.
7. **Mikes, O. and Coupek, J.,** Organic Supports, in *HPLC of Biomolecules*, Marcel Dekker, New York, 1990.
8. **Barth, H.G.,** A practical approach to steric exclusion chromatography of water-soluble polymers, *J. Chromatogr. Sci.*, 18, 409, 1980.
9. **Mant, C.T., Parker, J.M.R., and Hodges, R.S.,** Size-exclusion high-performance liquid chromatography of peptides. Requirement for peptide standards to monitor column performance and non-ideal behaviour, *J. Chromatogr.*, 397, 99, 1987.

SIZE-EXCLUSION CHROMATOGRAPHY OF PROTEINS

Heinz Engelhardt

I. PRINCIPLES OF SIZE-EXCLUSION CHROMATOGRAPHY (SEC)

In SEC, molecules are separated according to their molecular size in solution. The main applications of SEC are: separation of molecules differing in size, determination of average molecular weights, determination of hydrodynamic diameters, and separation of proteins from small molecules, e.g., desalting. Porous stationary phases with defined pore diameters are required. A prerequisite for SEC is that the eluent has to be adjusted in such a way that no interaction between sample molecules and the surface of the stationary phase takes place. By definition, the separation is finished when the smallest molecule, usually the eluent molecule, is eluted. The larger the size of a molecule, the smaller the amount of accessible pore volume and the earlier the molecule is eluted. Molecules which are larger than the largest pore diameter cannot penetrate into the pores and pass through the column fastest with the interstitial velocity (u_z) of the eluent. These molecules are eluted with the interstitial volume (V_z) of the column, i.e., the dead or void volume in SEC. The smallest molecule is eluted with V_m, the total volume (volume of the total mobile phase) of the column. Consequently, SEC is finished when sorptive chromatography starts. Due to the small separation volume, limited to the internal or pore volume (V_p) of the stationary phase, optimization strategies are extremely important in SEC. The volume of the mobile phase in the column combines V_z and V_p:

$$V_m = V_z + V_p$$

The elution volume (V_e) of a solute in SEC can be described as

$$V_e = V_z + K_{SEC} \cdot V_p$$

where K_{SEC} is the fraction of pore volume accessible to the solute, and is sometimes erroneously called partition coefficient in SEC. By definition, for a totally excluded molecule this "partition coefficient" is zero, and for a molecule penetrating all the pores, $K_{SEC} = 1$.

In analogy to the capacity ratio (k') in retentive chromatography, an exclusion chromatography ratio ($k*$) can be defined as

$$k^* = \frac{V_e - V_z}{V_z}$$

In contrast to the practically unlimited k' value (limited only by the decreasing detection sensitivity with increasing k'), $k*$ approaches a maximum value:

$$k^*_{max} = \frac{V_p}{V_z}$$

This demonstrates that k^*_{max} is determined by the pore volume of the stationary phase and the interstitial volume of the column. The latter is a factor of packing technology and particle geometry.

Elution in SEC can be described as

$$V_e = V_z\left(1 + k^*\right)$$

compared to adsorption (retentive) chromatography (retention volume V_R)

$$V_R = V_m\left(1 + k'\right)$$

Resolution in SEC is — as in adsorptive chromatography — a function of the differences in elution volume and the peak broadening of the two peaks. The differences in elution volume of two solutes with molecular weight M_1 and M_2 can be obtained from the calibration curve, where usually the logarithm of the solute molecular weight is plotted as a function of elution volume. The slope of the calibration curve within the separation range determines the elution volume difference. To obtain a large span of molecular weights, several columns packed with stationary phases with different average pore diameters can be coupled. For optimum separation, it has been found best to couple columns which differ in pore diameter by a factor of 10 to 20.[1]

The resolution equation in adsorption chromatography usually contains also the peak broadening factor by assuming that the peak widths of adjacent peaks are identical. This is not possible in SEC, because peak broadening is affected by two interacting and interdependent factors. The diffusion coefficient of the solute and, hence, its influence on peak broadening is dependent on its molecular weight and elution volume. The earlier the peaks are eluted, the lower their diffusion coefficients. However, the time these molecules spend in the mobile phase is smaller than that of smaller molecules with higher diffusion coefficients. Consequently, peak broadening is changing dramatically within the chromatogram. Discussion of this would be beyond the scope of this paper; details can be found in the literature.[1-3]

II. OPTIMIZATION STRATEGIES IN SEC

A. OPTIMIZATION OF CAPACITY FACTOR, k*

As demonstrated above, the relationship of pore volume and interstitial volume in the column influences the separation capacity of a stationary phase and the analysis time obtained with this stationary phase. A large separation capacity, k^*, means increased accuracy of determination of molecular mass distribution. A small interstitial volume, on the other hand, means a shorter analysis time, because the time needed for elution of the totally excluded peak (the dead time in SEC) is shorter. With classical swollen gels, such as Sephadex, this relationship can reach values of up to 2. With rigid materials, such as silica, values around one can usually be reached. To increase the ratio of V_p and V_z, two possibilities are available. First, silicas with larger specific pore volumes can be used. However, nature limits this possibility, because with increasing specific pore volume the silica becomes like a sponge and mechanical (pressure) stability decreases. On the other hand, it has been found that spherical silicas can be packed more densely than irregular ones of identical particle diameter. This also leads to a decrease of interstitial volume. Usually, with spherical particles the interstitial volume is 20% lower. However, owing to their preparation, the porosities of these materials are low. Depending on the shape of the silica and their pore volume, k^* values between 0.6 and 1.4 can usually be achieved in SEC.

B. REDUCTION OF PEAK DISPERSION

The only possibility of further increasing resolution in SEC — as in all types of HPLC — is to minimize peak dispersion by reduction of particle diameter and optimization of eluent velocity. To evaluate chromatographic dispersion in SEC is difficult because there is always a separation according to molecular weight superimposed. By using "monodisperse" polystyrene standards, it has been demonstrated that the peak capacity of SEC can be doubled by reducing the particle diameter from 10- to 3-μm.[2] Due to the small diffusion coefficients of proteins in aqueous buffers, which vary between 10^{-8} and 10^{-6} cm^2/sec within the usually accessible molecular weight range, the linear velocity has to be low to work close to the minimum of the van Deemter plot (H vs. u curve), where optimum (i.e., maximum) plate numbers are generated. Consequently, the problems encountered with very small particles (excessive pressure drop, heat of friction within the column) are negligible in size-exclusion chromatography. If small peak volumes create problems, the column diameter may be increased from the standard 4 mm to 6 or 8 mm.

In Figure 1, the increase of resolution in SEC with decreasing interstitial velocity is demonstrated with protein standards for a column packed with 10-μm particles. An interstitial velocity (u_z) of 2 mm/sec corresponds to the usual standard velocity of 1 ml/min (with 4 mm I.D. columns) in sorptive liquid chromatography. With an interstitial velocity of 0.13 mm/sec at this particle diameter, the optimum resolution has not yet been reached.

Figure 2 compares the separation achieved with 10-μm particles to that with 3-μm particles of the same silica at two different velocities. As expected, the resolution is better with the smaller particles, and at an interstitial velocity of 0.13 mm/sec the maximum efficiency should be achieved. At this velocity, one works close to the optimum of the H vs. u curve for both the low and the high molecular weight samples. It has been shown that working under optimal conditions and reducing the particle diameter from 10- to 3-μm yields a duplication of the separation efficiency obtained with a 10-μm particle diameter and reduces analysis time by a factor of 5.[2]

The optimum linear velocity is, of course, a function of the diffusion coefficient of the biopolymer and, hence, its molecular weight. An interstitial velocity around 0.1 mm/sec is optimum for solutes with molecular weights of up to 200,000 and particle diameters of 3- to 5-μm. For solutes with molecular weights of up to 1,000,000, optimum interstitial velocity has to be reduced further to 0.01 mm/sec and below.

C. OPTIMIZATION FOR SIZE EXCLUSION

Crosslinked hydrophilic gels have been used in classical SEC of proteins. These materials swell considerably in the presence of water, thus generating the pore structure required for SEC. Unfortunately, this swollen matrix is sensitive to pressure and also to changes in pH, ionic strength, and eluent composition. One problem with these gels is that they are not applicable to the conditions generally applied in HPLC to achieve high speed, high resolution, and high efficiency. Increasing the pressure stability by increased crosslinking is a limited option, because increased crosslinking results always in decreased pore diameter. Therefore, the advantages of porous silica with a rigid structure are obvious, especially because silica can easily be prepared with varying pore diameters in the range of 4 to 400 nm (40 to 4000 Å). Of course, the polar surface of silica with its acidic surface groups has to be modified to prevent irreversible adsorption and denaturation of proteins during chromatographic separation. Because proteins are polyelectrolytes with more or less hydrophobic centers at the molecule surface, they are able to interact with the stationary phase in many ways. Consequently, the eluent composition has to be adjusted properly to achieve a predominant SEC mechanism.

Polar bonded stationary phases with diol or N-acetylpropylamine groups show the weakest hydrophobic interaction,[5,6] allowing solute separation to be achieved without organic modifier in the eluent. Besides increased retention due to hydrophobic interaction, ionic effects can also obscure the separation according to a size-exclusion mechanism.

Depending on the pH of the buffer, its ionic strength, and the isoelectric point of the proteins, three different interactions of the protein with the stationary phase are possible and can be differentiated: proteins which are negatively charged in the buffered eluent are excluded much

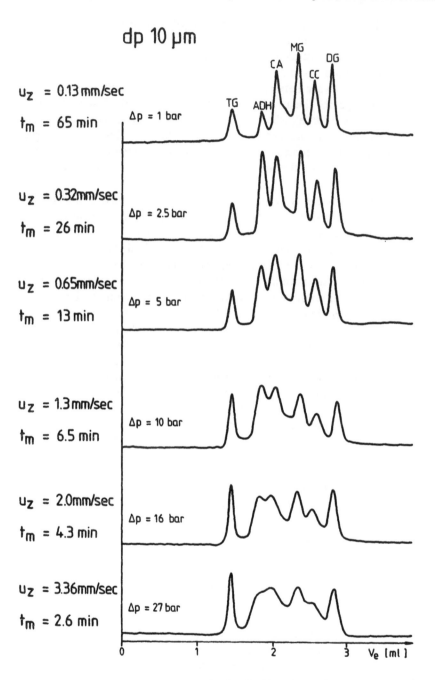

FIGURE 1. Influence of flow-rate on separation in SEC of proteins. Column: 250 × 4.1 mm I.D., packed with amide bonded to silica Grace 250 A, 10-μm particle size, 250-Å pore size (Grace GmbH, Worms, FRG). Mobile phase: 0.1 *M* Tris-buffer, pH 7, containing 0.4 *M* NaCl. t_m denotes analysis time. u_z denotes interstitial velocity of mobile phase; 2 mm/sec (u_z) corresponds to a standard velocity of 1 ml/min for a 4 mm I.D. column. The standard velocity, or flow-rate, is obtained by multiplying u_z by the interstitial (or dead) volume of the column, V_z, and then dividing by the length of the column. TG, ADH, CA, MG, CC, and DG denote, respectively, thyroglobulin, alcohol dehydrogenase, conalbumin, myoglobin, cytochrome c, and dinitrophenylglutamic acid. (From Engelhardt, H. and Schön, U. M., *Chromatographia*, 22, 388, 1986. With permission.)

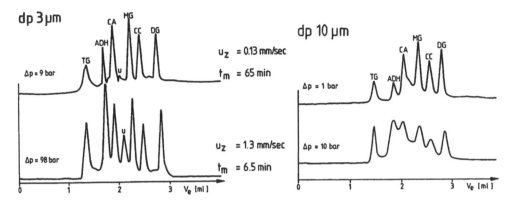

FIGURE 2. Influence of particle diameter on protein resolution. Column I (left): 250 × 4.1 mm I.D., packed with 10-μm particles. Column II: (right) 250 x 4.1 mm I.D., packed with 3-μm particles. Mobile phase: see Figure 1. Protein identities: see Figure 1. t_m denotes analysis time. u_z denotes interstitial velocity of mobile phase. From Engelhardt, H. and Schön, U. M., *Chromatographia*, 22, 388, 1986. With permission.)

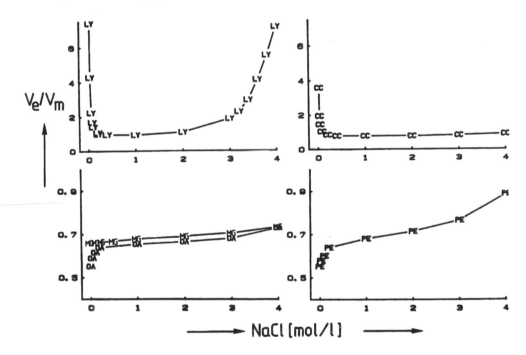

FIGURE 3. Elution behavior of proteins on a polar silica-based SEC column. Column packing: see Figure 1. Column dimensions: 150 × 4 mm I.D. Mobile phase: 0.01 M phosphate buffer, pH 7. Ionic strength adjusted by adding NaCl; flow-rate, 0.5 ml/min. LY, CC, MG, OA, and PE denote lysozyme, cytochrome c, myoglobin, ovalbumin, and pepsin, respectively. V_e denotes elution volume of protein; V_m denotes total volume (volume of the total mobile phase) of the column.

more than expected from their relative molecular mass, due to an anionic exclusion mechanism, sometimes described as Donnan potential. This effect is noticeable at low ionic strength (below 0.3), as seen in Figure 3 for ovalbumin and pepsin. Positively charged proteins can interact with acidic surface silanols, still present even after chemical modification, by an ion-exchange mechanism. In Figure 3, this is demonstrated for cytochrome C and lysozyme. This interaction can be decreased by increasing the ionic strength of the buffer up to 0.4 and 0.5. However, with this high ionic strength, the retention can increase again, as can be seen for pepsin and, at much

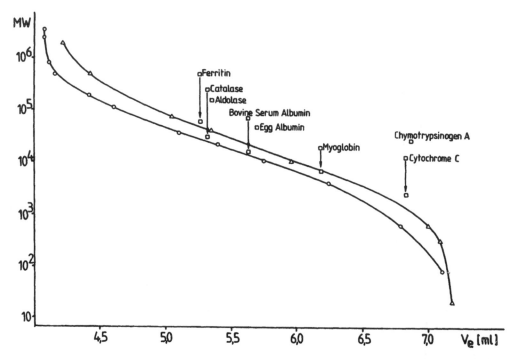

FIGURE 4. Calibration curve of a two-column system. Columns: each 300 × 4.2 mm I.D., packed with Lichrosorb Si 100 and Lichrosorb Si 500 modified with propylacetamide. Solutes: polystyrenes (circles), with dichloromethane as eluent; dextrans (triangles), with 1% aq. NaN_3 as eluent; proteins (squares), with 0.1 M Tris-buffer, pH 7.5 (ionic strength adjusted to 0.5 with NaCl) as eluent. Flow-rate: 1 ml/min.

higher ionic strength, also for lysozyme. In this case, the proteins are retarded on the bonded organic moiety on the surface by hydrophobic interaction and are eluted after V_m. This salting out effect depends , of course, on the hydrophobic character of the protein and can be used additionally for separation by hydrophobic interaction chromatography (HIC).

Summing up, the elution volume of proteins on chemically modified silica-based stationary phases is dependent solely on their size and the accessibility of the pore volume, if ionic interactions (ion-exchange or ion-exclusion) can be minimized and hydrophobic interactions are unimportant. This can be achieved by optimizing pH and the ionic strength of the buffer eluent. Hydrophobic interactions can, of course, be reduced by the addition of an organic modifier to the buffer. However, this may cause protein denaturation.

D. OPTIMUM PORE SIZE DISTRIBUTION

Silica is available with pore sizes ranging from 4 nm (40 Å) to 400 nm (4000 Å). In HPLC, the standard material has an average pore diameter around 10 nm (100 Å). With this silica, only small proteins with molecular masses below 50,000 daltons can be separated. For those with higher molecular masses, less than 20% of the pore volume is accessible. For the separation of larger proteins, silicas with pore diameters between 30 nm (300 Å) and 50 nm (500 Å) are required. With a silica of 50 nm pore diameter, about 50% of the pore volume is accessible for ferritin with a molecular mass of around 500,000 daltons.

To extend the separation capacity in SEC, the use of combined columns packed with silicas with different pore size distributions of exclusion limits is recommended. In Figure 4, the calibration curve for such a two column system is shown. Two columns packed with chemically modified silica with average pore diameters of 10 and 50 nm were coupled. Another advantage of silica-based materials is demonstrated additionally. These columns can be used with aqueous as well as organic eluents because the pore structure of the stationary phase and bed structure

TABLE 1
Molecular Weight and Size

	Molecular weight	Diameter or size (nm)	Molecular weight of a random coil polystyrene of identical size
Ferritin	480,000	15.6	57,000
Catalase	240,000	10.4	29,000
Bovalbumin	67,000	7	15,000
Myoglobin	17,800	$4.4 \times 4.4 \times 2.5$	7,000
Cytochrome C	12,400	$2.5 \times 2.5 \times 3.5$	2,500

of the columns are not altered by changing the eluent. Consequently, it is possible to calibrate the column system in an organic eluent with polystyrenes which can be easily obtained as narrow calibration standards in a wide range of available molecular weights.

The separation capacity of the combination shown in Figure 4 spans a relative molecular mass range of 10,000 for polystyrenes (circular) in dichloromethane and dextrans (triangles) in water. For proteins, the separation capacity of this column system also spans a wide molecular mass range and exceeds the 500,000-dalton range.

E. CORRELATION OF ELUTION VOLUME WITH MOLECULAR MASS AND SIZE

As stated above, a paramount advantage of silica-based stationary phases in SEC is the possibility to calibrate the columns with polystyrene standards in organic eluents and to use them in aqueous systems for protein separation. The size of the random coil of linear polymers is closely correlated to their molecular weight.[7] To each molecular mass of the polystyrene standard, a distinct pore diameter can be assigned at which the molecule is excluded from the pores.[8] The dextrans, which also form random coils in water, have a higher relative molecular mass per unit chain length. Therefore, they seem to be much more dense than corresponding polystyrene molecules; the relative volume of the random coil is smaller. This is the reason why they are eluted at larger elution volumes than polystyrenes of identical molecular weight.

Proteins exist in an ordered tertiary structure, which is much more dense than that of random coils. Consequently, they are eluted much later — as smaller molecules — than random coil polymers of identical molecular mass. Therefore, the elution volumes of proteins shown in Figure 4 are always displaced from that of polystyrenes and dextrans to higher values.

For some proteins, their molecular dimensions or hydrodynamic diameters have been determined by different ways. In Table 1, the molecular weights and the sizes of a few proteins are shown. Also included are the molecular weights of polystyrenes which have the same elution volume and, as random coils, the same size as the proteins. Bovalbumin with a molecular weight around 67,000 daltons has a Stoke's diameter of 7 nm. A random coil polystyrene molecule with a molecular weight around 15,000 daltons corresponds to this size. Similar calculations have been made for the other proteins. This shift of the molecular weight when going from ordered to random coil structure is indicated in Figure 4 by arrows. This new calibration curve for proteins now fits much better to that of polystyrenes and dextrans. This is also a confirmation of the absence of additional retention mechanisms other than size exclusion. By inversion of this procedure, SEC can be used to determine Stoke's radii for unknown proteins by correlating their elution volume to that of polystyrene standards.

III. CONCLUSIONS

Resolution in SEC can be improved significantly by applying surface modified silicas with large pore volumes and with small particle diameters. Peak dispersion is minimal if one is

1 Myoglobin
2 β-Lactoglobulin
3 Lysozyme
4 Ovalbumin
5 α-Chymotrypsin
6 Chymotrypsinogen A

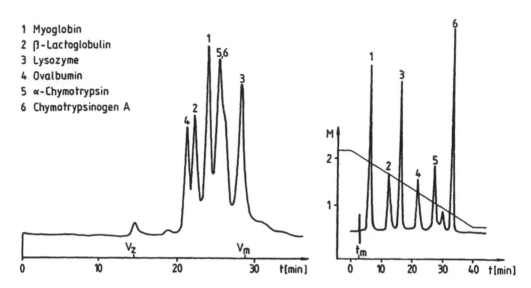

FIGURE 5. Two-modal column use: SEC and HIC. Column: 250 × 4.1 mm I.D., packed with amide on Grade Silica 250 A, 7-μm particle size, 250-Å pore size. SEC (left): mobile phase, Tris/HCl-buffer, pH 7.5, in 0.4 M NaCl; flow-rate, 0.1 ml/min. HIC (right): mobile phase, linear AB gradient (15 to 80% B in 40 min), where Eluent A is 2.5 M $(NH_4)_2SO_4$ in 0.1 M phosphate buffer, pH 7.0, and Eluent B is 0.1 M phosphate buffer, pH 7.0; flow-rate, 1 ml/min. V_m denotes total volume (volume of the total mobile phase) of the column. V_z denotes interstitial (or dead) volume of the column.

working close to the optimum value of linear velocity. The interstitial velocity in aqueous medium should be in the range of 0.1 to 0.01 mm/sec if high molecular weight proteins have to be separated. The pressure drop with aqueous eluents and 3-μm particles at these flow rates will always be below 50 bar. Here, an analysis time is still in an acceptable range (around 60 min) because with the small particle columns the plate numbers required are generated in shorter columns.

By optimizing the pH and the ionic strength of the aqueous eluent, secondary interactions of the proteins with surface groups can be minimized, and in many cases excluded. Because the eluent pH and ionic strength can be adjusted to optimum conditions, protein denaturation during chromatography is minimized and the proteins can be purified whilst retaining their biological activity.

The disadvantage of SEC is its limited separation capacity. However, proteins are adsorbed on SEC phases at high ionic strength. This hydrophobic interaction can be used to separate proteins with a polar SEC column according to a different mechanism.[9] The solutes are injected in an appropriate buffer at high ionic strength, usually containing 2.0 to 2.5 M ammonium sulfate. The proteins are salted out onto the stationary phase and are retarded. In a gradient elution mode, the ionic strength is reduced to that where SEC is possible. The proteins are eluted according to their hydrophobicity. Of course, the elution order is different from that on the same column in the SEC mode. In Figure 5, this is demonstrated for the separation of standard proteins in the SEC mode as well as in the HIC mode. The bimodal use of the same column, of course, is only possible because of the pressure stability, constant permeability and structure, as well as bed stability of silica-based stationary phases.

ACKNOWLEDGMENTS

This paper contains, in part, unpublished results from the Ph.D. Thesis of G. Ahr, M. Czok, D. Mathes, and U. M. Schön. Their work has been made possible by financial furtherance of the

Fonds der Chemischen Industrie, Frankfurt, and the Deutsche Forschungsgemeinschaft, Bonn. Their help is gratefully appreciated.

REFERENCES

1. **Yan, W.W., Kirkland, J.J., and Bly, D.D.,** *Modern Size Exclusion Liquid Chromatography,* John Wiley & Sons, New York, 1979.
2. **Engelhardt, H. and Ahr, G.,** Optimization of efficiency in size-exclusion chromatography, *J. Chromatogr.,* 282, 385, 1983.
3. **Guiochon, G.,** Theoretical investigation of the optimum particle size for the resolution of proteins by size-exclusion chromatography, *J. Chromatogr.,* 326, 3, 1985.
4. **Engelhardt, H. and Schön, U.M.,** Optimal conditions for size exclusion chromatography of proteins, *Chromatographia,* 22, 388, 1986.
5. **Pfannkoch, E., Lu, K.C., Regnier, F.E., and Barth, H.G.,** Characterization of some commercial high performance size-exclusion chromatography columns for water-soluble polymers, *J. Chromatogr. Sci.,* 18, 430, 1980.
6. **Engelhardt, H. and Mathes, D.,** High performance liquid chromatography of proteins using chemically-modified silica supports, *Chromatographia,* 14, 325, 1981.
7. **Vollmert, B.,** *Polymer Chemistry,* Springer, Berlin-Heidelberg New York, 1973.
8. **Halász, I. and Martin, K.,** Porengrößen von Festkörpern, *Angew. Chem.,* 90, 954, 1978.
9. **Engelhardt, H. and Schön, U.M.,** Separation of proteins on polar bonded phases by hydrophobic interaction chromatography, *J. Liq. Chromatogr.,* 9, 3225, 1986.

SIZE-EXCLUSION HIGH-PERFORMANCE LIQUID CHROMATOGRAPHY OF INTEGRAL MEMBRANE PROTEINS AND THE EFFECT OF DETERGENTS ON IMMUNOLOGICAL ACTIVITY

Gjalt W. Welling and Sytske Welling-Wester

I. INTRODUCTION

Since the introduction of chromatography, the recovery of protein bioactivity has been a major concern and the same is true for proteins subjected to HPLC. Recovery of bioactivity means that the conformation of a protein is still intact after chromatography or that the original structure is regained. Spectroscopic measurements of tyrosine and tryptophan exposure and second derivative spectra may provide data on conformational changes in a protein during chromatography.[1,2] After chromatography, biological activities, i.e., enzymatic activity, hormonal activity,[3] or immunological activities, may be determined to obtain information on the conformational state of a protein. There are differences to be considered in the type of immunological activity that is being investigated. Antibodies are directed against linear and conformational epitopes of a protein. A linear epitope comprises consecutive amino acids in a polypeptide chain and a conformational epitope consists of amino acids that are brought together in the native conformation from various parts of the polypeptide chain. Polyclonal antibodies against a protein can be used to detect the protein after purification. Since these antibodies are not directed exclusively against conformational epitopes but also against linear parts of the polypeptide chain, this may not give complete information on the conformational status of the protein. Similarly, monoclonal antibodies (Mabs) may be directed against linear epitopes and, as a consequence, reactivity with such Mabs gives less information on the possibility of conformational changes in a protein after chromatography than a reaction with Mabs directed against a conformational epitope. Polyclonal or monoclonal antibodies directed against linear epitopes may recognize the unfolded or denatured protein as well as the native protein. A second matter that deserves consideration is the method to determine the immunological activity. For example, in immunoblotting, which utilizes SDS-polyacrylamide gel electrophoresis (SDS-PAGE) under denaturing and reducing conditions, the predominant immunological reactivity measured involves linear epitopes and denatured proteins. This will provide data on the identity or purity of a protein but not on its native conformation. Native immunoblotting[4] (SDS-PAGE under non-reducing conditions and not more than 0.1% SDS) and utilization of Mabs directed against conformational epitopes may give information regarding the conformation of a particular protein. We have used antibodies, either polyclonal or monoclonal, directed against proteins to determine residual conformation following chromatography.[5,6] In the following sections, this is illustrated by the determination of the immunological activity of Sendai virus proteins following size-exclusion high-performance liquid chromatography (SEC) in different non-ionic detergents using Mabs directed against the intact proteins.

A. SENDAI VIRUS PROTEINS AS A MODEL MIXTURE OF INTEGRAL MEMBRANE PROTEINS

We have used Sendai virus proteins as model proteins to study various chromatographic methods for the separation of integral membrane proteins.[7-9] The envelope of Sendai virus is composed of a lipid bilayer in which two integral membrane proteins are embedded, the fusion protein F (Mr = 65,000) and the hemagglutinin-neuraminidase protein HN (Mr = 68,000). The F protein consists of two components F_1 (Mr = 50,000) and F_2 (Mr = 13 to 15,000) which are connected by disulfide bridges. The HN protein is found in a monomeric, a dimeric, and tetrameric form, HN, HN_2, and HN_4, respectively. HN_2 and HN_4 can be converted to the monomeric form by treatment with a reducing agent. Due to limited proteolysis, a truncated dimeric form (without the transmembrane end), HN_2- may be present in detergent extracts. The strategy to purify integral membrane proteins generally involves extraction of the integral membrane proteins from the lipid bilayer followed by chromatography. During chromatography, a denaturant, a detergent, or an organic solvent has to be present in the eluent to prevent aggregation. A wide variety of additives has been used for SEC of membrane proteins: sodium dodecyl sulfate (SDS),[10-12] sarkosyl,[13] triethylamine-Triton X-100,[14] Brij 35,[15] CHAPS,[16] Tween-20,[16] octylglucoside,[17] guanidine-HCl,[11,18] and deoxycholate.[19] We have used Sendai virus membrane proteins as a model mixture to investigate the effect of a number of these additives on the conformation of integral membrane proteins during SEC on two columns of Superose-6 coupled in series.[5,6]

Immunological activity profiles were determined using Mabs directed exclusively against the native conformation of Sendai virus F and HN protein in an enzyme-linked immunosorbent assay (ELISA). The methodology to determine such a profile is described using SEC of Sendai virus proteins with 0.1% SDS and 0.05% sarkosyl in the eluent as an example.

II. EXPERIMENTAL

A. VIRUS GROWTH; DETERGENT EXTRACTION; REMOVAL OF TRITON X-100

Virus was grown, isolated, and extracted with Triton X-100 as described.[6] Since non-ionic detergents interfere with the ELISA, Triton X-100 was removed by treatment with Amberlite XAD-2.[6] Prior to HPLC, the detergent-free extract was incubated for 20 min in an equal volume of the concentrated (2×) buffers used for elution.

B. SEC; SDS-PAGE

Chromatography was performed with an LKB 2150 pump (LKB, Bromma, Sweden), a Rheodyne 7125 injector (Inacom, Veenendaal, the Netherlands) and an LKB 2151 UV-monitor. The eluate fractions were analyzed with an LKB-midget system on 8% SDS-gels under non-reducing conditions. Polypeptide bands were visualized by silverstaining.

C. DETERMINATION OF IMMUNOLOGICAL ACTIVITY BY AN ELISA

Eluate fractions were collected in 70 × 11 mm minisorp-tubes (Nunc, Roskilde, Denmark). The tubes were covered with a square piece of dialysis tubing and closed by gently pressing 1.5 cm length of silicone tubing over it. After dialysis overnight against water in the cold, the fractions were lyophilized in the tubes. Demineralized water was added to each tube, the content was dissolved, and the fractions were diluted in 50 mM sodium carbonate buffer, pH 9.6, to a concentration of 1 µg/100 µl. The wells of polystyrene microtiterplates (Dynatech, Denkendorf, FRG) were coated with 100 µl from each tube. Wells were coated to determine the reaction with a positive antiserum containing either polyclonal (a population of antibodies directed against several epitopes) or monoclonal (directed against one epitope) antibodies, and a negative antiserum. Plates were kept for 2 hr at 37°C or overnight at 4°C. The plates were washed 3 times

(5 min) with phosphate-buffered saline containing 0.3% Tween 20, 0.2 M NaCl and 100 mg SDS/L (washing buffer). Then, 100 µl of a suitable dilution (generally 1:1000) of a monoclonal antibody directed against the protein or a positive polyclonal antiserum (dilution 1:100) in the same buffer supplemented with 0.5% BSA (dilution buffer) was added to each well and negative serum (dilution 1:100) to control wells. The plates were incubated for 1 hr at room temperature. The plates were washed 3 times (5 min) with washing buffer. Then, 100 µl of a suitable dilution of a conjugate (sheep antimouse IgG conjugated to horse-radish peroxidase) in dilution buffer was added. The plates were incubated for 1 hr at 37°C. After washing (3 times 5 min with washing buffer), 100 µl of a substrate solution containing 0.2 mg/ml orthophenylene diamine dihydrochloride (OPD) (Eastman Kodak, Rochester, NY), 0.006% H_2O_2 in 2% methanol, 0.05 M Na-K-phosphate, pH 5.6, was added to each well to visualize the enzyme reaction. The ELISA-plates were incubated for 30 min at room temperature. Then 50 µl of 2 M H_2SO_4 was added to terminate the peroxidase reaction. The optical density at 492 nm was read in a microplate photometer. The optical density was used as a measure of the immunological activity in each fraction. Background and control values should not exceed an OD_{492} of 0.200. Specific immunological activity of the fractions is expressed as OD_{492} per µg of protein.

Alternatively, if the amount of protein in the fractions is unknown, two-fold serial dilutions of the fractions are coated. The ELISA is performed as described above. Immunological activity is defined as the dilution resulting in an OD_{492} of 1.200.

Since ELISAs have a tendency to give aspecific positive reactions, the following controls were included: (a) a conjugate control in which the incubation step with the antisera was omitted to test the reaction between coat and conjugate; (b) aspecific coating of the antisera. To test the latter, the wells were treated with coating buffer and the possible reaction of antisera and conjugate can be investigated separately. Optical densities with the negative serum should always be less than 0.200. If the ELISA is aspecific positive, the composition of washing buffers should be modified by adding more salt, Tween-20 or occasionally 1 to 5% newborn calf serum.

III. RESULTS

A. SEC

The elution patterns obtained after chromatography of the mixture of Sendai virus proteins are shown in Figures 1 and 2. Figure 1 shows the separation obtained after elution with 0.1% SDS in 50 mM sodium phosphate, pH 6.5. SDS-PAGE shows that all fractions do contain the F protein, suggesting that large micellar complexes containing the F protein were separated on the column. When 0.05% sarkosyl in 10 mM Tris-HCl, pH 7.5, with 0.6 M NaCl was used in the eluent, the elution pattern shown in Figure 2 was obtained. The profile consists of four relatively sharp peaks which all contained the F protein. The HN_4 and HN_2 proteins were only present in peaks 2 and 3, respectively. This means that peaks 1 and 4 do contain relatively pure F protein.

B. IMMUNOLOGICAL ACTIVITY

We have used polyclonal antibodies[5,12] as well as monoclonal antibodies[6] to probe the structure of proteins. Polyclonal antibodies were obtained by infection of animals with a virus preparation. Part of such antibodies may be directed against linear epitopes of the protein that are insensitive to conformational changes, i.e., flexible linear amino acid sequences located on the outside of the protein molecule. As a consequence, the reactivity of such antibodies will be the average reactivity with a variety of structures and it may be difficult to detect subtle changes in conformation. Monoclonal antibodies are directed against one epitope. Epitopes may be linear or conformational. Mabs directed against conformational epitopes may be adequate probes to detect conformational changes in a protein. However, from the many Mabs available against a particular protein, those directed against these conformational epitopes have to be

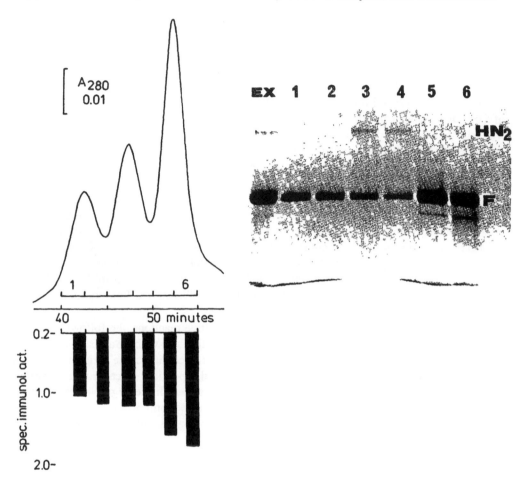

FIGURE 1. Size-exclusion chromatography of Sendai virus membrane proteins. Columns: two Superose-6HR 10/30 columns (300 × 10 mm I.D.; Pharmacia, Uppsala, Sweden) in tandem. Mobile phase: 0.1% SDS in 50 mM sodium phosphate, pH 6.5. Flow-rate, 0.5 ml/min. Absorbance monitored at 280 nm. Fractions 1 to 6 were analyzed on 8% SDS-gels. EX denotes the Sendai virus extract. The positions of the HN$_2$, and the F protein are indicated. The solid bars indicate reactivity of the monoclonal antibodies with F protein. HN protein does not react with monoclonal antibodies. (From Welling-Wester, S., Kazemier, B., Örvell, C., and Welling, G. W., *J. Chromatogr.*, 433, 255, 1988. With permission.)

selected. This was achieved by determining the reactivity of Mabs with Sendai virus proteins denatured by boiling for 2 min in 4% SDS and with untreated proteins.[6] Those that showed the largest difference in reactivity with intact and denatured protein were supposed to be directed against conformational epitopes. The application of such Mabs is illustrated by the reactivity of the eluate fractions in Figures 1 and 2. Polystyrene plates were coated with equal amounts of protein (1 μg/well) and the ELISA described in Section II.C was applied. The data were expressed as specific immunological activities i.e., OD$_{492}$ per μg of protein. Figure 1 shows that elution with 0.1% SDS results in F protein which is still reactive with a Mab directed against conformational epitopes, while the HN protein does not react. When 0.05% sarkosyl is present in the eluent (Figure 2), both F and HN proteins are reactive, indicating that under these conditions the conformation of both proteins remains largely intact. Sarkosyl differs only from SDS with respect to the hydrophilic part of the molecule, methylglycine instead of sulfate in SDS. It has been noted before that, in contrast to SDS, sarkosyl does not denature certain membrane proteins.[20] An additional advantage of Mabs is that the data from the ELISA can also be used as a measure of purity. For example, fraction 1 contains pure F protein with a high immunological activity (Figure 2).

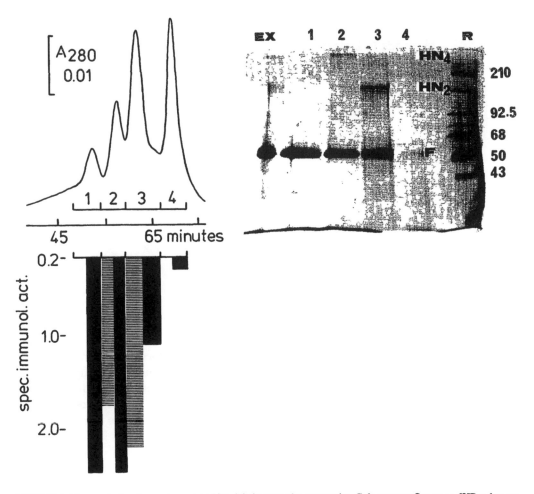

FIGURE 2. Size-exclusion chromatography of Sendai virus membrane proteins. Columns: two Superose-6HR columns (300 × 10 mm I.D.) in tandem. Mobile phase: 0.05% sarkosyl in 10 mM Tris-HCl, pH 7.5, containing 0.6 M NaCl. Flow-rate, 0.5 ml/min. The absorbance was monitored at 280 nm. Fractions 1 to 4 were analyzed on 8% SDS-gels. EX denotes the Sendai virus extract. The positions of the HN_4, HN_2, and the F protein are indicated. R denotes reference proteins, of which the molecular weight is indicated in kilodaltons. The solid bars and shaded bars indicate reactivity of the monoclonal antibodies against F protein and HN protein, respectively. (From Welling-Wester, S., Kazemier, B., Örvell, C., and Welling, G. W., *J. Chromatogr.*, 433, 255, 1988. With permission.)

C. OTHER METHODS TO DETERMINE IMMUNOLOGICAL ACTIVITY

In Table 1, a number of methods to determine immunological activity are listed. The methods differ in their capacity to detect conformational changes. We have used polyclonal antibodies from virus-infected animals[5] and Mabs selected for their reactivity with conformational epitopes[6] in an ELISA. The results were comparable. For example, the highest reactivity of the polyclonal antiserum was found with fractions 5 and 6 in the elution pattern of Figure 1. This indicates that the polyclonal antiserum contained a relatively large amount of antibodies directed against native structure. However, these antibodies do not allow a distinction to be made between the HN and F protein with regard to conformational disturbances. Immunoblotting is a useful method to detect proteins after HPLC. However, native immunoblotting (under non-reducing conditions and not more than 0.1% SDS) is necessary to obtain data on conformation. Since the antigen-antibody reaction occurs under physiological conditions, the other methods in Table 1 can be used to detect conformational changes in an HPLC-purified protein when appropriate antibodies are utilized. Immunization is also listed as a method to determine

TABLE 1
Methods to Determine Immunological Activity after HPLC

Method	Protein(s) (from)	Ref.
ELISA	Influenza virus	21
	Sendai virus	5, 6, 12
Immunization	Bovine viral diarrhea virus	22
	Epstein-Barr virus	23
	Measles virus	24
	Poliovirus	25
Immunoblotting	Bovine viral diarrhea virus	22
	Flavivirus	10
	Herpes simplex virus	26
	Measles virus	24
	Rubella virus	27
Immunodiffusion	Influenza virus	15
Immunoprecipitation	Adenovirus	28
Radioimmunoassay	Epstein-Barr virus	23
	Thyrotropin (TSH)	3
Time-resolved fluoroimmunoassay	Adenovirus	29

immunological activity after HPLC. HPLC-purified proteins may be used to immunize experimental animals to produce antibodies against the proteins. When the antiserum reacts with the intact protein or, in the case of viral proteins, is able to neutralize viral infectivity, this implies that a considerable part of intact structure must have been present, either consisting of linear or conformational epitopes. This has been shown for 90 to 150 kDa polypeptides of bovine viral diarrhea virus,[22] the Epstein-Barr virus membrane antigen gp340[23] and the H glycoprotein of measles virus.[24]

IV. CONCLUSIONS

Depending on how the HPLC-purified protein will be used, a method should be selected to determine immunological activity. When conformation is not essential for further study, any of the methods listed in Table 1 can be used. When conformation is essential, any of the methods can be used provided that the blotting procedure is carried out under non-reducing conditions and in the presence of not more than 0.1% SDS. In this case, the antibodies have to be selected for reactivity with the intact conformation of the protein. Polyclonal antibodies can be used, but the source of the protein is important. The antiserum may contain antibodies against impurities that may interfere with the immunoassay. Local changes may have long-range effects in a protein and may disturb a distantly located epitope consisting of, e.g., four different parts of the polypeptide chain. Therefore, one Mab directed against one conformational epitope is probably sufficient to detect conformational changes in a protein. However, a panel of properly selected Mabs directed against different conformational epitopes is probably the best choice.

REFERENCES

1. **Wu, S-L., Benedek, K., and Karger, B.L.,** Thermal behavior of proteins in high-performance hydrophobic-interaction chromatography; on-line spectroscopic and chromatographic characterization, *J. Chromatogr.*, 359, 3, 1986.

2. Hearn, M.T.W., Aguilar, M.I., Nguyen, T., and Fridman, M., High-performance liquid chromatography of amino acids, peptides and proteins; application of derivative spectroscopy to the study of column residency effects in the reversed-phase and size-exclusion liquid chromatographic separation of proteins, *J. Chromatogr.*, 435, 271, 1988.

3. Johnston, R.C., Stanton, P.G., Robertson, D.M., and Hearn, M.T.W., High-performance liquid chromatography of amino acids, peptides and proteins; separation of isoforms of the glycoprotein hormones from human pituitary extracts, *J. Chromatogr.*, 397, 389, 1987.

4. Cohen, G.H., Isola, V.J., Kuhns, J., Berman, P.W., and Eisenberg, R.J., Localization of discontinuous epitopes of Herpes simplex virus glycoprotein D: use of a nondenaturing ("native" gel) system of poly-acrylamide gel electrophoresis coupled with Western blotting, *J. Virol.*, 60, 157, 1986.

5. Welling, G.W., Kazemier, B., and Welling-Wester, S., Size-exclusion high-performance liquid chromatography of integral membrane proteins; effect of detergents on immunological activity, *Chromatographia*, 24, 790, 1987.

6. Welling-Wester, S., Kazemier, B., Örvell, C., and Welling, G.W., Effect of detergents on the structure of integral membrane proteins of Sendai virus studied with size-exclusion high-performance liquid chromatography and monoclonal antibodies, *J. Chromatogr.*, 443, 255, 1988.

7. Welling, G.W., van der Zee, R., and Welling-Wester, S., Column liquid chromatography of integral membrane proteins, *J. Chromatogr.*, 418, 223, 1987.

8. Welling, G.W., Slopsema, K., and Welling-Wester, S., Purification strategies for Sendai virus membrane proteins, *J. Chromatogr.*, 397, 165, 1987.

9. Welling, G.W., van der Zee, R., and Welling-Wester, S., High-performance liquid chromatography of Sendai virus membrane proteins, *Trends in Anal. Chemistry*, 5, 225, 1986.

10. Winkler, G., Heinz, F.X., Guirakhoo, F., and Kunz, C., Separation of Flavivirus membrane proteins by multistep high-performance liquid chromatography optimized by immunological monitoring, *J. Chromatogr.*, 326, 113, 1985.

11. Montelaro, R.C., West, M., and Issel, C.J., High-performance gel permeation chromatography of proteins in denaturing solvents and its application to the analysis of enveloped virus polypeptides, *Anal. Biochem.*, 114, 398, 1981.

12. Welling, G.W., Nijmeijer, J.R.J., van der Zee, R., Groen, G., Wilterdink, J.B., and Welling-Wester, S., Isolation of detergent-extracted Sendai virus proteins by gel-filtration, ion-exchange and reversed-phase high-performance liquid chromatography and the effect on immunological activity, *J. Chromatogr.*, 297, 101, 1984.

13. Kita, K., Murakami, H., Oya, H., and Anraka, Y., Quantitative determination of cytochromes in the aerobic respiratory chain of *Escherichia coli* by high-performance liquid chromatography and its application to analysis of mitochondrial cytochromes, *Biochem. Int.*, 10, 319, 1985.

14. McKean, D.J. and Bell, M., The use of size-exclusion and C_{18} reverse phase HPLC columns for separating the component polypeptides of Ia antigen membrane proteins, *Protides Biol. Fluids*, 30, 709, 1982.

15. Calam, D.H. and Davidson, J., Isolation of influenza viral proteins by size-exclusion and ion-exchange high-performance liquid chromatography: the influence of conditions on separation, *J. Chromatogr.*, 296, 285, 1984.

16. Matson, R.S. and Goheen, S.C., Use of high-performance size-exclusion chromatography to determine the extent of detergent solubilization of human erythrocyte ghosts, *J. Chromatogr.*, 359, 285, 1986.

17. DeLucas, L.J. and Muccio, D.D., Purification of bovine rhodopsin by high-performance size-exclusion chromatography, *J. Chromatogr.*, 296, 121, 1984.

18. Tempst, P., Woo, D.D.-L., Teplow, D.B., Aebersold, R., Hood, L., and Kent, S.B.H., Microscale structure analysis of a high-molecular weight, hydrophobic membrane glycoproteins fraction with platelet-derived growth factor-dependent kinase activity, *J. Chromatogr.*, 359, 403, 1986.

19. Lambotte, P., Van Snick, J., and Boon, T., Partial purification of a membrane glycoprotein antigen by high-pressure size-exclusion chromatography without loss of antigenicity, *J. Chromatogr.*, 297, 139, 1984.

20. Wroblewski, H., Burlot, R., and Johansson, K-E., Membrane protein solubilization with sarkosyl, *Biochimie*, 60, 389, 1978.

21. Phelan, M.A. and Cohen, K.A., Gradient optimization principles in reversed-phase high-performance liquid chromatography and the separation of influenza virus components, *J. Chromatogr.*, 266, 55, 1983.

22. Mohanty, J.G. and Elazhary, Y., High-performance liquid chromatographic separation and immunological characterization of soluble bovine diarrhea virus antigen, *J. Chromatogr.*, 435, 149, 1988.

23. David, E.M. and Morgan, A.J., Efficient purification of Epstein-Barr virus membrane antigen gp340 by fast protein liquid chromatography, *J. Immunol. Methods*, 108, 231, 1988.

24. Gerlier, D., Garnier, F., and Forquet, F., Haemagglutinin of Measles virus: Purification and storage with preservation of biological and immunological properties, *J. Gen. Virol.*, 69, 2061, 1988.

25. Heukeshoven, J. and Dernick, R., Characterization of a solvent system for separation of water-insoluble poliovirus proteins by reversed-phase high-performance liquid chromatography, *J. Chromatogr.*, 326, 91, 1985.

26. Gallo, M.L., Jackwood, D.H., Murphy, M., Marsden, H.S., and Parris, D.S., Purification of the Herpes simplex virus type 1 65-kilodalton DNA-binding protein: properties of the protein and evidence of its association with the virus-encoded DNA polymerase, *J. Virol.*, 62, 2874, 1988.

27. **Ho-Terry, L. and Cohen, A.,** Rubella virus haemagglutinin: association with a single virion glycoprotein, *Arch. Virol.,* 84, 207, 1985.
28. **Green, M. and Brackmann, K.H.,** The application of high-performance liquid chromatography for the resolution of proteins encoded by the human adenovirus type 2 cell transformation region, *Anal. Biochem.,* 124, 209, 1982.
29. **Waris, M. and Halonen, P.,** Purification of adenovirus hexon protein by high-performance liquid chromatography, *J. Chromatogr.,* 397, 321, 1987.

A REVERSED-PHASE TRACE ENRICHMENT TECHNIQUE FOR THE LOADING OF SIZE-EXCLUSION HPLC COLUMNS

Hugh P. J. Bennett

Some form of molecular sieving or gel filtration procedure is frequently an important component of a peptide or protein purification scheme. Size-exclusion high-performance liquid chromatography (SEC) is the high performance equivalent of this type of chromatography. SEC columns are packed with uniform, small, and highly porous gels. These properties make it possible to perform gel filtration procedures at 10 to 20 times the speed of conventional open column systems. Only a limited number of SEC columns are available. Toyo-Soda and Waters Associates manufacture SEC columns based upon controlled pore silica or polymer which has been treated with a hydrophilic bonded phase. Toyo-Soda columns are marketed by a variety of HPLC companies and can be recognized by the TSK prefix. Indeed, the high molecular weight column from Waters is a TSK column. An alternative to the silica or polymer based columns are the Superose series from Pharmacia. Superose is a gel filtration medium based upon cross-linked agarose.

All these systems provide rapid, high resolution SEC. However, their role is frequently confined to analytical applications. The high porosity of the gels means that quite small analytical columns can be used to resolve milligram quantitites of peptide and protein. A practical consideration presents a severe limitation in the use of analytical SEC columns for semi-preparative uses. Due to their small size (generally 300×7.8 mm I.D.), samples must be loaded onto these columns in very small volumes (i.e., generally of the order of 100 µl or less). The problem is how to load samples in these small volumes while maintaining satisfactory recoveries. Some form of trace enrichment procedure would obviate this problem. An electrophoretic trace enrichment procedure has been developed by Ofverstedt and Eriksson[1] for use with conventional column solvent systems based upon buffered salt solutions. We have recently developed a reversed-phase trace enrichment procedure for use with a SEC solvent system similar to ones frequently used for reversed-phase HPLC (RPC). We have previously shown that excellent SEC can be provided by a solvent system consisting of 40% aq. acetonitrile containing 0.1% TFA.[2] The combination of low pH and relatively high solvent concentration minimizes both polar and non-polar interactions with the column. This leads to excellent molecular sieving properties. Variations in this system have been used successfully by other laboratories.[3,4] Both Toyo-Soda and Waters columns have been used in these studies. As will become apparent later, the use of this solvent system is critical for the trace enrichment technique that we have developed. The only major limitation of this volatile solvent system is the bio-compatability of solutes with its relatively denaturing properties. However, we have noticed that for optimal performance it is not advisable to switch back and forth from this sytem to one containing conventional salt buffers. Columns should be dedicated to one solvent system or the other. In the studies outlined below we have used the Waters silica-based I-125 column. The trace enrichment technique provides excellent recoveries and has been fully justified elsewhere.[5] The practical aspects of the trace enrichment technique will be discussed in detail here. The basis of the technique is the trace enrichment of peptides and proteins onto a RPC pre-column prior to SEC.

STEP 1 - TRACE ENRICHMENT   STEP 2 - COLUMN LOADING

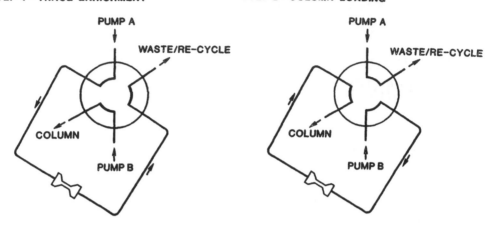

FIGURE 1. Diagrammatic representation of a Valco six port HPLC injector adapted for trace enrichment loading of SEC column. Pump A (attached to injection port) is used for trace enrichment onto the Guard-Pak (with ODS-silica insert) which is contained within the injection loop. Pump B meanwhile is used to elute solutes from the SEC column. Step 1 — trace enrichment of sample onto the Guard-Pak. Step 2 — the Guard-Pak is put in line with the column and the sample is loaded onto the column in a minimal volume.

The cartridge pre-column systems now available are ideal for trace enrichment. The Brownlee version consists of pre-packed stainless steel cartridges (30 × 2.1 mm I.D.) housed within a finger-tightened holder. An alternative pre-column is the Waters Guard-Pak system. This consists of plastic cartridges of loose ODS-silica (approximately 40 mg) contained within a hand-tightened holder. The holder is a radial compression device which, when tightened, converts the insert into a mini-column (approximately 5 × 6 mm I.D.). Indeed, the chromatographic efficiency of the insert is sufficiently high that simple procedures can be undertaken using the Guard-Pak alone. For instance, we have recently shown that a Guard-Pak can be used to prepare radio-iodinated peptides while minimizing contamination of the HPLC apparatus (i.e., pumps, injector, and columns) with radioactivity.[6] These experiences demonstrated that the chromatographic efficiency of the Guard-Pak inserts is comparable to that of the Brownlee cartridges. High chromatographic efficiency means low elution volumes. Small injection volumes are the major requirement for optimal performance of SEC. To use the Guard-Pak system for column loading, it must be attached to a Valco-type injector as indicated in Figure 1. The injection port has been removed and a HPLC pump attached in its place. A Guard-Pak with ODS-silica insert has been attached within the injection loop. Meanwhile, the SEC column is equilibrated isocratically with 40% aq. acetonitrile containing 0.1% TFA by pump B.

Step 1 - Trace enrichment: In this position, pump A is used to pump sample onto the Guard-Pak cartridge. Provided that the loading solution is aqueous or of low solvent strength, peptides and proteins enrich onto the ODS-silica cartridge. The volume of loading at this stage is not important and indeed can be quite large (see later). Samples can be enriched onto the Guard-Pak at relatively high flow rates owing to the low back pressure that is generated. The flow-rate is generally 2 or 3 ml/min. Like any form of ODS-silica, the Guard-Pak insert must be prewetted with solvent before use. Thus, prior to trace enrichment, 5 ml of 80% aq. acetonitrile containing 0.1% TFA is pumped through the cartridge followed by equilibration in 0.1% aq. TFA.

Step 2 - Column Loading: In this position, the Guard-Pak cartridge is put in line with the column. The column eluate now passes through the cartridge. The solvent strength has risen abruptly to that of the SEC solvent system (i.e., 40% aq. acetonitrile containing 0.1% TFA). Enriched peptides and proteins are eluted onto the SEC column in a very small volume. Interestingly, because the direction of the flow through the injection loop is reversed upon

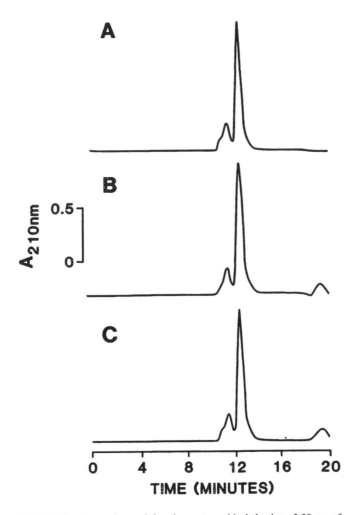

FIGURE 2. Comparison of the chromatographic behavior of 20 μg of ribonuclease loaded onto two Waters Protein-Pak I-125 columns (300 × 7.8 mm I.D.) connected in series. Increasing volumes were loaded as follows: Panel A (50 μl by injection), Panel B (5 ml), and Panel C (50 ml). For Panels B and C, samples were loaded by reversed-phase trace enrichment onto the Waters Guard-Pak in 0.1% aq. TFA. SEC was carried out in 40% aq. acetonitrile containing 0.1% TFA at a flow-rate of 1 ml/min.

switching from the trace enrichment position to the column loading position, the cartridge elution volume is minimized.

The effectiveness of the method can be illustrated best by considering the loading of 20 μg of ribonuclease onto the SEC column. Figure 2 shows a comparison of the elution profiles obtained after loading of the ribonuclease onto the SEC column in volumes of 50 μl (by injection, Panel A), 5 ml (Panel B), and 50 ml (Panel C). Column performance in terms of recoveries and peak shape is identical regardless of volume of loading when reversed-phase trace enrichment was used for the 5 and 50 ml sample volumes.

This technique is clearly applicable where a SEC step is needed at or towards the end of a purification scheme. Under these circumstances, the fractions in question would normally be dried and redissolved prior to injection onto the SEC column. Using the trace enrichment technique the sample can be loaded with a minimum of manipulations. When working in the submicrogram range, recoveries are dramatically improved. The method provides the user with quick final purification and/or estimation of purity and molecular weight.

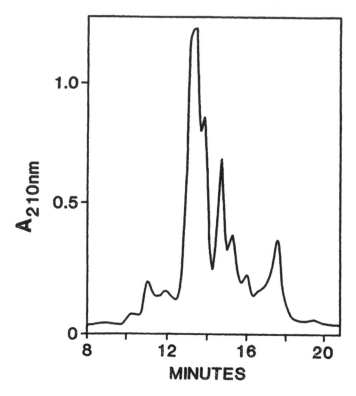

FIGURE 3. SEC of basic pool of peptides and proteins extracted from five posterior pituitaries. Sample was loaded by reversed-phase trace enrichment onto the Waters Guard-Pak and subsequently subjected to SEC on two Waters Protein-Pak I-125 columns (300 × 7.8 mm I.D.) in 40% aq. acetonitrile containing 0.1% TFA at a flow-rate of 1 ml/min.

The technique can also be used for semi-preparative applications when the trace enrichment is at or towards the beginning of the purification scheme. This is illustrated with the extraction and purification of basic peptides from the bovine posterior pituitary. Elsewhere in this volume, an ion-exchange batch fractionation technique is described. Using this technique to pre-fractionate tissue extracts into acidic, basic and neutral pools can greatly reduce column loading at subsequent reversed-phase HPLC steps. It can facilitate peptide purification. Furthermore, through the manipulation of the pH and molarity of the buffers it is possible to enrich peptides of interest even further. Five bovine posterior pituitaries (Pel-Freez Biologicals) were extracted in an acidic medium consisting of 1 M hydrochloric acid, containing 5% (v/v) formic acid, 1% (w/v) sodium chloride, and 1% (v/v) TFA.[7,8] The supernatant was subjected to reversed-phase batch extraction using a total of five C_{18} Sep-Pak cartridges as described previously.[8] The combined C_{18} Sep-Pak eluates (i.e., 20 ml of 80% acetonitrile containing 0.1% TFA) were lyophilized. The residue was redissolved in 5 ml of 50 mM Tris buffer (pH 7) containing 10% acetonitrile and subjected to batch ion-exchange fractionation using a CM-Accell Sep-Pak cartridge (see Reference 7 and article entitled "Use of Ion-Exchange Cartridges for Batch Extraction in the Purification of Pituitary Peptides" by Bennett). The CM-Accell cartridge was eluted with 5 ml 50 mM Tris (pH 7) containing 1 M sodium chloride and 20% acetonitrile. This was acidified with 1% TFA and subjected to reversed-phase trace enrichment followed by SEC. Fractionation of the basic pool derived from an extract of bovine posterior pituitaries is shown in Figure 3. It can be seen that quite a large mass of polypeptides has been enriched and loaded onto the Waters I-125 SEC system. The major component eluted at 13 and 14 min was subjected to RPC on a C_{18} μBondapak column where complete purification was achieved (not shown).

Amino acid analysis of this material revealed that it is tissue basic trypsin inhibitor or aprotinin.[9] Our results indicate that it is by far the greatest single basic peptide found in bovine posterior pituitary extracts. Some years ago, Li and Chung purified the same peptide from extracts of bovine anterior pituitaries.[10] The role of such high concentrations of this protease inhibitor in an endocrine organ remains to be determined.

The above example clearly shows that the trace enrichment technique can be used for semi-preparative applications. For the small scale application illustrated in Figure 2, the amounts of ribonuclease loaded were low. In the second application, a total of approximately 1.5 mg was loaded according to protein determination. Experience has revealed that the practical upper limit of loading is about 1 mg of total protein.[5] When the material which passed through the Guard-Pak during step 1 of the enrichment procedure (Figure 1) was recycled, it was found that a further 10% of protein could be extracted and chromatographed. A second re-cycling revealed no further extraction of protein. Since the composition of material to be loaded varies from sample to sample, repeated re-cycling of extracts is recommended. Fortunately, the efficacy of the enrichment procedure can be assessed easily and quickly by monitoring the SEC profiles. To simplify sample collection, the column fractions can be returned to the starting position on the fraction collector before beginning the next cycle. SEC is highly reproducible and this facilitates the re-cycling of column fractions in this manner.

REFERENCES

1. **Ofrerstedt, L.G. and Eriksson, K.O.,** Pre-column concentration of protein samples by displacement electrophoresis (isotachophoresis) followed by high performance molecular sieve chromatography on a packed agarose column, *Anal. Biochem.,* 137, 318, 1984.
2. **Bennett, H.P.J., Browne, C.A., and Solomon, S.,** α-N-Acetyl-β-endorphin$_{1-26}$ from the neurointermediary lobe of the rate pituitary: Isolation, purification and characterization by high-performance liquid chromatography, *Anal. Biochem.,* 128, 121, 1983.
3. **Swergold, G.D. and Rubin, C.C.,** High-performance gel permeation chromatography of polypeptides in a volatile solvent: rapid resolution and molecular weight estimations of proteins and peptides on a column of TSK-G3000PW, *Anal. Biochem.,* 131, 295, 1983.
4. **Eipper, B.A., Glembotski, C.C., and Mains, R.E.,** Selective loss of α-melanotropin-amidating activity in primary cultures of rat intermediate pituitary cells, *J. Biol. Chem.,* 258, 7292, 1983.
5. **Bennett, H.P.J. and James, S.,** A trace enrichment technique for the loading of gel-permeation high performance liquid chromatography column, *Anal. Biochem.,* 179, 222, 1989.
6. **Carriere, P. and Bennett, H.P.J.,** Reversed-phase liquid chromatography of radiolabelled peptides using a C$_{18}$ Guard-Pak pre-column system, *Peptides,* 10, 485, 1989.
7. **Bennett, H.P.J.,** Use of ion-exchange Sep-Pak cartridges in the batch extraction of pituitary peptides, *J. Chromatogr.,* 359, 383, 1986.
8. **Bennett, H.P.J., Browne, C.A., and Solomon, S.,** Purification of the two major forms of rat pituitary corticotropin using only reversed-phase liquid chromatography, *Biochemistry,* 20, 4530, 1981.
9. **Fritz, H. and Wunderer, G.,** Biochemistry and applications of aprotinin, the kallikrein inhibitor from bovine organs, *Drug Research,* 33, 479, 1983.
10. **Li, C.H. and Chung, D.,** Isolation of pancreatic trypsin inhibitor from bovine pituitary glands, *Proc. Natl. Acad. Sci. U.S.A.,* 80, 1204, 1983.

Section IV
Ion-Exchange Chromatography

THE USE OF PEPTIDE STANDARDS FOR MONITORING IDEAL AND NON-IDEAL BEHAVIOR IN CATION-EXCHANGE CHROMATOGRAPHY

Colin T. Mant and Robert S. Hodges

I. INTRODUCTION

High-performance ion-exchange chromatography (IEC) has become increasingly popular for the analysis of peptides in recent years.[1,2] Although IEC is used only sparingly for peptide separations compared to the widespread use of reversed-phase chromatography (RPC), the two techniques are often complementary, i.e., their combined use can provide optimal separation of a peptide mixture or assess the purity of a peptide separation. Both anion-exchange (AEX)[1-9] and cation-exchange (CEX)[1,2,7,9-17] columns have been utilized for peptide separations, with strong cation-exchange chromatography (strong CEX) being probably the most useful mode for IEC of peptide mixtures.[1,2,7,9,12,14-17] The utility of strong CEX packings, generally containing sulfonate functionalities, lies in their ability to retain their negatively charged character in the acidic to neutral pH range.[1,2,7,12] In contrast, weak cation-exchange (weak CEX) packings, usually containing carboxylate functionalities,[11] lose their negatively charged character below pH 4.0 to 4.5 and, thus, can only be utilized at, or close to, neutrality. At low pH (< pH 4.0 to 4.5), side chain carboxyl groups of acidic amino acid residues are protonated, emphasizing any positively charged character of the peptides. Thus, the advantages of strong cation-exchange chromatography lies in the fact that the pH of the mobile phase can be manipulated to change the net charge on the peptides without changing the properties of the column packing.

Prior to the use of an ion-exchange column, it is important to assess its performance capabilities (selectivity, efficiency), since the peptide resolving power of an ion-exchanger may vary from manufacturer to manufacturer or from batch to batch of ion-exchanger from the same manufacturer. The most logical approach to assessing column performance is to employ HPLC standards, preferably under a set of standard chromatographic run conditions, for a proper comparison of different HPLC packings.[1] It is preferable to use compounds that are structurally similar to the sample(s) of interest, and that presumably interact with the column packing in a similar manner to achieve the most precise and accurate analysis. Thus, ion-exchange peptide standards are most suited for monitoring peptide retention and resolution in IEC. The value of standards in monitoring the retention characteristics of ion-exchange columns is fourfold:

1. Standards will confirm that the column can, indeed, retain charged species (the weaker the charge that can be retained, the better the column). In addition, a mixture of standards with a range of net charge will determine whether the column is retaining charged species in a logical manner, i.e., a general increase in peptide retention time with increasing net charge.

2. Standards can confirm that a particular mobile phase will elute charged species from an ion-exchange column.

3. Standards can be used to assess the effect of pH variations on the resolving capability and loading capacity of an ion-exchange column. This is particularly important in strong CEX,

where the manipulation of mobile phases over the acidic to neutral pH range is frequently employed for peptide separations.

4.   Standards can be used to determine whether a particular ion-exchange packing exhibits any mechanism of separation other than one based on ion-exchange, i.e., non-ideal behavior. Although, as the name implies, the major separation mechanism of this mode of HPLC is electrostatic in nature, ion-exchange packings may also often exhibit significant hydrophobic characteristics, giving rise to mixed-mode contributions to solute separations.[18,19] As pointed out by Rounds et al,[20] a small amount of hydrophobic character in an ion-exchanger is not necessarily detrimental to solute separations, and may even enhance resolution by mixed-mode effects. Indeed, several researchers have exploited these mixed-mode effects to aid in peptide and protein separations on both AEX[5,19-22] and CEX[9,15,16] columns. However, when only the predominant, i.e., ionic, stationary phase-solute interaction is required (ideal ion-exchange behavior), the mobile phase must be manipulated so as to minimize non-specific interactions, e.g., by the addition of a non-polar organic solvent such as acetonitrile to the mobile phase buffers to suppress hydrophobic interactions between the solute and the ion-exchange packing. Although the hydrophobic character of high-performance ion-exchange packings has long been recognized, this article describes the first systematic approach to examining the effect and magnitude of the hydrophobicity of these packings during IEC of peptides.

In addition to overall net charge, other factors which may affect the retention behavior of peptides during IEC (under conditions ensuring ideal ion-exchange behavior) include peptide conformation, polypeptide chain length, charge distribution and charge density. To understand peptide retention behavior during IEC completely, it is not sufficient merely to demonstrate that these various factors have an effect on peptide retention, it is also necessary to quantitate the relative contribution each factor makes to retention behavior.

This article describes the design and application of a mixture of four synthetic undecapeptide HPLC standards for monitoring the performance of cation-exchange columns. From observation of the retention behavior of these standards on various strong cation-exchange columns, series of basic peptide polymers (5 to 50 residues) of varying hydrophobicity were designed and utilized to clarify the effects of hydrophobic interactions with ion-exchange packings and to gain a clearer understanding of the effect of both chain length and charge density on peptide retention behavior during strong CEX.

II. STRONG CEX COLUMNS

Peptide mixtures described in this article were separated on three strong cation-exchange columns:

1.   SynChropak S300, 250 × 4.1 mm I.D., 6.5-μm particle size, 300-Å pore size (SynChrom, Linden, IN). This is a silica-based column containing sulfonate groups as the negatively charged functionalities.[7,12]
2.   PolySULFOETHYL Aspartamide, 250 × 4.6 mm I.D., 5 μm, 300 Å (PolyLC, Columbia, MD). This is also a silica-based cation-exchanger with sulfonate functionalities, but the chemistry of sulfonate attachment to the silica support is different compared to that of the S300.[15,16]
3.   Mono S HR 5/5, 50 × 5 mm I.D., 10 μm (Pharmacia, Dorval, Quebec, Canada). The Mono S cation-exchanger consists of sulfonate groups attached to a polyether support.

III. MONITORING OF STRONG CATION-EXCHANGE COLUMN PERFORMANCE WITH SYNTHETIC PEPTIDES

An advantage of peptides over proteins as monitors of ion-exchange column characteristics lies in the fact that all charged groups in peptides are generally available to interact with the column packing even under benign (non-denaturing) conditions. The tendency for proteins under benign conditions to assume tertiary structures may cause shielding of charged residues, so that only a portion of the protein surface interacts with the ion-exchange matrix. Hence, the overall net charge on the protein will not be fully expressed unless highly denaturing conditions are employed. In contrast, peptides (< 50 residues) are less prone to conformational masking of charged residues and their overall net charge (as well as their overall hydrophobicity) is fully expressed under milder conditions. Thus, systematic changes in overall net charge and/or hydrophobicity of a peptide allow a high degree of precision in monitoring ion-exchange column performance.

A. REQUIREMENT FOR PEPTIDE STANDARDS IN STRONG CEX

1. Design of Peptide Standards

The sequences of four undecapeptide cation-exchange standards, C1 to C4 (+1 to +4 net charge, respectively) are shown in Table 1. IEC, generally in conjunction with RPC, has proved useful during peptide mapping of protein digests, and the 11-residue length of the standards was designed to approximate the average size of cleavage fragments from proteolytic (e.g., tryptic) digests of proteins. The uniform length of the peptides ensured that any potential chain length effect on peptide retention, if present, would be constant for all four peptide standards. Peptides C1 to C4 contain, respectively, 1 to 4 basic residues (lysine residues) with no acidic residues present. Thus, over the pH range used for the majority of cation-exchange separations (pH 3.0 to 7.0 [strong CEX] or pH > 4.5 to 7.0 [weak CEX]), the net charges of +1 to +4 for peptides C1 to C4, respectively, do not change. It has been observed[15] that the retention behavior of peptides on a strong CEX column may be affected by the distribution of basic residues in the peptide sequence. Thus, the retention effects of adjacent positively charged residues (Arg, in this case [15]) were less than additive on a PolySULFOETHYL Aspartamide column, whereas the retention behavior of peptides was in proportion to the number of basic residues if these residues were well separated. The distribution of lysine residues on peptides C1 to C4, with at least two neutral amino acid residues separating the positively charged lysine residues (Table 1), was designed to ensure full expression of the positively charged character of the peptides. The hydrophobicity of the standards increases from C1 to C4, with a concomitant increase in peptide sensitivity to potential non-ideal (i.e., hydrophobic, as opposed to ionic, column/solute interactions) retention behavior. Standards C2 and C4 contain tyrosine residues, enabling detection of these peptides at 280 nm and simplifying peak identification.

2. Effect of pH on Strong CEX Packings

Figure 1 demonstrates the elution profiles at pH 6.5 (top) and pH 3.0 (bottom) of the four peptide standards (C1 to C4) on the SynChropak S300 column. The standards were subjected to gradient elution (Buffer A = 5 mM KH$_2$PO$_4$ and Buffer B = Buffer A + 1 M NaCl) at 20 mM salt/min and a flow-rate of 1 ml/min, following 5 min isocratic elution with Buffer A. Linear gradient elution is generally the elution mode of choice when attempting to separate mixtures of peptides with a wide range of net charges.[1,7] Care should always be taken over the choice of ionic strength of the starting buffer (only 5 mM in the present case). If it is too high, weakly basic peptides which may otherwise be retained by cation-exchange columns (or weakly acidic

TABLE 1
Characteristics of Synthetic Peptides Used in this Study

Peptide designation	Peptide sequence[a]	No. of repeating Units (n)	No. of residues	Net charge	Hydrophobicity[b]
10X[c]	Ac-(Gly-Leu-Gly-Ala-Lys*-Gly-Ala-Gly-Val-Gly)$_n$-amide	1	10	+1	17.9
20X	Ac-(Gly-Leu-Gly-Ala-Lys*-Gly-Ala-Gly-Val-Gly)$_n$-amide	2	20	+2	35.8
30X	Ac-(Gly-Leu-Gly-Ala-Lys*-Gly-Ala-Gly-Val-Gly)$_n$-amide	3	30	+3	53.7
40X	Ac-(Gly-Leu-Gly-Ala-Lys*-Gly-Ala-Gly-Val-Gly)$_n$-amide	4	40	+4	71.6
50X	Ac-(Gly-Leu-Gly-Ala-Lys*-Gly-Ala-Gly-Val-Gly)$_n$-amide	5	50	+5	89.5
5A	Ac-(Leu-Gly-Leu-Lys*-Ala)$_n$-amide	1	5	+1	19.8
10A	Ac-(Leu-Gly-Leu-Lys*-Ala)$_n$-amide	2	10	+2	39.6
20A	Ac-(Leu-Gly-Leu-Lys*-Ala)$_n$-amide	4	20	+4	79.2
30A	Ac-(Leu-Gly-Leu-Lys*-Ala)$_n$-amide	6	30	+6	118.8
40A	Ac-(Leu-Gly-Leu-Lys*-Ala)$_n$-amide	8	40	+8	158.4
50A	Ac-(Leu-Gly-Leu-Lys*-Ala)$_n$-amide	10	50	+10	198.0
5L	Ac-(Leu-Gly-Leu-Lys*-Leu)$_n$-amide	1	5	+1	26.6
10L	Ac-(Leu-Gly-Leu-Lys*-Leu)$_n$-amide	2	10	+2	53.2
20L	Ac-(Leu-Gly-Leu-Lys*-Leu)$_n$-amide	4	20	+4	106.4
30L	Ac-(Leu-Gly-Leu-Lys*-Leu)$_n$-amide	6	30	+6	159.6
40L	Ac-(Leu-Gly-Leu-Lys*-Leu)$_n$-amide	8	40	+8	212.8
50L	Ac-(Leu-Gly-Leu-Lys*-Leu)$_n$-amide	10	50	+10	266.0
C1[c]	Ac-Gly-Gly-Leu-Gly-Gly-Ala-Gly-Gly-Leu-Lys*-amide	—	11	+1	18.6
C2	Ac-Lys*-Tyr-Gly-Leu-Gly-Gly-Ala-Gly-Gly-Leu-Lys*-amide	—	11	+2	23.4
C3	Ac-Gly-Gly-Ala-Leu-Lys*-Ala-Leu-Lys*-Gly-Leu-Lys*-amide	—	11	+3	30.2
C4	Ac-Lys*-Tyr-Ala-Leu-Lys*-Ala-Leu-Lys*-Gly-Leu-Lys*-amide	—	11	+4	35.0

a   Ac = N$^\alpha$-acetyl; amide = C$^\alpha$-amide. Positively charged residues (lysine) are marked with a star (*).

b   Peptide hydrophobicity is expressed as the sum of the hydrophobicity coefficients of the amino acid residues as described by Guo et al.[23] The coefficients were obtained by measuring the contribution of individual amino acid residues to the retention time of a model synthetic peptide (Ac-Gly-X-X-[Leu$_3$]-[Lys]$_2$-amide, where X was substituted by the 20 amino acids found in proteins) in reversed-phase chromatography at pH 7.0.

c   The "X" series of peptide polymers (a mixture of reversed-phase peptide standards) and the cation-exchange standards (C1 to C4) were obtained from Synthetic Peptides Incorporated, Department of Biochemistry, University of Alberta Edmonton, Canada.

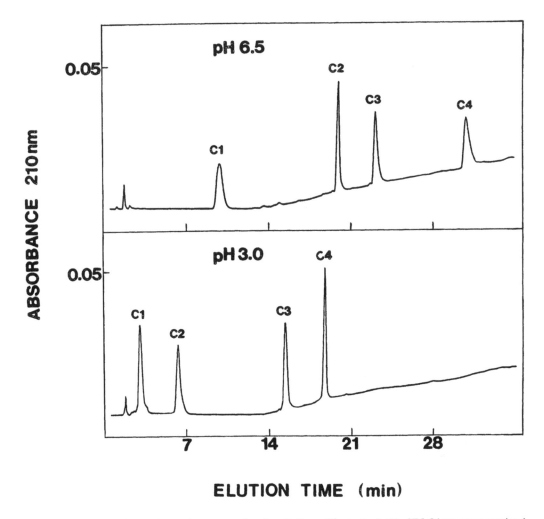

FIGURE 1. Strong cation-exchange chromatography of synthetic peptide standards. The HPLC instrument consisted of a Varian Vista Series 5000 liquid chromatograph (Varian, Walnut Creek, CA) coupled to a Hewlett-Packard (Avondale, PA) HP1040A detection system, HP85B computer, HP9121 disc drive, HP2225A Thinkjet printer and HP7470A plotter. Column: SynChropak S300 (250 × 4.1 mm I.D.). Conditions: linear AB gradient (20 mM salt/min, following 5 min isocratic elution with Buffer A), where Buffer A is 5 mM KH$_2$PO$_4$, pH 6.5 (top) or pH 3.0 (bottom), and Buffer B is Buffer A plus 1 M NaCl; flow-rate, 1 ml/min, temperature, 26°C. Peptide standards C1, C2, C3, and C4 contain net charges of +1, +2, +3, and +4 respectively (see Table 1).

peptides in the case of anion-exchange columns) may be eluted with unretained compounds. From Figure 1, at pH 6.5 (top), standards C2, C3, and C4 (+2, +3, and +4 net charge, respectively) were removed by the gradient, while peptide C1 (+1 net charge) was eluted during the initial isocratic elution. In contrast, at pH 3.0 (Figure 1, bottom), only peptides C3 and C4 were removed by the gradient, while both peptides C1 and C2 were eluted during the initial isocratic elution. In addition, the retention time of peptides C3 and C4 were reduced considerably. Ideally, there should have been no variation of elution time of these peptides with a change in pH of the buffer. The observed effects apparently resulted from a reduction in column capacity to retain charged species as the pH became more acidic.[7] A similar, but much smaller, effect was observed on the PolySULFOETHYL Aspartamide column (silica-based like the S300 but a different chemistry of sulfonate attachment). The polyether-based Mono S demonstrated no such pH effect, with the peptide standards exhibiting identical retention times at both pH 6.5 and 3.0.

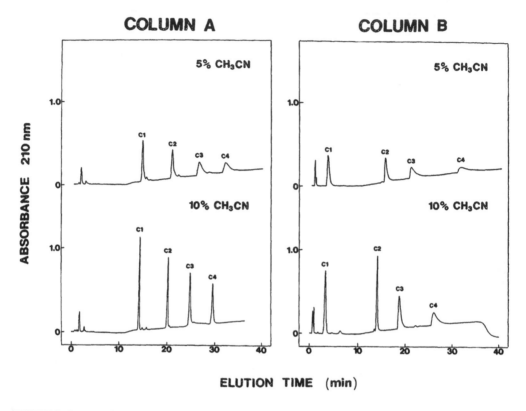

FIGURE 2. Strong cation-exchange chromatography of synthetic peptide standards. HPLC instrument : as Figure 1, except for a HP9000 Series 300 computer. Columns : Column A, PolySULFOETHYL Aspartamide (250×4.6 mm I.D.); Column B, Mono S (50×5 mm I.D.). Conditions : linear AB gradient (20 mM salt/min, following 5 min isocratic elution with Buffer A, where Buffer A is 5 mM KH$_2$PO$_4$, pH 6.5, and Buffer B is Buffer A plus 1 M NaCl, both buffers containing 5% (top profiles) or 10% (bottom profiles) acetonitrile (v/v); flow-rate, 1 ml/min; temperature, 26°C. Peptide standards C1, C2, C3, and C4 contain net charges of +1, +2, +3, and +4, respectively (see Table 1).

3. Effect of Hydrophobic Interactions on CEX of Peptide Standards

Figure 2 shows elution profiles of the peptide standards obtained on the PolySULFOETHYL Aspartamide column (Column A) and the Mono S column (Column B). The mobile phase buffers were identical to that shown for Figure 1, except for the addition of 5% (Figure 2, top profiles) or 10% (Figure 2, bottom profiles) acetonitrile (v/v) to both mobile phase buffers. The inclusion of acetonitrile in the mobile phase was found to be necessary to elute the four peptides from both columns within a reasonable time and with reasonable peak shape. In the absence of acetonitrile, neither peptide C3 nor C4 (+3 and +4 net charge, respectively) were seen to be eluted from the two columns, while C2 (+2 net charge) was eluted as a broad, skewed peak. The improvement in peptide elution profiles of the four standards on the two columns on addition of acetonitrile to the mobile phase indicated that in its absence both column packings were exhibiting hydrophobic, in addition to ionic, characteristics. In the presence of 5% acetonitrile (Figure 2, top profiles), all four peptide standards were eluted from the two columns in a similar time (~ 32 min). However, an increase in acetonitrile concentration to 10% in the mobile phase (bottom panels), substantially improved the peak shape of the peptides, particularly the more hydrophobic C3 and C4 (+3 and +4 net charge, respectively). In addition, peptide retention times generally decreased with increasing acetonitrile concentration, particularly those of C3 and C4. The improvement in peak shape of the peptides on the Mono S continued with a further increase in acetonitrile concentration to 40% (Figure 4B). Crimmins et al [16] also noted that the omission of 25% acetonitrile from their mobile phase buffers resulted in some peptide peak broadening on a PolySULFOETHYL Aspartamide column.

From Figure 2 (Column A, bottom profile; PolySULFOETHYL Aspartamide) and Figure 4B (Mono S), it is apparent that, when hydrophobic interactions were effectively suppressed, the retention times of the peptide standards increased linearly with increasing net charge. Thus, the differences in retention times for adjacent peptide peaks (eluted by the gradient and not during the initial isocratic elutron with Buffer A) were essentially equal. The greater difference in retention time between C3 and C4 compared to between C2 and C3 on the S300 column at pH 6.5 (Figure 1, top panel) indicated the presence of hydrophobic interactions even though all four peptides were removed from the column in the absence of acetonitrile in the mobile phase. This was confirmed when, following addition of acetonitrile to the mobile phase, standards C2, C3, and C4 were eluted in a linear fashion.

The value of the peptide standards to the researcher lies in the way they can quickly compare performance characteristics of different cation-exchange columns or to assess the performance characteristics of a specific cation-exchange column. The researcher can then decide on the column best suited to his needs and/or the most appropriate mobile phase conditions for the separation of a particular peptide mixture. Figures 1 and 2 demonstrated how the standards were used to evaluate the advantages and disadvantages of three very different types of strong CEX columns. Thus, an advantage of the PolySULFOETHYL Aspartamide column lay in its ability to retain the weakly basic peptide, C1 (+1 net charge) (Figure 2, Column A); this peptide was not retained by the S300 (Figure 1) or Mono S (Figure 2, Column B) columns, being washed through the columns during the initial isocratic elution with Buffer A. The standards were able to detect the undesirable effect on the S300 packing (a decrease in peptide retention times) on lowering the pH from 6.5 to 3.0 (Figure 1). This observation highlighted an advantage of the peptide standards over proteins as sensitive monitors of column performance. While the drop in pH had negligible effect on the charge characteristics of the peptide standards, the overall net charge on a protein may vary considerably with pH changes, thereby masking any potential pH effects on a column packing. The standards were also able to detect a similar but much smaller pH effect with the PolySULFOETHYL Aspartamide column, while the Mono S showed no such effect. The peptide standards detected hydrophobic characteristics in all three column packings; the magnitude of these hydrophobic interactions was smallest with the S300 column, where all four peptides were removed by a 100% aqueous mobile phase (Figure 1).

There is nothing particularly exceptional about the peptide standards in terms of the range of net charge, charge density or hydrophobicities they represent. In fact, the characteristics of these peptides probably reflect those of the great majority of peptides encountered during CEX of peptide mixtures. Thus, the sensitivity of the peptide standards to CEX column characteristics makes them eminently suitable for assessing accurately CEX column performance.

B. STRONG CEX OF PEPTIDE POLYMERS
1. Design of Peptide Polymers

In order to examine the effect of peptide chain length on peptide retention behavior during strong CEX, as well as carrying out a systematic examination of non-specific hydrophobic interactions between solute and column packing, it was necessary to design series of positively charged peptide polymers covering a similar range of chain length, but differing in overall hydrophobicity. The structures of three polymer sets subsequently used in this study are shown in Table 1. One set of polymers (the "X" series) is a mixture of five synthetic reversed-phase peptide standards (10, 20, 30, 40, and 50 residues; +1, +2, +3, +4, and +5 net charge, respectively). Two other sets of polymers were synthesized: (a) Ac-(Leu-Gly-Leu-Lys-Ala)$_n$-amide (the "A" series), where n = 1, 2, 4, 6, 8, 10 (5 to 50 residues; +1 to +10 net charge, respectively; (b) Ac-(Leu-Gly-Leu-Lys-Leu)$_n$-amide (the "L" series), where N = 1, 2, 3, 6, 8, 10 (5 to 50 residues; +1 to +10 net charge, respectively). The peptide polymers were synthesized on an Applied Biosystems peptide synthesizer Model L30A (Foster City, CA), using the general procedure for solid-phase synthesis described by Parker and Hodges[24] and Hodges et al.[25] The single lysine residue in each repeating unit of five residues in the "A" and "L" series of polymers ensured that the overall charge density of every peptide in these series was identical (+0.2/residue) and double that of the "X" series of peptides (+0.1/residue) which contained a single lysine residue in each repeating unit of 10 residues (Table 1). The hydrophobicity of the polymer series increased in the order, "X" series < "A" series < "L" series (see Table 1 for definition of peptide hydrophobicity). For the purposes of this article, each peptide is referred to by a number and letter which denote, respectively, the number of residues it contains and to which polymer series it belongs. Thus, 5A refers to the five-residue "A" series peptide; 30X refers to the 30-residue "X" series peptide, etc.

2. Effect of Hydrophobic Interactions on CEX of Peptide Polymers

Figure 3 shows elution profiles of the mixture of the "A" series of peptide polymers on the S300 column. Similar results were obtained on the PolySULFOETHYL Aspartamide and Mono S columns. The peptides were chromatographed using a linear sodium chloride gradient (20 mM NaCl/min, following 10 min elution with starting buffer, at a flow-rate of 1 ml/min) in 5 mM KH$_2$PO$_4$ buffer at pH 6.5. The mobile phase buffers also contained 0, 10, 20, 30, or 40% acetonitrile (v/v). In the absence of acetonitrile (results not shown), only peptides 5A (+1 net charge) and 10A (+2 net charge) were eluted, with reasonable peak shape, by a salt gradient up to a concentration of 0.5 M NaCl. Peptide 20A (+4 net charge) appeared as a late-eluted very broad, badly skewed peak. The more hydrophobic 30A, 40A, and 50A peptides (+6, +8, and +10 net charge, respectively) were not eluted by 0.5 M NaCl. It had previously been shown by Mant and Hodges[7] that a mixture of peptides of average hydrophobicity and a range of net charge from +2 to +8 at pH 6.5 was easily removed from the S300, in the absence of an organic solvent, by a salt gradient up to a concentration of only 0.4 M. Thus, it was apparent that, in addition to ionic interactions between the "A" series of peptides and the column packing, non-specific hydrophobic interactions were also affecting the retention behavior of the polymer series. With the addition of 10% acetonitrile to the mobile phase buffers (Figure 3A), peptides 5A, 10A, and 20A (+1, +2, and +4 net charge, respectively) were all now eluted with good peak shape within a concentration range of 0 to 0.3 M NaCl. As the concentration of acetonitrile was increased further to 20, 30, and 40% (Figure 3B, 3C, and 3D, respectively), the more hydrophobic peptides (30A, 40A, and 50A [+6, +8, and +10 net charge, respectively]) were also eluted from the column. In addition, the retention times of all peptides decreased with increasing levels of acetonitrile. At a level of 40% acetonitrile in the mobile phase buffers, the most hydrophobic peptide, 50A (+10 net charge), was eluted at a salt concentration of only ca. 0.2 M. It should be noted that the peptide with a net charge of only +1 (5A) was eluted from the S300 during the initial 10-min isocratic elution with the starting buffer and not by the subsequent gradient. However, the elution time of this peptide also decreased with increasing levels of acetonitrile.

The difference in hydrophobicity of the "A" and "L" series of peptide polymers ("A" < "L") was reflected in the intensity of their non-ionic interactions with the three ion-exchange packings. Peptides 20A, 30A, 40A, and 50A required at least, respectively, 10, 20, 20, and 30% acetonitrile in the mobile phase to be eluted from the S300 column (Figure 3); only the 10-residue peptide (10A) was eluted in the absence of acetonitrile. In contrast, in the case of the more hydrophobic "L" series (Table 1), peptides 10L, 20L, 30L, 40L, and 50L required at least, respectively, 10, 30, 30, 30, and 40% acetonitrile in the mobile phase to overcome hydrophobic interactions with the column packing. Again, similar results were obtained on the other two cation-exchange columns.

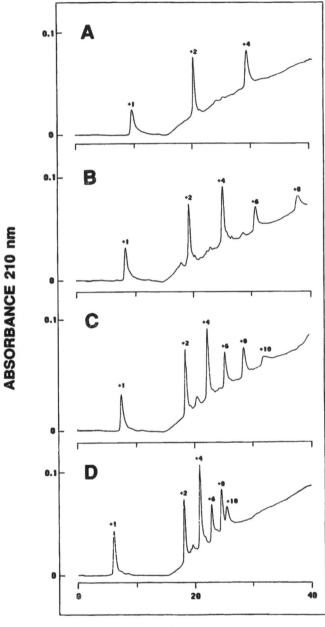

FIGURE 3. Strong cation-exchange chromatography of synthetic peptide polymers. HPLC instrument : as Figure 2. Column : SynChropak S300 (250 × 4.1 mm I.D.). Conditions: linear AB gradient (20 mM salt/min, following 10 min isocratic elution with Buffer A), where Buffer A is 5 mM KH$_2$PO$_4$, pH 6.5, and Buffer B is Buffer A plus 0.5 M NaC1, both buffers containing 10, 20, 30, or 40% acetonitrile (v/v) (Panels A, B, C, and D, respectively); flow-rate, 1 ml/min; temperature, 26°C. The sequence of the peptides was Ac-(Leu-Gly-Leu-Lys-Ala)$_n$-amide, where n = 1, 2, 4, 6, 8, 10 (+1, +2, +4, +6, +8, +10 net charge, respectively) ("A" series in Table 1). (From Burke, T. W. L., Mant, C. T., Black, J. A., and Hodges, R. S., *J. Chromatogr.*, 476, 377, 1989. With permission.)

C. EFFECT OF POLYPEPTIDE CHAIN LENGTH AND CHARGE DENSITY ON PEPTIDE RETENTION TIMES IN CATION-EXCHANGE CHROMATOGRAPHY

1. Effect of Polypeptide Chain Length and Charge Density on Peptide Retention Behavior

Figure 4 shows elution profiles of the "X" series of peptide polymers (10 to 50 residues; +1 to +5 net charge, respectively) (Table 1) and the four cation-exchange standards C1 to C4 (+1 to +4 net charge, respectively) on the Mono S strong CEX column. The peptides were eluted with a linear sodium chloride gradient (20 mM NaCl/min, following 10-min elution with starting buffer) at a flow-rate of 1 ml/min. Since only the effects of polypeptide chain length and/or charge density were being examined, it was important to minimize any non-specific, hydrophobic interactions of peptides with the ion-exchange packing. Thus, both the starting buffer (5 mM KH$_2$PO$_4$, pH 6.5) and the gradient buffer (5 mM KH$_2$PO$_4$, + 0.5 M NaCl, pH 6.5) contained 40% acetonitrile (v/v). From Figure 4, it can be seen that similarly charged species were not necessarily eluted at similar times. For instance, peptide 50X (+5 net charge) (Panel A) was not retained as long as C3 (+3 net charge) or C4 (+4 net charge) (Panel B). Similarly, peptide 40X (+4 net charge) (Panel A) was eluted prior to C3 (+3 net charge) (Panel B). The two series of peptide standards differed significantly in their range of both peptide chain length (10 to 50 residues for peptides 10X to 50X, respectively; eleven residues each for peptides C1 to C4) and charge density (+1 net charge/10 residues [+0.1/residue] for 10X-50X; +1 to +4 net charge/11 residues [+0.09 to + 0.36/residue] for C1 to C4). In order to rationalize the elution profiles shown in Figure 4, it was necessary to determine the relative contribution that polypeptide chain length and charge density individually make to peptide retention behavior during CEX.

2. Linearization of Peptide Retention Behavior

Figure 5, top panel, demonstrates the relationship between peptide elution time on the S300 column and peptide net charge for the "A" and "X" series of peptide polymers and the cation-exchange standards ("C"). The chromatographic conditions were the same as those described for Figure 4. The peptides containing a single net positive charge (5A, 10X, C1) were eluted during the initial 10-min isocratic elution with starting buffer and are not included in the plots. The plot for the remaining three cation-exchange standards, C2 to C4 (+2 to +4 net charge, respectively) demonstrated a linear relationship between peptide elution time and net charge. However, the plots for the two peptide polymer series, "X" and "A", showed a non-linear relationship, with the peptides eluted earlier than expected with increasing net charge and chain length. Plotting the elution times of the "X" and "A" series of peptides against the logarithm of the number of residues they contained (lnN) resulted in the straight-line relationships shown in Figure 5, middle panel. This exponential relationship between peptide retention time and peptide chain length reflected a similar relationship reported for reversed-phase chromatography of peptides.[1,26] A plot of elution time vs. lnN for C2 to C4 (all 11 residues in length) naturally produced a straight, vertical line. The divergence of the plots in Figure 5, top panel, and the difference in slopes in Figure 5, middle panel, appeared to reflect a difference in the charge densities of the peptides (+1 net charge/10 residues for the "X" series; +2 net charge/10 residues for the "A" series; +1 to +4 net charge/11 residues for C2 to C4).

From Figure 5, bottom panel, it can be seen that dividing the net charge of the peptides from the two polymer series ("X" and "A") and the mixture of cation-exchange standards (C2 to C4) by the logarithm of the number of residues they contain (net charge/lnN), and plotting this value against the observed elution time resulted in a single, straight-line plot with a correlation of 0.99 (determined by linear least-squares fitting).[27] The success of this simple linearization approach was particularly interesting considering the diversity of the peptides used in this study: the peptides varied substantially in charge density (+0.1 to +0.4/residue), net charge (+2 to +10), polypeptide chain length (10 to 50 residues) and overall hydrophobicity.

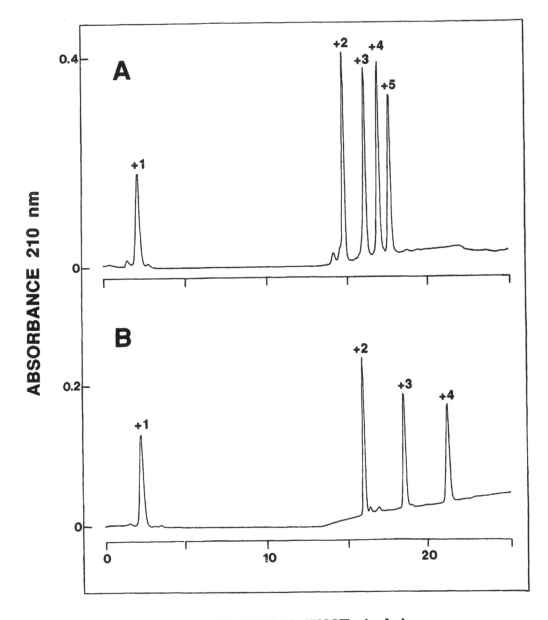

FIGURE 4. Strong cation-exchange chromatography of synthetic peptides. HPLC instrument : as Figure 2. Column: Mono S (50 × 5 mm I.D.). Conditions : linear AB gradient (20 mM salt/min, following 10 min isocratic elution with Buffer A), where Buffer A is 5 mM KH$_2$PO$_4$, pH 6.5, and Buffer B is Buffer A plus 0.5 M NaCl, both buffers containing 40% acetonitrile (v/v); flow-rate, 1 ml/min; temperature, 26°C. Panel A: mixture of five synthetic peptide size-exclusion standards (10 to 50 residues; +1 to +5 net charge) ("X" series in Table I). Panel B: mixture of four synthetic undecapeptide cation-exchange standards (+1 to +4 net charge; C1 to C4 in Table 1). (From Burke, T. W. L., Mant, C. T., Black, J. A., and Hodges, R. S., *J. Chromatogr.*, 476, 377, 1989. With permission.)

A simple relationship between peptide elution time and net positive charge during strong CEX was reported by Mant and Hodges (on a SynChropak S300 column)[7] and later by Crimmins et al. (on a PolySULFOETHYL Aspartamide column)[16], i.e., a satisfactory linear relationship was obtained without any correction for peptide chain length. The results shown in Figure 5 (top

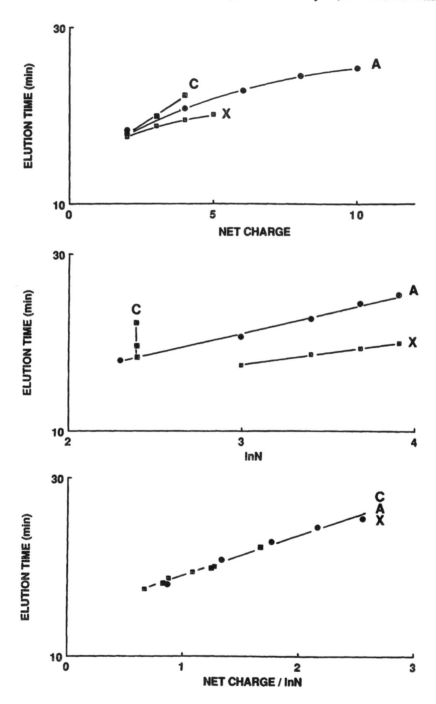

FIGURE 5. Relationship of peptide elution time to polypeptide chain length and charge density during strong cation-exchange chromatography of synthetic peptides. HPLC instrument: as Figure 2. Column: SynChropak S300 (250 × 4.1 min I.D.). Conditions: as Figure 3. Top panel: peptide elution time vs. peptide net charge. Middle panel: peptide elution time vs. logarithm of the number of residues (lnN). Bottom panel: peptide elution time vs. peptide net charge divided by the logarithm of the number of residues (net charge/lnN). The letters C, A, and X denote the cation-exchange standards, the "A" series of peptide polymers and the "X" series of size-exclusion standards, respectively. See Table 1 for sequences. (From Burke, T. W. L., Mant, C. T., Black, J. A., and Hodges, R. S., *J. Chromatogr.*, 476, 377, 1989. With permission.)

panel) suggested that the polypeptide chain length effect on peptide retention times becomes significant only beyond a length of ca. 20 residues. The peptides utilized by the previous workers ranged in chain length from only 12 to 21 residues[7] or 7 to 13 residues.[16] Thus, for these particular mixtures of peptides, any effect on peptide retention behavior due to chain length differences was probably fairly small. However, as shown in the present study, it is important to take polypeptide chain length into account when attempting to correlate net charge with the retention behavior of peptides > 20 residues in length.

It should also be noted that the choice of peptides employed by these researchers[7,16] may also have been fortuitous in terms of the absence and/or relative unimportance of any other factors (e.g., peptide conformation, charge distribution) which may potentially affect peptide retention behavior in CEX. Interestingly, conformational effects cannot explain the observed divergencies with polypeptide chain length demonstrated in this article. The peptide polymers used in this study do not have any unique tertiary structure, since the mobile phase conditions used to ensure ideal ion-exchange behavior (buffers containing 40% acetonitrile) are denaturing to tertiary structures, favoring the exposure of all charged residues. In addition, their secondary structure ranges from random coil to substantial α-helical content (as measured by circular dichroism), yet they still exhibited a similar polypeptide chain length effect.

IV. CONCLUSIONS

The peptides described in this article covered an extreme range of peptide hydrophobicities to values far exceeding those of most peptides encountered and it is not very likely, or desirable, that organic solvent concentrations of as high as 40% (v/v) in the mobile phase buffers will be regularly necessary for CEX of average peptide mixtures. However, the results described, particularly those showing the retention behavior of the cation-exchange standards (Figures 1 and 2), do suggest that the addition on a regular basis of a low level of acetonitrile (e.g., 10 to 20% [v/v]) to the mobile phase buffers would be worthwhile during CEX of peptides to suppress any hydrophobic interactions with the ion-exchange packing and thereby ensure complete elution of all peptides in a peptide mixture.

This article has also attempted to underline the importance of utilizing peptide standards to assess column performance, as well as to determine mobile phase effects on peptide retention behavior and/or the column packing, prior to application of the sample of interest. The development of a similar set of peptide standards (containing only acidic residues) for assessing anion-exchange column performance in peptide separations is a logical step and is presently underway in the authors' laboratory.

ACKNOWLEDGMENTS

This work was supported by the Medical Research Council of Canada and equipment grants from the Alberta Heritage Foundation for Medical Research.

REFERENCES

1. **Mant, C. T. and Hodges, R.S.**, HPLC of peptides, in *High-Performance Liquid Chromatography of Biological Macromolecules : Methods and Applications*, Gooding, K. and Regnier, F., Eds., Marcel Dekker, New York, 1989, 301.

2. **Mant, C. T. and Hodges, R. S.,** Optimization of peptide separations in high-performance liquid chromatography, *J. Liq. Chromatogr.,* 12, 139, 1989.
3. **Gariepy, J., Sykes, B. D., and Hodges, R. S.,** Lanthanide-induced peptide folding: variations in lanthanide affinity and induced peptide conformation, *Biochemistry,* 22, 1765, 1983.
4. **Welinder, B. S. and Linde, S.,** High performance ion-exchange chromatography of insulin and insulin derivatives, in *Handbook of HPLC for the Separation of Amino Acids, Peptides and Proteins,* Vol. 2, Hancock, W. S., Ed., CRC Press, Boca Raton, 1984, 357.
5. **Dizdaroglu, M.,** Weak anion-exchange high-performance liquid chromatography of peptides, *J. Chromatogr.,* 334, 49, 1985.
6. **Takahashi, N., Ishioka, N., Takahashi, Y., and Putnam, F. W.,** Automated tandem high-performance liquid chromatographic system for separation of extremely complex peptide mixtures, *J. Chromatogr.,* 326, 407, 1985.
7. **Mant, C. T. and Hodges, R. S.,** Separation of peptides by strong cation-exchange high-performance liquid chromatography, *J. Chromatogr.,* 327, 147, 1985.
8. **Patience, R. L. and Rees, L. H.,** Comparison of reversed-phase and anion-exchange high-performance liquid chromatography for the analysis of human growth hormones, *J. Chromatogr.,* 352, 241, 1986.
9. **Andrews, P. C.,** Ion-exchange HPLC for peptide purification, *Peptide Res.,* 1, 93, 1988.
10. **Takahashi, N., Takahashi, Y., and Putnam, F. W.,** Two-dimensional high-performance liquid chromatography and chemical modification in the strategy of sequence analysis. Complete amino acid sequence of the lambda light chain of human immunoglobulin D, *J. Chromatogr.,* 266, 511, 1983.
11. **Cachia, P. J., Van Eyk, J., Chong, P. C. S., Taneja, A. K., and Hodges, R. S.,** Separation of basic peptides by cation-exchange high-performance liquid chromatography, *J. Chromatogr.,* 266, 651, 1983.
12. **Mant, C. T. and Hodges, R. S.,** General method for the separation of cyanogen bromide digests of proteins by high-performance liquid chromatography: rabbit skeletal troponin I, *J. Chromatogr.,* 326, 349, 1985.
13. **Kumagaye, K. Y., Takai, M., Chino, N., Kimura, T., and Sakakibara, S.,** Comparison of reversed-phase and cation-exchange high-performance liquid chromatography for separating closely related peptides: separation of Asp[76]-human parathyroid hormone (1-84) from Asn[76]-human parathyroid hormone (1-84), *J. Chromatogr.,* 327, 327, 1985.
14. **Mychack, P. and Benson, J. R.,** Peptide separations using high-performance ion-exchange chromatography, *LC.GC Mag. Liq. Gas Chromatogr.,* 4, 462, 1986.
15. **Alpert, A. J. and Andrews, P. C.,** Cation-exchange chromatography of peptides on poly (2-sulfoethyl aspartamide)-silica, *J. Chromatogr.,* 443, 85, 1988.
16. **Crimmins, D. L., Gorka, J., Thoma, R. S., and Schwartz, B. D.,** Peptide characterization with a sulfoethyl aspartamide column, *J. Chromatogr.,* 443, 63, 1988.
17. **Burke, T. W. L., Mant, C. T., Black, J. A., and Hodges, R. S.,** Strong cation-exchange HPLC of peptides : (1) effect of non-specific hydrophobic interactions; (2) linearization of peptide retention behaviour, *J. Chromatogr.,* 476, 377, 1989.
18. **Kopaciewicz, W., Rounds, M. A., Fausnaugh, J., and Regnier, F. E.,** Retention model for high-performance ion-exchange chromatography, *J. Chromatogr.,* 266, 3, 1983.
19. **Kennedy, L. A., Kopaciewicz, W., and Regnier, F. E.,** Multimodal liquid chromatography columns for the separation of proteins in either the anion-exchange or hydrophobic-interaction mode, *J. Chromatogr.,* 359, 73, 1986.
20. **Rounds, M. A., Rounds, W. D., and Regnier, F. E.,** Poly(styrene-divinylbenzene)-based strong anion-exchange packing material for high-performance liquid chromatography of proteins, *J. Chromatogr.,* 397, 25, 1987.
21. **Kopaciewicz, W., Rounds, M. A., and Regnier, F. E.,** Stationary phase contributions to retention in high-performance anion-exchange protein chromatography: ligand density and mixed mode effects, *J. Chromatogr.,* 318, 157, 1985.
22. **Heinitz, M. L., Kennedy, L., Kopaciewicz, W., and Regnier, F. E.,** Chromatography of proteins on hydrophobic interaction and ion-exchange chromatographic matrices: mobile phase contributions to selectivity, *J. Chromatogr.,* 443, 173, 1988.
23. **Guo, D., Mant, C. T., Taneja, A. K., Parker, J. M. R., and Hodges, R. S.,** Prediction of peptide retention times in reversed-phase high-performance liquid chromatography. I. Determination of retention coefficients of amino acid residues of model synthetic peptides, *J. Chromatogr.,* 359, 499, 1986.
24. **Parker, J. M. R. and Hodges, R. S.,** I. Photoaffinity probes provide a general method to prepare synthetic peptide-conjugates, *J. Prot. Chem.,* 3, 465, 1985.
25. **Hodges, R. S., Heaton, R. J., Parker, J. M. R., Molday, L., and Molday, R. S.,** Antigen-Antibody interaction. Synthetic peptides define linear antigenic determinants recognized by monoclonal antibodies directed to the cytoplasmic carboxyl terminus of rhodopsin, *J. Biol. Chem.,* 263, 11768, 1988.

26. **Mant, C. T., Burke, T. W. L., Black, J. A., and Hodges, R. S.,** Effect of peptide chain length on peptide retention behavior in reversed-phase chromatography, *J. Chromatogr.,* 458, 193, 1988.
27. **Hodges, R. S., Parker, J. M. R., Mant, C. T., and Sharma, R. R.,** Computer simulation of high-performance liquid chromatographic separations of peptide and protein digests for development of size-exclusion, ion-exchange and reversed-phase chromatographic methods, *J. Chromatogr.,* 458, 147, 1988.

ION-EXCHANGE HIGH-PERFORMANCE LIQUID CHROMATOGRAPHY OF PEPTIDES

Andrew J. Alpert

I. INTRODUCTION

During the past 15 years or so, a scientist with peptides to purify or analyze has generally reached first for a reversed-phase chromatography (RPC) column. This reflects the general utility of the mode in peptide applications, the use of mobile phases that can be removed from the product by lyophilization, and, perhaps, the familiarity of this mode of chromatography to most users of high-performance liquid chromatography (HPLC) systems. It is unfortunate that such users are not better acquainted with alternatives to RPC, since there are peptide applications for which RPC fails. These include the obvious case of different peptides that happen to possess the same degree of hydrophobic character, as well as peptides too hydrophobic to be eluted from RPC columns, peptides too hydrophilic to bind at all, and polypeptides with important tertiary structure that could be lost in the denaturing RPC mobile phases. Moreover, no single mode of HPLC can now afford complete resolution of mixtures that contain many more than 30 peptides. This number can easily be exceeded in a crude tissue extract or in a protein digest generated for the purpose of peptide mapping. Thus, there is a need for modes of HPLC with selectivity complementary to that of RPC, either as alternatives or as supplements to RPC.

II. ION-EXCHANGE CHROMATOGRAPHY OF PEPTIDES

Perhaps the most promising alternative to RPC for peptide work is ion-exchange chromatography (IEC). This mode resolves solutes on the basis of charge, and most peptides are charged. Anion-exchange chromatography has been used for acidic peptides[1] and cation-exchange chromatography for basic ones,[2-5] generally involving salt gradients at a pH near neutrality. However, if a pH in the range 2.8 to 3.0 is employed, then cation-exchange becomes applicable to peptides in general.[6-8] At such a pH, the great majority of peptides are basic, having lost the negative charge at Asp- and Glu- residues and most of the negative charge at the C-terminus. At the same time, Lys-, Arg-, and His- residues and N-termini are positively charged. Peptide binding will, in principle, be governed by these basic residues rather than by the degree of hydrophobicity as is the case for RPC. Figure 1 demonstrates that standard peptides are eluted (with a salt gradient) in order of increasing absolute number of basic, positively charged, residues.

Converting anion-exchange to a mode of equally general utility with peptides would require a pH around 13. In practice this would not work, since all basic functional groups now used in anion-exchange chromatography packings would also lose their basicity in this range. Thus, the question of pH stability of silica vs. polymer-based supports would not arise. Polymer-based cation-exchange columns offer no advantage over silica-based columns for peptide applications, silica being quite stable at pH 3.0. However, with silica-based media, the pH should be

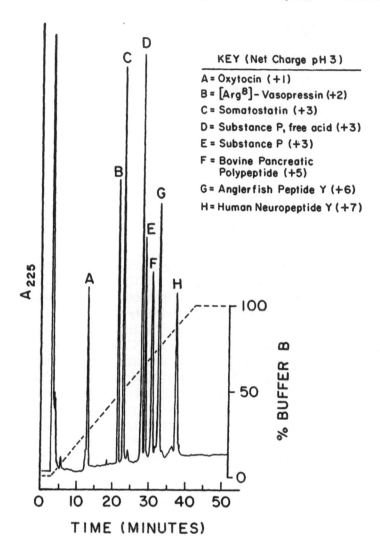

FIGURE 1. Elution of peptide standards from a strong cation-exchange column. Column: PolySULFOETHYL Aspartamide, 200 × 4.6 mm I.D., 5-μm particle size, 300-Å pore size (PolyLC, Columbia, MD). Conditions: linear AB gradient (2.5% B/min, equivalent to 6.25 m*M* salt/min), where Buffer A is 5 m*M* potassium phosphate, pH 3.0 and Buffer B is Buffer A containing 0.25 M potassium chloride, both buffers containing 25% (v/v) acetonitrile; flow-rate, 0.7 ml/min. Sample: 50 μl volume containing *ca.* 5 μg of each peptide. (From Alpert, A. J. and Andrews, P. C., *J. Chromatogr.*, 443, 85, 1988. With permission.)

kept above 2.5, since the Si-C bond of silane-based coatings starts to hydrolyze beginning around pH 2.0.

III. COLUMNS FOR CATION-EXCHANGE CHROMATOGRAPHY OF PEPTIDES

For cation-exchange chromatography at pH 3, weak cation-exchange (WCX) columns are not suitable since the functional carboxyl- groups (e.g., CM-type) are predominantly uncharged at pH 3. Appropriate functional groups include phosphates, phosphonic and sulfonic acids,

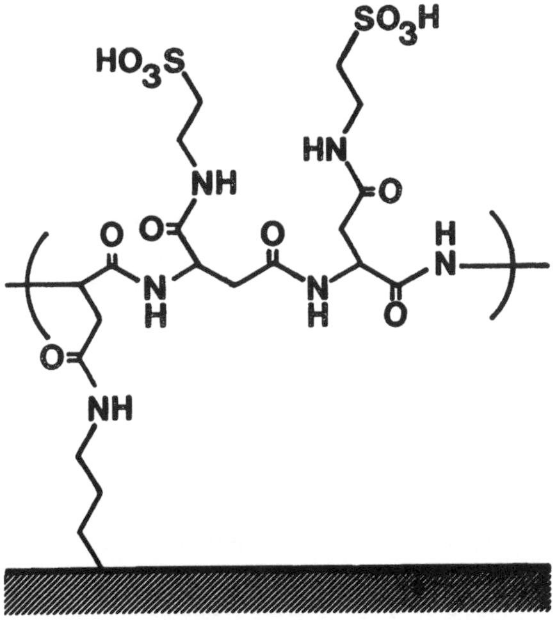

FIGURE 2. Structure of the coating of poly(2-sulfoethyl aspartamide)-silica.

which retain their negative charge at this low pH. Supports with such functional groups are strong cation-exchange (SCX) materials.

The most commonly-encountered SCX packings for HPLC contain sulfopropyl-(SP-) groups. This ligand possesses a degree of hydrophobicity due to the length of the propyl arm, conferring some mixed mode character upon peptide separations. Ion-exchange and hydrophobic effects are basically incompatible interactions in chromatography. Ionic interactions weaken as salt concentration increases, while hydrophobic interactions become stronger. These two opposing processes result in lower resolution and, in severe cases, lower yields. In an effort to eliminate such effects, a new SCX material was developed recently[6] that features sulfoethyl-groups. This is prepared by incorporating the amino acid taurine into a coating of the reactive polymer poly(succinimide) bound to silica. The result is a polypeptide coating, poly(2-sulfoethyl aspartamide), covalently attached to the silica (Figure 2). This material (trade name PolySULFOETHYL Aspartamide, or PolySULFOETHYL A) is extremely hydrophilic, and appears to be one of the best SCX materials currently available for peptide applications.

IV. FACTORS AFFECTING PEPTIDE RETENTION BY SCX

As Figure 1 shows, PolySULFOETHYL A does indeed resolve peptides on the basis of charge. Crimmins et al.[9] have confirmed this with a study involving 54 peptides of known sequence. Several tentative rules governing retention can be deduced from close inspection of the data in this study:

1. The most important factor is the absolute number of basic residues; charge to mass ratio had only a minor effect, if any, with the peptides studied.
2. In general, peptides differing by a single charge are completely resolved; however, peptides of the same charge but differing otherwise in composition can also be resolved, although the degree of separation is less than with peptides of different charge.

3. Concerning the data in Reference 9, juxtaposition of an acidic and a basic residue (or the N-terminus) appears to neutralize substantially the contribution of the basic residue to the retention of the peptide. While an isolated acidic residue would effectively be uncharged at pH 3.0, an adjacent basic residue would create a microenvironment of appreciably higher pH, lowering the perceived pKa of the acidic residue. Such conjugate acid-base pairs resemble those observed in the active sites of enzymes, where, for example, a His- residue can lower the pKa of a nearby Cys- residue by several pH units.[10] Another example is the appreciably greater retention of Val-Asp than Asp-Val in a related mode of chromatography.[11] Exceptions to rule 3 appear to be the lack of influence of the C-terminus and the lack of effect of Glu- on an adjacent His- residue.[9]

This ability to resolve peptides of the same charge could be attributed to some sort of mixed-mode interaction or to differences in peptide secondary structure, which could render some residues sterically unavailable for binding to the stationary phase surface. Figure 3 demonstrates the influence of mobile phase composition on selectivity; substance P and its free acid are coeluted when strictly aqueous buffers are used, but inclusion of 25% acetonitrile in both mobile phases affords complete resolution. This topic will be addressed later.

V. SCX AS A SUPPLEMENT TO RPC

The use of PolySULFOETHYL A in sequence with an RPC column has proved to be quite useful with mixtures too complex to be resolved by any single mode. In no case examined has this combination failed to yield, in two runs, peptides that are sequenceable (thus, presumably over 90% pure). Figure 4 illustrates the fractionation of peptides in a crude tissue extract, using a preparative-scale SCX column. Peaks were collected and rerun by RPC (e.g., Figure 5) to yield sequenceable products. It is most convenient to use the SCX column first. The capacity is several times that of an equivalent RPC column (typically, 3 to 5 mg per run on a 200×4.6 mm I.D. analytical column with retention of maximum resolution), and the salt present in the collected fractions is eliminated conveniently in the subsequent RPC run. The final product is thus collected in a volatile mobile phase.

The SCX-RPC combination has been used to purify synthetic peptides that could not be purified by RPC alone.[12,13] It has also been used in peptide mapping, [14] for which it represents a form of two-dimensional chromatography. It has been estimated[14] that this combination has the potential to resolve a mixture of up to 1800 peptides.

VI. SCX AS AN ALTERNATIVE TO RPC OF PEPTIDES

Selectivity in RPC is sensitive to the composition of most of the residues in peptides, since most residues possess some degree of hydrophobic character. In contrast, adsorption in SCX chromatography is effected by the relatively small number of basic residues. Any factor affecting these residues will have an inordinate effect on retention by SCX. This permits the specific analysis or isolation of classes of peptides by SCX for which RPC is not specific. An example is the analysis of peptides blocked at the N-terminus with a pyroglutamyl residue.[15] Enzymatic removal of the pGlu- residue produces a free N-terminus, increasing the net charge by +1. The substrate and product peptides are easily resolved by SCX.

Digestion of proteins with trypsin generates a mixture of peptides that lends itself readily to selective isolation of particular classes of peptides. The typical tryptic peptide has a net charge of +2 at pH 3, due to the N-terminus and the Lys- or Arg-residue at the C-terminus; peptides

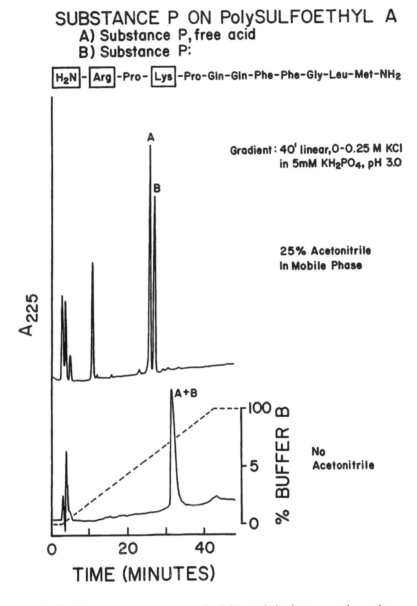

FIGURE 3. Effect of organic solvent on selectivity: resolution by strong cation-exchange chromatography of Substance P (B) and its free acid form (A). Column and conditions as described in Figure 1. (From Alpert, A. J. and Andrews, P. C., *J. Chromatogr.*, 443, 85, 1988. With permission.)

containing His- can have higher charges. The C-terminal peptide or a blocked N-terminal peptide will have a charge of +1, and is generally the first peptide to be eluted from an SCX column.[16]

A more intriguing application is the selective isolation of disulfide-linked peptides from a tryptic digest. Linking two peptides with charges of +2 produces a peptide with a charge of +4. One would therefore expect disulfide-linked peptides to be retained on an SCX column significantly longer than the rest of the digest, and such has proved to be the case.[12] Crimmins et al. have recently applied SCX to the analysis of V8 digests.[17]

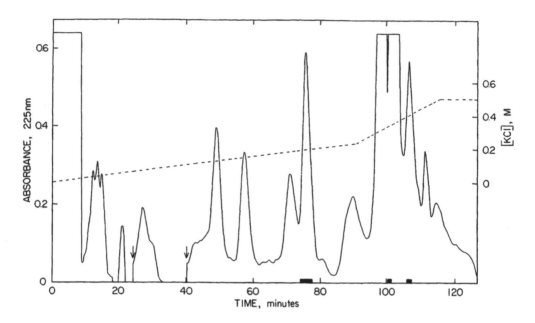

FIGURE 4. Preparative-scale purification of polypeptides from an extract of 111 g of guinea pig pancreas. Column: PolySULFOETHYL Aspartamide, 500 × 22 mm I.D., 15 to 20 μm particle size, 300-Å pore size (PolyLC). Conditions: mobile phase, 5 m*M* potassium phosphate, pH 3.0, with 25% (v/v) acetonitrile and a gradient to 0.5 *M* potassium chloride as shown; flow-rate, 30 ml/min. Solid bars indicate fractions that were collected and rerun by RPC. (From Alpert, A. J. and Andrews, P. C., *J. Chromatogr.*, 443, 85, 1988. With permission.)

VII. EFFECT OF THE MOBILE PHASE ON SELECTIVITY

Inclusion of an organic solvent in both mobile phases can have marked effects on selectivity in SCX, as was apparent in Figure 3. Increasing levels of organic solvent affect different peptides in different ways, decreasing the retention of some while increasing the retention of others.[6] For the separation of any two particular peptides, the optimal level of organic solvent must be determined empirically. A good level for general peptide SCX is 25% acetonitrile.

A set of five decapeptides has been developed for the calibration of RPC columns[18] (the structures of these peptides have been reported in Reference 19). Four of these peptides contain the same net positive charge (+2) and differ only in the hydrophobic character of two of the residues. These four peptides are not resolved by SCX at low levels of acetonitrile, but are partially or mostly resolved when the mobile phase contains 50% acetonitrile.[6] It has been found[20] that the order of elution of these four peptides from the PolySULFOETHYL A column is from least hydrophilic to most hydrophilic, the opposite of that in RPC. This suggests that the PolySULFOETHYL A sorbent becomes sensitive to the nonbasic residues of peptides with mobile phases containing high levels of organic solvent. With such mobile phases, retention is promoted by the hydrophilic residues in particular. A recent study[11] has established that this is indeed the case. The effect can be demonstrated with the individual amino acids, which are eluted in order of least to most hydrophilic.

These hydrophilic interactions dominate the chromatography at levels of acetonitrile above 70%; electrostatic effects become secondary under these conditions. For example, at low levels of acetonitrile, the presence of strongly acidic groups (e.g., phosphate or sulfate) in peptides speeds up their elution from PolySULFOETHYL A columns, due to electrostatic repulsion. At high levels of acetonitrile, such groups can actually promote retention, since a charged group, whether acidic or basic, is a very hydrophilic group. At sufficiently high levels, even uncharged peptides are readily retained and resolved on the basis of their hydrophilicity.[11]

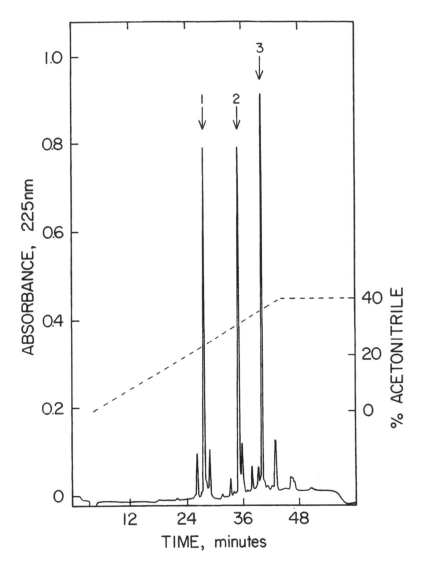

FIGURE 5. Reversed-phase chromatography of the peak eluted at 102 min from the preparative strong cation-exchange purification shown in Figure 4. Column: Vydac pH-stable RPC column, 250 × 4.6 mm I.D. (The Separations Group, Hesperia, CA). Conditions: linear AB gradient (1% B/min), where Eluent A is 0.1% aq. trifluoroacetic acid (TFA) and Eluent B is 0.1% TFA in acetonitrile; flow-rate, 0.7 ml/min. Peaks: 1 = unknown (blocked N-terminus); 2 = insulin; 3 = glucagon. (From Alpert, A. J. and Andrews, P. C., *J. Chromatogr.*, 443, 85, 1988. With permission.)

These effects would appear to account for much of the ability of PolySULFOETHYL A to resolve different peptides of the same charge. The mixed-mode effects in question reflect hydrophilic interactions, rather than the hydrophobic interactions encountered with less hydrophilic SCX materials. Thus, the PolySULFOETHYL A column can be employed in two alternate modes. At low levels of organic solvent, it functions in the SCX mode, with the chromatography governed by the basic residues. At high levels of organic solvent, it functions in a Hydrophilic Interaction Chromatography mode, sensitive to the composition of all the amino acids in a peptide but in the opposite sense from RPC. This new mode of chromatography can effect separations not possible with RPC or SCX.[11] Future development of these techniques will expand the alternatives available for peptide separations, facilitating progress in this field.

REFERENCES

1. **Dizdaroglu, M.,** Weak anion-exchange high-performance liquid chromatography of peptides, *J. Chromatogr.,* 334, 49, 1985.
2. **Mant, C.T. and Hodges, R.S.,** Separation of peptides by strong cation-exchange high-performance liquid chromatography, *J. Chromatogr.,* 327, 147, 1985.
3. **Isobe, T., Takayasu, T., Takai, N., and Okuyama, T.,** High-performance liquid chromatography of peptides on a macroreticular cation-exchange resin: Application to peptide mapping of Bence-Jones proteins, *Anal. Biochem.,* 122, 417, 1982.
4. **Takahashi, N., Takahashi, Y., and Putnam, F.W.,** Two-dimensional high-performance liquid chromatography and chemical modification in the strategy of sequence analysis, *J. Chromatogr.,* 266, 511, 1983.
5. **Kumagaye, K.Y., Takai, M., Chino, N., Kimura, T., and Sakakibara, S.,** Comparison of reversed-phase and cation-exchange high-performance liquid chromatography for separating closely related peptides: Separation of Asp^{76}-Human Parathyroid Hormone (1-84) from Asn^{76}-Human Parathyroid Hormone (1-84), *J. Chromatogr.,* 327, 327, 1985.
6. **Alpert, A.J. and Andrews, P.C.,** Cation-exchange chromatography of peptides on poly(2-sulfoethyl aspartamide)-silica, *J. Chromatogr.,* 443, 85, 1988.
7. **Imamura, T., Sugihara, J., Yokota, E., Kagimoto, M., Naito, Y., and Yanase, T.,** Analytical peptide mapping by ion-exchange high-performance liquid chromatography: application to haemoglobin variants, *J. Chromatogr.,* 305, 456, 1984.
8. **Kato, Y., Nakamura, K., and Hashimoto, T.,** New high-performance cation exchanger for the separation of proteins, *J. Chromatogr.,* 294, 207, 1984.
9. **Crimmins, D.L., Gorka, J., Thoma, R.S., and Schwartz, B.D.,** Peptide characterization with a sulfoethyl aspartamide column, *J. Chromatogr.,* 443, 63, 1988.
10. **Shipton, M. and Brocklehurst, K.,** Characterization of the papain active centre by using two-protonic-state electrophiles as reactivity probes. Evidence for nucleophilic reactivity in the uninterrupted cysteine-25-histidine-159 interactive system, *Biochem. J.,* 171, 385, 1978.
11. **Alpert, A.J.,** *J. Chromatogr.,* 499, 177, 1990.
12. **Andrews, P.C.,** Ion-exchange HPLC for peptide purification, *Peptide Res.,* 1, 93, 1988.
13. **Hong, A.L., Brasseur, M.M., and Kesuma, D.,** Purification of histidine-rich peptides, this publication.
14. **Schlabach, T.D., Colburn, J.C., Mattaliano, R.J., and Yuen, S.,** Meeting the challenge in peptide fragment purification for protein sequencing, in *Techniques in Protein Chemistry,* Vol. 1, Hugli, T.E., Ed., Academic Press, New York, 1989, ch. 48.
15. **Crimmins, D.L., McCourt, D.W., and Schwartz, B.D.,** Facile analysis and purification of deblocked N-terminal pyroglutamyl peptides with a strong cation-exchange sulfoethyl aspartamide column, *Biochem. Biophys. Res. Commun.,* 156, 910, 1988.
16. **Cheng, D.H.W., Lee, P.-J., and Watt, K.W.K.,** Selective isolation of carboxyl-terminal peptides from proteins by a strong cation-exchange, Sulfoethyl aspartamide column and an affinity-chromatography anhydrotrypsin column, *7th International Symposium on HPLC of Proteins, Peptides, and Polynucleotides,* Washington, D.C., Nov. 1987, Abstract 311.
17. **Crimmins, D.L., Thoma, R.S., McCourt, D.W., and Schwartz, B.D.,** Strong-cation-exchange sulfoethyl aspartamide chromatography for peptide mapping of *Staphylococcus aureus* V8 protein digests, *Anal. Biochem.,* 176, 255, 1989.
18. **Mant, C.T. and Hodges, R.S.,** Requirement for peptide standards in reversed-phase high performance liquid chromatography, *LC.GC Mag.,* 4, 250, 1986.
19. **Parker, J.M.R., Mant, C.T., and Hodges, R.S.,** A practical approach to the preparative purification of peptides using analytical instrumentation with analytical and semipreparative columns, *Chromatographia,* 24, 832, 1987.
20. **Hodges, R.S. and Mant, C.T.,** personal communication, 1988.

USE OF ION-EXCHANGE CARTRIDGES FOR BATCH EXTRACTION IN THE PURIFICATION OF PITUITARY PEPTIDES

Hugh P. J. Bennett

The aim of our approach to peptide purification has been to exploit the resolving power of reversed-phase HPLC (RPC) as much as possible. We have designed an extraction procedure which produces an essentially salt- and protein-free extract which can be loaded directly onto the column.[1] Using a variety of solvent systems, it is possible to exploit the hydrophobic, basic, and acidic characters of peptides and proteins in a systematic manner. One of the unique properties of RPC is that large and small molecules and also those of basic, neutral, and acidic charge can be run simultaneously. These properties are invaluable for certain analytical procedures. For instance, it is possible to analyze all the biosynthetic derivatives of pro-opiomelanocortin simultaneously.[2] With very few exceptions all peptides of endocrine origin can be expected to run on RPC columns. One can initiate an RPC isolation of an unknown compound from an endocrine tissue and confidently expect efficient chromatography without knowing anything about the overall character of the substance including charge and molecular weight.

When purifying a minor component from a tissue extract, the problem of column overloading arises. If the proportion of unwanted material is very large, the chromatographic behavior of the minor components becomes distorted. Peaks tend to streak and minor components are eluted in large volumes. Initial crude fractionation procedures are necessary to reduce column loading. Contaminants running close to or simultaneously with the peptide or protein in question are likely to have radically different pI values. Therefore, batch ion-exchange fractionation can be employed. This chapter describes how cartridges packed with weak ion-exchangers can be used, in a batch procedure, to fractionate peptides extracted from bovine posterior pituitaries into basic, acidic, and neutral pools. A full account of this technique has been published elsewhere.[3,4] Figures 1 and 2 and Table 1 are reproduced from Reference 4 with permission. The method greatly reduces column loading since a large proportion of unwanted material is removed prior to RPC. The Sep-Pak cartridges used in this study were obtained from Waters (Milford, MA). They are packed with either Accell quaternary methyl-ammonium (QMA) anion-exchange medium or Accell carboxy-methyl (CM) cation-exchange medium. In both instances, the functional group is linked to a polymer-coated silica base (37 to 55 μm particle size, 500 nm pore size).

It is interesting to compare these characteristics with those of the octadecylsilyl-silica (ODS-silica) Sep-Pak cartridges. ODS-silica or C_{18} Sep-Pak cartridges are packed with ODS-silica of 37 to 55 μm particle size and 100 nm pore size. This difference in pore size relative to that of the ion-exchange Accell packing is an important consideration. The small pore size of C_{18} Sep-Pak packing severely reduces the accessibility of large proteins (i.e. > 100 kDa) to the ODS-silica matrix. Since high molecular weight protein is essentially excluded in this way, ODS-silica extracts of raw tissue homogenates are largely protein-free.[1,5] Increasing the silica pore size to 500 nm greatly increases the silica surface surface available for adsorption of larger proteins. Thus, the Accell ion-exchanger will have a greater intrinsic capacity to bind higher molecular

TABLE 1

Summary of the Theoretical Charges (at pH 6 and 7) and the Actual Ion-Exchange Properties of Peptides Extracted from Bovine Posterior Pituitaries

Peptide no. (see Fig. 2)	Peptide identity	Overall charge pH 5	Overall charge pH 7	A/B/N pool 10 nM NH$_4$Ac pH 5	A/B/N pool 10 mM tris pH 7
1	AVP	2+	2+	B	N
2	POMC$_{80-103}$	7–	7–	A	A
3	Oxytocin	1+	1+	B/N	N
4	β-MSH	1+	0	B	N
5	Lys^{1}γ$_3$MSH	5+	4+	B	B
6	Des-acetyl-α-MSH	3+	2+	B	N
7	Mono-acetyl α-MSH	2+	1+	B	N
8	Lys^{1}γ$_1$MSH	4+	3+	B	B
9	Di-acetyl α-MSH	2+	1+	B	N
10	CLIP	2-	2–	A	A
11	16K$_{1-77}$	0	1–	B	A
12	16K$_{1-49}$	5–	5–	A	A
13	Posterior pituitary glycopeptide	2–	2–	A	A

Abbreviations: POMC$_{80-103}$ = the acidic joining peptide of pro-opiomelanocortin; AVP = arginine-vasopressin; α-, β-, and γ-MSH = α, β, and γ-melanotropin; CLIP = corticotropin-like intermediate lobe peptide (i.e., ACTH$_{18-39}$); 16K$_{1-77}$ and 16K$_{1-49}$ = amino terminal fragments of pro-opiomelanocortin; posterior pituitary glycopeptide = the 109 to 147 sequence of propressophysin. Using the sequence information from References 6 and 7, the overall charge for each peptide was determined. The overall charge represents the difference between the number of acidic residues (i.e. Asp, Glu, and free carboxyl termini) and the number of basic residues (i.e. Lys, Arg, His, and free amino termini). At pH 5, histidine residues are considered fully charged, while at pH 7, they are considered uncharged. A/B/N represent the acidic, basic or neutral pools into which peptides are found to be fractionated.

From Bennett, H. P. J., *J. Chromatogr.*, 359, 383, 1986. With permission. N.B. There is an error in Table 1 of Reference 4. AVP and POMC$_{80-103}$ are designated as peaks 2 and 1, respectively, in this previous article. The designations shown above are correct.

weight proteins than C$_{18}$ Sep-Paks through the greater surface area available for binding. However, the larger the pore size, the lower the total surface area for a given mass of silica. Thus, while ODS-silica of 100 nm pore size has limited capacity for larger proteins, it has a much larger total surface available for binding smaller molecules than the 500 nm Accell material. These are important considerations when using these materials.

An ideal system to test the performance of ion-exchangers, at least with regard to peptides and small proteins, are the products obtained through the biosynthetic processing of pro-opiomelanocortin (POMC). POMC is a biosynthetic precursor to a number of hormones and peptide fragments. POMC itself has a pI approaching 8. The hormonal sequences (i.e. corticotropin and endorphin), together with the pairs of basic residues that flank these sequences, contribute a very large number of positive charges. To arrive at a pI of 8, these basic components are balanced by components of acidic charge. Indeed, the charge distribution alternates between negative and positive as one progresses from the amino-terminus of POMC through to the carboxyl-terminus. Examination of the many cDNA sequences for pro-hormones that have been published shows that this charge distribution is a common feature.

To test the properties of the Accell cartridges, an extract of bovine posterior pituitaries (Pel-

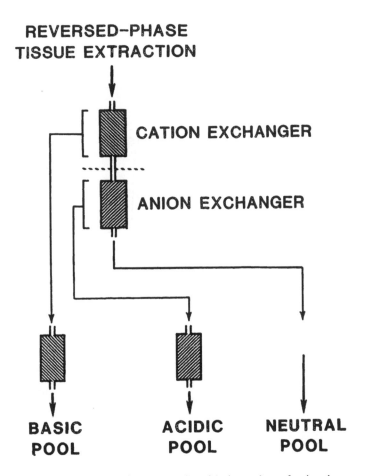

FIGURE 1. Diagrammatic representation of the ion-exchange fractionation procedure. The Sep-Pak ion-exchangers (approximately 1 × 1 cm, containing 350 mg packing) are connected in series for the fractionation of the tissue extract. The cartridges can conveniently be connected together using plastic tips used for pipette guns. The 250 μl size makes the best connector. Tissue extracts are passed through the Sep-Paks using a 10 ml syringe. They are then detached from each other and eluted separately. The material flowing through both cartridges is designated the neutral pool, while eluates from the cation and anion exchangers constitute the basic and acidic pools respectively.(From Bennett, H. P. J., *J. Chromatogr.*, 359, 383, 1986. With permission.)

Freez Biologicals) was prepared. Briefly, this involved extraction of 20 pituitaries with an extraction medium consisting of 1 M hydrochloric acid, containing 5% (v/v) formic acid, 1% (w/v) sodium chloride, and 1% (v/v) trifluoroacetic acid (TFA). This extract was subjected to reversed-phase extraction using ten C_{18} Sep-Pak cartridges.[1,5] The Sep-Pak eluates were combined, divided into ten equal portions (i.e. two pituitary equivalents), and taken to dryness. The dried residues were taken up in the appropriate buffer for ion-exchange fractionation with either ammonium acetate (pH 5) or Tris-HCl buffer (pH 7), containing 20% acetonitrile. Pilot experiments revealed that the inclusion of the organic solvent was important to obtain adequate recovery of peptides. All batch procedures were performed on a pair of Sep-Pak ion-exchangers, connected in series (Figure 1). For all studies, the cation-exchanger was connected ahead of the anion-exchanger. Pituitary extracts were passed through both cartridges in 3 ml of buffer. This was followed by a 6 ml wash with the same buffer. The unretained pool, now in 9 ml of buffer was designated the neutral pool (Figure 1). The cartridges were then disconnected and each

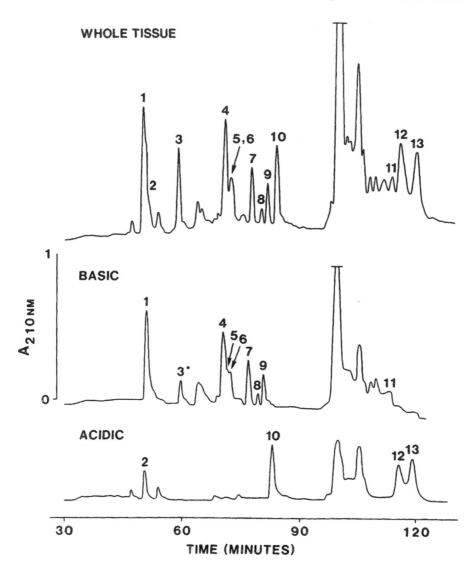

FIGURE 2. Analytical RPC of peptides extracted from bovine posterior pituitaries. The upper panel (whole tissue) shows the elution profile obtained from the RPC of two posterior pituitaries. A further portion of tissue extract, also corresponding to two posterior pituitaries, was dried, taken up in 3 ml of 10 mM ammonium acetate (pH 5), and subjected to ion-exchange batch extraction with cation- and anion-exchange Sep-Pak cartridges connected in series. The middle and lower panels show the RPC profiles obtained for the basic and acidic pools, respectively. The neutral pool was also analyzed (not shown), but only one peak was observed corresponding to oxytocin (Peak 3), denoted by a star in the basic pool. Peaks 1 to 13 are identified in Table 1. The column used was a Waters C$_{18}$ µBondapak, 300 × 3.9 mm I.D., 10-µm particle size, 125-Å pore size. Peptides were eluted with a linear AB gradient of 1.6 to 61.6% acetonitrile over 3 hr (0.33% acetonitrile/min), containing 0.1% trifluoroacetic acid throughout; flow-rate, 1.5 ml/min. (From Bennett, H. P. J., *J. Chromatogr.*, 359, 383, 1986. With permission.)

eluted with 5 ml of buffer containing 1 M sodium chloride. The cation-exchange eluate was designated the basic pool, while the anion-exchange eluate was designated the acidic pool (Figure 1). To get an impression of the way in which the pituitary extract had been distributed between the three pools, basic, acidic, and neutral pools were analyzed for peptide content by RPC.[1] These chromatograms were compared with one obtained with the starting material. All the major peaks observed correspond to fragments derived from the three precursor molecules, i.e., pro-opiomelanocortin (POMC), pro-pressophysin, and pro-oxyphysin.[6,7] The extract was derived from the whole bovine posterior pituitary consisting of the intermediate lobe and the neural lobe. POMC originates in the intermediate lobe, while the neural lobe gives rise to the oxytocin and vassopressin precursors. The various peaks were identified through their characteristic elution positions or a combination of amino acid analysis, radioimmunoassay and absorbance characteristics at 280 and 210 nm.

Figure 2 illustrates how bovine posterior pituitaries were fractionated at pH 5 using an ammonium acetate buffer. Interestingly, at this pH almost the entire tissue extract fractionates into either the acidic or basic pools (see also Table 1). The neutral pool contained a single peak of oxytocin accounting for about 50% of that found in the whole tissue extract (not shown). The almost complete distribution of peaks into the acidic and basic pools can be rationalized in terms of the charge distributions within the peptide hormone precursors. Furthermore, by considering the primary sequences of the pro-hormone fragments,[6,7] it can be seen that each fragment is behaving exactly as one might predict by adding up the positive and negative charges. Table 1 summarizes both the charge of each fragment considered and their observed distribution into the three pools.

The fractionation was repeated at pH 7 using a Tris-HCl buffer (Figure 3). A pattern markedly different from that at pH 5 (Figure 2) was observed. The weakly positively charged peptides were not retained on the cation-exchanger at pH 7. Figure 2 and Table 1 indicate that vasopressin, oxytocin, β-MSH and the α-MSHs were all recovered in the neutral pool. In contrast, all the acidic peptides, whether weakly charged or not, remained in the acidic pool. This behavior can be correlated well with the relative strengths of the two ion-exchangers. QMA-Accell is a relatively strong exchanger, while CM-Accell is a much weaker exchanger. All peptides having a histidine residue (pK 6) lose one positive charge when the buffer is changed from pH 5 to 7 and this contributes to the change in fractionation behavior.

These findings illustrate how ion-exchange Sep-Pak cartridges provide a useful technique for rapidly fractionating tissue extracts into acidic, basic or neutral pools prior to RPC. The technique has recently been used in the purification of two peptides that affect adrenal function. In one study, a thirty amino acid peptide with a net negative charge of 4 has been purified from a transplantable Leydig cell tumor. This polypeptide has been shown to regulate cholesterol-side chain cleavage activity in steroidogenic tissues including the adrenal.[8] In another study, four basic polypeptides have been purified from extracts of fetal and adult rabbit lung. These 33 or 34 amino acid peptides range in basic charge from +4 to +8. These polypeptides have been shown to inhibit corticotropin-stimulated steriodogenesis by rat adrenal cells. They appear to act as competitive inhibitors of corticotropin binding to its receptor and have been called corticostatins.[9] The use of batch ion-exchange fractionation has facilitated the purification of these two classes of polypeptide.[8,10] The ion-exchange fractionation technique is complementary to the C_{18} Sep-Pak extraction procedure.[1] Both procedures reduce column loading at the RPC steps.

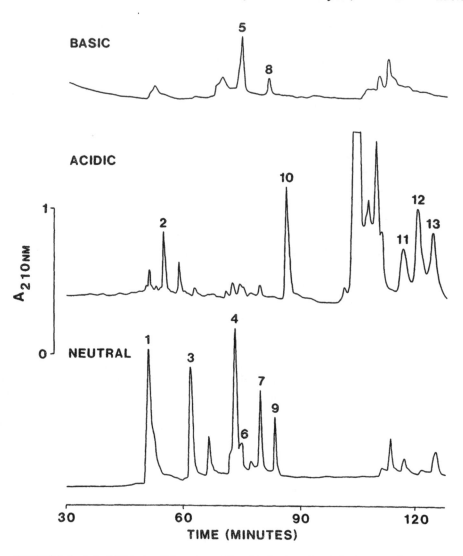

FIGURE 3 Analytical RPC of peptides extracted from bovine posterior pituitaries. A portion of tissue extract, corresponding to two posterior pituitaries, was dried, taken up in 3 ml of 10 m*M* Tris-HCl (pH 7), and subjected to ion-exchange batch extraction with cation- and anion-exchange Sep-Pak cartridges, connected in series. The upper, middle, and lower panel show the RPC profiles obtained for the basic, acidic, and neutral pools, respectively. Peaks 1 to 13 are identified in Table 1. For chromatography conditions, see legend to Figure 1. (From Bennett, H. P. J., *J. Chromatogr.*, 359, 383, 1986. With permission.)

REFERENCES

1. **Bennett, H.P.J., Browne, C.A., and Solomon, S.,** Purification of the two major forms of rat pituitary corticotropin using only reversed-phase liquid chromatography, *Biochemistry*, 20, 4530, 1981.
2. **Bennett, H.P.J.,** Biosynthetic fate of the amino-terminal fragment of pro-opiomelanocortin within the intermediate lobe of the mouse pituitary, *Peptides*, 7, 615, 1986.
3. **James, S. and Bennett, H.P.J.,** Use of reversed-phase and ion-exchange batch extraction in the purification of bovine pituitary peptides, *J. Chromatogr.*, 326, 329, 1985.

4. **Bennett, H.P.J.,** Use of ion-exchange Sep-Pak cartridges in the batch fractionation of pituitary peptides, *J. Chromatogr.*, 359, 383, 1986.
5. **Bennett, H.P.J., Hudson, A.M., McMartin, C., and Purdon, G.E.,** Use of octadecasilyl-silica for the extraction and purification of peptides in biological samples. Application to the identification of circulating metabolites of corticotropin-(1-24)-tetracosapeptide and somatostatin in vivo, *Biochem, J.*, 168, 9-13, 1977.
6. **Nakanishi, S. Inoue, A., Kita, T., Nakamura, M., Chang, A.C.Y., Cohen, S.N., and Numa, S.,** Nucleotide sequence of cloned cDNA for bovine corticotropin-β-lipotropin, *Nature (London)*, 278, 423, 1979.
7. **Land, H., Grez, M., Ruppert, S., Schmale, H., Rehbein, M., Richter, D., and Schutz, G.,** Deduced amino acid sequence from bovine oxytocin-neurophysin I precursor cDNA, *Nature (London)*, 302, 342, 1983.
8. **Pedersen, R. and Brownie, A.C.,** Steroidogenesis-activator polypeptide isolated from a rat Leydig cell tumor, *Science*, 236, 188, 1987.
9. **Zhu, Q., Hu, J., Mulay, S., Esch, F. Shimasaki, S., and Solomon, S.,** Isolation and structure of corticostatin peptides from rabbit fetal and adult lung, *Proc. Natl. Acad. Sci. U.S.A.*, 85, 592, 1988.
10. **Zhu, Q., Bateman, A., Singh, A., and Solomon, S.,** Isolation and biological activity of corticostatic peptides (anti-ACTH), *Endocrine Res.*, 15, 129, 1989.

HIGH-PERFORMANCE ION-EXCHANGE CHROMATOGRAPHY OF PROTEINS

Mark P. Nowlan and Karen M. Gooding

I. INTRODUCTION

Ion-exchange chromatography is a popular technique for the analysis and purification of proteins and peptides. This popularity is due both to the presence of surface charges on proteins which allows them to bind to ion-exchange sorbents and to the preservation of biological activity during the ion-exchange process. The easy transfer of methodologies from classical gels to high-performance columns has contributed to the latter's wide acceptance. High-performance ion-exchange chromatography (IEC) is more attractive than low pressure techniques because it yields better resolution than classical column methods in 10 to 20% of the time with equivalent or higher loading capacities and recoveries. Figure 1 aptly demonstrates these advantages for IEC of crystallins.[1]

II. CHARACTERISTICS OF THE COLUMN PACKING

A. COMPOSITION

All of the IEC packings which are currently available contain ionic functional groups grafted onto a base material composed of silica or a polymer which is suitable for HPLC. Although the ionic functional groups primarily determine the ion-exchange properties, the underlying composition can have retentive effects and can determine operational restrictions or considerations.

Silica and polymeric supports differ chiefly in mechanical and pH stability. Silica has excellent rigidity but dissolves in alkaline media and thus has a normal operating range of pH 2 to 8. The pH restrictions are not generally prohibitive because the bonded phases give the silica some protection from alkaline pH. Additionally, most proteins and peptides are run at biological or physiological pH where silica stability is not a problem. Polymeric supports have a wider pH stability range, especially on the basic side, but they have more limited mechanical stability than silica. Higher percentages of crosslinking agents (e.g., divinylbenzene in polystyrene matrices) are used to increase pressure and flow limits and diminish swelling effects; however, they also lower permeability and increase hydrophobicity. Generally, silica is available in more pore diameters and is less expensive than comparable polymeric matrices.

Complete surface coverage is necessary to avoid silanophilic or hydrophobic interactions of the protein or peptide with the support matrix. This coverage is usually provided by a polymeric layer which tends to be more stable than simple silane bonded phases. The ion-exchange properties are determined by the specific ionic moieties incorporated into the bonded phase. Weak anion-exchange sorbents contain diethylaminoethanol (DEAE), ethylamine (EA), or polyethyleneimine (PEI) functional groups, while strong anion-exchange sorbents are quaternized amines (Q). The tertiary amines of the DEAE or PEI sorbents assume charge according to mobile phase pH, whereas the Q sorbents have a relatively permanent positive charge which is independent of local pH.

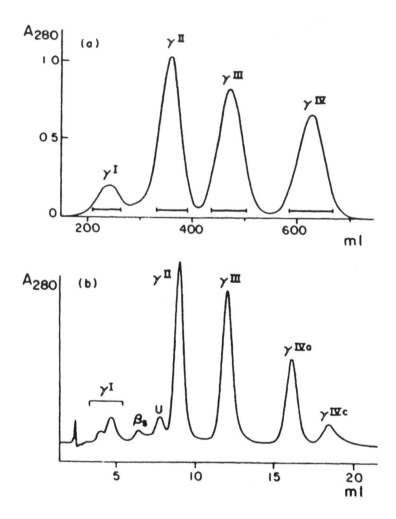

FIGURE 1. Comparison of classical (panel a) and high-performance (panel b) LC columns.[1] Columns: panel a, SP Sephadex C50 (400 × 15 mm I.D.); panel b, SynChropak CM 300 (250 × 4.1 mm I.D., 6.5-μm particle size, 300-Å pore size) with SynChropak CSC precolumn (50 × 4.1 mm I.D.). Mobile phase buffers: (A) 0.2 M sodium acetate, 0.02% NaN_3, pH 5.0; (B) A + 0.2 M sodium acetate; (C) 0.02 M Tris-acetate, 0.02% NaN_3, pH 5.0; (D) C + 0.5 M sodium acetate, pH 5.0. Gradients: panel a, 100% A to 100% B in 1600 min at 0.5 ml/min; panel b, 75% C, 25% D to 50% C, 50% D in 20 min at 1.0 ml/min. Temperatures: panel a, 4°C; panel b, 20°C. Samples: calf lens nuclear γ-crystallins at loads of 70 mg (panel a) or 40 μg (panel b).

Comparable charge and pH considerations also apply to cation-exchange columns. Weak cation-exchange functionalities are usually carboxymethyl (CM) and strong cation-exchange groups can be sulfonyl (S), sulfopropyl (SP), or phosphonyl (P) ionic groups. Table 1 lists characteristics of some commercial IEC packings.

Due to charge permanence, the ion-exchange capacity of strong ion-exchange sorbents remains high at extreme pH, whereas for weak anion-exchange and cation-exchange sorbents, it declines somewhat at high and low pH, respectively. Because the ionization of the composite amino acids of proteins is also affected by pH, some retentive and loading differences result from pH manipulation on both weak and strong ion-exchange columns.[2,7]

TABLE 1
Selected Microparticulate HPIEC Packings

Product	Chemistry[a]	Pore Diameter (Å)/support	Manufacturer[b]
Weak AEX			
Aquapore AX300	PEI	300/ silica	1
Bakerbond WP-PEI-WAX	PEI	300/ silica	3
SynChropak AX	PEI	100, 300, 1000/ silica	6
TSK DEAE-5-PW	DEAE	1000/ polystyrene divinylbenzene	7
TSK DEAE-3-SW	DEAE	250/ silica	7
Zorbax WAX	EA	300/ silica	2
Strong AEX			
Mono Q HR 5/5	Q	>1000/ styrene	4
Protein IEC QA-824	Q	1000/ polymer	8
SynChropak Q	Q	100, 300, 1000/ silica	6
Zorbax SAX	Q	300/ silica	2
Weak CEX			
Aquapore CX-300	CM	300/ silica	1
Bakerbond CEX	CM	300/ silica	3
Poly CAT-A	Polyaspartic acid	300/ silica	5
Protein IEC CM-825	CM	1000/polymer	8
SynChropak CM	CM	100, 300, 1000/silica	6
TSK CM-3-SW	CM	250/ silica	7
Zorbax WCX	CM	300/ silica	2
Strong CEX			
Mono P HR 5/20	P	>1000/ styrene	4
Mono S	S	>1000/ polymer	4
SynChropak S	S	300, 1000/ silica	6
TSK SP-5-PW	S	1000/ polystyrene divinylbenzene	7
Zorbax SCX	S	300/ silica	2

[a] PEI denotes polyethyleneimine; DEAE denotes diethylaminoethanol; EA denotes ethylamine; Q denotes quaternized amines; CM denotes carboxymethyl; P denotes phosphonyl; S denotes sulfonyl.

[b] Manufacturers: Applied Biosystems (1), DuPont (2), J.T. Baker (3), Pharmacia (4), PolyLC (5), SynChrom, Inc. (6), Toso Haas (7), Waters (8).

B. PARTICLE DIAMETER

Resolution, pressure and cost requirements determine which particle size is optimum. Microparticulate supports (≤ 10 micron) have back pressures from 500 to 1000 psi in standard analytical columns and serve best when subtle distinctions between sample components must be discerned, as in final preparation steps. The isoenzyme analysis in Figure 2 illustrates the excellent resolution and recoveries that can be obtained from columns with 6 micron particles.[3] Macroparticulate media (>15 micron) produce lower column pressures and offer cost-effective alternatives when premium resolution is unnecessary, or when there is high risk of fouling, as under high load conditions in crude, early stage purifications. Appropriate adjustment of operational parameters such as gradient slope and linear velocity can yield separations which are similar to those obtained on microparticulate columns, as can be seen in Figure 3.[4] Macroparticulate material (≥ 30 µm) can frequently be dry packed, in contrast to the requisite slurry packing for microparticulates.

For a given particle size option, sizing by the manufacturer within a narrow, uniform

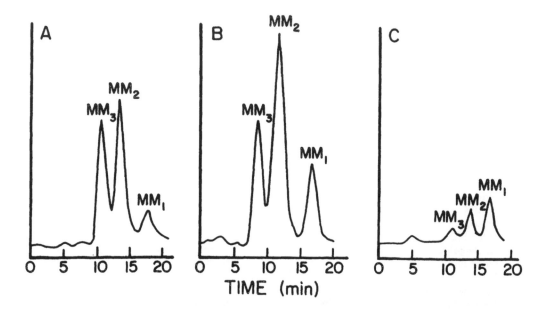

FIGURE 2. Resolution by weak AEX of creatine kinase (CK) MM isoenzyme subtypes in patients who experienced myocardial infarction. (A) 8 hr, (B) 12 hr, and (C) 84 hr after chest pain commences. Note changes in absolute amounts of subtypes, as well as in subtype ratio, MM_1/MM_3.[3] Column: SynChropak AX 300 (250×4.1 mm I.D., 6.5-μm particle size, 300-Å pore size) with 20×4.1 mm I.D. guard column. Mobile phase: 0.02 M Tris, pH 7.8 (isocratic). Flow-rate: 0.5 ml/min. Temperature: 45°C. Sample: diluted serum. Detection: post column reaction producing NADPH for fluorometric monitoring at 350 nm (excitation) and 450 nm (emission); reagent flow is 0.5 ml/min at 45°C.

distribution helps eliminate high pressure fines as well as larger particles that detract from resolution. Spherical particle shape is preferred over irregular on the basis of particle uniformity and flow characteristics but is more expensive and possibly inconsequential as particle size increases.

C. PORE DIAMETER AND LOADING CAPACITY

The molecular weight and shape of sample analytes dictate the pore diameter selection. Most of the surface area of a porous IEC support is within the pores and it is generally inversely related to the pore diameter. This surface area, or concomitantly the ion-exchange capacity, is not useful unless the pore is large enough to admit the solute protein. Table 2 illustrates that 300 Å is an ideal pore diameter for proteins smaller than 200 kDa in terms of maximizing loading capacity and resolution.[5] High molecular weight proteins may require larger pores to provide access to the surface. For peptides smaller than 5 kDa, 100 Å supports offer highest loadability.

Because denaturing mobile phases containing surfactants or organic solvents disrupt the tertiary structure of most proteins and linearize their shapes, they yield higher hydrodynamic volumes.[6] Such systems may thus require larger pore diameters than molecular weight alone dictates to permit access of the molecules to the ionic groups.

III. CONTRIBUTIONS OF THE MOBILE PHASE

A. GENERAL OPERATIONAL CONSIDERATIONS

Although the properties of the stationary phase and the solute define the separation

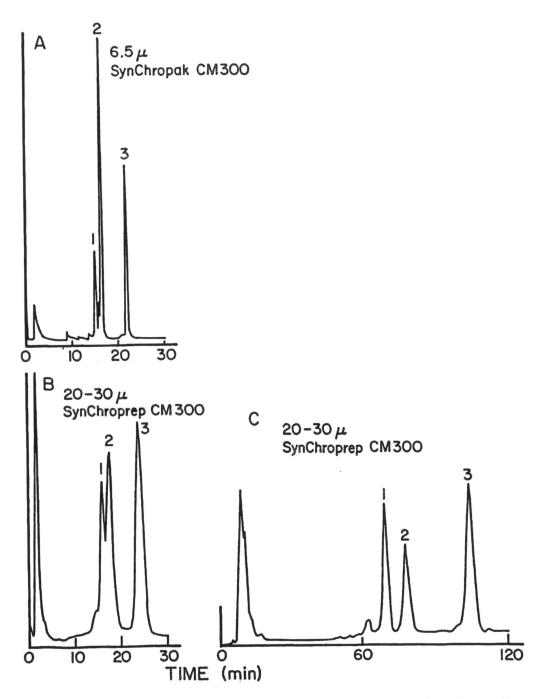

FIGURE 3. Comparative resolution of microparticulate (panel A) vs. macroparticulate weak cation-exchange packings without (panel B) and with (panel C) adjustments in gradient slope and linear velocity. Columns: panel A, SynChropak CM 300, 250 × 4.1 mm I.D., 6.5-μm particle size, 300-Å pore size; panel B, SynChroprep CM 300, 250 × 4.1 mm I.D., 20 to 30 μm, 300-Å; panel C, same as panel B. Mobile phase: 0.02 M Tris, pH 7.0; linear gradient of 0 to 0.5 M sodium acetate over 30 min (16.67 mM salt/min) (panels A and B) or 120 min (4.17 mM salt/min) (panel C). Flow-rate: 1 ml/min (panels A and B) or 1.5 ml/min (panel C). Sample: ribonuclease A (peak 1), α-chymotrypsinogen A (peak 2) and lysozyme (peak 3). Detection: absorbance at 254 nm.

TABLE 2
Ion-Exchange Capacities of Weak Anion-Exchange Packings with Different Pore Diameters

Pore diameter (Å)	Surface area (m²/g)	Ion exchange capacity		
		Picric acid (µmol/g)	Ovalbumin (mg/g)	Serum albumin (mg/g)
100	250	1415[a]	59[a]	64
300	100	656[a]	98[a]	130
500	50	308[a]	76[a]	59
1000	20	129	26	57
4000	6			

[a]   From Reference 5.

mechanism in ion-exchange chromatography, the mobile phase plays an equally vital role in the selectivity (i.e., the relative spacing of peaks) and resolution. Salt composition, pH, gradient slope, additives, and temperature can all determine the effectiveness of an IEC procedure.

B. EFFECT OF THE SALT

During elution in ion-exchange chromatography, salt counterions displace solute ions from the charged sites on the stationary phase. The ions may also affect retention by complexing with either the ion-exchange ligand or the solute to change their ionic properties. In the case of biological macromolecules, such as proteins, specific salts may alter tertiary structure and, consequently, ion-exchange interactions with the stationary phase. Taking into consideration this complexity, certain general guidelines can be formed:[7] divalent ions tend to be stronger displacers than monovalent species and thus produce lower retention; smaller ions tend to have a higher elution strength than larger ions of the same group. Arranged in order of decreasing solute retention, the approximate sequence for cations of commonly used salts is:

$$K^+ < Na^+ < NH_4^+ < Ca^{2+} < Mg^{2+}$$

and for anions is:

$$CH_3COO^- < Cl^- < HPO_4^{2-} \leq SO_4^{2-}$$

Discrepancies can exist between the elution strengths of ions for small monofunctional solutes and those for proteins due to the complex structure of the proteins. Both anions and cations associate with side chain functionalities of the proteins, as well as with the bonded phase ionic groups. This individualized salt effect can result in excellent selectivity and resolution for specific bonded phase/mobile phase/solute combinations.[7-9] Figure 4 illustrates this salt-induced selectivity during anion-exchange chromatography of a four-protein mixture.

C. EFFECT OF PH

The pI of a protein provides a guideline to selection of both an appropriate IEC column and an initial pH for the mobile phase. Unfortunately, the pI alone does not give a totally accurate prediction of ion-exchange behavior because it reflects all of the amino acids of a protein, whereas ion-exchange interactions only occur with surface amino acids on a protein. For

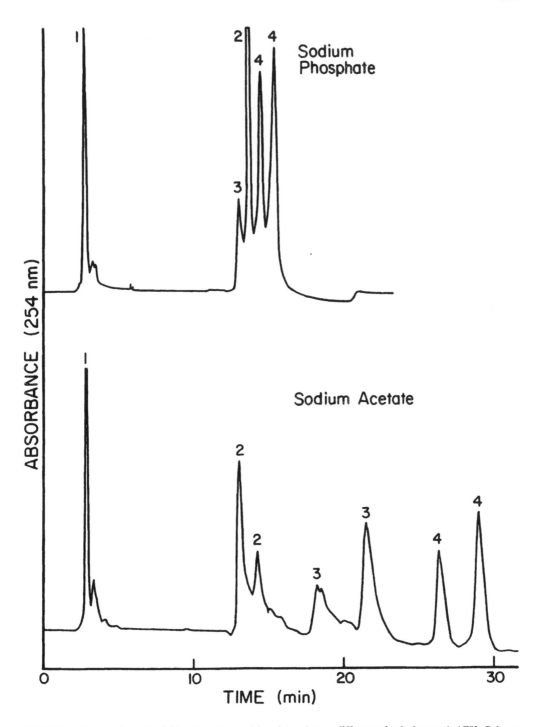

FIGURE 4. Comparative selectivities for a four-protein mixture in two different salts during weak AEX. Column: SynChropak AX 300, 250×4.1 mm I.D., 6.5-μm particle size, 300-Å pore size. Mobile phase: 0.02 M Tris, pH 7.0; linear gradient of 0 to 1 N salt over 30 min (33.3 mN salt/min); top panel, sodium phosphate; bottom panel, sodium acetate. Flow-rate: 1 ml/min. Sample: myoglobin (peak 1), conalbumin (peak 2), ovalbumin (peak 3), β-lactoglobulins B and A (peak 4). Detection: absorbance at 254 nm.

TABLE 3
Buffers Used for HPIEC

Buffer		pKa	pH range
Phosphate			
pK$_1$		2.1	1.5—2.7
pK$_2$		7.2	6.6—7.8
pK$_3$		12.3	11.7—12.9
Citrate			
pK$_1$		3.1	2.5—3.7
pK$_2$		4.7	4.1—5.3
pK$_3$		5.4	4.8—6.0
Formate		3.8	3.2—4.4
Acetate		4.8	4.2—5.4
MES	(2-[N-morpholino]ethanesulfonic acid)	6.1	5.5—6.7
BIS-TRIS	(bis[2-hydroxyethyl]iminotris-[hydroxymethyl]methane)	6.5	5.8—7.2
PIPES	(piperazine-N,N'-bis [2-ethanesulfonic acid])	6.8	6.1—7.5
BES	(N,N-bis[2-hydroxyethyl]-2-aminoethanesulfonic acid)	7.1	6.4—7.8
MOPS	(3-[N-morpholino]propanesulfonic acid)	7.2	6.5—7.9
HEPES	(N-[2-hydroxyethyl]piperazine-N'--ethanesulfonic acid])	7.5	6.8—8.2
TRIS	(tris[hydroxymethyl] aminomethane)	8.3	7.7—8.9
Ammonia		9.2	8.6—9.8
Borate		9.2	8.6—9.8
Diethylamine		10.5	9.9—11.1

example, proteins with disparate pI values may be co-eluted due to similar surface charge features. Conversely, proteins with nearly identical pI values may separate completely because their surface charges are differentially distributed.[10] Generally, to achieve adequate binding, the pH of the mobile phase should be at least 0.5 pH units above the pI of the solute for anion-exchange chromatography (AEX) and below the pI for cation-exchange chromatography (CEX).

A map of retention vs. pH for an individual protein will provide a better indication of optimum conditions when pI is insufficient or not predictive.[11] The intersection of anion and cation-exchange retention curves marks a pH, usually near the pI, where retention of the protein is relatively weak. At this pH and appropriate ionic strength, such a protein in a mixture with impurities can pass through a mixed anion- and cation-exchange bed unretained, while the impurities can be adsorbed for later removal. Such a selective protocol was described for IgG$_1$, in bovine serum by Stringham and Regnier,[12] where the low retention of the target protein, IgG$_1$, results in high recovery, purity, and throughput without the need for gradients, small particles, and scale-up concerns.

The critical effects of pH in IEC necessitate its control during a gradient for reproducibility. Buffer concentrations should range from 10 to 100 mM to supply enough buffering capacity without contributing excessive ionic strength. For preparative purposes, a volatile buffer, such as triethylammonium bicarbonate (TEAB)[13] or volatile salt, such as ammonium acetate,[14] can facilitate lyophilization while avoiding sample hydrolysis by acid. Table 3 lists some commonly used buffers for IEC, along with their optimum pH range.

In chromatofocusing, which has been successfully implemented on specific IEC columns which possess inherent buffering capacity (or broad titration curves), mixtures of electrolytes called polybuffers self-assemble on the ion-exchange matrix and generate a pH gradient in the column.[15] When a buffer of different pH is applied, the gradient proceeds down the column. Proteins desorb at pH $\leq$ pI and readsorb at pH > pI, thus accompanying the moving gradient at their pI values. Proteins of identical pI focus together because mobile phase flow exceeds pH progression down the column.

D. GRADIENT SHAPE

Due to the numerous charged sites on a protein's surface, a gradient must be used during IEC so that the increasing ionic strength or variance in pH can cause the release of the proteins in narrow bands. In a limited number of examples of IEC, proteins have been analyzed isocratically, but these are protein-specific methods and not generally applicable.[3] Gradient shape is less critical in IEC than it is in reversed-phase chromatography. Linear 15 to 60 min gradients have been successfully implemented for a multitude of proteins and columns. A shallow gradient slope may enhance resolution though it may concomitantly lower the sensitivity, which would be crucial in trace analyses. An isocratic hold or lower ionic concentration at the beginning of the gradient can improve resolution of early eluted peaks. Some gradient adjustments will usually be necessary when changing the manufacturer or ionic functionality of stationary phases.

IEC often utilizes gradients in ionic strength ranging to 1 M for monovalent salts or 0.5 M for divalent. Sodium acetate and sodium chloride have been used frequently, although halides are sometimes avoided due to their deleterious effects on stainless steel. Regardless of choice, salts for IEC should be free of UV-absorbing impurities which may interfere with detection or destabilize baselines.

E. ADDITIVES

Many substances can be added to the mobile phase to alter the selectivity, decrease retention, or bolster mass recovery through stabilization of the protein. Because ion-exchange packings possess varying degrees of hydrophobicity related to their composition, separations can be multimodal (ionic and hydrophobic) under given conditions.[17] Specifically, increased crosslinking in PEI sorbents may reduce the number of ion-exchange sites, while the nature of the crosslinker may affect hydrophobicity. Burke et al. found that certain peptide standards would not be eluted from most strong cation-exchange columns without the addition of acetonitrile to the mobile phase.[18] Andrews also observed that a low percentage of acetonitrile decreased the peak width of certain large peptides on cation-exchange columns.[19] Any addition of organic solvents like methanol, acetonitrile, or isopropanol to the mobile phase should be undertaken judiciously because they can adversely affect enzymatic or biological activity or cause salt precipitation.

Frequently, stabilizing agents are added to the mobile phase or sample extraction matrix for IEC of proteins and peptides. Chelating agents such as EDTA protect enzymes from metal impurities in the sample, system or mobile phase. Reducing agents, such as dithiothreitol (DTT) or β-mercaptoethanol,[20] are added to retard oxidation. Potassium cyanide stabilizes variant hemoglobins during their quantification by IEC, while molybdate performs a similar function in progestin receptor characterization.[16]

Surfactants and denaturants, such as urea and guanidinium hydrochloride, are added to a mobile phase to solubilize membrane proteins or to denature proteins. Nonionic surfactants, such as Triton X-100, Berol 172, Emulgen 911 and in the case of liver membrane proteins,[21] Genapol X-100, have been successfully utilized in IEC of proteins. These surfactants are especially useful for isolating individual chains or subunits of proteins; for example, urea has been used for troponin subunits.[22] The percentage of surfactant added may also affect resolution.[23] Generally, nonionic additives can be easily removed from IEC columns after use;

however, ionic (e.g., SDS) or zwitterionic (e.g., CHAPS) surfactants bind to the sorbents and are subsequently difficult to remove.

F. TEMPERATURE

The high speed of IEC analyses frequently precludes the necessity of using low temperatures to preserve the biological activity of the protein solutes. For very labile proteins such as the hormone receptors, however, subambient temperatures have been necessary.[24] The retention of some proteins, especially glycosylated hemoglobins, can be especially responsive to temperature variations. For instance, temperature effects on hemoglobin retention are high in phosphate buffers but low in Bis-Tris or malonate salts.[25]

If it is not deleterious to the protein, increased temperatures can yield lower retention and analysis times in ion-exchange chromatography. In post column reactors, elevated temperatures can boost the reaction rate and sensitivity. Temperature-mediated selectivity effects have been seen for isoenzymes of creatine kinase.[3]

IV. GUIDELINES TO ACHIEVING SUCCESSFUL IEC

In conclusion, IEC can be successfully implemented as a tool for protein analysis and purification using the following guidelines:

1.   Select the type of IEC column based on the pI or retention map of the solute.
2.   Select a pore diameter large enough to admit the protein and small enough to yield high ion-exchange capacity.
3.   Use a mobile phase which is compatible with preservation of biological activity.
4.   Run a 30-min linear salt gradient to 1 M at a moderate flow-rate (1 ml/min for 4.6 mm I.D. columns).
5.   Optimize the operating conditions by adjusting the salt, pH, or gradient.

REFERENCES

1.   **Siezen, R.J., Kaplan, E.D., and Anello, R.D.,** Superior resolution of γ-crystallins from microdissected eye lens by cation-exchange HPLC, *Biochem. Biophys. Res. Comm.,* 127(1), 153, 1985.
2.   **Gooding, K.M. and Schmuck M.N.,** A comparison of weak and strong high performance anion-exchange chromatography, *J. Chromatogr.,* 327, 139, 1985.
3.   **Wu, A.H.B. and Gornet, T.G.,** Measurement of creatine kinase MM sub-type by anion-exchange liquid chromatography, *Clin. Chem.,* 31(11), 1841, 1985.
4.   **Schmuck, M.N., Gooding, D.L., and Gooding, K.M.,** Comparison of porous silica packing materials for preparative ion-exchange chromatography, *J. Chromatogr.,* 359, 323, 1986.
5.   **Vanecek, G. and Regnier, F.E.,** Variables in high-performance anion-exchange chromatography of proteins, *Anal. Biochem.,* 109, 345, 1980.
6.   **Gooding, K.M. and Freiser, H.H.,** High performance size exclusion chromatography of proteins, this publication.
7.   **Gooding, K.M., Schmuck, M.N., and Nowlan, M.P.,** Correlation of salt effects on protein retention in high-performance anion- and cation-exchange chromatography, ACS Meeting, Miami, FL, September 1989.
8.   **Gooding, K.M. and Schmuck, M.N.,** Ion selectivity in the high-performance cation-exchange chromatography of proteins, *J. Chromatogr.,* 296, 321, 1984.
9.   **Kopaciewicz, W., Rounds, M.A., Fausnaugh, J., and Regnier, F.E.,** Retention model for high-performance ion-exchange chromatography, *J. Chromatogr.,* 266, 3, 1983.

10. **Rudolph, F.B., Cooper, B.F., and Greenhut, J.,** Enzyme purification by high-performance ion-exchange liquid chromatography, in *Progress in HPLC*, Parves et al, Eds., VNU Science Press, 1, 133, 1985.
11. **Regnier, F.E.,** High-performance ion-exchange chromatography, *Methods Enzymol.*, 104, 170, 1984.
12. **Stringham, R.W. and Regnier, F.E.,** Selective non-adsorption preparative chromatography of bovine IgG_1, *J. Chromatogr.*, 409, 305, 1987.
13. **Greenhut, J. and Rudolph, F.B.,** Preparative separations of nucleotides by high-performance ion-exchange liquid chromatography using a volatile buffer system, *J. Chromatogr.*, 319, 461, 1985.
14. **Jimeniz, M.A., Rico, M., Nieto, J.L., and Gutierrez, A.M.,** Separation and identification of ribonuclease S-peptide methyl esters by ion-exchange high-performance liquid chromatography and ^1H nuclear magnetic resonance spectroscopy, *J. Chromatogr.*, 360, 288, 1986.
15. **Wagner, G. and Regnier, F.E.,** Rapid chromatofocusing of proteins, *Anal. Biochem.*, 126, 37, 1982.
16. **Van Der Walt, L.A. and Wittliff, J.L.,** High-resolution separation of molybdate-stabilized progestin receptors using high-performance liquid chromatography, *J. Chromatogr.*, 425, 277, 1988.
17. **Heinitz, M.L., Kennedy, L., Kopaciewicz, W., and Regnier, F.E.,** Chromatography of proteins on hydrophobic interaction and ion-exchange chromatographic matrices: mobile phase contributions to selectivity, *J. Chromatogr.*, 443, 173, 1988.
18. **Burke, T.W.L., Mant, C.T., Black, J.A., and Hodges, R.S.,** Strong cation exchange HPLC of peptides: (1) Effect of non-specific hydrophobic interactions: (2) Linearization of peptide retention behaviour, *J. Chromatogr.*, 476, 377, 1989.
19. **Andrews, P.C.,** Ion-exchange HPLC as an orthogonal method for the purification of peptides, *Peptide Res.*, 1, 93, 1988.
20. **Murphy, R., Furness, J.B., and Costa, M.,** Measurement and chromatographic characterization of vasoactive intestinal peptide from guinea-pig enteric nerves, *J. Chromatogr.*, 336, 41, 1984.
21. **Josíc, D., Hofmann, W., and Reutter, W.,** Ion-exchange and hydrophobic-interaction HPLC of proteins — a practical study, *J. Chromatogr.*, 371, 43, 1986.
22. **Cachia, P.J., Van Eyk, J., McCubbin, W.D., Kay, C.M., and Hodges, R.S.,** Ion-exchange high performance liquid chromatographic purification of bovine cardiac and rabbit skeletal muscle troponin subunits, *J. Chromatogr.*, 343, 315, 1985.
23. **Funae, Y., Kotake, A.N., and Yamamoto, K.,** Multiple forms of cytochrome P-450 separable by HPLC. Progress in HPLC, Parves et al., Eds., VNU Science Press, 1, 59, 1985.
24. **Wittliff, J.L.,** HPLC of steroid-hormone receptors, *LC Magazine*, 4, 1092, 1986.
25. **Stenman, U.H.,** Determination of hemoglobin A_1c, in *HPLC of Biological Macromolecules: Methods and Applications*, Gooding, K.M. and Regnier, F.E., Eds., Marcel Dekker, New York, 1990.

HIGH-PERFORMANCE ION-EXCHANGE CHROMATOGRAPHY OF PROTEINS: APPLICATION TO THE PURIFICATION OF ENZYMES

Heidrun Matern and Siegfried Matern

I. INTRODUCTION

Chromatography of proteins on ion-exchange columns has a wide variety of practical applications to protein analysis and separation. Due to progress in the development of rigid chromatographic supports, conventional ion-exchange separations of proteins on gel-type columns may be replaced by high-performance separations with equal or superior resolution in a shorter separation time. The present review will discuss factors which have to be considered in high-performance ion-exchange chromatography (IEC) of proteins and will provide some representative applications of this technique to the purification of enzymes.

II. VARIABLES IN HIGH-PERFORMANCE ION-EXCHANGE CHROMATOGRAPHY (IEC) OF PROTEINS

Factors which have an influence on IEC of proteins have been systematically examined in the last years, e.g. column packing materials and support pore diameter, mobile phase composition, column length, separation time, and temperature.[1-5]

For high-performance ion-exchange separations of proteins, a number of packing materials are commercially available with different pore diameters.[1-3] Pore diameter plays an important role in IEC of proteins. Since more than 95% of the surface area of a support is contained within the pore, the diameters of pores must be sufficiently large to allow penetration of macromolecules. Supports with a 300-Å pore diameter have been shown to give highest loading capacity and resolution for proteins in the 50- to 100-kDa range.[2] Larger molecules may require greater pore diameters for optimum results.[2,3]

Two types of ionic stationary phases have been used in IEC of proteins: weak or strong exchanging groups. The classification depends upon the range of pH over which an ion-exchanger remains charged. Strong ion-exchangers are completely ionized over a broader range of pH than weak ion-exchangers and were found to be superior in high-performance separations of proteins.[4]

In ion-exchange chromatography of proteins, the nature of the mobile phase plays an important role in the control of protein retention and resolution. Basically, the same mobile phases used in the elution of conventional ion-exchange columns were used with high-performance ion-exchange packings. Three mobile phase variables are of importance in manipulating the selectivity in ion-exchange chromatography of proteins: (a) pH, (b) ionic strength, and (c) nature of the displacing ion.

The pH of a mobile phase has an effect on the charge of a protein. At any pH above its pI a protein will have a net negative charge as opposed to a positive charge below the pI. It has generally been assumed that a protein will be retained on an anion-exchange column within 1 pH unit above its pI whereas a cation-exchange column will retain a protein within 1 pH unit below its pI. Deviations, however, may occur due to charge asymmetry of proteins, since only a fraction of the protein surface interacts with the stationary phase.[6] Although anion- and cation-exchange columns are mostly operated at slightly basic and acidic pH, respectively, for each individual protein, the pH of the mobile phase which provides optimal resolution has to be found out empirically.

Elution of proteins from ion-exchange columns may be performed with gradients of ionic strength or pH. Application of an ionic strength gradient is the predominant mode of elution. The nature of the displacing ions can strongly influence selectivity of protein retention. Various ions have been characterized with regard to their displacing power in IEC of proteins.[4]

In contrast to conventional ion-exchange chromatography, column length has been shown to be of minor importance in the resolution of proteins in high-performance systems. Columns of 5-cm length have a similar resolution as those of 25-cm length.[1] With the use of short columns, operating pressures are diminished and eluents are more concentrated. A disadvantage of short columns is the lower loading capacity.

The velocity of the mobile phase has been shown to have an influence on the resolution of proteins in IEC.[1] An inverse relationship between resolution and mobile phase velocity was observed when the flow rate was decreased from 2 ml/min to a lower limit of about 0.5 ml/min.[1]

With biologically active proteins, separation procedures have to be performed in most cases with refrigerated high-performance columns due to lability of these proteins at high temperature. With more stable proteins, however, separations should be performed at room temperature, since in IEC of proteins a decrease in resolution has been observed with a decrease in column temperature.[1]

III. APPLICATIONS

IEC has been used for the separation of a variety of proteins for analytical or preparative purposes. In the following, the application of high-performance ion-exchange systems to the separation and purification of active enzymes will be demonstrated and the results of separations will be compared to those obtained with conventional ion-exchange columns.

A. SOLUBLE ENZYMES

Many examples already exist in the literature for the application of IEC to the purification of water soluble enzymes from various sources. In most cases, crude enzyme mixtures have been prepurified by several conventional techniques before application to a high-performance ion-exchange column.[7-14] The following reasons may be responsible for the preferred use of IEC in the final phases of purification procedures. First, crude enzyme extracts may contain compounds which adsorb irreversibly to column supports and reduce the lifetime of columns, e.g. phenols or tannins from plant extracts.[7] Second, preparative high-performance fractionation procedures have been mostly performed with short columns (5-cm length) of low protein load capacity. Except for the main disadvantage of lower loading capacity, short columns provide several advantages as compared to long columns (cf. above).

For purification of O-methyltransferases from plants by high-performance anion-exchange chromatography, crude plant extracts were subjected to pre-purification by ammonium sulfate precipitation and two conventional low-pressure column chromatographies.[7] A comparison between separation of partially purified plant enzymes on a conventional anion-exchange resin and the corresponding high-performance system shows the advantages of the high-performance system. The high-performance separation was achieved 60 times faster than on the conventional column with superior resolution and a recovery of enzyme activities ranging from 70 to 100%.[7] Furthermore, this study shows that in the high-performance anion-exchange system resolution could be improved when the slope of the salt gradient was reduced from 20 to 5 mM/min.[7]

In some cases, pre-purification of crude extracts by conventional techniques is not a prerequisite prior to IEC. Enzymes present in high amounts in biological materials, e.g. ribulose-1, 5-bisphosphate carboxylase from plant chloroplasts, can be prepared in highly purified form from crude extracts by IEC in one step.[15] In the case of ribulose-1,5-bisphosphate carboxylase from chloroplasts, the specific activity after one high-performance ion-exchange procedure was higher than obtained after several conventional purification steps.[15] Due to the short time required for the chromatographic separation in the high-performance system, proteolytic degradation was minimized and a stable enzyme preparation was obtained.[15]

IEC may also be applied for the separation of isoenzymes. With conventional techniques, e.g. size-exclusion chromatography or electrophoresis, these separations were time-consuming. In the case of citrate synthase and isocitrate dehydrogenase from microorganisms, efficient and rapid separation of soluble isoenzymes was achieved after fractionation of crude extracts on a high-performance anion-exchange column.[16] With a conventional anion-exchange resin, no separation into isoenzymes was observed.[16]

Application of a protein mixture in the presence of salt (0.1 to 0.4 M) to a high-performance ion-exchange column equilibrated at low ionic strength has been reported to improve resolution in some cases.[17] Injection of a small sample volume in the presence of salt may lead to separation of proteins during loading due to the development of a decreasing salt gradient. In subsequent elution of proteins with an increasing gradient of ionic strength, efficient separation may be obtained even from crude samples.[17]

B. MEMBRANE-BOUND ENZYMES

IEC of proteins is a rapidly expanding field but up to now only a few cases have been reported where it has been used used to purify membrane-bound enzymes in biologically active form; some examples are given in the references.[18-22] In contrast to soluble enzymes, particulate enzymes can be subjected to purification by high-performance liquid chromatography only after they have been solubilized. After solubilization, membrane-bound enzymes have to be kept in their monomeric form in an artificial lipid matrix, usually by the addition of a suitable detergent to the liquid medium. For each particular type of enzyme to be purified, the detergent system has to be carefully selected so as to be compatible with enzyme activity and yield optimal resolution in column chromatography. In ion-exchange chromatography, the detergents used should have either the same charge as the column packing or be zwitterionic or nonionic. Since membrane-bound enzymes are often unstable when removed from their natural lipid environment, stabilizers have to be included in the buffers in addition to detergents, e.g. glycerol in concentrations of 10 to 20% or more. These additives can cause considerable increase in mobile phase viscosity. Increased viscosity leads to higher column back-pressures in high-performance systems which might be detrimental to maintaining the native conformation of membrane-bound enzymes. Despite these difficulties in the utilization of high-performance techniques for purification of membrane-bound proteins, the feasibility of IEC to resolve membrane-bound enzymes has been demonstrated and will be shown in some examples.

Closely related cytochrome P-450 isoenzymes have been efficiently resolved on high-performance ion-exchange columns after pre-purification with conventional techniques and enzymes were recovered in a catalytically viable form.[21,22] In these cases, resolution in IEC was comparable or superior to that for electrophoresis.

A membrane-bound bile acid glucosyltransferase from human liver microsomes (molecular weight of 214,000 in size-exclusion chromatography) could be separated on a high-performance anion-exchange column with higher resolution and higher recovery of biological activity than on a conventional anion-exchange column.[18,19] Figure 1 shows the resolution of a partially purified preparation of this enzyme on a conventional DEAE-Trisacryl column. Bile acid glucosyltransferase activity was eluted as a single peak within a linear gradient of NaCl resulting in a ninefold purification with a recovery of about 60%.[18] For further purification of this partially

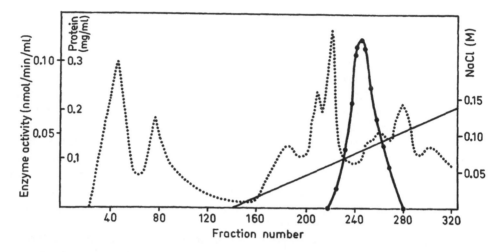

FIGURE 1. Purification of bile acid glucosyltransferase on DEAE-Trisacryl.[18] Column, 15 × 2.6 cm I.D.; flow-rate, 40 ml/hr; fraction volume, 3 ml; amount of protein applied, 129 mg; sample volume, 25 ml; buffer, 25 mM imidazole-HCl, pH 6.5, containing 10% (v/v) glycerol, 1 mM dithioerythritol and the following detergents: 0.5% (w/v) octyl glucoside and 0.1% (w/v) Zwittergent 3-12. ..., protein; •–•, enzyme activity; —— NaCl gradient.

purified enzyme preparation, rechromatography on a conventional DEAE-Trisacryl column was compared with high-performance anion-exchange chromatography on a Mono Q column.[18,19] It is apparent from Figure 2 that rechromatography of the enzyme peak in Figure 1 on a conventional DEAE-Trisacryl column did not lead to a resolution of proteins. However, chromatography of the enzyme peak in Figure 1 on a high-performance anion-exchange Mono Q column gave a further separation of membrane proteins and the resolution of bile acid glucosyltransferase into two partially resolved peaks (Figure 3). The purification of enzyme activity in this high-performance step was about 12-fold with a recovery of 80%. Complete resolution of bile acid glucosyltransferase into two peaks was achieved by subsequent high-performance chromatofocusing on an anion-exchange Mono P-column (Figure 4).[18,19]

To achieve an electrophoretically homogeneous preparation of bile acid glucosyltransferase, two conventional column chromatographic steps were combined with two high-performance techniques.[18,19] Pre-purification of the crude enzyme preparation on a conventional DEAE-Trisacryl column proved advantageous prior to high-performance Mono Q chromatography because a large amount of protein and microsomal phospholipids was removed by this conventional column and the remaining protein could be applied in one step to a short analytical Mono Q column (5-cm length). Concentration of the enzyme sample by Mono Q chromatography enabled the direct application of this preparation to a Mono P column after appropriate dilution of salt.[18,19] Since Mono P columns (20-cm length) exhibit higher operating pressures than short Mono Q columns, lower flow rates must be applied and small sample volumes are more practicable. Therefore, Mono P chromatography was performed after the Mono Q step where a more concentrated enzyme sample was available. Minor contaminants of protein present after the Mono P step were removed with conventional ethylagarose chromatography.[18]

IV. CONCLUSIONS

It may be concluded from the examples presented above that high-performance ion-exchange chromatography offers major advantages over conventional ion-exchange techniques in resolution, separation time and recovery of biologically active soluble and membrane-bound

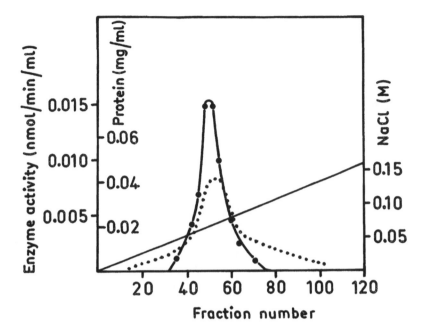

FIGURE 2. Rechromatography of bile acid glucosyltransferase on DEAE-Trisacryl. Column, 10 × 1 cm I.D.; flow-rate, 10 ml/hr; fraction volume, 1 ml; amount of protein applied, 1.9 mg; sample volume, 30 ml. Mobile phase and pH were the same as in Figure 1, except that the concentration of Zwittergent 3-12 was 0.05% (w/v). ..., protein; •‒•, enzyme activity; —NaCl gradient.

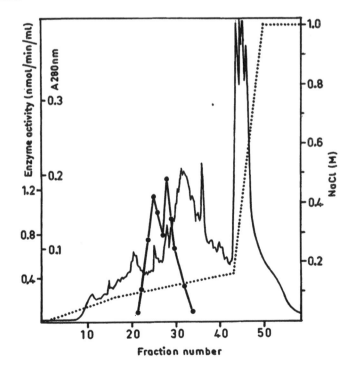

FIGURE 3. High-performance anion-exchange chromatography of bile acid glucosyltransferase after chromatography on DEAE-Trisacryl (Figure 1).[18] Mono Q column, 5 × 0.5 cm I.D. (Pharmacia); flow-rate, 1.5 ml/min; fraction volume, 1 ml; amount of protein applied, 8.6 mg; sample volume, 140 ml. For details of mobile phase, see legend to Figure 2. •‒•, enzyme activity; ....., NaCl gradient.

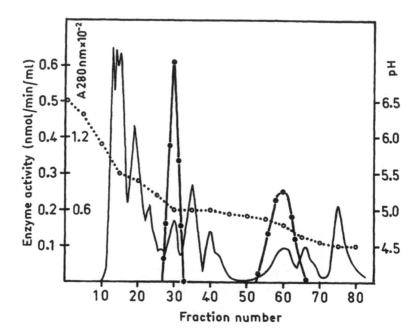

FIGURE 4. High-performance chromatofocusing of bile acid glucosyltransferase.[18] Mono P column, 20 × 0.5 cm I.D. (Pharmacia); flow-rate, 0.8 ml/min; fraction volume, 1 ml; amount of protein applied, 0.57 mg; sample volume, 27 ml. Mobile phase for column equilibration, see legend to Figure 2. Mobile phase for development of pH gradient, 10% (v/v) Polybuffer 74 (Pharmacia), pH 5.0 (40 ml) and pH 4.5 (40 ml) containing 10% (v/v) glycerol, 1 mM dithioerythritol, 0.5% (w/v) octyl glucoside and 0.05% (w/v) Zwittergent 3-12. •–•, enzyme activity; o....o, pH gradient.

enzymes. The same mobile phases used in conventional ion-exchange chromatography may be used in high-performance ion-exchange systems. In purification procedures, high-performance techniques are often applied to enzyme preparations which have been pre-purified by conventional methods. Thus, in combination with conventional purification techniques, high-performance ion-exchange chromatography makes an essential contribution to the availability of pure enzymes.

REFERENCES

1. **Vanecek, G. and Regnier, F. E.,** Variables in high-performance anion-exchange chromatography of proteins, *Anal. Biochem.*, 109, 345, 1980.
2. **Regnier, F. E.,** High-performance ion-exchange chromatography of proteins: The current status, *Anal. Biochem.*, 126, 1, 1982.
3. **Vanecek, G. and Regnier, F. E.,** Macroporous high-performance anion-exchange supports for proteins, *Anal. Biochem.*, 121, 156, 1982.
4. **Kopaciewicz, W. and Regnier, F. E.,** Mobile phase selection for the high-performance ion-exchange chromatography of proteins, *Anal. Biochem.*, 133, 251, 1983.
5. **Gupta, S., Pfannkoch, E., and Regnier F. E.,** High-performance cation-exchange chromatography of proteins, *Anal. Biochem.*, 128, 196, 1983.
6. **Kopaciewicz, W., Rounds, M. A., Fausnaugh, J., and Regnier F. E.,** Retention model for high-performance ion-exchange chromatography, *J. Chromatogr.*, 266, 3, 1983.

7. **Geoffroy, P., Legrand, M., Hermann, C., and Fritig, B.,** High-performance liquid chromatography of proteins. Purification of plant enzymes by ion-exchange chromatography, *J. Chromatogr.*, 315, 333, 1984.
8. **Rådmark, O., Shimizu, T., Jörnvall, H., and Samuelsson, B.,** Leukotriene A₄ hydrolase in human leukocytes. Purification and properties, *J. Biol. Chem.*, 259, 12339, 1984.
9. **Kovár, J. and Plocek, J.,** Purification of aldehyde reductase 1 from pig liver, *J. Chromatogr.*, 351, 371, 1986.
10. **Jeng, A. Y., Sharkey, N. A., and Blumberg, P. M.,** Purification of stable protein kinase C from mouse brain cytosol by specific ligand elution using fast protein liquid chromatography, *Cancer Res.*, 46, 1966, 1986.
11. **Josíc, D., Schütt, W., van Renswoude, J., and Reutter, W.,** High-performance liquid chromatographic methods for antibodies, glycosidases and membrane proteins, *J. Chromatogr.*, 353, 13, 1986.
12. **Warholm, M., Jensson, H., Tahir, M. K., and Mannervik, B.,** Purification and characterization of three distinct glutathione transferases from mouse liver, *Biochemistry*, 25, 4119, 1986.
13. **Skabrahová, Z., Turánek, J., Kovár, J., and Glatz, Z.,** Rapid chromatographic purification of the mitochondrial isoenzyme of beef heart malate dehydrogenase, *J. Chromatogr.*, 369, 426, 1986.
14. **Purwin, W., Laux, M., and Holzer H.,** Fructose 2-phosphate, an intermediate of the dephosphorylation of fructose 2,6-bisphosphate with a purified yeast enzyme, *Eur. J. Biochem.*, 164, 27, 1987.
15. **Salvucci, M. E., Portis, A., R., Jr., and Ogren, W. L.,** Purification of ribulose-1,5-bisphosphate carboxylase/oxygenase with high specific activity by fast protein liquid chromatography, *Anal. Biochem.*, 153, 97, 1986.
16. **Mitchell, C. G., O'Neil, S., Reeves, H. C., and Weitzman, P. D. J.,** Separation of isoenzymes of citrate synthase and isocitrate dehydrogenase by fast protein liquid chromatography, *FEBS Lett.*, 196, 211, 1986.
17. **Vardanis, A.,** Ion-exchange columns in high-performance liquid chromatography of proteins; a simple loading technique that improves resolution, *J. Chromatogr.*, 350, 299, 1985.
18. **Matern, H., Matern, S., and Gerok, W.,** Formation of bile acid glucosides by a sugar nucleotide-independent glucosyltransferase isolated from human liver microsomes, *Proc. Natl. Acad. Sci. U.S.A.*, 81, 7036, 1984.
19. **Matern, H.,** Application of HPLC to the purification of bile acid glucosyltransferase from human liver microsomes, *LC Magazine*, 3, 994, 1985.
20. **Muto, N. and Tan, L.,** Purification of oestrogen synthetase by high-performance liquid chromatography. Two membrane-bound enzymes from the human placenta, *J. Chromatogr.*, 326, 137, 1985.
21. **Funae, Y. and Imaoka, S.,** Simultaneous purification of multiple forms of rat liver microsomal cytochrome P-450 by high-performance liquid chromatography, *Biochim. Biophys. Acta*, 842, 119, 1985.
22. **Bornheim, L. M. and Correia, M. A.,** Fractionation and purification of hepatic microsomal cytochrome P-450 isoenzymes from phenobarbital-pretreated rats by h.p.l.c. A convenient tool for screening of isoenzymes inactivated by allylisopropylacetamide, *Biochem. J.*, 239, 661, 1986.

ANION EXCHANGE HIGH-PERFORMANCE LIQUID CHROMATOGRAPHY OF VIRAL PROTEINS

Gjalt W. Welling and Sytske Welling-Wester

I. INTRODUCTION

Ion-exchange chromatography is a relatively mild chromatographic method. Column materials are available to apply this type of chromatography in its high-performance version. In the following sections, the application of ion-exchange high-performance liquid chromatography (IEC) for the purification of viral proteins will be described.

II. VIRUSES

A. GENERAL PROPERTIES

Viruses differ from other microorganisms in: (1) size; virus particles vary in size from 10 to 300 nm and are relatively small, while bacteria have an approximate size of 1000 nm; (2) viruses have either DNA or RNA in their genome, and do not contain both. The DNA or RNA may be double-stranded, single-stranded and segmented; and (3) virus particles are metabolically inert and are completely dependent on susceptible cells or organisms for their multiplication. They do not possess ribosomes, mitochondria nor a protein-synthesizing apparatus.

A considerable number of viruses obtain a membrane or lipid bilayer by budding through the host cell surface. These viruses are called enveloped viruses. The envelope or the membrane surrounding the virus particle contains virus-specific glycoproteins. On electron micrographs of enveloped viruses these glycoproteins are visible as spikes, or outward-oriented projections from the surface. The viral membrane proteins are anchored in the lipid bilayer of the envelope by a membrane-spanning region, consisting of 20 to 27 predominantly uncharged, hydrophobic amino acid residues. The number of membrane glycoproteins present in the envelope varies for different virus groups. Because we have used Sendai virus proteins as a model mixture in the development of methods for HPLC of membrane proteins,[1-4] this virus will be described in more detail. The virion envelope is composed of a lipid bilayer with a matrix (M) protein (Mr 38,000) on the inside. Sendai virus is a member of the family of paramyxoviridea and of the genus of paramyxoviruses. Human parainfluenza virus, mumps virus, Newcastle disease virus, and Simian virus 5 are classified in the same genus. Two glycoproteins, the hemagglutinin-neuraminidase protein (HN) and the fusion protein (F) are present in the envelope of the virus particles. The HN protein, the larger of the two (Mr 68,000), carries both hemagglutinating and neuraminidase activities. The F protein (Mr 65,000) is involved in cell fusion, hemolysis, and virus penetration into the cell. During infection, a precursor of the F protein, designated F_o, is produced. This biologically inactive precursor is cleaved by proteolysis into two biologically active subunits, F_1 and F_2. The F_1 and F_2 subunits remain linked together by disulfide bridges. Both glycoproteins HN and F are able, like many other glycoproteins that are exposed on the outside of the virus particles, to induce the production of antibodies or other types of immune response. To study the contribution of individual glycoproteins to the immune response of an infected host, it is of importance to purify viral proteins. In addition, purified viral proteins can be used as vaccines.

B. PURIFICATION OF VIRAL PROTEINS

Depending on the virus protein of interest, the strategy for purification may be different. If it concerns a non-structural protein, a detergent extract of virus-infected cells can be used. For purification of structural proteins, virus particles are the preferred starting material. Therefore, the first step in a purification procedure involves separation of the virions from the cells in which they were produced. Excreted virus from infected cells can be used or virus can be released from infected cells by controlled homogenization or by repeated freeze-thawing. Usually, virus particles are separated from contaminating cell components by several cycles of centrifugation, i.e., low-speed centrifugations followed by rate zonal centrifugation through a gradient of sucrose or equilibrium gradient centrifugation in, for instance, cesium chloride or potassium tartrate. Besides these classical techniques, recombinant-DNA techniques are employed to clone genes of virus-specific proteins in order to produce large quantities of these proteins. Virus proteins are expressed and produced in mainly four systems, bacteria, yeasts, mammalian cells, and insect cells.

III. EXTRACTION AND SOLUBILIZATION OF VIRAL PROTEINS

Viral proteins are important as non-structural or structural components of the virion. Structural viral proteins are often tightly associated with either a lipid bilayer envelope or as part of the nucleocapsid. The strongly hydrophobic character of these proteins contributes to a large extent to this strong interaction. In addition, electrostatic interactions may play a role. In order to purify a protein by chromatographic methods, it has to be solubilized. Detergents (surfactants),[5-8] which are lipid-like substances, are often utilized for this purpose. Detergents are not only used to keep viral proteins in solution throughout a purification process, but also as a first step in the purification of integral membrane proteins. Detergent molecules are able to compete with the lipids in the lipid bilayer envelope. This results in detergent-protein complexes that are soluble in aqueous solutions. Non-ionic detergents generally do not affect the native conformation and the biological activity of the proteins and therefore they are often used for extraction as well as an additive in elution buffers. Triton X-100 and the structurally similar Nonidet P40 (NP-40) are the most popular non-ionic detergents although others, like the relatively expensive octylglucoside, are also frequently used. Advantages of octylglucoside over Triton X-100 and Nonidet P40 are: it does not show UV-absorbance at 280 nm and it can be removed easily by dialysis against water due to its relatively high critical micelle concentration (CMC). The CMC is the concentration of the detergent monomer at which micelles, i.e., spherical bilayer aggregates of detergent molecules, begin to form. Other useful non-ionic detergents that are now commercially available consist of a chain of 8 to 12 alkyl groups and a number of oxyethylene groups ranging from 5 to 12. Although these detergents do not or only slightly affect the biological activity of the proteins, they differ with respect to the amount of protein that can be extracted from a virus preparation. For example, extraction of Sendai virus (35 mg of protein) with octylglucoside resulted in 2.3 mg of protein, while with dodecylpentaethyleneglycol 4.5 mg of protein was extracted (van Ede et al., unpublished results). In addition to the non-ionic detergents, ionic detergents (e.g., sodium dodecylsulfate [SDS] and bile salts) and amphoteric detergents (e.g., CHAPS) can be used for extraction. However, ionic detergents are not compatible with ion-exchange chromatography and proteins solubilized by ionic detergents are preferably separated by size-exclusion chromatography. During detergent extraction, proteolytic enzymes may be present originating from the culture medium in which the virus was cultivated. These enzymes may degrade the viral proteins to small fragments or in some cases also to large stable fragments. Proteolytic activity can be inhibited by adding protease inhibitors to the extraction medium. However, limited proteolysis can also be utilized to advantage, i.e., to obtain a large soluble fragment of the protein without its membrane-anchoring segment. The fragment may still be biologically active. An example is the hemagglutinin obtained by bromelain digestion of the X-31 strain influenza A2/Hong Kong/68.[9] Bromelain (from pineapple) exhibits a proteolytic activity similar to that of papain (from Papaya latex) and is commercially available from Sigma (St. Louis, MO), Serva (Heidelberg, FRG), and Boehringer (Mannheim, FRG). It has a very wide specificity and its action in this particular case is probably limited proteolysis of an exposed loop in an otherwise tight globular structure. A similar treatment of the 1968 influenza virus Aichi/68 resulted in a soluble trimer of the hemagglutinin. This soluble trimer was crystallized and its three-dimensional structure could be determined.[10] In addition, organic solvents can be used to solubilize viral proteins or to extract them from a lipid bilayer. A two-phase system can be created, and proteins and viral lipids soluble in the organic phase can be separated from proteins in the aqueous phase.[11]

IV. ION-EXCHANGE CHROMATOGRAPHY

A. ION-EXCHANGE-CHROMATOGRAPHY (GENERAL)

In ion-exchange chromatography, differences in electrostatic interaction between column ligands and charged patches on the proteins are the basis for separation. Two methods are used for elution: pH change, which results in weaker binding and increase of ionic strength, resulting in decrease of electrostatic interaction between the protein and the column ligands. Elution conditions are generally mild since buffers of near neutral pH are used. They contain a mild detergent (non-ionic or zwitterionic) when membrane proteins or other hydrophobic proteins are subjected to chromatography.

B. ION-EXCHANGE HPLC OF VIRAL PROTEINS

The main problem encountered in the separation of viral proteins is to maintain complete dissociation of the proteins during the purification procedure. Because of their large tendency to aggregate, denaturants, strong detergents, or organic solvents often have to be used. Shire et al.[12] used guanidine hydrochloride and urea to solubilize the VP1 surface antigen of foot and mouth disease virus (FMDV) expressed as a fusion protein in *E. coli.*. A VP1-fusion protein pellet was dissolved in 7 M guanidine hydrochloride containing 0.1% 2-mercaptoethanol and 50 mM sodium phosphate, pH 7.0. After size-exclusion chromatography (SEC) on Sephacryl S-300, fractions containing the VP1 fusion protein were dialyzed against 8 M urea and 14 mM Tris-HCl, pH 8.5, containing 0.1% 2-mercaptoethanol. This solution was subjected to anion-exchange chromatography (AEC) on DE-52, resulting in 97% pure VP1 fusion protein in the flow-through fraction. The contaminating proteins from *E. coli* were eluted with a salt gradient. Direct chromatography on DE-52 in the presence of urea was not successful. VP1 fusion protein and *E. coli* proteins were then eluted as one large complex. Both guanidine hydrochloride and urea failed to dissociate a fusion protein produced in *E. coli* consisting of fragments of herpes simplex virus glycoprotein D and β-galactosidase of *E. coli* (Welling et al., unpublished results). The fusion protein contaminated with *E. coli* proteins was always eluted as one complex.

The high-performance mode of ion-exchange chromatography, IEC, has been used for purification of a number of viral proteins which are listed in Table 1, together with the columns and the buffer systems used for elution. Sendai virus proteins have been purified both by conventional and high-performance chromatography and their purification will be discussed in more detail. For conventional chromatography, an extract containing the HN, F, and M protein, was obtained by treatment of purified virions with KCl-Triton X-100[21] and applied to a DEAE-

TABLE 1
Viral Proteins Subjected to IEC

Protein(s)	Gradient system	Column	Ref.
Adenovirus	0.01—1 M NaCl in 10 mM HEPES, pH 7.6, 0.1% NP-40, 1 mM PMSF, 1 mM dithiothreitol, 5% glycerol	SynChropak AX300	13
	0—0.5 M NaCl in 50 mM Tris-HCl, pH 6.5	Mono Q	14
Bovine viral diarrhea virus	0—0.5 M NaCl in 20 mM ethanolamine, pH 9.2, 0.5% Berol 185	DEAE-TSK-5PW	15
	0—1 M NaCl in 50 mM Tris-HCl, pH 7.0, 0.18—0.20% Berol 172	Mono Q	16
	0—1 M NaCl in 20 mM piperazine-acetic acid, pH 6.0, 0.18—0.20% Berol 172		
Epstein-Barr virus	0.025—1 M NaCl in 20 mM piperazine-HCl, pH 5.2, 0.5% MEGA-9	Mono Q	17
Herpes simplex virus	0—0.5 M KCl in 10 mM Tris-HCl, pH 7.5, 50 mM NaCl, 1 mM EDTA, 2 mM 2-mercaptoethanol, 0.5 mM PMSF	Mono Q	18
Measles virus	0—1 M NaCl in 20 mM piperazine, pH 9.5, 0.03% Triton X-100 or 30 mM octylglucoside	Mono Q	19
Rubella virus (trypsin-treated)	0—0.5 M NaCl in 20 mM Tris-HCl, pH 8.6	Mono Q	20
Sendai virus	0.15—1.5 M NaCl in 20 mM sodium phosphate, pH 7.2, 0.1% Triton X-100	Mono Q	2
	0—0.5 M NaCl in 20 mM Tris-HCl, pH 7.8, 0.1% Triton X-100	Mono Q	3
	0—0.5 M NaCl in 20 mM Tris-HCl, pH 7.8, 0.1% decylpolyethyleneglycol-300	TSK-DEAE-NPR	This study

BioGel A column.[22] Proteins wereeluted stepwise with 0.1% Triton X-100 in 10 mM phosphate buffer, pH 7.2, containing 0.05 M NaCl or 0.5 M NaCl. The HN protein was eluted with phosphate buffer with and without 0.05 M NaCl. With high salt, F protein was eluted, contaminated with HN protein. Cation-exchange chromatography on CM-Sepharose CL-6B has been exploited to purify the F protein.[23] In this case, the F protein was found in the unadsorbed fraction, while the HN protein, contaminated with F protein, was eluted between 0.15 to 0.20 M NaCl in 10 mM sodium acetate, pH 6.0, containing 0.25% Triton X-100. We have used AEC on a Mono Q column for the purification of these proteins. They were eluted with a gradient from 0.15 to 1.5 M NaCl in 20 mM sodium phosphate, pH 7.2, containing 0.1% Triton X-100.[2] In contrast to the results with conventional anion-exchange chromatography, relatively pure F protein could be eluted with the NaCl gradient. The HN protein was not adsorbed to the column under these conditions. Therefore, in a later study,[3] elution conditions were changed. As starting buffer, 20 mM Tris-HCl, pH 7.8 with 0.1% Triton X-100 was used and the proteins were eluted with a gradient to 0.5 M NaCl. These elution conditions allowed chromatography of the HN protein. The F protein was eluted as a broad peak which might be due to different micellar forms or multimeric forms present in the extract, but probably also to differences in charge caused by the acidic carbohydrate chains attached to the protein. We have used other detergents in the eluent as well, e.g., 0.1% of Brij 35, octylglucoside, decylpolyethyleneglycol-300 (designated as decyl-PEG) containing 5 to 9 oxyethylene units, and a similar separation was obtained. However, there are differences in recovery. With 0.1% octylglucoside in the eluent, the yield was 18%; in contrast, with 0.1% decyl-PEG, yields were more than twice as high

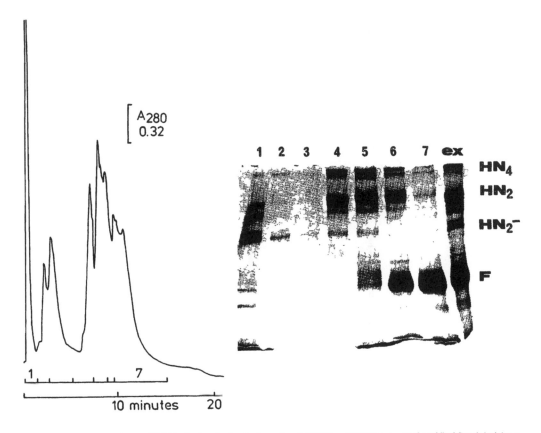

FIGURE 1. Anion-exchange HPLC of a decylpolyethyleneglycol-300 (decyl-PEG) extract of purified Sendai virions. Column: TSK DEAE-NPR, (35 × 4.6 mm I.D.) (Toyo-Soda, Tokyo, Japan). Conditions: linear AB gradient (0 to 0.5 M NaCl in 12 min), following 5 min isocratic elution with Buffer A, where Buffer A is 20 mM Tris-HCl, pH 7.8, and Buffer B is Buffer A containing 0.5 M NaCl; flow-rate, 1 ml/min; absorbance at 280 nm. Fractions (1 to 7) were analyzed by SDS-PAGE (8% gels). The positions of HN_4 (272 kDa), HN_2 (136 kD), HN_2^- (110 kDa) and the F protein (65 kDa) are indicated. ex = the decylPEG extract.

(Welling-Wester et al., unpublished results). As an example, the separation on a non-porous IEC column of Sendai virus proteins from a decylpolyethyleneglycol-300 extract is shown[24] (Figure 1). Except for the tetramer and dimer of the HN protein and the fusion protein F, the extract (lane ex) contains degraded Sendai virus proteins along with the proteolytically degraded HN dimer, HN_2^-. Preincubation of the detergent extract at 37°C results in an increasing amount of this fragment (not shown). Fractions 1 to 7 were analyzed by sodium dodecyl sulfate polyacrylamide gel electrophoresis (SDS-PAGE) under non-reducing conditions on 8% gels. After 5 min isocratic elution with 20 mM Tris HCl, pH 7.8, 0.1% decyl-PEG, the gradient to 0.5 M NaCl (in 12 min) was started. Most of the proteins are eluted with the gradient. The analysis by SDS-PAGE shows that fraction 7 contains relatively pure F protein and fraction 4 a mixture of the various forms of the HN protein and a small amount of F protein. For quantitation, we also analyzed the fractions by size-exclusion HPLC (SEC) on two Zorbax GF-450 columns in tandem. These columns were eluted with 0.1% SDS in 50 mM sodium phosphate, pH 6.5. Samples of fractions 1 to 7 (containing 4% SDS) were heated for 3 min in a boiling water bath prior to chromatography. As an example, the analysis of fractions 5 to 7 is shown in Figure 2. Fraction 5 mainly contains the tetramer and dimer of the HN protein (peak 2 and 3, respectively). Peak 1 is an aggregate which contains no protein and peak 4 contains the F protein. The amount of F protein (peak 4) increases in fraction 6, while in fraction 7 the ratio of HN_4 to HN_2 and F is 1:1.2:6.0. SEC is very useful in the quantitation of proteins, although the resolution of SDS-

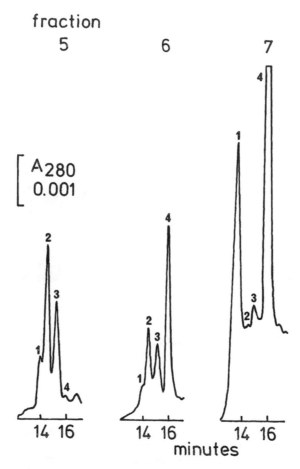

FIGURE 2. Size-exclusion HPLC of fractions 5, 6, and 7 from the anion-exchange HPLC shown in Figure 1. Columns: two Zorbax GF-450 columns (250 × 9.4 mm I.D. each) (Du Pont, Wilmington, DE) in tandem. Condition: elution buffer, 50 mM sodium phosphate, pH 6.5, containing 0.1% SDS; flow-rate, 1 ml/min; absorbance at 280 nm, 1, aggregate peak; 2, the tetramer of HN; 3, the dimer of HN; 4, the F protein.

PAGE is superior. It was argued that a non-porous IEC- column might be advantageous with regard to the recovery of large multimeric proteins, since they only interact with charged ligands on the outside of the column particles. However, we observed higher yields (at least 10% more) upon chromatography on a Mono Q column of which the particles have a pore size of 100 nm (1000 Å). Apparently, these large proteins do not enter the pores. A large Epstein-Barr virus membrane protein (Mr 340,000) was also successfully purified by IEC on a Mono Q column.[17]

In addition, the immunological activity of the fractions was investigated. A high reactivity with monoclonal antibodies directed against the intact proteins showed that the conformation of the proteins was not affected by the chromatographic procedure (data not shown).

V. CONCLUSIONS

IEC is probably one of the most versatile chromatographic methods that can be used for the purification of viral proteins. Elution conditions can be varied easily (pH, ionic-strength) and when hydrophobic membrane proteins are subjected to chromatography, a wide variety of detergents can be used as additives to maintain the proteins in solution. Moreover, the biological activity of the proteins is generally retained after IEC.

REFERENCES

1. **Welling, G.W., van der Zee R., and Welling-Wester, S.,** Column liquid chromatography of integral membrane proteins, *J. Chromatogr.*, 418, 223, 1987.
2. **Welling, G.W., Groen, G., and Welling-Wester, S.,** Isolation of Sendai virus F protein by anion-exchange high-performance liquid chromatography in the presence of Triton X-100, *J. Chromatogr.*, 266, 629, 1983.
3. **Welling, G.W., Nijmeijer, J.R.J., van der Zee, R., Groen, G., Wilterdink, J.B., and Welling-Wester, S.,** Isolation of detergent-extracted Sendai virus proteins by gel-filtration, ion-exchange and reversed-phase high-performance liquid chromatography and the effect on immunological activity, *J. Chromatogr.*, 297, 101, 1984.
4. **Welling, G.W., van der Zee, R., and Welling-Wester, S.,** High-performance liquid chromatography of Sendai virus membrane proteins, *Trends in Anal. Chem.*, 5, 225, 1986.
5. **Helenius, A. and Simons, K.,** Solubilization of membranes by detergents, *Biochim. Biophys. Acta*, 415, 29, 1975.
6. **Tanford, C. and Reynolds, J.A.,** Characterization of membrane proteins in detergent solutions, *Biochim. Biophys. Acta*, 457, 133, 1976.
7. **Helenius, A., McCaslin, D.R., Fries, E., and Tanford, C.,** Properties of detergents, *Methods Enzymol.*, 56, 734, 1979.
8. **Hjelmeland, L.M. and Crambach, A.,** Solubilization of functional membrane proteins, *Methods Enzymol.*, 104, 305, 1984.
9. **Brand, C.M. and Skehel, J.J.,** Crystalline antigen from the influenza virus envelope, *Nature*, 238, 145, 1972.
10. **Wilson, I.A., Skehel, J.J., and Wiley, D.C.,** Structure of the haemagglutinin membrane glycoprotein of influenza virus at 3 Å resolution, *Nature*, 289, 366, 1981.
11. **van Renswoude, J. and Kempf, C.,** Purification of integral membrane proteins, *Methods Enzymol.*, 104, 329, 1984.
12. **Shire, S.J., Bock, L., Ogez, J., Builder, S., Kleid, D., and Moore, D.M.,** Purification and immunogenicity of fusion VP1 protein of foot and mouth disease virus, *Biochemistry*, 23, 6474, 1984.
13. **Green, M. and Brackmann, K.H.,** The application of high-performance liquid chromatography for the resolution of proteins encoded by the human adenovirus type 2 cell transformation region, *Anal. Biochem.*, 124, 209, 1982.
14. **Waris, M. and Halonen, P.,** Purification of adenovirus hexon protein by high-performance liquid chromatography, *J. Chromatogr.*, 397, 321, 1987.
15. **Mohanty, J.G. and Elazhary, Y.,** High-performance liquid chromatographic separation and immunological characterization of soluble bovine diarrhea virus antigen, *J. Chromatogr.*, 435, 149, 1988.
16. **Kårsnäs, P., Moreno-Lopez, J., and Kristiansen, T.,** Bovine diarrhea virus: purification of surface proteins in detergent-containing buffers by fast protein liquid chromatography, *J. Chromatogr.*, 266, 643, 1983.
17. **David, E.M. and Morgan, A.J.,** Efficient purification of Epstein-Barr virus membrane antigen gp340 by fast protein liquid chromatography, *J. Immunol. Methods*, 108, 231, 1988.
18. **Gallo, M.L., Jackwood, D.H., Murphy, M., Marsden, H.S., and Parris, D.S.,** Purification of the Herpes simplex virus type 1 65-kilodalton DNA-binding protein: properties of the protein and evidence of its association with the virus-encoded DNA polymerase, *J. Virol.*, 62, 2874, 1988.
19. **Gerlier, D., Garnier, F., and Forquet, F.,** Haemagglutinin of Measles virus: Purification and storage with preservation of biological and immunological properties, *J. Gen. Virol.*, 69, 2061, 1988.
20. **Ho-Terry, L. and Cohen, A.,** Rubella virus haemagglutinin: association with a single virion glycoprotein, *Arch. Virol.*, 84, 207, 1985.
21. **Scheid, A. and Choppin, P.W.,** Identification of the biological activities of paramyxovirus glycoproteins. Activation of cell fusion, hemolysis, and infectivity by proteolytic cleavage of an inactive precursor protein of Sendai virus, *Virology*, 57, 475, 1974.
22. **Urata, D.M. and Seto, J.T.,** Glycoproteins of Sendai virus: purification and antigenic analysis, *Intervirology*, 6, 108, 1975/76.
23. **Fukami, Y., Hosaka, Y., and Yamamoto, K.,** Separation of Sendai virus glycoproteins by CM-Sepharose column chromatography, *Febs Lett.*, 114, 342, 1980.
24. **Kato, Y., Kitamura, T., Mitsui, A., and Hashimoto, T.,** High-performance ion-exchange chromatography of proteins on non-porous ion exchangers, *J. Chromatogr.*, 398, 327, 1987.

SAMPLE PREPARATION AND APPLICATION OF DIFFERENT HPLC METHODS FOR THE SEPARATION OF HYDROPHOBIC MEMBRANE PROTEINS

Djuro Josić and Werner Reutter

I. INTRODUCTION

In order to separate membrane proteins, they first have to be solubilized from the membrane structure. The more components the sample contains, the more care has to be taken in the solubilization process. Through the use of different reagents in different steps, the membrane proteins can be pre-separated according to their respective solubility and hydrophobic characteristics. This in turn allows further chromatographic separation. When working with comparatively simple systems, such as plasma membranes from the electric organ of the fish *Torpedo californica*, two extractions steps will suffice. By pre-treating the plasma membranes with diluted sodium hydroxide solution, the extrinsic membrane proteins are removed. In a second step, the main protein of the plasma membranes, acetylcholine-receptor, can be solubilized highly enriched by applying a detergent such as Triton or SDS.[1] However, with more complicated membrane structures, more complex extraction methods have to be applied. Figure 1 shows an extraction scheme for liver plasma membranes, which serves as a preparation for subsequent chromatographic separation. In the first step, freezing/thawing, the plasma membranes are mechanically destroyed, and the membrane associated, extrinsic membrane proteins are solubilized. No detergents have to be used with these proteins, and they do not precipitate when kept in diluted solutions. However, as they often aggregate, it is recommended to add mild detergents such as CHAPS or octylglucose before separation.

In the second step, the hydrophobic membrane proteins are extracted by the non-ionic detergent Triton X 114. This particular detergent is water soluble at +4°C, but forms water insoluble micelles at +25°C. These are detergent-rich and can easily be separated from the aqueous, detergent-poor phase. Phase separation of the detergent-protein solution can only be achieved if the concentration of the plasma membrane suspension taken for extraction is less than 2 mg of protein/ml. After extraction with a 1% Triton X 114 solution at +4°C and heating up to a temperature between 20 and 35°C for 3 min, the hydrophobic, intrinsic membrane proteins will be found in the detergent-rich phase, the less hydrophobic, mainly extrinsic membrane proteins in the aqueous, detergent-poor phase. This extraction method with subsequent phase separation was developed by Bordier[2] and has proved to be an efficient preparatory step for further chromatographic separation of membrane proteins. The proteins from the detergent-poor phase, as well as those from the detergent-rich phase, can be subjected after dilution with cold buffer to further HPLC separation, preparatively and analytically.

After treatment with Triton X 114, the pellet still contains about 20 to 30% of the initial protein quantity. A large part are membrane-associated proteins from the cytoskeleton. These proteins can be extracted by treatment with diluted sodium hydroxide solution at pH 11. The extraction should last only between 10 and 30 min, and is carried out at room temperature.

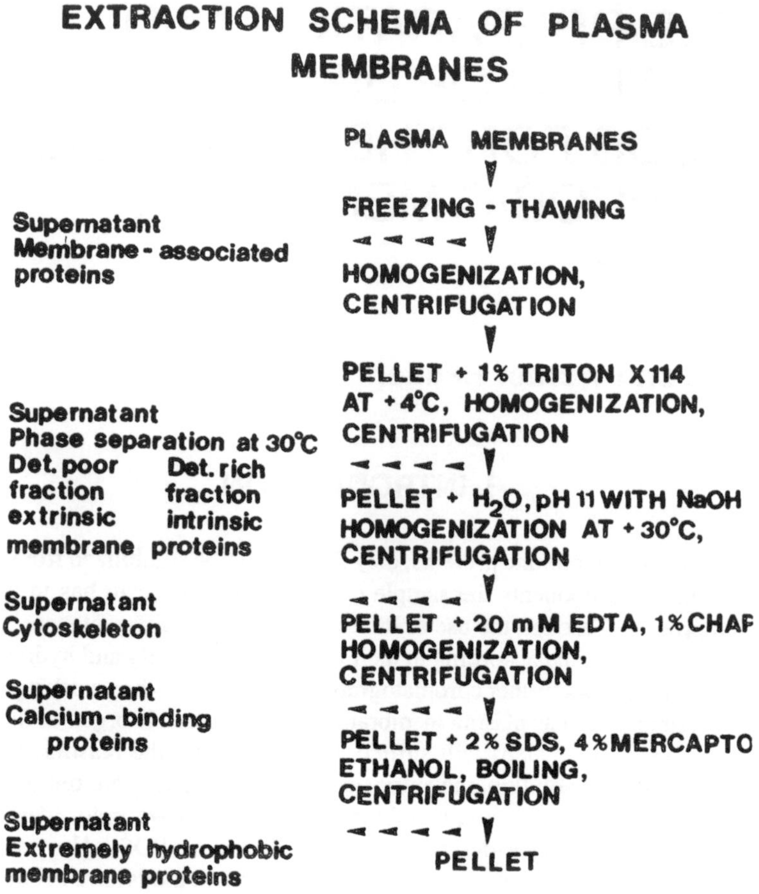

EXTRACTION SCHEMA OF PLASMA MEMBRANES

PLASMA MEMBRANES

↓

FREEZING - THAWING

Supernatant
Membrane - associated
proteins ◄ ◄ ◄ ◄ ↓

HOMOGENIZATION,
CENTRIFUGATION

↓

PELLET + 1% TRITON X 114
AT +4°C, HOMOGENIZATION,
Supernatant CENTRIFUGATION
Phase separation at 30°C
Det. poor   Det. rich ◄ ◄ ◄ ◄ ↓
fraction   fraction
extrinsic   intrinsic PELLET + H_2O, pH 11 WITH NaOH
membrane proteins HOMOGENIZATION AT + 30°C,
CENTRIFUGATION

Supernatant ◄ ◄ ◄ ◄ ↓
Cytoskeleton PELLET + 20 mM EDTA, 1% CHAP
HOMOGENIZATION,
CENTRIFUGATION

Supernatant ◄ ◄ ◄ ◄ ↓
Calcium- binding
proteins PELLET + 2% SDS, 4% MERCAPTO
ETHANOL, BOILING,
CENTRIFUGATION

Supernatant ◄ ◄ ◄ ◄ ↓
Extremely hydrophobic
membrane proteins PELLET

FIGURE 1. Stepwise extraction of membrane proteins.

Although they are not integral membrane components, these proteins will precipitate in water at pH 7. Therefore, 0.1% detergent should be added before neutralization in order to prevent association and precipitation.

In the next step, the pellet is treated with a solution of 1% zwitterionic detergent CHAPS and with 20 to 50 mM EDTA. In this way, calcium-binding proteins can be isolated. After pretreatment as described above, only a few proteins will remain, so that calcium-binding proteins can be isolated in an almost pure state in this step already, as shown in Figure 2.

The very hydrophobic membrane proteins, which have not been extracted despite the pretreatment, can be solubilized in the last step by 4% sodium dodecyl sulfate (SDS), to which mercaptoethanol is added, and by boiling. In the same way, larger amounts of lipids can be extracted. In order to remove them, it is recommended to treat the membranes first with ethanol-acetone.

II. HIGH-PERFORMANCE LIQUID CHROMATOGRAPHY — GENERAL ASPECTS

By selective extraction as described above, a pre-separation of the membrane proteins according to their hydrophobic characteristics is achieved. The pre-treatment facilitates subsequent chromatographic separation and provides a first guideline for the choice of detergent.

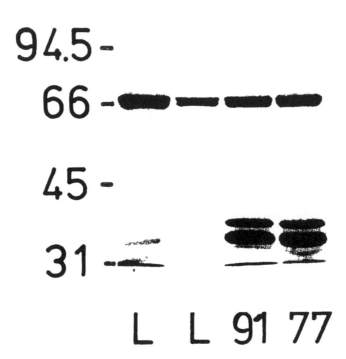

FIGURE 2. Extraction step of chelating calcium by EDTA (cf. Figure 1). Two polypeptides with apparent molecular masses of 65 and 67 kDa could be extracted from both liver and Morris hepatoma plasma membranes (line "L" for liver, "91" for Morris hepatoma 9121, and "77" for Morris hepatoma 7777). Two additional polypeptides with apparent molecular masses of 33 and 35 kDa appear in the extracts from both hepatoma plasma membranes. In the extracts from liver they appear in much smaller quantities. (The Figure was first published in Josić, Dj., et al., Purification of Liver and Hepatoma Membrane Proteins by HPLC, *FEBS Lett.*,185, 182—186, 1985; reprinted with the permission of Elsevier Science Publishers B.V.)

When choosing the detergent, one should stick to that with which the protein was solubilized, if this is at all possible. It is also advisable to add a less denaturing detergent such as CHAPS or octylglucose to those fractions that were obtained by freezing/thawing and extraction at pH 11, i.e., without the use of detergent. This applies to the sample as well as to the buffers.

Despite adequate sample preparation, difficulties can arise when the membrane proteins are separated, especially when they consist of complex mixtures with many different components. These difficulties are:

• Tendency towards association and aggregation

- Microheterogeneity, e.g., glycosylation
- Possibility of non-specific interaction with the support, which can lead to irreversible adsorption of the protein to the column, perhaps plugging it
- The application of buffers with high salt concentration and of detergents and other denaturing agents puts considerable strain on the HPLC system, column hardware, and support

Although practically all methods for separation of water soluble biopolymers are also used for HPLC separation of hydrophobic proteins, the latter field follows its own rules in some respects. Several methods such as size-exclusion HPLC (SEC), and quite often also reversed-phase (RPC) and ion-exchange HPLC (IEC), can be applied only under denaturing conditions. The addition of denaturing agents primarily prevents non-specific interaction with the support as well as association and aggregation of the sample.

The following HPLC methods can be used for preparative and analytical separation of hydrophobic membrane proteins:

A. SEPARATION UNDER DENATURING CONDITIONS
- *Size-exclusion HPLC* — its application requires the use of denaturing agents in the case of many membrane proteins.
- *Reversed-phase HPLC* — some hydrophobic proteins can bind to the column irreversibly. The application of formic acid, up to 60%, in both eluents is recommended.[3,4]
- *Ion-exchange HPLC* — the use of 6 *M* urea can sometimes be useful. In many cases, however, it can be avoided by using non-ionic or zwitterionic detergents.

B. SEPARATION UNDER NON-DENATURING CONDITIONS
- *Ion-exchange HPLC* — with many membrane proteins the problem of aggregation remains (see above). It can sometimes be overcome by using "mild", non-ionic detergents.
- *Hydrophobic-interaction HPLC* — a combination of salt solutions and detergent solutions as eluents, as has been recommended.[5]
- *Hydroxylapatite HPLC* — very effective, but in some cases the columns can easily be blocked.
- *Metal-affinity HPLC* — very effective under mild conditions. Some detergents, e.g. CHAPS, tend to impair column performance.[5]
- *Affinity chromatographic methods* (HPAC), such as lectin HPAC, immunoaffinity HPAC, heparin HPAC and other HPAC methods with immobilized, specific ligands.[6]

III. PROBLEMS CONCERNING THE HPLC SYSTEM AND THE COLUMN HARDWARE

This topic has recently been discussed on a wide scale.[3-6] In this paper, only a limited number of fundamental problems are given attention.

The difficulties listed above occur most frequently when concentrated salt solutions and detergents are applied. Some pumps cannot manage detergent solutions over 0.5% (w/v). Continuous use of concentrated salt solutions can cause leaks. In any case, only those pumps should be used which allow the rinsing of the pump heads, which in turn prevents the settling of salt crystals on the seals. Frequent switching of organic solvents to aqueous systems can also lead to difficulties.

It can be said, however, that pumps which have become available over the last 2 years meet these requirements. Still, aggressive solutions should not be allowed to stay in the system for long, e.g. overnight, otherwise they may cause corrosion. Whenever the system has to be free of metal ions, so-called biocompatible equipment and glass columns with connections made of

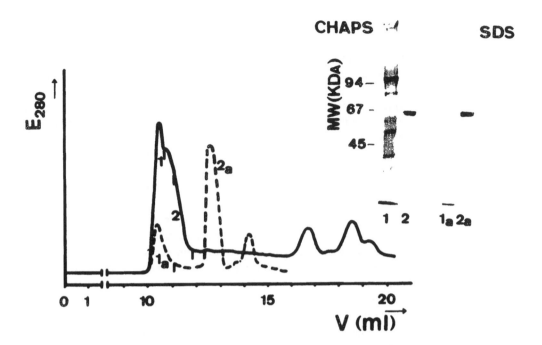

FIGURE 3. Influence of different detergents on the behavior of proteins during chromatographic separation on a size-exclusion TSK-3000 column, 600 × 7.5 mm I.D. The mixture consists of calcium-binding plasma membrane proteins of liver, pre-purified by stepwise extraction (cf. Figure 1). — The freeze-dried sample was dissolved in phosphate-buffered saline, PBS, pH 7.2, containing 1% (w/v) CHAPS; 0.1% CHAPS was added to the elution buffer. — The freeze-dried sample was dissolved in PBS, containing 1% (w/v) SDS. Then 0.1% SDS was added to the elution buffer. The fractions were collected and controlled electrophoretically. The 65 to 67 kDa calcium-binding protein was eluted with CHAPS as aggregate (fraction 2). With SDS it seems to appear as monomer(s) (fraction 2a). (The figure was first published in Josić, Dj., Baumann, H., and Reutter, W., *Anal. Biochem.*, 142, 473, 1984; reprinted with the permission of Academic Press, Inc.)

synthetic materials are recommended. It is to be expected that, in the near future, systems will generally be required which are free of metal ions, with columns made of glass or similar materials. Lacking these characteristics they will not be regarded as fit for biopolymer and membrane protein separation.

IV. THE CHOICE OF DETERGENT

Choosing the right detergent can be of vital importance in some experiments, with the extraction as well as with the separation of membrane proteins. Detergents keep the hydrophobic proteins in solution and prevent their aggregation and unwelcome interaction with the support during HPLC separation. In some processes, where separation occurs due to the hydrophobic characteristics of the sample components, the detergents can play an important role in sample separation itself, e.g., in hydrophobic-interaction HPLC[5] and in concanavalin A HPAC.[7] Basically, the detergents can be assigned to two different groups, the non-ionic and the ionic. The ionic detergents are subdivided into anionic, cationic, and zwitterionic. The "mild" detergents (CHAPS, octylglucose, and similar substances such as MEGA) can extract the proteins from the plasma membranes and keep them in solution without further denaturation or, e.g., destroying their enzymatic activity. The "strong" detergents (such as SDS or Triton) will destroy the plasma membrane more effectively, but will often denature the proteins and cause the loss of biological activity as a consequence of such treatment. After removing the strong detergent, that is after replacing it by a mild one, the protein's activity can sometimes be restored. Figure 3 shows the behavior of a strong detergent such as SDS in comparison to a mild detergent such as CHAPS during SEC separation. When SDS was applied, the protein appeared as a monomer during separation (cf. the corresponding control by SDS-PAGE), whereas self-aggregation could not be fully eliminated through the use of CHAPS. Although kept in solution, the protein formed aggregations with a markedly higher molecular weight when CHAPS was added.

In this short description of detergents, it has to be pointed out that there is no clear borderline between strong and mild detergents. Some proteins will stay fully active also in the presence of SDS, some proteins even need detergents for staying active.

Another important factor is the critical micelle-concentration (cmc) of the detergents. The higher the cmc, the easier it is to remove the detergent by dialysis. Some detergents with very low cmc, e.g. those of the Triton group, can hardly be removed by dialysis at all. Conclusive data on detergents are found in the literature.[8]

Several detergents, among them again those of the Triton group, absorb in the 280 nm range. Consequently, the detection of the protein is made almost impossible by the high background when these detergents are used. For this reason it is recommended to use corresponding derivatives, whose double-bonds have been saturated. Such detergents have recently become commercially available, and are sold under the name of, e.g., Triton X 100 or X 114, *reduced*. Their absorption in the 280 nm range is much lower, and they can be applied to HPLC separations without causing any difficulties with detection.

The choice of detergent has to ensure that interference during separation is avoided. Ionic detergents, e.g., can interfere with IEC. Some detergents such as CHAPS impair separation by metal-affinity HPLC.

V. HPLC SEPARATION OF MEMBRANE PROTEINS

As has been mentioned above, almost all methods can be used for membrane protein separation. Several particular problems will now be discussed. The use of silica-based column packings can sometimes present difficulties, as some detergents, e.g. Triton, may adsorb to the silica gel surface.[1] This, in turn, can lead to precipitation, the loss of the sample, and finally to the destruction of the column itself. Newly developed silica-based column packings, whose surface is protected by a polymer layer, widely neutralizes this effect. However, if columns with synthetic polymers are used, one has to expect certain hydrophobic characteristics on their surfaces, which may remain despite the introduction of hydrophobic groups.

The columns for SEC and RPC of membrane proteins have to meet particularly high standards. During SEC, in order to reduce to an absolute minimum the interaction between matrix and sample components, and also among the sample components themselves, very strongly denaturing agents have often to be added to the buffers. In order to reduce the disulfide bridges, the addition of mercaptoethanol is sometimes necessary. Such reagents can seriously damage some column materials. As a rule it can be said that the column life is severely shortened under such conditions.

Similar problems, though on the whole less severe, are encountered with reversed-phase HPLC. All columns will take normal eluents, such as 0.1% trifluoracetic acid and isopropanol. However, not all proteins can be isolated under these conditions.[4] The choice of less hydrophobic material can ease this problem, if a C_4 or even a C_1 sorbent is used instead of a C_{18} sorbent. Formic acid in concentrations up to 60% is still required from time to time. Many column materials will withstand these conditions for quite a while, but in some cases the performance will be impaired after only 1 or 2 runs.

There is no general rule for the durability of columns in membrane protein separations. Every column should be tested before being used. Thus, short columns should be used for budget reasons. On systems like this, separation can be carried out and possibly optimized on a small

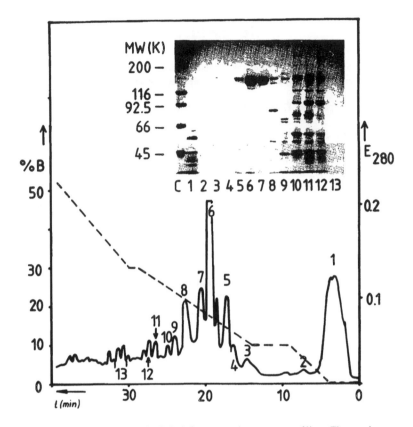

FIGURE 4. Anion-exchange HPLC of plasma membrane extract of liver. The membranes were solubilized by freezing/thawing (cf. Figure 1). Column, TSK-DEAE-5PW, 75 × 7.5 mm I.D. (LKB, Munich, F.R.G.); Buffer A, 10 m*M* HEPES, pH 6.1, with 0.05% CHAPS; Buffer B, 1 *M* sodium chloride in buffer A. Flow-rate, 1 ml/min; pressure, 12 bar. The gradient is marked. Samples were collected and aliquots (50 to 100 μg of protein) were used for electrophoresis (see upper part). (The Figure was first published in Josić, Dj., Hofmann, W., Wieland, B., Nuck, R., and Reutter, W., *J. Chromatogr.*, 359, 315, 1986; reprinted with the permission of Elsevier Science Publishers B.V.)

scale with a tiny amount of sample. For the control of recovery in the chromatographic run, radioactively labeled samples can be separated on such a small scale.[9] After selecting the most suitable column, the experiment can be upgraded to a larger scale.

Despite all the difficulties, whose discussion and better understanding has been the object of this report, membrane proteins can be successfully separated analytically as well as preparatively. The preconditions are adequate sample preparation and choice of detergent, as well as the selection of suitable column material. The separation results of such complex mixtures in the analytical field are comparable to those obtained with water-soluble proteins, as can be seen in Figure 4.

In the preparative field, the most favored method will be affinity HPLC. Different immobilized lectins, in most cases concanavalin A, and other ligands such as, e.g., heparin, will be used as well as monoclonal and polyclonal antibodies. This latter branch of affinity HPLC has seen a quick development over the last few years. Speed as well as minimal non-specific binding give this method superiority over conventional affinity chromatographic methods.

REFERENCES

1. **Josić, Dj., Baumann, H., and Reutter, W.,** Size-exclusion high-performance liquid chromatography and sodium dodecyl sulfate polyacrylamide gel electrophoresis of proteins: a comparison, *Anal. Biochem.*, 142, 473, 1984.
2. **Bordier, C.,** Phase separation of integral membrane proteins in Triton X 114 solution, *J. Biol. Chem.*, 256, 1604, 1981.
3. **Schwarz, W., Born, J., Tiedemann, H., and Molnar, I.,** Separation of proteins by SE- and RP-HPLC, in *Practical Aspects of Modern HPLC*, Molnar, I., Ed., Walter de Gruyter, Berlin, New York, 1982, 123.
4. **Heukeshoven, J. and Dernick, R.,** Separation of hydrophobic virus proteins, *Chromatographia*, 25, 230, 1988.
5. **Josić, Dj., Hofmann, W., and Reutter, W.,** Ion-exchange and hydrophobic-interaction high-performance liquid chromatography of proteins — a practical study, *J. Chromatogr.*, 271, 43, 1986.
6. **Josić, Dj., Hofmann, W., Habermann, R., and Reutter, W.,** High-performance liquid affinity chromatography of liver plasma membrane proteins, *J. Chromatogr.*, 397, 39, 1987.
7. **Josić, Dj., Hofmann, W., Habermann, R., and Reutter, W.,** Concanavalin A high-performance affinity chromatography of liver plasma membrane proteins, *J. Chromatogr.*, 444, 29, 1988.
8. **Helenius, A. and Simons, K.,** Solubilization of membranes by detergents, *Biochem. Biophys. Acta*, 415, 29, 1975.
9. **Josić, Dj., Hofmann, W., Wieland, B., Nuck, R., and Reutter, W.,** Anion-exchange high-performance liquid chromatography of membrane proteins from liver and Morris hepatomas, *J. Chromatogr.*, 359, 315, 1986.

CATION-EXCHANGE HIGH-PERFORMANCE LIQUID CHROMATOGRAPHY: GROUP SEPARATION OF ACIDIC AND BASIC PROTEINS USING VOLATILE SOLVENTS

A.J.M. van den Eijnden-van Raaij, I. Koornneef,
G.J.T. Zwartkruis, R. Nieuwland, and E.J.J. van Zoelen

I. INTRODUCTION

For the separation of proteins differing only slightly in charge, isoelectric focusing and ion-exchange HPLC (IEC) techniques are required. The disadvantage of isoelectric focusing is that this method is rather laborious for large-scale fractionation of proteins. Furthermore, highly basic proteins with isoelectric points above 9.5 cannot be separated because of the limited range of pH gradients generated by the commercially available ampholytes. In ion-exchange chromatography, the choice of the type of matrix (cationic or anionic) depends on the isoelectric point (pI) and the stability within a certain pH range of the proteins to be separated.[1] In general, an anion-exchange column should be used for the fractionation of acidic proteins, while a cation exchanger is most suitable for basic proteins. Protein separation can be achieved by choosing the right combination of pH and ionic gradient. For the separation of basic proteins on a cation exchanger, the starting pH and the ionic strength are normally chosen so that the proteins of interest are just bound to the matrix.[2] However, the purification of highly basic proteins (pI > 9.0) is often complicated by the instability of most proteins in the high alkaline pH range and by the difficulty in establishing satisfactory conditions for the separation of highly basic proteins with similar charge and size properties via cation-exchange chromatography.[3-7] The present paper shows that these problems can be overcome by cation-exchange HPLC (CEC) at low pH. We report the chromatographic properties of a silica-based weak cation-exchange HPLC column (Bio-Rad CM-2-SW) using acidic volatile buffer solutions. Depending on the eluting buffer system used, highly basic, neutral, or acidic proteins can be separated on this column. The advantages of volatile buffer solutions during chromatography and the applications of the ion exchanger for protein separation are discussed.

II. EXPERIMENTAL

The proteins used in this study were purchased from Sigma except chymotrypsinogen A which was obtained from Boehringer-Mannheim. Ammonium acetate (NH_4Ac) was obtained from Merck.

Cation-exchange high-performance liquid chromatography (CEC) was carried out on equipment from Millipore-Waters comprising a Model 680 automated gradient controller supplied with two 510 solvent pumps and a 481 variable wavelength detector linked to a BD41 chart recorder from Kipp. A combination of a Bio-Sil TSK HPLC guard column (75×7.5 mm,

TABLE 1
Polypeptide Standards

Protein	Source	pI[a]
Basic		
Trypsinogen	Bovine	9.3
Chymotrypsinogen A	Bovine	9.5[13]
Ribonuclease A	Bovine	9.6[14]
Cytochrome c	Horse	10.6[15]
Lysozyme	Chicken	11.0[16]
Myelin basic protein	Bovine	11.8
Acidic		
Trypsin inhibitor	Soybean	4.5—4.6
Ovalbumin	Chicken	4.6
Serum albumin	Bovine	4.9
β-Lactoglobulin	Bovine	5.1
Insulin	Bovine	5.3—5.7[17]
Neutral		
Carbonic anhydrase	Bovine	6.8
Myoglobin	Horse	6.8, 7.2
Lactic dehydrogenase	Bovine	8.1

[a]   The pI values of the proteins were obtained from the supplier, the *Merck Index* (9th Edition) and the *Handbook of Biochemistry* (2nd Edition), respectively. The pI value of myelin basic protein was calculated from the amino acid composition and the pK values of the individual amino acids.[12]

Bio-Rad) and a Bio-Sil TSK CM-2-SW column (250 × 4.6 mm, Bio-Rad) was used. Protein samples of 2 ml were applied to the column using a loop injector (U6K, Millipore-Waters).

Basic proteins (pI > 9) were eluted by a linear AB salt gradient (10 mM salt/min), where Eluent A was 1 M acetic acid (HOAc) (pH 2.4) and Eluent B was 1 M NH$_4$Ac in 1 M HOAc (pH 4.8). Acidic proteins (pI 4 to 6) were eluted by a linear AB salt gradient (3 mM salt/min), where Eluent A was 50 mM HOAc (pH 3.1) and Eluent B was 300 mM NH$_4$Ac (pH 6.7). Intermediate proteins (pI 6 to 9) were eluted by a linear AB salt gradient (10 mM salt/min), where Eluent A was 50 mM NH$_4$Ac and Eluent B was 1 M NH$_4$Ac, both eluents being adjusted to pH 5.0 by the addition of HOAc.

III. RESULTS AND DISCUSSION

In order to develop a general method for the separation of proteins differing slightly in charge, the chromatographic behavior of a number of different proteins (Table 1) was studied on a HPLC cation exchanger under various conditions. The proteins were divided into three groups representing the highly basic proteins (pI > 9.0), the acidic proteins (pI 4 to 6), and an intermediate class of proteins, called the neutral proteins (pI 6 to 9).

A mixture of highly basic proteins was dissolved in 1 M HOAc (pH 2.4) and injected on to a silica-based Bio-Sil CM-2-SW cation-exchange HPLC column equilibrated in this solvent. Figure 1A shows the elution profile of a mixture containing five basic proteins (pI 9.3 to 11.0) following application of a linear gradient of NH$_4$Ac in 1 M HOAc (10 mM NH$_4$Ac/min). The proteins emerged according to their isoelectric point as well-separated sharp peaks, in spite of the fact that they are all fully protonated under these conditions. The retention times of the proteins varied from 27.5 min for trypsinogen to 56.2 min for lysozyme.

During the linear NH$_4$Ac gradient, the pH is not constant. In Figure 1A (insert), the pH is

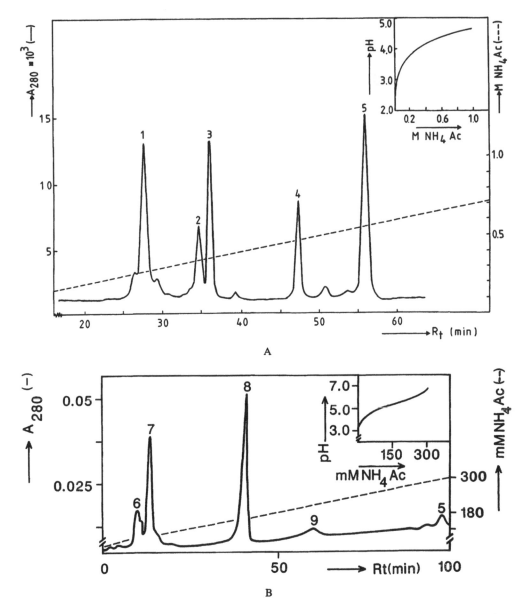

FIGURE 1. Separation of basic (A) and acidic (B) proteins on a Bio-Sil TSK CM-2-SW cation exchange HPLC column. (A) Linear gradient of 0 to 1 M NH$_4$Ac in 1 M HOAc (10 mM/min). (B) Linear gradient of 50 mM HOAc (pH 3.1) to 300 mM NH$_4$Ac (pH 6.7) (3 mM NH$_4$Ac/min). Flow-rate, 0.8 ml/min. Absorbance at 280 nm. Inserts: relationship between pH and molarity of ammonium salt in the eluting buffer during the linear salt gradient. (1) Trypsinogen (126 μg), (2) chymotrypsinogen A (50 μg), (3) ribonuclease A (100 μg), (4) cytochrome c (50 μg), (5) lysozyme (50 μg), (6) ovalbumin (25 μg), (7) trypsin-inhibitor (50 μg), (8) β-lactoglobulin (50 μg), and (9) insulin (25 μg).

plotted against the molarity of NH$_4$Ac in the eluting buffer. The pH of the starting buffer (1 M HOAc) is 2.4. Continuous addition of increasing amounts of NH$_4$Ac results in a rapid increase in pH from 2.4 to 3.6, followed by a gradual increase to pH 4.8 (1 M NH$_4$Ac/1 M HOAc). The convex pH gradient appeared to be very important for efficient separation of basic proteins. If the pH was kept constant, separation of the proteins was less effective or even absent.[8]

In the NH$_4$Ac system, with the combined pH and salt gradient, the proteins tested were eluted exactly according to their isoelectric point (Figure 2a). Proteins like chymotrypsinogen A and

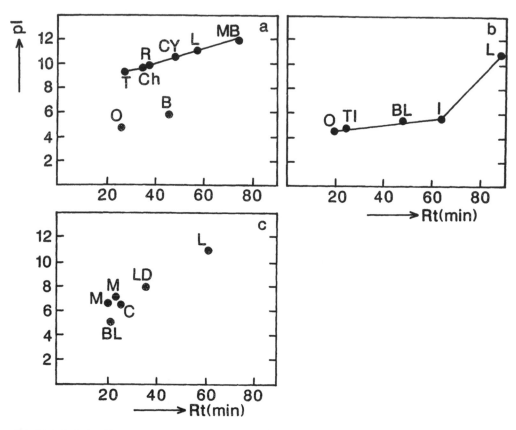

FIGURE 2. Relationship between isoelectric point (pI) and retention time (R$_t$) on a Bio-Sil TSK CM-2-SW HPLC column using as mobile phase a linear gradient of (a) 0 to 1 *M* NH$_4$Ac in 1 *M* HOAc, pH 2.4 to 4.8 (b) 50 m*M* HOAc to 300 m*M* NH$_4$Ac, pH 3.1 to 6.7, (c) 50 m*M* NH$_4$Ac to 1 *M* NH$_4$Ac, pH 5.0. Flow-rate, 0.8 ml/min. Absorbance at 280 nm. T = trypsinogen, Ch = chymotrypsinogen A, R = ribonuclease A, Cy = cytochrome c, L = lysozyme, MB = myelin basic protein, O = ovalbumin, B = bovine serum albumin, TI = trypsin inhibitor, BL = β-lactoglobulin, I = Insulin, C = carbonic anhydrase, M = myoglobin, LC = lactic dehydrogenase.

ribonuclease A with pI values that differ by only 0.1 unit are very well separated. Acidic proteins like ovalbumin (pI 4.6) and bovine serum albumin (pI 4.9) deviate from the correlation plot. These proteins were eluted from the column later than would be expected on the basis of their pI value. A possible explanation for the chromatographic behavior of the acidic proteins in the NH$_4$Ac system might be found in the net positive surface charge density and the composition of the basic residues involved in the interaction of proteins with the cation exchanger at low pH.[8] Smaller acidic proteins did not bind well to the column in HOAc and were eluted simultaneously shortly after initiation of the NH$_4$Ac gradient. The NH$_4$Ac system thus appeared to be useful exclusively for the separation of highly basic proteins.

As shown in Figure 1B, separation of acidic proteins (pI 4 to 6) on the CM-2-SW column is possible after lowering the ionic strength of the buffer solutions used during chromatography of basic proteins. The gradient system used was a linear gradient from 50 m*M* HOAc (pH 3.1) to 300 m*M* NH$_4$Ac (pH 6.7) resulting in a non-linear pH gradient (Figure 1B, insert). The proteins were eluted as well-separated sharp peaks, and even ovalbumin and trypsin inhibitor, which have near identical isoelectric points, were separated in this system. The broad peak of insulin might be explained by retardation of the protein after isoelectric precipitation during the gradient. In the system described above, the pH plays a more important role than ionic strength in the separation of these acidic proteins. Also, in this system there is a linear relationship for the acidic proteins between isoelectric point and retention time (Figure 2b), whereas the basic proteins deviate from this correlation plot. In contrast to the NH_4Ac system described for basic proteins, ovalbumin shows a normal chromatographic behavior in the low ionic strength system.

For proteins with isoelectric points between 6 and 9, the optimal gradient system for separation is a gradient of 50 mM NH_4Ac to 1 M NH_4Ac at constant pH (pH = 5.0). In Figure 2c, it is shown that proteins with pI values < 7.0 were hardly separated, whereas proteins with pI > 7.0 were clearly separated in this system.

Evaluating the results of our experiments, a linear gradient of NH_4Ac in HOAc at high ionic strength, inducing a convex pH gradient, appears to be the optimal condition for the separation of acid-stable basic proteins on a Bio-Sil CM-2-SW cation-exchange column. A combination of a linear gradient of NH_4Ac at low ionic strength and non-linear pH gradient is suited for the separation of acidic proteins, whereas an intermediate group of proteins (pI 6 to 9) can be optimally separated by an NH_4Ac gradient at constant pH. Changing the conditions of the salt and pH gradient therefore enables different groups of proteins to be separated on the same cation-exchange column.

Separation of a mixture of proteins with a wide pI range, e.g., a mixture of ribonuclease A, bovine serum albumin and myoglobin, can be achieved by a combination of two sequential gradient systems. First, a linear gradient of 50 mM HOAc (pH 3.1) to 100 mM NH_4Ac should be applied in order to elute the acidic bovine serum albumin; the pH should then be adjusted to 5.0 and kept constant during the following linear gradient from 100 mM NH_4Ac to 1 M NH_4Ac in order to separate myoglobin from ribonuclease A. The choice of the gradient systems is dependent on the proteins to be separated.

One of the applications of the CM-2-SW cation exchanger is the purification of acid-stable highly basic polypeptide growth factors such as platelet-derived growth factor (PDGF)-like polypeptides[9] and transforming growth factor β.[10] Detection of nanogram amounts of these proteins in column fractions is often only possible by testing them in a biological assay under sterile conditions.[9,10] For this reason, the use of the 1 M HOAc/NH_4Ac system (pH < 4) is very suitable during the chromatographic steps. The use of volatile buffer solutions like NH_4Ac facilitates the testing of large column fractions without the side effects of high salt concentrations. The CM-2-SW weak cation-exchange column has also proven to be very efficient in the identification of a heparin-binding growth factor from PC13 embryonal carcinoma cells as related to the basic form of fibroblast growth factor.[11] Considering the high recoveries of proteins from the column, even of sticky proteins such as PDGF-like growth factors,[10] as well as the advantages of volatile acidic buffer solutions, the present HPLC technique should find particular applications in the purification of highly basic, acid-stable proteins, as well as acidic and neutral biologically active proteins. The conditions determined in the present study for optimal separation of proteins on an analytical/semipreparative column can probably also be applied to preparative CM-2-SW cation exchangers.

ACKNOWLEDGMENTS

The authors wish to thank Dr. A. Teelken from the University Hospital Groningen, Department of Neurology, for the generous supply of myelin basic protein.

REFERENCES

1. **Himmelhoch, S.R.,** Chromatography of proteins on ion-exchange adsorbents, *Methods in Enzymology,* Vol. 22, Jakoby, W., Ed., pp 273, 1971.
2. **Lampson, G.P. and Tytell, A.A.,** A simple method for estimating isoelectric points, *Anal. Biochem.,* 11, 374, 1965.
3. **Dunkley, P.R. and Carnegie, P.R.,** Isolation of myelin basic protein, in *Research Methods in Neurochemistry,* Vol. 2, Marks, N. and Rodnight, R., Eds., 219, 1974.
4. **Hirshfeld, H., Teitelbaum, D., Arnon, R., and Sela, M.,** Basic encephalitogenic basic protein: a simplified purification on sulphoethyl-sephadex, *FEBS Lett.,* 7, 317, 1970.
5. **Barton, M.A., McPherson, T.A., and Martin, J.K.,** Purification of fully active human encephalitogenic basic protein on sulfoethyl-sephadex, *Can. J. Biochem,* 50, 684, 1972.
6. **Ohe, Y., Hayashi, H., and Iwai, K.,** Human spleen histone H2B, *J. Biochem.,* 85, 615, 1979.
7. **Senshu, T. and Iwai, K.,** Fractionation of calf thymus histone by a new method of CM[carboxymethyl] cellulose chromatography, *J. Biochem.,* 67, 473, 1970.
8. **Van den Eijnden-van Raaij, A.J.M., Koornneef, I., van Oostwaard, Th.M.J., de Laat, S.W., and van Zoelen, E.J.J.,** Cation-exchange high-performance liquid chromatography: separation of highly basic proteins using volatile acidic solvents, *Anal. Biochem.,* 162, 263, 1987.
9. **Van den Eijnden-van Raaij, A.J.M., Koornneef, I., van Oostwaard, Th.M.J., Feijen, A., Kruijer, W., de Laat, S.W., and van Zoelen, E.J.J.,** Purification of a PDGF-like growth factor and a type beta transforming growth factor secreted by mouse neuroblastoma cells: a general strategy utilizing exclusively volatile acidic solvents and resulting in a high protein recovery, *Biochem. J.,* 257, 375, 1989.
10. **Van den Eijnden-van Raaij, A.J.M., Koornneef, I., and van Zoelen, E.J.J.,** A new method for high yield purification of type beta transforming growth factor from human platelets, *Biochem. Biophys. Res. Comm.,* 157, 16, 1988.
11. **Van Veggel, J.H., van Oostwaard, Th.M.J., de Laat, S.W., and van Zoelen, E.J.J.,** PC13 embryonal carcinoma cells produce a heparin-binding growth factor, *Exp. Cell Res.,* 169, 280, 1987.
12. **Segel, I.H.,** in *Biochemical Calculations,* ch. III, 123, 1968.
13. **Long, C., King, E.J., and Sperry, W.M.,** *Biochemists Handbook,* 304, 1961.
14. **Tanford, C. and Hauenstein, J.D.,** Hydrogen ion equilibria of ribonuclease, *J. Am. Chem. Soc.,* 78, 5287, 1956.
15. **Tint, H. and Reiss, W.J.,** Studies on the purity and specificity of cytochrome c. I. Electrophoretic analyses, *J. Biol. Chem.,* 182, 385, 1950.
16. **Alderton, G., Ward, W.H., and Fevold, H.L.,** Isolation of lysozyme from egg white, *J. Biol. Chem.,* 157, 43, 1945.
17. **Malamud, D. and Drysdale, J.W.,** Isoelectric point of proteins: a table, *Anal. Biochem.,* 86, 620, 1978.

SEPARATION OF PROTEIN ISOFORMS: AN ION-EXCHANGE HPLC SEPARATION OF BOVINE AND CARDIAC TROPONIN COMPLEXES

Paul J. Cachia and Robert S. Hodges

I. INTRODUCTION

The separation of the troponin complex of muscle was first accomplished by Ebashi et al.[1] when they showed that native tropomyosin consisted of two molecules, tropomyosin and troponin, both of which were required to confer Ca^{2+} sensitivity to purified actomyosin preparations. It is now known that troponin is composed of three subunits which were first separated by employing sodium dodecyl sulfate (SDS) polyacrylamide gel electrophoresis (PAGE) and DEAE-Sephadex chromatography in the presence of 6 M urea.[2,3] Potter[4] has published two detailed descriptions of the purification of both cardiac and skeletal troponin complexes using Cibacron Blue-Sephacryl chromatography. This procedure, followed by various combinations of chromatography on DEAE-Sephadex and CM-Sephadex in the presence of 6 M urea and 1 mM EDTA, can also be used to prepare pure troponin subunits. In the past, our laboratory has used modifications[5] of the method of Straprans[6] for the purification of native rabbit skeletal troponin complex and its subunits, while the purification of bovine cardiac troponin complex and its subunits was performed using methods worked out by Burtnick and co-workers.[7,8] While these methods are successful at providing pure forms of the native troponin complexes, as well as the subunits, there are disadvantages. These methods are time consuming and, because they use open-column chromatography, the derived product(s) are dispersed in large volumes of column eluent.

With the advent of ion-exchange columns available for high-performance liquid chromatography (HPLC), we looked at the possibility of developing a method for the purification of the subunits using these columns.[9] The steps behind the development of this method are straightforward and we report them here to serve as a model for the development of other methods used to separate complex mixtures of proteins by ion-exchange chromatography (IEC). Furthermore, during development of this method it became apparent that HPLC columns were capable of not only resolving the different subunits of these complexes but also isoforms of the subunits themselves. This result was seen as a significant advantage in the purification and isolation of important biologically or synthetically modified proteins.

II. OPEN COLUMN METHOD

At the time we were developing our method, one report of the isolation of troponin T (TnT) isoforms by open-column IEC appeared in the literature. Gusev and co-workers [10] had identified two forms of TnT from bovine cardiac troponin on DEAE-cellulose support in the presence of 8 M urea. The heterogeneity that they observed was found to occur due to the presence of two forms of TnT, differing in their M_r values and amino acid content. It was further shown by amino acid analysis that the amino acid compositions of two homogenous forms of this protein isolated from the DEAE-cellulose column were significantly different in the observed content of glutamic acid/glutamine. They concluded that the elevated content of glutamic acid in one of the forms compared to the other was responsible for its increased net negative charge and, hence, explained both its tighter binding to the DEAE-cellulose column and the electrophoretic differences observed between the forms. These results, together with our own observations which indicated multiple forms of both troponin I (TnI) and TnT existed in both cardiac and rabbit skeletal muscle, prompted us to further develop our HPLC method for the isolation of these different forms of the subunits.

The traditional methods of open column chromatography used in our laboratory to purify bovine cardiac troponin subunits comprise a combination of DEAE-Sephadex A-25 and CM-Sephadex C-25, followed by gel filtration on Biogel P-200 and final desalting on Biogel P-2 for purification of the cardiac subunits, while the purification of the rabbit skeletal complex employs DEAE-Sephadex A-25 and SE-Sephadex. In the separation of the native cardiac subunits, it was found[7] that troponin C (TnC) isolated from crude troponin by chromatography on DEAE-Sephadex A-25 in 6 M urea is obtained directly as a homogeneous peak, as determined by SDS-PAGE. TnI isolated from this column is rechromatographed on CM-Sephadex C-25 in 8 M urea, followed by gel filtration on Biogel P-200 to remove high-molecular-weight contaminants.[8] TnT is isolated as crude material on DEAE-Sephadex A-25 and purified on CM-Sephadex C-25 in 8 M urea. The separation of rabbit skeletal subunits is carried out by initial chromatography of DEAE-Sephadex A-25 in 6 M urea.[5] TnI obtained in this way is 80 to 90% pure as judged by SDS-PAGE and may be used without further purification. TnT from this chromatography is 75 to 85% pure and is purified to homogeneity on SE-Sephadex at pH 8.0 in 6 M urea. TnC is obtained as a homogeneous product (SDS-PAGE) directly from DEAE-Sephadex A-25. The individual protein peaks are then dialyzed to remove any salts and urea.

III. ION-EXCHANGE HPLC METHOD

The initial purification of these troponin complexes by ion-exchange HPLC used a combination of both anion-exchange (AEX) and cation-exchange (CEX) analytical columns (Q300 and CM300, respectively; 250 × 4.1 mm I.D.; SynChrom, Linden, IN) These columns were selected for this application because their particle dimensions (6.5 µm) and pore size (300 Å) were suited for protein work. The same columns are also useful, however, in the separation of small to medium size peptides.[11] Two-dimensional gel electrophoresis was used to examine the purity of the fractions obtained during the separations.[12] We presumed that the isolectric points (pI) of the different protein components would govern the separation of the complex. The pI values of the different components can be estimated from 2D gels which separate the complex on the basis of pI and the molecular weights of the individual subunits (Figure 1). Non-eqilibrium pH gradient electrophoresis conducted in the first dimension was performed according to the method of O'Farrell and co-workers ,[12] while the second dimension molecular weight separation was carried out using the method of Laemmli.[13] Estimates of the pI values of the three subunits were obtained from the literature[14-17] and are as follows: TnC, an acidic protein with a pI of about 4; TnI, a basic protein with a pI of about 10; and TnT, a neutral protein with a pI of about 7. These agreed with our own estimates from 2D gels. Based on these values, it was predicted that TnT would not bind to either an anion- or cation-exchange column at pH 7. At the same pH, TnC would be eluted with the solvent front on a cation-exchange column or be retained by an anion-exchange column, while TnI would be eluted with the front from an anion-exchange column or be retained on a cation-exchange column. Examples of the behavior of these complexes at pH 6.5, using a linear salt gradient in 8 M urea on CM300 and Q300 ion-exchange columns, are shown in Figure 2. Contrary to what was expected, under these conditions TnT was resolved into multiple peaks on both the anion- and cation-exchange columns. As predicted, TnI was retained on the cation-exchange column and was also found to be resolved into multiple

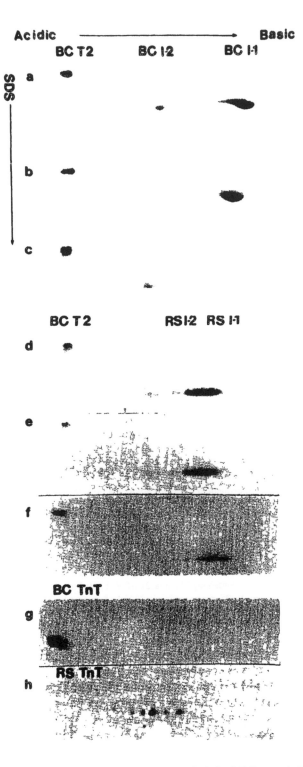

FIGURE 1. Separation of bovine cardiac (BC) and rabbit skeletal (RS) troponin T and troponin I on two-dimensional PAGE. In panels a to f, HPLC-purified bovine cardiac troponin T fraction, T2, was used as a standard marker. (a) mixture of HPLC-purified BC I1 and BC I2; (b) BC I1; (c) BC I2; (d) mixture of HPLC-purified RS I1 and RS I2; (e) RS I1; (f) RS I2; (g) native bovine cardiac troponin T; (h) native rabbit skeletal troponin T. (Reprinted from Cachia, P. J., Van Eyk, J., McCubbin, W. D., Kay, C. M., and Hodges, R. S., *J. Chromatogr.*, 343, 315, 1985. With permission.)

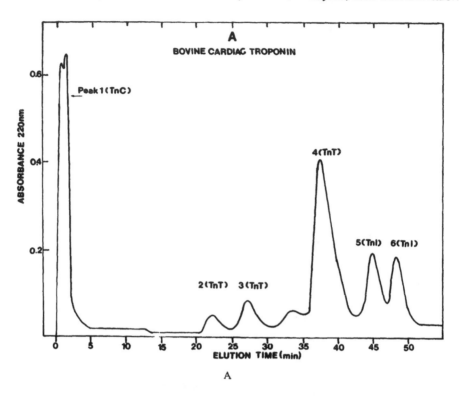

A

FIGURE 2. Separation of bovine cardiac troponin (A panels) and rabbit skeletal troponin (B panels) by ion-exchange HPLC. Programmed chromatographic runs were performed on a Varian Vista Series 5000 liquid chromatograph (Varian, Walnut Creek, CA) interfaced with a Varian CDS 401 data system and coupled to a Kratos SF769Z variable-wavelength UV spectrometer. (Figure A) column, Syn-Chropak CM300 weak cation-exchange column, 250 × 4.1 mm I.D., 6.5-μm particle size, 300-Å pore size (SynChrom, Linden, IN); conditions, linear AB gradient (90% A to 10% B at time 0 min to 78% A - 22% B at time 60 min, equivalent to 2 mM KCl/min) at a flow-rate of 1 ml/min, where Buffer A is 50 mM KH$_2$PO$_4$, pH 6.5, containing 8 M urea, 2 mM EGTA, 0.5 mM DTT, and Buffer B is Buffer A containing 1 M KCl; sample, 3 mg troponin per 100 μl of starting buffer; chart speed, 12 in./hr. (Figure B) column, SynChropak Q300 strong anion-exchange column, 250 × 4.1 mm I.D., 6.5 μm, 300 Å; conditions, linear AB gradient (80% A to 20% B at time 0 min to 60% A to 40% B at time 40 min for bovine cardiac troponin [panel A] or 90% A to 10% B at 0 min to 70% A to 30% B for rabbit skeletal troponin [panel B], both gradients equivalent to 5 mM KCl/min) at a flow-rate of 1 ml/min, where Buffer A is 50 mM Tris, pH 7.5, containing 8 M urea, 2 mM EGTA, 0.5 mM DTT, and Buffer B is Buffer A containing 1 M KCl; sample, 25 mg troponin per 200 μl of starting buffer; chart speed, 12 in./hr. (Reprinted from Cachia, P. J., Van Eyk, J., McCubbin, W. D., Kay, C. M., and Hodges, R. S., *J. Chromatogr.*, 343, 315, 1985. With permission.)

forms under these conditions on this column. TnC was retained on the anion-exchange column and was homogenous. Furthermore, examination of Figure 2A shows that, as expected, TnC is not retained on this column and the relative positions of TnI and TnT interchange on the cation-exchange matrix, CM300, during the separation of bovine cardiac and rabbit skeletal complexes. Figure 2B demonstrates that, on the anion-exchange column (Q300), only TnC and TnT are retained and interchange their relative positions with respect to each other, while TnI is eluted with the solvent front. Table 1 lists the size of the protein (i.e., the number of residues), the "net charge" on the protein at neutral pH, its pI, its relative hydrophobicity, and the salt concentrations required to elute the individual subunits. Examination of Table 1 indicates that there are anomalies when one compares the observed retention times of the individual subunits vs. their pI or "net charge". For example, when bovine cardiac troponin is separated on the cation-exchange column (Figure 2A), it would be predicted that both the TnC and TnT subunits would

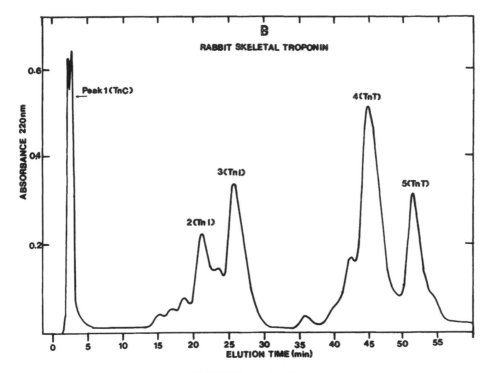

FIGURE 2A (continued)

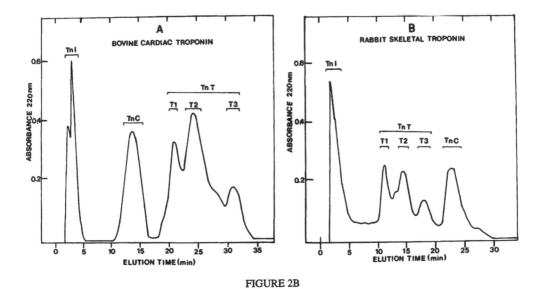

FIGURE 2B

be eluted with the solvent front as both have acidic "net charges". However, bovine cardiac TnT is retained by this column. Generally, it was assumed that the behavior of each subunit would be consistent with its pI as determined from 2D gel electrophoresis (Figure 1). However, it seems that the ionic contributions from the side chains of the constituent acidic and basic residues in TnT are fully expressed under the conditions of the chromatography. Therefore, it is reasonable for TnT to bind to both columns since an examination of the known sequences[18] of several TnT molecules show that they are comprised of distinct areas of positive or negative charge and it is these patches of charge (i.e., preferred binding domains) which bind either cation or anion

TABLE 1

Protein	Size	Net charge	Relative hydrophobicity[a]	pI	Salt conc (mM) Cation	Anion
AcRSTnC	159	−29	1.00	4		215
AcRSTnI	178	+6	1.06	10	250—270	
AcRSTnT	259	+5	1.32	7	300—330	150—200
AcBCTnC	161	−29	1.05	4		270
AcBCTnI	211	+15	1.31	10	185—200	
AcBCTnT1[b]	284	−8	1.29	7	140—180	290—370
AcBCTnT2[b]	279	−5	1.29	7	140—180	290—370

[a]   Relative hydrophobicity is calculated as the sum of the coefficients[28] of the component residues of the protein .This sum is then taken relative to that of RSTnC.

[b]   Taken from Reference 18.

matrices. Thus, the observed behavior of the TnT and TnI subunits cannot be correlated with intrinsic parameters such as net charge, pI, molecular weight, or hydrophobicity. Similar results were obtained by Kopaciewicz and co-workers [19] when they examined the suitability of the "net charge" concept in explaining chromatographic retention on high-performance ion-exchange columns. They observed that "zero" retention at or near the pI on both anion- or cation-exchange columns, in agreement with the "net charge" concept, was observed for only three of the fourteen proteins studied. Furthermore, they found that deviations from the predicted retention times based on the "net charge" effect did not occur in any systematic way. It was concluded then that retention times on ion-exchange columns could not be satisfactorily explained by the "net charge" concept alone and that the importance of protein pI had been over emphasised. Burke and co-workers [20] have demonstrated that there are hydrophobic interactions occurring between peptides and high-performance ion-exchange packings. Their results demonstrated that these interactions give rise to mixed-mode contributions in solute separations. In the separation of the troponin subunits shown here, there was no precaution taken to minimize non-specific hydrophobic interactions of this type. It is possible therefore that the unusual behavior observed for the I and T subunits may result in part from this kind of interaction with the column matrix.

Separation of the bovine and rabbit complexes exhibited other differences when the results from the weak cation-exchange column were analyzed by gel electrophoresis. Examination of the fractions from Figure 2A, for the separation of the bovine cardiac subunits, showed that peaks 2, 3, and 4 were TnT, while 5 and 6 were TnI. However, this analysis also showed that of peaks 2, 3, and 4, only peak 2 was pure TnT. While peaks 3 and 4 contained a mixture of both TnT and TnI, peaks 5 and 6 proved to be homogenous TnI peaks. The rabbit skeletal fractions did not exhibit this kind of cross-contamination under the same conditions on this column. Alternatively, when these complexes were separated by anion-exchange chromatography, it was found that no cross-contamination occurred in any of the fractions collected as judged by gel electrophoresis and by reinjection of the isolated peaks under the same conditions as they were collected. Therefore, to avoid problems of contamination in the purification of preparative amounts of these variants, the purification scheme devised employed (1) separation by anion-exchange chromatography to isolate pure C and the individual T variants, and (2) reinjection of the flow through peak from the anion-exchange column on a cation-exchange column with subsequent separation and isolation of two different forms of TnI from this column.

Since the publication of our HPLC method for the purification of the troponin subunits, two papers have appeared in the literature which report sequences of bovine cardiac TnT[18] and TnI.[21] However, only two forms of TnT were reported isolated. This was consistent with other recent publications delineating variants of TnT from chicken and rabbit cardiac muscle.[22,23] Furthermore, the bovine cardiac TnI which was isolated was found to be homogenous.

The sequencing results demonstrated that the variants of bovine cardiac TnT resulted from differences which reside in the N-terminal region of the sequence and was attributable to deletion of residues 15 to 19 (Glu-Ala-Ala-Glu-Glu) in T2. This deletion results in a change in the "net charge" from −8 to −5 when comparing T1 and T2, respectively. Our results show that we were able to isolate three distinct forms of bovine cardiac TnT as well as two forms of TnI; however, we did not identify the differences among the forms. Since our isolation method is based mainly on charge differences, we must assume that these differences are either sequence related or the result of post-translational modifications such as phosphorylation or deamidation. We cannot, however, rule out non-specific hydrophobic interactions which may contribute to the results we observe. However, as can be seen in the 2-D gel results in Figure 1g and 1h, bovine cardiac TnT consists of less well resolved components than rabbit skeletal T. It is possible, therefore, that the open column chromatographic methods used in the isolation of the variants for sequencing were of insufficient resolving power to separate the I and T variants isolated by our HPLC approach. Since the bovine cardiac TnT variants arise from the deletion of charged residues, the method best suited to the separation of such variants would be based on IEC. The methods used in the sequence papers of Leszyk et al.,[18,21] however, employ open-column IEC which may be of sufficient resolving power to isolate only the most abundant protein isoforms. This was followed by protein digestion and reversed-phase purification of the individual peptides used for the sequencing studies. It is possible, therefore, that peptides from isoforms other than the ones which are reported were missed, especially if these isoforms occurred as minor components (ca. 10%) in the overall mixture and failed to separate due to the poor resolving power of the method.

IV. APPLICATION OF ION-EXCHANGE HPLC SEPARATIONS TO OTHER PROTEIN ISOFORMS

The diagnosis of renal transplant rejection,[24] bilary obstruction,[25] and some forms of cancer,[25] are now possible by monitoring the presence of isoenzymes using ion-exchange HPLC methodology. These methods have proved very valuable for quick and reproducible analysis of these conditions in the clinical setting. One additional area where cation-exchange HPLC, in particular, has proved invaluable is in the study and diagnosis of hemoglobinopathies.

There are at present 485 (Sept., 1986) variants of hemoglobin of known structure.[26] A majority of the hemoglobin variants that are known were detected because of abnormal electrophoretic behavior. Electrophoretic screening has in the past been the most common method, along with isoelectric focusing and open-column chromatographic procedures, for the identification of variants. A comparative method using electrophoresis on citrate agar and cellulose acetate has been used to identify 102 variants. When electrophoresis on cellulose acetate (pH 8.6) and on citrate agar (pH 6.0 to 6.5) is coupled with electrophoresis of globin chains in 6 M urea, identification of most variants can be made. These electrophoretic methods, although sensitive, are labor intensive and require lengthy run times. A recent review by Huisman[27] covers the separation of hemoglobins and hemoglobin chains by HPLC. It has been found that separation of the hemoglobin variants by cation-exchange HPLC has become very popular in recent years. These HPLC ion-exchange methods have become attractive because they are rapid, accurate, and are easily automated so that multiple sample analysis is facilitated. The methods described employ CM300 analytical columns (250 × 4.1 mm I.D.) using a sodium acetate gradient (0 to 0.15 M) in 0.03 M Bis-Tris and 0.0015 M potassium cyanide, pH 6.4. It was interesting to note that variants with identical substitutions (Glu→Lys) but located in different positions of the β chain were eluted at different retention times, indicating that the chromatographic properties of these variants are greatly influenced by the local environment of the substitution. The method proved useful not only for the separation of variants based on point mutation but also variants resulting from post-translational modifications of the heme protein through glycosylation. It was shown, for instance, that high levels of glycosylated hemoglobin are found in older populations of red cells and that the level of glycosylation increases with age. The work presented demonstrates that separation and quantitation of many hemoglobin variants is best achieved on cation-exchange HPLC columns. The data presented also show that, as a general rule, the higher the positive charge on the protein, the longer the retention time. However, there are anomalies. These result from substitutions occurring within a helical structure. These isoforms are eluted faster than those which result from substitutions in non-helical portions of the hemoglobin molecule. This would be consistent with the argument posed earlier referring to the "preferred binding domains", since incorporation of the extra positive charge(s) in a helical region may not be as readily available for interaction with the cation-exchange surface as charge(s) placed in a more flexible region of the molecule. These arguments will be valid only when running the separation of the native hemoglobin structure under benign buffer conditions. Full expression of these extra charge(s) would occur in the presence of denaturants such as urea or guanidine hydrochloride. Consequently, it was shown that HPLC identification of hemoglobin isoforms is facilitated by locating the point mutations within the two structural units in the molecule (i.e., the α and β chains) by using denaturing solvent conditions to examine the retention times of the indavidual chains.

ACKNOWLEDGMENTS

This work was supported by the Medical Research Council of Canada and by equipment grants from the Alberta Heritage Foundation for Medical Research.

REFERENCES

1. **Ebashi, S., Kodama, A., and Ebashi, F.,** Troponin: preparation and physiological function, *J. Biochem. (Tokyo),* 64, 465, 1968.
2. **Greaser, M.L. and Gergely, J.,** Reconstitution of troponin activity from three protein components, *J. Biol. Chem.,* 246, 4226, 1971.
3. **Greaser, M.L. and Gergely, J.,** Purification and properties of the components from troponin, *J. Biol. Chem.,* 248, 2125, 1973.
4. **Potter, J.O.,** Preparation of troponin and its subunits, *Methods Enzymol.,* 85, 241, 1982.
5. **Chong, P.C.S. and Hodges, R.S.,** A new heterobifunctional cross-linking reagent for the study of biological interactions between proteins, *J. Biol. Chem.,* 256, 5064, 1981.
6. **Straparans, I., Takahashi, H., Russell, M.P., and Wantanabe, S.,** Skeletal and cardiac troponins and their components, *J. Biochem. (Tokyo),* 72, 723, 1972.
7. **Byers, D.M., McCubbin, W.D., and Kay, C.M.,** Hydrodynamic properties of bovine cardiac troponin, *FEBS Lett.,* 104, 106, 1979.
8. **Burtnick, L.D., McCubbin, W.D., and Kay, C.M.,** The isolation and characterization of the ATPase inhibitory protein (TN-I) from bovine cardiac muscle, *Can. J. Chem.,* 53, 1207, 1975.
9. **Cachia, P.J., Van Eyk, J., McCubbin, W.D., Kay, C.M., and Hodges, R.S.,** Ion-exchange high-performance liquid chromatographic purification of bovine cardiac and rabbit skeletal troponin subunits, *J. Chromatogr.,* 343, 315, 1985.
10. **Gusev, N.B., Barskaya, N.V., Verin, A.D., Duzhenkova, I.V., Khuchua, Z.A., and Zeltova, A.D.,** Some properties of cardiac troponin T structure, *Biochem. J.,* 213, 123, 1983.
11. **Cachia, P.J., Van Eyk, J., Chong, P.C.S., Taneja, A., and Hodges., R.S.,** Separation of basic peptides by cation-exchange high-performance liquid chromatography, *J. Chromatogr.,* 266, 651, 1983.
12. **O'Farrell, P.Z., Goodman, H.M., and O'Farrell, P.H.,** High resolution two-dimensional electrophoresis of basic as well as acidic proteins, *Cell,* 12, 1133, 1977.

13. **Laemmli, U.K.,** Cleavage of structural proteins during the assembly of the head of bacteriophage T4, *Nature,* 227, 680, 1970.
14. **Wilkison, J.M., Moir, A.J.G., and Waterfield, M.D.,** The expression of multiple forms of troponin in chicken-fast-skeletal muscle may result from differential splicing of a single gene, *Eur. J. Biochem.,* 143, 47, 1984.
15. **Giometti, C.S., Anderson, N.G., and Anderson, N.L.,** Muscle protein analysis 1. High resolution two-dimensional electrophoresis of skeletal muscle proteins for analysis of small biopsy samples, *Clin. Chem.,* 25, 1877, 1979.
16. **Murakami, U. and Uchida, K.,** Two-dimensional electrophoresis of troponin complex with nonequilibrium pH gradient-sodium dodecyl sulphate polyacrylamide slab gel, *J. Biochem. (Tokyo),* 95, 1577, 1984.
17. **Hirabayashi, T.,** Two-dimensional gel electrophoresis of chicken skeletal muscle proteins with agarose gels in the first dimension, *Anal. Biochem.,* 117, 443, 1981.
18. **Leszyk, J., Dumaswala, R., Potter, J.D., Gusev, N.B., Verin, A.D., Tobacman, L.S., and Collins, J.H.,** Bovine cardiac troponin T: amino acid sequences of the two isoforms, *Biochemistry,* 26, 7035, 1987.
19. **Kopaciewicz, W., Rounds, M.A., Fausnaugh, J., and Regnier, F.E.,** Retention model for high-performance ion-exchange chromatography, *J. Chromatogr.,* 266, 3, 1983.
20. **Burke, T.W.L., Mant, C.T., Black, J.A., and Hodges, R.S.,** Strong cation-exchange high-performance liquid chromatography of peptides: effect of non-specific hydrophobic interactions and linearization of peptide retention behaviour, *J. Chromatogr.,* 476, 377, 1989.
21. **Leszyk, J., Dumaswala, R., Potter, J.D., and Collins, J.H.,** Amino acid sequence of bovine cardiac troponin I, *Biochemistry,* 27, 2821, 1988.
22. **Cooper, T.A. and Ordahl, C.P.,** A single cardiac troponin T gene generates embryonic and adult isoforms via developmentally regulated alternate splicing, *J. Biol. Chem.,* 260, 11140, 1985.
23. **Pearlstone, J.R., Carpenter, M.R., and Smillie, L.B.,** Amino acid sequence of rabbit cardiac troponin T, *J. Biol. Chem.,* 261, 16795, 1986.
24. **Nicot, G., Lachatre, G., Gonnet, C., Dupuy, J.-L., and Valette, J.-P.,** Rapid assay of N-acetyl-β-D-glucosaminidase isoenzymes in urine by ion-exchange chromatography, *Clin. Chem.,* 33, 1796, 1987.
25. **Schoenau, E., Herzog, K.H., and Boehles, H.-J.,** Liquid chromatographic determination of isoenzymes of alkaline phosphatase in serum and tissue homogenates, *Clin. Chem.,* 32, 816, 1986.
26. **Lehmann, H.,** Human hemoglobin variants in *Hemoglobin: Molecular, Genetic and Clinical Aspects,* Dyson, J., W. B. Saunders, Philadelphia, 1986, chap. 10.
27. **Huisman, T.H.J.,** Separation of hemoglobins and hemoglobin chains by high-performance liquid chromatography, *J. Chromatogr.,* 418, 277, 1987.
28. **Guo, D., Mant, C. T., Taneja, A.K., Parker, J.M.R., and Hodges, R.S.,** Prediction of peptide retention times in reversed-phase high-performance liquid chromatography. 1. Determination of retention coefficients of amino acid residues of model synthetic peptides, *J. Chromatogr.,* 359, 499, 1986.

ASSESSMENT OF STEROID RECEPTOR POLYMORPHISM BY HIGH-PERFORMANCE ION-EXCHANGE CHROMATOGRAPHY

Ronald D. Wiehle and James L. Wittliff

I. INTRODUCTION

Separation of various forms of steroid hormone receptors on ion-exchange columns has been used widely to investigate the physical and functional parameters of these proteins. This methodologic approach is excellent in terms of distinguishing species of the receptor which are "activated", that is, able to bind nuclei or DNA tightly. Beyond this application, we have shown that ion-exchange chromatography using high-performance liquid chromatography techniques (HPIEC) also reveals multiple molecular forms (isoforms) of both the estrogen (ER) and progesterone (PR) receptors and can provide insights into ligand-receptor-complex structure and its stability. Our laboratory also has used HPIEC as a tool to study changes in these receptor isoforms in different tissues of the same species and during differentiation of a single tissue as well as in neoplastic tissues. Moreover, the speed, reproducibility, recovery, and ease of operation allow this method to be included in any scheme employing multiple methods of protein analysis and purification.

II. HORMONE BINDING ASSAYS

Steroid hormone receptor levels are usually determined by *in vitro* binding assays requiring the preparation of a crude cytosolic extract and its incubation with radiolabeled hormone analogs. The details of these methods are given elsewhere.[1,2] The precise quantitation of ER and PR sites in tissues and tumors is made by analyzing binding data from titrations by the method of Scatchard[3] or from competition/inhibition assays. These types of studies are excellent in terms of disclosing the number of binding sites, the affinity constant of the interaction, and even the homogeneity of the binding site, but provide little or no information about the polymorphism of the receptor molecule, i.e., the number and variety of the subunits themselves.

One early attempt to define more precisely steroid hormone receptor polymorphism simultaneously with receptor quantitation employed sedimentation velocity centrifugation on continuous sucrose density gradients (SDG). This discloses multiple isoforms of the receptors on the basis of their sedimentation coefficients.[4] However, this technique is not well suited to the analysis of large numbers of tumor samples which need to be analyzed for clinical purposes, chiefly because of the relatively long times required for the centrifugation step and the number of tumor samples which can be processed daily. The substitution of the vertical tube rotor technology for classical swinging bucket rotor centrifugation is an improvement in terms of speed and increased sample number at the expense of resolution.

High-performance liquid chromatographic methods of analysis were initially investigated to overcome limitations of SDG centrifugation for the determination of receptor polymorphism. We have demonstrated that HPLC methods have applications beyond this original intention in that these techniques provide new types of information previously not considered. The balance of this review will focus on the ways HPIEC has been used in our laboratory to explore certain aspects of estrogen and progesterone receptors mentioned above. HPIEC was one of the first modes we developed for steroid receptor characterization which retained many of the biological properties of these labile regulatory proteins. Important to all of these studies are the physical characteristics of the ion-exchange columns themselves.

III. PHYSICAL CHARACTERISTICS OF HIGH-PERFORMANCE ION-EXCHANGE COLUMNS

Our use of high-performance ion-exchange columns stemmed not only from other HPLC protein work but also from the need to enhance the resolution and applicability of ion-exchange chromatography as a method. We had long recognized that DEAE or DE52 column materials are capable of separating more than one species of the estrogen receptor[5] and that this method uncovered relationships between charged receptor isoforms not readily given by size-exclusion or sedimentation velocity centrifugation, both of which separate species essentially according to size.[4]

A typical separation of the ionic forms of the estrogen receptor from human breast cancer on an open column of DE52 indicated that, in general, three ionic species were detectable, a flow-through, a low salt peak, and a high salt species. The column recovery was 60 to 80% in our experience. When the same preparation of receptor was applied to an high-performance AX 1000 (weak anion-exchange) HPIEC column and eluted with a gradient of sodium phosphate,[6] the recovery was > 90% and the ionic isoform distribution was different (Figure 1). In this case, only a very small amount of radioactive material was eluted as flow-through and major peaks were recovered in a low salt region (50 to 100 mM phosphate) and in a high salt portion of the gradient. Moreover, the radioactivity in the low salt region appeared to correspond to multiple peaks. The same sample applied to the AX 300 HPIEC column, a column of similar packing material but different porosity (i.e., 300 rather than 1000 Å), gave a profile analogous to the other HPIEC column, except that the individual species were eluted later. The recovery given in Figure 2 was 94%.

In general, as the pore size of column matrix is reduced, the ionic strength required for the elution of the same species increases. Table 1 is a distillation of a study performed by Jung and Wittliff[7] concerned with the effect of pore size and column length on the separation parameters of a new type of ion-exchange resin from CliniChrom, Inc. (Lafayette, IN). In order to simplify the chromatogram, the estrogen receptor was preserved by the inclusion of 10 mM sodium molybdate in the incubation buffer; under these conditions, the receptor complex is eluted from the column as a single peak. The length of the column made little difference in terms of elution position since the molybdate-stabilized estrogen receptor isoforms were eluted at nearly the same position from each pair of ion-exchangers regardless whether a column of 10 or 25 cm in length was employed. In actuality, the 10 cm × 4.6 mm I.D. column showed somewhat broader peaks from that of the 25 cm × 4.6 I.D. column. Results obtained using molybdate-stabilized progesterone receptor were essentially identical (data not shown). These data are in contrast to traditional open column ion-exchange chromatography on supports such as DEAE-cellulose, where resolution increases as the diameter-to-length ratio of the matrix is increased.[8]

On the other hand, the pore size of column matrix material has a major impact on the elution characteristics of the molybdate-stabilized ER. The difference noted between columns of pore size 300 and 1000 Å was even more dramatic for the CliniChrom columns than that found earlier for the AX series.[6] A number of reports have indicated a relationship between protein elution characteristics and pore size.[9-11] Regnier[12] estimated that no more than 5% of the total surface area of porous column material is outside the pore. Thus, the macromolecule must enter the pore

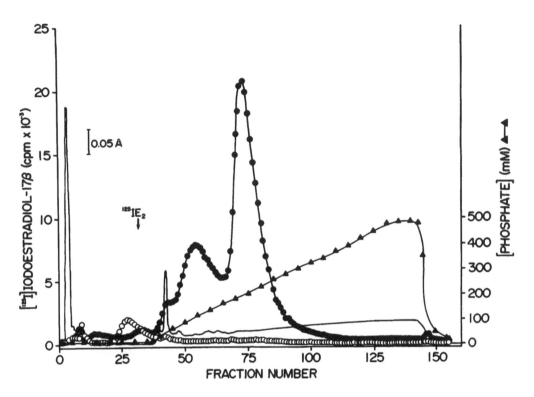

FIGURE 1. HPIEC separation of ionic forms of the estrogen receptor from human breast cancer tissue on a Synchropak AX 1000 weak anion-exchange column (250 × 4.6 mm I.D., 10-μm particle size, 1000-Å pore size; SynChrom, Lafayette, IN). Cytosol identical with that used in the experiments illustrated in Figures 1 and 3 was used. The cytosol was prepared and incubated in the presence (O) or absence (●) of 500-fold excess DES. DES (diethylstilbeserol) is an estrogen analog. Elution was performed at 1.0 ml/min using a gradient of potassium phosphate (▲) at pH 7.4. The elution of the labeled ligand alone was previously determined under identical conditions and is marked with an arrow. The recovery of radioactivity from the column was 91% for the aliquot of cytosol incubated in the absence of DES. A tracing of species absorbing at 280 nm is given by the continuous line. (From Wiehle, R. D. and Wittliff, J. L., *J. Chromatogr.*, 297, 313, 1984. With permission.)

to interact maximally with the reactive groups of the stationary phase. It seems reasonable that the greater number of contact points between the receptor proteins and the stationary phase afforded by the AX 1000 column leads to this enhanced interaction and affects the amount of salt needed for elution. Gooding and Schmuck[10] observed that proteins smaller than 100 kDa were well resolved on columns of 300-Å pore size, whereas proteins in excess of $M_r = 200,000$ are more suitable for analysis on columns with pores of under 1000 Å pore size. The receptor exists in a complex of 200,000 to 300,000 in molecular weight,[13] yet is accommodated within the 300-Å pore size column. This suggested to us[1] and to others[14] that, in certain cases, receptor dissociation *in vitro* is observed and it is the individual subunits that are being recovered.

An additional feature of the AX series is apparent from Figures 1 and 2. The highly lipophilic steroid itself is retained to different extents on the AX300 and AX1000 columns. This observation led us to question whether the separation afforded by the AX columns was influenced by the variable hydrophobic natures of the respective matrixes. This line of reasoning led us to explore high-performance hydrophobic interaction chromatography (HPHIC) as a method for the separation of receptor isoforms.[15-18] Interestingly, the CliniChrom columns do not retain steroid to any extent (unpublished results), suggesting that this separation is in no way mixed-mode and indicating that the differences observed between these column materials is due largely to the effects of pore size and ionic strength on the proteins.

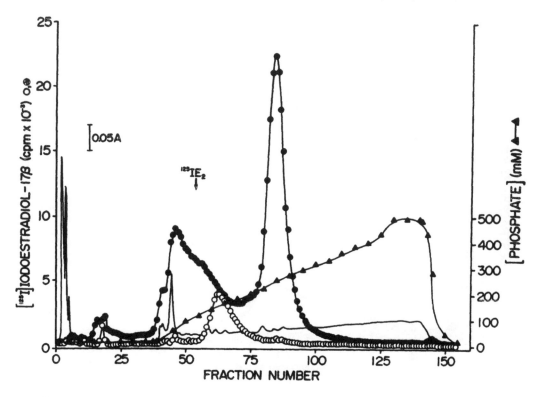

FIGURE 2. HPIEC separation of ionic forms of the estrogen receptors from human breast cancer tissue on a Synchropak AX 300 weak anion-exchange column (250 × 4.1 mm I.D., 6.5 μm, 300 Å). Cytosol was prepared from human breast cancer tissue and incubated in the presence (○) or absence (●) of 500-fold excess DES. Elution was performed at 1.0 ml/min using a gradient of potassium phosphate (▲) at pH 7.4. The elution of the labeled ligand alone was determined previously under identical conditions and is marked with an arrow. The recovery of radioactivity from the column was 94% for the aliquot of cytosol incubation in the absence of DES. A tracing of species absorbing at 280 nm is given by the continuous line. (From Wiehle, R. D. and Wittliff, J. L., *J. Chromatogr.*, 297, 313, 1984. With permission.)

IV. INVESTIGATION OF RECEPTOR COMPLEXITY BY HPIEC

HPIEC has been valuable in deciphering the meaning of the polymorphism detected in the estrogen receptor molecule. As indicated in Figures 1 and 2, the receptor is highly heterogenous in terms of the number of isoforms detected in the absence of molybdate. The multiplicity in the isoforms eluted in the low salt portion of the phosphate gradient is particularly evident in the two chromatograms shown. In both separations, a large peak and a shoulder were present.

The estrogen receptor from both the uterus and the lactating mammary gland of the rat exhibited polymorphism in the low salt portion of the phosphate gradient using the AX 1000 column (Figure 3). The results given in this figure were generated using on-line, flow-through monitoring of radioactivity and ionic strength[19] and demonstrates the reproducibility of the low salt portion of the gradient in the triplicate samples shown, particularly in the case of the uterus (area I). In contrast to data generated by HPIEC, the heterogeneity of receptors from rat uterus is not reflected in the profile as exhibited by sedimentation velocity centrifugation using sucrose density gradients (SDG), where a single nearly symmetrical peak is seen in extracts from uteri at 10 and 19 days postpartum (data not shown). To complicate matters further, estrogen receptors from mammary gland were clearly heterogeneous at these two stages of glandular development, demonstrating that receptor polymorphism was dependent on the day of lactation.

The number of isoforms detected in the high salt portion of the chromatogram is equally

TABLE 1
Elution of Estrogen Receptors in Cytosol from Human Breast Cancer Using Clinichrom Columns

Pore size (Å)		Column of 25 cm					Column of 10 cm		
	n	Phosphate concentration (mM)	Retention time (min)	Recovery (%)		n	Phosphate concentration (mM)	Retention time (min)	Recovery
300	5	150 ± 3	33 ± 2	71 ± 9		5	153 ± 3	32 ± 1	84 ± 9
500	5	102 ± 5	28 ± 1	78 ± 11		6	94 ± 13	26 ± 1	78 ± 19
1000	7	53 ± 4	25 ± 1	91 ± 16		5	50 ± 2	24 ± 1	85 ± 14

Note: Collection results are shown for the elution characteristics of the receptors using different pore sizes and column lengths. Data (mean ± S.D.) represent the phosphate concentrations and the retention times at which the receptors were eluted. The recovery of specific steroid-binding activity is expressed in percent of receptor applied to the column.

From Jung, S. S. and Wittliff, J. L., *J. Chromatogr.*, 425, 293, 1988. With permission.

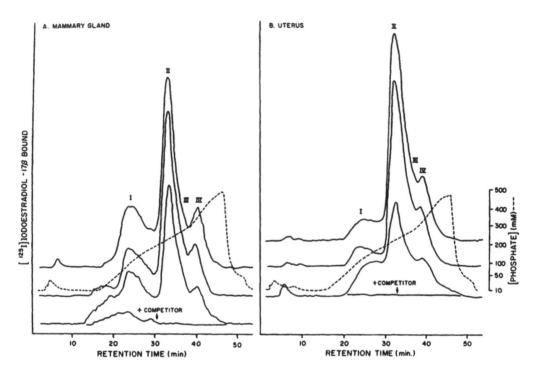

FIGURE 3. HPIEC elution profiles of labeled estrogen receptors from the lactating mammary glands (A) and uteri (B) of rats 19 days postpartum and their reproducibility. Labeled samples (100 µl) from three different rats were analyzed individually on a Synchropak AX 1000 weak anion-exchange column. The elution of radioactivity during chromatography was monitored with a Beckman Model 170 flow-through radioisotope detector. The recording for non-specific binding (i.e., plus competitor) is typical of other assays and therefore shown only once. (From Hyder, S. M., Wiehle, R. D., and Wittliff, J. L., *Comp. Biochem. Physiol.*, 91B, 517, 1988. With permission.)

interesting. Inspection of Figure 3 shows a number of isoforms which are numbered II, III, IV, and which correspond to species eluted at 200 to 250 m*M*, 250 to 300 m*M*, and 300 to 375 m*M* phosphate, respectively. In the uterus and mammary gland of lactating rats at 19 days post partum, species IV is very apparent. However, in both of these tissues at 10 days post partum, species IV is absent and a different peak appears as a shoulder in the area designated III (data not shown). Overall, it appears that, whereas the estrogen receptor from uterus and mammary vary considerably in the number of isoforms seen by SDG, this variability is dependent on the stage of lactation.[20] Moreover, HPIEC revealed that isoform II was always present in both tissues, and isoforms III and IV appeared at distinct stages of lactation, that is, isoform III appeared to be linked to 10 days of lactation and isoform IV to 19 days. Since this is true for both uterus and mammary gland, it seems reasonable to presume that estrogen receptors in both tissues are under the influence of some humoral factor rather than a tissue factor. Moreover, the protease inhibitor, diisopropylfluorophosphate (DFP) which preserves a large form of the estrogen receptor on SDG, does not substantially alter the distribution of isoforms in HPIEC. Thus, HPIEC discloses the tissue- and stage-dependent relationship between ER isoforms unencumbered by the influence of proteases, or at least those which are inhibited by DFP.

Clearly, HPIEC analyses of steroid receptors demonstrates they exhibit a far more complex profile, revealing details about the subunit organization in no way indicated by the SDG method which separates macromolecules essentially on the basis of their size, density, and the buoyancy of the solution. In general, we have observed that SDG and, to a lesser extent, high-performance size-exclusion chromatography (HPSEC) conserve the integrity of the larger ER complex, whereas HPIEC, HPHIC, and high-performance chromatofocusing (HPCF)[21] record the disruption of that complex.[1,2]

V. SUBUNIT COMPOSITION OF LARGE MOLECULAR WEIGHT SPECIES

The dissociation of the large isoforms detected by SDG and HPSEC may be characterized further by a multidimensional approach through the rechromatography of individual species by HPIEC. The progesterone receptor exhibited a definite lability when subjected to a multidimensional approach to isoform analysis using HPIEC and SDG.

As shown by van der Walt and Wittliff, (1986),[22] the 8 to 9 S (S = Svedberg units) form of the progestin receptor (Figure 4A) was isolated as two species from the AX 1000 column, one species in the flow-through region of the column at a position indistinguishable from the position of unbound steroid and the other species within the salt gradient at 60 mM phosphate (Figure 4C). The 4 to 5 S isoform was detected as a flow-through species and as a 60 mM phosphate peak upon rechromatography (Figure 4B). Upon rechromatography of the 60 mM phosphate eluted species, which was derived from the 8 S isoform, a peak appeared in the void volume as a 60 mM phosphate species (Figure 4D). This result is in sharp contrast to the profile seen when only HPIEC was used. In this latter case, cytosol fractionated by HPIEC exhibited a species in the void volume and an isoform eluted at 100 mM phosphate. Upon rechromatography, the moiety eluted at 100 mM phosphate was detected again along with a flow-through component (data not shown). Regardless of whether the void volume material represents free steroid dissociated from the progesterone receptor or distinct ligand-binding species, prepurification of the receptor on SDG appeared to remove or destabilize a non-hormone binding receptor subunit. This led to the appearance of a new isoform (60 mM phosphate peak) which was clearly distinguishable from one of those detected in the original profile. It is speculated that the form lost contained the 90 K heat shock protein known to be associated with nearly all classes of receptor proteins.[23,24] Another possibility is that the missing subunit is a receptor-associated protein kinase, preliminary evidence for which has been found.[25-27]

One surprising feature of this lability is that it has a time-dependent component, at least for the estrogen receptor. Normally, HPSEC profiles of extracts from normal human uterus or endometrial carcinoma exhibit at least two isoforms under low salt conditions, namely, a 60- and 30-Å isoform and the relative proportions of these species vary. We know that the 60-Å form can be converted to a smaller, (28 to 30 Å) form by the action of 400 mM KCl, presumably by the dissociation of the larger complex into subunits. The 30-Å species may also be converted to a (28 to 30 Å) component by the same physical process, although a salt-induced conformational change remains a possibility. In any case, the smaller forms are readily apparent by HPSEC. When the 60-Å species is reapplied to an AX 1000 HPIEC column within 15 min of isolation from the HPSEC system, three species were observed which were eluted at 180 mM, 50 to 70 mM, and in the wash-through (Figure 5C). This pattern was highly similar to the profile of estrogen receptor detected on HPIEC without the prepurification step. However, after 150 min at 4°C, the species recovered from the AX 1000 as a 60-Å moiety was altered, in that the 180 mM phosphate isoform clearly disappeared and the other two forms increased in amount even though the overall specific binding was conserved (Figure 5A). The 30-Å species does not appear to contain a 180 mM phosphate form but, nevertheless, a time-dependent shift in the distribution of isoforms remained (Figure 5B, 5D). Moreover, the heterogeneity of the 30-Å species was also observed in the 60-Å isoform given sufficient separation time. Collectively, we suggested that the receptor is composed of at least two types of hormone-binding subunits of the same size but differing surface charge and that their association is highly unstable.[28] The strict time-dependency of this lability is probably observable only due to the rapid analysis afforded by HPIEC.

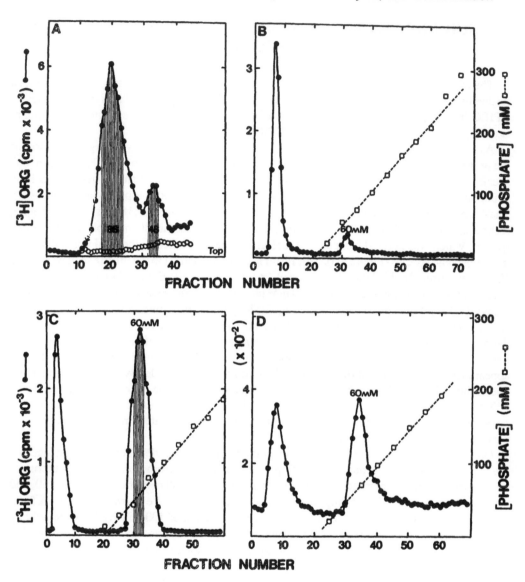

FIGURE 4. Characterization of progestin receptor isoforms from human uterus employing a combination of sucrose density gradient centrifugation and HPIEC. Cytosol prepared from human uterus was incubated with [³H]ORG 2058 in the presence (○) and absence (●) of excess unlabeled compound ORG 2058. Aliquots (100 μl) of incubate were applied to 18 to 20% sucrose density gradients (A). Subsequently, either the 8 to 9 S (fractions 17 to 24) or 4 to 5 S (fractions 32 to 35) receptor isoforms from gradients were pooled, injected and eluted on a Synchropak AX 1000 weak anion-exchange HPIEC column using a gradient of phosphate (□). The 8 to 9 S isoform profile (C); the 4 to 5 S isoform (B). Collection of the specific bound fraction from the first HPIEC run, i.e., fractions 30 to 33 in (C) was rechromatographed on HPIEC and fractionated (D). (From van der Walt, L. A. and Wittliff, J. L., *J. Steroid Biochem.*, 24, 377, 1986. With permission.)

VI. MULTIDIMENSIONAL ANALYSIS INVOLVING HPIEC

A typical flow diagram illustrating multidimensional microanalysis of estrogen receptor from human uterus in the presence and absence of the stabilizing agent, sodium molybdate, is given in Table 2 along with an abbreviated description of the results. By virtue of this sequencial approach, we are able to identify different isoforms in a variety of chromatographic systems. The

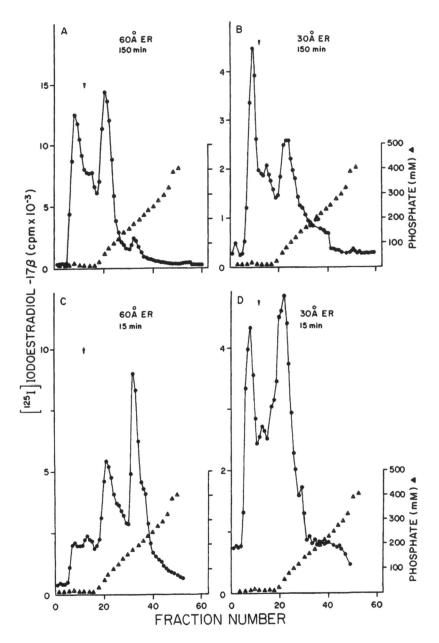

FIGURE 5. HPIEC analyses of estrogen receptor species eluted from HPSEC. Cytosol incubated with [¹²⁵I]iodoestradiol (●), was first eluted with phosphate buffer containing 100 or 400 mM KCl to allow the recovery of 60- or 30-Å species, respectively, from TSK-3000 SW size-exclusion columns. Aliquots (300 to 400 μl) of 60-Å species (A, C) or of 28- 30-Å form (B, D) was applied to AX 1000 weak anion-exchange columns within 15 min (C, D) or after 150 min (A, B) and eluted with a phosphate gradient (▲). The 60- and 28- to 30-Å components contained little nonspecific binding. The arrow indicates elution position of free [¹²⁵I]iodoestradiol. (From Shahabi, N. A., Hyder, S. M., Wiehle, R. D., and Wittliff, J. L., *Steroid Biochem.*, 24, 1151, 1986. With permission.)

inclusion of sodium molybdate in the homogenization incubation and in the running buffers has allowed us to determine that the stable species is 9 to 10 S, 65 Å, and corresponds to a component which is eluted from the AX 1000 column as a single peak at 110 m*M* phosphate. The stability of this entity through various chromatographic procedures is in contrast to the lability of estrogen

TABLE 2
Flow Diagram of Microanalysis of Estrogen Receptors by Multidimensional HPLC

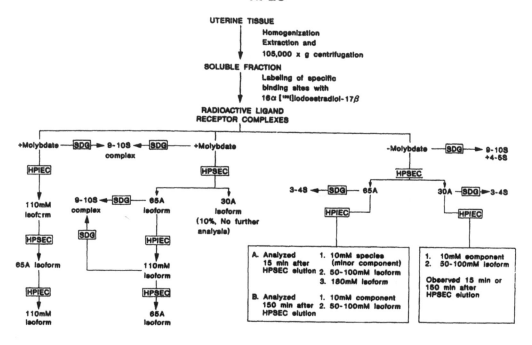

*SDG = Sucrose density gradient analysis
Recoveries in all cases ranged from 75% - 100%

From Hyder, S. M. and Wittliff, J. L., *BioChromatography* 2, 121, 1987. With permission.

receptors in the absence of sodium molybdate, where size and surface charge properties cannot be correlated.[29] Also apparent from Table 2 is the uniqueness of the stable 110 mM phosphate/ 65-Å component, in the sense that this moiety is *not* found in cytosols without molybdate. Thus, the 9 to 10 S form of the estrogen receptor seen in SDG in the presence and absence of molybdate is misleading — not all 9 to 10 S forms are created equal. The precise nature of the interaction of molybdate with steroid hormone receptors is not understood currently.

The multidimensional approach employing HPIEC in combination with other HPLC modes lends itself well to small-scale purification. Table 3 gives the results of two microanalysis schemes employing solely HPSEC and HPIEC. In each case, the original starting material was only 6 mg of a crude preparation of cytosol derived from a human uterus. The major drawback to the procedure is the vanishing small amounts of proteins recovered at the end of the last step. Since radioactive labeling is several orders of magnitude more sensitive than protein determination, the final purification factor is only an estimate. A conservative value for this estimation relies upon the limits of detection of silver stain procedure. Rationalizing that the efficiency of protein recovery in the final step would be the same as that for the first step of the scheme, that is, 1% for the HPSEC and 0.25% for the HPIEC method, would raise the final purification factors to between 7000 and 11,000× for the two schemes shown.

VII. EFFECTS OF LIGAND ON RECEPTOR ISOFORMS

The determination that estrogen and progestin binding activity is distributed over multiple isoforms suggested that species have individual characteristics which could be relevant in

TABLE 3
Purification and Microanalysis of Estrogen Receptors from Human Uterus
Using HPLC in the Presence of Sodium Molybdate

Purification step	Protein (μg/ml)	Total bound (fmol/ml)	Binding activity (fmol/mg protein)	Purification factor
Cytosol	6,000	1,754.8	292	1
HPSEC (65 Å)	70	661.0	9,442	32
HPIEC (100 m*M*)	1	57.5	57,500	197
HPSEC (65 Å)	a	21.1	700,000+	2,400+
Cytosol	6,000	1,754.8	292	+
HPIEC	15	197.0	13,133	45
HPSEC	1.5	30.0	19,800	68
HPIEC	a	12.3	400,000+	3,300+

a   Protein concentration was below the sensitivity of the Bio-Rad colorimetric method. Using an arbitrary value for protein estimated at 0.01 to 0.03 μg/ml based upon the limit of detection by the silver stain procedure and the percent recovery of protein from each of the first steps in the two approaches, a purification of greater than 2000-fold would have been achieved.

From Hyder, S. M., Shahabi, N. A., and Wittliff, J. L., *BioChromatography*, 3(5), 216, 1988. With permission.

hormone response. This has wide implications in the clinical setting through antihormones.[30] Compounds such as tamoxifen and toremifene are used widely in hormone treatment of breast cancer. Others have implied that these non-steroidal synthetic hormones bind to sites distinctly different from the steroid-binding site.[31,32] An important direction in this laboratory has been determining whether the isoforms isolated by virtue of their surface charge properties on HPIEC interact with various steroidal ligands to the same extent.

Using cytosol extracted from rat uteri, Myatt and Wittliff[33] showed that estrogen receptors bound to the synthetic estrogen [16alpha-^{125}I]iodoestradiol-17beta were fully displaceable by estradiol-17beta, the natural ligand (Figure 6). When tamoxifen was used in place of estradiol in a parallel competition series, a similar displacement was seen with the difference due primarily to the concentration of antihormone needed for ligand displacement. Another antiestrogen, 4-OH tamoxifen, gave a similar result but was effective at a lower concentration than tamoxifen (data not shown). Importantly, in no case was one isoform preferentially inhibited, indicating that, at this level of sensitivity, each receptor isoform identified appeared to bind estradiol, iodoestradiol, 4-OH-tamoxifen and tamoxifen, albeit with different affinities. Similarly, progesterone receptor bound with either R5020 or ORG 2058 was found to associate with a single isoform when eluted from AX 1000 columns by a gradient of phosphate.[22] Furthermore, HPCF of PR bound by [^{3}H]R5020 or [^{3}H]ORG 2058 revealed only one isoform eluted at pH 5.6 to 6.1.

VIII. INTERACTION WITH MONOCLONAL ANTIBODIES

Monoclonal antibodies against the estrogen receptor are an important tool in the clinical and research laboratory. The use of antibodies in EIA (enzyme-immunolinked assay) and ERICA (estrogen receptor immunocytochemical assay) kits (Abbott Laboratories, Inc.) has raised a number of questions with regard to their sensitivity and specificity. Equally important is the relevance of the results derived from these tests to those from the multipoint titration assay which has been used clinically for more than 15 years as predictive and prognostic indices of the course of breast cancer.[30] A number of reports have indicated that the numerical values derived from EIA and multipoint titration assays are not equivalent even though a clear correlation exists

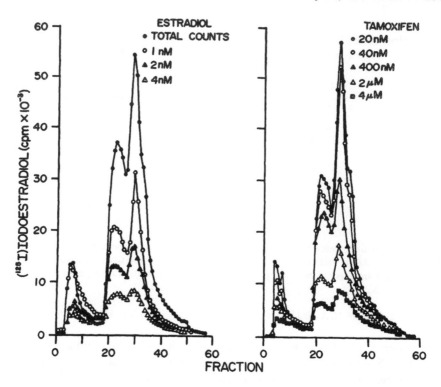

FIGURE 6.   Competition by estradiol or tamoxifen for [^{125}I]iodoestradiol binding to estrogen receptor species. Cytosol was labeled with 2 nM [^{125}I]iodoestradiol in the absence (O) or presence (●) of the indicated concentrations of estradiol or tamoxifen for 24 hr at 4°C prior to HPIEC and eluted with a 10 to 500 mM phosphate gradient. Fractions (1 ml) were collected and radioactivity counted. (From Myatt, L. and Wittliff, J. L., *J. Steroid Biochem.*, 24, 1041, 1986. With permission.)

between the two tests.[34-40] The treatment of patients with breast cancer depends in part upon the results derived from the analysis of the original tumor for the presence of estrogen and progesterone receptors;[4,11] therefore, there is a need to ensure that all ligand binding receptor isoforms are recognized by these antibodies and that those recognized are the biologically relevant ones. The answer to the first question is relatively more straightforward than the latter and has been addressed in our laboratory and elsewhere.[41] Whether the antibody-based assay is able to indicate a better cohort of patients for hormone therapy remains an open question.

The EIA for estrogen receptor is carried out using two monoclonal antibodies. One (D547) is coupled to a polystrene bead and a second (H222) is linked to horseradish peroxidase and utilized for the generation of a quantifiable color. Clearly, each receptor isoform must be recognized by both antibodies and the D547 coupled to the polystyrene bead should bind the receptor in a quantitative manner. We have used various HPLC methodologies to explore these and related questions with the goal of not only verifying the monoclonal assay kits but also using antibodies to discriminate among the various isoforms apparent upon separation.[42-44] To test the interaction of receptor and monoclonal antibody, the two were incubated under various conditions and the isoforms separated by HPLC. When the receptor was incubated with H222 under high salt conditions, the antibody recognized the receptor isoforms nearly quantitatively *under those conditions* as judged by the shift of the steroid-bound receptor to a larger sized species as shown by HPSEC.[43] However, under low salt conditions, the same monoclonal antibody showed selectivity in that one isoform of the receptor which was normally eluted from the HPIEC column with 150 m*M* phosphate is well recognized, whereas another isoform normally eluted at lower salt concentration (50 to 60 m*M*) is not recognized (Figure 7).

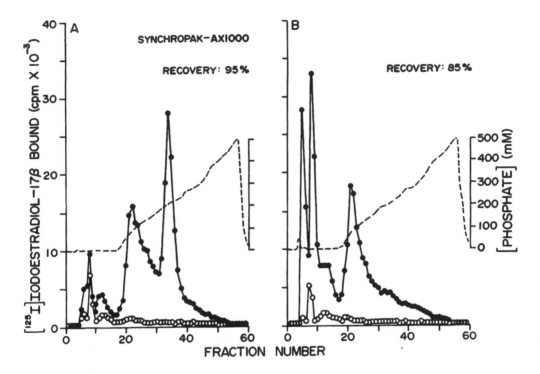

FIGURE 7. HPIEC evaluation of estrogen receptor interaction with H222Sp monoclonal antibody conjugate. In Panel A, [^{125}I]iodoestradiol-17 beta receptor complexes were incubated with non-specific rat IgG overnight in phosphate buffer, then separated on a Synchropak AX 1000 weak anion-exchange column. Panel B was parallel incubation with H222Sp conjugate and separation by HPIEC. (●) Total ligand binding; (○) non-specific binding. (From Brandt, D. W. and Wittliff, J. L., *J. Chromatogr.*, 397, 287, 1987. With permission.)

Interestingly, it is the former isoform which is induced by the action of phorbol esters in cells in culture[45] and may be hyperphosphorylated.[25,46] The D547 monoclonal antibody also recognized the majority of the 150 mM phosphate isoform as detected by HPIEC separation, although a proportion of the component does not appear to be reactive (data not shown). The ability of D547 to recognize the low phosphate moiety could not be evaluated in the same way, since this antibody interacted with both the AX 1000 and Altex DEAE-5PW weak anion-exchange columns.

Curiously, neither H222 nor D547 appeared to recognize the molybdate-stabilized estrogen receptor, since their profile detected by HPIEC in the presence molybdate was identical to control profiles employing non-specific rat IgG.[43] It was of further interest that the monoclonals AC88 or 4F3 which recognize HSP90, a non-steroid-binding subunit of the estrogen receptor, also had no effect on the profile of isolated isoforms. One implication of these data is that the two epitopes are blocked in the presence of molybdate, perhaps by their mutual interaction. Clearly, one can readily assay the estrogen receptor in the presence of molybdate in solution using components of the EIA method (Brandt and Wittliff, 1987).[43]

When estrogen receptor isoforms were first separated by HPIEC and then assessed for their ability to interact with the monoclonal antibodies, all isoforms separated now reacted to the same extent as judged by their association in the EIA assay. However, when the isoforms were labeled with [^{125}I]Iodoestradiol and separated by HPIEC and then reacted with the D547 on the polystyrene bead, the presence of radioactivity on the bead was only 30% of the total that had previously been retained on the bead in the absence of the isoform label.[42] This result is in sharp contrast to the finding that the amount of (mass) the receptor found by EIA is always higher than the amount detected by hormone binding assay (Figure 8).

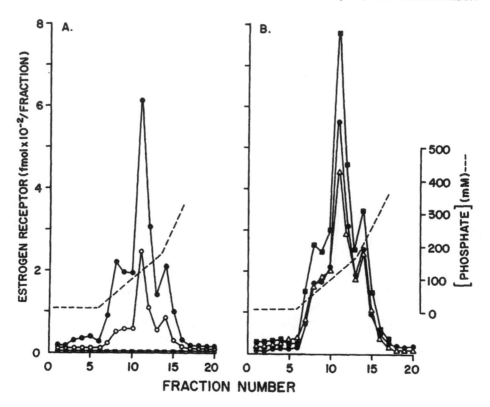

FIGURE 8. Interaction of estrogen receptors from human breast cancer with monoclonal antibodies, following HPIEC. The cytosol from breast cancer was incubated with 3 nM [^{125}I]iodoestradiol in the presence or absence of a 200-fold excess of DES. Elution of samples from the column was performed with a gradient of phosphate. After HPIEC separation, eluates from the column were used for determination of total receptor-bound radioactivity, radioactivity on the MAb-coated bead, and estrogen receptor concentration by ER-EIA. When using the unlabeled sample of an identical cytosol for HPIEC, estrogen receptor concentration in the eluate from the column was determined by ER-EIA. Panel A shows the number of steroid binding sites, determined by radioactivity measurements: (●) [^{125}I]iodoestradiol ratioactivity on the MAb-coated bead; (○) radioactivity of [^{125}I]iodoestradiol isoform complexes in solution; (■) radioactivity of DES-containing sample, labeled with [^{125}I]iodoestradiol. Panel B shows the estrogen receptor concentration determined by EIA: (■) measurements of the unlabeled sample; (●) sample previously labeled with [^{125}I]iodoestradiol; (Δ) the fractions labeled with DES and [^{125}I]iodoestradiol. (From Sato, N., Hyder, S. M., Chang, L., Thais, A., and Wittliff, J. L., *J. Chromatogr.*, 359, 475, 1986. With permission.)

Equally interesting is the finding of Sato et al.[42] that the amount of receptor found by monoclonal antibody in the ER EIA appears to be dependent upon the ligand employed. It was determined that binding capacity determined by EIA decreased in the following order: unliganded > DES >[^{3}H]estradiol 17B > iodoestradiol-bound receptor. These data suggest a number of things: (1) precursors or spent products of the estrogen receptor which do not bind steroid are also recognized by the antibody; (2) antibody recognizes additional subunits within the receptor complex; and (3) the ligand selected dictates epitope, suggesting receptor conformation is involved. A recent report[32] has indicated further that the binding of tamoxifen or 4-OH tamoxifen increased the immunoreactivity of the estrogen receptor for H222 but not for D547 or D75 monoclonal. The resolution of these questions and the significance of their answers for the clinic or understanding of hormonal response await further studies.

ACKNOWLEDGMENTS

Studies in the Hormone Receptor Laboratory related to the development of HPLC methods were supported in part by grants from the American Cancer Society (BC-514B), Phi Beta Psi Sorority, and USPHS grants CA-34211, CA-32102, and CA-42154.

REFERENCES

1. **Wittliff, J.L. and Wiehle, R.D.,** Analytical methods for steroid hormone receptors and their quality assurance, in, *Hormonally Responsive Tumors,* Hollander, V.P., Ed., Academic Press, New York, 1985, 383.
2. **Wittliff, J.L., Wiehle, R.D., and Hyder, S.M.,** HPLC as a means of characterizing the polymorphism of steroid hormone receptors, in *The Use of HPLC in Receptor Biochemistry,* Kerlavage, R.A., Ed., Alan R. Liss, New York, 1989, 155.
3. **Scatchard, G.,** The attractions of proteins for small molecules and ions, *Annals N.Y.A.S.,* 51, 660, 1949.
4. **Wittliff, J.L.,** Steroid-binding proteins in normal and neoplastic mammary cells, in *Methods in Cancer Research,* Vol. XI, Busch, H., Ed., Academic Press, New York, 1975, 293.
5. **Schmidt, T.J. and Litwack, G.,** Activation of the glucocorticoid-receptor complex, *Physiol. Rev.* 62, 1132, 1982.
6. **Wiehle, R.D. and Wittliff, J.L.,** Isoforms of estrogen receptors by high performance ion-exchange chromatography, *J. Chromatogr.,* 297, 313, 1984.
7. **Jung, S.S. and Wittliff, J.L.,** Influence of column length and pore size on high performance ion-exchange chromatography of estrogen and progestin receptors, *J. Chromatogr.,* 425, 293, 1988.
8. **Cooper, T.G.,** *The Tools of Biochemistry,* John Wiley & Sons, New York, 1977, 150.
9. **Unger, K.,** *Porous silica.,* Elsevier, Amsterdam, The Netherlands, 1979, 11.
10. **Gooding, K.M. and Schmuck, M.N.,** Comparison of weak and strong high-performance anion-exchange chromatography, *J. Chromatogr.,* 327, 139, 1985.
11. **Wittliff, J.L.,** HPLC of steroid-hormone receptors, *LC.GC Magazine of Liquid and Gas Chromatography,* 4, 1986, 1092.
12. **Regnier, F.E.,** High-performance ion-exchange chromatography of proteins: the current status, *Anal. Biochem.,* 126, 1, 1982.
13. **Miller, L.K., Tuazon, F.B., Niu, E.-N., and Sherman, M.R.,** Human breast tumor estrogen receptor: effects of molybdate and the electrophoretic analyses, *Endocrinolology,* 108, 1369, 1981.
14. **Madhock, T.C. and Leung, B.S.,** Characterization of uterine estrogen receptors by size exclusion and ion-exchange high-performance liquid chromatography, *Biochem. Biophys. Res. Commun.,* 115, 988, 1983.
15. **Hyder, S.M., Wiehle, R.D., Brandt, D.W., and Wittliff, J.L.,** High-performance hydrophobic interaction chromatography of steroid hormone receptors, *J. Chromatogr.,* 327, 237, 1985.
16. **Hyder, S.M., Sato, N., and Wittliff, J.L.,** Characterization of estrogen receptors and associated protein kinase activity by high-performance hydrophobic-interaction chromatography, *J. Chromatogr.,* 397, 251, 1987.
17. **Hyder, S.M. and Wittliff, J.L.,** High-performance hydrophobic interaction chromatography of a labile regulatory protein: the estrogen receptor, *BioChromatography* 2, 121, 1987.
18. **Wittliff, J.L., Folk, P., Dong, J., and Hyder, S.,** High-performance hydrophobic interaction chromatography of a labile regulatory protein, this publication.
19. **Boyle, D.M., Wiehle, R.D., Shahabi, N.A., and Wittliff, J.L.,** Rapid, high-resolution procedure for assessment of estrogen receptor heterogeneity in clinical samples, *J. Chromatogr.,* 327, 369, 1985.
20. **Hyder, S.M., Wiehle, R.D., and Wittliff, J.L.,** Alterations in estrogen receptor isoforms in the mammary gland and uterus of the rat during differentiation, *Comp. Biochem. Physiol.,* 91B, 517, 1988.
21. **Hutchens, T.W., Wiehle, R.D., Shahabi, N.A., and Wittliff, J.L.,** Rapid analysis of estrogen receptor heterogeneity by chromatofocusing with high performance liquid chromatography, *J. Chromatogr.,* 266, 115, 1983.
22. **van der Walt, L.A. and Wittliff, J.L.,** Assessment of progestin receptor polymorphism by various synthetic ligands using HPLC, *J. Steroid Biochem.,* 24, 377, 1986.
23. **Joab, I., Radong, C., Renoir, J., Bochou, T., Catnelli, M.G., Binart, N., Mester, J., and Baulieu, E.E.,** Common non-hormone binding component in non-transformed chick oviduct receptors of four steroid hormones, *Nature,* 308, 850, 1984.
24. **Sanchez, E.R., Toft, D.O., Schlesinger, M.J., and Pratt, W.B.,** Evidence that the 90 kDa phosphoprotein associated with the untransformed L-cell glucocorticoid receptor is a murine heat shock protein, *J. Biol. Chem.,* 260, 12398, 1985.

25. **Baldi, A., Boyle, D.M., and Wittliff, J.L.,** Estrogen receptors are associated with protein and phospholipid kinase activities, *Biochem. Biophys. Res. Commun.,* 135, 597, 1986.
26. **Garcia, T.T., Bouchou, J.M., Renoir, J., Mester, J., and Baulieu, E.E.,** A protein kinase co-purified with chick oviduct progesterone receptor, *Biochemistry,* 12, 3090, 1986.
27. **Migliaccio, A., DiDomenico, M., Green, S., deFalco, A., Kajtaniak, F., Blasi, F., Chambon, P., and Auricchio, F.,** Phosphorylation on tyrosine of *in vitro* synthesized human estrogen receptor activates its hormone binding, *Mol. Endo.,* 3, 1061, 1989.
28. **Shahabi, N.A., Hyder, S.M., Wiehle, R.D., and Wittliff, J.L.,** HPLC analysis of estrogen receptor by a multidimensional approach, *Steroid Biochem.,* 24, 1151, 1986.
29. **Shahabi, N.A., Hutchens, T.W., Wittliff, J.L., Helmo, S.D., Kirk, M.E., and Nisker, J.A.,** Physiochemical characterization of estrogen receptors from a rabbit endometrial carcinoma model, in *Progress in Cancer Research and Therapy,* Vol. 31, Raynaud, J.P., Bresciani, F., King, R.J.B., and Lippman, M.E., Eds., Raven Press, New York, 1984, 63.
30. **Wittliff, J.L.,** Steroid hormone receptors, in *Methods in Clinical Chemistry,* Pesce, A.J. and Kaplan, L.A., Eds., C.V. Mosby, St. Louis, 1987, 767.
31. **Faye, J.C., Jozan, S., Redueilh, G., Baulieu, E.E., and Bayard, F.,** Physico-chemical and genetic evidence for specific antiestrogen binding sites, *Proc. Natl. Acad., Sci. U.S.A.,* 80, 3158, 1983.
32. **Martin, P.M. Berthois, Y., and Jensen, E.V.,** Binding of antiestrogens exposes an occult and antigenic determinant in the human estrogen receptor, *Proc. Natl. Acad. Sci. U.S.A.,* 85, 2533, 1988.
33. **Myatt, L. and Wittliff, J.L.,** Characterization of non-activated and activated estrogen- and antiestrogen-receptor complexes by high performance ion exchange chromatography, *J. Steroid Biochem.,* 24, 1041, 1986.
34. **Mirecki, D.W. and Jorden, V.C.,** Steroid hormone receptor and human breast cancer, *Lab Med.,* 16, 287, 1985.
35. **Anonymous,** Estrogen receptor determination with monoclonal antibodies, *Cancer Res.,* (suppl) 46, 4231, 1986.
36. **Heubner, A., Beck, T., Grill, H.J., and Pollow, K.,** Comparison of immunocytochemical estrogen receptor assay, estrogen receptor enzyme immunoassay, and radioligand-labeled estrogen receptor assay in human breast cancer and uterine tissue, *Cancer Res.,* (suppl) 46, 4291, 1986.
37. **Nicholson, R.I., Colin, P., Francis, B., Keshra, R., Finlay, P., Williams, M., Elston, C.W., Blaney, R.W., and Griffins, K.,** Evaluation of an enzyme immunoassay for estrogen receptors in human breast cancer, *Cancer Res.,* (suppl) 46, 4299, 1986.
38. **Raam, S. and Urabel, D.M.,** Evaluation of an enzyme immunoassay kit for estrogen receptors measurements, *Clin. Chem.,* 32, 1496, 1986.
39. **Pausette, A., Gustafsson, S.A., Thornblad, A.M., Nordgren, A., Sallstrom, J., Lindgren, A., Sunderlin, P., and Gustafsson, J.A.,** Quantitation of estrogen receptors in seventy-five specimens of breast cancer: comparison between an immunoassay (Abbott ER EIA Monoclonal) and a [³H]estradiol binding assay based on isoelectric focusing in polyacrylamide gel, *Cancer Res.,* (suppl) 46, 4308, 1986.
40. **Pasic, R., Djulbegovic, B., and Wittliff, J.L.,** Comparison of enzymeimmunoassay with the multipoint titration assay for the determination of estrogen receptor and progestin receptor, *J. Clin. Lab. Anal.,* 4, 430, 1990.
41. **Moncharmont, B., Su, J.L., and Parikh, I.,** Monoclonal antibodies against estrogen receptors: interactions with different molecular forms and functions of the receptor, *Biochemistry,* 21, 6916, 1982.
42. **Sato, N., Hyder, S.M., Chang, L., Thais, A., and Wittliff, J.L.,** Interactions of estrogen receptor isoforms with immobilized monoclonal antibodies, *J. Chromatogr.,* 359, 475, 1986.
43. **Brandt, D.W. and Wittliff, J.L.,** Assessment of estrogen receptor-monoclonal antibody interaction by high-performance liquid chromatography, *J. Chromatogr.,* 397, 287, 1987.
44. **Hyder, S.M., Sato, N., Chang, L., Meyer, J.S., and Wittliff, J.L.,** Recognition of estrogen receptor isoforms from human breast cancer by immobilized monoclonal antibodies, *Tumor Diagnostic & Therapie,* 9, 233, 1988.
45. **Baldi, A., Ivanier, He, Y.J., and Wittliff, J.L.,** Influences of Phorbol Esters on Breast Tumor Cells, in *Recent Advances in Chemotherapy,* Anticancer Section 1, Iehigami, J., Ed., University of Tokyo Press, Japan, 1985, 233.
46. **Knabbe, C., Lippman, M.E., Greene, G.L., and Dickson, B.B.,** Phorbol ester induced phosphorylation of the estrogen receptor in intact MCF-7 human breast cancer cells, *Fed. Proc.,* 45, 1899, 1986.
47. **Hyder, S.M., Shahabi, N.A., and Wittliff, J.L.,** Microanalysis of estrogen receptors from human uteri by multidimensional HPLC, *BioChromatography,* 3(5), 216, 1988.

Section V
Reversed-Phase Chromatography

REVERSED-PHASE PACKINGS FOR THE SEPARATION OF PEPTIDES AND PROTEINS BY MEANS OF GRADIENT ELUTION HIGH-PERFORMANCE LIQUID CHROMATOGRAPHY

Ulrich Esser and Klaus K. Unger

I. INTRODUCTION

Reversed-phase HPLC (RPC) has grown to become a mature and well-established technique in the analysis, structure elucidation, and isolation of peptides and proteins.[1-3] Numerous studies have examined the influence of the mobile phase on the retention and selectivity of peptides and polypeptides in RPC.[1,2,4,5] In contrast, few systematic investigations have been performed to establish the role of the type of packing in RPC.[6,7] This paper aims at surveying the most relevant properties of reversed-phase (RP) packings and discusses the impact of these properties on the chromatographic performance of RP systems in peptide and protein separations.

II. THE ROLE OF THE STATIONARY PHASE IN GRADIENT ELUTION RPC OF PEPTIDES AND PROTEINS

A. THE PRINCIPLE OF SEPARATION

In gradient elution RPC, peptides and proteins are retained essentially according to their hydrophobic character. The retention process can be considered as adsorption of the solute at the hydrophobic stationary surface or as a partition between the mobile and the stationary phase.[8]

In the first case, the retention is related to the total interfacial surface of the RP packing and expressed by the adsorption coefficient K_A. The retardation is based on a hydrophobic association between the solute and the hydrophobic ligands of the surface and is described by the solvophobic theory, advocated by Horváth and co-workers.[9] By increasing the solvent strength of the mobile phase, the attractive forces are weakened and the solute is eluted. According to Hearn,[10] the process can be regarded as being entropically driven and endothermic, i.e., both ΔS and ΔH are positive. Through the relation between the solute capacity factor k' and the mobile and stationary phase properties, the solvophobic theory permits prediction of the effect of the organic modifier of the mobile phase, the ionic strength, the ion-pairing reagent, the type of ligand of the RP packing, and other variables on the chromatographic retention of peptides.

The second case assumes a partitioning of the solute between the mobile phase and the stationary phase, the latter being regarded as a hydrophobic bulk phase. This system resembles an n-octanol-water two phase system, where the selectivity is expressed by the partition coefficient P of the solute. There are pro and contra arguments for both retention principles.[8] In essence, the effective process is largely dependent on the molecular organization of the grafted n-alkyl chains of the RP packing in the solvated state and the size and conformation of the peptidic solute.

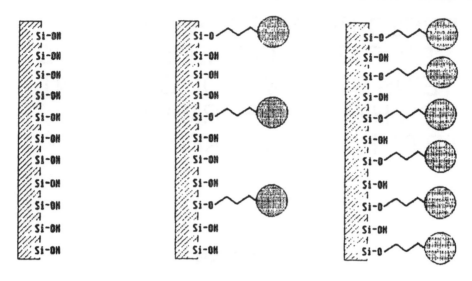

a) native silica   b) half coverage   c) maximum coverage

FIGURE 1. Scheme of a native and silanized silica surface.

B. THE CHARACTERISTIC FEATURES OF RP PACKINGS

RP packings based on silica are manufactured by reacting a suitable porous silica with a silane that contains a hydrophobic ligand, e.g., n-alkyl, phenyl, and cyclohexyl. Depending on the silane and the reaction conditions, the stationary surface is either of monolayer or polymeric type. In the case of a monolayer, the density of ligands varies from a low density up to a complete monolayer (cf. Figure 1). Due to steric restrictions, only half of the original hydroxyl group population of the silica surface is able to react even when maximum coverage is achieved. To avoid an effect of remaining hydroxyl groups on the chromatographic retention of solutes, the silanized materials are subjected to a second silanization with a small reactive silane. The products are termed as endcapped.

When in equilibrium with the mobile phase, the stationary phase of RP packing behaves as a dynamic system. The grafted n-alkyl chains are mobile in the solvated state. The mobility depends on the length of the n-alkyl chain and the ligand density. Furthermore, by the solvation power of the mobile phase, the stationary phase can expand in volume (by increasing the fraction of organic solvent of the mobile phase) and can be compressed (by increasing the water content of the mobile phase).

While silica-based RP packings are sensitive to hydrolytic cleavage of the siloxane bonds when using aqueous mobile phases, cross-linked poly(styrene-divinylbenzene),[11] and porous-graphitized carbon[12] packings offer a superior pH stability particularly at pH > 8. Another alternative is to coat silicas with a layer of a pH-stable polymer, e.g., poly(butadiene).[13] The retention behavior of polypeptides and proteins chromatographed on hydrophobic surfaces and using mobile phases rich in organic solvent might be influenced by the induction of conformational changes during separation.[14]

III. THE PROPERTIES OF THE RP PACKING

Silica-based RP packings are characterized by the type of hydrophobic ligand and by the carbon load. The latter can amount to about 20% (w/w). Typical ligands are n-octadecyl (C_{18}),

TABLE 1
Relation Between the Pore Diameter, pd, of Silica and the Specific Surface Area, a_s

pd (nm)	a_s (m²/ml)
5	300
10	150
30	50
100	15
400	4
1.5 μm non-porous	4

n-octyl (C_8), n-butyl (C_4), and phenyl. Cyanopropyl-grafted silicas have also been employed where the propyl spacer functions as the hydrophobic ligand. The pore size of the RP packing must be adapted to the size of solute. As a rule of thumb, the solute molecular diameter has to be 10 times smaller than the average pore diameter of the packing to avoid restricted diffusion of the solute and to provide an accessible stationary phase. Simply expressed, it means that 15 nm (150-Å) pore-size silicas enable the separation of peptides of < 10 kDa. For proteins, larger pore-size packings of a mean pore diameter of 50 or 100 nm (500 or 1000 Å, respectively) are required. By using nonporous particles the restricted diffusion is eliminated.

With increasing pore diameter, pd, the surface area, a_s, of the packing decreases as indicated in Table 1. As a consequence, the retention and the mass loadability decrease proportionally. While the mass loadability on a 10 nm (100 Å) RP packing amounts to about 100 mg/g, the loadability drops down to about a few milligrams per gram on a 400 (4000 Å) nm packing, provided the surface is totally accessible.

Another decisive variable of the RP packing is the particle size, dp. The smaller the dp, the better the resolution of the column, due to the inverse relationship of the column plate number, N, and the particle size:

$$N = \text{const} \frac{L}{dp} \tag{1}$$

where L is the column length. Thus, a trend to smaller particles can be observed. In practice, analytical separations are performed on columns 100 to 250 mm in length and 4 or 4.6 mm inner diameter packed with 5-μm particles.

IV. THE IMPACT OF STATIONARY PHASE PROPERTIES ON CHROMATOGRAPHIC RESOLUTION OF PEPTIDES AND PROTEINS

A. GENERAL ELUTION CONDITIONS

RPC of peptides and proteins is carried out on n-alkyl bonded silicas with an acidic low ionic strength mobile phase as a starting eluent and applying a gradient of organic solvent. The starting eluent is usually 10 mM acid, e.g., trifluoracetic acid (TFA), phosphoric acid, perchloric acid, and hepta-fluorobutyric acid (HFBA) of pH 2 to 5. The displacing eluent contains an organic solvent such as methanol, acetonitrile, 1- and 2-propanol. Linear gradients are run up to 70% (v/v) of the organic solvent within 10 to 60 min. Typically, flow rates are in the range of 0.5 to 2.0 ml/min with changes of the organic modifier of 0.2 to 5%/min.

For proteins, the lower range of flow-rate is best because of the low diffusion rate of the biomacromolecules in the chromatographic column. For peptides, higher flow rates can be used.

The retention of peptides and proteins decreases with increasing solvent strength. Retention is further regulated by the ionic strength when employing buffers such as triethylammonium phosphate, ammonium acetate and pyridinium acetate, and by ion pairing reagents. Separations are often reported to be carried out at elevated temperatures up to 80°C.[15] Higher temperatures favor a rapid mass transfer of the solute and result in an improved column efficiency. On the other hand, higher temperatures may not be favorable for a number of proteins with regard to their stability.

B. THE CHOICE OF THE RP PACKING

When selecting a RP-column for the analysis and isolation of peptides and proteins, several factors have to be taken into account. Retention of peptides and proteins follows their relative hydrophobicities and increases in the sequence of $C_{18} > C_8 > C_4 > C_2$ bonded silicas, at the same ligand density, for a given solute at a constant eluent composition. Hydrophobic peptides are preferably chromatographed on cyanopropyl bonded silicas. While peptides are capable of intercalating the stationary phase of the RP packing, proteins have restricted access due to their larger size and are assumed to interact at the outer interface. For a series of proteins, retention times were found to be lowest on a C_2 bonded silica and reached a plateau for the C_4, phenyl and C_8 phase. A slight increase in retention was seen for the C_{18} bonded silica.[16] Similar results are reported by Cohen et al.[17] When residual hydroxyl groups of the silica surface are accessible to peptide solutes, polar adsorption effects between the ionogenic groups of the peptide and the surface contribute to retention in addition to hydrophobic interactions. These often result in poor peak shape, poor resolution, and loss of recovery, particularly for basic solutes. In this aspect, exhaustively silanized silicas pose fewer problems than those with lower ligand density.

Retention studies of peptides and proteins on n-alkyl bonded silicas with various pore sizes in the range between 6 and 100 nm (60 and 1000 Å; respectively)[18,19] clearly indicated the phenomenon of limited access of solute to the stationary phase. Thus, it is mandatory to choose an appropriate pore size of the RP packing. Polypeptides bigger than 15 kDa are preferably resolved on RP silicas of 30 or 100 nm (300 or 1000 Å, respectively) pore size.[20] In order to avoid restricted access, slow mass transfer, and exclusion phenomena, nonporous silanized silicas or nonporous poly(styrene-divinylbenzene) gels are the RP packing of choice.[15,16] Although the external surface of such particles is relatively low (about 3 m²/ml of packing at 2-μm particle size), the retention power is sufficient for proteins and even for peptides.[15]

The type of n-alkyl silica is also fundamentally important with respect to column life-time and recovery. The shorter the chain length, the more sensitive the anchoring siloxane bonds are to hydrolytic cleavage. Furthermore, n-alkyl silicas with a polymer type of layer offer a superior stability compared to those of monolayer type. The problem of chemical stability of the stationary phase can be circumvented by using poly(stryrene-divinylbenzene)-based packings which are stable over the entire pH range.[15,20] Such packings are also available with graduated pore sizes. With longer n-alkyl chain of the RP packing, the strength of hydrophobic interaction increases which can result in significant losses of biorecovery for certain proteins. Putting all considerations together, it is advisable to use C_4 and C_8 bonded silicas for protein separations and C_8 and C_{18} bonded silicas for peptide separations.

Other important considerations include the choice of the particle diameter of the RP packing, the column length, and the flow rate. For analytical separations, 5-μm particles seem to be the best compromise between chromatographic resolution and speed. Larger particles are preferred for preparative isolation. 3-μm porous and 2-μm nonporous silicas are best suited for fast separations using columns of less than 50 mm in length.[15,16] When large plate numbers are required, columns in excess of 100 mm length are employed. Long columns together with low flow-rates, however, often yield poor biorecovery of proteins due to the long residence time the biopolymer spends in the column. Shorter columns should then be used. The relations between chromatographic resolution and particle diameter of packing, flow-rate of mobile phase, column length, etc., are discussed in depth by Snyder and Stadalius.[22]

V. CONCLUSION

This short review summarizes the most important aspects of RP packings for the separation of peptides and proteins. RP packings with a designed stationary phase, with graduated pore size and particle size are commercially available to adapt to specific requirements for rapid separation, analysis, and isolation of peptides and proteins. Models have been developed which permit prediction of the effect of stationary phase properties and column dimensions on a given separation problem and help to optimize a separation with regard to resolution, speed, and biorecovery.

REFERENCES

1. **Hancock, W.S. and Harding, D.R.K.**, Review of separation conditions, in *CRC Handbook of HPLC for the Separation of Amino Acids, Peptides and Proteins*, Vol 2, Hancock, W.S., Ed., CRC Press, Boca Raton, 1984, 3, 303.
2. **Hearn, M.T.W.**, High performance liquid chromatography and its application in protein chemistry, in *Advances in Chromatography*, Vol. 20, Giddings, J.C., Brown, P., and Cazes, J., Eds., Marcel Dekker, New York, 1982, 87.
3. **Hearn, M.T.W.**, High performance liquid chromatography of peptides, in *High Performance Liquid Chromatography: Advances and Perspectives*, Vol. 3, Horváth, Cs., Ed., Academic Press, New York, 1980, 1.
4. **Hearn, M.T.W. and Grego, B.**, Solvophobic considerations for the separation of unprotected peptides on chemically bonded hydrocarbonaceous stationary phases, *J. Chromatogr.*, 203, 349, 1981.
5. **Regnier, F.E.**, HPLC of proteins, peptides and polynucleotides, *Anal. Chem.*, 55, 1299 A, 1983.
6. **Cooke, N.H.C., Archer, B.G., O'Hare, M.J., Nice, and E.C., Capp, M.**, Effects of chain length and carbon load on the performance of alkyl-bonded silicas for protein separations, *J. Chromatogr.*, 255, 115, 1983.
7. **Nice, E.C., Capp, M.W., Cooke, N., and O'Hare, M.J.**, Comparison of short and ultrashort-chain alkylsilane bonded silicas for the high-performance liquid chromatography of proteins by hydrophobic interaction methods, *J. Chromatogr.*, 218, 569, 1981.
8. **Dill, K.A.**, The mechanism of solute retention in reversed-phase liquid chromatography, *J. Phys. Chem.*, 91, 1980, 1987.
9. **Melander, W. and Horváth, Cs.**, Reversed-phase chromatography, in *High Performance Liquid Chromatography-Advances and Perspectives*, Vol. 2, Horváth, Cs., Ed., Academic Press, New York, 1980, 114.
10. **Hearn, M.T.W.**, High performance liquid chromatography of peptides, in *High Performance Liquid Chromatography: Advances and Perspectives*, Vol. 3, Horváth, Cs., Ed., Academic Press, New York, 1980, 99.
11. **Su, S.-J., Grego, B., Niven, B., and Hearn, M.T.W.**, Pairing ion effects in the reversed-phase high performance liquid chromatography of peptides in the presence of alkylsulfonates, *J. Liq. Chromatogr.*, 4, 1745, 1981.
12. **Smith, N.W. and Brennan, D.**, The use of porous graphitic carbon in pharmaceutical analysis with particular reference to its stereoselectivity, 11th International Symposium on Column Liquid Chromatography, June 19—24, 1988, Washington, D.C., paper W-P-339.
13. **Schomburg, G.**, Stationary phases in high performance liquid chromatography. Chemical modification by polymer coating. *LC/GC International Magazine*, 1, 34, 1988.
14. **Hearn, M.T.W., Hodder, A.N., and Aguilar, M.I.**, Investigations on the effects of chromatographic dwell in the reversed-phase high performance liquid chromatographic separation of proteins, *J. Chromatogr.*, 327, 47, 1985.
15. **Maa, Y.F. and Horváth, Cs.**, Rapid analysis of proteins and peptides by reversed-phase chromatography with polymeric micropellicular sorbents, *J. Chromatogr.*, 445, 71, 1988.
16. **Jilge, G., Janzen, R., Giesche, H., Unger, K., Kinkel, J.N., and Hearn, M.T.W.**, Retention and selectivity of proteins and peptides in the gradient elution on non-porous monodisperse 1.5 μm reversed-phase silicas, *J. Chromatogr.*, 397, 71, 1987.

17. **Cohen, K.A., Grillo, S.A., and Dolan, J.W.,** A comparison of protein retention and selectivity on large-pore reversed-phase HPLC columns, *LC, Liq. Chromatogr. HPLC Mag.,* 3, 37, 1985.
18. **Hearn, M.T.W. and Grego, B.,** Evaluation of the effect of several stationary phase parameters on the chromatographic separation of polypeptides on alkylsilicas, *J. Chromatogr.,* 282, 541, 1983.
19. **Engelhardt, H. and Mueller, H.,** Optimal conditions for the reversed-phase chromatography of proteins, *Chromatographia,* 19, 77, 1984.
20. **Lewis, R.V., Fallon, A., Stein, S., Gibson, K.D., and Udenfriend, S.,** Supports for reversed-phase high performance liquid chromatography of large proteins, *Anal. Biochem.,* 104, 153, 1980.
21. **Tweeten, K.A. and Tweeten, T.N.,** Reversed-phase chromatography of proteins on resin-based wide-pore packings, *J. Chromatogr.,* 359, 111, 1986.
22. **Snyder, L.R. and Stadalius, M.A.,** High performance liquid chromatography separations of large molecules: a general model, in *High Performance Liquid Chromatography: Advances and Perspectives,* Vol. 4, Horváth, Cs., Ed., Academic Press, New York, 1980, 195.

COMMERCIALLY AVAILABLE COLUMNS AND PACKINGS FOR REVERSED-PHASE HPLC OF PEPTIDES AND PROTEINS

Kerry D. Nugent

I. INTRODUCTION

Reversed-phase high-performance liquid chromatography (RPC) has rapidly become the most widely used tool for the separation, purification, and analysis of peptides.[1] In the case of proteins, RPC has suffered from problems associated with denaturation including loss of activity, poor recoveries, broad misshapen peaks, and ghosting.[2,3] Although some of these problems can be directly related to the column being used,[4] many of the problems can be overcome by selective optimization of extra-column variables including sample pretreatment and mobile phase and hardware considerations.[5]

Loss of biological activity is a major problem for biochemists trying to purify proteins for further use and this limits the role of RPC in the preparative purification of large proteins. Recent work has shown that for smaller proteins (less than 30,000 daltons), some denaturation effects are reversible and therefore RPC can be successfully employed in the purification of many recombinant proteins.[6] RPC is also a very powerful tool for the micropreparative purification of proteins and peptides prior to microsequencing,[7] and is a widely used analytical tool in biotechnology for process monitoring, purity, and stability determinations and peptide mapping studies.[8]

In this chapter, we will try to review all of the column variables (packing support, bonded phase, pore size, particle size, and column dimensions) as well as the extra-column variables (sample pretreatment, mobile phase composition, pH, temperature, additives, gradient profile, and hardware) in the hope that users will be better able to utilize commercially available RPC columns to solve their current peptide and protein separations problems.

II. RPC COLUMNS

As with small molecule HPLC, there are a large number of reversed-phase columns for use with peptides and proteins. Since many of these columns have different characteristics, it is often difficult for the user to select the best column for his specific needs or to translate a separation developed on a column from one source to a similar column from a different manufacturer. In trying to select the most appropriate column for a new application or for broad applicability, the user should consider such column variables as support, bonded phase, pore size, particle size, and dimensions. Other factors to consider include loading capacity, reproducibility (run to run on one column as well as column to column variability from the same manufacturer), and overall column stability and lifetime for the application being considered.

A. PACKING SUPPORT

Most commercial RPC columns for peptides and proteins are silica based. Silica offers good mechanical stability and allows a wide range of selectivities through the bonding of various phases, but silica-based RPC columns can be troublesome depending on the source of silica, bonding chemistry employed, and the specific application conditions required. Although it is well known that silica-based RPC columns are not stable at basic pHs, recent reports have shown that in the acidic mobile phases typically used for protein and peptide separations, the bonded phase is slowly dissolved from the base silica.[9,10] This degradation process affects not only the reproducibility, stability, and lifetime of the column, but can also affect recoveries and modify selectivities due to uncovering of the silica surface. In preparative purification of recombinant proteins, this process can also affect final product with silica and bonded phase contaminants.

Because of the problems and limitations associated with silica-based columns, polymer-based columns have been gaining popularity for protein and peptide separations by RPC. Polymer-based supports are usually stable from pH 1 to 14 making them more widely applicable and easier to clean than silica-based supports. Although most polymer-based columns are still inferior to their silica-based counterparts in terms of mechanical strength, selectivity and efficiency, newer materials such as PLRP-S from Polymer Labs show superior performance in complex protein separations.[4] Several investigators have also recently shown that the use of polymer-based columns at high pH (8 to 11) can offer unique selectivities as a complement to classic acidic mobile phases.[11,12] These advantages, coupled with superior stability, longer column life, and better reproducibility make polymer-based columns a good first choice for complex protein separations.

B. BONDED PHASES

A wide variety of bonded phases exist for separation of peptides and proteins by RPC but the most common are C4, C8, and C18. In general, the shorter chain phases are preferable for more hydrophobic samples and the longer chain phases are better for hydrophilic samples. In well-controlled systems, such as the synthetic peptides shown in Figure 1,[13] only small differences can be seen in selectivity between columns from different sources. In more complex systems, such as very large hydrophobic proteins or very basic proteins, significant differences can be seen between different column sources.[4,14]

C. PORE SIZE

In the reversed-phase separation of peptides, a pore size of 300 Å has become the most popular, but the literature is full of good separations on 60 to 150 Å columns. For larger peptides and proteins between 10,000 and 100,000 daltons, 300 Å materials seem to offer the best overall performance.[4] For very large proteins, newer columns with 1000 and 4000 Å pores may offer better performance.[4] In working with complex samples, if all other column variables are equal, it is better to use a larger pore size to ensure that restricted diffusion or exclusion from the pores is not encountered for very large proteins.

D. PARTICLE SIZE

As with small molecules, theory predicts that performance should go up as particle size goes down. This is generally the case for most commercially available RPC columns being used for protein and peptide separations, but limitations may be encountered which will influence your choice of particle size. For analytical separations, most columns have 5 to 8 micron particles and little difference in performance can be seen over this range. Some manufacturers are now offering columns with particles of 2 to 3 microns, but small gains in efficiency may quickly be outweighed by higher column backpressures, greater susceptibility to plugging and shorter column life. For semipreparative applications (10 to 500 mg), particle sizes of 10 to 20 microns offer the best compromise between cost and performance, while for preparative separations (> 1 g), 20 to 40 micron particles are often the best choice.

281

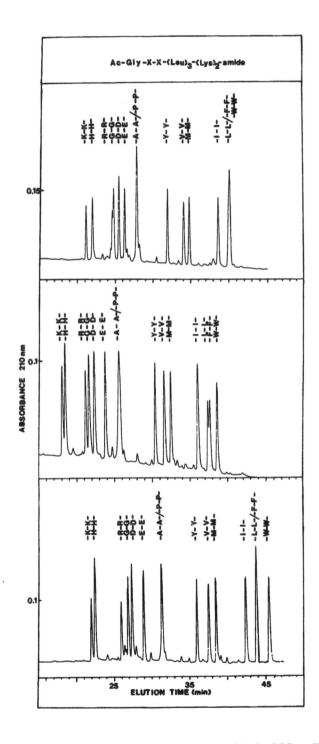

FIGURE 1. Separation of an identical mixture of synthetic model peptides by RPC at pH 2.0. Columns: Top, SynChropak RP-4 (C₄), 250×4.1 mm I.D., 6.5-μm particle size, 300-Å pore size, ca. 7.5% carbon loading (SynChrom, Linden, IN); middle, SynChropak RP-8 C₈, 250 × 4.1 mm I.D., 6.5 μm, 300 Å, ca. 7.5% carbon loading; bottom, Whatman Partisil 5 C₈, 250×4.6 mm I.D., 5 μm, 60 Å, ca. 9% carbon loading (Whatman, Clifton, NJ). Conditions: linear AB gradient (1% B/min), where Eluent A is 0.1% aq. TFA and Eluent B is 0.1% TFA in acetonitrile; flow-rate, 1 ml/min; 26°C. The model peptides are based on the sequence, Ac-Gly-X-X-(Leu)₃-(Lys)₂-amide, where position X is substituted by the 20 amino acids found in proteins. (From Guo, D., Mant, C. T., Taneja, A. K., Parker, J. M. R., and Hodges, R. S., *J. Chromatogr.*, 359, 499, 1986. With permission.)

E. COLUMN DIMENSIONS

The two criteria to consider when choosing a column size are the required efficiency and sample loading capacity. For peptides and small proteins (< 10,000 daltons), modest gains in efficiency can be seen with increasing column length.[4,22] For complex systems such as tryptic peptide maps, columns of 100 to 250 mm are routinely used. In the case of larger proteins (> 10,000 daltons), column length contributes very little to efficiency, but longer columns do have an adverse affect on protein recovery suggesting that a 20 to 50 mm column is preferable.[4,15]

Once the appropriate length has been chosen, the selection of column diameter is based on required sample capacity. Columns with internal diameters (I.D.) of 1 or 2 mm are best suited for submicrogram sample levels to minimize sample loss and maximize sensitivity (although commercial HPLC hardware is often difficult to use with these small columns). For analytical applications in the microgram to low milligram range, columns of 3 to 4.6 mm I.D. are best suited (although Hearn and co-workers have reported the possibility of loading up to 200 mg [3.3 μmol of a 60 kDa protein] of some proteins on a 250 × 4.6 mm I.D. column;[16] interestingly, Hodges et al. successfully loaded up to 84 mg [76 μmol] of a mixture of 10-residue peptides on a column of similar dimensions[19]). Preparative applications require column diameters of 10 to 100 mm.

III. EXTRA-COLUMN VARIABLES

Once the proper RPC column has been selected for a specific application, there are still several other areas which must be considered if one is to attain the best peptide or protein separation achievable. The effects of these extra column variables, which include sample pretreatment, mobile phase conditions and HPLC hardware considerations, are very dependent on the application in question. Since these variables may contribute as much or more than the column to the success of a particular separation problem, their impacts should be well understood.

A. SAMPLE PRETREATMENT

As stated earlier, one of the problems with RPC for peptides and proteins is the requirement for harsh conditions (low pH, organic solvents, high temperatures, hydrophobic stationary phases,[20] etc.) which are denaturing and not very comforting for most protein chemists. Except in the case of the user who requires preparative isolation of large proteins with recovery of biological activity, it is appropriate to take advantage of these and other denaturing conditions to aid in RPC of proteins.[5] By using an appropriate mixture of chaotropic agents (protein solubilizing reagents such as urea and guanidine hydrochloride), surfactants (detergents such as sodium dodecyl sulfate, Triton X-100, Zwittergen, etc.), ion-pairing acids, and organic solvents, one can solubilize and partially denature the sample components of interest and maintain their solubility when injected onto a reversed-phase column. This prevents on-column denaturation which leads to aggregation that can plug columns, lower recoveries, and increase ghosting. This sample pretreatment has also been found to improve dramatically peak shape for very large proteins, as they are no longer slowly denaturing and/or changing conformational states during a RPC separation.[5] These effects are less dramatic with peptides, but proper solubilization prior to analysis is important for any biological sample.

It should be noted that most new RPC columns require some "conditioning" prior to attaining reproducible performance. This conditioning is related to a variable loss of certain proteins and peptides that varies from column to column and application to application. It is suspected that these losses are due to reactive sites (metal surfaces or exposed silica) as well as partially restricted pores which need to be blocked prior to routine analysis. It has been found that two to three injections of a moderate level (0.1 μg/g of packing) of well solubilized peptides and proteins which cover the molecular size range of interest is sufficient to condition most columns.[5]

B. MOBILE PHASE

The proper selection of mobile phase conditions for the reversed-phase separation of peptides and proteins is even more critical than for small molecule HPLC. Most RPC separations carried out today use an acetonitrile/water/trifluoroacetic acid (TFA) gradient which is ideal for most types of samples. Figure 2 shows the dramatic impact of changing organic modifier in the separation of synthetic peptides,[13] and similar effects have been found with complex protein separations.[5]

Guo and co-workers found that the type and concentration of ion-pairing acid used as a modifier in peptide separations had the largest impact on selectivity.[17] This same observation has been verified for complex protein samples.[5] Other mobile phase modifiers including surfactants, chaotropes, and reducing agents have been explored with only marginal improvements seen except for the use of zwitterionic detergents when separating large hydrophobic proteins.[5]

C. HARDWARE CONSIDERATIONS

Although most commercially available HPLC systems can be used for reversed-phase separation of peptides and proteins, there are several variables under the user's control which can greatly improve performance. The HPLC system should be properly maintained and serviced to ensure reliable performance from the system components. The mobile phases should be of high purity, particulate free, and well degassed (ideally using a helium overlay rather than continuous sparging to ensure uniformity of mobile phase from run to run).

Temperature control is critical to good performance. The column should be maintained within 1°C for good retention time repeatability. Recent work has shown that the use of elevated temperatures is beneficial for many proteins, but caution must be used when working with silica-based columns where lifetime is reduced by elevated temperatures.[5] Elevated temperatures for peptide separations can improve resolution, but generally does not warrant the risk of peptide degradation or increased acceleration of column aging at elevated temperatures.[21]

Proper gradient conditions are also critical in optimizing separations for proteins and peptides by RPC. The gradient time, range, and shape are all important and must be optimized for a given sample. Often, methods development software such as Drylab G from LC Resources can be useful for this purpose. Other important parameters include gradient delay time and column reequilibration time. Gradient delay time is a function of system volume and should be minimized, because it will not only increase total analysis times but may have an adverse affect on protein recoveries.[5] Column reequilibration time is an extremely important variable. The time must be sufficient to allow the initial mobile phase to equilibrate in the internal pores of the column packing (at least 3 column volumes) and should be exactly repeatable (by using an auto sampler) if precise retention times are required.

IV. OPTIMIZED CONDITIONS

Once the user has optimized all of the hardware, mobile phase and sample preparation variables for a given system and selected the proper column for his application, he can truly enjoy the rewards of these efforts. In the case of complex proteins, this optimization process can result in dramatic improvements in performance as illustrated in Figure 3. One can see that the effects of optimization on peak shape are more significant for larger proteins, but this optimization process yielded more repeatable results with better recoveries for a wide range of proteins.[4,5] The bottom panel of Figure 3 shows that the polymer-based column gave much better peak shape and fewer conformers with β-galactosidase than the silica-based column.

V. FUTURE TRENDS

The role of RPC in the separation of peptides and proteins will continue to be an important

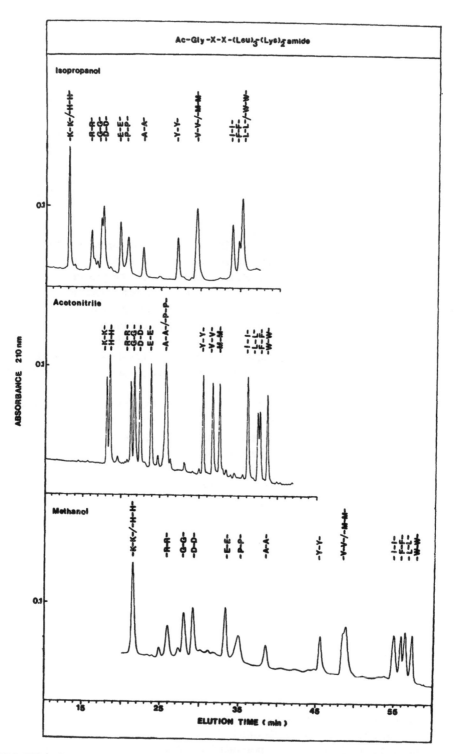

FIGURE 2. Effect of organic modifier on RPC of an identical mixture of synthetic model peptides at pH 2.0. Column: SynChropak RP-8 C_8, 250 × 4.1 mm I.D., 6.5-μm particle size, 300-Å pore size, ca. 7.5% carbon loading (SynChrom, Linden, IN). Conditions: linear AB gradient (1% B/min), where Eluent A is 0.1% aq. TFA and Eluent B is 0.1% TFA in isopropanol (top), acetonitrile (middle), or methanol (bottom); flow-rate, 1 ml/min; 26°C. Description of peptides is given in Figure 1. (From Guo, D., Mant, C. T., Taneja, A. K., Parker, J. M. R., and Hodges, R. S., *J. Chromatogr.*, 359, 499, 1986. With permission.)

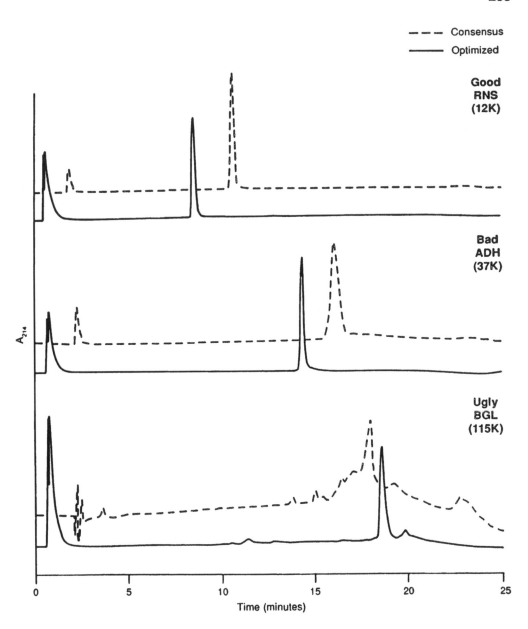

FIGURE 3. RPC of proteins. Columns: broken trace, 250 × 4.6 mm I.D. column containing silica-based Vydac C_4 packing (The Separations Group, Hesperia, CA), 5-μm particle size, 300-Å pore size; solid trace, 50 × 1.0 mm I.D. column containing polymer-based PLRP-S packing (Polymer Labs., Church Stretton, U.K.), 5 μm, 300 Å. Conditions (broken trace): linear AB gradient (3% B/min, from 5 to 80% B in 25 min), where Eluent A is 0.1% aq. TFA and Eluent B is 0.1% TFA in acetonitrile (pH 2.1); flow-rate, 0.6 ml/min; 21°C. Conditions (solid trace):linear AB gradient (2% B/min, from 10 to 60% B in 25 min), where Eluent A is 0.1% aq. TFA and Eluent B is 0.1% TFA in acetonitrile (pH 2.1); flow-rate, 30 μl/min; 60°C. The broken trace and solid trace show the results using "concensus" and "optimized" conditions, respectively (see Reference 4 for details). The chromatograms show the results obtained for (top) "good" (ribonuclease A), (middle) "bad" (alcohol dehydrogenase), and (bottom) "ugly" (β-galactosidase) proteins. "Good" denotes proteins giving symmetrical bandshapes; "bad" denotes proteins giving broad bands and/or somewhat asymmetric peaks; "ugly" denotes proteins giving broad, multiple and/or misshaped bands (see Reference 4). (From Nugent, K. D., Burton, W. G., Slattery, T. K., and Johnson, B. F., *J. Chromatogr.*, 443, 381, 1988. With permission.)

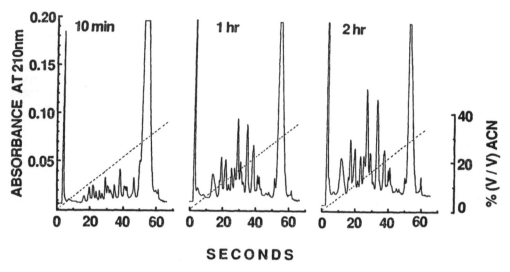

FIGURE 4. Chromatograms illustrating the time course of cytochrome c degradation by trypsin (10 min, 1 and 2 hr). Column: micropellicular C_8 silica, 30 × 4.6 mm I.D., 2 μm particle size. Conditions: Eluent A is 0.1% aq. TFA and Eluent B is 0.1% TFA in H_2O acetonitrile/water (95:5, v/v); flow-rate, 5 ml/min; 80°C. Aliquots (10 μl) taken from the reaction mixture (digest contains 5 μg reduced, S-carboxymethylated cytochrome c) at the incubation time indicated, were mixed with 10 μl of Eluent A and injected directly. (From Kalghatgi, K. and Horváth, Cs., *J. Chromatogr.*, 443, 343, 1988. With permission.)

one. In the area of recombinant protein process development and manufacturing, it will continue to grow as a powerful purification tool for smaller proteins. With the development of newer materials and associated HPLC hardware, it has the potential to be used as an on-line screening tool. An example of this was recently demonstrated by Horváth and co-workers[18] in the application of micropellicular columns to rapid peptide analysis as shown in Figure 4.

RPC will continue to be the workhorse for analytical separations of proteins and peptides. The newer polymeric materials should help to improve the reliability of separations and improve column life. New advances will be made in the microanalysis area where femtomole (10^{-15} mol) to picomole (10^{-12} mol) levels of peptides and proteins need to be separated for quantification and identification. This will undoubtedly involve the newer nonporous materials packed in microbore or capillary columns. Significant advances in HPLC hardware will be required to take full advantage of these applications.

REFERENCES

1. **Regnier, F.E.,** Peptide mapping, *LC/GC*, 5(5), 392, 1987.
2. **Pearson, J.D., Lin, N.T., and Regnier, F.E.,** in *High Performance Liquid Chromatography of Proteins and Peptides,* Hearn, M.T.W., Wehr, C.T., and Regnier, F.E., Eds., Academic Press, New York, 1982, 81.
3. **Hearn, M.T.W.,** in *High Performance Liquid Chromatography-Advance and Perspectives,* Vol. 3, Horváth, Cs., Ed., Academic Press, New York, 1986, 87.
4. **Burton, W.G., Nugent, K.D., Slattery, T.K., and Summers, B.R.,** Separation of proteins by reversed-phase HPLC: I. Optimizing the column, *J. Chromatogr.*, 443, 363, 1988.
5. **Nugent, K.D., Burton, W.G., Slattery, T.K., and Johnson, B.F.,** Separation of proteins by reversed-phase HPLC. II. Optimizing sample pretreatment and mobile phase conditions, *J. Chromatogr.*, 443, 381, 1988.
6. **Lu, X.M., Benedek, K., and Karger, B.L.,** Conformational effects in the HPLC of proteins: further studies on the reversed-phase chromatographic behavior of ribonuclease A., *J. Chromatogr.*, 359, 19, 1986.

7. **Yuen, S., Hunkapiller, M.W., Wilson, K.J. and Yuan, P.M.,** Applications of tandom microbore LC and sodium dodecyl sulfate-polyacrylamide gel electrophoresis/ electroblotting in microsequence analysis, *Anal. Biochem.*, 168, 5, 1988.

8. **Kunitani, M., Johnson, D., and Snyder, L.R.,** Model of protein conformation in the reversed-phase separation of interleukin-2-muteins, *J. Chromatogr.*, 371, 313, 1986.

9. **Glajch, J.L., Kirkland, J.J., and Köhler, J.,** Effect of column degradation on the reversed-phase HPLC separation of peptides and proteins, *J. Chromatogr.*, 384, 81, 1987.

10. **Sagliano, J., Flold, T.R., Hartwick, R.A., Dibussolo, J.M., and Miller, N.T.,** Studies on the stabilization of reversed-phases for liquid chromatography, *J. Chromatogr.*, 443, 155, 1988.

11. **Williams, R.C., Vasta-Russell, J.F., Glajch, J.L., and Golebiowski, K.,** Separation of proteins on a polymeric fluorocarbon HPLC column packing, *J. Chromatogr.*, 371, 63, 1986.

12. **Maa, Y. and Horváth, Cs.,** Rapid analysis of proteins and peptides by reversed-phase chromatography and polymeric micropellicular sorbents, *J. Chromatogr.*, 445, 71, 1988.

13. **Guo, D., Mant, C.T., Taneja, A.K., Parker, J.M.R., and Hodges, R.S.,** Prediction of peptide retention times in reversed-phase HPLC. I. Determination of retention coefficients of amino acid residues of model synthetic peptides, *J. Chromatogr.*, 359, 499, 1986.

14. **Ferris, R.J., Cowgill, C.A., and Traut, R.R.,** Separation of ribosomal proteins from *Escherichia coli* and rabbit reticulocytes using reversed-phase HPLC, *Biochemistry,* 23, 3434, 1984.

15. **Pearson, J.D.,** HPLC column length designed for submicrogram scale protein isolation, *Anal. Biochem.*, 152, 189, 1986.

16. **De Vos, F.L., Robertson, D.M., and Hearn, M.T.W.,** Effect of mass loadability, protein concentration and n-alkyl chain length on the reversed-phase HPLC behavior of bovine serum albumin and bovine follicular fluid inhibin, *J. Chromatogr.*, 392, 17, 1987.

17. **Guo, D., Mant, C.T., and Hodges, R.S.,** Effects of ion-pairing reagents on the prediction of peptide retention in reversed-phase HPLC, *J. Chromatogr.*, 386, 205, 1987.

18. **Kalghatgi, K. and Horváth, Cs.,** Rapid peptide mapping by HPLC, *J. Chromatogr.*, 443, 343, 1988.

19. **Hodges, R.S., Burke, T.W.L., and Mant, C.T.,** Preparative purification of peptides by reversed-phase chromatography. Sample displacement mode *versus* gradient elution mode, *J. Chromatogr.*, 444, 349, 1988.

20. **Lau, S.Y.M., Taneja, A.K., and Hodges, R.S.,** Effects of high-performance liquid chromatographic solvents and hydrophobic matrices on the secondary and quaternary structure of a model protein. Reversed-phase and size-exclusion high-performance liquid chromatography, *J. Chromatogr.*, 317, 129, 1984.

21. **Guo, D., Mant, C.T., Taneja, A.K., and Hodges, R.S.,** Prediction of peptide retention times in reversed-phase HPLC. II. Correlation of observed and predicted peptide retention times and factors influencing the retention times of peptides, *J. Chromatogr.*, 359, 519, 1986.

22. **Snyder, L.R. and Stadalius, M.A.,** HPLC separations of large molecules: a general model, *High Performance Liquid Chromatography,* 4, 195, 1988.

REQUIREMENT FOR PEPTIDE STANDARDS TO MONITOR COLUMN PERFORMANCE AND THE EFFECT OF COLUMN DIMENSIONS, ORGANIC MODIFIERS, AND TEMPERATURE IN REVERSED-PHASE CHROMATOGRAPHY

Colin T. Mant and Robert S. Hodges

I. INTRODUCTION

The excellent resolving power of reversed-phase chromatography (RPC) has resulted in its becoming the predominant mode of high-performance liquid chromatography (HPLC) for peptide separations in recent years. Prior to the use of a reversed-phase column, it is important to assess its performance capabilities and the logical approach to assessing reversed-phase column performance is to employ HPLC standards. Peptide standards are best suited for monitoring peptide retention in RPC, since it is preferable to use standards that are structurally similar to the sample of interest and that presumably interact with the reversed-phase packing in a similar manner. Reversed-phase peptide standards allow the researcher to:

1. Identify non-specific interactions of peptides with the reversed-phase packing (discussed later in this book)
2. Monitor column performance (efficiency, selectivity, resolution)
3. Monitor run-to-run reproducibility of peptide separations
4. Monitor column aging
5. Monitor the effect of mobile phase composition and pH;
6. Monitor the effect of gradient-rate, flow-rate (discussed later in this book) and temperature
7. Monitor the effect of changing column dimensions (length and diameter)
8. Monitor the effect of variations in ligand density and n-alkyl chain length of the bonded phase
9. Monitor the effect of particle size and pore size
10. Compare packing materials (identical bonded phases from different manufacturers; different batches of an identical packing from the same manufacturer)
11. Monitor sensitivity of detector response
12. Monitor effect of instrument variations

This article describes a series of five commercially-available synthetic decapeptide standards[1] and demonstrates their usefulness in monitoring the effects of column dimensions, column aging, organic modifier, and temperature on peptide separations in RPC.

TABLE 1
Sequences of Reversed-Phase Peptide Standards

Peptide standard	Peptide sequence
S1	Arg-Gly-│Ala-Gly│-Gly-Leu-Gly-Leu-Gly-Lys-amide
S2	Ac-Arg-Gly-│Gly-Gly│-Gly-Leu-Gly-Leu-Gly-Lys-amide
S3	Ac-Arg-Gly-│Ala -Gly│-Gly-Leu-Gly-Leu-Gly-Lys-amide
S4	Ac-Arg-Gly-│Val -Gly│-Gly-Leu-Gly-Leu-Gly-Lys-amide
S5	Ac-Arg-Gly-│Val -Val│-Gly -Leu-Gly-Leu-Gly-Lys-amide

Note: Ac denotes $N\alpha$-acetyl; amide = $C\alpha$-amide. Peptide S1 has a free α-amino group. Variations in composition of the peptide analogs are shown between the two vertical lines.

II. DESIGN OF PEPTIDE STANDARDS

The sequences of the peptide standards, S1 to S5, are shown in Table 1. RPC is often used during peptide mapping of proteins, and the 10-residue length of the standards was designed to reflect the average size of cleavage fragments from proteolytic (e.g., tryptic) digests of proteins. The hydrophobicity of the standards increases only slightly between S2 and S5 — between S2 and S3 there is a change from an α-H to a β-CH_3 group, between S3 and S4 there is a change from a β-CH_3 group to two methyl groups attached to the β-CH group, and between S4 and S5 there is a change from an α-H to an isopropyl group attached to the α-carbon — enabling very precise determination of the resolving power of a reversed-phase packing.

The peptide standards are available from Synthetic Peptides Incorporated (Department of Biochemistry, University of Alberta, Edmonton, Alberta, Canada), Pierce Chemical Co. (Rockford, Illinois) and SynChrom (Linden, Indiana).

III. MONITORING REVERSED-PHASE COLUMN PERFORMANCE WITH PEPTIDE STANDARDS

A. EFFECT OF COLUMN DIMENSIONS AND COLUMN AGING

Although excellent resolution of peptide mixtures can be obtained at acidic or neutral pH, the majority of researchers have carried out RPC at pH < 3.0.[2] Most reversed-phase columns employed for peptide separations are silica-based and, apart from the suppression of silanol ionization under acidic conditions (thereby suppressing undesirable ionic interactions with basic residues), silica-based columns are more stable at low pH.

Figure 1 shows the elution profiles of peptide standards S1 to S5 on four silica-based reversed-phase C_{18} columns of varying length, diameter, and degree of use.[3] The four columns contain identical column packings (6.5-μm particle size, 300-Å pore size, ca. 10% carbon loading). The separations were obtained using aqueous trifluoroacetic acid (TFA) to TFA-acetonitrile linear gradients (pH 2.0) at room temperature. In the authors' experience, this is the best initial approach to most peptide separations as well as the best standard approach to assessing the peptide-resolving power of a reversed-phase column.[4] Acetonitrile is the favored organic solvent for most purposes and the ion-pairing properties of TFA are effective in separating complex peptide mixtures. The excellent separations of the peptide standards on all four C_{18} columns demonstrate the columns to be eminently suitable for the resolution of peptide mixtures. In addition, the peptide elution profiles are similar for all four columns, the main difference being a shift in peptide retention times.

The effect of aging on the performance of a column is seen in Figures 1A and B, where the

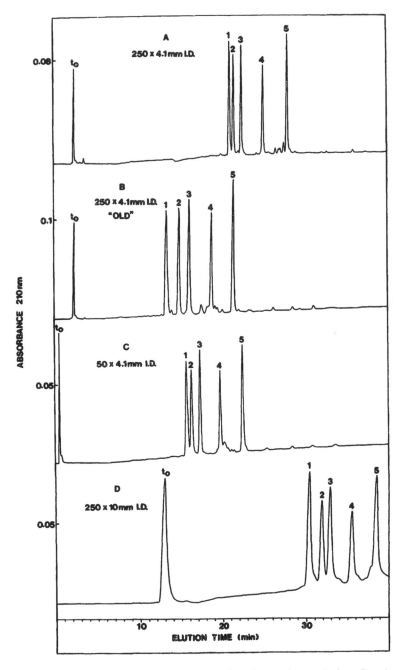

FIGURE 1. Separation of a mixture of five synthetic reversed-phase decapeptide standards on C_{18} columns at pH 2.0. The HPLC instrument consisted of a Spectra-Physics (San Jose, CA) SP8700 solvent delivery system and SP8750 organizer module, combined with a Hewlett-Packard (Avondale, PA) HP1040A detection system, HP3390A integrator, HP85 computer, HP9121 disk drive and HP7470A plotter. Columns: four SynChropak RP-P C_{18} columns (SynChrom, Linden, IN), 6.5-μm particle size, 300-Å pore size, ca. 10% carbon loading; 250 × 4.1 mm I.D. (A); 250 × 4.1 mm I.D., used extensively over a period of 4 months (at least 50 runs); (C) 50 × 4.1 mm I.D.; (D) 250 × 10 mm I.D. Mobile phase: linear AB gradient (1% B/min), where Eluent A is 0.1% aq. trifluoroacetic acid (TFA) and Eluent B is 0.1% TFA in acetonitrile; flow-rate, 1 ml/min; 26°C. Peptide mixture dissolved in 0.5% aq. TFA (A to C) or 2% aq. TFA (D); sample volume for D was double that for A to C. Excess TFA was used to dissolve the sample to enable measurement of t_o. The absorbance peak denoted t_o, produced by the excess of TFA concentrations in the sample, represents the elution time for unretained compounds. Peptides 1, 2, 3, 4, and 5 denote peptide standards S1, S2, S3, S4, and S5, respectively. (From Guo, D., Mant, C. T., Taneja, A. K., Parker, J. M. R., and Hodges, R. S., *J. Chromatogr.*, 359, 499, 1986. With permission.)

peptides are retained longer on a new column (Figure 1A, 250 × 4.1 mm I.D.) than a similar column used extensively over a period of months (Figure 1B). This decrease in peptide retention time on the older column is likely caused by a gradual loss of alkyl (octadecyl, in this case) chains with extensive column use, thereby reducing the amount of hydrophobic stationary phase available to interact with peptide solutes. Regular monitoring with the peptide standards enables the researcher to not only gauge the effect of aging on column performance, but also to decide whether the column is still suitable for peptide separations. Figure 1B clearly shows that although this "old" C_{18} column is not as retentive as a new column (Figure 1A), it is still able to separate peptide mixtures satisfactorily. It should be noted that the loss of a n-alkyl chain will have a much more serious effect on the resolution of a peptide mixture when the column is used in a preparative application.

A comparison of Figures 1A (25-cm column) and C (5-cm column) demonstrates that a five-fold difference in column length makes little difference to the separation of the peptide mixture. The excellent peptide resolution obtained on the 5-cm column (Figure 1C) suggests that, for many analytical applications, reversed-phase columns of this length would make perfectly acceptable substitutes for the 25- to 30-cm columns generally used for peptide separations, with subsequent cost-saving to the researcher.

Scale-up of an analytical reversed-phase peptide purification protocol to greater sample loads may require the use of larger columns with increased load capacity. It is then necessary to determine whether the peptide resolving capability of the larger column is sufficient for the researcher's needs. This is particularly important when peptide impurities are eluted close to the peptide of interest, as is often the case for synthetic peptide purifications. The effect of column diameter is demonstrated by comparing Figure 1A (4.1-mm I.D.) with Figure 1D (10-mm I.D.). The longer peptide retention times on the semipreparative column (Figure 1D) compared to the analytical column (Figure 1A) are due to the greater volume of column packing in the former. Despite the increased potential for solute diffusion with an increase in internal column diameter, with a concomitant increase in peak broadening, the peptide standards are still well resolved on the 10-mm I.D. column. The peptide standards allow the researcher to determine easily the change in peptide retention and resolution as the column dimensions or mobile phase flow-rate or gradient rate are changed.

B. EFFECT OF ORGANIC MODIFIER

The order of effectiveness of the three organic solvents most frequently used in RPC has been shown to be isopropanol > acetonitrile > methanol.[5-7] Acetonitrile is the favored organic solvent for most practical purposes. However, the alcohols can be used when a more nonpolar solvent (isopropanol) or a more polar solvent (methanol) is required for separations of very hydrophobic or very hydrophilic peptides, respectively.

Figure 2 demonstrates the usefulness of the peptide standards in comparing the effects of the above three organic modifiers on peptide resolution during linear gradient elution. The resolution of the five peptides is excellent when using isopropanol (top panel), acetonitrile (middle panel), or methanol (bottom panel) mobile phases. The increase in peptide retention times with the increase in polarity of the organic solvent (methanol < acetonitrile < isopropanol) confirms the order of effectiveness noted above. The optimum resolution of peptides is usually achieved between 15 and 40% of the organic solvent in the gradient,[7] and the five peptide standards are eluted at, or conveniently close to, the lower limit for all three solvents examined.

C. EFFECT OF TEMPERATURE

Peptide or protein retention times in RPC generally decrease with increasing temperature, due to increasing solubility of the solute in the mobile phase as the temperature rises.[5,8-15] In addition, improved peptide resolution, due to a more rapid transfer of the solutes between the stationary and mobile phases, generally accompanies a rise in temperature.

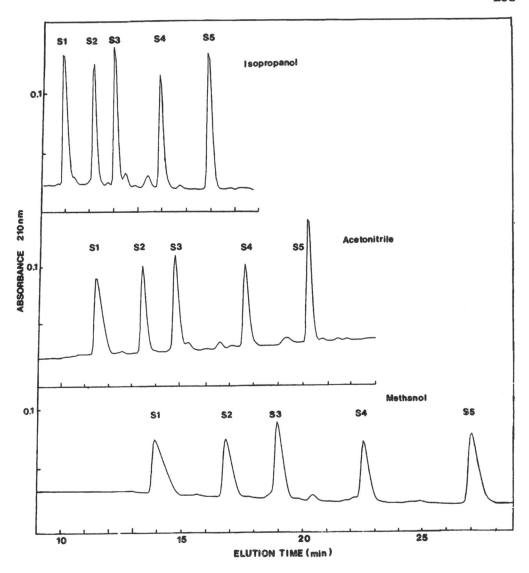

FIGURE 2. Effect of organic modifier on reversed-phase separation of a mixture of five synthetic decapeptide standards (S1 to S5). The HPLC instrument consisted of a Varian Vista series 5000 liquid chromatograph (Varian Associates, Walnut Creek, CA) combined with a Hewlett-Packard HP1040A detection system, HP85B computer, HP9121 disk drive, HP2225A Thinkjet printer and HP7470 plotter. Column: SynChropak RP-P C_{18}, 250 × 4.1 mm I.D., 6.5-μm particle size, 300-Å pore size, ca. 10% carbon loading. Mobile phase: linear AB gradient (1% B/min), where Eluent A is 0.1% aq. TFA and Eluent B is 0.1% TFA in isopropanol (top panel), acetonitrile (middle panel), or methanol (bottom panel) pH 2.0); flow-rate, 1 ml/min; 26°C.

The effect of temperature on peptide standards S1 to S5 during gradient elution RPC tends to confirm previous findings (Figure 3). The peptides exhibit a successive decrease in retention times on a C_8 column, coupled with slightly improved peptide resolution, as the temperature increases from 26 to 66°C.[15] It should be stressed that although enhanced peptide and protein resolution is advantageous, the improvement must be balanced against possible solute degradation at higher temperatures. Certainly, the slightly improved separation of the peptide standards obtained by raising the temperature from 26 to 66°C (Figure 3) does not justify the risk of peptide degradation or possible acceleration of column aging at elevated temperatures. It is worth noting, for example, that following the high temperature runs on the C_8 column used to produce

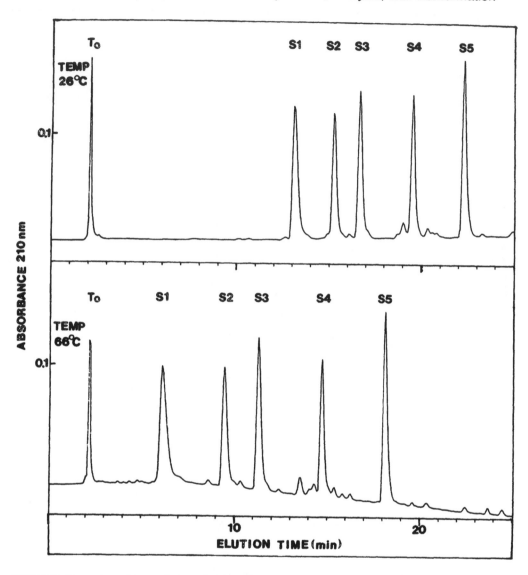

FIGURE 3. Comparison of reversed-phase elution profiles of a mixture of five synthetic decapeptide standards (S1 to S5) at 26 and 66°C. The HPLC instrument is the same as for Figure 2, except for the addition of a Varian column oven. Column: SynChropak RP-8 C_8, 6.5-μm particle size, 300-Å size, ca. 7.5% carbon loading. Mobile phase: linear AB gradient (1% B/min), where Eluent A is 0.1% aq. TFA and Eluent B is 0.1% TFA in acetonitrile (pH 2.0); flow-rate, 1 ml/min; 26 or 66°C. The absorbance peak denoted t_o, produced by an excess TFA concentration in the peptide sample, represents the elution time for unretained compounds. (From Guo, D., Mant, C. T., Taneja, A. K., and Hodges, R. S., *J. Chromatogr.*, 359, 519, 1986. With permission.)

the elution profiles shown in Figure 3, the column was found to have lost some retentiveness for the peptide standards when the temperature was lowered again to 26°C. Except where it cannot be avoided, it is probably unwise to employ reversed-phase columns at temperatures above 30 to 35°C, particularly for extended periods of time.

IV. CONCLUSIONS

For anyone involved in reversed-phase peptide separations, peptide standards are essential for accurate monitoring of column performance, as well as for gauging the possible effects of varying column and/or mobile phase parameters on the resolution of peptide mixtures. Indeed, the use of a universal set of peptide standards by chromatographers could greatly facilitate an accurate comparison of results between different researchers. In addition, the general applicability of 300 Å pore size packings to the resolution of small and large molecules also suggests that the monitoring of column performance for peptide standards can be extrapolated to separations obtained for polypeptides and proteins.

ACKNOWLEDGMENTS

This work was supported by the Medical Research Council of Canada and by equipment grants from the Alberta Heritage Foundation for Medical Research.

REFERENCES

1. **Mant, C.T. and Hodges, R.S.,** Requirement for peptide standards in reversed-phase high-performance liquid chromatography, *LC, Liq. Chromatogr. HPLC Mag.*, 2, 250, 1986.
2. **Mant, C.T. and Hodges, R.S.,** HPLC of peptides, in *High-Performance Liquid Chromatography of Biological Macromolecules: Methods and Applications,* Gooding, K. and Regnier, F., Eds., Marcel Dekker, New York, 1990, 301.
3. **Guo, D., Mant, C.T., Taneja, A.K., Parker, J.M.R., and Hodges, R.S.,** Prediction of peptide retention times in reversed-phase high-performance liquid chromatography. I. Determination of retention coefficients of amino acid residues of model synthetic peptides, *J. Chromatogr.*, 359, 499, 1986.
4. **Mant, C.T. and Hodges, R.S.,** Optimization of peptide separations in high-performance liquid chromatography, *J. Liq. Chromatogr.*, 12, 57, 1989.
5. **Mahoney, W.C. and Hermodson, M.A.,** Separation of large denatured peptides by reverse phase high performance liquid chromatography, *J. Biol. Chem.*, 255, 11199, 1980.
6. **Wilson, K.J., Honegger, A., Stötzel, R.P., and Hughes, G.J.,** The behaviour of peptides on reverse-phase supports during high-pressure liquid chromatography, *Biochem. J.*, 199, 31, 1981.
7. **Hermodson, M. and Mahoney, W.C.,** Separation of peptides by reversed-phase high-performance liquid chromatography, *Methods Enzymol.*, 91, 352, 1983.
8. **O'Hare, M.J. and Nice, E.C.,** Hydrophobic high-performance liquid chromatography of hormonal polypeptides and proteins on alkylsilane-bonded silica, *J. Chromatogr.*, 171, 209, 1979.
9. **Rubinstein, M.,** Preparative high-performance liquid partition chromatography of proteins, *Anal. Biochem.*, 98, 1, 1979.
10. **Hearn, M.T.W.,** The use of reversed phase high performance liquid chromatography for the structural mapping of polypeptides and proteins, *J. Liq. Chromatogr.*, 3, 1255, 1980.
11. **Barford, R.A., Sliwinski, B.J., Breyer, A.C., and Rothbart, H.L.,** Mechanism of protein retention in reversed-phase high-performance liquid chromatography, *J. Chromatogr.*, 235, 281, 1982.
12. **Cohen, K.A., Schellengerg, K., Benedek, K., Karger, B.L., Grego, B., and Hearn, M.T.W.,** Mobile-phase and temperature effects in the reversed phase chromatographic separation of proteins, *Anal. Biochem.*, 140, 223, 1984.
13. **Ingraham, R.H., Lau, S.Y.M., Taneja, A.K., and Hodges, R.S.,** Denaturation and the effects of temperature on hydrophobic-interaction and reversed-phase high-performance liquid chromatography of proteins, *J. Chromatogr.*, 327, 77, 1985.
14. **Hancock, W.S., Knighton, D.R., Napier, J.R., Harding, D.R.K., and Venable, R.,** Determination of thermodynamic parameters for the interaction of a lipid-binding peptide and insulin with a reversed-phase column, *J. Chromatogr.*, 367, 1, 1986.
15. **Guo, D., Mant, C.T., Taneja, A.K., and Hodges, R.S.,** Prediction of peptide retention times in reversed-phase high-performance liquid chromatography. II. Correlation of observed and predicted peptide retention times and factors influencing the retention times of peptides, *J. Chromatogr.*, 359, 519, 1986.

THE USE OF PEPTIDE STANDARDS TO MONITOR NON-IDEAL BEHAVIOR DUE TO UNDERIVATIZED SILANOLS IN REVERSED-PHASE CHROMATOGRAPHY

Colin T. Mant and Robert S. Hodges

I. INTRODUCTION

The majority of reversed-phase peptide separations are still performed on silica-based supports, chemically modified by the reaction of chloroalkylsilanes with silanols present on the surface of the silica. The success of a particular reversed-phase peptide separation is inextricably bound up with the correct choice of column(s) and chromatographic conditions. The proper choice of column will simplify optimization of the chromatographic conditions and any deviations from ideal column behavior, i.e., ionic, as opposed to hydrophobic, interactions between the column and peptide solutes must be identified and suppressed or eliminated. Residual silanol groups present on the surface of silanized silica can influence the reversed-phase chromatographic behavior of peptide solutes containing basic, positively charged amino acid residues, often producing undesirable effects such as peak tailing and non-reproducible peptide retention times.[1,2] A deterioration in column performance also becomes apparent after extensive use of a silica-based reversed-phase column, due to the gradual removal of hydrophobic ligands from the silica, resulting in an increased concentration of underivatized silanols.[3,4]

Researchers utilizing reversed-phase chromatography (RPC) for peptide separations would likely prefer not only to assess the level of silanol activity on a particular reversed-phase column, but also gauge the effect this activity may have on the retention behavior of the peptide solutes. It is preferable to use compounds that are structurally similar to the sample of interest, and that presumably interact with the reversed-phase sorbent in a similar manner, to achieve the most precise and accurate analysis of solute and stationary phase interaction.[5] Thus, although several empirical procedures have been proposed to monitor silanol activity on reversed-phase columns, generally employing normal- or reversed-phase separations of polar organic solutes,[6-10] peptide standards are best suited for monitoring peptide/protein interactions with reversed-phase packings.

In this article, we describe the design and application of a mixture of four synthetic peptide HPLC standards for monitoring free silanols on silica-based reversed-phase packings, demonstrating the versatility of the standards in monitoring silanol activity (over a pH range of 2.0 to 7.0) on commercial columns and columns prepared in the laboratory.

II. DESIGN OF PEPTIDE STANDARDS

The four undecapeptide standards (1 to 4), shown in Table 1, were synthesized on a Beckman Model 990 peptide synthesizer (Beckman Instruments, Berkeley, CA) using the general procedure for solid-phase synthesis described by Parker and Hodges,[11] and are available from

TABLE 1
Sequences of Peptide Standards

Peptide standard	Peptide sequence	No. of positively charged residues
1	Ac-Gly-Gly-Gly-Leu-Gly-Gly-Ala-Gly-Gly-Leu-Lys*-amide	1
2	Ac-Lys*-Tyr-Gly-Leu-Gly-Gly-Ala-Gly-Gly-Leu-Lys*-amide	2
3	Ac-Gly-Gly-Ala-Leu-Lys*-Ala-Leu-Lys*-Gly-Leu-Lys*-amide	3
4	Ac-Lys*-Tyr-Ala-Leu-Lys*-Ala-Leu-Lys*-Gly-Leu-Lys*-amide	4

Note: Ac denotes Nα-acetyl; amide denotes Cα-amide. Positively charged residues (lysine) are marked with a star (*).

Synthetic Peptides Incorporated, Department of Biochemistry, University of Alberta, Edmonton, Alberta, Canada. RPC is often used during peptide mapping of proteins, and the 11-residue length of the standards was designed to approximate the average size of cleavage fragments from proteolytic (e.g., tryptic) digests of proteins. Peptides 1 to 4 contain, respectively, 1 to 4 basic residues (lysine residues) with no acidic residues present. Thus, over the pH range used for the majority of reversed-phase separations (pH 2.0 to 7.0), the net charges of +1 to +4 for Peptides 1 to 4, respectively, do not change. The basic positively charged character of the peptides ensures their sensitivity to any ionic as opposed to hydrophobic stationary phase characteristics; this sensitivity increases with increasing basic character (Peptide 1 to 4) of the standards. The hydrophobicity of the standards also increases from Peptide 1 to 4.

The optimum resolution of peptides is usually achieved between 15 and 40% of the organic solvent in the gradient[12] and, through the use of amino acid residue hydrophobicity coefficients developed by Guo et al.,[13] the peptide standards were also designed to enable precise assessment of column selectivity at the limits of this range. Thus, Peptides 1 and 2 (+1 and +2 net charge, respectively) form a peptide pair, closely related in hydrophobicity, at the lower limit of the optimum organic solvent concentration range; Peptides 3 and 4 (+3 and +4, respectively) form a closely related peptide pair at the upper limit. Peptides 2 and 4 contain tyrosine residues, enabling detection of these standards at 280 nm and simplifying peak identification of the closely-related peptide pairs. In addition, the sequence of the standards was chosen to ensure the peptides would lack conformation on binding to the reversed-phase sorbent, since the non-polar environment of the matrix may induce α-helical structure in potentially α-helical molecules.[14,15]

III. MONITORING OF FREE SILANOLS ON REVERSED-PHASE PACKINGS

A. COLUMNS

The reversed-phase columns described in this article include two commercially purchased columns: analytical C_8 column, 220 × 4.6 mm I.D., 7-μm particle size, 300-Å pore size; analytical C_{18} column, 250 × 4.6 mm I.D., 5 μm, 300 Å. Two C_8 columns were also prepared in this laboratory[2,4] from underivatized silica from SynChrom (Linden, IN), 75 × 4.6 mm I.D., 6.5 μm, 300 Å; the "low silanol" C_8 column was maximally covered with C_8 ligands, while the "high silanol" C_8 column contained a significant level of underivatized silanols.

B. RPC OF PEPTIDE STANDARDS AT pH 2.0

The favored approach to most reversed-phase separations of peptides is to employ linear aqueous trifluoroacetic acid (TFA) to TFA/acetonitrile gradients at pH 2.0. Under these acidic conditions, silanol ionization is suppressed and, hence, ionic interaction of the hydrophobic stationary phase with positively charged peptide residues is minimized. Figure 1 demonstrates the elution profiles at pH 2.0 of the peptide standards obtained on three different reversed-phase columns.

The separation of the four standards on a commercial C_8 column (top panel) was excellent, with sharp, well-defined peaks and retention times close to those predicted by the method of Guo et al.[13] A similar result was obtained on a C_8 reversed-phase packing (middle panel) produced in this laboratory[2,4] and designed to contain a large amount of free, underivatized silanols ("high silanol"). Mant and Hodges[4] had shown previously that the conditions used to produce the elution profiles shown in Figure 1 were sufficient to suppress ionization of the silanols on the particular unmodified silica used to prepare this C_8 packing, i.e., even with this unmodified silica as the stationary phase, the predominant mechanism of peptide separation was hydrophobic in nature. Thus, it may be assumed that, at pH 2.0, non-ideal interactions of the peptides with the "high silanol" C_8 column (middle panel) and, indeed, the commercial C_8 column (top panel) were negligible. However, the elution profile of the standards on a commercial C_{18} column (bottom panel) demonstrated that a low pH does not necessarily eliminate silanol effects, since these effects as a function of pH will be dependent on the type of silica used. This C_{18} column exhibited significant ionic interactions with the basic peptide standards, apparent mainly from the retention behavior and peak deterioration of Peptides 3 and 4.

B. DEVELOPMENT OF PROCEDURE FOR MONITORING NON-IDEAL BEHAVIOR

For maximum practical value to the researcher, any method for monitoring free silanols on reversed-phase packings should be generally applicable to a wide range of columns exhibiting a wide range of silanol activities. Hence, for the development of a silanol monitoring procedure, chromatographic conditions were required which would allow significant, but controllable, ionic interactions of the peptide standards with the hydrophobic stationary phase. The generally effective suppression of silanol ionization at pH 2.0 precluded the gradient elution conditions demonstrated in Figure 1 for this method development. The unusually high silanol activity exhibited by the commercial C_{18} column at this low pH (Figure 1, bottom panel) is the exception rather than the rule.

Acting as weak acids, silanols are negatively charged above pH 3.5 to 4.0, resulting in strong ionic interactions with basic peptide residues. Thus, chromatographic conditions at pH values > 4.0 were deemed the most promising for development of a silanol monitoring procedure. In addition, it is important to assess the effect of pH values closer to neutrality on peptide and/or protein retention behavior as solubility problems may be encountered at acidic pH for many solutes, particularly proteins, precluding the pH 2.0 TFA/acetonitrile system. Also, the researcher can take advantage of cationic ion-pairing reagents (e.g., tetrabutylammonium phosphate) at pH values > 4.0 and manipulate the separation profiles of peptides and proteins containing negatively charged residues (glutamic and aspartic acid residues, free α-carboxyl groups).

1. Monitoring Silanol Activity at pH 7.0

At neutral pH, all free silanols should be ionized. Prior to examining the effects of silanol activity on the retention times of the peptide standards at pH 7.0, it was necessary to obtain their chromatographic elution profiles under conditions designed to minimize ionic interactions between the peptide standards and the stationary phase. Figure 2 (left panels) demonstrates elution profiles of the peptide standards on the "high silanol" C_8 column (top), the "low silanol" C_8 (middle), and the commercial C_8 column obtained with a linear AB gradient (1.67% B/min, equivalent to 1% acetonitrile/min), where Eluent A was 10 mM $(NH_4)_2HPO_4$, pH 7.0, and Eluent B was 60% aq. acetonitrile, both solvents containing 100 mM sodium perchlorate ($NaClO_4$).

pH 2.0

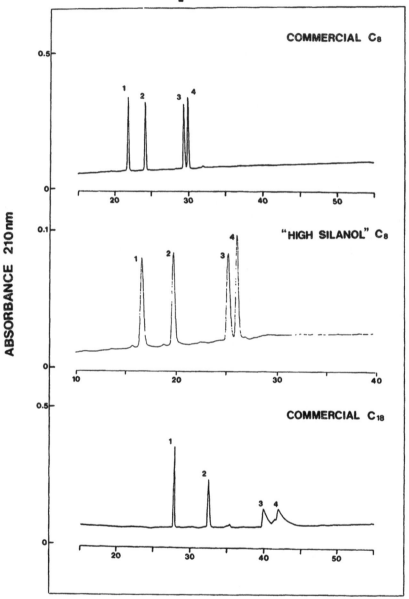

FIGURE 1. RPC of peptide standards at pH 2.0. Columns: top, commercial C_8, 220 × 4.6 mm I.D., 7-μm particle size, 300-Å pore size, middle "high silanol" C_8, 75 × 4.6 mm I.D., 6.5 μm, 300 Å; bottom, commercial C_{18}, 250 x 4.6 mm I.D., 5 μm, 300 Å. Mobile phase: linear AB gradient (1% B/min), where Eluent A is 0.05% aqueous trifluoroacetic acid (TFA) and Eluent B is 0.05% TFA in acetonitrile (pH 2.0); flow-rate, 1 ml/min; temperature, 26°C. The HPLC instrument consisted of a Spectra-Physics (Autolab Division, San Jose, CA) SP8700 solvent delivery system and SP8750 organizer module coupled to a Hewlett-Packard (Avondale, PA) HP1040A detection system, HP3390A integrator, HP85 computer, HP9121 disc drive and HP7470A plotter. The numbers 1 to 4 denote peptide standards 1 to 4, respectively (see Table 1 for details).

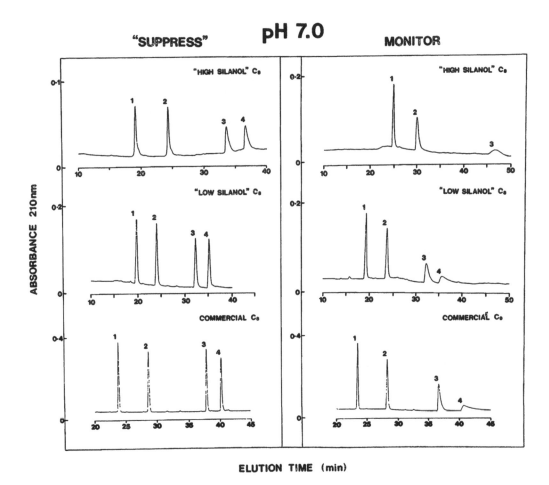

FIGURE 2. Monitoring of free silanols on reversed-phase columns at pH 7.0. Columns: top panels, "high silanol" C_8 (see Figure 1); middle panels, "low silanol" C_8, 75 × 4.6 mm I.D., 6.5-μm particle size, 300-Å pore size; bottom panels, commercial C_8 (see Figure 1). Mobile phases: left panels, linear AB gradient (1.67% B/min, equivalent to 1% acetonitrile/min), where Eluent A is aq. 10 mM $(NH_4)_2HPO_4$, pH 7.0, and Eluent B is 60% aq. acetonitrile, both eluents containing 100 mM $NaClO_4$; right panels, linear AB gradient (2% B/min, equivalent to 1% acetonitrile/min and 1 mM $NaClO_4$/min), where Eluent A is aq. 10 mM $(NH_4)_2HPO_4$, pH 7.0, and Eluent B is 50% aq. acetonitrile containing 50 mM $NaClO_4$; flow-rate, 1 ml/min; temperature, 26°C. Instrument details as for Figure 1. The numbers 1 to 4 denote peptide standards 1 to 4, respectively (see Table 1 for details).

Sodium perchlorate is a chaotropic reagent capable of suppressing ionic interaction between peptides and reversed-phase packing material, as was well demonstrated by the elution profiles obtained on the "low silanol" and commercial C_8 columns (middle and bottom panels, respectively). All four peptide standards were well resolved on these columns, exhibiting good peak shape. However, the peak broadening of the standards, particularly Peptides 3 and 4, on the "high silanol" column (top panel) compared to the pH 2.0 profile (Figure 1, middle panel), suggested some ionic interaction of the peptides with the "high silanol" C_8 due to incomplete suppression of silanol activity at neutral pH.

It was felt that adjustment of the sodium perchlorate concentration in the pH 7.0 mobile phase should enable control of the level of ionic interaction between the basic peptide standards and the negatively charged silanols. However, a single set of chromatographic conditions (i.e., a simple lowering of the level of $NaClO_4$ in the mobile phase) which would enable a convenient monitoring of the effect of ionic interactions on a wide range of columns exhibiting markedly different silanol activities was elusive. Peptides 3 and 4 (+3 and +4 net charge, respectively) proved to be extremely sensitive to ionic interactions with the hydrophobic stationary phase in terms of long retention times and peak broadening. In contrast, changes in retention times and peak shapes of Peptides 1 and 2 (+1 and +2 net charge, respectively) were much less dramatic. The best potential for a general silanol monitoring system at pH 7.0 applicable to a wide range of columns appeared to be a double gradient system with a gradient of increasing salt concentration from 0 mM NaClO$_4$ in Eluent A, concurrent with the acetonitrile gradient. The monitoring procedure developed involved a linear AB gradient (2% B/min, equivalent to 1% acetonitrile/min and 1 mM NaClO$_4$/min) at a flow-rate of 1 ml/min, where Eluent A was aq. 10 mM (NH$_4$)$_2$HPO$_4$, pH 7.0, and Eluent B was 50% acetonitrile containing 50 mM NaClO$_4$. The effect of this double gradient system on the retention times of the peptide standards is shown in Figure 2, right panels, for the "high silanol" C$_8$ (top), the "low silanol" C$_8$ (middle), and the commercial C$_8$ (bottom) columns. The difference in the elution profiles of the peptide standards in the double gradient system (right panels) compared to those obtained under conditions designed to minimize ionic interactions (left panels) is immediately obvious. All four peptides were retained longer on the "high silanol" C$_8$ (top) using the double gradient system (right panel), with a concomitant broadening of peak shape, compared to the conditions employing high levels of NaClO$_4$ in both eluents (left panel). The change in retention times became increasingly more pronounced with increasing net positive charge of the peptide standards. The most highly charged peptide (Peptide 4; +4 net charge) was, in fact, never observed to be eluted from the column. It was interesting to note that, even on columns considered to have maximum possible ligand coverage ("low silanol" C$_8$ and commercial C$_8$; middle and bottom, respectively), ionic interactions were detected by the peptide standards, again apparent from the retention behavior of Peptides 3 and 4. Steric effects preclude derivatization of 100% of available silanols during silanization.[16] Although the remaining free silanols are effectively shielded by the hydrophobic ligands on the silica support, some ionic interactions are inevitable, particularly at neutral pH. Even on well-covered columns ("low silanol" C$_8$ and commercial C$_8$), peptide standards 3 and 4 were extremely sensitive to the presence of residual free silanols. In contrast, the retention behavior of Peptides 1 and 2 was only seriously affected by the presence of high concentrations of free silanols ("high silanol" C$_8$).

2. Monitoring Silanol Activity at pH 4.5

Figure 3, top panel, shows the elution profiles of the peptide standards on the "high silanol" C$_8$ under conditions designed to minimize ionic interactions at pH 4.5. The peptides were eluted with a linear AB gradient (2% B/min, equivalent to 1% acetonitrile/min), where Eluent A was aq. 250 mM triethylammonium phosphate (TEAP), pH 4.5, and Eluent B was 50% Eluent A and 50% acetonitrile. The silanol masking properties of amines are well known[1,17] and the excellent separation profile obtained under these conditions was very similar to that obtained at pH 2.0 (Figure 1, middle panel). The slightly longer retention times of the peptides in the pH 2.0 mobile phase compared to the pH 4.5 system was probably due to the more hydrophobic nature of the TFA anionic counterion compared to the phosphate ion.

The peptide elution profiles shown in Figure 3, middle panel ("low silanol" C$_8$) and bottom panel ("high silanol" C$_8$) were obtained with a linear AB double gradient (2% B/min, equivalent to 1% acetonitrile/min and 1 mM TEAP/min), where Eluent A was aq. 10 mM TEAP, pH 4.5, and Eluent B was 50% aq. acetonitrile containing 60 mM TEAP. All four peptides exhibited excellent peak shape and reasonable run times on the "low silanol" C$_8$ column (middle panel), with no indication of the non-ideal behavior detected by the pH 7.0 double gradient system (Figure 2, middle left panel). In addition, although the peak shapes of the four standards have deteriorated on the "high silanol" C$_8$ column (bottom panel) compared to the "low silanol" C$_8$ (middle panel), their retention times have actually decreased from those obtained on the maximally covered column. Although the deterioration in peak shape is indicative of ionic

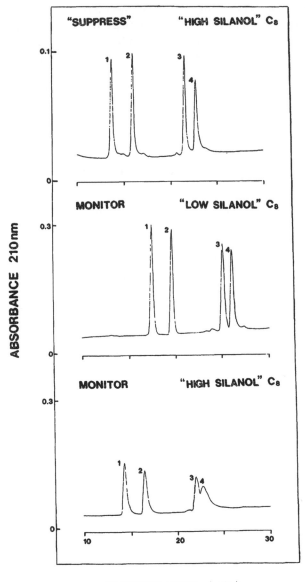

FIGURE 3. Monitoring of free silanols on reversed-phase columns at pH 4.5. Columns: top and bottom, "high silanol" C_8 (see Figure 1); middle "low silanol" C_8 (see Figure 2). Mobile phases: top, linear AB gradient (2% B/min, equivalent to 1% acetonitrile/min), where Eluent A is aq. 250 mM triethylammonium phosphate (TEAP), pH 4.5, and Eluent B is 50% Eluent A and 50% acetonitrile (Eluent A was prepared by adding sufficient triethylamine to 250 mM orthophosphoric acid until the desired pH was obtained); middle and bottom, linear AB gradient (2% B/min, equivalent to 1% acetonitrile/min and 1 mM TEAP/min), where Eluent A is aq. 10 mM TEAP, pH 4.5, and Eluent B is 50% aq. acetonitrile containing 60 mM TEAP (Eluent B was prepared by mixing equal volumes of acetonitrile and an aq. 120 mM solution of TEAP, pH 4.5); flow-rate, 1 ml/min; temperature, 26°C. Instrument details as for Figure 1. The numbers 1 to 4 denote peptide standards 1 to 4, respectively (see Table 1 for details).

interaction with the "high silanol" stationary phase (compare this profile to that obtained on this column under conditions designed to minimize ionic interactions [Figure 3, top panel]), the decrease in peptide retention times on this column compared with the more hydrophobic "low silanol" column (middle panel), suggested that these interactions had been effectively suppressed even by the relatively low level of TEAP used in the double gradient system.

The much reduced sensitivity of the double gradient system at pH 4.5 (Figure 3) compared to pH 7.0 (Figure 2, right) probably reflected a higher degree of silanol ionization (and, hence, a greater potential for ionic interaction) at the higher pH value. At pH 4.5, a pH value only slightly higher than the characteristic pKa range of free silanols (pH 3.5 to 4.0), there is a possibility that not all silanols will be ionized. Thus, these results confirmed that a pH 7.0 double gradient system, due to its excellent sensitivity to the presence of free silanols, had greater utility as a generally applicable monitoring procedure.

IV. CONCLUSIONS

This article describes the development of a silanol monitoring system based on the sensitivity of a mixture of four basic peptide standards (+1 to +4 net charge) to ionic interactions with reversed-phase packings. The monitoring procedure involves chromatographing the peptides at pH 7.0 with a combined acetonitrile (1% acetonitrile/min) and sodium perchlorate (1 mM NaClO$_4$/min) linear gradient. The impressive sensitivity of the peptide standards, particularly Peptides 3 and 4 (+3 and +4 net charge, respectively), toward non-ideal behavior on even maximally derivatized silica supports should prove of great value to researchers involved in reversed-phase separations of peptides and other polar solutes. Peptides with charges of +3 and +4 are not uncommon, and researchers can now demonstrate with the standards whether interactions are occurring and how the buffer system must be changed to suppress these interactions if they are observed in the monitoring system. In addition, the goal of manufacturers involved in the development of new and improved packings should be to prepare columns lacking interactions with, in particular, Peptides 3 and 4 using this pH 7.0 silanol monitoring system.

ACKNOWLEDGMENTS

This work was supported by the Medical Research Council of Canada and equipment grants from the Alberta Heritage Foundation for Medical Research.

REFERENCES

1. **Bij, K.E., Horváth, Cs., Melander, W.R., and Nahum, A.,** Surface silanols in silica-bonded hydrocarbonaceous stationary phases. II. Irregular retention behavior and effect of silanol masking, *J. Chromatogr.*, 203, 65, 1981.
2. **Mant, C.T. and Hodges, R.S.,** Monitoring free silanols on reversed-phase supports with peptide standards, *Chromatographia*, 24, 805, 1987.
3. **Mant, C.T., Parker, J.M.R., and Hodges, R.S.,** On-line derivatization to restore reversed-phase column performance in peptide separations, *LC.GC Mag. Liq. Gas Chromatogr.*, 4, 1004, 1986.
4. **Mant, C.T. and Hodges, R.S.,** On-line derivatization of silica supports for regeneration and preparation of reversed-phase columns, *J. Chromatogr.*, 409, 155, 1987.

5. **Mant, C.T. and Hodges, R.S.**, Requirement for peptide standards in reversed-phase high performance liquid chromatography, *LC.GC Mag. Liq. Gas Chromatogr.*, 4, 250, 1986.
6. **Majors, R.E. and Hopper, M.J.**, Studies of siloxane phases bonded to silica gel for use in high performance liquid chromatography, *J. Chromatogr. Sci.*, 12, 767, 1974.
7. **Karch, K., Sebestian, I., and Halász, I.**, Preparation and properties of reversed phases, *J. Chromatogr.*, 122, 3, 1976.
8. **Hemetsberger, H., Kellermann, M., and Ricken, H.**, Behavior of chemically bonded alkylmethyldichlorosilanes to silica gel in reversed-phase high-performance liquid chromatography, *Chromatographia*, 10, 726, 1977.
9. **Nondek, L., Buszewski, B., and Berek, D.**, Retention of pyridine and 2,6-dimethylpyridine on silanized silica. A simple test on residual silanols?, *J. Chromatogr.*, 360, 241, 1986.
10. **Walters, M.J.**, Classification of octadecyl-bonded liquid chromatography of columns, *J. Assoc. Off. Anal. Chem.*, 70, 465, 1987.
11. **Parker, J.M.R. and Hodges, R.S.**, I. Photoaffinity probes provide a general method to prepare synthetic peptide-conjugates, *J. Prot. Chem.*, 3, 465, 1985.
12. **Hermodson, M. and Mahoney, W.C.**, Separation of peptides by reversed-phase high-performance liquid chromatography, *Methods Enzymol.*, 91, 352, 1983.
13. **Guo, D., Mant, C.T., Taneja, A.K., Parker, J.M.R., and Hodges, R.S.**, Prediction of peptide retention times in reversed-phase high performance liquid chromatography. I. Determination of retention coefficients of amino acid residues of model synthetic peptides, *J. Chromatogr.*, 359, 499, 1986.
14. **Lau, S.Y.M., Taneja, A.K., and Hodges, R.S.**, Effects of high-performance liquid chromatographic solvents and hydrophobic matrices on the secondary and quaternary structure of a model protein. Reversed-phase and size-exclusion high-performance liquid chromatography, *J. Chromatogr.*, 317, 129, 1984.
15. **Mant, C.T. and Hodges, R.S.**, HPLC of peptides, in *High Performance Liquid Chromatography of Biological Macromolecules: Methods and Applications*, Gooding, K. and Regnier, F., Eds., Marcel Dekker, New York, 1990, 301.
16. **Roumeliotis, P. and Unger, K.K.**, Structures and properties of *n*-alkyldimethylsilyl bonded silica reversed-phase packings, *J. Chromatogr.*, 149, 211, 1978.
17. **Rivier, J.E.**, Use of trialkylammonium phosphate buffers in reversed-phase HPLC for high resolution and recovery of peptides and proteins, *J. Liq. Chromatogr.*, 1, 343, 1978.

THE EFFECT OF VARYING FLOW-RATE, GRADIENT-RATE, AND DETECTION WAVELENGTH ON PEPTIDE ELUTION PROFILES IN REVERSED-PHASE CHROMATOGRAPHY

T. W. Lorne Burke, Colin T. Mant, and Robert S. Hodges

I. INTRODUCTION

The flexibility of reversed-phase chromatography (RPC) makes it the obvious choice for an initial HPLC run to gauge the complexity of a peptide mixture. As described previously in this book,[1] the best initial approach to most analytical peptide separations is to employ a 0.1% aq. trifluoroacetic acid (TFA) to 0.1% TFA-acetonitrile linear gradient (pH 2) of 1% acetonitrile/min at a flow-rate of 1 ml/min. From the observed peptide elution profile obtained under these conditions, the need for further optimization of the separation, and the best approach to this optimization, may be assessed. Although changes in ion-pairing reagent will often offer the more powerful peptide resolving capability,[2,3] varying the flow- and gradient-rate can be very effective in optimizing peptide separations. The selection of optimal conditions may be critical in establishing an analytical method. This selection really depends on the individual researcher's requirements and is a compromise of run time (speed), resolution, sensitivity (detector response) and sample load. For example, if one desires maximum resolution and maximum sensitivity with minimal sample injected and is willing to sacrifice run time, what flow- and gradient-rate would be selected? On the other hand, if a total compromise of sensitivity (at a given sample load), resolution and run time is desired, totally different conditions would be selected. It should be remembered that alteration of run conditions (e.g., gradient- or flow-rate) to improve one aspect of a separation (e.g., resolution or sensitivity) may be at the expense of others. However, despite the often dramatic effect on peptide elution profiles when varying flow- and gradient-rate, particularly the latter, a rigorous step-by-step assessment of the effects of these parameters on the retention behavior of peptides during RPC is not easy to find in the literature.

This article examines the effect of flow- and gradient-rate on the reversed-phase elution profile of a mixture of closely-related peptide analogs. The effect on peptide retention times, peak height, peak width, and overall peptide resolution is clearly demonstrated. In addition, the effect of varying the detection wavelength on the sensitivity of peak detection is illustrated.

II. EXPERIMENTAL

A. INSTRUMENTATION

The HPLC instrument consisted of a Hewlett-Packard (Avondale, PA) HP1090 Series 300 computer, HP9133 disc drive, HP2225A Thinkjet printer, and HP7440A plotter.

TABLE 1
Peptide Sequences

Peptide	Peptide sequence[a]
I1	Ac-Arg-Gly-Gly-Gly-Gly-<u>Ile</u>-Gly-<u>Ile</u>-Gly-Lys-amide
I2	Ac-Arg-Gly-Gly-Gly-Gly-<u>Ile</u>-Gly-Leu-Gly-Lys-amide
S2	Ac-Arg-Gly-Gly-Gly-Gly-<u>Leu</u>-Gly-Leu-Gly-Lys-amide
S3	Ac-Arg-Gly-<u>Ala</u>-Gly-Gly-Leu-Gly-Leu-Gly-Lys-amide
S4	Ac-Arg-Gly-<u>Val</u>-Gly-Gly-Leu-Gly-Leu-Gly-Lys-amide
S5	Ac-Arg-Gly-<u>Val</u>-<u>Val</u>-Gly-Leu-Gly-Leu-Gly-Lys-amide

[a] Ac = N^α-acetyl; amide = C^α-amide The solid lines denote amino acid substitutions in the peptide S2 sequence.

B. COLUMN

Peptide separations were carried out on a SynChropak RP-P C_{18} column, 250×4.6 mm I.D., 6.5-μm particle size, 300-Å pore size (SynChrom, Lafayette, IN).

C. CONDITIONS

The peptide mixture was subjected to linear AB gradients, where Eluent A was 0.1% aq. TFA and Eluent B was 0.1% TFA in acetonitrile.

III. PEPTIDE STANDARDS

The six synthetic decapeptide analogs used in this study (I1, I2, S2, S3, S4, and S5) are shown in Table 1. The hydrophobicity of the peptides increases only slightly between S2 and S4 — between S2 and S3 there is a change from an α-H to a β-CH_3 group, between S3 and S4 there is a change from a β-CH_3 group to two methyl groups attached to the β-CH group, between S4 and S5 there is a change from an α-H to two methyl groups attached to the β-CH group. The hydrophobicity variations between I1, I2, and S2 are even more subtle. There is a change of only an isoleucine to a leucine residue between I1 and I2, and between I2 and S2. Guo et al.[4] demonstrated that leucine is slightly more hydrophobic than isoleucine, although these residues contain the same number of carbon atoms. Since isoleucine is β-branched, the β-carbon is close to the peptide backbone and not as available to interact with the hydrophobic stationary phase compared to the conformation of the leucine side-chain. Thus, this peptide mixture, in addition to being a very sensitive monitor for determining the effects of varying flow- and gradient-rate on peptide elution profiles, enables very precise determination of the resolving power of a reversed-phase column.

IV. RESULTS

A. EFFECT OF DETECTION WAVELENGTH ON SENSITIVITY OF PEAK DETECTION

Detection of peptides during HPLC is generally based on peptide bond absorbance at low UV wavelengths, usually over the 205 to 215 nm wavelength range. Although peptides may also contain UV chromophores, such as tryptophan, tyrosine or phenylalanine (side-chain absorbance in the range of 240 to 300 nm), these amino acid residues are not present in all peptides.

The RPC elution profiles shown in Figure 1 demonstrate the effect of varying detection wavelength (200 to 230 nm, in 10 nm steps) on the sensitivity of detection of the six decapeptide standards shown in Table 1. The sensitivity of peak detection decreases dramatically as the

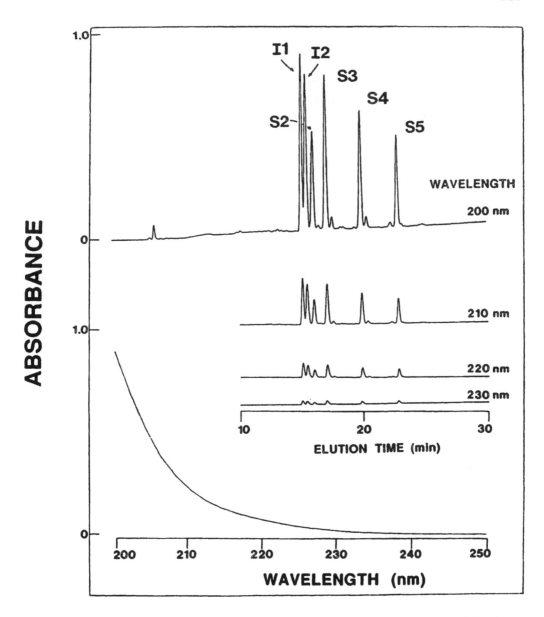

FIGURE 1. Effect of varying detection wavelength on sensitivity of UV detection of peptides during RPC. Conditions: linear AB gradient (1% B/min) at a flow-rate of 1 ml/min. Top and center: elution profiles of peptide standards I1, I2, S2, S3, S4, and S5 (sequences shown in Table 1) at wavelengths of 200, 210, 220, and 230 nm. Bottom: UV spectrum (200 to 250 nm) of peptide standard I1.

detection wavelength increases from 200 nm (~850 mAU for I1, the first eluted peak) to 230 nm (~15 mAU for I1). On each 10 nm increase in detection wavelength, there is an approximate fourfold decrease in sensitivity. Thus, there is a total change in detection sensitivity of ~64-fold between 200 and 230 nm.

The decrease in absorbance of the peptide bond with increasing detection wavelength is clearly apparent from the UV spectrum (200~to 250 nm) of peptide standard I1 shown at the bottom of Figure 1. Although not shown, the absorption maximum for the peptide bond is actually ~187 nm. However, several practical considerations preclude the routine use of wavelengths below 200 nm for solute detection; indeed, the use of wavelengths below 210 nm tends to be the exception rather than the rule. Detection below 210 nm can suffer from interference due to impurities present in buffers, solvents, or even the sample, which may obscure the peak(s) of interest. HPLC-grade solvents are the norm for most analytical purposes, especially where detectors are being used at high sensitivity. However, chemical groups or bonds in the pure solvents themselves may also contribute significantly to UV absorbance at wavelengths below 210 nm. The "UV cutoff" values — the wavelength in nm at which the absorbance of a 1-cm-long cell filled with the solvent has an absorbance of 1.0, measured against water as reference — of the popular RPC solvents methanol, acetonitrile and isopropanol are 205, 188, and 205 nm, respectively. Thus, the excellent sensitivity, with little detection interference, of the 200-nm wavelength peptide elution profile shown in Figure 1 is due partly to the low UV cutoff value of acetonitrile as well as the cleanliness of the sample. The quality of the detection system is, of course, also an important factor. The common use of 210 nm as the preferred detection wavelength for most analytical reversed-phase applications (and, indeed, for those of other HPLC modes) is a good compromise between detection sensitivity and potential detection interference.

It should be noted that wavelengths in the range of 220 to 230 nm are frequently employed in preparative applications,[5,6] where the use of more sensitive detection wavelengths would result in overloading of the detector response (usually above an absorbance of 2.0 to 2.5 AU).

B. EFFECT OF FLOW- AND GRADIENT-RATE ON SENSITIVITY OF PEPTIDE DETECTION
1. Effect of Varying Flow-Rate at Constant Gradient-Rate

Figure 2A shows the RPC elution profiles of the six decapeptide standards obtained following linear gradient elution at flow-rates of 2.0 ml/min (left profile) and 0.1 ml/min (right profile) at a constant gradient-rate of 1% B/min. The major effects of this considerable (20-fold) variation in flow-rate on detection sensitivity is immediately obvious, i.e., a substantial increase (8-fold) in peak height of I1 (the first eluted peak) on decreasing the flow-rate from 2.0 ml/min to 0.1 ml/min.

The effects of varying flow-rate at six fixed gradient-rates on peak height and area of peptide S5 (the last eluted peptide) are illustrated graphically in Figure 3A and C, respectively, and can be summarized as follows:

1.   From Figure 3A, peak height increases with decreasing flow-rate at any given gradient-rate, e.g., at a constant gradient-rate of 1% B/min, there is an eightfold increase in peak height of S5 on decreasing the flow-rate from 2.0 to 0.1 ml/min. It is worth noting that over the flow-rate range frequently employed with standard analytical columns (1.0 to 2.0 ml/min), the effect of flow-rate variation is quite small relative to that observed on decreasing the flow-rate below 1.0 ml/min. An increase in peak detection sensitivity with decreasing flow-rate has been reported by other researchers for both peptides[9-11] and proteins.[12]
2.   From Figure 3C, the peak area of S5 decreases with increasing flow-rate to the same extent at all gradient-rates. This decrease in peak area with increasing flow-rate is due to the detector response time and sample dilution.[9] In a similar manner to that noted above for the effect of flow-rate on peak height of S5, the effect on peak area over the 1.0 to 2.0 ml/min flow-rate range is small relative to that below 1.0 ml/min.

2. Effect of Varying Gradient-Rate at Constant Flow-Rate

From Figure 2B, at constant flow-rate (1.0 ml/min), there is a decrease in peptide peak height as the gradient-rate is decreased from 4.0% B/min (left profile) to 0.2% B/min (right profile). Thus, there is an eightfold decrease in the height of I1 with a 20-fold decrease in gradient-rate.

The effects of varying gradient-rate, at constant flow-rate, on the peak height and area of

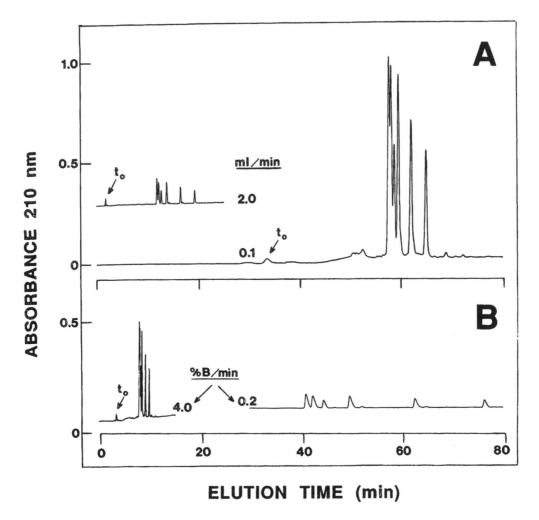

FIGURE 2. Effect of varying flow-rate or gradient-rate on sensitivity of peak detection in RPC of peptides. Conditions: panel A, linear AB gradient (1% B/min) at flow-rates of 2 ml/min (left profile) and 0.1 ml/min (right profile); panel B, linear AB gradient of 4.0% B/min (left profile) and 0.2% B/min (right profile) at a flow-rate of 1 ml/min. t_0 denotes the time for elution of unretained compounds. The sequences of the six peptide standards are shown in Table 1.

peptide S5 are illustrated graphically in Figure 3B and D, respectively, the results of which may be summarized as follows:

1. From Figure 3B, peak height increases with increasing gradient-rate at any given flow-rate[9,10]
2. From Figure 3D, in contrast to the effect of flow-rate (Figure 3C), peptide peak area is independent of gradient-rate

C. EFFECT OF FLOW- AND GRADIENT-RATE ON PEPTIDE RESOLUTION

Figure 4 demonstrates the effects of large changes in flow-rate (0.1 ml/min [panels A to C] and 2.0 ml/min [panels D to F], and gradient-rate (4.0% B/min [panels A and D], 1.0% B/min [panels B and E] and 0.2% B/min [panels C and F]) on the resolution of the six peptide standards. The somewhat extreme variations in flow-rate (a 20-fold change between 0.1 and 2.0 ml/min) and gradient-rate (a 20-fold change between 4.0 and 0.2% B/min) are designed to show as clearly as possible the effects of these parameters on peptide resolution.

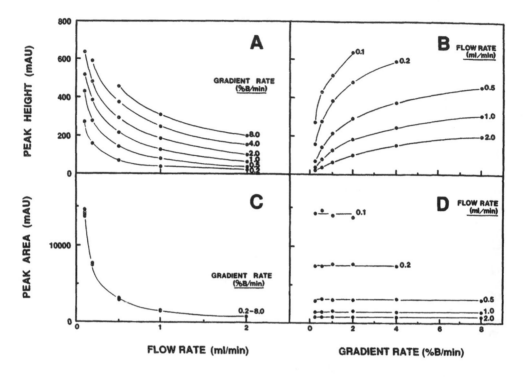

FIGURE 3. Graphical representation of effect of varying flow-rate or gradient-rate on sensitivity of peak detection in RPC of peptides. Conditions: flow-rates and linear AB gradients as shown. Panel A: effect of varying flow-rate, at constant gradient-rate, on peak height of peptide S5. Panel B: effect of varying gradient-rate, at constant flow-rate, on peak height of S5. Panel C: effect of varying flow-rate, at constant gradient-rate, on peak area of S5. Panel D: effect of varying gradient-rate, at constant flow-rate, on peak area of S5. The sequence of S5 is shown in Table 1.

As noted previously, linear AB gradient elution at 1.0% B/min at a flow-rate of 1.0 ml/min are very common starting conditions for many analytical RPC applications employing aq. TFA-TFA/acetonitrile mobile phases. Figure 5 demonstrates the effect on the resolution of the peptide analogs of manipulating these standard conditions. Thus, at a constant gradient-rate of 1.0% B/ min (left profiles), the flow-rate is varied from 0.1 to 1.0 ml/min; at a constant flow-rate of 1.0 ml/min (right profiles), the gradient-rate is varied from 4.0 to 0.2% B/min. In this figure, the time scales are constant for each vertical series of profiles (15 min for the left profiles; 40 min for the right profiles). Thus, the effect of varying flow-rate (left profiles) and gradient-rate (right profiles) on peptide retention times and the spread of the six peptides, as well as peptide resolution, is highlighted.

The term "resolution" in Figures 4 and 5 is qualitative, relying as it does on the observed separation of peptide peaks without attempting to quantify the effects of flow- and gradient-rate variation on the separation of the peptide analogs. For the purposes of the present study, the resolution between two peaks is described quantitatively by the expression $\Delta t/W_1 + W_2$, where Δt is the difference (in min) between the retention times of two retained components at their peak maxima, and W_1 and W_2 are the peak widths at half height (in min). Figure 6 illustrates graphically the effects of varying flow- and gradient-rate on peptide resolution. The effect of varying flow-rate (at constant gradient-rate) on the resolution of peptides S4 and S5 is shown in Figure 6A; the effect on the peak width at half height of S5 is shown in Figure 6B; the effect on Δt between S4 and S5 is shown in Figure 6C. The effect of varying gradient-rate (at constant flow-rate) on the resolution of S4 and S5 is shown in Figure 6D; the effect on the peak width at half height of S5 is shown in Figure 6E; the effect on Δt between S4 and S5 is shown in Figure 6F.

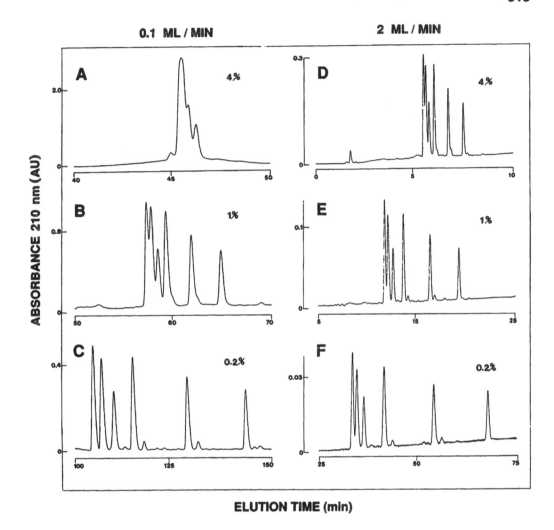

FIGURE 4. Effect of varying gradient-rate, at constant flow-rate, on separation of peptides in RPC. Conditions: flow-rates and linear AB gradients as shown. Left panels: effect of varying gradient-rate (4%, 1%, 0.2% B/min [panels A, B, and C, respectively]) at constant flow-rate (0.1 ml/min). Right panels: effect of varying gradient-rate (4%, 1%, 0.2% B/min [panels D, E, and F, respectively]) at constant flow-rate (2 ml/min). The sequences of the six peptide standards are shown in Table 1.

The results shown in Figures 4 to 6 may be summarized as follows:

1. From Figures 4, 5, and 6A; at constant gradient-rate, there is a general improvement in peptide resolution with increasing flow-rate.[10] One small exception to this can be seen in the slight worsening of the separation of the first three eluted peptides (I1, I2, and S2) obtained at a gradient-rate of 0.2% B/min (Figure 4C and F). Figure 6A also demonstrates that the major effects of flow-rate on peptide resolution occur below 1.0 ml/min. Thus, at a constant gradient-rate of 1.0% B/min, there is a twofold (i.e., 100%) increase in resolution of S4 and S5 on increasing the flow-rate from 0.1 to 0.5 ml/min; however, between 0.5 and 2.0 ml/min, there is only a further 7% increase in resolution (Figure 6A). The observation that the plots in Figure 6A all tend towards a plateau above a flow-rate of 0.5 ml/min suggests that the routine use of 1.0 ml/min ensures good reproducibility of peptide resolution during RPC on analytical columns (4.1 to 4.6 mm I.D.).

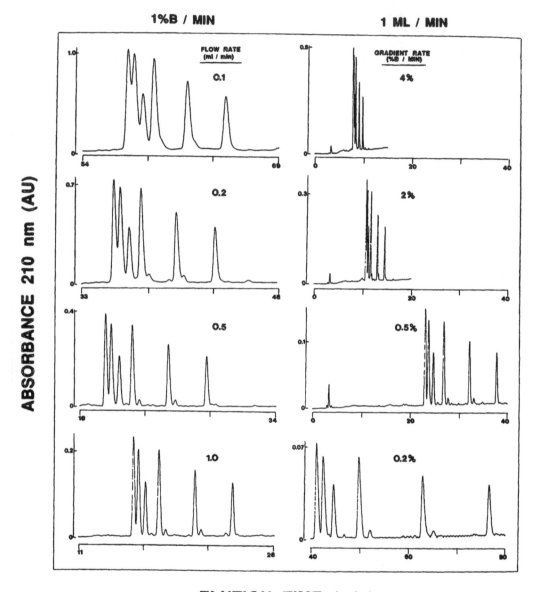

ELUTION TIME (min)

FIGURE 5. Effect of varying flow-rate or gradient-rate on separation of peptides in RPC. Conditions: flow-rates and linear AB gradients as shown. Left profiles: effect of varying flow-rate (0.1, 0.2, 0.5, and 1.0 ml/min) at constant gradient-rate (1% B/min). Right profiles: effect of varying gradient-rate (4, 2, 0.5, and 0.2% B/min) at constant flow-rate (1 ml/min). Sequences of the six peptide standards are shown in Table 1.

2.   From Figures 4, 5, and 6D, at constant flow-rate there is a general improvement in peptide resolution with decreasing gradient-rate.[9,10] This improvement is particularly marked at low flow-rates, e.g., in Figure 4A to C, the elution profile of the peptide standards progresses from negligible separation at 4.0% B/min (Figure 4A) to complete (baseline) separation of all six peptides at 0.2% B/min (Figure 4C). From Figure 6D, gradient-rate changes over the 2.0 to 8.0% B/min range have relatively little effect on peptide resolution compared to the large changes effected at gradient-rates < 2.0% B/min. The results of Figures 4, 5, 6A, and 6D also suggest that gradient-rate variations have a greater effect on

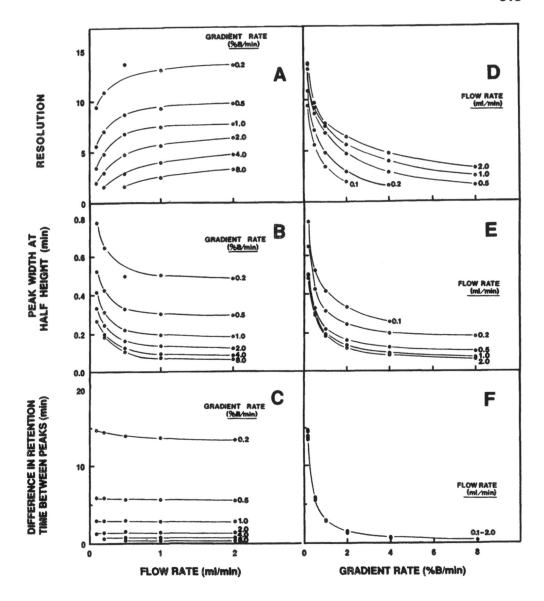

FIGURE 6. Graphical representation of effect of varying flow-rate or gradient-rate on peptide resolution in RPC. Conditions: flow-rates and linear AB gradients as shown. Left panels: effect of varying flow-rate, at constant gradient-rate, on resolution of peptides S4 and S5 (panel A), peak width at half height of S5 (panel B) and difference in retention times between S4 and S5 (panel C). Right panels: effect of varying gradient-rate, at constant flow-rate, on resolution of S4 and S5 (panel D), peak width at half height of S5 (panel E) and difference in retention times between S4 and S5 (panel F). Sequences of S4 and S5 are shown in Table 1.

peptide resolution than changes in flow-rate. Thus, at a constant flow-rate of 1.0 ml/min, there is an approximately threefold increase in resolution of S4 and S5 for a 20-fold (4.0 to 0.2% B/min) decrease in gradient-rate (Figure 6D); in contrast, at a constant gradient-rate of 1.0% B/min, there is just a twofold increase in resolution for a 20-fold (0.1 to 2.0 ml/min) increase in flow-rate.

3. From Figure 6B and E, there is an increase in peak width at half height of S5 with decreasing flow- and gradient-rate, respectively.

4. From Figures 4 and 5, decreasing the flow- or gradient-rate both lead to an increase in peptide retention times. For instance, from Figure 4, at a constant gradient-rate of 1.0% B/

min, the retention time of peptide S5 (last eluted peak) increases by 46 min on decreasing the flow-rate from 2.0 ml/min (Figure 4E) to 0.1 ml/min (Figure 4B); at a constant flow-rate of 2.0 ml/min, the retention time of S5 increases by 62 min on decreasing the gradient-rate from 4.0% B/min (Figure 4D) to 0.2% B/min (Figure 4F).

The increasing peptide retention times with decreasing flow-rate are due to the increasing time required for the gradient to reach the column from the pump via proportioning value and injector (gradient delay time). Once this delay time has been accounted for, flow-rate changes appear to have little overall effect on absolute peptide retention behavior, i.e., the distance between peptide peaks remains essentially constant with varying flow-rate.[10] This is clearly apparent in Figure 5 (left profiles) and Figure 6C. The improved resolution of S4 and S5 with increasing flow-rate (Figure 6A), is, thus, a result of a decrease in peptide diffusion as the flow-rate increases, producing smaller peak widths (Figure 6B).

In contrast to flow-rate changes, gradient-rate variations have a profound impact on peptide separations in terms of both peptide retention times (Figure 5, right profiles) and distances between peptides (Figure 5, right profiles, and Figure 6F) increasing with decreasing gradient-rate.[10,13-15] Thus, improved peptide resolution is obtained with decreasing gradient-rate (Figure 6D), despite the concomitant increase in peak widths (Figure 6E), since the resulting increase in Δt (Figure 6F) more than compensates for this peak broadening.

D. EFFECT OF FLOW- AND GRADIENT-RATE ON PEPTIDE RESOLUTION AND DETECTION SENSITIVITY

We have shown (Figures 2 and 3) that you can increase the sensitivity of peptide peak detection and peptide resolution by decreasing flow-rate and/or increasing gradient-rate. An equivalent change in flow-rate (at a constant, standard gradient-rate of 1.0% B/min) or gradient-rate (at a constant, standard flow-rate of 1.0 ml/min) has a similar effect on detection sensitivity. For example, from Figure 3, a 10-fold increase in flow-rate (e.g., 0.1 to 1.0 ml/min; Figure 3A) or 10-fold decrease in gradient-rate (e.g., 2.0 to 0.2% B/min; Figure 3B) both produce a 4-to 5-fold increase in peak height of peptide S5. However, if the objective of the researcher was to increase sensitivity while maintaining satisfactory peptide resolution, then a decrease in flow-rate should be considered rather than an increase in gradient-rate. The latter produces a marked deterioration in peptide resolution (Figure 6D), whereas the former, in comparison, has a much lesser effect.

V. CONCLUSIONS

This article has attempted to demonstrate the major effects of variations in flow-rate and gradient-rate on peptide elution profiles in terms of both detection sensitivity and peptide resolution, during RPC. In addition, it has been shown how manipulation of these parameters can be effective in optimizing the separation of closely-related peptides. Optimum conditions for a peptide separation will depend on the particular peptide mixture under investigation and the requirements of the separation, i.e., whether one component or, more demanding perhaps, all components of a peptide mixture need to be resolved. If the latter, then a compromise may have to be made between a short analysis time and the desire for maximum resolution and/or sensitivity.

ACKNOWLEDGMENTS

This work was supported by the Medical Research Council of Canada and by equipment grants from the Alberta Heritage Foundation for Medical Research.

REFERENCES

1. **Mant, C.T. and Hodges, R.S.**, Standard chromatographic conditions for SEC, IEC, RPC and HIC, this publication.
2. **Mant, C.T. and Hodges, R.S.**, The effects of anionic ion-pairing reagents on peptide retention in reversed-phase chromatography, this publication.
3. **Bennett, H.P.J.**, Manipulation of pH and ion-pairing reagents to maximize the performance of reversed-phase columns, this publication.
4. **Guo, D., Mant, C.T., Taneja, A.K., Parker, J.M.R., and Hodges, R.S.**, Prediction of peptide retention times in reversed-phase high-performance liquid chromatography. I. Determination of retention coefficients of amino acid residues of model synthetic peptides, *J. Chromatogr.*, 359, 499, 1986.
5. **Rivier, J., McClintock, R., Galyean, R., and Anderson, H.**, Reversed-phase high-performance liquid chromatography: preparative purification of synthetic peptides, *J. Chromatogr.*, 288, 303, 1984.
6. **Hoeger, C., Galyean, R., Boublik, J., McClintock, R., and Rivier, J.**, Preparative reversed-phase high performance liquid chromatography: effects of buffer pH on the purification of synthetic peptides, *BioChromatography*, 2, 134, 1987.
7. **Hodges, R.S., Burke, T.W.L., and Mant, C.T.**, Preparative purification of peptides by reversed-phase chromatography. Sample displacement mode *versus* gradient elution mode, *J. Chromatogr.*, 444, 349, 1988.
8. **Hong, A.L., Brasseur, M.M., and Kesuma, D.**, Purification of histidine-rich hydrophilic peptides, this publication.
9. **Meek, J.L. and Rossetti, Z.L.**, Factors affecting retention and resolution of peptides in high-performance liquid chromatography, *J. Chromatogr.*, 211, 15, 1981.
10. **Guo, D., Mant, C.T., Taneja, A.K., and Hodges, R.S.**, Prediction of peptide retention times in reversed-phase high-performance liquid chromatography. II. Correlation of observed and predicted peptide retention times and factors influencing the retention times of peptides, *J. Chromatogr.*, 359, 519, 1986.
11. **Stone, K.L., LoPresti, M.B., Crawford, J.M., DeAngelis, R., and Williams, K.R.**, Reversed-phase HPLC separation of sub-nanomole amounts of peptides obtained from enzymatic digests, this publication.
12. **Schlabach, T.D. and Wilson, K.J.**, Microbore flow-rates and protein chromatography, *J. Chromatogr.*, 385, 65, 1987.
13. **Wilson, K.J., Honegger, A., Stötzel, R.P., and Hughes, G.J.**, The behaviour of peptides on reverse-phase supports during high-pressure liquid chromatography, *Biochem. J.*, 199, 31, 1981.
14. **Sasagawa, T., Okuyama, T., and Teller, D.C.**, Prediction of peptide retention times in reversed-phase high-performance liquid chromatography during linear gradient elution, *J. Chromatogr.*, 240, 329, 1982.
15. **Hearn, M.T.W., Aguilar, M.I., Mant, C.T., and Hodges, R.S.**, High-performance liquid chromatography of amino acids, peptides and proteins. LXXXV. Evaluation of the use of hydrophobicity coefficients for the prediction of peptide elution profiles, *J. Chromatogr.*, 438, 197, 1988.

MANIPULATION OF pH AND ION-PAIRING REAGENTS TO MAXIMIZE THE PERFORMANCE OF REVERSED-PHASE COLUMNS

Hugh P. J. Bennett

Reversed-phase HPLC (RPC) has found wide acceptance in numerous peptide and protein laboratories around the world. RPC columns provide rapid, high resolution separations together with generally excellent recoveries. It is the technique of choice for peptide isolation and purification. RPC is particularly useful for mapping of peptide fragments following digestion with endopeptidases. Despite its high resolving power, RPC is frequently only used in the final stages of peptide and protein purification schemes. It is hoped that in this chapter the reader will gain insight into how to maximize the resolving power of RPC columns. Note that in discussing the concepts of ion-pairing and pH manipulation, I have taken what is a rather simplistic view of the mechanisms involved. The mechanisms involved are undoubtedly much more complex than those I have discussed. However, the interpretations in large part correlate well with the way peptides and proteins actually behave in the various solvent systems that are discussed. Hopefully, these interpretations are conceptually useful. The development and successful use of the solvent systems described has led to the availability of HPLC-grade reagents. For instance, the Pierce Chemical Co. (Rockford, IL) markets HPLC-grade versions of trifluoroactic acid, heptafluorobutyric acid, triethylamine, and tetrabutylammonium phosphate. Use of reagent grade chemicals leads to baseline problems and the appearance of spurious peaks. It is important to use high grade reagents to ensure optimal column performance. The preparation of solvents for RPC is discussed in detail elsewhere.[1]

The most important feature of a successful purification is that it be a convergent process. Each chromatographic step should ideally be based upon principles entirely or radically different from the preceding step. The use of size-exclusion chromatography (SEC), ion-exchange chromatography (IEC) and RPC is a common combination which exploits in a predictable manner the size, charge and hydrophobic character of the peptides and proteins in question. To maximize the contribution of RPC to a purification scheme, the use of the separation principle should also be a convergent process. RPC can be manipulated in a number of ways. First of all, the column itself can be varied. Numerous bonded-phase silica and polymer HPLC columns are available (e.g. C_2, C_8, C_{18}, CN, phenyl, etc.). The interaction of solutes with these bonded-phases varies from column to column often in quite subtle ways. The great disadvantage of using a selection of RPC columns for peptide and protein purification is that it is not a predictable process. That is, many reversed-phase columns often behave similarly and it may take several attempts before a column is found that enhances the purification scheme. Meanwhile, the losses incurred in running and re-running the sample could ultimately lead to failure. Similarly, altering the nature of the organic component of the solvent system can alter column behavior, but again not in a predictable manner. The most successful way in which column behavior is altered is by manipulating the ionic component of the solvent system. By altering the nature and strength of the ion-pairing reagent and the pH of the solvent system, it is possible to exploit both the acidic and basic character of solutes in a systematic and predictable manner.[1] Without knowing anything about the structure of a given peptide or protein, it is certain that it will have hydrophobic character through its content of the lipophilic amino acids leucine, isoleucine, valine, methionine, phenylalanine, tyrosine, and tryptophan. It will also have hydrophilic character dependent largely upon the presence of aspartic acid, glutamic acid, arginine, and lysine residues. The ratio between hydrophobic and hydrophilic amino acids determines the initial behavior observed for any peptide or protein in the most common RPC solvent system in use today, namely aqueous acetonitrile containing 0.1% trifluoroacetic acid (TFA).[2,3] To some extent the hydrophilic character of solutes is suppressed at the low pH of this system (i.e., approximately 2). All carboxylic acid groups are protonated under these conditions and their contribution to the hydrophilic character is reduced. This, in turn, improves peak shape. TFA is used in preference to other fully dissociated acids because it is completely volatile (c.f. phosphoric acid) and will not corrode stainless steel (c.f. hydrochloric acid). TFA also acts as a weak hydrophobic ion-pairing reagent. The behavior of solutes in this system serves as the starting point in the purification strategy outlined below (see Figure 3).

The concept of ion-pairing in RPC is very useful in both developing and understanding the mechanism of action of solvent systems. At any given pH, proteins and peptides will display ionized amino acids which are available for ion-pairing. For instance, in the pH range dictated by the stability of silica-based columns (i.e., pH 2 to 8) the basic amino acids (arginine and lysine) and the terminal amino group are fully ionized. It has been suggested that anions associate or ion-pair with these positive charges. Hydrophobic anions such as trifluoroacetate and alkylsulfonates form hydrophobic ion pairs with the basic charges. The resulting complex tends to mask the basic charge and tends also to increase the affinity of the peptide or protein for the column. This, in turn, leads to an increased retention time. The perfluorinated carboxylic acids trifluoroacetic acid, pentafluoropropionic acid (PFPA), and heptafluorobutyric acid (HFBA) are useful hydrophobic ion-pairing reagents. They are all fully dissociated acids and volatile. Figure 1 shows a comparison of a mixture of standard peptides in solvent systems containing 0.01 M TFA, PFPA or HFBA. Table 1 shows the structures of these model peptides. The chromatographic conditions are identical in each instance (i.e., column, molarity of acid, solvent gradient). It is immediately apparent that retention times for the various peptides are different in each of the solvent systems. More importantly, the relative retention times are different. In progressing from the TFA to the HFBA system, it can be seen that some peptides increase in retention time more than others. The retention times of $ACTH_{1-24}$ and $ACTH_{1-39}$ shift the most. This correlates well with the high number of basic residues found in these peptides (see Figure 1 and Table 1).

These findings demonstrate that by changing the ion pairing component of a solvent system, retention times of solutes can be altered in a predictable manner. The elements of cation-exchange chromatography are introduced into the purification scheme. Furthermore, the basic character of the peptides is exploited in isolation from the acidic character since the carboxylic acids are not ionized. The term "dynamic ion exchanger" has been used to describe an RPC column used in this way[4]. TFA and HFBA are weak and strong hydrophobic ion-pairing reagents, respectively. The use of solvent systems such as these in tandem is as far as the basic character of solutes can be exploited (see Figure 3). The sequential use of solvents containing TFA, HFBA, and finally TFA with the same RPC column is frequently sufficient to achieve purification of peptides.[3]

It is possible to use the resolving power of RPC further by exploiting the acidic character of peptides and proteins. To do this the pH must be raised well above the pK for the carboxylic acids (i.e., to at least pH 5). While it is unnecessary to buffer solvents at low pH, the presence of a buffer is very important to stabilize higher pHs. A convenient solvent of higher pH is provided by 0.01 M triethylamine acetate, pH 5.5 (TEA).[1,5] Like the perfluorinated acid systems, it is volatile. Like TFA, triethylamine is a weak hydrophobic ion-pairing reagent. Any changes in retention times found with the TEA system relative to the TFA system are due primarily to an increase in the pH and the resulting ionization of the carboxyl side chains of glutamic acid and aspartic acid and

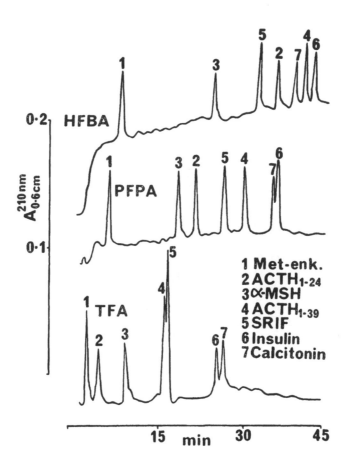

FIGURE 1. Comparison of the effectiveness of trifluoroacetic acid (TFA), pentafluoropropionic acid (PFPA) and heptafluorobutyric acid (HFBA) as hydrophobic ion-pairing reagents in RPC of seven natural and synthetic peptides. Peptides were eluted from the Waters Associates, μBondapak C_{18} column (300 × 3.9 mm I.D.) with a linear AB gradient of 20 to 40% acetonitrile containing 0.01 M concentrations of each acid over 1 hr (0.33% acetonitrile/min) at a flow-rate of 1.5 ml/min. The upper panel shows the elution behavior of the peptides with 0.01 M (0.13% v/v) HFBA, the middle panel shows the behavior with 0.01 M (0.10% v/v) PFPA and the lower panel shows the behavior with 0.01 M (0.07% v/v) TFA. Samples (2 μg) of the following peptides were injected onto the column (number of basic residues at pH 2, including histidine, is shown in brackets): (1) methionine enkephalin, (1); (2) $ACTH_{1-24}$, (9); (3) α-MSH, (3); (4) human $ACTH_{1-39}$,(9); (5) somatostatin, (3); (6) bovine insulin,(6); (7) human calcitonin (3).

the terminal carboxyl group. Peptides exhibit increased hydrophilic character in the TEA system and retention times are decreased dramatically. This is illustrated by considering the behavior of model peptides α-melanocotropin (α-MSH) and corticotropin-like intermediate lobe peptide (CLIP) in Figure 2. α-MSH and CLIP are biosynthetic derivatives of ACTH in the intermediate lobe of the pituitary. Each is subject to post-translational modification. α-MSH exists in mono- and di-acetylated forms, while CLIP exists in phosphorylated or non-phosphorylated forms[6]. At pH 5.5, α-MSH has one free carboxyl group (the carboxyl terminus is amidated). In contrast, rat CLIP has five free carboxyl groups (including the free carboxyl-terminus). The structures of α-MSH and CLIP are shown in Table 1. In switching from the TFA system (Left Panel, Figure 2) to the TEA system (Middle Panel, Figure 2) the polarity of CLIP increases greatly relative

TABLE 1
The Predicted Number of Positively and Negatively Charged Residues at pH 2 and pH 5.5 for the Model Peptides Used in Figures 1 and 2

Peptide identity[a]	Predicted no. of positively charged residues at pH 2	Predicted no. of negatively charged residues at pH 5.5	Peptide sequence (single letter code)
MET-enkephalin	1	1	YGGFM
Corticotropin (1–24)	9	2	SYSMEHFRWGKPVGKKRRPVKVYP
α-Melanotropin	3	1	Ac-SYSMEHFRWGKPV-NH$_2$
Corticotropin (1–39)	9	7	SYSMEHFRWGKPVGKKRRPVKVYPNGAEDESAEAAFPLEF
Somatostatin	3	1	AGCKNFFWKTFTSC
Insulin (bovine)	6	6	GIVEQCCASVCSLYQLENYCN FVNQHLCGSHLVEALYLVCGERGFFYTPKA
Calcitonin	3	1	CGNLSTCMLGTYTQDFNKFHTFPQTA–NH$_2$
Corticotropin-like intermediate lobe peptide (rat)	3	5	RPVKVYPNVAENESAEAAFPLEF

a The human sequence is shown unless indicated otherwise.

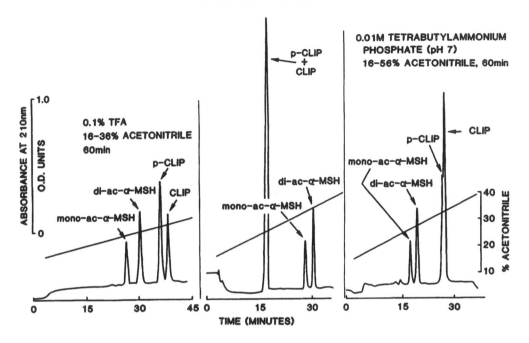

FIGURE 2. Comparison of the behavior of α-MSH (mono- and di-acetylated) and CLIP (phosphorylated and non-phosphorylated) in three different RPC systems. These peptide standards were purified from rat pituitaries as described previously.[3,6] Peptides were eluted from the μBondapak C_{18} column (300 × 3.9 mm I.D.) with linear gradients of acetonitrile at a flow-rate of 1.5 ml/min. The left panel shows the behavior of the peptides in a solvent system containing 0.1% TFA. The middle panel shows the behavior of the peptides in a system containing 0.01 M triethylamine acetate (pH 5.5). The right panel shows the behavior of the peptides in a system containing 0.01 M tetrabutylammonium phosphate (pH 7).

to α-MSH. The net result is that while the retention times of both peptides fall, they fall greater for the CLIP peptides.

These findings demonstrate that by increasing the pH of the solvent system the retention times of solutes can be altered in a predictable manner. The acidic character is now the basis of the separation. The use of the TEA system follows logically after the use of the HFBA although the order in which the solvent systems are used is not really important. What is important is that basic and acidic character are exploited in turn in the purification scheme (see Figure 3). The sequential use of TFA, HFBA, and TEA systems can be useful for more difficult separation problems.[1]

One more manipulation is possible in the maximization of the resolving power of RPC. At higher pHs, carboxylic acids are fully charged and available for ion-pairing. Unfortunately, there are no strong hydrophobic reagents that ion-pair with carboxylic acids and that are also volatile. However, tetrabutylammonium phosphate at pH 7 can be used with good effect in this role.[1,6] If α-MSH and CLIP are subjected to RPC in solvents containing 0.01 M tetrabutylammonium phosphate (pH 7), their relative retention times are again altered. As can be seen from Figure 2 (Right Panel), CLIP now emerges from the column after α-MSH. This can be rationalized in terms of ion-pairing of the tetrabutylammonium ion with free carboxyl-groups. The greater the number of carboxyl groups the greater the ion-pairing possible and the greater the increases in retention time. Thus, for the complete exploitation of the ionic character of a given peptide or protein the use of this solvent system is recommended.

The general approach to peptide purification is represented diagrammatically in Figure 3.

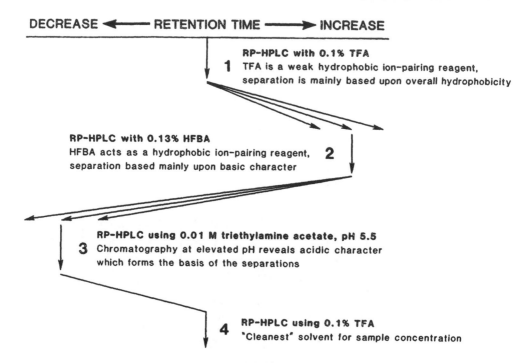

FIGURE 3.   Diagrammatic representation of a peptide purification scheme employing various solvent modifiers.

This diagram highlights the behavior of peptides and proteins through the recommended scheme. All the solvent systems used are volatile, making them compatible with bioassays, radioimmunoassays, and structural analysis procedures. As a general rule, however, solvents containing TFA leave the least residue upon evaporation. This is the system of choice for the preparation of samples for amino acid analysis or sequencing. Each solvent system exploits a different characteristic of the peptide of interest in a covergent process. The behavior of peptides and proteins is generally predictable. Occasionally separations are observed which are not accountable in the terms considered above. For instance, it is hard to rationalize the behavior of phosphorylated CLIP relative to that of CLIP (Figure 2). It is not obvious why they should tend to be co-eluted in the TEA and tetrabutylammonium phosphate systems. Similarly, it is not always possible to predict when a given chromatographic procedure will lead to poor yields. The scheme illustrated in Figure 3 has been used in the purification of all the biosynthetic derivatives of pro-opiomelanocortin purified from mouse neurointermediate pituitaries.[8] Yields were generally excellent throughout the purification scheme. Loading and re-loading samples onto the RPC column without taking them to complete dryness helps to maintain yields. In the past we have found that loading samples through the aqueous HPLC pump to be an invaluable procedure.[3] In these studies, a Waters Associates (Milford, MA) HPLC system consisting of two 6000A pumps was used. These pumps have a three-way port which facilitates loading of samples through the aqueous pump. Recently, we have connected a Milton-Roy Simplex pump (LDC, Riviera Beach, FL) to the system and dedicated it to sample loading.

REFERENCES

1. **Bennett, H.P.J.,** Isolation of pituitary peptides by reversed-phase high-performance liquid chromatography. Expansion of the resolving power of reversed-phase columns by manipulating pH and the nature of the ion-pairing reagent, *J. Chromatogr.*, 226, 501, 1983.
2. **Bennett, H.P.J., Hudson, A.M., McMartin, C., and Purdon, G.E.,** Use of octadecasilyl-silica for the extraction and purification of peptides in biological samples. Application to the identification of circulating metabolites of corticotropin-(1-24)-tetracosapeptide and somatostatin *in vivo*, *Biochem. J.*, 168, 9, 1977.
3. **Bennett, H.P.J., Browne, C.A., and Solomon, S.,** Purification of the two major forms of rat pituitary corticotropin using only reversed-phase liquid chromatography, *Biochemistry*, 20, 4530, 1981.
4. **Hearn, M.T.W. and Hancock, W.S.,** Ion-pair partition reversed-phase HPLC, *Trends in Biochemical Sciences*, 4, N58, 1979.
5. **Matrisian, L.M., Larsen, B.R., Finch, J.S., and Magun, B.F.,** Further purification of epidermal growth factor by high performance liquid chromatography, *Anal. Biochem.*, 125, 339, 1982.
6. **Browne, C.A., Bennett, H.P.J., and Solomon, S.,** Isolation and characterization of corticotropin and melanocortin-related peptides from the neurointermediary lobe of the rat pituitary by reversed-phase liquid chromatography, *Biochemistry*, 20, 4538, 1981.
7. **Hancock, W.S., Bishop, C.A., Battersby, J.E., Harding, D.R.K., and Hearn, M.T.W.,** High pressure liquid chromatography of peptides and proteins, XI the use of cationic reagents for the analysis of peptides by high pressure liquid chromatography, *J. Chromatogr.*, 168, 377, 1979.
8. **Bennett, H.P.J.,** Biosynthetic fate of the amino-terminal fragment of pro-opiomelanocortin within the intermediate lobe of the mouse pituitary, *Peptides*, 7, 615, 1986.

THE EFFECTS OF ANIONIC ION-PAIRING REAGENTS ON PEPTIDE RETENTION IN REVERSED-PHASE CHROMATOGRAPHY

Colin T. Mant and Robert S. Hodges

I. INTRODUCTION

Of the three major modes of HPLC (size-exclusion chromatography, ion-exchange chromatography, reversed-phase chromatography), reversed-phase chromatography (RPC) offers the widest scope for manipulation of mobile phase characteristics during peptide separations. Several mobile phase parameters, including pH, organic modifier, and choice of counterion, can be varied for optimum peptide resolution. Indeed, a major part of the flexibility and excellent peptide resolving power of RPC is derived from the availability and use of ion-pairing reagents. Peptides are charged molecules at most pH values and the presence of different counterions will influence their chromatographic behavior. Differences in the polarities of peptides in a peptide mixture can be maximized through careful choice of ion-pairing reagent. Favored models for the mechanism of ion-pair separations either involve formation of ion pairs with the sample solute in solution, followed by retention of the solute molecules on a reversed-phase column,[1,2] or a dynamic ion-exchange event in which the ion-pairing reagent is first retained by the reversed-phase column and then solute molecules exchange ions with the counterion associated with the sorbed ion-pair reagent.[3-6] Both models yield similar predictions concerning separation as a function of experimental conditions. Whatever the mechanism, the resolving power of ion-pairing reagents is effected by its interaction with the ionized groups of a peptide: anionic (negatively charged) counterions will interact with the protonated (positively charged) basic residues (arginine, lysine, histidine) of a peptide as well as protonated free N-terminal amino groups; cationic (positively charged) counterions will show an affinity for ionized (negatively charged) acidic residues (glutamic- and aspartic acid) as well as ionized free C-terminal carboxyl groups. In addition, the actual effect on retention time of a peptide will depend strongly on the hydrophobicity of the ion-pairing reagent and the number of oppositely charged groups on the peptide. The homologous series of alkylsulfonates (e.g., butane- and hexanesulfonic acid)[7-11] and perfluorinated carboxylic acids (e.g., trifluoroacetic acid [TFA] and heptafluorobutyric acid [HFBA])[12-22] offer a range of anionic ion-pairing reagents with gradually changing hydrophobicity. Commonly used cationic ion-pairing reagents include tertiary alkylamines,[23-25] particularly triethylammonium phosphate. Quaternary ammonium ions are also useful as cationic ion-pairing reagents;[9,10,12,26] tetrabutylammonium phosphate has shown particular use as a strongly hydrophobic cationic counterion.

The use of cationic ion-pairing reagents is, of course, limited to pH values above the pKa values of acidic side-chain groups (~ pH 4.0), generally necessitating the employment of non-volatile mobile phases to maintain the required pH, although a volatile triethylamine acetate system (triethylamine is a weak hydrophobic ion-pairing reagent) has seen some use in peptide separations.[12,13,27] The majority of researchers utilizing ion-pair RPC for peptide separations take advantage of the excellent resolving power and selectivity of the anionic ion-pairing properties of TFA, particularly in the use of TFA/water to TFA/acetonitrile gradients. The low pH (~ 2.0) of simple, unbuffered solutions (0.05 to 0.1% TFA [v/v]) ensures protonation of hydrophilic carboxylic acid groups, thereby increasing the interaction of peptides with the reversed-phase sorbent. In addition, the suppression of surface silanols in silica-based packings at low pH (<3.5 to 4.0) decreases ionic interactions with the hydrophobic stationary phase. Silica is also more stable at low pH. TFA is an excellent solvent for most peptides, is completely volatile and enables detection at wavelengths below 220 nm due to its low UV transparency.

If the presence of TFA is not sufficient to resolve a particular mixture of peptides efficiently, considerable flexibility in the degree of peptide retention and elution order may be achieved through use of a more hydrophilic (e.g., orthophosphoric acid) or a more hydrophobic (e.g., HFBA) anionic ion-pairing reagent. This article demonstrates how the effects of anionic ion-pairing reagents on peptide elution profiles in RPC may be manipulated in a highly predictable manner to achieve the desired separation. In addition, the effect of counterion concentration on peptide resolution and retention is discussed.

II. EFFECTS OF ANIONIC ION-PAIRING REAGENTS ON PEPTIDE RETENTION BEHAVIOR

A. EFFECT OF COUNTERION HYDROPHOBICITY ON PEPTIDE RETENTION

The labeling of a particular anionic counterion as hydrophobic or hydrophilic often tends to be somewhat arbitrary, relying as it does on the relative effectiveness of one ion-pairing reagent compared to another. TFA, for instance, has been described by various researchers as a hydrophilic, weakly hydrophobic, or a hydrophobic ion-pairing reagent. All anionic counterions are potentially capable of ion-pairing with the positively charged basic residues of a peptide, thereby reducing its overall hydrophilicity and increasing peptide retention. However, they differ in their ability to interact with the reversed-phase sorbent, thus producing a useful basis for defining the nature of a particular counterion. Hence, a hydrophobic counterion (e.g., trifluoroacetate, heptafluorobutyrate) is not only capable of ion-pairing with the positively charged solute, but, due to its hydrophobicity, can increase further the affinity of the peptide for the hydrophobic stationary phase. In contrast, a polar hydrophilic counterion (e.g., phosphate, chloride), following ion-pair formation with positively charged residues, would be unlikely to interact with the non-polar sorbent. The increase in peptide retention would only be due to reduction in hydrophilicity of positively charged residues by ion-pair formation. Several studies utilizing perfluorinated carboxylic acids as ion-pairing reagents have demonstrated increasing retention times of basic peptides with increasing hydrophobicity of the counterion.[15,18-21] HFBA has been used as the ion-pairing reagent of choice under circumstances where the resolving power of the TFA system has not been sufficient to separate satisfactorily a peptide mixture.[12,14,16-18,21,22] Apart from its effectiveness as an ion-pairing reagent, it shares with TFA the advantages of volatility and, at low concentrations, UV transparency to permit monitoring of column eluate at 210 nm. Despite being non-volatile, orthophosphoric acid (H_3PO_4) has proved useful as a hydrophilic ion-pairing reagent for hydrophobic peptides and proteins.[8,9,21,29-33] Its use permits a significant decrease in the concentration of organic solvent in the mobile phase required for elution of the protein, thus reducing the possibility of denaturation or precipitation.[30,31]

The effect of increasing anionic counterion hydrophobicity on the reversed-phase elution profile of a series of synthetic peptide cation-exchange standards is shown in Figure 1. The four standards, C1 to C4, contain net charges of +1 to +4, respectively (see Table 1 for peptide sequences). Peptides C2 and C4 (+2 and +4 net charge, respectively) contain tyrosine residues, enabling these peptides to be detected at 280 nm. The peptide standards were chromatographed on a SynChropak C_{18} column and an Aquapore C_8 column under linear AB gradient conditions (Eluent A = 0.1% [v/v] aq. H_3PO_4 [Panels A and D], TFA [Panels B and E] and HFBA [Panels

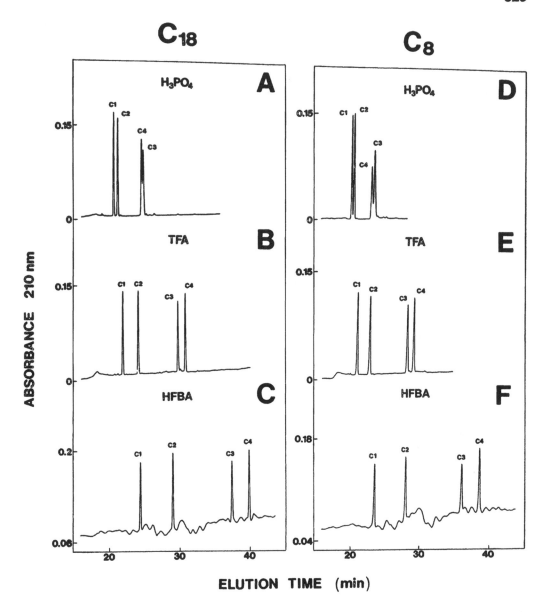

FIGURE 1. Effect of anionic ion-pairing reagents on the separation of a mixture of synthetic peptide HPLC standards in RPC. The HPLC instrument consisted of a Varian Vista Series 5000 liquid chromatograph (Varian, Walnut Creek, CA) coupled to a Hewlett-Packard (Avondale, PA) HP1040A detection system, HP9000 Series 300 computer, HP9133 disc drive, HP2225A Thinkjet printer and HP7470A plotter. Columns: Panels A to C, SynChropak C_{18}, 250×4.6 mm I.D., 6.5-μm particle size, 300-Å pore size (SynChrom, Linden, IN); Panels D to F, Aquapore C_8, 220×4.6 mm I.D., 7 μm, 300 Å (Brownlee Labs, Santa Clara, CA). Conditions: linear AB gradient (1% B/min), where Eluent A is water and Eluent B is acetonitrile, both eluents containing 0.1% H_3PO_4 (panels A and D), TFA (B and E) or HFBA (C and F); flow-rate, 1 ml/min; 26°C. Sequences of peptide standards C1 to C4 are shown in Table 1. Peptides C2 and C4 contain tyrosine residues and, thus, are detectable at 280 nm.

C and F]; Eluent B = 0.1% [v/v] of the respective ion-pairing reagents in acetonitrile; 1% B/min, 1 ml/min, 26°C). The rather poor baselines shown in Panels C and F are frequently observed when using HFBA as an ion-pairing reagent. The peptide elution profiles obtained on the two columns were very similar, with C1 and C2 forming one peptide pair and C3 and C4 being eluted as a second peptide pair. However, the retention behavior of the peptides varied quite dramatically between the three ion-pairing reagent systems. The retention times of all four

TABLE 1
Sequences of Peptides

Peptide	Peptide sequence[a]	Number of positively charged groups[b]
C1[c]	Ac-Gly-Gly-Gly-Leu-Gly-Gly-Ala-Gly-Gly-Leu-L*ys-amide	1
C2	Ac-L*ys-Tyr-Gly-Leu-Gly-Gly-Ala-Gly-Gly-Leu-L*ys-amide	2
C3	Ac-Gly-Gly-Ala-Leu-L*ys-Ala-Leu-L*ys-Gly-Leu-L*ys-amide	3
C4	Ac-L*ys-Tyr-Ala-Leu-L*ys-Ala-Leu-L*ys-Gly-Leu-L*ys-amide	4
S1[d]	*A*rg-Gly-Ala-Gly-Gly-Leu-Gly-Leu-Gly-L*ys-amide	3
S2	Ac-A*rg-Gly-Gly-Gly-Gly-Leu-Gly-Leu-Gly-L*ys-amide	2
S3	Ac-A*rg-Gly-Ala-Gly-Gly-Leu-Gly-Leu-Gly-L*ys-amide	2
S4	Ac-A*rg-Gly-Val-Gly-Gly-Leu-Gly-Leu-Gly-L*ys-amide	2
S5	Ac-A*rg-Gly-Val-Val-Gly-Leu-Gly-Leu-Gly-L*ys-amide	2
0	Ac-Thr-Asp-Leu-Leu-Gly-amide	0
1	Ac-Val-Ser-L*ys-Thr-Glu-Thr-Ser-Gln-Val-Ala-Pro-Ala-amide	1
2A	Ac-A*rg-Gly-Ala-Gly-Gly-Leu-Gly-Leu-Gly-L*ys-amide	2
2B	Ac-A*rg-Gly-Val-Gly-Gly-Leu-Gly-Leu-Gly-L*ys-amide	2
3	*A*rg-Gly-Ala-Gly-Gly-Leu-Gly-Leu-Gly-L*ys-amide	3
4	Ac-Ser-Asp-Gln-Glu-L*ys-A*rg-L*ys-Gln-Ile-Ser-Val-A*rg-Gly-Leu-amide	4
5	Ac-Gly-L*ys-Phe-L*ys-A*rg-Pro-Pro-Leu-A*rg-A*rg-Val-Gly-amide	5
6	Ac-Gly-L*ys-Phe-L*ys-A*rg-Pro-Pro-Leu-A*rg-A*rg-Val-A*rg-amide	6

[a]   Ac = Nα-acetyl; amide = Cα-amide.

[b]   Positively charged residues (lysine, arginine) are marked with a star (*); peptides 3 and S1 (which are identical) have a free α-amino group, which also contains a positive charge.

[c]   Peptides C1 to C4 are synthetic cation-exchange HPLC standards available from Synthetic Peptides Incorporated (Department of Biochemistry, University of Alberta, Edmonton, Alberta, Canada).

[d]   Peptides S1 to S5 are synthetic reversed-phase standards, also available from Synthetic Peptides Incorporated.

peptides increased with increasing counterion hydrophobicity ($H_2PO_4^- <$ TFA$^- <$ HFBA$^-$) (Table 2). However, the magnitude of this increase depended on the number of positively charged residues in the peptide; thus, the greater the net positive charge of the peptide, the greater the increase in retention time with increasing counterion hydrophobicity. For example, the retention times of peptide C1 (+1 net charge) on column 1 were 20.6, 21.8, and 24.6 min, respectively, for the ion-pairing reagents H_3PO_4, TFA, and HFBA, i.e., a difference of 4 min between the least hydrophobic (H_3PO_4) and most hydrophobic (HFBA) systems. In contrast, the retention times of peptide C4 (+4 net charge) were 24.5, 30.6, and 40.0 min, respectively, in the H_3PO_4, TFA and HFBA systems, i.e., a difference of 15.5 min between the H_3PO_4 and HFBA systems (Table 2). It is interesting to note that the retention order of peptides C3 and C4 (+3 and +4 net charge, respectively) changed between the H_3PO_4 (Figure 1, panels A and D and the TFA (panels B and E) systems, as confirmed by the detection of C4 at 280 nm. In the H_3PO_4 system, C4 was eluted before C3; in the TFA system (and, subsequently, the HFBA system [Panels C and F]), C4 was eluted after C3.

Figure 2 demonstrates the peptide resolving capability of RPC achievable by simply changing the hydrophobicity of the ion-pairing reagent. A mixture of basic peptides of similar size but containing varying numbers of positively charged groups (see Table 1 for peptide sequences) were subjected to linear gradient elution (same conditions as Figure 1) on a SynChropak C_{18} column in anionic ion-pairing systems of H_3PO_4, TFA and HFBA (Figure 2A, 2B, and 2C, respectively). As seen previously for the peptide standards (Figure 1, Table 2), the peptides in Figure 2 exhibited increasing retention times (Table 3) with increasing counterion

TABLE 2
Effect of Ion-Pairing Reagent on Retention Times of Synthetic Peptide HPLC Standards in RPC

Peptide standard	No. of positively charged residues (N)	Retention time (min)				Δ/N^a (HFBA-TFA)	Δ/N (TFA-orthophosphoric acid)
		Orthophosphoric acid	TFA	HFBA			
C$_{18}$ column							
C1	1	20.6	21.8	24.6		2.8	1.2
C2	2	21.2	24.0	29.1		2.6	1.4
C3	3	24.7	29.6	37.5		2.6	1.6
C4	4	24.5	30.6	40.0		2.4	1.5
						Average 2.5^b	Average 1.5
C$_8$ column							
C1	1	19.7	20.8	23.3		2.5	1.1
C2	2	20.0	22.6	27.8		2.6	1.3
C3	3	22.9	28.0	35.9		2.6	1.7
C4	4	22.5	29.0	38.4		2.4	1.6
						Average 2.5	Average 1.5

Note: Columns: SynChropak C$_{18}$ (250 × 4.6 mm I.D.); Aquapore C$_8$ (220 × 4.6 mm I.D.). HPLC instrument and conditions: same as Figure 1. Sequences of peptide standards C1 to C4 are shown in Table 1.

a Δ denotes difference in retention time of a peptide between two ion-pairing reagent systems; N denotes number of positively charged groups in peptide.

b The average values were obtained by summing the differences in peptide retention times of all four peptides between two ion-pairing systems and then dividing by the number of positive charges they possess in total ($1 + 2 + 3 + 4 = 10$).

COLUMN 1

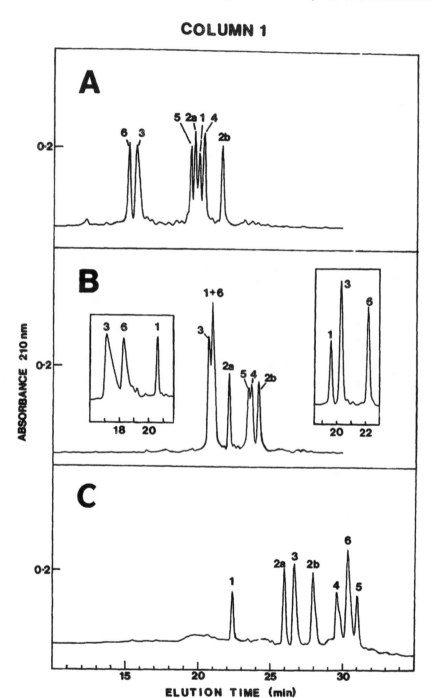

FIGURE 2. Effect of anionic ion-pairing reagent on the separation of a mixture of basic peptides in RPC. HPLC instrument: same as Figure 1, except a Hewlett Packard HP85B computer and HP9121 disc drive were used. Column: SynChropak C$_{18}$, 250 × 4.1 mm I.D., 6.5-µm particle size, 300-Å pore size. Conditions: linear AB gradient (1% B/min), where Eluent A is water and Eluent B is acetonitrile, both eluents containing 0.1% H$_3$PO$_4$ (Panel A), TFA (B) or HFBA (C); flow-rate, 1 ml/min; 26°C. Panel B, insets: left 0.01% TFA in Eluents A and B; right, 0.4% TFA in Eluents A and B. Sequences of peptides are shown in Table 1. (From Guo, D., Mant, C. T., and Hodges, R. S., *J. Chromatogr.*, 386, 205, 1987. With permission.)

TABLE 3
Prediction of Effect of Anionic Ion-Pairing Reagent on Retention Times of Basic Peptides in RPC

Peptide	t_R^{obs} [a] TFA	τ [b] H_3PO_4	t_R^{obs} H_3PO_4	τ HFBA	t_R^{obs} HFBA
C_{18} column					
1	21.0	19.9	20.1	23.0	22.4
2a	22.1	19.9	19.8	26.1	26.0
2b	24.1	21.9	21.7	28.1	28.0
3	20.8	17.5	15.8	26.8	26.7
4	23.6	19.2	20.4	31.6	29.6
5	23.5	18.0	19.5	33.5	30.4
6	21.0	14.4	15.3	33.0	31.0
C_8 column					
1	20.5	19.2	19.8	23.3	23.7
2a	22.5	19.9	19.3	28.1	28.9
2b	25.6	23.0	21.9	31.2	30.4
3	21.3	17.4	15.6	29.7	29.9
4	24.0	18.8	19.3	35.2	32.2
5	23.7	17.2	17.6	37.7	35.6
6	20.5	12.7	12.6	37.3	35.6

Note: Columns: SynChropak C_{18} (250×4.1 mm I.D.); Aquapore C_8 (220×4.6 mm I.D.). HPLC instrument and conditions: same as Figure 2. Sequences of peptides are shown in Table 1.

[a] t_R^{obs} denotes observed peptide retention time (min).

[b] τ denotes predicted peptide retention time (min), calculated as described in text. Retention times in the H_3PO_4 and HFBA systems were predicted from observed retention times in the TFA system. The values for Δ/N (see text) were obtained from the retention times of a mixture of synthetic decapeptide standards, S1 to S5 (Table 1): TFA $\Leftrightarrow H_3PO_4$, Δ/N = 1.1 min (C_{18}) or 1.3 min (C_8); TFA $\Leftrightarrow$ HFBA, Δ/N = 2.0 min (C_{18}) or 2.8 min (C_8).

hydrophobicity: $H_2PO_4^-$ (Figure 2A) < TFA⁻ (Figure 2B) < HFBA⁻ (Figure 2C). In addition, the elution order of the peptides changed from one counterion system to another, due to the difference in magnitude of the effect of counterion hydrophobicity on peptides containing different numbers of positively charged groups. Thus, as seen for the peptide standards (Figure 1, Table 2), the greater the net positive charge of a peptide, the greater the effect on peptide retention of varying the anionic counterion. For example, in Figure 2, the elution order of peptides 1, 3, and 6 (containing one, three, and six positively charged residues, respectively) was reversed as the counterion changed from $H_2PO_4^-$ (Figure 2A) to HFBA⁻ (Figure 2C).

B. EFFECT OF COUNTERION CONCENTRATION ON PEPTIDE RETENTION

Acidic ion-pairing reagents are generally used only at low concentrations (0.05 to 0.1% v/v) in the mobile phase. A number of researchers have demonstrated increasing peptide retention times with increasing concentrations of anionic counterions in the mobile phase.[9,18,20,21,34] Guo et al.[18] examined the effect of increasing concentrations (0.01 to 0.5% v/v) of H_3PO_4, TFA, and HFBA on the elution profiles of a mixture of five synthetic decapeptide reversed-phase standards on a C_8 column. Four of the peptides contained two positively charged groups, while the fifth peptide (peptide 3 in Table 1) contained three positively charged groups. The peptides all demonstrated increasing retention times with increasing concentrations of the acids. In a similar manner to the effect of counterion hydrophobicity on peptide retention (Figure 1 and 2), the increase in retention time of a peptide was related to the number of positively charged groups it contained: the higher the net positive charge of a peptide, the greater the increase in peptide retention time with increasing counterion concentration. Guo et al.[18] subsequently demonstrated that counterion concentration had an essentially equal effect on each positively charged group.

The potential resolving capability of variations in TFA concentration is shown in Figure 3. Several peptides of similar size but varying in their number of positively charged groups (Table 1) were chromatographed on a SynChropak C_{18} column under linear AB gradient conditions (1% B/min, 1 ml/min, 26°C), where Eluent A was water and Eluent B was acetonitrile, both eluents containing 0.01 to 0.8% (v/v) of TFA. As shown in Figure 3, the peptides exhibited a changing resolution profile as the concentration of TFA in the mobile phase was increased, producing the best separation (baseline resolution of all five peptides) at a 0.05% concentration of the ion-pairing reagent. It may have been expected, due to the absence of any positively charged residues, that the retention time of peptide 0 would be unaffected by changes in TFA concentration. This peptide was certainly affected least over the TFA concentration range examined, but still exhibited a small variation in retention time over the 0.1 to 0.8% TFA concentration range (Figure 3). As the concentration of the ion-pairing reagent increased from 0.01 to 0.8% in the mobile phase, the pH of the aqueous component varied from 3 to 1.3. Because organic solvents tend to suppress ionization, the pH of an aqueous solvent mixture may only be an apparent value.[15] Although the apparent pH of the mobile phase may not be a very critical parameter provided that it is well below the pK of the peptide carboxyl groups (the pk of Asp and Glu side-chain carboxyls generally varies between 3.9 and 4.4 for peptides and proteins, respectively; for C-terminal α-carboxyl groups, these values vary between 2.0 and 3.1 for peptides and proteins, respectively), it is possible that the combination of increasing concentration of the hydrophobic trifluoroacetate counterion, coupled with a significant drop in pH, is somehow affecting the interaction of the reversed-phase packing with the peptide.

A comparison of the effects of varying counterion concentration vs. counterion hydrophobicity is demonstrated in Figure 2. As noted above, the elution order of peptides 1, 3, and 6 (containing one, three, and six positively charged residues, respectively) was reversed as the counterion changed from $H_2PO_4^-$ (Figure 2A) to HFBA$^-$ (Figure 2C). In comparison, the peptide elution order in the right inset of Figure 2B was 1, 3, and 6 at 0.4%, while at 0.01% TFA (Figure 2B, left inset) the elution order was 3, 6, and 1. In many cases in the purification of synthetic peptides, contaminating peptides and the desired peptide can be very similar in hydrophobicity under the conditions used. If the contaminants vary in the number of positively charged residues they contain compared to the peptide of interest, changing the counterion hydrophobicity or concentration should resolve these contaminants from the desired peptide. This approach is probably more advantageous than searching for columns with different selectivities and, in addition, is a useful test of peptide homogeneity. It should be noted, however, that extensive use of high concentrations of these acids should be avoided as they may have a detrimental effect on the stability of the column packing.

III. PREDICTION OF EFFECTS OF ANIONIC ION-PAIRING REAGENTS ON PEPTIDE RETENTION TIMES

A. BASIS OF RETENTION TIME PREDICTION BETWEEN DIFFERENT ION-PAIRING REAGENT SYSTEMS

Guo et al.[18] compared the effects of H_3PO_4, TFA, and HFBA on the retention of a series of model synthetic peptides, originally used to determine sets of retention coefficients in a 0.1% TFA system: Ac-Gly-X-X-(Leu)$_3$-(Lys)$_2$-amide, where X is substituted by the 20 amino acids found in proteins.[35] The results suggested an essentially equal contribution by each positively charged residue to shifts in peptide retention when changing from one ion-pairing reagent to another. In addition, the results of Guo et al.[18] supported the premise that, at low pH, only positively charged residues need be taken into account when determining the effect of various anionic counterions. The negligible change in retention time of a neutral peptide in the three ion-pairing reagent systems further supported this view.

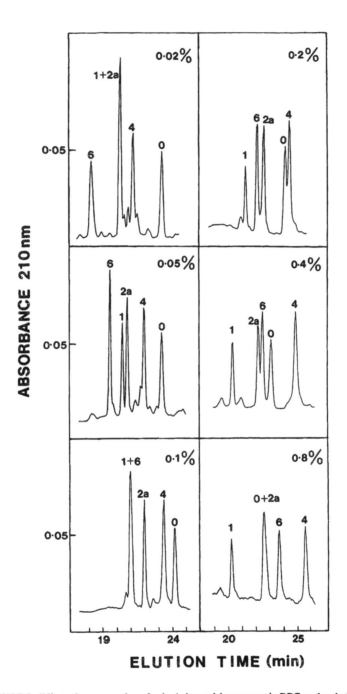

FIGURE 3. Effect of concentration of anionic ion-pairing reagent in RPC on the elution profile of a mixture of peptides with varying numbers of positively charged groups. HPLC instrument and column as Figure 2. Conditions: linear AB gradient (1% B/min), where Eluent A is water and Eluent B is acetonitrile, both eluents containing 0.02, 0.05, 0.1, 0.2, 0.4, or 0.8% TFA; flow-rate, 1 ml/min; 26°C. Sequences of peptides are shown in Table 1. (From Guo, D., Mant, C. T., and Hodges, R. S., *J. Chromatogr.*, 386, 205, 1987. With permission.)

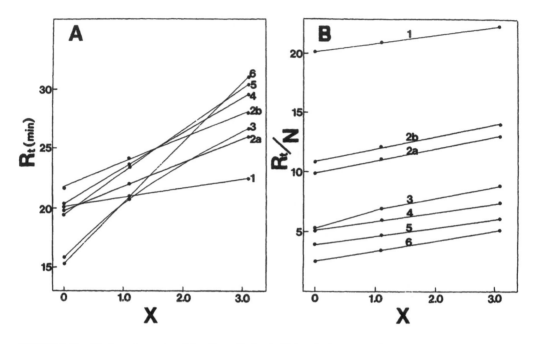

FIGURE 4. Graphical representation of the effect of hydrophobicity of anionic ion-pairing reagent on peptide retention in RPC. Data points were obtained from peptide elution profiles shown in Figure 2. Panel A: retention time (R_t) vs. the average increase in retention time (X) per positively charged group, obtained from a mixture of synthetic peptide standards, when TFA or HFBA was used as the mobile phase acid compared to the retention times of the peptide standards when orthophosphoric acid was used as the mobile phase acid; X = 1.1 (TFA) and 3.1 (HFBA) (Table 3). Panel B: R_t/N vs. X, where N denotes number of positively charged groups in peptide. The values for X were obtained from the retention times of a mixture of synthetic decapeptide standards (S1 to S5): average values of Δ/N (see Table 2 for details of these calculations) on the C_{18} column were 1.1 min (TFA $\Leftrightarrow$ H_3PO_4) and 2.0 min (TFA $\Leftrightarrow$ HFBA); thus, for $H_3PO_4 \rightarrow$ TFA, X = 1.1, and for $H_3PO_4 \rightarrow$ HFBA, X = 1.1 + 2.0 = 3.1. Sequences of peptides and peptide standards S1 to S5 are shown in Table 1. (From Guo, D., Mant, C. T., and Hodges, R. S., *J. Chromatogr.*, 386, 205, 1987. With permission.)

The observed retention times of the seven peptides shown in Figure 2 in the H_3PO_4, TFA, and HFBA systems were plotted against the average increase in retention time per positively charged residue (X) compared to the retention times of the peptide standards when H_3PO_4 is used as the mobile phase acid. The results are shown in Figure 4A. The values for X were obtained by chromatographing a mixture of synthetic reversed-phase decapeptide standards, S1 to S5, (S1 has three positively charged groups and S2 to S5 have 2 positively charged groups; Table 1) on the same C_{18} column used to obtain the elution profiles shown in Figure 2, under the same run conditions and using the same three ion-pairing reagent systems. As shown in Figure 4A, the slopes of the plots for each peptide are very different, depending as they do on the number of positive charges present in the peptides. With the exception of peptide 3, the plots illustrate a linear relationship between peptide retention time and counterion hydrophobicity. When X was then plotted against R_t/N (where R_t is peptide retention time and N is the number of positively charged residues in the peptides), the profiles shown in Figure 4B were obtained. The similar, if not absolutely parallel, slopes for most of the peptides support the view noted previously (see above) that anionic counterions exert their effects only through positively charged groups and that there is an essentially equal effect on each positively charged group.

This view is again supported by the retention behavior of peptide standards C1 to C4 on the C_8 and C_{18} columns (Figure 1). From Table 2, it can be seen that the change in retention time per positively charged residue (Δ/N) when changing from one ion-pairing reagent system to another is very similar for each peptide. The average change in retention times per positively charged residue between two ion-pairing systems was 2.5 min (TFA $\Leftrightarrow$ HFBA) and 1.5 min (TFA $\Leftrightarrow$ H_3PO_4) for both columns, i.e., there is a greater change in counterion hydrophobicity between TFA^- and $HFBA^-$ than between TFA^- and $H_2PO_4^-$.

The results of Guo et al.,[18] together with those described in this article, suggested that peptide retention times in different ion-pairing systems may be related by a simple correction factor.

B. PREDICTION OF PEPTIDE RETENTION TIMES

1. Requirement for Peptide Standards

Any correction factor allowing for the average contribution of each positive charge to changes in peptide retention time between different ion-pairing reagent systems would only be applicable to the particular column under investigation. Differences in peptide retention on different columns arise from a variety of factors, including column aging and variations in the ligand density or *n*-alkyl chain length of packing materials. It would be impractical and time-consuming to chromatograph a large number of peptides whenever a new column is used, but the results of Guo et al.[18] and of this article suggested that only a few standard peptides are needed to calculate the required retention time correction between different ion-pairing systems. Guo et al.[18] utilized a series of five decapeptide standards (S1 to S5; Table 1) which worked well for this purpose. However, peptide standards C1 to C4 (Figure 1, Tables 1 and 2) are probably even better suited for two reasons: (1) the wider variety of net charges exhibited by peptides C1 to C4 (+1 to +4, respectively) compared to S1 to S5 (S1 has three positively charged groups; S2 to S4 each have two) will likely allow a more accurate measurement of the average effect per positively charged residue of varying the counterion; (2) two of the three charged groups in S1 are in very close proximity (a free α-amino group on an arginine residue) and this may produce anomalous results. In fact, the close proximity of two of the charges in peptide 3 (Table 1), which is in fact, identical to S1, may explain its anomulous behavior in Figure 4.

2. Rules for Prediction of Peptide Retention Time when Varying Counterion

The degree of shift in peptide retention time with changes in counterion is dependent on counterion hydrophobicity, counterion concentration and the selectivity of the particular reversed-phase column used. In the past, researchers have stated that there are difficulties in predicting peptide retention values for a particular chromatographic system using coefficients derived from a different chromatographic system, particularly if the overall selectivities of the different systems diverge.[29,36] However, the following rules for prediction of peptide retention in one counterion system from the results of another system require only two basic assumptions: (1) only basic, positively charged residues contribute to shifts in peptide retention; (2) each positive charge, whether originating from a lysine, arginine or histidine side-chain, or from an N-terminal α-amino group, exerts an equal effect on peptide retention. These rules apply to linear AB gradients, where A and B are water and acetonitrile, respectively, each containing the ion-pairing reagent.

The contribution of each positively charged residue to shifts in peptide retention between two ion-pairing reagent systems is determined by chromatographing a basic peptide standard(s) with both counterions. The average contribution of each basic residue to a change in retention time is denoted by Δ/N, where Δ is the shift (in min) in retention time of the standard between the two counterion systems, and N equals the number of positively charged residues in the standard. The counterion correction factor (t_i) for a peptide of interest is then obtained by multiplying the number of positively charged residues of the peptide (n) by Δ/N for the standard,

$$t_i = n(\Delta / N)$$

When the retention time of a peptide of interest is known in the presence of one counterion, its predicted position in another counterion system is described by the expression,

$$\tau = t_R^{obs} + t_i$$

where τ is the predicted peptide retention in the desired counterion, t_R^{obs} is the observed retention time in another counterion system, and t_i is the counterion correction factor.

3. Accuracy of Peptide Retention Prediction

The above prediction rules were applied to the separation of the mixture of basic peptides of varying numbers of positively charged groups shown in Figure 2. The predicted retention times (τ) of the peptides in the H_3PO_4 and HFBA systems were calculated and compared to the observed retention values (t_R^{obs}) (Table 3). The values for Δ/N (retention time shift/basic residue) were obtained from the retention times of peptide standards S1 to S5 and represent the average value of Δ/N for all five synthetic peptide HPLC standards. The use of several peptide standards should ensure greater accuracy then the use of just one standard.

Sample calculations:

1. What is the predicted retention time in HFBA of peptide 2b on the C_{18} column from an observed retention time of 24.1 min in TFA?

    $$t_i = n(\Delta / N)$$

    where n = 2 (Table 1) and Δ/N = 2.0 min (Table 3);

    $$t_i = 2 \times 2.0 = 4.0 \ min$$

    Changing from TFA⁻ to the more hydrophobic HFBA counterion effects an increase in peptide retention time, i.e., t_i is positive;

    $$\tau(HFBA) = t_R^{obs}(TFA) + 4.0$$
    $$= 24.1 + 4.0$$
    $$= 28.1 \ min \ (t_R^{obs}[HFBA] = 28.0 \ min)$$

2. What is the predicted retention time in H_3PO_4 of peptide 4 on the C_8 column from an observed retention time of 24.0 min in TFA?

    $$t_i = n(\Delta / N)$$

    where n = 4 (Table 1) and Δ/N = 1.3 min (Table 3);

    $$t_i = 4 \times 1.3 = 5.2 \ min$$

    Changing from TFA⁻ to the less hydrophobic $H_2PO_4^-$ counterion effects a decrease in peptide retention time, i.e., t_i is negative;

    $$\tau(H_3PO_4) = t_R^{obs}(TFA) - 5.2$$
    $$= 24.0 - 5.2$$
    $$= 18.8 \ min \ (t_R^{obs}[H_3PO_4] = 19.3 \ min)$$

The results presented in Table 3 illustrate the accuracy of the predictive method for the peptides examined. The average deviations of predicted and observed values for the seven peptides on the C_{18} column were only 0.8 min (TFA → H_3PO_4) and 1.1 min (TFA → HFBA); for the C_8

column, the values were 0.7 min (TFA → H$_3$PO$_4$) and 1.3 min (TFA → HFBA). The accuracy of this predictive method is particularly impressive considering that the seven peptides utilized varied in both the numbers and types of basic residues they contained. Discrepancies between predicted and observed peptide retention times may be due to slightly unequal contributions of these residues.

IV. CONCLUSIONS

This article provides a clear understanding of the effect of changing anionic counterion hydrophobicity or concentration on peptide retention, and thus can be extremely beneficial in the purification of peptides and for providing proof of peptide homogeneity. In addition, the demonstration of a simple relationship between peptide retention time in different ion-pairing systems enabled the determination of rules for prediction of peptide retention times in one ion-pairing system from observed retention times in another system. This prediction method shows generally good accuracy and should certainly be beneficial in narrowing down the position of a peptide(s) of interest in the elution profile of a peptide mixture.

ACKNOWLEDGMENTS

This work was supported by the Medical Research Council of Canada and by equipment grants from the Alberta Heritage Foundation for Medical Research.

REFERENCES

1. **Horváth, Cs., Melander, W., Molnár, I., and Molnár, P.,** Enhancement of retention by ion-pair formation in liquid chromatography with nonpolar stationary phases, *Anal. Chem.,* 49, 2295, 1977.
2. **Horváth, Cs., Melander, W., and Molnár, I.,** Solvophobic interactions in liquid chromatography with non polar stationary phases, *J. Chromatogr.,* 125, 129, 1976.
3. **Kraak, J.C., Jonker, K.M., and Huber, J.F.K.,** Solvent-generated ion-exchange systems with anionic surfactants for rapid separations of amino acids, *J. Chromatogr.,* 142, 671, 1977.
4. **Hoffman, N.E. and Liao, J.C.,** Reversed-phase high performance liquid chromatographic separations of nucleotides in the presence of solvophobic ions, *Anal. Chem.,* 49, 2231, 1977.
5. **Kissinger, P.T.,** Comments on reverse-phase ion-pair partition chromatography, *Anal. Chem.,* 49, 883, 1977.
6. **Van de Venne, J.L.M., Hendrikx, J.L.H.M., and Deelder, R.S.,** Retention behavior of carboxylic acids in reversed-phase column liquid chromatography, *J. Chromatogr.,* 167, 1, 1978.
7. **Fransson, B.,** Reversed-phase ion-pair chromatography of basic, hydrophobic peptides, *J. Chromatogr.,* 361, 161, 1986.
8. **Hancock, W.S., Bishop, C.A., Meyer, L.J., Harding, D.R.K., and Hearn, M.T.W.,** High pressure liquid chromatography of peptides and proteins. VI. Rapid analysis of peptides by high-pressure liquid chromatography with hydrophobic ion-pairing of amino groups, *J. Chromatogr.,* 161, 291, 1978.
9. **Hearn, M.T.W., Grego, B., and Hancock, W.S.,** High-performance liquid chromatography of amino acids, peptides and proteins. XX. Investigation of the effect of pH and ion-pair formation on the retention of peptides on chemically-bonded hydrocarbonaceous stationary phases, *J. Chromatogr.,* 185, 429, 1979.
10. **Iskandarini, Z., Smith, R.L., and Pietrzyk, D.J.,** Investigation of the influence of hydrophobic ions as mobile phase additives on the liquid chromatographic separation of amino acids and peptides, *J. Liq. Chromatogr.,* 7, 111, 1984.
11. **Spindel, E., Pettibone, D., Fisher, L., Fernstrom, J., and Wurtman, R.,** Characterization of neuropeptides by reversed-phase, ion-pair liquid chromatography with post-column detection by radioimmunoassay. Application to thyrotropin-releasing hormone, substance P, and vasopressin, *J. Chromatogr.,* 222, 381, 1981.

12. **Bennett, H.P.J.,** Isolation of pituitary peptides by reversed-phase high performance liquid chromatography. Expansion of the resolving power of reversed-phase columns by manipulating pH and the nature of the ion-pairing reagent, *J. Chromatogr.,* 266, 501, 1983.

13. **Bennett, H.P.J.,** Manipulation of pH and ion-pairing reagents to maximize the performance of reversed-phase columns, this publication.

14. **Bennett, H.P.J., Browne, C.A., and Solomon, S.,** Purification of the two major forms of rat pituitary corticotropin using only reversed-phase liquid chromatography, *Biochemistry,* 20, 4530, 1981.

15. **Bennett, H.P.J., Browne, C.A., and Solomon, S.,** The use of perfluorinated carboxylic acids in the reversed-phase HPLC of peptides, *J. Liq. Chromatogr.,* 3, 1353, 1980.

16. **Browne, C.A., Bennett, H.P.J., and Solomon, S.,** The isolation of peptides by high performance liquid chromatography using predicted elution positions, *Anal. Biochem.,* 124, 201, 1982.

17. **Burgess, A.W., Knesel, J., Sparrow, L.G., Nicola, N.A., and Nice, E.C.,** Two forms of murine epidermal growth factor: Rapid separation by using reverse-phase HPLC, *Proc. Natl. Acad. Sci. U.S.A.,* 79, 5753, 1982.

18. **Guo, D., Mant, C.T., and Hodges, R.S.,** Effects of ion-pairing reagents on the prediction of peptide retention in reversed-phase high performance liquid chromatography, *J. Chromatogr.,* 386, 205, 1987.

19. **Harding, D.R.K., Bishop, C.A., Tarttellin, M.F., and Hancock, W.S.,** Use of perfluoroalkanoic acids as volatile ion-pairing reagents in preparative HPLC, *Int. J. Peptide Protein Res.,* 18, 314, 1981.

20. **Schaaper, W.M.M., Voskamp, D., and Olieman, C.,** Perfluoroalkanoic acids as lipophilic ion-pairing reagents in reversed-phase liquid chromatography of peptides including secretin, *J. Chromatogr.,* 195, 181, 1980.

.21. **Starratt, A.N. and Stevens, M.E.,** Ion-pair high performance liquid chromatography of the insect neuropeptide proctolin and some analogs, *J. Chromatogr.,* 194, 421, 1980.

22. **van der Rest, M., Bennett, H.P.J., Solomon, S., and Glorieux, F.H.,** Separation of collagen cyanogen bromide-derived peptides by reversed-phase high performance liquid chromatography, *Biochem. J.,* 191, 253, 1980.

23. **Hearn, M.T.W. and Grego, B.,** High performance liquid chromatography of amino acids, peptides and proteins. XL. Further studies on the role of the organic modifier in the reversed-phase high performance liquid chromatography of polypeptides. Implications for gradient optimization, *J. Chromatogr.,* 255, 125, 1983.

24. **Hearn, M.T.W. and Grego, B.,** High performance liquid chromatography of amino acids, peptides and proteins. XLVI. Selectivity effects of peptidic positional isomers and oligomers separated by reversed-phase high performance liquid chromatography, *J. Chromatogr.,* 266, 75, 1983.

25. **Tarr, G.E. and Crabb, J.W.,** Reverse phase high performance liquid chromatography of hydrophobic proteins and fragments thereof, *Anal. Biochem.,* 131, 99, 1983.

26. **Hancock, W.S., Bishop, C.A., Battersby, J.E., Harding, D.R.K., and Hearn, M.T.W.,** High-pressure liquid chromatography of peptides and proteins. XI. The use of cationic reagents for the analysis of peptides by high-pressure liquid chromatography, *J. Chromatogr.,* 168, 377, 1979.

27. **Matrisian, L.M., Larson, B.R., Finch, J.S., and Magun, B.F.,** Further purification of epidermal growth factor by high performance liquid chromatography, *Anal. Biochem.,* 125, 339, 1982.

28. **Fullmer, C.S. and Wasserman, R.H.,** Analytical peptide mapping by high performance liquid chromatography. Application to intestinal calcium-binding proteins, *J. Biol. Chem.,* 254, 7208, 1979.

29. **Gaertner, H. and Puigserver, A.,** Separation of some peptides and related isopeptides by high-performance liquid chromatography: structure-retention time relationships, *J. Chromatogr.,* 350, 279, 1985.

30. **Hancock, W.S., Bishop, C.A., Prestidge, R.L., Harding, D.R.K., and Hearn, M.T.W.,** Reversed-phase, high-pressure liquid chromatography of peptides and proteins with ion-pairing reagents, *Science (Washington, D.C.),* 200, 1168, 1978.

31. **Hancock, W.S., Bishop, C.A., Prestidge, R.L. Harding, D.R.K., and Hearn, M.T.W.,** High-pressure liquid chromatography of peptides and proteins. II. The use of phosphoric acid in the analysis of underivatized peptides by reversed-phase high-pressure liquid chromatography, *J. Chromatogr.,* 153, 391, 1978.

32. **Nice, E.C. and O'Hare, M.J.,** Simultaneous separation of β-lipotrophin, adrenocorticotropic hormone, endorphins and enkephalins by high-performance liquid chromatography, *J. Chromatogr.,* 162, 401, 1979.

33. **O'Hare, M.J. and Nice, E.C.,** Hydrophobic high-performance liquid chromatography of hormonal polypeptides and proteins on alkylsilane-bonded silica, *J. Chromatogr.,* 171, 209, 1979.

34. **Hearn, M.T.W. and Grego, B.,** High-performance liquid chromatography of amino acids, peptides and proteins. XXVII. Solvophobic considerations for the separation of unprotected peptides on chemically bonded hydrocarbonaceous stationary phases, *J. Chromatogr.,* 203, 349, 1981.

35. **Guo, D., Mant, C.T., Taneja, A.K., Parker, J.M.R., and Hodges, R.S.,** Prediction of peptide retention times in reversed-phase high-performance liquid chromatography. I. Determination of retention coefficients of amino acid residues of model synthetic peptides, *J. Chromatogr.,* 359, 499, 1986.

36. **Grego, B., Lambrou, F., and Hearn, M.T.S.,** High-performance liquid chromatography of amino acids, peptides and proteins. XLVIII. Retention behavior of tryptic peptides of human growth hormone isolated by reversed-phase high-performance liquid chromatography: a comparative study using different chromatographic conditions and predicted elution behavior based on retention coefficients, *J. Chromatogr.,* 266, 89, 1983.

THE BIOLOGICAL ACTIVITY OF POLYPEPTIDES AFTER REVERSED-PHASE HIGH-PERFORMANCE CHROMATOGRAPHY: A COMPARATIVE STUDY

Benny S. Welinder

I. INTRODUCTION

In the last decade, reversed-phase high-performance liquid chromatography (RPC) has been the preferred technique for the separation and purification of polypeptides on the analytical as well as preparative scale. The major reasons for this dominant position (RPC constitutes more than 70% of all HPLC-analyses[1]) are the speed and efficiency of RPC plus an almost unlimited degree of freedom in choice of mobile phase (buffer, pH, organic modifier) as well as stationary phase (at present more than 200 different reversed-phase columns are commercially available[2]).

Like all other chromatographic techniques, RPC utilizes the ability to manipulate the binding forces between sample molecules and a stationary phase. Although this manipulation in RPC is performed almost universally at acid pH in solvents containing various amounts of potential toxic or harmful organic solvents, the number of literature reports describing partial or total loss of biological activity is in no way overwhelming.

Several reasons may underline this point: only a limited number of bioactive samples are subjected to RPC due to a deeply rooted conviction that organic solvents are incompatible with the survival of the highly ordered three-dimensional structure necessary for the activity of many biopolymers (i.e., enzymes); a lack of proper reference compounds (i.e., sample of interest purified to a similar degree using techniques known for certain not to be harmful to the bioactivity); a general displeasure in western scientific literature towards reporting "negative" results.

However, reports on RPC of a number of different peptides/proteins (enzymes, hormones, growth factors) leading to partial or total loss of biological or immunological activity have been published (see Table 1) and from these reports it can be concluded that the major menaces to the bioactivity of polypeptides during RPC seem to be:

1. The commonly used mobile phases (highly acidic milieu, content of organic solvents).
2. Too strong binding forces between the sample and the stationary phase, leading to denaturation upon the column.
3. Binding of toxic impurities or degradation products in the mobile phase to the sample.
4. Binding of stationary phase degradation products to the sample.
5. Use of unqualified isolation procedures for regaining the sample from the column eluate.

Although these effects are somehow coherent, this review will discuss the individual critical points separately, followed by a conclusive strategy for reversed-phase-based purification and isolation of polypeptides retaining full bioactivity.

TABLE 1

Experimental Conditions (Stationary and Mobile Phase, Sample Isolation Procedure) for a Number of RPC Purifications of Polypeptides leading to Total or Partial Loss of Biological Activity

Sample	Buffer	pH	Organic modifier	Column	Sample isolation procedure
Interferon[3]	0.1 *M* formic acid	2.4	2-propanol	C$_{18}$	Lyophilization of the sample in the column eluate directly or after removal of the organic modifier with cyclohexane
Thyroid-stimulating hormone[4]	0.025 *M* sodium phosphate	7.0	Acetonitrile	C$_{18}$	BSA was added to the sample in the column eluate before lyophilization
Ribonuclease[5] BSA Horseradish peroxidase Ovalbumin	0.012 *M* hydrochloric acid	2	Ethanol/1-butanol	C$_{18}$	Neutralization of column eluate, partial evaporation, lyophilization
Papain[6]	0.010 *M* sodium phosphate	2.2 4.8	1-propanol	C$_4$	Lyophilization of the sample in the column eluate (pH 2.2) or after neutralization to pH 7.0
Chorionic gonadotropin[7]	0.1% TFA/3% ethyleneglycol dimethylether	2—3	Ethanol/1-butanol	C$_{18}$	Lyophilization of the sample in the column eluate (pH 2—3) or after neutralization to pH 7.5
Growth inhibitory glycopeptide[8]	Water adjusted with TFA	3.0	2-propanol	C$_4$	The sample in the column eluate was dialyzed against distilled water and lyophilized
Lutropin[9]	0.5% ammoniumtrifluoroacetate	3.3	Acetonitrile	C$_{18}$	The sample in the column eluate was lyophilized directly
Interleukin-2[10]	0.15 *M* sodium chloride, 0.027 *M* phosphate, 0.1% PEG 6000	2.5	1-propanol	C$_6$	1-propanol was removed using centrifugation *in vacuo* (Speed Vac)
Pituitary fibroblast growth factor[11]	0.1% TFA	2.1	Methanol Ethanol Propanol	C$_4$	The sample in the column eluate was diluted with Dulbeccos modified Eagle medium plus 0.5% BSA before the bioassay
β-glucisodase[12]	0.1% TFA	2	2-propanol	C$_8$	The sample in the column eluate was dialyzed against 0.01 *M* potassium phosphate before they were assayed
Trypsinogen[13] Proelastase Lipase Chymotrypsinogen I and II Amylase	0.1% TFA	2	Acetonitrile	C$_8$	Not reported

TABLE 1
Experimental Conditions (Stationary and Mobile Phase, Sample Isolation Procedure) for a Number of RPC Purifications of Polypeptides leading to Total or Partial Loss of Biological Activity (continued)

Procarboxypeptidase I and II Follitropin[14]	0.5% ammoniumtri-fluoroacetate	3.3	Acetonitrile	C_{18}	The sample in the column eluate was lyophilized directly
Growth hormone[15]	0.1% TFA	2	1-propanol	C_{18}	The sample in the column eluate was lyophilized directly
Monoiodoinsulins[16]	0.25 M trialkylam-monium phosphate		2-propanol Acetonitrile	C_{18}	The sample in the column eluate was lyophilized before bioassay
Prostatic acid phosphatase[17]	0.1% TFA	2	Acetonitrile	C_{18}	The sample in the column eluate was collected as such or into buffers containing various stabilizing agents

II. DISCUSSION

A. WHAT IS 'COMMON PRACTICE' IN RPC?

A number of literature reports describing total or partial loss of bioactivity of peptides and proteins following RPC have been collected in Tables 1 and 2. Sample identification, column, mobile phase, and isolation procedures are shown in Table 1; various critical events in the chromatographic process (from injection to sample isolation) leading to loss of activity are commented upon in Table 2. Before going into a detailed description of the individual reports, it might be advantageous to encircle the most common practice in RPC of polypeptides.

Insulin may be a well-qualified common denominator for this purpose since it was one of the first polypeptides subjected to analytical and preparative RPC (preparative meaning industrial scale as well); in addition, since insulin was found to be stable under harsh conditions (actually insulin is extracted from pancreatic glands on an industrial scale using acid ethanol), the resulting RPC analyses should reflect that the chromatographers could concentrate upon putting forward the most favorable chromatographic conditions — without any genuflections to 'physiological conditions'.

We have recently reviewed 129 RPC separations of insulin and insulin derivatives published from 1978 to 1988.[18] With respect to choice of buffer, organic modifier, pH and stationary phase, the results can be summarized in the following way:

- *Buffer substances*: Alkylammonium salts, acid phosphates, neutral salts or trifluoroacetic acid (TFA) were the major choices.
- *pH*: Almost all insulin analyses were performed at acid pH (pH 2.0 to 4.0).
- *Organic modifier*: More than 85% of all insulin analyses utilized acetonitrile as organic modifier.
- *Stationary phase*: Silica-based C_{18} columns were used in more than 75% of all insulin analyses.

TABLE 2
The Biological Activity of RPC-Purified Polypeptides and Attempts to Increase the Recovered Bioactivity

Ref. number	Biological Activity of the sample after RPC
3	Vacuum concentration (Speed Vac) of the sample in the column eluate leads to loss of 99% of the interferon activity. If the propanol is extracted with 4 vol. of cyclohexane, the residual cyclohexane removed under a nitrogen stream and the aqueous sample solution concentrated *in vacuo*, the biological activity is retained.
4	Column eluate fractions (0.5 ml) were collected into 2 ml phosphate-buffered saline containing 0 1% BSA followed by lyophilization. 57.7% of the bioactivity was recovered. No free subunits were observed.
5	Immunological reactivity after RPC: unchanged (BSA and ribonuclease), partly affected (horseradish peroxidase), total loss (ovalbumin). Biological activity after RPC: the enzymatic activity of ribonuclease was unchanged, whereas horseradish peroxidase lost 95 to 99% of its initial enzymatic activity.
6	Collection of the sample (separated at pH 2.2 at 4°C) at room temperature followed by lyophilization caused irreversible denaturation. This denaturation could be arrested if the sample immediately after collection was neutralized prior to lyophilization.
7	Human chorionic gonadotropin lost 40 to 90% of its biological activity after RPC at pH 2 to 3. The optimal conditions for retaining the activity were (1) neutralization of the sample before lyophilization; (2) ethanol/butanol instead of 1- and 2-propanol; (3) 0.1% TFA instead of 0.025% TFA.
8	When acetonitrile (HPLC grade) adjusted to pH 3 with TFA was lyophilized, the residue interfered with the bioassay (protein synthesis in 3T3 cells), leading to 20% inhibition. Such inhibition from the lyophilization residue was not observed when 2-propanol or ethanol were used as organic modifiers.
9	13% of the initial bioactivity was recovered after RPC.
10	Acetic acid or TFA in combination with acetonitrile or propanol made the elution of bioactive IL-2 impossible. Attempts to exchange the acidic eluate for a volatile or a physiological buffer (using HPLC gel chromatography) resulted in complete loss of activity of interleukin-2, whereas dialysis in the presence of human serum proteins resulted in an highly active IL-2.
11	RPC in 0.1% TFA containing organic solvents lead to 90 to 95% loss of activity of the isolated fibroblast growth factor. Incubation experiments showed that, after 2 hr in 0.1% TFA, only 2% of the activity remained, whereas addition of 50% ethylene glycol to the TFA resulted in full activity after incubation. Addition of 50% glycerol had a similar effect.
12	No enzymatic activity was recovered after RPC.
13	The biological activity of all enzymes in the pancreatic juice was totally lost after RPC. The isoelectric points were altered after chromtography.
14	96% of the bioactivity lost after RPC. Adjusting pH in the mobile phase to 4.3 and 5.3 resulted in comparable chromatograms, but recovery of bioactivity was still only approximately 4%.
15	Recovery of growth hormone immunoreactivity after RPC: 82% measured directly in the column eluate, 36% if the column eluate was dried by vacuum desiccation.
16	The biological and immunlogical activity of the four monoiodoinsulins was always reduced when the column eluate (containing nannogram amounts of the sample) was lyophilized directly. The immulogical and biological activity was retained if the column eluate was extracted with 4 vol. cyclohexane before lyophilization. Dilution of the sample in the column eluate with large volumes of albumin-containing buffers or removal of the mobile phase (gelchromatography in 40% ethanol) followed by lyophilization also resulted in samples with 100% bioactivity.
17	No enzymtic activity was found when the collected fractions were assayed directly or after removal of the acetonitrile followed by lyophilization. Addition of potassium phosphate, sodium acetate, ammonium bicarbonate, DMSO or PEG to the collected fractions had no effect upon the resulting activity. Hanks' Balanced Salt Solution or Delbeccos Phosphate Buffered Saline were moderate effective in stabilizing the enzymatic activity, and glycerol (45%) was found to be very active for protecting the enzymatic activity.

Note: See Table 1 for polypeptide identification.

This list of preferred conditions is in no way accidental. There are many good reasons for the favored buffer substances (ion-pairing agents minimize non-specific interaction with residual silanol groups, TFA has a chaotropic effect, is low-UV transparent, and is lyophilizable, etc.), the preferred pH (suppression of ionization of sample and stationary phase leading to very good peak shape), and the preferred organic modifier (acetonitrile has a very good elution strength, is low-UV transparent, has a low viscosity, and may be obtained ultrapure). In addition, the number of commercially available C_{18} stationary phases is probably higher than for any other bonded phase.

Parts of this 'common practice' can be recognized in Table 1: Acid pH (13 out of 15 reports); use of C_{18} columns (9 out of 15 reports); acetonitrile only was used as organic modifier in 6 cases.

From Tables 1 and 2, it can be concluded that RPC of polypeptides with biological activity can be performed in two different ways:

1. The chromatographic parameters are chosen with the purpose of obtaining the very best separation. Such parameters may lead to loss of biological activity, but renaturation may restore the activity.
2. The chromatographic parameters are modified to more 'physiological' conditions (alcohols instead of acetonitrile, neutral pH in the mobile phase) in order to prevent loss of activity during the chromatography.

B. MOBILE PHASE pH

As mentioned above, an acid (and even very acid) pH is preferred, although many polypeptides will lose their tertiary structure at pH 2 to 3. This may not necessarily be fatal for the resultant activity, but if the polypeptide consists of subunits stabilized by non-covalent bondings, the individual chains will most certainly be separated during the chromatography. This was found in the case of various glycopeptide hormones[4,7,9,14] composed of an alpha- and a beta-chain, where the best results were obtained after chromatography at neutral pH[4] or when chain recombination was performed after the separation[7]. Even then, RPC results in loss of bioactivity, and since lutropin can be purified using high-performance ion-exchange chromatography or chromatofocusing without loss of bioactivity,[19] it can be concluded that some polypeptides are too labile to allow RPC purification even when this technique is performed with great care.

It is puzzling that the majority of RPC separations of bioactive polypeptides supposed to result in isolation of active compounds is still carried out at pH 2 to 3. One reason may be the extreme popularity of trifluoroacetic acid, this popularity being more the result of uncritical copying of literature studies than considerations on preserving the three-dimensional structure, which in most cases is a necessary prerequisite for retaining the biological activity.

C. ORGANIC MODIFIER

Although acetonitrile has many good physico/chemical properties as organic solvent in RPC, it can hardly be described as a 'biocompatible' solvent. It is toxic, and even HPLC quality acetonitrile may contain contaminants that interfere with *in vivo* assays.[8] It is very destructive for the three-dimensional structure,[10,14] and for enzymes its use will almost invariably lead to total loss of activity. It is a striking observation that RPC of pancreatic juice in acetonitrile led to a complete loss of activity of all the enzymes as well as alterations of their isoelectric points;[13] in contrast, the same enzymes could be separated without any loss of enzymatic activity using high-performance hydrophobic interaction chromatography, a technique operating at neutral pH in purely aqueous milieu.[20]

Common physiological and biochemical knowledge will lead to the recommendation of alcohols (i.e., ethanol, 1- or 2-propanol) as modifiers in RPC of bioactive macromolecules, but it should be emphasized that no general rules can be given. HPLC purification of critical biopolymers will always demand some pre-experiments in order to explore the survival of the activity after incubation in a number of different buffer/organic modifier combinations.

D. COLUMN BLEEDING

When small amounts of the four monoiodoinsulins of porcine insulin were separated and isolated using RPC, it was found that the biological activity was reduced compared to similar substances purified and isolated after disc-electrophoresis and ion-exchange chromatography.[16] The reason for this reduced bioactivity was most likely column bleeding, i.e., contamination of the samples with stationary phase degradation products. The content of silicium in the column eluate measured for three different LiChrosorb RP-18 columns, following isocratic elution with 0·25 M triethylammonium formate/21.5% 2-propanol was found to be 216, 30, or 188 ng/ml, compared to 11 ng silicium/ml in the inlet buffer. After alkaline hydrolysis of extracts of three stationary phases (Lichrosorb RP-18, Spherisorb ODS-2, and Vydac 218 TPB 5) in a similar buffer, followed by gas chromatography, the presence of octadecane as well as octadecanol could be demonstrated.[21] Since only a few nanograms of each monoiodoinsulin were purified in each column cycle, a collection of an individual insulin tracer in 1 ml of column eluate might lead to contamination with 100 times more column degradation products, a problem which is common for all experiments where very small amounts of highly potent biopolymers are purified to homogeneity.

If the residue left after lyophilizing 2.5 ml of 0.25 M triethylammonium formate in 27% acetonitrile or 36% ethanol was added to tracers with 100% binding affinity, this binding affinity to isolated rat fat cells was reduced for all four monoiodoinsulins to 50 to 70% of the initial value.[16]

RPC-purified monoiodoinsulins with non-reduced biological activity could be obtained if the tracers in the column eluate were isolated following RPC under conditions where minimal hydrophobic interaction between the sample and silica-C_{18} derivatives released during RPC would occur:

1. Size-exclusion chromatography of the column eluate in 40% aqueous ethanol followed by lyophilization, or
2. Extraction of the column eluate with cyclohexane followed by lyophilization, or
3. Dilution with large volumes of albumin-containing buffers before the bioassay

These sample isolation procedures were found to be usable for nanogram as well as milligram amounts[21].

E. SAMPLE ISOLATION

From Tables 1 and 2, it can be seen that the sample isolation procedure applied after chromatography is of uttermost importance. Lyophilization of the column eluate containing sample, acidic buffer substances and organic modifier almost invariably leads to loss of bio-activity.[3,4,6,7,9,14-16] In the majority of these cases, recovery of bioactivity was found to increase if the isolation procedure was changed:

1. Removal of the organic modifier before lyophilization[11,16]
2. Adjusting the pH of the acidic mobile phase to approx. neutral before lyophilization[6,7]

Addition of various protecting substances to the column eluate before further treatment may be necessary in order to recover the bioactivity. Serum albumin[4,16], 45 to 50% glycerol[11,17] or 50% ethylene glycol[11] have been successfully employed for this protective purpose.

F. A RATIONAL STRATEGY

The main steps in RPC purification of polypeptides are binding of the sample to the stationary phase, shorter or longer fixation period with increasing amount of organic solvents and, finally, elution from the column followed by an isolation procedure in order to obtain the sample free of mobile phase contamination. During all these steps, the activity of a biopolymer may be reduced or lost due to the initial strong multipoint-binding to the stationary phase, treatment with solvents with high elution strength in this fixed state, elution with free mobility in organic solvents, and potential harmful liquid/solid surface interactions when the mobile phase is removed by lyophilization. Combined with the fact that each biological polypeptide is unique and demands careful mobile/stationary phase optimization before RPC purification, the general utility of any set of conditions claimed to improve recovery of bioactivity after RPC must necessarily be limited. However, some general advice based upon the previous reports may be given:

1. Avoid as far as possible extremely acid mobile phases. Use of TFA (in particular) and other strong acids (in general) is no physical necessity, and many reversed-phase stationary phases perform excellently at pH 5 to 7.
2. Avoid acetonitrile as an organic modifier. Propanol may be an equally good solvent and the increased back-pressure created by this solvent can be reduced using lower flow rates.
3. Realize that the eluted biopolymer may be in a cadaverous state with respect to structure. A renaturation procedure and/or addition of protective substances may restore the bioactivity. Glycerol, ethylene glycol and serum albumin have been used with good results, but specific advice cannot be given. An exchange of the mobile phase before further sample treatment (dialysis, desalting) may be advantageous.
4. Lyophilization of the sample directly from the mobile phase (containing organic solvent[s]) should be avoided. The organic modifier should always be removed and pH adjusted to neutral before lyophilization.
5. When very small amounts of polypeptides are purified to homogeneity under common RPC conditions, problems may arise due to self-aggregation, adhesion to surfaces (glass, plastic, stainless steel) or instability following removal of protective compounds, ligands, etc., present in the initial sample mixture. Addition of human serum albumin (0.05 to 0.3%) or exchanging the mobile phase with a solution of serum proteins may be helpful in these cases.

III. CONCLUSIONS

When the common conditions for RPC are evaluated, it is difficult to imagine that bioactive polypeptides can be chromatographed under these conditions with retained activity. However, several literature reports have documented extreme stability of peptides/proteins and only a limited number of RPC separations result in totally inactive biopolymers. Consequently, there is no reason for avoiding this separation principle, especially since an unrivalled separation capacity can be obtained. A few simple incubation experiments will show the necessity of adjustments in the conventional mobile phases and sample treatment, if any, and the actual separation can be performed very rapidly.

On the other hand, some polypeptides are too labile for this type of chromatography and there is not much point in time-consuming experiments with the purpose of extending the usability of RPC beyond the limits. In such cases, high performance ion-exchange chromatography or hydrophobic interaction chromatography will be the method of choice.

REFERENCES

1. **Majors, R.E.**, Advances in HPLC of relevance for biotechnology, *LC/GC Magazine*, 3, 774, 1985.
2. **Unger, K.K. and Lork, K.D.**, Stationary phases in reversed-phase chromatography, *Eur. Chromatogr. News*, 2, 14, 1988.
3. **Smith-Johannsen, H. and Tan, Y.H.**, Isolation of HuIFN-beta by immunosorbent and high-pressure liquid chromatography, *J. Interferon Res.*, 3, 473, 1983.
4. **Bristow, A.F., Wilson, C., and Sutcliffe, N.**, Reversed-phase high-performance liquid chromatography of human thyroid-stimulating hormone, *J. Chromatogr.*, 270, 285, 1983.
5. **Luiken, J., Van der Zee, R., and Welling, G.W.**, Structure and activity of proteins after reversed-phase high performance liquid chromatography, *J. Chromatogr.*, 284, 482, 1984.
6. **Cohen, S.A., Benedek, K.P., Dong, S., Tapuhi, Y., and Karger, B.L.**, Multiple peak formation in reversed-phase liquid chromatography of papain, *Anal. Chem.*, 56, 217, 1984.
7. **Wilks, J.W. and Butler, S.S.**, Biological activity of human chorionic gonadotropin following reversed-phase high performance liquid chromatography, *J. Chromatogr.*, 298, 130, 1984.
8. **Sharifi, B.G., Bascom, C.C., Khurana, V.Y., and Johnson, T.C.**, Use of urea and guanidine-HCl-propanol solvent system to purify a growth inhibitory glycopeptide by high-performance liquid chromatography, *J. Chromatogr.*, 324, 173, 1985.
9. **Hallin, P., Madej, A., and Edquist, L.-E.**, Subunits of bovine lutropin. Hormonal and immunological activity after separation with reversed-phase high performance liquid chromatography, *J. Chromatogr.*, 319, 195, 1985.
10. **Kniep, E.M., Kniep, B., Grote, W., Conradt, H.S., Monner, D.S., and Mühlradt, P.F.**, Purification of the T lymphocyte growth factor interleukin-2 from culture media of human peripheral blood leukocytes (buffy coats), *Eur. J. Biochem.*, 143, 199, 1984.
11. **Gospodarowicz, D., Massoglia, S., Cheng, J., Lui, G.-M., and Böhlen, P.**, Isolation of pituitary fibroblast growth factor by fast protein liquid chromatography (FPLC): Partial chemical and biological characterization, *J. Cell. Physiol.*, 122, 323, 1985.
12. **Fausnaugh, J.L., Kennedy, L.A., and Regnier, F.E.**, Comparison of hydrophobic-interaction and reversed-phase chromatography of proteins, *J. Chromatogr.*, 317, 141, 1984.
13. **Padfield, P.J., Grifin, M., and Case, R.M.**, Separation of guinea-pig pancreatic juice by reversed-phase high performance liquid chromatography, *J. Chromatogr.*, 369, 133, 1986.
14. **Hallin, P. and Khan, S.A.**, Preservation of biological LH and FSH activity after application on HPLC. Comparison between cation-exchange and reverse-phase high performance liquid chromatography, *J. Liq. Chromatogr.*, 9, 2855, 1986.
15. **Patience, R.L. and Rees, L.H.**, Comparison of reversed-phase and anion-exchange high performance liquid chromatography for the analysis of human growth hormones, *J. Chromatogr.*, 352, 241, 1986.
16. **Welinder, B.S., Linde, S.L., and Hansen, B.**, Reversed-phase high performance liquid chromatography of the four monoiodoinsulins: effect of column supports, buffers and organic modifiers, *J. Chromatogr.*, 298, 41, 1984.
17. **Strickler, M.P., Kintzios, J., and Gemski, M.J.**, The purification of prostatic acid phosphatase from seminal plasma by reverse phase high pressure liquid chromatography, *J. Liq. Chromatogr.*, 5, 1921, 1982.
18. **Welinder, B.S., Sorensen, H.H., Hejnaes, K.R., Linde, S., and Hansen, B.**, Stationary and mobile phase effects in high performance liquid chromatography of protein hormones, in HPLC of Proteins, Peptides and Polynucleotides, Hearn, M.T.W., Ed., VCH Publishers, New York, 1991.
19. **Hallin, P. and Khan, S.A.**, Biological characterization of a pituitary monkey lutropin preparation after chromatofocusing or after high performance anion exchange chromatography, *J. Liq. Chromatogr.*, 11, 1261, 1988.
20. **Padfield, P.J. and Case, R.M.**, Separation of the proteins present in pancreatic juice using hydrophobic interaction chromatography, *Anal. Biochem.*, 171, 294, 1988.
21. **Linde, S. and Welinder, B.S.**, RP-HPLC of insulin and iodinated insulin, this publication.

REVERSED-PHASE CHROMATOGRAPHY OF INSULIN AND IODINATED INSULIN

Susanne Linde and Benny S. Welinder

I. INTRODUCTION

Iodinated insulins are frequently used as biological tracers as well as in radioimmunoassays and insulin receptor studies. The biological and immunological activities of such iodoinsulins have been thoroughly studied, partly in order to utilize a labeled insulin derivative closely related to authentic insulin, and partly to obtain further knowledge about the relation between substitutions in the insulin molecule and the resulting biological and immunological activity.[1-4]

Iodination of insulin leads to a heterogeneous mixture of unsubstituted insulin and mono- and diiodinated insulins. Dependent upon the choice of iodination procedure, all four tyrosyl residues in insulin (A14, A19, B16, and B26) can be mono- or disubstituted and, in order to obtain a well-characterized tracer, it is necessary to isolate the individual iodoinsulins from the iodination mixture.

It has been shown that A14 monoiodoinsulin exhibits the same binding affinity and biological activity as native insulin, making it an ideal insulin-tracer.[1-5]

Previously, all four monoiodoinsulins were isolated using a combination of disc-electrophoresis and low-pressure ion-exchange chromatography, but these methods were very time consuming.[3,4] Several reports have described the applicability of reversed-phase high-performance liquid chromatography (RPC) for characterizing insulin and insulin derivatives (for a review, see Reference 6).

This review will deal with RPC fractionation of the heterogeneous iodination mixture leading to separation and isolation of very pure monoiodoinsulins. The effect of mobile phase (buffer substances, pH, organic modifier), stationary phase and separation temperature upon the separation will be discussed.

II. EXPERIMENTAL

A. IODINATION OF INSULIN

Iodinated insulin was prepared by lactoperoxidase iodination of highly purified porcine insulin using carrier-free $^{125}I^-$ or 0.1 N NaI containing trace amounts of $^{125}I^-$ (resulting in iodinated insulin with high or low specific radioactivity, respectively).[3,7] The average degree of iodination was chosen to approximately 0.15 I/mole insulin in order to minimize the content of diiodoinsulins.

B. HPLC INSTRUMENTATION

The RPC system consisted of a Spectra-Physics SP 8700 chromatograph, a Waters U6K sample injector, and a Pye Unicam UV detector. The column eluate was collected in 1-min fractions (Pharmacia, Frac 300 fraction collector) and the radioactivity was measured in a 16-channel γ-counter (Hydrogamma 16).

C. COLUMNS

LiChrosorb RP-18, 250 × 4.0 mm I.D., 5-μm particle size (Merck), LiChrosorb RP-18, 250 × 25 mm I.D., 7 μm (Merck), Spherisorb ODS-2 C_{18}, 150 + 100 × 4.0 mm I.D., 3 μm (Phase Separation), Vydac 218TPB5, 250 × 8.0 mm I.D., 5 μm, 300-Å pore size (Separation Group, Hesperia, CA), TSK ODS-120T C_{18}, 250 × 4.6 mm I.D., 5 μm (Toyo Soda), Techogel C_4, 250 × 4.0 mm I.D., 5 μm (HPLC technology), Techogel C_{18}, 250 × 4.0 mm I.D., 5 μm (HPLC Technology), PLRP-S, 150 × 4.0 mm I.D., 5 μm (Polymer Laboratories), PRP-1, 150 x 4.0 mm I.D., 10 μm (Hamilton).

D. MOBILE PHASES

1% triethylammonium trifluoroacetate (TEATFA), pH 3.0, was prepared by adjusting the pH of 1% trifluoroacetic acid (TFA) with triethylamine; 0.25 M triethylammonium phosphate (TEAP) was prepared by adjusting the pH of 0.25 M phosphoric acid with triethylamine to pH 3.0, 4.0, 5.0, or 6.0; 0.25 M triethylammonium formate (TEAF) was prepared by adjusting the pH of 0.25 M formic acid with triethylamine to pH 3.0, 4.0, 5.0, and 6.0. The volatile TEATFA system was used in preference to TFA, since isocratic elution with 0.1% TFA in acetonitrile often results in bad peak shapes (tailing) for insulin.

Acetonitrile, isopropanol or ethanol (Rathburn HPLC quality) were used as organic modifiers. All other chemicals were of analytical reagent grade. Distilled water was drawn from a Millipore Milli Q plant and all solvents were Millipore-filtered (0.45 μm) and degassed (vacuum/ultrasound) before use. During chromatography, helium was bubbled continuously through the mobile phases.

Solutes were eluted isocratically from the columns with a mixture of solution A (buffer) and solution B (equal volumes of buffer and organic modifier). The flow-rate was 1.0 ml/min (acetonitrile and ethanol as organic modifier) or 0.5 ml/min (isopropanol).

The UV-absorption of the column eluate was measured continuously at 210 nm (TEAP/acetonitrile), 230 nm (TEATFA/acetonitrile), or 240 nm (TEAF/isopropanol).

E. ISOLATION PROCEDURES

Following appropriate dilution, monoiodoinsulins with high specific radioactivity can be used directly from the RPC fractions in radioimmunoassays, receptor binding analysis, and for chemical analysis. Isolation for quantitative chemical and biological analyses was performed as follows:

1. A Sep-Pak C_{18} cartridge (Waters) was prewashed with 10 ml isopropanol-water (90:10, v/v) followed by 10 ml water. The pooled RPC fractions (containing the individual monoiodoinsulins) were loaded on the Sep-Pak after 1:1 dilution with water, and salts, etc., were washed out with 10 ml of water. The [^{125}I] insulin was then eluted with 2 ml isopropanol-water (90:10, v/v).
2. Removal of the organic modifier was performed by mixing the pooled RPC fractions or the Sep-Pak concentrate with 4 volumes of cyclohexane. The aqueous phase was isolated, flushed with nitrogen (until no isopropanol could be smelled), and lyophilized. Although other extraction solvents, e.g., hexane, could possibly be used, cyclohexane was used due to its great capacity for extraction of acetonitrile and isopropanol.

F. IDENTIFICATION AND CONTROL OF PURITY

The identification of the individual peaks obtained after RPC was performed by determination of the iodine distribution using oxidative sulfitolysis and enzymatic cleavage of the separated A- and B-chains as described previously.[2] The purity of the isolated monoiodoinsulins was determined using analytical RPC.

353

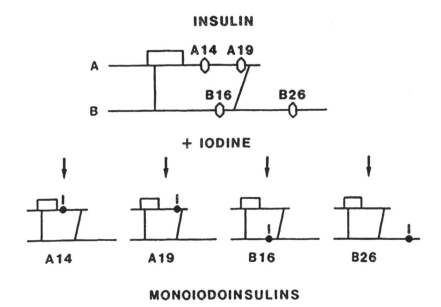

FIGURE 1. Structural formulas of insulin and the four monoiodinated porcine insulin isomers. A and B refer to the 21-residue A-chain and 30-residue B-chain, respectively, of the insulin molecule. A14 and A19 denote, respectively, Tyr^{14} and Tyr^{19} in the A-chain; B16 and B26 denote, respectively, Tyr^{16} and Tyr^{26} in the B-chain. I denotes monoiodinated tyrosine.

The overall content of diiodoinsulins was estimated by disc-electrophoresis.[2] The gel was sliced and the radioactivity in the slices containing diiodoinsulins was measured.

III. RESULTS AND DISCUSSION

The structures of the four monoiodinated insulins which were separated using RPC techniques is shown in Figure 1. The monoiodoinsulins are isomers with the same overall composition and their different retention times on the RPC columns must therefore reflect differences in hydrophobicity and/or charge distribution.

An evaluation of the quality of the separation of the four monoiodoinsulins and unsubstituted insulin using analytical size columns (4.0 to 4.6 to 8.0 mm I.D.) is summarized in Table 1.

The effect of scale-up from analytical LiChrosorb RP-18 columns (250 × 4 mm I.D.) to a preparative LiChrosorb RP-18 column (250 × 25 mm I.D.) upon the separation obtained with two selected mobile phases (TEAP, pH 4.0/acetonitrile and TEAF, pH 6.0/isopropanol) is shown in Figures 2 and 3.

The influence of individual parameters (column packing, mobile phase, temperature, and column load) upon the separation pattern for the mono- and diiodoinsulins as well as recovery and purity of the isolated monoiodoinsulins can be evaluated as follows:

A. COLUMN PACKINGS

The separation pattern obtained for an iodination mixture containing monoiodoinsulins (1 to 5 ng) and insulin (100 μg crystalline insulin added) on a number of reversed-phase column packings was compared using isocratic elution with 0.25 M TEAF, pH 6.0, containing approx. 20% isopropanol as the mobile phase.[8] Baseline separation of all five compounds could be obtained using LiChrosorb RP-18, TSK ODS-120T, or Spherisorb ODS-2 columns, but not on the Techogel C_4 and C_{18} columns nor the resin-based PLRP-S and PRP-1 columns. Substituting the C_{18}-ligand with C_4 (Techogel), furthermore, changed the elution behavior of A14 monoio-

TABLE 1
RPC Separation of Insulin and Monoiodinated Insulin Isomers

Stationary phase	Buffer	Mobile phase pH	Organic modifier	Remarks	Ref.
Spherisorb	TEATFA	3.0	Acetonitrile	Baseline separation	10
Spherisorb	TEAP	4.0	Acetonitrile	Baseline separation	11
LiChrosorb	TEAP	3.0	Acetonitrile	B26/B16 and A19/insulin partially resolved	11
LiChrosorb	TEAP	4.0	Acetonitrile	A19/insulin almost resolved	11
LiChrosorb	TEAP	5.0	Acetonitrile	Improved separation of A19/insulin, broad peaks	11
LiChrosorb	TEAP	6.0	Acetonitrile	Improved separation of A19/insulin, broad peaks	11, 12
LiChrosorb	TEAF	3.0	Acetonitrile	B16/A14 coeluted	12
LiChrosorb	TEAF	4.0	Acetonitrile	Baseline separation	12
LiChrosorb	TEAF	5.0	Acetonitrile	B16/A14 coeluted	12
LiChrosorb	TEAF	6.0	Acetonitrile	B16/A14 coeluted	12
LiChrosorb	TEAF	3.0	Isopropanol	Baseline separation	12
LiChrosorb	TEAF	4.0	Isopropanol	Baseline separation	12
LiChrosorb	TEAF	5.0	Isopropanol	B16/A14 coeluted	12
LiChrosorb	TEAF	6.0	Isopropanol	Baseline separation	7, 8, 12
LiChrosorb	TEAF	3.0	Ethanol	B26/A14 coeluted	12
LiChrosorb	TEAF	4.0	Ethanol	B26/A14 coeluted	12
LiChrosorb	TEAF	5.0	Ethanol	Poor separation of A14/B16	12
LiChrosorb	TEAF	6.0	Ethanol	Blurred peaks	12
Vydac	TEAP	4.0	Acetonitrile	Baseline separation, A19/insulin closely eluted	11, 12
Vydac	TEAP	5.0	Acetonitrile	Same	12
Vydac	TEAP	6.0	Acetonitrile	Same	12
Vydac	TEAF	4.0	Acetonitrile	Same	12
Vydac	TEAF	5.0	Acetonitrile	Same	12
Vydac	TEAF	6.0	Acetonitrile	Same	12
Vydac	TEAF	6.0	Ethanol	Nearly baseline separation, but broad peaks	12
Vydac	TEAF	6.0	Isopropanol	A14/B16 not resolved, broad peaks	12
Spherisorb	TEAF	6.0	Isopropanol	Baseline separation	8
TSK ODS-120T	TEAF	6.0	Isopropanol	Baseline separation	8
Techogel C4	TEAF	6.0	Isopropanol	B26/A14 coeluted	8
Techogel C18	TEAF	6.0	Isopropanol	A14/B16 coeluted	8
PLRP-S	TEAF	6.0	Isopropanol	A14/B16 coeluted	8
PRP-1	TEAF	6.0	Isopropanol	No separation	8

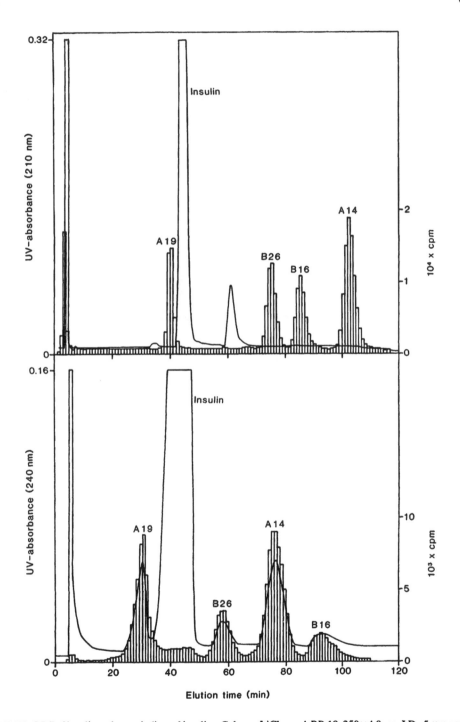

FIGURE 2. RPC of insulin and monoiodinated insulins. Column: LiChrosorb RP-18, 250 × 4.0 mm I.D., 5-μm particle size. Mobile phase: Upper panel, isocratic elution (flow-rate, 1 ml/min) with a 50:50 (v/v) mixture of Eluents A and B, where Eluent A is 0.25 M TEAP, pH 4.0 and Eluent B is equal volumes of Eluent A and acetonitrile; Lower panel, isocratic elution (flow-rate, 0.5 ml/min) with a 55:45 (v/v) mixture of Eluents A and B, where Eluent A is 0.25 M TEAF, pH 6.0 and Eluent B is equal volumes of Eluent A and isopropanol. The samples were a diluted iodination mixture containing 1 to 5 ng of monoiodoinsulins plus 100 μg of added crystalline porcine insulin (containing 5 to 10% deamidated insulin (upper panel) or 0.5 mg of iodinated insulin with low specific activity (lower panel). The continuous curve represents the UV absorption at 210 nm (upper panel) or 240 nm (lower panel). The second UV-peak in the top panel (eluted after insulin) is deamidated insulin. The histograms represent the radioactivity in the collected 1-min fractions.

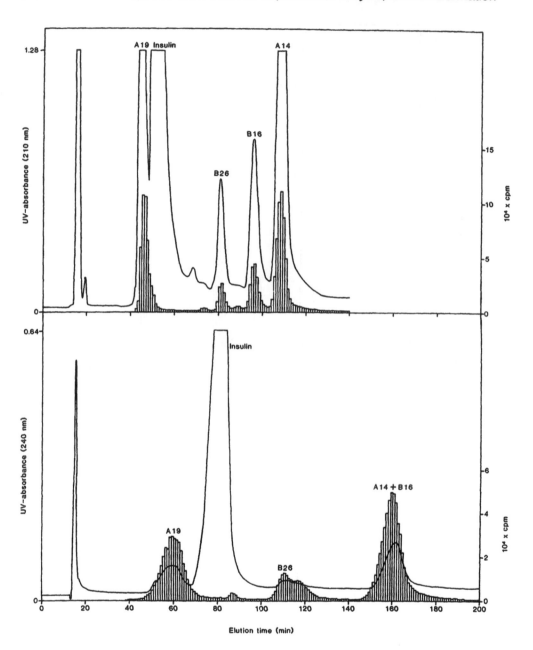

FIGURE 3. Preparative RPC of insulin and monoiodinated insulins. Column: LiChrosorb RP-18, 250 × 25 mm I.D., 7-μm particle size. Mobile phase: Upper panel, isocratic elution (flow-rate, 5 ml/min) with a 48:52 (v/v) mixture of Eluents A and B, where Eluent A is 0.25 *M* TEAP, pH 4.0 and Eluent B is equal volumes of Eluent A and acetonitrile; Lower panel, isocratic elution (5 ml/min) with a 55:45 (v/v) mixture of Eluents A and B, where Eluent A is 0.25 *M* TEAF, pH 6.0 and Eluent B is equal volumes of Eluent A and isopropanol. Samples were 29 mg (upper panel) and 25 mg (lower panel) iodinated insulin with low specific radioactivity. UV absorption and radioactivity registration as shown in Figure 2.

doinsulin, resulting in coelution with B16 monoiodoinsulin or B26 monoiodoinsulin, respectively.

The Vydac 218TPB5 column was unable to separate the four monoiodoinsulins, although wide-pore (300 Å) silica-based C_{18} columns have been reported to be advantageous for the separation of large polypeptides compared to similar 60- to 80-Å pore size columns. However, satisfactory separation was obtained when the insulins were eluted from the Vydac column with acetonitrile as organic modifier[12] (see Table 1).

Interestingly, the three columns capable of separating insulin and the four monoiodoinsulins were all silica-based C_{18} columns with 60- to 80-Å pore size, but with quite different carbon loads and degrees of end-capping. Although insulin and monoiodoinsulins were separated on these C_{18} columns, the individual elution order using a single mobile phase (TEAF, pH 6.0/isopropanol) was found to differ: A19-insulin-B26-A14-B16 (LiChrosorb and Spherisorb) and A19-insulin-B26-B16-A14 (TSK ODS-120T). For the sake of brevity, the monoiodoinsulins will be named only by the position of iodine.

B. MOBILE PHASES

The majority of mobile phases used for RPC separation of insulin and iodinated insulins were ion-pairing substances (TEAP, TEAF, TFA) in combination with acetonitrile. Several combinations of these buffer substances and alternative organic modifiers (isopropanol or ethanol), as well as column packings were investigated (Table 1). In general, mobile phases containing ethanol were not able to separate the monoiodinated insulins satisfactorily (broad peaks, only partial resolution). The use of acetonitrile in the mobile phase always resulted in sharper peaks compared to the use of isopropanol.

The elution order of the four monoiodoinsulins depended on the organic modifier: when the insulins were eluted from the LiChrosorb column with TEAF, pH 6.0, the elution order with acetonitrile was A19-insulin-B26-B16-A14 and with isopropanol was A19-insulin-B26-A14-B16. The mobile phase pH had a similar influence upon the elution order (the elution of A14 and B16 was reversed on the LiChrosorb column using TEAF/ethanol at pH 5.0 or 6.0).

In many respects the use of TEAP and TEAF buffers resulted in comparable chromatograms, and TEAF buffers were generally preferred because they are volatile.

C. SEPARATION TEMPERATURE

The isocratic elution of the four monoiodoinsulins was found to be very sensitive to minor changes in temperature. With the TEAF, pH 4.0/isopropanol system, all four monoiodoinsulins were eluted from the LiChrosorb column within 120 min at 28°C, whereas only A19 monoiodoinsulin was eluted within 140 min when the temperature was 19°C.[12]

Interesting observations were the reversal of the elution order of the B16 and A14 monoiodoinsulins after changing the separation temperature from 22 to 0°C (LiChrosorb column, TEAP, pH 4.0/acetonitrile mobile phase) and the coelution of A19 monoiodoinsulin and insulin at 0°C compared to reasonable separation at 22°C.[11] Such reversal was not observed when isopropanol was used as organic modifier (similar buffer substances and pH),[12] thus demonstrating the complex nature of the binding forces between these polypeptides and the C_{18}-bonded phase.

D. DIIODOINSULINS

The presence of diiodoinsulins in monoiodoinsulin tracers should always be avoided since their decreased binding affinities and decay to iodide and polymers with high non-specific binding can obscure the interpretation of receptor binding studies.[13]

The low degree of iodination used in these studies ensured minimal contents of diiodoinsulins (5%), which include insulin containing two iodine molecules substituted in different tyrosine residues as well as in the same tyrosine residue.[2] Diiodoinsulins isolated from disc-electrophoresis gel slices could be resolved into five components using RPC.[11] Two of the diiodoinsulins might contaminate the monoidoinsulins in some RPC systems described here, but the use of the LiChrosorb column with the TEAF, pH 6.0/isopropanol mobile phase resulted in virtually no contamination of the monoiodoinsulins.[12]

E. COLUMN LOAD

Satisfactory separation patterns were obtained when analytical columns (4.0 mm I.D.) were loaded with up to 500 μg iodinated insulin with low specific radioactivity containing insulin and the four monoiodoinsulins (Figure 2, lower panel). Higher loads resulted in reduced resolution.

The preparative column (25 mm I.D.) was loaded with 25 mg iodinated insulin using TEAF, pH 6.0/isopropanol as mobile phase (Figure 3, lower panel) and 29 mg iodinated insulin using TEAP, pH 4.0/acetonitrile as mobile phase (Figure 3, upper panel).

Although very good separation of the monoiodoinsulins was obtained when they were eluted from 4.0 mm I.D. columns with both mobile phases (Figure 2), only the TEAP, pH 4.0/ acetonitrile mobile phase resulted in a satisfactory separation on the 25 mm I.D. column[7] (Figure 3, upper panel). The reason for this discrepancy could be the different particle size (7 μm for the preparative column compared to 5 μm for the analytical column), different stationary phase batch number or a more pronounced influence of the reduced flow-rate using TEAF, pH 6.0/ isopropanol as mobile phase. Due to practical limitations, the flow-rate was only increased to 5.0 ml/min instead of the linearly calculated 20 ml/min.

F. RECOVERY

When an iodination mixture (high specific radioactivity) containing 1 to 5 ng iodoinsulins was applied to the analytical column (4.0 mm I.D.), the recovery varied from 50 to 100%, determined by measurement of radioactivity in the collected fractions.[10,11] The distribution of radioactivity among the four monoiodinated insulin isomers in the same iodination mixture was, nevertheless, constant, indicating non-specific losses of all monoiodoinsulins. It is well-known that insulin in low concentration has a tendency to adsorb to many surfaces (glass, plastic, etc.).[14] Addition of 50 to 100 μg insulin to the iodination mixture increased the recovery to 100%, indicating that the reduced recovery observed for nanogram amounts of iodoinsulins was caused by a non-specific binding to stainless steel and PTFE in the chromatographic equipment.[11]

The recovery of milligram amounts of monoiodoinsulins with low specific activity from the 25 mm I.D. column was quantitative. The recovery of monoiodoinsulins concentrated using Sep-Pak cartridges and in the water phase after extraction of the organic modifier with cyclohexane was similar (app. 90%).[7]

G. PURITY

Monoiodoinsulins were separated from an iodination mixture of 20 μg insulin iodinated with 1 mCi ^{125}I$^-$ on a LiChrosorb column (250 × 4.0 mm I.D.) by elution with TEAF, pH 6.0-isopropanol.[7] All isolated monoiodoinsulin tracers were more than 95% pure, containing less than 0.5% diiodoinsulin (B16 contained less than 2%). Monoiodoinsulins prepared from 30 mg iodinated insulin with low specific radioactivity by elution from a LiChrosorb column (250 × 25 mm I.D.) with TEAP, pH 4.0/acetonitrile (Figure 3, upper panel), were more than 94% pure.[7]

H. BIOLOGICAL ACTIVITY AND BINDING AFFINITY

The biological properties of the isolated monoiodoinsulins are discussed in detail in the previous chapter in this book.

IV. CONCLUSIONS

Insulin monoiodinated in Tyr A14, A19, B16, and B26 can be separated from insulin and diiodoinsulins using RPC. Several stationary and mobile phases were examined and, in conclusion, it is recommended to use TEAF, pH 6.0/isopropanol as mobile phase with analytical size columns (4.0 mm, 4.6 mm I.D.) such as LiChrosorb RP-18, Spherisorb ODS-2 and TSK ODS-120T. Baseline separation can easily be obtained and the mobile phase is lyophilizable.

A number of literature reports have described the fractionation of iodinated insulins using other column packings and buffers, but the separation of A19 monoiodoinsulin and insulin was always less satisfactory.[15-23]

With the preparative LiChrosorb column (25 mm I.D.), the use of TEAP, pH 4.0/ acetonitrile is recommended, since the use of only this mobile phase resulted in a separation pattern directly comparable to those obtained using the 4.0 mm I.D. column.

During these studies, it was observed that the desired separation in some cases was subject to batch-to-batch variations, at worst resulting in lack of separation of all four monoiodoinsulin isomers. This shows that the separation is only just within the capacity of the reversed-phase columns.

REFERENCES

1. **Gliemann, J., Sonne, O., Linde, S., and Hansen, B.,** Biological potency and binding affinity of monoiodoin-sulin with iodine in tyrosine A14 or tyrosine A19, *Biochem. Biophys. Res. Commun.,* 87, 1183, 1979.
2. **Linde, S., Hansen, B., Sonne, O., Holst, J.J., and Gliemann, J.,** Tyrosine A14 [^{125}I]monoiodoinsulin. Preparation, biologic properties, and long-term stability, *Diabetes,* 30, 1, 1981.
3. **Linde, S., Sonne, O., Hansen, B., and Gliemann, J.,** Monoiodoinsulin labelled in tyrosine residue 16 or 26 of the insulin B-chain. Preparation and characterization of some binding properties, *Hoppe-Seyler's Z. Physiol. Chem.,* 362, 573, 1981.
4. **Sonne, O., Linde, S., Larsen, T.R., and Gliemann, J.,** Monoiodoinsulin labelled in tyrosine residue 16 or 26 of the B-chain or 19 of the A-chain. II. Characterization of the kinetic binding constants and determination of the biological potency, *Hoppe-Seyler's Z. Physiol. Chem.,* 364, 101, 1983.
5. **Peavy, D.E., Abram, J.D., Frank, B.H., and Duckworth, W.C.,** Receptor binding and biological activity of specifically labelled [^{125}I]- and [^{127}I]monoiodoinsulin isomers in isolated rat adipocytes, *Endocrinology,* 114, 1818, 1984.
6. **Welinder, B.S., Sorensen, H.H., and Hansen, B.,** Reversed-phase high-performance liquid chromatography of insulin. Resolution and recovery in relation to column geometry and buffer components, *J. Chromatogr.,* 361, 357, 1986.
7. **Linde, S., Welinder, B.S., Hansen, B., and Sonne, O.,** Preparative reversed-phase high-performance liquid chromatography of iodinated insulin retaining full biological activity, *J. Chromatogr.,* 369, 327, 1986.
8. **Welinder, B.S., Linde, S., and Hansen, B.,** Reversed-phase high-performance liquid chromatography of insulin and insulin derivatives. A comparative study, *J. Chromatogr.,* 348, 347, 1985.
9. **Rabel, F.M., and Martin, D.A.,** Protein separation on octyl and di-phenyl bonded phases, *J. Liq. Chromatogr.,* 6, 2465, 1983.
10. **Welinder, B.S., Linde, S., and Hansen, B.,** Separation, isolation and characterization of the four monoiodi-nated insulin tracers using reversed-phase high-performance liquid chromatography, *J. Chromatogr.,* 265, 301, 1983.
11. **Welinder, B.S., Linde, S., Hansen, B., and Sonne, O.,** Binding affinity of monoiodinated insulin tracers isolated after reversed-phase high-performance liquid chromatography, *J. Chromatogr.,* 281, 167, 1983.
12. **Welinder, B.S., Linde, S., Hansen, B., and Sonne, O.,** Reversed-phase high-performance liquid chromatographic separation of the four monoiodoinsulins: effect of column supports, buffers and organic modifiers, *J. Chromatogr.,* 298, 41, 1984.
13. **Maceda, B.P., Linde, S., Sonne, O., and Gliemann, J.,** [^{125}I]Diiodoinsulins. Binding affinities, biologic potencies and properties of their decay products, *Diabetes,* 31, 634, 1982.
14. **Sato, S., Ebert, C.D., and Kim, S.W.,** Prevention of insulin self association and surface adsorption, *J. Pharm. Sci.,* 72, 228, 1983.
15. **Markussen, J. and Larsen, U.D.,** The application of HPLC to the analysis of radioiodinated tracers of glucagon and insulin, in *Insulin, Chemistry, Structure and Function of Insulin and Related Hormones,* Brandenburg, D., and Wollmer, A., Eds., Walter de Gruyter, New York, 1980, 161.
16. **Jorgensen, K.H. and Larsen, U.D.,** Homogeneous mono-^{125}I-insulin. Preparation and characterization of mono-^{125}I-(TyrA14)- and mono-^{125}I-(TyrA19)-insulin, *Diabetologia,* 19, 546, 1980.

17. **Frank, B.H., Peavy, D.E., Hooker, C.S., and Duckworth, W.C.,** Receptor binding properties of monoiodotyrosyl insulin isomers purified by high performance liquid chromatography, *Diabetes,* 32, 705, 1983.

18. **Frank, B.H., Beckage, M.J., and Willey, K.A.,** High-performance liquid chromatographic preparation of single-site carrier-free pancreatic polypeptide hormone radiotracers, *J. Chromatogr.,* 266, 239, 1983.

19. **Lioubin, M.N., Meier, M.D., and Ginsberg, B.H.,** A rapid, high-yield method of producing mono-[^{125}I]A14 iodoinsulin, *Prep. Biochem.,* 14, 303, 1984.

20. **Marchetti, P., Benzi, L., Cecchetti, P., and Navalesi, R.,** A rapid separation of A$_{14}$-^{125}I-insulin from heterogeneous iodination mixtures by high performance liquid chromatography (HPLC), *J. Nucl. Med. All. Sci.,* 28, 31, 1984.

21. **Rideout, J.M., Smith, G.D., Lim, C.K., and Peters, T.J.,** Comparative study of several phase systems for high-performance liquid chromatographic separation of monoiodoinsulins, *Biochem. Soc. Trans.,* 13, 1225, 1985.

22. **Benzi, L., Pezzino, V., Marchetti, P., Gullo, D., Cecchetti, P.,** Masoni, A., Vigneri, R., and Navalesi, R., A14-[^{125}I]monoiodoinsulin purified by different high-performance liquid chromatographic procedures and by polyacrylamide gel electrophoresis: preparation, immunochemical properties and receptor binding affinity, *J. Chromatogr.,* 378, 337, 1986.

23. **Stenz, F.B., Wright, R.K., and Kitabchi, A.E.,** A rapid means of separating A$_{14}$-^{125}I-insulin from heterogenous labelled insulin molecules for biologic studies, *Diabetes,* 31, 1128, 1982.

BATCH-EXTRACTION OF HUMAN PITUITARY PEPTIDES ON A REVERSED-PHASE MATRIX: APPLICATION TO THE PURIFICATION OF THE POMC-JOINING PEPTIDE

**Suzanne Benjannet, Claude Lazure,
James A. Rochemont, Nabil G. Seidah,
and Michel Chrétien**

I. INTRODUCTION

Despite the advent of molecular biology and the impact of biotechnology on the production of proteins and biologically active peptides and hormones, the extraction and isolation of peptides from vast quantities of tissues or fluids is still often the sole source of interesting novel material. Furthermore, even though the production of hormones and biologically active peptides from chemically engineered bacteria, viruses, or yeasts has been facilitated, purification from the organisms or from the production medium poses a serious and at times difficult problem.[1,2] Thus, studies aimed at improving the purity, the biological activity, and/or the yield of the biologically active moieties appear well warranted. To this end, the introduction of High-Performance Liquid Chromatography (HPLC) has proved pivotal since it enables rapid and efficient purification of minute amounts of proteins and peptides from the vast array of extracts. Thus, the convenience, rapidity, sensitivity, and versatility of HPLC-based techniques offer the protein chemist very powerful tools. Considering that HPLC instrumentation is also the basis for numerous analytical procedures, such as amino acid analysis and peptide sequencing (as reviewed recently[3]), it is easy to understand its popularity and wide distribution.

Throughout the years, our laboratory has relied heavily on this approach in order to purify or analyze various novel peptides or proteins. In particular, a program devoted many years ago to the purification of biologically active human growth hormone (hGH, 191 residues) from frozen pituitary glands obtained at autopsy, was initiated. This endeavor provided us with some challenging problems related to the amount of material (a thousand glands had to be extracted), the complexity in terms of peptides present, the conservation of the biological activity of the isolated hGH, and our desire to recover as much as possible the secondary products which might be related to our studies on Pro-opiomelanocortin (POMC). This chapter will summarize results obtained while using a purification protocol based entirely upon reversed-phase HPLC (RPC) and, in particular, will focus on peptides present in the non-precipitating fraction of a neutral salt precipitation procedure.

II. ISOLATION OF HUMAN POMC (79 TO 108) PEPTIDE

A. THEORETICAL CONSIDERATIONS

Considering that the initial purpose of the study was the purification of biologically active hGH for clinical uses, it was necessary to adopt a protocol enabling us to obtain this product with a reasonable yield from a material both precious and difficult to obtain. Various means by which growth hormone (GH) could be purified from pituitary extracts have been proposed over the years and suffered in many instances either in poor recovery and/or low biological activity. These aspects have recently been discussed in detail and the purified material appeared to be homogenous, was obtained in good yields and displayed high potency when assayed according to the GH standard obtained from the National Institute of Arthritis, Metabolism, and Digestive Diseases (NIAMDD).[4] It should be mentioned that, following RPC, hGH was slightly soluble and almost inactive, a behavior alleviated by including a mandatory dialysis step after the final purification.

As mentioned previously, we were very much interested in obtaining the other fractions which could be derived from the extraction of sometimes as many as 2000 frozen pituitaries. Starting from an homogenate of finely powdered frozen pituitaries, the initial step involved a neutral salt precipitation yielding two fractions. For both fractions, the purification was carried out using a protocol incorporating the following steps:

- Batch-wise adsorption on a reversed-phase matrix to permit either preliminary fractionation or desalting of the fractions
- A step or serial elution scheme was used on the batch-adsorbed material
- RPC was accomplished using an HPLC instrument incorporating simple devices allowing processing of large volumes in a highly repetitive and automatic fashion
- In the case of hGH, dialysis of the purified fraction against neutral or alkaline buffers was obligatory

Whereas it can be seen that this protocol relied entirely for its success on the attainment of a preliminary enriched fraction and on the resolving power of RP-matrices or RPC, other approaches have been described. Thus, manipulation of pH and ion-pairing reagents, use of C_{18} Sep-Pak and of ion-exchange Sep-Pak cartridges were described and applied to the separation of pituitary peptides following tissue extractions.[5-7]

B. PRACTICAL CONSIDERATIONS
1. Preparation of the Extract and Batch-Wise Extraction
Here, we will summarize only the steps involved in the analysis of the non-precipitable neutral salt material, since other conditions used were described previously.[4]

The 250 frozen pituitaries (individually kept in liquid nitrogen) were first reduced to a fine powder and then homogenized using a Polytron PT35 in the presence of 0.1 M Tris.HCl and 2.5 mM EDTA (ethylenediaminetetraacetic acid) adjusted to pH 8.5. Following centrifugation to remove cellular debris, the proteins contained in the supernatant were then precipitated at pH 8.5 using ammonium sulfate to a 70% saturation level. After centrifugation at 4°C for 60 min, the pellet and the supernatant were separated and kept for further analysis.

The supernatant was acidified to pH 3.0 by adding trifluoroacetic acid (TFA), centrifuged for 30 min and manually mixed with 50 g of C_{18}-silica (50-µm particle size; Waters Associates, Milford, MA) previously equilibrated with 5% aq. acetonitrile containing 0.1% TFA; the addition of a low amount of organic solvent facilitates manipulation of the matrix. After washing the matrix with the equilibrating buffer, the adsorbed material was eluted directly with 2 × 250 ml of 50% aq. acetonitrile containing 0.1% TFA (in the case of the salt-precipitated fraction a serial elution scheme was preferred[4]).

2. HPLC Purification
a. Equipment Required
Depending upon the scale at which the extraction procedure was done, the initial purification

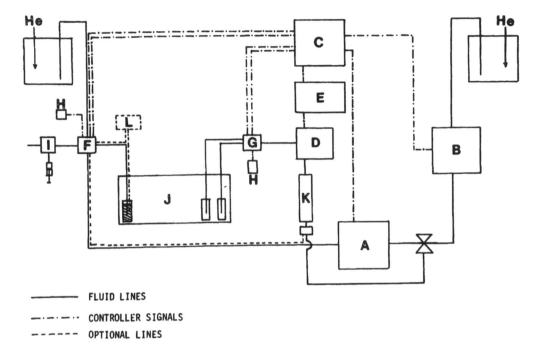

FLUID LINES

CONTROLLER SIGNALS

OPTIONAL LINES

FIGURE 1. Block diagram of the HPLC system used for automatic preparative separations. The various letters used represent the following devices: solvent pumps (A, B), programmable controller (C), UV-detector (D), and integrator/plotter (E). This basic HPLC was interfaced with a 4-port valve with electric valve actuator to serve for injection (F) and with a 16-port valve also electrically activated for collection of the fractions (G). Both valves can also be manually operated by manual controllers (H). A 2-port valve (I) is solely used to prime the sample up to unit F during the initial run. A thermoregulated bath (J) and a custom made reversed-phase column (1 × 12 in., K) complete the system. The eluting buffers are continuously kept under Helium (He) to prevent bubble formation. An optional auxiliary pump (L) to deliver the sample directly to the column is also shown.

can be accomplished with basic HPLC equipment with large volume injection loops or using an instrument dedicated solely to this endeavor (represented in Figure 1). This instrument incorporates a number of low cost devices which render its use more convenient in the context of large or numerous samples. Here, the sample was loaded on the column by using either the aqueous buffer pump (in which case, it must be cleaned thoroughly with 50% nitric acid after use) or an auxiliary inexpensive pump (Figure 1, part L). Switching from the "inject" position to the "run" position was carried out automatically by the 4-port valve activated by a programmed event from the HPLC controller. Thus, all operations can be accomplished in an entirely automated fashion by judicious programming of the relay activating the 4-port valve responsible for the sample application and the 16-port valve responsible for fraction collection. Over the years, this low-cost system has proved extremely reliable and versatile; furthermore, the chromatographic reproducibility was excellent while affording unattended operation.

b. Assessment of Methodology

It was possible to follow the course of our hGH purification by routinely monitoring its biological activity using a radioreceptor assay. In other cases, purification can be conveniently monitored directly on aliquots (0.005 to 0.1% of total amount) by radioimmunoassays, gel electrophoresis in denaturing conditions or analytical HPLC. Finally, the purity as well as the identification of the material so obtained was carried out by amino acid analysis or, when warranted, by peptide sequencing according to previously described procedures.[3,8,9]

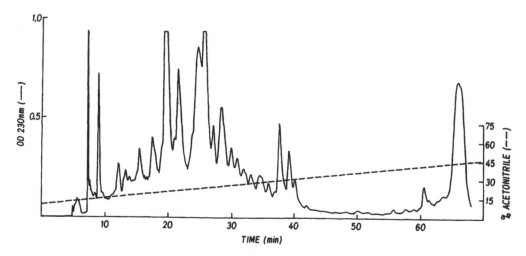

FIGURE 2. Reversed-phase separation of the pituitary extract supernatant obtained following ammonium sulfate precipitation and batch adsorption. Column: μBondapak C$_{18}$, 300 x 7.8 mm I.D., 10-μm particle size (Waters Associates, Milford, MA). Mobile phase: linear AB gradient (0.5% B/min, starting from 10% B at 0 min), where Eluent A is 0.1% aq. trifluoracetic acid (TFA) and Eluent B is 0.1% TFA in acetonitrile; flow-rate, 2 ml/min; absorbance at 230 nm in view of the semi-preparative separation mode. Fractions were manually collected according to the UV-absorbance profile.

III. RESULTS AND DISCUSSION

As mentioned above, the use of ammonium sulfate precipitation yielded two fractions which were both analyzed using the protocol described herein. The neutral salt precipitate obtained from whole pituitary extract was extremely rich in various peptides. A partial list of identified human peptides includes the POMC N-terminal fragment (1 to 76), adrenocorticotropin (ACTH, 39 residues), β-lipotropin (β-LPH, 89 residues) the C-terminal part of propressophysin (copeptin, 39 residues) and three novel peptides denoted GAWK (74 residues), CCB (57 residues), and 7B2 (186 residues).[8-10] Missing from that list as far as the POMC molecule is concerned was the POMC (79 to 108) peptide, also called joining peptide. This peptide, whose existence was first predicted on the basis of the gene sequence encoding the whole POMC molecule,[11] was later purified from pituitary extracts by carboxymethylcellulose chromatography and RPC.[12]

The absence of the (79 to 108) peptide from the salt precipitated fraction prompted us to investigate the supernatant fraction. The use of batch adsorption on a RP-matrix allowed us to circumvent the desalting procedure normally accomplished by gel permeation (thus minimizing the loss of small molecular weight peptides) and to concentrate the rather diluted supernatant. Whereas serial elution with different acetonitrile concentrations had proved beneficial in the case of the salt precipitated fraction, a single eluting buffer was now used. The eluted fraction was either suitably diluted or lyophilized before being separated on the automated preparative HPLC. A representative chromatogram obtained at a semi-preparative scale, identical to the one using the preparative HPLC, is shown in Figure 2. As can be seen, surprisingly numerous products are not precipitated by using such a high ammonium sulfate concentration. One of the major UV-absorbing peaks eluted at 26.6 min on the C$_{18}$ column (Figure 2) was further purified on a wide-pore C$_4$ column (Figure 3) and, through analysis, was identified as the human joining peptide. Indeed, the repurified peptide yielded the expected amino acid composition (data not shown) and its identity was confirmed by automated Edman degradation using an Applied Biosystems (Foster City, CA; ABI Model 470A) Gas-phase sequenator as shown in Figure 4. The last residue identified is a Glu which, according to the absence in the amino acid composition

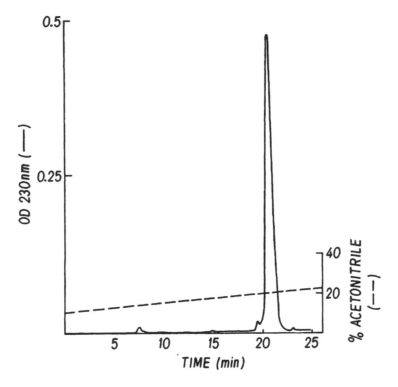

FIGURE 3. Final repurification of the human POMC (79 to 108) joining peptide. Column: Vydac C$_4$, 250 × 10 mm I.D., 5-μm particle size, 300-Å pore size (The Separations Group, Hesperia, CA). Mobile phase: linear AB gradient (0.5% B/min, starting from 10% B at 0 min), where Eluent A is 0.1% aq. TFA and Eluent B is 0.1% TFA in acetonitrile; flow-rate, 2 ml/min; absorbance at 230 nm. Fractions were manually collected according to the UV-absorbance profile.

of the C-terminal Gly predicted by the gene sequence, is amidated. The isolated peptide thus represents the region encompassing position 79 to 108 of the POMC precursor.

Further analysis of other fractions from the supernatant yielded numerous other peptides which were identified by either sequencing or amino acid analysis. Thus, it was found that the joining peptide is actually quite heterogeneous, since numerous fractions in the region between 25 to 30 min (Figure 2) contained joining peptide related material. For example, the peptide eluted at 27.3 min corresponded to POMC (79 to 98) and probably arises from cleavage after the single Arg residue; possibly, this could explain the previously reported low yield of intact peptide.[12] Other peptides were also identified, such as the nonapeptide Arg-Vasopressin and a fragment of copeptin (position 27 to 37) in the same eluting region. Considering the number of UV-absorbing peaks present in the supernatant which are still awaiting further characterization, their identification will present a formidable task.

It is obvious that the techniques presented herein, while not novel, are extremely powerful since they allow rapid purification of peptides (present in either the precipitate or supernatant following precipitation with ammonium sulfate) from large amounts of material and from intrinsically complex mixtures. The use of batch-adsorption allows concentration of peptides and at the same time the necessary desalting without loss of smaller peptides (as substantiated here by the recovery of Arg-Vasopressin). The use of the automated HPLC, which can be adapted without costly modifications in any laboratory, permits separation of large extract volumes, thus minimizing lyophilization steps, and of numerous samples in a completely automated and highly repetitive fashion. Thus, even though this methodology was intended

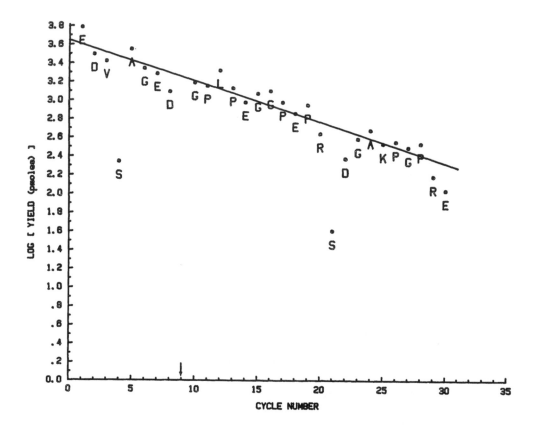

FIGURE 4. Yields of phenylthiohydantoin(PTH)-amino acids as a function of sequenator cycle number obtained during the NH₂-terminal sequence analysis of the human POMC (79 to 108) joining peptide. The arrow at cycle 9 corresponds to the position of the sole Cys residue which was not identified here. The initial and repetitive yields were 4.43 nmol and 90.3%, respectively, as estimated by linear regression using the yield of selected stable PTH-amino acids corrected according to an internal standard, PTH-NorLeu, (correlation coefficient of 0.90).

primarily for the purification of biologically active hGH, it has proved to be extremely useful by permitting the isolation of numerous interesting peptides.

ACKNOWLEDGMENTS

This study was made possible by financial support from the Medical Research Council of Canada. C.L. is a "chercheur-boursier" of the "Fonds de la Recherche en Santé du Québec".

REFERENCES

1. **Marston, F.A.O.,** The purification of eukaryotic polypeptides synthesized in *E. coli, Biochem. J.* 240, 1, 1986.
2. **Mattaliano, R.J., Rosa, J.J., Foeller, C., Woodard, J.P., and Bertolini, M.J.,** Analysis of recombinant proteins: current trends and practical limits in analytical stringency, in *Methods in Protein Sequence Analysis,* Walsh, K.A., Ed., Humana Press, Clifton, 79, 1987.
3. **Lazure, C., Rochemont, J.A., Seidah, N.G., and Chrétien, M.,** Amino acids in protein sequence analysis, in *HPLC of Biological Macromolecules,* Regnier, F. and Gooding, K.M., Eds., Marcel Dekker, New York, 1990, 263.

4. **Lazure, C., Benjannet, S., Rochemont, J.A., Seidah, N.G., and Chrétien, M.,** High-performance liquid chromatography (HPLC): A versatile tool in peptide and protein chemistry, in *Laboratory Methodology in Biochemistry,* Fini, C., Floridi, A., Finelli, V.N., and Wittman-Liebold, B., Eds., CRC Press, Boca Raton, FL, 1990, 83.

5. **Bennett, H.P.J.,** Isolation of pituitary peptides by reversed-phase high performance liquid chromatography: expansion of the resolving power of RP-columns by manipulating pH and the nature of the ion-pairing reagent, *J. Chromatogr.,* 266, 501, 1983.

6. **James, S. and Bennett, H.P.J.,** Use of reversed-phase and ion-exchange batch extraction in the purification of bovine pituitary peptides, *J. Chromatogr.,* 326, 329, 1985.

7. **Bennett, H.P.J.,** Use of ion-exchange Sep-Pak cartridges in the batch fractionation of pituitary peptides, *J. Chromatogr.,* 359, 383, 1986.

8. **Benjannet, S., Leduc, R., Lazure, C., Seidah, N.G., Marcinkiewicz, M., and Chrétien, M.,** GAWK, a novel human pituitary polypeptide: isolation, immunocytochemical localization and complete amino acid sequence, *Biochem. Biophys. Res. Comm.,* 126, 602, 1985.

9. **Benjannet, S., Leduc, R., Adrouche, N., Falgueyret, J.P., Marcinkiewicz, M., Seidah, N.G., Mbikay, M., Lazure, C., and Chretien, M.,** Chromogranin B (Secretogranin I), a putative precursor of two novel pituitary peptides through processing at paired basic residues, *FEBS Lett.,* 224, 142, 1987.

10. **Mbikay, M., Grant, S.G.N., Sirois, F., Tadros, H., Skowronski, J., Lazure, C., Seidah, N.G., Hanahan, D., and Chrétien, M.,** cDNA sequence of neuroendocrine protein 7B2 expressed in beta cell tumors of transgenic mice, *Int. J. Peptide Protein Res.,* 33, 39, 1989.

11. **Chang, A.C.Y., Cochet, M., and Cohen, S.N.,** Structural organization of human genomic DNA encoding the proopiomelanocortin peptide, *Proc. Natl. Acad. Sci. U.S.A.,* 77, 4890, 1980.

12. **Seidah, N.G., Rochemont, J., Hamelin, J., Benjannet, S., and Chrétien, M.,** The missing fragment of the pro-sequence of human Pro-opiomelanocortin: sequence and evidence for C-terminal amidation, *Biochem. Biophys. Res. Comm.,* 102, 710, 1981.

AUTOMATED DETERMINATION OF AMINO ACID ENANTIOMERS USING DERIVATIZATION WITH 1-(9-FLUORENYL)ETHYL CHLOROFORMATE AND REVERSED-PHASE LIQUID CHROMATOGRAPHY

Stefán Einarsson and Göran Hansson

I. INTRODUCTION

The important role of chirality in nature has long been recognized. Large groups of biologically important molecules such as amino acids are chiral, and for understanding the properties of these substances it is necessary to separate and quantitate the enantiomers individually. The diverse vital function of amino acids in living organisms make it important to determine the ratios of the enantiomers in various areas, such as dating,[1] food chemistry,[2] and studies of extraterrestrial material.[3] For peptide synthesis, it is important to determine the enantiomeric purity of the amino acids used and the extent of racemization which may occur during the synthesis. For analytical procedures which involve liberation of the amino acids from peptides and proteins, it is necessary to check the extent of racemization in the hydrolysis procedure used. Studies have shown that the degree of racemization during hydrolysis varies from one type of peptide to another, and that peptide-bound amino acids generally racemize faster than free amino acids.[4]

Besides the resolution of the enantiomers, quantitation of amino acid enantiomers normally requires separation of the enantiomers of interest from other amino acids in the sample. It is, moreover, usually necessary to form derivatives of the amino acids for the purpose of enhancing sensitivity in detection. Methods for amino acid analysis based on pre-column derivatization with fluorescence reagents and separation by reversed-phase chromatography (RPC) have gained widespread use, because of the low detection limits attainable and the flexibility of the separation system.[5-7] Automated methods have been developed using the non-chiral reagents o-phthaldialdehyde/mercaptoethanol[8] (OPA) for primary amino acids and 9-fluorenylmethyl chloroformate[9,10] (FMOC-C1) for primary and secondary amino acids. Derivatization of chiral amino acids with a chiral reagent results in diastereomers, which are, in principle, separable on non-chiral columns. By using this approach, it has been possible to perform analysis of the enantiomers of the protein amino acids in a single run. Since the chromatographic separation normally takes 1 to 2 hr, it is important to be able to automate fully the analytical procedure. A prerequisite is a simple derivatization reaction which proceeds to completion in a short time at room temperature. Reagents based on different chiral thiols in combination with OPA have been used for separation of amino acid enantiomers,[11-13] and an automated method has been presented.[14]

The chiral reagent 1-(9-fluorenyl)ethyl chloroformate (FLEC) which has been used for resolution of amino acid enantiomers, has the advantages of forming stable derivatives and being reactive towards both primary and secondary amino acids.[15]

This paper describes an automated procedure for determination of amino acid enantiomers using the FLEC reagent. The method was applied for determination of amino acid enantiomers after hydrolysis of peptides.

II. EXPERIMENTAL

A. APPARATUS

The chromatographic system consisted of a gradient delivery system (Varian 5500) and an autosampler (Varian 9090) equipped with a gas-actuated Valco injector with a fixed 10-μl sample loop. A Shimadzu RF-535 fluorescence detector was used for monitoring the derivatives, with the excitation and emission wavelengths set at 260 and 315 nm, respectively. A Varian DS 651 data system was used for automatic control of the chromatography and the autosampler, and for sampling and storage of chromatograms. Autosampler vials and microvial inserts were obtained from Varian (Solna, Sweden).

B. CHEMICALS

The FLEC reagent was obtained from Eka-Nobel (Surte, Sweden). Amino acid standards and peptides were purchased from Sigma (St. Louis, MO), boric acid (suprapur) and the OPA reagent from Merck (Darmstadt, FRG), and iodoacetic acid (sodium salt) and sodium azide from Fluka (Buchs, FRG). Acetonitrile, THF, acetone, pentane and ethyl acetate (all HPLC grade) were all obtained from Rathburn (Walkerburn, UK).

C. DERIVATIZATION PROCEDURES
1. Derivatization of Primary and Secondary Amino Acids

The reaction and the extraction procedures were performed in 190 μl microvial inserts placed in vials with screwcaps with teflon-lined membranes. The autosampler was programmed to mix 25 μl of the buffered sample (in 0.2 M borate buffer, pH 9.0) and 25 μl of the FLEC reagent (5 mM in acetone) in the microvial (Figure 1A). The solution was mixed by driving 80 μl of air through the sample, and then the reaction was allowed to proceed for 10 min. After completion of the reaction, 60 μl of the extraction solution (pentane:ethyl acetate, 85:15) were added, followed by 6 cycles of air mixing. After equilibration of the solution for 10 min, the lower water phase was sampled for injection. Six needle wash cycles (85:15, acetone:water) were included between all pipetting steps.

2. Selective Derivatization of Secondary Amino Acids

To 80 μl of buffered sample (in 0.1 M borate buffer, pH 9.5), the following solutions were added in turn (Figure 1B): 8 μl OPA reagent (50 mg OPA and 25 μl mercaptoethanol/ml, in acetonitrile), 8 μl iodoacetate (1 M in 0.1 N NaOH) and 24 μl FLEC reagent (5 mM in acetone). After addition of each reagent, the solution was mixed (air, 80 μl), and the needle washed (5 wash cycles). The solution was then extracted by addition of 50 μl of diethyl ether followed by 5 air-mix cycles. After equilibration for 10 min, the lower water phase was sampled for injection on the column. Actual reaction times (i.e., including the time taken for washes) were 4.5 min for the OPA reagent and the iodoacetic acid reagent and 7 min for the FLEC reagent.

The autosampler was programmed to derivatize a sample while a previous sample was being chromatographed, thus minimizing the total analysis time.

D. HYDROLYSIS PROCEDURES

Culture tubes (Corning) with teflon-lined screw caps were used for the liquid-phase hydrolysis. The samples were hydrolyzed in constant boiling hydrochloric acid (~6 N), under nitrogen (4 cycles of evacuation and purging).

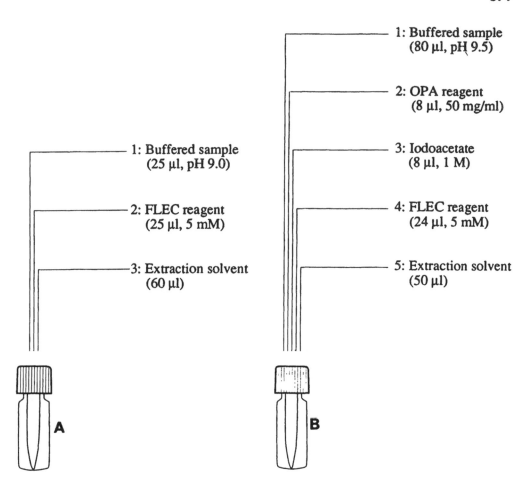

FIGURE 1. (A) Procedures used for derivatization of primary and secondary amino acids, and (B) selective derivatization of secondary amino acids.

For gas-phase hydrolysis, the samples were pipetted into 0.7 ml tubes, evaporated to dryness under vacuum and then placed in a vessel (115 ml) equipped with a valved cap (Waters). Constant boiling hydrochloric acid (70 μl) was pipetted into the vessel beside the sample tubes, and then the vessel was evacuated and placed in an oven for hydrolysis.

E. CHROMATOGRAPHY

The system was equipped with a scavenger column (C_{18}, 36 × 4.5 mm I.D., 20-μm particle size, Rsil) placed between the pump and the injector, and a guard column (RP-8, 15 × 3.2 mm I.D., 7-μm particle size, Brownlee) between the injector and the analytical column. NaN_3 (100 mg/l) was added in the aqueous mobile phase buffers as a bactericidal agent.

1. Gradient Separation of the Protein Amino Acids

For the separation (Figure 2), a column (300 × 4.6 mm) packed with a Kromasil octyl material (5 μm) was used. The ternary gradient was composed of, A: THF, B:acetate buffer (1 ml glacial acetic acid/l, pH adjusted to 7.0) and C: acetate buffer (1.8 ml glacial acetic acid/l, pH adjusted to 4.24). Flow-rate, 1 ml/min.

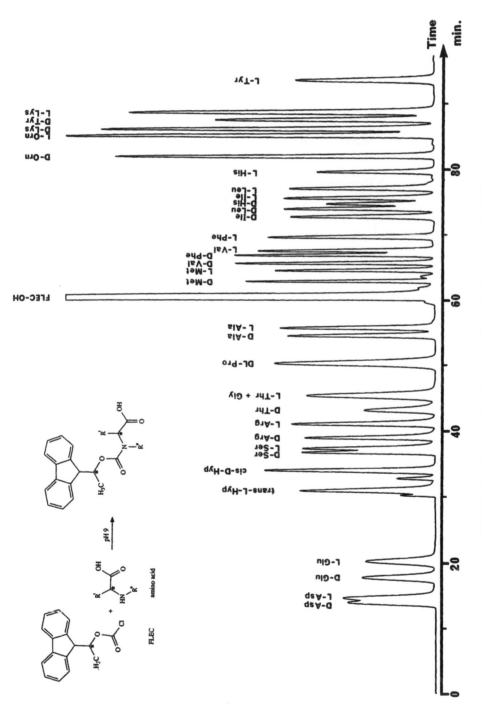

FIGURE 2. Gradient separation of an amino acid standard derivatized with (+)-FLEC.

Time (min)	A (%)	B (%)	C (%)
0	15	85	0
16.99	16	84	0
17	28	0	72
28	28	0	72
37	29	0	71
50.99	38	0	62
51	38	31	31
61	40	30	30
75	44	28	28
82	44	28	28
89.99	46	27	27
90	55	45	0
94.99	55	45	0
95	15	85	0

2. Isocratic Separation of Secondary Amino Acids

The same column was used as for the gradient separation.

Mixtures of acetonitrile and phosphoric acid (0.1 N) were used as mobile phases for the separations of both FMOC-derivatives (acetonitrile:phosphoric acid, 39:61) and FLEC-derivatives (44:56). The flow-rate was 1.5 ml/min.

III. RESULTS AND DISCUSSION

The rates at which amino acids react with FLEC at a given temperature are primarily dependent on the concentration of the reagent and the pH of the reaction solution. Tyrosine, which has a phenolic side group, may be determined either as a mono- or as a bis-derivative, depending on the reaction conditions. A procedure based on quantitation of the bis-derivative of tyrosine makes it possible to use the derivatization of this amino acid as a control of the reaction procedure, as formerly used in methods based on the FMOC-C1 reagent.[16] Incidents such as insufficient excess of the reagent, or reduction in pH by highly buffered samples, both causing less than complete reaction yields, will thus be detected by an increase in the mono-derivative of tyrosine. Under the conditions described in the experimental section, 90% of tyrosine is converted to the bis-derivative. After completion of the derivatization, an extraction with an organic solvent is used to separate the excess of the reagent and part of its hydrolysis product from the amino acid derivatives, which remain in the aqueous phase.

The two different automatic procedures used for the derivatization of the amino acids are shown in Figure 1. Procedure A was used for derivatization of both primary and secondary amino acids. Procedure B, however, was used for selective derivatization of the secondary amino acids. The selectivity is based on reaction of the primary amino acids with the OPA reagent, followed by derivatization of the secondary amino acids with FLEC. Owing to the different fluorescence properties of the OPA derivatives, the secondary amino acids can be detected without disturbances from the primary amino acids.[17]

The linearity of the response for 32 amino acid enantiomers was checked in the concentration range 1 to 100 µmol/l. Good linear response was obtained (mean correlation coefficient = 0.9995, range 0.9988 to 0.9998, n = 14) for all the amino acids except D- and L-tyrosine, which exhibited a linear response up to 50 µmol/l. It should be noted that the ratio of the total amino acid concentration to the reagent was only 1:1.3 at the highest level (100 µmol/l).

The fluorescence responses of the amino acid derivatives were very similar under the elution conditions used. The lowest response (78% of the mean response for the mono derivatives) was obtained for threonine, while higher responses were obtained for the bis-derivatives of lysine (190%), ornithine (170%), and tyrosine (150%).

The procedure for selective determination of the secondary amino acid enantiomers (Figure 1B) involves a sequence of three rapid reactions. The first two reactions serve as a clean-up before the derivatization of the secondary amino acids with FLEC. Selective blocking of substances containing primary amines is accomplished by reaction with OPA/mercaptoethanol. The following reaction with iodoacetic acid serves to eliminate the excess of mercaptoethanol in order to prevent the FLEC reagent from being consumed by the thiol. Less than 1 min is needed for each of the first two reactions, and the reaction with FLEC is complete in less than 5 min. The actual time needed to carry out the whole procedure is therefore determined by the number of wash-cycles between the pipetting steps.

Calibration curves in the concentration interval 0.1 to 100 μmol/l showed excellent linearity of response for all the secondary amino acids (mean correlation coefficient 0.9995, range = 0.9991 to 0.9999, n = 14).

The mean standard deviation of the retention times, evaluated for 35 amino acids over a time period of 7 days, was 0.2 min (range 0.125 to 0.32 min, 40 runs of different samples). The smallest variation was in the late eluted derivatives. Systematic variation in the retention of groups of closely eluted peaks made it possible to use relative retention times for increasing the reliability of the peak identification.

Samples such as peptide hydrolyzates typically contain amino acid enantiomers at widely different concentration levels. The repeatability of peak areas was evaluated from calibration curves in the range 20 to 80 μmol/l for the L-amino acids and 0.3 to 2.8 μmol/l for the D-amino acids. The mean relative standard deviation (RSD) was 3.6% for the L-amino acids and 6.2% for the D-amino acids. Normalizing the area using one amino acid as a reference resulted in a decreased mean RSD (2.2% for the L-amino acids and 5.3% for the D-amino acids).

For the isocratic separation of the secondary amino acids (Figure 3), the range in the standard deviation (n = 10) of the retention times was 0.0082 to 0.040 min.

The mean RSD (concentration 5 μmol/l, n = 10) of peak areas for the secondary amino acids was 1.1% (range 0.75 to 1.44%). An excess of L- over D-proline by a factor of 100 (50 μM vs. 0.5 μM) was not found to influence the variation in the peak area of the D-enantiomer (RSD 1.1%).

The autosampler used was equipped with a fixed 10 μl injection loop. In order to avoid imprecision due to the injection, complete filling of the sample loop should be ensured. To obtain steady response in the peak areas, at least 25 μl was needed to be sampled for injection due to nonlinearity in loop-filling.[18]

The mobile phase was based on THF,[15] and the separation was tuned by changing the composition of the aqueous acetate buffer (Figure 2). Starting the gradient with a buffer at pH 7 separates aspartic acid and glutamic acid from the polar amino acids.[19] The pH is then switched to 4.24 for optimal separation of the non-ionic hydrolysis product, and the ionic strength of the buffer chosen so that arginine is eluted between serine and threonine. For optimized separation of the D-Phe/L-Val and D-His/L-Ile pairs, the pH is adjusted (4.78) by mixing the two buffers. Optimal elution conditions can be expected to vary between different stationary phases. The presence of residual silanol groups strongly affects the retention of arginine, which can be counteracted by addition of an amine to the mobile phase (e.g., triethyl amine, tetramethyl ammonium range 1 to 20 mM).

A mobile phase with a low pH is needed to resolve the enantiomers of hydroxyproline and proline. A complete separation was obtained by simple isocratic elution, as shown in Figure 3. Similar chromatographic conditions allow separation of the cis and trans isomers of hydroxyproline after derivatization with the achiral reagent FMOC-Cl.

Measurements of the enantiomeric purity of amino acids and the degree of racemization in protein hydrolyzates require a system capable of determining the amino acids in a wide concentration range. Detection of traces of the D form in peptide hydrolyzates can be problematic

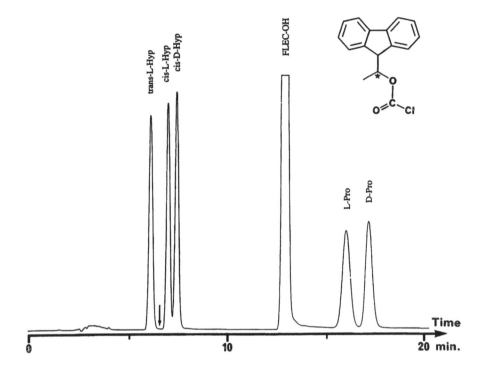

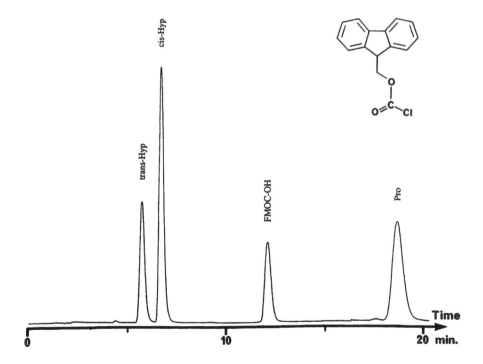

FIGURE 3. Isocratic separations of an amino acid standard containing 36 primary and secondary amino acids (25 μmol/l each) after derivatization using procedure B with (+)-FLEC (upper chromatogram) and the achiral reagent FMOC-Cl (lower chromatogram). The arrow in the upper chromatogram indicates the position of trans-D-Hyp, determined separately after derivatization of trans-L-Hyp with (–)-FLEC.

because of possible interference, such as contamination by other amino acids and the reagent. High resolution in the separation is therefore advantageous. Access to both enantiomers of the reagent is valuable, since it increases the reliability of the identification by enabling reversal in the elution order of the amino acid enantiomers (Figure 4). Another advantage is the possibility to determine the retention of the derivative of an amino acid enantiomer which is not available, by chromatographing its enantiomerically related derivative. This was utilized for trans-D-hydroxyproline (Figure 3).

Racemization of amino acids was determined after acid hydrolysis of different peptides (neurotensin, α-MSH, substance P, oxytocin) and free amino acids. A chromatographic separation of a peptide hydrolysate is shown in Figure 4. The extent of racemization after hydrolysis was comparable with previously published data.[4] Comparison of procedures based on liquid and gas-phase hydrolysis in 6 N HCl at 110°C), did not reveal any differences in the racemization of the amino acids.

ACKNOWLEDGMENTS

We thank the Varian Instrument Group (Walnut Creek, California) for loan of instruments. This work was supported by the Swedish National Board for Technical Development and the Erna and Victor Hasselblad Foundation.

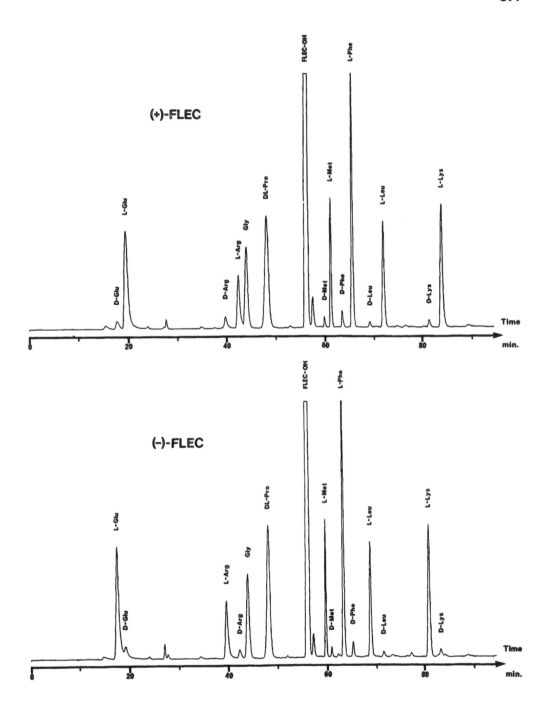

FIGURE 4. Gradient separations of a hydrolysate of substance P (145°C, 4 hr in 6 N HCl) after derivatization with (+)-FLEC (upper chromatogram) and (−)-FLEC (lower chromatogram).

REFERENCES

1. **Hare, P.E., Hoering, T.C., and King, K., Jr., Eds.,** *Biogeochemistry of Amino Acids,* John Wiley & Sons, New York, 1980.
2. **Man, E.H. and Bada, J.L.,** Dietary D-amino acids, *Ann. Rev. Nutr.,* 7, 209, 1987.
3. **Cronin, J.R. and Pizzarello, S.,** Amino acids in meteorites, *Adv. Space Res.,* 3, 5, 1983.
4. **Frank, H., Woiwode, W., Nicholson, G., and Bayer, E.,** Determination of the rate of acidic catalyzed racemization of protein amino acids, *Liebigs Ann. Chem.* 354, 1981, and references cited therein.
5. **Einarsson, S., Josefsson, B., and Lagerkvist, S.,** Determination of amino acids with 9-fluorenylmethyl chloroformate and reversed-phase high-performance liquid chromatography, *J. Chromatogr.,* 282, 609, 1983.
6. **Lindroth, P. and Mopper, K.,** High performance liquid chromatographic determination of subpicomole amounts of amino acids by precolumn fluorescence derivatization with o-phthaldialdehyde, *Anal. Chem.,* 51, 1667, 1979.
7. **Taphui, Y., Schmidt, D.E., Lindner, W., and Karger, B.L.,** Dansylation of amino acids for high-performance liquid chromatography analysis, *Anal. Biochem.,* 115, 123, 1981.
8. **Smith, R.J. and Panico, K.A.,** Automated analysis of o-phthalaldehyde derivatives of amino acids in physiological fluids by reverse phase high performance liquid chromatography, *J. Liq. Chromatogr.,* 8, 1783, 1985.
9. **Cunico, R., Mayer, A.G., Wehr, C.T., and Sheehan, T.L.,** High sensitivity amino acid analysis using a novel automated precolumn derivatization system, *BioChromatography,* 1, 1986.
10. **Betnér, I. and Földi, P.,** The FMOC-ADAM approach to amino acid analysis, *LC.GC,* 6, 832, 1988.
11. **Aswad, D.W.,** Determination of D- and L-aspartate in amino acid mixtures by high-performance liquid chromatography after derivatization with a chiral adduct of o-phthaldialdehyde, *Anal. Biochem.,* 137, 405, 1984.
12. **Buck, R.H. and Krummen, K.,** High-performance liquid chromatographic determination of enantiomeric amino acids and amino alcohols after derivatization with o-phthaldialdehyde and various chiral mercaptans, *J. Chromatogr.,* 387, 255, 1987.
13. **Einarsson, S., Folestad, S., and Josefsson, B.,** Separation of amino acid enatiomers using precolumn derivatization with o-phthalaldehyde and 2,3,4,6-tetra-O-acetyl-1-thio-β-glucopyranoside, *J. Liq. Chromatogr.,* 10, 1589, 1987.
14. **Buck, R.H. and Krummen, K.,** Resolution of amino acid enatiomers by high-performance liquid chromatography using automated pre-column derivatization with a chiral reagent, *J. Chromatogr.,* 315, 279, 1984.
15. **Einarsson, S., Josefsson, B., Möller, P., and Sanchez, D.,** Separation of amino acid enatiomers and chiral amines using precolumn derivatization with (+)-1-(9-fluorenyl)ethyl chloroformate and reversed-phase liquid chromatography, *Anal. Chem.,* 59, 1191, 1987.
16. **Mayer, A.G. and Sheehan, T.L.,** Fully automated HPLC analysis of amino acids using precolumn derivatization with fluorenylmethyl chloroformate, poster paper II, P72, 17th International Symposium on Chromatography, Vienna, 1988.
17. **Einarsson, S.,** Selective determination of secondary amino acids using precolumn derivatization with 9-fluorenylmethyl chloroformate and reversed-phase high-performance liquid chromatography, *J. Chromatogr.,* 348, 213, 1985.
18. **Dolan, J.W.,** Troubleshooting autosamplers, part II, *LC.GC,* 5(3), 224, 1987.
19. **Lagerkvist, S.,** Personal communication, 1988.

REVERSED-PHASE HIGH-PERFORMANCE LIQUID CHROMATOGRAPHY FOR THE SEPARATION OF PEPTIDE DIASTEREOISOMERS

Victor J. Hruby, Andrew Kawasaki, and Geza Toth

I. INTRODUCTION

Diastereoisomers of peptides with biological activities often exhibit pronounced antagonistic or agonistic properties which differ greatly from each other. This fact, as well as the difficulty in obtaining many uncommon amino acids in optically pure form, underlies the importance of having at hand effective chromatographic methods for the purification and characterization of diastereoisomeric mixtures of peptides. Diastereoisomeric peptides, isomers in which one or more asymmetric centers have opposite configuration, may possess similar or different physical-chemical or biological properties. There exists a number of ways a diastereoisomeric mixture of peptides can be generated: (1) formation of a diastereoisomeric mixture may result from a peptide synthesis methodology which leads to racemization of one or more amino acid residues; (2) non-eukaryotic amino acids are often synthesized as a racemic mixture and are utilized as such for peptide synthesis; (3) commercially available D-amino acids usually contain small amounts of the L isomer; and (4) incorporation of isotopes into a peptide is usually accomplished by synthesis of a racemic modification of the amino acid followed by separation of the resulting diastereoisomeric peptides. This review will focus on the practical aspects of utilizing reversed-phase high-performance liquid chromatography (RPC) as a method for the purification and characterization of peptide hormone diastereoisomers. Since aspects of this subject were last reviewed,[1,2] additional successful separations of peptide diastereoisomers have been reported. These recent advances will be addressed in this review, as will work from our laboratory which presents a practical approach to the separation of four closely related diastereoisomers of [D-Pen^2,D-Pen^5]enkephalin (DPDPE) which differ from each other only stereochemically at a single amino acid residue.

II. RESULTS AND DISCUSSION

The more recent reports on the separation of biologically active peptide diastereoisomers (Table 1, Figure 1) have served to confirm the importance of the factors involved in RPC separation of peptide diastereoisomers, e.g., the choice of the solid support, the mobile phase, and the pH of the mobile phase. A survey of the literature shows that octadecylsilane (ODS, C$_{18}$, RP-18) bonded solid supports have emerged as the most popular stationary phase for RPC for the separation of diastereoisomeric or nondiastereoisomeric peptides. Though the resolving power and efficiency of the ODS sorbent is widely acknowledged, there have been examples reported in the literature where RPC utilizing an ODS bonded phase resolved peptide diastereoisomers, but failed to separate certain associated impurities of the peptide diastereoisomers.

TABLE 1
Separations of Biologically Active Peptide Diastereoisomers

Diastereoisomer sample	Column	Mobile phase	pH	Detector	Ref.
Enkephalin and endorphin	μBondapak phenyl	CH_3CN/NH_4OAc	4.5	UV	5
Met- and Leu-enkephalin	Ultrasphere ODS	CH_3CN/phosphate	2.1	UV	6
Met- and Leu-enkephalin	Spherisorb ODS	CH_3CN/NH_4OAc or $CH_3CN/NH_4OAC/AcOH$ or $CH_3CN/Cu(OAc)_2$	NR	UV	7
Enkephalin	Spherisorb CN	CH_3CN/TEAP	2.5—5.4	UV	
Oxytocin	μBondapak C_{18}	CH_3CN/NH_4OAc	4.0	UV	8
Oxytocin	μBondapak C_{18}	CH_3CN, THF, or dioxane/NH_4OAc	4.0 / 6.0	UV	9
Oxytocin	Partisil 10 ODS	CH_3CN/NH_4OAc	4.0	UV	10
Oxytocin	Separon SI-C_{18}	CH_3OH/NH_4OAc	7	UV	11
		CH_3OH/TFA	2	UV	
Cyclic-α-MSH analogues	Vydac C18 or Altech	CH_3CN/TFA or CH_3OH/TFA/TMAP	2.2	UV	12
α-MSH analogues	Vydac C_{18}	CH_3CN/TFA or CH_3CN/HFBA or CH_3OH/TFA	2.0	UV	13
Angiotensin I	MicroPak AX-10	CH_3OH/TEAP	2.2	UV	14
Angiotensin I and II	ODS	CH_3CN/TEAA	6	UV	15
AVP	μBondapak C_{18}	CH_3CN/TEAP	3.5	UV	16
	Spherisorb ODS LiChrosorb RP-18	CH_3OH/CH_3CN, or THF/TEAA	4.0	UV	
LH-RH, TRH, somatostain	μBondapak	CH_3CN/NH_4OAc	4.0	UV	17
LH-RH	μBondapak C_{18} μBondapak CN	CH_3CN/TEAP TEAP	3.0	UV	18
LH-RH	μBondapak CN μBondapak phenyl μBondapak C_{18}	CH_3CN/TEAP	3.0	UV	19
LH-RH	ODS-hypersil	CH_3CN/phosphate	6.5	UV	20
LH-RH	Vydac C_{18}	CH_3CN/TEAP	3.7	UV	21
Somatostatin	μBondapak C_{18}	CH_3CN/NH_4OAc	4.1	UV	22

Secretin	LiChrosorb RP-18	$CH_3OH/H_2O/AcOH$	NR	UV, RI	23
Secretin	LiChrosorb RP-18 Nucleosil C_{18} Polygosil CN Polygosil phenyl	CH_3OH or CH_3CN/TFA or phosphate buffer	NR	UV	24
hPTH	Nucleosil C_{18}	$CH_3CN/K_2HPO_4/H_3PO_4/Na_2SO_4$	2.6	UV	25
hGRF	Nucleosil C_{18}	$CH_3CN/K_2HPO_4/H_3PO_4$	2.6	UV	26
	TSK gel CM-25W	CH_3CN/Na_3PO_4	6	UV	
		$CH_3CN/Na_3PO_4/NaCl$	7	UV	
Neurotensin	MicroPak AX-10 or LiChrosorb-Si-60	$CH_3CN/TEAA$	6	UV	27
Gramicidin S and Gratisin	Finepak SIL C_{18}	$CH_3OH/NaClO_4$	NR	UV	28

Tyr-Gly-Gly-Phe-Met   Tyr-Gly-Gly-Phe-Leu   Pyr-Leu-Tyr-Glu-Asn-Lys-Pro-Arg-Arg-Pro-Tyr-Ile-Leu

Met-enkephalin   Leu-enkephalin   Neurotensin

Tyr-Gly-Gly-Phe-Met-Thr-Ser-Glu-Lys-Ser-Gln-Thr-Pro-Leu-Val-Thr-Leu-Phe-Lys-Asn-Ala-Ile-Ile-Lys-
Asn-Ala-Tyr-Lys-Lys-Gly-Glu

β-Endorphin

Cys-Tyr-Ile-Gln-Asn-Cys-Pro-Leu-Gly-NH₂   Ac-Ser-Tyr-Ser-Met-Glu-His-Phe-Arg-Trp-Gly-Lys-
Pro-Val-NH₂

Oxytocin   α–MSH

Ac-Ser-Tyr-Ser-Cys-Glu-His-Phe-Arg-Trp-Cys-Lys-Pro-Val-NH₂

cyclic α–MSH   Asp-Arg-Val-Tyr-Ile-His-Pro-Phe

Angiotensin II (human)

Asp-Arg-Val-Tyr-Ile-His-Pro-Phe-His-Leu   Ala-Gly-Cys-Lys-Asn-Phe-Phe-Trp-Lys-Thr-Phe-Thr-Ser-Cys

Angiotensin I   Somatostatin

Cys-Tyr-Phe-Gln-Asn-Cys-Pro-Arg-Gly-NH₂   Pyr-His-Trp-Ser-Tyr-Gly-Leu-Arg-Pro-Gly-NH₂

AVP   LH-RH

His-Ser-Asp-Gly-Thr-Phe-Thr-Ser-Glu-Leu-Ser-Arg-Leu-Arg-Glu-Gly-Ala-Arg-Leu-Gln-Arg-
Leu-Leu-Gln-Gly-Leu-Val-NH₂

Secretin (human)

Tyr-Ala-Asp-Ala-Ile-Phe-Thr-Asn-Ser-Tyr-Arg-Lys-Val-Leu-Gly-Gln-Leu-Ser-Ala-Arg-Lys-
Leu-Leu-Gln-Asp-Ile-Met-Ser-Arg-Gln-Gln-Gly-Glu-Ser-Asn-Gln-Glu-Arg-Gly-Ala-Arg-
Ala-Arg-Leu-NH₂

h-GRF

Cyclo-(-Val-Orn-Leu-D-Phe-Pro)₂   Cyclo-(-Val-Orn-Leu-Phe-Pro-Tyr)₂

Gramicidin S-1   Gratisin 1

FIGURE 1. Sequences of various biologically active peptides referred to in this article.

For example, in the case of angiotensin I, weak anion-exchange HPLC (AEC) was utilized to resolve the impurities in question (Table 1, Reference 14). It was concluded by the workers that complementary use of both RPC and AEC offers a superior approach for the determination of the diastereoisomeric purity of peptides. In another report, RPC was compared with cation-exchange HPLC (CEC) for the separation of diastereoisomers and closely related structural isomers of human parathyroid hormone (hPTH) and human growth hormone-release factor (hGRF) (Table 1, Reference 26). The method of RPC allowed for more efficient resolution of the peptide diastereoisomers, while certain closely related structural peptide isomers could only be separated by CEC. The cyanopropylsilyl bonded phase has been utilized to resolve various peptide diastereoisomers including certain enkephalin analogs (Table 1, Reference 7). In this work, the cyanopropyl support exhibited decreased affinity for the peptides compared to ODS sorbents, but showed a higher efficiency of resolution with retention. Important factors for optimizing resolution with the cyanopropyl bonded phase were the pH of the mobile phase and temperature at which the separation was performed.

Other reversed-phase packings containing bonded phases such as butyl, octyl, or diphenylsilanes are now commercially available, although there are not many examples in which these sorbents have been utilized for the separation of peptide diastereoisomers. In addition, changes in the organic modifier (acetonitrile, methanol, etc.) and the ionic strength of the mobile phase can be used in RPC and can become critical factors affecting the separation of peptides. On a practical note, the volatility of the buffer used is also a consideration particularly because it affects the ease of work-up. In regard to the separation of peptide diastereoisomers, the use of perfluoro-alkanoic acids other than trifluoroacetic acetic acid (TFA), e.g., heptafluorobutyric acid (HFBA), and of triethylammonium phosphate (TEAP) warrants further investigation, especially if the counter anion is involved in an ion-pairing mechanism. With certain α-MSH analogs, it has been reported that the use of TEAP in comparison to TFA results in an increased capacity factor (k') and an improved selectivity (α) (Table 1, Reference 13), but interestingly the TFA-containing mobile phase had a greater influence on selectivity than the HFBA-containing mobile phase with either acetonitrile or methanol as the organic modifier.

A. A PRACTICAL APPROACH TO THE SEPARATION OF DIASTEREOISOMERS OF DPDPE

DPDPE (H-Tyr-D-Pen-Gly-Phe-D-Pen-OH), a cyclic enkephalin analog with extraordinary δ opioid receptor selectivity, was developed in our laboratory.[3] In our efforts to understand the relationship of the structure and conformation of DPDPE to its biological activity, we have prepared a series of diastereoisomeric analogs based on DPDPE.[4,29] In this work, the chromatographic behavior of six DPDPE analogs, which differ only in the configuration of one individual amino acid, was analyzed. We present here a semi-preparative separation of [erythro-D,L-β-Me-p-NO$_2$Phe^4]DPDPE. The structure of one of the diastereoisomers of [β-Me-p-NO$_2$Phe^4]DPDPE (the erythro-L-isomer) is presented in Figure 2, as are the four optical isomers (D,L-threo and D,L-erythro) of β-Me-p-NO$_2$Phe which, when incorporated into DPDPE, give rise to the different diastereoisomers of DPDPE. The peptides were synthesized by solid-phase methodology as described previously.

The crude [erythro-D,L-β-Me-p-NO$_2$Phe^4]DPDPE was purified by gel-filtration chromatography (Sephadex G-10) using 30% aq. acetic acid as eluent. Final separation of the diastereoisomers and purification of the lyophilized solid was achieved by semi-preparative RPC. The diastereoisomeric peptides were separated during the initial isocratic elution conditions over a 20-min period (Figure 3). A subsequent rapid linear gradient effectively rids the column of the more strongly adsorbed impurities.

After peaks 1 and 2 (Figure 3) were collected and lyophilized, the composition of amino acids in the samples was determined by amino acid analysis. The configuration of β-Me-p-NO$_2$Phe in the two peptides was determined by enzymatic assay using L-amino acid oxidase. On the basis of this assay, the first peak corresponds to [erythro-L-β-Me-p-NO$_2$Phe^4]DPDPE, and the [erythro-D-β-Me-p-NO$_2$Phe^4]DPDPE was eluted as the second peak. A similar approach was used to resolve and characterize the [threo-L-] and [threo-D-β-Me-p-NO$_2$Phe^4]DPDPE.

The chromatographic behavior of DPDPE and DPLPE and the four pure diastereoisomers of [β-Me-p-NO$_2$Phe^4] DPDPE was studied utilizing RPC with six different mobile phases (Table 2).

In examining the capacity factor (k') for the various analogs, it was observed that a change in the stereochemistry of one amino acid invariably resulted in different retention times for that particular diastereoisomer. The elution order of an analog within a diastereomeric series depends on the position of the stereochemically altered amino acid in the peptides. In the case of DPDPE and DPLPE, DPDPE was eluted first and DPLPE second. In the case of the four isomers of [β-Me-p-NO$_2$Phe^4]DPDPE, the elution order depends not only on the configuration

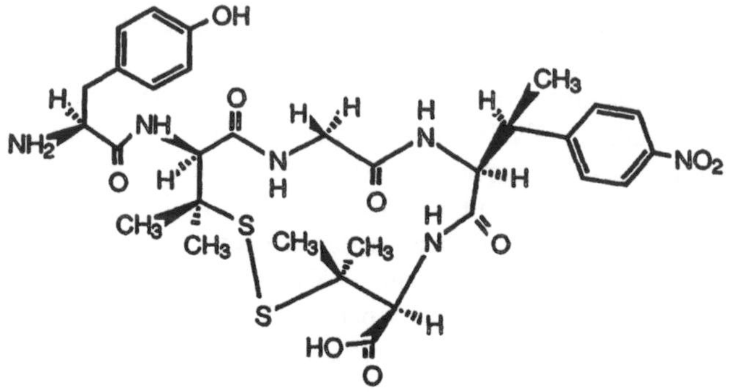

Tyr-D-Pen-Gly-(2S,3S)-β–Me-p-NO$_2$Phe-D-Pen

erythro-L-β-Me-p-NO$_2$Phe[4]-DPDPE

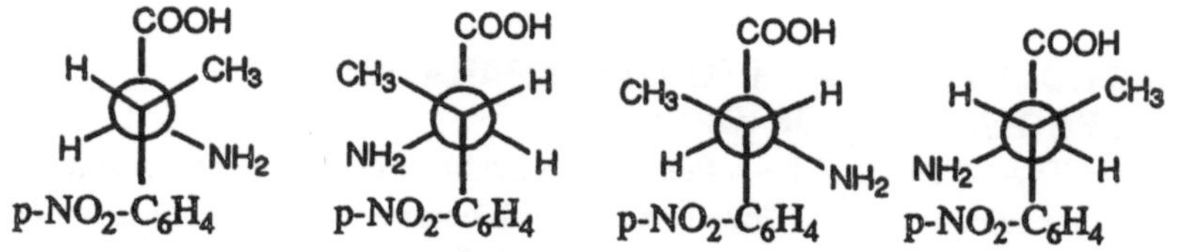

2S,3R	2R,3S	2S,3S	2R,3R
L-threo	D-threo	L-erythro	D-erythro

FIGURE 2. The structure of [erythro-L-β-Me-p-NO$_2$Phe[4]] DPDPE (top) and of the four optimal isomers of β-Me-p-NO$_2$Phe (bottom).

of β-Me-p-NO$_2$Phe, but also on the mobile phase. In acidic conditions, the elution order is [erythro-L-β-Me-p-NO$_2$Phe[4]]DPDPE, [threo-L-β-Me-p-NO$_2$Phe[4]]DPDPE, [threo-D-β-Me-p-NO$_2$Phe[4]]DPDPE, and [erythro-D-β-Me-p-NO$_2$Phe[4]]DPDPE (conditions A, B, C, and D, Table 2). In the neutral water-CH$_3$OH mobile phase system, the elution order was changed relative to the order obtained under acidic conditions to that shown in Table 2 (conditions E and F).

The separation of DPDPE and DPLPE was optimal in the water-methanol system (condition E, Table 2), where the peptides were in a zwitterionic form. The optimal mobile phase for separation of the four isomers of [β-Me-p-NO$_2$Phe[4]]DPDPE was solvent system B (Figure 4, Table 2), where there is essentially a baseline separation of all four isomers. Although a greater difference between the k′ values for the four isomers was observed with solvent systems A and D, the isomers were eluted as broad peaks which were not practically separable on a semi-preparative scale.

ACKNOWLEDGMENTS

This research was supported by grants from the U.S. Public Health Service and the National Science Foundation.

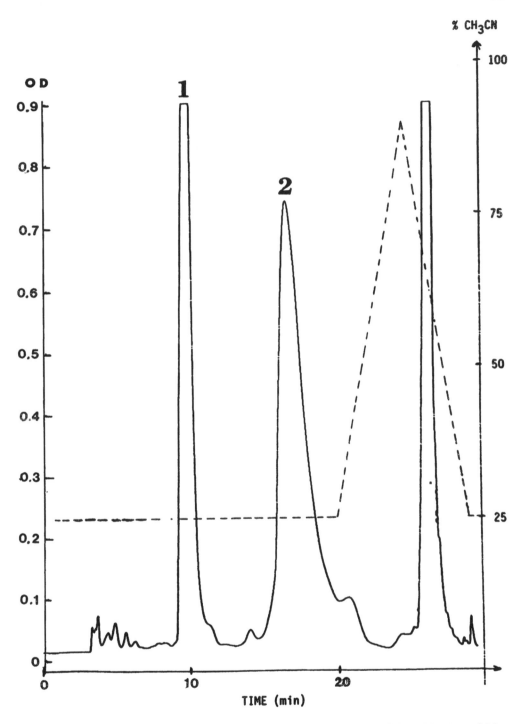

FIGURE 3. Reversed-phase separation of [erythro-L-β-Me-p-NO₂Phe⁴]DPDPE (Peak 1) and [erythro-D-β-Me-p-NO₂Phe⁴]DPDPE (Peak 2) from a crude synthetic peptide mixture. Column: Vydac 218TP1010 C_{18} (Vydac, The Separations Group, Hesperia, CA), 250 × 10 mm I.D., 10-μm particle size, 300-Å pore size. Mobile phase: isocratic elution for 20 min with 75% A-25% B, followed by a linear gradient from 25% B to 90% B in 5 min and a decreasing linear gradient from 90% B to 25% B in 5 min; finally, reequilibration to 75% A-25% B over 5 min, where Eluent A is 0.1% aqueous trifluoroacetic acid (TFA) and Eluent B is acetonitrile (CH_3CN); flow-rate, 4 ml/min; absorbance at 280 nm. A sample of the crude peptide (5 to 7 mg) was dissolved in 25% acetonitrile in 0.1% aq. TFA (200 μl), filtered through a 0.4 μm filter and injected onto the column. The dotted line denotes the concentration of acetonitrile in the mobile phase.

TABLE 2
Capacity Factor Values (k') of DPDPE Analogs with Various Mobile Phases

Peptide	Conditions					
	A	B	C	D	E	F
H-Tyr-D-Pen-Gly-Phe-D-Pen-OH	1.85	1.29	1.95	3.68	2.42	1.44
H-Tyr-D-Pen-Gly-Phe-L-Pen-OH	1.83	1.29	2.03	4.07	3.62	2.25
[e-L-β-Me-p-NO$_2$Phe⁴]DPDPE	2.91	1.91	3.00	5.36	3.78	2.37
[t-L-β-Me-p-NO$_2$Phe⁴]DPDPE	3.61	2.29	3.38	6.17	3.78	2.18
[t-D-β-Me-p-NO$_2$Phe⁴]DPDPE	5.14	3.05	4.00	7.68	10.54	6.94
[e-D-β-Me-p-NO$_2$Phe⁴]DPDPE	7.52	4.20	5.50	11.12	9.15	5.13

Note: Column: Vydac 218TP104 C$_{18}$, 250 × 4.6 mm I.D., 10-μm particle size, 300-Å pore size.
Conditions: A, 0.1% aq. TFA-CH$_3$CN (77:23); B, 0.1% aq. TFA-CH$_3$CN (75:25); C, 0.1% aq. TFA-CH$_3$OH (55:45); D, 0.1% aq. TFA-CH$_3$OH (60:40); E, 45% aq. CH$_3$OH; F, 50% aq. CH$_3$OH. Flow-rate, 1.5 ml/min. The capacity factor, k', is calculated by the equation: $k' = (t_R - t_o)/t_o$, where t_R is the peptide retention time and t_o is the column dead time (retention time of unretained compounds).

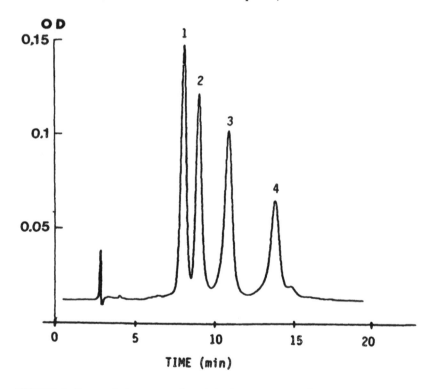

FIGURE 4. Reversed-phase separation of the four isomers of [β-Me-p-NO$_2$Phe⁴]DPDPE. Column: Vydac 218TP104 C$_{18}$, 250 × 4.6 mm I.D., 10-μm, 300-Å. Mobile phase: 25% acetonitrile in 0.1% aq. TFA; flow-rate, 1.5 ml/min; absorbance at 280 nm. The peaks are erythro-L-, threo-L-, threo-D- and erythro-D-β-Me-p-NO$_2$Phe⁴-DPDPE in order of elution.

REFERENCES

1. Blevins, D.D., Burke, M.F., and Hruby, V.J., Diastereoisomer separations, in *CRC Handbook of HPLC Sep. Amino Acids, Peptides and Proteins*, Vol. 2, Hancock, W.S., Ed., CRC Press, Boca Raton, 1984, 137.

2. Krummen, K., HPLC in the analysis and separation of pharmaceutically important peptides, *J. Liq. Chromatogr.*, 3, 1243, 1980.

3. Mosberg, H.I., Hurst, R., Hruby, V.J., Gee, K., Yamamura, H.I., Galligan, J.J., and Burks, T.F., Bispenicillamine enkephalins possess highly improved specificity toward δ opioid receptors, *Proc. Natl. Acad. Sci. U.S.A.*, 80, 5871, 1983.

4. Hruby, V.J., Toth, G., Prakash, O., David, P., and Burks, T.F., Cyclic enkephalins which are optically pure isomers of [β-Me-p-NO$_2$-Phe^4]DPDPE possess extraordinary δ opioid receptor selectivities, *Peptides*, 1988, Proceedings 20th European Peptide Symposium, Bayer, E. and Jung, G., Eds., Walter de Gruyter, Berlin, 1989, 597.

5. Currie, B.L., Chang, J.K., and Cooley, R., High performance liquid chromatography of enkephalin and endorphin peptide analogs, *J. Liq. Chromatogr.*, 3, 513, 1980.

6. Mousa, S., Mullet, D., and Couri, D., Sensitive and specific high performance liquid chromatographic method for methionine and leucine enkephalins, *Life Sci.*, 29, 61, 1981.

7. Hunter, C., Sugden, K., Lloyd-Jones, J.G., HPLC of peptides and peptide diastereoisomers on ODS- and cyanopropyl-silica gel column packing materials, *J. Liq. Chromatogr.*, 3, 1335, 1980.

8. Larsen, B., Viswanatha, V., Chang, S.V., and Hruby, V.J., Reversed-phase high pressure liquid chromatography for the separation of peptide hormone diastereoisomers, *J. Chromatogr. Sci.*, 16, 207, 1978.

9. Larsen, B., Fox, B.L., Burke, M.F., and Hruby, V.J., The separation of peptide hormone diastereoisomers by reversed phase high pressure liquid chromatography. Factors affecting separation of oxytocin and its diastereoisomers-structural implications, *Int. J. Peptide Protein Res.*, 13, 12, 1979.

10. Viswanatha, V., Larsen, B., and Hruby, V.J., Synthesis of DL-[2-^{13}C]Leucine and its use in the preparation of [3-DL-[2-^{13}C]leucine]oxytocin and [8-DL-[2-^{13}C]leucine]oxytocin. Preparative separation of diastereoisomeric peptides by partition chromatography and high pressure liquid chromatography, *Tetrahedron*, 35, 1575, 1979.

11. Lebl, M., Amino acids and peptides CLXXXI. Separation of diastereoisomers of oxytocin analogues, *J. Chromatogr.*, 264, 459, 1983.

12. Lebl, M., Cody, W.L., and Hruby, V.J., Cyclic melanotropins. Part VI. Reverse phase HPLC studies, *J. Liq. Chromatogr.*, 7, 1195, 1984.

13. Cody, W.L., Wilkes, B.C., and Hruby, V.J., Reversed-phase high-performance liquid chromatography studies of α-MSH fragments, *J. Chromatogr.*, 314, 313, 1984.

14. Margolis, S.A. and Dizgaroglu, M., Separation and purification of diastereomers of angiotensin I by weak anion-exchange high-performance liquid chromatography, *J. Chromatogr.*, 322, 117, 1985.

15. Margolis, S.A. and Konash, P.L., The high-performance liquid chromatographic analysis of diastereoisomers and structural analogs of angiotensins I and II, *Anal. Biochem.*, 134, 163, 1983.

16. Blevins, D.D., Burke, M.F., Hruby, V.J., and Larsen, B.R., Factors affecting the separation of arginine vasopressin peptide diastereoisomers by HPLC, *J. Liq. Chromatogr.*, 3, 1299, 1980.

17. Burgus, R. and Rivier, J., Use of high pressure liquid chromatography in the purification of peptides, in *Peptides 1976*, Loffet, C.A., Ed., Editions de l'Universite de Bruxelles, Belgium, 1976, 85.

18. Rivier, J., Spiess, J., Perrin, M., and Vale, W., Application of HPLC in the isolation of unprotected peptides, in *Biological/Biomedical Applications of Liquid Chromatography II*, Hawk, G.L., Ed., Marcel Dekker, New York, 1979, 223.

19. Rivier, J.E., Use of trialkylammonium phosphate (TAAP) buffers in reversed phase HPLC for high resolution and high recovery of peptides and proteins, *J. Liq. Chromatogr.*, 1, 343, 1978.

20. Sertl, D.C., Johnson, R.N., and Kho, B.T., An accurate, specific HPLC method for the analysis of a decapeptide in a lactose matrix, *J. Liq. Chromatogr.*, 4, 1134, 1981.

21. Parnes, H., Shelton, E.J., and Huang, G.T., Synthesis of ^{14}C-labelled peptides by asymmetric reduction of dehydroamino acids, *Int. J. Peptide Protein Res.*, 28, 403, 1986.

22. Meyers, C.A., Coy, D.H., Huang, W.Y., Schally, A.V., and Redding, T.W., Highly active position eight analogues of somatostatin and separation of peptide diastereomers by partition chromatography, *Biochemistry*, 17, 2326, 1978.

23. Bakkum, J.T.M., Beyerman, H.C., Hoogerhout, P., Olieman, C., and Voskamp, D., Reversed-phase high performance liquid chromatography of protected peptides in the sequential synthesis of secretin and analogues, *Recueil, J.R. Netherlands Chem. Soc.*, 96, 301, 1977.

24. Voskamp, D., Olieman, C., and Beyerman, H.C., The use of trifluoroacetic acid in the reversed-phase liquid chromatography of peptides including secretin, *Recueil, J.R. Netherlands Chem. Soc.*, 99, 105, 1980.

25. **Kimura, T., Yoshizawa, K., Takai, M., and Sakakibara, S.,** Problems in the synthesis of human parathyroid hormone, in *Frontiers in Biochemical and Biophysical Studies of Proteins and Membranes,* Liu, T.Y., Sakikibara, S., Schechter, A.N., Yagi, K., Yajima, H., and Yasunobu, K.T., Eds., Elsevier, New York, 1983, p. 81—89.

26. **Kumagaye, K., Takai, M., Chino, N., Kimura, T., and Sakakibara, S.,** Comparison of reversed-phase and cation-exchange high-performance liquid chromatography for separating closely related peptides: separation of Asp^{76}-human parathyroid hormone (1-84) from Asn^{76}-human parathyroid hormone (1-84), *J. Chromatogr.,* 327, 327, 1985.

27. **Dizdaroglu, M., Simic, M.G., Rioux, F., and St. Pierre, S.,** Separation of diastereomers and analogues of neurotensin by anion-exchange high-performance liquid chromatography, *J. Chromatogr.,* 245, 158, 1982.

28. **Tamaki, M., Takimoto, M., Nozaki, S., Muramatsu, I.,** HPLC behavior of peptides related to Gratisin and Gramicidin S., *Peptide Chemistry,* 1985, Kiso, Y., Ed., Protein Research Foundation, Osaka, 1986, p. 327—332.

29. **Hurst, R.,** Synthesis and comparative biological activities of conformationally restricted enkephalin analogs, Master of Science Thesis, University of Arizona, 1984, pp. 53.

SEPARATION OF INTRACHAIN DISULFIDE BRIDGED PEPTIDES FROM THEIR REDUCED FORMS BY REVERSED-PHASE CHROMATOGRAPHY

Kok K. Lee, James A. Black, and Robert S. Hodges

I. INTRODUCTION

The presence of intrachain disulfide bridges in peptides and proteins is a common feature. Examples of these bridges are found in peptide hormones such as insulin and vasopressin, in enzymes such as lysozyme, ribonuclease, and serine proteases and in small protease inhibitors such as trypsin inhibitors and kallikrein inactivators. The main function of these disulfide bridges is to stabilize the protein conformation. With the technological advances in peptide synthesis, many active fragments of peptides and proteins are being synthesized with their native disulfide bonds or the addition of disulfide bonds to confer to the fragments a structure similar to that in the native protein.[1]

The main problem encountered in the formation of the intrachain disulfide bridge is the availability of a rapid and simple monitoring method to follow the extent of oxidation of the cysteine residues to cystine. In addition, a sensitive and quantitative method is required to resolve the disulfide bridged (oxidized) peptide from the corresponding cysteine containing (reduced) peptide. Assessment of the purity of the oxidized conformer would then be possible by quantitative determination of the undesired reduced conformer. Reversed-phase high-performance liquid chromatography (RPC) could be utilized as such a method if baseline separation between the reduced and oxidized conformers of the peptide could be achieved using standard conditions of chromatography (for example, a linear AB gradient, where A is 0.05% aqueous TFA and B is 0.05% TFA/acetonitrile; a gradient rate of 1.0% B/min; a flow rate of 1.0 ml/min). In this study, we have used a group of fourteen peptides to examine the separation of the reduced and oxidized peptides by RPC. In addition to using RPC to monitor the extent of oxidation of peptides with time, we have also improved the resolution of the reduced and oxidized conformers by modification of the free sulfhydryl (-SH) groups of the reduced peptides with a -SH specific reagent, giving rise to a better separation between the two conformers.

II. EXPERIMENTAL

Trifluoroacetic acid (TFA) was redistilled prior to use. HPLC-grade acetonitrile was obtained from J.T. Baker (Phillipsburg, NJ). Deionized water was purified by reverse-osmosis using an HP661A water purifier (Hewlett Packard, Avondale, PA). All other chemicals used were reagent grade. Purified peptides were obtained from the Alberta Peptide Institute. The HPLC instrument consisted of a Spectra-Physics (San Jose, CA) SP8700 solvent delivery system and a SP8750 organizer module combined with a Hewlett Packard HP1040A detection system, HP3390A integrator, HP85 computer, HP9121 disc drive, and HP7470A plotter.

Samples were injected with a 2.0 ml injection loop (Model 7125, Rheodyne, Cotati, CA). Air oxidations were performed by dissolving the reduced peptide in 0.1 M ammonium bicarbonate to a concentration of 0.1 mg/ml and stirring the solution in an erlenmeyer flask open to the air. The extent and/or rate of oxidation was monitored by removing 100 µl samples at various time points and mixing them with 10 µl of an aqueous solution of 1 mg/ml N-ethylmaleimide (NEM). This mixture was allowed to sit at room temperature for at least 15 minutes before it was injected onto a RP column. The RP column used was an Aquapore RP-300 (C_8) analytical column (220 × 4.6 mm I.D., 7-µm particle size, 300-Å pore size (Brownlee Labs, Santa Clara, CA). In all cases, a linear AB gradient was used where solvent A is 0.05% aqueous TFA and solvent B is 0.05% TFA/acetonitrile at a gradient rate of 1% B/min and a flow rate of 1 ml/min.

III. RESULTS AND DISCUSSION

A. SEPARATION OF REDUCED AND OXIDIZED PEPTIDES BY RPC

A series of 14 peptides in their reduced and oxidized forms was chromatographed by RPC and the results are summarized in Table 1. The number of amino acid residues within the disulfide loop of the peptides used in this study varied between 4 and 15 residues, and the length of these peptides ranged from 9 to 29 residues. Some of these peptides are closely related analogs varying by a single amino acid substitution (Peptides 13 and 14), multiple substitutions (Peptides 4 to 7), and others varying in polypeptide chain length (Peptides 4 and 8; 10 to 14).

Representative chromatograms of the separation of the reduced and oxidized forms of three peptides are shown in Figure 1. The difference in retention time between reduced and oxidized forms of Peptides 7, 13, and 14 was 3.25 min, 0.63 min, and 1.38 min, respectively. In fact, as shown in Table 1, the separation of the reduced and oxidized forms of most of these peptides was possible with the exception of Peptides 2 and 9. These two peptides showed no clear separation between the conformers. In addition, no dimeric interchain peptides were observed when a low concentration (0.1 mg/ml) was used for air oxidation. Whenever a difference in retention time was observed between the two forms of the peptide, the reduced conformer was always eluted later than the disulfide bridged peptide. This result suggested that, on formation of the disulfide loop, some residues in the peptide were not as accessible to interact with the stationary phase as in the reduced peptide. In other words, the loop formation decreased the overall hydrophobicity of the peptide.

It has been noted by Creighton[2] that cystine has a lower solubility in water than cysteine, thus making the disulfide group relatively more hydrophobic than -SH groups. The solubility for the cystine in water was 0.01g/100 ml, less than 1/1000 that of cysteine. An observation made by Schulze-Gahmen et al.[3] showed that the reduced form of their 16 amino acid residue peptide (NH$_2$-YGCTVGGGGGGGGVTCG-OH) was eluted 1 to 2 min earlier than the oxidized form. The column and conditions used by these authors were similar to those used in this study (a Baker C_{18} column [250 × 4.6 mm I.D., 300-Å pore size] was run at ambient temperature with a flow rate of 1.0 ml/min and a linear gradient applied of 2% B/min equivalent to 1.2% acetonitrile where Eluent A is 0.1% aqueous TFA and Eluent B is 0.1% TFA in 60% acetonitrile). This observation could be explained by the increase in hydrophobicity of the peptide due to the presence of cystine vs. two cysteine residues. Another explanation for the appearance of the disulfide bridged peptide prior to, unresolved from or later than the reduced conformer is simply differences in conformation of the loop peptides as a result of sequence variability. For example, with the series of peptides (4 to 9), which have 12 amino acid residues between the cysteine residues, the ΔR_t between reduced and oxidized conformers ranged from 0 to 3.25 min. Similarily, for peptides with 4 amino acid residues between cysteine residues, the ΔR_t for Peptide 1 was 1.88 min while Peptide 2 showed no noticeable difference between the two conformers. In addition, Peptides 13 and 14 differed by only a single amino acid change within the loop (aspartic acid to arginine) and yet ΔR_t between reduced and oxidized conformers varied from

TABLE 1

Peptide no.	Amino acid sequence	No. of residues within disulfide loop	Retention time (min)		
			(-SH)	(-S-S-)	(ΔR$_t$)
1	NH₂-GIVECSTSICSLY-amide	4	33.22	31.34	1.88
2	desamino-CYFQNCPRG-amide	4	26.82	26.82	0.00
3	NH₂-AGCKNFFWKTFTSC-OH	10	35.44	34.44	1.00
4	Ac-KCTSDQDEQFIPKGCSK-OH	12	23.23	21.24	1.99
5	Ac-ACAADQDEQFIPKGCSK-OH	12	25.84	24.09	1.75
6	Ac-ACAAAADEQFIPKGCSK-OH	12	27.56	25.19	2.37
7	Ac-ACAAAAAQFIPKGCSK-OH	12	31.54	28.29	3.25
8	Ac-ITLTRTAADGLWKCTSDQDEQFIPKGCSK-OH	12	34.89	33.02	1.87
9	Ac-ACKSTQDPMFTPKGCDN-OH	12	24.76	24.76	0.00
10	Ac-CFGGRMDRIGAQSGLGC-amide	15	29.80	29.17	0.63
11	Ac-CFGGRMDRIGAQSGLGCNSFRY-OH	15	32.87	32.07	0.80
12	NH₂-SSCFGGRIDRIGAQSGLGCNSFRY-OH	15	28.82	28.07	0.75
13	NH₂-SLRRSSCFGGRMDRIGAQSGLGCNSFRY-OH	15	29.36	28.73	0.63
14	NH₂-SLRRSSCFGGRMRRIGAQSGLGCNSFRY-OH	15	29.45	28.07	1.38

Note: ΔR$_t$ denotes the difference in retention time between reduced (-SH) and (-S-S-) oxidized peptides. The symbols ▲ and ○ denote single amino acid sequence changes between peptides; ▽ denotes position of cysteine or cystine residues; Ac-, acetyl; -OH, C-terminal carboxyl group; NH₂, N-terminal amino group.

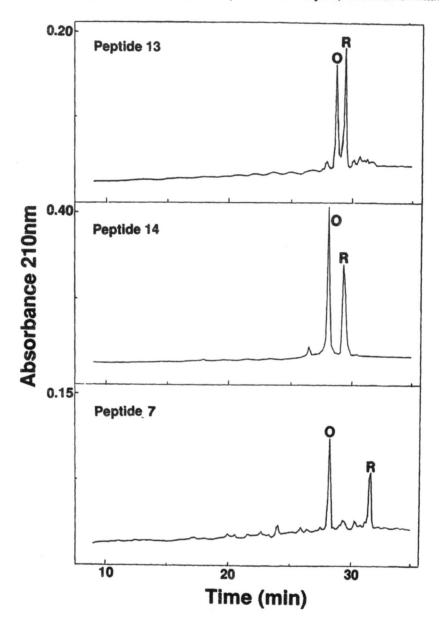

FIGURE 1. Separation of reduced (R) and oxidized (O) peptides by RPC. Column: Aquapore RP-300 C$_8$, 220 × 4.6 mm I.D., 7-µm particle size, 300-Å pore size. Mobile phase: linear AB gradient (1% B/min), where Eluent A is 0.05% aq. TFA and Eluent B is 0.05% TFA in acetonitrile (pH 2.0); flow-rate, 1 ml/min.

0.63 min for Peptide 13 to 1.38 min for Peptide 14. These results suggested that there is no direct correlation between the size of the loop and the differences in ΔR_t obtained on RPC.

B. RPC OF THE NEM-MODIFIED AND OXIDIZED PEPTIDES

Peptides 2 and 9 used in these studies did not show any noticeable separation between the reduced and oxidized forms, and were eluted off the column as a single peak. In order to improve the resolution between the reduced and oxidized conformers, we have used a thiol specific reagent, N-ethylmaleimide (NEM), to modify the reduced peptide. This modification results in

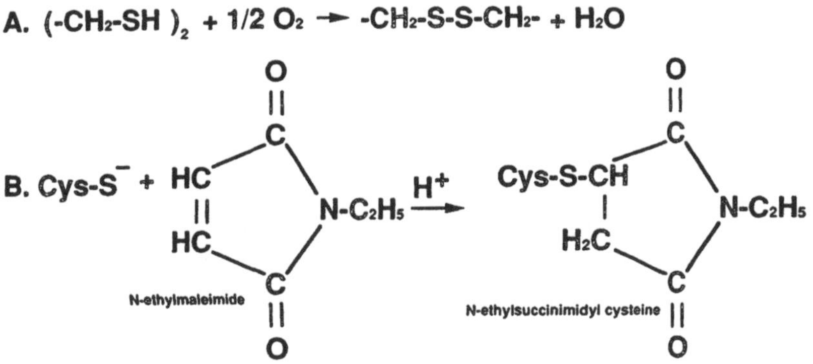

A. $(-CH_2-SH)_2 + 1/2 O_2 \rightarrow -CH_2-S-S-CH_2- + H_2O$

FIGURE 2. Panel A: Air oxidation of cysteine side-chains to form a disulfide bond. Panel B: Reaction of cysteine with N-ethylmaleimide (NEM).

the formation of a stable covalent linkage between the thiol and NEM (the thiolate anion attacks one of the double bonded carbon atoms forming N-ethyl succinimidyl cysteine) as shown in Figure 2B. The covalently modified reduced peptide becomes more hydrophobic and is eluted off the column later than the reduced peptide. Thus, baseline resolution can now be obtained between the oxidized conformer and the NEM-modified peptide. The time course of the oxidation can be followed by the disappearance of the NEM-modified peptide with time and the concomitant appearance of the peak corresponding to the oxidized peptide The extent of oxidation can also be monitored by titrating the unreacted -SH groups with DTNB (5,5'-dithiobis [2-nitrobenzoic acid]), commonly referred to as Ellman's reagent[4]. This method measures the quantity of SH groups by following the increase in absorbance at 412 nm of the brightly yellow colored aromatic thiol liberated by the reagent on disulfide exchange with the cysteine residues in the peptide. Larger amounts of peptides are required in this procedure and only the free sulfhydryls are monitored. The product of the oxidation process, i.e., the oxidized peptide, is not quantitated or monitored. The use of electrochemical detection of thiol and disulfide containing peptides has also been documented (Allison and Shoup[5] and Jacobs[6]). This method is still relatively new and requires additional specialized detectors which may not be easily available or accessible.

The modification of the -SH groups with NEM was easily accomplished by reacting the reagent with the reduced peptide (10 µg) with excess NEM for 15 min in 0.1 M NH$_4$HCO$_3$, pH 8.3 or phosphate buffer saline, pH 7.4, at room temperature. This procedure has previously been used by Lunte and Kissinger[7] to modify free thiol groups in liver samples for LC/EC studies. The reagent reacts only with the reduced peptide (data not shown). When Peptide 2 or 9 was NEM-modified, the peptides were eluted off the column 3 min or more after the reduced peptide. This result is illustrated in Figure 3. The reduced and oxidized forms of Peptide 9 were eluted off the column at 24.8 min and a 1:1 mixture of the two forms resulted in a single peak. When the same mixture was reacted with NEM, it was found that the modified reduced peptide was eluted off the column at 28.6 min. The reagent itself is quite hydrophilic, and was eluted off the RP column at 12 min. It would therefore not interfere or coincide with most peptides which are usually eluted off the column later in the 15 to 40% acetonitrile range of the gradient. We have also used N-phenylmaleimide (data not shown). This particular reagent gave a larger separation (8 min) between modified and unmodified peptides. However, the more hydrophobic phenyl group resulted in the reagent being eluted off the column at approximately 22 min. It was not considered the reagent of choice since the increase in retention time may interfere with a larger number of peptides.

The results shown in Figure 4 illustrated the use of RPC to monitor the oxidation of peptides

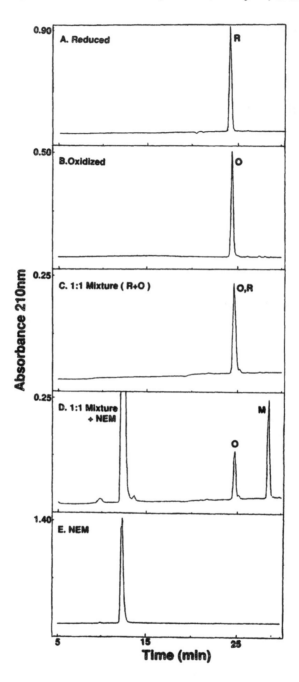

FIGURE 3. The separation of reduced and oxidized forms of Peptide 9 following modification of the reduced peptide with NEM. The NEM-modified peptide is denoted by M. The conditions of chromatography are shown in the legend of Figure 1. Conditions for air oxidation and chemical modification with NEM are given in the Experimental section.

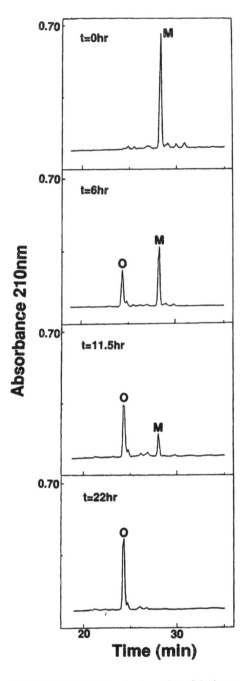

FIGURE 4. Monitoring the formation of the intra-chain disulfide bridge with time during air oxidation of Peptide 9 by RPC. The reduced peptide was modified with NEM prior to injection onto the column to increase the separation between the oxidized and reduced peptides. The modified peptide is denoted by M. The conditions of chromatography are shown in the legend of Figure 1. Conditions for air oxidation and chemical modification with NEM are given in the Experimental section.

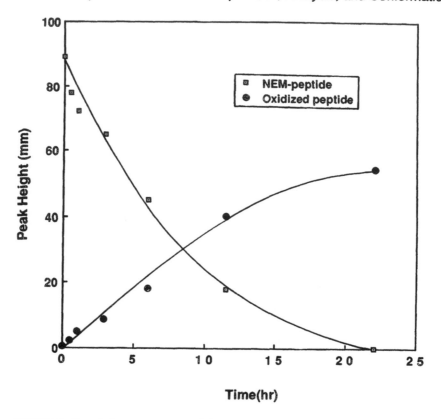

FIGURE 5.   Quantitative measurements of the reduced and oxidized Peptide 9 during an air oxidation process. The peak heights of the two conformers obtained from RPC chromatograms (from Figure 4) were used.

that show no separation between their reduced and oxidized conformers. Such a monitoring procedure can be used for any peptides whether they show good or poor resolution. At various time points, samples can be removed and the NEM reaction will stop any further oxidation by covalently modifying the free -SH groups of the reduced peptide, denoted by M in Figure 4. Monitoring the intrachain disulfide bridge formation by RPC allows us to quantitate simultaneously the two conformers, unlike the Ellman procedure which titrates free sulfhydryls of the reduced peptides. The extent of the increase in the quantity of oxidized peptide and the decrease in the reduced conformer can be quantitated and plotted out over the air oxidation period as shown in Figure 5. Although these modified peptides cannot be retrieved, only a small amount of a peptide is required because of the high sensitivity of the UV detection. The modified peptide was also shown to have an enhanced absorption at 210 nm due to the succinimidyl derivative covalently attached to it. The peak height of the modified peptide at the beginning of the oxidation process was 89 mm and the completely oxidized peptide was only 54.5 mm. This increase in the sensitivity would be useful for detecting any small quantity of reduced peptide left in the air oxidation mixture.

C. RATES OF PEPTIDE OXIDATION

The rates of oxidation of several peptides were compared during this study. These peptides were air oxidized by stirring 0.1 mg/ml in 0.1 M NH_4HCO_3 (pH 8.3) at room temperature. The reaction for the formation of the disulfide bond is shown in Figure 2A. A more detailed discussion on the mechanism of oxidation is given by Friedman.[8] This pH is commonly employed by many researchers to reoxidize denatured proteins. We have tried using phosphate buffer, pH 7.4, to air oxidize some of our peptides and have found that the rate of oxidation was slower (data not shown). Aeration can be used to speed up the oxidation. This method of oxidation was found to be very convenient because the oxidized peptide can be obtained by repeated freeze-drying since the ammonium bicarbonate is volatile. There is no need for further purification or desalting steps, unlike the potassium ferricyanide oxidation procedure of Hope et al.[9] The use of metal catalysts such as Cu^{2+}, Fe^{2+}, Co^{2+}, and Mn^{2+} is supposed to enhance markedly air oxidation of peptides. Cowgill[10] has shown that addition of 10 to 50 μM metal catalyst and aeration greatly enhanced the oxidation process of a paramyosin protein. The metal-catalyzed mechanism of disulfide bond formation is thought to occur via metal complexes of thiol with molecular oxygen.[8] We did not add any trace metal catalysts into the oxidation mixture. However, small traces of iron are present (0.002%) in the ammonium bicarbonate from the supplier. This might be sufficient to catalyze disulfide bond formation. Peptides 2, 5, and 6 were completely oxidized after 1.5, 3, and 6 hr, respectively. Peptide 9 took about 20 hr to be completely oxidized. It was interesting to note that Peptide 2, with the smallest number of residues within the loop (4 residues), had the fastest oxidation rate. Air oxidation studies by Heaton et al[11] using Cys-(Gly)$_n$-Cys peptides (n = 0 to 4) showed that at least four glycine residues between cysteines were required to obtain a good yield of the oxidized peptide (90%). Their results suggested that loop formation was hindered sterically when cysteines were separated by less than 4 residues, resulting in lower yields of the oxidized peptide. Based on the conformational entropy associated with ring closure, the rate of oxidation of Peptide 2 should be faster than Peptides 5, 6, and 9 (12 residues between cysteines). Statistical considerations of loop size by Kauzmann[12] would have suggested that all three of these peptides have the same probability of loop formation and should have similar rates of oxidation. Since the rates are different, conformational considerations must be involved. This conformational factor arises from the inherent properties of the peptide sequence as suggested by White.[13] This consideration was confirmed in a recent study by Milburn and Scheraga.[14] A 30 residue fragment from ribonuclease A (50 to 79) with three cysteine residues at positions 58, 65, and 72 was used. Three intramolecular disulfide bonds 58 to 65, 65 to 72, and 58 to 72 were possible. Reoxidation in a Tris-HCl buffer at pH 8.0 resulted in the 65 to 72 disulfide bond being favored over the other two. This is the same disulfide bond found in the native protein, suggesting that native conformation persisted in the peptide and this modulates the bridge formation between 65 to 72.

IV. CONCLUSIONS

From this study, it was found that most reduced and oxidized peptides can be separated by RPC. The ΔR_t was dependent on the amino acid sequences of the peptides and was not related to the number of residues within the disulfide loop or the length of the peptides. Whenever a separation was observed between the two conformers, the reduced peptide always seemed to be retained longer than the oxidized peptide. However, it is possible for the peptides to be coeluted (this study) or for the reduced peptide to be eluted earlier as shown by other researchers. When the two conformers are coeluted, we could obtain baseline resolution of the reduced peptide from the oxidized peptide by modifying the reduced conformer with NEM. This makes the reduced peptide more hydrophobic because of two extra covalently attached succinimidyl groups. The modified reduced peptides are retained 3 min or more on the column than the unmodified reduced peptide. Thus, the oxidation process of any peptide, whether the reduced and oxidized conformers can be resolved or not on RPC, can be followed with time using RPC after modification of the reduced conformer with NEM. This allows the reduced and oxidized peptides on a HPLC chromatogram to be easily quantitated.

ACKNOWLEDGMENTS

We are indebted to the Medical Research Council of Canada and the Alberta Heritage Foundation for Medical Research for their support in the form of equipment grants (RSH) and an AHFMR Studentship (KKL).

REFERENCES

1. **Atassi, M.Z., McDaniel, C.S., and Manshouri, T.,** Mapping by synthetic peptides of the binding sites for acetylcholine receptor on α-bungarotoxin. *J. Protein Chem.* 7, 655, 1988.
2. **Creighton, T.E.,** *Proteins: Structure and Molecular Properties,* W.H. Freeman, New York, 1983, 23.
3. **Schulze-Gahmen, U., Klenk, H-D., and Beyreuther, K.,** Immunogenicity of loop-structured short synthetic peptides mimicking the antigenic site A of imfluenza virus hemagglutinin, *Eur. J. Biochem.,* 159, 283, 1986.
4. **Ellman, G.L.,** Tissue sulfhydryl groups, *Arch. Biochem. Biophys.,* 82, 70, 1959.
5. **Allison, L.A. and Shoup, R.E.,** Dual electrode liquid chromatography detector for thiols and disulfides, *Anal. Chem.,* 55, 8, 1983.
6. **Jacobs, W.,** Monitoring disulfide peptides via LCEC, *Curr. Sep.,* 8, 44, 1987.
7. **Lunte, S.M. and Kissinger, P.T.,** Detection of thiols and disulfides in liver samples using liquid chromatography/electrochemistry, *J. Liq. Chromatogr.,* 8, 691, 1985.
8. **Friedman, M.,** *Oxidation Reactions in the Chemistry and Biochemistry of the Sulfhydryl Group in Amino Acids, Peptides and Proteins,* Pergamon Press, Oxford. 1973, chap. 3
9. **Hope, D.B., Murti, V.V.S., and Vigneaud, V.D.,** A highly potent analogue of oxytocin, desamino-oxytocin, *J. Biol. Chem.,* 237, 1563, 1962.
10. **Cowgill, R.W.,** Location and properties of sulfhydryl groups on the muscle protein paramyosin from *Mercenaria mercenaria, Biochemistry,* 13, 2467, 1974.
11. **Heaton, G.S., Rydon, H.N., and Schofeld, J.A.,** Polypeptides. III. The oxidation of some peptides of cysteine and glycine, *J. Chem. Soc.,* 3157, 1956.
12. **Kauzmann, W.,** Relative probabilities of isomers in cystine-containing randomly coiled polypeptide, in *Sulfur in Proteins,* Benesch, R., Benesch, R.E., Boyer, P.D., Klotz, M., Middlebrook, W.R., Szent-Gyorgi, A.G., and Schwarz, D.R., Eds., Academic Press, New York, 1959, 93.
13. **White, F.H.,** Regeneration of native secondary and tertiary structures by air oxidation of reduced ribonuclease, *J. Biol. Chem.,* 236, 1353, 1961.
14. **Milburn, P.J. and Scheraga, H.A.,** Local interactions favor the native eight residue disulfide loop in the oxidation of a fragment corresponding to the sequence Ser-50 to Met-79 derived from bovine pancreatic ribonuclease A, *J. Protein Chem.,* 7, 377, 1988.

CHROMATOGRAPHY OF PROTEINS AT HIGH ORGANIC SOLVENT CONCENTRATIONS: AN INVERSE-GRADIENT REVERSED-PHASE HPLC METHOD FOR PREPARING SAMPLES FOR MICROSEQUENCE ANALYSIS

Richard J. Simpson and Robert L. Moritz

I. SURVEY OF METHOD

It is not generally recognized that a skewed U-shaped, or bimodal, dependency exists between retention times and concentration of organic solvent during reversed-phase chromatography (RPC) of low-M_r organic compounds, peptides and small proteins.[1-6] Interestingly, those reversed-phase packings which best exhibit this behavior for proteins (Figures 1B, C, and D) (i.e., strong retention at high concentrations of organic solvent) are characterized by large surface areas (200 to 400 m^2/g) and high carbon content (7 to 15%). Such packings (e.g., Beckman Ultrasphere-ODS, [Figure 1B] Zorbax-ODS [Figure 1C], ODS-Hypersil [Figure 1D], Brownlee C_8 VeloSep [data not shown]) are characterized by small pore sizes (6 to 12 nm, 60 to 120 Å) and are commonly utilized for RPC of low-M_r substances (e.g., peptides). By contrast, those packings which exhibit only weak retention for proteins at high organic solvent concentrations (e.g., Brownlee RP-300 [Figure 1A]; Beckman RPSC and Toya-Soda TSK-Phenyl-5PW [data not shown]) are those designed specifically for RPC of proteins (typically, packings with large pore sizes [30 to 100 nm, 300 to 1000 Å] and small surface areas [approx. 60 m^2/g]).

Proteins bound to the matrix under conditions of high organic solvent concentration (e.g., 90% n-propanol/10% water) can be recovered by introducing an inverse gradient of decreasing organic modifier concentration; high sample recovery (> 85%) and reduced peak volumes (100 to 200 μl) can be achieved by introducing an ion-pairing agent (e.g., trifluoroacetic acid [TFA]) into the eluent solvent.[4]

In this chapter, we describe a number of applications of inverse gradient RPC for preparing subnanomole quantities of protein for microsequence analysis.

II. APPLICATIONS OF "INVERSE GRADIENT" REVERSED-PHASE CHROMATOGRAPHY

It is now well-established that sample preparation is one of the major limitations to obtaining amino acid sequence information from subnanomole quantities of peptides and proteins.[7-9] Although RPC is an established tool for purifying a large range of proteins and peptides, it does have limitations with certain classes of proteins (e.g., large M_r hydrophobic proteins). For these proteins, electrophoretic methods (e.g., one- and two-dimensional polyacrylamide gel electrophoresis) provide important alternative high-resolution techniques.

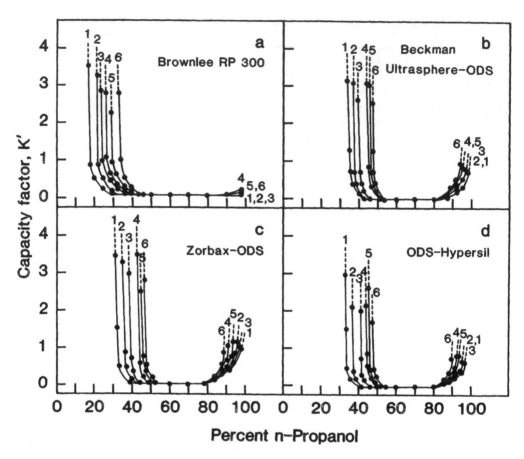

FIGURE 1. Dependence of protein capacity factor, K′, on mobile-phase n-propanol concentration using different reversed-phase bonded alkylsilica columns. The retention data were obtained isocratically for various proteins. All protein samples (20 μl) contained 0.1% SDS. Capacity factors were calculated from $K' = (t_R\text{-}t_o/t_o)$; where t_R is the sample retention time and t_o is the non-retained time for the particular column. Key to protein samples: 1, ribonuclease; 2, cytochrome c; 3, α-lactalbumin; 4, bovine serum albumin; 5, trypsin inhibitor; 6, ovalbumin. n-Alkylsilica columns used: (a) Brownlee RP-300; (b) Beckman Ultrasphere-ODS; (c) Zorbax ODS; (d) ODS-Hypersil. Instrument: chromatography was performed using either: (a) a Perkin-Elmer liquid chromatograph (Model LC4) equipped with a variable wavelength spectrometer (Model LC95) and a scanning fluorimeter (Model LS4) connected in series or (b) a Hewlett Packard liquid chromatograph (Model 1090) equipped with a diode-array detector (Model 1040A). Samples were injected via a Rheodyne injection valve (Model 7125) equipped with a 2-ml sample loop. (From Simpson, R.J. et al., *Eur. J. Biochem.*, 165, 21, 1987. With permission.)

One of the drawbacks in the past with electrophoretically purified proteins has been the difficulty in isolating them in sufficient quantities and in a form suitable for microsequence analysis. The most commonly used procedures for recovering proteins from SDS-PAGE (sodium dodecylsulfate-polyacrylamide gel electrophoresis) gels, after visualization (e.g., staining with Coomassie blue) are electroelution and, more recently, direct electrophoretic transfer (electroblotting) onto an immobilizing matrix (see Reference 9 for review).

A. RECOVERY OF PROTEINS FROM SDS-PAGE ELECTROELUATES
From the numerous reports in the literature describing procedures and equipment for electroeluting proteins from gel slices, it is obvious that no single method has proven to be entirely satisfactory. A common problem frequently encountered with protein electroeluates is that they contain excessively large amounts of SDS (> 1% w/v), buffer salts and acrylamide-

related contaminants which interfere with the Edman degradation procedure and subsequent HPLC identification of PTH-amino acids.

Although proteins can be recovered from SDS-PAGE electroeluates by selective precipitation with organic solvent or ion-pairing agents, these procedures often result in poor recoveries.[9] Unfortunately, conventional high-performance liquid chromatography (HPLC) methods (e.g., reversed-phase or size-exclusion) are of limited value, since numerous artifactual peaks (depending on the quality of detergents and electrophoresis reagents used) seriously interfere with the interpretation of the chromatograms (Figures 2A and B). However, if protein electroeluates are applied to a small-pore reversed-phase packing (e.g., ODS-Hypersil, Brownlee VeloSep C_8) under inverse-gradient RPC conditions (i.e., 95% aqueous n-propanol), proteins are retained on the column while SDS and acrylamide-gel artifacts wash through the column (Figure 2C). To ensure reproducible chromatography, we routinely pre-equilibrate the column at high flow rate (1 ml/min for a 2.1 mm ID column; back pressure, 250 bar) with at least 50 column volumes of primary solvent (> 90% aqueous n-propanol). Proteins can be recovered from the column in high yield (> 85%) by decreasing the concentration of organic modifier to < 60% and by simultaneously adding TFA 0.4% to the mobile phase.[4] Using a steep gradient at a low flow rate (20 to 200 µl/min), proteins are typically recovered in < 150 µl in a form suitable for direct sequence analysis.

Typically, small amounts of protein (1 to 50 µg) can be recovered from SDS-PAGE electroeluates in high yield (> 85%) using a 2.1 mm I.D. column, provided the assayed SDS concentration[19] does not exceed 1.2% (w/v). It should be stressed that the sample SDS concentration affects protein retention at high organic solvent concentrations. Samples (≤ 300 µl) with concentrations of SDS > 1.2% require dilution with n-propanol (up to 1.5 ml) in the sample-loading syringe; in such cases, a small bubble should be introduced in the syringe to ensure thorough mixing prior to loading the sample via a 2-ml sample loop. Conventional columns (4.6 mm I.D.) can be used for isolating larger quantities of protein (500 µg) from electroeluates of preparative gel bands.[4]

The general utility of this chromatographic strategy has been demonstrated by its successful use in recovering a wide variety of proteins (10 to 500 pmol range) from SDS-PAGE electroeluates for N-terminal sequence analysis.[4,10-12]

B. INTERNAL SEQUENCE ANALYSIS OF ELECTROBLOTTED PROTEINS

Electrotransfer (electroblotting) of SDS-PAGE-resolved proteins onto immobilizing matrices, e.g., PVDF (poly[vinylidine difluoride]) for the purpose of obtaining N-terminal sequence data has received considerable attention in recent times due to the simplicity and speed of the technique as well as its potential for handling proteins refractive to conventional RPC technology (see Reference 9 and references therein). However, further characterization (e.g., immunological studies, peptide mapping, internal sequence analysis) often necessitates recovering the protein from the immobilizing matrix.

Internal sequence data is important for several reasons, namely, (1) N-terminally blocked proteins are refractory to Edman degradation, (2) construction of DNA probes, (3) confirming cDNA-deduced protein sequences, (4) determination of post-translational modification sites, (5) fingerprinting recombinant DNA-derived proteins, and (6) epitope mapping.

A number of different regimens have been reported for eluting proteins from PVDF membranes after their transfer from SDS-polyacrylamide gels.[13,14] In our experience, detergent mixtures are the most efficient agents. Typically, we use 2% SDS/1% Triton X-100/0.1% dithiothreitol; to a lesser extent, mixtures of organic solvent and TFA can be employed.[14]

For proteolytic digestion of detergent-eluted proteins, it is necessary to reduce the detergent concentration since most proteolytic enzymes are inactive at concentrations in excess of 0.1%. Attempts to lower the detergent concentration by simply diluting the eluate are inappropriate

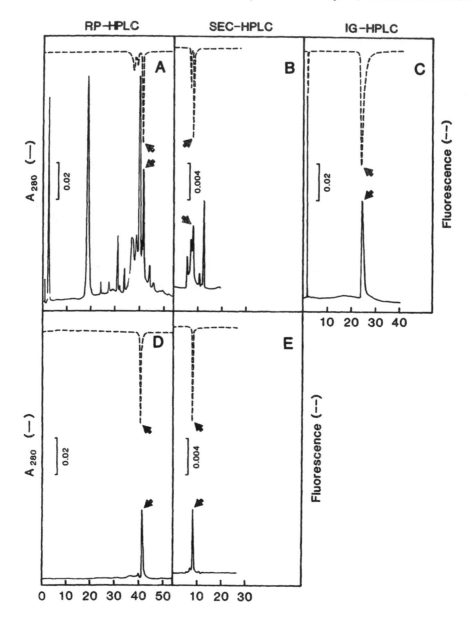

FIGURE 2. Inverse-gradient, reversed-phase and size-exclusion HPLC of β-lactoglobulin recovered from SDS-PAGE gels by electrophoresis. Electroeluted Coomassie-blue stained β-lactoglobulin (5 μg in 50 μl) was chromatographed as follows. (A) Reversed-phase HPLC (RP-HPLC). Column: Brownlee RP-300 C_8 (100 × 2.1 mm I.D.) Chromatographic conditions: a linear 60-min gradient was applied from 0 to 100% B, where Eluent A was water containing 0.1% (v/v) TFA, pH 2.0, and Eluent B was 60% aq. acetonitrile; flow-rate, 0.1 ml/min. Proteins were detected by their absorbance at 280 nm and by their endogenous tryptophan fluorescence using excitation and emission wavelengths of 295 and 360 nm, respectively. (B) Size-exclusion chromatography (SEC-HPLC). Column: TSK-2000SW (30 × 7.5 mm I.D.). Chromatographic conditions: solvent, 100 m*M* sodium phosphate buffer, pH 7.0; flow-rate 0.5 ml/min. (C) Inverse-gradient chromatography (IG-HPLC). Column: ODS-Hypersil (5 μm). 100 × 2.1 mm I.D. Chromatographic conditions: a linear 50-min gradient was applied from 0 to 50% B, where Eluent A was 90% n-propanol/10% water and Eluent B was water containing 0.4% (v/v) TFA; flow-rate, 20 μl/min. β-Lactoglobulin recovered from (C) was re-chromatographed by reversed-phase HPLC (D), using the same conditions described in (A) and size-exclusion chromatography (E) using the same conditions described in (B). Arrows indicate position of β-lactoglobulin. (From Simpson, R.J. et al., *Eur. J. Biochem.*, 165, 21, 1987. With permission.)

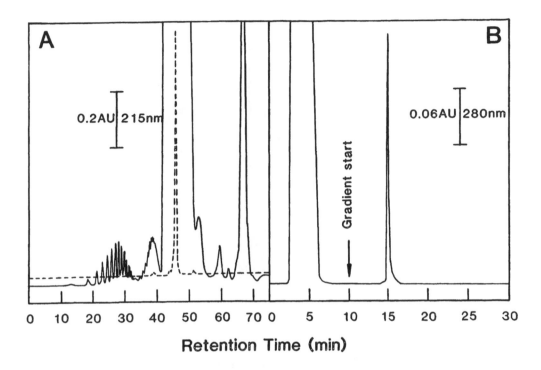

FIGURE 3. Chromatography of protein/Triton X-100 mixtures using conventional and inverse-gradient RP-HPLC. Column: Brownlee C_8 VeloSep, 5-μm particle size, 40 × 3.2 mm I.D. (A) Reversed-phase elution mode. The elution position of plant glycoprotein S_7[10], determined in the absence of detergent, is denoted by the broken line (– – –); (B) Inverse-gradient elution mode. Chromatographic conditions: the column was developed with a linear 50-min gradient from 0 to 100% B, where Eluent A was 100% n-propanol and Eluent B was 50% n-propanol/50% water containing 0.4% TFA; flow-rate; 400 μl/min; detection, UV at 280 nm; column temperature, 40°C. Sample: 200 μl of aqueous 1% Triton X-100 containing 10 μg of glycoprotein S_7. (From Simpson, R. J., et al., *J. Chromatogr.*, 476, 345, 1989. With permission.)

since Coomassie blue and ultraviolet-absorbing detergents trace-enrich on reversed-phase supports and interfere with peptide map interpretation (Figure 3). To overcome this problem, proteins can be recovered from detergent eluates (free from Coomassie blue and detergents) by inverse-gradient RPC (Figure 4C) and then subjected to proteolytic digestion and reversed-phase separation (Figure 4D). Before addition of enzyme, samples are diluted (to 1.0 to 1.5 ml) with an appropriate buffer containing 0.01% Tween 20 to lower the organic solvent concentration (and also to buffer-exchange) since most proteases are inactive above 15% organic solvent.[14] Resultant peptides can be purified by microbore column (1 to 2 mm I.D.) RPC[7-9,16] employing a low-pH mobile phase buffer/acetonitrile system (Figure 4D) (for review see Reference 9).

For peptide mapping on microbore columns, we prefer to use short columns (< 10 cm) since they permit the use of high flow rates (typically, 400 to 1000 μl/min for a 1 mm I.D. column and 500 to 2000 μl/min for a 2.1 mm I.D. column) which allow, (1) rapid trace-enrichment of sample onto an interactive sorbent and (2) rapid column re-equilibration.[17] Previously, we have demonstrated that short microbore columns do not seriously compromise the chromatographic separation of peptides and proteins on reversed-phase columns.[9,17]

The use of multiple-wavelength detection (e.g., diode-array) is quite useful for peptide mapping since it provides a facile procedure for rapidly targeting aromatic-amino-acid-containing peptides for sequence analysis and design of DNA probes.[18] For example, tryptophan, with its unique codon, is readily identifiable by this means. An inspection of the spectral

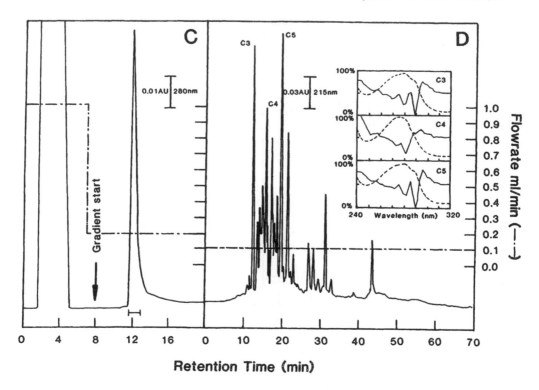

Retention Time (min)

FIGURE 4. Peptide mapping of Coomassie blue-stained S-carboxymethyl-lysozyme (g-type) (7 μg) from the Black Swan[15] recovered from detergent mixtures using inverse-gradient RPC. Conditions for SDS-PAGE electrophoresis, electroblotting onto PVDF membrane and visualization with Coomassie blue are described elsewhere.[14] Coomassie-blue stained lysozyme was eluted with 200 μl of SDS/1% Triton X-100/0.1% dithiothreitol in 50 m*M* Tris-HCl buffer, pH 9.0. (C) Inverse-gradient RPC of Coomassie blue-stained lysozyme. Chromatographic conditions are given in Figure 3. (D) Separation of peptic peptides of S-carboxymethyl lysozyme recovered from Figure 4C. Column: Brownlee RP-300 (50 × 1.0 mm I.D.). Chromatographic conditions: the column was developed with a linear 60-min gradient from 0 to 100% B, where Eluent A was 0.1% (v/v) TFA and Eluent B was 60% acetonitrile/40% water containing 0.09% TFA. Peptides selected for sequence analysis are indicated along with their spectral analysis (see inset). The absorption spectra have been normalized to relative absorbance on a scale of 0 to 100%. Zero-order derivative spectra (- - -); second-order derivative spectra (—). (From Simpson, R. J., et al., *J. Chromatogr.*, 476, 345, 1989. With permission.)

data in Figure 4D indicates that peptides C3, C4, and C5 contain aromatic amino acids as judged by their absorbance in the range 270 to 300 nm. Enhancement of resolution by second-order derivative spectroscopy predicts the presence of a tyrosine residue(s) in peptide C4 (extremum at 280 ± 2 nm), and the extremum at 290 ± 2 nm in peptides C3 and C5 is characteristic of tryptophan residues (these assignments were subsequently confirmed by sequence analysis).

Thus, by combining gel electrophoresis (either SDS-PAGE or two-dimensional), electroblotting and RPC (both conventional and inverse-gradient), it is possible to obtain internal sequence data for electroblotted proteins isolated in the low microgram range.[14]

C. FRACTIONATION OF PROTEINS AT HIGH ORGANIC SOLVENT CONCENTRATIONS

Using the inverse-gradient RPC procedure, we have recently demonstrated that proteins can be chromatographically fractionated at high organic solvent concentrations (Figure 5).[20] In this study, we compared the retention behavior of ten proteins (of known primary structure) on different reversed-phase columns operated in both the classical reversed-phase as well as the inverse-gradient elution mode. Unlike the chromatographic behavior of small peptides, where

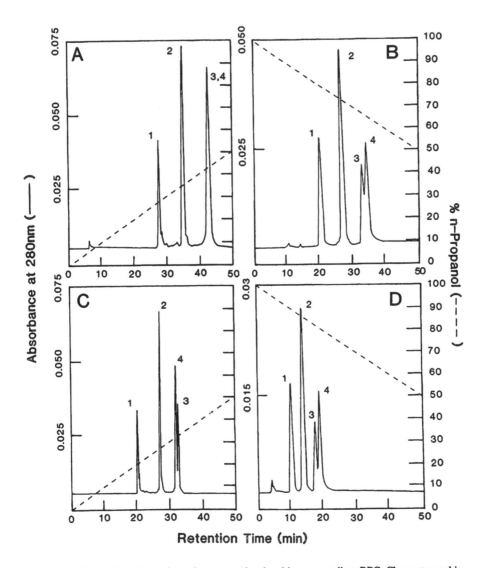

FIGURE 5. Separation of proteins using conventional and inverse-gradient RPC. Chromatographic conditions. Panel A: ODS-Hypersil (100 × 2.1 mm I.D.) in conventional reversed-phase elution mode; linear 50-min gradient from 0 to 100% B, where Eluent A was 0.1% (v/v) TFA, pH 2.0, and Eluent B was 50% aqueous n-propanol containing 0.1% (v/v) TFA. Panel B: ODS-Hypersil (100 × 2.1 mm I.D.) in inverse-gradient reversed-phase elution mode; linear 50-min gradient from 0 to 100% B, where Eluent A was 100% n-propanol and Eluent B was 50% aqueous n-propanol containing 0.1% (v/v) TFA. Panel C: Brownlee C₈ VeloSep (40 × 3.2 mm I.D.) in conventional reversed-phase elution mode (same as Panel A). Panel D: Brownlee C₈ VeloSep (40 x 3.2 mm I.D.) in inverse-gradient reversed-phase elution mode (same as Panel B). Flow rates, Panels A and B, 200 μl/min; Panels C and D, 400 μl/min. Column temperature, 40°C. Proteins: 1, bovine insulin; 2, α-lactalbumin; 3, α-amylase; 4, carbonic anhydrase. Sample load: 10 μg in 20 μl water. (From Simpson, R. J. and Moritz, R. L., *J. Chromatogr.*, 474, 418, 1989. With permission.)

it is well documented[3] that a linear relationship exists between the polarity of a peptide and its retention order, we (and others[21]) observed no clear correlation between the calculated hydrophobicities[22] of the proteins examined and their elution times with n-propanol. Interestingly, for a particular reversed-phase column, although the protein retention order was essentially the same in both elution modes, some reversals in protein selectivity pattern were apparent. For example, the retention order of insulin and ribonuclease as well as ovalbumin and transferrin are reversed in the two chromatographic modes (date not shown; also, note α-amylase and carbonic anhydrase in panels C and D of Figure 5, peaks 3 and 4, respectively). The reversals in polypeptide retention during the two chromatographic modes were also shown to occur during elution with other organic solvents such as methanol and acetonitrile.[3]

These findings are in accord with previously reported observations[1-3] that multiple retention processes, including both solvophobic (hydrophobic) and silanophilic interactions are involved in the binding of peptides and proteins to silica-based hydrocarbonaceous supports. Since solvophobic theory[23] alone cannot explain the skewed U-shaped dependence between solute retention time and concentration of organic solvent, it has been suggested[1-5] that silanophilic interactions (due to residual silanol groups remaining on a support) may be responsible for this irregular behavior. This hypothesis is supported by Bij and co-workers[2] who demonstrated that the addition of n-butylamine (an attenuator of silanophilic interactions) abrogated this U-shaped dependence. However, the fact other alkylamines (e.g., triethylammonium phosphate[1,3,5]) do not dramatically influence this behavior suggests that more than one mechanism may be involved.

The protein selectivity pattern changes observed in the inverse-gradient RPC[4,20] (see also Figure 5, panels C and D) are indicative of quasi-polar-phase (or normal phase) behavior due, presumably, to silanophilic interactions. These selectivity changes may reflect different protein conformations being induced in the "reversed-phase" and "inverse-gradient" modes, since it is well documented[21,24,25] that conformational transitions in proteins can be induced by apolar compounds such as detergents and organic solvents. For instance, using physiochemical techniques such as circular dichroism, fluorescence, and visible spectroscopy, it has been shown for a number of proteins that n-propanol can induce a reversible conformational change to an apparently ordered helical form.[21] Consistent with this hypothesis is the recent report by Karger and coworkers[24] that protein conformation can have a marked influence on protein retention behavior on reversed-phase columns. In these studies, "native" and "denatured" proteins were resolved on reversed-phase packings and it was established that the kinetics of protein unfolding was a function of both organic modifier employed and the incubation time that a protein spent on the bonded-phase surface prior to column development. Likewise, it has been demonstrated for small peptides that their retention behavior on reversed-phase columns can be strikingly altered if the peptide can be induced to form an amphipathic helix.[26,27] Thus, protein selectivity changes at high organic solvent concentrations may result from the exposure of buried surfaces within the protein (due to unfolding of the protein) and their interaction with surfaces of the hydrocarbonaceous bonded-phase. The nature of this interaction, yet to be fully defined, is presently under investigation.

III. SUMMARY

Inverse-gradient RPC exploits the U-shaped dependency between protein retention behavior and organic solvent concentration. This technique provides a facile method for recovering proteins from detergent mixtures such as those used to elute electroblotted proteins from PVDF membranes as well as SDS-polyacrylamide gel electroeluates. Proteins recovered in this manner are in a form suitable for microsequence analysis, i.e., free from detergent and gel-related artifacts, and in a volatile solvent. Moreover, inverse-gradient RPC provides the potential for larger-scale fractionation of proteins at high organic solvent concentrations.

REFERENCES

1. Sander, L.C. and Field, L.R., Effect of eluent composition on thermodynamic properties in high-performance liquid chromatography, *Anal. Chem.*, 52, 2009, 1980.
2. Bij, K.E., Horváth, C., Melander, W.R., and Nahum, A., Surface silanols in silica-bonded hydrocarbonaceous stationary phases. II. Irregular retention behavior and effect of silanol masking, *J. Chromatogr.*, 203, 65, 1981.
3. Hearn, M.T.W. and Grego, B., High-performance liquid chromatography of amino acids, peptides and proteins. Organic solvent modifier effects in the separation of unprotected peptides by reversed-phase liquid chromatography, *J. Chromatogr.*, 218, 497, 1981.
4. Simpson, R.J., Moritz, R.L., Nice, E.C., and Grego, B., A high-performance liquid chromatography procedure for recovering subnanomole amounts of protein from SDS-gel electroeluates for gas-phase sequence analysis, *Eur. J. Biochem.*, 165, 21, 1987.
5. Armstrong, D.W. and Boehm, R.E., Gradient LC separation of macromolecules: theory and mechanism, *J. Chromatogr. Sci.*, 22, 378, 1984.
6. Guinebault, P., Broquaire, M., Colafranceschi, C., and Thenot, J.P., High-performance liquid chromatographic determination of alfuzosin in biological fluids with fluorimetric detection and large-volume injection, *J. Chromatogr.*, 353, 361, 1986.
7. Simpson, R.J. and Nice, E.C., Preparation of subnanomole amounts of protein and polypeptides for microsequencing, in *Integration Control of Metabolic Processes: Pure and Applied Aspects: FOAB Conference (4th:1986:Singapore)*, Kon, O.L., Ed, ICSU Press, Miami, 1987, 473.
8. Simpson, R.J. and Nice, E.C., The role of microbore HPLC in the purification of subnanomole amounts of polypeptides and proteins for gas-phase sequence analysis, in *Methods in Protein Sequence Analysis*. Walsh, K.A., Ed, Humana Press, Clifton, NJ, 1987, 213.
9. Simpson, R.J., Moritz, R.L., Begg, G.S., Rubira, M.R., and Nice, E.C., Micropreparative procedures for high sensitivity sequencing of peptides and proteins, *Anal. Biochem.*, 177, 221, 1989.
10. Anderson, M.A., Cornish, E.C., Mau, S-L., Williams, E.G., Hoggart, R., Atkinson, A., Bonig, I., Grego, B., Simpson, R.J., Roche, P.J., Haley, J.D., Penschow, J.D., Niall, H.D., Tregear, G.W., Coglan, J.P., Crawford, R.J., and Clarke, A.E., Cloning of cDNA for a stylar glycoprotein associated with expression of self-incompatability in *Nicotiniana alata*, *Nature*, 321, 38, 1986.
11. Mau, S-L., Williams, E.G., Atkinson, A., Anderson, M.A., Cornish, E.C., Grego, B., Simpson, R.J., Kheyr-Pour, A., and Clarke, A.E., Style proteins of a wild tomato (Lycopersicon peruvianum) associated with self-incompatibility, *Planta*, 169, 184, 1986.
12. Ramasamy, R., Simpson, R.J., Dexter, A., Keeghan, M., Reed, C., Bushell, G., Ingram, L.T., Henderson, T., Moloney, M.B., Moritz, R.L., Rubira, M.R., and Kidson, C., Isolation and partial characterization of a 26 kilodalton antigen from plasmodium falciparum recognized by an inhibitory monoclonal antibody, *Mol. Biochem. Parasitol.*, 29, 25, 1988.
13. Szewczyk, B. and Summers, D.F., Preparative elution of proteins blotted to immobilon membranes, *Anal. Biochem.*, 168, 48, 1988.
14. Simpson, R.J., Ward, L.D., Batterham, M.P., and Moritz, R.L., Peptide mapping and internal sequencing of proteins electroblotted from poly(vinylidene difluoride) membranes: a chromatographic procedure for separating proteins from detergents, *J. Chromatogr.*, 476, 345, 1989.
15. Simpson, R.J., Begg, G.S., Dorow, D.S., and Morgan, F.J., Complete amino acid sequence of the goose-type lysozyme from the egg white of the Black swan, *Biochemistry*, 19, 1814, 1980.
16. Simpson, R.J., Moritz, R.L., Rubira, M.R., and Van Snick, J., Murine hybridoma/plasmacytoma growth factor: complete amino acid sequence and relation to human interleukin-6, *Eur. J. Biochem.*, 176, 187, 1988.
17. Nice, E.C., Grego, B., and Simpson, R.J., Application of short microbore HPLC guard columns for the preparation of samples for protein microsequencing, *Biochem. International*, 11, 187, 1985.
18. Grego, B., Van Driel, I.R., Goding, J.W., Nice, E.C., and Simpson, R.J., Use of high-performance liquid chromatography for purifying subnanomole levels of polypeptides for microsequencing: structural studies on the murine plasma cell antigen PC-1, *Int. J. Peptide Protein Res.*, 27, 201, 1986.
19. Waite, J.H. and Wang, C.Y., Spectrophotometric measurement of dodecyl sulfate with basic fuchsin, *Anal. Biochem.*, 70, 279, 1976.
20. Simpson, R.J. and Moritz, R.L., Chromatographic fractionation of proteins at high solvent modifier concentrations, *J. Chromatogr.*, 474, 418, 1989.
21. Sadler, A.J., Micanovic, R., Katzenstein, G.E., Lewis, R.V., and Middaugh, R.C., Protein conformation and reversed-phase high-performance-liquid chromatography, *J. Chromatogr.*, 317, 93, 1984.
22. Bigelow, C.C., On the average hydrophobicity of proteins and the relation between it and protein structure, *J. Theoret. Biol.*, 16, 187, 1967.
23. Horváth, C., Melander, W., and Molnar, I., Solvophobic interactions in liquid chromatography with nonpolar stationary phases, *J. Chromatogr.*, 125, 129, 1976.

24. **Benedek, K., Dong, S., and Karger, B.L.,** Kinetics of unfolding of proteins on hydrophobic surfaces in reversed-phase liquid chromatography, *J. Chromatogr.*, 317, 227, 1984.
25. **Mattice, W.L., Riser, J.M., and Clark, D.S.,** Conformational properties of the complexes formed by proteins and sodium dodecyl sulfate, *Biochemistry*, 15, 4624, 1976.
26. **Houghten, R.A. and Ostresh, H.M.,** Conformational influences upon peptide retention behavior in reverse phase high performance liquid chromatography, *BioChromatography*, 2, 80, 1987.
27. **Hancock, W.S., Knighton, D.R., and Harding, D.R.K.,** The use of reversed-phase HPLC to study the lipid associating properties of polypeptides, in *Peptides 1984*, Rognarsson, U., Ed., in Proc. of the 18th European Peptide Symposium, Stockholm, Sweden, 1985, 145.

DISCONTINUOUS REVERSED-PHASE CHROMATOGRAPHY: A SEPARATION METHOD FOR COMPLEX PROTEIN MIXTURES APPLICABLE IN BOTH ANALYTICAL AND LARGE SCALE HIGH-PERFORMANCE LIQUID CHROMATOGRAPHY

Joachim R. Grün and Richard Reinhardt

I. INTRODUCTION

Since the problem of fast protein separations at high resolution has been solved by the application of reversed-phase high-performance liquid chromatography (RPC), the question has now become how to separate large amounts of biological material with good recovery. Until recently, RPC seemed to be a technique which was only capable of separating minimal amounts of biological sample per injection or per gram of stationary phase, as has been reported for load capacities using small molecules (e.g., 2 mg/g).[1] Ion-exchange high-performance liquid chromatography (IEC) seemed to be the only alternative for the separation of large amounts of, e.g., proteins. In the meantime, it has been shown that reversed-phase (RP) material can bind amounts of protein that are similar to those which IEC can separate. Reported maximum load capacities are, e.g., 3.7 to 6.4 mg/g in Reference 2, 5.35 or 20 mg/g in Reference 3 and up to 100 mg protein per gram of RP material.[4] Although there is no longer any doubt that RP material can bind proteins with a high load capacity, the question remains as to how to utilize this high load capacity for practical separations of complex biological material. We are of the opinion that discontinuous reversed-phase HPLC (Disc RPC) offers a solution to this problem.

The separation problem with which we are concerned is the isolation of the complete set of ribosomal proteins from the archaebacterium *Sulfolobus acidocaldarius* (DSM 1616).[5] If isolated ribosomal proteins from *S. acidocaldarius* are N-terminally sequenced, this sequence information can be used as a basis for synthesizing oligonucleotide probes to locate and isolate the genes for the proteins, and hence to deduce complete primary structures for the proteins from the corresponding DNA sequences. Comparative studies of ribosomal protein sequences from *Sulfolobus* and other archaebacteria should help to shed light on the relative evolutionary position of *Sulfolobus* within the class of archaebacteria, which represent a third evolutionary line of descent, showing characteristics of both the eubacterial and eukaryotic classes.

We have isolated ribosomal proteins of *Sulfolobus*, utilizing two different methods: (1) the use of non-denaturing conditions (cf. Reference 6), which should permit subsequent crystallization of the proteins and the determination of their tertiary structure; and (2) the use of high-performance liquid chromatography (HPLC) for sequence determination, peptide mapping and amino acid analysis (cf. Reference 7). The major advantage of HPLC with volatile eluents is that sequence and amino acid analysis, as well as gel electrophoresis, can be undertaken directly after evaporation of the salt-free eluents. In pilot experiments, we attempted to separate the total protein mixtures from the small and the large subunit (TP30 and TP50) with advanced HPLC methods, e.g., gel-permeation or size-exclusion chromatography (SEC) and IEC[8,9] or with RPC.[9] In *Escherichia coli,* which is the best characterized member of the eubacteria,[10] the 30S subunit consists of 21 "S" proteins and the 50S subunit of 34 "L" proteins. The number of proteins seems to be higher in the subunits of *Sulfolobus.* Our total protein mixtures of *S. acidocaldarius* consist of up to 60 individual proteins,[11] including both ribosomal and non-ribosomal proteins (e.g., DNA-binding proteins, cf. Reference 12). Most ribosomal proteins of *S. acidocaldarius* cannot be isolated in pure form in a single chromatography run, even by RPC methods. Often, several proteins are coeluted within a single peak, as revealed by sodium dodecylsulfate (SDS) or two-dimensional (2D) polyacrylamide gel electrophoresis (PAGE). A more advanced method for detecting cross-contamination is peak analysis using correlation matrix contour plots,[13] which for any one peak shows correlations of at least eight complete spectra against each other. Spectrum data sampling for this new peak purity check is accomplished with a photodiode array detector.

Purification of the cross-contaminated protein fractions is achieved by re-chromatography under different RPC conditions,[14,15] or with IEC, using volatile buffers,[14,16] under conditions similar to those reported in Reference 17. To obtain identical samples of starting material for several re-chromatography runs from the same pool, we increased the load capacity of our pre-fractionation technique by developing the Disc RPC method.[11,18] The Disc RPC pre-fractionation method provided groups of one to six ribosomal proteins. Since they were eluted in volatile solvents, they could either be re-chromatographed in another RPC run without concentrating, or could be subjected to an IEC run in a volatile buffer[14] after evaporation and subsequent solution in 1 M acetic acid or after simple dilution of the pre-fraction with water.

II. EXPERIMENTAL

A. SAMPLE PREPARATION

Sulfolobus acidocaldarius cells were grown in a 150-l glass fermenter at 80°C and a pH of 4.5, with aeration at a flow-rate of 130 l/min as described in Reference 5. Harvesting and breakage of cells, preparation of ribosomes, ribosomal subunits, and total protein mixtures was carried out as in References 5 and 9. The dialyzed acetic acid extract containing the total protein mixture was divided into aliquots in Eppendorf tubes, frozen in liquid nitrogen, and either lyophilized and stored at –20°C[18] or stored at –70°C without lyophilization.[15] For RPC, 850 µl water and 150 µl formic acid was added to each of the desired number of tubes containing 9.2 A_{230} (5.7 mg) lyophilized TP30. Alternatively, formic acid was added to a final concentration of 15% to the tubes containing thawed TP50 dialysate. The final concentration of TP50 was 3.4 A_{230}/ml.[15] Tests with more concentrated TP30 solutions gave similar separations, but were not advantageous because of the limited solubility of the proteins. After centrifuging for 5 min at 9000 $g,$ the TP solution was filtered through a 0.45 or 0.22 µm pore size filter[15,18] and injected directly onto the HPLC system.

B. REVERSED-PHASE CHROMATOGRAPHY

Proteins with different affinities for the hydrophobic stationary phase of a reversed-phase column cannot be separated effectively under isocratic conditions,[19] requiring instead gradient elution. For sample injection, either a free programmable autoinjector from Gilson (Villier le Bel, France) or an injection valve, Model 7125, from Rheodyne (Cotati) is used. The size of the sample loops ranges from 0.1 up to 22.7 ml.[11,15]

We usually monitor column eluates at 230 nm. This wavelength is only half as sensitive for proteins as 220 nm and one fourth as sensitive as at 210 nm. Thus, one can monitor larger amounts of proteins (cf. Figure 6) without problems. Acetonitrile, 2-propanol and mixtures of both solvents are used as the organic modifier (Eluent B) in RPC. Organic eluents were of Lichrosolv grade from Merck (Darmstadt, FRG), and the water for Eluent A was deionized and doubly distilled in a quartz system. All eluents contained 0.001% (w/v) sodium azide, 0.001% (v/v) 2-mercaptoethanol, and 0.1% trifluoroacetic acid (TFA) as ion-pairing reagent (cf. Reference 11).

The organic modifier 2-propanol has a sixfold higher viscosity than acetonitrile. This fact alone should exclude 2-propanol as a useful solvent for HPLC of proteins. On the other hand, 2-propanol is a stronger RP solvent than acetonitrile and is reported to increase the recovery of proteins (e.g. Reference 20). More important for our purposes, Cooke et al.[21] pointed out that propanol interacts with accessible silanol sites, which exist in uncapped phases or in smaller amounts in all silica gel based phases as unreacted silanol groups.[22] This interaction probably reduces silanol effects with "more acidic"[22] column materials, whereas Stadalius et al.[23] used triethylamine or morpholine to reduce this effect.

Silanol sites lead to irreversible protein binding and peak tailing, and have the strongest influence on proteins containing many basic amino acids,[23] as is the case in the ribosomal proteins of *S. acidocaldarius*: To date, 45 N-terminal sequences of these proteins have been determined (8 to 86 steps), giving a total of 1694 sequenced amino acids (20 S- and 25 L-proteins[24]). Of these sequenced amino acids, 22.0% (cf. Reference 11) are basic (232 lysine, 120 arginine, and 20 histidine residues), whereas only 9.6% are acidic residues (94 glutamic and 69 aspartic acid residues).

It is clear from our data that TP samples are eluted in fewer peaks with 0.1% TFA in 2-propanol as Eluent B than with 0.1% TFA in pure acetonitrile. In the latter case, the same protein is present in several peaks[11] (corresponding to elution of different conformers of one protein, cf., e.g., Reference 25). We are not yet sure whether this multiple peak effect is induced by the acetonitrile itself, or whether it is due to sample preparation with the different conformers not being resolved in 2-propanol. To reduce the extent of multiple peak elution of the individual proteins, and to exploit the advantages described above, we use a low viscosity mixture of 9 parts acetonitrile and 1 part (v/v) 2-propanol as Eluent B for pre-fractionation of large amounts of TP. Nevertheless, for re-chromatography with end-capped Vydac column material (cf. Table I in Reference 22), we routinely use acetonitrile[15] as well as 2-propanol and the 9:1 mixture of both. A 1:2 mixture of acetonitrile and 2-propanol gave better resolution than pure 2-propanol (in accordance with Reference 26), but did not allow the isolation of as many single proteins as the 9:1 mixture, as judged by SDS PAGE.

C. FRACTION COLLECTION

For sample fractionation in preparative runs, we use SuperRac fraction collectors from Pharmacia-LKB. When separating highly complex protein mixtures, it is advantageous to run the SuperRac in peak mode, where it detects the beginning and end of a peak by calculating the first and second derivatives of the detector signal independently of any increase of the baseline, i.e., not merely like a level sensor. It is therefore not possible for two peaks, just separated by the HPLC system, to be inadvertently re-combined in a single fraction tube, as can occur in the time mode of conventional fraction collectors.

III. PRINCIPLES OF DISCONTINUOUS REVERSED-PHASE CHROMATOGRAPHY

Discontinuous RPC is a type of multi-column chromatography and is referred to as "Disc RPC" because, (1) all of the injected sample runs through a discontinuous bed of stationary phases, comprised of at least two different column materials and (2) when the columns are combined in optimal sequence (i.e., if the proteins are eluted from the first column material at lower concentrations of the B eluent than from the next), one observes an enrichment effect, which, by analogy with discontinuous gel electrophoresis, minimizes band broadening.[18] Disc RPC is distinct from other kinds of multi-column chromatography, as detailed in References 11, 14, and 18. Our goal with this technique was to increase the load capacity of RP columns, used as a pre-fractionation technique for complex ribosomal protein mixtures instead of SEC,[11] without too much loss of chromatographic resolution and without the need for a more sophisticated multi-column system. We know that discontinuous chromatography is applicable not only to RPC (in both analytical and semi-preparative HPLC), but also to open column liquid chromatography with IEC.[27] With Disc RPC, the single column is replaced with different coupled columns, that is, two or more columns connected in series.

A. PROTEIN ELUTION/REBINDING MECHANISM IN DISC RPC

To investigate the mechanism of protein elution in Disc RPC, we combined three different mini-columns (5 × 4 mm I.D., Knauer, Bad Homburg, FRG), dry-packed with SynChropak C$_1$ 1R-103 (Bischoff, Leonberg, FRG), uncapped Vydac C$_{18}$ 201-TP (Machery-Nagel, Düren, FRG), and C$_{18}$ TSK ODS 120T (Labotron Instrument AG, Zürich, Switzerland) reversed-phase materials. The use of these extremely short columns ensured that the observed effects would also be seen with longer columns. Observation of the elution behavior of the TP30 proteins of *S. acidocaldarius* on the combined mini-columns (sequence C$_1$ + C$_{18}$ + C$_{18}$) as shown in Figure 1 led us to the following conclusions:

1. The injected protein binds onto the first column material as a relatively broad band (dependent on the amount of injected protein, sample volume injected and the binding capacity of the material in the first column).
2. Binding forces between the proteins and the first stationary phase are counteracted by the protein-eluent interaction at increasing organic modifier concentrations during gradient elution. Consequently, only selected portions of the individual proteins are eluted from the first column at each point of the gradient (the proteins exhibiting in binding or elution mode, a quasi on-off mechanism, but cf. Reference 26).
3. Proteins in the elution mode migrate along the column (with only weak interactions with the stationary phase), with peak broadening dependent on the column length.
4. Ideally, those proteins which are eluted from the first column will bind again to the top of the second column. Because of the reduced quantity of protein bound to the second column, this results in sharper bands.
5. Again, proteins bound to the second column will be eluted fractionally with increasing concentrations of Eluent B during gradient elution, and will bind again to the top of the third column.
6. At each point of the gradient, the greater part of the non-eluted proteins remains bound to the first column.
7. At any given time, only differential fractions of the total amount of injected protein are passing through the last column, but nevertheless the entire amount will have passed through this column by the time the gradient is finished.
8. At each step, where eluted proteins are rebound to the following column, band broadening effects due to the length of the previous column are eliminated, so that only the length of the final column has any influence on this type of band broadening.

Some properties of Disc RPC have been described several times (e.g., References 29 and 30, reviewed in more detail in References 11 and 18). Combinations of different SEC columns (quasi Disc SEC[31]), as well as the combination of a cation- and anion-exchange column in series[32] (referred to as tandem columns) have been reported, and show that discontinuous HPLC is not restricted to RPC. It should be noted that Disc chromatography not only represents a combination of different stationary phases, but also has the effect that groups of solutes, eluted from the first column can bind to or be retained by the following column, as shown in Figure 1 (an effect which is not possible, in principle, with SEC). These groups are ideally separated on the following column with a different selectivity, leading to the formation of less complex sub-groups. Otherwise, increasing the column length by a factor of 2 would lead to a broadening of the peak width by a factor of $\sqrt{2}$.

Rubinstein[33] already emphasized in 1979 that analytical RP columns could be used for preparative applications with gradient elution, and that even large injection volumes (up to 100 ml, applied with a pump) caused no loss of resolution. He concluded that the loss of resolution with increased protein amounts is due to inefficient binding and not to overloading. Our initial studies with single column RPC support this conclusion. These studies showed that the quality of the separation was dependent on the solvent in which the TP30 was applied to the column,[14] and the best results were obtained when the proteins were dissolved in 15% formic acid.[9] In addition to this effect, we think that Disc RPC overcomes the effect of inefficient binding, because the last column is not greatly influenced by the binding conditions on the first column.

B. DETERMINATION OF THE COUPLING SEQUENCE IN DISC RPC

With untried samples and untried columns, it is necessary to determine the optimal coupling sequence of the columns. The goal should be to find those columns from which the sample is eluted at the lowest and highest organic modifier compositions, respectively. For Disc RPC, it is better to use different columns from different manufacturers, because the silica gel (its carbon content, porosity, degree of free silanol groups, metal impurities, etc.), and not only the bonded phase, has a significant influence on the retention times and on the quality of the separation. We have tested SynChropak C_1, C_4, and C_8 stationary phases for use in Disc RPC,[11] but the proteins were eluted with only slight or no differences in the retention times. This is in contrast to C_1 and C_{18} stationary phases from Toyo Soda (TSK TMS-250 and TSK ODS 120T). Proteins are eluted from these columns with the largest difference in organic modifier concentration we have observed to date (considering the different column dimensions, but note that the support silica gels, carbon content, etc., are different!).

The simplest way to determine the column sequence is shown in Figure 2. The unknown sample is eluted from each individual column with a gradient. Next, either the Eluent B concentration is determined at which the last proteins are eluted from the columns, or the Eluent B concentration is determined at which the same protein is eluted from the different columns. The columns are then coupled in the sequence of increasing B concentration values (in this case: 1. C_1 TSK TMS-250, 2. C_{18} Vydac 201 TP, 3. C_{18} TSK ODS 120T). It is also possible, but more time-consuming, to elute one protein isocratically from the different columns with various % B concentrations. Barford et al.[34] eluted bovine serum albumin (BSA) isocratically from different manufactured C_1, C_8 or C_{18} single columns with 2-propanol as Eluent B. With a retention volume of 6 ml (i.e., always with the same capacity factor [k'] value), BSA was eluted from the C_1 column at 27.5% B, from the C_8 column at 30% B and from the C_{18} column at 37.5% B. The coupling sequence for Disc RPC would be: 1. C_1, 2. C_8, 3. C_{18}.

IV. APPLICATION OF DISC RPC TO RIBOSOMAL 30S TOTAL PROTEIN MIXTURES OF *SULFOLOBUS ACIDOCALDARIUS*

We tested Disc RPC in several combinations of 14 different stationary phases, 7 different column lengths, 6 different diameters, and even coupled columns with different diameters, summarized in five different categories of Disc RPC.[11] This technique has so far been applied to mixtures of DNA binding proteins, tryptic peptides, membrane proteins (cf. Reference 14)

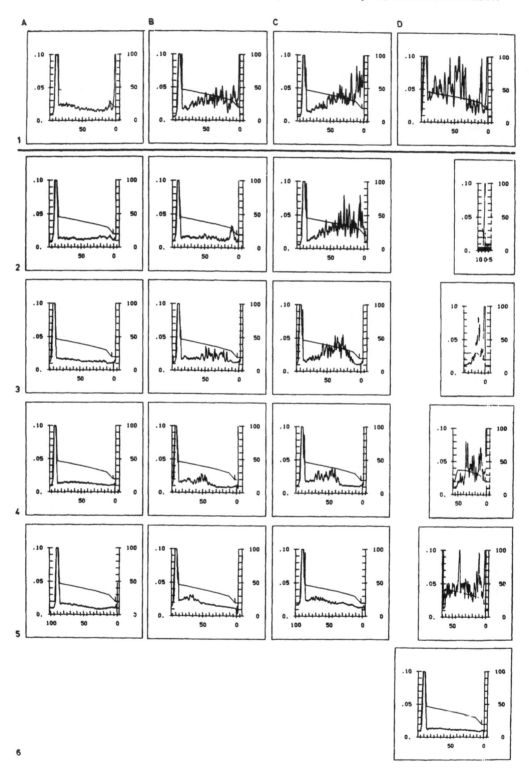

FIGURE 1

FIGURE 1 (on page 414). Investigation of protein elution and rebinding in Disc RPC with coupled mini-columns. The figure is subdivided into 4 rows (A to D) and six lines (1 to 6). Columns: Synchropak 1R-103 C_1, Vydac 201 TP C_{18}, TSK ODS 120T C_{18}, all 5 × 4 mm I.D. HPLC instruments: a 420 HPLC controller, two 114 M pumps, a 340 organizer (with injection valve and dynamic mixing chamber), all from Beckman, and an Uvicord SII fixed wavelength UV detector (Pharmacia-LKB). HPLC conditions: AB gradient, 20 to 47% B in 190 min, in 3 segments; whole gradient used in chromatograms of row A, B, and C and line 1 and 6, parts thereof used in chromatograms D/2 to D/5. D/2: isocratic elution with 20% B; D/3: 20 to 30% B and additional isocratic elution at 30% B, D/4: 20 to 37% B and additional isocratic elution at 37% B; D/5: 20 to 42% and subsequent reversed gradient to 20% B, as in D/3 and D/4, too. Absorbance, 226 nm; detector range 0.1 (with 2.5 mm cell). Injected only in chromatograms of line 1 and of row D (not in D/6) were 2.85 mg (4.6 A_{230}) of TP30, dissolved in 500 μl 15% formic acid. Eluent A is 0.1% aq. TFA (pH 2.0) and Eluent B is 0.1% TFA in acetonitrile/2-propanol (9:1, v/v); flow-rate, 0.5 ml/min; recorder, 5 ml (or 10 min)/cm. Row A shows in each case chromatograms obtained with the 5 × 4 mm TSK ODS 120T single column, row B those obtained with the Vydac 201 TP single column, and row C those with the Synchropak 1R-103 C_1 single column. Row D shows in each case chromatograms with mini-column Disc RPC in the combination sequence 1. Synchropak C_1, 2. Vydac C_{18}, 3. TSK C_{18}. Line 1 shows chromatograms with an injection of TP30 onto each single column (A-C/1) or onto the combination (D/1), respectively. Lines 2 to 5 show chromatograms where only one injection onto the combination was carried out per line (in the chromatograms of row D), with subsequent partial gradient elution of the proteins, as given under HPLC conditions. After each partial elution had been stopped, the combination was disassembled and the remaining proteins were eluted from the three single columns with always the same gradient. A subsequent increase of Eluent B to 100% and re-equilibration completed the gradient, which is indicated in the chromatograms. Protein recovery is inversely proportional to the retention time, and therefore the remaining proteins were always eluted in the following order from the disassembled columns: (a) TSK C_{18}, (b) Vydac C_{18}, and last (c) Synchropak C_1. Line 6 shows a chromatogram of the combination without TP30 injection (D/6). It is obvious from the figure, that nearly all of the injected proteins bind to the first column (this corresponds to a load capacity of *ca.* 39 mg lyophilized TP30/g stationary C_1 phase) and that proteins are rebound after elution from the first column. The results of this figure led us to the conclusions described in the text. These results were first presented at the 2nd Würzburger Chromatographie Gespräche, 1987.

and a test mixture of seven aromatic compounds (eluted isocratically).[14] Ribosomal proteins from *E. coli* ,[14] *Bacillus stearothermophilus*,[13] and *S. acidocaldarius*[11,15,18] have also been separated with Disc RPC. Here, we summarize our results with Disc RPC separations of ribosomal TP30 from *S. acidocaldarius* under various conditions.

Reversed-phase single column HPLC with the Vydac 214 TP 54 column (from Machery-Nagel) is shown in Figure 3. This C_4 column works best in combination with a C_4 Nucleosil (from Macherey-Nagel) pre-column for re-chromatography of pre-fractionated groups. In Figure 3, 5.25 A_{230} of TP30 dissolved in 5% acetic acid were injected and eluted with a multi-segment gradient of 2-propanol from 10 to 55% over a period of 280 min (gradient as indicated in Figure 3). The TP preparations separated in Figures 2 to 6 yielded *ca.* 40 proteins.

The combination of a C_1 (TSK TMS-250) and a C_{18} (TSK ODS 120T) column, run with 0.1% TFA in acetonitrile as Eluent B is illustrated in Figure 4. This separation of 36.4 A_{230} TP30 in 3.95 ml 15% formic acid was completed within 265 min (20 to 55% B).

The three-column combination (C_1, C_{18}, C_{18}) was tested with acetonitrile, 2-propanol and a mixture of both (9:1). Figure 5 shows the chromatogram obtained with 2-propanol. Because of the high viscosity of 2-propanol, the flow-rate was reduced to 0.25 ml/min, which increased the gradient time to 718 min (30 to 55% B). Disc RPC with 2-propanol and this particular three-column combination did not show better resolution than the two-column combination of Vydac 201 TP (C_{18}) and TSK ODS 120T (C_{18}) (data not shown).

Finally, the same three-column combination (TSK TMS-250, Vydac 201 TP and TSK ODS 120T) was run with a mixture of acetonitrile: 2-propanol (9:1), as shown in Figure 6. 72 A_{230} of TP30 in up to 8 ml of 15% formic acid could be injected onto this analytical column combination. Under these conditions, it was possible to separate the TP30 into approximately 20 peaks, most of them baseline separated.

The recovery of TP30 was determined by weight to be 78.1%: 20.17 mg of lyophilized TP30 were injected in 3.7 ml of 15% formic acid onto a combination of columns with different lengths and diameters (16 × 30 + 8 × 60 + 4.6 × 250 mm) with a total volume of 13.2 ml[11] of stationary

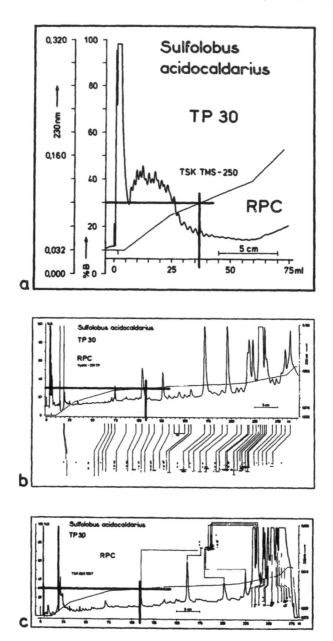

FIGURE 2. Determination of the coupling sequence in Disc RPC of ribosomal TP30 proteins from *S. acidocaldarius*. Columns: (a) TSK TMS-250 C_1, 75 × 4.6 mm I.D., 10-μm particle size (LKB, Bromma, Sweden); (b) Vydac 201 TP C_{18}, 250 × 4.0 mm I.D., 10 μm, (packing material obtained from Macherey, Nagel and Co., Düren, F.R.G.); (c) TSK ODS 120T C_{18}, 250 × 4.6 mm I.D., 5 μm (LKB). HPLC instrument: Pharmacia-LKB system with a 2152 HPLC controller, two 2150 pumps and a 2151 variable-wavelength monitor. HPLC conditions: (a) AB gradient (10 to 55% B in 145 min, in two segments, as indicated on the chromatogram) with a detector range of 0.32 AU and 18 A_{230} units of TP30 injected; (b) AB gradient (0 to 55% B in 526 min, in different segments, as indicated on the chromatogram) with a range of 0.16 AU and 5 A_{230} units injected; (c) same gradient as (b) with a range of 0.08 AU and 2.5 A_{230} units injected; Eluent A is 0.1% aq. trifluoroacetic acid (TFA) (pH 2.0) and Eluent B is 0.1% TFA in 2-propanol; flow-rate, 0.5 ml/min; absorbance, 230 nm; recorder, 5 ml (or 10 min)/cm. Gradients are indicated on each chromatogram. The 30% B value is marked on each chromatogram. This composition of Eluent B eluted nearly all proteins from the TSK C_1 column (a), the first larger peak from the Vydac C_{18} column (b) and only the first contaminant proteins from the TSK C_{18} column (c). The coupling sequence was therefore determined to be: (1) TSK C_1, (2) Vydac C_{18}, and (3) TSK C_{18}. All three chromatograms are reproduced, with slight changes, from Reference 18.

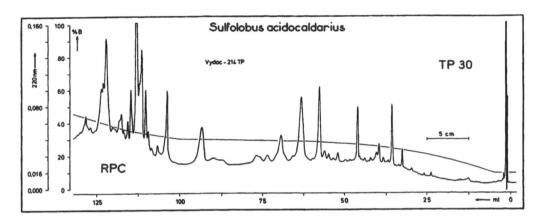

FIGURE 3. Single column RPC of ribosomal TP30 proteins. Column: Vydac 214 TP54 C$_4$, 250 × 4.6 mm I.D., 5-μm particle size (column obtained from Machery, Nagel and Co.). HPLC instrument: dual pump HPLC system from Beckman as in Figure 1, detector as in Figure 2; HPLC conditions: AB gradient (10 to 55% B in 280 min, multi-segment gradient as indicated on the chromatogram), where Eluent A is 0.1% aq. TFA (pH 2.0) and Eluent B is 0.1% TFA in 2-propanol; flow-rate, 0.5 ml/min; temperature, 35°C; absorbance, 220 nm; detector range 0.16 AU; recorder, 2.5 ml (or 5 min)/cm; 5.25 A$_{230}$ units of TP30 injected. This end-capped C$_4$ column was the best column (besides the Vydac C$_{18}$, 218 TP 54) for RP re-chromatography of pre-fractionated total protein mixtures under the conditions we have used.

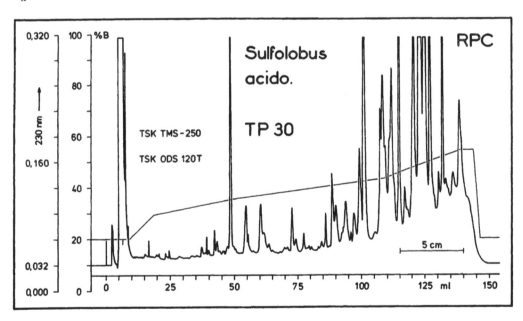

FIGURE 4. Two-column Disc RPC of ribosomal TP30 proteins. Columns: TSK TMS-250 C$_1$ and TSK ODS 120T C$_{18}$; other details as shown in Figure 2. HPLC instrument: as in Figure 2. HPLC conditions: AB gradient (20 to 55% B in 265 min, multi-segment gradient as indicated on the chromatogram), where Eluent A is 0.1% aq. TFA (pH 2.0) and Eluent B is 0.1% TFA in acetonitrile; flow-rate, 0.5 ml/min; temperature, 20°C; absorbance, 230 nm; detector range, 0.32 AU; recorder, 5 ml (or 10 min)/cm; 36 A$_{230}$ units of TP30 injected. With acetonitrile, this two-column combination gave improved separations relative to single column RPC with the TSK ODS 120T column. (From Grün, J. R. and Reinhardt, R., *J. Chromatogr.*, 397, 327, 1987. With permission.)

phase (*ca.* 13.9 g). After elution with a 120 ml gradient (eluent B was acetonitrile), 15.75 mg protein were recovered[11] (following lyophilization to constant weight).

To scale up the results obtained with analytical columns, we combined materials similar to

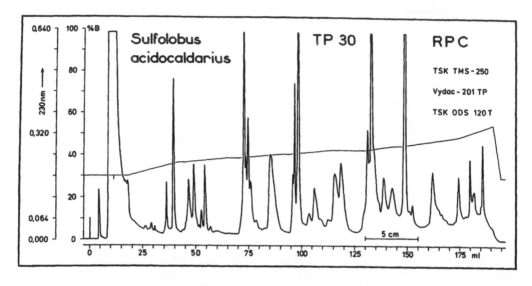

FIGURE 5. Three-column Disc RPC of ribosomal TP30 proteins. Columns: same as in Figure 2; column sequence, TSK C_1 + Vydac C_{18} + TSK C_{18}. HPLC instrument: same as in Figure 2. HPLC conditions: AB gradient (30 to 55% B in 718 min, multi-segment gradient as indicated on the chromatogram), where Eluent A is 0.1% aq. TFA (pH 2.0) and Eluent B is 0.1% TFA in 2-propanol; flow-rate, 0.25 ml/min; absorbance, 230 nm; detector range, 0.64; recorder, 5 ml (or 20 min)/cm; 36.4 A_{230} units of TP30 injected. This chromatogram demonstrates that 2-propanol is not a favorable eluent for this column combination. Because of the low flow-rate necessary with 2-propanol, this separation was only completed after more than 718 min. Nevertheless, many groups of peaks are not baseline separated.

those shown in Figure 6 (although using SynChropak C_1 material instead of the TSK C_1 material), packed in columns with inner diameter of 8 mm.[11,15,18] This semi-preparative column combination was loaded with 57 mg TP30 in 10 ml 15% formic acid[11,18] or with 109 mg TP50 in 58 ml 15% formic acid.[15] TP30 separations were repeated 20 times[11] and TP50 separations 10 times[35] to give sufficient pre-fractionated material for purification of all the proteins sequenced to date.

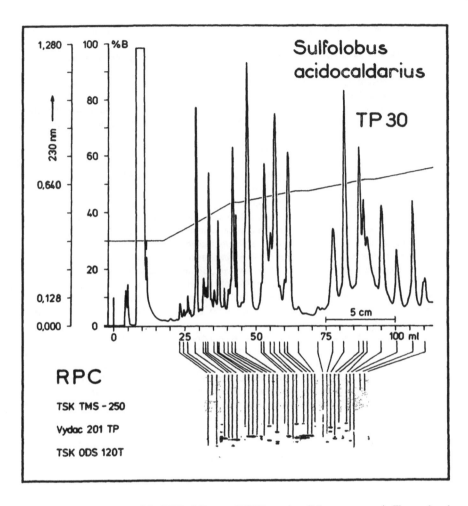

FIGURE 6. Three-column Disc RPC of ribosomal TP30 proteins. Columns: same as in Figures 2 and 5; same column sequence as in Figure 5. HPLC instrument: same as in Figure 2. HPLC conditions: AB gradient (30 to 56% B in 212 min, + 26 min delay time from mixing chamber to detector cell;[11] multi-segment gradient as indicated on the chromatogram), where Eluent A is 0.1% aq. TFA (pH 2.0) and Eluent B is 0.1% TFA in acetonitrile/2-propanol (9:1, v/v); absorbance, 230 nm; detector range, 1.28; 72 A$_{230}$ units of TP30 injected. These were the best separation conditions we could obtain for TP30 of *Sulfolobus*. About 10 proteins were eluted sufficiently pure for sequencing. Aliquots of the peaks (dependent on the peak heights), which were collected in peak mode by the SuperRac fraction collector, were evaporated in a vacuum centrifuge, redissolved in 25 μl sample buffer[18] and applied to an SDS gel. There are ten lanes of TP30 as marker between the separated fractions. (From Grün, J. R. and Reinhardt, R., *J. Chromatogr.*, 397, 327, 1987. With permission.)

REFERENCES

1. **Karch, K., Sebestian, I., and Halász I.,** Preparation and properties of reversed phases, *J. Chromatogr.*, 122, 3, 1976.
2. **Pearson, J.D., Lin, N.T., and Regnier, F.E.,** The importance of silica type for reverse-phase protein separations, *Anal. Biochem.*, 124, 217, 1982.
3. **Pohl, T. and Kamp, R.M.,** Desalting and concentration of proteins in dilute solution using high-performance liquid chromatography, *Anal. Biochem.*, 160, 388, 1987.

4. Unger, K.K., Phase systems and operation — HPLC of biopolymers (German), *GIT Supplement Chromatographie*, 3, 14, 1987.
5. Grote, M., Dijk, J., and Reinhardt, R., Ribosomal and DNA binding proteins of the thermoacidophilic archaebacterium *Sulfolobus acidocaldarius, Biochim. Biophys. Acta,* 873, 405, 1986.
6. Dijk, J. and Littlechild, J.A., Purification of ribosomal proteins from *Escherichia coli* under nondenaturing conditions, *Methods Enzymol.,* 59, 481, 1979.
7. Kossmann, B., Richter, R., Grün, J.R., and Reinhardt, R., Phenylisothiocyanate amino acid determination — a strategy for routine analyses, in *UltroChrom 88, 3. Würtburger Chromatographie-Gespräche, Symposium,* Vol. 1—3, Pharmacia LKB GmbH, Freiburg, 1989, 67.
8. Grün, J.R., Kamp, R.M., and Reinhardt, R., Separation of 30S ribosomal proteins from *Sulfolobus acidocaldarius* by ion-exchange and gel-permeation chromatography, *Chromatographia,* 22, 432, 1986.
9. Grün, J.R., Trennung und Identifizierung ribosomaler 30S Proteine von *Sulfolobus solfataricus* mit Einsatz der HPLC-Technik, *Thesis,* Freie Universität Berlin, 1986.
10. Wittmann, H.G., Components of bacterial ribosomes, *Annu. Rev. Biochem.,* 51, 155, 1982.
11. Grün, J.R., Kossmann, B., and Reinhardt, R., Discontinuous reversed-phase high-performance liquid chromatography increases load capacity of analytical columns. Separation of ribosomal proteins from the archaebacterium *Sulfolobus acidocaldarius, Chromatographia,* 25, 189, 1988.
12. Dijk, J. and Reinhardt, R., The structure of DNA-binding proteins from eu- and archaebacteria, in *Bacterial Chromatin,* Gualerzi, C.O. and Pon, C.L., Eds, Springer, Berlin, Heidelberg, 1986, 185.
13. Grün, J.R., Beck, A., and Reinhardt, R., The Correlation Matrix Contour Plot — a new peak purity test in photodiode array detected HPLC — applied to semi-preparative separations of ribosomal proteins, in *Ultro-Chrom 88, 3. Würtburger Chromatographie-Gespräche, Symposium,* Vol. 1—3, Pharmacia LKB GmbH, Freiburg, 1989, 14.
14. Grün, J.R., Isolierung ribosomaler 30S-Proteine aus dem Archaebakterium Sulfolobus acidocaldarius mit her Dislcontinnierlichen Reversed-Phase Hochleistungsflüssigkeits chromatographie sowie Entwicklung einer Peakreinheitskontrolle für Proteine, Ph.D. Thesis, Freie Universität, Berlin, 1990.
15. Kossmann, B., Grün, J.R., and Reinhardt, R., Separation of 50S ribosomal proteins from *Sulfolobus acidocaldarius* by discontinuous reversed-phase chromatography, *Chromatographia,* 25, 215, 1988.
16. Grün, J.R., Kossmann, B., and Reinhardt, R., Discontinuous reversed-phase chromatography, a separation method for highly complex protein mixtures: HPLC purification of ribosomal proteins for sequence analysis, *J. Protein Chem.,* 7, 230, 1988.
17. van den Eijnden-van Raaij, A.J.M., Koornneef, I., van Oostwaard, Th. M.J., de Laat, S.W., and van Zoelen, E.J.J., Cation-exchange high-performance liquid chromatography: separation of highly basic proteins using volatile acidic solvents, *Anal. Biochem.,* 163, 263, 1987.
18. Grün, J.R. and Reinhardt, R., Purification of ribosomal 30S proteins from the archaebacterium *Sulfolobus acidocaldarius* by ion-exchange and discontinuous reversed-phase high-performance liquid chromatography, *J. Chromatogr.,* 397, 327, 1987.
19. Hearn, M.T.W. and Grego, B., HPLC of amino acids, peptides and proteins. XL. Further studies on the role of the organic modifier in the reversed-phase high-performance liquid chromatography of polypeptides. Implications for gradient optimization, *J. Chromatogr.,* 255, 125, 1983.
20. Cohen, K.A., Schellenberg, K., Benedek, K., Karger, B.L., Grego, B., and Hearn, M.T.W., Mobile-phase and temperature effects in the reversed-phase chromatographic separation of proteins, *Anal. Biochem.,* 140, 223, 1984.
21. Cooke, N.H.C., Archer, B.G., O'Hare, M.J., Nice, E.C., and Capp, M., Effects of chain length and carbon load on the performance of alkyl-bonded silicas for protein separations, *J. Chromatogr.,* 255, 115, 1983.
22. Leach, D.C., Stadalius, M.A., Berus, J.S., and Snyder, L.R., Reversed-phase HPLC of basic samples, *LC·GC Int.,* 1(5), 23, 1988.
23. Stadalius, M.A., Quarry, M.A., and Snyder, L.R., Optimization model for the gradient elution separation of peptide mixtures by reversed-phase high performance liquid chromatography. Application to method development and the choice of column configuration, *J. Chromatogr.,* 327, 93, 1985.
24. Grün, J.R., Kossmann, B., Kruft, V., Choli, T., van den Broek, R., Reinhardt, R., and Wittmann-Liebold, B., to be published.
25. Cohen, S.A., Benedek, K., Tapuhi, Y., Ford, J.C., and Karger, B.L., Conformational effects in the reversed-phase liquid chromatography of ribonuclease A, *Anal. Biochem.,* 144, 275, 1985.
26. Meek, J.L. and Rossetti, Z.L., Factors affecting retention and resolution of peptides in high-performance liquid chromatography, *J. Chromatogr.,* 211, 15, 1981.
27. Grün, J.R., Richter, J., and Reinhardt, R., Purification of ribosomal components with gradient elution on discontinuous stationary phases. Application in LC and HPLC mode, presented at the *3rd Würzburger Chromatographie Gespräche,* 1988.
28. DiBussolo, J.M. and Gant, J.R., Vizualization of protein retention and migration in reversed-phase liquid chromatography, *J. Chromatogr.,* 327, 67, 1985.

29. **Wulfson, A.N. and Yakimov, S.A.,** HPLC of nucleotides. II. General methods and their development for analysis and preparative separation. An approach to selectivity control, *J. High Resolut. Chromatogr. Chromatogr. Commun.,* 7, 442, 1984.

30. **Jurenitsch, J., Kopp, B., Bamberg-Kubelka, E., Kern, R., and Kubelka, W.,** High-performance liquid chromatographic separation of the cardiac glycosides of *Convallaria majalis L.* by coupling "reversed-phase" columns of different polarity (German), *J. Chromatogr.,* 240, 125, 1982.

31. **Meyer, V.R.,** *Practical High-Performance Liquid Chromatography,* John Wiley & Sons, Chichester, New York, 1988, 189.

32. **El Rassi, Z. and Horváth, Cs.,** Tandem columns and mixed-bed columns in high-performance liquid chromatography of proteins, *J. Chromatogr.,* 359, 255, 1986.

33. **Rubinstein, M.,** Preparative high-performance liquid partition chromatography of proteins, *Anal. Biochem.,* 98, 1, 1979.

34. **Barford, R.A., Sliwinski, B.J., Breyer, A.C., and Rothbart, H.L.,** Mechanism of protein retention in reversed-phase high-performance liquid chromatography, *J. Chromatogr.,* 235, 281, 1982.

35. **Kossmann, B.,** personal communication, 1987.

Section VI
Hydrophobic Interaction Chromatography

HYDROPHOBIC INTERACTION CHROMATOGRAPHY OF PROTEINS

Richard H. Ingraham

I. INTRODUCTION

Both hydrophobic interaction chromatography (HIC) and reversed-phase chromatography (RPC) rely on hydrophobic interactions with the stationary phase to effect separation. The special utility of HIC lies in its ability to separate proteins on the basis of their surface accessible hydrophobicity without, in general, causing denaturation and loss of enzymatic or other activity.[1-5] HIC can be used to accomplish this because it differs from RPC in three areas: (1) mobile phase solvents utilized; (2) ligand density on the column support; and (3) hydrophobicity of the column matrix.

II. CHARACTERISTICS OF HIC

A. MOBILE PHASE

In HIC, high salt concentrations are used to promote binding of proteins to matrices which are relatively weak hydrophobes. Proteins adsorb to the stationary phase because of hydrophobic interactions between non-polar patches on the polypeptide surface and the hydrophobic ligands of the stationary phase. Solute species are eluted in order of increasing hydrophobicity either isocratically or, more generally, by a descending salt gradient which weakens hydrophobic interactions and causes the solute to be released from the column. These solvent conditions generally stabilize protein tertiary and quaternary structure;[6] in contrast, the acidic conditions and organic solvents used for sample elution in RPC tend to promote denaturation of proteins.

B. FUNCTIONAL LIGAND

A variety of weakly hydrophobic ligands have been shown to be useful for HIC and are now commercially available. In the SynChrom series of HIC columns, alkyl and aryl groups are incorporated into a silica-bound hydrophilic polymeric matrix. Figure 1, adapted from Gooding et al.[7] illustrates the effects of different hydrophobic groups on the retention times of five proteins. These results, in conjunction with earlier findings,[8] demonstrate that, for the ligands in this series of packings, protein retention generally increases in the order: hydroxypropyl < methyl < benzyl = propyl < isopropyl < phenyl < pentyl. Similarly, Alpert, et al.[9] prepared a series of silica-bonded poly(alkylaspartamide) columns and found that hemoglobin binding capacity increased in the order: methyl (PolyMETHYL A column) < ethyl (PolyETHYL A column) < propyl (PolyPROPYL A column). A number of other approaches to HIC packings have also been successful. Examples include: silica-based ether-bonded alkyl phases[10] such as the Spherogel CAA-HIC methyl-polyether column available from Beckman; neopentyl-agarose (Alkyl-Superose HR 5/5), and phenyl-agarose (Phenyl-Superose HR 5/5) FPLC columns from Pharmacia; a silica-based polymer-bound polyethylene glycol column (Hydropore HIC) from Rainin; and the silica-based diol phase column (LC-HINT) from Supelco. Additional examples of HIC columns or mixed-mode columns which can be used in HIC mode can be found in the catalogs of most column suppliers. The ligands used for HIC are inherently

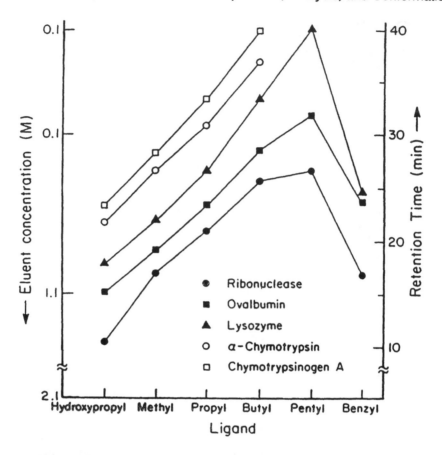

FIGURE 1.   Effect of ligand arm on protein retention in HIC. Columns: SynChropak Hydroxypropyl, Propyl, Methyl, Butyl, Pentyl, and Benzyl HIC columns, 250 × 4.1 mm I.D., 6.5-μm particle size, 300-Å pore size (SynChrom, Linden, IN). Instrumentation: the HPLC instrument consisted of a Varian Model 5000 liquid chromatograph (Varian, Walnut Creek, CA) coupled to a Chem Research Model 2020 UV detector (ISCO, Lincoln, NE) with a Linear Model 1200 recorder (Linear Instruments, Irvine, CA); samples were injected with a Model 7125 injector (Rheodyne, Cotati, CA). Mobile phase: 0.1 *M* potassium phosphate (pH 6.8); 30-min gradient from 2 to 0 *M* ammonium sulfate; flow-rate, 1 ml/min. (Adapted from Gooding, D. L., Schmuck, M. N., Nowlan, M. P., and Gooding, K. M., *J. Chromatogr.*, 359, 331, 1986. With permission.)

less hydrophobic than the C-8 and C-18 alkyl chains routinely used for RPC. Consequently, they are both less denaturing and allow for elution with aqueous rather than organic solvents.

In HIC, the choice of bonded-phase ligand has a profound effect on both protein retention time and resolution.[1,8,9,11] For example, in one study,[1] the retention time of β-glucuronidase increased from 11.3 to 24.2 min when the alkyl chain length of the ligand was increased from one to five. Conalbumin, on the other hand, was eluted in 5.3 min from the first column, but was not eluted from the second. Such results stand in sharp contrast with RPC, where the effects of chain length variation on protein retention are relatively insignificant.[12]

Retention time and selectivity are so ligand-dependent in HIC because the ligands generally interact with hydrophobic amino acid residues which are accessible on the surface of the folded native molecule. In contrast, with RPC, bonding interactions occur throughout the length of the denatured polypeptide chain, including those normally buried in the hydrophobic protein core. When such groups are located primarily within hydrophobic pockets on the protein surface, HIC columns with longer ligands will form more binding interactions than those with ligands which cannot contact these surfaces.

C. LIGAND DENSITY

The primary effect of increasing ligand density in HIC is enhanced protein retention. Fausnaugh et al.[2] prepared a series of columns derivatized with various mixtures of phenyl and acetyl ligands and examined the retention of twelve proteins as a function of increasing phenyl ligand density. Based on the results, the proteins were grouped into three categories. The smallest proteins exhibited decreases in retention time with decreasing phenyl ligand density until the latter was reduced twofold. At this point, retention was minimal and no further decreases were observed. The leveling-off of retention was attributed to the relatively small number of phenyl groups available to interact with the protein within a given area. Somewhat larger proteins exhibited a continuous, though non-linear, decrease in retention with phenyl ligand density. Presumably, this group of proteins have a surface area which is large enough to ensure a significant number of binding contacts even on the low-density columns. The highest molecular weight proteins were not eluted from the high phenyl density columns under HIC gradient conditions, but were eluted from the lower density columns.

Choosing which column to use for purifying a given protein can be difficult, since the selectivities involved in resolving the target protein from contaminants will vary. Consequently, it is useful to examine the chromatographic behavior of a sample on at least two columns of differing hydrophobicities before finalizing a purification protocol. In choosing which column(s) to try, an important guideline is to use more hydrophobic sorbents for proteins having little surface hydrophobicity and sorbents with either shorter alkyl chains or lower ligand density for more hydrophobic or labile samples.[11] The former guideline should be followed to ensure retention of a protein during loading so that it may be gradient-eluted, while the latter should minimize irreversible binding and denaturation.

III. OPERATIONAL PARAMETERS IN HIC

A. EFFECTS OF TYPE AND INITIAL CONCENTRATION OF SALT IN THE MOBILE PHASE

The selectivity and retention of a HIC column can be modified by adjusting mobile-phase variables such as salt type, salt concentration, pH, temperature, and mobile-phase modifiers.[2,8,13,14]

Protein retention in HIC is highly dependent upon the chemical nature of the salt used to promote binding as has been shown by a number of investigators.[2,8,13-15] Salts useful in this regard are high in the lyotropic series of ions[14,15] and are termed antichaotropic ("structure-making"). They are also known for their ability to salt out proteins. The ability of such salts to enhance hydrophobic interactions can be understood in terms of their effects on the free-energy, ($\Delta G°$) of the system, consisting of the solute, the solvent, and the stationary phase. This system contains two sets of hydrophobic groups, those of the protein and those incorporated into the stationary phase. An important property of water is that, in the vicinity of non-polar groups, it becomes significantly more ordered than in the bulk solvent. The decrease in solvent entropy which results increases the free-energy of the system as shown by Equation 1:

$$\Delta G° = \Delta H° - T\Delta S° \qquad (1)$$

When the hydrophobic ligands of the column bind to the protein, structured water is released from both and entropy ($\Delta S°$) increases, resulting in a lower $\Delta G°$ and a more stable system. Within this context, antichaotropic salts enhance hydrophobic interactions by making structured solvent more energetically unfavorable.

Horváth, et al.[14,16,17] have made use of these concepts in explaining and attempting to quantify the effects of different salts on protein retention within the framework of the solvophobic theory.

TABLE 1
Molal Surface Tension Increments of Selected Salts

Salt	$(\sigma \times 10^3$ dyn-g/cm-mol)
KCl	1.50
NaCl	1.64
Na_2HPO_4	2.03
Mg_2SO_4	2.06
$(NH_4)_2SO_4$	2.17
Na_2SO_4	2.74
Na_3PO_4	2.88
Sodium citrate	3.12

Adapted from Horváth, C., Melander, W., and Molnár, I., *J. Chromatogr.*, 125, 129, 1976.

According to this theory, the primary parameters responsible for retention are the surface tension of the mobile phase and the exposed surface area of the protein which is hydrophobic in nature. These parameters are important because the free energy involved in forming the ordered solvent cavity in which a molecule resides is a direct function of them. Consequently, an increase in surface tension is predicted to lead to increased sample retention and vice-versa. The surface tension, γ, of an aqueous solution is estimated as the product of the molal salt concentration, m, times the molal surface tension increment of the salt, σ, plus the surface tension of water, γ° (72 dyne/cm). Therefore, increased protein retention should be achievable by either increasing the salt concentration in the mobile phase or changing to a salt with a greater molal surface tension increment. Table 1 lists values for various salts which may be useful in HIC and is adapted from Horváth, et al.[16]

Melander et al.[14] compared the effects of four different salts on HIC of proteins with their predicted effects and found that the data corroborated the general predictions made by their theory. Similarly, Fausnaugh et al.[2] examined the effects of five different salts on the retention of cytochrome c, conalbumin, and β-glucuronidase. The salts used were: sodium citrate, Na_2SO_4, $(NH_4)_2SO_4$, NaCl, and the chaotropic salt NH_4SCN ($\sigma = 0.45 \times 10^3$ dyn-g/cm-mol). In accordance with expectations, proteins were most highly retained when the starting buffer contained 1.0 M sodium citrate, followed in order of effectiveness by Na_2SO_4 and $(NH_4)_2SO_4$. Virtually no retention occurred when NaCl or NH_4SCN were used (however, see Reference 8, where NaCl was effective). Significantly, optimal resolution did not necessarily correlate with greatest retention as cytochrome c and conalbumin were best resolved in the sodium citrate gradient, while conalbumin and β-glucuronidase were best resolved in $(NH_4)_2SO_4$. From this, it is concluded that more hydrophilic proteins are best resolved by salts high in the lyotropic series, e.g., Na_2SO_4, while relatively more hydrophobic proteins are better resolved when salts with lower molal surface tension increments, e.g., $(NH_4)_2SO_4$, are used. From such observations, it is apparent that surface tension plays a major role in HIC. However, it has also been shown that specific interactions can occur between proteins and various ions and lead to results which are incongruous with solvophobic theory.[14,18]

Several practical considerations have made $(NH_4)_2SO_4$, with a value of 2.160×10^3 dyn-g/cm-mol, the most commonly used salt for HIC. Among these are its high solubility (4 M at 25°C), lack of significant ultraviolet absorbance or specific-ion effects, relative intransigence to microbial growth, and commercial availability of high-purity or HPLC grade preparations. In contrast, the solubility of Na_2SO_4 is limited to 1.5 M at 25°C, while sodium citrate has high ultraviolet absorbance and, in a similar manner to phosphate salts, promotes microbial growth. Despite the apparent advantages of $(NH_4)_2SO_4$, the selectivity gains which may result from utilization of other salts indicates they should not be neglected in developing a purification procedure.

Initial salt concentrations used in HIC generally range from 1 to 3 M. In accordance with expectations, proteins are eluted earlier when lower initial salt concentrations are used. Fausnaugh et al.[1] reported that initial salt ionic strength can significantly effect selectivity, suggesting that, by altering initial loading conditions, protein resolution may be enhanced. Moreover, the effect of this parameter is greatest on the resolution of the more hydrophilic early eluted components of a mixture.[8,13] Because the effect on resolution is less on more hydrophobic proteins, a potential benefit of using a lower initial ionic strength is that such proteins will be eluted earlier, resulting in shorter analysis times.

B. EFFECT OF pH

Mobile phases used in HIC are generally in the neutral pH range, pH 5 to 7, buffered with sodium or potassium phosphate. Melander et al.[14] predicted that retention would be increased or decreased, respectively, as the pH is shifted towards or away from the isoelectric point of a protein. The effects of pH choice on retention are indeed protein dependent according to studies by Fausnaugh et al.[2,18] and Kato et al.[13] Variation of this parameter may therefore result in enhanced selectivity. However, there is no clear correlation between pH effects and isoelectric point. An alternative explanation is that the ionization states of amino acid residues located within the contact surface area of the protein are involved. Evidence for this explanation comes from a study of HIC of avian lysozyme variants which suggested that the presence of a histidine (His) residue in the contact area was responsible for lesser retention at pH 6 (His positively charged) than at pH 8 (His uncharged).[18]

C. EFFECT OF MOBILE PHASE MODIFIERS

An additional method of adjusting retention is the addition of modifiers to the mobile phase. The presence of varying concentrations of the non-ionic detergent CHAPS [3([3-cholamidopropyl]dimethylammonio)-1-propanesulfonate] was found to affect retention of lysozyme, ribonuclease, and trypsin inhibitor differentially, indicating selectivity as well as solubility may be enhanced by the presence of detergents.[19,20] Addition of 5 to 10% methanol or isopropanol or even 20% ethylene glycol to the mobile phase promotes desorption of proteins without necessarily causing denaturation and loss of activity.[13,15,21,22]

Similar effects are observed with the chaotropic agents urea and guanidine hydrochloride.[13,21] Because these modifiers can affect proteins differentially, their utilization may yield altered selectivities, as well as enhanced resolution, due to narrower peaks. However, these reagents are excellent protein denaturants at high concentrations (6 to 8 M) and must be used at lower concentrations (1 to 2 M) to prevent denaturation while promoting the desired selectivity effects.

D. EFFECT OF TEMPERATURE

Temperature plays an important role in HIC, since the strength of hydrophobic interactions increases with temperature, and a number of studies have demonstrated that protein retention times generally increase with temperature.[11,21,23,24] Figure 2 shows the effect of increasing temperature on the HIC elution profiles of cytochrome c (Figure 2A) and myoglobin (Figure 2B). In contrast to RPC, increasing the temperature at which HIC is performed generally enhances protein binding to the matrix without change in conformation.[25] It is clear from Figure 2B that elevated temperatures enhance the binding of myoglobin to the hydrophobic matrix; at 15°C, for example, the retention time of the peak is 9.3 min, whereas at 45°C it is 12.4 min. However, as the temperature increases above 40°C thermal unfolding of myoglobin is observed, as evidenced by the appearance of a very broad envelope between 12 and 24 min as the original myoglobin peak at 12 min disappears between 40 and 50°C. When the elution profile obtained at 400 nm (where heme absorbs) was compared to that observed at 210 nm, it was found to be

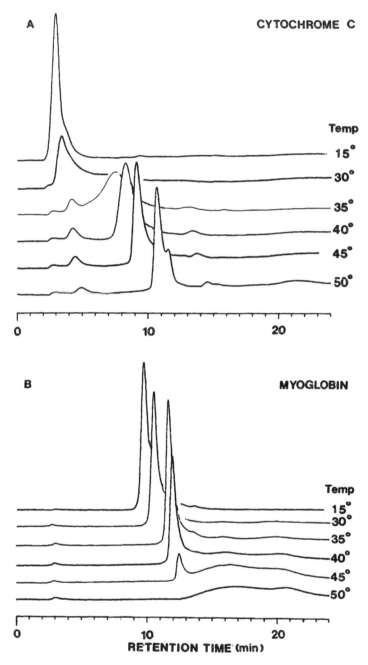

FIGURE 2. Effects of temperature on the HIC elution profile of cytochrome c and myoglobin. Column: Bio-Gel TSK-Phenyl-5-PW column, 75 × 7.5 mm I.D., 10-μm particle size, 1000-Å pore size (BioRad Lab., Richmond, CA). Instrumentation: the HPLC instrument consisted of a Spectra-Physics (San Jose, CA) SP8700 solvent delivery system and SP8750 organizer module in conjunction with a Hewlett-Packard (Avondale, PA) HP1040A detection system, HP3390A integrator, HP85 computer, HP9121 disc drive and HP7470A plotter. Conditions: linear AB gradient (100% A to 100% B in 15 min, followed by an additional 5 min at 100% B), where Buffer A was 0.1 M sodium phosphate (pH 7.0) containing 1.7 M ammonium sulfate and Buffer B was 0.1 M sodium phosphate (pH 7.0); flow-rate, 1 ml/min. A circulating Haake water bath with variable temperature control was used to adjust the temperature of the column, which was enclosed by a 500-ml jacket, constructed from a plastic graduated cylinder and rubber stoppers. A 50% aq. ethanol solution was used as the circulating solvent. The protein profiles were monitored at both 210 and 400 nm. The normalized absorbances are identical for the two wavelengths. (From Ingraham, R. H., Lau, S. Y. M., Taneja, A. K., and Hodges, R. S., *J. Chromatogr.*, 327, 77, 1985. With permission.)

identical, i.e., the heme group is not released from the polypeptide chain. Thus, unlike the situation where this protein is run on the same column under RPC conditions,[23] denaturation of myoglobin at these elevated temperatures is not total. Instead, it appears that myoglobin can exist in various partially unfolded states under these conditions.

From Figure 2A, cytochrome c at 15°C is eluted as a sharp peak following the column dead volume and, hence, is not retained by the column at this temperature. Increasing the temperature in increments of 5°C results in an increase in protein retention time together with peak broadening up to 35°C, and then a reversal of this trend with peak sharpening up to 50°C. The observed progression in peak shape, from sharp to broad to sharp, suggests that cytochrome c may undergo a localized conformational change between 15 and 40°C, which increases its hydrophobicity and, hence, increases protein retention time. Such an occurrence would be expected to contribute to the increase in retention time and also to produce peak broadening at intermediate temperatures, where the portion of the molecule undergoing conformational change would be in equilibrium between differing nearly isoenergetic states. Alternatively, peak broadening could be due to slow kinetic interchange between native and denatured forms of the protein. Ingraham et al.[23] observed that there was very little effect of temperature (15 to 45°C) on the HIC retention time of lysozyme, reflecting similar results by Goheen and Engelhorn.[21]

The above results suggested that the degree to which temperature affects retention time may vary significantly between proteins. Hence, utilizing temperature variation during HIC may be used to promote better resolution of protein peaks as well as to sharpen individual peaks. Lowering column temperature, particularly, often results in peak sharpening and resolution. Coupling of this factor with the decreased lability of proteins at lower temperatures suggests that operation at sub-ambient temperatures may often prove beneficial.[23,24]

E. EFFECT OF GRADIENT-RATE

Gradient rate effects on protein resolution have been examined using the TSK gel Phenyl-5PW column.[13] Linear gradients of $(NH_4)_2SO_4$ from 1.8 to 0 M in 0.1 M phosphate buffer were run using gradient times from 30 min (a typical value in analytical HIC) to 240 min. Resolution of six sample proteins increased with gradient time up to 120 min, indicating that longer runs are generally unwarranted. Interestingly, Hjertén and Liao[26] found that gradient times longer than 12 min could actually decrease resolution when an experimental agarose-based HIC column was used. At the opposite extreme, studies have demonstrated that extremely fast gradients can yield useful separations[26-28] as shown in Figure 3 from Reid and Gisch.[28] In panel A, a linear gradient from 3 M $(NH_4)_2SO_4$ in 0.5 M NH_4OAc to 0.5 M NH_4OAc is used to separate four sample proteins using a gradient time of 20 min. In panel B, shortening of the gradient time to 0.5 min, i.e., a 40-fold increase in gradient-rate, only reduced the retention time for α-chymotrypsin from 24 to 11 min. Since the 0.5 min gradient is essentially a step gradient, all four proteins were eluted isocratically after binding under high salt conditions. Significantly, Janzen et al.[27] have succeeded in resolving eight proteins in 3.0 min on an experimental pellicular HIC packing (1.5-μm particle diameter) utilizing a 2.5 min gradient. Clearly, HIC has great potential as an analytical tool with relatively high throughput.

F. PREPARATIVE HIC

HIC columns generally have reasonably high sample capacities of between 1 to 10 mg/ml of column packing. Consequently, HIC is very amenable to large sample loadings. Preparative-scale versions of analytical HIC columns are available from a number of manufacturers. For example, the Phenyl 5-PW preparative column (150 × 21.5 mm I.D.) gave excellent resolution of several enzymes at sample loads between 50 and 200 mg.[29] Similarly, SynchroPrep Propyl columns (250 × 10 mm I.D.) from SynChrom have a dynamic load capacity for ovalbumin of 170 mg.[7] Pharmacia states that a typical loading capacity for its Phenyl Superose HR 10/10 FPLC column (100 × 10 mm I.D.) is 80 mg. For scale-up, Gooding et al.[7] recommend a fourfold increase in gradient time over that used on a corresponding analytical column.

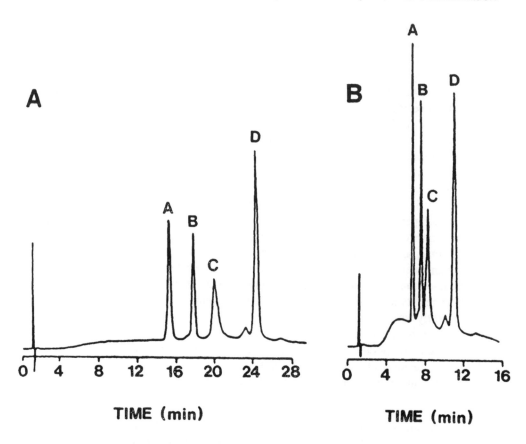

FIGURE 3. Effect of gradient steepness on HIC separation of protein standards. Column: Supelcosil LC-HINT, diol bonded phase, 100 × 4.6 mm I.D., 5-μm particle size (Supelco Inc., Bellefonte, PA). Instrumentation: the HPLC instrument consisted of a Spectra-Physics (San Jose, CA) SP8700 solvent delivery system, SP8500 dynamic mixer, SP4270 recorder-integrator and a Spectroflow 773 UV detector (Kratos, Ramsey, NJ); samples were injected with a Model 7125 injector (Rheodyne, Cotati, CA). Mobile phase: 3.0 M ammonium sulfate in 0.5 M ammonium acetate with pH adjusted to 6.0 with glacial acetic acid (Solvent A), 0.5 M ammonium acetate with pH adjusted to pH 6.0 with glacial acetic acid (Solvent B). Gradient: Panel A, 100% A to 0% A over 20 min, then 10-min hold, then to 100% A over 5 min; Panel B, 100% A to 0% A over 0.5 min, then 10-min hold, then to 100% A over 5 min. Flow-rate: 1 ml/min. Detection: 280 nm at 0.5 AUFS. Peaks A, B, C, and D denote cytochrome c, ribonuclease A, ovalbumin and α-chymotrypsinogen A, respectively. (From Reid, T. S. and Gisch, D. J., *LC.GC*, 5, 986, 1987. With permission.)

IV. HIC OF PEPTIDES

Generally, the method of choice for peptide purification is RPC due to excellent selectivity, volatility and ease of preparation of mobile phases, ease of operation, and the ability of peptides to retain biological activity after denaturation. Alpert[30] subjected a number of peptides to RPC and HIC and compared their chromatographic behavior. In most cases, selectivities as evaluated by elution order were similar or identical for the two methods, but with better efficiency exhibited with RPC. However, for some peptides, HIC may be a better choice. HIC performed much better than RPC for purification of an extremely hydrophobic peptide, a synthetic analog of a lipoprotein fragment. HIC may also give better resolution for small proteins, or peptides large enough to possess secondary or tertiary structure as has been found with synthetic calcitonin analogs[31] as well as with semi-purified snake venom.[30]

Another area where HIC can supplement RPC is purification of hydrophilic peptides. Alpert[30] reports that several mixtures of glycopeptides were resolved on a PolyPROPYL

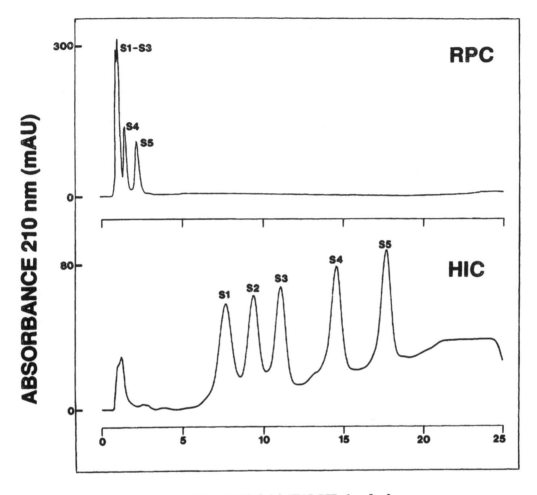

FIGURE 4. Elution profiles of peptide standards on a reversed-phase column run under reversed-phase (top) or HIC (bottom) conditions. Column: BioGel TSK Phenyl RP+, 75 × 4.6 mm I.D., 10-μm particle size, 1000-Å pore size (BioRad Lab., Richmond, CA). Instrumentation: the HPLC instrument consisted of a Varian Vista Series 5000 liquid chromatograph (Varian, Walnut Creek, CA) coupled to a Hewlett-Packard (Avondale, PA) HP1040A detection system, HP85B computer, HP9121 disc drive, HP2225A Thinkjet printer and HP7470A plotter; samples were injected with a Model 7125 injector (Rheodyne, Cotati). Conditions for RPC (top): linear AB gradient (1% B/min), where Eluent A is 0.1% aq. trifluoroacetic acid (TFA), pH 2.0, and Eluent B is 0.1% TFA in acetonitrile; flow-rate, 1 ml/min. Conditions for HIC (bottom): See Figure 2. S1-S5 are synthetic decapeptide reversed-phase standards obtained from Synthetic Peptides Incorporated, Department of Biochemistry, University of Alberta, Edmonton, Alberta, T6G 2H7, Canada.

Aspartamide HIC column which did not bind to various RPC columns. Similarly, Figure 4 shows the separations of five peptide standards obtained on a Bio-Gel TSK Phenyl-RP+ reversed-phase column run in both RPC, panel A, and HIC, panel B, modes. In RPC mode, the five peptides exhibit only slight retention in 0.1% TFA and are only partially resolved (Figure 4A). In contrast, in panel B (HIC mode), the peptides bind to the column and are well resolved upon gradient elution. Consequently, reversed-phase columns may be used in HIC mode to effect separation of peptides which are too weakly hydrophobic to be resolved under standard RPC conditions.

REFERENCES

1 Fausnaugh, J.L., Pfannkoch, E., Gupta, S., and Regnier, F.E., High-performance hydrophobic interaction chromatography of proteins, *Anal. Biochem.*, 137, 464, 1984.

2. Fausnaugh, J.L., Kennedy, L.A., and Regnier, F.E., Comparison of hydrophobic-interaction and reversed-phase chromatography of proteins, *J. Chromatogr.*, 317, 141, 1984.

3. Pavlu, B., Johansson, U., Nyhlén, C., and Wichman, A., Rapid purification of monoclonal antibodies by high-performance liquid chromatography, *J. Chromatogr.*, 359, 449, 1986.

4. Hyder, S.M. and Wittliff, J.L., High-performance hydrophobic interaction chromatography of a labile regulatory protein: the estrogen receptor, *BioChromatography*, 2, 121, 1987.

5. Withka, J., Moncuse, P., Baziotis, A., and Maskiewicz, Use of high-performance size-exclusion, ion-exchange, and hydrophobic interaction chromatography for the measurement of protein conformational change and stability, *J. Chromatogr.*, 398, 175, 1987.

6. Ahmad, F. and Bigelow, C.C., Thermodynamic stability of proteins in salt solutions: a comparison of the effectiveness of protein stabilizers, *J. Prot. Chem.*, 5, 355, 1986.

7. Gooding, D.L., Schmuck, M.N., Nowlan, M.P., and Gooding, K.M., Optimization of preparative hydrophobic interaction chromatographic purification methods, *J. Chromatogr.*, 359, 331, 1986.

8. Gooding, D.L., Schmuck, M.N., and Gooding, K.M., Analysis of proteins with new, mildly hydrophobic high-performance liquid chromatography packing materials, *J. Chromatogr.*, 296, 107, 1984.

9. Alpert, A.J., High-performance hydrophobic-interaction chromatography of proteins on a series of poly(alkyl aspartamide)-silicas, *J. Chromatogr.*, 359, 85, 1986.

10. Miller, N.T., Feibush, B., and Karger, B.L., Wide-pore silica-based ether-bonded phases for separation of proteins by high-performance hydrophobic-interaction and size-exclusion chromatography, *J. Chromatogr.*, 316, 519, 1985.

11. Ueda, T., Yasui, Y., and Ishida, Y., Study on optimal packing materials for high-performance hydrophobic interaction chromatography of proteins, *Chromatographia*, 24, 427, 1987.

12. Pearson, J.D. and Regnier, F.E., The influence of reversed-phase n-alkyl chain length on protein retention, resolution, and recovery: implications for preparative HPLC, *J. Liq. Chrom.*, 6, 497, 1983.

13. Kato, Y., Kitamura, T., and Hashimoto, T., Operational variables in high-performance hydrophobic interaction chromatography of proteins on TSK Gel Phenyl-5PW, *J. Chromatogr.*, 298, 407, 1984.

14. Melander, W.R., Corradini, D., and Horváth, Cs., Salt-mediated retention of proteins in hydrophobic-interaction chromatography: applications of solvophobic theory, *J. Chromatogr.*, 317, 67, 1984.

15. Strop, P., Hydrophobic chromatography of proteins on semi-rigid gels: effect of salts and interferents on the retention of proteins by Spheron P 300, *J. Chromatogr.*, 294, 213, 1984.

16. Horváth, C., Melander, W., and Molnár, I., Solvophobic interactions in liquid chromatography with non-polar stationary phases, *J. Chromatogr.*, 125, 129, 1976.

17. Melander, W. and Horváth, C., Salt effects on hydrophobic interactions in precipitation and chromatography of proteins: an interpretation of the lyotropic series, *Arch. Biochem. Biophys.* 183, 200, 1977.

18. Fausnaugh, J.L. and Regnier, F.E., Solute and mobile phase contributions to retention in hydrophobic interaction chromatography of proteins, *J. Chromatogr.*, 359, 131, 1986.

19. Wetlaufer, D.B. and Koenigbauer, M.R., Surfactant-mediated protein hydrophobic-interaction chromatography, *J. Chromatogr.*, 359, 55, 1986.

20. Buckley, J.J. and Wetlaufer, D.B., Use of the surfactant 3-(3-cholamidopropyl)-dimethylammoniopropane sulfonate in hydrophobic interaction chromatography of proteins, *J. Chromatogr.*, 464, 61, 1989.

21. Goheen, S.C. and Engelhorn, S.C., Hydrophobic interaction high performance liquid chromatography of proteins, *J. Chromatogr.*, 317, 55, 1984.

22. Heinitz, M.L., Kennedy, L., Kopaciewicz, W., and Regnier, F.E., Chromatography of proteins on hydrophobic interaction and ion-exchange chromatographic matrices: mobile phase contributions to selectivity, *J. Chromatogr.*, 443, 173, 1988.

23. Ingraham, R.H., Lau, S.Y.M., Taneja, A.K., and Hodges, R.S., Denaturation and the effects of temperature on hydrophobic-interaction and reversed-phase high-performance liquid chromatography of proteins. Bio-Gel TSK-Phenyl-5-PW column, *J. Chromatogr.*, 327, 77, 1985.

24. Wu, S.-L., Benedek, K., and Karger, B.L., Thermal behavior of proteins in high-performance hydrophobic-interaction chromatography. On-line spectroscopic and chromatographic characterization, *J. Chromatogr.*, 359, 3, 1986.

25. Scopes, R., in *Protein Purification: Principles and Practice*, Cantor, C.R., Ed., Springer-Verlag, New York, 1982, 43.

26. Hjertén, S. and Liao, J.L., High-performance liquid chromatography of proteins on compressed, non-porous agarose beads. I. Hydrophobic interaction chromatography, *J. Chromatogr.*, 457, 165, 1988.

27. **Janzen, R., Unger, K.K., Giesche, H., Kinkel, J.N., and Hearn, M.T.W.,** Evaluation of advanced silica packings for the separation of biopolymers by high-performance liquid chromatography. V. Performance of non-porous monodisperse 1.5-μm bonded silicas in the separation of proteins by hydrophobic interaction chromatography, *J. Chromatogr.,* 397, 91, 1987.
28. **Reid, T.S. and Gisch, D.J.,** Fast protein separations by high-performance hydrophobic interaction chromatography, *LC.GC,* 5, 986, 1987.
29. **Kato, Y., Kitamura, T., and Hashimoto, T.,** Preparative high-performance hydrophobic interaction chromatography of proteins on TSK gel phenyl-5PW, *J. Chromatogr.,* 333, 202, 1985.
30. **Alpert, A.J.,** Hydrophobic interaction chromatography of peptides as an alternative to reversed-phase chromatography, *J. Chromatogr.,* 444, 269, 1988.
31. **Heinitz, M.L., Flanigan, E., Orlowski, R.C., and Regnier, F.E.,** Correlation of calcitonin structure with chromatographic retention in high-performance liquid chromatography, *J. Chromatogr.,* 443, 229, 1988.

EFFECTS OF HPLC SOLVENTS AND HYDROPHOBIC MATRICES ON DENATURATION OF PROTEINS

Colin T. Mant and Robert S. Hodges

I. INTRODUCTION

The protein resolving capabilities of both hydrophobic interaction chromatography (HIC) and reversed-phase chromatography (RPC) are based on the strength of hydrophobic interactions between the protein and a non-polar matrix. However, the two techniques differ considerably in the solvent conditions utilized and the relative hydrophobicity and density of the bonded ligand. The development of HIC was spurred by the tendency of proteins to become denatured, with subsequent loss of biological activity, during RPC due to the somewhat harsh mobile phase conditions (organic solvents, low pH, low ionic strength) and strongly hydrophobic stationary phases typical of the latter technique.[1-6] For HIC, ligands of lower hydrophobicity, lower ligand densities and non-denaturing aqueous solvents are used, all with the intention of maintaining a protein in its native conformation. Wu et al.[7] noted that protein structural changes, i.e., changes in secondary, tertiary and/or quaternary structure, can occur on any surface, depending on the protein, the mobile phase composition and pH, and the column temperature. Thus, in spite of the fact that HIC is conducted with weakly hydrophobic surfaces and, in general, with high concentrations of stabilizing salts, protein conformational effects can be exhibited during utilization of even this relatively mild technique.[8] For an overview of the monitoring of protein conformational changes during HIC and RPC, including the effect of temperature, the reader is directed to the article in this book by Wu and Karger ("On-Line Conformational Monitoring of Proteins in HPLC") and references therein.

It is important for the researcher to be aware of the potential effects of HIC and RPC mobile phase solvents and/or the stationary phase matrix on the conformation of a protein of interest, particularly as this may have a direct bearing on the conformational state (random coil or degree of ordered structure) of the protein following elution from the column. This article provides a brief examination of the effects of the more common HIC and RPC HPLC solvents on protein conformation (secondary, tertiary and quaternary structure), together with a demonstration of protein denaturation by hydrophobic stationary phases.

II. DISCUSSION

A. PROTEIN STANDARDS

The protein denaturing tendencies of HPLC solvents and/or hydrophobic matrices can be best assessed by examination of their effects on standard proteins.

1. Synthetic Model Proteins

Lau et al.[9] reported the synthesis of a series of synthetic analogs of tropomyosin, a much-studied two-stranded α-helical coiled-coil. The sequence of the synthetic peptide polymers was Ac-(Lys-Leu-Glu-Ala-Leu-Glu-Gly)$_n$-Lys-amide, where n = 1 to 5 (8, 15, 22, 29, and 36 residues, respectively; denoted TM-8, TM-15, TM-22, TM-29, and TM-36, respectively). The

29- and 36-residue peptides (TM-29 and TM-36, respectively) form extremely stable two-stranded α-helical coiled-coils (referred to as TM-24 dimers and TM-36 dimers), where the subunits are held together by hydrophobic interactions, providing an excellent model for evaluating chromatographic effects on quaternary structure. In addition, since the secondary and quaternary structure of these model proteins is so stable[9], (the TM-36 dimer is 30% denatured by temperature at 74°C and maintains 70% of its α-helicity in 6 M urea), a demonstration of complete denaturation by HPLC solvents and/or hydrophobic matrices would then be representative of the situation for most proteins.

2. Non-Synthetic Protein Standards

The study of monomeric proteins with various degrees of lability can provide a criterion for evaluating changes in tertiary structure (i.e., protein unfolding) resulting from solvent or matrix effects. Proteins such as lysozyme,[5-7,11,12,14] cytochrome c,[6,7,10-14] and myoglobin,[6,10-14] for instance, have frequently been utilized for such purposes. Myoglobin is particularly useful in demonstrating solvent and/or stationary phase effects on tertiary structure due to the presence of a non-covalently bound heme group.

B. EFFECT OF HPLC SOLVENTS
1. HIC vs. RPC Starting Solvents

A high ammonium sulfate concentration (1.5 to 2.0 M) in a phosphate buffer at pH 7.0 are typical starting conditions for HIC of proteins and are designed to maintain the native state of the protein as well as inducing interactions between hydrophobic amino acid residues on the protein surface and the mildly hydrophobic stationary phase of the HIC column. Protein elution is then effected by a decreasing ammonium sulfate gradient.

The most commonly used solvent systems for RPC of peptides and proteins involve linear increasing gradients, starting with water and increasing concentrations of organic solvent (usually methanol, acetonitrile, or isopropanol). These solvent systems generally employ low concentrations of perfluorinated organic acids (e.g., trifluoroacetic acid [TFA]) at a concentration of 0.05 to 0.1% (v/v) in both the water and the organic solvent, resulting in a pH value of ~2.0. Apart from the instability of many proteins at such acidic pH, TFA is a chaotropic agent and is known to be capable of destabilizing the native protein state.[5]

The monomeric-dimeric structure of TM-8, TM-15, TM-22, TM-29, and TM-36 at pH 2.0 was examined[3] by size-exclusion HPLC (SEC) on a TSK G2000SW column (Figure 1). Profile C of Figure 1 shows the elution profile of the five peptides when employing 0.1% aq. TFA as the mobile phase at a flow-rate of 0.4 ml/min. Figure 2 (closed circles) shows the results obtained when plotting the retention times of the five peptides in Figure 1 (profile C) against the logarithm of their molecular weights. The results of Figures 1 and 2 demonstrated that TM-8, TM-15, and TM-22 were monomers under these conditions; in contrast, TM-29 and TM-36 maintained their dimeric structure under these conditions, i.e., the quaternary two-stranded α-helical coiled-coils structure of these two peptides was not disrupted by the acidic nature of the mobile phase.

It is interesting to note that, whereas TM-22 is a monomer in 0.1% aq. TFA (Figures 1 and 2), the presence of 1.7 M ammonium sulfate in a benign buffer system (100 mM sodium phosphate, pH 7.0) induced α-helix formation of the peptide and promoted formation of TM-22 dimers.[6] In contrast, this peptide was monomeric and only moderately helical in a benign buffer system containing 1.1 M potassium chloride and 50 mM sodium phosphate at pH 7.0. Thus, these results demonstrated that, in the case of TM-22, these HIC starting conditions not only maintained the conformation of the peptide but actually promoted its secondary and quaternary structure.

The effects of the starting solvents of HIC and RPC on denaturation of three protein standards (cytochrome c, myoglobin, and lysozyme) is demonstrated in Figure 3. The proteins were run

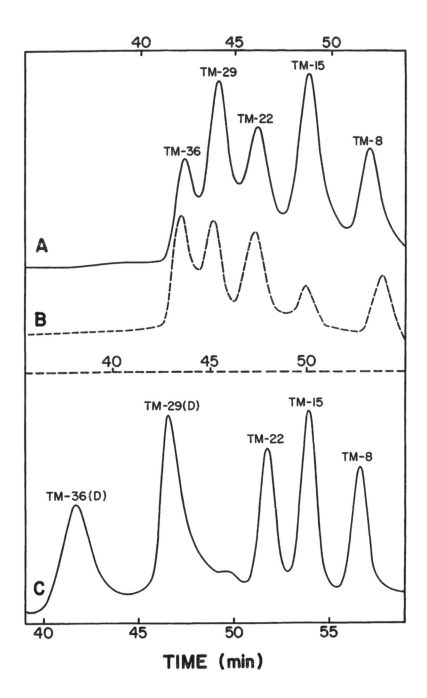

FIGURE 1. Comparison of the SEC elution profiles of a peptide sequence in three different solvent systems. The synthetic peptide mixture consisted of five peptides (TM-8, TM-15, TM-22, TM-29, and TM-36) of the sequence Ac-(Lys-Leu-Glu-Ala-Leu-Glu-Gly)$_n$-Lys-amide, where n = 1 to 5. Column: TSK G3000SW (600 × 7.5 mm I.D.) with a guard column (60 × 7.5 mm I.D.) (Toyo Soda, Tokyo, Japan). Instrumentation: the HPLC instrument consisted of a Spectra-Physics (San Jose, CA) SP8700 solvent delivery system and SP8750 organizer module, combined with a Hewlett-Packard (Avondale, PA) HP1040A detection system, HP3390A integrator, HP85 computer, HP9121 disc drive and HP7470A plotter. Profile A (solid line): solvent, trifluoroethanol (TFE) — 0.1% aq. TFA (1:1). Profile B (dotted line): solvent, acetonitrile — 0.1% aq. TFA (1:1). Profile C (solid line): solvent, 0.1% aq. TFA. Flow-rate, 0.4 ml/min. Absorbance measured at 210 nm. The time axes have been shifted to align the separations. The time axis for profile A is on the top; for profile B, it is represented by the dashed line in the center, and for profile C, it is on the bottom of the figure. The symbol (D) denotes the dimeric form of the peptide. (From Lau, S. Y. M., Taneja, A. K., and Hodges, R. S., *J. Chromatogr.*, 317, 129, 1984.)

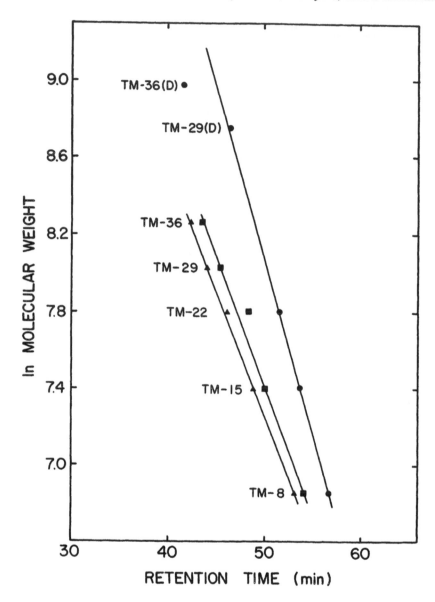

FIGURE 2. Plots of ln molecular weight vs. retention time of the five peptides separated by SEC in three different solvent systems (see Figure 1). The symbols are ● = 0.1% aq. TFA, ▲ = TFE — 0.1% aq. TFA (1:1), and ■ = acetonitrile — 0.1% aq. TFA (1:1). The symbol (D) denotes the dimeric form of the peptide. (From Lau, S. Y. M., Taneja, A. K., and Hodges, R. S., *J. Chromatogr.*, 317, 129, 1984.)

on a TSK Phenyl-5-PW HIC column in both HIC mode (starting solvent of 1.7 *M* ammonium sulfate in 100 m*M* sodium phosphate at pH 7.0) (panel A) and RPC mode (starting solvent of 0.1% aq. TFA, pH 2.0) (panel B). In HIC mode (top), the proteins were eluted in the order cytochrome c, myoglobin, and lysozyme, suggesting that their relative surface hydrophobicities increase in the same order. In contrast, in RPC mode, the elution order changes to lysozyme, cytochrome c, apomyoglobin and heme. The altered elution order results from lysozyme being the only one of the three proteins retaining a native conformation under starting RPC conditions

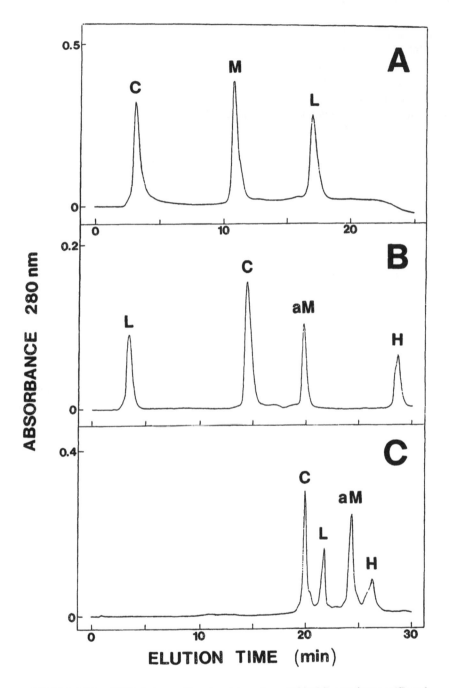

FIGURE 3. HIC and RPC elution profiles of a mixture of myoglobin (M), cytochrome c (C), and lysozyme (L). Columns: panels A and B, Bio-Gel TSK Phenyl-5PW HIC column, 75 × 7.5 mm I.D., 10-μm particle size, 1000-Å pore size (Bio-Rad Labs.,Richmond, CA); panel C, Bio-Gel TSK Phenyl-RP+ RPC column, 75 × 4.6 mm I.D., 10 μm, 1000 Å (Bio-Rad Labs). HPLC instrumentation: see Figure 1. Conditions: panel A (HIC mode), linear descending gradient of 1.7 M ammonium sulfate to 0 M ammonium sulfate in 100 mM sodium phosphate buffer, pH 7.0, in 15 min at a flow-rate of 1 ml/min, followed by continued isocratic elution with the buffer for 5 min; panels B and C (RPC mode), linear AB gradient (2% B/min), at a flow-rate of 1 ml/min where Eluent A was 0.1% aq. TFA and Eluent B was 0.05% TFA in acetonitrile. H denotes heme and aM denotes apomyoglobin.

in the presence of the hydrophobic stationary phase. Similar observations of several proteins (insulin [chain B], ribonuclease A, avidin, lysozyme, concanavalin A, α-chymotrypsinogen A) maintaining native or only partially unfolded states in 0.1% aq. TFA were also reported by Mant et al.[14] (see also Figure 5). The low temperature (4°C) used by these authors[14] also favored maintenance of native protein conformation. Ingraham et al.[6] demonstrated by SEC that the heme group is not dissociated from myoglobin when using 0.1% aq. TFA as the mobile phase, i.e., the protein did not unfold under these acidic conditions. The observed dissociation of the heme group from the myoglobin in Figure 3 (panel B) was likely due to the hydrophobicity of the HIC stationary phase causing some partial unfolding in the region of the heme binding site.

2. Organic Solvents

From the results described above, it is clear that the acidic nature of the RPC starting solvent, 0.1% aq. TFA, does not necessarily denature all proteins. However, since the organic modifiers used in RPC are protein denaturants, it is important to know the effects of these solvents on protein conformation to understand better the behavior of proteins under RPC conditions.

Of the three most commonly used organic solvents in RPC, the order of effectiveness in eluting peptides and proteins has been found to be propanol > acetonitrile > methanol.[15-17] In general, acetonitrile is appropriate for most peptides and proteins, methanol being used for hydrophilic molecules and propanol for more hydrophobic ones.

Panel B of Figure 1 shows the elution profile of TM-8, TM-15, TM-22, TM-29, and TM-36 on the TSK G2000SW SEC column when employing a 1:1 mixture (v/v) of 0.1% aq. TFA — acetonitrile as the mobile phase.[3] From the retention time vs. ln molecular weight plot shown in Figure 2 (closed squares), it can be seen that all five peptides were eluted as monomers. This result was in contrast to the profiles obtained when using 0.1% aq. TFA as the mobile phase, where only TM-8, TM-15, and TM-22 were eluted as monomers and TM-29 and TM-36 were eluted as dimers. A similar result was obtained when using trifluoroethanol (TFE), an α-helix-inducing solvent, in place of acetonitrile (Figure 1, profile A; Figure 2, closed triangles). Lau et al.[3] observed that TM-15, which was a monomer in 0.1% aq. TFA with little α-helical structure, increased its molar ellipticity substantially on the addition of organic solvents, including acetonitrile. Thus, these results suggested that the addition of acetonitrile to the starting solvent for RPC, 0.1% aq. TFA, resulted in denaturation of the quaternary structure of TM-29 and TM-36, while stabilizing the secondary structure (α-helix) of the individual polypeptide chains. Disruption of the tertiary and quaternary structure of a protein in the non-polar medium of RPC is not surprising considering that the major stabilizing forces are hydrophobic interactions. However, since hydrogen bonds which stabilize the α-helix are unstable in the presence of water, it would be expected that, as the non-polarity of the medium increases, the stability of the secondary structure (α-helix) in single-stranded polypeptides would increase.[18,19]

The denaturing effects of acetonitrile on various proteins has often been noted.[1,3,5,6] Benedek et al.,[5] for instance, noted the strongly denaturing effect of this solvent on α-chymotrypsinogen. Ingraham et al.[6] observed that, while the heme is not dissociated from myoglobin in 0.1% aq. TFA, it is dissociated from the protein molecule when 25% acetonitrile is added to this RPC starting solvent, i.e., the acetonitrile disrupted the tertiary structure of myoglobin.

Aqueous solutions of alcohols are known to denature proteins, and Bull and Breese[10] observed that the rate of denaturation increases with the increasing alkyl chain length of the alcohol. These authors suggested that this indicated monohydric alcohols such as methanol, ethanol, butanol, and propanol interacted principally with the hydrocarbon parts of the amino acid residues in the protein, thereby destabilizing them. This report also suggested that proteins in general do not differ greatly in their susceptibility to denaturation by alcohols. In their investigation of the effects of aqueous solutions (at pH 4.0) of propanol on the structures of several proteins (e.g., hemoglobin, β-lactoglobulin, cytochrome c, myoglobin, ribonuclease, lysozyme, and α-lactalbumin), Sadler et al.[12] deduced that, while the solvent destabilized tertiary protein structure, there was a concomitant induction of secondary structure (α-helix) with increasing propanol concentration. These propanol-induced changes were very similar to that described above for acetonitrile and TFE.

C. EFFECT OF HYDROPHOBIC MATRICES

The overall hydrophobicity of a HIC or RPC stationary phase, and its subsequent effect on protein structure, is a combination of the hydrophobicity of the functional ligand and the ligand density. Acyl and short-chain alkyl groups are commonly used in HIC packings, whereas RPC functional ligands are generally long-chain (C_8 or C_{18}) alkyl groups. The ligand density of HIC packings is also generally considerably lower than that of RPC matrices; a reversed-phase ligand density of 10 to 100 times the density of hydrophobic groups in HIC packings has been quoted.[11] Since the stability of protein tertiary and quaternary structures are heavily dependent on hydrophobic interactions, there will always be competition between hydrophobic groups, either buried within the protein core or at a subunit interface and the hydrophobic stationary phase of the column. This competition will necessarily weaken the structural stability of the protein.

1. Functional Ligand

Several research groups[8,13,20-22] have demonstrated an increase in protein retention times during HIC with increasing length and, thus, hydrophobicity of alkyl functional ligands, e.g., pentyl > butyl > propyl > methyl.[22] Similarly, Kato et al.[23] demonstrated longer protein retention times on HIC stationary phases containing phenyl compared to ether ligands, these sorbents having been prepared by introducing phenyl or oligoethylene glycol, respectively, into a resin-based hydrophilic support. More strongly hydrophobic ligands may be useful for the separation of more hydrophilic proteins, but care must be taken that the hydrophobicity of the ligand does not lead to protein denaturation, even under benign HIC conditions. Ueda et al.[13] investigated the effect of temperature and alkyl chain length, at constant ligand density, on the HIC retention behavior of a number of proteins. Figure 4 shows the HIC elution profiles of myoglobin (left profiles) and α-lactalbumin (right profiles) on an n-propyl (C3) phase (top profiles) and an ethyl phase (C2) (bottom profiles) at temperatures of 25, 35, and 45°C. Apart from the greater retention times of myoglobin on the propyl phase (C3) compared to the ethyl phase (C2) at each temperature, the most striking difference between the two sets of myoglobin elution profiles was the appearance of a second peak observed at higher temperatures on the propyl column. The first peak was ascribed to the native form of myoglobin and the second peak to the denatured form. Dramatic effects of temperature and stationary phase were observed on the elution profiles of α-lactalbumin (Figure 4, right profiles). This protein was eluted as a sharp peak from the C2 phase at 25 and 35°C, but in a broader peak at 45°C. In contrast with the result on the C2 phase, α-lactalbumin was eluted as a heavily tailing peak from the C3 phase at 25°C, but not eluted above 35°C. This behavior suggested that the contact of α-lactalbumin with the propyl (C3) stationary phase would lead to partial denaturation of this protein even at room temperature. These results, and others in the same study, indicated that conformational, i.e. denaturing, changes were induced by contact with the more hydrophobic functional ligand as well as by elevated temperatures.

2. Ligand Density

Ligand density effects on protein structure and retention can be demonstrated by comparing the chromatograms of cytochrome c, lysozyme and myoglobin run in reversed-phase mode on the Bio-Gel TSK Phenyl-5-PW column (Figure 3B) and Bio-Gel TSK Phenyl-RP+ column (Figure 3C). The two columns differ in the density of the phenyl functional ligands. The HIC packing (Phenyl-5-PW) contained phenyl groups only sparsely distributed across the polymer-based support ("low phenyl"). The second column (Phenyl-RP+) was based on the above HIC

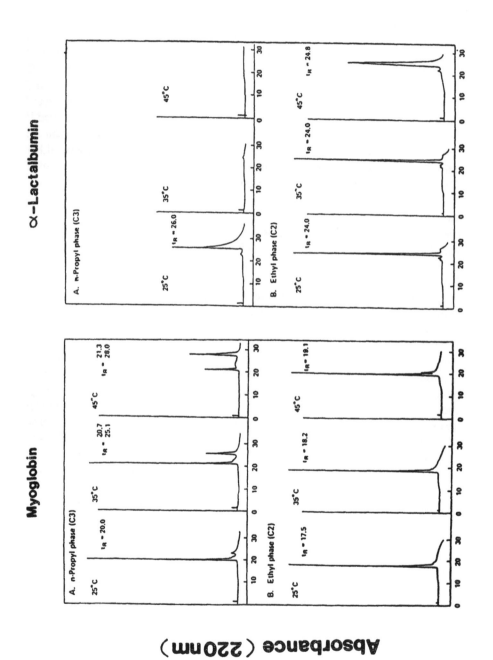

Elution Time (min)

445

FIGURE 4. Effects of temperature and functional ligand on the HIC elution profiles of myoglobin (left profiles) and α-lactalbumin (right profiles). Columns: the columns (50 × 4 mm I.D.) contained silica-based supports (5-μm particle size, 300-Å pore size) derivatized with ethyl (C2, bottom profiles) or n-propyl (C3, top profiles) functional ligands. HPLC instrumentation: the HPLC instrument was an LC-6A liquid chromatograph system (Shimadzu Corporation, Kyoto, Japan), consisting of two Model LC-6A solvent delivery pumps, an SCL-6A system controller, an SIL-6A autoinjector, a CTO-6A column oven, an SPD-6A spectrophotometric detector, and a C-R3A data processor. Conditions: a 20-min descending linear gradient was run from 3.0 M ammonium sulfate in 100 mM phosphate buffer, pH 7.0, to 100 mM phosphate buffer, pH 7.0, at a flow-rate of 0.5 ml/min. Column temperatures are indicated to the left of each profile. (Adapted from Ueda, T., Yasui, Y., and Ishida, Y., *Chromatographia*, 24, 427, 1987. With permission.)

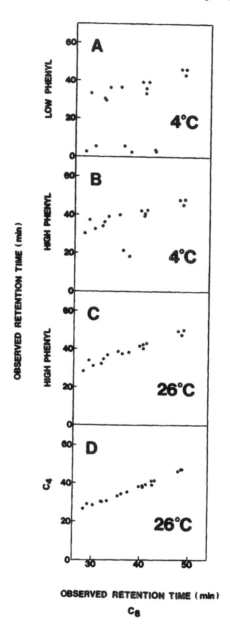

FIGURE 5. Effect of hydrophobicity and ligand density of the stationary phase on the retention times of proteins in RPC. Columns: panel A, Bio-Gel TSK Phenyl-5PW HIC column, 75 × 7.5 mm I.D., 10-µm particle size, 1000-Å pore size (Bio-Rad Labs., Richmond, CA); panels B and C, Bio-Gel TSK Phenyl-RP+ RPC column, 75 × 4.6 mm I.D., 10 µm, 1000 Å (Bio-Rad Labs); panel D, SynChropak RP-4 C_4 RPC column, 250 × 4.1 mm I.D., 6.5 µm, 300 Å (SynChrom, Linden, IN); the C_8 column on the abscissa is an Aquapore RP300 C_8 RPC column, 220 × 4.6 mm I.D., 7 µm, 300 Å (Brownlee Labs, CA). HPLC instrumentation: the HPLC instrument consisted of a Hewlett-Packard (Avondale, PA) HP1090 liquid chromatograph, including an HP1040A detection system, coupled to an HP9000 Series 300 computer, HP9133 disc drive, HP2225A Thinkjet printer and HP7440A plotter. A circulating Haake water bath with variable temperature control was used to adjust the temperature of the column, which was enclosed by a 500-ml jacket constructed from a plastic graduated cylinder and rubber stoppers; a 50% aq. ethanol (v/v) solution was used as the circulating solvent. Conditions: linear AB gradient (1% B/min) at a flow-rate of 1 ml/min, where Eluent A was 0.1% aq. TFA and Eluent B was 0.1% TFA in acetonitrile; temperatures as shown; absorbance at 210 nm. The proteins (listed in Table 1) were chromatographed on all four columns. Where the occasional protein produced a badly skewed peak, thereby preventing accurate retention time measurement, this data point was not used. (From Mant, C. T., Zhou, N. E., and Hodges, R. S., *J. Chromatogr.*, 476, 363, 1989. With permission.)

TABLE 1
Proteins Used in Figure 5

	Protein	Molecular weight	N[a]
1	Insulin (chain B)	3500	30
2	Insulin	6030	51
3	Cytochrome c	11700	104
4	α-Lactalbumin	14180	123
5	Ribonuclease A	13690	124
6	Avidin	14330	128
7	Lysozyme	14310	129
8	Myoglobin	17200	153
9	RsTnC	17960	159
10	Turkey TnC	18000	162
11	RsTnI	20700	178
12	Prolactin	22550	198
13	Papain	23430	212
14	Conconavalin A	25570	237
15	Elastase	25900	240
16	α-Chymotrypsinogen A	25670	245
17	RsTnT	30520	259
18	RcTnT	32880	276
19	RcTM	32000	284

[a] Number of amino acid residues.

packing, but contained a much greater density of phenyl groups ("high phenyl") and was intended to function as a reversed-phase column. The proteins were chromatographed on both columns using a linear gradient (1% B/min) at a flow-rate of 1 ml/min, where Eluent A was 0.1% aq. TFA and Eluent B was 0.1% TFA in acetonitrile. On the column designed for RPC (Figure 3C), all three proteins, including lysozyme which was not retained on the HIC column under these conditions (Figure 3B), appeared to be bound in their denatured form and are also retained significantly longer than on the HIC column.

A similar comparison of the two column packings can be seen in Figure 5 which compares the reversed-phase retention times of 19 proteins (Table 1) obtained on the Phenyl-5-PW column (Figure 5A) and Phenyl-RP+ column (Figure 5B) with those obtained on a C_8 reversed-phase column which had been shown to denature proteins effectively.[14] It would be expected that, if two hydrophobic sorbents denatured proteins to the same extent, and allowing for slight selectivity differences between the packings, then a plot of observed protein retention times on one column against observed retention times on the other would produce a linear plot with little scatter of data points. This linear relationship is clearly lacking for the low phenyl column at 4°C (Figure 5A), where there is a wide scatter of data points. In fact, as noted previously, several proteins were barely, if at all, retained by the column, suggesting that these proteins have retained their native state or have only been partially unfolded. In contrast, the effect of the more hydrophobic nature of the high phenyl column on the retention times of the proteins is clearly apparent in Figure 5B, where the scatter of data points has lessened considerably. All proteins were now retained by the column, with just lysozyme and avidin still exhibiting only partial unfolding. The temperature for the low phenyl and high phenyl runs shown in Figure 5A and B, respectively, remained constant (4°C); thus, the increase in hydrophobicity of the column packing, due to increased phenyl ligand density, was solely responsible for the improvement in correlation of the data points shown in Figure 5B. The further improvement in correlation between observed protein retention times on the high phenyl and C_8 columns on increasing the temperature from 4 (Figure 5B) to 26°C (Figure 5C) was a result of the denaturing effect of increasing temperature.

The above results are consistent with the ligand density studies of Fausnaugh et al.[4] and Ueda et al.[13] Significantly, these researchers also demonstrated that if ligand densities were too high, some proteins would not be eluted from the columns under standard HIC conditions. Hence, there are practical restrictions on the ligand densities which can be used in HIC.

3. Effect of Overall Matrix Hydrophobicity on Protein Denaturation

Lau et al.[3] subjected the five model protein analogs (TM-8, TM-15, TM-22, TM-29, and TM-36) to RPC (using linear aq. TFA to TFA-acetonitrile gradients) on three n-alkyl matrices of varying chain length (C_3, C_8, and C_{18}). A subsequent plot of ln molecular weight *versus* retention times of the five synthetic peptides indicated that all five peptides were eluted from all three columns as a function of their monomeric molecular weight. These results, in conjunction with the size-exclusion data described in Figures 1 and 2 (where TM-29 and TM-36 were eluted as dimers in 0.1% aq. TFA), suggested that the quaternary structure of TM-29 and TM-36 was disrupted upon binding to the hydrophobic matrices. Both TM-29 and TM-36 are extremely stable dimers, as indicated by temperature and urea denaturation studies in 0.1% aq. TFA.[9] Thus, it was the hydrophobic matrix that caused the disruption of the hydrophobic interactions between the two subunits of these peptides. This denaturation occurred even on the ultra-short (C_3) matrix with low carbon loading. Thus, although the organic mobile phases used in RPC were shown to cause denaturation and disruption of the dimers (Figures 1 and 2), the hydrophobicity of the matrix was the important factor in denaturation on binding to the matrix in 0.1% aq. TFA, the starting conditions of RPC.

Ingraham et al.[6] noted that, when subjected to HIC on a TSK Phenyl-5-PW (low density phenyl) HIC column, TM-22 was eluted in its monomeric form, while TM-36 was eluted as a dimer. It was mentioned previously that the starting conditions for HIC (1.7 M ammonium sulfate in 0.1 M sodium phosphate, pH 7.0) actually promoted formation of TM-22 dimers; thus, the dimeric quaternary structure of this peptide was disrupted by the HIC column despite the relatively mild nature of the matrix hydrophobicity. This destabilizing effect of the matrix on TM-22 under HIC conditions was reminiscent of the same effect on TM-36 on the same column under RPC conditions.[6] It was presumed by the authors[6] that TM-36 did not lose its dimeric structure under HIC conditions as it did in RPC mode due to stabilization brought about by the presence of ammonium sulfate and the fact that the hydrophobicity of the HIC column is not sufficient to denature the more stable TM-36 dimer.

The effect of increased sorbent hydrophobicity on the reversed-phase retention behavior of 19 proteins (Table 1) at constant temperature (26°C) can be seen by comparing Figure 5C and 5D. The correlation of observed protein retention times on a C_4 and C_8 column (Figure 5D) showed an improvement over that obtained between the less hydrophobic Phenyl RP + ("high phenyl") and C_8 columns (Figure 5C). In fact, the linearity of the plot shown in Figure 5D indicated that the C_4 and C_8 sorbents denatured the proteins to the same extent.

III. CONCLUSIONS

This article has attempted to demonstrate potential effects of HPLC solvents and hydrophobic matrices on the secondary, tertiary, and quaternary structures of proteins. The observed effects on a protein subjected to HIC or RPC depend on a combination of various parameters. These parameters include matrix hydrophobicity (functional ligand and/or ligand density), mobile phase, temperature and protein lability. In fact, it has been observed by several researchers[6,14,24,25] that changes in sorbent hydrophobicity and/or temperature could be used to improve the resolution of proteins with similar retention times. Thus, the controlled denaturation of a protein(s) in a protein mixture may occasionally be a useful separation tool.

ACKNOWLEDGMENTS

This work was supported by the Medical Research Council of Canada and equipment grants from the Alberta Heritage Foundation for Medical Research.

REFERENCES

1. Regnier, F.E., High-performance liquid chromatography of proteins, *Methods Enzymol.*, 91, 137, 1983.
2. Luiken, J., van der Zee, R., and Welling, G.W., Structure and activity of proteins after reversed-phase high-performance liquid chromatography, *J. Chromatogr.*, 284, 482, 1984.
3. Lau, S.Y.M., Taneja, A.K., and Hodges, R.S., Effects of high-performance liquid chromatographic solvents and hydrophobic matrices on the secondary and quaternary structure of a model protein. Reversed-phase and size-exclusion high-performance liquid chromatography, *J. Chromatogr.*, 317, 129, 1984.
4. Fausnaugh, J.L., Kennedy, L.A., and Regnier, F.E., Comparison of hydrophobic-interaction and reversed-phase chromatography of proteins, *J. Chromatogr.*, 317, 141, 1984.
5. Benedek, K., Dong, S., and Karger, B.L., Kinetics of unfolding of proteins on hydrophobic surfaces in reversed-phase liquid chromatography, *J. Chromatogr.*, 317, 227, 1984.
6. Ingraham, R.H., Lau, S.Y.M., Taneja, A.K., and Hodges, R.S., Denaturation and the effects of temperature on hydrophobic-interaction and reversed-phae high-performance liquid chromatography of proteins. Bio-Gel TSK-Phenyl-5-PW column, *J. Chromatogr.*, 327, 77, 1985.
7. Wu, S.-L., Benedek, K., and Karger, B.L., Thermal behavior of proteins in high-performance hydrophobic interaction chromatography. On-line spectroscopic and chromatographic characterization, *J. Chromatogr.*, 359, 3, 1986.
8. Wu, S.-L., Figueroa, A., and Karger, B.L., Protein conformational effects in hydrophobic interaction chromatography. Retention characterization and the role of mobile phase additives and stationary phase hydrophobicity, *J. Chromatogr.*, 371, 3, 1986.
9. Lau, S.Y.M., Taneja, A.K., and Hodges, R.S., Synthesis of a model protein of defined secondary and quaternary structure. Effect of chain length on the stabilization and formation of two-stranded α-helical coiled-coils, *J. Biol. Chem.*, 259, 13253, 1984.
10. Bull, H.B. and Breese, K., Interaction of alcohols with proteins, *Biopolymers,* 17, 2121, 1978.
11. Goheen, S.C. and Engelhorn, S.C., Hydrophobic interaction high-performance liquid chromatography of proteins, *J. Chromatogr.*, 317, 55, 1984.
12. Sadler, A.J., Micanovic, R., Katzenstein, G.E., Lewis, R.V., and Middaugh, C.R., Protein conformation and reversed-phase high-performance liquid chromatography, *J. Chromatogr.*, 317, 93, 1984.
13. Ueda, T., Yasui, Y., and Ishida, Y., Study on optimal packing materials for high-performance hydrophobic-interaction chromatography of proteins, *Chromatographia,* 24, 427, 1987.
14. Mant, C.T., Zhou, N.E., and Hodges, R.S., Correlation of protein retention times in reversed-phase chromatography with polypeptide chain length and hydrophobicity, *J. Chromatogr.*, 476, 363, 1989.
15. Mahoney, W.C. and Hermodson, M.A., Separation of large denatured peptides by reversed-phase high-performance liquid chromatography. Trifluoroacetic acid as a peptide solvent, *J. Biol. Chem.*, 255, 11199, 1980.
16. Wilson, K.J., Honegger, A., Stötzel, R.P., and Hughes, G.J., The behaviour of peptides on reversed-phase supports during high-pressure liquid chromatography, *Biochem. J.*, 199, 31, 1981.
17. Hermodson, M. and Mahoney, W.C., Separation of peptides by reversed-phase high-performance liquid chromatography, *Methods Enzymol.*, 91, 352, 1983.
18. Klotz, I.M. and Franzen, J.S., The stability of interpeptide hydrogen bonds in aqueous solution, *J. Am. Chem. Soc.*, 82, 5241, 1960.
19. Talbot, J.A. and Hodges, R.S., Tropomyosin: a model for studying coiled-coil and α-helix stabilization, *Acc. Chem. Res.*, 15, 224, 1982.
20. Fausnaugh, J.L., Pfannkoch, E., Gupta, S., and Regnier, F.E., High-performance hydrophobic interaction chromatography of proteins, *Anal. Biochem.*, 137, 464, 1984.
21. Alpert, A.J., High-performance hydrophobic-interaction chromatography of proteins on a series of poly(alkyl aspartamide)-silicas, *J. Chromatogr.*, 359, 85, 1986.
22. Schmuck, M.N., Nowlan, M.P., and Gooding, K.M., Effects of mobile phase and ligand arm on protein retention in hydrophobic interaction chromatography, *J. Chromatogr.*, 371, 55, 1986.

23. **Kato, Y., Kitamura, T., and Hashimoto, T.,** New resin-based hydrophilic support for high-performance hydrophobic-interaction chromatography, *J. Chromatogr.,* 360, 260, 1986.
24. **Snyder, L.R.,** Temperature-induced selectivity in separations by reversed-phase liquid chromatography, *J. Chromatogr.,* 179, 167, 1979.
25. **Barford, R.A., Sliwinski, B.J., Breyer, A.C., and Rothbart, H.L.,** Mechanism of protein retention in reversed-phase high-performance liquid chromatography, *J. Chromatogr.,* 235, 281, 1982.

HIGH-PERFORMANCE HYDROPHOBIC INTERACTION CHROMATOGRAPHY OF A LABILE REGULATORY PROTEIN

Salman M. Hyder, Jing Dong, Petr Folk, and
James L. Wittliff

I. INTRODUCTION

A. PRINCIPLE AND BACKGROUND TO HPHIC

High-performance hydrophobic interaction chromatography (HPHIC) is now recognized as a valuable technique for the separation of proteins under non-denaturing conditions.[1-4] Earlier, Arfmann and Shaltiel[5] made the important observation that the affinity of the proteins for hydrophobic matrices was proportional to the chain length of the derivative. With the advent of defined silica particles, studies from Regnier's group firmly established that hydrophobic interaction chromatography of proteins was feasible in the high-performance liquid chromatography (HPLC) mode. Reports from other laboratories confirmed these initial findings.[6,7]

Principles of HPHIC involve an initial weak hydrophobic interaction of the protein with the bonded phase of the stationary silica particles. This interaction is promoted through the use of a high ionic strength mobile phase. Most commonly used salts for the promotion of this initial interaction include sodium sulfate, ammonium sulfate, sodium chloride, and sodium citrate. The initial high salt conditions enhance the hydrophobic interaction by removing water molecules from the vicinity of the protein surface, thus allowing their interaction with the apolar bonded phase. The retained proteins are then selectively eluted during a descending salt gradient which allows the proteins to rehydrate selectively and be eluted from the column with retention of their biological activity.[1-4,8] This is in sharp contrast to reversed-phase liquid chromatography (RPC), where the stronger interactions are neutralized with organic solvents, thus denaturing the proteins of interest. This infers that the mobile phase employed for HPHIC generally allows the retention of biological activities of the respective proteins.

B. APPLICATION OF HPHIC TO SEPARATE STEROID RECEPTOR ISOFORMS

Steroid receptors, which belong to a family of labile transcriptional factors, appeared to be candidates for separation by hydrophobic interaction chromatography in the HPLC mode. We undertook these studies from two points of view: first, to purify the steroid receptors rapidly and with good recoveries (discussed below); and second, to understand the on-column behavior of oligomeric proteins for which the steroid receptor is ideal, since it forms complexes with several other macromolecules.[3]

To ascertain the molecular mechanisms of steroid hormone actions, considerable emphasis has been placed on the purification of the receptor proteins responsible for their physiological effects. Some success has been achieved in this respect (e.g., References 9 to 12). The majority of current procedures are lengthy and lead to a considerable loss of receptor during the purification process, presumably due to degradation or ligand dissociation.[12] In addition, analysis of purified estrogen receptors by centrifugation followed by DEAE-cellulose chromatography (e.g., Reference 13) or by sodium dodecyl sulfate-polyacrylamide gel electrophoresis indicated a heterogeneous population. The latter point has been supported by the finding that steroid receptors exhibit polymorphism, as demonstrated by HPLC in a variety of modes.[15-19] These HPLC methods include size-exclusion,[16] ion-exchange,[17] and chromatofocusing.[18] Separation is rapid, efficient, and reproducible, giving recoveries of 75 to 100%.[12] Similar polymorphism of steroid receptors is found in the hydrophobic interaction mode with near 100% recovery of biological activity of steroid receptors as determined by either ligand binding[2,3,20,21] or by immunorecognition of the receptor protein eluted from hydrophobic matrices.[22] A distinct advantage in the study of steroid receptors was the availability of iodinated steroids with very high specific activity. This has allowed us to monitor the receptor elution pattern related to one of its biological activities, i.e., ligand binding.[23] Here we describe our experience with HPHIC of estrogen receptors (ER) utilizing two separate mildly hydrophobic columns. Some applications of HPHIC are also discussed which emphasizes the usefulness of HPLC in general for rapid detection, quantitation, and extension of the chromatographic observation to possible reactions *in vivo*. With minimal modifications, this methodology may be applied to other biologically-active labile proteins. A complete step-by-step guide to HPHIC procedures is described in a recent set of protocols.[4]

Any discussion of HPLC of steroid receptors will be incomplete without first giving a description of the preparation of soluble estrogen receptors from target-tissues. In addition, general procedures of chromatography and the columns used are also briefly discussed below.

II. EXPERIMENTAL

A. PREPARATION AND LABELING OF SOLUBLE ESTROGEN RECEPTOR

All procedures were performed at 4°C. Rat uterine tissue (1 ml per uterus) or human breast tumors (ca. 200 to 400 mg/ml) were homogenized in $P_{10}EDG$ [10 mM phosphate — 1.5 mM EDTA — 1 mM DTT — 10% (v/v) glycerol, pH 7.4]. Homogenization was performed with two 10-sec bursts with a Brinkman Polytron homogenizer (Westbury, NY).

Soluble fractions were prepared by centrifugation of the homogenate for 30 min at 40,000 rpm in a Beckman Ti 70.1 rotor using an L8-7D ultracentrifuge. Supernatant was removed carefully, avoiding the layer of fat at the top. Soluble fractions were labeled with 2 to 3 nM [16α-^{125}I]iodoestradiol-17β] in the presence and absence of a 200-fold excess of diethylstilbestrol for 2 to 4 hr at 4°C, unless otherwise stated. The reaction was terminated by removing unbound steroid with a pellet, derived from dextran-coated charcoal suspension (1% charcoal, 0.05% dextran). Labeled cytosol was applied to the charcoal pellet, mixed, and allowed to stand for 5 min at 4°C. Dextran-coated charcoal was then removed by centrifuging the sample for 5 min at 1000 *g*. Cytosol protein concentrations were determined by the method of Bradford,[24] using bovine serum albumin as the standard. Protein concentrations generally ranged from 4 to 8 mg/ml.

B. CHROMATOGRAPHY

All chromatographic procedures were performed at 4°C in a Puffer-Hubbard (Ashville, NC) cold box. Buffers were filtered under vacuum through Millipore 0.45 μm HAWP filters (Bedford, MA) before use. Filters of 0.5 μm pore size (Type FHLP, also from Millipore) were substituted when buffers containing organic solvents were filtered. Free steroid or estrogen-receptor complexes were applied to the columns with an Altex Model 210 sample injection valve. Elution was carried out with a Beckman 112 or 114 solvent delivery module, including a Model 421 system controller. Further details on HPHIC of steroid receptors can be found in References 3 and 4.

C. HPHIC COLUMNS

Although HPHIC columns with different bonded phases are available from various sources

(Toyo Soda, Beckman, SynChrom, Pharmacia, The Nest Group, J.T. Baker, Poly LC), we utilized both 300- and 500-Å pore size silica-based propyl columns (SynChrom, Lafayette, IN) and a 300-Å silica-based polyether bonded phase column (CAA-HIC, Beckman/Altex, San Ramon, CA) which is non-ionic in nature. These two types of stationary phases are treated separately in the following sections.

D. HPHIC ON SYNCHROPAK PROPYL COLUMNS

In our first study[25] of separation of steroid receptors using HPHIC, we used SynChrom propyl 500 columns (250×4.6 mm I.D.). Subsequently, we used 300- and 500-Å columns of shorter length (100×4.6 mm I.D.). The gradient conditions described in this section apply to the longer 250 mm column and the gradient conditions (elution conditions) described in the next section (Section E), also apply to the shorter propyl columns.

Several combinations of buffers were tested, and two different column elution programs were used. The column (Propyl 500 or 300 Å, 250×4.6 mm I.D.) was equilibrated with high-ionic strength buffer composed of 500 mM phosphate, 1.5 mM EDTA, 1 mM DTT, 10% v/v glycerol, pH 7.4 (P_{500}EDG), and eluted with a reverse salt gradient (500 to 10 mM phosphate). Initially, a long program (105 min) of the reverse phosphate gradient was used with a flow-rate of 0.2 ml/min to allow greater contact time of the receptor with the stationary phase. After 10 min, flow-rate was increased to 1 ml/min for the next 75 min, during which the reverse salt gradient was developed. The column was washed for 20 min with a similar buffer, except that the phosphate concentration was 10 mM. Later in the study, the elution program-time was reduced to 60 min, consisting of an initial flow-rate of 1 ml/min for 5 min and then a descending salt gradient-elution for 20 min. At the end of the elution program, the column was washed with either water or P_{10}-EDG buffer (10 mM phosphate, 1.5 mM EDTA, 1 mM DTT, and 10% v/v glycerol) for an additional 35 min. This latter step was followed by washing the column with methanol-P_{10}EDG buffer (50:50). Eluted steroid (free and protein bound) was collected as 1-ml fractions and detected radiometrically in a Micromedics 4/600 gamma counter (Rohm & Haas, Cleveland, OH). The counting efficiency was 75 to 80%. On occasion, continuous recordings were made with a Beckman Model 170 flow-through radioisotope detector.

E. HPHIC ON CAA-HIC POLYETHER BONDED COLUMN

Unless otherwise stated, the gradient program for the CAA-HIC column (300 Å, 100×4.6 mm I.D.) consisted of an initial elution with eluent A (P_{10}EDG, containing 2 M ammonium sulfate, pH 7.4) at a flow-rate of 1 ml/min. Following sample injection, a descending salt gradient was developed to 100% eluent B (P_{10}EDG) over 30 min before stopping and reequilibrating to 100% eluent A. There was a gradient delay period of ca. 5 min. This time period was not subtracted from the reported t_R values. The above described gradient elution program was used in most cases, but the nature of some experiments dictated use of other gradient elution conditions. These are described in the various figure legends.

Following chromatography, the eluted steroid (free and protein-bound) was collected as described earlier. Since the non-specific binding (radioactivity eluted from cytosols labeled in the presence of DES) showed mainly baseline levels and represented only 5 to 10% of the total binding, these are usually not shown in the figures. Recoveries of total radioactivity and injected protein were 75 to 100%.

F. IMMUNOPRECIPITATION AND DETECTION OF PROTEIN KINASE ACTIVITY

Following HPHIC, fractions which contained protein-bound radioactivity and fractions from other elution positions within the gradient were incubated with a monoclonal antibody (MAb D547) to purified estrogen receptors which was immobilized on polystyrene beads (Abbott Laboratories, Chicago, IL). Two MAb-coated beads were added directly to each of the test tubes, containing labeled receptor proteins and other fractions (controls), and incubated for 18 hr at

4°C. The beads were removed and one of these was developed for quantification (measurement of mass) of ER associated with MAb, as recommended by the manufacturer. The second bead was used for protein kinase activity measurements and was processed exactly as described in our previous publications.[26-28]

Briefly, beads were washed first with distilled water, then with P_{10}EDG containing 0.05% NP-40, followed by P_{10}EDG, and incubated at 30°C for 30 min with 5 to 10 μCi of ^{32}P-labeled ATP in the presence of 10 μg phosvitin (Sigma), which served as exogenous substrate for transfer of ^{32}P phosphate from ATP to a polypeptide. The phosphorylated polypeptides were eluted and analyzed by sodium dodecylsulfate-polyacrylamide gel electrophoresis (SDS-PAGE) under denaturing conditions in 7.5% slab gels as described by Laemmli.[29] Molecular weight markers for SDS-PAGE were obtained from Sigma.

III. ANALYSIS OF STEROID RECEPTORS ON SYNCHROPAK PROPYL COLUMNS

A. STUDIES WITH 250 × 4.6 mm I.D. COLUMN

We were interested in the evaluation of HPHIC as an alternative purification tool mainly due to the labile characteristics of steroid receptors. However, another primary interest is to characterize the origin of receptor polymorphism that is observed when steroid receptors are separated by their properties of charge, shape, size and hydrophobicity.[1-4, 15-19, 30-32]

Our experience indicated that it is essential to characterize the behavior of unbound steroid ligands on an HPLC column to identify steroid-receptor complexes. For initial studies, we used an elution program of 105 min. The starting buffers, pH 7.4, were either P_{10}EDG buffer containing 3 or 1 M potassium chloride, or P_{500}EDG with 10 or 40% glycerol. The final buffer concentration at the end of the gradient was always P_{10}EDG containing 10 or 40% glycerol, depending on the nature of the experiment (low ionic strength buffers).

A primary consideration is to evaluate the influence of the mobile phase on the chromatographic behavior of unbound ligand. When various reverse salt gradients were applied, no steroid was released.[25] Furthermore, only a small quantity of steroid was released when the column was washed with either water or P_{10} buffer for 1 hr. A 100% recovery of steroid was achieved when the column was washed with methanol. These data suggest that if labeled steroid was observed in fractions eluted during the reverse salt gradients, it may be attributed to protein-bound [^{125}I]iodoestradiol-17β.

When ER complexes were applied to the Propyl 500 column in the absence of organic solvents, no labeled steroid, either free or protein-bound, was eluted. It was unclear whether (a) the stationary phase stripped the labeled steroid from the receptor, allowing the unliganded species to be eluted, (b) the steroid-receptor complexes were tightly bound to the hydrophobic column, or (c) the receptor protein was denatured as a result of high column pressure, releasing the steroid to interact with the column matrix.

The high pressure does not seem to be the cause since, following other modes of HPLC, virtually all the receptor activity was recovered. In this respect, other modifications of the gradient were tested, including a combination of descending salt concentration and ascending glycerol concentrations. In all cases, no release of protein bound steroid was observed during the elution time.

We reasoned that if the steroid was dissociated from the receptor by interaction with the hydrophobic column during chromatography, perhaps a program consisting of a shorter reverse salt gradient would elute the receptor isoforms on the basis of their hydrophobicity. Thus, the elution program was reduced to 60 min. The free [^{125}I]iodoestradiol behaved somewhat differently under these conditions. Again, no free steroid was eluted from the column during the reverse salt gradient, which took approximately 30 min. However, immediately after the gradient, labeled iodoestradiol-17β began to be eluted gradually. Thus, if [^{125}I]-iodoestradiol-17β appeared during the reverse salt gradient when a receptor preparation was applied to the SynChropak Propyl column, it should be attributed to receptor-bound steroid.

To evaluate hydrophobic-interaction chromatography, tissue extracts labeled with [^{125}I]iodoestradiol-17β were applied to the SynChropak Propyl column and eluted with a 60-min program described earlier. Two peaks of radioactivity were detected, each of which appeared to contain specific protein-bound iodoestradiol-17β (see Reference 25). As noted, when unlabeled diethylstilbestrol (inhibitor) was incubated with labeled steroid and tissue extracts, the binding of [^{125}I] was greatly diminished, indicating the specificity of the interaction. One of the components (peak 1) was eluted in the void volume, while the others were eluted at 350 to 450 mM phosphate (peak 2). Also, it was noted that peak 1 was eluted at a position where the majority of proteins were detected by their absorption at 280 nm.

Using the same column (300 Å) and similar receptor preparation, further study was conducted to show the hydrophobic nature of protein in peak 2.[25] Labeled estrogen receptors were applied to the stationary phase and the column was eluted with a continuous wash of P$_{500}$EDG buffer without phosphate gradient. Only volume-dependent peak 1 was eluted, i.e., no steroid-binding components identified as peak 2 by phosphate gradient elution were observed. This finding substantiates the suggestion that estrogen-binding components in peak 2 exhibit hydrophobic properties. Radioactivity bound by components in peaks 1 and 2 represented 3 to 12% of that applied, suggesting that the SynChropak Propyl matrix is either stripping the ligand from the receptor or has retained other estrogen-binding components.

Using conventional open column hydrophobic chromatography, other workers have recognized that certain estrogen receptors from carcinogen-induced mammary tumors of rats, as well as androgen receptors from rat prostate, display hydrophobic characteristics. Maggi et al.[33] have used the latter procedure to study affinity-labeled progestin receptors from chick oviduct. Currently, the specific chemical basis for the hydrophobic interaction common to these receptors is unknown. However, the report of Tenenbaum and Leclercq[34] may shed light upon this question, since they suggest the estrogen receptor in the liganded and unliganded state behaves differently on a hydrophobic column (Blue Sepharose CL-6B).

Moderate concentrations of organic solvents, such as methanol and acetonitrile, have been used successfully in reversed-phase HPLC for the purification of various proteins. In this study, we have evaluated the influence of various organic solvents in retarding the interaction of iodoestradiol-17β with the stationary phase and in isolating labeled receptor or intact isoforms. Our results with phosphate:acetonitrile buffer (80:20) show that iodoestradiol was eluted with a retention time of approximately 15 min. It was observed that free steroid was now eluted within the phosphate gradient in the presence of acetonitrile.

After establishing the elution position of free iodoestradiol-17β, we investigated the separation of labeled estrogen receptors by HPHIC. Two [^{125}I]iodoestradiol-binding components were observed in the presence of 20% acetonitrile. Both components appeared to bind labeled estrogen specifically since unlabeled diethylstilbestrol diminished association (Figure 1). The first peak of radioactivity was eluted in the same position as unbound steroid. However, inhibition of binding by diethylstilbestrol suggests that peak 1 may consist of both protein-bound and free iodoestradiol-17β. The peak appearing at approximately 50 mM phosphate appeared to consist of specific estrogen-binding components, since ligand association was diminished by inhibitor, and the elution position was different from that of free steroid. Separation profiles of receptor isoforms should reflect differences in the intrinsic properties (such as amino acid sequence) of individual estrogen-binding components. The physiological significance of the molecular heterogeneity observed in a variety of tissues (2, 31, 32) will have to await progress in the application of HPHIC to steroid hormone receptors.

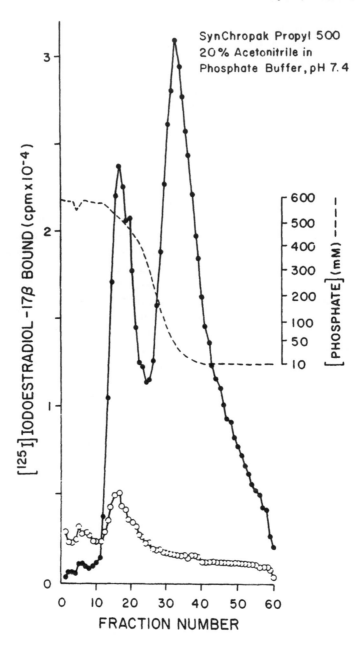

FIGURE 1. HPHIC separation of estrogen receptors from human breast cancer. All phosphate buffers contained 20% acetonitrile. Of the applied radioactivity, 90% was recovered, of which 30% was located in the first peak (fractions 11 to 24) and 70% in the second peak (fractions 24 to 60). Total binding is indicated by (●) while non-specific binding is indicated by (○). (From Hyder, S. M., Wiehle, R. D., Brandt, D. W., and Wittliff, J. L., *J. Chromatogr.*, 327, 237, 1985. With permission.)

B. STUDIES WITH 100 × 4.6 mm I.D. COLUMN

All the previously described studies on HPHIC of steroid receptors were performed on a SynChrom propyl column with dimensions of 250 × 4.6 mm I.D. SynChrom has marketed Propyl 300 and 500 Å columns of shorter length (100 mm) with similar packing material to their

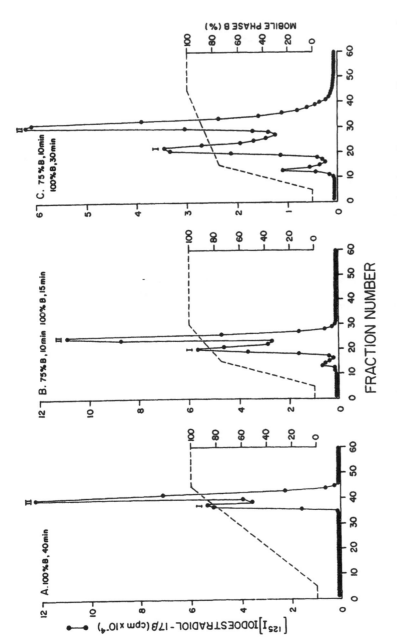

FIGURE 2. Influence of gradient development time on HPHIC separation of estrogen receptor isoforms from human breast cancer. Dextran-coated charcoal-treated human breast tumor cytosol was injected into the SynChropak propyl column and chromatographed with gradient times reaching (A) 100% eluent B in 40 min, (B) 75% B in 10 min followed by 100% B in next 15 min and (C) 75% B in 10 min, followed by 100% B in 30 min. All samples were adjusted to 1.5 M ammonium sulfate prior to injection. Eluent A in this experiment was P$_{10}$EDG, containing 2 M ammonium sulfate; eluent B was P$_{10}$EDG. For clarity, only total cpm/fraction (●) are shown. (From Hyder, S. M. and Wittliff, J. L., $J.$ $Chromatogr.$, 476, 455, 1989. With permission.)

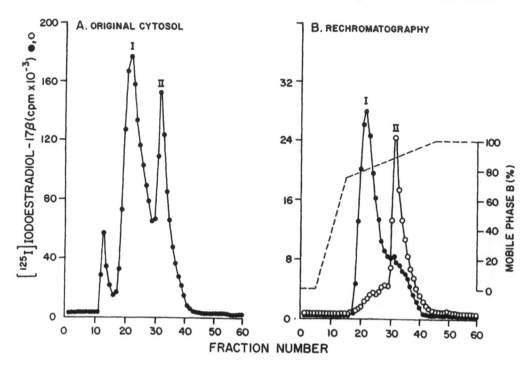

FIGURE 3. Rechromatography of Isoforms I and II by HPHIC. (A) Cytosol (8.1 mg/ml) was first injected onto the SynChropak Propyl 300 column and eluted with buffers lacking sodium molybdate. Each of the eluted isoforms I and II was then reinjected separately onto the SynChropak Propyl Column 300 (B) and eluted under similar conditions. Recoveries were approximately 100% in all cases. (From Hyder, S. M. and Wittliff, J. L., *J. Steroid Biochem.*, 33, 965, 1989. With permission.)

original columns. We next undertook an extensive study of estrogen and progesterone receptor under varying conditions of the mobile phase. One should note that the mobile phase was changed from that of phosphate buffer to that containing ammonium sulfate for initial hydrophobic bonding of receptor with the stationary phase.

When HPHIC was performed on the SynChrom propyl column of shorter length, with a linear gradient of 2 to 0 M ammonium sulfate in 40 min, estrogen receptor was eluted with a retention time of 38 min for peak I and 41 min for peak II (Figure 2A). Although the resolution on the propyl column was not optimal under these conditions, the results were in agreement with previous experiments where two receptor isoforms were observed.[20,25]

To enhance resolution of the estrogen receptor isoforms, the gradient conditions were modified to reach 75% B in 10 min and, in a second phase, to reach 100% B in 15 min (Figure 2B). This increased the relative resolution slightly. Also, a minor third peak (t_R = 13 min) was observed, which had been ignored initially. However, the majority of molybdate-stabilized receptor, which we have termed MI, was eluted at this retention time (see below). A better resolution of receptor isoforms was observed when the second phase of the gradient was extended to reach 100% eluent B in 30 min (Figure 2C).

Since the propyl column was hydrophobic, the effect of starting the gradient at a lower ionic strength was analyzed on the basis of the retention of the estrogen receptor molecule. When the initial ionic strength of ammonium sulfate was 1 M, the separation of estrogen receptor was comparable to that with 2 M ammonium sulfate as the starting buffer.[20] The sample concentration was also adjusted to 1 M ammonium sulfate prior to injection. However, when the initial concentration of eluent A was lowered to 0.5 M ammonium sulfate, a considerable proportion of receptor protein was not associated with the bonded phase. This occurred regardless of whether the sample concentration was adjusted to 0.5 *M* ammonium sulfate or to 1.0 *M* ammonium sulfate prior to sample application. Therefore, it is imperative that a critical ammonium sulfate concentration be reached both in eluent A and in the receptor preparation to ensure immobilization of the proteins on the bonded phase. This probably reflects the removal of water molecules associated with the receptor protein, which is known to be ionic,[26,61] or water molecules from associated proteins.

The hydrophobic properties of the estrogen receptor from human breast cancer changed when this protein was incubated overnight in the presence of iodoestradiol at 4°C. When analyzed after a short incubation with the steroid, the receptor was eluted as isoforms I and II. The longer incubation (17 to 24 hr) at 4°C converted isoform II into isoform I. Since the receptor may have been proteolyzed with time, resulting in the elution of receptors as the single peak I, receptor size was monitored by HPSEC. Trypsinized or proteolyzed receptor was eluted as a single, sharp peak at 25 to 30 Å which retained the steroid-binding domain.[35] The receptor is known to dissociate from other complex macromolecules, such as HSP-90[36] with time, to produce a non-proteolyzed form which is also eluted in the 25 to 30 Å range in HPSEC and is difficult to separate from the proteolyzed form. This process can be accelerated with potassium chloride in buffers. Although some of the 65-Å receptor complex was transformed to the 30-Å species during a 24-hr incubation, this conversion cannot be responsible for the extensive change in receptor hydrophobicity. This suggests that receptor size and hydrophobicity are unrelated.

Estrogen receptors from both human uterus and human breast cancer cells in culture were analyzed for their hydrophobic properties. Estrogen receptor separated into two isoforms, whether obtained from uterus or from breast cancer cells.[20] The retention times of the isoforms were the same as those of receptors separated from human breast tumor cytosols. This indicates that estrogen receptor in different tissues undergoes similar post-translational modifications and associates with similar proteins. A protein known to interact with estrogen receptor is HSP-90, which has been detected in many different tissues.[37] Using monoclonal antibody raised against HSP-90 from chick oviduct, HSP-90 was detected only in the M1 isoform.[37]

Steroid receptors are known to be labile proteins, found in very small amounts in target cells. This makes their isolation and study extremely difficult.[3,38] Properties of the hydrophobic stationary phase were investigated for separation and characterization of estrogen receptor isoforms and as a means of storing receptor protein without denaturation for short and long periods. Estrogen receptor that was eluted as isoforms I and II in control cytosol retains its chromatographic characteristics when immobilized on the stationary phase for 90 min at 4°C.[20] However, when the receptor was in contact with the stationary phase for 16 hr at 4°C, the peak characteristics and elution patterns were altered. Curiously, both isoform I and II were eluted with t_R values that were 3 min longer than those observed for the control. Since the experimental conditions were identical, except for the holding time, these results suggest unfolding of the protein with an increase in hydrophobicity. It is intriguing that both receptor forms showed a similar change in hydrophobicity, suggesting a common event. Both receptor isoforms retained their ligand-binding properties, indicating that extensive denaturation did not occur.

Our most current analysis of estrogen receptor using HPHIC has shown that the two distinct receptor species (I and II) are high molecular weight proteins (> 60 Å) as determined by high-performance size-exclusion chromatography.[39] The hydrophobic properties of individual receptors were conserved when HPHIC-eluted ER isoforms were reanalyzed (Figure 3). This indicates that individual receptor isoforms (I and II) have distinct contact points with the bonded phase and there was no interconversion of the two within the experimental time required to complete the assay.[39] To confirm that the high molecular weight components of ER were polymorphic, we partially purified ER from human breast cancers by HPSEC both in the absence and presence of molybdate. Each of the high molecular weight protein complexes was then analyzed by HPHIC. The results (Figures 4A and 4C) showed that each homogeneous HPSEC peak was actually composed of heterogeneous hydrophobic forms, with the elution profile being identical

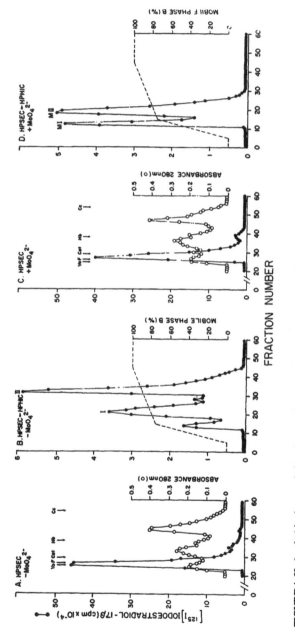

FIGURE 4. Hydrophobic characteristics of high molecular weight estrogen receptors separated first by size-exclusion chromatography. (A) Cytosol (8.6 mg/ml) was injected onto the TSK-3000 SW column both in the absence (A) and presence (C) of sodium molybdate and eluted with P_{10}EDG + 100 mM KCl. The receptor which was eluted in the void volume was then analyzed by HPHIC in the absence (B) or presence (D) of sodium molybdate. Recoveries in all cases were approximately 100%. (From Hyder, S. M. and Wittliff, J. L., *J. Steroid Biochem.*, 33, 965, 1989. With permission.)

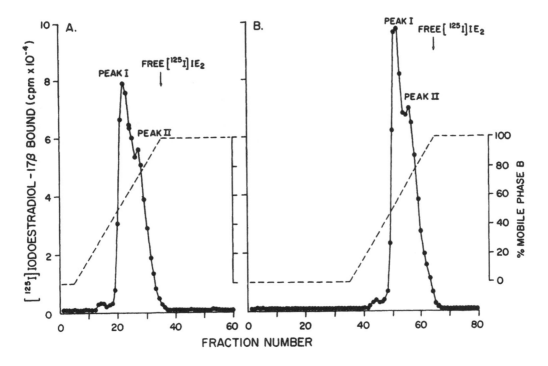

FIGURE 5. Influence of stationary phase contact time with ER on HPHIC separation profile of ER isoforms from rat uterus. Labeled cytosol was cleared of excess steroid and chromatographed on a CAA-HIC column with a 30 min linear gradient of ammonium sulfate from 2 to 0 M (A, control). A second sample of the same cytosol was injected and 2 M ammonium sulfate was first delivered isocratically for 30 min, followed by a 30-min gradient to 0 M ammonium sulfate (B). For clarity, only total cpm/fraction (●) is shown. (From Hyder, S. M., Sato, N., and Wittliff, J. L., *J. Chromatogr.*, 397, 251, 1987. With permission.)

to that obtained with original cytosol.[39] These data indicate that the two types of hydrophobic peaks arise due to highly specific surface characteristics of the receptor complexes. Reasons for this phenomenon include (1) direct interaction of the receptor with the bonded phase, (2) interactions of the associated nonreceptor component with the bonded phase, or (3) an alteration of either of the above interactions due to post-translational modifications in the receptor molecule itself, such as protein phosphorylation.

Collectively, our results indicate that the two hydrophobic isoforms of ER represent distinct molecular complexes. The physiological significance of these high molecular weight forms of receptor remains to be explored. The human estrogen receptor gene encodes a single protein of molecular weight 65000 daltons.[40,41] However, we have clearly shown the presence of polymorphism within this protein, which most probably arises as a result of post-transcriptional modification of proteins *in vivo*.

IV. ANALYSIS OF STEROID RECEPTORS ON A CAA-HIC POLYETHER BONDED HYDROPHOBIC COLUMN

When ER from rat uterine cytosol was analyzed on the CAA-HIC column with a linear gradient of 1 to 0 M ammonium sulfate, developed in 30 min, the majority of the receptor was eluted in the void volume (t_R = 3 to 4 min); a minor portion was eluted as two peaks with t_R = 16 min and t_R = 25 min.[27] However, when the sample was first adjusted to the ionic strength of the initial mobile phase, a relatively lower proportion of receptor was found in the void volume, while the remainder was found at t_R = 16 and 25 min. All the receptor isoforms were specific, as judged by DES inhibition of radioactivity, associated with the ER. This provided the first clue that either unfolding of receptor structure was important for promoting its hydrophobic bonding with the stationary phase or that the hydrophilic groups on the surface of the protein were neutralized and only the hydrophobic groups were left to interact with the column matrix.

To confirm the above observation, receptor was separated by using an elevated salt concentration ($2\,M$) in the initial mobile phase, with increasing ionic concentration in the sample injected. When cytosol was injected without altering its ionic strength, some specific receptor was eluted in the void volume as a result of the lack of interaction with the stationary phase. One other distinct component was present at $t_R = 25$ min together with a trailing edge. However, increasing the ionic strength of the sample to 2.0 to 1.5 M ammonium sulfate prior to injection completely eliminated the specific bound radioactivity in the void volume and resolved the receptor into two peaks of $t_R = 22$ min (peak I, $t_R = 22 \pm 1$ min, $n = 16$) the $t_R = 28$ min (peak II, $t_R = 27 \pm 0.5$ min, $n = 14$). Further increase of ionic strength in the sample to 2 M ammonium sulfate prior to injection led to a loss in resolution. Our data suggest that this loss of resolution is better described as a slow interconversion of peak I to peak II.[27] We chose to adjust all of our cytosol preparation to 1.5 M ammonium sulfate prior to injection. Ionic strength adjustment of the sample with ammonium sulfate does not denature the receptor as evidenced by its retention of biological activity (ligand binding), as indicated. We do not know, however, if high ionic strength alters the receptor protein structure in any way. One must presume that some alteration of the receptor with high ionic strength occurs since inclusion of ammonium sulfate exposes hydrophobic groups.[27] However, this alteration is not significant enough since both receptor ligand binding properties and immuno-recognition properties are intact.[27]

Steroid receptors are known to undergo structural changes which are ligand and time-dependent.[42] It has been known, for example, that liganded receptors are much more stable than unliganded receptors. In addition, we have found that the unliganded estrogen receptor is more prone to structural alterations following HPLC in the ion-exchange mode, leading to reduced association of receptors with specific monoclonal antibodies.[43] Some of the time-dependent alterations are related to receptor activation, i.e., the receptor affinity for binding to DNA increased.[44] We used HPHIC to study the time-dependent alteration of receptor structure to ascertain isoform conversion.

Because cytosols from uteri exhibit different proportions of peak I and peak II, following HPHIC, we analyzed different types. Regardless of the initial distribution of receptor on the hydrophobic column, overnight incubation always resulted in the conversion of peak I into peak II. We believe that peak II represents the active isoform of rat uterine cytosolic receptor which has an affinity for DNA. It is known that activation of receptor may be carried out following a brief exposure of cytosol to elevated temperatures, such as 25 to 30°C, for 20 to 30 min.[44] This temperature renders the receptor more hydrophobic, leading to a delayed elution of receptor peaks ($t_R = 27$ min) from the column. These results agree with time-associated changes indicating the active isoform of the receptor is more hydrophobic.[24] Such an increase in hydrophobicity may also result from aggregation of proteins during incubation, providing greater surface area for protein stationary phase interaction. In contrast to the results described for the rat uterine cytosol,[27] human breast cancer cytosol shows an opposite effect on overnight incubation, isoform II converting to isoform I.[28] Peak I may also represent the active isoform since the receptor in the peak is shown to bind DNA also.[21]

Steroid receptors are prone to aggregation and/or thermal degradation when incubated with or without steroid or chromatographed on strong hydrophobic matrices.[11] We were not sure whether the molecular heterogeneity observed following elution from the CAA-HIC column was due to time-dependent conformational changes taking place while the receptor was in contact with the stationary phase. However, we found that if the salt gradient was started immediately after sample injection or if mobile phase was first released isocratically for 30 min after sample injection and prior to the start of the gradient, the separation profiles obtained were the same (Figure 5). Therefore, we concluded that the stationary phase itself does not contribute to receptor heterogeneity. Importantly, the elution of receptor isoforms was dependent upon ionic strength and not time.

We have not evaluated receptor stability after longer time periods because these do not appear necessary in purification procedures. Because of the absence of stationary phase-induced conformational changes, one may inject multiple volumes of protein to increase sample loads. When the gradient developing time was increased from 30 to 50 min, it led to improved resolution of the two peaks and band broadening, as expected. No further distinct receptor species were seen. However, in other experiments where the two receptor isoforms were less well resolved, increasing the gradient time did not improve separation.

A. INFLUENCE OF RECEPTOR MODIFYING REAGENTS ON HYDROPHOBIC PROPERTIES OF ESTROGEN RECEPTOR

1. Sodium Molybdate

The use of sodium molybdate (MoO_4^{2-}) is gaining wide support, mainly because the steroid receptors are labile regulatory proteins which become even more unstable in the partially purified stage. However, MoO_4^{2-} protects the ligand binding property of the receptor and keeps the protein in the high molecular weight, non-transformed state.[45] The actual mechanism of action of MoO_4^{2-} is unknown, although several alternatives such as interaction directly with cysteine residues or sulfhydryl groups on the receptor protein[46] or with RNA[47] have been suggested.

Our laboratory and those of others have found that MoO_4^{2-} reduced receptor polymorphism and eluted the receptor as a single entity on weak anion-exchange columns.[31] This was not the case, however, when strong anion-exchange columns were used; in the presence of MoO_4^{2-}, two ionic isoforms were detected.[48]

Two molecular forms of estrogen receptors separated based upon hydrophobicity were observed either in the absence or presence of MoO_4^{2-}. Figure 6 shows the influence of MoO_4^{2-} and time of incubation on the separation of ER by HPHIC. When the receptor was extracted in the absence of MoO_4^{2-} and chromatographed, two receptor peaks with $t_R = 25$ to 27 min (peak I) and $t_R = 34$ to 36 min (peak II) were detected (Figure 6A). In a previous publication, the appearance of these two peaks was not time-dependent but rather dependent upon ammonium sulfate.[27] When receptor from the same tissue was extracted with buffers containing 10 mM MoO_4^{2-} and eluted with buffers containing 10 mM MoO_4^{2-}, two isoforms were again detected (Figure 6B). Peak MI was eluted at $t_R = 15$ to 17 min and peak MII was eluted at $t_R = 24$ to 26 min. Peaks I (Figure 6A) and MII (Figure 6B) exhibited similar retention behavior and appear to represent the same receptor isoforms. However, within experiments, MII was consistently eluted one or two fractions (1 to 2 ml) earlier. We attribute this minor variation to the presence of the oxyanion in eluting buffers. In any event, peak MI was the most hydrophilic compared to all other receptor isoforms detected by HPHIC. This increase in the hydrophilicity of the receptor proteins in the presence of MoO_4^{2-} may be responsible for receptor stabilization and maintenance of the regulatory protein in its non-activated form.

When the receptor was first extracted in buffers lacking MoO_4^{2-} and then made 10 mM with respect to MoO_4^{2-} (Figure 6C), the chromatographic behavior was similar to the results described in Figure 6B. Therefore, it appears that cytosol contains components responsible for promoting the MoO_4^{2-} effect in reducing the hydrophobicity of receptor protein. It is interesting that Toft's group[49] detected two different types of 8 S (Svedberg units) progesterone receptor (PR) when analyzed on sucrose density gradients. Other workers have shown that cytosol contains both 90 K heat-shock proteins and actin[50,51] components suggested to be involved in promoting MoO_4^{2-} effects on steroid receptors. This implies that certain receptor heterogeneity (polymorphism) may arise due to association with nonreceptor components. The extent of heterogeneity is likely to be dependent upon the stoichiometry of the interaction. To ascertain

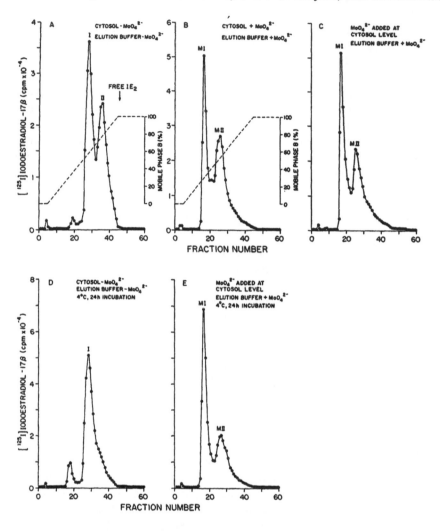

FIGURE 6. Influence of sodium molybdate and time of incubation on the separation of estrogen receptor isoforms by HPHIC. Breast cancer cytosol was injected onto the CAA-HIC column and eluted with buffers either lacking (A) or containing 10 mM sodium molybdate (B and C). Another aliquot of cytosol was incubated overnight (24 hr) and then chromatographed with buffers lacking (D) or containing (E) 10 mM sodium molybdate. Recovery of iodoestradiol in these experiments ranged between 85 to 100%. Arrow indicates position of free steroid. (From Hyder, S. M., Sato, N., Hogancauup, W., and Wittliff, J. L., *J. Steroid Biochem.*, 29, 197, 1988. With permission.)

this possibility, the composition of the two hydrophobic species of receptor is being analyzed by our laboratory exploiting their properties by various HPLC modes and affinity chromatography (manuscript in preparation).

MoO_4^{2-} conserved the specific ligand binding capacity of steroid receptors separated by HPHIC to an extent similar to that reported by Ratajczak et al.,[52] using sucrose density gradient centrifugation. Importantly, it also preserved the two molecular forms of receptor separated based on their property of hydrophobicity (Figures 6D and 6E). In contrast, in the absence of MoO_4^{2-}, the non-stabilized isoform II converted into peak I (Figure 6D). This does not appear to be a result of proteolysis, since trypsin-treated receptor consistently separated with a shorter retention time (t_R = 24 min; [Reference 53]). In certain cases where the receptor activity was reduced by 75% in the absence of MoO_4^{2-}, addition of MoO_4^{2-} stabilized both ligand binding capacity and preserved the hydrophobic forms of receptor (not shown). Without an understanding of the mode of action of MoO_4^{2-}, absence of the oxyanion appears to contribute to the loss of certain components in the high-molecular weight species leading to a more hydrophobic protein. This suggestion is supported by the observation of others.[52] Interestingly, when the molybdate concentration was increased from 0 to 5 or 10 mM, formation of a new highly acidic form of progestin receptor was detected by high performance chromatofocusing.[54]

In order to assess the interrelationship of the various hydrophobic forms of these receptors, the non-stabilized receptors were eluted with buffers containing MoO_4^{2-}. Elution revealed a shift of peak II into isoform MI, while peak I remained unaffected.[28] This alteration in migration position suggested a selective interaction of MoO_4^{2-} with receptors in peak II. This does not rule out the possibility that MoO_4^{2-} interacts with components in peak I without altering their hydrophobic properties.

To assess whether an extended (15 min) wash of immobilized, non-stabilized receptor with MoO_4^{2-}-containing buffers would totally convert peak II into MI (for example, Figure 6C), buffer containing 10 mM MoO_4^{2-} and 2 M ammonium sulfate was delivered for 15 min. The decreasing salt gradient was then initiated. Even after the extended wash with MoO_4^{2-}, peak II was not completely converted into MI. This result indicates that once bound to the stationary phase, only a portion of the receptor in peak II is under the influence of MoO_4^{2-}. One possibility is that a non-receptor macromolecule, which is less hydrophobic, has a higher affinity for receptor isoform II than the receptor has for the stationary phase. This results in elution of the macromolecular complexes. As this factor becomes limiting with time, proportion of receptor would be left associated with the stationary phase for elution later with its expected retention time.

To probe further the molecular interconversion of the hydrophobic species of receptor and the on-column behavior of these proteins, we analyzed MoO_4^{2-}-stabilized receptors with elution buffers lacking MoO_4^{2-}. Under these conditions, isoform MI converted to peak II. Component I again remained constant, implying that inter-conversion occurred primarily between peak II and MI. These results indicate that isoforms MII and I most probably represent a common species of ER within the experimental conditions used.

The effect of on-column behavior of protein was evaluated by using an extended wash of MoO_4^{2-}-stabilized receptor with buffers lacking MoO_4^{2-}. The receptor was injected onto the column and eluted with a salt gradient which was started either immediately or after a 15 min wash with high salt buffer.[28] There was a greater conversion of MI into peak II than of peak II into MI, again implying that, in the latter case, some component(s) were limiting. Even after extensive washing with these buffers, a small percentage (< 5%) of receptor remained as MI. Furthermore, a small proportion of MI receptor was retained even when the receptor was extracted from tissues with non-MoO_4^{2-}-containing buffer. This suggests that one of the native forms of the receptor is MI and that the interconversion into other hydrophobic species occurs on-column. These experiments also indicate the usefulness of HPHIC columns as an apparatus for buffer-exchange (dialysis) as well as its usefulness in showing that the MoO_4^{2-} effect on receptor is reversible, in agreement with Muller et al.[55]

MoO_4^{2-} selectively influenced the transformation of isoform II into MI and vice-versa. This suggests there would be a loss of peak II with a concomitant increase in component MI when cytosol was first made 10 mM with respect to MoO_4^{2-} and then chromatographed with MoO_4^{2-}-containing buffers. In general, isoform II converted into MI but, in a few experiments, an increase of isoform I also occurred. This observation suggests that receptors in MII/I may serve as an intermediate species for conversion of II to MI. Secondly, the presence of three hydrophobic species of receptors suggest that the conversion of isoforms in the presence of MoO_4^{2-}-containing buffers (Figure 6B) is not due to the oxyanion effect only. If this was the case, elution of non-MoO_4^{2-}-stabilized receptor with MoO_4^{2-}-containing buffers would have resulted in only two instead of the three isoforms observed. MoO_4^{2-} has been implicated to act directly on receptors but it appears that it may be involved in multiple mechanisms.

Previously, our group reported that ER from rat uterine cytosol also separated into similar isoforms as those shown here for human breast cancer.[27] However, unlike the interconversion of peak II to I after overnight incubation (Figures 6D and E), uterine receptor showed an opposite time-dependent conversion of isoform I into peak II. Based on these collective results, it appears that peak II represents the activated form of receptor, although recent analysis has shown that receptors from both peaks I and II may bind DNA.[21] Since the proteolyzed (trypsinized) receptor from human breast cancer was more hydrophilic than either peak I or II,[54] this suggested that peak I is not a proteolyzed form. Another interesting comparison between the HPHIC profiles of rat and human ER is that certain rat uterine cytosols (approximately 15%) exhibited only isoform I and none represented exclusively peak II.[27] In contrast, all samples analyzed from breast cancer cystosols exhibited both peaks I and II with variations in their relative quantities. We are now evaluating the origin of the two hydrophobic species.

2. Effect of N-Ethylmaleimide (NEM) on Separation Characteristics of Estrogen Receptors

Protection of sulfhydryl groups on the estrogen receptor are known to preserve certain ligand-binding properties. The DNA binding domain of ER is rich in cysteine residues.[41] In the absence of sulfhydryl-modifying reagents, the receptor was eluted at $t_R = 26$ min (peak I) and $t_R = 34$ (peak II).[35] Some free ligand was present ($t_R = 46$ min). The presence of free ligand suggests that the stationary phase may promote dissociation of steroid and receptor; this phenomenon varied from column to column. The Spherogel CAA-HIC column used in our earlier study[27] did not show this effect. In the presence of the sulfhydryl-modifying reagent, NEM, there was a dose-dependent loss of peak resolution.[35] There appeared to be a larger loss of peak I than of peak II with 10 mM NEM. This produced a broad peak, possibly owing to multiple conformations of receptor. Peak II appeared to be due to the interaction of the DNA-binding domain of the receptor with the stationary phase.[28] These results suggest that receptors in peak II are less susceptible to the NEM effect than to those in peak I. This is despite the fact that peak II may contain more sulfhydryl groups, which may be modified. Receptor activation, which exposes the DNA-binding domain of the receptor, is known to be a temperature-, time-, and ionic-strength-dependent process. In our experiments, the samples were adjusted to a high ionic strength just prior to injection, and this is likely to expose the DNA-binding domain. Prior to this treatment, sulfhydryl residues are unable to react with NEM, since they appear to be buried within the protein molecule. The period between increasing the ionic strength of the sample and injection (< 2 min) probably was insufficient for NEM to modify the sulfhydryl groups of the DNA-binding domain. In addition to loss of peak resolution in the presence of NEM, there was a reduction in the quantity of receptors, based on steroid-binding domains. Surprisingly, this loss was greater when sodium molybdate was present in the cytosol (20 to 30%) than when sodium molybdate was absent (5 to 10%). In addition, the presence of NEM in cytosols also led to increased stripping of ligand during chromatography. This appears to reflect the importance of sulfhydryl groups in maintaining the high affinity of the receptor for the ligand.[35]

3. Effects of Trypsin on Hydrophobic Properties of Receptor

Limited trypsin treatment of ER reduces the receptor size to *ca.* MW 35,000 with a loss in its DNA-binding properties.[22,56] The hydrophobic characteristics of ER in the intact and trypsin-treated (or mero-receptor) states were compared. Trypsinized receptor (I′) was consistently eluted earlier ($t_R = 23$ to 24 min) but similarly to isoform I ($t_R = 26$ to 28 min).[35] This suggests that the steroid-binding domain near the C-terminus region of the receptor must contribute significantly to the interaction with the stationary phase. Further evidence for this suggestion is provided by the trypsin study conducted in the presence of sodium molybdate. Although receptor was resolved into two isoforms in the presence of sodium molybdate after trypsin treatment, a single isoform (MII′, $t_R = 24$ min) was observed, similar to isoform I, in the absence of sodium molybdate. These results indicate that the chromatographic behavior of trypsin-treated receptor was independent of the sodium molybdate effect and reaffirm that isoform II involves interaction of the DNA-binding domain with the stationary phase. Furthermore, sodium molybdate does not prevent proteolysis of ER under the conditions used and should not be considered an inhibitor of receptor-modifying proteases.

4. Effect of RNase A on Receptor Hydrophobicity

It has been suggested that steroid receptors are associated with RNA.[57,58] However, it is now clear that RNA is not associated with receptor, as it is normally extracted from the cell. Rather, the receptor protein complex first undergoes dissociation, followed by reassociation with RNA molecules(s).[59] This may be observed when receptors are fractionated according to their size. Removal of RNA with RNase led to a form of glucocorticoid receptor migrating more slowly in sucrose density gradients.[57] We demonstrated that the high-molecular weight forms of ER do not contain RNA molecules[35] and that RNase treatment does not alter the hydrophobic properties of the receptor. When the effect of RNase A of the highest purity from two different sources was studied in concentrations up to 10 mg/ml, no effect was observed on receptor hydrophobic or size characteristics compared to a control. In addition, the presence of sodium molybdate in the reaction mixture also did not influence the hydrophobic properties of ER.[35] These data rule out the possibility that the appearance of isoform MI is due to interaction of receptor RNA complexes with the stationary phase via the nucleic acid. However, it cannot be ruled out that selective complexes, e.g., receptor-RNA-receptor (or receptor-RNA-another protein), interact with the stationary phase. Collectively, our results show that high molecular weight forms of ER exist independent of exposed RNA. It is unknown whether the high molecular weigh ER complex contains RNA buried within the protein structure. Results from other studies with glucocorticoid receptors have indicated that this is not the case.[59,60]

V. ELUTION OF PROTEIN KINASE ACTIVITY WITH THE LIGAND BINDING FORM OF ESTROGEN RECEPTOR

Current research in our laboratory has investigated a Mg^{2+}-dependent protein kinase activity, associated with immunopurified ER from human breast cancer cells (MCF-7)[26,61,62] and rat uterus.[27,62] It was shown that receptors eluted by HPIEC retained this kinase activity (unpublished observation). We investigated whether retention of kinase activity was also possible following HPHIC. Figure 7A illustrates a typical isoform chromatogram of ER, separated from rat uteri and used for analysis of protein kinase activity associated with ER. These separations resulted in a five- to twentyfold purification for each isoform, depending upon the relative proportion present. In this experiment, both components were purified *ca.* fifteenfold following a single pass. Karger's group[8] has had similar success in resolving two labile enzymes on the CAA-HIC column with retention of their activities, confirming its mild nature.

To demonstrate protein kinase activity, fractions from the receptor peaks (fractions 22 and 28) and two control points at fractions 12 and 50 were incubated directly with polystyrene beads linked to D547 monoclonal antibodies against ER.[63] A separate peak was incubated with non-fractionated receptor in P_{10}EDG buffer for each experiment. After an overnight incubation, these antibodies were washed (see Experimental) and then one bead was analyzed for ER content by the EIA procedure (Abbott Labs) and the other was used for protein kinase assay with phosvitin as the exogenous substrate. Histones were also used successfully as substrates. Figure 7B is an auto-radiogram demonstrating that only the ER isoform eluted in fraction 22 was immunoprecipitated with monoclonal antibody D547 and exhibited protein kinase activity. Importantly, no reaction was observed when the monoclonal antibody was allowed to interact with fraction 12, where most of the nonreceptor proteins were eluted.

Unlike previous studies with human breast cancer cells,[26,61] we have not been able to

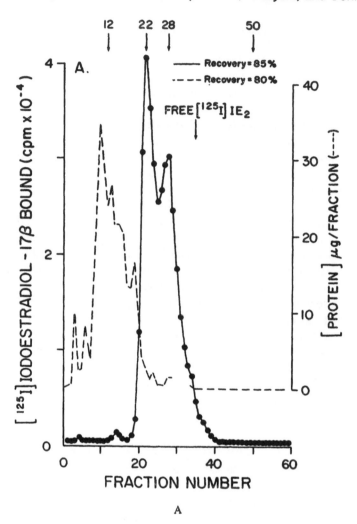

A

FIGURE 7. Protein kinase activity, associated with ER isoforms, separated by HPHIC. (A) Rat uterine cytosol was chromatographed on the CAA-HIC column as described in Experimental. (●) Total cpm/fraction, (- - -) protein profile, as determined by the Bradford procedure.[13] For clarity, the non-specific binding profile which was virtually undetectable is omitted. (B) Fractions 12, 22, 28, and 50 from the HPHIC-separated sample (shown in A) were directly incubated with monoclonal antibody (D547), which was coated on polystyrene beads. A non-fractioned control sample was also incubated with the monoclonal antibody complex bead. Following an 18-hr incubation and subsequent washing, one bead was analyzed for ER content (mass) by an EIA procedure and the second bead was tested for protein kinase activity, as described in Experimental. The receptor content associated with the mono-clonal antibody in fmol/bead was 0 in fraction 12, 1.4 in fraction 22, 2.1 in fraction 28, and 0 in fraction 50. The control bead contained 7 fmol of receptor from the unfractioned cytosol in this representative experiment. (From Hyder, S. M., Sato, N., and Wittliff, J. L., *J. Chromatogr.*, 397, 251, 1987. With permission.)

demonstrate auto-phosphorylating activity of ER from rat uterus. In the present experiments (Figure 7), both isoforms were purified to the same extent (ca. fifteen- to sixteenfold) and yet only isoform I (peak I) exhibited protein kinase activity. Our current analysis indicates that some

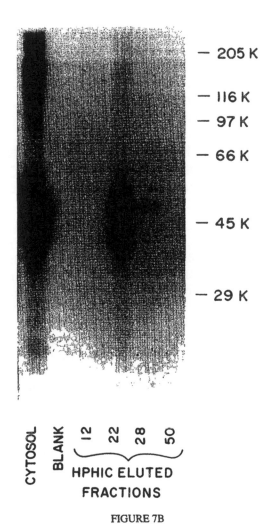

— 205 K

— 116 K

— 97 K

— 66 K

— 45 K

— 29 K

CYTOSOL   BLANK   12   22   28   50

HPHIC ELUTED
FRACTIONS

FIGURE 7B

of the bead-associated kinase may be a non-specific component(s) which was precipitated simultaneously with the receptor monoclonal antibody complexes.[28] Furthermore, immunoprecipitation of kinase(s) occurred when either receptor positive or negative human breast cancer tissue was analyzed with the monoclonal antibody D-547 linked to polystyrene beads.

The extent of this kinase activity varied from tissue to tissue; certain ER positive tissues showed a negative result, while some ER negative tissues showed a positive result.[28] Receptor measurements were made by the multipoint titration assay[65] and confirmed by a single-point saturation (3 nM) assay. Receptor negativity or positivity was also confirmed with a second monoclonal antibody H222 which was peroxidase-labeled (Abbott Labs) in an EIA assay. Therefore, unless there is a non-ligand binding form of the receptor which is not recognized by the H222 monoclonal antibody bead, these results indicate non-specific interaction of certain ATP/Mg^{2+}-dependent kinases with either the D-547 monoclonal antibody or with the polystyrene beads.

We have not excluded the possibility that receptor may be associated with protein kinase activity as reported by others.[65] However, the present methodology, employing monoclonal antibody on a solid matrix is capable of precipitating other kinase(s) in a non-specific fashion (also, see Reference 66). On the other hand, recent reports indicate that what initially appeared to be a steroid receptor associated kinase actually turned out to be a separate entity when more stringent conditions were used.[67] In our study, the origin of the protein kinase which is precipitated with the polystyrene matrix is unknown. A recent report[66] has shown that other non-receptor proteins were capable of binding to the immunomatrix (antibody beads). We have reported similar observations.[22,68] Nevertheless, if this enzymatic activity is not associated directly with the ER, the methodology of HPHIC still describes a successful separation of protein kinase(s) which themselves are labile regulatory components of the cell.

VI. A NOVEL METHOD OF SIMULTANEOUS IDENTIFICATION OF ESTROGEN AND PROGESTERONE RECEPTORS BY HPLC USING A DOUBLE ISOTOPE ASSAY

Polymorphism of ER and progestin receptors (PR) was analyzed simultaneously using HPHIC to characterize their isoforms.[69] ER and PR were prepared from human breast cancer and labeled with 4 nM of either [^{3}H]estradiol-17β (^{3}HE) or [^{125}I]iodo-estradiol-17β (^{125}IE), while PR was associated with 5 nM of either [^{3}H]R5020 (^{3}HR) or [^{125}I]iodovinylnortesterone (^{125}IVNT).[69] ER was resolved by HPHIC into isoforms MI (t_R = 11 min), I (t_R = 16 min), and II (t_R = 24 min). PR separated into isoforms MI (t_R =14 min) and I (t_R = 21 min, 80% of specific binding) when eluted with the same gradient used for ER chromatography. Upon inclusion of 10 mM molybdate, ER resolved into isoforms MI and MII (t_R = 16 min) and PR into isoforms MI and I (here, however, isoform MI represented 80 to 95% of specific binding). HPHIC profiles of ER isoforms labeled with either ^{125}IE or ^{3}HE were identical (Figure 8) as were PR isoform profiles labeled with either ^{3}HR or ^{125}IV.[69] Pairs of ^{125}I and ^{3}H labeled ligands were used in either combination to monitor ER and PR profiles simultaneously. Isoforms analyzed in 50 biopsies gave reproducible retention times; however, the ratio between I and II for ER and MI and I for PR varied. This method allows the rapid, simultaneous monitoring of chromatographic behavior of ER and PR using various HPLC modes. It is particularly useful in combination with HPHIC to establish the interrelationships of receptor isoforms and the native composition of these labile, regulatory proteins.

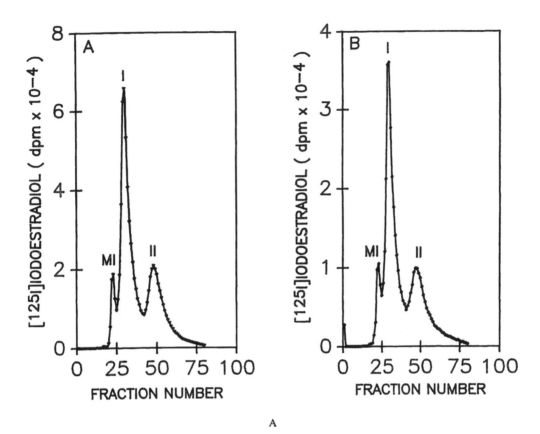

A

FIGURE 8. HPHIC profiles of ER from a human breast cancer cytosol using a double-isotope protocol. Cytosol was prepared in P_{10}EDG buffer, as described earlier, labeled with either 4 nM ^{3}HE or 2 nM ^{125}IE. After removing unbound ligand, labeled cytosols were applied to the HPHIC column. ER isoform profiles from either type of ligand using a single isotope labeling approach (A and C) were virtually identical to those detected by the double-isotope assay (B and D). As shown, the ER profile using ^{125}IE from the double isotope assay (B) gave a better resolution compared to that using ^{3}HE (D). (From Folk, P., Dong, J., and Wittliff, J. L., 9th Int. Symp. on HPLC of Proteins, Peptides and Polynucleotides, Philadelphia, Nov. 6—8, 1989.)

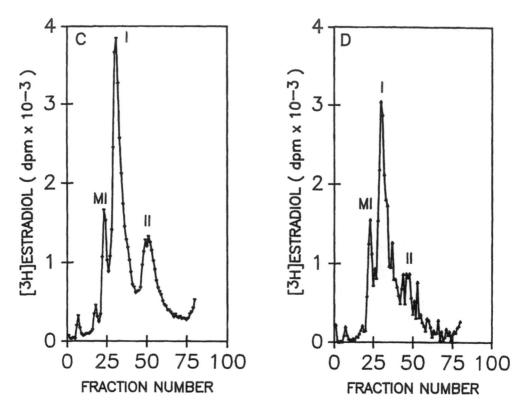

FIGURE 8 (continued)

REFERENCES

1. **Fausnaugh, J.L., Pfannkoch, E., Gupta, S., and Regnier, F.E.,** High-performance hydrophobic interaction chromatography of proteins, *Anal. Biochem.*, 137, 464, 1984.
2. **Wittliff, J.L., Wiehle, R.D., and Hyder, S.M.,** HPLC as a means of characterizing the polymorphism of steroid hormone receptors, in *The Use of HPLC in Receptor Biochemistry*, Kerlavage, A.R., Ed., Alan R. Liss, New York, 1989, 155.
3. **Hyder, S.M. and Wittliff, J.L.,** High-performance hydrophobic interaction chromatography of a labile regulatory protein: the estrogen receptor, *BioChromatography*, 2, 121, 1987.
4. **Hyder, S.M. and Wittliff, J.L.,** High-performance hydrophobic interaction chromatography, in *Current Protocols in Molecular Biology*, Ansusel, F., Brent, R., Moore, D., Smith, J.A., Seidman, J., and Struhl, K., Eds., Greene Publishing, New York, 1987, 10.15.1.
5. **Arfmann, H.A. and Shalteil, S.,** Resolution and purification of histone on homologous series of hydrocarbon-coated agarose, *Eur. J. Biochem.*, 70, 269, 1987.
6. **Gooding, D.L., Schmuck, M.N., and Gooding, K.M.,** Analysis of protein with new, mildly hydrophobic high-performance liquid chromatography packing material, *J. Chromatogr.*, 296, 107, 1984.
7. **Kato, Y., Kitamura, T., and Hashimoto, T.,** Operation variables in high-performance hydrophobic interaction chromatography of proteins on TSK Gel Phenyl-5PW, *J. Chromatogr.*, 298, 407, 1984.
8. **Miller, N.T., Feibush, B., Corina, K., Lee, S.P., and Karger, B.L.,** High-performance hydrophobic interaction chromatography: purification of rat liver carbamoylphosphate synthetase I and ornithine transcarbonylase, *Anal. Biochem.*, 148, 510, 1985.
9. **Sica, V. and Bresciani, F.,** Estrogen binding proteins of calf uterus. Purification to homogeneity of receptor from cytosol by affinity chromatography, *Biochemistry*, 19, 2369, 1979.

10. Schrader, W.T., Birnbaumer, M.E., Hughes, M.R., Weigel, N.L., Grody, W.W., and O'Malley, B.W., Studies on the structure and function of chicken progesterone receptor, *Rec. Prog. Hormone Res.*, 37, 583, 1981.

11. Bruchovsky, N., Rennie, P.S., and Comeau, T., Partial purification of nuclear androgen receptor by micrococcal nuclear digestion of chromatin and hydrophobic interaction chromatography, *Eur. J. Biochem.*, 120, 399, 1981.

12. Wittliff, J.L. and Wiehle, R.D., Analytical methods for steroid hormone receptors and their quality assurance, in *Hormonally Sensitive Tumors*, Hollander, V.P., Ed., Academic Press, New York, 1985, 383.

13. Kute, T.E., Heidemann, P., and Wittliff, J.L., Molecular heterogeneity of cytosolic forms of estrogen receptors from human breast tumors, *Cancer Res.*, 38, 4307, 1978.

14. Puca, G.A., Nola, E., Sica, V., and Bresciani, F., Estrogen-binding protein of calf uterus. Partial purification and preliminary characterization of two cytoplasmic proteins, *Biochemistry*, 10, 3769, 1971.

15. Hutchens, T.W., Wiehle, R.D., Shahabi, N.A., and Wittliff, J.L., Rapid analysis of estrogen receptor heterogeneity by chromatofocusing with high performance liquid chromatography, *J. Chromatogr.*, 266, 115, 1983.

16. Wiehle, R.D., Hofmann, G.E., Fuchs, A., and Wittliff, J.L., High performance size exclusion chromatography as a rapid method for the separation of steroid hormone receptors, *J. Chromatogr.*, 39, 307, 1984.

17. Wiehle, R.D. and Wittliff, J.L., Isoforms of estrogen receptors by high performance ion-exchange chromatography, *J. Chromatogr.*, 297, 313, 1984.

18. Shahabi, N.A., Hutchens, T.W., Wittliff, J.L., Halmo, S.D., Kirk, M.E., and Nisker, J.A., Physicochemical characterization of estrogen receptors from a rabbit endometrial carcinoma model, in *Progress in Cancer Research and Therapy*, Vol. 31, Raynaud, J.P., Bresciani, F., King, R.J.B., and Lippman, M.E., Eds., Raven Press, New York, NY, U.S.A., 1984, 63.

19. Van de Walt, L. and Wittliff, J.L., High-resolution separation of molybdate-stabilized progestin receptors using high-performance liquid chromatography, *J. Chromatogr.*, 425, 277, 1988.

20. Hyder, S.M. and Wittliff, J.L., Separation of two molecular forms of human estrogen receptor by hydrophobic interaction chromatography: Gradient optimization and tissue comparison, *J. Chromatogr.*, 476, 455, 1989.

21. Folk, P., Hyder, S.M., and Wittliff, J.L., Hydrophobic properties of activated estrogen and progestin receptors, The Endocrine Society Meeting, Seattle, WA, June 21—24, 1989.

22. Sato, N., Hyder, S.M., Chang, L., Thais, A., and Wittliff, J.L., Interaction of estrogen receptor isoforms with immobilized monoclonal antibodies, *J. Chromatogr.*, 359, 475, 1986.

23. Boyle, D.M., Wiehle, R.D., and Wittliff, J.L., Rapid high-resolution procedure for assessment of estrogen receptor heterogeneity in clinical samples, *J. Chromatog.*, 327, 369, 1985.

24. Bradford, M.M., A rapid and sensitive method of quantitation of microgram quantities of protein utilizing the principle of protein dye binding, *Anal. Biochem.*, 72, 248, 1976.

25. Hyder, S.M., Wiehle, R.D., Brandt, D.W., and Wittliff, J.L., High-performance hydrophobic interaction chromatography of steroid hormone receptors, *J. Chromatogr.*, 327, 237, 1985.

26. Baldi, A., Boyle, D.M., and Wittliff, J.L., Estrogen receptor is associated with protein and phospholipid kinase activities, *Biochem. Biophys. Res. Commun.*, 135, 597, 1986.

27. Hyder, S.M., Sato, N., and Wittliff, J.L., Characterization of estrogen receptors and associated protein kinase activity by high-performance hydrophobic-interaction chromatography, *J. Chromatogr.*, 397, 251, 1987.

28. Hyder, S.M., Sato, N., Hogancamp, W., and Wittliff, J.L., High-performance hydrophobic interaction chromatography of estrogen receptors and magnesium dependent protein kinase(s): detection of two molecular forms of estrogen receptor in the presence and absence of sodium molybdate, *J. Steroid Biochem.*, 29, 197, 1988.

29. Laemmli, U.K., Cleavage of structural proteins during the assembly of the head of bacteriophase T4, *Nature*, 227, 680, 1970.

30. Shahabi, N.A., Hyder, S.M., Wiehle, R.D., and Wittliff, J.L., HPLC analysis of estrogen receptor by a multidimensional approach, *J. Steroid Biochem.*, 24, 1151, 1986.

31. Wittliff, J.L., HPLC of steroid-hormone receptors, *LC.GC*, 4, 1092, 1986.

32. Wittliff, J.L., Shahabi, N.A., Hyder, S.M., van der Walt, L.A., Myatt, L., Boyle, D.M., and He, Y.J., High-performance liquid chromatography as a means of characterizing isoforms of steroid hormone receptor proteins, in *Protein: Structure and Function*, L'Italien, J.J., Ed., Plenum Publishing, New York, 1987, 61.

33. Maggi, A., Compton, J.G., Fahnstock, M., Schrader, W.T., and O'Malley, B.W., Purification of chick oviduct progesterone receptor apoprotein, *J. Steroid Biochem.*, 15, 63, 1981.

34. Tenenbaum, A. and Leclercq, G., Different chromatographic behavior on blue Sephanse CL-6B of free and estradiol complexed forms of the estrogen receptor from DMBA-induced rat mammary tumors, *J. Steroid Biochem.*, 13, 829, 1980.

35. Hyder, S.M. and Wittliff, J.L., High performance hydrophobic interaction chromatography as a means of identifying estrogen receptors expressing different binding domains, *J. Chromatogr.*, 44, 225, 1988.

36. Sullivan, W.P., Sullivan, B.T., Vroman, V.J., Bauer, R.K., Puri, R.M., Pearsdon, G.R., and Toft, D.O., Immunological evidence that the non-hormone binding component of avian steroid receptors exist in a wide range of tissues and species, *Biochemistry*, 24, 6586, 1985.

37. Toft, D., personal communication, 1988.
38. Wittliff, J.L., Steroid receptor analysis. Quality control and clinical significance, in *Cancer of the Breast*, Donegan, W. and Spratt, J., Eds., W.B. Saunders, Philadelphia, 1988, ch. 11, 303.
39. Hyder, S.M. and Wittliff, J.L., Detection of two high molecular weight hydrophobic forms of the human estrogen receptor, *J. Steroid Biochem.*, 33, 965, 1989.
40. Evans, R.M., The steroid and thyroid hormone receptor superfamily, *Science*, 240, 889, 1988.
41. Green, G., Walter, P., Kumar, V., Krust, A., Burnest, J.M., Argus, P., and Clambon, P., Human estrogen receptor DNA: Sequence, expression and homology to versa, *Nature*, 320, 134, 1986.
42. Katzenellenbogen, J.A., Ruth, T.S., Carlson, K.E., Iwamoto, H.S., and Gorski, J., Ultraviolet photosensitivity of the estrogen binding protein from rat uterus. Wavelength and ligand dependent photocovalent attachment of estrogens to protein, *Biochemistry*, 14, 2120, 1975.
43. Sato, N., Hyder, S.M., Chang, L., Thais, A., and Wittliff, J.L., Interaction of estrogen receptor isoforms with immobilized monoclonal antibodies, *J. Chromatogr.*, 359, 475, 1986.
44. Hyder, S.M., Murdocl, E., Lim, L., and Myatt, L., The interaction of estrogen receptor with oligodeoxynucleotides, *Biochem. Soc. Trans.*, 12, 322, 1984.
45. Redeuilh, G.C., Baulieu, E.E., and Richard-Foy, H., Calf uterine estradiol receptor: effects of molybdate on salt induced transformation process and characterization of a non-transformed receptor state, *J. Biol. Chem.*, 256, 11496, 1981.
46. de Boer, W., Bolt, J., Brinkmann, A.O., and Mulder, E., Differential effects of molybdate on the hydrodynamic and DNA-binding properties of the non-activated and activated form of the androgen receptor in calf uterus, *Biochem. Biophys. Acta*, 889, 240, 1986.
47. Feldman, M., Kallos, J., and Hollander, V.P., RNA inhibits estrogen receptor binding to DNA, *J. Biol. Chem.*, 256, 1145, 1980.
48. Ruh, M.F. and Ruh, T.S., Analysis of two forms of the molybdate-stabilized estrogen receptor, *Endocrinology*, 115, 1341, 1984.
49. Daugherty, J.J., Puri, R.K., and Toft, D.O., Polypeptide components of two 8S forms of chicken oviduct progesterone receptor, *J. Biol. Chem.*, 259, 8004, 1984.
50. Schuh, S., Yonemento, W., Brugge, J., Bauer, V.J., Riehl, R.M., Sullivan, W.R., and Toft, D.O., A 90,000-dalton binding protein common to both steroid receptors and the rous sarcoma virus transforming protein, pp60[V-SRC], *J. Biol. Chem.*, 260, 14292, 1985.
51. Koyasu, S., Nishida, E., Kadowaki, T., Matsuzaki, F., Iida, K., Harada, F., Kasuga, M., Sakai, H., and Yahara, I., Two mammalian heat shock proteins, HSP90 and HSP100, are actin-binding proteins, *Proc. Natl. Acad. Sci., U.S.A.*, 83, 8054, 1986.
52. Ratajczak, T., Sounec, A.M., and Hahnel, R., Requirement for a reduced sulfhydryl entity in the protection of molybdate-stabilized estrogen receptor, *FEBS Lett.*, 149, 80, 1982.
53. Hyder, S.M., Hogancamp, W.H., and Wittliff, J.L., Hydrophobic characteristics of estrogen receptor (ER) isoforms in the presence of protein modifying reagents. 7th Int. Symp. HPLC Abstr. 816, Washington, D.C., Nov. 2—4, 1987.
54. Mujaji, W.B., van der Walt, L., Shahabi, N., Hyder, S.M., and Wittliff, J.L., Assessment of progestin receptors in the presence of sodium molybdate using HPLC, unpublished data.
55. Muller, R.E., Traish, A.M., Beebe, D.A., and Wotiz, H.H., Reversible inhibition of estrogen receptors transformation by sodium molybdate, *J. Biol. Chem.*, 257, 1295, 1982.
56. Greene, G., Sobel, N., King, W., and Jensen, E.V., Immunochemical studies of estrogen receptor, *J. Steroid Biochem.*, 20, 51, 1984.
57. Tymoczko, J. and Phillips, M.M., Dexamethasone receptor: increased affinity for deoxyribonucleic acid and altered sedimentation profile, *Endocrinology*, 112, 142, 1983.
58. Feldman, M., Kallos, J., and Hollander, V.P., RNA inhibits estrogen receptor binding to DNA, *J. Biol. Chem.*, 256, 1145, 1981.
59. Vedeckis, W.V., LaPointe, M.C., and Kovacic-Milivojevic, B., Glucocorticoid receptor analysis using high performance size exclusion liquid chromatography, *BioChromatography*, 2, 121, 1987.
60. Kasayama, S., Noma, K., Sato, B., Nakao, M., Nishizawa, Y., Matsumoto, K., and Kishimoto, S., Sodium molybdate converts the RNA-associated transformed oligomeric form of the glucocorticoid receptor into the transformed, monomeric form, *J. Steroid Biochem.*, 28, 1, 1987.
61. Baldi, A., Hyder, S.M., Sato, N., Boyle, D.M., and Wittliff, J.L., Monoclonal antibody recognition of estrogen receptor isoforms, in *Advances in Gene Technology: Molecular Biology of the Endocrine System*, ICSU Short Reports, Vol. 4, Puett, D. et al., Eds., Cambridge University Press, Cambridge, England, 1986, 264.
62. Wittliff, J.L., Hyder, S.M., and Baldi, A., Association of protein kinase activities with estrogen receptor in breast and uterine cells, in *Receptor Phosphorylation*, Moudgil, V.K., Ed., CRC Press, Boca Raton, FL, 1988, 357—371.
63. Anonymous, Estrogen receptor determination with monoclonal antibodies, *Cancer Res.* (Suppl.) 42, 4231, 1986.

64. **Wittliff, J.L.,** Steroid hormone receptors in breast cancer, *Cancer,* 53, 630, 1984.
65. **Singh, V.B. and Moudgil, V.K.,** Protein kinase activity of purified rat liver glucocorticoid receptor, *Biochem. Biophys. Res. Commun.,* 125, 1067, 1984.
66. **Sato, B. and Matsumoto, K.,** Effects of estrogen and vanadate on the proliferation of newly established transformed mouse Leydig cell line *in vitro, Endocrinology,* 120, 1112, 1987.
67. **Garcia, T., Buchou, T., Renoir, J.M., Mester, J., and Baulieu, E.E.,** A protein kinase co-purified with chick oviduct progesterone receptor, *Biochemistry,* 25, 7937, 1986.
68. **Hyder, S.M., Sato, N., Chang, L., Meyer, J. and Wittliff, J.L.,** Recognition of estrogen receptor isoforms from human breast cancer by immobilized monoclonal antibodies, *Tumor Diagnostic and Therapie,* 9, 233, 1988.
69. **Folk, P., Dong, J., and Wittliff, J.L.,** A novel method of simultaneous identification of estrogen and progesterone receptors by HPLC using a double isotope assay, 9th Int. Symp. on HPLC of Proteins, Peptides and Polynucleotides, Philadelphia, Nov. 6—8, 1989.

Section VII
Affinity and Immunoaffinity Chromatography

HIGH-PERFORMANCE AFFINITY CHROMATOGRAPHY OF PEPTIDES AND PROTEINS

Jennifer E. Van Eyk, Colin T. Mant, and Robert S. Hodges

I. INTRODUCTION

All methods of chromatography exploit physical or chemical differences between molecules in order to achieve their separation. In the case of affinity chromatography, separation is based on the biological properties of the molecule(s), in particular their relative binding affinities for a specific ligand bound to an inert column support. The choice of ligand is critical, and is the main determinant for achieving the extreme specificity of which this chromatographic method is capable. The advantage of affinity chromatography over other chromatographic methods is this inherent biospecificity, coupled with the option of employing nondenaturing conditions during the separation.

Affinity chromatography is a powerful technique for both investigating the molecular properties of protein-protein or peptide-protein interactions and for purifying proteins. Since the advent of high-performance affinity chromatography (HPAC),[1] which dramatically decreases the time required for each chromagraphic separation compared to conventional affinity chromatography, the routine identification and quantitation of minute amounts of protein(s) from complex mixtures of proteins and peptides has become possible. Recent reviews of HPAC can be found in References 2 and 3. This article will deal with the optimization of HPAC when peptides and proteins are used as the solute and ligand molecules, respectively.

II. GENERAL PRINCIPLES OF HPAC

The overall processes involved in affinity chromatography are based on the bioselective interaction between the molecule(s) of interest in the sample mixture and the ligand which is attached to an inert column support. Figure 1 is a schematic representation of the basic procedures involved in achieving separation of molecules by affinity chromagraphy. A sample which contains both the molecule of interest and impurities is loaded onto the affinity column (Step 1); adsorption of the solute molecule occurs upon binding to the ligand, while the impurities are washed through the column (Step 2); the solute molecules are desorbed from the ligand by a competing desorption agent present in the dissociating buffer (Step 3); finally, the column is then re-equilibrated (Step 4) for the next separation. At each step in the separation, there are numerous conditions which must be optimized for each particular separation: thus, the choice of ligand and its chemical attachment to the inert column support will dictate the type of column that can be used; the type of desorption agent and whether it is applied as a step gradient or as a linear gradient will determine the efficiency of the separation; the choice of flow-rate will affect peak width, peak height and peak elution volume.

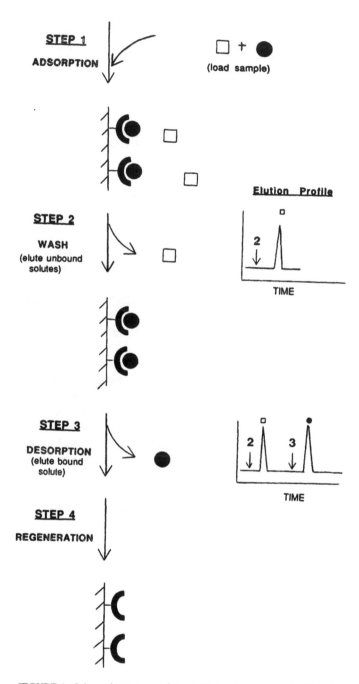

FIGURE 1. Schematic representation of affinity chromatography. *Step 1*: Sample is loaded onto the affinity column. The solute molecule is biospecifically adsorbed onto the column when it binds to the ligand-support. *Step 2*: The sample impurities are washed through the column and detected as the first peak. *Step 3*: The solute molecule is released from the ligand by the dissociation (or desorbing) buffer containing a desorbing agent. The second peak will contain the pure solute molecule. *Step 4*: The affinity column is regenerated and is ready for the next separation.

III. ADVANTAGES OF HPAC OVER CONVENTIONAL AFFINITY CHROMATOGRAPHY

The main advantage of HPAC over conventional affinity chromatography is the increased speed during the loading, elution, and regeneration steps of the separation. This is achieved as a result of the small particle size of the silica or polymer packing used in HPAC compared to that used in conventional soft gels. The small diameter of the particles (~10 μm) results in increased mass transfer rates that allow the solute to move in and out of the matrix more rapidly. This means that the dynamic sample capacity of the column (the capacity of the column when flowing) is maintained, even at high flow rates. Hence, the small particle size used in HPAC increases the efficiency of the column, not by increasing the number of theoretical plates as with other chromatographic methods, but rather by increasing the capacitance of the column. Flow-rates of up to 140-fold faster than conventional columns can be used with HPAC. However, since the capacity of the column and efficiency of the ligand-molecule interaction depends on the diffusion rate of the solute molecule(s) plus its association rate with the ligand, the capacity of the column may be reduced at very high flow-rates. Excessively high flow-rates are not normally used because of large column backpressures which may damage the immobilized ligand. Hence, for most affinity chromatography, flow-rates of 0.5 to 1.0 ml/min are utilized for analytical columns, where sample capacity is less affected by flow-rate. It is especially important during purification or quantitation of proteins that the sample capacity of the column be greater than the quantity of the solute molecule in the sample. Though lower flow-rates can increase sample capacity, it should be ensured that the total quantity of sample is able to bind to the ligand and not go through the column along with the impurities. The sample capacity of a column can also be affected by the quantity of the ligand derivatized on the column.

Due to the reduced time for each separation in HPAC, multiple runs can be carried out in a single day, thus eliminating any day-to-day fluctuations in results. This is a particular concern when comparing quantities isolated from different samples or investigating the elution times between various related solute molecules. The reproducibility between runs can be very high in HPAC; for example, three consecutive runs of peptide binding to a protein affinity column resulted in peptide elution within a 0.2-min margin over a 30-min run time.[4]

IV. CHOICE OF LIGAND AND LIGAND ATTACHMENT

There are two classes of ligands that can be used in affinity chromatography: ligands can be specific for a particular molecule or may be multifunctional and bind a number of related molecules. Employing the second class of ligand depends on both the sample and experimental application. A ligand must possess the following properties: it must bind reversibly to the solute molecule of interest, be stable under conditions required to dissociate the molecule, and possess high specificity for the solute, i.e., it must not bind impurities that might be present in the sample. Equally important, the ligand must contain a functional group that can be used to couple it to the column support such that it does not interfere with the solute binding site. Thus, when the ligand is bound to the support, its binding site for the solute is unaffected.

The chemistry of ligand immobilization to the stationary support is critical, since the interaction with the solute molecule may be affected by conformational changes or steric hindrance induced in the ligand upon attachment to the support. Unfortunately, at the present time there is not a large choice of matrix functional groups on commercially available underivatized HPAC supports. Most researchers are restricted to using the commercially prepacked HPAC columns that are available as either activated supports (ready for coupling the ligand of choice) or as ready-to-use matrixes that have commonly used ligands (e.g., Protein A) already coupled to the gel. Table 1 lists several manufacturers and underivatized columns which are available.

There are a number of different matrix functional groups available which will bind amino, hydroxyl or sulfhydryl groups of amino acids on the ligand. The most versatile is the epoxy functional group. Figure 2 illustrates the chemistry of epoxide linkages of a ligand to a silica support. Although epoxy groups bind preferentially to sulfhydryl groups, there is some selectivity towards various other functional groups (e.g., hydroxy and amino) on the ligand, depending on the pH of the derivatizing buffer. Thus, when derivatizing a column with a protein ligand, a certain percentage of the protein ligands will be attached to the support by different reactive groups. In addition, since hydroxy (OH) and amino (NH_2) groups are prevalent in an amino acid sequence, a protein would be attached at different sites along its sequence. This is not usually a problem, even if some ligand-support linkages occur at the ligand-solute binding site, since the majority of attachment sites will be at sites other than that of the ligand-solute interaction. Under acidic conditions, the protonated epoxide group is extremely reactive and will react with nucleophiles. The relative reactivity is SH > OH > NH_3^+. To promote amino linkages, the pH is increased and/or a non-aqueous solvent (e.g., acetonitrile) is used so that the epoxy group is not protonated; also, NH_2 is a stronger nucleophile than NH_3^+. It should be noted that, since supports that couple only to primary amines will bind at numerous sites located throughout the protein, the number of ligands on the column able to interact with the molecule of interest may be lowered.

Ligands of small molecular weights, such as peptides, pose additional problems during ligand-support attachment. Locating the functional group involved in the binding of the ligand to the support away from the actual binding site for the solute molecule is difficult due to the small size of the peptide. Ideally, it should be at either the amino or carboxyl terminus of the peptide, whichever is located furthest away from the amino acid residues involved in solute binding. Thus, if synthesizing a peptide to be used as an affinity chromatographic ligand, a cysteine linker could be added at either terminus of the sequence, if there is no cysteine already in the sequence, for reaction with the epoxy functional group on the support. Even so, the closeness of the ligand attachment site to the solute binding site can lead to the disruption or inaccessibility of the solute for the ligand. The use of spacer arms to expose the small peptide to the solvent is commonly used in conventional affinity chromatography. Although the number of commercially available HPAC underivatized columns is limited, Nu-Gel H-AF polyepoxy support, manufactured by Separation Industries, has a long (12 Å) spacer arm attached to a silica support. When dealing with a peptide and a protein, it is recommended to use the protein as the ligand and the peptide as the solute molecule. In so doing, attachment problems are minimized. However, if it is essential for a peptide to be a ligand, the above concerns should be kept in mind. An example where a peptide was successfully used as a ligand is with a synthetic peptide of a region of troponin I (TnI) which was used to bind troponin C (TnC; MW = 17,960 daltons).[5] The sequence of the TnI peptide is Ac-G^{104}-K-F-K-R-P-P-L-R-R-V-R^{115}-amide. The conventional CNBr Sepharose matrix, which binds to primary amines, was used, but an HPAC column could have been employed. The coupling of the peptide to the matrix was through the ε-NH_2 of the Lys residues at positions 105 and 107. Since the peptide when bound to the support still bound TnC, the residues of the peptide important for binding this protein must be situated distal to the residues involved in ligand-support attachment. Ni and Klee[6] also reported an interesting example of using peptide ligands, when they coupled calmodulin tryptic fragments to Sepharose 4B. This affinity packing was then used to test the ability of different calmodulin-regulated enzymes to recognize different domains of calmodulin.

Another concern when derivatizing a ligand is that its concentration on the support should be reduced if either the ligand or the solute molecule has a large molecular weight, since the

TABLE 1
Examples of Underivatized HPAC Columns

Manufacturer	Column name	pH range	Pore size (Å)	Particle size (μm)	Support material	Support functional group	Ligand specificity
Beckman	Ultraffinity-EP	2–7	—	10	Silica	Epoxy	SH NH$_2$ OH
Biorad	Affi-Prep 10	2–10	N/A	N/A	Polymer	N-hydroxysuccinimide	NH$_2$
Pierce	SelectiSpher-10	<8.5	300	10	Silica	Tresyl	NH$_2$ SH
Separation Industries	NuGel H-AF polyepoxy	2.5–8.5	300	10	Silica	Epoxy	NH$_2$
Separation Industries	NuGel H-AF polybond	2.5–8.5	300	10	Silica	Hydroxyl	NH$_2$

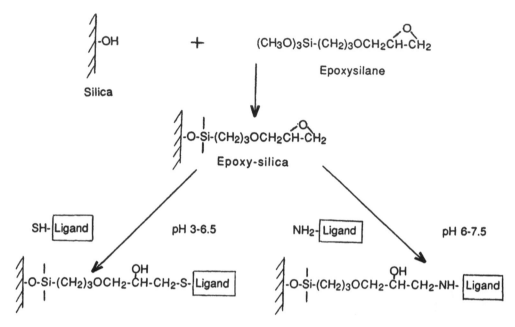

FIGURE 2. The chemistry of ligand derivatization to epoxy-silica matrix. Silica is reacted with epoxysilane to produce epoxy-silica, which is available commercially. Epoxides are highly reactive due to the ease of opening the highly strained 3-membered rings. A nucleophilic group on the ligand attaches the epoxide. At low pH, the epoxide is protonated and is very reactive. In contract, at more alkaline pH, the nonprotonated epoxide requires a stronger nucleophilic agent (i.e., NH_2) for the reaction to occur. Hence, some degree of selectivity of ligand functional group attachment to the support can be achieved.

larger physical size of the proteins increases the chance of steric hindrance between the ligands bound on the support, or the solute molecules once bound, if the ligands are situated close to each other. The neighboring ligands will interfere with the solute binding and, hence, lower the efficiency of the column. Reduction of ligand concentration may also be necessary if the ligand-solute dissociation constant is very low ($k_d < 10^{-9}$), necessitating a higher concentration of desorbing agent to elute the solute molecule. This is a consideration only if the desorption agent is a peptide or protein and a lower concentration of this agent is desired (especially if the counter ligand is expensive or in short supply). In addition, if the interaction between the solute and ligand is strong, the dissociation of the solute from the column will occur slowly. This leads to broader peaks regardless of whether step gradient or linear gradient elution is employed. Hence, the volume of eluting buffer required for elution of the solute would increase. By reducing the ligand concentration, elution volumes will be minimized, a particular advantage during purification. Unreacted derivatizing functional groups on the support may be blocked by β-mercaptoethanol or glycerol, which will increase the column hydrophilicity but not alter the net charge of the ligand-support. Blocking functional groups with ethanolamine, which adds a cationic charge, and glycine, which is an ampholyte, will contribute unwanted ionic interactions. It should be noted that reducing the ligand concentration on the support will also reduce the maximum sample capacity of the column. Thus, for preparative purifications, larger amounts of derivatized support will be required compared to analytical applications. Finally, if a solute-ligand complex has a high dissociation constant ($> 10^{-7}$), it is desirable to maximize ligand concentration on the support.

V. RUN CONDITIONS IN HPAC

A. CHOICE OF BUFFERS

All the equilibrating, running or adsorption buffers must be nondenaturing. This will ensure that the ligand and/or solute molecule(s) do not lose their biospecificity for each other.

B. SAMPLE RECOVERY TECHNIQUES

1. Choice of Desorption Agent and Elution Conditions

Optimization of the desorption step in the affinity separation offers the greatest flexibility in customizing HPAC for the system being studied. The elution of the protein or peptide bound to the ligand involves breaking the same types of bonds that are involved in maintaining the native conformation of proteins. Thus, the conditions chosen to disrupt ligand-solute interactions must be mild to minimize the denaturation of the ligand and solute molecules. Depending on the ligand and solute molecules, either a general (non-specific) or specific agent can be used to dissociate the solute molecule.

There are numerous nonspecific desorption agents that disrupt the interaction between the ligand and solute molecule, including low pH or chaotropic ions. Reducing the pH of the buffer will eliminate ionic interactions between the ligand and solute molecule. It must be remembered, of course, that low pH may denature proteins. A common example of using low pH in the desorption step is in the purification of antibodies on a protein A derivatized column (many companies can supply a ready-to-use protein A column). The bound IgG is eluted by dropping the pH of the same buffer used in the loading step from pH values above neutrality to pH values in the range of 2 to 3.[7] It has been reported that low pH reduces the life of the column.[8] Also, sudden drops in pH can be harmful to column life and it has been suggested that a pH gradient is less harmful. There are many reviews that deal with the complex area of purification of IgG molecules from sera, ascites fluid or cell culture media.[9] Chaotropic agents are used to destabilize ligand-solute interactions by disrupting the structure of water and hence, reducing hydrophobic interactions between the ligand and solute. Unlike during pH manipulation, there is no denaturation of ligand or solute, except at very high concentrations of the chaotropic agent, and, thus, this is the desorption method of choice. The effectiveness of chaotropic salts is as follows: $CCl_3COO^- > SCN^- > CF_3COO^- > ClO_4^- > I^- > Cl^-$. The use of a chaotropic agent was described by Van Eyk and Hodges,[4] in which the relative binding affinities of a series of analogs of the TnI peptide (region 104 to 115, as described above) were determined on a TnC HPAC column (Figure 3). A linear KCl gradient was applied and the concentration of KCl required to elute the various peptide analogs was a reflection of the binding strength of the peptide-TnC interaction.

Specific desorbing agents that compete with the solute molecule for the same binding site on the ligand can be very useful in purification of a protein that is extremely sensitive to denaturation. Also, the binding specificity of the solute for the ligand can be determined by its competition with other solute molecules for the ligand and monitoring the elution of a solute molecule first attached to the ligand. For example, a 28-residue synthetic peptide from the N-terminal region of actin binds to a TnI HPAC column (MW TnI = 20,700 daltons). The actin peptide can be eluted with a KCl gradient, or by injecting a competing desorbing molecule, such as holo actin (unpublished data). A slight twist to this approach is to have the same molecule as the ligand in the dissociation (desorbing) buffer so that it is able to compete with the ligand for the bound solute molecule. In this case, the molecule will be eluted as a complex with the competing ligand. For example, the synthetic actin peptide could be eluted off the TnI HPAC

PEPTIDE SEQUENCE

A   Ac-G-(G)-F-(G)-(G)-P-P-L-R-R-V-R-amide

B   Ac-G-(G)-F-(G)-R-P-P-L-R-R-V-R-amide

C   Ac-G-K-F-K-(G)-P-P-L-R-R-V-R-amide

D   Ac-G-K-F-(G)-R-P-P-L-R-R-V-R-amide

E   Ac-G-(G)-F-K-R-P-P-L-R-R-V-R-amide

F   Ac-G-K-F-K-R-P-P-L-R-R-V-R-amide

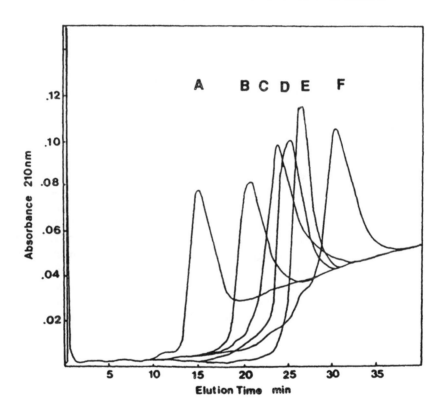

FIGURE 3. Comparison of retention times of single and multiple glycine-substituted troponin I peptide analogs on a troponin C HPAC column. Individual peptide analogs were chromatographed on a Beckman Ultraffinity-EP column (50 × 4.6 mm I.D.), derivatized with rabbit skeletal troponin C. The HPLC instrument consisted of a Varian Vista Series 5000 liquid chromatograph (Varian, Walnut Creek, CA), coupled to a Schoeffel GM770 single wavelength detector and Technical Marketing plotter. Linear AB gradient elution was employed to elute the analogs, where buffer A is 20 mM Tris, 0.1 mM EGTA, and 1 mM CaCl$_2$, pH 6.9, and buffer B is buffer A plus 1 M KCl. Flow-rate: 0.5 ml/min. Analogs: single substituted, Gly 105, Gly 107, Gly 108, double substituted Gly 105-107 and triple substituted Gly 105-107-108. NS is the native troponin I peptide [104 to 115] with the sequence Ac-G-K-F-K-R-P-P-L-R-R-V-R-amide.

column by injecting TnI (unpublished data). Under certain circumstances, a solute protein may bind so tightly that denaturation of the ligand is required to elute the bound solute molecule. For example, TnI could not be eluted from the TnC affinity column in 4 M KCl, and it has been reported that the TnI-TnC complex in the presence of calcium is not dissociated by 6 M urea.[10] A solution of TnC (2 mg/ml) was pumped through the HPAC column at 0.2 ml/min for 1 hr. The TnI that was bound to the column was eluted as the TnC-TnI complex.

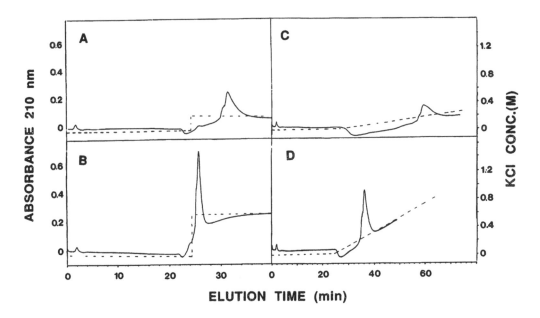

FIGURE 4. Comparison of step gradient elution and linear gradient elution of troponin I peptide from a troponin C HPAC column. The column, HPLC instrument and mobile phases are described in Figure 3. Panels A and B: step gradient elution using 200 or 400 mM KCl, respectively, as the desorbing agent. Panels C and D: linear gradient elution at 5 or 15 mM KCl/min, respectively. The dotted line shows the concentration of KCl with time under the various elution conditions. Flow-rate: 0.3 ml/min. All runs had an initial 10-min delay time following sample application (10 nmol peptide) and prior to the start of the linear gradient or step gradient KCl front.

2. Application of Non-Specific Desorbing Buffers

Non-specific desorbing agents, such as H$^+$ ions or chaotropic salts, can be applied either as a step gradient or as a linear gradient. Figure 4 shows the effect of the desorbing buffer (containing KCl) (buffer B) on the elution profile of the synthetic TnI peptide during step gradient elution from a TnC HPAC column. An initial delay time after loading the sample in the starting (or loading) buffer (buffer A) was required to wash impurities through the column prior to isocratic application of the desorbing buffer (buffer B). In panels A and B, there was an initial 10-min delay, followed by an isocratic step of either 200 or 400 mM KCl, respectively. In panel B, the concentration of KCl in the isocratic step was greater than needed to elute the peptide from the column. Therefore, when the 400 mM KCl front reached the column [at 23.3 min = t_d + 10 min delay; t_d, the gradient delay time, is the time required for the gradient (or KCl front, in this case) to reach the top of the column from the solvent proportioning valve] the peptide was immediately eluted as a sharp peak. In panel A, the concentration of KCl in the desorbing buffer (200 mM) was low enough to effect a slow dissocation of the solute with the ligand and/or reassociation of the solute with the ligand as the solute molecule travelled down the column. This resulted in peak broadening and a longer retention time after the KCl front had reached the column.

Panels C and D in Figure 4 illustrate the effect of linear gradient elution of the TnI peptide from the TnC HPAC column at gradient-rates of 5 and 15 mM KCl/min, respectively, following a 10-min delay. Regardless of the gradient-rate, the peptide was eluted at the same KCl concentration (~ 180 mM), so that, at the slower gradient-rate, the retention time increased. Also, at the slower gradient-rate, the peak was broadened, since the gradient was not steep enough to eliminate the reassociation of peptide as it travelled down the column.

Either type of gradient can be easily programmed on an HPLC instrument, by altering the amount of B buffer with respect to A buffer over time, where A is the starting (or loading) buffer and B is the desorbing buffer. Linear gradients are easier from the standpoint that one does not

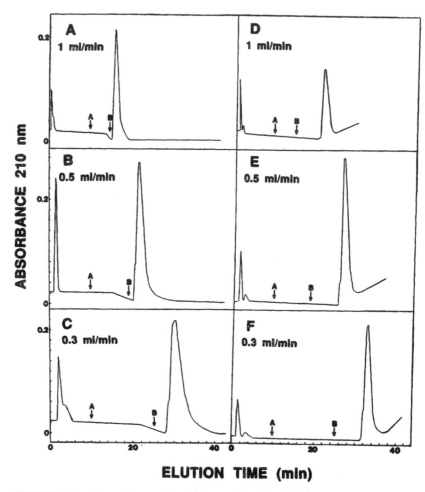

FIGURE 5. The effect of flow-rate on the elution of the troponin I peptide from a troponin C HPAC column. The column, HPLC instrument and mobile phases are described in Figure 3. Panels A-C: step gradient elution (200 mM KCl). Panels D-F: linear gradient elution (20 mM KCl/min). Flow-rate: 1 ml/min (panels A and D), 0.5 ml/min (panels B and E) or 0.3 ml/min (panels C and F). A denotes the gradient delay time, t_d (described in the text); B denotes the start of the linear gradient or step gradient KCl front, following a 10-min delay (B minus A) after sample application (5 nmol peptide).

need to have an approximate idea of the conditions required for elution of the molecule, as one does with a step gradient. Therefore, the use of linear gradients is a more universal approach and can be used very effectively for any application.

C. FLOW-RATES

Flow-rates affect the mass transfer properties of the column and, hence, affect the retention time and peak shape. At reduced flow-rates, the retention time of a peak will be greater due to the increase in the gradient delay time (t_d). The value of t_d can be determined for a particular HPLC instrument by noting the time required to observe an off-scale absorbance, due to the addition of B-mercaptoethanol (3 mM) to buffer B in the absence of a column. The value of t_d at any flow-rate can be calculated as a proportion of the observed t_d at a particular flow-rate. For instance, the HPLC system used to obtain the elution profiles shown in Figure 4 had a t_d value of 4.6 min at 1 ml/min; thus, at flow-rates of 0.5 and 0.3 ml/min, the t_d values will be 9.2 and 15.3 min, respectively.

Peak shape of the eluted peptide is altered by varying the flow-rate. Figure 5 shows the effect of various flow-rates (0.3, 0.5, and 1 ml/min) on the elution profile of the TnI peptide from a TnC HPAC column. These profiles were obtained by step gradient (panels A to C) or linear gradient (panels D to F) elution. Peak characteristics (retention time, peak width, etc.) obtained from these runs are reported in Table 2. A number of observations can be made concerning the results presented in Figure 5 and Table 2. Thus, at constant gradient-rate (20 mM KCl/min) during linear gradient elution, or constant ionic strength (200 mM KCl) of the desorption buffer during step gradient elution, peptide retention time increased, as expected, with decreasing flow-rate in both elution modes; peak width remained constant with varying flow-rate during linear gradient elution, but increased with decreasing flow-rate during step gradient elution; there was a general increase in peak height with decreasing flow-rate in both elution modes; finally, peak elution volume decreased with decreasing flow-rate during linear gradient elution, but remained essentially constant during step gradient elution. It should be noted that, when a protein, troponin T (TnT; MW = 30,500 daltons), was used as the solute molecule on the TnC HPAC column, there was a small increase in peak width at low flow-rates during linear gradient elution. This is in contrast to the results shown in Table 2 for the TnI peptide, where peak width was unaffected by flow-rate variations during linear gradient elution. This increase in peak width observed for TnT was most likely due to enhanced diffusion of the protein at lower flow-rates. With either a peptide or protein as the solute molecule, peak heights are larger at slower flow-rates because the eluted solute molecule spends more time in the detector than at faster rates, i.e., the detector has more time to "see" the molecule at slower flow-rates. Thus the observed absorbance is increased. The constant elution volume, with varying flow-rate, of the peptide during step gradient elution, indicated that peak width increased in direct proportion to the decrease in flow-rate.

Table 2 also reports the effect on peptide characteristics of elution at constant flow-rate, but varying gradient-rate during linear gradient elution, or varying ionic strength of the desorbing buffer during step gradient elution. Thus, peptide retention time increased with both decreasing gradient-rate and decreasing ionic strength of the desorbing buffer, although the latter decrease was not as dramatic as the former; peak width increased with both decreasing gradient-rate and decreasing ionic strength of the desorbing buffer; peak height decreased with both decreasing gradient-rate and decreasing ionic strength of the desorbing buffer; finally, as expected from the peak width results, peak elution volume increased with both decreasing gradient-rate and decreasing ionic strength of the desorbing buffer.

The relationship between dynamic sample capacity (capacity under flow) and flow-rate depends on both the diffusion rate of the solute and the chemical association rate of the solute/ligand interaction.[11] When maximum capacity is required, the flow-rate can be decreased until the dynamic capacity approaches the equilibrium capacity. On an HPAC column, this point is reached at a higher linear velocity than for conventional soft gel columns. This is because the mass transfer rate possible with the small particles of HPAC supports is much higher than that for large particles. As a result, HPAC packings allow more efficient use of all the immobilized ligands at any flow-rate. Thus, at higher flow-rates, the dynamic capacity of the Ultraffinity-EP columns (10-μm particle size) used in the present study is typically one order of magnitude higher than conventional (100-μm particle size) affinity support material. This results from a combination of the higher mass transfer rate of the solute into and out of the affinity matrix and the high ligand concentration possible. It is only at extremely high flow-rates that the capacity of an HPAC column decreases to any degree. At these high flow-rates, the back pressure is great, perhaps damaging the ligand, and should not be used. An average analytical HPAC column can be successfully used at a flow-rate of 0.5 to 1.0 ml/min.

VI. CARE AND MAINTENANCE OF COLUMNS

After derivatization of the support, the HPAC column should be stored in a low salt buffer

TABLE 2
Effect of Run Conditions on Peak Characteristics — Peptide Elution from Protein HPAC Column

Step Gradient Elution

Panel[a]	Flow-rate (ml/min)	t_r[b] (min)	Buffer B[c] ionic strength (mM KCl)	$W_{1/2}$[d] (min)	Peak height (mAU)	Peak elution[e] volume (ml)
A	1	17.5	200	0.8	220	0.80
B	0.5	22.5	200	1.6	272	0.80
C	0.3	30.0	200	2.9	224	0.87
—	0.3*	25.7	600	0.8	600	0.24
—	0.3*	26.7	400	1.0	487	0.31
—	0.3*	32.0	200	2.05	200	0.62

Linear Gradient Elution

Panel	Flow-rate (ml/min)	t_r (min)	Gradient-rate (mM KCl/min)	$W_{1/2}$ (min)	Peak height (mAU)	Peak elution volume (ml)
D	1	21.5	20	1.4	133	1.40
E	0.5	26.8	20	1.4	276	0.70
F	0.3	33.2	20	1.4	224	0.42
—	0.3*	37.2	15	1.2	259	0.36
—	0.3*	45.8	10	2.7	160	0.81
—	0.3*	61.3	5	4.5	103	1.35

a   Refers to panels in Figure 5.
b   Denotes retention time; all runs had an initial 10-min delay time following sample application and prior to the start of the linear gradient or step gradient KCl front.
c   The desorbing buffer during step gradient elution.
d   Peak width at half height.
e   Peak elution volume is calculated by multiplying $W_{1/2}$ by the flow-rate.
*   Runs marked with a star were carried out on a Varian Vista Series 5000 liquid chromatograph (Varian, Walnut Creek, CA), coupled to a Hewlett-Packard (Avondale, PA) HP1040A detection system, HP9000 Series 300 computer, HP9133 disc drive HP2225A Thinkjet printer and HP7440A plotter. The unmarked runs were carried out on a Vista Series 5000 LC coupled to a Schoeffel GM770 single wavelength detector and Technical Marketing plotter.

with 0.02% bacterioside (NaN$_3$) at 4°C. The column should not be frozen. Also, care must be taken to ensure that the pH of the storage buffer does not go above pH 7 if the column support is silica. It is extremely important to use preinjection filters and in-line filters to minimize plugging of the column. The stability and lifetime of HPAC columns may be extensive, as long as care is taken with them.

ACKNOWLEDGMENTS

This work was supported by research grants from the Medical Research Council of Canada and by equipment grants, studentship (J.E.V.) and a research allowance from the Alberta Heritage Foundation for Medical Research.

REFERENCES

1. Ohlson, S., Hansson, L., Larsson, P.O., and Mosbach, K., High performance liquid affinity chromatography (HPLAC) and its application to the separation of enzymes and antigens, *FEBS Lett.*, 93, 5, 1978.
2. Cooke, N., High performance affinity chromatography, *LC.GC.*, 5, 866, 1987.
3. Ernst-Cabrera, K. and Wilchek, M., High-performance affinity chromatography, *Trends in Anal. Chem.*, 7, 58, 1988.
4. Van Eyk, J.E. and Hodges, R.S., The biological importance of each amino acid residue of the troponin I inhibitory sequence 104-115 in the interaction with troponin C and tropomyosin-actin, *J. Biol. Chem.*, 263, 1726, 1988.
5. Van Eyk, J.E. and Hodges, R.S., Calmodulin and troponin C: affinity chromatographic study of divalent cation requirements for troponin I inhibitory peptide [104-115], mastoparan and fluphenazine binding, *Biochemistry and Cell Biology*, 65, 982, 1987.
6. Ni, W.-C. and Klee, C.B., Selective affinity chromatography with calmodulin fragments coupled to Sepharose, *J. Biol. Chem.*, 260, 6974, 1985.
7. Ohlson, S. and Wieslander, J., High-performance liquid affinity chromatographic separation of mouse monoclonal antibodies with protein A silica, *J. Chromatogr.*, 397, 207, 1987.
8. Josíc, D., Hofmann, W., Habermann, R., Becker, A., and Reutter, W., High-performance liquid affinity chromatography of liver plasma membrane proteins, *J. Chromatogr.*, 397, 39, 1987.
9. Nau, D.R., Chromatographic methods for antibody purification and analysis, *BioChromatography*, 4, 4, 1989.
10. Chong, P.C.S. and Hodges, R.S., A new heterobifunctional cross-linking reagent for the study of biological interactions between proteins. II. Application to the Troponin C-Troponin I interaction, *J. Biol. Chem.*, 256, 5071, 1981.
11. Technical brochure on Fast Affinity Chromatography, Bulletin 5933, Beckman Instruments, Berkeley, CA.

HPLC PURIFICATION OF PLASMA SERINE PROTEINASES: APPLICATION TO PLASMA KALLIKREIN AND FACTOR XII

Nabil G. Seidah, Claude Lazure, and Michel Chrétien

I. INTRODUCTION

The purification of proteolytic enzymes is usually performed using classical protocols of ion-exchange and size-exclusion chromatography, often coupled to affinity procedures. One of the difficulties encountered in obtaining a homogenous proteinase preparation, especially if the amount of material is low, is the decomposition of the enzyme during the purification steps, probably due to autolysis and degradation by other contaminating proteinases. In our laboratory, we have been interested in the purification and structural characterization of a Serine proteinase originally isolated from porcine pituitary and rat heart.[1-3] Subsequent experiments demonstrated that this enzyme is also found in plasma and was purified from that source in three species including human, porcine, and rat.[4] This proteinase was found to be highly selective for cleavage of a number of polypeptide hormone precursors at pairs of basic amino acids,[1-5] and hence was considered a serious candidate as a processing enzyme of certain pro-proteins.[6] In order to obtain this enzyme in sufficient purity for sequence determination, an overall purification factor of about 50,000 was deemed necessary.

The present paper describes the procedure ultimately developed for the purification of the tissue and plasma forms of this proteinase, and its separation from contaminating proteinases of similar chromatographic properties. This involved the use of three affinity columns, a size-exclusion column and a reversed-phase high-performance liquid chromatography (HPLC) column. This methodology allowed the isolation of a sufficiently pure enzymatic preparation for determination of the NH_2-terminal sequence of the two chains of this enzyme. The resulting sequence permitted its identification as plasma kallikrein and the contaminating proteinase as factor XII.

II. EXPERIMENTAL

A. AFFINITY PROCEDURES

The procedure involving the initial purification of the IRCM-Serine protease 1, as it was originally called[1-3], consisted of:

1. Homogenization of the tissue (e.g. pituitary or heart) or plasma, followed by centrifugation at 10,000 and then at 100,000 g to remove all debris. The ultracentrifugation supernatant was then concentrated and dialyzed against 50 mM HEPES (N-[2-hydroxyethyl]piperazine-N'-[2-ethanesulfonic acid]), 1 mM EDTA (ethylenediaminetetraacetic acid), 10% DMSO (dimethyl sulfoxide) using a Millipore Minitan tangential flow system equipped with 30 kDa cut off filters.

2. Injection of the retained solute onto a Phenyl Boronate-Agarose (Amicon, Oakville,

Ontario, Canada) (200 ml/100 g of tissue) column (400 × 25 mm I.D.), eluted with the same buffer, then followed by an overnight wash with the same buffer containing 100 mM sorbitol to elute all proteins which contain 5% carbohydrate or less. After this extensive wash (10 × column volume), a 1 M guanidine hydrochloride (GnHCl) step is then applied and the enzyme activity is found to be eluted with the major protein peak (absorbance at 280 nm).

3. The pooled 1 M GnHCl peak is then dialyzed and concentrated against 100 mM MES (2-[N-morpholino]ethanesulfonic acid), 1 mM EDTA, pH 6 (buffer A), and then applied either to a Benzamidine-Agarose (Sigma, St. Louis, MO) column (125 × 25 mm I.D.; 50 ml of gel) or to two columns in tandem of Lys-Sepharose (Pharmacia, Baie Du'Urfé, Québec, Canada; 125 × 25 mm I.D.; 50 ml of gel) followed by Soya Bean Trypsin Inhibitor-(SBTI)-Agarose (Sigma; 5 ml of gel). The material retained on the Benzamidine column was eluted with a linear gradient of 0.1 to 1 M benzamidine, and the fraction exhibiting enzyme activity (as followed by the release of 7-amino-4-methylcoumarin [AMC] from the tripeptide fluorogenic substrate Z-Ala-Lys-Arg-AMC[1]) was then applied directly onto a C_4 reversed-phase column, as described below, resulting in the copurification of both plasma kallikrein and factor XII. In contrast, proteins bound non-specifically to the SBTI column were eluted during a wash step with 1 M KCl. This wash step was then followed by a 5 mM NaOH (pH 10.8) step to elute the material specifically attached to the SBTI column. Whereas the benzamidine procedure resulted in a mixture of plasma kallikrein and factor XII, the SBTI procedure provided an enzymatically pure preparation of plasma kallikrein. However, as judged by sodium dodecylsulfate/polyacrylamide gel electrophoresis (SDS/PAGE) and microsequencing, the protein is barely 10% pure, even though an overall purification factor of about 18,000 was achieved at that stage.

B. HPLC PROCEDURES AND SEQUENCING

The material obtained from the SBTI affinity column was dialyzed (against buffer A), and concentrated with a Pierce model 320 ProDiCon microconcentrator. The concentrated enzyme preparation was then applied 500 μl at a time to a Bio-Sil TSK-250 size-exclusion HPLC column, (600 × 7.5 mm I.D.; Bio-Rad, Mississauga, Ontario, Canada), where the plasma kallikrein was eluted as a dimer with an apparent Mr of 160,000.[1-5] This final procedure allows us to keep the enzyme in an active state such that we can follow the proteolytic activity with the release of a fluorescent coumarin leaving group in the tripeptide Z-Ala-Lys-Arg-AMC.[1]

Since the enzyme is composed of two chains (a heavy chain called regulatory chain and a light chain called catalytic chain) held together by a disulfide bond, the microsequencing of the protein results in the identification of two amino acids at each cycle, each belonging to its respective polypeptide chain. In order to confirm our assignment, the Cys residues were S-carboxymethylated and the protein chains were then separated by reversed-phase chromatography (RPC) on a Vydac C_4 column (250 × 4.6 mm I.D., 300-Å pore size; The Separations Group, Hesperia, CA).[4] The eluted peptides were then subjected to microsequencing on an Applied Biosystems (Foster City, CA) gas-phase sequenator (Model 470A), as described previously.[7]

III. RESULTS AND DISCUSSION

The affinity purification procedure presented in the Experimental section resulted in an 18,000-fold increase in specific activity of the purified enzyme.[5] In Figure 1, we compare the elution profiles of two preparations of porcine plasma kallikrein on a reversed-phase C_4 column obtained following linear gradient elution with a water/acetonitrile (CH_3CN) system containing 0.13% (v/v) heptafluorobutyric acid (HFBA).[2-5] The first preparation was obtained following elution from a Benzamidine-Agarose affinity column (Figure 1A), whereas the second was obtained after elution from a SBTI-Agarose affinity column (Figure 1B). Since acetonitrile

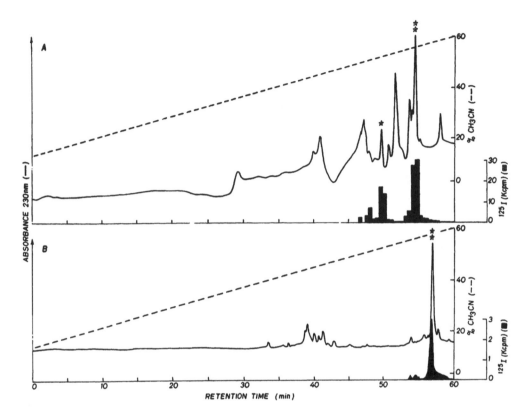

FIGURE 1. Reversed-phase purification of plasma kallikrein preparations obtained following Benzamidine-Agarose (A) or SBTI-Agarose (B) affinity chromatography. Column: Vydac C_4, 250 × 4.6 mm I.D., 300-Å pore size. Mobile phase: linear AB gradient (0.8% B/min, starting from 12% B at 0 min), where Eluent A is 0.13% (v/v) aq. heptafluorobutyric acid (HFBA) and Eluent B is 0.13% HFBA in acetonitrile; flow-rate, 1 ml/min; absorbance at 230 nm (0.1 AUFS) and 280 nm (not shown); radioactivity in fractions was determined using a Model 1274 RIA gamma counter (LKB) (A) or detected using an on-line radioactivity detector (Radiomatic Instruments and Chemicals, Tampa, FL; FLO-ONE/Beta Model IC) at the outlet of the UV-detector following mixing of the effluent with aquasol-2 in a ratio of 1:3.4 (v/v) (B). The dark peaks represent the elution positions of markers (active site radiolabeled enzymes within each preparation). Clearly, plasma coagulation factor XII (*) is only found in the preparation obtained from the Benzamidine-Agarose column, whereas plasma kallikrein (**) is obtained by both procedures, albeit in a much more pure form from the SBTI-Agarose column.

completely inactivates the enzyme, we identified the elution position of the proteinases by the elution positions of active-site radiolabeled enzymes included (less or equal to 0.04% on a molar basis) within each preparation ($[^{125}I]$ radiolabeled pentapeptide chloromethyl ketone [D-Tyr-Glu-Phe-Lys-Arg-COCH₂Cl] was used to label the active site), as described previously.[1] Clearly, the benzamidine procedure revealed the presence of two enzymatic activities, which, when sequenced, gave the sequences of porcine plasma kallikrein and factor XII (Figure 2).

The separation of the catalytic and regulatory chains of plasma kallikrein was achieved by RPC of the reduced and carboxymethylated enzyme. As shown in Figure 3, such a separation permitted the correct identification of each amino acid in the sequence of the porcine plasma kallikrein, as shown in Figure 2. A similar methodology was also applied for the isolation of the homologous enzyme from both human and rat plasma. This powerful characterization methodology permitted the unequivocal identification of plasma kallikrein as the candidate processing enzyme, specific for cleavage post-paired-basic amino acids and C-terminal to certain single Arg- residues, all known to be cleaved in vivo.[1-5] Furthermore, the sequence obtained for the rat preparation allowed the identification of the signal peptidase cleavage site and zymogen activation site of the rat pre-pro-plasma kallikrein, as deduced from its cDNA structure.[8]

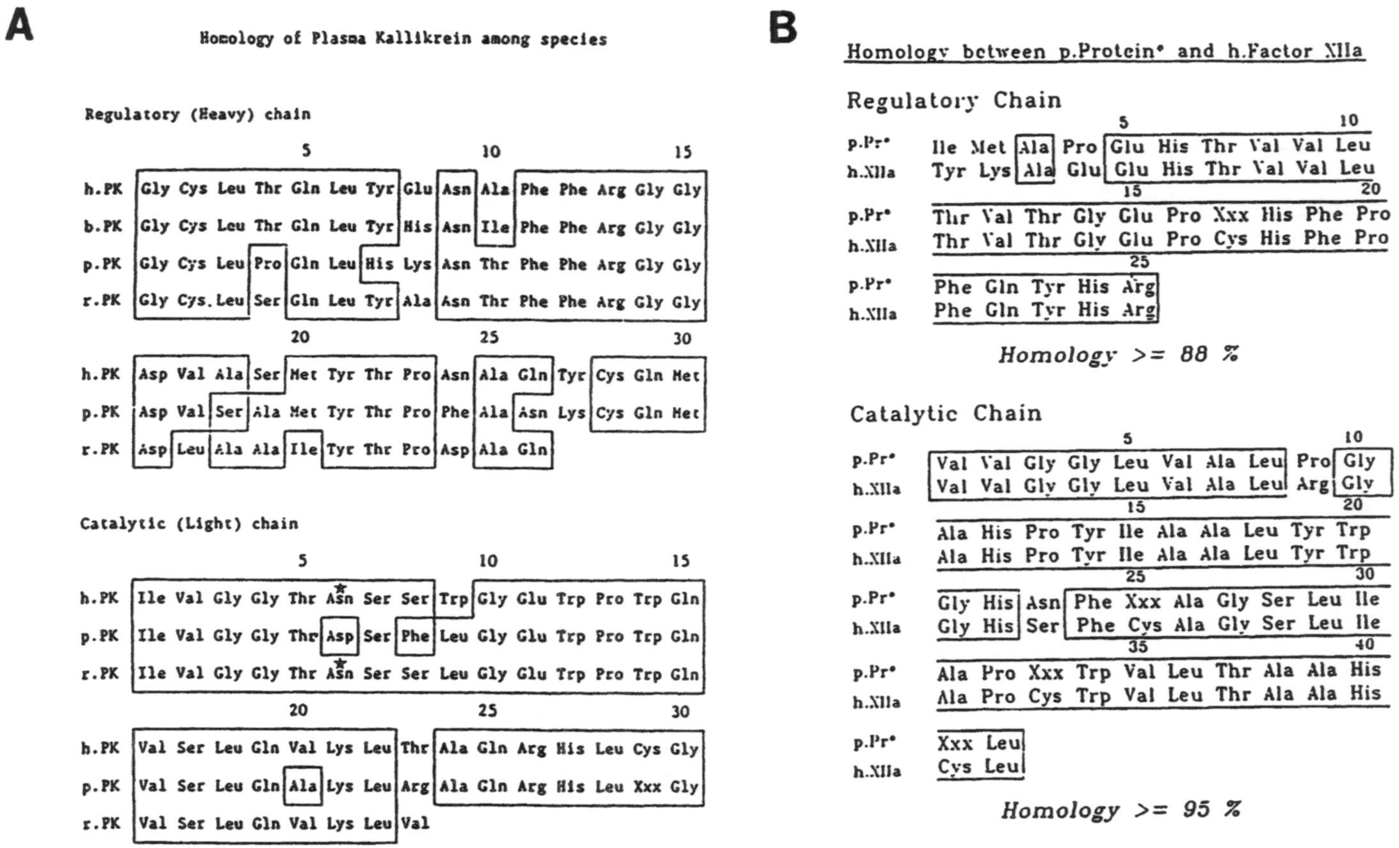

FIGURE 2. (A) N-terminal sequence deduced for plasma kallikrein (PK) obtained from three species including human (h.PK), rat (r.PK), and porcine (p.PK) following the procedures employed in this work. These are also compared to the sequence reported for bovine plasma kallikrein (b.PK). (From Heimark, R. D. and Davie, E. W., in *Methods in Enzymology*, Lorand, L., Ed., Academic Press, 80, 157, 1981. With permission.) (B) Deduced N-terminal sequence of porcine factor XII (p.Pr) and comparison with the reported human sequence (h.XIIa). (From McMullen, B. A. and Fujikawa, K., *J. Biol. Chem.*, 260, 5328, 1985. With permission.)

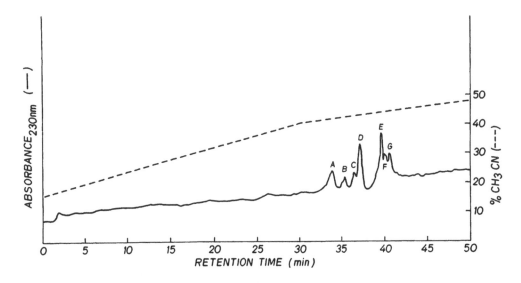

FIGURE 3. Reversed-phase separation of the 2-chains of the reduced and carboxymethylated plasma kallikrein isolated from extracts of anterior lobes of porcine pituitaries. Column: as Figure 1. Mobile phase: linear AB gradient (1% B/min up to 30 min, starting from 20% B at 0 min, then 0.5% B/min), where Eluent A is 0.1% aq. trifluoroacetic acid (TFA) and Eluent B is water/acetonitrile (20/80, v/v) containing 0.1% TFA; flow-rate, 1 ml/min; absorbance at 230 nm (0.1 AUFS) and 280 nm (not shown). The sequence of peaks A and B demonstrated that they represent the regulatory subunit, whereas peaks C and D contained the catalytic chain of plasma kallikrein (see Figure 2); peaks E, F, and G represent contaminating proteins unrelated to plasma kallikrein. The dashed line represents the CH₃CN elution gradient used.

One of the drawbacks of RPC is the complete inactivation of the proteinase once purified. As an alternative approach to obtain pure and active proteinase, we used instead a Bio-Sil TSK-250 molecular sieve column for the purification of the enzyme obtained from the SBTI-affinity column. Here, we exploited the fact that plasma kallikrein is a dimer in solution at pH 6[1-3] and used this HPLC molecular sieve column to obtain a homogenous preparation of active plasma kallikrein, with a final overall purification factor of about 50,000 (Figure 4).

The developed procedure exploited both the powerful affinity procedures for selectively purifying proteins and the high resolving power of HPLC for the final purification of both plasma kallikrein and factor XII. We have found that a similar protocol would also be applicable to other plasma proteinases such as plasmin. The biochemist involved in the identification of proteins by microsequencing usually needs an 80 to 95% pure protein in order to assign unambiguously the correct amino acid sequence at the N-terminus of the polypeptide chain. The data presented in this paper demonstrated that affinity procedures alone do not provide such a high degree of purity, especially for those proteins present in very minute amounts. However, the combination of affinity and HPLC procedures allows such a goal to be reached. Furthermore, one can also exploit other modes of HPLC, such as molecular sieving or ion-exchange chromatography, in order to obtain homogenous and active enzyme preparations. The methods developed allow the complete purification in a preparative mode of plasma kallikrein in good yield and excellent purity within 1 week of the start of the extraction. We have now repeated this procedure at least 10 times with different species and consistently obtained reproducible results. It must be emphasized that, although the method developed is used to purify the active form of the enzyme, its zymogen can also be purified by replacing the SBTI-column with another affinity column binding the regulatory subunit of plasma kallikrein.[9,10]

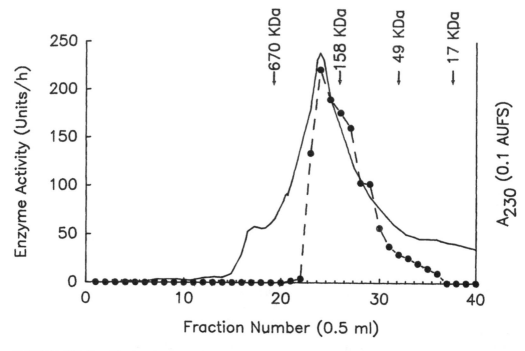

FIGURE 4. HPLC purification of plasma kallikrein by size-exclusion chromatography. Column: Bio-Sil TSK-250, 600 × 7.5 mm I.D. Mobile phase: 100 mM MES, pH 6; flow-rate, 1 ml/min; absorbance at 230 nm. Enzyme activity was detected following the hydrolysis of the fluorogenic tripeptide substrate Z-AlaLysArg-AMC, as described previously. (From Cromlish, J. A., Seidah, N. G., and Chrétien, M., *J. Biol. Chem.*, 261, 10850, 1986. With permission.)

ACKNOWLEDGMENTS

This study was made possible thanks to the financial support from the Medical Research Council of Canada and the National Institutes of Health U.S.A. and from the "Succession J.A. de Sève". C.L. is a "chercheur boursier" of the "Fonds de la Recherche en Santé du Québec".

REFERENCES

1. **Cromlish, J.A., Seidah, N.G., and Chrétien, M.,** A novel Serine Protease (IRCM-Serine Protease 1) from Porcine Neurointermediate and Anterior Pituitary Lobes, *J. Biol. Chem.*, 261, 10850, 1986.
2. **Cromlish, J.A., Seidah, N.G., and Chrétien, M.,** Selective Cleavage of human ACTH, β-Lipotropin, and the N-terminal Glycopeptide at Pairs of basic Residues by IRCM-Serine Protease 1, *J. Biol. Chem.*, 261, 10859, 1986.
3. **Seidah, N.G., Cromlish, J.A., Hamelin, J., Thibault, G., and Chrétien, M.,** Homologous IRCM-serine protease 1 from pituitary, heart atrium and ventricle: a common pro-hormone maturation enzyme?, *Biosci. Rep.*, 6, 835, 1986.
4. **Seidah, N.G., Paquin, J., Hamelin, J., Benjannet, S., and Chrétien, M.,** Structural and immunological homology of human and porcine pituitary and plasma IRCM-serine protease 1 to plasma kallikrein: marked selectivity for pairs of basic residues suggests a widespread role in pro-hormone and pro-enzyme processing, *Biochimie*, 70, 33, 1988.
5. **Metters, K.M., Rossier, J., Paquin, J., Chrétien, M., and Seidah, N.G.,** Selective cleavage of proenkephalin-derived peptides (< 23,300 Daltons) by plasma kallikrein, *J. Biol. Chem.*, 263, 12543, 1988.

6. **Lazure, C., Seidah, N.G., Pélaprat, D., and Chrétien, M.,** Proteases and posttranslational processing of prohormones: a review, *Can. J. Biochem. Cell Biol.,* 61, 501, 1983.

7. **Seidah, N.G., Donohue-Rolfe, A., Lazure, C., Auclair, F., Keusch, G.T., and Chrétien, M.,** Complete amino acid sequence of shigella toxin B-chain, *J. Biol. Chem.,* 261, 13928, 1986.

8. **Seidah, N.G., Mbikay, M., Lazure, C., Chrétien, M., Ladenheim, R., Lutfalla, G., and Rougeon, G.,** The cDNA structure of rat plasma kallikrein: a candidate processing proteinase of pro-hormones and pro-enzymes, in *Progress in Endocrinology,* Imura, H. et al. eds., Elsevier Science Publishers B.V. (Biomedical Division), 1988, pp. 319.

9. **Chung, D.W., Fujikawa, K., McMullen, B.A., and Davie, E.W.,** Human plasma kallikrein, a zymogen to a Serine Protease that contains four tandem repeats, *Biochemistry,* 25, 2410, 1987.

10. **Tait, J.F. and Fujikawa, K.,** Primary structure requirements for the binding of human high molecular weight kininogen to plasma prekallikrein and factor XI, *J. Biol. Chem.,* 262, 11651, 1987.

11. **Heimark, R.D. and Davie, E.W.,** Bovine and human plasma prekallikrein, in *Methods in Enzymology,* Lorand, L., Ed., Academic Press, 80, 157, 1981.

12. **McMullen, B.A. and Fujikawa, K.,** Amino acid sequence of the heavy chain of human a-factor XIIa (activated Hageman Factor), *J. Biol. Chem.,* 260, 5328, 1985.

THE USE OF HEPARIN AFFINITY HIGH-PERFORMANCE LIQUID CHROMATOGRAPHY FOR THE ISOLATION AND CHARACTERIZATION OF FIBROBLAST GROWTH FACTORS

Niggi Iberg and Michael Klagsbrun

I. INTRODUCTION

A wide range of proteins bind strongly to heparin, a highly anionic glycosaminoglycan, and affinity chromatography on heparin covalently linked to a matrix (e.g., Sepharose) has often been used as a purification step for biologically active heparin-binding proteins.[1] In 1983, it was demonstrated that an 18,000 molecular weight endothelial cell growth factor isolated from rat chondrosarcoma had a strong affinity for heparin and that this growth factor could be purified to homogeneity using heparin affinity chromatography as the critical purification step.[2] Soon thereafter, it was found that virtually all endothelial cell growth factors could bind tightly to heparin and that the purification of these growth factors was greatly facilitated by heparin affinity chromatography. Two of these heparin-binding endothelial cell growth factors turned out to be acidic fibroblast growth factor (aFGF) and basic fibroblast growth factor (bFGF), polypeptides originally isolated from brain.[3-8] aFGF, also known as heparin-binding growth factor I (HBGF-I) or endothelial cell growth factor (ECGF), has a pI of 5 to 7, a molecular weight of 16,000 to 18,000, and is found mainly in neural tissue. bFGF, also known as heparin-binding growth factor II (HBGF-II), has a pI of 8 to 10, a molecular weight of 18,000 and is fairly ubiquitous, being found in neural tissue, skeletal tissue, reproductive tissue and tumors among others (for reviews see References 9 to 12).

Heparin affinity chromatography is a technique which is fairly specific for the purification of aFGF and bFGF. Most other growth factors, e.g., epidermal growth factor (EGF), transforming growth factor alpha (TGF-alpha) and transforming growth factor beta (TGF-beta) don't bind at all. Platelet derived growth factor (PDGF) binds to heparin but is eluted at about 0.5 M NaCl, a salt concentration which is typical for a cationic protein. Conventionally, heparin-Sepharose affinity chromatography has been used for the purification of aFGF and bFGF. However, these two growth factors don't always separate that well on these columns. Accordingly, we have used heparin affinity chromatography in high-performance liquid chromatography (HPLC) as a new method using TSK Heparin-5PW columns. Heparin affinity HPLC has several advantages. Since the HPLC column matrix generally is more regularly packed and the gradient is more accurate, resolution is higher and results are highly reproducible. In addition, since the volume of the column is smaller, the fractions are less diluted. aFGF and bFGF can be readily separated by heparin affinity HPLC. The fractions obtained are more concentrated so that much lower amounts of aFGF or bFGF can be applied and detected in a bioassay. This method therefore seems to be the ideal tool for quantitative studies of aFGF and bFGF in tissues or cells.

In this chapter, we describe the use of heparin affinity HPLC in the purification of fibroblast growth factors from tissues and cells.

II. MATERIALS

Dithiothreitol (DTT) and 3[(3-cholamidopropyl)-dimethylammonio]-1-propanesulfonate (CHAPS) were purchased from Pierce (Rockford, IL). Heparin-Sepharose was purchased from Pharmacia (Uppsala, Sweden). 'Endothelial Mitogen', an extract of bovine hypothalamus which contains aFGF and bFGF,[13] was purchased from Biomedical Technologies Inc. (Stoughton, MA). It is useful to pretreat the 'Endothelial Mitogen' with DTT (25 mM, 37°C, 1 hr) in order to restore the full activity of bFGF. Stripped mouse brains were purchased from Pel Freeze (Rogers, AR). The cells used were ras-transformed rat fibroblasts (EJ-RAT).[14]

III. METHODS

A. CONVENTIONAL HEPARIN AFFINITY CHROMATOGRAPHY

In conventional heparin affinity chromatography, a bed of 3 to 5 ml heparin-Sepharose in a column is equilibrated with starting buffer (0.6 M NaCl, 0.02 M Tris, pH 7.3); the sample is then applied and the column washed with 40 ml starting buffer. The bound proteins are eluted with an 80 ml gradient of 0.6 to 3 M NaCl in 0.02 M Tris, pH 7.3. Fractions of 2.5 ml are collected and assayed for the ability to stimulate DNA synthesis in 3T3 cells.[2]

B. HEPARIN AFFINITY HPLC
1. HPLC Instrumentation

HPLC is performed on a Beckman system, consisting of two 110 A pumps and a 421 A controller. The sample is applied with a 10 or a 50 ml sample loop (FPLC system, Pharmacia).

2. Column

The analytical (75 mm × 8 mm I.D.) and guard columns used are TSK Heparin-5PW glass columns, manufactured by TOSOH corporation (Japan). In the U.S., these columns can be purchased directly from the distributor TOSOHAAS (Philadelphia, PA) as well as from Novex (Encinitas, CA), The Nest Group (Southboro, MA), and Supelco (Bellafonte, PA). The column matrix — TSKgel 5PW — consists of a hydrophilic vinyl polymer-based material with a particle size of 10 μm and with a large pore size of 1000 Å, which is especially suited for proteins.[15] The amount of heparin covalently bound to this matrix is 5 mg/ml of wet gel. The column has a high capacity for heparin-binding growth factors and samples of milligram amounts can be applied. The column matrix is stable over a pH range of 5.5 to 10.0, and can be exposed above pH 10 only for a short time. TSK-heparin 5PW is also available in stainless steel columns. However, the use of glass columns is recommended, since, (1) metal ions might decrease the biological activity of some proteins, and (2) the high salt concentration in the gradient is corrosive to the stainless steel. The glass column should not be operated at pressures higher than 225 psi (15 Atm.), and the suggested maximum flow-rate is 1.2 ml/min.

In order to clean the column, it should be washed with 0.1 to 0.2 N NaOH, which is applied by several injections of 1 to 2 ml via the sample loop. Alternatively, the column is washed with 1 to 2 ml of 20 to 40% aq. acetic acid. Should this procedure not be sufficient, washing with 6 M urea, or nonionic detergent or up to 20% organic solvent in Tris-buffer, pH 7.5, is recommended. The manufacturer suggests storage of the column in water. In order to prevent bacterial growth upon prolonged storage, we keep our columns in 20% aq. methanol. To prolong the life of the analytical column, the use of a guard column is recommended. In our laboratory, TSK-heparin columns have been used for at least 50 runs.

The buffers used are (1) 0.6 M NaCl, 1 mM DTT, 0.1% CHAPS in 0.02 M Tris, pH 7.5, and (2) 3 M NaCl, 1 mM DTT, 0.1% CHAPS in 0.02 M Tris, pH 7.5. Both buffers are filtered through a nylon-filter of 0.45 μm. Since DTT slowly oxidizes, the buffers have a limited stability and are routinely used for 1 week only. Inclusion of DTT and CHAPS in the buffers gives a much better recovery, up to 80% compared to 30 to 40% in the conventional protocol.

3. Sample Preparation

The starting materials for our samples are tissues or cell cultures. The preparation of the samples is carried out as follows: The sample is homogenized in 2 to 10 ml ice cold 1 M NaCl, 0.02 M Tris, pH 7.5 with a Polytron (3 to 5 g tissue), or by sonication (10^7 cells/ml). All steps are performed on ice. After homogenization, DTT is added to a final concentration of 25 mM. The cell extracts are diluted with 1 mM DTT, 0.1% CHAPS, in 0.02 M Tris, pH 7.5 to a salt concentration of about 0.25 M NaCl, and the cell debris is centrifuged at 25,000 g for 30 min. In the case of tissue extracts, centrifugation might not be sufficient to obtain a clear solution. The remaining turbidity is not filterable and conventional filters immediately clog. Therefore, the homogenates are clarified by ultracentrifugation at 250,000 g for 1 hr. Prior to application, the sample is filtered through a 0.2 μm filter; it is very important that the sample is completely clarified before application. Otherwise, the column will have a relatively short life.

4. Column Chromatography

The filtered samples are applied with a sample loop of 10 or 50 ml. The flow-rate is 1 ml/min. After application, the column is washed for 10 min with buffer A. The adsorbed proteins are eluted with a gradient of 0 to 60% B (0.6 to 2.0 M NaCl) over a 30-min period. Fractions of 0.5 ml are collected in microfuge tubes and aliquots of 1 to 5 μl are assayed for growth factor activity in a bioassay which measures the stimulation of DNA synthesis in quiescent BALB-3T3 cells by acidic or basic FGF.[2] The amounts of aFGF and bFGF are too low to be detected by UV.

IV. RESULTS

Both tissue extracts and cell sonicates were analyzed by heparin affinity HPLC. About 20 mg of 'Endothelial Mitogen', a commercially available hypothalamus extract, was applied to an HPLC-heparin column (Figure 1). Two peaks of activity were eluted with the NaCl gradient. The first peak, eluted at 1.1 to 1.3 M NaCl, is aFGF and the second peak, eluted at 1.5 to 1.6 M NaCl, is bFGF, as ascertained by Western blot analysis using specific anti-aFGF and anti-bFGF antibodies (not shown).[16] The separation of aFGF and bFGF is excellent and the two peaks are separated to baseline.

An extract of mouse brain was applied to the HPLC-heparin column (Figure 2). Two peaks of activity were eluted with the NaCl gradient. Again, the first peak, eluted at 1.1 to 1.3 M NaCl, is aFGF and the second peak, eluted at 1.5 M NaCl, is bFGF. The two peaks, brain-derived aFGF and bFGF, were eluted at salt concentrations similar to hypothalamus-derived aFGF and bFGF, indicating that the heparin affinity HPLC gives highly reproducible results. Estimation of the protein content in filtered extracts and in the pooled peak-fractions indicates a purification of about 20,000-fold using HPLC-heparin columns. The yield of aFGF and bFGF is about 10 to 100 pmol.

An extract of ras-transformed rat fibroblasts was applied to the HPLC-heparin column (Figure 3). One peak of growth factor activity, shown to be bFGF by Western blotting (result not shown), was eluted at 1.5 M NaCl. The recovery of bFGF in this application is 80%. In general, the recovery is somewhere between 50 to 80%. Good recovery depends on the presence of CHAPS and is also enhanced by the presence of DTT.

V. DISCUSSION

Heparin affinity HPLC on TSK Heparin-5PW is a valuable tool for the purification and

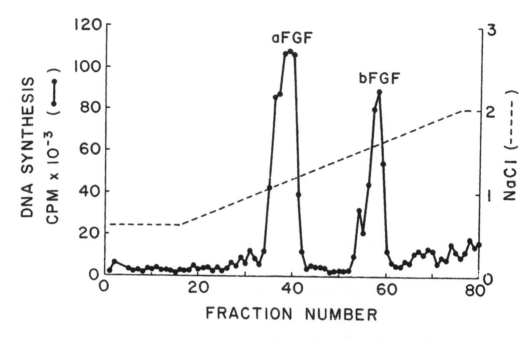

FIGURE 1. Heparin affinity HPLC of Hypothalamus Extract. 20 mg of a lyophilized extract of bovine hypothalamus, 'Endothelial Mitogen', were dissolved in 0.6 M NaCl, 25 mM DTT, 0.02 M Tris, pH 7.5, and incubated for 1 hr at 37°C. The sample was filtered and applied to the column. After washing the column for 10 min with buffer A, the column was eluted with a linear AB gradient of 0 to 60% Buffer B (0.6 to 2.0 M NaCl) over a 30-min period (46.7 mM NaCl/min), where Buffer A was 0.02 M Tris, pH 7.5, containing 0.6 M NaCl, 1 mM DTT, and 0.1% CHAPS, and Buffer B was the same as Buffer A except for an increase in the NaCl concentration to 3 M. The flow-rate was 1 ml/min. Fractions of 0.5 ml were collected and aliquots analyzed for the ability to stimulate DNA synthesis in 3T3 cells.

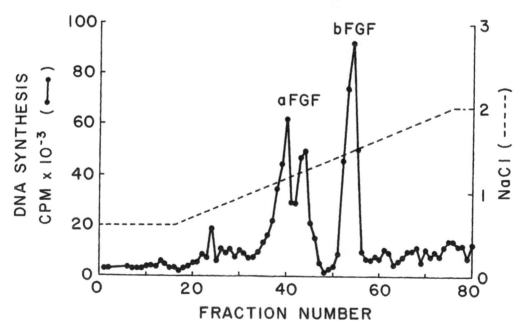

FIGURE 2. Heparin affinity HPLC of aFGF and bFGF of Mouse Brain. Five mouse brains (about 2.5 g) were homogenized in 2 ml ice cold 1 M NaCl, 0.02 M Tris, pH 7.5. 1.5 mg DTT (final concentration of 5 mM) was added, and the solution was sonicated. All steps were performed at 4°C. The homogenate was diluted with 0.1% CHAPS, 0.02 M Tris, pH 7.5 to 10 ml and centrifuged for 1 hr at 250,000 g. The supernatant was filtered and applied to the column. Chromatographic conditions and bioassay are described in Figure 1.

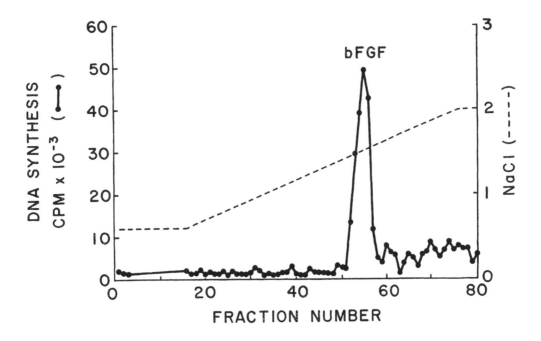

FIGURE 3. Heparin affinity HPLC of bFGF of EJ-RAT cells. EJ-RAT cells (1.2 × 10⁷ cells) were sonicated 3 times for 20 sec in 1.5 ml 1 *M* NaCl, 0.02 *M* Tris, pH 7.5. All steps were performed at 4°C; 1.2 mg DTT (final concentration of 5 m*M*) was added, and the extract was diluted to 10 ml and centrifuged at 25,000 *g* for 30 min. The sample was filtered and applied to the column. Chromatographic conditions and bioassays are described in Figure 1.

separation of aFGF and bFGF. By this method, aFGF and bFGF can be separated in one step from other mitogenic factors such as PDGF and EGF. Separation is possible with high resolution and excellent reproducibility. The HPLC-heparin column is particularly useful for purification of aFGF and bFGF from small samples and with high recovery. The advantages of using heparin affinity HPLC vs. heparin-Sepharose are as follows: (1) TSK Heparin-5PW gives a very good resolution of aFGF and bFGF with sharp peaks and baseline separation of the two peaks, (2) the elution profiles are highly reproducible allowing good comparison between applications of samples of different tissues or cells, (3) small sample volumes can be assayed; the pooled fractions in the peak have a volume of 3 to 4 ml, whereas in a similar isolation using heparin-Sepharose the pooled fractions add up to about 15 to 25 ml, and (4) there is no loss of biological activity of bFGF unlike reversed-phase chromatography where up to 95% of the activity is lost. There are some disadvantages with heparin affinity HPLC: (1) The application of large volumes is time-consuming, (2) the volume and size of the column are fixed, limiting the sample size, and (3) samples have to be cleared and filtered before application. For large samples, ultracentrifugation and filtration of sample might not be possible. But when large samples have to be analyzed, one strategy is to use conventional heparin-Sepharose with batch elution as a first step, followed by a subsequent heparin affinity HPLC.

Since aFGF and bFGF are eluted at high salt concentrations (1.1 and 1.5 *M*, respectively), it is possible that this high ionic strength might increase the hydrophobic interaction between these polypeptides and the affinity matrix.[17] To counteract this interaction, 0.1% CHAPS, a weak detergent, was included in the buffer system. CHAPS is zwitterionic and does not affect subsequent purification steps. In addition, it does not adversely affect the bioassay and, in any case, can be removed by dialysis.

Since both aFGF and bFGF have free -SH groups, which are at least in part necessary for full activity, it is important during isolation, purification and storage to prevent disulfide formation or oxidation. In order to protect the free cysteine groups, 1 mM DTT, a reducing agent, was included in all buffers. DTT is often useful as well to restore the biological activity of bFGF lost during storage for longer periods.

REFERENCES

1. Heparin-Sepharose CL-6B for affinity chromatography, *Handbook of Pharmacia*, Uppsala, Sweden.
2. Shing, Y., Folkman, Y., Sullivan, R., Butterfield, C., Murray, J., and Klagsbrun, M., Heparin affinity: purification of a tumor-derived capillary endothelial cell growth factor, *Science*, 223, 1296, 1984.
3. Gospodarowicz, D., Cheng, J., Lui, G.-M., Baird, A. and Böhlen, P., Isolation of brain fibroblast growth factor by heparin-Sepharose affinity chromatography: Identity with pituitary fibroblast growth factor, *Proc. Natl. Acad. Sci. U.S.A.*, 81, 6963, 1984.
4. Maciag, T., Mehlman, T., Friesel, R., and Schreiber, A.B., Heparin binds endothelial cell growth factor, the principal endothelial cell mitogen in bovine brain, *Science*, 225, 932, 1984.
5. Lobb, R.R. and Fett, J.W., Purification of two distinct growth factors from bovine neural tissue by heparin affinity chromatography, *Biochemistry*, 23, 6295, 1984.
6. Gimenez-Gallego, G., Rodkey, J., Bennett, C., Rios-Candelore, M., DiSalvo, J., and Thomas, K., Brain-derived acidic fibroblast growth factor: complete amino acid sequence and homologies, *Science*, 230, 1385, 1985.
7. Esch, F., Baird, A., Ling, N., Ueno, N., Hill, F., Denoroy, L., Klepper, R., Gospodarowicz, D., Boehlen, P., and Guillemin, R., Primary structure of bovine pituitary basic fibroblast growth factor (FGF) and comparison with the amino-terminal sequence of bovine brain acidic FGF, *Proc. Natl. Acad. U.S.A.*, 82, 6507, 1985.
8. Lobb, R., Sasse, J., Sullivan, R., Shing, Y., D'Amore, P., Jacobs, J., and Klagsbrun, M., Purification and characterization of heparin-binding endothelial cell growth factors, *J. Biol. Chem.*, 261, 1924, 1986.
9. Folkman, J. and Klagsbrun, M., Angiogenic factors, *Science*, 235, 442, 1987.
10. Lobb, R.R., Harper, W., and Fett, J.W., Purification of heparin-binding growth factors, *Anal. Biochem.*, 154, 1, 1986.
11. Thomas, K.A. and Gimenez-Gallego, G., Fibroblast growth factors: broad spectrum mitogens with potent angiogenic activity, *TIBS*, 11, 81, 1986.
12. Gospodarowicz, D., Ferrara, N., Schweigerer, L., and Neufeld, G., Structural characterization and biological functions of fibroblast growth factors, *Endocrine Reviews*, 8, 95, 1987.
13. Klagsbrun, M. and Shing, Y., Heparin affinity of anionic and cationic capillary endothelial cell growth factors: analysis of hypothalamus-derived growth factors and fibroblast growth factors, *Proc. Natl. Acad. Sci. U.S.A.*, 82, 805, 1985.
14. Land, H., Parada, L.F., and Weinberg, R.A., Cellular and viral oncogenes cooperate to achieve tumorigenic conversion of rat embryo fibroblasts, in *Cancer Cells 2 / Oncogenes and Viral Genes*, Cold Spring Harbor Laboratory, Cold Spring Harbor, 1984, 473.
15. Nakamura, K., Toyoda, K., and Kato, Y., High-performance affinity chromatography of proteins of TSKgel Heparin-5PW, *J. Chromatogr.*, 445, 234, 1988.
16. Wadzinski, M.G., Folkman, J., Sasse, J., Devey, K., Ingber, D., and Klagsbrun, M., Heparin-binding angiogenesis factors: detection by immunological methods, *Clin. Physiol. Biochem.*, 5, 200, 1987.
17. Scopes, R.K., *Protein Purification: Principles and Practice*, Springer Verlag, New York, 1987, 140.

THEORY AND PRACTICAL ASPECTS OF HIGH-PERFORMANCE IMMUNOAFFINITY CHROMATOGRAPHY

Terry M. Phillips

I. INTRODUCTION

Immunoaffinity chromatography is a unique separation procedure which depends on immunological reactions to isolate the material of interest. Although the majority of immunoaffinity techniques center around low or medium pressure systems, the development of silica particles and glass beads[1-3] with reactive side-chains suitable for protein immobilization provided suitable packing material which has allowed this technique to be applied to high-performance liquid chromatographic (HPLC) systems. This new technique has been called high-performance immunoaffinity chromatography (HPIAC). Recently, protein A and avidin-coated glass beads have been developed as suitable HPIAC immobilization supports.[4,5] These beads have greatly simplified the procedure for antibody immobilization and have widened the application of immobilized antibody separation technology to high-performance liquid chromatography.

II. THEORY OF IMMUNOAFFINITY REACTIONS

The efficiency of HPIAC separations depends on the selectivity of immobilized antibodies to perform the separation. The antibody is bound to an inert matrix, which forms the packing material of the column. The HPIAC separation is performed in two phases: a primary isolation phase, followed by a secondary elution phase. In the primary phase, a solution containing the material to be isolated (the antigen) is passed over the immobilized antibody, which reacts with the antigen forming an immobilized antibody-antigen complex. This complex retains the antigen while the unreactive (non-antigenic) material passes through the column and forms the primary peak of the chromatogram. The bound antigen is recovered during the second or elution phase, by breaking the antibody/antigen complex and allowing the freed antigen to be flushed through the column, to form the second peak of the chromatogram. The procedure of HPIAC is diagrammatically outlined in Figure 1.

III. COLUMN PACKING MATERIAL

The packing material of choice for HPIAC is silica, either in the form of irregular particles or as glass beads. Both forms are commercially available, although some element of basic chemistry is required to immobilize the ligand to the particle or bead surface. Table 1 summarizes the available side-chains used to immobilize proteins to silica or bead surfaces.

A. PARTICLES
Irregular silica particles which have been chemically modified to accept reactive side-chains

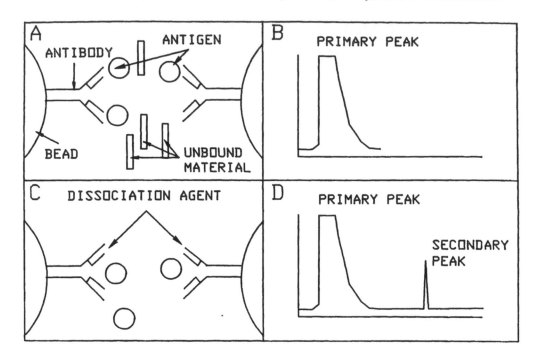

FIGURE 1. The principles of HPIAC separations. A. The antibody-coated beads react with and retain the antigen, while the unbound material is washed through the column. B. The unbound material forms the primary peak of the chromatogram. C. The bound antigen is recovered by applying a gradient containing a dissociation agent, which breaks the antibody-antigen complex. D. The recovered antigen passes through the column to form the second peak of the chromatogram.

TABLE 1
Reactive Side-Chains Used for Protein Immobilization

Side-chain	Ligand coupling conditions
N-hydroxysuccinimide	Coupling through primary amino groups on the ligand by incubation at pH 9.0
Carbonyldiimidazole	Coupling through primary amino groups on the ligand by incubation at pH 9.0
Epoxy or epoxide	Coupling through amino, hydroxyl, and sulfhydryl groups on the ligand by incubation at pH 8.0—9.0
Thiol	Coupling through carbonyl or alkyl groups on the ligand in the presence of a water-soluble carbodiimide at pH 9.0
Carboxyl	Coupling through primary amino groups on the ligand in the presence of a water-soluble carbodiimide at pH 9.0

are commercially available from several sources (Diagnostic Specialties, Metuchen, NJ; Serva, Westbury, NY; Beckman Instruments, Palo Alto, CA). The reactive side-chains are usually epoxy or thiol groups, although other more novel side-chains are now being introduced. For those not wishing to deal with the labors of column packing, Beckman Instruments (Palo Alto, CA) have developed a prepacked epoxy-silica column, in which the ligand immobilization is performed in situ, as per the manufacturers instructions.

B. GLASS BEADS

Chemically modified controlled pore glass beads, coated with a hydrophilic, nonionic carbohydrate surface — called glycophase, which decreases the non-specific interactions between glass and biological materials — are suitable for HPIAC. The beads are commercially available with a wide variety of different side-chains attached to the bead surface (Fisher Scientific, Columbia, MD; Spectrum Medical Industries, Los Angeles, CA: Pierce Chemical Co., Rockford, IL). Derivatized glass beads can easily be coated with specialized proteins to enable efficient binding of antibodies to their surfaces.

1. Protein A or Protein G-Coated Glass Beads

A modification of the glass bead is binding a coating of bacterial coat proteins to the surface of the beads. This coating enables IgG class antibodies to be easily immobilized to the bead surface.[6] Both protein A and G have the ability to bind IgG antibodies via their tails or Fc portions. Binding via the Fc portion helps to orient the antigen-binding sites of the antibodies into the mobile phase of the column, thereby producing the maximum efficiency from the ligand.[5] Antibodies naturally bind to the IgG receptors on the protein A or G molecules and can be permanently immobilized by crosslinking the antibody to the bacterial protein with a solution of 50 mM carbodiimide.

2. Avidin and Streptavidin-Coated Glass Beads

Another modification of the glass bead is to add a coating of either egg white avidin or the more refined streptavidin. Both of these proteins have high binding affinity for materials labeled with the vitamin, biotin.[7,8] Biotin is commercially available in many forms, ranging from N-hydroxysuccinimide derivatized biotin to the specialized biotin hydrazine.[9] The former is the easiest of the biotin derivatives to use when biotinylating proteins; the binding takes place when the protein and the biotin are incubated at room temperature for 2 hr at pH 9.0 in carbonate buffer. The biotin is crosslinked randomly to exposed amino groups on the surface of the protein of interest (Scheme 1). The latter is used for biotinylating carbohydrates and has been used to label the carbohydrate content of IgG antibodies.[5,9] The advantage of this technique is that the biotin is attached to the Fc portion of the IgG antibody and when incubated with the avidin-coated beads, becomes attached to the biotin receptors via its Fc portion (Scheme 2). In this way, biotinylated antibodies become oriented in a similar way to those bound to protein A-coated glass beads. A disadvantage of this form of biotinylation is that the antibodies have to be modified by treatment with periodate prior to linkage of the hydrazide-biotin. Due to the extremely high affinity exhibited by either avidin or streptavidin for biotin-labeled antibodies, no chemical modification is required, such as chemical cross-linking, once the antibody is attached to the coated bead.

C. ANTIBODIES USED AS LIGANDS IN HPIAC

The antibodies used as the immobilized ligand in HPIAC can be obtained from many different commercial sources, clinical samples, or from animal immunizations. The IgG class of antibody is the most useful for HPIAC techniques although the introduction of the hydrazide-biotin labeled procedure has allowed all classes of antibodies to be immobilized on avidin-coated glass beads. The IgE class of antibodies is useful in isolating antigens involved in allergy and the IgA antibodies are useful against some forms of bacterial and viral antigens. Apart from the different classes of antibody, there are also several different types of antibody which can be used, each with its own properties.

1. Polyclonal Antibodies

Animals immunized with specific antigens produce a type of antibody which is called a polyclonal antibody. This antibody is usually the major immunological component of hyperimmune serum and is present in company with other less or non-reactive antibodies. Therefore, the reactive IgG has to be isolated before the antibody can be used in HPIAC. This type of antibody is a natural product of an induced immune response and often reacts with common determinants on the antigen.

A. BIOTINYLATION OF A PROTEIN OR ANTIBODY

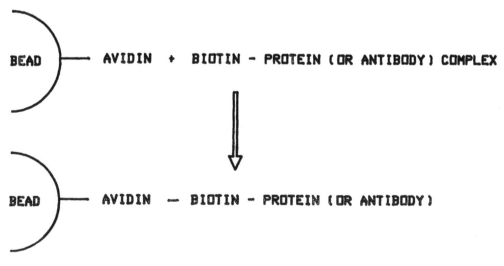

B. FORMATION OF AVIDIN – BIOTIN – PROTEIN LIGAND

SCHEME 1

A. BIOTINYLATION OF ANTIBODY WITH BIOTIN HYDRAZINE

(I) R – C – H $\xrightarrow{\text{PERIODATE}}$ R – C – H

CARBOHYDRATE ON
Fc (OR TAIL)
OF ANTIBODY

(II) R – C – H + H_2N – N – C – $(CH_2)_4$ ...

$\Downarrow$ H_2O

R – C = N – N – $(CH_2)_4$ ... BIOTIN-HYDRAZINE (BH)

B. FORMATION OF AVIDIN – BIOTINYLATED ANTIBODY LIGAND

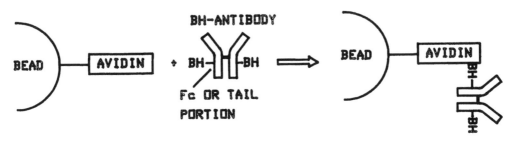

SCHEME 2

2. Autoantibodies

This type of antibody is a polyclonal antibody produced by the animal or person against components of its own body.[10] These antibodies are often made against specific entities such as cell membrane receptors and subcellular fractions. In addition, they are often made against hormones and cell secretory by-products. When used to study certain disease states these antibodies are useful in defining the materials considered antigenic by the host.

3. Monoclonal Antibodies

The introduction of laboratory-designed antibodies in 1975[11] greatly increased the use of antibodies as specific separation reagents. The advantage of monoclonal antibodies is that they are the product of cellular fusions, performed at the discretion of the investigator, and all of the antibody molecules are reactive against the same defined region of the antigen. These antibodies often exhibit greater specificity than their polyclonal counterparts and are well suited for isolating subcomponents of proteins and polypeptides.

4. Idiotypic Antibodies

Idiotypes and their counterparts, anti-idiotypes are specialized antibodies which are part of the normal immune system regulatory circuits.[12] They have distinct properties not exhibited by other antibodies, such as being able to act as receptor substrates and to this end they are valuable reagents for HPIAC isolations. Specific anti-idiotypes can be produced by monoclonal antibody technology and these antibodies exhibit such defined specificity that they will react only with a defined area on the antigen and with very little else.

IV. COLUMN CONSTRUCTION

Coating of the silica particles or glass beads can easily be achieved by activation of any of the reactive side-chains by the conditions given in Table 1. Coating of glass beads with either a bacterial coat protein or with one of the two types of avidin can easily be achieved by incubating either N-hydroxysuccinimide or carbonyldiimidazole derivatized glass beads with the protein at pH 9.0 in 50 m*M* carbonate buffer for 2 hr at room temperature on an overhead mixer.[4,5] This incubation is followed by a further 1 hr incubation with the antibody solution. Following this, the antibody-coated beads can be packed into either short, wide-bore or long, narrow-bore stainless steel columns. Conventional 4.6 mm I.D. columns in lengths ranging from 5 to 25 cm make excellent HPIAC columns.[13] The length depends on the amount of available antibody and/ or the amount of antigen in the test solution. Long, narrow columns often yield higher amounts of antigen from dilute solutions than do short, wide columns.

Packing can be performed in the same way as for conventional packing materials although the pumped-slurry packing apparatus has been shown to be gentler on the protein-coated HPIAC packing material than the gas-activated packing apparatus.[13] Whichever technique is used, care must be exercised to ensure that excessive packing pressures are not encountered, which can shear the protein coat off the packing materials. Aqueous solvents must be used when packing HPIAC columns, but these solvents have been shown to produce more friction than organic solvents during the packing process. When aqueous buffers such as 0.01 *M* phosphate are used, packing pressures up to 500 psi can be obtained without undue damage occurring to the packing material. When using glass beads, dry-packing of freeze-dried coated beads can be successfully performed but this technique does not work well with protein-coated irregular silica particles.[13]

When the column is packed it can be used with any conventional HPLC apparatus, from a simple isocratic system to a complicated gradient system. The latter is preferable because it gives greater control over the formation of the elution gradients.[14]

V. RUNNING CONDITIONS

Flow-rates of 0.5 to 1.5 ml/min are commonly used in HPIAC separations,[15,16] although the higher flow-rates can interfere with the interactions between the immobilized antibody and the antigen. When labile antigens are to be separated, fast flow-rates coupled with long, narrow-bore columns will maintain the maximum efficiency of the system. In contrast, when the sample is small and contains a large number of different materials, a slow flow-rate on a short, wide-bore column will often prove to be the most efficient separation strategy.

The life of the column will be affected by the pressure and temperature employed during runs. Running pressures should not exceed 1000 psi as stripping of the bound antibody/antigen complex can occur above this pressure. The use of coated glass beads helps to keep the running pressure low as the shape of the beads allows for greater flow of the mobile phase.[5]

Temperature is very important both for maintaining the integrity of the immobilized antibody and for maintaining the activity of the isolated antigen. Most HPIAC runs should be performed at 4°C, to ensure maximum column life and to maintain the activity of the isolated antigens. This temperature can easily be maintained by using a column jacket and a circulating ice bath.[13] The addition of a thermocouple, inserted into the jacket, will help maintain the correct temperature.

VI. ELUTION AGENTS AND RECOVERY OF ISOLATED MATERIAL

The second phase of HPIAC is the elution or recovery phase which entails breaking the intermolecular forces which are involved in the antibody/antigen complex. This can be achieved by several different strategies, but it must be borne in mind that delicate proteins can easily be denatured and require delicate elution protocols.[17] Also, the forces which hold antibody/antigen complexes together are identical to the forces which are responsible for maintaining the tertiary structure of proteins, and protein and glycoprotein antigens can easily lose their activity during prolonged or extremely harsh elution. The two most common techniques for recovering isolated materials from HPIAC columns are acid or chaotropic ion elution, although other techniques have been described.[18] Table 2 summarizes these elution buffers.

A. ACID ELUTION

Lowering of the pH of the running buffer to pH 2 or lower will effectively break antibody/antigen binding forces and a variety of different acidic elution buffers have been used in HPIAC. When using acid elution, it is advisable to use a steep controlled gradient to perform the elution. Silica is susceptible to continued exposure to low pH and even though the protein coat will help to protect the packing, eventually the acid will cause deterioration of the packing material. In addition, prolonged exposure to acid will also damage the antigen and reduce its activity. Figure 2 is an example of an HPIAC isolation using a linear acid elution gradient.

B. CHAOTROPIC ION ELUTION

Chaotropic ion elution is becoming the most popular technique for recovering antigens from HPIAC columns.[4,5,18] These salts interfere with the organization of the ionic interactive forces and cause a dissociation of the antibody/antigen complex. The effective dissociation parameter of chaotropic ions is as follows:

$$Cl^- < I^-, ClO_4^- < CF_3COO^- < SCN^- < CCl_3COO^-$$

Chaotropic ion gradients ranging from 0 to 2.5 to 3.0 M have been reported to be very effective in recovering bound antigens from HPIAC columns. All of these reports have used steep, controlled gradients to recover the bound materials.[4,5,19,20]

TABLE 2
Elution Buffers Used in HPIAC

pH manipulation
   Acidic buffers
      0.1—1 M glycine   pH range 1.0—1.5
      0.1—1 M Tris/HCl   pH range 1.0—2.0
      0.33 M citric acid   pH range 1.5—2.5
      0.5—1 M acetic acid   pH range 1.0—1.5
   Alkaline buffers
      0.5—2 M sodium hydroxide   pH range 10.0—14.0
      0.5—2 M potassium hydroxide pH range 10.0—14.0
Chaotropic ions
   2—3 M sodium thiocyanate
   2—3 M sodium iodide
   2.5—6 M sodium chloride
   2—4 M polyvinylpyrrolidone/iodide complex

Note: All of the values, e.g., 2—3 M, pH 1.0—1.5, are effective ranges
      for each elution buffer and are not gradients. These ranges are
      gleaned from several different reference sources.

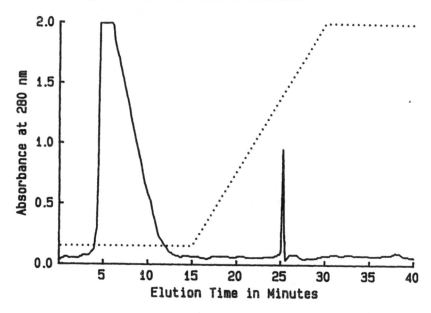

FIGURE 2. HPIAC isolation of human transferrin from normal human serum on a 150 × 4.6 mm
I.D. HPIAC column run at 1 ml/min in 0.01 M phosphate buffer, pH 7.2. HPIAC column
constructed of protein A — coated glass beads to which monoclonal anti-transferrin antibodies
have been attached. Sample size: 100 μl; detection: 280 nm at 0.08 AUFS. Column maintained
at 4°C throughout the run. Linear acid gradient from pH 7.2 to pH 1.5 started at 15 min into the
run and maintained for a further 15 min (dotted line). Transferrin isolated as the sharp second
peak seen at 25 min.

C. OTHER TECHNIQUES

Agents such as urea, guanidine hydrochloride and the polarity-reducing agents, dioxane and
ethylene glycol have been used to recover materials from HPIAC columns. However, because
of their denaturing effects, they have not gained the popularity of either acid or chaotropic
elution techniques.[13] When used, most of these agents have been applied in stepwise elution
techniques.

D. TREATMENT OF RECOVERED ANTIGENS

Once the antigen has been eluted from the column and collected, the elution agent has to be removed to protect the antigen from denaturation. Several techniques are available for performing this task, the simplest being dialysis which is either performed by on-line continuous flow dialysis or following the chromatographic separation. Another technique involves desalting the isolated antigen on small size-exclusion columns such as the P10 column from Pharmacia (filled with Sephadex G25). This technique is advantageous when the isolated antigen is of large molecular weight and can easily be separated from the low molecular weight elution agents by size-exclusion chromatography.

REFERENCES

1. **Weetall, H.H. and Filbert, A.M.,** Porous glass for affinity chromatography applications, *Methods Enzymol.,* 34, 59, 1974.
2. **Scouten, W.H.,** *Affinity Chromatography,* John Wiley & Sons, New York, 1981, 42.
3. **Mohr, P. and Pommerening, K.,** *Affinity Chromatography,* Marcel Dekker, New York, 1985, 7.
4. **Phillips, T.M., Queen, W.D., More, N.S., and Thompson, A.M.,** Protein A-coated glass beads: a universal support for high-performance immunoaffinity chromatography, *J. Chromatogr.,* 327, 213, 1985.
5. **Babashak, J.V. and Phillips, T.M.,** Use of avidin-coated glass beads as a support for high-performance immunoaffinity chromatography, *J. Chromatogr.,* 444, 21, 1988.
6. **Boyle, M.D.P.,** Applications of bacterial Fc receptors in immunotechnology, *Biotechniques,* November, 334, 1984.
7. **Wilchek, M. and Bayer, E.A.,** The avidin biotin complex in immunology, *Immunol. Today,* 5, 39, 1984.
8. **Buckle, J.W. and Cook, G.M.W.,** Specific isolation of surface glycoproteins from intact cells by biotinylated concanavalin A and immobilized streptavidin, *Anal. Biochem.,* 165, 463, 1986.
9. **O'Shannessy, D.J. and Quarles, R.H.,** Labelling of the oligosaccharide moieties of immunoglobulins, *J. Immunol. Methods,* 99, 153, 1987.
10. **Blecher, M.,** Speculations on potential anti-receptor autoimmune disease, in *Receptors, Antibodies and Disease,* Evered, D. and Whelan, J., Eds., Pitman, London, 1982, 279.
11. **Kohler, G. and Milstein, G.,** Continuous cultures of fused cells secreting antibody of predefined specificity, *Nature* (London), 256, 495, 1975.
12. **Bona, C.A. and Kohler, H.,** Anti-idiotypic antibodies and internal images, in *Monoclonal and Anti-Idiotypic Antibodies: Probes for Receptor Structure and Function,* Venter, J.C., Fraser, C.M., and Lindstrom, J., Eds., Alan R. Liss, New York, 1984, 141.
13. **Phillips, T.M.,** High-performance immunoaffinity chromatography, *Adv. Chromatogr.,* Giddings, J.C., Grushka, E., and Brown, P.R., Eds., Marcel Dekker, New York, 1989, 133.
14. **Phillips, T.M.,** High-performance immunoaffinity chromatography, *LC.GC,* 3, 962, 1985.
15. **Walters, R.R.,** High-performance affinity chromatography: pore size effects, *J. Chromatogr.,* 249, 19, 1982.
16. **Roy, S.K., McGregor, W.C., and Orichowskyj, S.T.,** Automated high-performance immunosorbent assay for recombinant leukocyte A interferon, *J. Chromatogr.,* 327, 189, 1985.
17. **Lapanje, S.,** *Physiochemical Aspects of Protein Denaturation,* John Wiley & Sons, New York, 1978, 56.
18. **Rouslahti, E.,** Antigen-antibody interaction, antibody affinity and dissociation of immune complexes, in *Immunoadsorbents in Protein Purification,* Rouslahti, E., Ed., University Park Press, Baltimore, 1976, 3.
19. **Sica, V., Puca, G.A., Molinari, M., Buonaguro, F.M., and Bresciani, F.,** *Biochemistry,* 19, 83, 1980.
20. **Dandliker, W.B., Alonso, R., de Sausserre, V.A., Kierszenbaum, F., Levison, S.A., and Shapiro, H.C.,** The effect of chaotropic ions on the dissociation of antigen antibody complexes, *Biochemistry,* 6, 1460, 1967.

ISOLATION AND MEASUREMENT OF MEMBRANE PROTEINS BY HIGH-PERFORMANCE IMMUNOAFFINITY CHROMATOGRAPHY

Terry M. Phillips

I. INTRODUCTION

High-performance immunoaffinity chromatography (HPIAC) can be used to isolate any biological material or chemical to which an antibody can be made. It is a new technique, however, and HPIAC has been used in only a limited number of applications since its introduction. Immobilized monoclonal antibodies, directed against substance P, have been used to isolate substance P receptors from solubilized lymphoblastic cell membranes[1] and other workers have used HPIAC to isolate liver membrane proteins, similar to those isolated by concanavalin A lectin chromatography.[2] Immunoaffinity has been used by several groups as a sample preparation technique prior to conventional high-performance liquid chromatography (HPLC)[3,4] or in tandem with reversed-phase analysis of the immunoaffinity isolated materials.[5]

Although HPLC columns containing protein A-coated packings are available from several commercial sources, these columns have been used almost exclusively for the isolation of IgG antibodies from either ascites fluids of animals producing monoclonal antibodies or from the supernatants of antibody-producing hybridomas.[6]

However, laboratory-made streptavidin-coated glass beads have been used to isolate both lymphocyte histocompatibility antigens[7] from normal human lymphocytes and specific lymphocyte receptors from immunized animals.[8,9]

II. ISOLATION OF COMPLEMENT COMPONENT 3b (C3b) RECEPTORS

A. LYMPHOCYTE RECEPTORS (SCHEME 1)

1. Coating Glass Beads for HPIAC

Five milligrams of protein A (Pharmacia, Piscataway, NJ) was dissolved in 10 ml of 50 mM carbonate buffer, pH 9.0, by gentle agitation, before adding 10 g of carbonyldiimidazole-derivatized glass beads (Spectrum Medical Industries, Los Angeles, CA) and incubating for 18 hr at 4°C. Following this incubation, the beads were extensively washed by sedimentation in 0.01 M phosphate buffer, pH 7.0 and used within 1 week.

2. Preparation of HPIAC Column

Monoclonal antibodies (MAb) raised against lightly fixed human B lymphocytes were tested for reactivity by performing an antigen binding assay, using I^{125}- labeled C3b. The reactive clones were expanded and the MAb immobilized onto protein A-coated glass beads by incubating 250 µg MAb with 2 g of protein A-coated beads for 2 hr at room temperature on an

A.   COATING GLASS BEADS WITH PROTEIN A

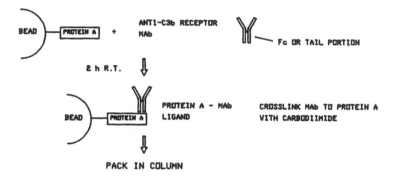

B.   PREPARATION OF HPIAC COLUMN

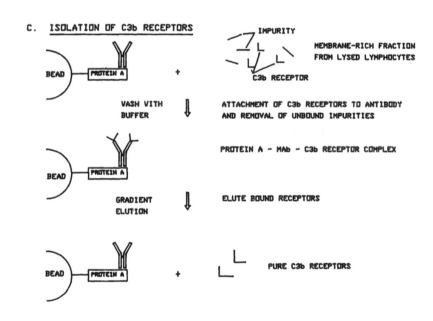

SCHEME 1

overhead mixer. The beads were washed five times in 0.01 *M* phosphate buffer, pH 7.0, and the MAb lightly cross-linked to the protein A by incubating the beads in 10 m*M* 1-cyclohexyl-3-(2-morpholinethyl)metho-p-toluene sulfate (Pierce Chemical Co., Rockford, IL) for 30 min before slurry-packing into a 100 × 4.6 mm I.D. stainless steel column and extensively flushing the column with the phosphate buffer.

3. Preparation of Solubilized Membranes

A cell pellet containing 1×10^6 B lymphocytes was hypotonically lysed, sonicated at full power for 2 min and the membrane-rich fraction isolated by centrifugation at 10,000 *g* for 30

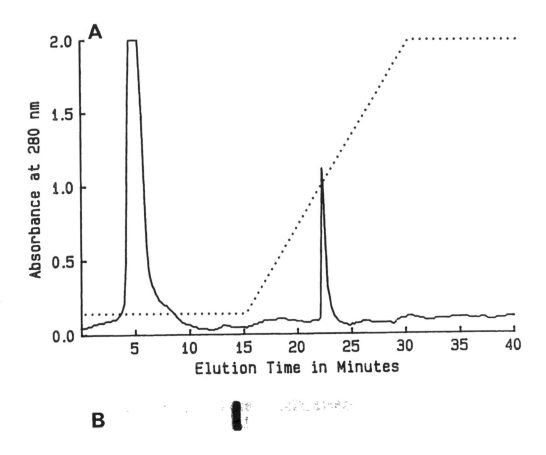

FIGURE 1. HPIAC isolation of C3b receptors from human B lymphocytes. (A) Chromatogram produced by passing 100 µl of the solubilized membrane material through a 100×4.6 mm I.D. HPIAC column, containing immobilized anti-C3b receptor MAb. The column was developed at 1 ml/min with a running buffer of 0.01 M phosphate, pH 7.0, containing 0.01% NP40. The dotted line represents the 0 to 3.5 M sodium chloride elution gradient. (B) Silver stained SDS/PAGE gel showing that the material recovered in the second peak contained a single protein band at approximately 120 kDa.

min. The membrane-rich fraction was solubilized in 1% Nonidet NP 40 and injected onto the column.

4. HPIAC Isolation of C3b Receptors

The chromatogram was developed by running the column in 0.01 M phosphate buffer (pH 7.0), to which 0.01% NP 40 was added, for 15 min at 1 ml/min, before starting a 0 to 3.5 M sodium chloride gradient. The gradient was developed over 15 min and maintained at the upper level of the gradient for a further 10 minutes before recycling the column by reversing the gradient.

The chromatogram illustrated in Figure 1A shows that the unreacted material formed a large primary peak followed by the HPIAC isolated material, which was eluted as a sharp, well defined peak at 23 min into the chromatography run. No other peaks were detected during the run.

B. MACROPHAGE RECEPTORS

Identical experiments using human macrophage membranes, isolated by nitrogen cavitation and solubilization in Triton X-100 were performed. The membranes were passed through the same HPIAC column under identical conditions and demonstrated similar results, to those found with the lymphocytes, although the second peak developed during this run was smaller.

III. ISOLATION OF SPECIFIC LYMPHOCYTE RECEPTORS

A. MONOCLONAL ANTI-P2 ISOLATION (SCHEME 2)
1. Production of Antibodies

In these studies, Balb/C mice were immunized with the P2 antigenic determinant of myelin basic protein (MBP) by subcutaneous injection of MBP in complete Freund's adjuvant. Five days following booster injection, the animals were splenectomized and the spleen cells fused and cloned and used for the production of MAb against P2.[10]

2. Biotinylation of Antibodies

Reactive anti-P2 MAb were isolated by ion-exchange chromatography and biotinylated by the technique of O'Shannessy and Quarles.[11] Briefly, this technique uses biotin hydrazine to label the carbohydrate portion of the IgG molecule, which is located on the Fc or tail portion. In this way, IgG molecules can be immobilized via their Fc portions onto avidin-coated beads in a similar fashion as that used for protein A immobilization. One milliliter containing 250 μg of MAb was incubated for 20 min with 1 ml of 10 mM sodium metaperiodate, followed by the addition of 20 ml of 5% ethylene glycol and overnight dialysis against 0.01 M phosphate buffer, pH 7.0. The MAb solution was added to 1 ml of phosphate buffer containing 1 mg of sodium cyanoborate and 1 mg of hydrazine-biotin and incubated for 1 hr in a tightly capped tube. The reaction was stopped by a further overnight dialysis against the phosphate buffer. All incubations and dialysis were performed at 4°C.

3. HPIAC Isolation of Specific P2 Receptors

The HPIAC column was prepared by incubating 250 μg of the biotinylated anti-P2 MAb with 2 g of streptavidin-coated glass beads (coated in the same way as described for protein A), on an overhead mixer for 1 hr. The beads were washed in 0.01 M phosphate buffer and slurry-packed into a 100 × 4.6 mm I.D. stainless steel column. Lymphocytes from P2 immunized animals were allowed to react with MBP (which possesses several P2 antigen sites), adsorbed onto plastic plates. The adherent cells were recovered by gentle sonication and the membrane-rich fraction obtained in a similar manner to that described in the section on C3b receptors. One hundred microliters of membrane preparation were injected into the column, which was developed in a analogous way to that described above (C3b section) except that the elution gradient used was 0 to 2.5 M sodium thiocyanate.

B. ANTI-IDIOTYPIC ISOLATION OF P2 RECEPTORS (SCHEME 3)

Similar experiments were performed on the isolation of P2 lymphocyte receptors, using a monoclonal anti-idiotypic antibody as the immobilized ligand.

1. Preparation of the Monoclonal Anti-Idiotypic Antibodies

This antibody was prepared by forming immune complexes between equal amounts of the anti-P2 MAb and murine rheumatoid factor (a naturally occurring antibody which reacts with the Fc portions of IgG antibodies, forming immune complexes in which the antigen receptors of the target IgG antibody become recognized as antigens when the complexes are processed by macrophages) and injecting these complexes into unimmunized mice as described by Phillips et al.[12] MAb were prepared from these animals and screened for the presence of antibodies which were able to inhibit the binding of P2 to the original anti-P2 MAb. The reactive hybrids were cloned and expanded to produce the desired monoclonal anti-idiotypic antibodies.

521

A. ISOLATION OF P2 ANTIGEN - P2 RECEPTOR COMPLEX

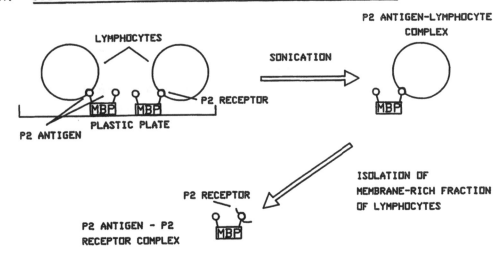

B. ISOLATION OF P2 RECEPTORS BY HPIAC

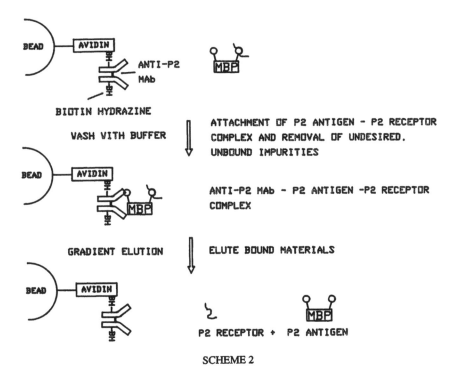

SCHEME 2

2. HPIAC Isolation of Receptors

The anti-idiotypes were isolated and biotinylated as described for the anti-P2 MAb and immobilized on streptavidin-coated glass beads, as the HPIAC ligand.

A sample of lymphocyte membranes prepared from P2 immunized animals, not reacted with MBP, but solubilized in the same way, was injected into the anti-idiotypic HPIAC column and the run developed as described for the anti-P2 column. Figure 2 shows examples of the chromatograms produced by these two different HPIAC columns.

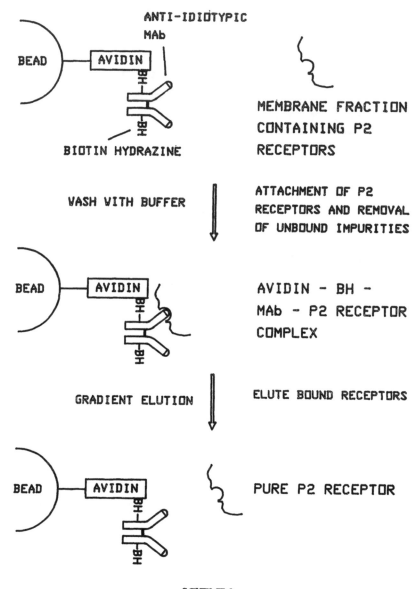

ANTI-IDIOTYPIC
MAb

BEAD ─ AVIDIN

BIOTIN HYDRAZINE

MEMBRANE FRACTION
CONTAINING P2
RECEPTORS

WASH WITH BUFFER

ATTACHMENT OF P2
RECEPTORS AND REMOVAL
OF UNBOUND IMPURITIES

BEAD ─ AVIDIN

AVIDIN - BH -
MAb - P2 RECEPTOR
COMPLEX

GRADIENT ELUTION

ELUTE BOUND RECEPTORS

BEAD ─ AVIDIN

PURE P2 RECEPTOR

SCHEME 3

IV. PURITY OF ISOLATED PROTEINS

Membrane proteins isolated by the different HPIAC techniques were analyzed by sodium dodecyl sulfate/polyacrylamide gel electrophoresis (SDS/PAGE).[13] Briefly, 25 µl of the second HPIAC peak was mixed 1:1 with 0.1% SDS and incubated at 100°C for 10 min. The solution was cooled to 4°C and a 10 µl sample run on a 10 to 30% linear PAGE gel at a constant voltage of 150 V for 3 hr. The gel was fixed in 4:1 methanol:acetic acid for 30 min and the isolated bands visualized by silver staining.[14] The specificities of the isolated bands were checked by specific immunostaining[15] with the MAb used as the HPIAC ligand, following Western blotting of unfixed SDS/PAGE protein maps of the isolated bands.[16]

A. C3b RECEPTORS

Isolation of the C3b receptors showed the presence of a single band of approximately 120

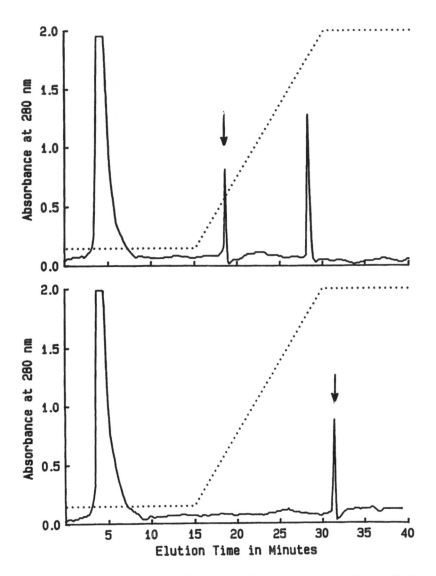

FIGURE 2. HPIAC isolation of P2 antigen receptors from primed murine lymphocytes. Both chromatograms were produced by passing 100 μl of solubilized membrane over either immobilized anti-P2 MAb (upper chromatogram) or immobilized anti-P2 anti-idiotypic MAb (lower chromatogram). Running conditions for both chromatograms as described in Figure 1, except that the dotted line represents a sodium thiocyanate elution gradient from 0 to 2.5 *M* (upper level). The isolated receptors are present in the second peak of both chromatograms (indicated by the arrow). In the upper chromatogram, the receptors are dissociated from the P2 antigen at a lower ionic strength and are eluted as a discrete peak in the early phase of the elution gradient. In both the upper and lower chromatograms, higher ionic strengths are required to dissociate either the MPB (which was used to probe the specific receptor and appears as the third peak in the upper chromatogram) or the P2 receptors (lower chromatogram) from the immobilized antibody ligands.

kDa, when compared to the molecular weight markers. This band was the only protein detected by silver staining the gel and reacted with the anti-C3b MAb when Western blots of unfixed SDS/PAGE gels were immunostained. The size of the isolated receptor was the same in both the lymphocyte and macrophage membrane preparations. Figure 1B shows a typical silver stained gel of the second HPIAC peak recovered from the anti-C3b column.

B. LYMPHOCYTE RECEPTORS

SDS/PAGE analysis of the receptors isolated from the primed lymphocytes of P2 immunized animals demonstrated the presence of a single band in the second HPIAC peak, from both the anti-P2 MAb and the anti-idiotypic MAb columns. The band isolated from the anti-P2 column was shown to be approximately 90 kDa when compared to the molecular weight standards. The band isolated by the anti-idiotype column was slightly larger (approximately 95 kDa), but similar to the band isolated by the anti-P2 MAb. No explanation for the size difference in the P2 receptors isolated by the two different columns has been found, although it is easy to postulate that binding the antigen to the receptor prior to membrane solubilization produces conformational changes which result in a smaller membrane fragment. No other bands were seen in either isolations in silver stained gels and immunostaining of Western blots of the SDS/PAGE gels confirmed that the bands isolated in the second peaks were reactive with the antibodies used as the HPIAC ligand. Figure 3 is a silver stained gel showing the bands isolated by the two HPIAC columns as compared to the unfractionated, solubilized membrane preparation.

V. BIOACTIVITY OF ISOLATED RECEPTORS

All of the isolated HPIAC fractions were tested for their ability to bind radiolabeled substrate. This ability was tested against a comparable number of intact cells and against receptors isolated by standard biochemical and lectin chromatographic techniques. 1×10^6 intact cells or the equivalent amount of receptor isolated from the same number of cells were incubated with 50 µg of I^{125}-labeled C3b or with labeled antigen. Samples were taken at 30, 60, and 120 min and the levels of bound radioactivity measured in a gamma counter.

In all cases, the highest level of binding was exhibited by the intact cells, although the HPIAC isolated receptors generally exhibited between 75 to 80% of the binding demonstrated by the intact cells. The binding levels for the MAb isolated C3b and P2 receptors were approximately the same, both receptors demonstrating the ability to bind 75 and 77% of the intact cell level, respectively. The P2 receptors isolated by anti-idiotypic HPIAC showed a slightly higher binding level, which was 82% of the intact cell binding. Although the anti-idiotypic HPIAC column produced the more active receptors, the results were not outstandingly superior to those produced by the anti-P2 column. Considering the technical skill required to produce anti-idiotypic antibodies, most investigators will obtain satisfactory results by using standard antibodies as the HPIAC ligand. Biochemical and lectin isolated receptors varied greatly, ranging from 30 to 65% of the intact cell activity. Table 1 summarizes these results.

VI. OTHER APPLICATIONS

Work in the author's laboratory has also focused on other applications for HPIAC. The technique has been used to quantitate the levels of IgG in cerebrospinal fluid samples from pediatric patients[17] and IgE levels in patients with allergies.[18] The technique compared well with more conventional procedures but could be performed in under 1 hr on microsamples. HPIAC has also been used to measure the levels of transferrin in brain tissue of normal and demyelinated mice.[19] The levels measured by integration of the HPIAC isolated peaks corresponded to levels measured by biochemical techniques and correlated with immunocytochemical localization in the same tissue.

Recently, idiotypic antibodies isolated from cancer patients have been immobilized and used as HPIAC ligands to measure the levels of inhibitory anti-idiotypic antibodies in the same patients during immunotherapy.[20] The levels of anti-idiotypic antibodies as measured by HPIAC correlated with the loss of anti-tumor activity in those patients not responding to the treatment.

FIGURE 3. SDS/PAGE gel of the materials isolated in the second peaks of the chromatograms illustrated in Figure 2. Lane 1, the whole solubilized lymphocyte membrane preparation; lane 2, the HPIAC isolated P2 receptor, using MBP, containing P2 antigen as the receptor probe; lane 3, HPIAC isolated P2 receptor using an immobilized anti-idiotypic MAb.

These anti-idiotypic antibodies have been used to isolate specific receptors on the membranes of tumor-reactive lymphocytes from the patients exhibiting the presence of the anti-idiotypes. In this way it is hoped that valuable insights into the pathways of immune regulation in cancer patients may be gained.

VII. CONCLUSIONS

High-performance immunoaffinity chromatography is a useful separation technique which can be applied to the isolation of any biochemical material. The technique combines the

TABLE 1
Bioactivity of the HPIAC Isolated Receptors Test Material

Test material	Amount of labeled substrate bound (cpm × 10³)		
	30[a]	60[a]	120[a]
Labeled C3b			
Intact lymphocytes	8.4	13.7	23.8
Intact macrophages	5.0	9.2	16.3
HPIAC isolated receptors (lymph)	6.3	9.8	18.1
HPIAC isolated receptors (macro)	3.8	6.6	12.1
Lectin isolated receptors (lymph)	4.3	8.0	11.5
Lectin isolated receptors (macro)	2.0	4.6	9.1
Biochemically isolated receptors (lymph)	3.0	4.8	8.8
Biochemically isolated receptors (macro)	1.5	3.5	5.9
Labeled P2 antigen			
Intact lymphocytes	10.4	18.8	27.1
HPIAC anti-idiotype isolated receptors	8.2	15.2	22.2
HPIAC MAb isolated receptors	7.8	14.1	20.9
Lectin isolated receptors	5.3	10.9	16.3
Biochemically isolated receptors	3.9	5.6	8.1

Note: All values minus background.

[a] Sample testing time in minutes.

specificity of immunological reactions with the precision of HPLC and can be applied to the detection and analysis of minute quantities of test material. The availability of commercially produced derivatized silica and protein A prepacked HPLC columns makes this technique easier to perform and more appealing to all types of chromatographers.

REFERENCES

1. McGillis, J.P., Organist, M.L., and Payan, D.G., Immunoaffinity purification of membrane protein constituents of the IM-9 lymphoblast receptor for substance P, *Anal. Biochem.*, 164, 502, 1987.
2. Josic, D., Hofmann, W., Habermann, R., Becker, A., and Reutter, W., High-performance liquid affinity chromatography of liver plasma membrane proteins, *J. Chromatogr.*, 397, 39, 1987.
3. Roy, S.K., Weber, D.V., and McGregor, W.C., High-performance immunosorbent purification of recombinant leukocyte A interferon, *J. Chromatogr.*, 303, 225, 1984.
4. Rybacek, L., D'Andrea, M., and Tarnowski, J.S., Rapid dual-column chromatographic assay for recombinant leukocyte interferon, *J. Chromatogr.*, 397, 355, 1987.
5. Janis, L.J. and Regnier, F.E., Immunological-chromatographic analysis, *J. Chromatogr.*, 444, 1, 1988.
6. Hammen, R.F., Pang, D., Remmington, K., Thompson, H., Judd, R.C., and Szuba, J., Rapid quantification of mouse IgG by protein A: high performance affinity chromatography, *BioChromatogr.*, 3, 54, 1988.
7. Babashak, J.V. and Phillips, T.M., The use of avidin-coated glass beads as a support for high-performance immunoaffinity chromatography, *J. Chromatogr.*, 444, 21, 1988.
8. Phillips, T.M. and Frantz, S.C., Isolation of specific lymphocyte receptors by high-performance immunoaffinity chromatography, *J. Chromatogr.*, 444, 13, 1988.
9. Phillips, T.M., Frantz, S.C., and Chmielinska, J.J., Isolation of bioactive lymphocyte receptors by high-performance immunoaffinity chromatography, *BioChromatogr.* 3, 149, 1988.
10. Campbell, A.M., Fusion procedures, in *Laboratory Techniques in Biochemistry and Molecular Biology. Monoclonal Antibody Technology*, Burdon, R.H. and van Knippenberg, P.H., Eds., Elsevier, Amsterdam, 1984, 120.

11. **O'Shannessy, D.J. and Quarles, R.H.,** Labeling of the oligosaccharide moieties of immunoglobulins, *J. Immunol. Methods*, 99, 153, 1987.
12. **Phillips, T.M., Frantz, S.C., and Holohan, T.V.,** Production of anti-DNA antibodies in normal mice after immunization with anti-idiotypic antibodies, *Ann. N.Y. Acad. Sci.*, 529, 228, 1988.
13. **Chrambach, A. and Rodbard, D.,** Gel electrophoresis, in *Gel Electrophoresis of Proteins: A Practical Approach*, Hames, B.D. and Rickwood, D., Eds., IRL Press, Washington, D.C., 1981, 93.
14. **Sammons, D.W., Adams, L.D., and Nishizawa, E.E.,** Ultrasensitive silver-based color staining of polypeptides in polyacrylamide gels, *Electrophoresis*, 2, 135, 1981.
15. **Tsang, V.C.W., Peralta, J.M., and Simons, A.R.,** The enzyme-linked immuno-electro-transfer blot technique (EITB) for studying the specificities of antigens and antibodies separated by gel electrophoresis, *Methods Enzymol.*, 92, 377, 1983.
16. **Towbin, H., Staehelin, T., and Gordon, J.,** Electrophoretic transfer of proteins from polyacrylamide gels to nitrocellulose sheets: procedure and some applications, *Proc. Natl. Acad., Sci. U.S.A.*, 76, 4350, 1979.
17. **Phillips, T.M., More, N.S., Queen, W.D., Holohan, T.V., Kramer, N.C., and Thompson, A.M.,** High-performance immunoaffinity chromatography: a rapid technique for the isolation and quantitation of IgG from cerebral spinal fluid, *J. Chromatogr.*, 317, 173, 1984.
18. **Phillips, T.M., More, N.S., Queen, W.D., and Thompson, A.M.,** The isolation and quantitation of serum IgE levels by high-performance immunoaffinity chromatography, *J. Chromatogr.*, 327, 213, 1985.
19. **Connor, J.R., Phillips, T.M., Lakshman, M.R., Barron, K.D., Fine, R.E., and Csiza, C.K.,** Regional variation in the levels of transferrin in the central nervous system of normal and myelin-deficient rats, *J. Neurochem.*, 49, 1523, 1987.
20. **Phillips, T.M.,** High-performance immunoaffinity chromatographic detection of immunoregulatory anti-idiotypic antibodies in cancer patients receiving immunotherapy, *Clin. Chem.*, 34, 1698, 1988.

Section VIII
Column Packing Techniques

PACKING LIQUID-CHROMATOGRAPHY COLUMNS

M. Verzele and C. Dewaele

I. INTRODUCTION

Chromatography is used for both analytical and preparative purposes. The mass of compound mixture processed can therefore vary enormously, e.g., from picogram (pg) to ton quantities. This is reflected in the size of the columns holding the chromatographic stationary phase or packing material. The Internal Diameter (ID) of packed columns ranges therefore from 50 μm to 2 m. In the earlier days of chromatography, columns were made of glass. While glass is still used in some cases, stainless steel has taken over for the most part. Filling these columns with packing material has become a complex problem with the advent of smaller particles and the very wide range of different column sizes and packing materials now available. These problems are the subject matter of the present contribution.

II. COLUMN HARDWARE

The size of a packed liquid chromatography (LC) column should be determined by the sample size that has to be handled. However, the following facts should be considered:

- Analytical detection in LC (UV, fluorescence, electrochemical, mass spectrometry) is at its best with amounts in the 1 to 10 ng range. This is quite straightforward for mass measuring detectors. This is true also for concentration measuring detectors, because the cell path length has to be adapted to the column size. The smaller cell volume of miniaturized systems (less dispersion, higher concentration) has to be compensated by a shorter path length. These two effects, therefore, more or less cancel each other out.
- LC columns can, without showing efficiency loss, handle samples that are 10^5 to 10^6 times smaller than their mass of stationary phase. This depends, of course, on the nature of the stationary phase sample, but all phases will handle 1 μg sample/g phase easily, and most will not be overloaded by 10 μg/g.

Packed columns for purely analytical purposes should therefore contain between 1 to 10 ng $\times 10^5$ to 10^6 or 0.1 to 10 mg of stationary phase. Over a column length of 10 to 30 cm, the most applicable column length, the larger mass (10 mg) implies an ID of about 250 to 320 μm. This is the size of current packed fused silica capillary columns, which accordingly seem best suited for analytical LC. In practice, however, at the present time, analytical columns have a much larger ID in the range of 2 to 6 mm or about 100 times the above larger mass. This has much to do with packing and operating problems. It is (or it was) easier to pack these somewhat larger columns efficiently than is possible with smaller capillary columns. Columns with much larger IDs, mainly used for preparative purposes, are again more difficult to pack. These differences in packing propensity of columns as a function of the ID are mainly due to wall effects as explained later. Still, column ID determines the nature of the column hardware. Since packing and the actual chromatography are best carried out at a high pressure, sometimes far exceeding 100 Bar, it is obvious that the column must be able to withstand these pressures. While a fused silica capillary of 250 to 320 μm ID can be packed at 500 to 1000 Bar, this is not possible with fused silica or glass columns of higher ID. Stainless steel is then the preferred material and the wall thickness must be adapted to the column dimensions. A stainless steel cylinder of 20 cm ID with a wall thickness of 1 cm can maybe stand up to 100 Bar, but not much more. While bursting of an analytical column with an ID of a few millimeters would not seem to be especially dangerous, it is obvious that this is not the case for much larger preparative systems. Columns with IDs up to about 20 mm can be made with a wall thickness of 1 mm and can be closed with the nut and ferrule approach. Columns of higher ID have to be terminated by other means, e.g. flanges. Columns of larger ID are unstable because of the thermal wall effect (see further). For good results, they have to be compressed during the actual chromatography as reviewed recently.[1]

By far the larger number of LC columns are currently made of stainless steel. In some cases in the biotechnology field, inertness of stainless steel is not good enough. In this case, glass, ceramic, or a plastic material like PTFE (polytetrafluoroethene) or PEEK (polyethereth-erketone) can be preferred. This influences the packing procedure and the choice of packing materials. Since pressure, especially packing pressure, cannot be high with these column materials, it is useless to choose small rigid particles. Furthermore, the first biocompatible stationary phases were rather soft and could not stand high packing pressures. Soft or gel-like packing materials packed at low pressure do not lead to efficient chromatography. There is therefore a strong tendency to switch to new, rigid, small, but also biocompatible stationary phases in the biotechnological field.

III. PACKING MATERIALS

Reviewing the fascinating history of packing materials is outside the scope of this contribution. Packing materials today are characterized by two trends: packing materials are generally becoming smaller and the number of phases is growing ever more rapidly. This phase diversification has actually become a proliferation problem.

Smaller packing materials lead to higher chromatographic efficiency. But smaller also means higher back pressure during chromatography and greater packing difficulties. In analytical LC, the tendency is towards 5 μm particles, while in preparative LC it is towards 10 to 20 μm. If this is the case, it is because current technology can handle the problems with these smaller particles.

Silica gel and silica gel-derivatized phases are by far the most used in LC at present. Dextran-based and dextran-derivatized phases play a relatively large role in peptide and protein chromatography. Great efforts are made to develop satisfactory alternatives also based on silica gel, so that proteins can be chromatographed on pressure-resistant packings. Stationary phase diversification is very conspicuous in LC. Every Journal issue brings new suggestions. Organic polymers and derivatized materials have interesting possibilities. Coated and/or derivatized materials have already proved this. Polymethacrylates and polyvinylalcohols of suitable bead size, rigidity, and surface area have been proposed.

Often the case for a particular packing material is presented as if it were the only or best choice for a specific problem. This is rarely the case. Most separation problems can be handled by several different stationary phases. Alkaloids, for example, can be separated on plain silica gel as well as on reversed-phase materials or on silica gel derivatized in other ways (cyanopropyl-ated, polyphenol derivatized, etc.).

The choice of a particular material has much to do with the fact that a well-developed packing procedure is available. This is, for example, the case for octadecylated silica gels which readily give the highest plate counts. Other phases do not pack so well or so readily. This is an area where much progress can be expected.

Phases are characterized by their surface chemistry, mean pore size, particle size and specific surface area:

- *Surface chemistry* is controlled by the nature of the bulk material, or by derivatization or by coating with a specific polymer layer. The number of different surface chemistries available is therefore unlimited. Discussing this aspect is outside the scope of the present paper. Excellent review papers by Schomburg[2] and Unger[3] on the subject are recommended.

- *Mean pore size* of the phases is important as this should be adapted to the size of the molecules that are subjected to chromatography. Small and large molecules are best chromatographed on, respectively, small and large mean pore size materials. Biomolecules such as proteins are nowadays chromatographed mostly on packing materials with at least 30 nm as mean pore size. Packing materials of larger pore size tend to be not as strong as materials of smaller pore size. Silica gel with a large pore size is, for example, usually very brittle. Brittleness is not the same for all silica gels. This is therefore a quality criterion since the packing pressure has to be adapted to the resistance of the particles. Particle crushing leads not only to fines and unwanted permeability problems, but also to changed surface chemistry for surface derivatized materials. Crushing, indeed, exposes new surfaces to the medium. In the context of this paragraph, the recent introduction of non-porous silica gel and derivatized materials must be mentioned[4]. The absence of pores in this material seems to favor the chromatography of biomolecules.

- *Particle size* has a tendency to become smaller as already mentioned. Today it is possible to synthesize almost all particles with uniform spherical size (monodisperse). The importance of this is often exaggerated. Non-monodisperse spherical and even irregularly shaped particles with some spread of the average size mostly give equal, if not better, chromatographic results.

- *Specific surface area* is, of course, related to the mean pore size and the apparent density of the material. A wide pore silica gel has necessarily a small specific surface area, and therefore a reduced sample capacity. Silica gel of 50 nm pore size will have a specific surface area of only about 30 m²/g. The number of adsorption sites is reduced and only little of a small molecule may be adsorbed. For a large molecule like a protein this is not such a drawback, as a reduced number of molecules is still a large mass. This is particularly true for the non-porous silica gel beads mentioned above. For Paraffin-Olefin-Naphthalene-Aromatic (PONA analysis), for example, silica gel of very large surface area (600 to 800 m²/g) may well be best suited. Such material has a mean pore size of about 3 nm and can only chromatograph small molecules.

IV. PACKING COLUMNS

The purpose of packing a column is to obtain a chromatographically efficient and stable bed of stationary phase particles. A strong interaction-attraction between the particles would therefore be a positive factor. Strong attraction of the particles in the packing slurry leads, however, to particle aggregation and this is a negative factor, since aggregates lead to inhomogeneities in the packed bed. This contradiction points to a necessary compromise.

Wall effects in LC are important as explained above. We may consider three different wall effects: the direct wall effect, the bridging effect and the thermal wall effect.[5] The first two effects have much to do with efficient packing. The *direct wall effect* is the physical and chemical attraction between the wall and the particles in direct contact with the wall. This effect stabilizes the packed bed since it holds the particles in position. All factors intensifying the direct wall effect have therefore a positive influence on column stability. Chemical wall modification can have an influence. Wall roughening or an elastic polymer layer on the wall can also hold and stabilize particles.[5] On the other hand, of course, a roughened wall will counter the forces pushing down the particles in the column and this will lead to a less densely packed bed which is in principle less stable. This contradiction again points to a compromise. The nature of an eventual elastic polymer layer on the wall and the packing solvent can be such that, in the first instance, the wall layer is relatively hard and does not hinder the particles sliding down along the wall in the column. After a few minutes, the polymer may have swollen somewhat, as is often the case. In this swollen condition the interaction between wall and particles would be especially effective.[5] The *bridging effect* becomes gradually more important as the column ID goes down. In Micro-LC columns (packed capillary columns with an ID of 50 to 500 μm; at the present time mostly in fused silica capillaries) the bridging effect is very important. It stabilizes the packed bed but opposes dense packing. This controversial situation has to be mastered to attain efficient Micro-LC. The *thermal wall effect* is most important in preparative LC columns of very wide ID. Packing such columns is very difficult if a non-compressed system is used. Column stability will be poor anyway in such a system. To the contrary, packing preparative LC columns with compressed instrumentation is easy and is usually done in the solvent of the first intended separation.[1]

Several review papers cover this extensive field.[6,7] A general recipe for packing analytical size columns cannot be given. All packing materials are different and may require adapted packing procedures. This is even the case for similar materials from different manufacturers. Some general remarks are as follows:

- Packing pressure should be as high as compatible with column hardware, packing material and packing device. This is especially true as the column ID becomes smaller.
- Efficient packing is more difficult as column ID and packing materials particle size become smaller.
- Apolar derivatized silica gel more readily gives high efficiency packing results than polar silica gel phases and the latter more readily than polystyrene based phases.
- The slurry should have a homogeneous aspect. Slurries which settle out quickly will not produce good columns. Good drying of the packing materials and of the solvents is most important in this respect.
- The packing material quality is more important than the packing procedure. Good quality packing materials pack more easily and will do so readily with a variety of packing procedures. Sophisticated packing procedures are more necessary for packing materials of lesser quality.
- High quality solvents should be used.
- Balanced density slurries, high viscosity slurries, adding salts or detergents to the slurries, are all not as effective as the number of papers mentioning these would indicate.
- Some ammonia (added as water solution when this is possible — added as gas to, for example, ether) is beneficial to harden packed beds.
- Slurry concentration is usually not so important and can be anything between 1 and 30%. We mostly use 15%.
- For smaller particles, up-tube packing[8] (with a tube of the same diameter as the ID of the column as packing vessel) is superior to downward packing.
- Sonication of the slurries before packing is recommended.

More specifically, procedures to start with could be as follows:

- Slurry in carbon tetrachloride and acetonitrile (CH_3CN) as follow-up solvent for reversed-phase and apolar phases in general
- Slurry in CCl_4 and hexane as follow-up solvent for underivatized silica gel and for polar phases in general (PONA phases, NH_2-propyl silica gel, Polyol phases, etc.)
- Slurry in acetone and acetone as follow-up solvent as first alternative for reversed phase packing materials (also for some special phases like Polyphenol-RSiL)

- Slurry in methanol/water 90/10 and methanol as follow-up solvent as alternative for normal phase packing materials
- Slurry in aqueous buffers with 0.2 M NaCl and follow with water for water compatible phases and ion-exchange phases
- Completely different solvent systems may also give satisfactory results.

If an octadecylated silica gel material does not give a reduced plate height of 2 to 2.5, something is probably wrong with the packing material. Whether this is the case or that, indeed, the packing procedure or chromatographic instrumentation is at fault, can be deduced from the h/u curve. The expression h/u stands for a plot of reduced plate height (h) vs. linear eluent velocity (u). The reduced plate height (h) is obtained by measuring the efficiency of the column, converting this to height equivalent to a theoretical plate and then dividing by the particle size of the packing material. The linear eluent velocity (u) is calculated in min/sec from the eluent flow-rate and the column dimensions. The h/u curve plots the values for this ratio at increasing eluent flow-rates. This curve (the van Deemter plot) reveals the quality of the packing technique and packing material. It is indeed most worthwhile to establish the h/u curve for evaluating packing procedures. A description of this approach and further discussion of our views on column packing can be found in the references mentioned.[8-11] Often, disappointing chromatography (low efficiency, poor reproducibility) are imputed to faulty packing material manufacturing, while the trouble is in fact due to bad packing and deficient instrumentation and insufficient chromatographic understanding.

V. REMARKS

Packing LC columns is based on know-how. Packing is commercially important. Packing know-how is therefore not readily given away and the really important tricks are kept secret. Ten to twenty years ago only a restricted number of laboratories could pack columns well. "Secrets" do leak out, however, and the expertise to pack good columns is available to most labs today. We believe, however, that this is mostly due to the ever-improving quality of the packing materials. What is described in the present contribution is practically common knowledge. There are, of course, new things which are not divulged yet, especially with the commercial firms. This is mostly the case for phases which are notably difficult. Spectacular improvement may, for example, be expected for polystyrene phases. Packing LC columns is still not completely understood and final solutions or answers do not yet exist. Hence, it is worthwhile to study packing technology further.

REFERENCES

1. **Verzele, M., De Coninck, M., Vindevogel, J., and Dewaele,** C., Column hardware in preparative liquid chromatography with axial flow, *J. Chromatogr.*, 450, 47, 1988.
2. **Schomburg, G.,** Stationary phases in high performance liquid chromatography, *LC & GC Magazine*, 6, 36, 1988.
3. **Unger, K.,** *Porous Silica, Journal of Chromatography Library*, Vol. 16, Elsevier, Amsterdam, 1982.
4. **Unger, K., Jilge, G., Janzen, R., Gieske, H., and Kinkel, J.,** Non-porous microparticulate supports in high performance liquid chromatography (HPLC) of biopolymers. Concepts, realization and prospects, *Chromatographia*, 22, 379, 1986.
5. **Verzele, M., Dewaele, C., De Weerdt, M., and Abbott, S.,** Inner wall coating of Micro-LC columns. Wall effects in LC, *JHRC & CC*, 12, 164, 1989.

6. **Elgass, M., Engelhardt, H., and Halasz, I.,** Reproduzierbare Methode zum Packen von Trennsäulen für die Chromatographie mit Kieselgel (5—10 μm), *Z. Anal. Chem.,* 294, 97, 1977.
7. **Martin, M. and Guiochon, G.,** Revue et discussion a propos des diverses techniques de preparation des colonnes pour la chromatographie en phase liquide a haute performance, *Chromatographia,* 10, 194, 1977.
8. **Verzele, M. and Dewaele, C.,** Low-viscosity solvent packing of HPLC columns using the "Up-tube" packing procedure, *LC & GC Magazine,* 4, 614, 1986.
9. **Verzele, M., Dewaele, C., and Duquet, D.,** Observations and ideas on slurry pa king of liquid chromatography columns, *J. Chromatogr.,* 391, 111, 1987.
10. **Dewaele, C., De Coninck, M., and Verzele, M.,** Preparative liquid chromatography and the h/u curve, *Sep. Sci. Technol.,* 22, 1919, 1987.
11. **Verzele, M. and Dewaele, C.,** *Preparative High-Performance Liquid Chromatography.* A practical guideline, 163 p. 1986, Alltech.

PACKING COLUMNS IN THE MICROBORE TO SEMI-PREPARATIVE RANGE

Richard Henry and Matthew V. Piserchio

I. INTRODUCTION

The objective of all column preparation methods is to produce a reproducible, uniform packing structure that is stable under reasonable use conditions. Some variables which affect column preparation are shown in Table 1. Although packed columns cannot be expected to last indefinitely, a useful lifetime of several months and several hundred sample injections is a reasonable objective. It is difficult to justify the preparation of packed columns in-house on the basis of cost alone. Some other factors which should be weighed in the decision to establish a safe column preparation area include:

1. The need to use special packings not available in prepacked form
2. Slow delivery to a remote location
3. As back-up to a commercial source

Some users also prefer to prepare columns in-house during method development where several packings and column geometries may be explored quickly and economically. Obviously, this strategy may only be employed to evaluate packings which are available in bulk. Once a method has been established, most users prefer to arrange for an outside commercial supply that meets specification. This article will describe methods for column preparation, testing and maintenance. Other useful information can be obtained from packing manufacturers, pump manufacturers, and from several recent articles.[1-3]

II. DRY-PACKING

The choice of a method of column preparation depends upon particle size and distribution, and upon column geometry. In general, packings having an average particle size greater than 20 μm can be dry-packed with simple procedures.

A. GENERAL DRY-PACK METHOD
A variety of methods have been reported with the following selected as preferred conditions:

1. Always prepare column in a vertical orientation.
2. Use guide funnel or flared attachment to avoid spilling.
3. Add packing in small increments, approximately equal to an increase in bed height of a few millimeters (e.g., a 250-mm column may require the addition of up to a hundred small increments for best results).
4. After each addition, light tapping with a spatula on the side of the tube near bed-level, accompanied by rotation, has produced good results. Both vertical tapping and vibration have been reported to be less effective in producing a stable, uniform bed structure. In any case, columns prepared by dry-packing are not highly reproducible.

TABLE 1
Variables which Affect Column Performance

Type of packing
    Nature of base silica, polymer, etc.
    Nature of bonded phase modification
    Pore-size and pore-size distribution
    Particle-size and particle-size distribution
    Particle shape
    Dimensional stability to pressure and solvents
Preparation method
    Dry-pack or slurry-pack
    Slurry solvent and slurry concentration
    Pushing solvent
    Pressure
    Up-fill vs. down-fill
    Geometry of slurry reservoir and column
    Time and column-volumes of solvent
    Equilibration solvent or rest time prior to removal and testing
Column hardware
    Length
    Internal diameter
    Wall material and wall finish
    Method of termination (frit, screen, etc.)
    Types of end fitting

III. SLURRY-PACKING

Packings having average particle sizes below 20 μm usually must be packed from a dilute liquid slurry. Solvent at high pressure is used to force the slurry from a reservoir into a column where it rapidly concentrates into a uniform bed. Both constant-volume, motor-driven, metering pumps and constant-pressure, pneumatic pumps have been employed. Figure 1 shows a schematic diagram of a high pressure, slurry-packing system using the upward-displacement method. While both up- and down-fill methods have been employed successfully, the subject is very controversial. The safer and more convenient down-fill method is strongly recommended as the initial method to be attempted. A strong warning must be issued to anyone wishing to attempt column preparation using slurry displacement methods with high-pressure, pneumatic pumps. The procedures can be very dangerous and should be left to commercial suppliers of packed columns whenever possible. Use of the high-pressure apparatus in a hood with pull-down safety shield is strongly recommended. Two examples of commercial equipment for preparing columns by high-pressure slurry methods are shown in Figures 2 and 3, while some commercial reservoirs are shown in Figures 4 and 5. Empty columns vary greatly in dimension and design, and must be adapted to the slurry reservoir with fittings that imitate the final column end-fitting. Although not usually necessary for column performance, a short precolumn and union is often employed between the column and reservoir to facilitate removal of excess packing and minimize wear on the reservoir fitting.

A. GENERAL SLURRY-PACK METHOD

Some general rules which can be applied to column preparation by the high-pressure slurry method are:

1. Slurry solvents should wet and disperse the packing particles. Poor solvents permit agglomeration which results in preparation of a loose bed with poor efficiency. Good slurry solvents can be screened using test tube experiments. The term "sedimentation quotient" has

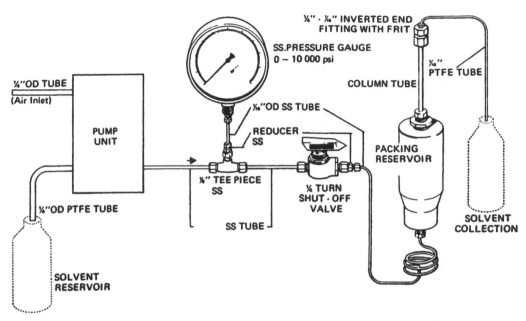

FIGURE 1. Schematic diagram of a slurry-packing system (Phase Separations, Ltd.).

FIGURE 2. Example of commercial slurry-displacement pump (Haskel, Inc.).

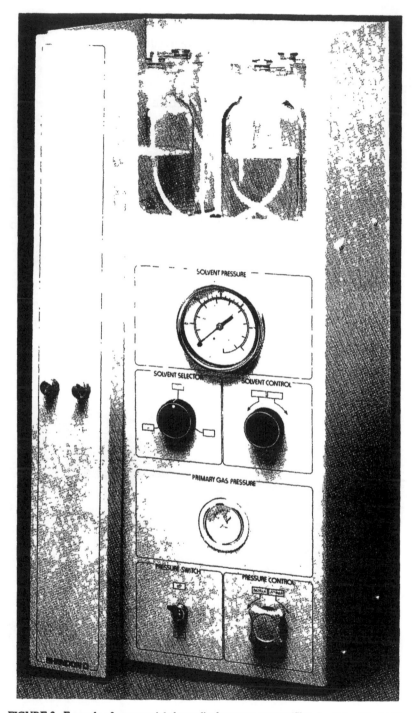

FIGURE 3. Example of commercial slurry-displacement system (Shandon Scientific, Ltd.).

been developed to measure dispersing ability of a solvent.[4] For best results, a slurry should
be uniform and free of clumps, and not show evidence of settling during the time required
for column preparation (typically a few minutes).

2. Slurry density can vary greatly; however, a good starting point is about 10% (w/v) for silica-
based packings. This results in a slurry reservoir which must be about ten times the volume
of the column to be prepared. The appropriate amount of packing to fill the column plus about

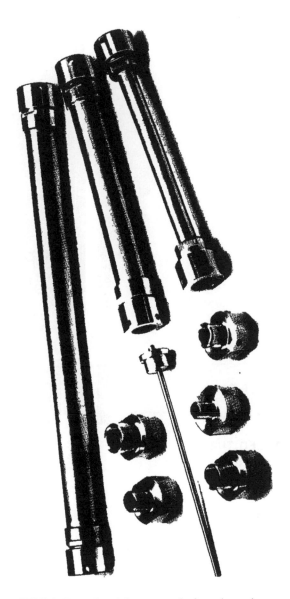

FIGURE 4. Examples of slurry reservoirs for various column sizes (Shandon Scientific, Ltd.).

10% excess should be weighed and added to a measured amount of slurry solvent. Either manual stirring with a spatula or sonic bath have been used to disperse and suspend the packing.

3. Down-fill methods with common solvents such as water, alcohols, and acetone are most safe and most common. Stable slurries which produce good columns can be prepared with narrow-distribution 3 or 5 μm packings. In difficult cases, mixed solvents may be employed. Packings with broad size distribution or with narrow distribution in the 10 μm size range may settle rapidly and therefore require the consideration of balanced-density slurry, high-viscosity slurry, or stirred reservoir. All have been used successfully to counteract the effect of slurry segregation due to settling of larger particles. The balanced-density method employs dense solvents (typically chlorinated or brominated hydrocarbons) to create slurries that do not settle rapidly; however, dense solvents are often toxic, impure, and

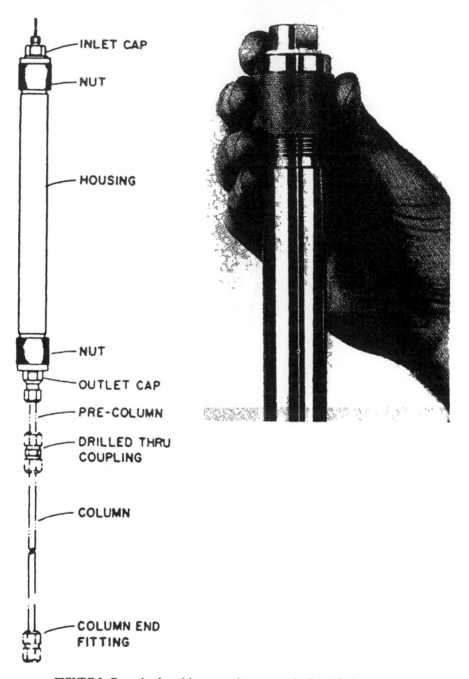

FIGURE 5.   Example of a quick connect slurry reservoir (Scientific Systems, Inc.).

expensive. A reasonable compromise may be the use of chlorinated solvents mixed with other organic solvents. The viscosity method usually employs ethanol or 2-propanol to slow the settling process. The use of surfactants and electrolytes in the slurry has also been reported. In any case, the filling process must be carried out as quickly as possible to minimize settling which creates column non-uniformity.

4. The pushing solvent can be the same as the slurry solvent or different; however, it is usually selected to be miscible with the slurry solvent. There have been many variations in the selection of pushing solvent, and this variable does not seem to be as important as the

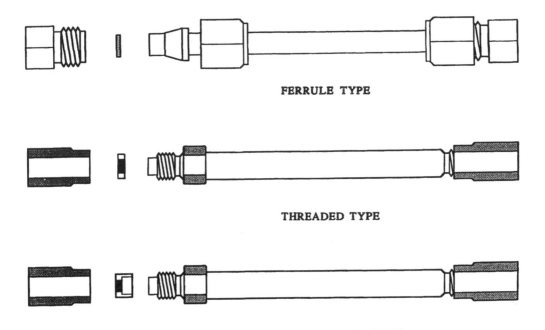

FERRULE TYPE

THREADED TYPE

CAPPED CARTRIDGE TYPE

FIGURE 6. Examples of ferrule-type and threaded-type column assemblies (Keystone Scientific, Inc.).

selection of slurry solvent. Some have even reported an equilibration step with mobile phase testing solvent (usually a mixture) before the column is removed from the reservoir and capped with a frit and fitting. This latter procedure may be especially important if the packing contracts or swells even slightly in different solvents. Multiple solvent equilibration is time-consuming and should not be employed unless absolutely necessary.

5. After final equilibration, pressure should be shut off and allowed to dissipate through the packed column while still attached to the slurry reservoir. A relaxation period of 10 min is usually adequate, but more may be needed if the packing or solvent is very compressible. The packed column should be loosened at the precolumn or reservoir and lowered with a twisting motion to minimize disturbance of the column inlet. If packing extrudes when the column is removed, longer equilibration may be needed. The top of the column should be squared-off with a spatula and cleaned thoroughly of all loose packing before installation of a fitting with frit or screen. Well-prepared columns may be tested and used in either the same direction as prepared or in the reverse direction.

Extreme caution must be employed when using slurry packers or any air-driven pump which can generate high velocity solvent streams and other projectiles. Only experienced personnel should attempt to prepare LC columns by the high-pressure slurry displacement method. Safety glasses, safety shields, ventilation, and protection from inhalation of airborne dust should be part of any protocol. If possible, install the column preparation apparatus in a fume hood with pull-down safety shield.

IV. COLUMN HARDWARE

Columns may be purchased as complete assemblies, as shown in Figure 6, or assembled from component parts that are designed to match. Also, columns that have lost performance often may be unpacked and refilled several times. A column that has been unpacked should be thoroughly

TABLE 2
Column[a] Terminology and Dimensions

Category	Internal diameter[b] (I.D.) (mm)	Approximate volume (ml)	Approximate wt. packing (grams)
Preparative	20	80	80
Large-Bore	10	20	20
Standard-Bore	4.6	4	4
Small-Bore	2	0.8	0.8
Micro-Bore	1	0.2	0.2
Capillary	0.5	0.05	0.05

[a]   Figures shown are for columns 250 mm in length.
[b]   Check upper pressure limit before preparing any column by high-pressure slurry displacement.

cleaned with solvents and passivated with concentrated nitric acid before refilling. The use of new frits or screens that exactly match the old ones is strongly recommended in all cases. Column hardware may not be reused indefinitely without leaks and loss of performance. Threaded columns and fittings can generally be reused more often than ferrule-type columns and fittings. Threaded columns are also less likely to slip during the high pressure filling operation, especially with the larger diameter columns used for preparative sample collection. A listing of volume calculations and approximate amounts of packing required to fill the column is shown in Table 2. Each packing will require different amounts due to different densities. When an empty column has been cleaned and assembled, it may be desirable to leak and pressure test it by connecting to the slurry reservoir with a plug or capillary restrictor in the column outlet. Valuable packing may be wasted if the assembly leaks or loosens during the packing operation. Material of construction is most commonly 316 stainless steel that has been specially cleaned and polished to remove burrs and chemical residue. Glass-lined and other inert materials have also been employed, but are not recommended for the inexperienced person. The column should be attached to the slurry reservoir by an adapter or union that matches the desired column fitting. The most common internal (female) type end fitting is compared to the older external (male) type end fitting in Figure 7. The internal fitting is strongly recommended for strength and low internal volume. Unfortunately, internal fittings from different manufacturers are not perfectly inter- changeable, so it is very desirable to standardize on one brand. If this is impossible, the use of adjustable connectors at the 1/16-in. end can avoid damage to end fittings and poor test results.

If a preassembled empty column has not been purchased, end fittings with frits or screens that are smaller in porosity than the smallest packing particles should be assembled and tested for tightness and safety in a shielded, ventilated environment. Before attaching to the slurry reservoir, one fitting and frit must be removed and set aside for later replacement when the filled column has been removed and cleaned to be free of packing particles around the column exterior.

V. METHOD FOR ODS SILICA

While techniques for all types of packings cannot be covered, the approach used for the most popular bonded phase, ODS (C_{18}) silica can serve as an example. Silica and its bonded phases are often dried at 100°C for several hours to remove excess moisture that can become an unwanted variable in the packing process. Once dried, packings should be capped tightly until used or they will pick up moisture from the atmosphere. Packing should be weighed without introducing dust, stirred into a beaker and inspected for clumps or rapid settling. A slurry must not settle appreciably prior to applying high pressure or the final column will not be uniform.

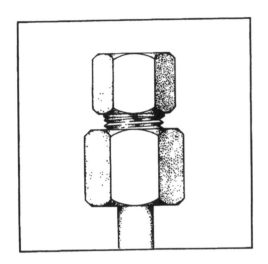 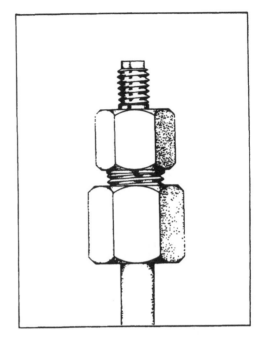

INTERNAL TYPE   EXTERNAL TYPE

FIGURE 7. Examples of internal (female) and external (male) column end fittings (Upchurch Scientific, Inc.).

Generally, the slurry should not change appearance for at least 5 min for best results. The following method has been reported[5] for a very common packing, ODS Hypersil[R].

A. SPECIFIC STEPS FOR ODS HYPERSIL: 5-μm PARTICLE SIZE, 250 × 4.6 mm I.D. COLUMN

1. Place 4.0 g of dry 5 μm ODS Hypersil into a clean beaker containing 40 ml of ethanol.
2. Stir or sonicate for 5 min while preparing the slurry reservoir with an empty 250 × 4.6 mm I.D. column tube. A short precolumn between the slurry reservoir and analytical column is optional.
3. Quickly transfer the 40 ml slurry into the top of the reservoir, close the reservoir, and pressurize with methanol at 7000 psi for 10 min or until at least 150 ml of solvent has passed through the column.
4. Shut off the pressure and allow the flow to drop to zero. A rest period of 10 min should be adequate in this case.
5. Loosen the analytical column from the reservoir with a twisting motion to minimize disturbance of the column inlet. A clean, dry-looking surface at the top of the column is a good sign that a stable, uniform bed has been prepared. The short plug of excess packing remaining in the precolumn or reservoir may be removed and dried for later use. A column that has not been filled due to incorrect weighing or estimating, must be extruded, cleaned, and repacked. Adding more packing to the partially-full column in a second step generally will not produce good results.
6. Square-off the top of the column with a spatula and clean all traces of packing from the edges and sides of the column. Replace the frit and end fitting, then install the column in an LC system for testing.

A great many variations may be made on this procedure to allow columns to be prepared with

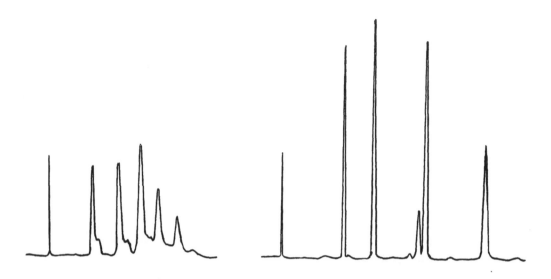

FIGURE 8. Test chromatograms of poorly-prepared (A) and well-prepared (B) columns (250×4.6 mm I.D.) containing the same ODS packing material (Keystone Scientific, Inc.). The test sample contained theophylline, *p*-nitroaniline, methyl benzoate, phenetole and *o*-xylene. Eluent: 60% aq. acetonitrile. Flow-rate: 1.25 ml/min.

other reservoirs and packings. Acetone is often added to or employed in place of ethanol, and mixed solvents containing chloroform or carbon tetrachloride are popular for packings which are difficult to disperse and suspend.

Microbore and smallbore columns can be prepared by scaling down the amount of packing, volume of slurry solvent, and size of slurry reservoir. Similar, or somewhat higher, pressures are often employed with pneumatic slurry-displacement systems. Flow rate must be scaled down also if a constant-volume, metering pump is used to displace the slurry.

Semi-preparative and preparative columns can be prepared by scaling-up the amount of packing, volume of slurry solvent, and size of slurry reservoir. Somewhat lower pressures are often employed in pneumatic pump systems for safety reasons, while flow rate must be scaled up in metering pump systems. Preparative columns are often prepared with large-particle packings due to the high cost of small-particle packings. Increasing the particle diameter may alter the slurry method very significantly or even permit dry packing.

VI. COLUMN TESTING

Equilibration time with the testing mobile phase is usually only 15 min or less if the test solvent is miscible with the pressurizing solvent. A multicomponent test mix of well-behaved compounds is recommended so that selectivity and efficiency can be measured. The packing manufacturer can make recommendations. More difficult test mixes may be attempted once the easy test mix has confirmed that a uniform bed structure has been obtained. Figure 8 shows test chromatograms for both a poor column and good column. In this case, the difference is so dramatic that an efficiency calculation on the poor column is not necessary, but calculation of plates per meter (N/m) by an approved method should always be carried out on a later peak and recorded in a log book for future comparison. As a rule of thumb, the N/m values in Table 3 should be achievable with good hardware, packing material, and procedure.

The most common formula for calculating efficiency is $N = 16 (t_R / W_{1/2})^2$, where $W_{1/2}$ is width of peak at half-height. A recent review article has described the half-height and other methods

TABLE 3
Typical Efficiency Values for Various Particle Sizes

Particle size (μm)	Expected efficiency range (N/m)
3	100,000—150,000
5	60,000—100,000
7	40,000—60,000
10	25,000—40,000

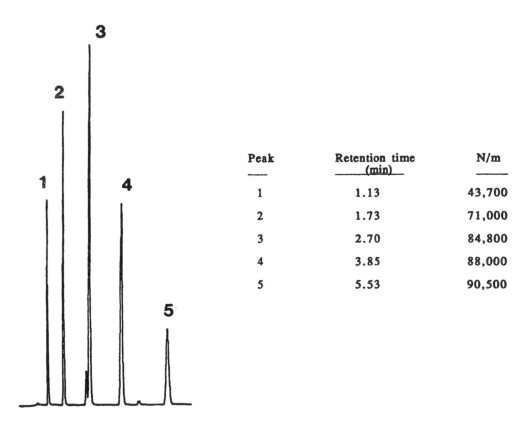

Peak	Retention time (min)	N/m
1	1.13	43,700
2	1.73	71,000
3	2.70	84,800
4	3.85	88,000
5	5.53	90,500

FIGURE 9. Test chromatogram showing effect of retention time on theoretical plate calculation (Keystone Scientific, Inc.). Column and conditions same as Figure 8. The test sample contained theophylline, *p*-nitroaniline, methyl benzoate, phenetole and *o*-xylene (peaks 1 to 5, respectively).

for calculating column efficiency.[6] The chromatogram in Figure 9 shows that the N/m calculation is closest to the true column value when later eluted peaks are measured. Peaks eluted near the solvent front experience significant bandspreading due to extra-column volume in the injector, tubing, end fittings and detector. The magnitude of this effect varies greatly from system to system and can be minimized by making efficiency measurements on late-eluted peaks with capacity factor, k', of five or greater. Also, it is very important to make careful connections to injector and detector.

As shown in Figure 10 (right), a connecting tube can be installed incorrectly leaving excessive volume in the column fitting at either end. This problem is common when fittings from different manufacturers are used with capillary tubing and permanent, stainless steel ferrules. End fitting standardization or adjustable connecting methods are recommended to minimize and control extra-column bandspreading.

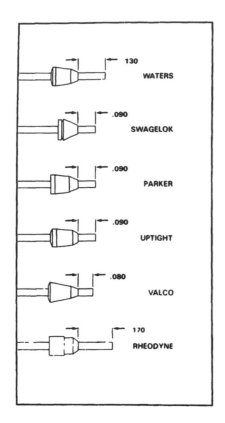

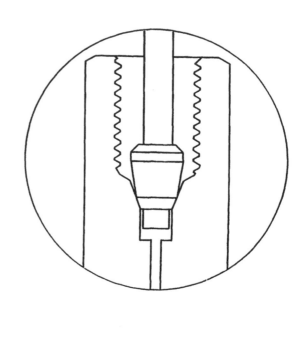

FIGURE 10.   Examples of non-interchangeable capillary tubing connections that can affect column performance (Upchurch Scientific, Inc.). Only Swagelok and Parker fittings are interchangeable.

VII. COLUMN MAINTENANCE AND TROUBLE SHOOTING

The use of high quality, filtered solvents and guard devices such as filters or short guard columns is recommended for maximum column life. A further discussion of guard devices is given in the article entitled "HPLC Guard Columns" by Henry.

After each use, the column should be flushed with solvent that is free of additives and not reactive to the packing. Most commonly employed are aqueous-organic mixtures or pure organic solvents, after inorganic buffers and salts have been removed. Columns should be tightly capped when not in use, especially if the packing changes dimension when exposed to different solvents. Many organic polymer packings fall into this category and can be damaged permanently if allowed to dry out.

Column inlet pressure and performance factors such as k', α(selectivity), N, and symmetry should be monitored to give warning that a column is changing or deteriorating. Any changes in flow or mobile phase composition should be made gradually to avoid local surges in pressure or temperature that might disturb the packed bed. If a column inlet frit becomes plugged, a frit change can usually be accomplished without great difficulty, especially when the column has been prepared in-house and exact spare parts are available. If a sudden loss of efficiency or the appearance of shoulders and split peaks turns out to be caused by a void under the inlet frit, column performance can usually be improved by filling the void with the same packing by wet tamping, or by adding chemically similar, dry solid-core material. Usually, the development of a void is serious, and the column will have to be repacked. The method of preparation may require modification or the operating conditions changed to avoid recurrence of the voiding problem.

VIII. CONCLUSION

In spite of many efforts to define and control critical variables, the preparation of stable, high-performance packed columns remain a combination of scientific method and art. Although the information and guidelines given here may be helpful for many packings and column geometries, there will always be room for a new approach and a fresh idea.

REFERENCES

1. **Shelly, D.C., Antonucci, V.L., Edkins, T.J., and Dalton, T.J.,** Insights into the slurry packing and bed structure of capillary liquid chromatographic columns, *J. Chromatogr.,* 458, 267, 1989.
2. **Shelly, D.C. and Edkins, T.J.,** Contribution to the packing of liquid chromatographic microcolumns, *J. Chromatogr.,* 411, 185, 1987.
3. **Verzele, M. and Dewaele, C.,** Low-viscosity solvent packing of HPLC columns using the "Up-tube" packing procedure, *LC.GC Magazine,* 4, 614, 1986.
4. **Meyer, R.F. and Hartwick, R.A.,** Ph.D. Dissertation, Rutgers University, 1986.
5. Keystone Scientific, Inc., Internal column preparation method.
6. **Bidlingmeyer, B.A. and Warren, Jr., F.V.,** Column efficiency measurement, *Anal. Chem.,* 56, 1583A, 1984.

Section IX
Methods of Detection

THE PRACTICAL APPLICATION OF DIODE ARRAY UV-VISIBLE DETECTORS TO HIGH-PERFORMANCE LIQUID CHROMATOGRAPHY ANALYSIS OF PEPTIDES AND PROTEINS

Paul Zavitsanos and Heinz Goetz

I. INTRODUCTION

One major objective of chemical analysis is the discovery of the molecular composition of a given sample. Complete information about this sample involves both qualitative and quantitative components, and generally qualitative chemical analysis encompasses the processes of sampling, separations, and identification. High-performance chromatographic techniques greatly enhance the separation and quantitation abilities of the analyst, but the measures of retention or mobility derived from these techniques typically yield insufficient information to positively identify a compound or elucidate its structure. While identification strategies built on retention time information are conceivable and available in limited application areas, positive identification and structure elucidation are still largely tasks for spectroscopic techniques.

Three Dimensional Ancillary Spectroscopic techniques have made great contributions to the effectiveness of chromatographic systems in a variety of application areas. Capillary Gas Chromatography/Mass Spectroscopy (CAP GC/MS), Capillary Gas Chromatography/Fourier Transform Infra-Red (CAP GC/FT-IR) and Capillary Gas Chromatography/Atomic Emission (CAP GC/AE) have added new high levels of confirmation in the fields of environmental, forensic, and drug analysis.

The protein-peptide chemist works primarily in the liquid phase for chromatography and consequently has fewer ancillary spectroscopic techniques at his disposal. Liquid Chromatography/UV-Visible Diode Array Detection Spectroscopy (LC/DAD), like the techniques mentioned above, is a three-dimensional technique that extends many of the advances of ancillary spectroscopy to the worker in Liquid Chromatography.

Extensive treatments of the theoretical aspects of reverse optics diode array UV-Visible Spectrophotometry can be found in References 1 to 6. The objective of this article is to present a practical summary of the potential impact an ancillary UV-Vis Diode Array Detector (DAD) can have on the HPLC analysis of proteins and peptides.

II. AREAS IN PROTEIN AND PEPTIDE ANALYSIS WHERE ANCILLARY UV-VISIBLE DIODE ARRAY SPECTROSCOPY WOULD BE AN AID TO THE ANALYST

The protein/peptide chemist has extensive challenges on all fronts with most aspects of protein/peptide analysis.

The separation task is formidable when compared to separations of conventional small molecules. This is not generally because of any separation problem with a target protein, but by the fact that there are so many largely similar molecules in a typical sample matrix. With the limited resolution of Liquid Chromatographic techniques, there is always the danger of an incomplete separation going undetected and contaminating an isolated product. Subsequent results from sequencing and amino acid analysis would be unsatisfactory or misleading. Had the contamination been detected during or shortly after the chromatographic run, a fair amount of sample, time and labor will have been saved. As will be explained later in detail, Diode Array Detectors allow extensive application of a process commonly referred to as Peak Purity Verification to aid in detecting such situations.

The primary structure elucidation (amino acid sequence analysis) of proteins is a highly involved process when compared to conventional small molecule analysis. Techniques such as NMR and High Resolution Mass Spectroscopy are not generally applicable to the sequence determination of proteins and peptides of weights even as low as 2000 Daltons. The structures of these compounds are derived by a determination of their sequence of amino acids. Before such a sequencing can take place on large protein molecules, there are many steps required to break the task into manageable portions. Digests and isolations of fragments are followed by amino acid analysis and sequencing and possibly further digests of the larger fragments and so on. In all these steps, the separation and isolation of fragments from complex chromatograms can be subjected to peak purity verification.[7] In addition, as will be shown later, UV-spectroscopy and Second Order UV Spectroscopy can be used to determine the possible aromatic amino acid contents of the proteins and peptides after separation. These techniques allow some degree of spectroscopic structure elucidation to be performed ancillary to the separation technique.[7,10-12]

III. THE VALIDITY OF UV/DAD SPECTRA IN PROTEIN AND PEPTIDE ANALYSIS

Traditionally, the UV spectra of proteins and peptides were thought to yield little information. This belief stems from the fact that while the UV spectra of proteins and peptides do contain information, they typically require instruments with absolute wavelength accuracy to visualize the differences between compounds. This ability was not readily available until the introduction of Diode Array UV Spectrometers.[8]

Due to the mechanical nature of the scanning mechanisms, conventional UV spectrophotometers had wavelength accuracy and repeatability such that only low slope regions of the absorption curve (minima and maxima) could be reliably used for identification purposes. Compounds that had no discernable peaks were said to have little UV information.

The reverse optics system yields several operational advantages over other UV-Vis Detection Systems. While this article emphasizes the use of UV-Vis Diode Array systems as HPLC detectors, the validity of diode array detection UV techniques in protein structure elucidation has been established. The parallel detection nature of the reverse optics diode array detection systems provides the potential for absolute wavelength accuracy; therefore, repetitive spectra of the same sample should yield identical spectra at all points. This means that the UV spectra of even compounds with no prominent peaks in the UV spectra can be compared and the analyst can be confident that spectral differences are not artifacts of instrument design, but true differences resulting from the sample under study.[13,14]

To the protein and peptide chemist, this result is invaluable.[7] The spectra of these compounds are no longer devoid of UV information but are comparable to other spectra. Because all points on the spectrum itself are valid information, they can be manipulated mathematically to extract even more information.

On-line spectra of very high quality and sensitivity can be obtained. Due to the high speed of spectral acquisition, there is little distortion of the spectrum across the entire wavelength range. Moreover, high quality spectra can generally be taken down to below the 0.001 AU (Absorbance Units) range, depending on the compound.

Multiple wavelengths can be extracted and processed from the spectra either during a run or post-run, depending on the design, without a corresponding increase in noise level. Typically, the noise level for all signals across the entire wavelength range can be low as 2×10^{-5} AU. In part, this signal performance is due to the fact that signal to noise ratio on a DAD can be optimized via variable band widths in both the wavelength and time domains.

These characteristics of DADs are foreshadowed by the fact that the protein-peptide chemist generally does his analytical separations under gradient conditions and at low concentrations. The resulting chromatograms often have a high degree of background drift, wander and noise. The greater part of this drift and wander is caused not by true mobile phase absorbance changes, but by refractive index effects. Most UV-Vis detectors present an output that is the sum of both Absorbance and Refractive Index (RI) components. A detector that does not measure its reference signal through the same cell and at the same time as the sample generally cannot differentiate between RI effects and absorbance effects. While the magnitude of these two variables may be different depending on cell design, at low concentration, RI effects will hinder UV-Vis absorbance detection and quantitation under gradient conditions. The DAD has the potential to use a non-absorbing portion of the UV-Vis spectrum to act as a true flowing reference. Assuming that RI effects are largely independent of wavelength, then the resulting signal should reflect only the true absorbance changes in the gradient chromatogram. Some DAD designs require the user to store the reference signal and perform a post-run subtraction to remove RI effects. Other designs incorporate a true non-absorbing reference wavelength as part of the signal definition and the chromatogram subtraction is carried out during a run with no post-run process required. Lower detection limits in gradient elution in LC/DAD are the result of the DADs potential.

The balance of this article will deal with the new techniques and their applications that have developed around diode array spectroscopy.

IV. THE APPLICATION OF DIODE ARRAY DETECTOR DATA TO THE ANALYSIS AND STRUCTURE ELUCIDATION OF PROTEINS

There are two main applications in protein and peptide analysis in which the capabilities of DAD technology are important, if not essential: (1) peak purity verification and (2) aromatic amino acid content of proteins and peptides.

A. PEAK PURITY VERIFICATION

As previously discussed, one of the major concerns that the protein chemist faces is the worry that a chromatographically isolated protein is contaminated by a co-eluted compound. The diode array detector configuration can help minimize these concerns. Figure 1A shows a chromatogram of several pure fibrino-peptides. The spectra of fibrino-peptide A and fibrino-peptide Des-

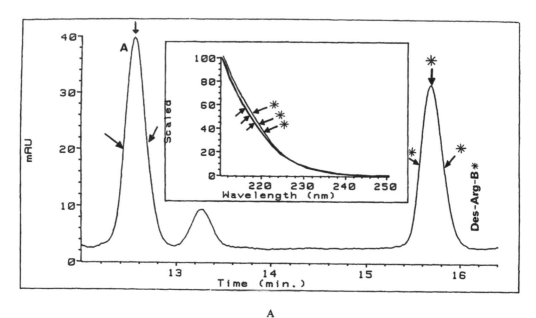

FIGURE 1. (A) Comparison of spectra for fibrino-peptide A and fibrino-peptide Des-Arg-B. Spectra collected at the apex and up and down slopes of the peptide peaks (collection points designated by arrows, together with a * for Des-Arg-B) are overlayed and normalized. (B) Spectra showing Des-Arg-B impurity in Peptide A. The symbols indicate the location of the compared spectra and show that the apex and downward slope spectra are identical but different from the spectra taken on the upward slope. The spectra illustrated in all the figures in this article were obtained from a Hewlett-Packard HP1040A diode-array detector. Peptide sequences: Peptide A, Ala-Asp-Ser-Gly-Glu-Gly-Asp-Phe-Leu-Ala-Glu-Gly-Gly-Gly-Val-Arg; Des-Arg-B, Glx-Gly-Val-Asn-Asp-Asn-Glu-Glu-Gly-Phe-Phe-Ser-Ala.

Arg-B are shown in the center of Figure 1A. Three spectra are overlayed and normalized from each component peak. The precision and absolute wavelength accuracy of the DAD allows comparison of normalized spectra from different points on a chromatographically pure peak that are indistinguishable from one another. The illustration shows that the spectra of fibrino-

TABLE 1
Advantages and Limitations of Various Peak Purity Techniques

	Signal ratio	Spectra	Retention time	Area ratio
Compatible with gradients	—	+	+	+
Detection of small impurities	—	—	+	+
Compatible with small signals	—	—	+	+
Detection of co-eluted impurities	+	+	—	+
Documentation/automation	+	—	+	+

peptides A and Des-Arg-B are in fact different. The difference in the spectra would be considered suspect, even with conventional high-performance spectrophotometers. The design compromises inherent in moving monochromator scanning LC detectors further increases spectral wavelength and absorbance error, thereby rendering the two spectra virtually indistinguishable. Given the great precision, absolute wavelength accuracy and reproducibility of spectral acquisition with DAD units, even such small differences in the spectra become very significant.[7-9] These characteristics of the DAD can be used to advantage in peak purity questions. Should the composition of a given peak not be consistent along the peak width, then the DAD spectra still reflect that fact. For example, Figure 1B demonstrates a situation that all chromatographers routinely face. Figure 1B shows a chromatogram of some fibrino-peptides. In the center of the figure, we see normalized overlayed spectra across the width of peptide A. All spectra are not the same, indicating that composition is not constant along the peak width. Spectral overlays of peptide peak A reveal an inconsistency in the composition of peak A. Experienced chromatographers could detect the purity problem with peak A by its general peak shape. Such interpretation is difficult and verification can only be made by changing the selectivity of the chromatographic system. The spectral analysis is less ambiguous and instantly verifiable. Other techniques exist for peak purity measurement and Table 1 summarizes the advantages and limitations of the various techniques.

As a general rule, the greater the spectral difference between major component and impurity, the lower the level of the impurity that can be detected. The more similar the spectra, the more gross the contamination has to be detected. Spectral overlay in proteins is usable for moderate levels of contamination at relatively high concentrations. The various peak purity parameters that presently exist show the same limitations as spectral overlay in that its efficacy largely depends on the magnitude of the spectral difference between major component and impurity. The signal ratio method is frequently proposed as a peak purity detection method. However, in gradient mode, interpretation of signal ratio data can be arbitrary, especially at low levels.[14]

Since DAD signals are all acquired at the same time, all retention time data should be identical for all wavelengths. A spectrally different impurity would skew the retention time of the peak at different wavelengths. A very sensitive integration algorithm is required for this method to be applied as inter-signal. Retention time differences are usually in the range of 0.001 to 0.010 min. Combined with peak compression techniques made possible by the internal reference wavelengths of the diode array systems, this mode of peak contamination detection becomes very powerful. The technique is unaffected by gradients and does not require a prior knowledge of the sample.

A definitive way of determining whether or not a peak is contaminated is by quantitating at two different wavelengths at the same time. A contaminated peak would show up as two different quantitative results. The advantage of this mode is that purity estimations can be made at low levels, the process easily automated and the results easily documented. The major disadvantage is that pure material is required for DAD calibration at the multiple wavelengths shown.

Taken together, these techniques often spell the difference between isolating a pure or contaminated product. As will be shown later in this article, even with the precision of the DAD, similarity of spectra in protein and peptide work often precludes the use of a simple peak purity detection methods. Often, spectral overlay and manual interpretation is the only technique that can be relied upon. In addition, spectral manipulation in the form of first or second derivative spectroscopy is required to enhance and visualize the differences in protein and peptide work.

B. AROMATIC AMINO ACID CONTENT OF PROTEINS AND PEPTIDES: NON-DESTRUCTIVE IDENTIFICATION OF AROMATIC AMINO ACIDS IN PROTEINS AND PEPTIDES

The previous sections outlined the reasons why the DAD UV-Vis spectra of proteins were precise and why they could be used as determinants of chromatographic peak purity. As mentioned earlier, there are areas in the analysis of proteins and peptides where even an estimation of the types of aromatic amino acids and their relative abundances in a given protein would aid in protein structure elucidation, sequencing and peptide mapping.

The UV spectra of proteins and peptides are greatly dependent on the spectra of the amino acids of which they are composed. Except for the aromatic amino acids phenylalanine, tyrosine and tryptophan, the other amino acids have very uninformative spectra. Only the aromatics have absorption maxima between 250 and 300 nm, but the spectra are broad and they overlap. As a consequence, trying to use the unenhanced spectrum of a protein to determine the aromatic amino acids that it contains becomes very difficult. To estimate the relative ratios of these aromatic amino acids becomes almost impossible. Therefore, there is an increasing interest in the use of derivative spectroscopy of proteins and peptides as an aid to aromatic amino acid identification and estimation. Second order derivations transform peaks and shoulders into minima. Second order derivative spectroscopy is widely used under non-chromatographic conditions. The properties of derivative spectra have been studied for some time and their use is becoming more prevalent. Various orders of derivatives have been used in an attempt to enhance the rather featureless zero order spectra of proteins and peptides. Odd order spectra are characterized by their bi-polar form; in even order derivatives, the centroid peak typically corresponds to the original zero order band maximum. The resolution of overlapping spectral bands is enhanced with every successive derivative, but there is also a corresponding increase in the noise and in the complexity of the resultant spectrum. Consequently, the second order derivative provides a good compromise between resolution enhancement and decreased noise.

The zero and second order derivative spectra of the aromatic amino acids have previously been presented.[11,12] Figure 2 illustrates the zero and second order spectra of aromatic amino acids. The most characteristic feature of these second derivative spectra is that they show a minimum at every maximum in the zero order spectra. From Table 2, it can be seen that the most prominent minima for tyrosine (282 nm) and the secondary minimum for tryptophan (278 nm) overlap. This overlap makes the identification of tyrosine in the presence of tryptophan difficult. The ratios of the various minima to one another are also roughly diagnostic of the particular aromatic amino acids. Grego[7] found the identification of tyrosine ambiguous at ratios of TYROSINE/TRYPTOPHAN of 1:1 or greater. At ratios greater than 3:1, the tyrosine minimum at 282 adds to the tryptophan minimum at 278 nm to produce a peak greater than the major tryptophan minimum at 290 nm.

Figure 3A shows an example of this second order analysis applied to a tetrapeptide GLY-PRO-TRP-LEU. The second order spectrum clearly indicates the presence of tryptophan by the second order minima at 288 and 278 nm. From the ratio of the two minima, it can be concluded that the TYROSINE/TRYPTOPHAN ratio is 1:1 or smaller. The absence of phenylalanine is indicated by the lack of a minimum at 257 nm.

The non-destructive identification of tryptophan in peptide maps is of great interest. Tryptophan is specified in DNA by a single codon and it is therefore most convenient to

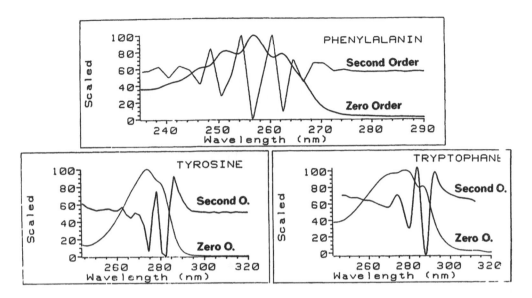

FIGURE 2. Zero and second order derivative spectra of free aromatic amino acids. Note: UV chromophores do not vary greatly in peptides compared to the free amino acids.

TABLE 2
Second Order Derivative UV Spectra for Aromatic Amino Acids: Major Minima and Side Minima

Aromatic amino acid	Minima (± 2 nm)
Phenylalanine	240, 246, 250, <u>257</u>, 266, 267
Tyrosine	274, <u>282</u>
Tryptophan	278, <u>290</u>

Note: Major minima are underlined.

construct oligo-nucleotide probes for gene fishing with data derived from tryptophan-containing peptides. Identification of tryptophan also alerts the amino acid analysts to its presence. Special amino acid hydrolysis must be performed to quantitate tryptophan from proteins. Ratios of tyrosine to tryptophan from the amino acid analysis can confirm the results derived from sequencing and PTH AA analysis. Such analysis can be performed on as little as one nanomole of peptide eluted from an HPLC column and this presents a useful analytical tool to the biochemist for peptide mapping.

The previous illustration showed an analysis performed on a simple molecule. The question is asked whether it is possible to extract similar information from much larger and more complex molecules. The chromatogram in Figure 3B illustrates the separation of three fibrinogen chains α, β, and γ (38, 44, and 48 aromatic amino acid residues, respectively).

The zero and second order spectra of the γ chain clearly identifies the presence of both tryptophan and phenylalanine. Tyrosine is possibly present, but the ratio of tyrosine to tryptophan is less than 1:1 as indicated by the relative heights of the minima at 282 and 290 nm. Figure 3B shows an example of the zero and second order spectra of the γ chain. In this case, tryptophan is clearly identified by the 290 nm minimum. Phenylalanine is also identified by the 257 minimum. The ratio of tyrosine to tryptophan in this case can be judged to be greater than 1:1 because the ratios of the 282 nm to 290 nm minima (indicative of tyrosine and tryptophan,

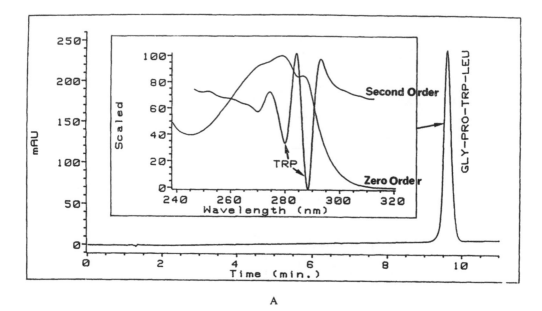

A

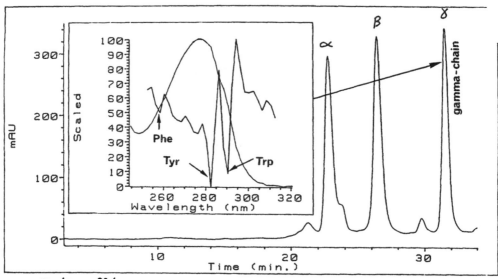

Ratio: Tyr/Trp = $^{20}/_{10}$

B

FIGURE 3. (A) Characteristic minima in second order derivative UV spectrum of a small peptide indicating the presence of tryptophan. (B) Characteristic minima in second order derivative UV spectrum of fibrinogen chain protein indicating content of aromatic amino acids. The 20/10 Tyr/Trp ratio is known from amino acid analysis and is used here to compare spectral results as a function of Tyr/Trp ratio.

respectively) show a larger contribution of the 282 minimum. Further examples of such analysis can be found.

Another use of spectral data from DADs is the confirmation of compound identification. Given the precision of DAD spectra and the existing computer programs that allow visualization of features, even the relatively featureless spectra of proteins and peptides can yield adequate information to identify compounds based on their retention time for a given chromatographic

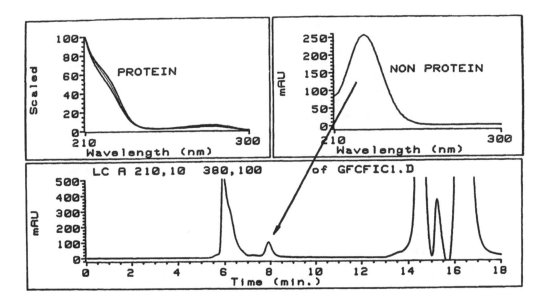

FIGURE 4. UV spectra indicating presence of a non-protein component of a protein mixture separated by size-exclusion chromatography.

system and their zero and second order derivative UV spectra. Figure 4 shows a size-exclusion chromatogram of a group of eluted proteins. The main peak eluted at approximately 6 min has the characteristic spectrum of a protein. However, the peak at approximately 8 min clearly does not have a conventional protein spectrum. With some systems, the zero and second order spectra can be searched against a library of spectra and a fit match is calculated. The contribution of the aromatic amino acids to the second order spectrum aids in identifying the protein from the spectral library.

V. CONCLUSION

This article has attempted to highlight some of the important analytical implications of Diode-Array UV-Visible spectrophotometric detectors to the HPLC analysis of proteins. These detectors and the software that helps visualize the data are an invaluable tool in the hands of the protein analyst.

REFERENCES

1. **Hopkins, W.H., Nordman, R.G., and Willis, B.G.,** Optical and Optomechanical design of a high through put ultraviolet-visible spectrophotometer, *Proc. Soc. Photo-optical Instrumentation Engineers,* 191, 48, 1979.
2. **Strong III, F.C.,** Correlation of measurements of absorbance in the ultraviolet and visible regions at different spectral slitwidths, *Anal. Chem.* 48, 2155, 1976.
3. **Willis, B.G., Fustier, D.A., and Bonelli, E.J.,** A high-performance parallel access spectrophotometer, *Am. Lab.,* June 1981.
4. *Chemometrics,* Chemical Analysis, Vol. 82, Sharaf, Illman, and Kowalski, Eds., John Wiley & Sons, New York, 1986.
5. **Garden, J.A., Mitchel, D.G., and Mills, W.N.,** Non constant variance regression techniques for calibration-curve-based analysis, *Anal. Chem.,* 52, 2310, 1980.

6. **Owen, A.J.**, The Diode-Array Advantage in UV/Visible spectroscopy; Hewlett-Packard Primer, publication number 12-5954-8912.

7. **Grego, B., Nice, E.C., and Simpson, R.J.**, Use of scanning diode array detector with reversed-phase microbore columns for the real-time spectral analysis of aromatic amino acids in peptides and proteins at the sub microgram level, applications to peptide and protein microsequencing, *J. Chromatogr.*, 352, 359, 1986.

8. **Fell, A.F., Scott, H.P., Gill, R., and Moffat, A.C.**, Applications of rapid-scanning multichannel detectors in chromatography, *J. Chromatogr.*, 273, 3, 1983.

9. **Fell, A.F., Scott, H.P., Gill R., and Moffat, A.C.**, Novel Techniques for peak recognition and deconvolution by computer-aided photodiode array detection in high-performance liquid chromatography, *J. Chromatogr.*, 282, 123, 1983.

10. **Fell, A.F., Clark, B.J., and Scott, H.P.**, Analysis and characterization of aromatic amino acids, metabolites and peptides by rapid scanning photodiode array detection in high-performance liquid chromatography, *J. Chromatogr.*, 297, 203, 1984.

11. **Balestrieri, C., Colonna, C.T., Giovane, A., Irace, C., and Servillo, L.**, Second-derivative spectroscopy of proteins, *Anal. Biochem.*, 106, 49, 1980.

12. **Balestriere, C., Colonna, C.T., Giovane, A., Irace, C., and Servillo, L.**, Second-derivative spectroscopy of proteins, *Eur. J. Biochem.*, 90, 433, 1978.

13. **Milano, M.J., Lam, S., Savonis, M., Pautler, D.B., Pav, J.W., and Grushka, E.**, Characterization of photodiode array detector in liquid chromatography, *J. Chromatogr.*, 149, 599, 1978.

14. **Clark, B.J., Fell, A.F., Scott, H.P., and Westerlund, P.**, Rapid-scanning, multi-channel high-performance liquid chromatographic detection of Zimeldine and metabolites with three-dimensional graphics and contour plotting, *J. Chromatogr.*, 286, 261, 1984.

PEAK IDENTIFICATION IN HPLC OF PEPTIDES USING SPECTRAL ANALYSIS AND SECOND ORDER SPECTROSCOPY OF AROMATIC AMINO ACID RESIDUES

James A. Black and Robert S. Hodges

I. INTRODUCTION

The advent of ultraviolet-visible (UV-VIS) diode-array detection systems in high- perform-ance liquid chromatography (HPLC) has made the real time spectral analysis of peptide components (which are eluted from an HPLC column) a practical and useful tool to the peptide chromatographer. While it is known that UV-VIS spectrophotometers are capable of producing wavelength scans of peptides and proteins, their useful application is limited to fairly pure samples. Diode-array detection systems, coupled with digital computing equipment, allow the near instantaneous acquisition of spectra at any time during a chromatographic separation.[1] Using HPLC and spectral analysis, a complex mixture of peptides, together with any possible non-peptide components, can be separated and identified. The acquisition of spectra at the apex and on the upslope and downslope of each peak can yield valuable information concerning peak purity.

Upon viewing the spectrum of any particular peak, the absence of a strong absorption band in the 200 to 230 nm region for the peptide bond (wavelength maximum $\approx$188 nm) will immediately inform the analyst that the peak is a non-peptide component. For a further review using diode array detection systems for peak purity analysis and other uses, see the chapter by Zavitsanos and Goetz[2] in this book.

The aromatic amino acids (phenylalanine [Phe], tyrosine [Tyr] and tryptophan [Trp]) each have their own unique near UV spectrum and, since there is no appreciable peptide bond absorbance above 240 nm, analysis of a spectrum from 240 to 300 nm will show whether an aromatic amino acid (or multiple aromatic amino acids) is present. The assignment of the aromatic amino residues in the peptide is complicated since the spectra of the three aromatic amino acid residues overlap. Since the molar extinction of tryptophan is higher than that of tyrosine, which is higher than that of phenylalanine, the spectrum will roughly resemble the strongest chromophore present (assuming an equimolar ratio of each chromophore).[3] In this situation, one can use the digital computer (which has all raw spectral and wavelength data stored) attached to the diode array detection system to apply the parameters developed by Savitzky and Golay to obtain second derivative spectra.[4] To obtain a second derivative spectrum, one first applies the Savitzky Golay parameters to a UV or zero order spectrum. This results in a plot of change in slope (of the zero order spectrum) vs. wavelength, resulting in a first order or first derivative spectrum. Derivatization of the first order spectrum results in a second derivative spectrum (rate of change of slope vs. wavelength). The non-mathematician may comprehend this concept by comparing these orders of spectrum to a one dimensional motion. If zero order represents distance vs. time, first order is a change in velocity vs. time and the second order is a change in acceleration vs. time. The end result of second order derivatization of a UV spectrum is an enhancement of the differences in a broad absorption band. Absorption minima appear which correspond to approximate wavelength maxima for the aromatic amino acids. Derivatives of the second order are normally used in preference to higher orders of derivatization since they represent a compromise between selectivity (intensity of absorption minima and maxima increase with each derivative) and interference from false absorptions (due to noise factors).[4]

Since the absorption spectrum for phenylalanine is different from that of tyrosine, which is again different from tryptophan, it is easy to determine which aromatic amino acid(s) is present. Also, assuming absorptions of each amino acid are additive, by measuring the absorption minima at a particular wavelength for each amino acid, one can sometimes determine a ratio of one aromatic amino acid to another.[4,5]

Real time spectral analysis is extremely useful in the separation of complex mixtures of peptides. Any resolved peaks in an HPLC chromatogram can be scanned and the resulting spectra viewed. These spectra can then determine whether or not the peak is a peptide or protein component or some contaminant. It can also be determined whether an aromatic amino acid is present and if it is in some derivatized form. This is extremely important in peptide chemistry where aromatic side-chain protecting groups are used during synthesis and must be completely removed from the peptide of interest after completion of synthesis. These undesired modifications can often go undetected since the protecting groups may be cleaved from the peptide during acid hydrolysis, and amino acid analysis would indicate the desired peptide. Aromatic amino acid content and identification in peptides is particularly useful in protein digest applications where a known protein sequence is fragmented and certain predicted fragments can be identified solely by their UV spectral data.[6]

II. EXPERIMENTAL PROCEDURES

A. MATERIALS

Reagent grade anisole and trifluoroacetic acid (TFA) were distilled prior to use. Dimethylsulfide (DMS) was 'gold label' quality from Adrich Chemical (Milwaukee, WI). Deionized water was purified by reverse-osmosis using a Hewlett-Packard (Avondale, PA) model HP661A water purifier. HPLC grade acetonitrile was obtained from J.T. Baker Chemical (Phillipsburg, NJ). All other chemicals were reagent grade.

B. HPLC APPARATUS

All peptides were separated on an Aquapore RP-300 C$_8$ column (220 × 4.6 mm I.D., 7-μm particle size, 300-Å pore size; Brownlee Labs., Santa Clara, CA). The HPLC instrument consisted of a Hewlett-Packard (Avondale, PA) HP 1090 liquid chromatograph coupled to an HP 1040A detection system, HP 9000 series 300 computer, HP 9133 disk drive, HP 2225A Thinkjet printer and HP 7440A plotter.

C. METHODS

Peptide synthesis and reversed-phase chromatography (RPC) purification was performed by the Alberta Peptide Institute. Peptides were synthesized on an Applied Biosystems (Foster City, CA) model 430A peptide synthesizer using the general procedure for solid-phase synthesis described by Erickson and Merrifield[7] with modifications by Hodges et al.[8] Synthesis of peptide 7 (Figure 1) used N-formyl protection for the indole side chain of the tryptophan residue.

The peptide mixture, 7a and 7b (Figure 1), resulted from using the following hydrogen fluoride (HF) cleavage conditions: the protected peptide resin (from solid phase synthesis) was stirred at –5°C for 1 hr in HF:anisole:DMS:*p*-thiocresol:resin (10 ml:1 ml:0.5 ml:0.2 ml:1 g).[9]

All peptides were chromatographed by RPC using a linear AB gradient (1%B/min) and a

PEPTIDE   SEQUENCE

	1	2	3	4	5	6	7	8	9	10	11
1	NH₂ - Ser	- Ser	- (Trp)	- Ala	- Val	- Leu	- Glu	- Val	- Ala	- amide	
2	Ac - Nle	- (Tyr)	- Lys	- Leu	- Gly	- Pro	- Lys	- Thr	- Gly	- Pro	- Gly - amide
3	Ac - Gly	- Ala	- (Phe)	- Gly	- Gly	- Ser	- Glu	- Asn	- Ala	- Thr	- amide
4	Ac - Arg	- Gly	- Val	- Gly	- Gly	- Leu	- Gly	- Leu	- Gly	- Lys	- amide
5	(Bb) - Nle	- Leu	- Glu	- Gln	- Pro	- Val	- Ser	- Asn	- Asp	- Leu	- Ser - amide
6	NH₂ - Arg	- Gly	- (Phe)	- (Phe)	- (Tyr)	- Thr	- amide				
7a	Ac - (Phe)	- Met	- His	- Asn	- Leu	- Gly	- Lys	- His	- Leu	- Ser	- Ser -
	Met	- Glu	- Arg	- Val	- Glu	- (Trp)	- Leu	- Arg	- Lys	- Lys	- Leu -
	Gln	- Asp	- Val	- His	- Asn	- (Phe)	- amide				
7b	As 7a, but containing - Trp(For) - instead of - Trp -										

FIGURE 1. Amino acid sequences of the peptides used in this study. The circled residues indicate the aromatic acids in each peptide.

flow-rate of 1 ml/min, where Eluent A was 0.05% aq. TFA and Eluent B was 0.05% TFA in acetonitrile.

D. DERIVATIVE CALCULATIONS

Slope is calculated by using the least squares method. Although the computer uses the Savitzky-Golay parameters, the following explanation is equivalent. A line equation is generated for each point on the plot using an equidistant number of points on either side of the one of interest. The slope of the line is then calculated from the equation. This procedure is repeated for all points on the plot to generate the derivative spectrum. As a note of interest, a smoothed line plot is performed in the same manner where each particular point on the plot has an equation generated and the point is re-adjusted to fit the equation. The total number of points used in calculating a particular line equation is called the filter length. The Hewlett-Packard computer system used to generate these plots allows for a filter length between five (two points on either side) and twenty-five. The longer the filter length, the less noise in the resulting derivative spectrum. However, the filter length chosen should be such that only one inflection is observed in the set of points used.[10] Even when using a filter of five points, some fine spectra can be lost (due to multiple inflections within the filter length) and therefore spectral resolution plays an important part in the quality of derivative spectrum plots obtained.

III. DISCUSSION

The peptides used for all the UV spectra were chosen not only because of their aromatic side-chain chromophores but also for their similar chain lengths. Thus, all peptides, with the exceptions of numbers six and seven, were of lengths between nine and eleven amino acid residues (Figure 1). This was done to show the relative absorbance intensities between the amino acid chromophore(s) and the peptide bonds (λ max $\approx$ 188 nm).[11] The lowest wavelength of the spectra is 200 nm. Although this is not at the wavelength maximum, it is the practical limit of the detection system due to the UV cut off of the mobile phase used in the HPLC of peptides. The maximum absorbances shown at 200 nm in Figure 2 are mainly on the down slope of the additive peptide bond absorbances. Knowing this, one can now appreciate the relative intensities between the aromatic side chain chromophores and the multiple peptide bonds.

The control peptide used in Figure 2 (peptide 4) not only demonstrates the lack of any appreciable absorbance beyond 230 nm in peptides with no aromatic side chains or other

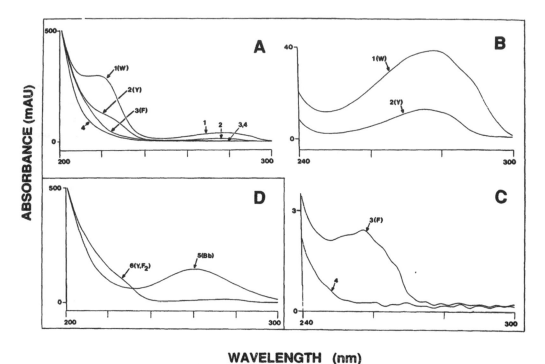

FIGURE 2. Comparison of the ultraviolet spectra of peptides 1 to 6 (Figure 1). Panel A: The spectra of peptides 1, 2, and 3 which contain a single aromatic amino acid chromophore, Trp, Tyr and Phe, respectively. Peptide 4 does not contain an aromatic amino acid. Panel B and panel C show the expansion of the wavelength region 240 to 300 nm of the peptide spectra shown in panel A. Panel D shows the spectra of a 4-benzoylbenzoyl (Bb) containing peptide and a peptide containing multiple aromatic amino acids (one tyrosine and two phenylalanine residues). The letter in brackets indicates the aromatic amino acid residue in the peptide (W, tryptophan; Y, tyrosine and F, phenylalanine).

chromophores, but also illustrates how aromatic amino acid containing peptides will add to the overall absorbance measurements in the HPLC elution profiles monitored in the 210 to 220 nm range. Cystine does have an absorption band above 240 nm (λ max $\approx$ 250 nm), but its contribution is fairly small and has been ignored for the purpose of this discussion. The peptide containing the 4-benzoylbenzoyl amino terminal group (peptide 5, Figure 1) was chromatographed (Figure 2, panel D) to show the relative absorbance of a highly aromatic absorbing group (benzophenone) to that of the aromatic amino acids. It was used to show the simple aromaticity of this organic group (smooth single gaussian curve) vs. the aromatic amino acid side chain chromophores which are not simple smooth curves but a complex mixture of overlapping gaussian curves.

Panel B (Figure 2) demonstrates the similarity of the absorption curves of tyrosine and tryptophan in the 280 nm range. However, the two differ in the magnitude of absorbance (tryptophan absorbing more strongly) and have differences in their fine spectral absorbances, as well as a small shift in their absorption maxima in the 280 nm range (this is more clearly demonstrated in the second order derivative spectra to be discussed below).

Panels B and C of Figure 2 clearly demonstrate the relative absorbance between all three chromophores. To fully appreciate this, the differences in the y axis absorbance values in panels B and C must be noted (a 0 to 40 scale for the Trp and Tyr peptide spectra panel B and a 0 to 3 scale for the Phe containing peptide [panel C]). Panel C also shows the limitations of the detection system in that the sine wave absorption pattern in the control and phenylalanine-containing peptides (peptides 3 and 4, respectively) can be attributed to some unknown form of instrument noise.

Panel D (Figure 2) shows a peptide containing both a tyrosine and a phenylalanine residue. Due to the stronger molar extinction of tyrosine, the spectrum resembles that of the single tyrosine-containing peptide (peptide 2, Figure 1 and panel A of Figure 2) and is insufficient to identify qualitatively all aromatic residues present.

While it is true that wavelength scans or zero order spectra do yield some information as to the composition of an eluted peak during HPLC, they are limited to the following: first, they distinguish peptides and proteins vs. organic eluents; second, they determine whether any aromatic side chromophore is present and tentatively identify the strongest absorbing chromophore; and third, they can be employed to monitor loss of specific side chain protecting groups (to be discussed later). To extract the most useful information from these aromatic amino acid side chain absorption spectra, one must look at a second order derivatization plot. As noted previously, this plot enhances the absorption maxima by plotting rate of change of slope instead of wavelength vs. absorbance. The results enhance the complex gaussian curves present in the spectra.

The second order spectra from 210 to 300 nm (Figure 3) reaffirm the observation that the additive absorptions of the peptide bond are generally much stronger in intensity than a single side chain absorbing chromophore. This demonstrates that to appreciate the capabilities of second order derivative spectrum, one must enlarge the region where the chromophore itself absorbs and to eliminate effects from peptide bond absorbances. The region between 250 to 300 nm works well for this purpose.

The second order spectrum of the 4-benzoylbenzoyl peptide (Figure 3, panel A) further illustrates its uncomplicated absorption curve. The minimum at 262 nm is consistent with the maximum of the zero order spectrum. When looking at the second order derivative tryptophan spectrum from 250 to 300 nm (compared to the zero order 240 to 300 nm spectrum; Figure 2, panel B), the spectrum is much smoother and simpler with basically two minima at 280 and 288 nm (Figure 3, panel B). This demonstrates that when using a five data point filter length some fine structure is lost. The limitations are related to the spectral resolution of the diode array detector (2 nm for the spectra generated in Figures 2 to 4). Spectral resolution becomes an important consideration when evaluating diode array detection systems for HPLC.

Another very valuable piece of information derived from the second derivative spectrum of tryptophan (Figure 3, panel B) is that the 280 and 288 nm minima demonstrate that the tryptophan side-chain has not undergone chemical modification, since modifications drastically affect the spectrum (to be discussed below).

The spectrum of tyrosine (Figure 3, panel C) shows the absorption maxima for this residue at 272 and 280 nm. One important limitation in identifying which aromatic amino acids are present in any second derivative spectrum is that the weaker absorption maxima of tyrosine overlap with those of tryptophan, thereby making tyrosine determination more difficult in the presence of tryptophan. The much more complex second derivative spectrum of phenylalanine in the 260 nm region (Figure 3, panel D) makes it possible to identify qualitatively its presence in most peptides which contain this chromophore. Panel E of Figure 3 resembles the second derivative spectrum of tyrosine. However, closer comparison with panel C shows that the minima in the 260 nm region indicates the presence of phenylalanine in the peptide. Although it is not shown, one can actually measure the difference between certain minima and maxima to determine ratios of aromatic amino acids present in a particular peptide or protein. This procedure is well described by Balestrieri et al.[4,5]

Figure 4 shows how UV spectra can be applied to analysis of peptides resulting from chemical synthesis. A 28-residue peptide containing one tryptophan residue was prepared by solid-phase peptide synthesis. Hydrogen fluoride cleavage conditions used resulted in an approximate equimolar amount of the native peptide and the same peptide still containing a formyl protecting group (from synthesis) on the indole side chain of the tryptophan residue [Trp(For)]. Panel A of Figure 4 shows the impressive resolving power of RPC in separating two 28-residue peptides

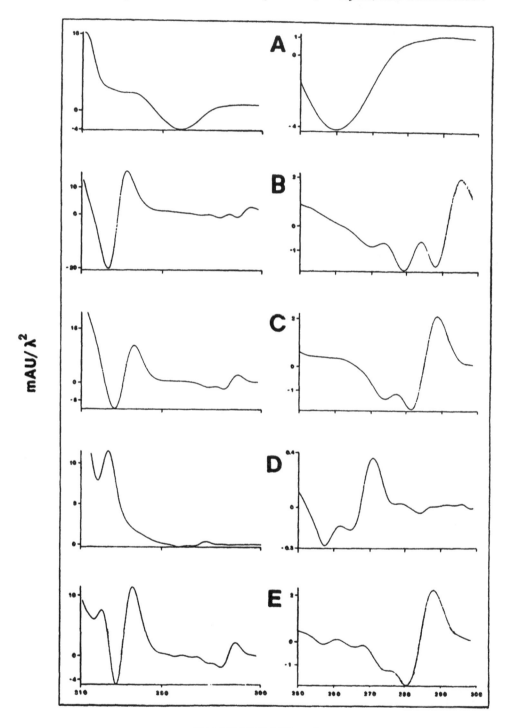

WAVELENGTH , λ (nm)

FIGURE 3. Second derivative ultraviolet spectra of peptides containing various aromatic chromophores. Panels A, B, C, and D correspond to spectra of peptides containing the single aromatic chromophores 4-benzoylbenzoyl (peptide 5), tryptophan (peptide 1), tyrosine (peptide 2) and phenylalanine (peptide 3), respectively. Panel E shows the spectra of peptide 6 which contains one tyrosine and two phenylalanine residues. The left panels are the spectra from 210 to 300 nm and the right panels are the spectra in the 250 to 300 nm range.

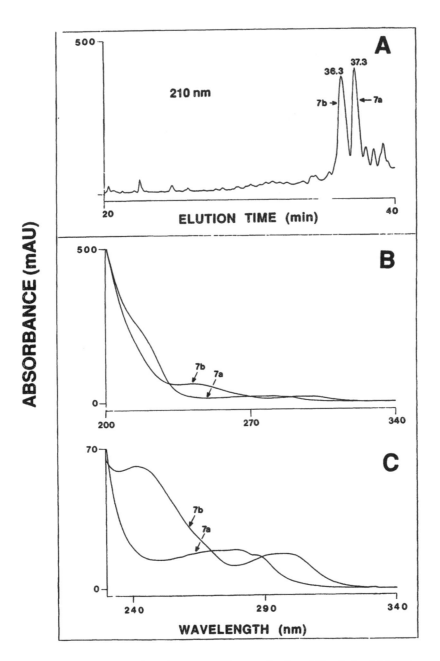

FIGURE 4. The reversed-phase chromatographic separation (panel A) and ultraviolet spectra (panels B and C) of two peptides which differ only in the presence of tryptophan (peptide 7a) or N-formyltryptophan (peptide 7B). The numbers above the peaks in panel A refer to the elution times of peptides 7a and 7b (Figure 1). The chromatographic conditions are described in the Experimental section.

differing by only a formyl group. One interesting observation is that, although the Trp(For) peptide is of higher molecular weight, it is eluted off the RPC column first due to an overall lower net hydrophobicity attributed to the hydrophilic formyl group.

This is an excellent example of how UV spectra can differentiate between these two peaks eluted off an analytical RPC column. These UV spectra also differentiate these two components from any others in the separation that do not contain either the tryptophan or Trp (For) residues.

Monitoring at 280 nm would not reveal that the first major peak is actually an analog of the desired peptide, since Trp(For) has little absorbance at this wavelength (Figure 4, panel B).

Panel C (Figure 4) shows the effect that the formyl group has on the absorbance maxima of the tryptophan side chain.

Although not shown, it must be remembered that if these two peaks (Figure 4, panel A) are not resolved, the spectrum of both peaks together (as the case would be if scanning the complete mixture in a UV-Vis spectrophotometer) is additive and the resulting net spectrum does not appear similar to either spectrum (Figure 4, panel B). As a result, even rough quantitation of the amount of each peptide would be difficult.

Although it has not been demonstrated by the authors, it is our belief that diode-array detection can be applied to monitor successful maintenance or removal of other protecting groups such as the dinitrophenyl group used for histidine protection in solid-phase peptide synthesis.

In conclusion, this chapter was written to provide an easy and short explanation of the value and limitations of spectral analysis during HPLC. For further and more complete discussions of any of the areas dealt with, the references cited should be consulted.

ACKNOWLEDGMENTS

I would like to thank personally my co-workers whose extra effort allowed me the time to prepare this chapter. I would also like to thank Dr. Paul Cachia of Synthetic Peptides Inc. and T.W. Lorne Burke for their help in the mathematical interpretations explained.

REFERENCES

1. **Fell, A.F. and Scott, H.P.,** Applications of rapid-scanning multichannel detectors in chromatography, *J. Chromatogr.,* 273, 3, 1983.
2. **Zavitsanos, P. and Goetz, H.,** The practical application of diode array UV-visible detectors to the HPLC analysis of proteins and peptides, this publication.
3. **Wetlaufer, D.B.,** Ultraviolet Spectra of proteins and amino acids, *Advances in Protein Chemistry,* Anfinsen, C.B., Jr., et al. Eds., Academic Press, N.Y., 1962, 314.
4. **Balestrieri, C., Colonna, G., Giovane, A., Irace, G., and Servillo, L.,** Second derivative spectroscopy of proteins; a method for the quantitative determination of aromatic amino acids in proteins, *Eur. J. Bioch.,* 90, 433, 1978.
5. **Balestrieri, C., Colonna, G., Giovane, A., Irace, G., and Servillo, L.,** Second-derivative spectroscopy of proteins: studies on tyrosyl residues, *Anal. Biochem.,* 106, 49, 1980.
6. **Grego, B., Nice, E.C., and Simpson, R.J.,** Use of scanning diode array detector with reversed-phase microbore columns for the real-time spectral analysis of aromatic amino acids in peptides and proteins at the sub-microgram level: applications to peptide and protein microsequencing, *J. Chromatogr.,* 352, 359, 1986.
7. **Erickson, B.W. and Merrifield, R.B.,** Solid-phase peptide synthesis, in *The Proteins,* Neurath, H. and Hill, R.L., Eds., Academic Press, New York, 1976, 255.
8. **Hodges, R.S., Semchuk, P.D., Taneja, A.K., Kay, C.M., Parker, J.M.R. and Mant, C.T.,** Protein design using model synthetic peptides, *Peptide Res.,* 1, 19, 1988.
9. **Hong, A.L. and Culwell, A.R.,** An introduction to the use of hydrogen fluoride in peptide synthesis, Peptide Synthesizer Model 430 User Bulletin, Applied Biosystems Inc., 1987.
10. **Savitzky, A. and Golay, M.J.E.,** Smoothing and differentiation of data by simplified least squares procedures, *Anal. Chem.,* 36, 1627, 1964.
11. **Donovan, J.W.,** *Ultraviolet Absorption in Physical Principles and Techniques of Protein Chemistry,* Part A, Leach, S.J., Ed., Academic Press, N.Y., 1969.

DETECTION ALTERNATIVES FOR HPLC ANALYSIS OF LHRH AND LHRH ANALOGS

Karen Lockhart, Cecilia Nguyen, and Maryann Lee

I. INTRODUCTION

The decapeptide luteinizing hormone-releasing hormone (LHRH) and its analogs are a major area of current research. As such, convenient, rugged analytical methods are needed to not only resolve the analogs but also provide sensitive detection. Based on our comparison of ultraviolet (UV), electrochemical (EC), and fluorescence detectors for the analysis of an LHRH analog that contains a good ultraviolet chromaphore,[1] we would like to extend this comparison to other LHRH analogs.

The amino acid content of a peptide is a major factor in determining whether UV, EC, or fluorescence detection will be the most sensitive. Hundreds of amino acids are found in nature, but the most common naturally occurring amino acids are alanine (Ala), arginine (Arg), asparagine (Asn), aspartic acid (Asp), cysteine (Cys), glutamic acid (Glu), glutamine (Gln), glycine (Gly), histidine (His), isoleucine (Ile), leucine (Leu), lysine (Lys), methionine (Met), phenylalanine (Phe), proline (Pro), serine (Ser), threonine (Thr), tryptophan (Trp), tyrosine (Tyr), and valine (Val).[2] Man has created some additional amino acids such as naphthylalanine (Nal), parachlorophenylalanine (pClPhe) and diethylhomoarginine (hArg(Et$_2$)). Although all peptides contain a weak chromaphore, the peptide bond -HN-C=O, only a few amino acids contain significant UV chromaphores, including Nal, Trp, Phe, pClPhe, and Tyr. Amino acids which contain UV chromaphores generally also show fluorescence. However, substituents strongly affect fluorescence, with electron-donating groups increasing fluorescence (such as -OH in Tyr) and electron-withdrawing groups decreasing fluorescence (such as -Cl in pClPhe). In addition, fluorescence is often observed for amines adjacent to extended conjugation (such as in Trp).[3] Finally, amino acids which respond to EC detection include Tyr, Trp, and Cys (the electroactive groups are phenol, indole, and thiol, respectively).[4] Peptides with a terminal NH$_2$ group exhibit a weak response to EC detection.[5]

For the purposes of comparison, we chose LHRH, D-Trp[6] LHRH, D-Phe[2], D-Ala[6] LHRH, nafarelin, and detirelix:

pGlu-His-Trp-Ser-Tyr-Gly-Leu-Arg-Pro-Gly-NH$_2$	LHRH
pGlu-His-Trp-Ser-Tyr-Trp-Leu-Arg-Pro-Gly-NH$_2$	D-Trp[6] LHRH
pGlu-Phe-Trp-Ser-Tyr-Ala-Leu-Arg-Pro-Gly-NH$_2$	D-Phe[2], D-Ala[6] LHRH
pGlu-His-Trp-Ser-Tyr-Nal-Leu-Arg-Pro-Gly-NH$_2$	nafarelin

pGlu-His-Trp-Ser-Tyr-Nal-Leu-Arg-Pro-Gly-NH$_2$
  1   2   3   4   5   6   7   8   9   10
Ac-Nal-pClPhe-Trp-Ser-Tyr-hArg(Et$_2$)-Leu-Arg-Pro-Ala-NH$_2$ detirelix
    1   2   3   4   5   6   7   8   9   10

LHRH contains two amino acids, Trp and Tyr, which are responsive to all three detectors. The other peptides in this study contain, in addition to Trp and Tyr, a third responsive amino acid and, therefore, would be expected to exhibit higher detector responses than LHRH. D-Trp[6] LHRH and D-Phe[2], D-Ala[6] LHRH in particular would be expected to show a higher fluorescence than LHRH since they contain an additional Trp and Phe, respectively. The extra Trp in D-Trp[6] LHRH would also be expected to produce a better EC response than LHRH. Nafarelin and detirelix would be expected to exhibit a higher UV than LHRH since they both contain Nal.

To compare detector sensitivity, the "figure of merit" (FOM) suggested by Roe[6] will be used. As previously shown,[1] the FOM is a good measure of detector sensitivity, comparable to a lower limit of detection (LLD) plot. For the purpose of choosing the best detector, it is an advantage to make all calculations from one low-concentration sample injection rather than the half-dozen injections required for an LLD plot. Valuable sample and time will be saved and one LLD plot can be done later after the best detector and analysis conditions are selected.

To optimize detector sensitivity:

- The choices of mobile phase and gradient or isocratic conditions should be carefully considered to meet the needs of the separation and sensitivity
- Spectroscopic and voltammetric determinations should be carried out for each peptide
- The settings of each detector should be optimized
- The FOM can be used to select the most sensitive detector for the analysis

Selectivity may also be a consideration. The careful choice of a detector can increase the selectivity of the analysis. If two peptides coelute during HPLC and one is electroactive while the other is not, the compounds can be "resolved" by determining one by EC and both by UV or fluorescence. Also, the response of some compounds can be eliminated by appropriate selection of the wavelength or electrode potential, although sensitivity may also be affected. Nal, Trp, Phe, and Tyr absorb weakly at about 280 nm; UV detection at this wavelength would increase selectivity but decrease sensitivity for peptides containing these amino acids.[5]

If more sensitivity or selectivity is needed than can be obtained from the unaltered peptide, derivatization followed by fluorescence[7] or EC detection[8] is an option. However, this technique requires additional work and will not be discussed here.

The convenience of the detector is also important. UV detectors are commonly available and generally more reliable, even though they may not be as sensitive or selective as fluorescence or EC detectors. Since less optimization work is required for the UV detector, UV should be considered first for routine analysis. If special needs arise, such as detecting compounds in a complex matrix (e.g., blood) where greater selectivity is needed or detecting compounds at trace levels where greater sensitivity is needed, try the fluorescence or EC detectors.

Generally, there is not one universal detection mode for all needs. Different detectors will supply better sensitivity, selectivity, or convenience.

II. EXPERIMENTAL

A. MATERIALS

LHRH, D-Trp[6] LHRH, and D-Phe[2], D-Ala[6] LHRH were obtained as acetate salts from Sigma Chemicals. Nafarelin acetate[9] and detirelix acetate[10] were prepared by the Syntex Institute of Organic Chemistry. Acetonitrile and methanol (Burdick and Jackson Labs) were HPLC grade. Trifluoroacetic acid (TFA, Aldrich) was 99% pure. Potassium dihydrogen phosphate (Mallinkrodt) was analytical reagent grade. Distilled water for mobile phase and sample preparation was further purified with a Barnstead Nanopure filtration system.

B. HPLC SYSTEM

The following components comprised the HPLC instrumentation: model SP8800 ternary pump (Spectra-Physics), Goldenfoil column heater (Systec), model 710B WISP autosampler (Waters Assoc.), and Cals Peak Pro data system (Beckman). The analytical reversed-phase column was a C_{18} "Protein and Peptide" column from Vydac (250×4.6 mm I.D., 5 µm particle size, 300 Å pore size). A precolumn (70×2.1 mm I.D.) packed with CO:Pell ODS (Whatman, Inc.) was installed before the autosampler to protect the analytical column.

Mobile phases were mixtures of (A) water with 0.1% TFA and (B) acetonitrile with 0.1% TFA (v/v, 0.013 M).

- *Isocratic conditions*: 70% A/30% B
- *Gradient conditions*: 81% A/19% B to 50% A/50% B in 12 min; hold at 50% A/50% B for 8 min; reequilibrate at 81% A/19% B for 15 min.

At a flow-rate of 1.0 ml/min and 45°C column temperature, the backpressure was approximately 1300 psi.

C. DETECTORS

All detectors and accessories are commercially available and have not been modified to improve sensitivity. UV HPLC traces were recorded with a LDC Spectromonitor III variable wavelength detector at 210 and 225 nm. A model LC-4B (Bioanalytical Systems) detector with glassy carbon working electrode at + 0.9 V vs. an old Ag/AgCl reference electrode and 10 nA range was used for EC detection. The fluorescence detector was a model FL-749 (McPherson) equipped with a 24-µl flow cell, xenon lamp, arc stabilizer, high sensitivity accessory, a model CF-320 flow cell filter, excitation wavelength set at 289 nm, emission wavelength set at 344 nm, slit width 16, range 0.03, and lamp current 7.5 amps.

D. PROCEDURES

Samples were prepared as 3.5×10^{-7} M solutions (content expressed as free base) in a dilution solvent of 25% methanol in 0.025 M KH_2PO_4. This dilution solvent was used rather than mobile phase, because acetonitrile injected onto the HPLC system sometimes results in artifactual peaks and because phosphate buffer helps prevent adsorption of the peptides to the glassware. Injection volume was 20 µl (approximately 10 ng of each peptide).

Nafarelin in the isocratic mobile phase was used to optimize all three detectors.

Detector sensitivity was assessed with the "figure of merit" described by Roe.[6] For each detector, peak response in mV (S), baseline noise in mV (N), and peak width at half height in µl (W) were measured. The peak width in µl is obtained by multiplying the peak width in minutes by the flow rate (ml/min) and by 1000 µl/ml. For a given amount of compound injected (ng) and injection volume in µl (V), the figure of merit (FOM) is equal to (S × V)/N × W × ng), and has units of reciprocal nanograms. Multiplying the FOM by the molecular weight of each compound yields the FOM with units of reciprocal nanomoles. Thus, a higher FOM indicates better detector sensitivity.

III. RESULTS AND DISCUSSION

Using LHRH and the chosen analogs, the goal is to develop and optimize a simple, reliable, and sensitive analytical method. To accomplish our goal, the method development procedure is broken into steps.

First, a mobile phase must be chosen. Water and an organic solvent are generally used for reversed-phase separation of peptides. For the purposes of UV and fluorescence detection, the organic solvent should have a low fluorescence and UV cut-off. Acetonitrile (UV cut-off 188 nm) is the usual choice. Organic solvents cannot be used with some EC electrodes; however, a glassy carbon electrode can tolerate nonaqueous conditions. Also, for the purposes of EC detection, an electrolyte must be included in the mobile phase, usually a buffer. If the peptides are going to be isolated from the HPLC eluent, a volatile buffer or solvent such as TFA should be used. The concentration of the buffer generally needs to be at least 0.01 *M*.

A choice between isocratic and gradient HPLC conditions must be made. Gradient analysis can separate compounds of widely varying hydrophobicity (in RPC) in a single chromatogram. This ability is not matched by isocratic analysis and, in our particular case, only three of the five chosen peptides could be analyzed under the isocratic conditions (LHRH was unretained and detirelix did not elute). Peaks eluted using gradient analysis are sharper than those eluted under isocratic conditions, but it may be difficult to accurately measure peak responses due to the baseline drift from the changing mobile phase composition. Isocratic HPLC instrumentation is less expensive and isocratic conditions may yield better resolution of very similar compounds. For running gradients, all solvents and reagents used in mobile phase preparation should be of high purity to avoid elution of mobile phase impurities during the gradient. Also, it should be confirmed that the organic solvent will not precipitate the buffer during the gradient run.

The best approach in determining the optimal detector settings is to record the UV and fluorescence spectra and the cyclic voltammograms of each peptide in the selected mobile phase. The solvent is critical since it can shift the maximum wavelength or potential. Various mobile phases should be investigated; however, the composition may be limited by the requirements of the separation. The maximum response obtained is the optimal setting for each detector. If sample is limited or the instrumentation is not available, a literature search for the peptide, related peptide or the amino acid components of the peptide may provide the necessary information. Injecting the sample onto the HPLC using settings which bracket the reported maximum will confirm or provide new optimal settings. In our case, all three detectors were optimized using nafarelin.

For the UV detector, further optimization work was minimal. The detector was set to its lowest sensitivity setting and only a few minutes were required for lamp warm-up. Sensitivity was somewhat limited by baseline drift during gradient runs.

Optimizing the EC detector required a bit more work. First, the glassy carbon working electrode should be polished. Over time, the signal becomes smaller due to fouling of the electrode with oxidation reaction products; therefore, the working electrode needs to be polished every day for maximum sensitivity. The detector should also be warmed-up overnight. Injecting nafarelin and bracketing the potential settings at + 0.8, 0.9, 1.0, and 1.1 V, the best signal-to-noise (stn) ratio was found at + 0.9 V. However, the potential and stn ratio depended on the condition of the Ag/AgCl reference electrode. An older electrode which gave its best response at + 0.9 V actually had a better stn ratio than a brand new reference electrode which required a potential of + 1.2 V. Next, using the old reference electrode with an applied potential of + 0.9 V, different range settings were tried. The 10 nA setting gave the best stn ratio.

Running a gradient on the EC detector caused a substantial amount of baseline drift (1200 mV). This presented no problem for the Beckman data system; however, a chart recorder with a smaller voltage range would go off-scale. The sharp angle of the baseline drift also makes it difficult to detect small peaks, causing large uncertainty in peak height and width measurements and effectively reducing accuracy.

EC detectors can be destructive if the reaction is not reversible or a second electrode is not used downstream to reverse the oxidation that has occurred. This should be kept in mind, if the peptides are to be isolated from the eluate. For the same reason, if UV, fluorescence and EC detectors are used in series, the EC detector should be last in the series.

Optimizing the fluorescence detector also required some work. The lamp should be warmed-up overnight. This detector has many variables which are critical to its sensitivity. Two wavelengths are required for operation, making the optimization of both settings more important — deviation of the excitation and emission wavelengths by 10 nm from the maxima can result

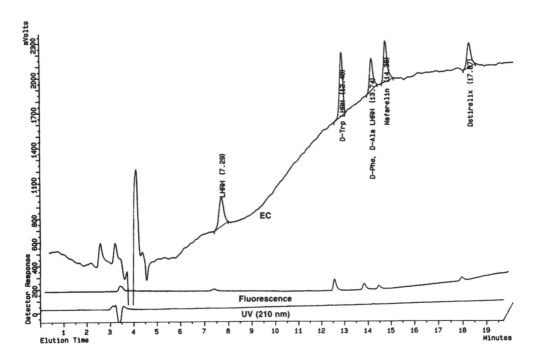

FIGURE 1. Gradient chromatograms of a mixture of LHRH and some LHRH analogs using ultraviolet (UV) (210 nm), fluorescence and electrochemical (EC) detectors. Column: Vydac C$_{18}$ (250 × 4.6 mm I.D., 5-μm particle size, 300-Å pore size). Conditions: linear AB gradient (81% A/19% B to 50% A/50% B in 12 min) followed by isocratic elution in 50% A/50% B for 8 min, where Eluent A is 0.1% TFA in water and Eluent B is 0.1% TFA in acetonitrile; flow-rate, 1 ml/min; temperature, 45°C.

in no peak detection at all. Once the excitation and emission wavelengths are determined from fluorescence scans or bracketing experiments, a flow cell (emission) filter is chosen which will reduce transmission of wavelengths other than the emission wavelength. A higher lamp current (up to a limit of 7.5 amps) and a larger slit width both yield better sensitivity. The optimum gain and range settings need to be determined by bracketing experiments. Gradient baseline shift was minor using this detector.

After optimizing all three detectors, we were ready to compare their sensitivity. Figure 1 shows gradient chromatograms following injection of a solution that is equimolar in LHRH and each chosen analog. As expected, the peptides are eluted in order of increasing hydrophobicity — LHRH, D-Trp^6 LHRH, D-Phe^2, D-Ala^6 LHRH, nafarelin and detirelix. The Beckman data system allows all three detector responses to be plotted on the same scale, despite the drifting baseline of the EC detector; however, the peaks in the fluorescence and UV chromatograms are small or not observable. To obtain the needed FOM parameters, the chromatograms were expanded until the peaks were measurable. Figure 2 shows an expanded scale plot of Figure 1 — the fluorescence chromatogram now has measurable peaks, but further expansion was required for the UV chromatogram (not shown). Figure 3 shows isocratic chromatograms obtained following injection of a solution that is equimolar in D-Trp^6 LHRH, D-Phe^2, D-Ala^6 LHRH, and nafarelin. As for the gradient chromatograms, the peptides are again eluted in order of increasing hydrophobicity and expanded plots were generated to obtain measurable peaks.

For both gradient and isocratic conditions, the FOM of each peak on each detector was calculated and the results are shown in Table 1. The gradient conditions resulted in better sensitivity in all cases due to sharper peaks. In particular, the latest-eluted peak under isocratic conditions, nafarelin, gave a much lower FOM.

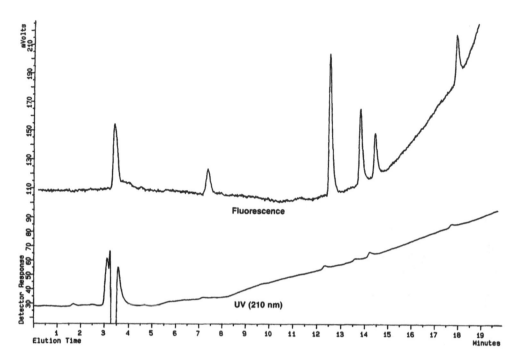

FIGURE 2. Expanded gradient chromatograms (from Figure 1) of a mixture of LHRH and some LHRH analogs using ultraviolet (UV) (210 nm) and fluorescence detectors.

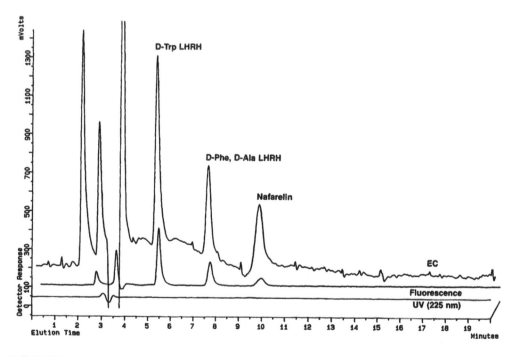

FIGURE 3. Isocratic chromatograms of a mixture of some LHRH analogs using ultraviolet (UV) (225 nm), fluorescence and electrochemical (EC) detectors. Column: as in Figure 1. Conditions: isocratic elution with 70% A/30% B, where Eluent A is 0.1% TFA in water and Eluent B is 0.1% TFA in acetonitrile; flow-rate, 1 ml/min; temperature, 45°C.

TABLE 1
Comparison of Detector Sensitivity for LHRH and LHRH Analogs
Using a Figure of Merit (FOM)

Peptide	UV 225	UV 210	EC	Fluorescence
Gradient conditions FOM (per nmole)				
LHRH	50	40	70	110
D-Trp^6 LHRH	100	90	220	800
D-Phe^2, D-Ala^6 LHRH	80	70	120	470
Nafarelin	190	120	150	230
Detirelix	120	90	90	210
Isocratic conditions FOM (per nmole)				
D-Trp^6 LHRH	60	30	180	440
D-Phe^2, D-Ala^6 LHRH	30	20	50	130
Nafarelin	40	20	30	30

In general, for all the peptides under both isocratic and gradient conditions, the fluorescence detector was the best, ranging from only slightly better than UV detection at 225 nm for nafarelin to eight times better than UV detection for D-Trp^6 LHRH (Table 1). EC detection was second best, ranging from slightly to three times better than UV detection at 225 nm, except in the case of nafarelin and detirelix where UV detection was second best. UV detection at 225 and 210 nm was approximately the same, again with the exception of nafarelin and detirelix, where UV detection at 225 nm was better. Response and background noise were both higher at 210 than at 225 nm, resulting in approximately the same signal-to-noise ratio for UV at 210 and 225 nm.

Comparing the FOM of the peptides using each detector under the gradient conditions, nafarelin and detirelix showed the highest UV response followed by D-Trp^6 LHRH, D-Phe^2, D-Ala^6 LHRH, and LHRH (third responsive amino acids are Nal, Trp, Phe, and none, respectively). Generally, the degree of UV detector response is dependent on the chromaphore of the third responsive amino acid, with Nal stronger than Trp which in turn is stronger than Phe. The contribution of pClPhe to the total UV response of detirelix appears to be minimal. Using the EC detector, D-Trp^{6}LHRH, with an additional electroactive Trp, clearly exhibited the highest response, with the rest, all containing one Trp and one Tyr, roughly the same. Using the fluorescence detector, D-Trp^6 LHRH exhibited the highest response, followed by D-Phe^2, D-Ala^6 LHRH, nafarelin, detirelix, and LHRH. Again, the third responsive amino acid determines the fluorescence of the peptide, with Trp greater than Phe which is greater than Nal. Nal may be lower than Phe due to steric crowding. The contribution of pClPhe is again minimal.

Under isocratic conditions, D-Trp^6 LHRH gave the highest FOM on all detectors. Using the UV detector, D-Trp^6 LHRH was higher than nafarelin because it is eluted earlier, resulting in a smaller bandwidth and therefore a larger FOM.

Even though the UV detector was the least sensitive, the other detectors were only about twice as sensitive, with the exception of fluorescence for D-Trp^6 LHRH and D-Phe^2, D-Ala^6 LHRH. For these peptides, fluorescence detection yields a five- to seven-fold increase in sensitivity. However, in general, not much of a sensitivity increase was gained for the additional optimization work required to use the EC or fluorescence detectors. Using the FOM, the most sensitive detector can be selected for each analysis. Complete validation of the method, including a lower limit of detection plot, can then be done under the optimized conditions.

IV. CONCLUSIONS

The requirements of each analysis will determine whether UV, EC, or fluorescence is the best detector to use. For most assays where sensitivity is not critical, the ease and reliability of UV detection is the best. If the sensitivity or selectivity of UV detection is not sufficient, further work must be done to carefully optimize EC or fluorescence detection. Since fluorescence was found to be the most sensitive, it is a good second choice. The FOM can be used to assist in the detector selection process using minimal sample and time.

REFERENCES

1. **Lockhart, K.L., Kenley, R.A., and Lee, M.O.**, Comparing electrochemical, fluorescence, and ultraviolet detectors for HPLC analysis of the decapeptide, nafarelin, *J. Liq. Chromatogr.*, 10, 2999, 1987.
2. IUPAC-IUB Commission on Biochemical Nomenclature Symbols for Amino-Acid Derivatives and Peptides, Recommendations (1971), *J. Biol. Chem.*, 247, 977, 1972.
3. **Willard, H.H., Merritt, Jr., L.L., Dean, J.A., and Settle, Jr., F.A.**, Structural factors, in *Instrumental Methods of Analysis*, 6th ed., Willard, H.H., Merritt, Jr., L.L., Dean, J.A., and Settle, Jr., F.A., Eds., Wadsworth Publishing, Belmont, 1981, chap. 4.1, 107.
4. **Bennett, G.W., Brazell, M.P., and Marsden, C.A.**, Electrochemistry of neuropeptides: a possible method for assay and *in vivo* detection, *Life Sci.*, 29, 1001, 1981.
5. **White, M.W.**, High-performance liquid chromatography of tyrosine-related peptides with electrochemical detection, *J. Chromatogr.*, 262, 420, 1983.
6. **Roe, D.K.**, Comparison of amperometric and coulometric electrochemical detectors for HPLC through a figure of merit, *Anal. Lett.*, 16(A8), 613, 1983.
7. **Schlabach, T.D. and Wehr, T.C.**, Fluorescent Techniques for the selective detection of chromatographically separated peptides, *Anal. Biochem.*, 127, 222, 1982.
8. **Joseph, M.H. and Davies, P.**, Electrochemical activity of o-phthalaldehyde-mercaptoethanol derivatives of amino acids, *J. Chromatogr.*, 277, 125, 1983.
9. **Nestor, Jr., J.J., Ho, T.L., Simpson, R.A., Horner, B.L., Jones, G.H., McRae, G.I., and Vickery, B.H.**, Synthesis and biological activity of some very hydrophobic superagonist analogues of luteinizing hormone-releasing hormone, *J. Med. Chem.*, 25, 795, 1982.
10. **Nestor, Jr., J.J., Ho, T.L., Tahilramani, R., Horner, B.L., Simpson, R.A., Jones, G.H., McRae, G.I., and Vickery, B.H.**, LHRH agonists and antagonists containing very hydrophobic amino acids, in *LHRH and its Analogs: Contraceptive and Therapeutic Applications*, Vickery, B.H., Nestor, J.J., and Hafez, E.S.E., Eds., MTP Press, Boston, 1984, 22.

POST COLUMN REACTION SYSTEMS FOR FLUORESCENCE DETECTION OF POLYPEPTIDES

C. Timothy Wehr

I. INTRODUCTION

Post column reaction (PCR) detection employs an on-line derivatization system to convert chromatographically-separated polypeptides to fluorescent products prior to transfer to a fluorimetric detector. In its simplest form, a post column reaction system would include a post column pump to deliver the derivatization reagent, a mixing tee to combine the reagent with the column eluent, and a reactor with sufficient delay time for adequate product formation. Specific post column chemistries may require additional components.

Post column reaction detection offers advantages when compared to detection of polypeptides by native absorbance, by native fluorescence, or by precolumn derivatization with fluorogenic reagents. Detection of polypeptides by UV absorbance in the 205 to 220 nm region is the most common detection method because of its simplicity and because the characteristic absorbance of the peptide bond at these wavelengths enables universal detection of all polypeptides. However, since UV absorbance detection at short wavelengths is very nonselective, nonpeptide solutes will generate peaks and many of the commonly-used mobile phase components will produce baseline drift. In contrast, fluorescence detection is inherently more selective by virtue of employing excitation and emission wavelengths which discriminate against fluorescence of groups other than the derivatized peptide. Fluorescence detection can enable use of buffers, ion-pairing agents, and organic modifiers which would be impractical with UV absorbance detection. In addition, the common post column reagents produce derivatives with high fluorescence yield, so sensitivity is typically better compared to UV absorption detection.

The amino acids tyrosine and tryptophan exhibit native fluorescence, and detection of peptides based on fluorescence of these residues should provide the benefits described above without the nuisance of adding a post column reaction system. However, sensitivity is dependent on the tyrosine or tryptophan content of the polypeptide, and the technique will not detect species which do not contain these amino acids. In practice, detection of native florescence is generally used in concert with UV absorbance detection as a means of identifying which components in a complex peptide mixture contain aromatic amino acids.

Precolumn derivatization of the sample prior to injection can provide good detection selectivity and sensitivity. Unfortunately, precolumn derivatization may yield multiple products for individual peptides which are resolved during chromatography, complicating peak identification and quantitation. In addition, some derivatization products may be labile, with half lives within the range of typical analysis times. In this case, precise quantitation is difficult even with automated derivatization procedures.

Although post column reaction detection can provide improved sensitivity and selectivity, the limitations of the technique must be considered as well. First, the PCR pump, heater, and hydraulics increase the total cost of the analytical system. In actuality, this cost must be compared to the labor costs incurred in manual precolumn derivatization or hardware costs for

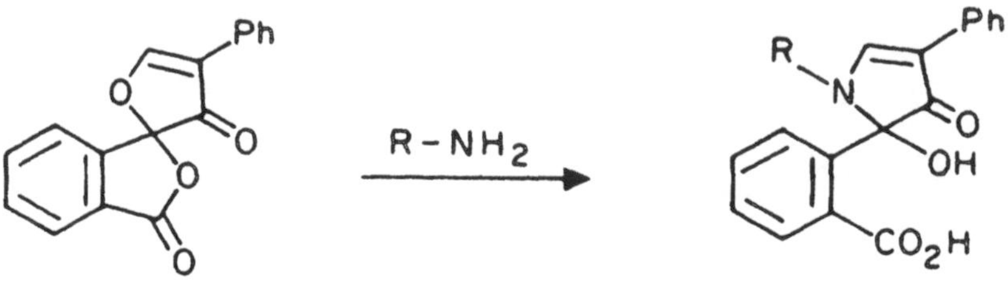

FIGURE 1. Reaction of fluorescamine with primary amines.

automation of precolumn chemistries. The performance requirements for post column reagent pumps are sufficiently modest to permit the use of inexpensive single-piston reciprocating pumps. A more serious limitation of PCR systems is the additional cost in maintenance and downtime. Not only are additional maintenance steps such as pump seal replacement necessary, but some PCR chemistries may increase the risk of plugging or rupture of reaction coils. Post column reaction systems are most cost-effective when operated routinely on a daily basis with a rigorous preventive maintenance program. A third limitation of PCR detection is the requirement for a nonfluorescent reagent which reacts rapidly with polypeptides under mild conditions to yield a fluorescent product. Only a few reagents fill these requirements; the two most widely used are fluorescamine and o-phthalaldehyde (OPA).[1] Both of these reagents react with peptide amino groups, and are commercially available in suitable purity from a number of vendors. A final limitation of post column reaction detection is common to both of these reagents: they will react with nonpeptide amine-containing contaminants in the sample, mobile phase solvents, and PCR buffers to produce spurious peaks or high backgrounds. Application of these reagents to high-sensitivity detection requires use of high purity chemicals and maintenance of a clean laboratory environment.

II. EXPERIMENTAL

Chromatography was performed with a Varian Model 5060 ternary liquid chromatograph (Varian, Palo Alto, CA) equipped with a Varian UV-5 (Figure 4) or Varian UV-100 (Figures 2 and 5) detector.

Post column reactions were performed with the Varian System I (Figure 4) or Varian System II (Figures 2 and 5) post column reaction systems, including the Fluorichrom fluorescence detector in both systems.

III. RESULTS

A. POST COLUMN REACTION DETECTION USING FLUORESCAMINE

Fluorescamine reacts with primary amines under alkaline conditions to yield a highly fluorescent product[2] (Figure 1); the reaction is complete in less than 1 min at ambient temperature. The product is optimally excited at 390 nm with emission collected at 490 nm.[3] Fluorescence yield is not influenced by pH between 4 and 10, but may be affected by solvent.[4] Fluorescamine also reacts with secondary amines to yield a product which is nonfluorescent,[5] and forms hydrolysis by-products which are nonfluorescent as well.[1]

Since fluorescamine is insoluble in water, it must be dissolved in an organic solvent such as acetone (typically at a concentration of 0.02%) and combined with an alkaline buffer (typically 0.3 M boric acid adjusted to an appropriate pH with lithium hydroxide) prior to introduction of the column eluent.[6] This requires the use of two post column reagent pumps, one for the fluorescamine/acetone solution and a second for the borate buffer. Fluorescamine, buffer, and column eluent are mixed in a ratio of 1:1:2 with a total flow-rate of 1 to 2 ml/min. The

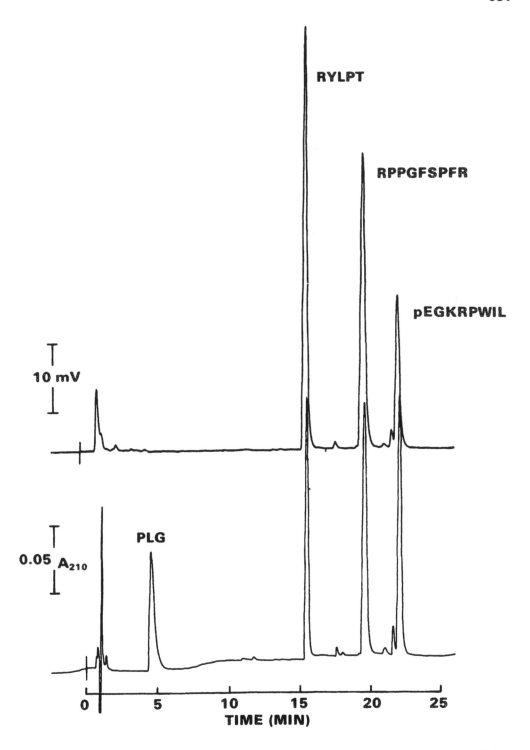

FIGURE 2. Dual detection of peptides with UV (bottom) and fluorescamine reaction (top) after reversed-phase separation. Column: Varian Micro Pak-SP-C$_{18}$, 150 × 2 mm ID, 3-μm particle size, 100-Å pore size. Conditions: Linear AB gradient (5% B to 45% B in 10 min and then to 60% B by 25 min), where Eluent A is 0.1% aq. trifluoroacetic acid (TFA) and Eluent B is 0.1% TFA in acetonitrile/H$_2$O (80:20); flow-rate, 0.4 ml/min. Peptides in order of elution are MSH-releasing inhibitor factor (PLG), proctolin (RYLPT), bradykinin (RPPGFSPFR) and xenopsin (pEGKRPWIL). (From Schlabach, T. D., *J. Chromatogr.*, 266, 427, 1983. With permission from Elsevier Science Publishers.)

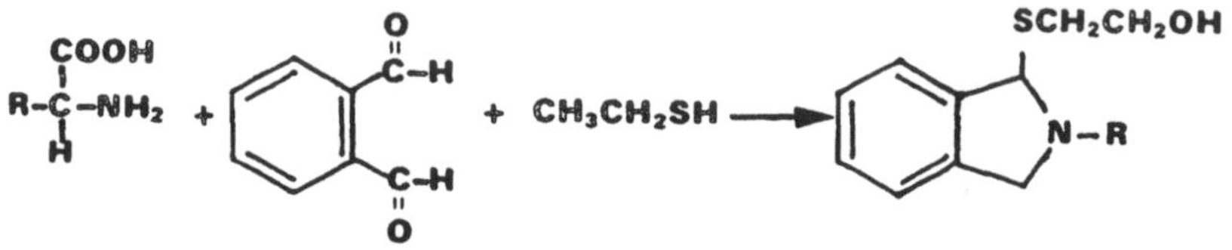

FIGURE 3. Reaction of O-phthalaldehyde with primary amines.

fluorescamine response has been shown to be linear to 1 μg of injected bovine serum albumin, and should provide sensitivity for detection of peptides at the low picomole level and proteins with normal lysine content at the low nanogram level.[6] An important feature of the fluorescamine chemistry is the fact that reaction with ammonia produces 1000-fold less fluorescence compared to primary amines. Hence the technique is much less sensitive to reagent and environmental contamination than the OPA system.

Reaction of fluorescamine with polypeptide amino groups is pH-dependent.[2,7] The N-terminal amino group exhibits maximum reactivity at lower pH, while the side chain amino group of lysine is more reactive at higher pH. This is evident from the data presented in Table 1, in which the fluorescence/absorbance (F/A) ratios of triglycine and serum thymic factor (which has the sequence pEAKSQYGGSN) are compared at several pH values[8]. Peptides were separated by reversed-phase chromatography and detected by absorbance at 210 nm followed in series by fluorescamine derivatization at the designated pH with fluorescence monitoring. The F/A ratio of triglycine (which possesses only a free N-terminus) decreases 25% as pH is increased from 8.4 to 10.4, while the F/A ratio of serum thymic factor (which has a reactive lysine side chain but an unreactive N-blocked pyroglutamic acid group as the amino terminus) increases 10-fold over the same range. This suggests that derivatization at pH ranges towards neutrality will be selective for N-terminal amino groups, while lysine will contribute to fluorescence only at alkaline pH values. The effect of amino acid sequence on fluorescamine reactivity is also demonstrated by dual absorbance/fluorescence detection in Figure 2 (fluorescamine reaction carried out at pH 9.0). MSH-releasing inhibitor factor (PLG) has proline as the N-terminus and hence yields no signal in the fluorescence trace (top panel); xenopsin (pEGKRPWIL) also has a blocked amino terminus, but a lysine in position 3 which is fluorescamine-reactive. From Figure 2, although equal amounts of RPPGFSPFR and pEGKRPWIL were applied to the column (according to the peak heights in the UV profile and the number of peptide bonds in the peptide), the greater peak height of the former in the fluorescent profile compared to the latter peptide reflects greater reactivity of an N-terminal group compared to the side-chain amino group of lysine.

B. POST COLUMN REACTION DETECTION USING *o*-PHTHALALDEHYDE (OPA)

Like fluorescamine, OPA forms a fluorescent product only with primary amino groups. In the reaction, OPA combines with the free amino group and stoichiometric amounts of a thiol such as mercaptoethanol to form a fluorescent isoindole (Figure 3).[9] Optimal sensitivity is achieved by excitation at 340 nm and fluorescence collection at 450 nm. The reaction is rapid, producing quantitative derivative yields at ambient temperature in about 30 sec. The OPA reagent is soluble in aqueous solution, and can be prepared directly in the borate reaction buffer (typically at a concentration of 0.05% in 1 M potassium borate, pH 10.4). OPA in solution is subject to oxidation, and the reagent should be prepared fresh daily. The OPA system is less complex than the fluorescamine system since only a single reagent pump and mixing tee is required. Column eluent and OPA reagent are mixed in a 1:1 to 1.5:1 ratio, with total flow of

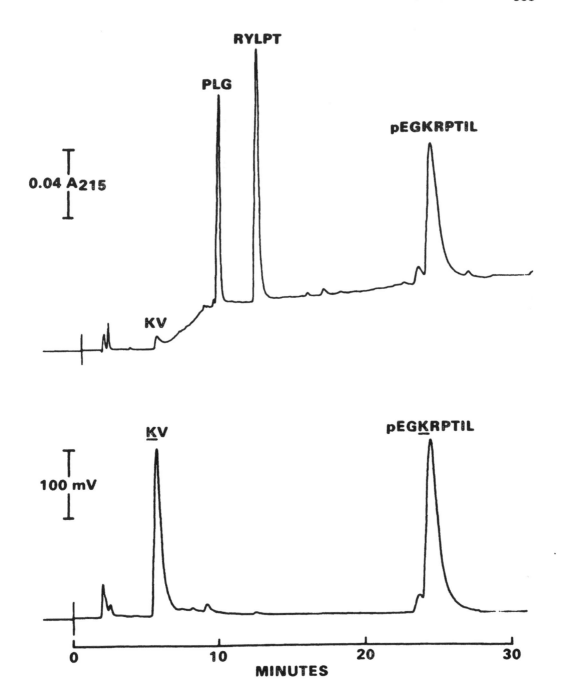

FIGURE 4. Dual detection of peptides with UV (top) and OPA reaction (bottom) after reversed-phase separation. Column: Varian Micro Pak-SP-C$_{18}$, 150 × 4.6 min ID, 3-μm particle size, 100-Å pore size. Conditions: linear AB gradient (0% B to 20% B in 5 min, up to 40% B by 30 min, and, finally, up to 60% B by 40 min), where Eluent A is 0.05% aq. TFA and Eluent B is 0.05% TFA in acetonitrile/water (90:10); flow-rate, 0.8 ml/min. Peptides in order of elution were lysyl valine, MSH-releasing inhibitor factor, proctolin and xenopsin. (From Schlabach, T. D. and Wehr, C. T., *Anal. Biochem.*, 127, 222, 1982. With permission from Academic Press.)

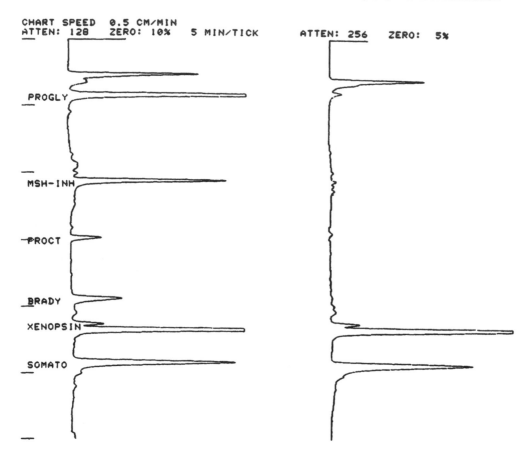

FIGURE 5. Comparison of peptide detection after reversed-phase separation using post column OPA reaction (upper chromatogram) and combined hypochlorite/OPA reaction (lower chromatogram). Column: Varian Micro Pak-SP-C$_{18}$, 150 × 4.6 mm ID, 3-μm particle size, 100-Å pore size. Conditions: linear AB gradient (2% B to 20% B in 5 min and then to 55% B by 20 min), where Eluent A is 0.1% aq. TFA and Eluent B is 0.1% TFA in acetonitrile/water (80:20); flow-rate, 0.7 ml/min. (From Schlabach, T. D., *J. Chromatogr.*, 266, 427, 1983. With permission from Elsevier Science Publishers.)

TABLE 1
Effect of pH on Detection of Peptides by Fluorescamine

Reagent pH	Peptide	Fluorescence (mV)	Absorption 210 nm	Fluorescence/ absorbance
10.4	GGG	127	0.170	747
10.4	pEAKSQGGSN	25.8	0.143	180
9.4	GGG	131	0.146	897
9.4	pEAKSQGGSN	16	0.143	112
8.4	GGG	129	0.129	1000
8.4	pEAKSQGGSN	3.6	0.187	19

Note: The absorbance at 210 nm is used as a measure of the quantity of peptide observed on the column. The fluorescence/absorbance ratio is a measure of the extent of reaction with fluorescamine.

From Schlabach, T. D., *J. Chromatogr.*, 266, 427, 1983. With permission from Elsevier Science Publishers.

Table 2
Fluorescence Response of Peptides in the OPA Post Column Reaction

Peptide sequence	Fluorescence (mV/nmol)
KV	180.0
YG	3.2
WG	3.0
GGG	12.0
PLG	1.0
TKPR	240.0
TPRK	210.0
YGGFM	3.3
KFIGLM	42.0
pEGPKRPWIL	190.0
pEHWSYGLRPG	2.5

From Schlabach, T. D. and Wehr, C. T., *Anal. Biochem.*, 127, 222, 1982. With permission from Academic Press.

about 1 ml/min. Reaction of OPA with ammonia produces a fluorescence signal comparable to that for primary amines, indicating that high reagent purity and reduced environmental contamination will be critical for high sensitivity OPA detection.

Although OPA reacts with the alpha amino group of free amino acids,[9] it does not react appreciably with the N-terminal amino group in polypeptides.[10,11] As seen in Table 2 and Figure 4, only peptides with lysine residues yield fluorescence response with OPA, suggesting that the lysine side chain amino is far more reactive than the N-terminus. In spite of this, the fluorescence intensity of the derivatized lysyl residues is sufficient to provide good sensitivity. However, OPA detection limits are in general inferior when compared to fluorescamine detection of polypeptides.

The OPA chemistry can be modified to make it selective for both lysine and proline residues.[12] This is accomplished by converting proline to an OPA-reactive species by oxidative decarboxylation. The reaction opens the proline ring to generate a primary amine which appears to be as reactive as the lysyl side chain amine. Hypochlorite (0.5% in 0.1 M sodium hydroxide) is the most commonly used reagent, with elevated temperature used to accelerate the reaction. An example of this approach is shown in Figure 5; note that peptides containing proline (prolylglycine, MSH releasing inhibitor factor, proctolin, and bradykinin) exhibit responses with the OPA/hypochlorite system (bottom panel) but not the OPA system alone (top panel). On-line proline oxidation requires a second post column pump to deliver the hypochlorite solution, and a heated reactor.

C. POST COLUMN REACTION SYSTEM COMPONENTS

The basic elements of a post column reaction system are the pump, mixing tee, and reactor. Inexpensive single-piston reciprocating pumps are satisfactory for this purpose and are available from a number of manufacturers, such as Milton Roy, Eldex, and Scientific Systems. The reagent pump should be capable of delivering flows up to 1 to 2 ml/min at operating pressures up to 1500 psi. Since both the fluorescamine and OPA reactions are quite rapid, open tubular reactors constructed from Teflon or stainless steel tubing are adequate. Delay times of 30 sec at total flows of 1 ml/min are provided by tubing lengths of 2.5 and 10 m for tubing diameters of 0.5 and 0.25 mm (0.02 and 0.01 in.), respectively.

In addition to the pump and reactor, several additional components are desirable for ease of use and reliability. A pressure gauge between the reagent pump and mixing tee is almost essential for diagnosing system problems such as PCR pump seal failure or loss of prime, and plugging or rupture of the reactor tubing. In addition, the Bourdon-tube type pressure gauge will act as a pulse damper when filled with fluid, improving system sensitivity. A bypass valve installed downstream from the pressure gauge will enable convenient system priming. Finally, a wash reservoir connected to a solvent selector valve installed at the PCR pump inlet will permit convenient flushout of the system after operation. Removing reagents from the system at shutdown, and periodic flushing of the reaction coil with a wash solvent will prevent buildup of crystalline deposits and help insure trouble-free operation. It should be noted that the reciprocating pumps usually employed for PCR reagent delivery are difficult to prime and prone to cavitation; this can be minimized by pressurizing reagent and wash reservoirs at 2 to 5 psi with nitrogen.

If recovery of intact sample following chromatography is required, a stream splitting device may be installed to divert a portion of column eluent to a fraction collector prior to addition of the post column reagent. A stream splitter employing a sampling valve has been described by Stein and Moschera[6] and by Lewis.[1] This approach requires an additional pump to deliver a makeup solvent during the sample collection cycle.

REFERENCES

1. **Lewis, R.V.,** Post-Column Fluorometric Detection, in *CRC Handbook of HPLC for the Separation of Amino Acids, Peptides, and Proteins*, W.B. Hancock, Ed., CRC Press, Boca Raton, pp. 193—196, 1984.
2. **Udenfriend, S., Stein, S., Bohlen, P., Doirman, W., Leimgruber, W., and Weigele, M.,** Fluorescamine: a reagent for assay of amino acids, peptides, proteins, and primary amines in the picomole range, *Science*, 178, 871, 1972.
3. **Weigele, M., DeBernardo, S.L., Tengi, J.P., and Leimbruber, W.,** A novel reagent for the fluorometric assay of primary amines, *J. Am. Chem. Soc.*, 94, 5927, 1972.
4. **DeBernardo, S., Weigele, M., Toome, V., Manhart, K., Leimgruber, W., Bohlen, P., Stein, S., and Udenfriend, S.,** Studies on the reaction of fluorescamine with primary amines, *Arch. Biochem., Biophys.*, 163, 390, 1974.
5. **Toome, V. and Manhart, K.,** A simple colorimetric determination of primary and secondary amines with fluorescamine, *Anal. Lett.*, 8, 441, 1975.
6. **Stein, S. and Moschera, J.,** High-performance liquid chromatography and picomole-level detection of peptides and proteins, *Methods. Enzymol.*, 79, 7, 1981.
7. **Nakai, N., Lai, C.Y., and Horecker, B.L.,** Use of fluorescamine in the chromatographic analysis of peptides and proteins, *Anal. Biochem.*, 58, 563, 1974.
8. **Schlabach, T.D.,** Dual-detector methods for selective identification of prolyl residues and amide-blocked N-terminal groups in chromatographically separated peptides, *J. Chromatogr.*, 266, 427, 1983.
9. **Roth, M.,** Fluorescence reaction for amino acids, *Anal. Chem.*, 43, 880, 1971.
10. **Joys, T.M. and Kim, H.,** O-phthalaldehyde and the fluorogenic detection of peptides, *Anal. Biochem.*, 94, 371, 1979.
11. **Schlabach, T.D. and Wehr, C.T.,** Fluorescent techniques for the selective detection of chromatographically separated peptides, *Anal. Biochem.*, 127, 222, 1982.
12. **Felix, A.M. and Terkelsen, .G.,** Total fluorometric amino acid analysis using fluorescamine, *Arch. Biochem. Biophys.*, 175, 177, 1973.

Section X
Analysis of Peptide and Protein Conformation by HPLC

HPLC ANALYSIS OF PROTEIN FOLDING

Albert Light

I. INTRODUCTION

In the last step in the biosynthesis of proteins, the completed polypeptide chain folds into the three-dimensional structure of a functional protein. The amino acid sequence dictates the folding process and accounts for the packing of the chain into a compact globular structure.[1] Present research efforts are directed toward elucidating the pathway of folding. To reach this goal, the *in vitro* refolding of proteins of known sequence and structure is a useful experimental system to understand the *in vivo* process and to explain how the informational content of the amino acid sequence dictates the folding pathway.

In the *in vitro* refolding of a well-characterized protein, the initial step is to unfold the molecule in the presence of denaturing agents; then renaturation is initiated on removal of denaturants. Refolding to the native protein is a spontaneous process.[1] Proteins containing disulfide bonds are fully reduced by reagents that specifically cleave disulfides. The restoration of native structure requires correct folding and pairing of half-cystine residues.[2-4] In refolding disulfide-containing proteins, intermediate species may be detected because they are stabilized by disulfide bonds between regions of the polypeptide chain.

The native structure is recognized from the change in properties of the molecule in undergoing the transition from the denatured to the refolded state. If an enzyme, or a precursor to an enzyme, is studied, the restoration of enzymatic activity serves as a powerful probe of structure. Changes in circular dichroism, hydrodynamic volume, isoelectric point, resistance to proteolysis, NMR properties, immunological identity, and other measures of structure are used when possible. Clearly, studies with more than one probe provide useful information on the secondary and tertiary structures of the folded protein. The restoration of native structure is conveniently followed kinetically and the properties of the final structure are related to the native molecule.

In the experimental studies described in this chapter, the proteins that were refolded are members of the serine protease family. These are the pancreatic proteolytic enzymes which are well-characterized molecules — the enzymatic properties, amino acid sequence, and three-dimensional structures are known.[5-7] In this family, bovine chymotrypsin, trypsin, and elastase are the best characterized proteins. They have identical mechanisms of enzyme action, homologous amino acid sequences, and essentially the same three-dimensional structures. However, trypsin contains six disulfide bridges, chymotrypsin five, and elastase four.[5] These differences influence the stability of intermediate species, and comparative studies of the folding process reveal similarities as well as differences.[4] In refolding serine proteases, the inactive zymogens, trypsinogen and chymotrypsinogen, were refolded since they were enzymatically inactive and autolysis was avoided.[4,8,9] Once the renatured molecules regained structure, the zymogens were activated and the enzyme activity measured.[8]

We followed the refolding process using size-exclusion high performance liquid chromatography (SEC).[10] SEC separates molecules differing in size, based on hydrodynamic volumes (Stokes radius). Although trypsinogen does not change size during folding, the hydrodynamic volume undergoes large changes. The unfolded molecule has a large Stokes radius because it occupies a large volume. As the polypeptide chain folds and forms secondary structure, it becomes more compact and the Stokes radius decreases. Indeed, the Stokes radius continually decreases as the folding progresses until the compact stable native structure is reached. Therefore, the order of elution of components during SEC is directly related to the order of their formation in the folding process.[10] SEC is almost unique as a separation tool in providing such useful information. Unfortunately, the resolution from SEC is not sufficiently high to resolve the many intermediates present at any one time during the folding process. It is therefore necessary to collect a series of sequential fractions and to rechromatograph each.[10] On rechromatography using the same conditions as in the initial separation, each fraction had the same retention time as it had in the first separation. Therefore, these fractions represented stable species of a given conformation and with a unique Stokes radius. Each fraction was a partly refolded molecule on the folding pathway.

II. EXPERIMENTAL

A. PREPARATION OF MIXED DISULFIDE OF FULLY REDUCED TRYPSINOGEN

Trypsinogen (15 mg/ml), in 6 M guanidine hydrochloride (Gdn.HCl), at pH 8.5, was reduced with 0.01 M dithioerythritol (or dithiothritol) for 2 hr at 37°C under a nitrogen atmosphere.[4,8,10] The high concentration of denaturant unfolded the protein and the reducing agent converted all cystine residues to half-cystines. The reduced protein and a large excess of gluthathione disulfide (0.1 M final concentration), in 6 M Gdn.HCl, were kept at pH 8.6, at a final protein concentration of 1.5 mg/ml, for 3 hr at room temperature, under a nitrogen atmosphere. The glutathione disulfide reacted with thiols forming a glutathione-protein mixed disulfide (Protein-S-SG) in high yield. The protein derivative was separated from small molecules on chromatography on a Sephadex G-25 column equilibrated with 0.15 M acetic acid.

B. REFOLDING OF TRYPSINOGEN-S-SG

Trypsinogen-S-SG (20 to 30 μg/ml) dissolved in 0.05 M Tris, pH 8.6, containing 3 mM cysteine, and 1 mM cystine, was kept at 4°C under strictly anaerobic conditions.[10] The mixed disulfide was soluble at these dilute concentrations, and the derivative prevented aggregation and precipitation.[4] Cysteine was the disulfide-interchange catalyst, making and breaking disulfides during folding.[8] If insolubility of an unfolded protein is a problem, 0.25 M Gdn.HCl can be added to improve the solubility of partially folded structures.[9] The temperature of 4°C decreased the rate of folding, increased the yield, compared to 37°C, and made it relatively easy to monitor the folding with time. Samples removed at timed intervals were immediately quenched with 2 M iodoacetate (25 × molar excess over the total thiol concentration) at pH 8.6, for 1 hr, at 25°C. The large excess of iodoacetate reacted within seconds with free thiols inhibiting further disulfide interchange.[11] The folded samples were dialyzed, lyophilized, and kept at –20°C until analysis on HPLC.

C. HPLC INSTRUMENTATION

HPLC analyses were performed on a Varian (Walnut Creek, CA) 5020 HPLC system.

The columns used were Toya Soda G-2000SW, (500 × 7.5 mm I.D.), with two columns connected in series for separation of intermediate species. The separations were at ambient temperatures, and the effluent was continuously monitored at 280 nm with a flow-through chromatography cell in a Gilson Model HM Holochrome detector.

III. SIZE-EXCLUSION CHROMATOGRAPHY

A. HPLC ANALYSIS OF REFOLDING

The size-exclusion columns were equilibrated and the proteins eluted at 0.5 ml/min with 0.1

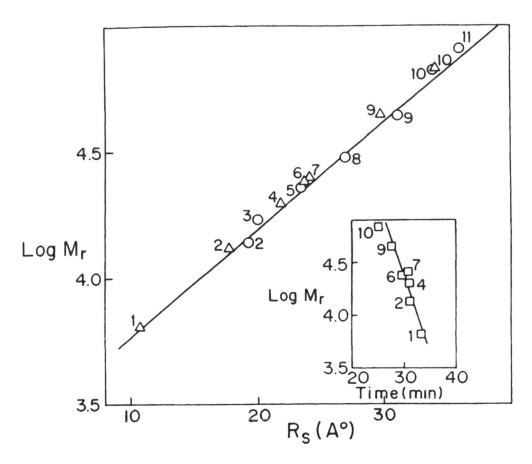

FIGURE 1. The linear relationship between the log molecular weight (Mr) and Stokes radius (Rs) and the log molecular weight and the retention time (inset). The samples are (1) bovine pancreatic trypsin inhibitor, (2) bovine ribonuclease, (3) sperm whale myoglobin, (4) soybean trypsin inhibitor, (5) carbonic anhydrase, (6) bovine trypsinogen, (7) bovine chymotrypsinogen, (8) β-lactoglobulin, (9) ovalbumin, (10) bovine serum albumin, and (11) transferrin.

M ammonium formate, pH 4.0, containing 2 M urea. The low concentration of urea improved the recovery of partially refolded intermediates. In refolding trypsinogen-S-SG, early intermediates were stable in 2 M urea as shown by finding a constant retention time on repeated rechromatography.[12]

Sequential fractions of equal volume (approximately 0.5 ml) were collected manually from the initial separation. After dialysis and lyophilization, these fractions were rechromatographed on the columns at the same flow-rate with the same buffer. The retention time of separated components was recorded from the time of injection to the maximum concentration of the peak. The retention time was related to a standard curve of log Mr vs. retention time for a set of standard proteins, and the log Mr was used to estimate the Stokes radius (Rs) from known values for these proteins (Figure 1).

The TSK G2000SW columns lacked the resolution to separate directly intermediate species (Figure 2). However, the presence of intermediates was obvious from the shape of the effluent curve, which was broad and occasionally had a shoulder. Rechromatography of a series of sequential fractions collected in the initial separation gave the elution time and relative yield of the principal intermediates (Figure 3). Essentially, this experimental approach deconvoluted the peak and provided useful information on the properties of the intermediates and how they changed with the time of refolding (Figure 4).

HPLC separated intermediates because they were stabilized by disulfide bonds. No further changes in conformation (Stokes radius) were detected during rechromatography of the

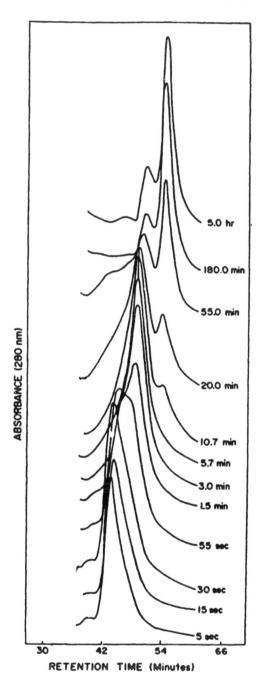

FIGURE 2. The time course for folding the mixed disulfide of trypsinogen. Samples quenched as a function of time were separated on two TSK G2000SW columns connected in series. The columns were protected with a GSWP guard column. Mobile phase: 0.1 M ammonium formate (pH 4.0) containing 2 M urea; flow-rate, 0.5 ml/min.; ambient temperature; absorbance at 280 nm. The mixed disulfide derivative had a Stokes radius of 33.2 Å (at 5 sec) and the native trypsinogen was 22.4 Å (at 5 hr).

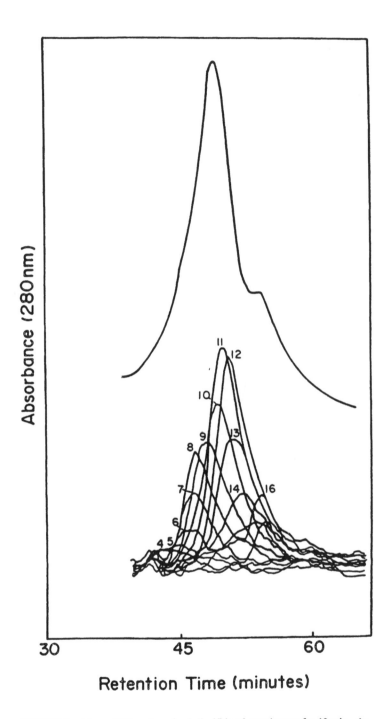

FIGURE 3. After refolding the mixed disulfide of trypsinogen for 10 min., the quenched sample was separated by SEC (upper curve) (conditions as described in Figure 2). Sequential fractions were collected and rechromatographed and shown as overlapping peaks (lower curves). The numbers refer to sequential fractions collected from the initial separation.

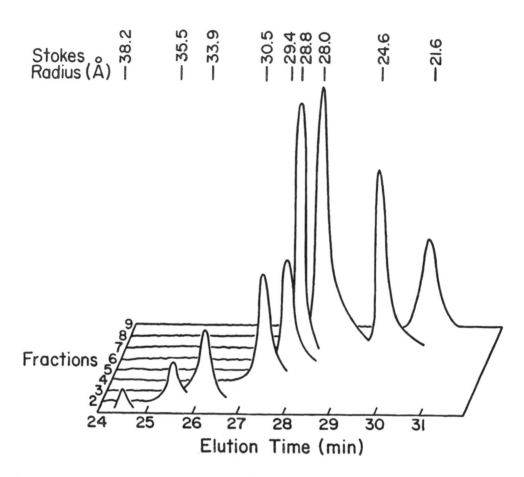

FIGURE 4.   A representation of the principal intermediates detected at 10 min folding of the mixed disulfide of trypsinogen. The area of the peaks is the absorbancy at 280 nm and the Stokes radius was estimated from the plot shown in Figure 1.

fractions; the elution time was the same on repeated chromatography.[12] Each fraction was unique and had a specific Stokes radius (Table 1).

At a 10-min refolding time, we observed intermediate species ranging from loose globular structures (large Stokes radius), to partly folded structures, to molecules that were close to the native molecule (Figure 4). Native protein molecules were present in low yield. Clearly, some molecules of denatured trypsinogen-S-SG folded rapidly on one pathway while other molecules followed alternate pathways that were slower.

At longer folding times, fewer intermediates were present and the yield of native molecules (Figure 5, fractions 6, 7, 8, and 9) increased. If the refolding time was increased to 8 hr, few intermediates were detected and the yield of native trypsinogen was close to 60%, based on the amount of active trypsinogen used in the experiment. We also observed some intermediates that were dead-end structures, molecules trapped in stable conformations or molecules covalently modified during the folding process.

The detection of intermediates with different Stokes radii suggested that folding followed a multiple pathway before the globular structures formed.[13] Apparently, a large number of pathways and early intermediates were initially possible (Table 1). The partly folded structures had fewer and fewer pathways to follow as folding continued (Figure 6). An almost native-like structure accumulated (R_s of 29.2 Å) as the immediate precursor of the native protein (Table 1). Apparently, the last step in the formation of the native protein was the rate-limiting step.

TABLE 1
Characterization of the Principal
Intermediate Species by
Size-Exclusion Chromatography[a]

Time interval	Hydrodynamic volume[b] (apparent Rs)
0	33.9 Å
0 to 5 sec	33.2 Å
15 to 30 sec	33.0 Å
55 to 90 sec	30.8 Å
3 to 10 min	29.2 Å
10 min to 5 hr	22.4 Å

[a] Trypsinogen-S-SG was folded at a concentration of 20µg/ml at pH 8.6, in the presence of 2 M urea, 3 mM cysteine, 1 mM cystine, at 4°C, under a nitrogen atmosphere. Samples were removed at timed intervals, quenched with 2 M iodoacetate, dialyzed, lyophilized, and analyzed by size-exclusion chromatography on columns of TSK G2000SW.

[b] The hydrodynamic volume is given as the Stokes radius (Rs) in angstroms. The Rs was estimated from a calibration curve of proteins of known Rs plotted against their retention times (Figure 1).

IV. SUMMARY

SEC is an important analytical tool to study the refolding of disulfide-containing proteins. The methodology developed in these studies involved rechromatography of a series of sequential fractions collected in an initial separation. The rechromatography identified intermediate species that ranged from partly folded molecules to the structure of the native protein. The hydrodynamic volume and amount of the intermediates present as a function of time are necessary for a kinetic analyses. The rate of formation and disappearance of intermediates are helpful in proposing a folding pathway for the renaturation process. The intermediates are stable structures because of disulfide bonds, and the principal intermediates can be further characterized by still other analytical techniques and by physical-chemical measurements.

ACKNOWLEDGMENTS

This investigation was supported partly by funds from the Army Research Office, the Immunex Corp. of Seattle, Washington, and the Department of Chemistry, Purdue University. The author is indebted to the excellent work of J. N. Higaki and A. M. Al-Obeidi whose contributions are cited in this chapter. Judith Heine has been most helpful in the preparation of the manuscript and preparing the final copy.

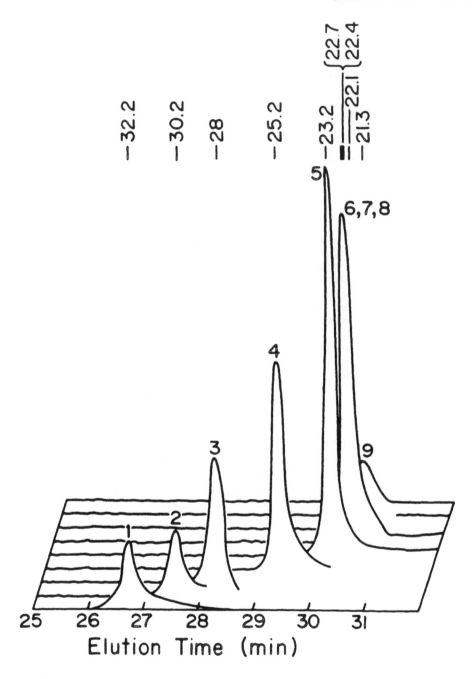

FIGURE 5. A representation of the principal intermediates detected at 2 hr folding of the mixed disulfide of trypsinogen. See Figure 4 for further details.

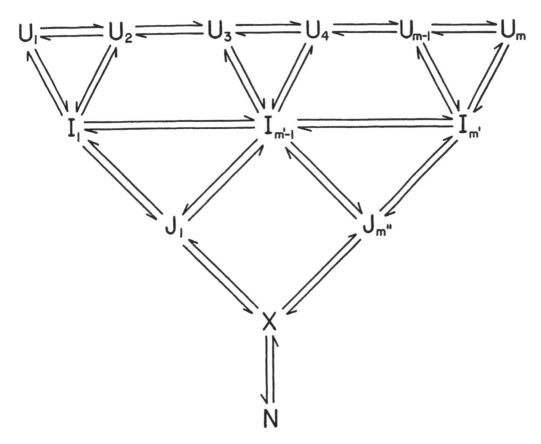

FIGURE 6. A proposed folding pathway for serine proteases. Initially, a large number of loose structures (large Rs) were formed (U_1 to U_m) that fold further to J structures then form a near native molecule (X), which changes in a two-state transition to the native protein (N).

REFERENCES

1. **Anfinsen, C.B.**, Principles that govern the folding of protein chains, *Science,* 181, 223, 1973.
2. **Anfinsen, C.B. and Haber, E.**, Studies on the reduction and reformation of protein disulfide bonds, *J. Biol. Chem.,* 223, 1361, 1969.
3. **Creighton, T.E.**, Experimental studies of protein folding and unfolding, *Prog. Biophys. Mol. Biol.,* 33, 231, 1978.
4. **Light, A.**, Protein solubility, protein modifications and protein folding, *BioTechniques,* 3, 298, 1985.
5. **Dayhoff, M.D.**, *Atlas of Protein Sequence and Structure,* Vol. 5, National Biomedical Research Foundation, Silver Springs, MD, 1972.
6. **Bode, W., Fehlhammer, H., and Huber, R.**, Crystal structure of bovine trypsinogen at 1.8 Å resolution. I. Data collection, application of Patterson search techniques and preliminary structural interpretation, *J. Mol. Biol.,* 106, 325, 1976.
7. **Keil, B.**, *The Enzymes,* Vol. 3, Academic Press, New York, 249, 1971.
8. **Odorzynski, T.W. and Light, A.**, Refolding of the mixed disulfide of bovine trypsinogen and glutathione, *J. Biol. Chem.,* 254, 4291, 1979.
9. **Duda, C.T. and Light, A.**, Refolding of bovine threonine-neochymotrypsinogen, *J. Biol. Chem.,* 257, 9866, 1982.

10. **Light, A. and Higaki, J.N.,** Detection of intermediate species in the refolding of bovine trypsinogen, *Biochemistry,* 26, 5556, 1987.
11. **Creighton, T. E.,** Disulfide bond formation in proteins, *Methods Enzymol.,* 107, 305, 1984.
12. **Al-Obeidi, A.M. and Light, A.,** Size-exclusion high-performance liquid chromatography of native trypsinogen, the denatured protein, and partially refolded molecules, *J. Biol. Chem.,* 263, 8642, 1988.
13. **Harrison, S.C. and Durbin, R.,** Is there a single pathway for the folding of a polypeptide chain?, *Proc. Natl. Acad. Sci. U.S.A.,* 82, 4028, 1985.

ANALYSIS OF PROTEIN FOLDING BY SIZE-EXCLUSION CHROMATOGRAPHY

Earle Stellwagen and William Shalongo

I. PRINCIPLES AND PRACTICE

The change in hydrodynamic volume accompanying the reversible folding of a protein into its native structure can be easily observed by high-performance size-exclusion chromatography (SEC) as a change in the elution time of the protein. Experimental conditions can usually be identified so that the dynamics of the folding process can be observed. Inspection of such profiles can identify the participation of intermediate structures in the folding process while analysis by computer simulation can establish the refolding mechanism and the equilibrium and kinetic parameters of each component in the mechanism.

Three kinds of profiles are necessary for a complete analysis. Folding profiles are obtained by injection of an unfolded protein in a SEC column equilibrated with buffer or a low concentration of a protein denaturant. Unfolding profiles are obtained by injection of a native protein in a column equilibrated with a high concentration of denaturant. Equilibrium profiles are obtained by equilibration of both the protein and the column with the same concentration of denaturant prior to injection. Typical chromatographic profiles observed using the protein ribonuclease and the denaturant guanidine hydrochloride (GuHCl) are illustrated in Figure 1.

It is useful to select a SEC column for these measurements having a pore size such that both the native and unfolded forms of the protein of interest are fractionally included. Each chromatographic measurement is obtained at a constant denaturant concentration, pH, temperature, and flow-rate with the denaturant concentration the variable between measurements. Any protein denaturant which does not damage the matrix may be employed. Since the product of the observed peak width and flow-rate is independent of flow-rate over the range 0.2 to 1.0 ml/min, it is experimentally advantageous to use the fastest flow-rate recommended by the manufacturer. Each profile should be monitored by an absorbance flow detector set at a wavelength at which the native and unfolded forms of the protein of interest have the same extinction coefficient, i.e., an isosbestic wavelength. Additional flow detectors may be added in series to monitor parameters unique to individual structures such as circular dichroism, fluorescence or biofunction.

II. EQUILIBRIUM PROFILES

The effect of guanidine hydrochloride (GuHCl) on the equilibrium profile of a small protein containing a single polypeptide chain is illustrated in Figure 1. The protein is eluted as a single nearly symmetrical peak in a low [GuHCl] in which the native protein is the dominant structure, panel A, and in a high [GuHCl] in which the unfolded protein is the dominant structure, panel G. The unfolded protein is eluted before the native protein since unfolding increases the hydrodynamic volume of a globular protein.

The dependence of the elution time of several proteins on the [GuHCl] is illustrated in Figure 2. The proteins include a persistent globular protein, pancreatic trypsin inhibitor, a persistent denatured protein, performic acid oxidized ribonuclease, and a globular protein displaying a reversible unfolding transition, ribonuclease. Surprisingly, the elution times for each protein

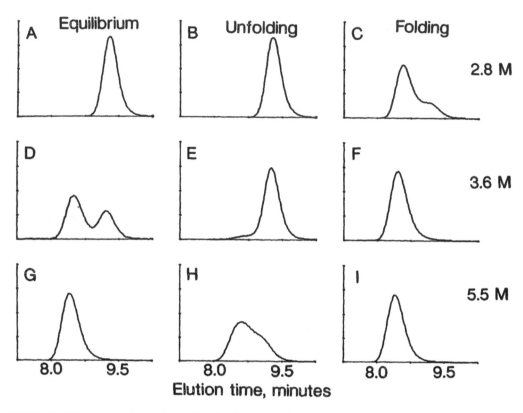

FIGURE 1. Chromatographic elution profiles for ribonuclease. Column: Toyo Soda G2000 SW, 300 × 7.5 mm ID, containing 300 plates effective for proteins; flow-rate, 1 ml/min; absorbance at 225 nm. All profiles were obtained at 4°C using solvents containing 50 mM cacodylate buffer, pH 6.0 , and the following concentrations of GuHCl: panels A to C, 2.8 M; panels D to F, 3.6 M; and panels G-I, 5.5 M. The ordinate is the absorbance at 225 nm expressed in millivolts and is scaled identically in each panel. Panels A, D, and G are equilibrium profiles; panels B, E, and H are unfolding profiles; and panels C, F, and I are folding profiles.

display a common curvilinear dependence on [GuHCl]. Modeling this curvilinear dependence as well as peak shape suggests that the native and unfolded proteins both bind weakly and reversibly with the chromatographic matrix. Additional modeling indicates that the extent and kinetics of this binding does not significantly perturb the analysis of the protein structural change described below.

The equilibrium profile obtained in the transition zone is shown in panel D of Figure 1. The transition zone is defined as the range of denaturant concentrations in which both the native and unfolded structures are significantly populated at equilibrium. The shape of an equilibrium profile indicates the exchange time between the native and unfolded structures. The exchange time is defined as the reciprocal sum of the rates of folding and of unfolding. The equilibrium profile illustrated in panel D has two peaks flanking a well defined valley. Such a profile is characteristic for slow exchange, which is greater than 1000 sec for the measurement illustrated. It is advantageous to place a protein in slow exchange in the transition zone, in order to extend the kinetic analysis of folding over a wide range of denaturant concentrations. Most proteins can be placed in slow exchange in the transition zone by simply lowering the temperature of the SEC column to 4°C.

Inspection of the equilibrium profiles of a protein in slow exchange, such as illustrated in Figure 1D, will provide an estimate of the midpoint and width of the structural change. The midpoint is defined as the [GuHCl] at which the native and unfolded structures are equally populated while the width is defined as the change in [GuHCl] required to change the relative

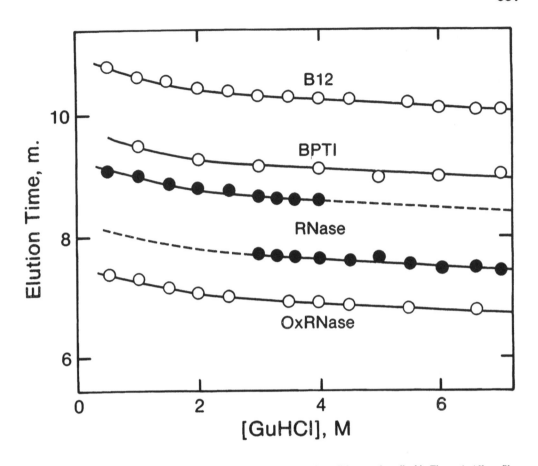

FIGURE 2. Dependence of elution time on the [GuHCl]. Column and conditions as described in Figure 1. All profiles were obtained at 4°C using solvents containing 50 mM cacodylate buffer, pH 6.0. The elution times for vitamin B12; bovine pancreatic trypsin inhibitor, BPTI; and oxidized ribonuclease, OxRNase, are indicated by open circles. The elution times for native and unfolded ribonuclease, RNase, are indicated by filled circles. The elution time for the blue dextran was 5.44 ± 0.02 min and for water, observed as a refractive discontinuity, was 10.32 ± 0.04 min in each of these solvents.

population of a structure from 10 to 90%. For example, the structural change labeled N/UF in Figure 3A has a midpoint at 3 M and a width of 1 M.

III. KINETIC PROFILES

The presence of transiently stable structures involved in folding and unfolding can often be detected by inspection of folding profiles. This can be most simply illustrated using the four-state reaction mechanism shown in Equation 1, which is representative of the refolding of proteins in which a cis/trans peptide isomerization involving a proline residue occurs in the unfolding of the native protein, N, which rapidly generates UF, an unfolded protein containing all-native peptide isomers. The cis/trans isomerization of a single peptide bond in UF populates US, an unfolded protein containing a single non-native peptide isomer. In moderate [GuHCl], UF folds rapidly while US folds slowly since the folding of US must be preceded by a slow peptide isomerization to UF. However, in low [GuHCl], a folding intermediate containing the non-native peptide bond, IS, can be transiently populated. Population of this intermediate facilitates folding prior to peptide isomerization.

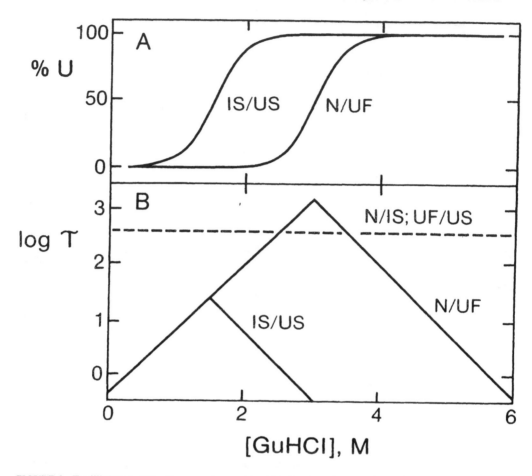

FIGURE 3. Equilibrium and kinetic parameters assumed for the two-, three-, and four-state mechanisms. Panel A illustrates the transition curves for each of the two structural changes expressed as the percentage of the protein in the unfolded form at equilibrium. Panel B is a semilogarithmic plot of the dependence of the exchange times on the concentration of denaturant. The dependence of the exchange times for the N/UF and the IS/US structural changes on denaturant concentration are related as if IS were an equilibrium mutant of N. The exchange time for peptide isomerization is assumed to be independent of denaturant concentration and to have a value of 400 sec in both the native and unfolded structures. At equilibrium, the N/IS ratio is 1000 while the UF/US ratio is 1.

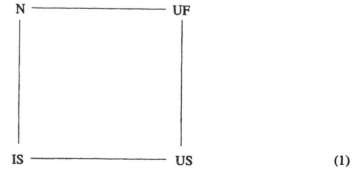

$$(1)$$

Figure 4 illustrates folding profiles which can be used to distinguish a simple two-state mechanism, N-UF, a three-state mechanism not involving a transient intermediate, N-UF-US, and the four-state mechanism given in Equation 1. These profiles were constructed using the equilibrium and kinetic parameters for each reaction which are given in Figure 3. In [GuHCl]

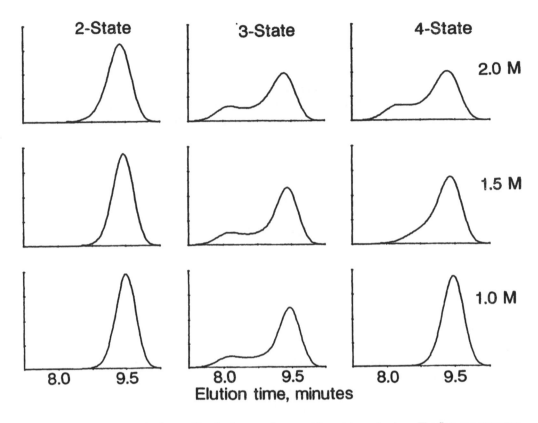

FIGURE 4. Representative folding profiles for the two-, three-, and four-state mechanisms. The first row represents folding profiles observed in 2.0 *M* GuHCl, the second row in 1.5 *M* GuHCl, and the third row in 1.0 *M* GuHCl. The left column represents the two-state mechanism, the middle column the three-state mechanism, and the right column the four-state mechanism.

of 2 *M* or less, all the protein in the two-state mechanism appears to have refolded to N since the folded protein is stable and folding is rapid. By contrast, a significant fraction of the protein in the three state mechanism appears to be persistently unfolded. This persistently unfolded protein represents US whose folding is limited by the slow peptide isomerization. The three- and four-state mechanisms display similar folding profiles in 2 *M* GuHCl. However, in lower [GuHCl], the transient intermediate IS becomes populated and the persistently unfolded US observed in the three-state mechanism folds to IS in the four-state mechanism.

IV. QUANTITATIVE ANALYSIS BY SIMULATION

An expandable menu-driven computer program has been written to simulate folding, unfolding, or equilibrium profiles based on the plate theory transfer equations of Endo and associates.[1,2] The program first requires coupling a desired number of structural changes and peptide isomerizations into an explicit mechanism. The reaction parameters, which include the midpoint, transition width and exchange time for each peptide isomerization, the composition of the sample injected, and the [GuHCl] in the isocratic chromatographic solvent, are then specified. Finally, the chromatographic parameters, which include the volume and protein concentration in the sample injected, the chromatographic flow-rate, the number of effective chromatographic plates, the elution time for each component in the mechanism and the elution time for an excluded molecule such as blue dextran and a small included molecule such as sodium dichromate, are specified. Simulation of a simple two-state mechanism, such as N-UF, with a Vax 11/780 computer requires less than a minute while simulation of a seven component mechanism requires several minutes. The number of effective plates is determined by fitting an equilibrium profile in a [GuHCl] distant from a structural change with a two-state transition in rapid exchange using the plate number as the variable.

It is most convenient to display simulated and observed profiles on a graphics terminal and to adjust the parameters until the two profiles are superimposed. Small changes in the specified parameters, e.g., a 0.1 *M* variation in the midpoint value for a structural change, a 25% variation in an exchange time or a 0.05 min change in elution time, are visually evident. Exchange times ranging from 50 to 10,000 sec, which encompasses most protein structural changes at 4°C, can be determined by simulation.

Considering the number of parameters to be specified, any given profile could be simulated equally well by numerous combinations of mechanisms and parameters. In order to obtain a unique solution, it is necessary to simulate the folding, unfolding, and equilibrium profiles obtained at different [GuHCl] using a common minimal mechanism and a consistent set of chemical and chromatographic parameters. The number of effective theoretical plates, the midpoint and width of each structural change and the equilibrium distribution, and exchange time for each peptide isomerization must be the same in all simulations. The elution times employed must construct a regular curvilinear denaturant dependence on [GuHCl] as illustrated by the solid and dashed lines for ribonuclease in Figure 2. The exchange times for each structural change must display an inverted triangular dependence on the [GuHCl] with the apex for each triangle at the midpoint of the equilibrium transition as illustrated in Figure 3. These constraints severely limit the number of mechanisms which can simulate a given set of SEC profiles.

The SEC profiles for thioredoxin[3] and ribonuclease A[4] have each been analyzed by simulation using a seven component reaction mechanism including two transiently stable intermediate conformations. The parameters used in these simulations predict the equilibrium and kinetic results obtained from simultaneous analysis using spectral measurements. The SEC measurements extend the spectral measurements by defining the hydrodynamic volume and stability of each component in the reaction mechanism. Surprisingly, the transient population of folding intermediates having hydrodynamic volumes distinct from those of the native or the unfolded protein has not been observed in the folding of either thioredoxin or ribonuclease. However, such intermediates may be populated during the folding of other proteins, particularly those having more than one domain or those possessing a quaternary structure.

REFERENCES

1. **Endo, S., Hayachi, H., and Wada, A.,** Affinity chromatography without immobilized ligands; a new method for studying macromolecular interactions using high-performance liquid chromatography, *Anal. Biochem.*, 124, 372, 1982.
2. **Endo, S., Saito, Y., and Wada, A.,** Denaturant-gradient chromatography for the study of protein denaturation: principle and procedure, *Anal. Biochem.*, 131, 109, 1983.
3. **Shalongo, W., Ledger, R., Jagannadham, M. V., and Stellwagen, E.,** Refolding of denatured thioredoxin observed by size-exclusion chromatography, *Biochemistry* 26, 3155, 1987.
4. **Shalongo, W., Jagannadham, M. V., Flynn, C., and Stellwagen, E.,** Refolding of denatured ribonuclease observed by size-exclusion chromatography, *Biochemisry*, 28, 4820, 1989.

COMPLEMENTARY APPLICATION OF SIZE-EXCLUSION CHROMATOGRAPHY AND CIRCULAR DICHROISM TO THE STUDY OF PROTEIN FOLDING

Wayne J. Becktel and Margaret A. Lindorfer

I. INTRODUCTION

Two techniques which have been used in the characterization of the native and denatured states of proteins are size-exclusion chromatography (SEC) and circular dichroism spectroscopy (CD).[1,2] Both techniques are, by themselves, useful in studying the properties of proteins because of their precision, need for relatively small amounts of protein for each experiment, and ease of use. Joint application of CD and SEC provides the additional advantage of measuring completely different physical properties. SEC monitors the radius of gyration or mean end-to-end distance of the macromolecule.[3] CD, in the peptide backbone region (190 to 250 nm), detects retention of secondary structure such as the α-helix.

Use of more than one physical parameter is also valuable in studies of the transition from folded to unfolded conformations. Lumry and co-workers proposed that agreement between more than one measure of thermal denaturation, where the techniques monitor different facets of protein conformation, was one way of establishing the existence of simple, two-state thermal transitions.[4] This concept may also be used for solvent denaturation.

It is the intent of this article to provide the reader with examples of the utility of joint application of SEC and CD to the study of protein stability. The examples given are drawn from the literature and from our own recent studies. The reader is referred to Ackers,[5] Pfannkoch et al.[6] and Yau et al.[3] for comprehensive theoretical treatments of SEC of proteins, and to Pace for a review of solvent denaturation.[7] This article will first describe the experimental conditions and calibrations involved in the use of SEC in solvent denaturation. Urea denaturation studies of myoglobin and T4 lysozymes in high and low salt by both CD and SEC will also be described.

II. MEASUREMENT TECHNIQUES

CD measurements were made as previously reported.[8,9] Stock protein solutions were diluted with the desired urea concentration to a final protein concentration of 0.02 mg/ml. CD spectra were measured with a Jasco J500-C in a 10 mm path cell from 220 to 230 nm. The UV absorption of urea prevented CD measurement below 220 nm.

SEC was performed on a Beckman System Gold HPLC with a Model 166 detector at 230 nm. Twenty to forty micrograms of protein were injected in two to twenty microliters of 10 mM potassium phosphate, pH 3, and 0.15 M KCl. No difference in elution volume was observed when protein solutions were first equilibrated in urea before SEC. The column was a TSK 3000 SW (7.5×300 mm) with a TSK SW guard column (7.5×75 mm) (Phenomenex). Flow-rate was normally 0.4 ml/min. At this flow-rate, the back pressure varied from 200 to 280 psi, depending upon urea concentration. We chose these chromatographic conditions because: (1) the resolution at 0.4 ml/min was a compromise between adequate resolution and practical time constraints,

(2) at higher flow-rates and pressures, the effect of pressure on the partial excluded volume of the macromolecule and the buffer can have significant effects such as changing the pH of phosphate solutions,[10] and (3) thermodynamic properties of proteins are much more sensitive to changes in pressure within the region of a conformational transition than outside those regions.

Eight molar urea (Amresco) stock solutions were prepared, deionized on AG 501-X3 mixed bed ion exchange resin (BioRad), diluted to desired concentrations with 10 mM potassium phosphate buffer, pH 3, and 0.15 or 0.50 M KCl. Solutions were filtered (Gelman, FP Vericel 200, 0.2 µm) immediately before use and sparged with helium during use.

Elution times were reproducible to within 0.2% (± 0.05 min/25 min). In fact, this criterion was employed to judge pump performance. When replicate, consecutive injections failed this test, the necessary steps (pump priming, check valve cleaning, or replacement) were taken to restore performance. The precision of the elution times compare well with that of CD spectroscopy.

Corbett and Roche[1] describe in detail the determination of protein molecular weight calibration curves for TSK columns. By using a set of standard proteins whose intrinsic viscosity is known, one can determine the dependence of the Stokes radius on elution volume under the solvent conditions (pH, ionic strength, temperature) of interest.[1,5] Corbett and Roche found a salt concentration of 0.2 was sufficient to suppress ionic interactions between the proteins and stationary phase. Pfannkoch et al.[6] and Unger et al.[11] have determined the ionic and solvophobic interaction of commercially available SEC columns.

III. UREA DENATURATION IN LOW AND HIGH SALT

Urea denaturation of myoglobin, as measured by both CD and SEC was reported by Corbett and Roche.[1] They noted that, as the urea concentration in the mobile phase increased, early and late elution volume peaks were observed. The ratio of these peak areas constitute the equilibrium constant of the protein unfolding process. These authors showed that reasonably linear plots of ΔG vs. [urea] can be constructed and $\Delta G°$ (the extrapolated value of ΔG at zero urea) determined. The C_m (urea concentration where the concentrations of unfolded and folded protein are equal) determined by SEC agrees well with CD measurements. However, whereas the CD signal shows no change prior to the transition, the SEC measurements show a gradual increase in elution volume prior to the transition.

Urea denaturation was carried out for T4 lysozyme at pH 3 in 0.15 and 0.50 M KCl. A pH of 3 was chosen because WT T4 lysozyme is fully unfolded in 6 M urea and because ionic interactions with the stationary phase are minimized. Figure 1 shows several of the chromatograms for wild type T4 lysozyme as a function of urea concentration. Before the transition, the elution volume decreases gradually with increasing urea concentration. There is a slight peak broadening through the transition, but no evidence of resolution of folded and unfolded species. Decreasing the flow-rate by one half to 0.2 ml/min did not appreciably change the peak shape. In contrast to myoglobin, the unfolding and refolding rates for T4 lysozyme under these conditions are fast relative to the chromatography time scale, so the equilibrium "average" is observed. Beyond the transition range, the denatured elution volume continues to decrease, in a non-cooperative manner, with increases in denaturant concentration.

Denaturation by urea of T4 lysozyme, as observed by means of CD in the peptide backbone region of the spectrum, is also sigmoidal in shape. Comparison of urea denaturation, as measured by CD and SEC can be made more quantitative by first converting the individual dichroism or elution volume to fraction of folded protein and then calculating the apparent equilibrium constants for each to determine the free energy of unfolding (ΔG). The procedures for doing this are well known.[6-9] The results, in 0.15 M KCl, are shown in Figure 2A. As the figure indicates,

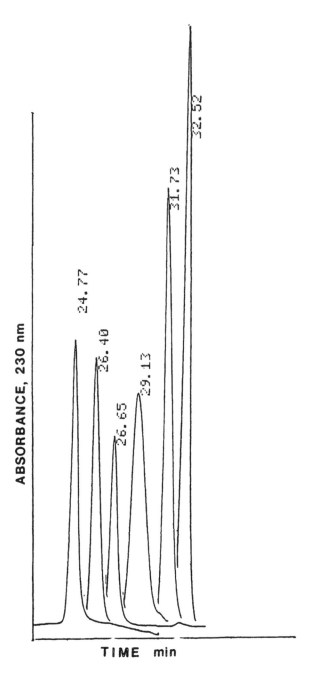

FIGURE 1. SEC chromatograms of WT T4 lysozyme as a function of urea concentration. Retention time 32.32 min, 1 *M* urea, 31.73 min, 2.5 *M* urea, 29.13 min, 3 *M* urea; 26.65 min, 3.5 M urea; 26.40 min, 4 *M* urea; 24.77 min, 6.5 *M* urea. Column: TSK 3000 SW (7.5 × 300 mm) plus TSK SW guard column (7.5 × 75 mm). Mobile phase: 0.15 *M* KCl, 10 m*M* potassium phosphate, pH 3, urea concentration as indicated. Absorbance monitored at 230 nm. Flow-rate 0.4 ml/min. Approximately 20 μg were injected. Notice the peak broadening at 3 *M* urea.

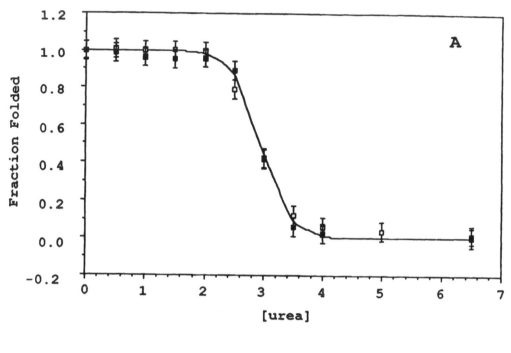

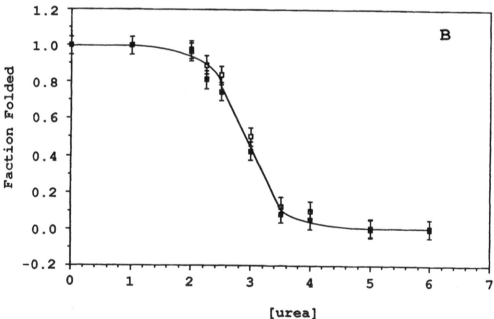

FIGURE 2. (A) Urea denaturation of wild type T4 lysozyme as determined by CD spectroscopy and SEC. The open squares are derived from CD and the filled squares from SEC. The denaturations were carried out at pH 3.0, 0.15 M KCl, 10 mM potassium phosphate and varying amounts of urea. CD measurements were made at 229 nm at 21°C. The size-exclusion measurements were made at 21°C (see Section II. Measurement Techniques). The error bars represent the standard deviation of the data and are nominally ± 0.05. The raw data were transformed to the fraction folded using the dependence of the native and denatured dichroism or retention volume at concentrations of urea outside of the transition region. (B) Same as Figure 2A, except 0.5 M KCl, 10 mM potassium phosphate, pH 3, 27.5°C. Adapted from Reference 2.

the fraction of native protein, as a function of urea concentration, is the same when measured by CD and SEC. The coincidence and cooperativity of the two denaturation curves is consistent with, but not proof of, a two-state transition. The C_m (urea concentration where $K = [U]/[F] = 1$) is $2.9 \pm 0.1 \ M$.

Urea denaturation was also performed in 0.5 M KCl, pH 3. Native state elution volumes increased from 13.2 ml in 0.15 M KCl to 14.0 ml for 0.5 M KCl. Increasing the ionic strength of the mobile phase increases the elution volume at all concentrations of urea at pH 3. The implication is either that the mean end-to-end distance decreases with increased ionic strength, i.e. the macromolecule becomes more compact, or the macromolecule interacts differently with the column material under the different salt conditions. Pfannkoch et al. studied hen lysozyme under similar conditions.[6] These authors concluded that, at pH below 4, electrostatic interactions with silica were eliminated. In addition, at ionic strengths between 0.1 and 0.6, hydrophobic interactions between the protein and column, particularly in the presence of a nonaqueous component of the mobile phase such as urea, are not significant. Use of CD under similar conditions also suggests that it is a decrease in the radius of gyration and not a column interaction. At pH 3, 0.15 M KCl and 6 M urea, the wild type protein exhibits practically no dichroism and an elution volume of 12 ml. If the KCl concentration is increased to 0.50 M, the retention volume increases along with the dichroism. A plot of the dichroism of the wild type at pH 3 and 6 M urea vs. [KCl], however, is completely non-cooperative, indicating a lack of a "transition" introduced by salt. The combination of the two techniques establishes that a reduction in the volume of the native and denatured states of T4 lysozyme accompanies an increase in ionic strength. The SEC and CD results are shown in Figure 2B. Again the two techniques give superimposable unfolding curves, within experimental error. The C_m is $2.9 \pm 0.1 \ M$ urea, in agreement with the 0.15 M KCl, pH 3 results.

Knowing the fraction of folded protein, as a function of urea concentration, also allows us to determine the energetic effects of urea denaturation.[12,13] The free energy of unfolding as determined by the two techniques is plotted as a function of urea concentration in Figure 3. First of all, the points lie along a series of straight lines which are practically coincident to one another. The extrapolated value of $\Delta G°$ ranges from 6.65 to 7.25 kcal/mole. Since the low (top panel) and high (bottom panel) salt measurements were carried out at 21 and 27.5°C, respectively, we expect the higher temperature values to have smaller slopes — which is what is observed. All in all, the two types of measurements yield essentially the same straight line. This, in turn changes slightly with temperature. We note furthermore that the concentration of urea employed is significantly less than at neutral pH, so that comparison to thermal denaturation data may entail less of an extrapolation and hence, less error. It is known from thermal denaturation of the protein at pH 3 and 0.15 M KCl (Becktel, unpublished) that $\Delta G(21°C)$ is 7.5 kcal/mole, while at 27.5°C it is 6.9 kcal/mole.

Thus, the free energies determined by extrapolating to zero urea concentration using either CD or SEC are in good agreement with thermal denaturation data. A combination of two disparate techniques not only increases the probability of detection of stable intermediates, should they exist, but also helps confirm the validity of thermal measurements and our treatment of protein stability curves.[14]

IV. OUTLOOK AND CONCLUSIONS

The studies reported here indicate the utility and some of the scope of combining SEC with spectroscopic measures of the state of proteins. Urea denaturation of T4 lysozyme at low and high salt yields unfolding curves and energies. The increase in elution volume and slight increase

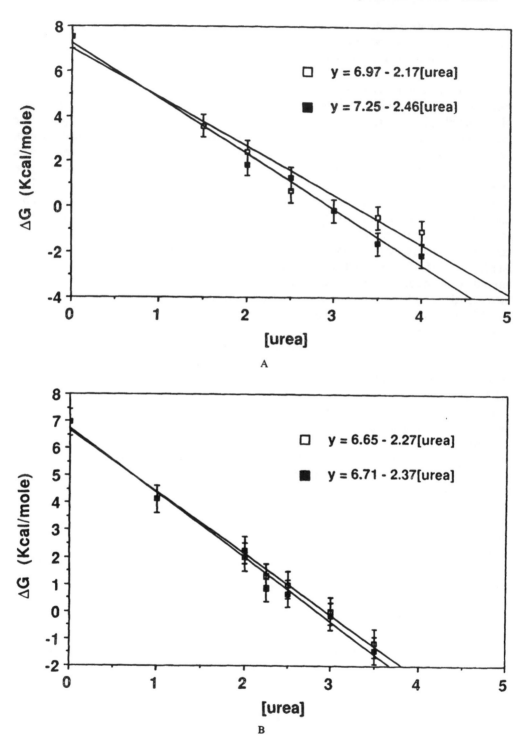

FIGURE 3. (A) Plot of ΔG(294 K, urea) derived from CD and SEC data (Figure 2A), 0.15 *M* KCl. The open squares are for CD measurements and closed squares are for SEC measurements. The values of ΔG were derived from the fraction of folded protein at each concentration of urea. ΔG°(CD) = 6.97 kcal/mole, ΔG°(SEC) = 7.25 kcal/mole. (B) Plot of ΔG(300 K, urea) derived from CD and SEC data (Figure 2B), 0.5 *M* KCl. ΔG°(CD) = 6.65 kcal/mole, ΔG°(SEC) = 6.71 kcal/mole. Adapted from Reference 2.

in dichroism with increasing ionic strength indicates that a thermodynamically stable, solvent accessible intermediate is not formed in high salt. In general, addition of SEC to optical or calorimetric measurements adds an additional dimension of insight into the area of protein stability. Further, since the operation of HPLC (in terms of accuracy and precision) depends upon the mechanical performance of its constituent parts, there is every reason to assume that these properties will continue to improve. SEC as a quantitative analytical tool will continue to increase in utility and popularity in protein chemistry.

ACKNOWLEDGMENTS

The authors are pleased to acknowledge the support of the National Institutes of Health (J.A. Schellman and B.W. Matthews) and the National Science Foundation (J.A. Schellman).

REFERENCES

1. **Corbett, R.J.T. and Roche, R.S.,** Use of high-speed size-exclusion chromatography for the study of protein folding and stability, *Biochemistry*, 23, 1888, 1984.
2. **Lindorfer, M.A. and Becktel, W.J.,** Spectroscopic and chromatographic studies of native and denatured states of T4 lysozymes, in *Current Research in Protein Chemistry*, Villafranca, J.J., Ed. Academic Press, San Diego, 1990, 309.
3. **Yau, W.W., Kirkland, J.J., and Bly, D.D.,** *Modern Size-Exclusion Liquid Chromatography*, John Wiley & Sons, New York, 1979.
4. **Lumry, R., Biltonen, R., and Brandts, J.F.,** Validity of the "two-state" hypothesis for conformational transitions of proteins, *Biopolymers*, 4, 917, 1966.
5. **Ackers, G.K.,** Analytical gel chromatography of proteins, *Advan. Protein Chem.*, 24, 343, 1970.
6. **Pfannkoch, E., Lu, K.C., Regnier, F.E., and Barth, H.G.,** Characterization of some commercial high performance size exclusion chromatography columns for water-soluble polymers, *J. Chrom.*, 18, 430, 1980.
7. **Pace, C.N.,** Determination and analysis of urea and guanidine hydrochloride denaturation curves, *Methods Enzymol.*, 131, 266, 1986.
8. **Elwell, M.L. and Schellman, J.A.,** Stability of phage T4 lysozymes. I. Native properties and thermal stability of wild-type and two mutant lysozymes, *Biochim. Biophys. Acta*, 494, 367, 1977.
9. **Becktel, W.J. and Baase, W.A.,** Thermal denaturation of bacteriophage T4 lysozyme at neutral pH, *Biopolymers*, 26, 619, 1987.
10. **Zipp, A. and Kauzmann, W.,** Pressure denaturation of metmyoglobin, *Biochemistry*, 12, 4217, 1973.
11. **Unger, K.K., Janzen, R., and Jilge, G.,** Packings and stationary phases for biopolymer separations by HPLC, *Chromatographia*, 24, 144, 1987.
12. **Schellman, J.A.,** Solvent denaturation, *Biopolymers,* 17, 1305, 1978.
13. **Schellman, J.A.,** Selective binding and solvent denaturation, 26, 549, 1987.
14. **Becktel, W.J. and Schellman, J.A.,** Protein stability curves, *Biopolymers*, 26, 1859, 1987.

ON-LINE CONFORMATIONAL MONITORING OF PROTEINS IN HPLC

Shiaw-lin Wu and Barry L. Karger

I. INTRODUCTION

It is well known that the conformation of a peptide or protein being eluted from a chromatographic column may not be the same as that injected. Any time a protein comes in contact with an environment that is non-physiological, some structural alteration can be anticipated. This alteration can be in the secondary, tertiary, or quaternary structure. The more removed the conditions from physiological, the more likely an alteration will take place. In this regard, it is well accepted that in reversed-phase liquid chromatography (RPC), the highly dense hydrophobic stationary phase and the low pH and organic solvent of the mobile phase makes it likely that conformational changes occur in the column. Denaturation of papain,[1] disaggregation,[2] and an increase in α-helicity for oligopeptides[3] have all been documented in RPC.

It is important to understand and control these structural alterations for a variety of reasons. First, the conformation(s) on the chromatographic surface clearly controls retention and separation. Indeed, many protein separations in LC arise from conformational manipulation of sample, often unbeknownst to the worker. Second, chromatographic artifacts can result from these changes, particularly if the rate of conversion from one state to another is comparable to the rate of migration of the species through the column. Thus, multiple or significantly broadened peaks can arise from pure materials.[4] It is essential to be able to recognize the difference between chromatographic artifacts and real impurities of the sample. Third, the form in which samples are collected during preparative LC may well be related to what has happened in the column. It is important to distinguish column induced conformational changes from the form of the species injected.

The characterization of collected fractions is an obvious approach to assess the state and "conformational purity" of species eluted from the column. However, it is likely that changes in protein structure may occur in solution after the stress of the adsorbed state is removed by desorption. Depending on the rate of change for the desorbed species, appropriate information may or may not be obtained from collected fractions. Characterization of the eluted fractions, however, will not likely allow conclusions as to the causes of chromatographic artifacts appearing on the recorder trace.

An important diagnostic tool in this regard is the use of on-line UV spectroscopy to assess conformational states and purities of chromatographic peaks. This approach can either involve the examination of the absorbance ratio at two separate wavelengths or the second derivative spectrum, in both cases at the apex or across the peak. For both types of measurements rapid wavelength scanning is required, and this conventionally means the use of a photodiode array spectrometer. The application of on-line rapid spectral scanning for characterization of chromatographic peaks is the subject of this article.

Before proceeding, it needs to be recognized that the spectral examination of the conformational state of a species eluted from a column does not necessarily mean that that specific form existed on the chromatographic surface. If there is a rapid conformational change upon desorption (half-life ~ msec), then the eluted state will not represent the adsorbed species.[5] Nevertheless, with this caveat in mind, quite useful information can be obtained by on-line UV spectroscopy.

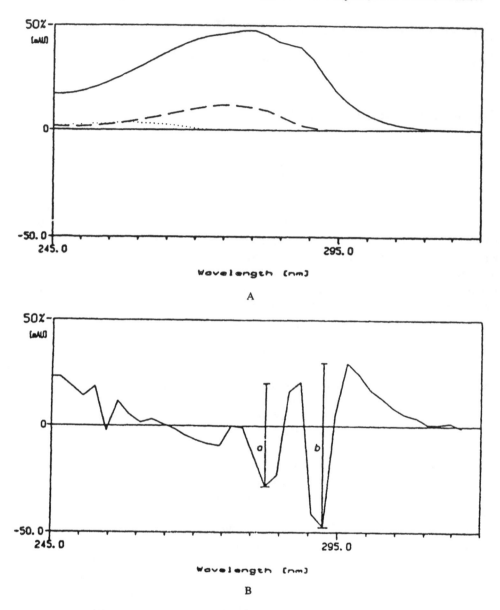

FIGURE 1. (A) Normal UV spectra of an equimolar solution of N-AcTrpNH$_2$ (—), N-AcTyrNH$_2$ (- - -) and N-AcPheOEt (···), dissolved in 0.5 *M* ammonium acetate. The detector was a Hewlett-Packard Model 1040A photodiode array detector (Palo Alto, CA). (B) On-line second-derivative spectrum of lysozyme eluted from a hydrophobic interaction chromatography column. HPLC instrument: Series 8800 gradient controller with a three head chromatographic pump (DuPont Instruments Product Division, Wilmington, DE) coupled to a Hewlett-Packard (HP) Model 1040A detector, HP85B personal computer, DPU multichannel integrator, HP 9121 P/S disc drives, HP 7470 A graphics plotter and HP think-jet printer. Column: ether-HIC column (100 × 4.6 mm I.D.); a precolumn (66 × 4.6 mm I.D.) containing the same phase was inserted between the pump and injector. Conditions: linear AB gradient (5% B/min), where Solvent A is 0.5 *M* ammonium acetate containing 2 *M* ammonium sulfate (pH 6.0) and Solvent B is 0.5 *M* ammonium acetate (pH 6.0); flow-rate, 1 ml/min. The peak-to-peak distance between the maximum at 288.5 nm and the minimum at 282.5 nm (a), and between the maximum at 296.5 nm and the minimum at 292.5 nm (b) defines γ (= a/b).

Classically, UV spectroscopy has been employed to assess conformational changes of proteins as a function of alteration in conditions, e.g., temperature, pH, organic solvent.[6] Increases in the intensities of the aromatic amino acids — phenylalanine (Phe), tyrosine (Tyr) and tryptophan (Trp) — occur upon exposure of these residues to the aqueous solvent. Figure 1A shows the UV spectrum of the three aromatic amino acids between 240 and 300 nm; in this region there is almost no interference caused by buffer solutions. The maximum absorbances for Trp, Tyr and Phe occur at wavelengths of 280, 274, and 254 nm, respectively. The extent of the total UV spectrum resulting from each of these amino acids will be dependent on their relative number and exposure to the solvent. Note the very low extinction coefficient of Phe, which means this amino acid does not generally contribute to the overall spectrum in the presence of the other two. Note also that Trp has a significantly higher extinction coefficient than Tyr.

While absorbent ratios can be employed to characterize bands, another useful approach is second derivative spectroscopy.[7] Figure 1B shows the second derivative spectrum of a solution of lysozyme in the region of 245 to 320 nm. As generally observed, the spectral region between 280 and 300 nm consists of two major peaks. The peak at the lower wavelength can be ascribed to a combination of the aqueous solvent exposures of Tyr and Trp, whereas the peak above 290 nm is predominantly due to the solvent exposure of Trp alone. It has been shown that the a/b (or γ) ratio in Figure 1B is an index of relative Tyr exposure.[8] Thus, the γ ratio can be used as a measure of the microenvironment for the aromatic residues, and changes in γ will relate to changes in conformation or structure. It should be noted that the content of the Tyr and Trp ratio in a protein is critical in the use of the second derivative γ ratio. If the ratio of Tyr vs. Trp is more than 5, the error in the γ ratio will be large. The optimal range of the content of Tyr and Trp ratio is thus between 1 and 5.

II. ABSORBANCE RATIO

As noted, it is important to characterize bands as they are eluted from the column, since dynamic processes can continue to occur in solution upon fraction collection. This is particularly true in the case of RNase A (ribonuclease A) where the refolding process is known to have rates comparable to the time of migration through the column.[9]

In RPC using low pH-propanol gradients and room temperature, chromatographic peak shapes as displayed in Figure 2 are observed.[10] It was suspected that the odd peak shape was due to refolding of the protein in the mobile phase upon desorption from the column. Collection of the sample and subsequent characterization could not provide definitive evidence of this behavior since the protein would be completely folded before analysis was complete. However, collection and reinjection of various portions of the chromatographic band did show that the behavior arose from the protein itself and not impurities. We therefore turned to on-line spectroscopy.

Difference spectroscopy revealed that the thermally denatured protein had a peak maximum difference relative to the native species at 287 nm due to exposure of Tyr amino acids to the solvent (RNase A has no Trp's.) On the other hand, there was no difference between the native and unfolded species at 254 nm. Therefore, the absorbance ratio for 287/254 could be used to assess the conformation of the species eluted from the column.

Figure 2 shows that the broad band represents one state and the sharp late-eluted band another conformational state. Indeed, comparison of the absorbance ratio values to the difference spectroscopy revealed that the broad band is the folded state and the sharp late-eluted band the unfolded state. Thus, the chromatographic behavior arises from the slow refolding of RNase A in solution upon desorption. Indeed, by varying the mobile phase flow rate, while maintaining the gradient volume constant, it was possible to measure the rate constant for refolding.[10] Figure 2 illustrates how on-line spectroscopy (in conjunction with reinjection) can be useful in elucidating conformational processes in chromatographic behavior.

III. SECOND DERIVATIVE SPECTROSCOPY

Figure 3 shows another type of chromatographic behavior due to conformational changes.[11]

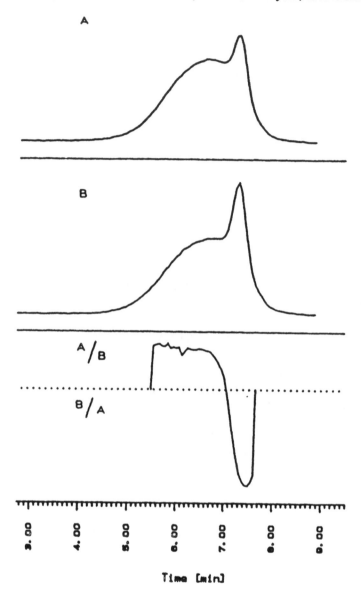

FIGURE 2. Reversed-phase chromatography behavior of ribonuclease A as a function of detection wavelengths at 287 nm (A) and 254 nm (B). HPLC instrument: see Figure 1. Column: C₄ bonded phase prepared from Vydac (Separations Group, Hesperia, CA) spherical silica (7-μm particle size, 300-Å pore diameter) and packed into a 100 × 4.6 mm I.D. stainless steel column. Gradient elution conditions: segment 1, 15 min, isocratic, Solvent A; segment 2, 25 min, 0 to 85% Solvent B (3.4% B/min); segment 3, 5 min, 85% Solvent B, where Solvent A is 10 m*M* orthophosphoric acid (pH 2.2) and Solvent B is 1-propanol/10 m*M* orthophosphoric acid (pH 2.2) (55:45, v/v); flow-rate, 1 ml/ min. (From Lu, X. M., Benedek, K., and Karger, B. L., *J. Chromatogr.*, 359, 19, 1986. With permission.)

Here, calcium-depleted α-lactalbumin (α-LACT) is eluted as a function of temperature (10 to 50°C) under hydrophobic interaction chromatography (HIC) conditions on a weakly hydrophobic stationary phase. It can be seen that the peak is relatively sharp at 10 and 50°C. In the intermediate region of 25 to 40°C, an enhanced broadening of the band is observed. Such

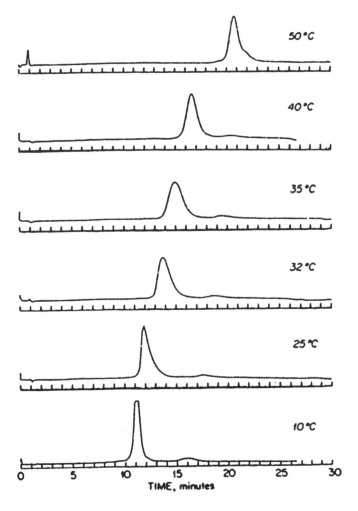

FIGURE 3. Effect of temperature on the chromatographic peak width of α-lactalbumin in hydrophobic interaction chromatography. HPLC instrument, column and conditions: see Figure 1. (From Wu, S.-L., Benedek, K., and Karger, B. L., *J. Chromatogr.*, 359, 3, 1986. With permission.)

broadening of intermediate regions has also been found in HIC for other proteins.[12] Furthermore, reinjection of collection fractions revealed a similar broadened band in this temperature region, indicating that the phenomenon is due to the protein itself and not impurities.

The sigmoidal increase in retention with temperature found for α-LACT is highly suggestive of unfolding of the labile species on the chromatographic stationary phase as a function of increasing temperature. As a general rule, unfolded species are eluted later than folded species in a chromatogram in RPC and HIC due to the greater contact area of the former species with the adsorbent surface. Note also from Figure 3 that column temperature can be used to manipulate separation in this case. Indeed, as we have noted, selectivity changes as a function of column conditions are often due to conformational changes in the protein which alter binding to the chromatographic stationary phase.

Further proof of this conformational behavior could be obtained by on-line spectroscopy of the band as a function of temperature. Table 1 presents the on-line absorbance ratios and γ values at the peak maximum for α-LACT from 0.5 to 50°C. The amino acid residue ratio for Trp/Tyr is 1:1 for this protein. At 32°C the peak is much broader than at 25°C (Figure 3), and at this

TABLE 1

Chromatographic and Spectroscopic Characteristics of α-Lactalbumin in HIC as a Function of Column Temperature

Temperature (°C)	Absorbance ratio			Second derivative ratio (γ)
	292/254	274/292	287/254	
0.5	1.41	1.43	1.78	0.82
5	1.42	1.42	1.79	0.83
10	1.43	1.43	1.80	0.85
18	1.40	1.44	1.77	0.89
25	1.41	1.43	1.77	0.87
32	1.45	1.41	1.87	0.70
50	1.40	1.46	1.84	0.85

Conditions: Ether-HIC column;[11] flow-rate, 1 ml/min; 20-min linear gradient. Mobile phase A: 2 M ammonium sulfate, 0.5 M ammonium acetate, pH 6; mobile phase B: 0.5 M ammonium sulfate, pH 6

temperature the spectroscopic characteristics are also altered. The 292 nm/254 nm and 287 nm/254 nm absorbance ratios increase from 25 to 32°C, suggesting exposure of Trp. This indication is strengthened by the decrease in the γ value from 25 to 32°C. Interestingly, solution studies have demonstrated exposure of Trp during unfolding.[13] At 50°C, the decrease in the 292 nm/254 nm ratio coupled with the increase in the 274 nm/292 nm ratio and the increase in γ in Table 1, now suggests a further exposure of Tyr. Thus, the spectroscopic data complements other pieces of evidence of conformational change and provides further characterization of the eluted species.

Finally, as another example, Figure 4 shows the HIC elution as a function of column temperature of α-LACT on a more hydrophobic stationary phase — C2-ether.[13] Two peaks are observed, with the earlier one decreasing with increasing temperature. Reinjection of collected fractions of both bands reproduced the chromatogram at any temperature, and this result confirmed that the behavior was due to the protein itself and not to impurities. As in other examples,[4] we are looking at the unfolding of the protein on the surface, with the first peak being in a more folded state.

The second peak can be observed to be very broad. In order to examine the causes of this broadening, we determined the UV characteristics at positions 1 to 4 in Figure 4B, iii. These characteristics are shown in Table 2. It can be seen that the broad second peak is inhomogeneous, probably due to the refolding of α-LACT in solution. Note that positions 2 and 3 are similar to the first peak, further suggesting a refolding process. It can be seen that both the absorbance and γ ratios provide a useful index of homogeneity of a single eluted band.

IV. CONCLUSION

The use of on-line spectroscopy to characterize eluted bands is valuable. Moreover, many instrument companies have available photodiode array detectors, and any LC laboratory should consider such a detector for their use. This article has suggested several ways that UV spectroscopy can be utilized for characterization. In the right set of circumstances, much can be learned from the use of such an approach. Several further examples can be found in References 14 and 15.

ACKNOWLEDGMENT

B.L. Karger gratefully acknowledges NIH under grant GM 15847 and Genentech, Inc., for support of this work. This is contribution number 379 from the Barnett Institute.

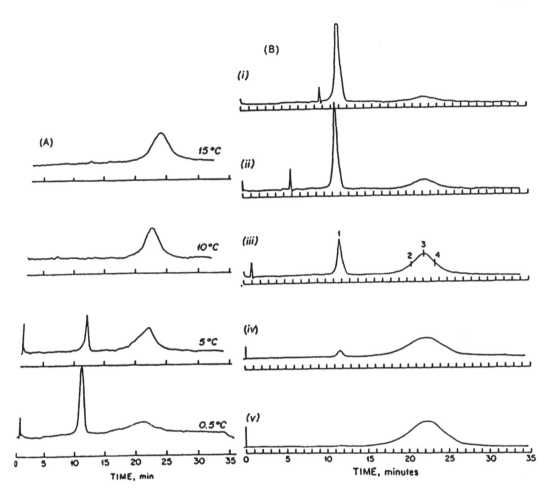

FIGURE 4. (A) Effect of temperature on the chromatographic behavior of α-lactalbumin in hydrophobic interaction chromatography (HIC). (B) Effect of contact time of α-lactalbumin on the HIC column at 5°C; (i) injection 9 min after start of gradient, (ii) injection 5 min after start of gradient, (iii) injection at start of gradient (standard conditions), (iv) 15 min incubation prior to start of gradient, (v) 30 min incubation prior to start of gradient (incubation refers to an isocratic hold under Solvent A conditions). HPLC instrument, column and conditions: see Figure 1. (From Wu, S.-L., Figueroa, A., and Karger, B. L., *J. Chromatogr.*, 371, 3, 1986. With permission.)

TABLE 2
Spectral Characteristics of the Two Peaks of α-Lactalbumin During HIC on a C$_2$-Ether Phase

Position	Absorbance ratio			Second derivative ratio (γ)
	274/292	292/254	287/254	
1	1.42	1.43	1.81	0.82
2	1.34	1.45	1.70	0.65
3	1.34	1.45	1.69	0.64
4	1.77	1.18	1.45	1.38

Note: See Figure 4B (iii) for definition of retention positions.

Conditions: C$_2$-ether phase[13]; temperature, 5°C; flow-rate, 1 ml/min; 20-min linear gradient. Mobile phase A: 2 *M* ammonium sulfate, 0.5 *M* ammonium acetate, pH 6; mobile phase B: 0.5 *M* ammonium acetate, pH 6.

REFERENCES

1. **Benedek, K., Dong, S., and Karger, B.L.,** Kinetics of unfolding of proteins on hydrophobic surfaces in reversed-phase liquid chromatography, *J. Chromatogr.,* 317, 227, 1984.
2. **Ingraham, R.H., Lau, S.Y.M., Taneja, A.K., and Hodges, R.S.,** Denaturation and the effects of temperature on hydrophobic-interaction and reversed-phase high performance liquid chromatography of proteins. Bio-Gel TSK-Phenyl-5-PW column, *J. Chromatogr.,* 327, 77, 1985.
3. **Heinitz, M.L., Flanigan, E., Orlowski, R.C., and Regnier, F.E.,** Correlation of calcitonin structure with chromatographic retention in high performance liquid chromatography, *J. Chromatogr.,* 435, 271, 1988.
4. **Cohen, S.A., Benedek, K., Tapuhi, Y., Ford, J.C., and Karger, B.L.,** Conformational effects in the reversed-phase liquid chromatography of ribonuclease A, *Anal. Biochem.,* 144, 275, 1985.
5. **Karger, B.L. and Blanco, R.,** Protein conformational changes upon adsorption to chromatographic surfaces, in *Protein Recognition of Immobilized Ligands,* UCLA Symposia, Pub., Alan R. Liss, New York, in press.
6. **Donovan, J.W.,** Ultraviolet difference spectroscopy — new techniques and applications, *Methods Enzymol.,* 27, 497, 1973.
7. **Balestrieri, C., Colonna, G., Giovane, A., Irace, G., and Servillo, L.,** Second-derivative spectroscopy of proteins — a method for the quantitative determination of aromatic amino acids in proteins, *Eur. J. Biochem.,* 90, 433, 1978.
8. **Ragone, R., Colonna, G., Balestrieri, C., Servillo, L., and Irace, G.,** Determination of tyrosine exposure in proteins by second-derivative spectroscopy, *Biochemistry,* 23, 1871, 1984.
9. **Hagerman, P.J., Schmid, F.X., and Baldwin, R.L.,** Refolding behavior of a kinetic intermediate observed in the low pH unfolding of ribonuclease A, *Biochemistry,* 18, 293, 1979.
10. **Lu, X.M., Benedek, K., and Karger, B.L.,** Conformational effects in the HPLC of proteins: Further studies of the reversed-phase chromatographic behavior of ribonuclease A, *J. Chromatogr.,* 359, 19, 1986.
11. **Wu, S.-L., Benedek, K., and Karger, B.L.,** Thermal behavior of proteins in high performance hydrophobic-interaction chromatography: on-line spectroscopic and chromatographic characterization, *J. Chromatogr.,* 359, 3, 1986.
12. **Strop, P.,** Hydrophobic chromatography of proteins on semi-rigid gels: Effect of salts and interferents on the retention of proteins by Spheron P300, *J. Chromatogr.,* 294, 213, 1984.
13. **Wu, S.-L., Figueroa, A., and Karger, B.L.,** Further studies into protein conformational effects in hydrophobic interaction chromatography: retention characterization and the role of mobile phase additives and stationary phase hydrophobicity, *J. Chromatogr.,* 371, 3, 1986.
14. **Hearn, M.T.W., Aguilar, M.I., Nguyen, T., and Fridman, M.,** High performance liquid chromatography of amino acids, peptides and proteins. LXXXIV. Application of derivative spectroscopy to the study of column residency effects in the reversed-phase and size-exclusion liquid chromatographic separation of proteins, *J. Chromatogr.,* 435, 271, 1988.
15. **Hancock, W.S., Canova-Davis, E., Chloupek, R.C., Wu, S.-L., Baldonado, I.P., Battersby, J.E., Spellman, M.W., Baba, L.J., and Chakel, J.A.,** Characterization of degradation products of recombinant human growth hormone, *Banbury Report 29: Therapeutic Peptides and Proteins: Assessing the New Technologies,* Cold Spring Harbor, 95, 1988.

REVERSED-PHASE HPLC SEPARATION OF HOMOLOGOUS PROTEINS: INTERLEUKIN-2 MUTEINS

Michael Kunitani

I. INTRODUCTION

Several methods have been developed for correlating polypeptide structures with reversed-phase high-performance liquid chromatography (RPC) retention in response to the growth of RPC as an efficient and popular method of separating polypeptides. One approach sums empirically derived retention coefficients representing the hydrophobic contribution of each amino acid residue.[1] The assumption behind this approach is that each amino acid residue contacts the sorbent surface and that the total hydrophobicity of the solute determines reversed-phase retention. This approach works well for peptides of 20 residues or less, but generally is not an accurate method of predicting reversed-phase retention for larger proteins. Studies with peptides containing amphiphilic helices (hydrophobic and hydrophilic residues on opposite sides of an α-helix) indicate that secondary protein structure plays a large role in determining the surface of a polypeptide which is exposed to the hydrophobic sorbent.[2] Other studies on retention of proteins during RPC have shown that proteins are less sensitive to changes in the HPLC support or bonded phase than are corresponding small molecule separations.[3] These results imply that protein retention in RPC is governed largely by the protein surface and not by the support surface chemistry.

RPC is inherently denaturing, in that it induces the protein to unfold to expose its more hydrophobic interior surfaces in order to bind the hydrophobic stationary phase. Spectroscopic studies have shown that some proteins undergo tertiary structural perturbations while bound to the reversed-phase sorbent and regain their original tertiary structure after elution.[4] This denaturing effect may lead to the formation of two peaks; an earlier peak of native conformation and a later eluted peak of denatured conformation.[5] Although some proteins show a loss of biological activity (irreversible denaturation) during RPC, others retain full biological activity. Thus, the hydrophobic forces necessary for reversed-phase binding compete with those required to maintain the protein's secondary and tertiary structure. Clearly, many proteins retain significant secondary and tertiary structural features during their reversed-phase separation. Through steric, hydrophobic, and ionic constraints, these structural features define a chromatographic surface or "footprint" of a protein, which in turn determines its reversed-phase binding. The concept of multipoint attachment of a protein to the sorbent surface is consistent with a relatively large chromatographic footprint for a protein over that of small molecule separations.

Geng and Regnier have developed the stoichiometry factor, Z, which represents the number of solvent molecules displaced during the binding of the protein to the column packing. Thus, the value of Z is a measure of the size of the effective chromatographic footprint of that protein.[6] For RPC, Kunitani et al. have derived Equation 1 which shows that Z is proportional to the term S:

$$Z = 2.3\Phi S \qquad (1)$$

where Φ is the volume fraction of the organic in the mobile phase.[7] Under defined gradient conditions, every protein will have an S value described by Equation 2.

$$\log k' = \log k_w - S\Phi \tag{2}$$

Here k' is the retention of the protein (capacity factor) and k_w is the value of k' when water is the mobile phase ($\Phi = O$). Although empirically derived, Equation 2 is a good approximation of RPC retention of solutes in both isocratic and gradient modes and is a useful relationship, as the term S is experimentally much easier to derive than Z.[8] For a series of related proteins of nearly identical composition, the hydrophobic contribution to reversed-phase retention will be similar, and thus observed changes in Z (and S) should reflect changes in the structure of the protein which contacts the sorbent surface.

Examination of Equation 2 reveals why small changes in the organic composition of reversed-phase eluents result in large changes in protein retention, as opposed to small molecule separations. Proteins will have a much larger chromatographic footprint (Z and S) than small molecules. Thus, changes in Φ (% organic) will be amplified by large values of S resulting in even larger changes in retention (log k').

Studies with RPC of proteins have shown that selectivity is only modestly affected by chromatographic variables such as bonded phase chain length, particle pore size, ligand density, etc.[9] Generally, the most successful approach to optimizing the separation of proteins in RPC is to vary the solvent pH or gradient shape.[10] In this chapter, the strategy employed is to vary the slope of a linear acetonitrile gradient and to utilize the relationship of Equation 2 to relate changes in reversed-phase retention to differences in protein structure for a series of interleukin-2 variants. These interleukin-2 (IL-2) proteins were named muteins in that they consisted of a series of genetically engineered proteins expressed from a nucleic acid sequence which had been altered by site-specific mutagenesis.[11] This resulted in one or more substitutions, additions, or deletions of amino acids from that of the parent IL-2 protein. Thus, these IL-2 muteins yielded a series of proteins of essentially identical molecular weight (ca. 15,500) and composition but which vary in their structure and reversed-phase behavior. These changes in RPC retention and their correlation to changes in the protein's chromatographic footprint via changes in the terms Z and S provide a model for predicting chromatographic retention and for optimizing the separation of related proteins in RPC.

For this gradient optimization scheme to be effective, the tertiary protein structure (chromatographic footprint) must be stable over the range of solvent conditions explored. Chromatographic distortions such as peak tailing, peak fronting, or multiple peak formation are indicative of changes in protein structure during RPC. However, changes in protein structure during RPC may not be readily visible. The reversible on-column structural changes seen for lysozyme while bound to the reversed-phase sorbent probably accounts for the lack of RPC resolution within a series of lysozyme variants.[4,12]

II. EXPERIMENTAL

Reversed-phase separations were performed on an Altex 322 chromatograph (Beckman, Berkeley, CA) with a WISP 710B injector (Waters, Milford, MA), a Nelson 6000 data system (Nelson Analytical, Cupertino, CA) and a Jones column heater (Jones Chromatography, Columbus, OH). The column dead time, t_o, and system dwell time, t_D, were 1.7 and 3.1 min, respectively, at 2 ml/min.

IL-2 muteins were produced and purified at Cetus Corporation. After preliminary purification, the cystine disulfide was generated by a proprietary process.[11] Oxidation of methionines to methionine sulfoxides in IL-2 was performed by the addition of a 1 to 50 molar excess of chloramine T (Sigma, St. Louis, MO) buffered with 1.5 M Tris, pH 8.8, and incubated for 2 to 60 min at room temperature. To reduce cystine to cysteines, dithiothreitol (DTT; Sigma, St. Louis, MO) was added to 10 mM, buffered with Tris to pH 8 and incubated for 30 min at 60°C. The calculations of Z, S, and anticipated gradient retention times are described in References 13 and 14 and are commercially available as a personal computer software program called Dry Lab G (LC Resources, Orinda, CA). Two linear gradients, similar in beginning and ending organic composition but varying in run time, were used to calculate the terms Z and S and predict the retention and resolution of the solutes for an optimized gradient. Other details of the apparatus, materials and methods are described in Reference 11.

A. PROTEIN CHEMISTRY OF THE IL-2 MUTEINS

IL-2 is a lymphokine, very hydrophobic in composition and with a high degree of helical structure.[15] Earlier studies on native IL-2 established that a single cystine disulfide (S^{58}-S^{105}) is present in the biologically active native IL-2.[16] Peptide mapping and refolding studies of recombinant IL-2 have confirmed the identity of the (S^{58}-S^{105}) cystine disulfide as the major species produced from cysteine (SH^{59}, SH^{105}, SH^{125}) containing IL-2.[17] Since only two of the three cysteines are necessary for cystine formation, muteins of IL-2 were created in which the cysteines were replaced, one at a time, by serine. Oxidation of these muteins resulted in the desAla^1(S^{58}-S^{105})Ser^{125} IL-2 mutein retaining full biological activity, while the desAla^{1}Ser^{58}(S^{105}-S^{125}) and desAla^1(S^{58}-S^{125})Ser^{105} IL-2 muteins were much less active.[18] A mutein of IL-2 in which all of the cysteines were replaced by serine, desAla^{1}Ser^{58,105,125} IL-2, also shows greatly reduced biological activity. The absence of biological activity of these muteins confirms that the (S^{58}-S^{105}) cystine is a conformational feature necessary for biological activity. From these results, as well as the crystal structure of IL-2, it is presumed that cystine formation of the (S^{58}-S^{125}) or (S^{105}-S^{125}) disulfides are expected to result in large perturbations in the IL-2 structure.[15]

Other studies have established that recombinant IL-2 of the native sequence, as expressed from *E. coli*, retains the N-terminal methionyl residue as a result of incomplete post-translational processing.[17] A series of IL-2 muteins was prepared that omitted the N-terminal alanyl residue at position one of the native sequence. These desAla^1 IL-2 muteins were all expressed with a homogeneous N-terminal sequence beginning with proline, the second residue in the native IL-2 sequence. The desAla^1 deletion retains full biological activity and presumably results in minimal tertiary structural changes.[19] Serine represents a single-atom structural perturbation from cysteine, but other IL-2 muteins, such as desAla^{1}Ala^{125} have been constructed to test conformational integrity.

Following extended liquid storage, many of these IL-2 preparations contain low levels of an earlier eluted RPC species (Figure 1.1). This protein (Peak A) has an amino acid composition, N-terminal amino acid sequence and biological activity identical to that of the primary IL-2 component (Peak B). By a process of semi-preparative isolation of Peak A, followed by tryptic peptide mapping or by sequencing of the mixture of the cyanogen bromide fragments, it has been established that Peak A is due to oxidation of the methionine solely at position 104.[11] Chemical oxidation of methionine with chloramine T to the corresponding methionine sulfoxide provided a positive control to the authentic Peak A material in the peptide map experiment (Figure 1.2).

III. RESULTS AND DISCUSSION

The nine IL-2 muteins used for this study are listed in Table 1. All of the IL-2 muteins were chromatographed in the cysteine containing (reduced) as well as the cystine containing (oxidized) forms, except the desAla^{1}Ser^{58,105,125} IL-2, which contains no cysteine and cannot oxidize. The retention times in Table 1 are for the 40 min gradients, while the Z values are calculated from a combination of 20 and 80 min gradient runs. Peak B$_{red}$ retention times and Z

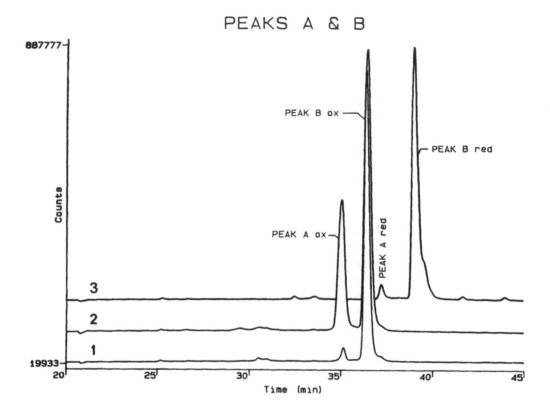

FIGURE 1. RPC of Peaks A and B in desAla[1]Ser[125] IL-2. Column: Vydac 214TP C$_4$ (Vydac, Hesperia, CA), 250 × 4.6 mm I.D., 5-μm particle size, 300-Å pore size. Mobile phase: linear AB gradient (10% acetonitrile at 0 min, 40.6% at 12 min, 70.3% at 45 min, and 100% at 50 min, followed by a 7 min reequilibration with 10% acetonitrile), where Eluent A is 10% aq. acetonitrile and Eluent B is acetonitrile, both eluents containing 0.1% trifluoroacetic acid (TFA); flow-rate, 1 ml/min; 30°C. (1) Disulfide-containing (S[58]-S[105]) IL-2 showing the location of Peak A$_{ox}$ and Peak B$_{ox}$; (2) a 2 min equilmolar chloramine T oxidation of the IL-2 shown in Figure 1.1 rapidly converts Peak B$_{ox}$ into Peak A$_{ox}$; (3) reduction of cystine (S[58]-S[105]) containing IL-2 to cysteine (SH[58],SH[105]) containing IL-2 shown in Figure 1.1 with DTT, showing the location of Peak A$_{red}$ and Peak B$_{red}$ and the integrity of this ratio during disulfide reduction. (From Kunitani, M., Hirtzer, P., Johnson, D. Halenbeck, R., Boosman, A., and Koths, K., *J. Chromatogr.*, 359, 391, 1986. With permission.)

values for all of the muteins are similar although most of the IL-2 muteins can be resolved with the 40 min gradient. The similarity in Z values for the column of Peak B$_{red}$ muteins is indicative of gross similarities in chromatographic footprint size and in protein structure among these reduced proteins.

A. STRUCTURAL FEATURES, PEAK A

For both the oxidized and reduced muteins, the Peak A retention is consistently less than that of Peak B, which corresponds to a larger Z value for Peak A over that of Peak B by 6% to 3%. Since this difference is close to the accuracy of the Z measurement itself, the Peak A values are divided by 1.06 and averaged with the Peak B values in Table 1.[7] This consistent difference in Z value suggests that the methionine sulfoxide at position 104 gives rise to a similar structural change in all those muteins containing a Peak A.

The absence of Peak A among some of the IL-2 muteins is consistent with their expected protein chemistry. The desAla[1]Ala[104]Ser[125] IL-2 has an alanyl residue at the position where the methionine sulfoxide producing Peak A would normally be. Although the other three methionines in the IL-2 sequence can be oxidized to methionine sulfoxide, they are not involved in the hydrophobic shift and structural change necessary to create Peak A. The desAla[1]Ser[58,105,125] IL-2 cannot form a Peak B$_{ox}$ or Peak A$_{ox}$ as it contains no cysteines to form the cystine disulfide.

TABLE 1
Retention Times and Z Values for IL-2 Muteins in RPC

	Peak B$_{red}$ retention	(Z)	Peak A$_{red}$ retention	Peak B$_{ox}$ retention	(Z)	Peak A$_{ox}$ retention
					Native disulfide bridged	
Ala¹Cys¹²⁵ (parent)	39.0 min	(19.4)	36.3 min	35.0 min	(26.4)	33.2 min
Ala¹Ser¹²⁵	34.6	(23.2)	31.9	31.4	(29.6)	29.6
desAla¹Cys¹²⁵	38.8	(21.1)	36.1	34.9	(23.8)	33.0
desAla¹Ser¹²⁵	34.6	(19.0)	32.1	31.4	(25.8)	29.7
desAla¹Ala¹²⁵	40.9	(19.8)	38.1	36.6	(23.4)	34.6
desAla¹Ala¹⁰⁰Ser¹²⁵	32.8	(25.5)	—	30.0	(27.4)	—
					Unnatural disulfides	
desAla¹Ser⁵⁸	38.9	(23.2)	36.4	19.9	(37.0)	—
des Ala¹Ser¹⁰⁵	38.0	(21.0)	35.2	20.3	(48.7)	—
					No disulfide possible	
desAla¹Ser⁹⁶,¹⁰⁵,¹²⁵	34.4	(21.5)	33.7	—	—	—

Column: Vydac 214TP C$_4$ (Vydac, Hesperia, CA), 250 x 4.6 mm I.D., 5-µm particle size, 300-Å pore size. Mobile phase: linear AB gradients (35 to 60% acetonitrile over 20, 40, or 80 min, followed by 10 min reequilibration with 35% acetonitrile), where Eluent A is 10% aq. acetonitrile and Eluent B is acetonitrile, both eluents containing 0.1% trifluoroacetic acid (TFA); flow-rate, 2 ml/min. The retention times shown were derived from the 40 min gradients; the Z values were calculated from a combination of the 20 and 80 min runs.

The Z values for Peak B$_{red}$ and Peak A$_{red}$ covary and, thus, the Peak A values are divided by 1.06 and averaged in with the Peak B values. The Z values for Peak B$_{ox}$ and Peak A$_{ox}$ are averaged in a similar manner.[7]

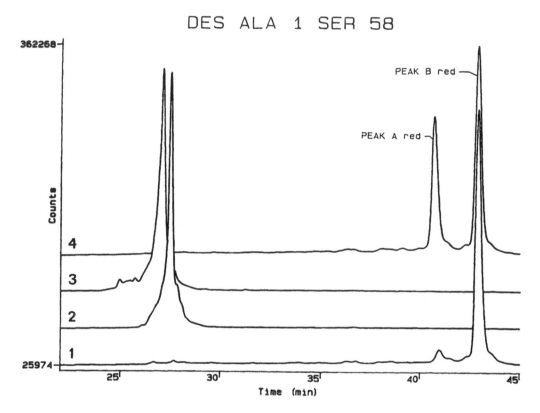

FIGURE 2. Oxidation of cysteines and methionines in desAla[1]Ser[58] IL-2. Gradient conditions: see Figure 1. (1) Reduced desAla[1]Ser[58] (SH[105],SH[125]) IL-2 containing 6% Peak A$_{red}$; (2) cystine-containing desAla[1]Ser[58] (S[105]-S[125]) IL-2 prepared from the protein in chromatogram 2.1; (3) chromatogram of the IL-2 shown in Figure 2.2 which has been oxidized with chloramine T to generate methione sulfoxides; (4) chromatogram of the IL-2 in Figure 2.3 which has been reduced with DTT to yield desAla[1]Ser[58](SH[105],SH[125]) IL-2 containing 38% Peak A$_{red}$. (From Kunitani, M., Hirtzer, P., Johnson, D. Halenbeck, R., Boosman, A., and Koths, K., *J. Chromatogr.*, 359, 391, 1986. With permission.)

B. NATIVE DISULFIDE BRIDGED MUTEINS

The muteins in the first six rows of Table 1 can all form the native (S[58]-S[105]) disulfide when oxidized. The Peak B$_{ox}$ retentions for these six muteins are consistently less than those for the corresponding Peak B$_{red}$ IL-2. This decrease in retention can also be represented by an average increase in Z value of 4.7 ± 2.2. This consistent increase in Z value suggests that the conformational changes among these muteins during the formation of the native (S[58]-S[105]) cystine are similar.

C. UNNATURALLY OXIDIZED MUTEINS

Two muteins, desAla[1]Ser[58] and desAla[1]Ser[105] IL-2, produce unusually large retention shifts when oxidized. These muteins cannot form the native (S[58]-S[105]) disulfide and appear to distort their tertiary conformation to form the unnatural disulfides (S[58]-S[125]) and (S[105]-S[125]). The absence of biological activity in these muteins also suggests significant conformational changes. Under the strongly denaturing condition of 6 *M* guanidine hydrochloride, other workers have found that the unnatural (S[58]-S[125]) and (S[105]-S[125]) disulfides can be formed from IL-2 of native sequence (Ala[1]Cys[125]), although the native (S[58]-S[105]) disulfide is favored under less denaturing conditions.[20]

Also very distinctive is the lack of a Peak A for the two unnaturally oxidized IL-2 muteins (desAla[1]Ser[58](S[105]-S[125]) IL-2 is shown in Figure 2.2).[11] Although reduced versions of these proteins may contain large percentages of Peak A$_{red}$ (methione sulfoxide 104), the corresponding cystine containing proteins do not demonstrate a corresponding Peak A$_{ox}$ (Figures 2.3 and 2.4). Presumably, in the unnaturally oxidized muteins, the methionine sulfoxide 104 does not occupy the same exterior position in the protein tertiary structure and the resulting hydrophobic contribution and chromatographic footprint do not result in a corresponding Peak A$_{ox}$. The large increase in Z value for these unnaturally oxidized IL-2 muteins, ($\Delta Z = + 13.8$ and $+ 27.7$, Table 1) upon oxidation, correlates with the distorted structure of these proteins. The "denaturation" of the tertiary structure upon unnatural oxidation results in a much larger Z value and indicates a substantial increase in the chromatographic footprint size.

Unlike the denaturation of most proteins, these unnatural disulfide bridged proteins have a shorter RPC retention under the gradient conditions of Table 1. Apparently, these particular structural distortions decrease the hydrophobicity available to the reversed-phase sorbent. However, in the hypothetical case where water is the mobile phase, the unnaturally oxidized IL-2 muteins, consistent with their larger Z values, are more strongly retained by several orders of magnitude.[7] Thus, the model of protein denaturation as a process of conformational distortions leading to exposure of a higher degree of interior hydrophobicity and therefore greater RPC retention is somewhat oversimplified. This example demonstrates the principle that RPC retention is a balance of both protein surface hydrophobicity and the size of the chromatographic footprint. These two factors are generally presumed to increase with the degree of protein denaturation, but for these unnaturally oxidized IL-2 muteins this relationship does not hold true.

D. STRUCTURAL FEATURES, SUBSTITUTIONS

The IL-2 muteins also illustrate exquisite structural sensitivity (one atom differences in a protein of ca. 15,500 molecular weight), and emphasize the potent selectivity of protein RPC separations due to the multipoint interaction of the protein surface with the reversed-phase sorbent. Substitutions of different amino acid residues into the IL-2 protein have a significant effect upon RPC retention and presumably upon protein conformation. Given that these substitutions alter the hydrophobicity and structure of the chromatographic footprint of the protein, differences in reversed-phase retention will be readily apparent, while changes far from the binding surface of the protein will not alter RPC retention appreciably. Compare the retention times of the oxidized and reduced forms of Peaks A and B for the three muteins: desAla^{1}Ser^{125}, desAla^{1}Cys^{125}, and desAla^{1}Ala^{125} IL-2 in Table 1. This series of muteins represents changes on the β-carbon at position 125 of an -OH, -SH, and -H, yet all four peaks for each of these three muteins are clearly separated on this gradient. In contrast, removal of the hydrophobic (Met) Ala^1 residues from the N-terminal of either the Cys^{125} or Ser^{125} proteins does not change the reversed-phase retention of Peak A or B in Table 1. Presumably, the hydrophilic N-terminal end of the protein is distant from the hydrophobic face of the IL-2 protein responsible for RPC retention.

E. GRADIENT OPTIMIZATION

Thus, even for proteins of similar molecular weight and amino acid sequence, subtle differences in structure may yield differences in Z (and S) which allow prediction of band spacing for various gradient conditions. To demonstrate the feasibility of this approach, the reversed-phase retention for desAla^{1}Ser^{125} was investigated and found to be linear (log $\bar{k}$ vs. $\bar{\Phi}$) over a broad gradient range ($\bar{k} = 1.5$ to 30).[7] Thus, the linearity of the relationship in Equation 2 allows accurate calculations of protein retention and implies that the tertiary conformation of these muteins is stable over this gradient range. The retention graphs for part of this range are shown in Figure 3 for Peak A$_{red}$, Peak B$_{red}$, Peak A$_{ox}$, and Peak B$_{ox}$. According to Equation 2, the slopes of the lines in Figure 3 are equivalent to -S.

It is seen that retention plots in Figure 3 for Peaks A and B of either the cystine or cysteine-containing molecules are parallel. This means that a change in gradient conditions will have little

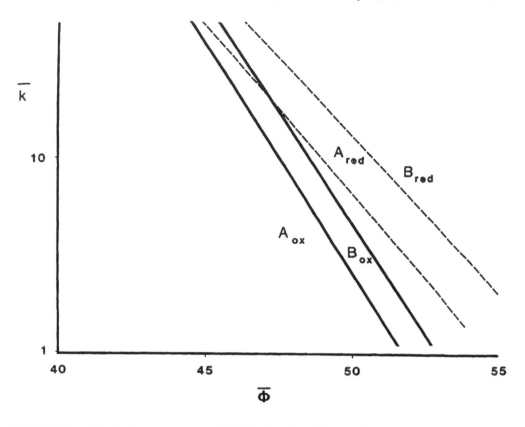

FIGURE 3. Retention (log k) vs. percent acetonitrile Φ for Peaks A and B of cystine-containing desAla[1](S^{58}-S^{105})Ser[125] IL-2 (solid lines) and cysteine-containing desAla[1](SH^{58},SH^{125})Ser[125] IL-2 (dashed lines). Column: see Figure 1. Mobile phase: linear AB gradients (35 to 60% acetonitrile over 20, 40, or 80 min, followed by 10 min reequilibration with 35% acetonitrile), where Eluent A is 10% aq. acetonitrile and Eluent B is acetonitrile, both eluents containing 0.1% TFA; flow-rate, 2 ml/min; 30°C. (From Kunitani, M., Johnson, D., and Snyder, L. R., *J. Chromatogr.*, 371, 313, 1986. With permission.)

effect on band-spacing for either the cystine- or cysteine-containing pair (Peaks A and B) of compounds. However, the plots for the cystine-containing Peak B and the cysteine-containing Peak A (MetSO[104]) are not parallel, so that the spacing of these peaks will be sensitive to experimental conditions. Using the plots of Figure 3, we would predict that at lower values of $\overline{\Phi}$ (very shallow gradients) the cystine-containing Peak B will reverse its order of elution with the cysteine-containing Peak A. Conversely, at higher values of $\overline{\Phi}$ (steeper gradients), the separation of cystine-containing Peak B and cysteine-containing Peak A will be much greater. The chromatograms of Figure 4 confirm this prediction. With the steeper gradients of Figure 4a, the cystine-containing Peak B (B_{ox}) and cysteine-containing Peak A (A_{red}) are clearly separated, while with the shallow gradient conditions of Figure 4b they reverse their order of elution. At both gradient extremes, the separation of the Peak A/B pairs remains similar. Thus, by changing flow-rate and gradient time, we can improve the separation of the oxidized vs. reduced pairs of Peaks A and B. However, we cannot improve the separation of a cysteine or cystine-containing IL-2 from its corresponding methionine sulfoxide 104 counterpart. This gradient optimization approach has led us to an unintuitive result: that for this particular protein separation, a steeper gradient yields better resolution than a shallower gradient. This example illustrates both the possibilities and limitations of a change in gradient conditions to affect the resolution of peptide and protein mixtures in RPC.

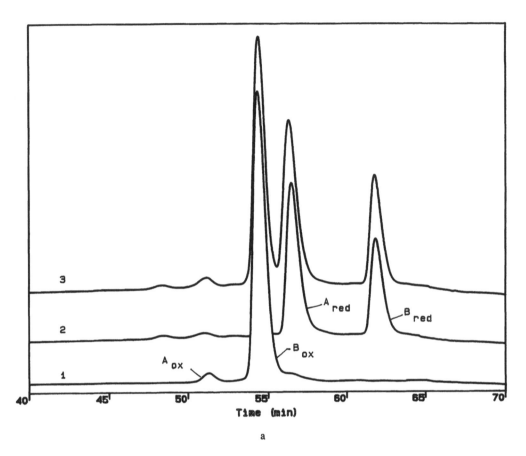

a

FIGURE 4. Reversed-phase chromatograms of desAla¹Ser¹²⁵ IL-2. Column and mobile phase eluents: see Figure 1. (a) linear AB gradient of 41 to 60% acetonitrile in 60 min (0.32% acetonitrile/min) at a flow-rate of 0.5 ml/min and a temperature of 30°C; (1) Peaks A_{ox} and B_{ox}; (2) Peaks A_{red} and B_{red}; (3) mixture of all four compounds in (1) and (2). Panel. (b) Linear AB gradient of 44.5 to 50.8% acetonitrile in 80 min (0.079% acetonitrile/min) at a flow-rate of 2 ml/min and a temperature of 30°C; peak identities same as Panel a. (From Kunitani, M., Johnson, D., and Snyder, L. R., *J. Chromatogr.*, 371, 313, 1986. With permission.)

IV. CONCLUSIONS

This chapter has shown that by changing the gradient slope one can optimize the RPC separation of closely related proteins. The changes in Z and S values generated by this predictive approach can be related to changes in tertiary protein structure. The RPC separation of IL-2 muteins established that reversed-phase retention is a combination of both protein surface hydrophobicity and protein surface contact size. However, there is poor correlation of many single residue changes in IL-2 proteins to corresponding changes in RPC retention.[11] This lack of correlation is largely due to the complex interaction of protein primary structure to its secondary and tertiary structure. Proteins containing a high degree of structural rigidity (structural elements such as α-helices or cystine disulfides) are most likely to reflect these primary structural changes as changes in reversed-phase retention. The structural differences among this series of IL-2 muteins exemplify the utility of a gradient optimization scheme.

Without details on protein secondary and tertiary structure during RPC retention, prediction of protein retention will continue to be an empirical process. However, we now have a basic model relating the role of the protein structure to RPC retention and a simple method of optimizing protein resolution via change in the gradient slope.

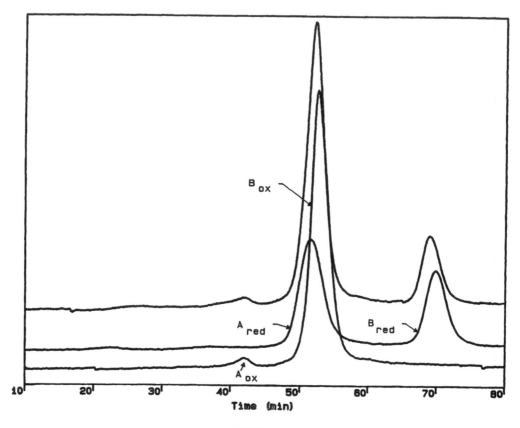

FIGURE 4b

REFERENCES

1. **Guo, D., Mant, C.T., Parker, J.M.R., and Hodges, R.S.,** Prediction of peptide retention times in reversed-phase high-performance liquid chromatography. I. Determination of retention coefficients of amino acid residues using model synthetic peptides, *J. Chromatogr.,* 359, 499, 1986.
2. **Heintz, M.L., Flanigan, E., Orlowski, R.C., and Regnier, F.E.,** Correlation of calcitonin structure with chromatographic retention in high-performance liquid chromatography, *J. Chromatogr.,* 443, 229, 1988.
3. **O'Hare, M.J., Capp, M.W., Nice, E.C. Cooke, N.H., and Archer, B.G.,** Factors influencing chromatography of proteins on short alkylsilane-bonded large pore size silicas, *Anal. Biochem.,* 126, 17, 1982.
4. **Lu, X-M., Figueroa, A., and Karger, B.L.,** Intrinsic fluorescence and HPLC measurement of the surface dynamics of lysozyme absorbed on the hydrophobic silica, *J. Am. Chem. Soc.,* 110, 1978, 1988.
5. **Benedek, K., Dong, S., and Karger, B.L.,** Kinetics of unfolding of proteins on hydrophobic surfaces in reversed-phase liquid chromatography, *J. Chromatogr.,* 317, 227, 1984.
6. **Geng, X. and Regnier, F.E.,** Retention model for proteins in reversed-phase liquid chromatography, *J. Chromatogr.,* 296, 15, 1984.
7. **Kunitani, M., Johnson, D., and Snyder, L.R.,** Model of protein conformation in the reversed-phase separation of interleukin-2 muteins, *J. Chromatogr.* 371, 313, 1986.
8. **Stadalius, M.A., Gold, H.S., and Snyder, L.R.,** Optimization model for the gradient elution separation of peptide mixtures by reversed-phase high-performance liquid chromatography. Verification of retention relationships, *J. Chromatogr.,* 296, 31, 1984.
9. **Burton, W.G., Nugent, K.D., Slattery, T.K., Summers, R.B., and Snyder, L.R.,** Separation of proteins by reversed-phase high-performance liquid chromatography. Optimizing the column. *J. Chromatogr.,* 433, 363, 1988.

10. Glajch, J.L., Quarry, M.A., Vasta, J.F., and Snyder, L.R., Separation of peptide mixtures by reversed-phase gradient elution. Use of flow rates changes for controlling band spacing and improving resolution, *Anal. Chem.*, 58, 280, 1986.

11. Kunitani, M., Hirtzer, P., Johnson, D., Halenbeck, R., Boosman, A., and Koths, K., Reversed-phase chromatography of interleukin-2 muteins, *J. Chromatogr.*, 359, 391, 1986.

12. Fausnaugh-Pollitt, J., Thevenon, G., Janis, L., and Regnier, F.E., Chromatographic resolution of lysozyme variants, *J. Chromatogr.*, 443, 221, 1988.

13. Quarry, M.A., Grob, R.L., and Snyder, L.R., Prediction of precise isocratic data from two or more gradient elution runs. Analysis of some associated errors., *Anal. Chem.*, 58, 907, 1986.

14. Dolan, J.W., Snyder, L.R., and Quarry, M.A., Computer simulation as a means of developing an optimized reversed-phase gradient-elution separation, *Chromatographia*, 24, 261, 1987.

15. Bandhuber, B.J., Boone, T., Kenney, W.C., and McKay, D.B., Three-dimensional structure of interleukin-2, *Science*, 238, 1707, 1987.

16. Robb, R.J., Kutny, R.M., Panico, M., Morris, H.R., and Chowdhy, V., Amino acid sequence and post-translational modification of human interleukin-2, *Proc. Natl. Acad. Sci. U.S.A.*, 81, 6486, 1984.

17. Lahm, H.-W. and Stein, S., Characterization of recombinant human interleukin-2 with micromethods, *J. Chromatogr.*, 326, 357, 1985.

18. Wang, A., Lu, S.-D., and Mark, D.F., Site-specific mutagenesis of the human interleukin-2 gene: structure-function analysis of the cysteine residues, *Science*, 224, 1431, 1984.

19. Rosenberg, S.A., Grimm, E.A., McGrogan, M., Doyle, M., Kawasaki, E., Koths, K., and Mark, D.F., Biological activity of recombinant human interleukin-2 produced in *Escherichia coli*, *Science*, 223, 1412, 1984.

20. Tsuji, T., Nakagawa, R., Sugimoto, N., and Fukcuhara, K.-I., Characterization of disulfide bonds in recombinant proteins: reduced human interleukin-2 in inclusion bodies and its oxidative refolding, *Biochemistry*, 26, 3129, 1987.

REVERSED-PHASE CHROMATOGRAPHY: THE EFFECT OF INDUCED CONFORMATIONS ON PEPTIDE RETENTION

John M. Ostresh, Klaus Büttner, and Richard A. Houghten

I. INTRODUCTION

Reversed-phase high-performance liquid chromatography (RPC) is currently the method of choice for the rapid estimation of the homogeneity of synthetic peptides. The possibility of using RPC "retention coefficients" for the prediction of peptide retention times is based generally upon the assumption that peptide retention is due primarily to the amino acid composition of the peptide.[1-7] We have shown, however, that the use of amino acid retention coefficients to predict peptide retention times can be unreliable. This appears to be due to the induction of secondary structure in the peptides by the hydrophobicity of the n-alkyl or hydrocarbon chains of the solid support, or by the increasingly organic component of the mobile phase,[8-12] although we believe the latter to be less likely. Previous investigators, using sets of unrelated peptides, have found correlations as high as 0.99 between predicted and observed retention times.[1-5,13-15] Deviations have been attributed to anomalous conformational effects and/or stationary phase interactions which, as yet, have not been well defined. We believe that these variations from theory can be better understood through the systematic study of closely related peptides to narrow the range of conformational possibilities.

In our initial studies involving 260 closely related substitution analogs ("flu" analogs) of a 13 amino acid sequence (YPYDVPDYASLRS) from the influenza virus hemagglutinin (HA1:98-110),[16-19] peptides were synthesized in which each position of the peptide was substituted with each of the twenty common naturally occuring amino acids (yielding 260 peptides, 247 of which were related analogs and 13 of which were the native control sequence). In this set of peptides, there were subsets of peptides of the same composition but differing sequence ("SCDS" analogs), which were shown to have retention times which differed from one another by as much as 6 min using a 0.5%/min gradient. While the experimental retention times found were in relative agreement with the predicted retentions, the agreement was not sufficient such that one would be able to distinguish the desired peptide from a related contaminating sequence in a natural or synthetic mixture, nor would it be possible to verify or deny a peptide's composition based on its elution position. These earlier studies[8-10] have led us to believe that the observed retention times of peptides in RPC are due predominantly to a combination of three factors: (1) the amino acid composition;[1-7,13-15] (2) ionic and other intramolecular interactions between neighboring amino acids;[26-28] and (3) induced conformational effects.[8-10] While amino acid composition is, overall, the most substantial factor influencing retention, intramolecular ionic effects and induced conformational effects can predominate for a significant number of peptides. While it is intuitively obvious, and experimentally well established, that more hydrophobic peptides will be retained longer than less hydrophobic peptides, and that more hydrophilic peptides will be retained for shorter times than less hydrophilic peptides, retention coefficients must be more accurate than those currently available to have a general utility.

Our earlier work prompted us to synthesize several sets of "SCDS" analogs in order to investigate more fully the possible conformational nature of these interactions, generally hypothesizing the induction of an α-helical conformation in these peptides by this two phase separation system. The first "SCDS" analogs were a set of 14 peptides derived from the positional variation resulting when a single lysine is "walked" along a 13-mer alanine chain amidated at its C-terminal (Ala-Lys/14-mers). Another set of 30 "SCDS" analogs was prepared in which the variable sequences contained seven lysines and seven leucines (Lys-Leu/14 mers), with 14 of these analogs having an intrinsic potential for the formation of amphipathic α-helices. These particular peptide compositions were chosen to give more extreme variation in retention than would peptides of more "average" composition, thus exaggerating the observed differences. One final set of "SCDS" analogs was produced containing nine lysines and nine leucines (Lys-Leu/18-mers) to determine whether "gaps" in the axial symmetry of the Lys-Leu/14-mers were influencing peptide retention behavior. "Gaps" refer to those regions, corresponding to positions 15, 16, 17, and 18 of a complete 5 turn 18 amino acid stretch of α-helix, which are not present in the 14 amino acid peptides (see Figure 2 axial projections). We synthesized these peptides to consider what effect this lack of hydrophilic/hydrophobic symmetry had on the retention of our peptides. All of the Lys-Leu analogs were acetylated at the N-terminal and amidated at the C-terminal to rule out possible ionic interactions due to these end groups. The lack of N- or C-terminal charge also ensures that the predominant effect studied was an alteration in the interaction between the induced α-helical conformation and the stationary phase caused by the variation of the position of the alkyl group of the leucine, or the protonated ε-amino group of the lysine, as they were moved through the Lys-Leu sequences.

II. EXPERIMENTAL

The methodology of simultaneous multiple peptide synthesis[16,20,21] was used to synthesize approximately 100 peptides at a time. Hydrogen fluoride (HF) was used to cleave 24 peptides at a time using a multiple vessel cleavage apparatus (Multiple Peptide Systems, San Diego, CA). Analytical RPC of crude HF extracts showed average purities for these peptides of 85%. Amino acid analysis determinations were within 10% of theory. No further purification or analysis other than analytical RPC was carried out.

A Beckman gradient HPLC system consisting of two Altex Model 110A pumps, a Beckman Model 421 microprocessor (Beckman Instruments, Anaheim, CA), a Hitachi Model 100-20 variable wavelength spectrophotometer (Baxter Scientific Products, Los Angeles, CA), a Shimadzu CR3A integrator (Cole Scientific, Calabasas, CA) and a Biorad Model AS-48 autosampler (Biorad Laboratories, Richmond, CA) was used. Samples (20 μl, 0.2 mg/ml) were analyzed on Vydac 218TP54 C_{18} columns (Alltech Associates, Los Altos, CA) (250 × 4.6 mm I.D., 5-μm particle size). Further analyses were carried out on a Vydac pH stable 208TP104 C_8 column (250 × 4.6 mm I.D., warranteed by the manufacturer to be exhaustively endcapped). Both types of columns were used at both pH 2.1 (0.1% TFA) and 7.2 (0.1% TFA/NH_3). The 260 "flu" analogs were analyzed using the aforementioned gradient with a buffer system consisting of Eluent A (0.1% trifluoroacetic acid/water) and Eluent B (0.1% trifluoroacetic acid/acetonitrile). All other analogs were analyzed using a 1% gradient with the same buffer system. In a separate experiment, the addition of 0.1 M $(NH_4)_2SO_4$ to Eluent A[22] was used to saturate further any uncapped silanol groups with ammonium groups.

Retention times have been normalized in these studies with the use of an internal standard to eliminate deviations in the results due to fluctuations in injection time and possible degradation of the columns used. The internal standard was the 13 residue sequence of influenza hemagglutinin (HA1:98-110)[16] Samples were injected without the internal standard, as well as simultaneously with the standard to preclude the possibility of interactions between the analogs and the internal standard affecting the elution. No differences were observed except those seen

TABLE 1

Y	P	Y	D	V	P	D	Y	A	S	L	R	S
1	2	3	4	5	6	7	8	9	10	11	12	13

RPC Retention Times for SCDS Peptide Analogs

Position:	1 Y TYR	3 Y TYR	8 Y TYR	2 P PRO	6 P PRO	4 D ASP	7 D ASP	10 S SER	13 S SER
1 LYS	17.60	15.01	12.71	21.09	20.80	20.40	19.01	23.61	23.75
2 HIS	18.00	14.21	12.95	20.85	20.85	21.41	20.41	23.72	23.04
3 GLN	19.61	16.61	14.26	21.46	23.12	21.29	22.01	24.78	24.84
4 ARG	19.60	16.21	14.77	20.11	24.09	23.11	19.61	23.81	24.16
5 ASN	19.71	17.11	14.80	21.56	21.73	22.31	22.31	24.50	24.68
6 SER	21.41	16.71	15.36	20.91	22.52	23.90	22.71	24.01	24.21
7 ASP	20.01	17.91	15.41	21.91	22.01	24.01	24.01	25.05	25.16
8 GLY	20.81	17.71	16.22	21.91	23.04	23.91	22.21	24.82	25.26
9 GLU	20.11	17.31	15.21	22.61	24.52	24.46	23.21	26.56	24.90
10 ALA	20.51	19.51	17.81	22.21	21.76	24.33	24.61	25.87	27.29
11 THR	19.81	18.24	16.73	22.11	25.05	24.12	22.91	25.71	27.49
12 PRO	21.47	21.46	19.61	24.01	24.01	25.33	24.51	27.03	28.68
13 CYS	21.81	21.21	20.56	23.91	26.72	25.76	25.91	27.53	29.00
14 VAL	23.22	23.73	23.99	25.11	25.85	25.43	27.11	28.53	28.94
15 TYR	24.01	24.01	24.01	25.81	28.37	26.66	27.51	28.58	32.59
16 MET	28.59	24.01	24.89	26.21	29.76	27.16	28.11	29.69	32.69
17 ILE	24.91	26.06	23.56	25.85	31.05	29.01	28.92	30.23	35.18
18 LEU	26.33	28.66	21.69	28.31	32.26	28.96	29.51	32.03	34.27
19 PHE	29.71	28.92	24.96	28.03	32.75	31.51	31.51	32.50	36.22
20 TRP	32.51	30.42	27.18	30.62	35.06	33.06	33.03	33.53	36.93

between identical injections. These small differences were due to random differences between injection and gradient start time, ± 0.1 min, caused by the use of an autosampler.

III. RESULTS AND DISCUSSION

Previous results[8-10] (partially summarized in Table 1) lead us to believe that accurate prediction of peptide retention times will only be accomplished when those factors involving induced secondary structure have been ascertained and are included as a factor in existing retention coefficients. For example, Table 1 shows that, depending on which of the three tyrosines in our native "flu" sequence is substituted with lysine, a change in retention time is found which can vary by nearly 5 min.

The variation in the retention times of these and other analogs has been suggested by others as being due to the interaction of the peptides with the uncapped silanols of the solid support by hydrogen bonding interactions and/or ion exchange mechanisms[22] The addition of $(NH_4)_2SO_4$ to Eluent A increased the overall retention time, but left the relative results identical to those without $(NH_4)_2SO_4$. We believe that these results negate the supposition by these authors that the variation in retention is substantially due to interactions with uncapped silanols. Since at pH 2.1 ionization of silanols is suppressed, similar results when these peptides were run at both pH 2.1 and 7.2 further support the conclusion that ionic interactions of the protonated amino groups of the peptides being studied with unreacted silanols are not significantly affecting our results.

More recent data are shown in Figure 1, which illustrates the variation of retention time vs. the position of a lysine in a 14 amino acid peptide containing 13 alanines. This periodicity of retention times, as well as the high potential of alanine to form α-helices according to Chou/

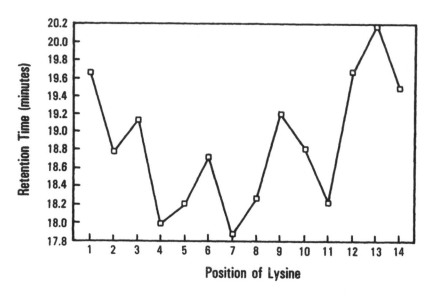

FIGURE 1. Periodicity of RPC retention times of Ala/Lys analogs in which the lysine is "walked" through the sequence, i.e., 1 = KAAAAAAAAAAAAA-NH$_2$, 2 = AKAAAAAAAAAAAA-NH$_2$, etc). NH$_2$ denotes C-terminal amide.

Fasman,[23] correlates well with the assumption that these peptides exist in an induced α-helical conformation which is amphipathic depending on the position of the protonated ε-amino group of the inserted lysine relative to the protonated α-amino group of the N-terminal amino acid. To substantiate further this assumption, we prepared another set of peptides which were identical to the above set, except for the acetylation of the N-terminal amino groups. An overall variation of approximately 2.4 min between the earliest and latest retention times for these analogs was observed; however, the retention time increased linearly with distance from the amino terminus with no appreciable periodicity found. We believe that the peptides in this N-acetylated series are, in fact, also in an induced α-helical conformation due to the high α-helix potential of alanine. One possible explanation for the exaggerated periodicity of the non-acetylated peptides is that the periodicity is caused by a distortion of the α-helix, which more fully exposes the amino groups to the aqueous mobile phase and also causes a variation in the hydrophobic contact area of the peptide with the solid support. This depends, in turn, simply on the relative positions of the two protonated amino groups if these peptides exist as ideal α-helices. When the amino groups are on opposite sides of an ideal α-helix, the contact area is lessened in order to permit exposure of both protonated amines to the aqueous phase, thus causing the peptide to be more weakly retained. This effect should be a periodic effect dependent on the relative position in an α-helix, as was demonstrated in Figure 1. The lack of a second protonated amine in the N-acetylated analogs allows these analogs, in an α-helical arrangement, to rotate essentially around their horizontal axes, permitting the single hydrophilic charged ε-amino group to be optimally exposed to the aqueous mobile phase. These peptides were found to have increasing retention depending on the position of the lysine in the sequence. Other investigators have found that an α-helical peptide has a dipole which causes a field difference equal to one half an elementary charge at each end of the α-helix.[29,30] The effective partial negative charge is located at the carboxyl terminal and a protonated lysine would stabilize the helix the closer it was to this effective charge. This increased stability, which should give a more ideal helix, would in turn cause increasing retention as the lysine moved from the N- to C-terminal, as is found for this series of peptides (data not shown).

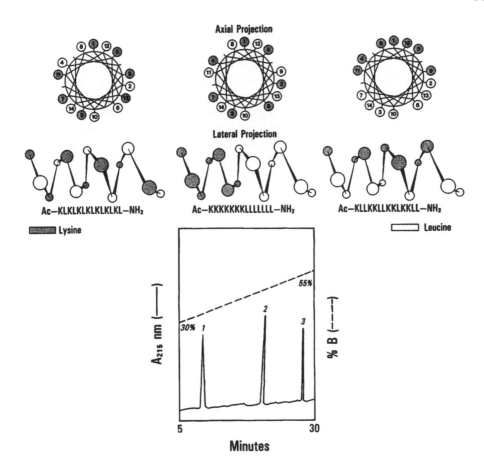

FIGURE 2. The lower figure shows the varying retention times for three Lys/Leu analogs (Peak 1: Ac-KLKLKLKLKLKLKL-NH$_2$, Peak 2: Ac-KKKKKKKLLLLLLL-NH$_2$, Peak 3: Ac-KLLKKLLKKLKKLL-NH$_2$) having the same composition but differing sequence. The upper figures illustrate the axial and horizontal projections of the potential α-helical conformations of these peptides. Ac denotes N-terminal acetyl; NH$_2$ denotes C-terminal amide.

Figure 2 graphically illustrates the major effect secondary structure can have on retention time for Lys-Leu/14mers with the same composition but differing sequences. These peptides have a strong potential for helix formation according to the scale of Chou/Fasman.[23] Figure 2 shows the potential α-helical comformations of the peptides and demonstrates that overall polarity, whether linear (Figure 2, central projections; lower graph, peak 2) or classically amphipathic (Figure 2, rightmost projections; lower graph, peak 3) causes substantially increased retention in RPC. The hydrophobic face (and large hydrophobic contact area) of the amphipathic helix of Peptide 3 causes this peptide to be retained for a much longer time than Peptide 1, which, while it has an ordered linear sequence composition, cannot exist in a classical amphipathic α-helical arrangement. Peptide 2 is segmentally amphipathic with a linear polarity and has an intermediate retention time relative to Peptide 3. Peptide 1 has the least polarity of the three peptides when viewed either as a linear sequence or as an α-helix. However, when viewed as an α-helix, Peptide 1 has lines of symmetry that spiral around the horizontal axis of its potential α-helical conformation (Figure 2, leftmost figure of lateral projection). A narrow β-sheet type of interaction with the C$_{18}$ chains cannot be excluded.

The assumption of α-helicity led us to believe that peptides could be found which would have even lower retention times if the sequences were more random and had less overall symmetry such as spiralling lines of symmetry possible in the leftmost peptide of Figure 2. We, therefore, prepared peptides having the same composition as Peptide 1, but which have more "random" sequences and less recognizable symmetry. Several of these analogs (i.e., Ac-LKLLLKKKLLKKKKLLKL-amide) were retained for shorter periods of time than Peptide 1, as would be anticipated if our assumptions were correct. The reduction in retention was approximately 5 min. Related studies using the corresponding Lys-Leu/18-mers show similar results to the Lys-Leu/14-mers found in Figure 2, although as expected they have longer overall retention times due to the greater overall hydrophobic contact area. These results suggest that the "gaps" (positions 15 to 18) in a 14-mer amphipathic helix have only a "subtractive" effect on retention when compared to a complete amphipathic 18-mer having a complete five turn α-helix cycle. We believe that the results obtained for the shorter peptides are very probably also applicable for the longer peptides.

Further studies were carried out with 18 analogs of Ac-LKLLKKLLKKLKKLLKKL-amide, nine of which were analogs composed of eight lysines and ten leucines, and nine of which were analogs composed of ten lysines and eight leucines, each of which were acetylated at their N-terminal and amidated at their C-terminal. These analogs permit the study of the effect of substituting a single leucine for lysine on the hydrophobic/hydrophilic surface area. Figure 3 shows the linear plot of retention time vs. amino acid substitution. This plot again shows the aforementioned periodicity. An identical periodicity is seen when the hydrophobic moment of the peptides is plotted vs. the retention time (data not shown).[24] These data again suggest that the variation is due to conformational effects. The results are better demonstrated if one makes the assumption of an α-helical configuration and the data from Figure 3 is replotted as in Figure 4 in which one plots the position of a single leucine residue as it is "walked" along the hydrophilic face of an amphipathic helix. The retention of this series of peptides first decreases and then increases, as the leucine moves around the hydrophilic face of the α-helix, which is, in effect, altering the hydrophobic area available to the stationary phase. This further substantiates the assertion that these peptides exist in an α-helical arrangement. Similar, but slightly greater changes in retention are seen when "walking" a lysine on the leucine side (data shown in Figure 3). In solution, the α-helical content, as measured by circular dichroism spectra of these peptides, was found to increase with the CH_3CN concentration; however, the percent helix determined was low (data not shown and may not be relevant to results found in this two-phase system).

These results continue to support our hypothesis that induced conformational factors, especially α-helical arrangements, influence the retention behavior of peptides in RPC. We believe that every peptide has an energetically favored conformation which is induced by the interaction of the peptide with the hydrophobic groups of the stationary phase. Other than the specific conformations of the peptide induced by its dynamic interation with the stationary phase, we believe that the ionic and hydrophobic interactions between amino acids within a sequence, as well as the actual overall amino acid composition of the peptide, are the major influences on RPC retention. While we expect that other factors will be found which influence the retention times of peptides in RPC, the above factors seem to predominate in our studies. The results found in our studies utilizing synthetic peptides may be useful in understanding the interactions of specific peptide domains in proteins, both as these interactions pertain to other areas of the protein, and as specific domains of proteins interact with the stationary phase in protein separation techniques.[25] Our results also suggest that, with further study, it may be possible to predict secondary structural properties of peptides relevant to biological systems with the use of RPC. The information may be of value for a number of biological and immunological studies.

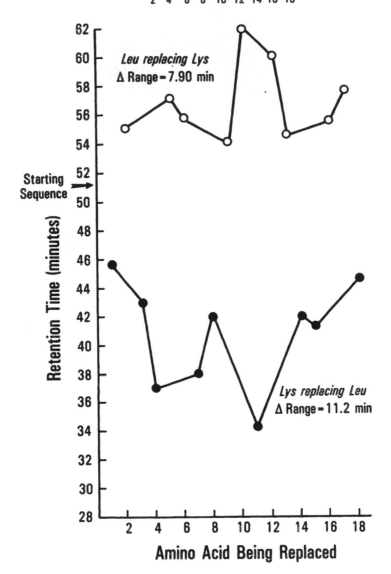

Ac—LKLLKKLLKKLKKLLKKL—AMIDE

FIGURE 3. Plot of retention times vs. position of inserted amino acid (Lys or Leu) in the sequence Ac-LKLLKKLLKKLKKLLKKL-NH₂. Ac denotes N-terminal acetyl; NH₂ denotes C-terminal amide.

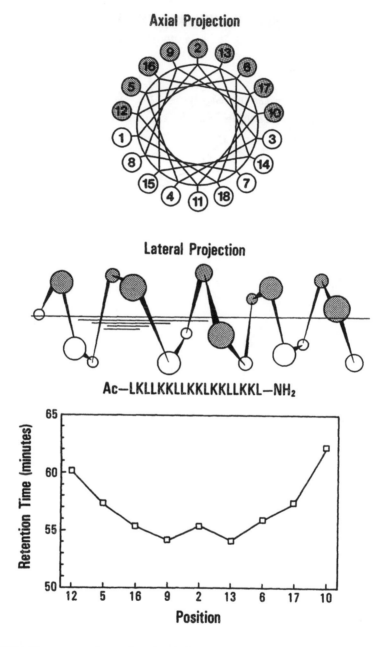

FIGURE 4. The upper two figures show the axial and horizontal projections of the potential α-helical conformation of the native sequence, Ac-LKLLKKLLKKLKKLLKKL-NH$_2$. Ac denotes N-terminal acetyl; NH$_2$ denotes C-terminal amide. The bottom figure shows a plot of peptide retention time vs. the position of a leucine residue as it is "walked" along the hydrophilic face of the amphipathic helix shown in the top figure. Shaded circles denote lysine residues; clear circles denote leucine residues.

ACKNOWLEDGMENT

The authors would like to thank Dr. Oded Arad for his critical reading of this manuscript and for helpful discussions. This work was supported in part by NIDA R01DA04491 (RAH) and NIDA Contract 271878122.

REFERENCES

1. Meek, J. L., Prediction of peptide retention time in high-pressure liquid chromatography on the basis of amino acid composition, *Proc. Natl. Sci. Acad. U.S.A.*, 77, 1632, 1980.
2. Meek, J. L. and Rossetti, Z. L., Factors affecting retention and resolution of peptides in high-performance liquid chromatography, *J. Chromatogr.*, 211, 15, 1981.
3. Su, J. J., Grego, B., and Hearn, M. T. W., High performance liquid chromatography of amino acids, peptides and proteins: pairing ion effect in RP-HPLC of peptides in the presence of alkylsulfonates, *J. Liq. Chromatogr.*, 4, 1745, 1981.
4. Wilson, K. J., Honegger, A., Stötzel, R.P., and Hughes, G. J., The behaviour of peptides on reverse-phase supports during high-pressure liquid chromatography, *Biochem. J.*, 199, 31, 1981.
5. Browne, C. A., Bennett, H. P. J., and Solomon, S., The isolation of peptides by high-performance liquid chromatography using predicted elution positions, *Anal. Biochem.*, 124, 201, 1982.
6. Guo, D., Mant, C. T., Taneja, A. K., Parker, J. M. R., and Hodges, R. S., Prediction of peptide retention times in reversed-phase high-performance liquid chromatography, *J. Chromatogr.*, 359, 499, 1986.
7. Guo, D., Mant, C. T., Taneja, A. K., and Hodges, R. S., Prediction of peptide retention times in reversed-phase high-performance liquid chromatography, *J. Chromatogr.*, 395, 519, 1986.
8. Houghten, R.A. and DeGraw, S.T., Effect of positional environmental domains on the variation of high-performance liquid chromatographic peptide retention coefficients, *J. Chromatogr.*, 386, 223, 1987.
9. Houghten, R. A. and Ostresh, J. M., Conformational influences upon peptide retention behavior in reverse phase high performance liquid chromatography, *Biochromatography*, 2, 80, 1987.
10. Houghten, R. A. and Ostresh, J. M., *Peptide Chemistry 1987*, Protein Research Foundation, Osaka, Japan, 1988, pp. 101—104.
11. Pietrzyk, D. J., Smith, R. L. and Cahill, W. R., Jr., The influence of peptide structure on the retention of small chain peptides on reverse stationary phases, *J. Liquid Chromatogr.*, 6(9), 1645, 1983.
12. Lau, S. Y. M., Taneja, A. K., and Hodges, R. S., Effects of high-performance liquid chromatographic solvents and hydrophobic matrices on the secondary and quaternary structure of a model protein, *J. Chromatogr*, 317, 129, 1984.
13. O'Hare, M. J. and Nice, E. C., Hydrophobic high-performance liquid chromatography of hormonal polypeptides and proteins on alkylsilane-bonded silica, *J. Chromatogr.*, 171, 209, 1979.
14. Parker, J. M. R., Guo, D., and Hodges, R. S, New hydrophobicity scale derived from high performance liquid chromatography: correlation of predicted surface residues with antigeniety and x-ray derived accessible sites, *Biochemistry*, 25, 5424, 1986.
15. Sasagawa, T., Okuyama, T., and Teller, D. C., Prediction of peptide retention times in reverse phase high performance liquid chromatography during linear gradient elution, *J. Chromatogr.*, 240, 329, 1982.
16. Houghten, R. A., General method for the rapid-solid phase synthesis of large numbers of peptides: specificity of antigen-antibody interaction at the level of individual amino acids, *Proc. Natl. Acad. Sci. U.S.A.*, 82, 5131, 1985.
17. Houghten, R. A., Facile determination of exact amino acid involvement in peptide antigen monoclonal antibody interactions, in *Modern Approaches to New Vaccines*, Channock, R., Ginsberg, H., Lerner, R., and Brown, F., Eds., Cold Spring Harbor Press, New York, 1987, pp. 1—7.
18. Niman, H. L., Houghten, R. A., Walker, L. E., Reisfeld, R. A., Wilson, I. A., Hogel, J. M., and Lerner, R. A., Generation of protein-reactive antibodies by short peptides is an event of high frequency: implications for the structural basis of immune recognition, *Proc. Natl. Acad. Sci. U.S.A.*, 80, 4949, 1983.
19. Wilson, I. A., Niman, H. L., Houghten, R. A., Cherenson, A. R., Connolly, M. L., and Lerner, R. A., The structure of antigenic determinant in a protein, *Cell*, 37, 767, 1984.
20. Houghten, R. A., Bray, M. K., DeGraw, S. T., and Kirby, C. J., Simplified procedure for carrying out simultaneous multiple hydrogen flouride cleavages of protected peptide resins, *Inter. J. Pept. Prot. Res.*, 27, 673, 1986.
21. Houghten, R. A., DeGraw, S. T., Bray, M. K., Hoffman, S. R., and Frizzell, N. D., Simultaneous multiple peptide synthesis: the rapid preparation of large numbers of discrete peptides for biological, immunological, and methodological studies, *BioTechniques*, 4, 522, 1986.

22. **Snider, R. H., Moore, C. F., Nylén, and Becker, K. L.,** Prediction of peptide retention times on reversed-phase HPLC: prospects and limitations, *BioChromatography,* 3, 100, 1988.

23. **Chou, P. Y. and Fasman, G. D.,** Conformational parameters for amino acids in helical, β-sheet, and random coil regions calculated from proteins, *Biochemistry,* 13, 211, 1974.

24. **Büttner, K., Ostresh, J. M., and Houghten, R. A.,** Prediction of immunodominant helper T-cell antigenic site using RP-HPLC, in preparation.

25. **Regnier, F. E.,** The role of protein structure in chromatographic behavior, *Science,* 238, 319, 1987.

26. **Bierzynski, A., Kim, P. S., and Baldwin, R. L.,** A salt bridge stabilizes the helix formed by isolated C-peptide of RNase A, *Proc. Natl. Acad. Sci. U.S.A.,* 79, 2469, 1982.

27. **Shoemaker, K. R., Kim, P. S., Brems, D. N., Marqusee, S., York, E. J., Chaiken, I. M., Stewart, J. M., and Baldwin, R. L.,** Nature of the charged-group effect on the stability of the C-peptide helix, *Proc. Natl. Acad. Sci. U.S.A.,* 82, 2349, 1985.

28. **Shoemaker, K. R., Kim, P. S., York, E. J., Stewart, J. M., and Baldwin, R. L.,** Tests of the helix dipole model for stabilization of α-helices, *Nature,* 326, 563, 1987.

29. **Hol, W. G. J.,** The role of the α-helix dipole in protein function and structure, *Prog. Biophys. Molec. Biol.,* 45, 149, 1985.

30. **Shoemaker, K. R., Kim, P. S., York, E. J., Stewart, J. M., and Baldwin, R. L.,** Tests of the helix dipole model for stabilization of α-helices, *Prog. Biophys. Molec. Biol.,* 45, 149, 1985.

THE EFFECT OF HYDROPHOBIC RESIDUE DISTRIBUTION IN α-HELICAL PEPTIDES ON PEPTIDE RETENTION BEHAVIOR IN REVERSED-PHASE HPLC

Nian E. Zhou, Paul D. Semchuk, Cyril M. Kay, and Robert S. Hodges

I. INTRODUCTION

The application of reversed-phase high-performance liquid chromatography (RPC) to the separation of peptides and proteins has increased significantly in recent years (for recent reviews, see References 1 to 3). A major factor governing the retention behavior of peptides and proteins during RPC is the relative hydrophilic/hydrophobic contribution that the side-chains of individual amino acid residues make to the overall hydrophobicity of the molecule. In fact, several research groups[4-14] have determined retention coefficients for the prediction of peptide retention time in RPC on the assumption that peptide retention is primarily related to the amino acid composition of the peptide. However, some researchers[15-19] have presented examples of peptides with identical composition and different sequences which have different retention times during RPC. These deviations are generally explained in terms of sequence-specific conformational differences, leading to preferential interaction sites, or anomalous stationary phase interactions.[6,16,18-21] For example, a non-polar environment, such as a hydrophobic stationary phase, may induce helical structures in potentially helical molecules. If a molecule becomes helical on binding and contains a preferred binding domain, as in the case of an amphipathic α-helix, then obviously some residues may not be interacting with the reversed-phase sorbent. This would then result in a deviation from a predicted retention time calculated assuming complete accessibility of all residues to interact with the stationary phase - an example of a sequence-dependent conformational effect on RPC retention behavior.

In order to study the effect of peptide conformation on peptide retention behavior in RPC, we synthesized two sets of peptides of varying lengths (7, 14, 21, 28, and 35 residues) having the same composition but different sequence (Table 1). Retention behavior of these peptides was studied by RPC and size-exclusion chromatography (SEC). The α-helicity of the ten peptides was determined by circular dichroism in the presence of trifluoroethanol (TFE). The helix-inducing solvent, TFE, was used to mimic the induction of α-helix in potentially α-helical peptides by the hydrophobicity of the stationary phase. The results of this study support the conclusion that amphipathic helices increase retention time because of the preferred binding of the hydrophobic face of the helix to the matrix and the inequivalent expression of all residues to the overall hydrophobicity of the peptide.

II. EXPERIMENTAL

A. MATERIALS

Unless otherwise stated, chemicals and solvents were reagent grade. Diisopropylethylamine

TABLE 1
Peptide Polymers of same Composition but Different Sequence

Peptide	Amino acid sequence
A7	Ac-Lys-Cys-Ala-Glu|Gly|Glu|Leu|amide
B7	Ac-Lys-Cys-Ala-Glu|Leu|Glu|Gly|amide
A14	Ac-Lys-Cys-Ala-Glu|Gly|Glu|Leu|-[Lys-Leu-Glu-Ala|Gly|Glu|Leu|-]$_1$-amide
B14	Ac-Lys-Cys-Ala-Glu|Leu|Glu|Gly|-[Lys-Leu-Glu-Ala|Leu|Glu|Gly|-]$_1$-amide
A21	Ac-Lys-Cys-Ala-Glu|Gly|Glu|Leu|-[Lys-Leu-Glu-Ala|Gly|Glu|Leu|-]$_2$-amide
B21	Ac-Lys-Cys-Ala-Glu|Leu|Glu|Gly|-[Lys-Leu-Glu-Ala|Leu|Glu|Gly|-]$_2$-amide
A28	Ac-Lys-Cys-Ala-Glu|Gly|Glu|Leu|-[Lys-Leu-Glu-Ala|Gly|Glu|Leu|-]$_3$-amide
B28	Ac-Lys-Cys-Ala-Glu|Leu|Glu|Gly|-[Lys-Leu-Glu-Ala|Leu|Glu|Gly|-]$_3$-amide
A35	Ac-Lys-Cys-Ala-Glu|Gly|Glu|Leu|-[Lys-Leu-Glu-Ala|Gly|Glu|Leu|-]$_4$-amide
B35	Ac-Lys-Cys-Ala-Glu|Leu|Glu|Gly|-[Lys-Leu-Glu-Ala|Leu|Glu|Gly|-]$_4$-amide

Note: Ac denotes N^α-acetyl; amide denotes C^α-amide. The residues (Gly and Leu) shown between the vertical lines are responsible for the sequence variation of the peptides.

(DIEA), dichloromethane, anisole and trifluoroacetic acid (TFA) were redistilled prior to use. *N,N*-dimethylformamide (DMF) and all the above solvents were obtained from General Intermediates of Canada. HPLC-grade water, methanol and acetonitrile were obtained from J.T. Baker Chemical (Phillipsburg, NJ). Dicyclohexylcarbodiimide (DCC) was obtained from Applied Biosystems (Foster City, CA), 1,2-ethanedithiol (EDT) was purchased from Aldrich Chemical (Milwaukee, WI), and acetic anhydride was obtained from Fisher Scientific (Fairlawn, NJ). Co-poly (styrene, 1% divinyl benzene) benzhydrylamine-hydrochloride resin (0.8 mmol of NH$_2$/g of resin) was purchased from Institute Armand Frappier (Laval, Quebec, Canada). tert-Butyloxycarbonyl (Boc) amino acids were purchased from Institute Armand Frappier and Bachem Fine Chemicals (Torrance, CA).

B. PEPTIDE SYNTHESIS

Peptides were synthesized using the general procedure for solid-phase synthesis described by Hodges et al.[22,23] on an Applied Biosystems (Foster City, CA) peptide synthesizer Model 430A. All amino groups were protected at the α-amino position with the Boc-group and the following side-chain protecting groups were used: Benzylester (Glu), α-chlorobenzyloxycarbonyl (Lys), 4-methylbenzyl (Cys). All amino acids were coupled using symmetrical anhydride couplings (2:1 equivalent, protected amino acid/DCC). Boc-groups were removed at each cycle by an 80-sec reaction with 33% TFA/dichloromethane (v/v), followed by a second reaction with 50% TFA/dichloromethane (v/v) for 18.5 min. Neutralizations were carried out using 2 × 1 min washes with 10% DIEA/DMF (v/v). The completed peptides were deprotected as described previously and acetylated with 25% acetic anhydride/dichloromethane (v/v) for 10 min. The peptides were cleaved from the resin support by treatment with hydrogen fluoride (20 ml/g resin) containing 10% anisole and 2% 1,2-ethanedithiol for 1 hr at −5 to 0°C. The solvents were removed under reduced pressure at 0°C. The resin was washed with ether, and extracted with 30% aqueous acetic acid.

C. PEPTIDE PURIFICATION AND ANALYSIS

Crude peptides were purified by RPC on either a SynChropak RP-P C$_{18}$ column (250 × 10 mm I.D., 6.5-μm particle size, 300-Å pore size; SynChrom, Lafayette, IN) or an Aquapore RP-300 C$_8$ column (220 × 4.6 mm I.D., 7 μm, 300 Å; Brownlee Labs., Santa Clara, CA). The HPLC instrument consisted of a Spectra-Physics (San Jose, CA) SP8700 solvent delivery system and SP 8750 organizer module, combined with a spectraflow monitor SF770, Monochrometer GM770 (Schoeffel Instrument-Corp), and Kratos SF 769 (Analytic Instruments) detector.

Peptides were eluted with a linear AB gradient (ranging from 0.2 to 1.0% B/min, depending on the peptide) at flow-rates of 2 ml/min (C_{18} column) or 1 ml/min (C_8 column), where Eluent A was 0.05% aq. TFA and Eluent B was 0.05% TFA in acetonitrile.

Amino acid analyses were performed on a Durram D-500 amino acid analyzer by conventional methods. The results indicated that all of the synthetic peptides had the expected amino acid composition.

III. RESULTS AND DISCUSSION

A. PEPTIDE DESIGN

We believe that the use of model synthetic peptide analogs offers a very promising approach to examining the effect of peptide conformation on retention behavior in RPC. In the present study, we designed two sets of model peptides of 7, 14, 21, 28, and 35 residues with a high potential to form α-helical structure. The seven residue repeat in the A and B series was [Lys-Leu-Glu-Ala-Gly-Glu-Leu] and [Lys-Leu-Glu-Ala-Leu-Glu-Gly], respectively (the peptide sequences are shown in Table 1). The amino acid sequence of the heptapeptide repeat in the B series was chosen based on the criteria described by Hodges et al.[24] Previous studies by our laboratory[23-26] and others[27-30] have demonstrated that hydrophobic interactions can play a dominant role in secondary structure formation and contribute in a major way to the stability of α-helices in aqueous solution. All of the synthetic peptides in the present study were acetylated at the N-terminal and amidated at the C-terminal to rule out possible ionic interactions due to the presence of these end groups. The amino acid sequences of A and B series peptides are viewed as axial projections of α-helices in a helical wheel (Figure 1). The hydrophobic amino acid residues, e.g., leucine residues, are more evenly distributed on the two sides of the α-helix in the A series peptides and are segregated on one side of the α-helix in B series peptides to form amphipathic α-helices. If we use the difference in the number of leucine residues between the two sides of the α-helix as an expression of amphipathicity then the A and B peptides can represent non-amphipathic helices and amphipathic helices, respectively. For example, in the A peptides (Figure 1), the difference in the number of leucine residues between the two sides (indicated by the solid and open arcs on the outside of the helical wheel) is 1 for A14 (2 minus 1), A21 (3 minus 2), A28 (4 minus 3) and A35 (5 minus 4). By comparison, all the leucine residues are present on one side of the α-helix in the B peptides (solid arcs, Figure 1) where B14, B21, B28, and B35 contain 3, 5, 7, and 9 leucine residues, respectively. If we compare the difference between the A and B peptides with regard to the number of leucine residues in the most hydrophobic side of the helix, the B peptides are more hydrophobic (0 between A7 and B7, 1 between A14 and B14, 2 between A21 and B21, 3 between A28 and B28, and a difference of 4 leucine residues between A35 and B35). Thus, as the polypeptide chain length increases, there is an increase in hydrophobicity of the hydrophobic face of the B series peptides compared to the most hydrophobic face of the A series peptides. In other words, the amphipathicity of the B series peptides increases with increasing polypeptide chain length and one would expect an increase in Δt (retention time difference between peptides of the A and B series of the same polypeptide chain length) if this hydrophobic face binds preferentially to the hydrophobic stationary phase.

B. HELICITY OF THE MODEL PEPTIDES

The α-helicity of the model peptides was determined by circular dichroism (CD). All of the CD spectra were measured in 0.1% aqueous TFA (pH = 2) containing trifluoroethanol (TFE) 1:1 (v/v), a solvent that induces helicity in single-chain potentially α-helical peptides and 1 mM DTT to prevent the formation of an interchain disulfide bridge between two peptide monomers. TFE is considered to be a non-interacting (inert) solvent.[31,32] Lau et al.[33] have shown previously

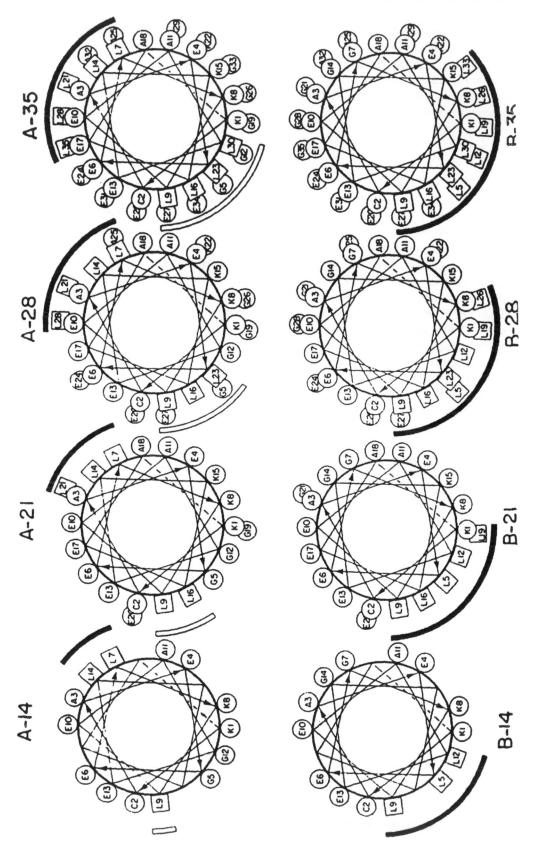

FIGURE 1. The amino acid sequences for the A and B series peptides (Table 1) are represented as helical wheels. The perimeter of the wheel corresponds to the polypeptide backbone and the external circles and squares represent the individual amino acid side-chains. The leucine residues are denoted by squares and all other side-chains by circles. Since the α-helix has 3.6 residues per turn, adjacent side-chains in the sequence are separated by 100° of arc on the wheel. Two hydrophobic surfaces are observed on opposing sides of the helix for the peptides in the A series as indicated by the solid and hatched bars. The peptides of the B-series are amphipathic containing one dominant hydrophobic surface on one side of the α-helix as indicated by the solid bars. The A and B series peptides are denoted by the letters A and B, respectively, followed by a number which denotes the number of residues in the polypeptide chain.

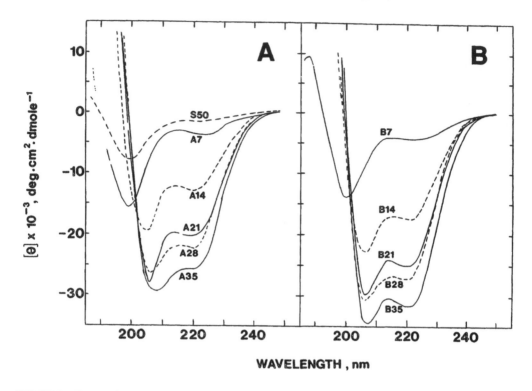

FIGURE 2. Circular dichroism (CD) spectra of the model peptide polymers. The CD spectra of the peptides were measured in 0.1% aq. TFA (pH 2.0) containing trifluoroethanol (1:1, v/v) and 1 mM DTT. Circular dichroism spectra were recorded on a JASCO J-500C spectropolarimeter attached to a JASCO DP-500N data processor. The instrument was routinely calibrated with an aqueous solution of recrystallized d-10-camphorsulphonic acid. Constant N$_2$ flushing was employed. The sequences of the A and B series of peptides are shown in Table 1. S50 is a synthetic peptide standard with the sequence Ac-(Gly-Leu-Gly-Ala-Lys-Gly-Ala-Gly-Val-Gly)$_5$-amide.

that 50% TFE disrupts the quaternary structure of α-helical coiled-coils, producing single-stranded α-helices, i.e., TFE is a denaturant of tertiary and quaternary structure stabilized by hydrophobic interactions. This solvent system was used to mimic the hydrophobicity of the reversed-phase matrix in inducing α-helical structure and disrupting the tertiary and quaternary structure of the model peptides or disrupting any non-specific aggregation.

The CD spectra of A and B series peptides are shown in Figure 2A and B, respectively, and the molar ellipticity values at 220 nm are summarized in Table 2. The heptapeptides (A7 and B7) do not show significant sign of helicity ([θ]$_{220}$ = –3600 and –4300°, respectively). The molar ellipticity values of the model peptides at 220 nm increase with increasing peptide chain length and the α-helical conformation dominates the model peptides of both series when chain length is longer than 14 residues (Table 2). Lau et al.[26] have obtained similar results at pH 7 in 50% aqueous TFE.

As explained by Lau et al.,[26] the predicted molar ellipticities for 100% α-helical peptides of varying polypeptide chain lengths (Table 2) were calculated based on theoretical values as described by Chen et al.[38] Although the α-helical segments in proteins are generally short, ranging from 3 to about 20 peptide units, the ends of the helices are conformationally restricted by the tertiary structure of the remaining polypeptide chain. Thus, it is not surprising that the molar ellipticities observed for these small polyheptapeptides in solution (end effects would be maximal compared to the same segment length in a protein) deviate substantially from the predicted values of Chen et al.[38] which were derived from protein data. Interestingly, as the peptides increase in polypeptide chain length, their helicity in trifluoroethanol increases; in

TABLE 2
Circular Dichroism Results of the A and B Series Synthetic Peptide Polymers

Peptide	$[\theta]_{220}$[a]	%-helix[b]	Peptide	$[\theta]_{220}$	%-helix[b]	Predicted[c]
A7	−3600	16	B7	−4300	19	−22,800
A14	−13000	44	B14	−17400	59	−29,600
A21	−20400	64	B21	−25100	79	−31,800
A28	−22300	68	B28	−27200	82	−32,900
A35	−25700	76	B35	−31700	94	−33,600

[a] A 1:1 (v/v) mixture of 0.1% aq. TFA with trifluoroethanol (TFE) at pH 2.0. Temperature was 25°C for ellipticity measurements.
[b] The % α-helix was calculated based upon the predicted value taken as 100% α-helix at a given polypeptide chain length.
[c] The predicted molar ellipticity (X_H^n) is based on the theoretical values as described by Chen et al.[38] of $[\theta]_{220}$ values of −36,300 degree cm²dmol⁻¹ for a helix of infinite length (X_H^∞). The equation used is $X_H^n = X_H^\infty (1-k/n)$ where n is the number of residues/helix and k is a wavelength dependent constant (2.6 at 220 nm).

addition, the more amphipathic the peptide, the higher the helicity (compare A and B series, Table 2). This suggests that hydrophobic interactions along the helix are offering stability to these single stranded α-helices (Figure 1).

C. RPC AND SEC

The retention behavior of the model peptides was investigated by RPC on a C_8 column under conditions identical with those employed by Guo et al.[11] who obtained retention coefficients for all 20 amino acid residues found in peptides and proteins (linear AB gradient of 1% B/min at a flow-rate of 1 ml/min, where Eluent A is 0.1% aq. TFA and Eluent B is 0.1% TFA in acetonitrile at pH 2). The acidic nature of the TFA-containing mobile phase suppresses the ionization of surface silanols on silica-based columns, thereby overcoming undesirable ionic interactions between basic solutes and the column packing.[39] Figure 3A shows the elution profiles of five peptide mixtures A7 and B7, A14 and B14, A21 and B21, A28 and B28, A35 and B35 where each mixture contains two peptides of the same chain length. It can be seen clearly that four out of five of the mixtures of the two peptides can be separated by RPC, the exception being A7 and B7, although the two peptides in each case have the same amino acid composition and similar secondary structure. The difference in retention time between peptides of the same length increases with the peptide chain length (Table 3).

Lau et al.[33] reported a linear relationship between ln molecular weight and peptide retention time during RPC for a series of five peptide polymers of 8 to 36 residues. Mant and Hodges[1] demonstrated a similar exponential relationship for a series of five peptide polymers of 10 to 50 residues. Figure 3B demonstrates a linear relationship between ln molecular weight and RPC retention time of the A and B series of peptides. These results indicate that all of the A and B peptides are bound to the reversed-phase column in their monomeric form. The different slopes for the A and B peptides (Figure 3B) could be explained by the increasing difference in hydrophobicities of the most hydrophobic faces (the most hydrophobic sides of the α-helices) between the A and B peptides as their polypeptide chain length increases.

To provide further support that these model peptides are monomeric when bound to the reversed-phase column, their structure was examined by SEC (Figure 4). Two solvent systems were used to mimic the conditions of RPC and CD: system 1 was acetonitrile — 0.1% aq. TFA (1:1) (v/v) containing 0.1 M KCl and 1 mM DTT, pH 2 (Figure 4A and B); system 2 was trifluoroethanol — 0.1% aq. TFA (1:1) (v/v) containing 0.1 M KCl and 1 mM DTT, pH 2 (Figure 4D and E). The nonpolar solvents (acetonitrile and trifluoroethanol) at a 1:1 ratio in aqueous solution were used to mimic the hydrophobicity of the reversed-phase column. Since the

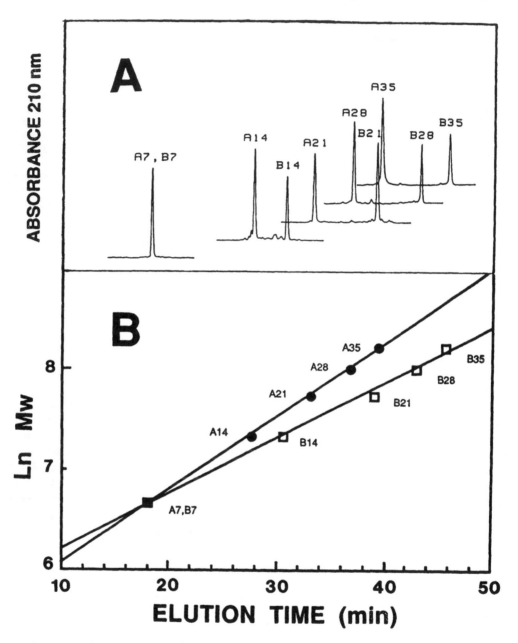

FIGURE 3. RPC of the A and B series of peptides (Table 1). In panel A are the elution profiles of 5 mixtures of peptides where each mixture contains two peptides of different sequences but the same polypeptide chain length. In panel B, the retention times of the peptides are plotted against ln molecular weight. Column: Aquapore RP-300 C_8 (220 × 4.6 mm I.D., 7-μm particle size, 300-Å pore size; Brownlee Labs., Santa Clara, CA). HPLC Instrument: the instrumentation consisted of a Varian Vista Series 5000 liquid chromatograph (Varian, Walnut Creek, CA) coupled to a Hewlett-Packard (Avondale, PA) HP1040A detection system, HP 9000 series 300 computer, HP 9153 B disc drive, HP 2225 A Thinkjet printer and HP 7440 A plotter. Samples were injected with a 500 μl injection loop (Model 7125, Rheodyne, Cotati, CA). Conditions: linear AB gradient, where Eluent A is 0.1% aq. TFA and Eluent B is 0.1% TFA in acetonitrile; flow-rate, 1 ml/min; 26°C. The peptides were dissolved in 0.1% aq. TFA containing 1 mM DTT prior to application.

trifluoroethanol was used for the CD measurements, it was important to verify the monomeric structure of the peptides in this medium. The DTT was added to the eluent to keep the peptides in their reduced form and prevent interchain disulfide bridge formation between two peptide monomers. The 0.1 M KCl was necessary to overcome the undesirable ionic interactions

TABLE 3
Comparison of Predicted and Observed Retention Times in Reversed-Phase HPLC for Three Series of Peptide Polymers

Peptide	t_A^{obs*} (min)	Peptide	t_B^{obs} (min)	$\Delta t_{(B-A)}^{obs}$† (min)	$\tau_{A\ and\ B}$‡ (min)	$\Delta t_A^{obs-pred.}$# (min)	$\Delta t_B^{obs-pred.}$ (min)	Peptide	t_B^{obs} (min)	τ_S (min)	$\Delta t_S^{obs-pred.}$ (min)
A7	19.3	B7	19.3	0	20.0	0.7	0.7	S10	19.0	20.0	1.0
A14	29.6	B14	32.5	2.9	27.3	2.3	5.2	S20	24.9	24.1	0.8
A21	34.8	B21	40.4	5.6	32.5	2.3	7.9	S30	27.2	26.8	0.4
A28	38.3	B28	45.0	6.7	36.3	2.0	8.7	S40	28.5	28.5	0
A35	40.7	B35	48.0	7.3	39.2	1.5	8.8	S50	29.3	29.7	0.4
Average difference						1.7	6.3				0.5

* The observed retention times (min) were obtained on an Aquapore RP-300 C$_8$ (220 x 4.6 mm I.D.) column under conditions of linear AB gradient elution (1% B/min), where Eluent A is 0.1% aq. TFA and Eluent B is 0.1% TFA in acetonitrile (pH 2.0); flow-rate, 1 ml/min; 26°C; absorbance at 210 nm.

† The difference in observed retention times between A and B peptides ($t_B^{obs} - t_A^{obs}$).

‡ τ is the predicted retention time of a peptide - calculated as described in the text.

$\Delta t^{obs-pred.}$ is the difference between the predicted (τ) and observed (t^{obs}) peptide retention time.

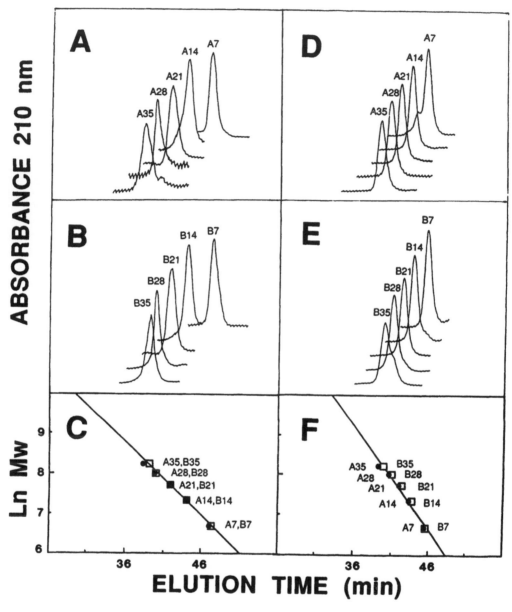

FIGURE 4. SEC of the A and B series of peptides (Table 1). Column: Altex Spherogel TSK G2000SW (300×7.5 mm I.D., 10-μm particle size, 130-Å pore size; Beckman Instruments, Berkeley, CA). HPLC instrument: as Figure 3. Mobile phase: panels A and B, 0.1% aq. TFA containing acetonitrile (1:1, v/v), 0.1 *M* KCl and 1 m*M* DTT (system 1 in the text; panels D and E, 0.1% aq. TFA containing trifluoroethanol (1:1, v/v), 0.1 *M* KCl and 1 m*M* DTT (system 2). Flow-rate 0.2 ml/min. Panels C and F show plots of ln molecular weight vs. peptide retention time in mobile phase systems 1 and 2, respectively. The peptides were dissolved in 0.1% aq. TFA containing 0.1 *M* KCl and 1 m*M* DTT prior to application.

between solutes and the column matrix,[34] since ideal SEC can only occur when non-specific hydrophobic and ionic interactions between the peptides and the matrix are eliminated. These effects have been observed on all commercial size-exclusion columns tested to date, and it is generally recommended that SEC columns should be operated at ionic strengths greater than 0.1 to 0.2 *M* to overcome or minimize these effects.[34,35] All ten model peptides in both solvent systems are monomers as observed by the linear plots of ln monomeric molecular weight vs.

retention time (Figure 4, panels C and F). These results support the conclusion that the peptides are monomeric under the conditions of RPC (Figure 3) and the conditions used to determine their α-helical content by circular dichroism (Figure 2). The observation that peptides of the same polypeptide chain length in the A and B series have essentially identical retention times in SEC also suggests that the peptides have similar conformations.

Disruption of the tertiary and quaternary structure of a polypeptide or non-specific peptide aggregation in the nonpolar medium of RPC is not surprising when one considers that the major stabilizing forces are hydrophobic interactions. Since hydrogen bonds which stabilize the α-helices are exceedingly unstable in the presence of water, one would expect that, as the non-polarity of the medium increases, the stability of the secondary structure (α-helices) in single-stranded polypeptides would increase.[25]

D. EFFECT OF α-HELICAL CONFORMATION AND HYDROPHOBIC RESIDUE DISTRIBUTION ON PEPTIDE RETENTION BEHAVIOR IN RPC

Our laboratory has demonstrated that, if a peptide is not subject to sequence-dependent conformational or nearest-neighbor effects, its reversed-phase chromatographic behavior can be correlated with its amino acid composition and the number of the residues in the polypeptide chain;[36,37] thus, peptide retention time in RPC can be predicted by the following equation:

$$\tau_c = \Sigma R_c - \left[m \Sigma R_c \cdot \ln N + b \right] \tag{1}$$

where τ_c is a predicted retention time, ΣR_c is the sum of the retention coefficients of Guo et al.[11] for the amino acid residues in a peptide, N is the number of residues in the peptide, and m and b are constants dependent on the column and instrumentation used and are obtained from running a set of peptide standards of varying polypeptide chain lengths. A series of five synthetic peptide polymers with the sequence Ac-(Gly-Leu-Gly-Ala-Lys-Gly-Ala-Gly-Val-Gly)$_n$-amide (where n = 1 to 5, denoted S10, S20, S30, S40, and S50, respectively) was used as a set of standards in the present study. This series of peptides was chosen since the high glycine content of the peptides minimizes or eliminates any tendency towards secondary structure formation.[34] In addition, the observed retention times in RPC of this series peptides and the predicted retention times calculated from their retention coefficients and the number of residues in the peptide chain have an excellent correlation.[37] The similar molecular weight range of this series of peptides (826 to 3894 dalton) compared with the molecular weight of A and B series peptides (789 to 3749 dalton) and the similar positive charges of this series of peptides (+1 to +5 for S10 to S50) compared with A and B series peptides (+1 to +5 for A7, B7 to A35, B35) make them excellent standards.

The CD results have shown that the peptide standards do not have any α-helicity even in the presence of trifluoroethanol, an α-helix inducing solvent (Figure 2A, peptide S50). By comparing the retention behavior of these peptide standards, which have no α-helical structure, with the A and B series peptides, where the α-helical structure dominates their conformation, one should be able to understand the effect of α-helical structure and the distribution of hydrophobic residues in the α-helical peptides on their retention behavior in RPC (Figure 5A to 5C). A plot of $\Sigma R_c - t^{obs}$ vs. $\Sigma R_c \ln N$, where t^{obs} is observed peptide retention time, for the five peptide standards is shown in Figure 5D. The slope (m) and intercept (b) values were determined and subsequently substituted into Equation 1 to produce the predicted peptide retention values shown in Figure 5E and Table 3. The effects of peptide chain length and overall peptide hydrophobicity have been taken into account in the predicted retention time. Deviation of the observed retention times of the A and B series peptides from predicted retention time can be considered indicative of the effect of α-helical structure or the distribution of hydrophobic residues in the α-helical peptide (preferred binding domains) on peptide retention behavior in RPC.

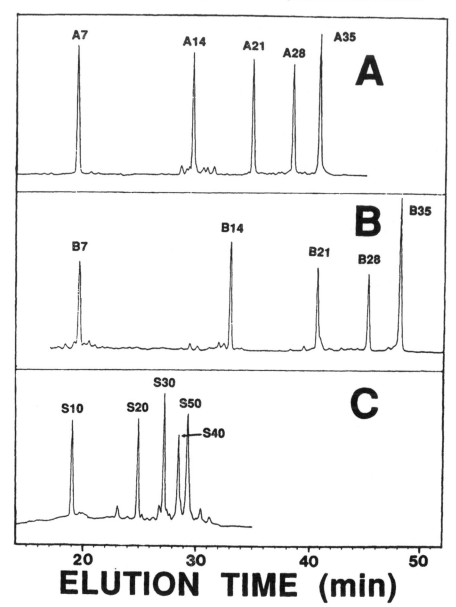

FIGURE 5. RPC elution profiles and correlation of predicted vs. observed retention times of synthetic peptide polymers. Column, HPLC instrumentation and conditions as Figure 3. Panels A, B, and C show elution profiles of the A series, B series and S series of peptides, respectively. Panel D: plot of $\Sigma R_c\text{-}t^{obs}$ vs. $\Sigma R_c \ln N$ for the S series of peptides (details shown in text); panel E: correlation of predicted and observed peptide retention times for the A series (●), B series (□) and S series (▲) of peptides. The sequences of the A and B series of peptides are shown in Table 1. The sequence of the S series of peptides is Ac-(Gly-Leu-Gly-Ala-Lys-Gly-Ala-Gly-Val-Gly)$_n$-amide, where n = 1 to 5 (10, 20, 30, 40, and 50 residues); S10 denotes the 10-residue peptide, etc.

The average deviation between observed and predicted retention times for the five standards (S series) used to provide the linear prediction equation (Figure 5D) is only 0.5 min (Table 3). These standards allow one to correct for differences in retention behavior between columns and to correct for the polypeptide chain length effect on peptide retention.[36,37] The average deviation between observed and predicted retention times for the A series peptides was only 1.7 min (Table 3). This small deviation indicates that the predictive method which utilizes peptide standards

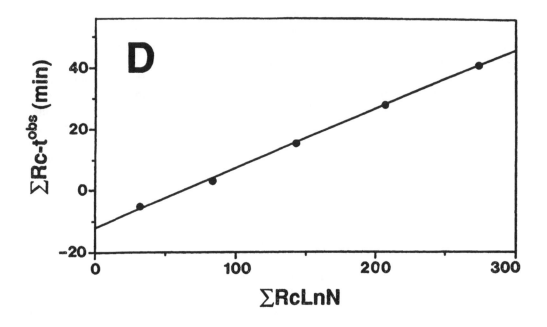

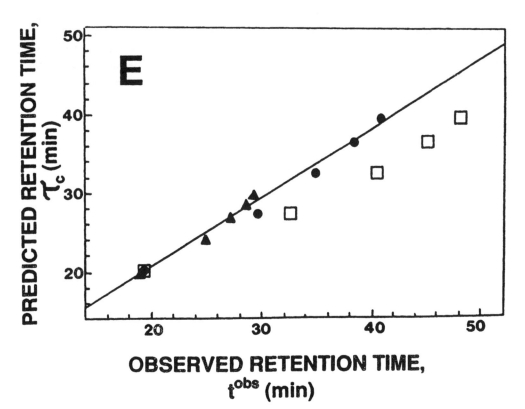

FIGURE 5 (continued)

with no observed secondary structure can still be used to predict the retention times of peptides with a high content of α-helical structure. This accuracy is highlighted by the high correlation ($r = 0.99$) of predicted vs. observed peptide retention times for the A and S series peptides (Figure 5E). These results suggest that if the hydrophobic residues of a peptide are well distributed around the α-helical structure of the peptide, the α-helical conformation of the peptide does not have any significant effect on peptide retention behavior in reversed-phase chromatography and that all residues in the α-helix are contributing to the overall hydrophobicity of the peptides on interacting with the hydrophobic stationary phase.

The deviation between the observed and predicted retention times for the B series peptides ranged from 0.7 (7-residue peptide) to 8.8 min (35-residue peptide) with an average deviation of 6.3 min (Table 3). Similarly, the difference in observed retention times between the A peptides, which are predicted reasonably well, and the B peptides increases with polypeptide chain length. This difference can only be explained by the difference of hydrophobic residue distribution in the α-helical structure of the peptides (Figure 1), since the two series of peptides (A and B series) have the same composition and similar secondary structure. In other words, peptides which form amphipathic α-helices cannot be predicted using the present method (Figure 5E).

IV. CONCLUSION

Our results clearly demonstrate that if the hydrophobic residues of a peptide are well distributed around the α-helical structure of a peptide, the α-helical conformation of the peptide *per se* does not have a significant effect on peptide retention behavior in RPC and the predicted retention behavior is still related to overall amino acid composition. On the other hand, our results clearly indicate that the distribution of hydrophobic residues is a significant factor affecting retention behavior of peptides in RPC and must be taken into account for prediction of peptide retention behavior for peptides which form amphipathic α-helices. Amphipathic helices result in an increase in the retention time over that predicted because of the preferred binding of the hydrophobic face of the helices with the hydrophobic surface of the matrix and the inequivalent expression of all residues to the overall hydrophobicity of the peptide. Preliminary results suggest that it is possible to adapt our equations for peptide retention prediction to handle amphipathic α-helices and that it is possible to predict amphipathic α-helical structure in peptides based upon peptide retention data.[40]

ACKNOWLEDGMENTS

This work was supported by research grants from the Medical Research Council of Canada and by equipment grants, postdoctoral fellowship (N.E.Z.) and research allowance from the Alberta Heritage Foundation for Medical Research. We thank K. Oikawa in the laboratory of Dr. C.M. Kay for his skilled technical assistance in performing the circular dichroism measurements.

REFERENCES

1. **Mant, C.T. and Hodges, R.S.,** HPLC of Peptides, in *High Performance Liquid Chromatography of Biological Macromolecules: Methods and Applications*, Gooding, K. and Regnier, F., Eds., Marcel Dekker, New York, 1990, 301.
2. **Mant, C.T. and Hodges, R.S.,** Optimization of peptide separations in high-performance liquid chromatography, *J. Liq. Chromatogr.*, 12, 139, 1989.
3. **Mant, C.T. and Hodges, R.S.,** Optimization and prediction of peptide retention behavior in reversed-phase chromatography, in *HPLC of Proteins, Peptides and Polynucleotides*, Hearn, M.T.W., Ed., VCH Publishers, 1991, 277.

4. **Meek, J.L.**, Prediction of peptide retention time in high-pressure liquid chromatography on the basis of amino acid composition, *Proc. Natl. Acad. Sci. U.S.A.*, 77, 1632, 1980.
5. **Meek, J.L. and Rossetti, Z.L.**, Factors affecting retention and resolution of peptides in high-performance liquid chromatography, *J. Chromatogr.*, 211, 15, 1981.
6. **Su, S.-J., Grego, B., and Hearn, M.T.W.**, Analysis of group retention contributions for peptides separated by reversed-phase high performance liquid chromatography, *J. Liq. Chromatogr.*, 4, 1745, 1981.
7. **Wilson, K.J., Honegger, A., Stötzel, R.P., and Hughes, G.J.**, The behaviour of peptides on reversed-phase supports during high-pressure liquid chromatography, *Biochem. J.*, 199, 31, 1981.
8. **Browne, C.A., Bennett, H.P.J., and Solomon, S.**, The isolation of peptides by high-performance liquid chromatography using predicted elution positions, *Anal. Biochem.*, 124, 201, 1982.
9. **Sasagawa, T., Okuyama, T., and Teller, D.C.**, Prediction of peptide retention times in reversed-phase high-performance liquid chromatography during linear gradient elution, *J. Chromatogr.*, 240, 329, 1982.
10. **Sasagawa, T., Ericsson, L.H., Teller, D.C., Titani, K., and Walsh, K.A.**, Separation of peptides on a polystyrene resin column, *J. Chromatogr.*, 307, 29, 1984.
11. **Guo, D., Mant, C.T., Taneja, A.K., Parker, J.M.R., and Hodges, R.S.**, Prediction of peptide retention times in reversed-phase high-performance liquid chromatography. I. Determination of retention coefficients of amino acid residues using model synthetic peptides, *J. Chromatogr.*, 359, 499, 1986.
12. **Guo, D., Mant, C.T., Taneja, A.K., and Hodges, R.S.**, Prediction of peptide retention times in reversed-phase high-performance liquid chromatography. II. Correlation of observed and predicted retention times and factors influencing the retention times of peptides, *J. Chromatogr.*, 359, 519, 1986.
13. **Sakamoto, Y., Kawakami, N., and Sasagawa, T.**, Prediction of peptide retention times, *J. Chromatogr.*, 442, 69, 1988.
14. **Jinno, K. and Tanigawa, E.**, Retention prediction of small peptides in reversed-phase liquid chromatography, *Chromatographia*, 25, 613, 1988.
15. **Iskandarini, Z. and Pietrzyk, D.J.**, Liquid chromatographic separation of amino acids, peptides, and derivatives on a porous polystyrene-divinylbenzene copolymer, *Anal. Chem.*, 53, 489, 1981.
16. **Terabe, S., Konaka, R., and Inouye, K.**, Separation of some polypeptide hormones by high performance liquid chromatography, *J. Chromatogr.*, 172, 163, 1979.
17. **Houghten, R.A. and DeGraw, S.T.**, Effect of positional environmental domains on the variation of high-performance liquid chromatographic peptide retention coefficients, *J. Chromatogr.*, 386, 223, 1987.
18. **Hearn, M.T.W. and Aguilar, M.I.**, High-performance liquid chromatography of amino acids, peptides and proteins. LXIX. Evaluation of retention and bandwidth relationships of myosin-related peptides separated by gradient elution reversed-phase high-performance liquid chromatography, *J. Chromatogr.*, 392, 33, 1987.
19. **Houghten, R.A. and Ostresh, J.M.**, Conformational influences upon peptide retention behavior in reversed-phase high performance liquid chromatography, *BioChromatography*, 2, 80, 1987.
20. **Ostresh, J.M., Büttner, K., and Houghten, R.A.**, Reversed-phase chromatography: the effect of induced conformational effects on peptide retention, this publication.
21. **Heinitz, M.L., Flanigan, E., Orlowski, R.C., and Regnier, F.E.**, Correlation of calcitonin structure with chromatographic retention in high-performance liquid chromatography, *J. Chromatogr.*, 443, 229, 1988.
22. **Hodges, R.S., Heaton, R.J., Parker, J.M.R., Molday, L., and Molday, R.S.**, Antigen-antibody interaction. Synthetic peptides define linear antigenic determinants recognized by monoclonal antibodies directed to the cytoplasmic carboxyl terminus of rhodopsin, *J. Biol. Chem.*, 263, 11768, 1988.
23. **Hodges, R.S., Semchuk, P.D., Taneja, A.K., Kay, C.M., Parker, J.M.R., and Mant, C.T.**, Protein design using model synthetic peptides, *Peptide Res.*, 1, 19, 1988.
24. **Hodges, R.S., Saund, A.K., Chong, P.C.S., St-Pierre, S.A., and Reid, R.E.**, Synthetic model for two-stranded α-helical coiled-coils. Design synthesis and characterization of an 86-residue analog of tropomyosin, *J. Biol. Chem.*, 256, 1214, 1981.
25. **Talbot, J.A. and Hodges, R.S.**, Tropomyosin: a model protein for studying coiled-coil and α-helix stabilization, *Acc. Chem. Res.*, 15, 224, 1982.
26. **Lau, S.Y.M., Taneja, A.K., and Hodges, R.S.**, Synthesis of a model protein of defined secondary and quaternary structure: effect of chain length on the stabilization and formation of two-stranded α-helical coiled coils, *J. Biol. Chem.*, 259, 13253, 1984.
27. **Kanehisa, M.I. and Tsang, T.Y.**, Local hydrophobicity stabilizes secondary structures in proteins, *Biopolymers*, 19, 1617, 1980.
28. **Eisenberg, D., Weiss, R.M., and Terwilliger, T.C.**, The hydrophobic moment detects periodicity in protein hydrophobicity, *Proc. Natl. Acad. Sci., U.S.A.*, 81, 140, 1984.
29. **DeGrado, W.F. and Lear, J.D.**, Induction of peptide conformation at apolar/water interfaces. 1. A study with model peptides of defined hydrophobic periodicity, *J. Am. Chem. Soc.*, 107, 7684, 1985.
30. **DeGrado, W.F., Wasserman, Z.R., and Lear, J.D.**, Protein design, a minimalist approach, *Science*, 243, 622, 1989.

31. **Lotan, N., Berger, A., and Katchalski, E.,** Conformation and conformational transitions of poly-α-amino acids in solution, *Ann. Rev. Biochem.,* 41, 869, 1972.
32. **Nelson, J.W. and Kallenback, N.R.,** Stabilization of the ribonuclease S-peptide α-helix by trifluoroethanol, *Proteins: Structure, Function, and Genetics,* 1, 211, 1986.
33. **Lau, S.Y.M., Taneja, A.K., and Hodges, R.S.,** Effects of HPLC solvents and hydrophobic supports on the secondary and quaternary structure of a model protein: reversed-phase and size-exclusion high performance liquid chromatography, *J. Chromatogr.,* 317, 129, 1984.
34. **Mant, C.T., Parker, J.M.R., and Hodges, R.S.,** Size-exclusion HPLC of peptides: requirement for peptide standards to monitor non-ideal behaviour, *J. Chromatogr.,* 397, 99, 1987.
35. **Regnier, F.E.,** High-performance liquid chromatography of proteins, *Methods Enzymol.,* 91, 137, 1983.
36. **Mant, C.T., Zhou, N.E., and Hodges, R.S.,** Correlation of protein retention times in reversed-phase chromatography with peptide chain length and hydrophobicity, *J. Chromatogr.,* 476, 363, 1989.
37. **Mant, C.T., Burke, T.W.L., Black J.A., and Hodges, R.S.,** The effect of peptide chain length on peptide retention behaviour in reversed-phase chromatography, *J. Chromatogr.,* 458, 193, 1988.
38. **Chen, Y.-H., Yang, J.T., and Chan, K.H.,** Determination of the helix and β-form of proteins in aqueous solution by circular dichroism, *Biochemistry,* 13, 3350, 1974.
39. **Mant, C.T. and Hodges, R.S.,** Monitoring free silanols on reversed-phase supports with peptide standards, *Chromatographia,* 24, 805, 1987.
40. **Zhou, N.E., Mant, C.T., and Hodges, R.S.,** Effect of preferred binding domains on peptide retention behavior in reversed-phase chromatography: amphipathic α-helices, *Peptide Res.,* 3, 8, 1990.

Section XI
Introduction to Narrow Bore, Microbore,
and Rapid HPLC Analysis Techniques

MICRO-HPLC TECHNIQUES FOR THE PURIFICATION AND ANALYSIS OF PEPTIDES AND PROTEINS

Timothy D. Schlabach, Lynne R. Zieske, and Kenneth J. Wilson

I. INTRODUCTION

Although the advantages of microbore techniques for protein purification were clearly illustrated more than 6 years ago,[1,2,3] most protein chemists still use conventional columns with internal diameters of 4 mm or greater. Microbore is associated commonly with columns of 1-mm diameters, but even 2-mm I.D. columns offer substantial advantages over conventional columns. Generally, all column formats less than 4 mm in diameter are referred to as microcolumns,[4] and we shall adopt Micro-LC for describing this technique. The belated acceptance of Micro-LC contrasts sharply with the rapid proliferation of conventional HPLC techniques in protein chemistry. Perhaps the benefits of Micro-LC are not as apparent, while the perceived costs are exaggerated.

Great advances were made over the last ten years in protein and peptide separations by HPLC.[5] The early eighties welcomed major advances in the purification of polypeptides and peptide fragments for structural analysis. The introduction of wide-pore supports,[6] and the discovery of trifluoroacetic acid (TFA) as a mobile phase modifier[7] improved the efficiency and recovery of proteins to an extent that revolutionized protein purification technology.[8] Conventional 4 mm I.D. columns were quite adequate for most, or all, protein/peptide purification problems until the advent of micro-sequencing,[9] which allowed primary sequence information to be obtained from as little as 5 pmol of polypeptide. Such minute quantities of materials are not detectable with conventional systems, which require 20 to 50 ng of polypeptide or protein to produce a detectable signal under favorable conditions.[3] Assuming that the average tryptic peptide has a mass of about 1000 daltons, it is clear that new techniques must be developed to work at the 5 to 25 pmol level.

The most striking advantage of Micro-LC is the higher sensitivity which enables purification of samples containing less than 10 ng of material.[10] For example, a 50 pmol sample may be barely detectable above the baseline noise on a 4 mm column, but will have a peak height more than 20 times higher on a 1 mm column and 5 times higher on a 2 mm column. The sensitivity of 2 mm columns can be further improved by lowering the flow rate below 100 μl/min.[11]

Micro-LC has other advantages as well. It circumvents the need for subsequent concentration steps when preparing samples for further analysis, because peak volumes are typically 60 μl or less.[11] Lower flow rates demand less solvent so less time is spent preparing mobile phases. Less solvent means less waste which translates into fewer trips for solvent disposal.

The most onerous drawback to Micro-LC is the need to acquire new equipment to reap fully the sensitivity benefits associated with 1-mm I.D. columns. Most modern HPLC hardware, however, is compatible with 2-mm I.D. columns. By simply replacing a conventional column with a 2-mm I.D. column, a substantial improvement can be made in sensitivity. Until recently the selection of 2-mm columns was fairly limited, but now many manufacturers offer them with a wide range of materials for protein separations.

A lingering misconception about Micro-LC is that the smaller diameter plumbing required by this technique fouls more frequently. The widespread use of microbore tubing (0.007 in. I.D.

or less) in many commercial LC systems indicates that fouling is not a problem, at least in laboratories that conscientiously filter or centrifuge their samples and use high quality solvents. Most fouling problems relate to salt precipitates which occur during improper solvent change-overs that omit the necessary cleansing of the system with distilled water.

Peptide mapping also followed the same trend towards analyses at the low picomole level. This technique compares the pattern of a standard protein digest with a potentially modified sample which should exhibit a shift in retention time for one or more peptide fragments.[12] Micro-LC techniques improve the sensitivity of peptide mapping by as much as twenty-fold,[10] allowing multiple analytical runs to be made with smaller amounts of material. Because many proteins can only be isolated in trace amounts, there is a growing need for peptide mapping at the low-to-mid picomole level.

Practical aspects of micro-purification and analysis are stressed in this study as they relate to the characterization of proteins and protein digests. Particular emphasis is given to sample requirements for micro-sequencing, so as to convey the specific benefits and costs of Micro-LC in determining protein structure at the mid-to-low picomole level.

II. EXPERIMENTAL

A. MATERIALS

Protein and protein digests were obtained from Applied Biosystems, Inc. (Foster City, CA), except for transferrin obtained from Sigma Chemical (St. Louis, MO) and TPCK-trypsin from Worthington Biochemical Laboratories (Freehold, NJ). Samples were dissolved in 0.1% trifluoroacetic acid in distilled water and refrigerated when not in use.

B. INSTRUMENTATION

Applied Biosystems supplied the Model 130 HPLC for flow rates less than 1 ml/min and the Model 150 for conventional flow rates.

C. TRYPTIC DIGESTION

Approximately 5 mg of transferrin was dissolved in 1 ml 0.2 M ammonium bicarbonate (pH 8.0 to 8.5). A similar amount of trypsin was dissolved first in 1 ml of 0.1% trifluoroacetic acid, which solubilizes the enzyme and prevents any autodigestion. Later, a 50 µl aliquot was added to the transferrin solution, followed by incubation overnight at 37°C.

This same procedure was employed for sperm whale apomyoglobin. Note that proteins derived from the sperm whale are now restricted from sale in the U.S.

III. RESULTS AND DISCUSSION

Sensitivity or minimal detectable quantity varies inversely with column diameter.[13] The effect can be understood as a scaling down of volumes, particularly peak volumes, while holding the mass constant. Thus, components are eluted in smaller volumes on micro-LC columns, resulting in higher concentrations. The separation in Figure 1 clearly shows that the tryptic peptides of apomyoglobin produce peaks that are 20- to 30-fold higher on the 1-mm column. The inferior sensitivity of the 4.6-mm column is so poor that many peptides observed in the upper trace were not even discernable in the lower profile. The apparent sensitivity limit was above 100 pmol for the 4.6-mm column and as low as 20 to 40 pmol on the 1-mm column. Lower detection limits can be obtained on conventional columns with the aid of baseline subtraction or other signal enhancing techniques.[14]

Quality peptide separations on 1-mm columns require HPLC equipment that delivers smooth gradients at flow rates of 50 µl/min or less. Such gradients can be achieved with high-pressure,

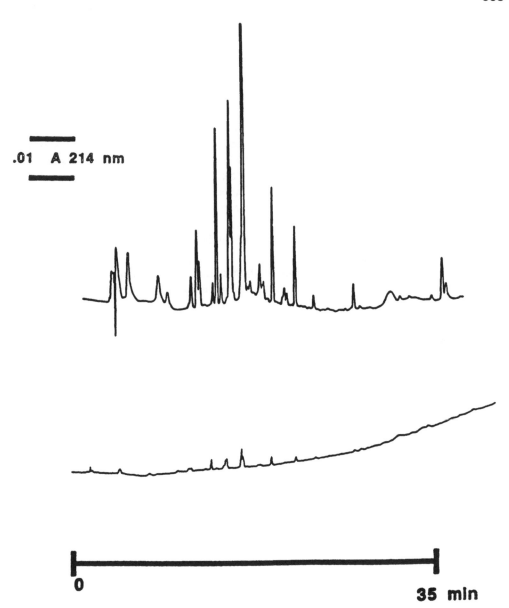

.01 A 214 nm

0   35 min

FIGURE 1. Comparative peptide mapping of apomyoglobin-tryptic digest on conventional and microbore columns. Columns: lower chromatogram, Aquapore RP300 C$_8$, 220 × 4.6 mm I.D., 7-μm particle size, 300-Å pore size (Applied Biosystems); upper chromatogram, Aquapore RP300 C$_8$, 250 × 1 mm I.D., 7 μm, 300 Å. Conditions: lower chromatogram, linear AB gradient (Eluent A = 0.1% aq. TFA; Eluent B = 0.088% TFA in acetonitrile/water [70:30]) from 0 to 60% acetonitrile in 45 min at a flow-rate of 1 ml/min; upper chromatogram, same conditions as above except for a flow-rate of 50 μl/min. The sample size was 100 pmol on both columns.

micro-syringes that deliver almost pulseless flow at rates as low as a few microliters per minute.[15] Our HPLC system produced sharp, linear gradients at flows as low as 40 μl/min.[16] This system has a dead volume of less than 300 μl, which includes both a static and dynamic mixer. Dead volume refers to the internal volume between the output of the pumps and the injection valve. The larger the dead volume, the longer the delay before a gradient reaches the column. At 50 μl/min, this delay is approximately 6 min with our system, which may have the smallest dead volume of any commercial LC system.

Tubing connections from the column to the injector and to the detector are critical. Tubing diameters should be 0.007 in. or smaller, because 0.01-in. tubing can produce noticeable losses in efficiency in Micro-LC systems.[17] Teflon™ tubing is now available in 0.005-in. diameters and is an attractive alternative to stainless-steel tubing, because Teflon™ is easily cut with a razor, compatible with both steel and plastic fittings, and can withstand pressures up to about 2000 psi in the presence of up to 50% acetonitrile.

Smaller volume flow cells can improve the performance of Micro-LC systems. A 2.4-μl flow cell was installed in the Micro-LC detector, replacing the conventional 12-μl flow cell. The smaller cell results in less band broadening, but more importantly it facilitates direct and convenient collection of fractions, because a peak can be collected as it appears on the detector. With a 12-μl cell, there would be a time delay of about 14 sec at a flow of 50 μl/min before the peak leaves the detector, which could lead to collection errors and missed peaks. Tubing from the flow cell to the collection vial can also produce significant delays at a flow rate of 50 μl/min. A 20-cm length of 0.010-in. tubing adds an additional delay of 12 sec, whereas the same length of 0.005-in. tubing adds only 3 sec.

The sensitivity gains inherent with smaller bore columns are not limited to reversed-phase separations. Figure 2 shows that the smallest diameter, anion-exchange column has about ten-fold higher sensitivity than the 4.6-mm column. The difference in sensitivity is even larger when the 4.6-mm column is operated at 1 ml/min instead of 0.5 ml/min. Similar increases in sensitivity were found for proteins separated by hydrophobic interaction chromatography.[10]

Smaller peak volumes are even more important in ion-exchange separations because concentration of column fractions by vacuum drying can result in salt cakes and loss of material. Because ion-exchange is often used as an early fractionation step in a multi-stage purification process, smaller fraction volumes permit direct transfer to the next purification stage. Two-dimensional HPLC purifications are thus becoming common where a crude protein extract[18] or complex protein digest[19] is fractionated on an ion-exchange column, followed by further purification of each fraction by injection and gradient elution on a reversed-phase column. Small fraction volumes (50 to 200 μl) from the ion-exchange column allow for direct, loop-injection of the entire fraction.

A major limitation of 2-mm or smaller diameter columns was the limited selection of column types and the inferior efficiency of these columns. Many major suppliers of columns for biochemical separations now offer several types of 2-mm diameter columns which appear comparable to conventional columns in terms of their resolution and efficiency. Figure 3 demonstrates the high efficiency and resolution of a commercially available 2-mm column in separating dozens of peptides in the tryptic digest of transferrin. Only about 150 pmol of material was injected, suggesting that the minimal detectable level should be less than 30 pmol. The elevated column temperature (45°C) produced about 10% better resolution than the separation at room temperature. We believe that a higher temperature reduces the tendency for peptide fragments to aggregate.

The performance advantages and systems requirements for micro-LC columns are listed in Table 1. In general, relationships between columns are proportional to the square of their diameters. The high and low sensitivity values for each column diameter demonstrate this proportionality. The range between high and low sensitivity reflects the effect of flow rate on sensitivity.[11] The high and low capacity data for each diameter, however, exhibited greater variation from one column size to another. The cause is not understood at present, but may reflect problems in distributing large protein masses uniformly over smaller columns.

Whereas capacity depends linearly on column length, peak volume varies logarithmically

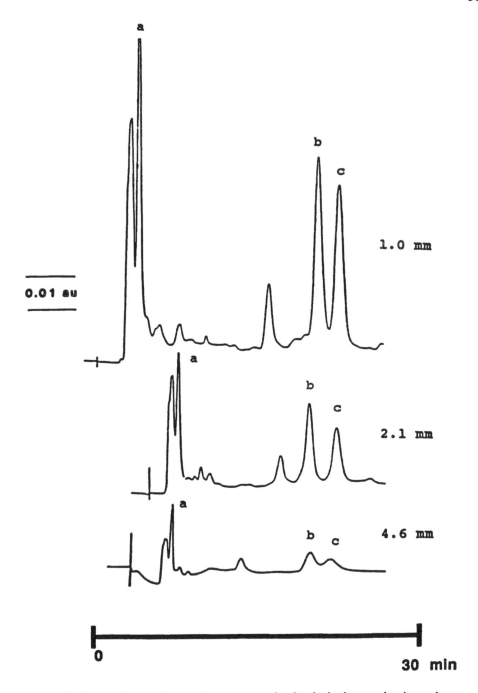

FIGURE 2. Comparative protein separations on conventional and microbore weak anion-exchange columns. Columns: Aquapore AX-300, 250 × 1.0 mm I.D., 100 × 2.1 mm I.D., and 100 × 4.6 mm I.D., 7-μm particle size, 300-Å pore size (Applied Biosystems). Conditions: linear AB gradient (5% B to 60% B in 20 min), where Buffer A is 0.02 M Tris buffer, pH 7.5, and Buffer B is Buffer A containing 0.5 M sodium chloride; flow-rate, 80 μl/min (1.0 mm I.D. column), 200 μl/min (2.1 min), 500 μl/min (4.6 min). Identical samples containing 2.5 μg carbonic anhydrase (a), 1.0 μg α-lactalbumin (b) and 2.5 μg ovalbumin (c) were separated.

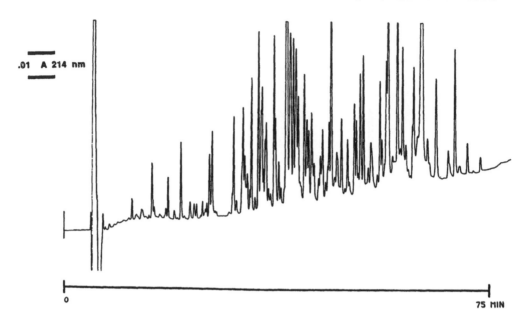

FIGURE 3. High efficiency peptide mapping of transferrin-tryptic digest on a narrow bore column. Column: Vydac TP52 C_{18}, 250 × 2 mm I.D., 5-μm particle size, 300-Å pore size (Vydac, Hesperin, CA). Conditions: linear AB gradient (1% B/min, starting from 5% B), where the eluents were the same as shown in Figure 1; flow-rate, 150 μl/min; temperature, 45°C; sample volume, 20 μl.

TABLE 1
Comparison of Column Performance and System Requirements

	Column diameter		
	1 mm	2 mm	4 mm
Column performance			
Upper capacity (mg)	0.04—0.30	0.25—1.8	1.4—10
Sensitivity (pmol)[a]	2—6	8—25	25—75
Flow rate range (μl/min)	20—100	50—300	400—2000
Peak volumes (μl)	10—40	20—100	200—800
System requirements			
System dead volume (ml)	0.1—0.4	0.3—1.0	1.0—3.0
Flow cell volume (μl)	0.5—5	2—10	5—20
Tubing diameters (inch)	0.005	0.007	0.010

[a] Protein digest that yields more than ten peptide fragments.

with flow rate.[11] Peak volumes of 25 to 50 μl are ideal for micro-sequencing and micro-amino acid analysis, and are readily achievable on either 1- or 2-mm I.D. columns. Even smaller peak volumes are desirable for capillary electrophoresis.[20] This technique consumes but a fraction of a microliter for injection, and smaller peak volumes would provide higher concentrations and improved sensitivity.

The most rigorous requirement in Table 1 is the system dead volume constraint, because this is solely at the discretion of the HPLC manufacturer. Currently, only HPLC systems that employ micro-syringe pumps meet the dead-volume requirements for 1-mm columns. Not surprisingly, only a few percent of chromatographers use 1-mm columns. Therefore, 1-mm column development has lagged behind, and the variety and performance of these columns is generally inferior.

When advanced HPLC hardware for Micro-LC is not available, the best alternative is to use 2-mm columns with modern HPLC hardware that can produce sharp gradients at flows from 100 to 200 µl/min. The resulting gains in sensitivity and performance will enhance the quality of the protein sequence and composition data which is the ultimate aim of micro-purification techniques. Eliminating the need to concentrate fractions in multi-step purifications not only saves time but also improves recoveries. For genetic studies on protein variants, less material is required for comparative mapping techniques so less time is spent preparing samples. Unless a laboratory is constrained to use outdated HPLC equipment, the newer 2-mm columns will become a must in laboratories at the leading edge of protein chemistry.

REFERENCES

1. **Nice, E.C., Lloyd, C.J., and Burgess, A.W.,** The role of short microbore high-performance liquid chromatography columns for protein separations and trace enrichment, *J. Chromatogr.* , 296, 153, 1984.
2. **Knize, M.G., Wuebbles, B.J., and Felton, J.A.,** The advantages of a semimicro HPLC system for the purification of small amounts of biologically active material, *LC, Liq. Chromatogr. HPLC Mag.*, 2, 222, 1984.
3. **Van Der Zee, R. and Welling, G.W.,** Microbore reversed-phase chromatography of proteins with conventional gradient equipment for high-performance liquid chromatography, *J. Chromatogr.*, 325, 187, 1985.
4. **Kucera, P., Ed.,** *Microcolumn High-Performance Liquid Chromatography*, Elsevier, Amsterdam, 1984.
5. **Chang, S.-H., Gooding, K.M., and Regnier, F.E.,** High performance liquid chromatography of proteins, *J. Chromatogr.*, 125, 103, 1976.
6. **Wilson, K.J., Van Wieringen, E., Klauser, S., and Berchtold, M.W.,** Comparison of the high-performance liquid chromatography of peptides and proteins on 100- and 300- Angstrom reversed-phase supports, *J. Chromatogr.*, 237, 407, 1982.
7. **Mahoney, W.C. and Hermodson, M.A.,** Separation of large denatured peptides by reverse-phase high performance liquid chromatography, *J. Biol. Chem.*, 255, 11199, 1980.
8. **Regnier, F.E.,** HPLC Revolutionizes the life sciences: report on the 2nd international symposium on HPLC of protein, peptides, and polynucleotides, *LC, Liq. Chromatogr. HPLC Mag.* 1, 18, 1983.
9. **Hunkapiller, M.W., Strickler, J.E., and Wilson, K.J.,** Contemporary methodology for protein structure determination, *Science,* 226, 304, 1984.
10. **Wilson, K.J., Dupont, D.R., Yuan, P.M., Hunkapiller, M.W., and Schlabach, T.D.,** Practical and theoretical aspects of microbore HPLC of proteins, peptides and amino acid derivatives, in *Modern Methods in Protein Chemistry*, L'Italien, J., Ed., Plenum, New York, 1987, 21.
11. **Schlabach, T.D. and Wilson, K.J.,** Microbore flow-rates and protein chromatography, *J. Chromatogr.*, 385, 65, 1985.
12. **Ebbert, R.F. and Schmetzer, C.H.,** Isolation and comparative peptide mapping of fibrinogen subunits by reversed-phase high-performance liquid chromatography, *J. Chromatogr.*, 443, 309, 1988.
13. **Novotny, M.,** Microcolumn liquid chromatography, *LC, Liq. Chromatogr. HPLC Mag.* 3, 876, 1985.
14. **Stone, K.L. and Williams, K.R.,** High performance peptide mapping and amino acid analysis in the sub-nanomole range, *J. Chromatogr.*, 359, 203, 1986.
15. **Schwartz, H.E. and Brownlee, R.G.,** Comparison of dynamic and static mixing devices for gradient micro-HPLC, *J. Chromatogr. Sci.*, 23, 402, 1985.
16. **Schwartz, H.E. and Berry, V.V.,** Gradient elution for micro HPLC, *LC, Liq. Chromatogr. HPLC Mag.*, 3, 110, 1985.
17. **Scott, R.P.W. and Kucera, P.,** The design of column connection tubes for liquid chromatography, *J. Chromatogr. Sci.*, 9, 641, 1971.
18. **Zieske, L.R. and Schlabach, T.D.,** A novel approach to protein fraction isolation using direct on-column collection, *Int. Labmate*, 13, 37, 1988.
19. **Takahashi, N., Takahashi, Y., Ishioka, N., Blumberg, B.S., and Putnam, F.W.,** Application of an automated tandem high-performance liquid chromatographic system to peptide mapping of genetic variants of human serum albumin, *J. Chromatogr.*, 359, 181, 1986.
20. **Lauer, H.H. and McManigill, D.,** Capillary zone electrophoresis of proteins in untreated fused silica tubing, *Anal. Chem.*, 58, 166, 1986.

REVERSED-PHASE HPLC SEPARATION OF SUB-NANOMOLE AMOUNTS OF PEPTIDES OBTAINED FROM ENZYMATIC DIGESTS

Kathryn L. Stone, Mary B. LoPresti, J. Myron Crawford, Raymond DeAngelis, and Kenneth R. Williams

I. INTRODUCTION

Since the percentage of eukaryotic proteins that have blocked NH_2-termini may be as high as 80 to 90%,[1,2] alternative procedures other than direct sequencing of the intact protein are frequently required to obtain primary structure information. One reasonable approach is to carry out a complete enzymatic digestion and then separate the resulting peptides by reversed-phase HPLC (RPC). If a reasonably specific enzyme such as trypsin, endoproteinase Lys-C, or *Staphylococcus aureus* V8 protease is used, the resulting peptides will have an average length of 9 (trypsin) to 17 (*Staphylococcus aureus* V8 protease) residues, based on the average occurrence of arginine, lysine, and glutamic acid in sequenced eukaryotic proteins.[3] In our experience, most peptides in or near this size-range resolve well and are recovered in high yield from RPC columns.

Despite the relative simplicity of RPC, there are many choices that have to be made regarding the manufacturer of the instrument and column, the column dimensions, and appropriate gradients and sample loading conditions. Each of these questions will be dealt with in this chapter with the goal of providing practical suggestions that will work well with most proteolytic digests. Although most of the studies that will be described were carried out on 50 to 250 pmol (picomole) aliquots derived from a large-scale transferrin digest, similar conditions have been used in the Yale School of Medicine Protein and Nucleic Acid Chemistry Facility to separate tryptic peptides derived from over 50 different proteins. Previous publications[4-7] provide suggested protocols for carrying out enzymatic digestions on sub-nanomole amounts of proteins.

II. EXPERIMENTAL

The HPLC system used was a Hewlett Packard (Avondale, PA) Model 1090 that was equipped with a diode array detector and a 200 µl injection loop. Data was acquired on a Nelson Analytical (Cupertino, CA) Model 4416X Multi-Instrument Data System.

All the comparative HPLC experiments that follow utilized 50 pmol aliquots derived from a large scale trypsin digest of carboxamidomethylated transferrin. After extensive dialysis, 8.7 mg of carboxamidomethylated transferrin was dried *in vacuo* and then redissolved in 1.0 ml of 8 *M* urea. Approximately 0.35 ml of this transferrin solution, containing 3.0 mg of carboxamidomethylated transferrin, was heated at 65°C for 10 min prior to adding 0.92 ml of 0.1 *M* NH_4HCO_3 and 0.12 ml of a 1.0 mg/ml solution of trypsin. The resulting transferrin to trypsin ratio was 25:1 on a weight basis, which corresponds to a molar ratio of approximately 8:1. After incubating at 37°C for 24 hr, suitable dilutions of this digest were made so that the appropriate amounts would be injected in a total volume of 20 µl of 0.05% aq. trifluoracetic acid (TFA).

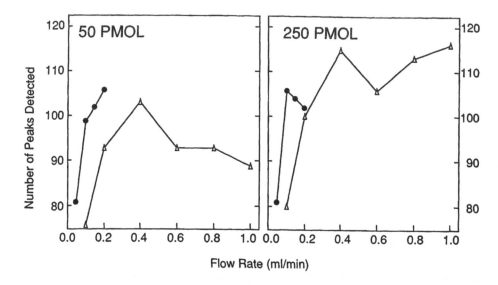

FIGURE 1. Influence of flow rate on peptide resolution in RPC. Sample: 50 pmol (left panel) or 250 pmol (right panel) of a tryptic digest of carboxamidomethylated transferrin. Columns: Delta Pak C_{18}, 15 cm × 2 mm I.D. (•—•) or 15 cm × 3.9 mm I.D. (Δ—Δ). Mobile-phase: linear AB gradient (2% at 0 min to 37% B at 63 min, to 75% B at 95 min and, finally, to 98% B at 105 min), where Eluent A is 0.060% aq. trifluoroacetic acid (TFA) and Eluent B is 0.056% TFA in acetonitrile/water (80:20); flow-rate, 0.04 to 0.2 ml/min (15 cm × 2 mm I.D.) or 0.1 to 1.0 ml/min (15 cm × 3.9 mm I.D.); absorbance at 210 nm. Small adjustments, on the order of ± 0.002%, were made in the concentration of TFA in either eluent as appropriate to balance the apparent absorbances at 210 nm. The gradient was followed by an isocratic wash with 2.5 ml of 98% Eluent B prior to re-equilibrating the column with approximately 15 column volumes of Eluent A. The extent of resolution was quantitated in terms of the number of peaks detected by a Nelson Analytical Model 4416X Multi-Instrument Data System using an area threshold of 50 μ V-sec (100 μAU-sec) and a noise threshold of 1 μV. Using these parameters, peak "shoulders" will not be counted as individual peaks, but even a barely resolved doublet would be scored as two peaks.

III. RESULTS

A. GUIDELINES FOR CHOOSING A NARROW BORE (2.0 TO 2.1 mm I.D.) VS. A STANDARD BORE (3.9 TO 4.6 mm I.D.) COLUMN

The major theoretical advantage of using a narrow bore instead of a standard bore column results from the fact that when solutes are eluted from both columns at the same linear flow velocity (which is proportional to the square of the cross-sectional area), the flow rate on the narrow bore column will be only 20 to 25% of that on the standard bore column. In addition to decreasing the amount of mobile phase used, the sensitivity of detection is markedly increased since peak volume is directly related to the flow rate when the gradient times are kept constant.[5-8] Many of the advantages of narrow bore columns can be achieved on standard bore columns simply by reducing the flow rates (while maintaining the same gradient times) to a range (such as from 0.2 to 0.4 ml/min) that is closer to the range of flow rates used on narrow bore columns. As shown in Figure 1, reducing the flow rate on a 4.6 mm I.D. column from 1.0 to 0.2 ml/min resulted in less than a 15% decrease in resolution, as measured by the number of detected peaks, even though the gradient volume had been decreased by 80%. This decrease in flow rate was accompanied by a three- to four-fold increase in sensitivity of detection.[6,7]

Based on the data shown in Figure 1, it would appear that 250 pmol might serve as an approximate guideline for the lower limit of the amount of protein digest that should be applied to a standard bore (3.9 to 4.6 mm I.D.) column. For this amount of digest, optimum resolution was obtained with a flow rate of 0.4 ml/min (Figure 1), which is sufficiently low to yield several peptide peaks with an absorbance at 210 nm (using a 6-mm path length flow cell) above 0.05 AU (see Figure 14 in Stone et al.[6]). When the amount of protein digest is decreased to 50 pmol, Figure 1 indicates that optimum resolution and sensitivity of detection are obtained on a narrow bore column at a flow rate of 0.15 to 0.2 ml/min. Under these conditions, chromatography of as little as 50 pmol of an enzymatic digest yields several peptide peaks that have an absorbance above 0.025 AU at 210 nm (see Figure 3). As the flow rate is decreased below 0.15 ml/min, peak resolution decreases (Figure 1) and problems are encountered with the use of commercially available automatic peak detectors.[7] As a general guide, therefore, it would appear that amounts of protein digest in the 50 to 250 pmol range are best chromatographed on narrow bore columns at flow rates of 0.15 to 0.20 ml/min.

In terms of the actual HPLC system used, we have previously obtained similar resolution when 50 pmol amounts of a digest were chromatographed at 0.15 ml/min on a narrow bore column on a Hewlett-Packard (Avondale, PA) Model 1090, an Applied Biosystems (Foster City, CA) Model 130, and a Waters Assoc. System (Milford, MA) equipped with two model 510 pumps. The advantages of the HP system were that it had a considerably shorter "gradient delay" than the Waters System (see Figure 6 in Stone et al.[7]) and it is completely automated, whereas the Applied Biosystems HPLC System required manual sample injection.

B. PEPTIDE RESOLUTION IS DIRECTLY RELATED TO COLUMN LENGTH

In contrast to protein HPLC, where column length has been reported to play a negligible role in determining resolution,[8] it is evident from Figure 2 that, in the case of RPC of peptides, peak resolution is directly related to column length. As shown in Figure 2, decreasing the length of a standard bore column from 25 to 5 cm resulted in a 28% decrease in resolution. That is, the number of absorbance peaks detected decreased from 121 on the 25-cm column to 87 on the 5-cm column. Since a similar result was previously seen on a narrow bore column,[6] it would appear that, in general, a 25-cm column should be used. One exception is the Delta Pak column which is only available in lengths up to 15 cm. Based on its ability to resolve complex enzymatic digests, the effective height-equivalent plate count on the Delta Pak column appears to be above average. Both standard and narrow bore versions of this column have previously been observed to provide resolution that is higher than that obtained on a 22- or 25-cm Aquapore C_8 column and is nearly equivalent to that obtained on a 25-cm Vydac C_{18} column.[6]

C. INFLUENCE OF COLUMNS FROM DIFFERENT MANUFACTURERS ON PEPTIDE RESOLUTION AND SELECTIVITY

The resolution limits of the five commercially available narrow bore columns listed in Table 1 were compared by chromatographing 50 pmol aliquots of a tryptic digest of transferrin on each column and then determining the number of absorbance peaks that were detected by our Nelson Analytical Multi-Instrument Data System. As shown in Table 1, approximately 25% fewer absorbance peaks were detected on the particular SynChropak RP-P column tested than on the Alltech Macrosphere column, which, among the individual columns that were tested, appeared to give the best resolution. As was found previously,[6] the 15-cm Delta Pak C_{18} column gave resolution comparable to, or better than, that obtained on a 25-cm Vydac C_{18} column, or on a 22-cm Aquapore C_8 column. In analyzing the data in Table 1 with respect to the number of peaks detected, it should be mentioned that replicate HPLC runs give a standard deviation of approximately ±3 absorbance peaks detected. In addition, testing of multiple Vydac C_{18} and Delta Pak C_{18} standard bore columns gave a standard deviation of ±4 to ±5 peaks. Based on these data, a difference of 10 or more peaks detected would seem to be significant.

A visual comparison of the resulting chromatograms (Figure 3) suggests that, of the columns tested, the Vydac and Delta Pak C_{18} columns are most similar to each other in terms of their selectivity. In most instances, it is relatively easy to identify a peak on the Delta Pak chromatogram that corresponds to any given peak in the Vydac chromatogram. In contrast, it

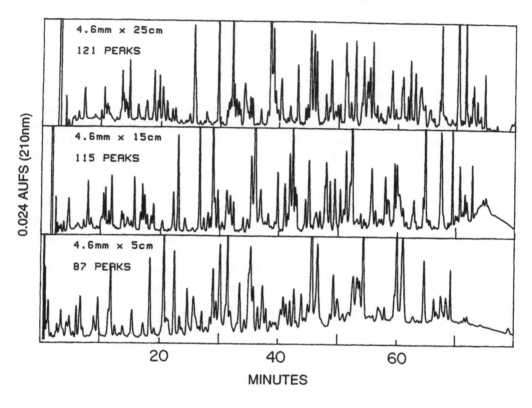

FIGURE 2. Effect of reversed-phase column length on resolution of peptides. Columns: Vydac C_{18}, 25 cm × 4.6 mm I.D. (top), 15 cm × 4.6 mm I.D. (middle), 5 cm × 4.6 mm I.D. (bottom). Mobile phase: see Figure 1; flow-rate, 1 ml/min. Aliquots of a large scale tryptic digest of carboxamidomethylated transferrin were injected onto the three columns. The number of absorbance peaks detected by the Model 4416 Multi-Instrument Data System are indicated for each column.

is considerably more difficult to identify corresponding peaks in the Aquapore C_8 and SynChropak RP-P chromatograms. Most of the differences seen in these two chromatograms probably do not relate simply to chain length of the bonded phase. A previous study obtained similar peptide maps with Vydac C_{18} and C_4 columns.[9] The major difference was that the elution profile was shifted earlier in the case of the C_4 column and, as expected, short peptides (that were generally eluted in the first one-third of the gradient) were better resolved on the C_{18} sorbent.[9] By "screening" small aliquots or actually re-purifying peptides that were originally isolated on a Vydac C_{18} column on an Aquapore C_8 column, it is possible to utilize the different selectivity of the Aquapore C_8 column to detect peptides that were co-eluted on a Vydac C_{18} column.[5] In this way, it is frequently possible to detect peptide mixtures prior to amino acid sequencing. Although other approaches, such as changing the pH of the mobile phase or using ion-exchange HPLC (IEC), can be used to detect peptide mixtures, it is generally easier to change the column rather than the mobile phase, particularly when only one HPLC system is available.

D. GRADIENT TIME — AN IMPORTANT PARAMETER THAT IS OFTEN OVERLOOKED

Having selected an HPLC system and an appropriate column, the next decisions relate to choice of mobile phase and gradient conditions. With respect to mobile phase, the water/acetonitrile solvent system containing 0.05 to 0.10% TFA has gained widespread acceptance because of its excellent solubilizing properties, low ultraviolet absorbance, volatility, and resolving power. In terms of the actual elution conditions, the two major variables are the flow

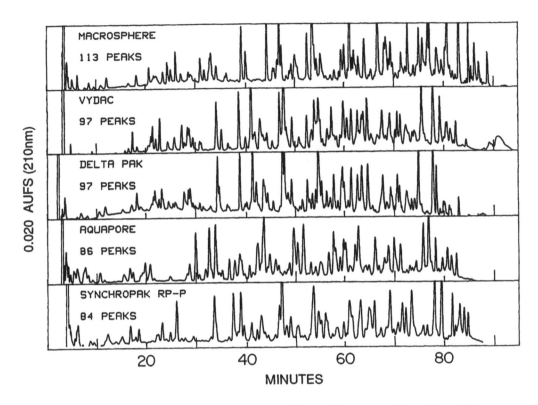

FIGURE 3. Comparison of reversed-phase separations of peptides on different narrow bore columns. Columns: see Table 1 for details. Mobile phase: see Figure 1; flow-rate, 0.2 ml/min. Aliquots (50 pmol) of a large-scale trypsin digest of carboxamidomethylated transferrin were injected onto the columns. The number of absorbance peaks detected by the Model 4416 Multi-Instrument Data System are indicated for each column.

rate and gradient program, which together determine the gradient slope and the number of column volumes over which the elution will occur. Figure 1 demonstrates that, with a standard bore column, the flow rate can be decreased from the usual value of 0.7 to 1.0 ml/min to a value such as 0.2 ml/min, a flow rate usually reserved for narrow bore columns, with only minimal loss in resolution. Since the gradient times were held constant in Figure 1, this result implies that, within reasonable limits, the flow rate and gradient slope (in terms of percent increase in Eluent B/ml) are less important determinants of the resolution that will be achieved than might have been expected. Hence, in Figure 1, decreasing the total gradient volume used to elute 250 pmol of a tryptic digest from a 15 cm x 3.9 mm I.D. column from approximately 60 column volumes (at 1.0 ml/min) to only about 12 column volumes (at 0.2 ml/min) resulted in only a 14% decrease in resolution, as measured by the number of peaks detected. Decreasing the flow rate to 0.1 ml/min, which corresponds to only a six-column volume gradient, resulted in approximately a 31% loss in resolution.

Figure 4 suggests that the gradient time may be a more important parameter than the flow rate. In this figure, the gradient volume has been kept constant at about six column volumes but the flow rate has been changed from 0.1 ml/min (bottom chromatogram) to 1.0 ml/min (top chromatogram), thereby increasing the gradient slope ten-fold in terms of percent increase in Eluent B/min.. This ten-fold decrease in run time resulted in a 77% decrease in resolution, which suggests that specialized packings and perhaps higher temperatures[10] are indeed needed in order to decrease significantly reversed-phase run times. The decrease in peptide resolution shown in Figure 4 was effectively caused by the ten-fold increase in gradient slope (in terms of percent increase in Eluent B/min) of each part of the segmented gradient. This more marked effect on

TABLE 1
Comparative Peptide Resolution on Narrow Bore Columns[a]

Column[b]	Bonded phase	Dimensions (cm × mm)	Particle size (microns)	Pore size (angstroms)	Number of absorbance peaks detected[a]
Alltech Macrosphere	C_{18}	25 × 2.1	5	300	113
Vydac	C_{18}	25 × 2.1	5	300	97
Waters Delta Pak	C_{18}	15 × 2.0	5	300	97
Brownlee Aquapore	C_8	22 × 2.1	7	300	86
SynChropak RP-P	C_{18}	25 × 2.1	6.5	300	84

[a] Number of absorbance peaks detected at 210 nm by the Nelson Analytical Model 4416X Multi-Instrument Data-System following injection of a 50 pmol tryptic digest of transferrin on an HP1090 HPLC. Elution was at 0.2 ml/min following the gradient program described in Experimental. Although the extent of possible variability in the number of peaks detected on two otherwise identical columns from the same manufacturer is not known, repetitive analyses on the same column result in a standard deviation of approximately ± 3 peaks detected.

[b] Column manufacturers: Alltech Macrosphere (Alltech Associates, Inc., Deerfield, IL); Vydac (The Separations Group, Hesperia, CA); Waters Delta Pak (Milford, MA); Brownlee Aquapore (Brownlee Labs., Santa Clara, CA); SynChropak RP-P (SynChrom, Linden, IN)

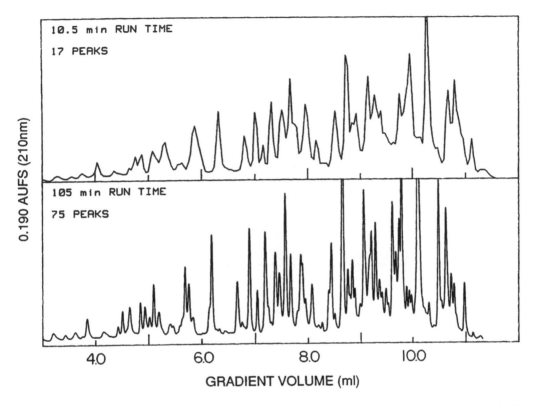

FIGURE 4. Effect of run time on RPC separation of tryptic peptides from 250 pmol of carboxamidomethylated transferrin. Column: Delta PAK C$_{18}$, 15 cm × 3.9 mm I.D. Mobile phase: eluents shown in Figure 1; the peptides were eluted with a total gradient volume of 10.5 ml at a flow-rate of either 0.1 ml/min and a run time of 105 min (lower panel) or a flow-rate of 1 ml/min and a run time of 10.5 min (top panel). In the latter case, the gradient slope (in terms of % increase in Eluent B/min) of each part of the segmented gradient used in Figure 1 was increased ten-fold. In each case, the number of absorbance peaks detected by the Model 4416 Multi-Instrument Data System are as indicated. It should be noted that a column with a different serial number was used for the lower panel in the above experiment than was used for the corresponding experiment in Figure 1, where 80 peaks were detected when 250 pmol of a transferrin digest was eluted at 0.1 ml/min with the same gradient program.

peptide resolution of changes in gradient slope (Figure 4) compared to flow-rate changes within the optimal flow-rates for a particular column (Figure 1) has been observed by other researchers in the peptide field. As a general guide, it would appear that approximately a 100-min gradient program, coupled with a flow rate of 0.4 to 0.5 ml/min for a standard bore and 0.15 to 0.2 ml/min for a narrow bore column, represents a reasonable compromise in terms of both optimum resolution, sensitivity, and analysis time.

E. EFFECT OF SODIUM DODECYLSULFATE (SDS) AND OTHER SAMPLE PARAMETERS ON RESOLUTION

Because of its high resolving power, SDS polyacrylamide gel electrophoresis is frequently used as a final purification step and, as a result, SDS is a common contaminant of proteins. A previous study has shown that as little as 10 µg SDS is sufficient to interfere with and 100 µg SDS is sufficient to prevent completely trypsin digestion from occurring when the digest is carried out in 0.2 ml of 2 M urea.[7] Figure 5 demonstrates that similar amounts of SDS also interfere with RPC. In this figure, small changes can be seen (note especially the absorbance peaks eluted near 47, 54, and 61 min) upon the addition of only 5 µg SDS to the sample prior to sample injection. A significant decrease in resolution, as well as an apparent loss of early-

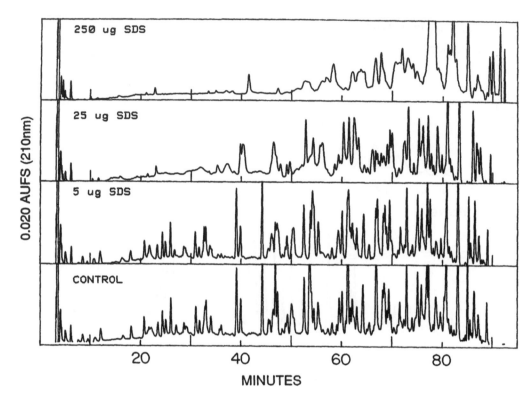

FIGURE 5. Effect of sodium dodecylsulfate (SDS) on RPC of peptides. Column: Alltech Macrosphere C_{18} (25 cm × 2.1 mm I.D.). Mobile-phase: see Figure 1; flow-rate, 0.2 ml/min. The indicated amounts of SDS were added to 50 pmol aliquots of a large scale trypsin digest of carboxamidomethylated transferrin that were then injected onto the column in a volume of 20 µl.

eluted peaks, occurs as the amount of SDS is increased to 25 µg and then to 250 µg. In contrast to the above results with SDS, nearly identical chromatograms were obtained when 50 pmol amounts of a tryptic digest of transferrin were dissolved in 20 µl of 0.05% aq. TFA, 2 or 8 M urea and 6 M guanidine hydrochloride (data not shown). Under these various loading conditions, the only differences noted were slightly decreased resolution or actual loss of a few weakly retained peptides that were eluted prior to 15 min (about 8% acetonitrile).

IV. CONCLUSIONS

Our results demonstrate that several different commercially available HPLC systems can be used to isolate sub-nanomole amounts of peptides from enzymatic digests of proteins. In terms of deciding between a 3.9 to 4.6 mm I.D. (standard) or a 2.0 to 2.1 mm I.D. (narrow bore) column, 250 pmol represents an approximate lower limit to the amount of an enzymatic digest that can be optimally separated on a standard bore column. The sensitivity of detection on standard bore columns can be increased, with minimal or no loss in resolution, by decreasing the flow rate from the usual 0.7 to 1.0 ml/min to 0.4 to 0.5 ml/min. Amounts of peptides in the 50 to 250 pmol range can easily be detected following elution at 0.15 to 0.20 ml/min from narrow bore columns. Although the sensitivity of detection obtained from narrow bore columns can be further increased by decreasing the flow rate to 0.05 to 0.10 ml/min, while maintaining the gradient times, these latter flow rates are too low to permit commercially available peak detectors to fractionate and collect complex chromatograms accurately. Of the several different narrow bore columns tested, highest resolution was obtained on three C_{18} reversed-phase columns (Alltech Macrosphere, Vydac, and Delta Pak) that were packed with 5 μm material. In contrast to reversed-phase protein isolations,[8] where column length appears to be relatively unimportant, a direct relationship was found between column length and peptide resolution in the case of peptide separations. The resolution achieved on reversed-phase columns was found to depend more strongly on the rate of increase of the acetonitrile concentration that occurred per minute rather than per milliliter of eluting solvent. That is, the actual gradient time appears to be a more important determinant of peptide resolution than the flow rate or gradient volume. This finding suggests that the kinetics of interaction between a peptide and the bonded phase play a major role in determining the extent of resolution that can be achieved by RPC. In general, at least 100-min gradients should be used with either the standard or narrow bore columns used in this study. Since even small amounts (i.e., 5 to 25 μg) of SDS interfere with RPC, this and other similar detergents should be removed prior to HPLC. Our results demonstrate that complex enzymatic digests from as little as 50 pmol protein can be easily fractionated and the resulting peptides individually and automatically collected[5] using commercially available HPLC systems and narrow bore, reversed-phase columns.

ACKNOWLEDGMENTS

The authors wish to thank the Sep/a/ra/tions Group as well as the Waters Chromatography Division, Millipore Corporation for providing some of the HPLC columns used in this study. We also thank A. Heckendorf of the Nest Group and R. Pfeifer of the Millipore Corporation for their advice and continued interest in these studies. K. Williams is supported by the Howard Hughes Medical Institute.

REFERENCES

1. **Brown, J.L. and Roberts, W.K.,** Evidence that approximately eighty per cent of the soluble proteins from Ehrlich ascites cells are N^{α}-acetylated, *J. Biol. Chem.,* 251, 1009, 1976.
2. **Brown, J.L.,** A comparison of the turnover of α-N-acetylated and nonacetylated mouse L-cell proteins, *J. Biol. Chem,* 254, 1447, 1979.
3. **Barker, W.C., Hunt, L.T., George, D.G., Yeh, L.S., Chen, H.R., Blomquist, M.C., Johnson, G.C., Seibel-Ross, E.I., Hong, M.K., Bair, J.K., and Ledley, R.S.,** *Protein Sequence Database,* Release 18.0, 1988.
4. **Williams, K.R., Stone, K.L., Fritz, M.K., Merrill, B.M., Konigsberg, W.H., Pandolfo, M., Valentini, O., Riva, S., Reddigari, W., Patel, G.L., and Chase, J.W.,** Use of HPLC comparative peptide mapping in structure/function studies, in *Proteins: Structure and Function,* L'Italien, J.J., Ed. Plenum Press, New York, 1987, 45.
5. **Stone, K. and Williams, K.,** Small bore HPLC purification of peptides in the sub-nanomole range, in *Macromolecular Sequencing and Synthesis,* Schlesinger, D., Ed. Alan R. Liss, New York, 1988, 7.
6. **Stone, K.L., LoPresti, M.B., and Williams, K.R.,** Enzymatic digestion of proteins and HPLC peptide isolation in the sub-nanomole range, in *Focus on Laboratory Methodology in Biochemistry,* Fini, C., Floridi, A., and Finelli, V., Ed. CRC Press, Boca Raton, 1989, 1990, 181.
7. **Stone, K.L., LoPresti, M.B., Williams, N.D., Crawford, J.M., DeAngelis, R., and Williams, K.R.,** Enzymatic digestion of proteins and HPLC peptide isolation in the sub-nanomole range, in *Techniques in Protein Chemistry.* T. Hugli, Ed. Academic Press, 1989, 377.
8. **Schlabach, T.D. and Wilson, K.J.,** Microbore flow rates and protein chromatography, *J. Chromatogr.,* 385, 65, 1987.
9. **Stone, K.L. and Williams, K.R.,** High performance liquid chromatographic peptide mapping and amino acid analysis in the sub-nanomole range, *J. Chromatogr.,* 359, 203, 1986.
10. **Kalghatgi, K. and Horváth, C.,** Rapid peptide mapping by high-performance liquid chromatography, *J. Chromatogr.,* 443, 343, 1988.

COMPARISON OF PEPTIDE RESOLUTION ON CONVENTIONAL, NARROWBORE, AND MICROBORE REVERSED-PHASE COLUMNS

T. W. Lorne Burke, Colin T. Mant, and Robert S. Hodges

I. INTRODUCTION

The advantages of employing HPLC columns with internal diameters smaller than the conventional 4 to 4.6-mm internal diameter (I.D.) for specific peptide and protein applications are now well established.[1-8] These advantages include increased detection sensitivity of the solute(s) of interest, higher concentrations of eluted solutes, and decreased solvent consumption. A good general review of microcolumn liquid chromatography can be found in reference 9.

A number of researchers[1,2,4,5] have compared the resolution of peptides and/or proteins obtained on standard reversed-phase chromatography (RPC) analytical columns (4 to 4.6 mm I.D.) with that on narrower-bore columns (1 to 2.5 mm). As expected, detection sensitivity increased and peak elution volumes decreased with decreasing column I.D. In addition, solute recovery, particularly for proteins, is often improved when using small-bore columns.[4,7]

Our experience from reviewing the literature on separations of peptides and proteins on narrowbore (2 to 3 mm, I.D.) and microbore ($\leq$ 1 mm I.D.) columns is that assessment of the performance of these columns, as well as comparing the effectiveness of columns of different diameters, is generally carried out using mixtures of unrelated peptides (synthetic or protein fragments) or proteins. This laboratory tends to the view that a more consistent and accurate assessment of the peptide separation performance of a column, or a comparison of different columns, can best be achieved through the use of well-defined and structurally-related synthetic peptide standards.[10-12] Thus, the main objective of the present article was to compare the separation of a set of reversed-phase peptide standards on conventional (4.6 mm I.D.), narrowbore (2.1 mm I.D.) and microbore (1.0 mm I.D.) columns of the same length (100 mm) and containing the identical C_8 reversed-phase packing. By comparing peptide resolution of the same peptide mixture on these three columns, the researcher has the information required to make the decision of when to choose narrowbore or microbore chromatography over the use of conventional columns and what the comparable recommended flow-rate range is for each type of chromatography to maintain the desired resolution.

II. EXPERIMENTAL

A. INSTRUMENTATION

All runs described in this article were carried out on a Hewlett-Packard (Avondale, PA) HP1090 liquid chromatograph coupled to an HP1040A detection system, HP9000 Series 300 computer, HP9133 disc drive, HP2225A Thinkjet printer, and HP7440A plotter.

B. COLUMNS

Runs were carried out on three Aquapore RP300 C_8 columns (7-μm particle size, 300-Å pore

TABLE 1
Sequences of Peptide Standards

Peptide	Peptide sequence[a]
I1	Ac-Arg-Gly-Gly-Gly-Gly-**Ile**-Gly-**Ile**-Gly-Lys-amide
I2	Ac-Arg-Gly-Gly-Gly-Gly-**Ile**-Gly-Leu-Gly-Lys-amide
S2	Ac-Arg-Gly-Gly-Gly-Gly-Leu-Gly-Leu-Gly-Lys-amide
S3	Ac-Arg-Gly-**Ala**-Gly-Gly-Leu-Gly-Leu-Gly-Lys-amide
S4	Ac-Arg-Gly-**Val**-Gly-Gly-Leu-Gly-Leu-Gly-Lys-amide
S5	Ac-Arg-Gly-**Val**-**Val**-Gly-Leu-Gly-Leu-Gly-Lys-amide

[a]   Ac = N^α-acetyl; amide = C^α-amide.

Note: Residues in bold denote amino acid substitutions in the peptide S2 sequence.

These standards were obtained from Synthetic Peptides Incorporated, Department of Biochemistry, University of Alberta, Edmonton, Alberta, Canada.

size; Brownlee Labs., CA): microbore (100 × 1 mm I.D.), narrowbore (100 × 2.1 mm I.D.) and conventional analytical (100 × 4.6 mm I.D.).

C. PEPTIDE STANDARDS

The six synthetic decapeptide analogs used in this study (I1, I2, S2, S3, S4 and S5; Table 1) are described in detail in this volume by Burke et al.[13] The subtle hydrophobicity variations of these peptides enables a very precise determination of the resolving power of a reversed-phase column.

III. RESULTS AND DISCUSSION

The best initial approach to most analytical peptide separations is to employ 0.1% aq. trifluoroacetic acid (TFA) to 0.1% TFA-acetonitrile linear gradients.[14] Varying flow- and gradient-rates can be very effective in optimizing peptide separations,[13] with the selection of optimal conditions on a specific column and on a specific LC instrument depending on the individual researcher's requirements. The development of such optimal conditions is generally a compromise between run time (speed), resolution, and sensitivity. Burke et al.,[13] elsewhere in this volume, demonstrated the major effects of varying flow- and gradient-rate on separation of the peptide standards shown in Table 1 on an analytical (4.6 mm I.D.) C_8 column and, thus, will not be repeated here. The results described below illustrate that column I.D. is also a potential variable to consider when attempting to develop optimal conditions for a particular separation problem.

A. EFFECT OF COLUMN I.D. ON PEPTIDE ELUTION PROFILES USING STANDARD RUN CONDITIONS

Figure 1 shows the elution profiles of the peptide standards on the analytical (top), narrowbore (middle) and microbore (bottom) C_8 columns. The elution order of the standards is (from left to right) I1, I2, S2, S3, S4, and, finally, S5. The run conditions were chosen to reflect those typically used for RPC of peptides,[14] i.e., linear AB gradient elution (1% B/min, where eluent

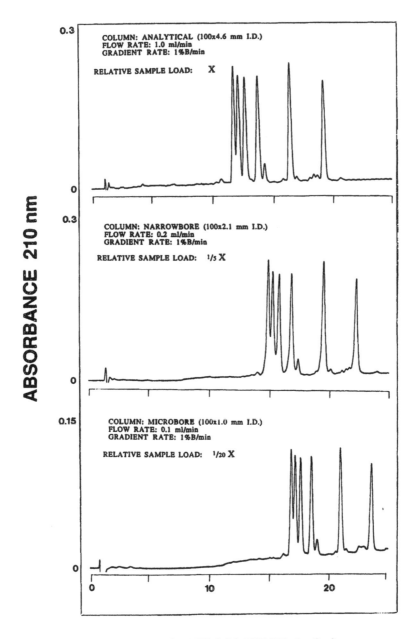

ELUTION TIME (min)

FIGURE 1. Reversed-phase separation of synthetic peptide standards on columns of varying internal diameter. Conditions: linear AB gradient (1% B/min) at flow-rates of 1 ml/min (analytical column, top profile), 0.2 ml/min (narrowbore column, middle profile) or 0.1 ml/min (microbore, bottom profile), where Eluent A is 0.1% aq. TFA and Eluent B is 0.1% TFA in acetonitrile. Sequences of the six peptide standards (I1, I2, S2, S3, S4, and S5) are shown in Table 1. The elution order of the peptides is (from left to right) I1, I2, S2, S3, S4, and, finally, S5.

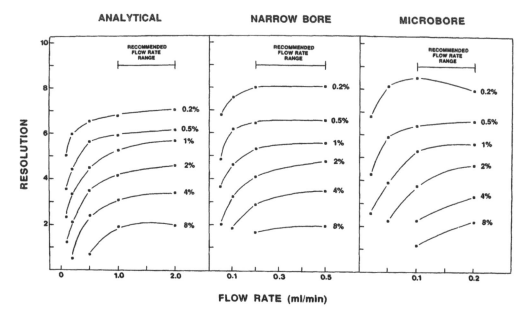

FIGURE 2. Graphical representation of the effect of varying flow-rate and gradient-rate on peptide resolution on RPC columns of varying internal diameter. Conditions: flow-rates and linear AB gradients as shown; mobile phase eluents same as Figure 1. The profiles show the effect of varying run conditions on the resolution of peptide standards S4 and S5 (sequences shown in Table 1).

A is 0.1% aq. TFA and eluent B is 0.1% TFA in acetonitrile) at flow-rates of 1 ml/min (analytical), 0.2 ml/min (narrowbore) and 0.1 ml/min (microbore). The increase in detection sensitivity with decreasing column I.D. is immediately apparent, with the relative sample load necessary to ensure strong peak detection decreasing significantly with decreasing I.D. Apart from this increase in sensitivity, the elution profiles are all quite similar.

B. EFFECT OF VARYING FLOW-RATE AND GRADIENT-RATE ON PEPTIDE RESOLUTION

Schlabach and Wilson[4] noted that little attention is usually given to selection of flow-rates, stating that flow-rates of 1, 0.2, and 0.05 ml/min tend to be automatically selected for 4-, 2-, and 1-mm I.D. columns. Burke et al.[13] in this volume demonstrated how both flow-rate and, particularly, gradient-rate could be manipulated to improve the separation of a peptide mixture on an analytical column. Figure 2 demonstrates the effect of varying gradient- and flow-rate on peptide resolution on the three C_8 columns of varying I.D. Resolution (R_s) was calculated using $R_s = \Delta t/W_1 + W_2$, where W_1 and W_2 are the peak widths at half height of peptides S4 and S5, respectively, and Δt is the difference in retention time between the two peptide peaks. In all cases, increasing flow-rate increases peptide resolution, which levels off in what we refer to as the recommended range of flow-rate for each column: analytical, 1 to 2 ml/min; narrowbore, 0.2 to 0.5 ml/min; and microbore (0.1 to 0.2 ml/min). In all cases, resolution improves with decreasing gradient-rate, as expected; in addition, at any given gradient-rate, the resolution is comparable whether using analytical, narrowbore, or microbore columns. It should be noted that improved resolution with decreasing gradient-rate is at the expense of run time. For instance, the increase in run time when comparing 4% B/min to 0.2% B/min is approximately 10-fold (see later in Figure 4). As discussed in detail by Burke et al.,[13] the improved resolution with decreasing gradient-rate is due to an improved Δt relative to the corresponding increase in peak widths.

Figure 3A shows the effect of varying linear flow velocity (at a constant gradient-rate of 1%

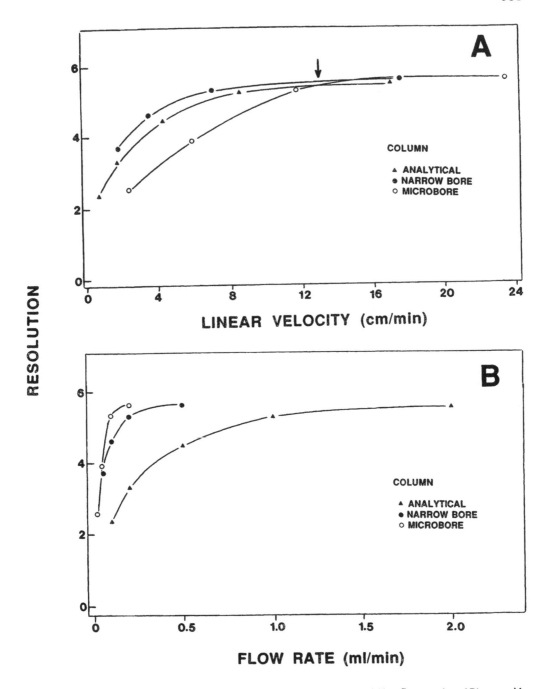

FIGURE 3. Graphical representation of the effect of varying linear velocity (panel A) or flow-rate (panel B) on peptide resolution on RPC columns of varying internal diameter. Conditions: linear AB gradient (1% B/min) at flow-rates as shown; mobile phase eluents same as Figure 1. The profiles show the effect of varying linear velocity and flow-rate on the resolution of peptide standards S4 and S5 (sequences shown in Table 1).

B/min) on resolution of peptides S4 and S5 on the three columns. Figure 3B provides a comparison of the effect of varying flow-rate on peptide resolution. Since the three columns contain the same packing material, it would be expected that each column should provide the same resolution if the same linear flow velocity is used. From Figure 3A, this only appears to be true for all three columns at linear velocities above 13 cm/min, as indicated by the arrow. The

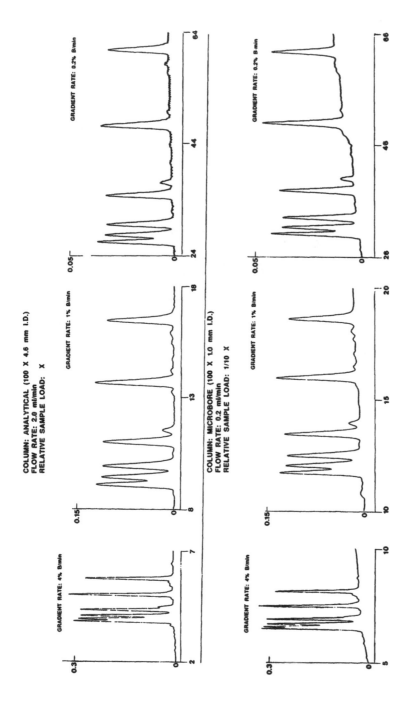

ABSORBANCE 210 nm

ELUTION TIME (min)

FIGURE 4. Effect of varying gradient-rate, at constant flow-rate, on reversed-phase elution profiles of synthetic peptide standards. Conditions: linear AB gradient [4% B/min (left profiles), 1% B/min (middle profiles) or 0.2% B/min (right profiles)] at flow-rates of 2 ml/min (analytical column, top profiles) or 0.2 ml/min (microbore column, bottom profiles); mobile phase eluents same as Figure 1. Sequences of the peptide standards are shown in Table 1. The elution order of the peptides is described in Figure 1.

TABLE 2
Resolution vs. Flow-Rate on Microbore, Narrowbore, and Conventional Analytical Columns

Column	Resolution[a]	Flow-rate	
		ml/min	cm/min[b]
Microbore	5.3	0.1	12
Narrowbore	5.6	0.4	14
Analytical	5.4	1.5	13

[a] Resolution was calculated using $R_s = \Delta t/w_1 + w_2$, where w_1 and w_2 are the peak widths at half height and Δt is the difference in retention time between the two peaks of interest (peptides S4 and S5; see Table 1).

[b] Linear flow velocity (v) is related to the flow rate in ml/min (F) by the equation $v = F(1)/v_o$, where l is the column length and v_o is the void column; $v_o = t_o F$, where t_o is the elution time for unretained compounds.

results on the narrowbore and analytical columns compare reasonably well over the entire range of linear velocities. The reason the microbore results are so different at very low flow-rates may be related to solute diffusion, leading to loss of resolution, as a result of mixing in the 8-μl volume detector flow cell. Table 2 shows the essentially identical peptide resolution obtained at flow-rates of 0.1, 0.4, and 1.5 ml/min (which result in very similar linear flow velocities) on the microbore, narrowbore and analytical columns, respectively.

C. COMPARISON OF PEPTIDE ELUTION PROFILES ON ANALYTICAL AND MICROBORE COLUMNS AT VARYING FLOW-RATES AND GRADIENT-RATES

Figure 4 displays the elution profiles of the peptide mixture at gradient-rates of 4% B/min (left hand profiles), 1% B/min (middle profiles), and 0.2% B/min (right hand profiles) on the analytical (top profiles) and microbore (bottom profiles) C_8 columns. The flow-rates were chosen on the basis of the results shown in Figure 2, where, at any given gradient-rate, flow-rates of 2 ml/min (analytical column) and 0.2 ml/min (microbore column) produced optimum peptide resolution. At these optimum flow-rates, the peptide elution profiles are essentially equal in all chromatographic parameters (retention time, resolution, etc.). In fact, the results indicate that the only major advantages of the microbore column over the conventional analytical column are the 10-fold decrease in solvent consumption and the 10-fold increase in detection sensitivity (the sample quantity injected on the microbore column was 1/10 that of the analytical column).

IV. CONCLUSIONS

The selection of a microbore column (1 mm I.D.) will provide an increase in sensitivity over that of conventional analytical columns, resulting in a requirement for less sample per injection. In addition, there is a significant decrease in solvent consumption compared to conventional analytical chromatography. Depending on the number of analytical runs carried out/ annum this can result in a substantial reduction in operating costs. However, LC instrumentation with the necessary performance characteristics is required to take full advantage of the benefits of these small diameter columns. Thus, the LC instrument must be able to generate gradients accurately and reproducibly at flow-rates of 10 to 100 μl/min. The dead volume between the gradient mixer and the top of the column should be < 100 μl. The instrumentation must have an efficient micromixer for preparing gradients, and the post-column volume, including detector cell volume, must be minimized to prevent post-column mixing and band broadening. If autosamplers are used with microbore chromatography, they must be able to inject accurately and reproducibly microliter volumes and 80 to 90% of the total sample volume so as not to lose the advantage of increased sensitivity.

The need for specialized LC instrumentation may be a financial burden for many researchers who wish to take advantage of columns with internal diameters smaller than conventional analytical columns. However, most modern HPLC hardware is compatible with narrowbore columns (2 to 3 mm I.D.).[6] The results of the present article support the observation by Schlabach et al.[6] that by simply replacing a conventional column with a narrowbore column, a substantial improvement can be made in terms of both increased sensitivity and decreased solvent consumption.

ACKNOWLEDGMENTS

This work was supported by the Medical Research Council of Canada and by equipment grants from the Alberta Heritage Foundation for Medical Research.

REFERENCES

1. **Van der Zee, R. and Welling, G.W.,** Microbore reversed-phase chromatography of proteins with conventional gradient equipment for high-performance liquid chromatography, *J. Chromatogr.*, 325, 187, 1985.
2. **Stone, K.L., Williams, K.R., Rabin, B.A., Murphy, J.B., Chase, J.W., and Pfeifer, R.,** HPLC purification and identification of peptides using microbore columns, *J. Anal. and Purification*, October, 20, 1986.
3. **Wilson, K.J., Hong, A.L., Brasseur, M.M., and Yuan, P.M.,** Microbore high performance liquid chromatography: an example of its application in peptide characterization, *BioChromatography*, 1, 106, 1986.
4. **Schlabach, T.D. and Wilson, K.J.,** Microbore flow-rates and protein chromatography, *J. Chromatogr.*, 385, 65, 1987.
5. **Götz, H.,** An evaluation of new microbore and high-speed columns for protein separations using HPLC and derivative spectroscopy, *BioChromatography*, 4, 156, 1989.
6. **Schlabach, T.D., Zieske, L.R., and Wilson, K.J.,** Mirco-HPLC techniques for the purification and analysis of peptides and proteins, this publication.
7. **Stone, K.L., LoPresti, M.B., Crawford, J.M., DeAngelis, R., and Williams, K.R.,** Reversed-phase HPLC separation of sub-nanomole amounts of peptides obtained from enzymatic digests, this publication.
8. **Nugent, K.D.,** Ultrafast HPLC for protein analysis, this publication.
9. **Novotny, M.,** Microcolumn liquid chromatography. Biomedical and pharmaceutical applications, *LC.GC*, 3, 876, 1985.
10. **Mant, C.T. and Hodges, R.S.,** Requirement for peptide standards to monitor ideal and non-ideal behavior in size-exclusion chromatography, this publication.
11. **Mant, C.T. and Hodges, R.S.,** The use of peptide standards for monitoring ideal and non-ideal behavior in cation-exchange chromatography, this publication.
12. **Mant, C.T. and Hodges, R.S.,** Requirement for peptide standards to monitor column performance and the effect of column dimensions, organic modifiers and temperature in reversed-phase chromatography, this publication.
13. **Burke, T.W.L., Mant, C.T., and Hodges, R.S.,** The effect of varying flow-rate, gradient-rate and detection wavelength on peptide elution profiles in reversed-phase chromatography, this publication.
14. **Hodges, R.S. and Mant, C.T.,** Standard chromatographic conditions for size-exclusion, ion-exchange, reversed-phase and hydrophobic interaction chromatography, this publication.

RAPID HPLC OF PEPTIDES AND PROTEINS

Krishna Kalghatgi and Csaba Horváth

I. INTRODUCTION

Traditionally, high-performance liquid chromatography (HPLC) of biopolymers is carried out by using stationary phases and chromatographic systems similar to those originally developed for the separation of small molecules. Thus, in HPLC of biological macromolecules, such as proteins and nucleic acids, microparticulate stationary phases made with macroporous supports are used and their separation is accomplished by reversed-phase, ion-exchange, or other types of chromatography under gradient elution conditions. However, a major drawback of using porous stationary phases stems from the mass transfer resistances with concomitantly long analysis times in the analysis of peptides and proteins. Further, poor recovery is frequently encountered due to entrapment of large molecules in the cavernous interior of porous sorbents. Recent applications of biopolymer HPLC such as measurement of purity, determination of sample composition, as well as process monitoring in fermentation or preparative chromatography call for introduction of rapid analytical methods.

Historically, enhancement of separation efficiency, increase in detection sensitivity, and reduction of analysis time has always been associated in the development of HPLC. The role and importance of mass transfer in chromatographic processes is well understood.[1-3] Particle size of the stationary phase is an important parameter in determining the efficiency of the column and reduction in particle size usually results in greater column efficiency and higher speed of analysis. However, this approach is generally restricted by the concomitant decrease in column permeability so that high column inlet pressures are required to attain even moderate flow velocities. Another important factor is the nature of the stationary phase proper. In the earlier days of column liquid chromatography of biopolymers, stationary phases were made from relatively large particles of polysaccharide matrices such as cellulose or agarose which are mechanically weak and have low efficiency due to diffusional resistances in the particles. Attempts to improve the mechanical stability and mass-transfer properties of the stationary phase were already made with limited success in the 1950s by the use of surface-coated packings. For instance, Celite was coated with a layer of ion-exchange resins for the separation of proteins[4,5] or cellulose was treated with polyethylenimine for the separation of nucleic acid constituents by thin-layer chromatography.[6] With the advent of HPLC, pellicular sorbents made by the coating of 40-μm glass diameter beads with a retentive layer, were used to reduce diffusional resistances by reducing the diffusional path length in the stationary-phase particles. Horváth[7] first demonstrated the merits of pellicular stationary phases for rapid HPLC for analysis of nucleic acid constituents. However, this approach was superseded in the late 1960s by the introduction of totally porous microparticulate bonded phases which exhibited greater efficiency and loading capacities in the chromatography of small molecules. As a result, the first generation of pellicular stationary phases lost its significance and since then the use of such column packings has been restricted mainly to guard columns.

II. MICROPELLICULAR STATIONARY PHASES

Recent advances in fine particle technology have led to the development of novel micropel-

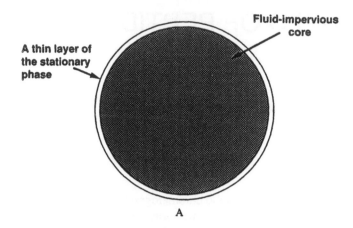

FIGURE 1. (A) A schematic illustration of the pellicular configuration. (B) Electron micrograph of 2-μm micropellicular silica microspheres.

licular supports for chromatography of biological macromolecules. In the middle of the 1980s, Unger[8] revived the concept of pellicular configuration and introduced non-porous reversed-phase packings made of monodisperse, 1.5-μm spherical silica particles for the chromatography of proteins. In the course of our investigations aimed at the development of rapid HPLC methods for biopolymer analysis, our laboratory has studied the use of columns packed with micropellicular packings made from 2-μm fused silica microspheres for rapid chromatography of peptides and proteins. A schematic illustration and a scanning electron micrograph of these types of sorbents are shown in Figure 1. The microspheres have a mean particle diameter of 2-μm and a specific surface area of 1.2 m²/g which is close to the calculated value of the geometric surface. Due to the solid, fluid-impervious core of these micropellicular packings, the columns are generally more stable at high pressures and elevated temperature than conventional HPLC packings made from porous silica or polymeric supports. Since with micropellicular sorbents the chromatographic interactions are limited to the outer surface, and the diffusional distances in the mobile phase are short, mass transfer resistances are also small with respect to those in columns packed with conventional stationary phases. This is particularly significant in the chromatography of large molecules which exhibit improved recovery in the absence of porous interior which may serve as a trap for macromolecules. Thus, micropellicular stationary phases offer the potential of both rapid analysis and high recovery of biopolymers. Additional benefits of micropellicular supports stem from rapid column regeneration after gradient elution and ease of maintenance.

III. REVERSED-PHASE CHROMATOGRAPHY OF PEPTIDES

Reversed-phase chromatography (RPC) of peptides is a powerful technique for peptide mapping, for isolation of peptides, to determine amino acid sequence,[9,10] for studies on microheterogeneity,[11] mutational variants,[12,13] and other isomers,[14,15] as well as for study of glycosylation sites[16] in proteins. Peptide mapping is now widely used in biotechnology for analysis of purity and quality control of recombinant proteins.[11,17,18]

Columns packed with a suitable micropellicular stationary phase are eminently suitable for rapid chromatographic separation of peptides, including peptide mapping. This is illustrated by peptide maps obtained by HPLC of a tryptic digest of recombinant tissue plasminogen activator (r-tPA) and a chymotryptic digest of β-lactoglobulin A in Figure 2. In each case, the components of the enzymic digests are separated with satisfactory resolution in 15 min, which is much shorter

FIGURE 1B

than that required for such peptide maps by using conventional porous stationary phases. The high resolution and high speed of separation exhibited by these maps is attributed not only to the pellicular configuration of the stationary phase but also to the use of elevated temperature which results in improved sorption kinetics.[19,20] The high column temperature did not compromise the stability of the micropellicular reversed-phase column.

IV. OPERATIONAL VARIABLES

Due to their relatively high stability, micropellicular reversed-phase columns can be used under a wide range of operating conditions. The small particle size of the stationary phase in the

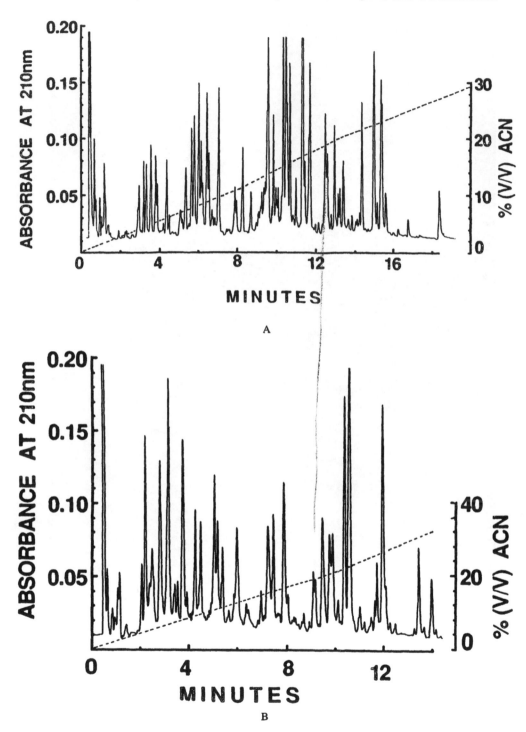

FIGURE 2. (A) Tryptic map of r-tPA. Sample: ≈ 10 μg of reduced and *S*-carboxymethylated r-tPA. Column: micropellicular C$_{18}$ silica, 75 × 4.6 mm I.D., Conditions: Eluent A, 50 m*M* phosphate buffer containing 1 mM sodium octyl sulfate, pH 2.8; Eluent B, 60% (v/v) acetonitrile (ACN) in 50 m*M* phosphate buffer, pH 2.8. Flow-rate, 1.5 ml/min; temperature, 80°C. (B) Chymotryptic map of β-lactoglobulin A. Sample: 3.5 μg of reduced and *S*-carboxymethylated β-lactoglobulin A. (Both chromatograms from Kalghatgi, K. and Horváth, Cs., *J. Chromatogr.*, 443, 343, 1988. With permission.)

columns has low permeability but this limitation can be overcome by carrying out the separation at elevated temperatures where the viscosity of the mobile phase is lower. We have shown previously[19] that an increase in temperature from 20 to 80°C results in a three-fold decrease in column inlet pressure due to a corresponding decrease in eluent viscosity which is also associated with an increase in the rate of diffusion. By virtue of favorable mass transfer properties of pellicular configuration, the columns can be operated at elevated temperature and at high flow velocities without significant loss in resolution. However, the phase-ratio of columns packed with micropellicular sorbents is lower than that of conventional columns packed with porous stationary phases used for separation of small molecules. As a result, short, polar peptides in protein digests are not retained sufficiently when the separation is carried out at high temperature. This is remedied by addition of suitable ion-pairing agents such as sodium hexyl- or octyl sulfate to the mobile phases. Nevertheless, the benefits of micropellicular packings manifest themselves in faster analyses, even at room temperature, and lower flow rates than those used in Figure 2.

V. RAPID HPLC OF PROTEINS

The chromatography of large molecules requires conventional stationary phases having correspondingly large pores in order to provide access to the binding sites and to facilitate diffusion in the particle interior. It follows that when only a fraction of the pores is explored by the polymeric eluites, the stationary phase is poorly utilized. Furthermore, restricted diffusion in pores having insufficiently large diameter can reduce the efficiency of separation and entrapment of solutes can result in poor recovery.

With micropellicular stationary phases, these problems are eliminated by virtue of the missing porous interior. Due to the small particle size, columns packed with these stationary phases exhibit a phase-ratio for biopolymers which is commensurate with that found with conventional columns. Consequently, pellicular sorbents are particularly suitable for chromatography of large molecules such as proteins and, due to the considerations outlined previously, they offer an efficient means for protein analysis. Since HPLC is rapidly assuming a leading role in analytical biotechnology, an increase in the speed of protein HPLC is of considerable importance.

Previous results from our laboratory have demonstrated the separation of five standard proteins by gradient elution in 20 sec and the two β-lactoglobulins by isocratic elution in 15 sec by using short columns packed with silica-based micropellicular non-polar stationary phases.[19] In order to reach such a high speed of analysis, high column temperature and high flow rates were used. Since most commercially available liquid chromatographs are not designed for such high analyses, certain modifications may be necessary to take full advantage of the micropellicular sorbent configuration. Such analytical speed may be necessary in monitoring the effluent of a preparative column or some other process of biotechnological significance. For most applications that require protein separations, an analysis time of 2 to 5 min may be quite appropriate at present. The chromatograms of some commercial samples of insulin and trypsin (Figure 3) illustrate the potential of micropellicular stationary phases in this regard.

Rapid protein analysis has been carried out at elevated column temperatures according to the considerations made above about the effect of temperature on the speed of separation. RPC of proteins employs conditions which result in denaturation. In fact, the unique selectivity of this technique is due to the interaction of unfolded protein molecules with the non-polar chromatographic surface. At elevated temperatures, we may expect further denaturation, the extent of which, however, is mitigated by the short residence time in the column. Due to their stability, micropellicular sorbents allow us to take advantage of elevated column temperature in analytical applications of RPC where the protein does not have to be recovered in the native form for further

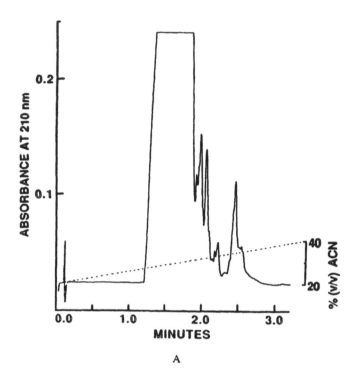

FIGURE 3. Chromatograms of commercial preparations of insulin (A) and trypsin (B). Column, micropellicular C_{18} silica, 30 × 4.6 mm I.D., Eluent A, 0.1% (v/v) TFA in water; Eluent B, 95% (v/v) ACN in water containing 0.1% (v/v) TFA; 20 to 40% B in 4 min gradient. Flow-rate, 3 ml/min; temperature, 80°C. Samples, 25 μg of insulin or 15 μg of trypsin both from Sigma. The off-scale peak in each chromatogram represents the protein of interest.

investigations. Fast separation of protein mixtures is of interest mainly in routine analysis in order to utilize better the capacity of the instrument and to obtain expeditiously the analytical information. In the future, we may expect an increase in the use of multi-column analytical HPLC systems. In order to make this approach practicable in routine work, a reduction in the time needed for the individual runs will be necessary so that the total time of analysis involving multiple runs will remain within practical limits.

The major shortcoming of silica-based bonded phases, which are most widely used in HPLC, is the hydrolytic decay of the silica support in contact with aqueous mobile phases having alkaline pH. This precludes the use of such materials at pH values higher than 9 or treatment of the column with caustic soda for cleaning purposes. Under such circumstances, suitable polymer- based micropellicular sorbents, which have already been made from various supports,[21-23] are commercially available. A detailed description of such a column material for RPC has been provided recently.[24] The support was made of highly cross-linked polystyrene and the column was stable in the pH range from 2 to 11 at temperatures up to 80°C.

Best results in RPC of proteins were obtained either at low pH with 0.1% trifluoroacetic acid (TFA) or at pH 11 with 1.5 mM Na_3PO_4 in the hydro-organic eluent. The selectivity of the chromatographic system was significantly different at the two pH extremes and the separation of a mixture of five proteins was completed in less than 30 sec in both cases. Such a column may find application in alkaline RPC of proteins when striving for alteration of selectivity. They are expected to be useful also in the analysis of "dirty" samples directly since the column can be

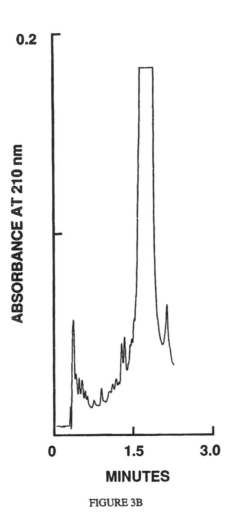

FIGURE 3B

cleaned with a sodium hydroxide solution to remove strongly bound impurities. In this way, elaborate and time-consuming sample preparation can be avoided.

VI. BIOSPECIFIC INTERACTION CHROMATOGRAPHY OF PROTEINS

Affinity chromatography, which was developed primarily as a preparative technique, is receiving increasing attention for analytical separations. It is anticipated that analytical affinity techniques will play a major role in monitoring and control of bioreactors and various bioprocesses. Affinity chromatography has been carried out by stationary phases made of porous matrices and often at subambient temperature. Biospecific pellicular sorbents appear to be ideal for analytical affinity separations, since they allow maximum exposure of the affinity ligate at the surface to interact with the molecules to be separated. We have recently demonstrated that micropellicular affinity sorbents are indeed suitable for rapid biospecific interaction chromatography.[25] Examples are shown in Figure 4A for quantitative analysis of IgG from human serum (I) and tissue culture supernatant (II) within 3 min by using a short column packed with silica-based micropellicular Protein A. In a similar fashion, a commercial sample of horseradish peroxidase was analyzed in 1 min by using a micropellicular lectin column with immobilized concanavalin A (Con A) (Figure 4B).

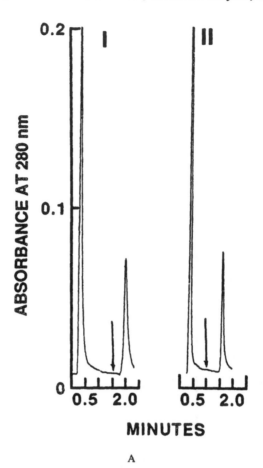

A

FIGURE 4. (A) Biospecific interaction chromatography of human IgG from serum (I) and mouse monoclonal antibody (II) on a Protein A column[25] with stepwise elution. Column: 30×4.6 mm I.D., 2-μm silica-based micropellicular Protein A. Binding buffer, 100 mM citrate (pH 7.4); debinding buffer (introduced at the arrow), 100 mM citrate (pH 2.2). Flow-rate, 1 ml/min; temperature, 25°C; sample, 20 μl of tissue culture supernatant or serum, diluted with binding buffer (1:40). (B) Biospecific interaction chromatography of horseradish peroxidase on Con A column by stepwise elution. Column: 30×4.6 mm, 2-μm silica based micropellicular Con A. Binding buffer, 25 mM Tris-HCl (pH 7.0) containing 150 mM sodium chloride, 1 mM manganese chloride and 1 mM calcium chloride; debinding buffer (introduced at the arrow) 50 mM α-methyl-D-glucopyranoside in the binding buffer. Flow-rate, 1 ml/min; temperature, 25°C. Detector sensitivities, 0.1 and 0.05 AUFS at 280 and 405 nm. Sample, 20 μg of horseradish peroxidase from Sigma. (Both chromatograms from Várady, L., Kalghatgi, K., and Horváth, Cs., *J. Chromatogr.*, 458, 207, 1988. With permission.)

VI. CONCLUSIONS

Short columns packed with micropellicular reversed-phase sorbents and eluted under gradient elution conditions are used at relatively high flow rates and at elevated temperature for rapid separation of peptides and proteins. In process monitoring applications, samples containing a few components are analyzed within seconds and, in less demanding routine analyses, within a few minutes. On the other hand, highly complex mixtures like enzymatic digests of proteins are separated in less than 20 min, i.e., a much shorter time than that required when columns packed with conventional porous stationary phases are used. Since rapid separations

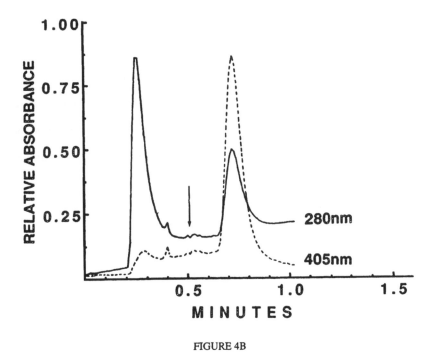

FIGURE 4B

are possible only at relatively high flow rates for high speed analysis, micropellicular columns are preferably operated at elevated temperature where the viscosity of the mobile phase is reduced. However, these columns also provide excellent performance at ambient temperature and may offer higher sensitivity due to lower flow rates under these conditions.

ACKNOWLEDGMENTS

This work was supported by Grants CM 20993 and CA 21948 from the National Institute of Health, U.S. Department of Health and Human Services.

REFERENCES

1. **Van Deemter, J.J., Zuiderweg, F.J., and Klingenberg, A.,** Longitudinal diffusion and resistance to mass transfer as causes of non-ideality in chromatography, *Chem. Eng. Sci.*, 5, 271, 1956.
2. **Horváth, Cs. and Lin, H.-J.,** General plate height equation and a method for the evaluation of the individual plate height contributions, *J. Chromatogr.*, 149, 43, 1978.
3. **Snyder, L.R. and Kirkland, J.J.,** in *Introduction to Modern Liquid Chromatography*, John Wiley & Sons, New York, 1979, pp. 168—268.
4. **Feitlson, J. and Partridge, S.M.,** An ion-exchange reagent for use in the chromatography of large peptides, *Biochem. J.*, 64, 607, 1956.
5. **Boardman, N.K.,** Chromatography of proteins on celite ion-exchange resins, *J. Chromatogr.*, 2, 388, 1959.
6. **Randerath, K.,** in *Thin Layer Chromatography*, 2nd ed., Academic Press, New York, 1956, pp. 229—243.
7. **Horváth, Cs.,** High-performance ion-exchange chromatography with narrow-bore columns: rapid analysis of nucleic acid constituents at the subnanomole level, in *Methods of Biochemical Analysis*, Vol. 21, Glick, D., Ed., John Wiley & Sons, New York, 1973, pp. 79—154.

8. **Unger, K.K., Jilge, G., Kinkel, J.N., and Hearn, M.T.W.,** Evaluation of advanced silica packings for the separation of biopolymers by high-performance liquid chromatography, *J. Chromatogr.*, 359, 61, 1986.

9. **Fullmer, C.S. and Wasserman, R.H.,** Routine peptide mapping by high-performance liquid chromatography, in *Methods in Protein Sequence Analysis*, Elzinga, M., Ed., Humana Press, Clifton, N.J., 1982, 489.

10. **L'Italien, J.J. and Stickler, J.E.,** Applications of high performance liquid chromatographic peptide purification to protein sequencing by solid-phase Edman degradation, in *High Performance Liquid Chromatography of Proteins and Peptides*, Hearn, M.T.W., Regnier, F.E., and Wehr, C.T., Eds., Academic Press, New York, 1983, 195.

11. **Hancock, W.S.,** Significance of purity in the manufacture of recombinant DNA-derived proteins, *Chromatogr. Forum* , 2, 57, 1986.

12. **Leadbeater, L. and Ward, F.P.,** Analysis of tryptic digests of bovine β-casein by reversed-phase high-performance liquid chromatography, *J. Chromatogr.* 397, 435, 1987.

13. **Takahashi, N., Takahashi, Y., Ishioka, N., Blumberg, B.S., and Putnam, F.W.,** Application of an automated tandem high-performance liquid chromatography system to peptide mapping of genetic variants of human serum albumin, *J. Chromatogr.*, 359, 181, 1986.

14. **Pohl, G., Kallastrom, M., Bergsdorf, N., Walless, P., and Jornvall, H.,** Tissue plasminogen activator: peptide analyses confirm an indirectly derived amino acid sequence, identify the active site serine residue, establish glycosylation sites, and localize variant residue, establish glycosylation sites, and localize variant differences, *Biochemistry*, 23, 3701, 1984.

15. **Browning, J.L., Mattaliano, R.J., Chow, E.P., Liang, S-M., Allet, S.M., Rosa, J., and Smart, J.E.,** Disulfide scrambling of interleukin-2: HPLC resolution of the three possible isomers, *Anal. Biochem.* 155, 123, 1986.

16. **L'Italien, J.J.,** Microscale elucidation of an N-linked glycosylation site by comparative high-performance liquid chromatography peptide mapping, *J. Chromatogr.*, 359, 213, 1986.

17. **Borman, S.,** Analytical biotechnology of recombinant products, *Anal. Chem.*, 59, 969A, 1987.

18. **Garnick, R.L., Solli, N.J., and Papa, P.A.,** The role of quality control in biotechnology: an analytical perspective, *Anal. Chem.* 60, 2546, 1988.

19. **Kalghatgi, K. and Horváth, Cs.,** Rapid analysis of proteins and peptides by reversed-phase chromatography, *J. Chromatogr.*, 398, 335, 1987.

20. **Kalghatgi, K. and Horváth, Cs.,** Rapid peptide mapping by high-performance liquid chromatography, *J. Chromatogr.*, 443, 343, 1988.

21. **Burke, D.J., Duncan, J.K., Dunn, L.C., Cummings, I., Siebert, C.J., and Ott, G.S.,** Rapid protein profiling with novel anion-exchange material, *J. Chromatogr.*, 353, 425, 1986.

22. **Kato, Y., Kitamura, T., Mitsui, A., and Hashimoto, H.,** High-performance ion-exchange chromatography of proteins on non-porous ion exchangers, *J. Chromatogr.*, 398, 327, 1987.

23. **Lee, D.P.,** Chromatographic evaluation of large-pore and non-porous polymeric reversed phases, *J. Chromatogr.*, 443, 143, 1987.

24. **Maa, Y.-F. and Horváth, Cs.,** Rapid analysis of proteins and peptides by reversed-phase chromatography with polymeric micropellicular sorbents, *J. Chromatogr.*, 445, 71, 1987.

25. **Várady, L., Kalghatgi, K., and Horváth, Cs.,** Rapid high-performance affinity chromatography on micropellicular sorbents, *J. Chromatogr.*, 458, 207, 1988.

ULTRAFAST HPLC FOR PROTEIN ANALYSIS

Kerry D. Nugent

I. INTRODUCTION

Liquid chromatography (LC) has been a very useful tool to biochemists for several decades. It is used extensively in the isolation and purification of biomolecules and its use as an analytical tool to characterize samples continues to grow. With the significant improvements in column packings and instrumentation over the past few decades, LC has evolved into a rapid and reliable tool for the analysis of biomolecules (Table 1).

In the past few years, there has been an increasing number of papers showing examples of very fast (less than 5 min) separations of proteins by reversed-phase chromatography (RPC) and ion-exchange chromatography (IEC) using both nonporous and macroporous supports.[1-6] Although these separations are impressive, the real criterium for ultrafast protein analysis is not time to resolve the components but rather, the total analysis time from injection to injection with accurate and reproducible results. When using these new columns, one must carefully optimize the instrumentation with respect to system volumes and flow dynamics in order to achieve ultrafast protein analyses. This is verified by Horváth who concludes: "Appropriate changes in instrumental design are necessary to take full advantage of this approach".[2]

In this paper, a new tool (the Ultrafast Microprotein Analyzer-UMA) will be described which allows rapid (1 to 3 min) gradient separations by both RPC and IEC. The UMA has been specifically designed to allow rapid re-equilibration by minimizing volumes and optimizing flow dynamics, and uses a variety of new fast flow HPLC packings in a 50 × 1 mm I.D. column at 5 to 10 times the normal flow velocities for these microbore columns with standard packing materials. This allows highly efficient, reproducible separations with a total analysis time of under 5 min. Application of the UMA to ultrafast analysis of recombinant proteins by RPC will be demonstrated.

II. EXPERIMENTAL

A. INSTRUMENTATION

The analyses described in this article were performed on a prototype Ultrafast Microprotein Analyzer (Michrom BioResources Inc., Pleasonton, CA). This unit was equipped with a custom microbore binary gradient pumping system optimized for ultrafast analysis, a manual Valco 10 port biocompatible sample injector, a column oven, a variable wavelength UV/Vis detector equipped with a custom flow cell and a PC based controller/data system.

B. COLUMNS

The column packing material used in this study consisted of polymeric macroporous spheres obtained from Polymer Laboratories, Inc., (Amherst, MA). This reversed-phase (PLRP-S, 8-μm particle size, 4000-Å pore size) material was packed into special column cartridges (50 × 1.0 mm I.D.) in our laboratory.

TABLE 1
Evolution of Biochemical Liquid Chromatography

Year	Analysis mode	Typical run times
1950s	Open column LC	Days
1960s	Low pressure LC	Many hours
1970s	Medium pressure LC	A few hours
1980s	HPLC	Less than 1 hr
1990s	Ultrafast LC	Less than 5 min

III. CHARACTERISTICS OF ULTRAFAST PROTEIN ANALYSIS

A. INSTRUMENT REQUIREMENTS

Most modern analytical HPLC instrumentation is designed to be versatile for a broad range of applications using conventional HPLC column technology. Compromises are often made in flow-rate and compositional accuracy and precision at the extremes of the system capabilities to allow a broader range of usability. Flow dynamics, mixing volumes, and extra column volumes are generally compromised or left for the user to optimize for specific applications. For these reasons, a conventional HPLC must be extensively modified if any of the potential gains in ultrafast analysis are to be achieved.

In this study, a dedicated UMA was constructed to take full advantage of the power of rapid protein separations. A dual microbore binary gradient pumping system was chosen which allowed reproducible, rapid (1 to 3 min) gradient formation with precise flows in the 2 to 1000 μl/min range. A positive displacement helium degas module ensured precise solvent composition and reproducible volumes of degassed solvents displaced by each piston stroke at pressures up to 6000 psi. By using a dual pump high pressure mixing system, the volumes and flow dynamics of the pumps had no effect on re-equilibration time between runs.

Since the main criteria for accessing the usefulness of a system for ultrafast protein analysis is the total analysis time from injection to injection, it is critical that the flow dynamics and system volumes be carefully optimized. In working with conventional HPLC systems, it was found that, although several could generate good gradients in 2 to 5 min, the total analysis time was often 10 to 20 min because of the inability of the system to re-equilibrate rapidly for the next sample injection. Attempts to inject subsequent samples prior to proper gradient re-equilibration would cause extremely poor retention time repeatability.

In the UMA, the total system volume was reduced to less than 100 μl using a custom 3 μl mixing tee and a 80 μl static mixing chamber, both designed to ensure adequate solvent mixing with all volumes well swept. A Valco 10 port injector was used to ensure precise injections in an inert valve, while minimizing dispersion effects. The use of a special column cartridge coupled to a custom detector flow cell reduced the extra column volume to less than 1 μl. The combination of these designs, together with strict attention to solvent flow dynamics, created a system which could be re-equilibrated in less than 1 min.

The UMA also contained a column oven capable of controlling temperature to ± 0.5°C from 30 to 99°C. This oven is important not only to ensure reliable retention time repeatability, but also because elevated temperatures help to improve the partitioning of proteins at these fast flows and reduce solvent viscosity which lowers the pressure drop across these fast-flow columns.[2] A variable wavelength UV/Vis detector was used to monitor the separations at 214 nm and excellent sensitivity was achieved using the 1.0 mm I.D. columns.

B. COLUMN REQUIREMENTS

In selecting a column for ultrafast protein analysis, the most important factor to consider is the packing material. The best place to start is to choose a packing material which was specifically developed for fast protein separations. With conventional HPLC packings, maximum efficiency is obtained at slow flow-rates and efficiency drops off as the flow velocity increases.[7] With the new fast flow materials, efficiency remains constant (or actually increases) as the flow velocity increases.

For ultrafast protein analysis, two main classes of materials have demonstrated this type of phenomenon. The first class are extremely small (1 to 3 μm) nonporous particles, which allow rapid partitioning of proteins on the surface of the particles.[1-4] The second class of materials are small (3 to 10 μm) macroporous particles (usually greater than 1000 Å pores) which maximize permeability and allow rapid partitioning of the proteins within these large pores at high flow velocities.[5,6]

Although both silica-based and polymeric resin-based materials are commercially available, the common criterium is that they must have good mechanical stability to withstand the higher flows and subsequent higher pressures required by the ultrafast analysis. In this study, the polymeric resins from Polymer Labs were found to give the best overall performance when considering such factors as efficiency, pressure drop, stability (mechanical, chemical and thermal), mass loading, ghosting, and recovery.

The second factor in selecting a column for ultrafast protein analysis is the column length. When using RPC with proteins larger than 10 kD, very short columns (2 to 50 mm) can give adequate resolution for most separations;[8] however, for complex samples and smaller proteins and peptides, a 50-mm long column offers the best compromise for efficiency and recovery.[9] In the current study, a 50-mm column length was also found to be optimal for these ultrafast protein analyses, since this gave adequate resolution of most peptide and protein samples by RPC and IEC. Very short columns (less than 20 mm), although better for pressure drop, recovery, and void volume, gave inadequate resolution and were susceptible to sample breakthrough for many of these fast separations. Longer columns (100 to 250 mm) gave better resolution and peak capacity for complex samples and small peptides and proteins, but the greater pressure drop, void volume, and sample losses made them unsuitable for routine ultrafast protein analysis.

Although not as important as length, the column I.D. must also be considered. Most reports to date have used a conventional 4.6-mm I.D. column for these rapid separations. Although useful, they require flows of 5 to 10 ml/min to achieve optimum results. These high flow-rates can create very large pressure drops, difficulties in solvent mixing, significant waste of expensive solvents and dramatic decreases in detection sensitivity. For these reasons, a smaller column I.D. is recommended, provided that suitable instrumentation is available to take full advantage of the gains which can be achieved. This study used 1.0-mm I.D. columns at flows of 0.25 to 0.50 ml/min in the UMA, which was optimized for this column size.

An example of the performance of the UMA in a RPC separation of 1 μg each of 11 protein standards (5 to 97 kD) is shown in Figure 1.

IV. APPLICATION OF ULTRAFAST RPC TO RECOMBINANT PROTEIN PROCESS MONITORING

Over the past decade, the field of biotechnology has grown from a few R&D start-ups to well over a thousand companies involved in the research, development, and production of recombinant protein products. Since the production of recombinant proteins is quite complex, a significant burden has been placed on the analyst trying to monitor both yield and purity through a multi-step process.

Unlike conventional chemical processing, which is generally a continuous stream, recombinant protein production is generally carried out in a batch mode. Although a process is generally developed prior to the start of production of a new recombinant protein, each batch must be carefully monitored from fermentation through purification to final product to maximize yield

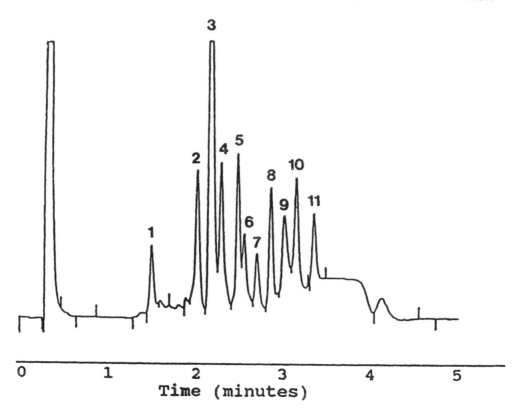

FIGURE 1. Separation of aprotinin (1), ribonuclease A (2), insulin (3), lysozyme (4), human glycoprotein (5), α-lactalbumin (6), trypsin inhibitor (7), carbonic anhydrase (8), bovine serum albumin (9), ovalbumin (10), and amyloglucosidase (11) by ultrafast RPC. Column: PLRP-S 4000 Å. Conditions: a 3-min linear gradient from 10 to 60% acetonitrile in 0.1% aq. trifluoroacetic acid (TFA) at a flow-rate of 250 μl/min.

and minimize contaminants.[10] Since throughput is the most important consideration in production, techniques such as overloading columns coupled with heartcutting and recycling are often employed. An example of this can be seen in Figure 2.

An anion-exchange separation was developed as an intermediate step in the purification of a proprietary recombinant protein. In the upper trace (A) of Figure 2, a semipreparative separation was developed to isolate the protein of interest from its major contaminants with good resolution. The actual production run was then operated in an overload condition to maximize throughput and the resulting separation is shown in the lower trace (B) of Figure 2. From the preparative profile, it is much more difficult to see how to collect fractions to maximize yield and purity. Since the process group had been using an HPLC assay which required 90 min/run, they had decided to collect only five fractions during this run, as shown in Figure 2. These fractions were subsequently analyzed by RPC on the UMA, and the results of these analyses can be seen in Figure 3. It is easy to see from these analytical runs that fractions 2 and 3 contain significant levels of impurity and that the protein of interest is still being eluted well into fraction 5. By combining fractions 1 and 2, and recycling fraction 3 (10% of product), a 70% yield with a 85% purity was taken on to the next step in the process. In light of these results, a subsequent preparative run was fractionated into 48 samples (every 5 min for 4 hr) and analyzed off-line on the UMA. By combining the proper fractions from the 48 sample run, a 55% yield with a 97%

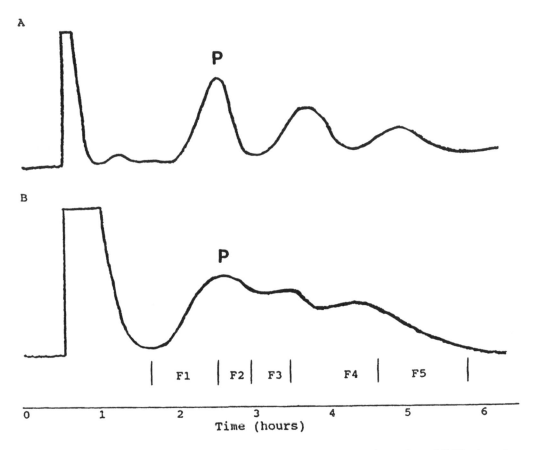

FIGURE 2. (A) Semipreparative purification of recombinant protein sample by anion-exchange LC (20 × 1 cm glass column packed with DEAE Sepharose) with normal column load (5 mg). (B) Preparative purification of recombinant protein sample by anion-exchange LC (20 × 5 cm glass column packed with DEAE Sepharose) with column overloading (1 g). (P) designates the protein of interest. F1-F5 denote five fractions.

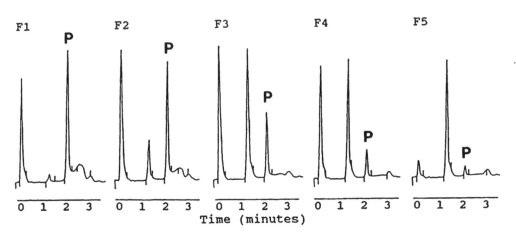

FIGURE 3. Ultrafast RPC analysis of the five fractions (F1-F5) from Figure 2. Conditions: as Figure 1. (P) designates the protein of interest.

purity was achieved. Additional fractions containing significant amounts of the protein of interest (40% of product) were recycled for separation in a subsequent run. The UMA allowed this process group to increase both total yield and purity without impacting their processing time or analytical support requirements.

V. CONCLUSIONS

Although new columns have recently been introduced to allow rapid separation of protein mixtures, a great deal of care is required to optimize all of the instrumental parameters which will impact this technique. Since the important criterium in ultrafast protein analysis is the total analysis time from inject to inject with reproducible results, the column and HPLC system must be properly integrated. A dedicated analyzer has been discussed (Ultrafast Microprotein Analyzer — UMA) which minimizes all of the system volumes and provides the other optimum system parameters to allow reproducible separations of complex protein mixtures in under 5 min.

Ultrafast protein analysis by RPC has been shown to be extremely useful for recombinant protein process monitoring. By employing a UMA at line or directly on line, process groups will be able to improve both yields and purity while saving analytical support.

Although this article has focused on protein analysis by RPC, additional materials for ultrafast protein analysis covering all modes of HPLC are currently under development by several companies and these will complement the reversed-phase mode described here. A preliminary look at a commercial fast flow anion-exchange material (PLSAX 4000 from Polymer Labs) has already been reported.[11]

In conclusion, the UMA should be useful for anyone desiring rapid analysis of large numbers of protein samples.

REFERENCES

1. **Kalghatgi, K. and Horváth, Cs.,** Rapid analysis of proteins and peptides by reversed-phase chromatography, *J. Chromatogr.,* 398, 335, 1987.
2. **Kalghatgi, K. and Horváth, Cs.,** Rapid peptide mapping by high-performance liquid chromatography, *J. Chromatogr.,* 443, 343, 1988.
3. **Rounds, M.A. and Regnier, F.E.,** Synthesis of a non-porous, polystyrene-based strong anion-exchange packing material and its application to fast high-performance liquid chromatography of proteins, *J. Chromatogr.,* 443, 73, 1988.
4. **Maa, Y. and Horváth, Cs.,** Rapid analysis of proteins and peptides by reversed-phase chromatography with polymeric micropellicular sorbents, *J. Chromatogr.,* 445, 71, 1988.
5. **Kitagawa, N.,** New hydrophilic polymer for protein separations by HPLC, in *Techniques in Protein Chemistry,* Hugli, T.E., Ed., Academic Press, San Diego, CA, 1989, 348.
6. **Regnier, F.E., Rounds, M.A., Thevenon, J., and Yang, Y.B.,** Macroporous Styrene-Divinyl Benzene-Based Media for Proteins, lecture presented at Twelfth Int. Symp. on Column Liquid Chromatography, Washington, D.C., June 19—24, 1988.
7. **Snyder, L.R. and Kirkland, J.J.,** *Introduction to Modern Liquid Chromatography,* 2nd ed., John Wiley & Sons, New York, 1979.
8. **Pearson, J.D.,** High-performance liquid chromatography column length designed for submicrogram scale protein isolation, *Anal. Biochem.,* 152, 189, 1986.
9. **Burton, W.G., Nugent, K.D., Slattery, T.K., Summers, B.R., and Snyder, L.R.,** Separation of proteins by reversed-phase high-performance liquid chromatography. I. Optimizing the column, *J. Chromatogr.,* 443, 363, 1988.
10. **Garnick, R.L., Solli, N.J., and Papa, P.A.,** The role of quality control in biotechnology: an analytical perspective, *Anal. Chem.,* 60, 2546, 1988.
11. **Nugent, K.,** Ultrafast protein analysis: a powerful tool for recombinant protein process monitoring, in *Current Research in Protein Chemistry,* Villafronca, J.J., Ed., Academic Press, San Diego, CA,, 1990.

Section XII
Prediction and Computer Simulation of
Peptide and Protein Separations and
Structure for Research and Teaching

DEVELOPMENT AND APPLICATION OF A COMPUTER SIMULATION PROGRAM FOR METHOD DEVELOPMENT AND TEACHING IN HPLC: ProDigest-LC

Colin T. Mant and Robert S. Hodges

I. INTRODUCTION

The efficient isolation of peptides has become increasingly important for an ever-widening range of research disciplines in recent years and high-performance liquid chromatography (HPLC) has shown immense versatility in the separation and purification of peptides from a great variety of sources. Peptide mixtures derived from different sources differ widely in complexity and quantity, and the approach to their separation must be tailored to the separation goals:

- Peptides obtained from biological tissues, for instance, are often found in only very small quantities and may require extensive purification. Thus, prior knowledge of the location of a biologically active peptide from various tissue sources in chromatograms obtained by different HPLC modes would be extremely beneficial.
- Separation of peptides from a chemical and/or proteolytic digest of a protein for subsequent peptide characterization is vital in structure-function studies of proteins. The complexity of the resulting peptide mixture will depend on the particular digesting agent as well as the properties (size, amino acid composition and sequence, etc.) of the protein of interest.
- During biosynthesis of proteins for therapeutic purposes, impurities very similar to the desired protein will be present. Separation systems are required that can detect small changes in the polypeptide chain; thus, peptide mapping, following protein digestion by chemical or proteolytic agents, is one way to verify the structure of a genetically engineered protein.
- The wide use of automated solid-phase peptide synthesis in recent years has also necessitated efficient isolation of peptides from various impurities, usually closely related to the peptide of interest (deletion, terminated, or chemically modified peptides) and perhaps missing only one amino acid residue.

Peptides derived from various sources differ widely in size, net charge and hydrophobicity, and purification of a single peptide from a complex mixture will require an approach different from that necessary for separating all components of a mixture. The former approach may require the application of only a single HPLC mode; in contrast, the latter approach will require a combination of separation modes for efficient resolution of all desired peptides.[1-3] The three main modes of HPLC used for peptide separations utilize differences in peptide size (size-exclusion HPLC or SEC), net charge (ion-exchange HPLC or IEC) or hydrophobicity (reversed-phase HPLC or RPC). Within these modes, mobile-phase conditions may be manipulated to maximize the separation potential of a particular HPLC column. Although a desired peptide separation may be obtained by trial and error, this may take many attempts, with subsequent loss of time and valuable peptide sample in cases where only limited quantities are available. Thus, any methodology that can aid the researcher in selecting a purification protocol without using precious sample or requiring an extensive method development time is invaluable.

A computer software program, ProDigest-LC,[4,5] has been developed to assist scientists in devising methods for the analytical separation and purification of biologically active peptides and peptide fragments from enzymatic and chemical digests of proteins by SEC, cation-exchange chromatography (CEC) and RPC. The experiments simulated on the computer eliminate the time-consuming trial-and-error methods involved in obtaining suitable separation or purification methods. In addition, ProDigest-LC is also a teaching aid for chromatographers, designed to help the student or researcher to select the correct conditions for chromatography (HPLC mode, column and mobile phase) and allowing him or her the option of examining the effect of varying flow-rate, gradient-rate, sample size and sample volume on the separation. No prior information about the sample peptides, except their amino acid composition, is required for the simulation performed by ProDigest-LC.

This article summarizes the main features of ProDigest-LC, outlines the concepts which led to its development and demonstrates some of its prediction and simulation capabilities.

II. GENERAL FEATURES OF THE PROGRAM

The most important requirement of any computer program is that it be user-friendly. Each menu is self-explanatory in ProDigest-LC, providing simple instructions and one-letter keying to access any particular section of the program. A full description of the various program menus can be found in References 4 and 5.

A summary of the major features of ProDigest-LC is as follows:

- Predicts retention behavior and simulates elution profiles of peptides of known composition, containing 2 to 50 amino acid residues, in SEC, CEC and RPC.
- Simulates the effect of varying sample size, sample volume (SEC only), flow-rate, and gradient-rate (CEC and RPC only) on peak height, peak width, peak retention time, and resolution of adjacent peaks.
- Carries out chemical and enzymatic cleavage of proteins and simulates elution profiles of protein fragments in SEC, CEC and RPC.
- Aids the researcher in deciding which HPLC mode or combination of HPLC modes is most suitable for resolution of a particular protein digest.
- Allows the researcher, through a peptide mapping option, to identify and isolate mutant peptide fragments from digests of mutant proteins by CEC or RPC.

ProDigest-LC also contains an extensive information menu, from which the operator can select the following features:

- A *definitions* section provides definitions of HPLC parameters used in the program.
- A *standards information menu* describes peptide standards that are available for SEC, CEC, and RPC, together with a list of references. For each mode of chromatography, the design features of the standards are described, together with examples of separations of the standards by means of various mobile phases to demonstrate the importance of standards in selecting mobile phase conditions.
- Under the *columns and conditions* section, reasons are given for using a particular HPLC mode for the separation of peptides and for choosing a particular column within this mode; the value of peptide standards for monitoring the resolving power of a column under various conditions and for detecting non-specific peptide/packing interactions are described. In addition, column maintenance and storage conditions are provided. All relevant references are listed, so that the researcher can quickly access more detailed information.

ProDigest-LC is available from Synthetic Peptides Incorporated, Department of Biochemistry, University of Alberta, Edmonton, Alberta, Canada. The minimum requirements for operating the program is an IBM-AT or compatible computer with 256K memory, equipped with two floppy disk drives and a monitor with graphics capability. The use of a math coprocessor is also strongly recommended.

III. IMPORTANCE OF PEPTIDE STANDARDS

The use of peptide standards has played a key role in the evolution of ProDigest-LC as a valuable analytical tool and chromatography teaching aid. Their importance is highlighted in four major areas:

1. Predictable retention behavior of peptides on HPLC requires that their separation is based on only one separation mechanism. Thus, predictable peptide retention times in SEC requires that they are separated based on their molecular size only; peptides should be separated solely on the basis of their net charge in IEC; and the mechanism by which peptides interact with reversed-phase sorbents should be based solely on peptide hydrophobicity. Any non-ideal interactions between peptides and the column packing (e.g., ionic interactions in SEC and RPC, or hydrophobic interactions in SEC and IEC) must be identified and suppressed by manipulation of the mobile phase. Peptide standards for SEC, CEC, and RPC are commercially available for identifying non-specific interactions and are supplied with ProDigest-LC. The use of these standards for SEC, CEC and RPC is discussed in detail in the relevant sections of this book.

2. Peptide standards were required during development of the simulation capabilities of ProDigest-LC, in terms of both prediction of peptide retention time and prediction of the effects of varying chromatographic parameters (sample load, sample volume, gradient-rate, flow-rate) on peptide retention time, peptide resolution, peak height, and peak width. For further details, see Section IV.B below.

3. Standards enable calibration of a researcher's HPLC column and instrumentation. After chromatographing a set of standards (suppled with ProDigest-LC) on a particular column, the quantity injected, peak heights, peak widths, retention times, and other parameters necessary for the program to adjust the predicted elution profiles to the researcher's particular column can be entered. The data generated by this single standard run calibrate the researcher's column to allow for column-to-column differences and instrumentation variations.

4. Model peptides have proved vital during rigorous assessment of the simulation capabilities of ProDigest-LC. The most logical approach to testing the accuracy of the program is to chromatograph, under varying conditions, well-defined mixtures of peptides differing in sequence and conformation from those used to calibrate the column [see point 3 above] and comparing observed and simulated results. Examples of such program assessments are shown below in Sections V.A and V.B.

IV. DEVELOPMENT OF SIMULATION CAPABILITIES OF ProDigest-LC

A. PREDICTION OF PEPTIDE RETENTION TIMES

Development of ProDigest-LC required accurate methods of retention time prediction in SEC, CEC, and RPC of peptides. Thus, this section outlines the principles of predictive behavior of peptides in HPLC from which this program was evolved.

1. SEC

To achieve its full potential as a useful, and predictable mode of HPLC for peptide separations, perhaps as one dimension or step in a multidimensional protocol, it is important to achieve a linear relationship between logMW and retention time or volume over a wide molecular weight range. This, in turn, requires ideal size-exclusion behavior, i.e., no non-specific (hydrophobic and/or ionic) peptide/packing interactions. In addition, the tendency of peptides or protein fragments to maintain or retain a particular conformation in non-denaturing media, as opposed to a random coil configuration in denaturing media, will complicate retention time prediction and, hence, must be prevented. The development and use of synthetic peptide SEC standards to identify non-specific interactions, and methods to suppress these interactions, are dealt with in detail by Mant and Hodges,[6] elsewhere in this book. Furthermore, if the conformational character of a peptide/protein mixture in a particular mobile phase is uncertain and ideal size-exclusion behavior is desired, SEC should always be carried out under highly denaturing conditions. Thus, Mant et al.,[7] for instance, could obtain a linear logMW vs. peptide retention time relationship for myoglobin and its cyanogen bromide fragments (2500 to 17,000 daltons), together with the above mentioned peptide standards (10, 20, 30, 40, and 50 residues), only under highly denaturing conditions (Figure 1).

2. CEC

Once any non-specific (hydrophobic) interactions have been identified (through the employment of peptide standards[8,9]) and suppressed, the retention times of peptides may be correlated with their amino acid compositions. Factors which affect the retention behavior of peptides during CEC (including net charge, polypeptide chain length, and charge density) are described extensively by Mant and Hodges,[9] elsewhere in this book.

The method of linearizing peptide retention behavior in CEC, described in detail in References 8 and 9, is summarized in Figure 2. Figures 2A and 2B show elution profiles of, respectively, a mixture of five size-exclusion standards (10, 20, 30, 40, and 50 residues; +1, +2, +3, +4, and +5, respectively) and a mixture of five synthetic peptide cation-exchange standards (11 residues; +1, +2, +3, and +4 net charge) on a strong cation-exchange column. The polypeptide chain lengths of these peptides vary considerably (10 to 50 residues), together with charge density (+1 net charge per 10 residues to +4 net charge per 11 residues). This leads to such observations as the 50-residue peptide (+5 net charge; Figure 2A) not being retained as long as 11-residue peptides with net charges of +3 or +4 (Figure 2B); similarly, the 40-residue peptide (+4 net charge; Figure 2A) was eluted prior to an 11-residue peptide with a net charge of +3 (Figure 2B). These results are rationalized by a linear peptide elution time vs. net charge/lnN relationship (where N is the number of residues the peptide contains)[8,9] (Figure 2C).

3. RPC

Development of the ability of ProDigest-LC to predict accurately peptide retention times in RPC required a reliable means of assigning a value to the hydrophobicity of a peptide. Specifically, a means of expressing peptide hydrophobicity in terms of HPLC-derived parameters was required. Mant and Hodges[10] have reviewed methods of determining sets of amino acid residue hydrophobicity coefficients for predicting peptide retention times in RPC. These methods are based on the assumption that the chromatographic behavior of a peptide is mainly or solely dependent on amino acid composition, and this assumption holds well enough for small peptides (up to ca. 15 residues).[10]

The most precise set of coefficients currently available were reported by Guo et al.,[11] and these formed the basis of peptide retention prediction in RPC by ProDigest-LC. These coefficients, listed elsewhere in this book,[12] were obtained by measuring the contribution (at pH 2.0 and 7.0, the former pH being employed for ProDigest-LC) of individual amino acids to the retention time of a synthetic model peptide at a fixed chain length. The octapeptide sequence,

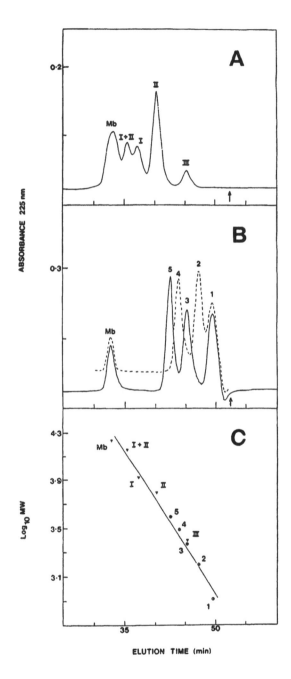

FIGURE 1. Ideal SEC of protein fragments and a mixture of synthetic peptide standards. Column: Altex Spherogel TSK G2000SW (300 × 7.5 mm I.D.; Beckman Instruments, Berkeley, CA). HPLC instrument: Varian Vista Series 5000 liquid chromatograph (Varian, Walnut Creek, CA), coupled to a Hewlett-Packard (Avondale, PA) HP1040A detection system, HP85B computer, HP9121 disc drive, HP2225A Thinkjet printer and HP7470A plotter. Mobile phase: 50 mM KH$_2$PO$_4$/0.5 M KCl/8 M urea, pH 6.5; flow-rate, 0.2 ml/min; temperature, 26°C. Panel A: Elution profile of horse-heart myoglobin (Mb) and its cyanogen bromide cleavage fragments (I, II, I + II, III). Panel B: elution profile of horse-heart Mb and synthetic peptide standards 1 to 5. Panel C: plot of log MW *versus* retention time of Mb, cyanogen bromide fragments of Mb and the five synthetic peptide standards. The peptide standards have the sequence Ac-(G-L-G-A-K-G-A-G-V-G)$_n$-amide, where n denotes 1, 2, 3, 4, and 5 (10, 20, 30, 40, and 50 residues, respectively). The standards were obtained from Synthetic Peptides Incorporated, Department of Biochemistry, University of Alberta, Edmonton, Alberta, Canada. The arrows denote the elution time for the total permeation volume of the column. (From Mant, C. T. and Parker, J. M. R., and Hodges, R. S., *J. Chromatogr.*, 397, 99, 1987. With permission.)

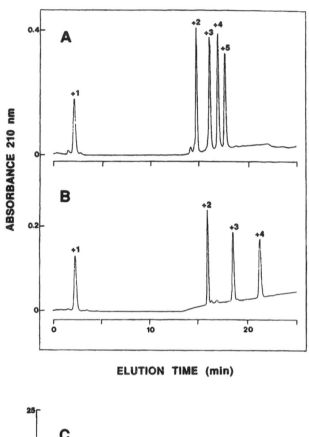

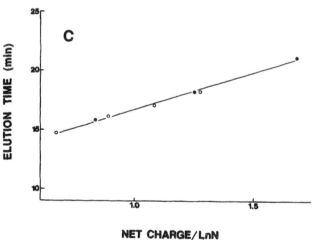

FIGURE 2. Strong cation-exchange chromatography of synthetic peptides. Column: Mono S HR 5/5 (50 × 5 mm I.D., 10-μm particle size, Pharmacia, Dorval, Canada). HPLC instrument: Varian Vista Series 5000 liquid chromatograph coupled to a Hewlett-Packard HP1040A detection system, HP9000 Series 300 computer, HP9133 disc drive, HP2225A Thinkjet printer and HP7440A plotter. Conditions: linear AB gradient (20 mM salt/min, following 10 min isocratic elution with buffer A), where buffer A is 5 mM KH$_2$PO$_4$ (pH 6.5) and buffer B is buffer A plus 0.5 M NaCl, both buffers containing 40% acetonitrile (v/v); flow-rate, 1 ml/min, temperature, 26°C. Panel A: mixture of five synthetic size-exclusion standards (10 to 50 residues; +1 to +5 net charge, respectively). The sequences of the peptides are shown in Figure 1. Panel B: Mixture of four synthetic peptide cation-exchange standards (+1 to +4 net charge). Sequences of the peptides are Ac-G-G-G-L-G-G-A-G-G-L-K-amide (C1), Ac-K-Y-G-L-G-G-A-G-G-L-K-amide (C2), Ac-G-G-G-A-L-K-A-L-K-G-L-K-amide (C3) and Ac-K-Y-A-L-K-A-L-K-G-L-K-amide (C4). Panel C: plot of observed peptide elution time *versus* peptide net charge divided by the logarithm of the number of residues (net charge/LnN). Both sets of peptide standards were obtained from Synthetic Peptides Incorporated.

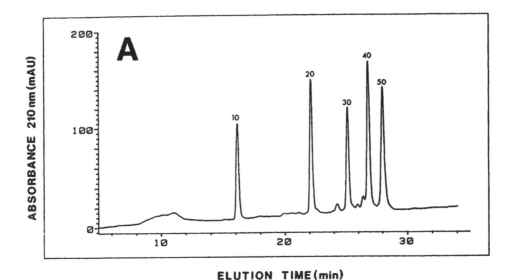

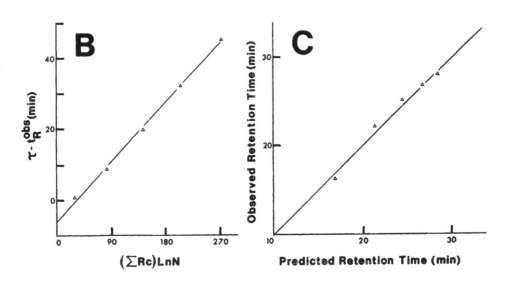

FIGURE 3. RPC of a mixture of synthetic peptide polymers. Column: Synchropak RP-P C_{18} (250 × 4.6 mm I.D., 6.5-μm particle size, 300-Å pore size; SynChrom Inc., Lafayette, IN). HPLC instrument: same as Figure 1. Conditions: linear AB gradient (1% B/min), where eluent A is 0.1% aq. TFA and eluent B is 0.1% TFA in acetonitrile; flow-rate, 1 ml/min; temperature, 26°C. Panel A: elution profile of five peptide polymers (10 to 50 residues). Panel B: plot of predicted minus observed retention time (τ - t_r^{obs}) vs. the sum of the retention coefficients (ΣR_c) of Guo et al.[11] times the logarithm of the number of residues (LnN). Panel C: correlation of predicted and observed retention times of the peptide polymers (derived from the plot shown in panel B as described by Mant et al.)[14] The sequences of the peptide polymers are shown in Figure 1. (From Hodges, R. S., Parker, J. M. R., Mant, C. T., and Sharma, R. R., *J. Chromatogr.*, 458, 147, 1988. With permission.)

Ac-Gly-X-X-(Leu)$_3$-(Lys)$_2$-amide, was substituted at position X by all 20 amino acids found in proteins. The predicted retention time (τ) of a peptide in RPC was then equal to the sum of the retention coefficients (ΣR_c) for the amino acid residues in the peptide, plus the time correction for an internal standard.

There is an exponential relationship between peptide chain length and peptide retention time in RPC.[2,4,5,13,14] Thus, in Figure 3A, the effect on peptide retention time of increasing length of five synthetic peptide polymers (10, 20, 30, 40, and 50 residues) decreased progressively with each 10-residue addition. The intimate relationship between peptide hydrophobicity and chain length and their combined effect on peptide retention behavior in RPC was detailed by Mant et al.,[14] who demonstrated a linear relationship between predicted (τ) minus observed (t_r^{obs}) retention time *versus* the product of peptide hydrophobicity [expressed as ΣR_c, the sum of the coefficients of Guo et al.[11]] and the logarithm of the number of residues, lnN (Figure 3B). Using the slope and intercept of such a plot, the retention behavior of peptides of up to 50 residues in length can be predicted. Figure 3C shows the good correlation between predicted and observed retention times for the peptides shown in Figure 3A, once peptide chain length has been taken into account.

B. PREDICTION OF EFFECT OF VARYING CONDITIONS ON PEPTIDE ELUTION PROFILES

Manipulation of parameters such as flow-rate in SEC, and flow-rate and (especially) gradient-rate in IEC and RPC are common approaches to optimizing peptide separations. Thus, knowledge of the effect of varying run conditions on peptide elution profiles simulated under one particular set of conditions was vital for the development of ProDigest-LC as a flexible and practical aid to the researcher.

Mixtures of synthetic peptide standards were subjected to SEC, CEC, and RPC at varying flow-rates, gradient-rates (CEC and RPC only), sample loads and sample volumes (SEC only).[4] The mobile phases employed were as follows: SEC, 50 mM KH_2PO_4 (pH 6.5), containing 0.1 M KCl; CEC, linear AB gradient at pH 3.0 or 6.5, where buffer A was 5 mM KH_2PO_4 and buffer B was buffer A containing 1 M NaCl; RPC, linear AB gradient, where eluent A was 0.1% aq. TFA and eluent B was 0.1% TFA in acetonitrile. These mobile phases represent HPLC conditions commonly employed in the peptide field, an important consideration for ProDigest-LC to be of widespread practical use.

Data from these chromatographic experiments were subsequently used to derive empirical equations predicting the effects of experimental parameters (sample size, sample volume, flow-rate, and gradient-rate) on peptide retention times, peak heights, peak widths, and resolution. Simulated elution profiles generated by the program are based on the mobile phases used to derive the equations.

V. SIMULATION OF PEPTIDE ELUTION PROFILES IN RPC

A demonstration of all the simulation capabilities of the ProDigest-LC software would require considerable space. For this reason, the present article will only focus on simulating elution profiles obtained on the most widely-used mode of HPLC at present, namely RPC. The ability of this technique to separate peptides of closely-related structures has made it an extremely powerful analytical and preparative tool. Whatever the source of a particular peptide sample, the resolving capability of RPC makes it the obvious choice for the initial HPLC run to gauge the complexity of the peptide mixture and help design the best approach for its resolution. In addition, RPC is frequently employed for peptide mapping of proteins.

A. PREDICTION OF PEPTIDE RETENTION TIMES AT VARYING GRADIENT-RATES AND FLOW-RATES

The sample peptide mixture shown in Figure 4 represented a stringent test of the retention time prediction capabilities of ProDigest-LC.[5] This mixture contained peptides varying significantly in size (11 to 50 residues), charge (+1 to +8), and hydrophobicity. In addition, the peptides varied considerably in their degree of secondary structure (α-helix). The peptide mixture was subjected to linear AB gradient elution on a C_8 column at all combinations of gradient-rates of 0.5, 1.0, and 2.0% B/min and flow-rates of 0.5, 1.0, and 2.0 ml/min (eluent A was 0.1% aq. TFA

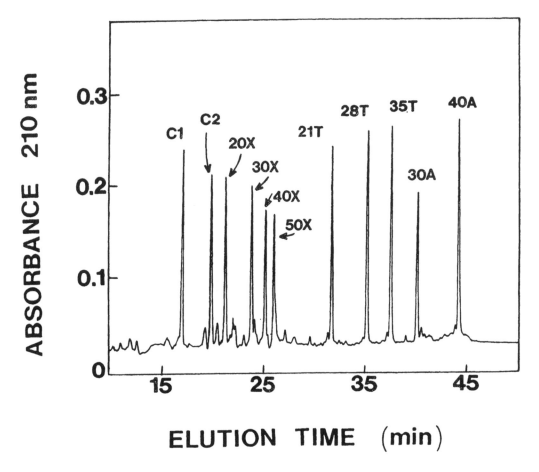

FIGURE 4. RPC of a mixture of synthetic peptides. Column: Aquapore RP300 C_8 (220 × 4.6 mm I.D.) HPLC instrument: same as Figure 2. Conditions: same as Figure 3. The sequences of peptides C1 and C2 are shown in Figure 2; the sequences of peptides 20X, 30X, 40X, and 50X are shown in Figure 1 (20, 30, 40, and 50 residue peptides, respectively); the sequences of peptides 21T, 28T, and 35T are Ac-K-C-A-E-G-E-L-(K-L-E-A-G-E-L)$_n$-amide, where n = 2, 3, and 4, respectively; the sequence of peptides 30A and 40 A is Ac-(L-G-L-K-A)$_n$-amide, where n = 6 and 8, respectively. This peptide mixture was subjected to variations in both gradient- and flow-rate; subsequent observed vs. predicted (by ProDigest-LC) retention times are shown in Table 1. (From Mant, C. T., Burke, T. W. L., Zhou, N. E., Parker, J. M. R., And Hodges, R. S., *J. Chromatogr.*, 485, 365, 1989. With permission.)

and eluent B was 0.1% TFA in acetonitrile). These parameters were chosen as being typical of the optimum range of conditions used for analytical peptide separations in RPC.

A comparison of observed peptide retention times and retention times predicted by ProDigest-LC (Table 1) demonstrates excellent predictive accuracy by the program. The average errors at each combination of gradient-rate and flow-rate was small, with the error range for individual peptides essentially representing maximum errors likely to be experienced by the researcher for most applications. The results shown in Table 1 are even more impressive when one considers what is required from the program, i.e., to predict peptide retention times over a range of gradient- and flow-rates with no prior information about the peptides except their amino acid composition. It is considerably more straightforward to predict the effects of variation in gradient- and flow-rate from observed peptide retention times.[3]

The results shown in Table 1 also make the important point that peptide conformation, *per se*, does not necessarily preclude accurate prediction of peptide retention behavior in RPC. In fact, the predicted retention times of peptides which are known to exhibit considerable α-helicity[5] (e.g., 30L, 40A, 21T-35T in Figure 4) showed very good accuracy. These results

TABLE 1
Predicted vs. Observed Peptide Retention Times when Varying Flow-Rate and Gradient-Rate

Peptide[a]	Flow-rate: 0.5 ml/min[b]									Flow-rate: 1.0 ml/min									Flow-rate: 2.0 ml/min								
	0.5% B/min			1.0% B/min			2.0% B/min			0.5% B/min			1.0% B/min			2.0% B/min			0.5% B/min			1.0% B/min			2.0% B/min		
	τ[c]	t_R^{obs}[d]	Δt[e]	τ	t_R^{obs}	Δt	τ	t_R^{obs}	Δt	τ	t_R^{obs}	Δt	τ	t_R^{obs}	Δt	τ	t_R^{obs}	Δt	τ	t_R^{obs}	Δt	τ	t_R^{obs}	Δt	τ	t_R^{obs}	Δt
C1	32.2	33.8	1.6	19.7	22.6	2.9	13.4	16.0	2.6	28.6	26.2	2.4	16.1	17.0	0.9	9.8	11.4	1.6	26.8	22.9	3.9	14.3	13.1	1.2	8.0	8.5	0.5
C2	35.2	39.4	4.2	21.2	25.1	3.9	14.2	17.1	2.9	31.6	32.2	0.6	17.6	19.8	2.1	10.6	12.6	2.0	29.8	27.1	2.7	15.8	16.1	0.3	8.8	9.9	1.1
20X	40.4	42.1	1.7	23.8	26.2	2.4	15.5	17.5	2.0	36.8	35.6	1.2	20.2	21.1	0.9	11.9	13.2	1.3	35.0	31.1	3.9	18.4	17.9	0.5	10.1	10.6	0.5
30X	46.2	47.3	1.1	26.7	28.5	1.8	17.0	18.5	1.5	42.6	41.2	1.4	23.1	23.7	0.6	13.4	14.3	0.9	40.8	37.3	3.5	21.3	20.7	0.6	11.6	11.9	0.3
40X	50.2	50.1	0.1	28.7	29.7	1.0	18.0	19.0	1.0	46.6	44.4	2.2	25.1	25.1	0	14.4	14.9	0.5	44.8	40.8	4.0	23.3	22.3	1.0	12.6	12.6	0
50X	52.9	51.9	1.0	30.0	30.5	0.5	18.6	19.3	0.7	49.3	46.4	2.9	26.4	26.0	0.4	15.0	15.3	0.3	47.5	43.0	4.5	24.6	23.3	1.3	13.2	13.0	0.2
21T	57.7	62.0	4.3	32.4	36.6	4.2	19.8	22.5	2.7	54.1	57.0	2.9	28.8	31.7	2.9	16.2	18.3	2.1	52.3	52.8	0.5	27.0	28.5	1.5	14.4	15.9	1.5
28T	65.8	70.3	4.5	36.5	39.9	3.4	21.9	24.0	2.1	62.2	64.3	2.1	32.9	35.2	2.3	18.3	20.0	1.7´	60.4	60.5	0.1	31.1	32.2	1.1	16.5	17.7	1.2
35T	72.2	74.9	2.7	39.7	42.1	2.4	23.4	25.1	1.7	68.6	69.2	0.6	36.1	37.5	1.4	19.8	21.1	1.3	66.8	65.6	1.2	34.3	34.7	0.4	18.0	18.8	0.8
30A	84.6	80.2	4.2	45.9	44.9	1.0	26.5	26.5	0	81.0	76.3	4.7	42.3	40.1	2.2	22.9	22.4	0.5	79.2	74.5	4.7	40.5	37.2	3.3	21.1	20.2	0.9
40A	93.6	88.0	5.6	50.4	48.6	1.8	28.8	28.3	0.5	90.0	84.3	5.7	46.8	44.0	2.8	25.2	24.3	0.9	88.2	83.8	4.4	45.0	41.3	3.7	23.4	22.2	1.2
Average error		2.8			2.2			1.6			2.4			1.5			1.2			3.0			1.3			0.7	

[a] Peptide sequences are shown in Figure 1 (20X-50X), Figure 2 (C1 and C2) and Figure 4 (21T-35T, 30A and 40A). [b] The peptide mixture was chromatographed on an Aquapore RP300 C8 column (220 x 4.6 mm I.D.) under linear AB gradient elution conditions, where Eluent A was 0.1% aq. TFA and Eluent B was 0.1% TFA in acetonitrile, pH 2.0. [c] τ denotes peptide retention times (min) predicted by ProDigest-LC. [d] t_R^{obs} denotes observed peptide retention times (min). [e] Δt denotes error (min) between predicted and observed peptide retention times.

TABLE 2
Peptide Sequences

Peptide	Peptide sequence[a]
I1	Ac-Arg-Gly-Gly-Gly-Gly-**Ile**-Gly-**Ile**-Gly-Lys-amide
I2	Ac-Arg-Gly-Gly-Gly-Gly-**Ile**-Gly-Leu-Gly-Lys-amide
S2[b]	Ac-Arg-Gly-Gly-Gly-Gly-Leu-Gly-Leu-Gly-Lys-amide
S3	Ac-Arg-Gly-**Ala**-Gly-Gly-Leu-Gly-Leu-Gly-Lys-amide
S4	Ac-Arg-Gly-**Val**-Gly-Gly-Leu-Gly-Leu-Gly-Lys-amide

[a] Ac = N^α-acetyl; amide = C^α-amide.
[b] Residues in bold denote substitutions in the sequence of peptide S2.

suggested that, if the conformation of a peptide does not present a preferred binding site, such as that found in the unusual situation of an amphipathic helix, its RPC retention behavior should be predictable.

B. SIMULATION OF PEPTIDE ELUTION PROFILES AT VARYING GRADIENT-RATES AND FLOW-RATES

ProDigest-LC was applied to predicting the effect of varying gradient-rate and flow-rate on the elution profile of a mixture of five decapeptide analogs closely related in hydrophobicity: I1, I2, S2, S3, and S4 (Table 2).[5] The hydrophobicity of the peptides increases only slightly between S2 and S4-between S2 and S3 there is a change from an α-H to a β-CH_3 group, between S3 and S4 there is a change from a β-CH_3 group to two methyl groups attached to the β-CH group. The hydrophobicity variations between I1, I2, and S2 are even more subtle. There is a change of only an isoleucine to a leucine residue between I1 and I2, and between I2 and S2. Guo et al.[11] demonstrated that leucine is slightly more hydrophobic than isoleucine, although these residues contain the same number of carbon atoms. Since isoleucine is β-branched, the β-carbon is close to the peptide backbone and not as available to interact with the hydrophobic stationary phase compared to the conformation of the leucine side-chain.

1. Effect of Gradient-Rate

Figure 5 shows the observed (left) and simulated (right) elution profile of the peptide mixture on a C_{18} column at a fixed flow-rate of 1 ml/min and gradient-rates of 4.0, 2.0, and 0.5% B/min (top, middle, and bottom elution profiles, respectively). The similarity of the observed and simulated profiles is immediately apparent, with the program successfully simulating the major effects of varying gradient-rate on peptide elution profiles, i.e., an increase in peptide retention times and peak widths, a decrease in peak height, and improved peptide resolution with decreasing gradient-rate.

The shaded peaks shown in the simulated peptide elution profiles (Figure 5, right) serve to highlight another feature of ProDigest-LC, namely being able to simulate the required minimum resolution between adjacent peptide peaks. The required minimum peptide resolution was set at 0.8, meaning that adjacent peaks that are not separated to at least this degree of resolution will be shaded. Hence, the simulated elution profile at 0.5% B/min (Figure 5, bottom right) suggested that all five peptides would be separated to at least a resolution of 0.8, since all peptide peaks are unshaded. In contrast, with an increase in gradient-rate to 2.0% B/min (middle right) and 4.0% B/min (top right), I1, I2, and S2 were shaded, denoting that a minimum resolution of 0.8 between these peptides had not been achieved.

2. Effect of Flow-Rate

Figure 6 shows the observed (left) and simulated (right) elution profile of the five peptide mixture on the C_{18} column at a fixed gradient-rate of 1.0% B/min and flow-rates of 1.0, 0.5, and

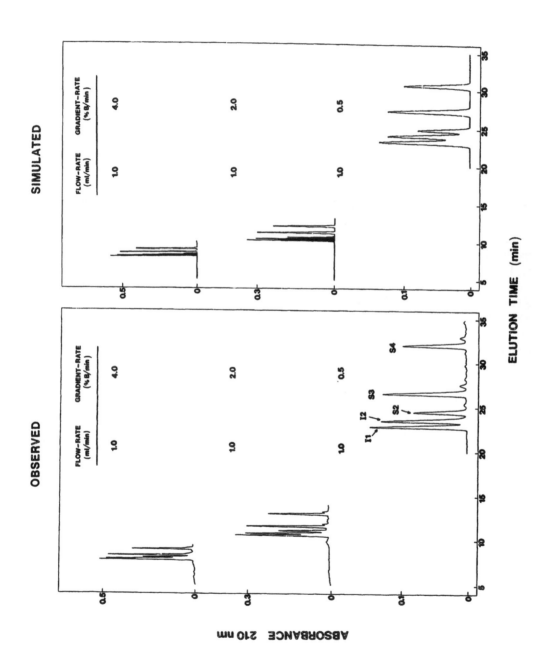

FIGURE 5. Computer simulation of the effect of varying gradient-rate on the reversed-phase elution profile of a mixture of synthetic peptide standards. Column: same as Figure 3. HPLC instrument: same as Figure 2. Left: observed RPC elution profiles, obtained with a linear AB gradient (4, 2, or 0.5% B/min), where eluent A is 0.1% aq. TFA and eluent B is 0.1% TFA in acetonitrile; flow-rate, 1 ml/min; temperature, 26°C. Right: simulated peptide elution profiles. Shaded peak areas in the simulated profiles at 2 and 4% B/min denote unresolved peptides, based on the specified peptide resolution of 0.8. The sequences of the five peptides are shown in Table 2. (From Mant, C. T., Burke, T. W. L., Zhou, N. E., Parker, J. M. R., and Hodges, R. S., *J. Chromatogr.*, 485, 365, 1989. With permission.)

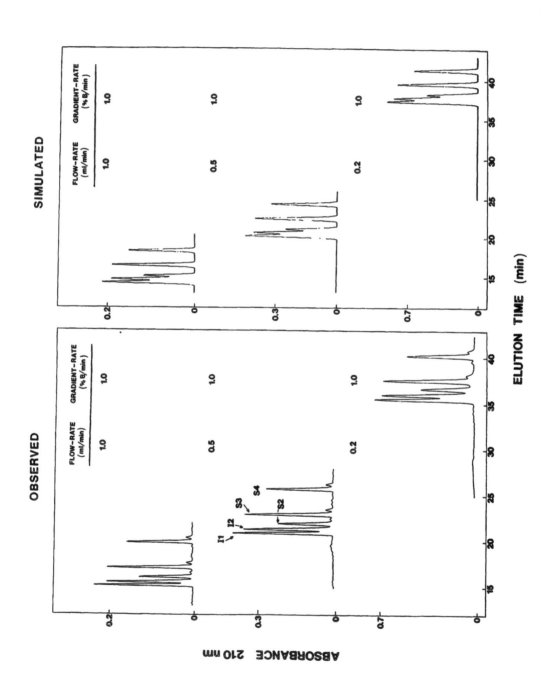

FIGURE 6. Computer simulation of the effect of varying flow-rate on the reversed-phase elution profile of a mixture of synthetic peptide standards. Column: same as Figure 3. HPLC instrument: same as Figure 2. Left: observed RPC elution profiles, obtained with a linear AB gradient (1% B/min), where eluent A is 0.1% aq. TFA and eluent B is 0.1% TFA in acetonitrile; flow-rate, 1.0, 0.5, or 0.2 ml/min; temperature, 26°C. Right: simulated peptide elution profiles. Specified peptide resolution is 0. The sequences of the five peptides are shown in Table 2. (From Mant, C. T., Burke, T. W. L., Zhou, N. E., Parker, J. M. R., and Hodges, R. S., *J. Chromatogr.*, 485, 365, 1989. With permission.)

0.2 ml/min (top, middle, and bottom elution profiles, respectively). Similar observed and simulated profiles are again demonstrated, with the program accurately simulating the major effects of varying flow-rate on peptide elution profiles, i.e., an increase in peptide retention times and peak heights, and a decrease in peptide resolution with decreasing flow-rate. Unlike gradient-rate (Figure 5), the effect of varying flow-rate on peak widths is small. The minimum peptide resolution required was set at 0, so that all three simulated elution profiles contained unshaded peptide peaks.

Although the observed peptide elution profiles in Figures 5 and 6 exhibited a little better resolution than the simulated profiles tended to suggest, this is a minor point considering that the program has to simulate the effects of a wide variation in flow- and gradient-rate during RPC of a mixture of closely-related peptide analogs, with the only information about these peptides being their amino acid composition. The major effects on peptide retention times, resolution, peak heights, and widths were certainly well predicted, and refinements to the program will increase this predictive accuracy even further. It is also important to note that the researcher may simulate the effects of parameters such as gradient-rate, flow-rate, and sample load on RPC of a particular mixture without ever actually having to run the sample of interest or peptide standards to calibrate the column. This is of major advantage to the researcher and can be achieved simply by using the Standards default file already in the program.

C. PEPTIDE MAPPING

Troponin I (TnI) is a 178-residue protein, involved in the regulation of muscle contraction. It has been shown that the region of amino acid residues 105 to 114 are responsible for the inhibitory activity of TnI,[15,16] and differences in relative inhibitory activity of rabbit skeletal fast and cardiac TnI can be at least partially and perhaps solely explained by a single amino acid insertion, a leucine residue, between positions 112 and 113 of the skeletal sequence. To investigate this point further, it would be appropriate to engineer, by site-specific mutagenesis, a skeletal protein containing the cardiac leucine insertion. Following site-specific mutagenesis, peptide mapping would be carried out on the engineered protein to verify that the change has indeed been made. It would also be of interest to isolate the inhibitory peptides from the mutant and native protein for biological activity measurements. Hence, being able to predict the retention behavior of these peptides relative to the other peptides in the digest would be advantageous.

The mutant protein was created by inserting a leucine residue between positions 112 and 113 of the native TnI sequence and saving this sequence in the program as TnI-179 (for 179 residues, as opposed to 178 residues in the native sequence). Cyanogen bromide cleavage was chosen and, hence, the program was instructed to cleave the mutant protein on the C-terminal side of all the methionine residues (9) in the protein. This was followed by a request to simulate the RPC elution profile (SEC and CEC are the other options) at a gradient-rate of 1% acetonitrile/min and a flow-rate of 1 ml/min (Figure 7). The entire cyanogen bromide digest profile of 10 peaks is not shown in Figure 7. Instead, the program zoom option has been selected to narrow the simulated profile down to the section of interest (16 to 32 min). The cursor, shown by the bar below the peak at 29.6 min, marks the inhibitory fragment (residues 96 to 117) of the mutant protein. Rabbit skeletal fast TnI was then loaded, followed by cyanogen bromide cleavage and selection of the RPC option (1% acetonitrile/min and 1 ml/min) (Figure 7B). The cursor marks the inhibitory fragment (residues 96 to 116) of the native protein. The differences in the retention times of the native fragment and the mutant fragment (26.8 and 29.6 min, respectively) clearly show that a change in the native sequence by site-specific mutagenesis can be easily detected. To highlight this, the peptide mapping option, following cyanogen bromide cleavage of the native TnI, was used to produce the simulated chromatogram shown in Figure 7C. This elution profile is an overlay of the cyanogen bromide cleavage fragments of both proteins (i.e., an overlay of Figure 7A and B). Any mutant fragment not observed in the native sequence is identified by an inverted

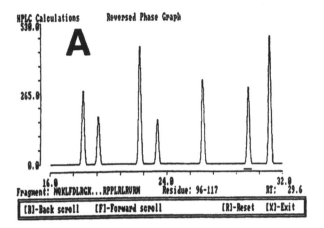

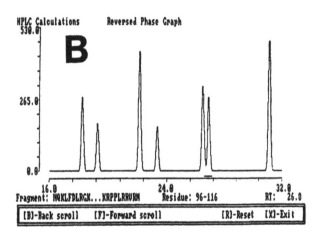

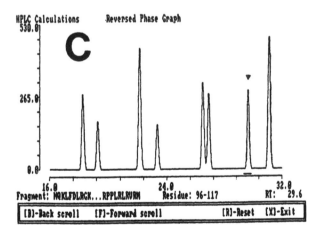

FIGURE 7. Computer simulation of peptide mapping by ProDigest-LC. All panels represent RPC elution profiles, obtained with a linear AB gradient (1% B/min) at a flow-rate of 1 ml/min, where eluent A is 0.1% aq. TFA and eluent B is 0.1% TFA in acetonitrile. Panel A: cyanogen bromide digest of a mutant protein, obtained by site-specific mutagenesis of rabbit skeletal fast troponin I (TnI); this mutant protein, the result of a leucine insertion between positions 112 and 113 is denoted as TnI-179 in the text. Panel B: cyanogen bromide digest of the native TnI protein. Panel C: overlay of cyanogen bromide digests of a mutant protein (TnI-179) and the native TnI protein; the inverted triangle identifies the mutant peptide fragment. The zoom option of the program has been selected to show only sections of the protein digests.

triangle above the peak (Figure 7C). In addition, if required, the program will list the retention times of the fragments, as well as the residue numbers and sequences of the peptides, the mutant fragments being denoted as such.

D. FURTHER OPTIMIZATION OF PEPTIDE ELUTION PROFILES

As stated previously, apart from being an HPLC method-development tool for the researcher, ProDigest-LC is also a teaching aid for chromatographers, designed to help the student or researcher to select the correct conditions for chromatography (HPLC mode, column and mobile phase) and allowing him or her the option of examining the effect of varying flow-rate, gradient-rate, sample size, and sample volume on the separation. Thus, ProDigest-LC simulates peptide elution profiles without even having to carry out an actual chromatographic run with the sample of interest. Having manipulated the program until the desired separation has been simulated, the researcher may then carry out the run. From the observed peptide elution profile, the researcher may then decide that further optimization of the chromatographic conditions, perhaps through the use of a complementary optimization program, may be required for the desired separation to be achieved.

It is always more accurate to simulate the effect of changes in run parameters on observed elution profiles than to base these simulations on previously predicted profiles, and a commercially-available optimization program, DryLab G, has been designed specifically for this purpose.[17]

For a recent review of optimization programs, in addition to other aspects of computer-assisted method development in chromatography, the reader is directed to Reference 18.

VI. SUMMARY AND FUTURE PROSPECTS

This article has described the development and use of a computer software program, ProDigest-LC, designed to assist researchers in devising methodologies for the analytical separation and purification of biologically active peptides and peptide fragments from enzymatic and chemical digests of proteins by SEC, CEC, and RPC. The flexibility and ease-of-use of ProDigest-LC has also ensured its value both as a teaching aid and as an analytical tool for workers involved in peptide and protein research. Future prospects for ProDigest-LC include refinements in the program to improve further simulations of peptide elution profiles and to take into account the presence of preferred binding domains.[19] In addition, the IEC predictive facility of the program will be extended to anion-exchange chromatography.

ACKNOWLEDGMENTS

This work was supported by the Medical Research Council of Canada, S.P.I. Synthetic Peptides Incorporated and by the Alberta Heritage Foundation for Medical Research.

REFERENCES

1. **Mant, C.T. and Hodges, R.S.**, General method for the separation of cyanogen bromide digests of proteins by high-performance liquid chromatography. Rabbit skeletal Troponin I., *J. Chromatogr.*, 326, 349, 1985.

2. **Mant, C.T. and Hodges, R.S.**, HPLC of peptides, in *High-Performance Liquid Chromatography of Biological Macromolecules: Methods and Applications*, Gooding, K. and Regnier, F., Eds., Marcel Dekker, New York, 1990, p. 301.

3. **Mant, C.T. and Hodges, R.S.**, Optimization of peptide separations in high-performance liquid chromatography, *J. Liq. Chromatogr.*, 12, 139, 1989.

4. **Hodges, R.S., Parker, J.M.R., Mant, C.T., and Sharma, R.R.**, Computer simulation of high-performance liquid chromatographic separations of peptide and protein digests for development of size-exclusion, ion-exchange and reversed-phase chromatographic methods, *J. Chromatogr.*, 458, 147, 1988.

5. **Mant, C.T., Burke, T.W.L., Zhou, N.E., Parker, J.M.R., and Hodges, R.S.**, Reversed-phase chromatographic method development for peptide separations using a computer simulation program: ProDigest-LC, *J. Chromatogr.*, 485, 365, 1989.

6. **Mant, C.T. and Hodges, R.S.**, Requirements for peptide standards to monitor ideal and non-ideal behavior in size-exclusion chromatography, this publication.

7. **Mant, C.T., Parker, J.M.R., and Hodges, R.S.**, Size-exclusion high-performance liquid chromatography of peptide. Requirement for peptide standards to monitor column performance and non-ideal behavior, *J. Chromatogr.*, 397, 99, 1987.

8. **Burke, T.W.L., Mant, C.T., Black, J.A., and Hodges, R.S.**, Strong cation-exchange high-performance liquid chromatography of peptides. Effect of non-specific hydrophobic interactions and linearization of peptide retention behavior, *J. Chromatogr.*, 476, 377, 1989.

9. **Mant, C.T. and Hodges, R.S.**, The use of peptide standards for monitoring ideal and non-ideal behavior in cation-exchange chromatography, this publication.

10. **Mant, C.T. and Hodges, R.S.**, Optimization and prediction of peptide retention behavior in reversed-phase chromatography in *HPLC of Proteins, Peptides and Polynucleotides*, Hearn, M.T.W., Ed., VCH Publishers, Weinheim, 1991, 277.

11. **Guo, D., Mant, C.T., Taneja, A.K., Parker, J.M.R., and Hodges, R.S.**, Prediction of peptide retention times in reversed-phase high-performance liquid chromatography. I. Determination of retention coefficients of amino acid residues of model synthetic peptides, *J. Chromatogr.*, 359, 499, 1986.

12. **Hodges, R.S. and Mant, C.T.**, Properties of peptides/proteins and practical implications, this publication.

13. **Lau, S.Y.M., Taneja, A.K., and Hodges, R.S.**, Effects of high-performance liquid chromatographic solvents and hydrophobic matrices on the secondary and quaternary structure of a model protein. Reversed-phase and size-exclusion high-performance liquid chromatography, *J. Chromatogr.*, 317, 129, 1984.

14. **Mant, C.T., Burke, T.W.L., Black, J.A., and Hodges, R.S.**, Effect of peptide chain length on peptide retention behavior in reversed-phase chromatography, *J. Chromatogr.*, 458, 193, 1988.

15. **Talbot, J.A. and Hodges, R.S.**, Synthesis and biological activity of an icosapeptide analog of the actomyosin ATPase inhibitory region of troponin I, *J. Biol. Chem.*, 254, 3720, 1979.

16. **Talbot, J.A. and Hodges, R.S.**, Synthetic studies on the inhibitory region of rabbit-skeletal troponin I, *J. Biol. Chem.*, 256, 2798, 1981.

17. **Snyder, L.R., Dolan, J.W., and Lommen, D.C.**, Computer simulation as a tool for optimizing gradient separation, this publication.

18. **Giese, R.W., Haken, J.K., Macek, K., and Snyder, L.R.**, Eds., Computer assisted method development in chromatography, *J. Chromatogr.*, 485, 1989.

19. **Zhou, N.E., Mant, C.T., and Hodges, R.S.**, Effect of preferred binding domains on peptide retention behavior in reversed-phase chromatography: amphipathic α-helices, *Peptide Res.*, 3, 1, 1990.

COMPUTER SIMULATION AS A TOOL FOR OPTIMIZING GRADIENT SEPARATIONS

Lloyd R. Snyder, John W. Dolan, and D. C. Lommen

I. INTRODUCTION

Samples of peptides, proteins and other large biomolecules are commonly separated by means of gradient elution. When an effort is made to optimize the various experimental conditions, the resulting separations can be very impressive. Samples that contain 20 or more bands can sometimes be completely resolved within run times of a few hours or less. However, developing a final (successful) separation can require weeks or months of effort by a trained chromatographer. There are several reasons for this:

1. Any sample with 20 or more components presents a real challenge for HPLC separation; achieving the proper spacing of bands within the limits of the chromatogram is not easily effected.
2. High-molecular-weight samples require longer run times, so that individual exploratory runs may require two or more hours each.
3. There are a large number of experimental variables to consider, and most of these must be explored in the case of difficult separations; variables in gradient conditions alone include: gradient shape, initial and final mobile-phase compositions, gradient time, flow rate, and column dimensions.
4. The use of the most popular mobile phase conditions (low pH, e.g., 0.1% trifluoroacetic acid [TFA]/water/acetonitrile) can lead to column degradation and changes in retention during use; this in turn can cause considerable difficulty in method development if a larger number of exploratory runs are required before a final method is achieved.

An alternative to traditional trial-and-error method development in the laboratory is the use of *computer simulation*. This approach[1-5] is based on the predictability of gradient separations, when certain information about individual sample compounds is available. The required information can be obtained from two initial experimental runs with the sample of interest, where gradient time is varied and all other conditions are maintained constant. Once these two starting runs have been carried out, the resulting data can be entered into a personal computer (PC) and subsequent experiments can be carried out by computer simulation. Since each computer simulation requires only a few seconds, a large number of experiments can be carried out within a short time — without worry that the column performance will change, and without wasting any sample. Computer simulation offers a number of further advantages, in that the computer can furnish additional information in various forms; e.g., graphs, tables, gradients superimposed on chromatograms, etc.

The use of computer simulation with a number of representative samples — including peptides, proteins, and compounds of nonbiological origin — has also uncovered several new opportunities for the improved use of gradient elution for virtually every sample. For example, it has been found that changes in gradient steepness often have a profound effect on band spacing and separation.[1-7] This in turn means that *intermediate* gradient times often provide optimum separation, as opposed to simply decreasing gradient steepness without limit. This also suggests tailoring the overall gradient so as to optimize gradient steepness for individual groups of bands within the chromatogram, i.e., the use of multi-segment gradients.

In the remainder of this chapter, we will examine a few examples of the use of computer simulation in the design of a final gradient elution separation. These representative cases involve reversed-phase separation, but the same approach can be used for ion-exchange or hydrophobic-interaction chromatography.

II. THEORY

The theory underlying the use of computer simulation has been described in detail.[1,3-5,8-11] Gradient separations can be considered equivalent to stepwise elution, where a large number of (very small) isocratic steps of increasing %-organic solvent are used. In each isocratic step, the solute capacity factor k' is given as

$$\log k' = \log k_w - S\emptyset \tag{1}$$

where $\emptyset$ refers to the volume-fraction of organic solvent B in the mobile phase, k_w is the (extrapolated) value of k' for water as mobile phase ($\emptyset = 0.00$), and S is a constant that is characteristic for the solute. For water/acetonitrile mobile phases, S is given (very approximately) as

$$S = 0.48 \, (\text{mol. wt.})^{0.44} \tag{2}$$

If two experimental gradient separations are carried out with a given sample (only gradient time varying), it is possible to calculate exact values of k_w and S for each component of the mixture. The resulting values of these parameters then allow us to predict separation for any change in gradient conditions: gradient shape, gradient time, initial and final %B (in an A/B gradient), etc.

It is impractical to calculate predicted separations manually, but quite feasible to do this using a computer. Commercial software (DryLab G, LC Resources) is available for use with IBM-compatible PCs. This software was used in the following examples.

III. SEPARATION OF A MIXTURE OF 23 PEPTIDES

A sample containing 23 synthetic peptides with molecular weights between 500 and 4000 Da was prepared as a means of evaluating the present and other[12] software. Method development was begun by carrying out two gradient separations with the conditions shown in Figure 1 (gradient times of 45 min [A] and 180 min [B]). A total of 22 bands can be seen in each chromatogram, meaning that there is a band overlap in each run. The use of band areas and relative retention allows the various peaks to be matched between these two runs, including the identification of overlapping bands.

The run conditions, retention times and band areas from Figures 1A and B were next entered into the computer and computer-simulation was begun. It is usually advantageous to begin computer simulation by comparing a simulated run with one of the starting experimental runs. The DryLab G program allows the user to select a value of the column plate number N so as to match resolution between experimental and simulated runs. The resulting simulation for the 45-min gradient is shown in Figure 1C (N = 10,000). At this point, one can then attempt trial-and-error improvement of the separation via computer simulation, just as one might do in the laboratory.

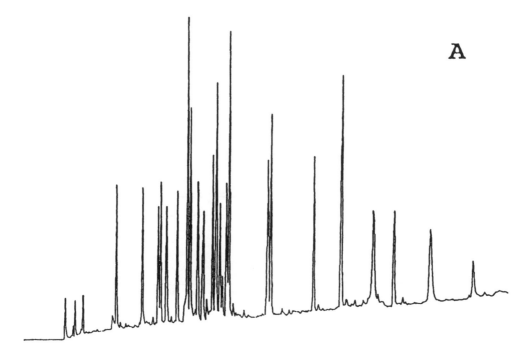

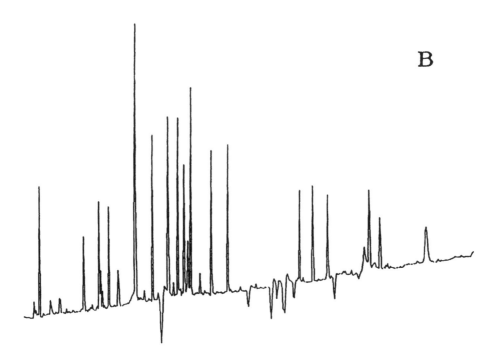

FIGURE 1. Separation of sample containing 23 synthetic peptides. Conditions: 250 × 4.6 mm I.D. column of Zorbax[R] Rx (DuPont, Wilmington, DE); 5 to 50%B gradients at 1.0 ml/min (0.1% TFA/water/acetonitrile); 30°C.[15] (A) Experimental 45-min run; (B) experimental 180-min run; (C) simulated 45-min run (same conditions as Figure 1A; N = 10,000); (D) relative resolution map.

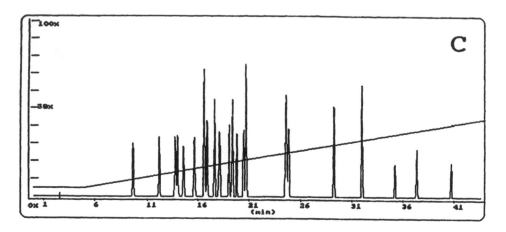

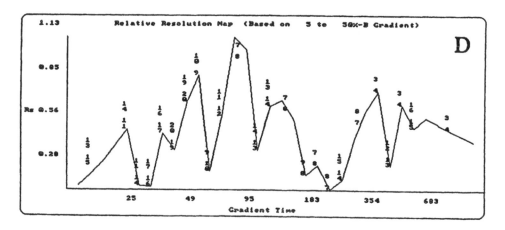

FIGURE 1 (continued)

A better approach, however, is to request a *relative resolution map:* a plot of resolution, R_s, (for the poorest-resolved band pair) as a function of gradient time.[13] This is shown for the present sample in Figure 1D. Here we see that resolution varies markedly with gradient time, even for small changes in this variable. The numbers in Figure 1D correspond to *critical* band pairs: those which are closest together (least resolved) for a given gradient time. Every value in the plot of Figure 1D corresponding to $R_s = 0$ means that, for this gradient time, two bands are completely overlapped, e.g., bands 7 and 8 for a gradient time of ~ 200 min; for greater or smaller gradient times, the resolution of the critical band pair improves in the chromatogram and the two bands making up the critical band pair changes. So we see that band spacing changes markedly with gradient time for this sample. For this reason, resolution is critically dependent on a particular gradient time. From Figure 1D, we also see that a gradient time of about 85 min provides maximum resolution, where bands 7 and 8 are the least resolved pair at a resolution of 1.1.

The selection of a linear gradient from 5 to 50% in 85 min represents a reasonable choice of separation conditions for this sample. The predicted separation is shown in Figure 2A. In this case, two experimental runs (Figures 1A and B) plus about 5 min with the computer provide a reasonable method for this sample. However, it is possible to improve on this choice considerably by further computer simulations — as described in References 4 and 5. For the present sample, it was found (computer simulation) that a 3-segment gradient (5/21/26/50%B in 0/29/

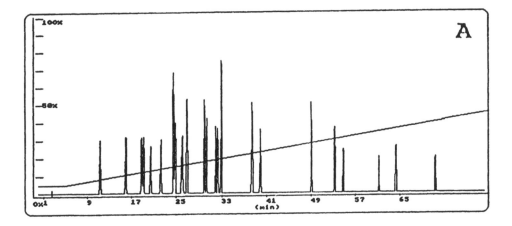

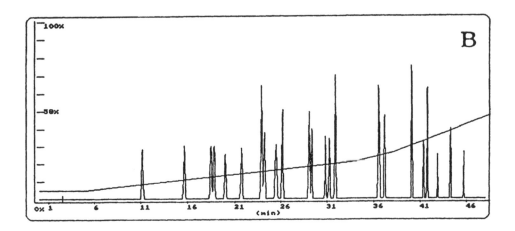

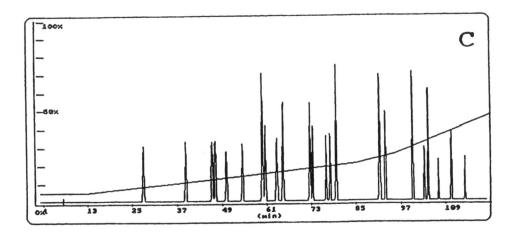

FIGURE 2. Optimized separations of synthetic peptide sample of Figure 1; conditions the same unless otherwise noted. (A) Simulated separation for optimum gradient time of 85 min (linear 5 to 50%B gradient); (B) simulated separation for multi-segment gradient (5/21/26/50% in 0/29/33/45 min); (C) simulated separation for gradient of Figure 2B, except flow rate equals 0.4 ml/min (5/21/26/50% in 0/72.5/82.5/112.5 min); (D) experimental run using conditions of Figure 2C.

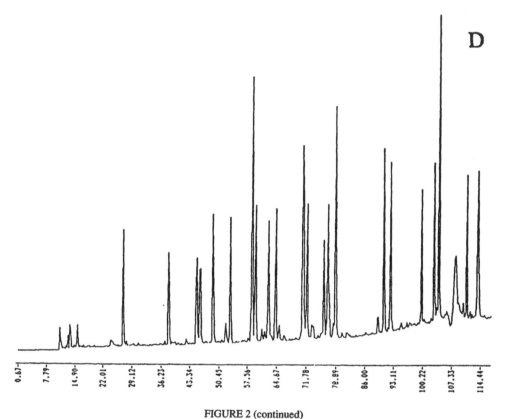

FIGURE 2 (continued)

33/45 min) gives a resolution of $R_s = 1.1$ in only 45 min (Figure 2B), i.e., half the time for a linear gradient in 85 min (Figure 2A).

Drylab G also allows the user to change so-called "column conditions" (column dimensions, particle size, or flow rate) so as to vary column plate number. This option was used to optimize this separation further; for a flow rate of 0.4 ml/min (instead of 1 ml/min originally), it was predicted that resolution could be increased to $R_s = 1.4$ in a total time of 112 min.* The simulated chromatogram for these conditions is shown in Figure 2C, and the experimental chromatogram is compared with it in Figure 2D. The comparison between predicted and actual chromatograms is seen to be quite good. Retention time predictions agree with experimental values within ± 1.4% B (1 SD), and resolution is predicted within ± 8% (1 SD). These are typical results for experimental data vs. predictions based on computer simulation; this degree of predictional accuracy is also quite adequate for the purposes of method development.

IV. SEPARATION OF 50S RIBOSOMAL PROTEINS FROM *E. COLI*

The ribosomal proteins comprise 53 proteins which can be separated by zonal centrifugation into two fractions: 21 30S and 32 50S proteins. Although several groups have attempted the separation of these fractions by reversed-phase HPLC, until recently no one has reported the complete separation of either sample in a single run (the best previous separations leave several band-pairs unresolved). The use of computer simulation[3,5] has changed this situation dramati-

---

* Following the optimization of gradient conditions (shape, time, initial and final %B in an A/B gradient), band spacing can be maintained constant (while column conditions are varied) by changing gradient time. In the present example, the flow rate was reduced from 1.0 to 0.4 ml/min, which required an increase in overall gradient time by a factor of (1.0/0.4) = 2.5.

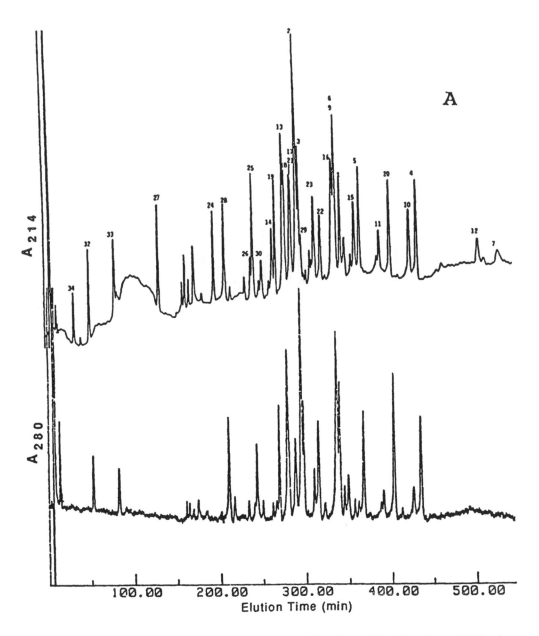

FIGURE 3. Separation of 50S ribosomal proteins. Conditions: 250 × 4.6 mm I.D. Zorbax Protein PLUS column (DuPont); 18 to 66%B (acetonitrile/water gradient with added triethylamine plus TFA); flow rate, 0.7 ml/min; 25°C.[16] (A) Gradient time of 192 min (214 nm detection upper, 280 nm lower); (B) gradient time of 720 min.

cally, permitting the separation of all 21 30S proteins and 31 of the 32 50S proteins. Here we will summarize the application of computer simulation to the separation of the 50S ribosomal proteins.

Figure 3 shows the initial two experimental runs for the 50S proteins: gradient times of 192 min (Figure 3A) and 720 min (Figure 3B). Detection at both 214 and 280 nm was used as an aid in matching bands between the two runs. The column plate number was adjusted to N = 900, in order to match the resolution of experimental and predicted runs, and a relative resolution map

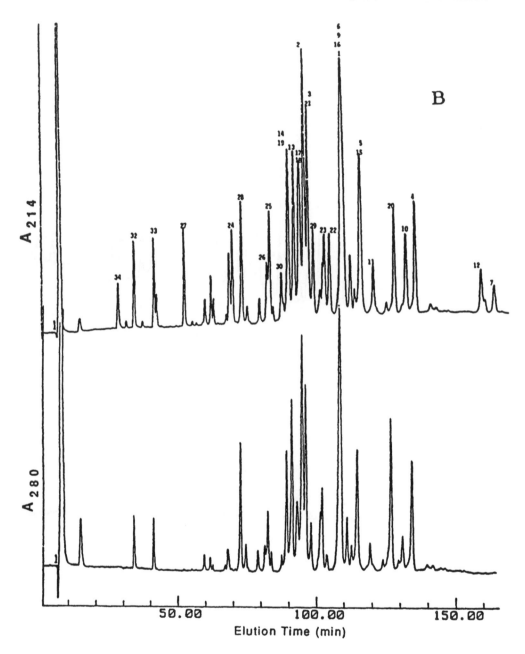

FIGURE 3 (continued)

was requested (Figure 4A). This suggests that the maximum possible resolution is $R_s = 0.8^*$ for an 8-hr run, assuming a linear gradient from 18% to 66%B. The use of a multi-segmented gradient was found to improve this separation, with $R_s = 0.9$ possible in a time of 5 hr. This predicted separation is shown in Figure 4B. Finally, the experimental run corresponding to the conditions of Figure 4B is shown in Figure 4C. Agreement between the two chromatograms is quite good: retention times, ± 0.5% B (1 SD); resolution, ± 8% (1 SD).

V. CONCLUSIONS

The use of computer simulation is now well established as an aid in method development for the HPLC separation of various samples.[14] This approach is particularly useful for the case of large molecules of biological origin separated by gradient elution. These separations often benefit from the use of complex, multi-segmented gradients, and the development of such gradients generally requires a large number of trial-and-error runs. Computer simulation requires only two experimental runs, following which separation can be predicted as a function of various separation conditions: gradient time, initial and final %B, gradient shape (multi-segmented gradients), flow rate, column dimensions and particle size.

Computer simulation reduces the time spent in method development, allows much improved separations, avoids the problem of column changes during method development, conserves sample and solvents, and has other advantages. In many cases, a practical separation can only be obtained by means of this approach.

---

* One band-pair was inseparable for any choice of gradient time; therefore data for 31 different bands were entered into the program for computer simulation, and values of R_s were calculated on the basis of 31 different sample components.

Relative resolution map (based on 18 to 66% B gradient)

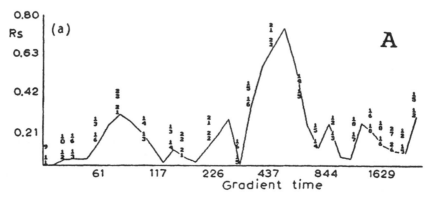

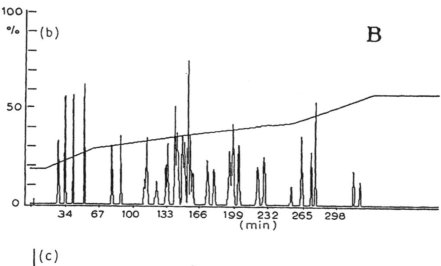

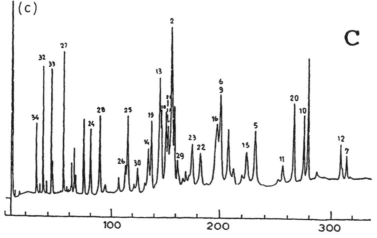

FIGURE 4. Separation of 50S ribosomal proteins. Conditions as in Figure 3, unless noted otherwise. (A) Relative resolution map (N = 900); (B) predicted optimum separation (18/29/37/43/58%B in 0/46/142/241/ 320 min); (C) experimental run corresponding to that of Figure 4B.

REFERENCES

1. **Dolan, J.W., Snyder, L.R., and Quarry, M.A.,** Computer simulation as a means of developing an optimized reversed-phase gradient-elution separation, *Chromatographia*, 24, 261, 1987.
2. **Dolan, J.W. and Snyder, L.R.,** Using computer simulation to develop a gradient elution method for reversed-phase HPLC, *LC/GC Mag.*, 5, 970, 1987.
3. **Ghrist, B.F.D., Cooperman, B.S., and Snyder, L.R.,** Design of optimized HPLC gradients for the separation of either small or large molecules. I. Minimizing errors in computer simulations, *J. Chromatogr.*, 459, 1, 1989.
4. **Ghrist, B.F.D. and Snyder, L.R.,** Design of optimized HPLC gradients for the separation of either small or large molecules. II. Background and theory, *J. Chromatogr.*, 459, 25, 1989.
5. **Ghrist, B.F.D. and Snyder, L.R.,** Design of optimized HPLC gradients for the separation of either small or large molecules. III. An overall strategy and its application to several examples, *J. Chromatogr.*, 459, 43, 1989.
6. **Glajch, J.L., Quarry, M.A., Vasta, J.F., and Snyder, L.R.,** Separation of peptide mixtures by reversed-phase gradient elution. Use of flow rate changes for controlling band spacing and improving resolution, *Anal. Chem.*, 58, 280, 1986.
7. **Kunitani, M., Johnson, D., and Snyder, L.,** Model of protein conformation in the reversed-phase separation of interleukin-2 muteins, *J. Chromatogr.*, 371, 313, 1986.
8. **Snyder, L.R.,** Gradient elution, in *High-Performance Liquid Chromatography. Advances and Perspectives*, Vol. 1, Horváth, Cs., Ed., Academic Press, New York, 1980, p. 207.
9. **Jandera, P. and Churacek, J.,** *Gradient Elution in Column Liquid Chromatography — Theory and Practice*, Elsevier, Amsterdam, 1985.
10. **Stadalius, M.A. and Snyder, L.R.,** HPLC separations of large molecules: a general model, in *High-performance Liquid Chromatography, Advances and Perspectives*, Vol. 4, Horváth, Cs., Ed., Academic Press, New York, 1986, p. 195.
11. **Quarry, M.A., Grob, R.L., and Snyder, L.R.,** Prediction of precise isocratic retention data from two or more gradient elution runs. An analysis of some associated errors, *Anal. Chem.*, 58, 907, 1986.
12. **Hodges, R.S., Parker, J.M.R., Mant, C.T., and Sharma, R.R.,** Computer simulations of HPLC separations of peptide and protein digests for development of size-exclusion, ion-exchange and reversed-phase chromatographic methods, *J. Chromatogr.*, 458, 147, 1988.
13. **Dolan, J.W.,** Method optimization and maintenance using a resolution map, *LC/GC Mag.*, 6, 1052, 1988.
14. **Dolan, J.W., Lommen, D.C., and Snyder, L. R.,** DryLab^R computer simulation for high-performance liquid chromatographic method development, *J. Chromatogr.*, 485, 91, 1989.
15. **Hodges, R.S., Parker, J.M.R., Mant, C.T., Dolan, J.W., and Snyder, L.R.,** A novel approach to HPLC peptide separations based on computer simulation, Abstract No. 701, presented at 8th Int. Symp. on HPLC of Proteins, Peptides and Polynucleotides, Copenhagen, Denmark, October 31 to November 2, 1988.
16. **Ghrist, B.F.D.,** The use of linear-solvent-strength theory to understand protein behavior in reversed-phase HPLC, Ph.D. thesis from University of Pennsylvania, Chemistry Department, 1989.

HPLC HYDROPHOBICITY PARAMETERS: PREDICTION OF SURFACE AND INTERIOR REGIONS IN PROTEINS

J. M. Robert Parker and Robert S. Hodges

I. INTRODUCTION

The significance of hydrophobicity in stabilizing protein structure was first described by Kauzman[1] and has been reviewed by Tanford[2] and Ben-Naim.[3] Data on the hydrophobicity of 11 of the 20 naturally occurring amino acids derived from the transfer between octanol/water were first reported by Tanford.[4,5] Since then, well over 200 sets of parameters have been reported as a measure of the biochemical and biological properties of amino acids.[6] Of these, over 80 parameters describe hydrophobicity. However, a quantitative description and understanding of hydrophobicity and its role in protein folding is still being investigated.

Recently, we described a new set of hydrophobicity parameters[7] derived from the retention time of peptides separated by reversed-phase high-performance liquid chromatography (RPC) (Figure 1). This is the first set of hydrophobicity/hydrophilicity parameters where model synthetic peptides were studied in a systematic manner. Each of the 20 naturally occurring amino acids were substituted in position X of the sequence, Ac-Gly-X-X-(Leu)$_3$-(Lys)$_2$-amide. The HPLC retention coefficients for each side chain, derived from the retention data, were assumed to be a measure of the partitioning of the amino acid residue between the octadecylsilane stationary phase and the buffered aqueous phase at pH 7.0. These retention coefficients would then be a measure of hydrophobicity similar to other parameters derived from the partitioning of amino acids between octanol and water. In a following paper,[8] these parameters were used to predict surface regions in proteins using only primary sequence information.

In this article, these HPLC parameters will be compared to other hydrophobicity parameters. In addition, new results of an improved version of a computer program, SurfacePlot, using the HPLC parameters to predict surface regions in proteins will be presented.

II. HYDROPHOBICITY PARAMETERS

Table 1 shows a listing of the HPLC parameters and their correlation to 18 other parameters. The HPLC parameters are more closely related to those parameters listed in Tables 1A and B than to the parameters listed in Table 1C.

Generally, the largest variation between parameters was observed for the residues cysteine, lysine, and arginine. In most of the other scales, the value of cystine rather than cysteine is reported. Only the value of cysteine is reported for the HPLC parameters because of anomalous results obtained when disulfide-bridged peptides were investigated. It was assumed that the disulfide bridge restricts the conformation of the peptide and therefore the accessibility of side chains to interact with the hydrophobic stationary phase. It is reasonable to expect that the hydrophilicity of cysteine (free thiol) and cystine (disulfide bridge) will be different and that cystine values cannot be used to represent cysteine. For this reason, when cystine values were

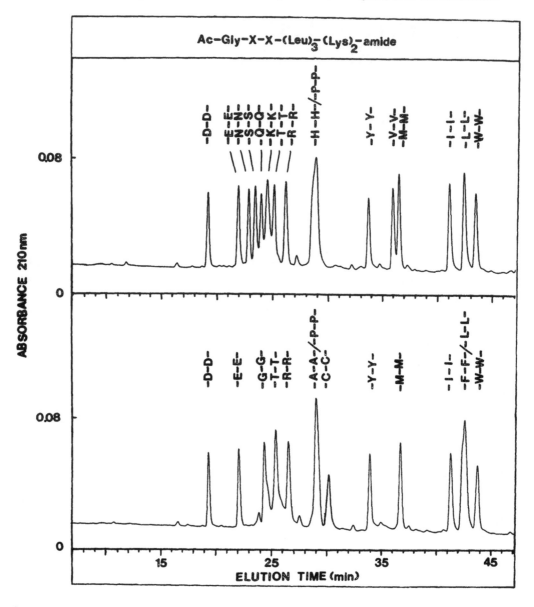

FIGURE 1. Representative RPC profile at pH 7.0 of synthetic peptides with the sequence Ac-Gly-X-X-(Leu)$_3$-(Lys)$_2$-amide, where X is substituted by the 20 amino acids found in proteins. The HPLC instrumentation consisted of a Spectra-Physics (San Jose, CA) SP8700 solvent delivery system and SP8750 organizer module, combined with a Hewlett-Packard (Avondale, PA) HP1040A detection system, HP3390A integrator, HP85 computer, HP9121 disc drive and HP7470A plotter. Column: SynChropak RP-P C$_{18}$ (250 × 4.1mm I.D., 6.5-μm particle size, 300-Å pore size; SynChrom, Linden, IN). Conditions: linear AB gradient (1.67% Eluent B/min, equivalent to 1% acetonitrile/min), where Eluent A is aq.10 m*M* (NH$_4$)$_2$HPO$_4$, pH 7.0, and Eluent B is water/acetonitrile (40:60, v/v), both eluents containing 0.1 *M* NaClO$_4$; flow-rate, 1 ml/min; 26°C. (From Parker, J. M. R., Guo, D., and Hodges, R. S., *Biochemistry*, 25, 5425, 1986. With permission.)

removed from all sets of parameters, most parameter sets displayed a marked improvement in the correlation to the HPLC parameters. In addition, removing lysine and arginine also improved the correlation (see bottom row in Table 1 A, B, and C). In this case, variations in salt conditions or pH conditions used to determine the parameters could lead to large variations in the hydrophobicity assignments for charged residues.

TABLE 1A

	Parker et al.[8] HPLC	Pliska et al.[19] π		Miger[20]		Abraham/Leo[21] π + F_BP		Guy[22]		Rose[17] Å-A/Å		Bull Breese[16]	
Trp	-10.0	2.25	(-10.0)	4.42	(-5.5)	1.88	(-10.0)	3.90	(-4.4)	0.85	(-6.9)	-2010	(-6.5)
Phe	-9.2	1.79	(-7.2)	6.33	(-10.0)	1.87	(-9.9)	6.33	(-10.0)	0.88	(-8.5)	-2330	(-9.0)
Leu	-9.2	1.70	(-6.6)	5.55	(-8.2)	1.81	(9.7)	4.90	(-6.7)	0.85	(-6.9)	-2460	(-10.0)
Ile	-8.0	1.80	(-7.2)	6.05	(-9.3)	1.81	(-9.7)	5.21	(-7.4)	0.88	(-8.5)	-2260	(-8.5)
Met	-4.2	1.23	(-3.7)	6.22	(-9.7)	1.05	(-6.9)	5.08	(-7.1)	0.85	(-6.9)	-1470	(-2.4)
Val	-3.7	1.22	(-3.7)	4.38	(-5.4)	1.27	(-7.8)	4.54	(-5.9)	0.86	(-7.4)	-1560	(-3.1)
Tyr	-1.9	0.96	(-2.1)	1.96	(0.3)	1.20	(-7.5)	2.24	(-0.6)	0.76	(-2.3)	-2240	(-8.3)
Cys	1.4	1.54	(-5.6)	4.80	(-6.4)	1.05	(-6.9)	5.36	(-7.7)	0.91	(-10.0)	-450	(5.3)
Ala	2.1	0.31	(1.9)	1.18	(2.2)	0.32	(-4.3)	1.06	(2.2)	0.74	(-1.3)	-200	(7.2)
Pro	2.1	0.72	(-0.6)	-0.85	(7.0)	0.95	(-6.6)	-0.88	(6.6)	0.64	(3.8)	-980	(1.3)
His	2.1	0.13	(3.0)	0.86	(2.9)	0.34	(-4.4)	2.72	(-1.7)	0.78	(-3.3)	-120	(7.8)
Arg	4.2	-1.01	(10.0)	-0.42	(6.0)	-1.25	(1.4)	-1.30	(7.6)	0.64	(3.8)	-120	(7.9)
Thr	5.2	0.26	(2.2)	0.01	(4.9)	0.33	(-4.3)	0.42	(3.6)	0.70	(0.8)	-520	(4.8)
Lys	5.7	-0.99	(9.9)	-2.13	(10.0)	-1.32	(1.7)	-2.33	(10.0)	0.52	(10.0)	-350	(6.1)
Gly	5.7	0.00	(3.8)	0.00	(4.9)	0.00	(-3.1)	0.00	(4.6)	0.72	(-0.3)	0	(8.8)
Gln	6.0	-0.22	(5.2)	-0.85	(7.0)	-0.91	(0.18)	-0.97	(6.8)	0.62	(4.9)	160	(10.0)
Ser	6.5	-0.04	(4.0)	-0.54	(6.2)	0.01	(-3.2)	-0.27	(5.2)	0.66	(2.8)	-390	(5.8)
Asn	7.0	-0.60	(7.5)	-1.07	(7.5)	-0.34	(-1.9)	-0.21	(5.1)	0.63	(4.4)	80	(9.4)
Glu	7.8	-0.64	(7.7)	-1.15	(7.7)	-3.60	(10.0)	-1.09	(7.1)	0.62	(4.9)	-300	(6.5)
Asp	10.0	-0.77	(8.5)	-1.24	(7.9)	-2.55	(6.2)	-1.18	(7.3)	0.62	(4.9)	-200	(7.2)
Correlation to Parker		0.82		0.81		0.72		0.74		0.68		0.83	
Correlation to Parker (minus Cys, Arg and Lys)		0.96		0.88		0.87		0.86		0.85		0.83	

Note: Amino acid hydrophobicity parameters. Each column of parameters lists the orginal scale on the left. The scale on the right (in brackets) has been recalculated with the most hydrophobic amino acid scaled to –10 and the most hydrophilic scaled to 10.

TABLE 1B

	Parker et al.[9] HPLC	Hopp/Woods[11]		Eisenberg[24]		Hopp[25]		Krig[18]		Cornette[19]		Meek[14]		Ponnus[23]		Kyte/ Dolittle[22]	
Trp	-10.0	-3.4	(-10.0)	0.37	(-7.1)	-3.0	(-10.0)	2.78	(-6.6)	1.04	(0.6)	14.9	(-10.0)	12.95	(-0.1)	-0.9	(2.0)
Phe	-9.2	-2.5	(-7.2)	0.61	(-9.0)	-2.7	(-9.0)	2.59	(-7.2)	4.44	(-7.2)	13.2	(-8.9)	13.43	(-2.3)	2.8	(-6.2)
Leu	-9.2	-1.8	(-5.0)	(0.53)	(-8.4)	-2.5	(-8.3)	3.93	(-2.9)	5.66	(-10.0)	8.8	(-6.2)	14.10	(-5.4)	3.8	(-8.4)
Ile	-8.0	-1.8	(-5.0)	(0.73)	(-10.0)	-2.5	(-8.3)	2.31	(-8.1)	4.77	(-8.0)	13.9	(-9.4)	14.77	(-8.6)	4.5	(-10.0)
Met	-4.2	-1.3	(-3.4)	(0.26)	(-6.2)	-1.8	(-6.0)	2.44	(-7.7)	4.23	(-6.7)	4.8	(-3.6)	14.33	(-6.5)	1.9	(-4.2)
Val	-3.7	-1.5	(-4.1)	(0.54)	(-8.4)	-1.7	(-5.6)	3.31	(-4.9)	4.67	(-7.7)	2.7	(-2.3)	15.07	(-10.0)	4.2	(-9.3)
Tyr	-1.9	-2.3	(-6.6)	(0.02)	(-4.4)	-2.0	(-6.6)	3.58	(-4.0)	3.23	(-4.4)	6.1	(-4.5)	13.29	(-1.6)	-1.3	(2.9)
Cys	1.4	-1.0	(-2.5)	(0.04)	(-4.5)	-2.6	(-8.6)	1.73	(-10.0)	4.07	(-6.4)	-6.8	(3.6)	14.93	(-9.3)	2.5	(-5.5)
Ala	2.1	-0.5	(-0.9)	(0.25)	(-6.2)	-0.5	(-1.6)	4.32	(-1.6)	0.22	(2.4)	0.5	(-0.9)	12.28	(3.1)	1.8	(-4.0)
Pro	2.1	0.0	(0.6)	(-0.07)	(-3.7)	2.6	(8.6)	7.19	(7.6)	-2.23	(8.1)	6.1	(-4.5)	11.19	(8.2)	-1.6	(3.5)
His	2.1	-0.5	(-0.9)	(-0.40)	(-1.0)	-0.4	(-1.3)	5.66	(2.7)	0.46	(1.9)	-3.5	(1.6)	12.84	(0.4)	-3.2	(7.1)
Arg	4.2	3.0	(10.0)	(-1.80)	(10.0)	0.3	(1.0)	6.55	(5.6)	1.42	(-0.3)	0.8	(-1.1)	11.49	(6.8)	-4.5	(10.0)
Thr	5.2	-0.4	(-0.6)	(-0.18)	(-2.8)	-0.1	(-0.3)	5.16	(1.1)	-1.90	(7.3)	2.7	(-2.3)	11.65	(6.0)	-0.7	(1.5)
Lys	5.7	3.0	(10.0)	(-1.10)	(4.4)	1.4	(4.6)	7.92	(10.0)	-3.04	(10.0)	0.1	(-0.7)	10.8	(10.0)	-3.9	(8.7)
Gly	5.7	0.0	(0.6)	(0.16)	(-5.5)	3.0	(10.0)	6.09	(4.1)	0.00	(3.0)	0.0	(-0.6)	12.01	(4.3)	-0.4	(0.9)
Gln	6.0	0.2	(1.2)	(-0.69)	(1.2)	-0.2	(-0.6)	6.13	(4.2)	-2.81	(9.4)	-4.8	(2.4)	11.28	(7.7)	-3.5	(7.8)
Ser	6.5	0.3	(1.6)	(-0.26)	(-2.2)	1.8	(6.0)	5.37	(1.8)	-0.45	(4.0)	1.2	(-1.4)	11.26	(7.8)	-0.8	(1.7)
Asn	7.0	0.2	(1.2)	(-0.64)	(0.8)	2.3	(7.6)	6.24	(4.6)	-0.46	(4.0)	0.8	(-1.1)	11.00	(9.1)	-3.5	(7.8)
Glu	7.8	3.0	(10.0)	(-0.62)	(0.7)	0.5	(1.6)	6.17	(4.3)	-1.81	(7.1)	-16.9	(10.0)	11.19	(8.2)	-3.5	(7.8)
Asp	10.0	3.0	(10.0)	(-0.72)	(1.5)	2.1	(7.0)	6.04	(3.9)	-3.08	(10.0)	-8.2	(4.5)	10.97	(9.2)	-3.5	(7.8)
Correlation to Parker		0.67		0.54		0.69		0.58		0.65		0.65		0.60		0.53	
Correlation to Parker (minus Cys, Arg, and Lys)		0.76		0.76		0.74		0.73		0.72		0.70		0.70		0.56	

Note: Amino acid hydrophobicity parameters. Each column of parameters lists the orginal scale on the left. The scale on the right (in brackets) has been recalculated with the most hydrophobic amino acid scaled to −10 and the most hydrophilic scaled to 10.

TABLE 1C

Parker et al.[8] HPLC		Janin[27] f		Chothia[28]		Wolfenden[26]		Welling[30]	
Trp	−10.0	1.6	(3.2)	0.27	(1.2)	−5.88	(−2.6)	−0.114	(−2.2)
Phe	−9.2	2.2	(0.5)	0.50	(−6.6)	−0.76	(−7.2)	−0.141	(−3.0)
Leu	−9.2	2.4	(−0.3)	0.45	(−4.9)	2.28	(−9.9)	0.075	(3.2)
Ile	−8.0	3.1	(−3.4)	0.60	(−10.0)	2.15	(−9.8)	−0.292	(−7.3)
Met	−4.2	1.9	(1.8)	0.40	(−3.2)	−1.48	(−6.5)	−0.385	(−10.0)
Val	−3.7	2.9	(−2.5)	0.54	(−7.9)	1.99	(−9.6)	−0.013	(0.7)
Tyr	−1.9	0.5	(8.0)	0.15	(5.2)	−6.11	(−2.4)	0.013	(1.4)
Cys	1.4	4.6	(−10.0)	0.50	(−6.6)	−1.24	(−6.7)	−0.120	(−2.4)
Ala	2.1	1.7	(2.7)	0.38	(−2.5)	1.94	(−9.6)	0.115	(4.3)
Pro	2.1	0.6	(7.5)	0.18	(4.2)	2.00	(−9.6)	−0.053	(−0.5)
His	2.1	0.8	(6.7)	0.17	(4.6)	−10.27	(1.3)	0.312	(10.0)
Arg	4.2	0.1	(9.8)	0.01	(10.0)	−19.92	(10.0)	0.058	(2.7)
Thr	5.2	0.7	(7.1)	0.23	(2.5)	−4.88	(−3.5)	−0.045	(−0.2)
Lys	5.7	0.05	(10.0)	0.03	(9.3)	−9.52	(0.7)	0.206	(6.9)
Gly	5.7	1.8	(2.3)	0.36	(−1.8)	2.39	(−10.0)	−0.184	(−4.2)
Gln	6.0	0.3	(8.9)	0.07	(7.9)	−9.38	(0.5)	−0.011	(0.7)
Ser	6.5	0.8	(6.7)	0.22	(2.9)	−5.06	(−3.3)	−0.026	(0.3)
Asn	7.0	0.4	(8.5)	0.12	(6.3)	−9.68	(0.8)	−0.077	(−1.1)
Glu	7.8	0.3	(8.9)	0.18	(4.2)	−10.20	(1.3)	−0.071	(−0.9)
Asp	10.0	0.4	(8.5)	0.15	(5.2)	−10.95	(1.9)	0.065	(2.9)
Correlation to Parker		0.34		0.44		0.26		0.116	
Correlation to Janin				0.83		0.47		0.19	
Correlation to Janin (minus Cys)				0.95					

Note: Amino acid hydrophobicity parameters. Each column of parameters lists the orginal scale on the left. The scale on the right (in brackets) has been recalculated with the most hydrophobic amino acid scaled to −10 and the most hydrophilic scaled to 10.

Previously, Rose[9] suggested that hydrophobicity parameters could be divided into two groups. The first group of hydrophobicity parameters resemble our HPLC parameters and those of Nozaki and Tanford[5] which are a measure of the amino acid residue partitioning between a hydrophobic and aqueous medium. This group is represented by those parameters listed in Tables 1A and B. As expected, one of the highest correlations was observed between the HPLC parameters and the π parameters of Pliska[10] which were determined from the partitioning of N-acetyl amino acid amides between octanol and water (correlation 0.96 when cysteine, lysine and arginine were removed). In this case the difference in hydrophobicities for charged residues could be due to different counterions used in the experiments (ClO_4^{-2} and PO_4^{-2} for the HPLC and π parameters, respectively). It should be noted that perchlorate was used in the mobile phase when determining the HPLC parameters, to suppress silanol interaction at pH 7.0[7] with the positively charged groups in the peptides.

Two other sets of parameters, Hopp-Woods[11] and Kyte-Dollitle,[12] which are probably the most widely used set of hydrophobicity parameters, show a very poor correlation to the HPLC parameters (0.67 and 0.53, respectively). In these two sets of parameters, the original scale of Nozaki and Tanford[5] was used. It is important to note that, in the original work, no attempt was made to control the pH or ionic strength of the aqueous solution when determining these parameters and therefore their reliability must be questioned. The difference in the hydrophobicity scales is mainly due to the values assigned for all charged residues. The effect of pH, counterion, self association of free amino acids, and solubility in organic solvents on hydrophobicity has been previously discussed by several authors.[9,10,13]

The Meek parameters,[14] which were derived from the retention time of 25 peptides in RPC, showed a very poor correlation (0.65) with our HPLC parameters. This can be explained by the fact that the small sample size does not give an equal representation of amino acid residues and the computer calculated regression analysis results in large errors to those residues of low occurrence. For example, the largest deviations were observed for cysteine and glutamic acid which have low occurrence in the peptides studied. Also, it has been documented that there is a polypeptide chain length effect on peptide retention in RPC.[15] The use of a random sampling of peptides of varying chain lengths would result in errors in the derived parameters when compared with our HPLC parameters which were determined from peptides at a fixed chain length.

The remaining parameters in this group show a moderate to low correlation to the HPLC parameters. The variations in assignments and unusual values, relative to the concensus for most of the other parameters, are discussed below.

The Bull and Breese[16] parameters, derived from surface tension measurement, assign unusual hydrophilic values to glutamine, asparagine, and glycine relative to amino acids with charged side chains. Both the Rose[17] parameters, derived from X-ray accessibility data, and the Krigbaum[18] parameters, which are an estimate of interaction parameters from X-ray data, were the only parameters of this group to assign the highest hydrophobic value to cysteine. Although several other parameter sets[9] assigned lysine the most hydrophilic value, the unusual exposure of lysine in the Rose parameters was attributed to the uncertainty of the side chain atom values. Previously, Cornette[19] described the correlation of 40 hydrophobicity parameters. The highest correlation was observed between the Miyazawa[20] parameters, which were contact energies derived from X-ray data, the Abraham and Leo[21] parameters, which are the fragmental (π) constants derived from partition coefficients of amino acids, and the Guy[22] parameters, which were the average of four different scales. Large variations were observed in these three scales for lysine, glutamine, asparagine, and tyrosine. Using these 40 parameters, Cornette derived a new scale which assigned unusual values to tryptophan and arginine. The Ponnusway[23] parameters, derived from the average surrounding hydrophobicity in X-ray data, also assigned an unusual value for tryptophan. The Eisenberg parameters,[24] which are the average of five scales, assign a hydrophilic value to arginine that is unusually high relative to the other charged residues in that set. The acrophilicity parameters of Hopp,[25] derived from X-ray data, assigned the highest hydrophilic values to glycine and proline, while the values of arginine, glutamine, and glutamic acid were much less hydrophilic.

The second group of hydrophobicity parameters (Table 1C), according to Rose,[9] resemble the parameters of Wolfenden[26] which were derived from the transfer of organic molecules (representing side chains of amino acids) from water to the vapor phase. The group of parameters in Table 1C resemble those of Wolfenden and all show a low correlation with the HPLC parameters. However, a moderate correlation (0.83) is observed between the Janin[27] and Chothia[28] parameters, both derived from X-ray structures, which is improved by removing cysteine (cystine) (0.95). This discrepancy is probably due to the low frequency of cysteine (cystine) residues in X-ray structures. The Wolfenden parameters do not correlate well with the Janin or Chothia parameters. It has been reported that the Wolfenden parameters correlated very well with the Janin-free energy of transfer parameters.[29] The Welling parameters,[30] which describe the antigenicity of residues, were derived from immunological and X-ray data and show the lowest correlation to all of the above parameters.

III. HYDROPHOBICITY MODULATION

Several workers have suggested that no single parameter can describe hydrophobicity.[9,10,19,21,31]

This has been attributed to "proximity effects" of polar functional groups.[32-34] Very little is known about the aqueous environment surrounding different amino acid residues. The fact that the hydration of polar amino acids is different from apolar residues is readily accepted. But what has not been described previously is that there may be a repulsive free energy between these two processes.[35] Recently, Urry and workers[35-37] have suggested a new and very interesting mechanism to describe hydrophobicity based on the water of hydration for apolar amino acid residues. Their work is centered on the concept of an ordered layer or "clathrate cage"[38,39] of water molecules surrounding hydrophobic residues. This, in turn, results in changes in the entropy of the system due to changes in water structure. Experimentally, hydrophobicity can be modulated by controlling the enthalphy and entropy of a system with changes in pH, ionic strength, and temperature.[36,37] This has been called the "vicinal chemical modulation of the hydrophobic effect".[35] This implies hydrophobicity is not a constant and that hydrophobicity is closely related to the hydration of amino acids in a changing environment.

IV. HYDROPHOBICITY CORRELATES WITH SURFACE AREA

Considering the above results — (a) that few of the 18 parameters derived for hydrophobicity correlate well, (b) that many hydrophobicity scales of amino acids determined previously did not take into account variations in pH, ionic strength, or self association and their effect on hydrophobicity, and (c) that hydrophobicity appears to change depending on residue environment — we suggest that the best approximation for hydrophobicity of amino acids in proteins would be a scale based on amino acid hydrophobicity in peptides (HPLC parameters) in solution under conditions of controlled pH and ionic strength. Further support for the use of HPLC-derived hydrophobicity parameters to describe the hydrophobicity/hydrophilicity of amino acids in proteins comes from the observation that there is a good correlation between hydrophobicity and aqueous cavity surface area or residue surface area[40-44] for amino acids. This is related to the observation by Urry[35-37] that hydrophobicity depends on the hydration or degree of water solvation of amino acids. Figure 2 shows a plot of the HPLC parameters vs. the surface area[17,28,45] calculated for amino acid residues in peptide side chains. A linear relationship is observed between the increasing hydrophobicity and increasing surface area for the apolar residues G, A, V, I, L, F, for the residues C, P, M, W, for the polar residues S, T, H, Y and N, Q and for the charged residues D, E, K, R. It is interesting that the amino acids appear to fall into four groups of hydrophobicity, i.e., very hydrophobic (I, L, F, W) hydrophobic (V, M, Y), hydrophilic (A, C, P, H), and very hydrophilic (G, S, T, N, Q, D, E, K, R).

V. HYDROPHOBICITY PARAMETERS PREDICT SURFACE AND INTERIOR REGIONS OF PROTEINS

Previously,[8] it was shown that one set of parameters could not correctly predict all antigenic regions or surface regions for several different proteins. Qualitatively, however, most parameters describe the same general features of hydrophilic and hydrophobic patterns[25] with the HPLC parameters generally scoring the highest[8] in predicting surface regions of proteins. After investigating several possible combinations of parameters, the best correlation with X-ray defined surface regions and immunologically-defined antigenic sites was demonstrated with a combination of the HPLC[7,8] hydrophilicity, Janin[27] accessibility and Karplus and Schulz[46] flexibility parameters. A computer program, called SurfacePlot,[8] was written to predict linear surface regions in proteins using these three sets of parameters. These three parameters were chosen because of their low correlation to each other. Since it was shown that all parameters qualitatively describe surface regions, each parameter was assumed to predict some different property that described hydrophobicity. More recently, the 18 parameters listed in this work

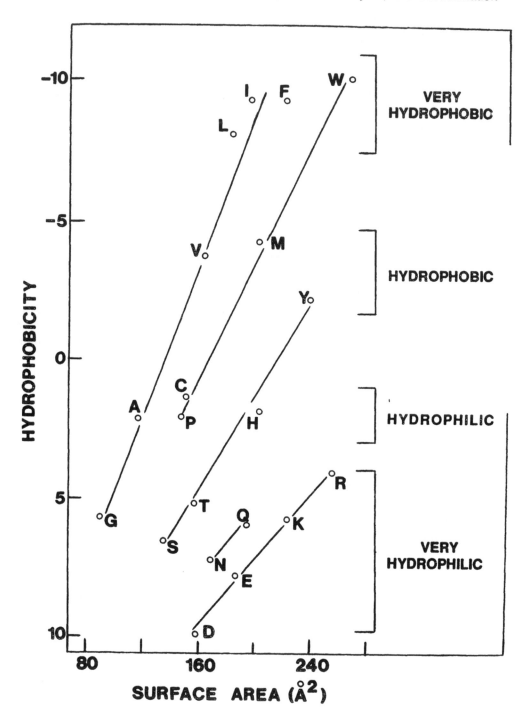

FIGURE 2. Correlation of HPLC hydrophobicity parameters (see Table 1) with surface area of amino acid residues. Brackets indicate four groupings of amino acids into very hydrophobic, hydrophobic, hydrophilic and very hydrophilic.

were also tested and found to lead to no improvement in the successful prediction of antigenic sites or surface regions. It is noteworthy that the Acrophilicity[25] and Welling[30] parameters showed very good agreement with the HPLC parameters in predicting surface regions in proteins but were still not as effective as SurfacePlot.

TABLE 2A
Myoglobin
(α class)

Surface		Interior	
Predicted	X-ray	X-ray	Predicted
5	3—6	7—17 (4—17H	6—20
21—26	18—20	21—33 (21—35H)	27—35
36—39	34—38 (37—42H)		40—43
44—45	41—45	46—49 (44—47H)	46—52
53—62	50—54 (52—57H)	55—76 (59—76H)	63—77
78—82	77—79	86—94 (83—95H)	83—92
	83—85	104—115 (101—118H)	96—118
93—95	95—98	123—146 (125—149H)	130—147
	116—118		
119—129	120—122		
148—150	147—153		

Predicted vs. X-ray surface and interior regions in myoglobin. Predicted and X-ray surface regions are listed in the two left columns. The interior regions are listed in the two right columns. Residue numbers in brackets are secondary structural regions of helix (H).

Using the SurfacePlot computer program (a higher cut-off value than described in the previous work[8] was employed), and comparing the predictions to 15 X-ray-defined structures, a very good correlation was achieved for surface or highly accessible regions, and interior regions or regions of low accessibility. These predicted surface regions have high hydrophilicity, accessibility or flexibility properties. This new improvement has been incorporated into the latest version of SurfacePlot.*

Examples of predicted (outer lanes) and X-ray (inner lanes) surface and interior regions in proteins using SurfacePlot are shown in Table 2. One example is given for each class of α, β, and α/β proteins (Tables 2A, B, and C, respectively). Linear X-ray surface regions are defined as three or more consecutive residues which individually have 40 Å² or more of exposed surface. The value of 40 Å² is approximately 25% of the standard state surface area of amino acid residues calculated by Rose.[17] This definition was derived after inspecting the accessibility pattern for several proteins. Single residues with high accessibility did not appear to define structural regions. It was interesting to note that lysine was the only residue that consistently gave high accessibility values as was reported earlier by Rose.[17] Again, two consecutive residues with high accessibilities did not appear to define structural regions. In fact, pairs of highly accessible residues were frequently observed in α-helical structures. However, the pattern of three or more consecutive residues with high accessibility was found to define surface structural regions (β-turns, loops, bends) of high accessibility. The regions of low accessibility in between these surface structural regions (helix, sheet, coil) were defined as interior regions. This pattern of linear accessible regions rarely occurred in helical or extended secondary structure regions. X-ray surface areas were calculated using the program of Kabsch and Sanders.[47] All secondary structure regions (these regions are listed in brackets in Table 2; H = helix, S = sheet) are from the assignments by Kabsch and Sanders.[47] Additional regions found in the Brookhaven Protein Data Bank[48] are marked with a superscript 1. Interior regions with no secondary structure are marked by open brackets ( ). In general, the correlation between X-ray and predicted regions is very high. Although exact residues cannot be identified, predicted regions generally correspond

---

* SurfacePlot version 1.2 is available from S.P.I. Synthetic Peptides Inc., University of Alberta, 355 Medical Sciences Building, Edmonton, Alberta, Canada, T6G 2H7.

TABLE 2B
Concanavalin A
(β class)

Surface		Interior	
Predicted	X-ray	X-ray	Predicted
	1—3	4—34 (4—10S)	1—10
11—21		4—34 (24—29S)	22—32
33—38	35—37 (36—39S)	38—40 (36—39S)	39—45
46	41—44	45—56 (47—55S)	47—56
57	57—60	61—81 (60—66S)	58—70
71		61—81 (73—78S)	72—76
77		89—98 (88—96S)	82—94
80—81	82—84 (81—84H)		96—101
	86—88	102—117 (103—106S)[1]	103—114
95		102—117 (105—116S)	
102	99—101	128—133 (124—130S)	123—133
115—122	118—122	137—157 (140—144S)	138—145
	124—127	137—157 (147—148S)	
134—137	134—136	137—157 (154—155S)	153—159
146—152		163—175 (170—175S)	170—183
160—169	158—162	179—181 (179—180S)	
	176—178	186—201 (188—198S)	188—202
184—187	182—185	206—233 (209—216S)	207—220
203,205—206	202—205		227—237
221—226			
	234—237		

Predicted vs. X-ray surface and interior regions in concanavalin A. Predicted and X-ray surface regions are listed in the two left columns. The interior regions are listed in the two right columns. Residue numbers in brackets are secondary structural regions of helix (H) and β-sheet or extended regions (S) defined by Kabsch and Sander.[47] Additional regions marked with a superscript 1 are listed in the Brookhaven Protein Data Bank.[48]

TABLE 2C
Lysozyme
(α/β class)

Surface		Interior	
Predicted	X-ray	X-ray	Predicted
	5—7	1—4 (1—3S)[1]	
16—21	18—23	8—17 (5—14H)	1—15
36—51	43—48 (43—45S)	24—42 (26—36H)	22—35
67—72	72—75	24—42 (38—40S)[1]	
	77—79	49—71 (51—53S)	52—66
85—88	85—87	49—71 (58—59S)	
103—104	101—103	80—84 (80—84H)[1]	73—84
	112—114	88—100 (89—98H)[1]	89—102
116—117	116—119	104—111 (109—114H)	105—115
		120—129 (—)	118—129

Predicted vs. X-ray surface and interior regions in lysozyme. Predicted and X-ray surface regions are listed in the two left columns. The interior regions are listed in the two right columns. Residue numbers in brackets are secondary structural regions of helix (H) and β-sheet or extended regions (S) defined by Kabsch and Sander.[47] Additional regions marked with a superscript 1 are listed in the Brookhaven Protein Data Bank.[48]

TABLE 3
Comparing X-Ray-Defined and Predicted Surface and Interior Regions

Protein	Surface			Interior		
	X-ray	Predicted	Over	X-ray	Predicted	Over
Alpha						
Myoglobin	12 (6)	9 (5)	0	7	7	1
Hemoglobin-α	8 (8)	4 (6)	6	7	6	3
Hemoglobin-β	7 (5)	5 (3)	3	8	7	1
Beta						
Elastase	16 (17)	12 (13)	5	14	14	2
Bovine Trypsin	15 (15)	11 (8)	7	17	11	0
Concanavalin A	15 (19)	10 (13)	5	14	14	1
Chymotrypsin	12 (10)	9 (6)	8	16	15	3
Alpha-Beta						
Thermolysin	20 (16)	16 (13)	6	18	16	2
Subtilisin	15 (17)	13 (13)	5	14	13	1
Carboxypeptidase						
A	14 (23)	9 (16)	11	19	16	3
Rat Mast Cell						
Protease	13 (17)	10 (12)	4	11	11	2
Ribonuclease	13 (4)	10 (3)	0	9	8	1
Papain	12 (7)	6 (5)	7	9	8	2
Staphloccocal						
Nuclease	11 (8)	6 (7)	3	7	6	3
Lysozyme	9 (11)	6 (7)	0	8	7	0
Total	192(183)	136(130)	70	178	160	25
% X-ray		71 (71)	36		90	14

Note: Comparison of X-ray, predicted and overpredicted surface and interior regions in 15 proteins in three classes of proteins (α, β, α/β). The numbers in brackets represent β-turn regions for each protein.

to X-ray structural regions. It is interesting that this simple search for interior and surface X-ray regions also appears to define secondary structure regions.

Table 3 shows the comparison of predicted and X-ray surface and interior regions for 15 proteins. Predicted regions which were contained in, or overlapped with the X-ray surface or interior region were considered successful predictions. Predicted regions which were outside X-ray regions were listed as over-predictions. As can be seen in Table 3, a very good prediction (90%) of interior regions was obtained with little overprediction (14%). A reasonable prediction (71%) was obtained for surface regions with 36% over-prediction. Although the overpredicted surface regions did not correspond to the definition of X-ray surface regions of three or more consecutive residues, these predicted surface regions did correspond to breaks or boundaries of secondary structure regions. An example of these over predicted surface regions is shown in conconavalin A (Table 2B) for predicted surface regions 11 to 21, 71, 77, 95, 146 to 152, 221 to 226. Also shown in Table 3 are the number of β-turns in each protein as defined by Kabsch/Sander[47] and Chou/Fasman.[49] These are listed in brackets in the X-ray Surface column. The number of predicted surface regions that correspond to these X-ray defined β-turns is shown in brackets in the predicted column. Approximately 71% of the X-ray defined turns were correctly identified within four residues.

VI. SUMMARY

We consider the HPLC parameters derived from amino acid residues in model synthetic peptides to be the most reliable hydrophobicity parameters yet determined which take into account the short range interaction of residues with other amino acids and the solvent. However, no one set of parameters will successfully predict surface regions in proteins from the amino acid sequence since medium and long range interaction with other residues in the sequence will also affect conformations. SurfacePlot, a computer program which combines hydrophilicity, accessibility and flexibility predictions, provides a good description of surface, interior and break regions in protein secondary structures.

ACKNOWLEDGMENTS

This work was supported by the Medical Research Council of Canada and Synthetic Peptides Incorporated, and by equipment grants from the Alberta Heritage Foundation for Medical Research.

REFERENCES

1. **Kauzmann, W.,** Some factors in the interpretation of protein denaturation, *Adv. Prot. Chem.* 14, 1, 1959.
2. **Tanford, C.,** *The Hydrophobic Effect,* 2nd ed., Wiley-Interscience, John Wiley & Sons, New York, 1980.
3. **Ben-Naim, A,** *Hydrophobic Interactions,* Plenum Press, New York, 1980.
4. **Tanford, C.,** Contribution of hydrophobic interactions to the stability of the globular conformation of proteins, *J. Am. Chem. Soc.,* 84, 4240, 1962.
5. **Nozaki, Y. and Tanford, C.,** The solubility of amino acids and two glycine peptides in aqueous ethanol and dioxane solutions. Establishment of a hydrophobicity scale, *J. Biol. Chem.,* 246, 2211, 1971.
6. **Nakai, K., Kidera, A., and Kanehisa, M.,** Cluster analysis of amino acid indices for prediction of protein structure and function, *Prot. Engineering,* 2, 93, 1988.
7. **Guo, D., Mant, C.T., Taneja, A.K., Parker, J.M.R., and Hodges, R.S.,** Prediction of peptide retention times in reversed-phase high performance liquid chromatography 1. Determination of retention coefficients of amino and residues of model synthetic peptides, *J. Chromatogr.,* 359, 499, 1986.
8. **Parker, J.M.R., Guo, D., and Hodges, R.S.,** New hydrophilicity scale derived from high performance liquid chromatography peptide retention data: correlation of predicted surface residues with antigenicity and X-Ray derived accessible sites, *Biochemistry,* 25, 5425, 1986.
9. **Rose, G.D., Gierasch, L.M., and Smith, J.A.,** Turns in peptides and proteins, *Adv. Prot. Chem.,* 37, 1, 1985.
10. **Fauchere, J.L. and Pliska, V.,** Hydrophobic parameters π of amino acid side chains from the partioning of N-acetyl-amino acid amides, *Eur. J. Med. Chem.* 18, 369, 1983.
11. **Hopp, T.P. and Woods, K.R.,** Prediction of protein antigenic determinants from amino acid sequences, *Proc. Natl. Acad. Sci. U. S. A.,* 78, 3824, 1981.
12. **Kyte, K and Doolittle, R.F.,** A simple method for displaying the hydropathic character of a protein, *J. Mol. Biol.* , 157, 105, 1982.
13. **Younger, L.M. and Cramer, R.D.,** Measurement and correlation of partition coefficients of polar amino acids, *Mol. Pharm.,* 20, 602, 1981.
14. **Meek, J.L.,** Prediction of peptide retention times in high pressure liquid chromatography on the bases of amino acid composition, *Proc. Natl. Acad. Sci. U. S. A.,* 77, 1632, 1980.
15. **Mant, C.T., Burke, T.W.L., Black, J.A., and Hodges, R.S.,** Effect of peptide chain length on peptide retention behavior in reversed phase chromatography, *J. Chromatogr.,* 458, 193, 1988.
16. **Bull, H.B. and Breese, K.,** Surface tension of amino acid solutions: a hydrophobicity scale of the amino acid residues, *Arch. Biochem. Biophys.,* 161, 665, 1974.
17. **Rose, G.D., Geselowitz, A.R., Lesser, G.J., Lee, R.H., and Zehfus, M.H.,** Hydrophobicity of amino and residues in globular proteins, *Science,* 229, 834, 1985.
18. **Krigbaum, W.R. and Komoriya, A.,** Local interactions as a structure determinant for protein molecules: II, *Biochim. Biophys. Acta,* 576, 204, 1979.
19. **Cornette, J.L., Cease, K.B., Margalit, H., Spouge, J.L., Berzofsky, J.A., and Delisi, C.,** Hydrophobicity scales and computational techniques for detecting amphipathic structures in proteins, *J. Mol. Biol.,* 195, 659, 1987.

20. **Miyazawa, S. and Jernigan, R.L.,** Estimation of effective interresidue contact energies from protein crystal structures: quasi-chemical approximations, *Macromolecules,* 18, 534, 1985.

21. **Abraham, D.J. and Leo, A.J.,** Extension of the fragment method to calculate amino acid zwitterion and side chain partition coefficients, *Proteins,* 130, 1987.

22. **Guy, H.R.,** Amino acid side-chain partition energies and distribution of residues in soluble proteins, *Biophys. J.,* 47, 61, 1985.

23. **Ponnuswamy, P.K., Prabhakaran, M., and Manavalan, P.,** Hydrophobic packing and spatial arrangement of amino acid residues in globular proteins, *Biochim. Biophys. Acta,* 623, 301, 1980.

24. **Eisenberg, D., Weiss, R.M., Terwilliger, T.C., and Wilcox, W.,** Hydrophobic moments and protein structure, *Far. Symp. Chem. Soc.,* 17, 109, 1982.

25. **Hopp, T.P.,** Protein surface analysis, methods for identifying antigenic determinants and other interaction sites, *J. Immunol. Methods,* 88, 1, 1986.

26. **Wolfenden, R.W., Cullis, P.M., and Southgate, C.C.F.,** Water, protein folding, and the genetic code, *Science,* 206, 575, 1979.

27. **Janin, J.,** Surface and inside volumes in globular proteins, *Nature,* 277, 491, 1979.

28. **Clothia, C.,** The nature of the accessible and buried surfaces in proteins, *J. Mol. Biol.,* 105, 1, 1976.

29. **Wolfenden, R., Andersson, L., Cullis, P.M., and Southgate C.C.B.,** Affinities of amino acid side chains for water, *Biochemistry,* 20, 849, 1981.

30. **Welling, G.W., Weijer, W.J., van der Zee, R., and Welling-Wester, S.,** Prediction of sequential antigenic regions in proteins, *FEBS,* 188, 215, 1985.

31. **Charton, M. and Charton, B.I.,** The structural dependence of amino acid hydrophobicity parameters, *J. Theor. Biol.,* 99, 629, 1982.

32. **Hansch, C. and Leo, A.,** *Substituent Constants for Correlation, Analysis in Chemistry and Biology,* Wiley Interscience, New York, 1979.

33. **Rekker, R.F.,** *The Hydrophobic Fragmental Constant,* Elsevier, Amsterdam, 1977.

34. **Roseman, M.,** Hydrophobicity of the peptide C0..NH hydrogen bonded group, *J. Mol. Biol.,* 210, 621, 1988.

35. **Urry, D.W.,** Developing biophysical priciples leads to realizing the potential of biotechnology, *Research and Development,* Nov., 72, 1988.

36. **Urry, D.W., Harris, R.D., and Prasad, K.U.,** Chemical potential driven contraction and relaxation by ionic strength modulation of an inverse temperature transition, *J.A.C.S.,* 110, 3303, 1988.

37. **Urry, D.W.,** Entropic elastic processes in protein mechanisms. 1. Elastic structure due to an inverse temperature transition and elasticity due to internal chain dynamics., *J. Prot. Chem.,* 7, 1, 1988.

38. **Swaminathan, S., Harrison, S.W., and Beveridge, D.L.,** Monte Carlo studies on the structure of a dilute aqueous solution of methane, *J.A.C.S.,* 100, 5705, 1978.

39. **Scheraga, H.A.,** Interactions in aqueous solutions, *Acc. Chem. Res.,* 12, 7, 1979.

40. **Chothia, C.,** Hydrophobic bonding and accessible surface area in proteins, *Nature,* 248, 338, 1974.

41. **Reynolds, J.A., Gilbert, D.B., and Tanford, C.,** Empirical correlation between hydrophobic free energy and aqueous cavity surface area, *Proc. Natl. Acad. Sci. U. S. A.,* 71, 2925, 1974.

42. **Harris, M.J., Higachu, T., and Rytting, J.H.,** Thermodynamic group contributions from ion pair extraction equilibria for use in the prediction of partition coefficients. Correlation of surface area with group contributions. *J. Phys. Chem.,* 77, 2694, 1973.

43. **Hermann, R.B.,** Theory of hydrophobic bonding. II. The correlation of hydrocarbon solubility in water with solvent cavity surface area, *J. Phys. Chem.,* 76, 2754, 1972.

44. **Hermann, R.B.,** Theory of hydrophobic bonding. I. The solubility of hydrocarbons in water, within the context of the significant structure theory of liquids, *J. Phys. Chem.,* 75, 363, 1971.

45. **Shrake, A. and Rupley, J.A.,** Environment and exposure to solvent of protein atoms. Lysozyme and Insulin, *J. Mol. Biol.,* 79, 351, 1973.

46. **Karplus, P.A. and Schulz, G.E.,** Prediction of chain flexibility in proteins, *Naturwissenschaften,* 72, 5, 212, 1985.

47. **Kabsch, W. and Sander, C.,** Dictionary of protein secondary structure: Pattern recognition of hydrogen-bonded and geometrical features, *Biopiolymers,* 22, 2577, 1983.

48. **Bernstein, F.C., Koetzle, T.F., Williams, G.J.B., Meyer, Jr. E.F., Brice, M.D., Rodgers, J.R., Kennard, O., Shimanouchi, T., and Tasumi, M.,** The protein data bank: A computer based archival file for macromolecular structures, *J. Mol. Biol.,* 112, 535, 1977.

49. **Chou, P.Y. and Fasman, G.D.,** β-Turns in proteins, *J. Mol. Biol.,* 115, 135, 1977.

Section XIII
Preparative Chromatography of Peptides and Proteins

PRACTICAL ASPECTS OF PREPARATIVE REVERSED-PHASE CHROMATOGRAPHY OF SYNTHETIC PEPTIDES

Carl A. Hoeger, Robert Galyean, Richard A. McClintock, and Jean E. Rivier

I. INTRODUCTION

The application of high-performance liquid chromatography (HPLC) technology in biochemistry has been instrumental in the isolation and eventual structural elucidation of a plethora of new peptides and proteins.[1] The need for scalar extrapolation from analytical to semipreparative and ultimately to the preparative separation of purified biologically active peptide and protein hormones, originally recognized by Burgus and Rivier,[2] is now well documented.[3-7] As new peptides were isolated and characterized, the need for their duplication by total synthesis became imperative, and it is here that the power and usefulness of preparative reversed-phase high-performance liquid chromatography (RPC) was unmistakably realized. While crude extracts from biological sources may contain a large number of components, most will be dissimilar enough (in molecular weight, hydrophobicity, and ionic charge) to allow their separation. The characteristic scarcity of the starting materials can, in most cases, further minimize the need for preparative separation. Crude synthetic peptides, on the other hand, can be generated in large quantities and when obtained using a solid-phase peptide synthesis approach (SPPS),[8] for example, can contain fewer undesired components, yet these impurities are more likely to be, by the very nature of their genesis, very closely associated (in structure) to the desired peptide. As a result, homogeneity in most cases cannot be achieved or be unequivocally proven. In fact, the crude peptidic preparations may contain the desired product as only a small fraction (1 to 70%) of the total. Since peptidic impurities cannot be tolerated for most physico-chemical and biological studies, we have spent considerable effort on the design of chromatographic methods directed at the separation of these impurities. Variations in pH of the buffers, chromatographic supports, gradient shapes, and organic modifiers have been examined in this laboratory over more than a decade on hundreds of compounds. We have derived from this experience general approaches that are delineated below.

II. EXPERIMENTAL

A. APPARATUS

The analytical chromatographic system consisted of two Waters Associates M-45 pumps and an Automated Gradient Controller; Houston Instruments Omniscribe strip chart recorder; Shimadzu Chromatopak-EIA integrator; Rheodyne 7125 injector and a Kratos 769Z variable-wavelength UV detector.

The preparative chromatographic systems consisted of:

1. Waters Associates: PrepLC/System 500 which was altered to allow flow-rate increments of

25 ml/min; Model 450 variable-wavelength UV detector; Linear instruments Model 455 strip chart recorder and an Eldex Laboratories Chromat-A-Trol Model II gradient maker.
2. Waters Associates: DeltaPrep Model 3000 instrument; Houston Instruments Omniscribe strip chart recorder and a Kratos Model 757 variable wavelength UV detector.

B. ANALYTICAL COLUMNS

The analytical columns (250×4.6 mm I.D.) were packed with Vydac 5-μm particle size, 300-Å pore size, C_{18} silica,[3] obtained from the Separations Group (Hesperia, CA).

C. PREPARATIVE CARTRIDGES

Empty polyethylene cartridges (30 cm × 5 cm I.D.) and frits (part numbers 50411 and 50421) obtained from Waters Associates were dry-packed in our laboratory with Vydac C_{18} derivatized silica (The Separations Group, Hesperia, CA), 15- to 20-μm particle size, and 300-Å pore size. For a discussion regarding packing materials for preparative chromatography, the reader is directed to Reference 3.

D. SOLVENT SYSTEMS

The aqueous TEAP 2.25, 5.2, 7.0 (an abbreviation for triethylammonium phosphate buffer at pH 2.25, 5.2 and 7.0, respectively) and the trifluoroacetic acid (TFA) buffer systems have been described earlier.[9-11] The TEAP 2.25 buffer was made from 0.9% phosphoric acid (v/v), 0.9% triethylamine; the TEAP 5.2 buffer was made from 0.4% phosphoric acid (v/v), 0.85% triethylamine (v/v); the TEAP 7.0 buffer was made from 0.1% phosphoric acid (v/v), 0.3% triethylamine (v/v). The 0.1 M ammonium acetate buffer was prepared by dissolving the appropriate amount of ammonium acetate (Fluka, puriss. p.a.) in distilled water. Solvent A was always the aqueous buffer; solvent B contained the organic modifier and was usually composed of 60% acetonitrile, 40% buffer A. Gradients are reported as changes in the percent B buffer linearly mixed with the A buffer. Analytical flow rates were 2 ml/min and preparative flow rates 85 to 100 ml/min. The wavelength at which the eluent was monitored is shown on the ordinate of the figures and the gradient shape is indicated as an overlay on figures.

E. PEPTIDES

Peptides were synthesized by the solid-phase approach either manually or on a Beckman 990 synthesizer using classical SPPS techniques.[12] In brief, peptides were assembled on either chloromethylated or methyl benzhydrylamine resins depending on whether the desired final product had a C-terminus carboxylic acid or carboxamide, respectively. Hydrogen fluoride cleavage and deprotection in the presence of scavengers yielded, after lyophilization, the crude peptidic preparations which were submitted to preparative chromatography.[13] The composition of the final purified products was determined by 4 N methanesulfonic acid (Pierce Chemical, Rockford, IL) hydrolysis at 110°C for 24 hr, followed by amino acid analysis.

F. SAMPLE PREPARATION AND LOADING

Approximately 1 to 3 g of the crude peptide was dissolved in a buffer (250 ml) whose strength in organic modifier was equal to or below that used for the equilibration of the cartridge prior to the run and filtered through a plug of glass wool. This solution, often slightly opalescent, was then loaded onto the radially compressed column through the pumps and chromatographed using the gradient conditions given in the individual figure legends.

III. DISCUSSION

For the purposes of this manuscript we will concern ourselves with only the preparative scale reversed-phase purification of 10 to 40 residue long peptides prepared by SPPS.[12] Therefore, crucial to the discussion that follows is the definition of the term *preparative scale*. In the context of this work, we define a preparative scale purification as one involving a crude peptide load of > 100 mg. In practice, we have used the instrumentation and techniques described to purify crude peptide preparations on amounts as low as 10 mg and as high as 3.5 g per load. For comparative purposes, analytical scale typically ranges from 10 to 100 µg and is used in this laboratory almost exclusively as a method for defining preparative scale conditions and the subsequent screening of fractions obtained during the purification. Semi-preparative separations (crude peptide loads of a few milligrams up to 100 mg *utilizing a semipreparative column*) suffer in our view from a number of shortcomings: (1) prepacked columns 1 in. in diameter containing either 5 or 10 µm C_{18} particles are expensive; (2) seem to be highly sensitive to the nature of the repetitive load; finally, (3) loss of resolution due to alteration of the top of the columns occurs relatively quickly, thus increasing the expense dramatically unless proper guard columns are used. In this report, we will concern ourselves exclusively with the use of octadecyl (C_{18}) reversed-phase silica-based sorbents. This is not due to a lack of experience with other sorbents, as this aspect has been reported elsewhere;[3] it is simply because the vast majority of peptides synthesized or isolated in our laboratory can be purified quite effectively on a sorbent of this type.

A. INITIAL APPROACH

A typical analytical HPLC chromatogram of a crude peptide toxin run in 0.1% TFA/CH_3CN is presented in Figure 1 and is indicative of the problems one must address in a purification. This peptide is the tricyclic ω-conotoxin GVIA (see Table 1). Our synthesis of this peptide, first isolated and characterized from the fish-hunting snail *Conus geographus*, has been published.[10] The starred peak corresponds to the desired product (determined by coelution with a sample of the native peptide). Quite often the major component of a peptide mixture has associated with it a large number of closely related impurities detected as a broad absorbance envelope, underlying that of the desired major component. The challenge that arises is to separate the desired peptide from those undesired species accounting for the envelope absorption. Even then, minor products may be still associated with the desired peptide. Thus, the initial task is to find a solvent system and associated conditions that will allow maximization of purity to be achieved.

B. CHROMATOGRAPHIC SOLVENT SELECTION: TEAP 2.25/CH_3CN AND 0.1% TFA/CH_3CN; DIFFERENCES IN SELECTIVITY

Early work in this laboratory on the preparative purification of peptides dealt with the identification and application of solvent systems and conditions that would maximize purity; this strategy has been described elsewhere[4] (see Experimental for a brief description). For utmost efficiency, analytical isocratic conditions that will yield maximum information on the composition of the mixture around the desired product have to be determined; generally this is obtained when the desired product is eluted with a solute capacity factor (k′) between 4 and 8. We have found that analytical HPLCs run using 0.1% TFA/CH_3CN buffers are particularly convenient; this is in part due to the transparency of this buffer system at 210 nm allowing for high sensitivity, extended column life span, and reproducibly good separations. Once the isocratic analytical conditions have been determined, successive and rapid assessment of the identity and purity of the fractions obtained from the preparative HPLC purification can be carried out. Preparative scale purification is run first using TEAP 2.25/CH_3CN buffers, since we have often found TEAP to give higher resolution and different selectivity than the corresponding TFA buffers. Gradient conditions are selected on the basis of our findings with the TFA buffer. Due to the strong elutropic characteristics of the TEAP buffer, preparative gradient conditions are generally started at about 10% lower acetonitrile concentration than the isocratic analytical conditions determined in a TFA system. Coupling this condition with a linear AB gradient slope of 1% CH_3CN increase per 300 ml of eluting solvent (20% increase in B buffer in 60 min with

TABLE 1

Triethylammonium Phosphate (TEAP) Purification[a] of Selected Synthetic Peptides and Gradient Conditions Used for their Isolation

Peptide	Molecular weight	TEAP buffer pH; gradient[b]
GnRH[e] and analogs		
GnRH: Glp-His-Trp-Ser-Tyr-Gly-Leu-Arg-Pro-Gly-NH₂[f]	1181	2.25; 16—60′—32% B[d]
[DTrp⁶,Pro⁹-NHEt]-GnRH	1282	2.25; 30—40′—50% B[d]
"Nal-Glu" GnRH Antagonist: Ac-DNal-D4Cpa-D3Pal-Ser-Arg-D-4-p-methoxybenzoyl-2-aminobutyric-acid-Leu-Arg-Pro-DAla-NH₂[f]	1486	2.25; 40—60′—60% B[d]
hpGRF (1-40)OH[e] Tyr-Ala-Asp-Ala-Ile-Phe-Thr-Asn-Ser-Tyr-Arg-Lys-Val-Leu-Gly-Gln-Leu-Ser-Ala-Arg-Lys-Leu-Leu-Gln-Asp-Ile-Met-Ser-Arg-Gln-Gln-Gly-Glu-Ser-Asn-Gln-Glu-Arg-Gly-Ala-OH	4538	2.25; 25—60′—55% B[d]
oCRF[e] Ser-Gln-Glu-Pro-Pro-Ile-Ser-Leu-Asp-Leu-Thr-Phe-His-Leu-Leu-Arg-Glu-Val-Leu-Glu-Met-Thr-Lys-Ala-Asp-Gln-Leu-Ala-Gln-Gln-Ala-His-Ser-Asn-Arg-Lys-Leu-Leu-Asp-Ile-Ala-NH₂	4670	7.0; 22—60′—39% B
α-Helical CRF (9-41) Asp-Leu-Thr-Phe-His-Leu-Leu-Arg-Glu-Met-Leu-Gly-Met-Ala-Lys-Ala-Glu-Gln-Glu-Ala-Gln-Ala-Ala-Leu-Asn-Arg-Leu-Leu-Leu-Glu-Glu-Ala-NH₂	3755	6.5; 30—60′—42.5% B
Inhibin fragments H-Pro-Inhibin α(40-58)G-Y-OH: Gly-Ser-Glu-Pro-Glu-Glu-Glu-Glu-Asp-Val-Ser-Gln-Ala-Ile-Leu-Phe-Pro-Ala-Thr-Gly-Tyr-OH	2206	5.2; 20—60′—32.5% B
porcine Inhibin βB(13-35): Arg-Gln-Gln-Phe-Phe-Ile-Asp-Phe-Arg-Leu-Ile-Gly-Trp-Ser-Asp-Trp-Ile-Ile-Ala-Pro-Thr-Gly-Tyr-OH	2830	7.0; 35—60′—50% B
Misc. human atrial natriuretic factor: Ser-Leu-Arg-Arg-Ser-Ser-Cys-Phe-Gly-Gly-Arg-Met-Asp-Arg-Ile-Gly-Ala-Gln-Ser-Gly-Leu-Gly-Cys-Asn-Ser-Phe-Arg-Tyr-OH	3077	5.2; 10—60′—25% B
Conotoxin GVIA: Cys-Lys-Ser-Hyp-Gly-Ser-Ser-Cys-Ser-Hyp-Thr-Ser-Tyr-Asn-Cys-Cys-Arg-Ser-Cys-Asn-Hyp-Tyr-Thr-Lys-Arg-Cys-Tyr-NH₂[f]	3035	2.25; 0—60′—15% B
[Glp¹]-pHGnRH(1-26)[e,c]: Glp-His-Trp-Ser-Tyr-Gly-Leu-Arg-Pro-Gly-Gly-Lys-Arg-Asp-Ala-Glu-Asn-Leu-Ile-Asp-Ser-Phe-Gln-Glu-Ile-Val-OH[f]	3014	7.0; 25—60′—45% B[d]
Neuromedin K[c]: Asp-Met-His-Asp-Phe-Phe-Val-Gly-Leu-Met-OH	1211	7.0; 40—30′—60% B[d]
salmon MCH[c]: Asp-Thr-Met-Arg-Cys-Met-Val-Gly-Arg-Val-Tyr-Arg-Pro-Cys-Trp-Glu-Val-OH	2099	2.25; 25—60′—45% B[d]

a Unless noted, all compounds were initially purified in TEAP 2.25 prior to TEAP buffer indicated in Table 1: chromatographic sorbent used was C_{18}, 300 Å, 15 to 20 μm Vydac; the fractions from this treatment that were deemed "acceptable" were pooled and desalted using either the 0.1% TFA or NH_4OAc system.

b 2.25 = TEAP 2.25; 7 = TEAP 7.0 = A; 5.2 = TEAP 5.2 = A; 6.5 = TEAP 6.5 = A; B = acetonitrile unless otherwise noted (see footnote d).

c compound insoluble at acidic pH, therefore, TEAP 2.25 purification could not be done; the fractions from this treatment that were deemed "acceptable" were pooled and desalted using NH_4OAc.

d B = acetonitrile in A (60:40).

e hpGRF: human pancreatic (tumor) growth hormone releasing factor; oCRF: ovine corticotropin releasing factor; pHGnRH: gonadatropin releasing hormone associated protein; GnRH: gonadatropin releasing hormone; MCH: melanin concentrating hormone.

f -NH_2 denotes C-terminal amide; -OH denotes C-terminal carboxyl group; Ac-, acetyl; Glp, pyroglutamic acid; Hyp, hydroxyproline; DNal, 3-(2-naphthyl)-D-alanine; DCpa, 3-(4 chlorophenyl)-D-alanine; DPal, 3-(3-pyridyl)-D-alanine.

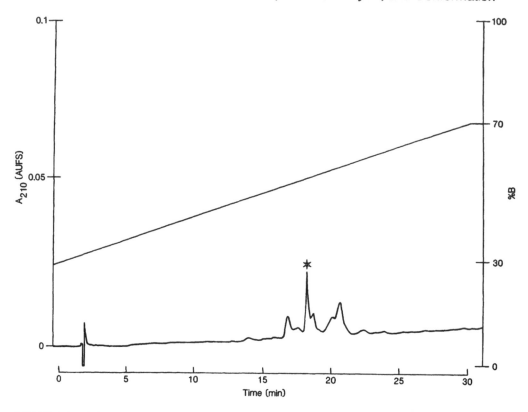

FIGURE 1. Load: Crude lyophilized, Conotoxin GVIA (15 μl, ca. 2 μg) obtained after cyclization.[14] Column: Vydac 5 μm, C_{18}, 250 × 4.6 mm I.D. Buffers: 0.1% TFA in water = A; 0.1% TFA in acetonitrile-water (60:40) = B. Linear AB gradient: starting from 30% B at time 0 min and ending at 70% B at 30 min (30%-30'-70% B). Flow rate: 2.0 ml/min, 3000 p.s.i. back pressure. Detector: 0.1 AUFS (210 nm). Chart speed: 1.0 cm/min. Star denotes peak of interest.

a solvent flow of 90 to 100 ml/min) will provide the chromatographer with preparative conditions that are essentially optimized. While this practice works very well for most separations, individual chromatographers will modify slightly the procedure by altering the gradient slope; some examples of the diversity of gradients employed can be found in Table 1. The fractions obtained from this purification step are screened analytically and the appropriate fractions pooled for desalting with 0.1% TFA/CH_3CN. Although a high degree of purity can be achieved with the TEAP 2.25 system, the TFA step is required to free the peptide from any TEAP salt. Purification in two systems (TEAP followed by TFA) and analysis in the second (TFA) was found in general to minimize the probability of missing any impurities and maximize the probability of obtaining a pure final product while offering good recovery.

The application of this strategy can best be illustrated by examining the purification of ω-Conotoxin GVIA (Figure 1). The analytical isocratic conditions were determined for this peptide and the crude material was applied to a C_{18} cartridge. The compound was then eluted using TEAP 2.25/CH_3CN buffers; the profile for this purification is given in Figure 2. Based on the absorbance profile from the previously obtained analytical HPLC profile (Figure 1), it was expected that fractions 9, 10, and 11 would contain the desired peptide, as we have found that relative elution order does not generally change dramatically in going from TFA to TEAP for linear 10- to 50-residue peptides; however, analysis of these fractions by isocratic HPLC indicated that the peptide was not present in these fractions, but rather was located in fractions 4 and 5 (Figure 3). These two fractions were pooled and preparatively desalted using the volatile 0.1% TFA/CH_3CN buffer system (Figure 4); the fractions obtained from this procedure that contained the desired peptide (as determined by analytical HPLC; Figure 5) were then pooled

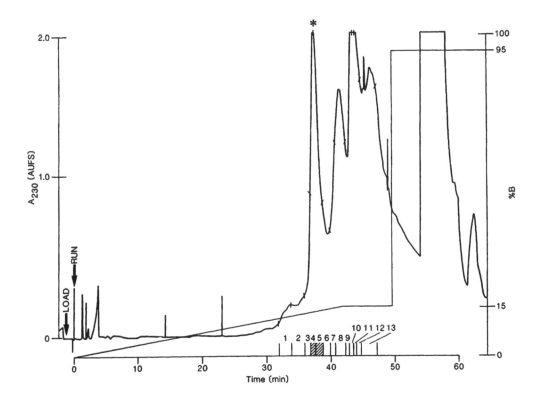

FIGURE 2. Load: lyophilized crude Conotoxin GVIA (100 ml, 930 mg). Column: 30 × 5 cm I.D., packed with Vydac 15 to 20 μm, C_{18} material. Buffers: TEAP 2.25 = A; acetonitrile = B. Gradient: 0%-42'-15%-8'-15% B, followed by 95% B wash (to remove hydrophobic materials from the cartridge). Flow-rate: 100 ml/min; 25 atm column pressure. Detector 2.0 AUFS (230 nm). Chart speed: 0.5 cm/min. Hatched areas denote fractions containing peak of interest (starred peak), as determined by analytical HPLC (Figure 3).

and lyophilized to give purified ω-Conotoxin GVIA. The analytical HPLC of the final product obtained is illustrated in Figure 6; the synthetic material was demonstrated to be coeluted with the natural product and had the anticipated biological activity.[14] The reason for the unusual behavior of this peptide in TEAP 2.25 (i.e. the early elution of the desired product out of the peptide by-product envelope) is not important to the ensuing discussion; however, it does point out the importance of isocratic screening of fractions and the advantages of using a two buffer system purification scheme, even though such unique solvent selectivity (which seems to be characteristic for highly constrained peptide structures) is not commonly seen for linear peptides.

C. THE USE OF pH EFFECTS IN THE PURIFICATION OF SYNTHETIC PEPTIDES

While the purification of the vast majority of peptides encountered can be achieved readily by the use of the aforementioned strongly acidic solvent system, occasionally one is confronted with a peptide, the solubility, and/or stability of which are marginal under these conditions. In general, peptides will fall into one of three categories, as arbitrarily determined by their isoelectric point (pI): acidic (pI < 6), neutral (6.5 < pI < 7.5), or basic (pI > 7.5).[15] Basic peptides usually can be purified with excellent results in the acidic mileu previously described. Neutral and acidic peptides (as well as acid-sensitive peptides), however, present more of a challenge as their solubility (or stability) can be poor at the pHs described above. When these situations arise, we have turned to the use of the TEAP solvent system at various pHs. One such group of

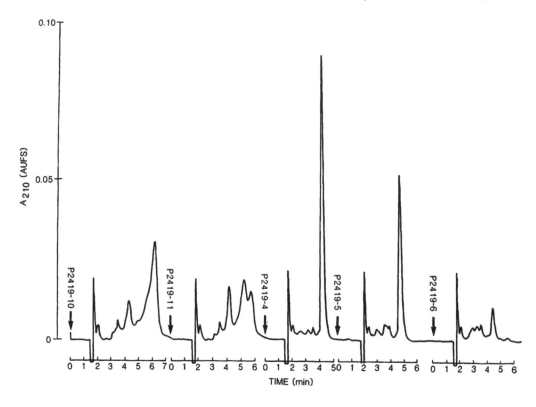

FIGURE 3. Isocratic analytical HPLC analysis of fractions 4, 5, 6, 10, and 11 of Conotoxin GVIA from preparative purification shown in Figure 2. Column: Vydac 15 to 20 μm, C_{18}, 250 × 4.6 mm I.D. Buffers: 50% aqueous acetonitrile containing 0.1% TFA. Flow-rate: 2 ml/min; 3000 p.s.i. back pressure. Detector: 0.1 AUFS (210 nm). Chart speed: 1 cm/min.

peptides are those related to the gonadotropin releasing hormone (GnRH) precursor protein, pHGnRH, a 69-residue peptide,[16] a specific homologous fragment of which is [Glp¹]-pHGnRH(1-26) (Table 1). This peptide, extremely insoluble in TEAP 2.25/CH_3CN, could be easily dissolved and chromatographed using TEAP at pH 7.0/CH_3CN (data not shown). The sharp, well-resolved HPLC-UV absorbances that were obtained using this buffer system indicated that TEAP 7.0 is a viable alternative for purification of acidic, hydrophobic peptides. With the knowledge that these purified peptides would probably not be soluble under the acidic conditions usually employed in the desalting process, we reinvestigated the use of ammonium acetate $(NH_4OAc)^2$ as an elutropic, volatile salt for desalting these purified fractions. Using an identical gradient to that employed for basic or neutral peptides with 0.1% TFA/CH_3CN (10% to 90%B in 15'), we can conclude on the basis of recovery studies and the purity of the final product that these higher pH buffer systems are equally effective (see Table 1 for other TEAP 7.0 examples).

The use of pH permutations to achieve purification is not limited simply to neutral and acidic peptides; this procedure can also be applied to basic peptides to realize enhanced purity levels. Adjusting the pH of the TEAP buffer from 2.25 to 5.3 or 6.5 can have marked effects on the purification of a number of peptides; some examples are given in Table 1. For example, CRF[17] and GRF[18] analogs are routinely subjected to both a TEAP 2.25 *and* a TEAP 5.2/6.5 step prior to desalting. This extra step aids considerably in the purification of these peptide analogs in that the main component can be separated, under these conditions, from the impurities which belong to the underlying envelope of compounds detectable at 210 nm. Whether some of the closely associated impurities result from deletions of acidic and/or basic residues has not been documented. Adjustments in the pH employed in the purification will change the net charge on

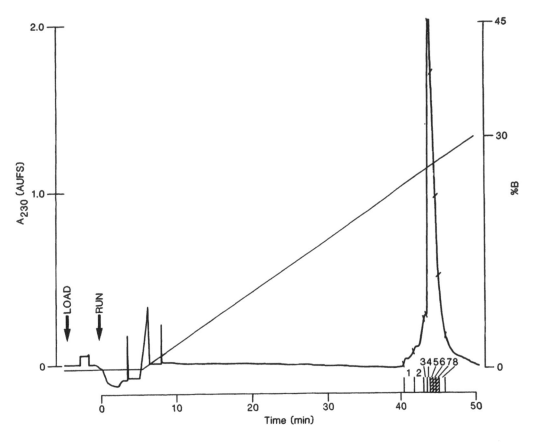

FIGURE 4. Load: Preparative desalting of Conotoxin GVIA fractions 4 and 5 from preparative purification shown in Figure 2. Column: 30 × 5 cm I.D., packed with Vydac 15 to 20 μm, C_{18} material. Buffers: 0.1% TFA in water = A; acetonitrile = B. Gradient: 0% - 5' - 0% - 40' - 30% B. Flow-rate: 100 ml/min; 25 atm column pressure. Detector 2.0 AUFS (230 nm). Chart speed: 0.5 cm/min. Hatched areas denote fractions containing peak of interest as determined by analytical HPLC (Figure 5).

the individual species present, thus possibly altering both their overall hydrophobic character and/or their overall conformations and accordingly their interaction at the reversed-phase sorbent/mobile phase interface.

IV. SUMMARY

We have endeavored to provide here a simple guide to the preparative scale purification of peptides. While the purpose of this manuscript is to aid the novice in his initial search for HPLC conditions that will achieve adequate separations, it must be remembered that there is no substitute for experience. If all peptides were alike, purification would be a rather straightforward task; alas, that is not the case. New and different peptides present new and different challenges for their purification and while the methods presented herein make a good foundation from which to start, innovation and observation will always prove to be an important factor in the final success as measured by the degree of purity of the final product.

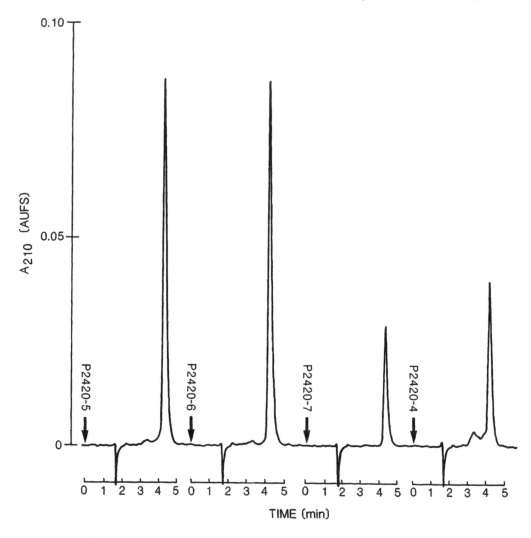

FIGURE 5. Isocratic analytical HPLC analysis of fractions 5 to 7 of Conotoxin GVIA from preparative desalting shown in Figure 4. Column: Vydac 15 to 20 μm, C_{18}, 25 × 0.46 cm I.D. Buffers: 50% aqueous acetonitrile containing 0.1% TFA. Flow-rate: 2 ml/min; 3000 p.s.i. back pressure. Detector: 0.1 AUFS (210 nm). Chart speed: 1 cm/min.

ACKNOWLEDGMENTS

Research supported by NIH Grants AM26741 and HD13527 and NIH contract NO1-HD-7-2907. We thank John Dykert, Ron Kaiser, and Charleen Miller for their expert technical assistance.

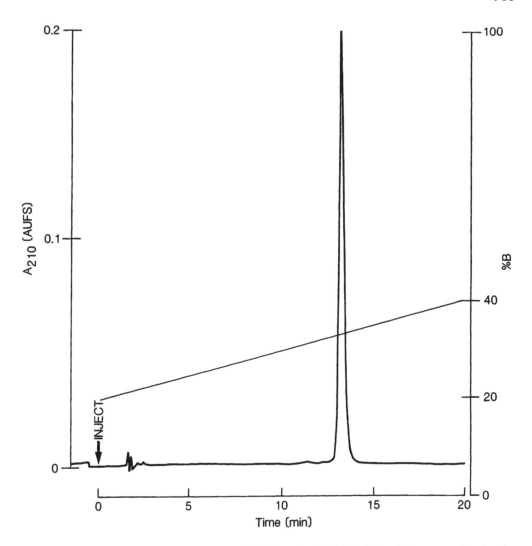

FIGURE 6. Load: lyophilized pool of fractions 5 to 7 of Conotoxin GVIA (15 μl, 15 μg) from preparative desalting shown in Figure 4. Column: Vydac 15 to 20 μm, C_{18}, 250 × 4.6 mm I.D. Buffers: TEAP 2.25 = A; acetonitrile in A (30:70) = B. Gradient: 20% B-20'-40% B. Flow-rate: 2 ml/min; 3000 p.s.i. back pressure. Detector: 0.2 AUFS (210 nm). Chart speed: 1 cm/min. Overall yield: 25.6 mg of > 98% purity from 930 mg.

REFERENCES

1. **Rivier, J., and McClintock, R.,** Purification and characterization of biologically active peptides, in *The Use of HPLC in Protein Purification and Characterization,* Vol. 14, Kerlavage, A.R., Ed., John Wiley & Sons, New York, 1989, 77—105.
2. **Burgus, R. and Rivier, J.,** Use of high pressure liquid chromatography in the purification of peptides, in *Peptides 1976,* Loffet, A., Ed., Editions de l'Universite de Bruxelles, Belgium, 1976, 85.
3. **Rivier, J. and McClintock, R.,** Reversed-phase high-performance liquid chromatography of insulins from different species, *J. Chromatogr.,* 268, 112, 1983.
4. **Rivier, J., McClintock, R., Galyean, R., and Anderson, H.,** Reversed-phase high-performance liquid chromatography: preparative purification of synthetic peptides, *J. Chromatogr.,* 288, 303, 1984.

5. **Majors, R.,** Practical aspects of preparative liquid chromatography, *LC, Liquid Chromatography HPLC Magazine,* 3, 862, 1985.

6. **Sitrin, R., DePhillips, P., Dingerdissen, J., Erhard, K., and Filan, J.,** Preparative liquid chromatography, a strategic approach, *LC-GC, Magazine of Liquid & Gas Chromatography,* 4, 530, 1986.

7. **Verzele, M. and Dewaele, C.,** Preparative liquid chromatography, *LC, Liquid Chromatography HPLC Magazine,* 3, 22, 1985.

8. **Merrifield, R.B.,** Solid phase peptide synthesis. I. The synthesis of a tetrapeptide, *J. Am. Chem. Soc.,* 85, 2149, 1963.

9. **Bennett, J.P.J., Hudson, A.M., McMartin, C., and Purdon, G.E.,** Use of octadecasilyl-silica for the extraction and purification of peptides in biological samples, *Biochem. J.,* 168, 9, 1977.

10. **Hoeger, C., Galyean, R., Boublik, J., McClintock, R., and Rivier, J.,** Preparative reversed-phase high-performance liquid chromatography: effects of buffer pH on the purification of synthetic peptides, *BioChromatography,* 2, 134, 1987.

11. **Rivier, J.,** Use of trialkylammonium phosphate (TAAP) buffers in reversed-phase HPLC for high resolution and high recovery of peptides and proteins, *J. Liq. Chromatogr.,* 1, 343, 1978.

12. **Stewart, J.M. and Young, J.D.,** *Solid Phase Peptide Synthesis,* 2nd ed., Pierce Chemical Co., Rockford, IL, 1984.

13. Peptides containing disulfide bridges were, after acidolytic treatment, allowed to oxidize to the desired Cys-Cys species. See Reference 14 for a typical example of this procedure.

14. **Rivier, J., Galyean, R., Gray, W.R., Azimi-Zonooz, A., McIntosh, J.M., Cruz, L.J., and Olivera, B.M.,** Neuronal calcium channel inhibitors. Synthesis of ω-conotoxin GVIA and effects of ^{45}Ca uptake by synaptosomes, *J. Biol. Chem.,* 262, 1194, 1987.

15. In practice, it is more convenient to count the acidic (i.e. Asp, Glu, etc.) and basic (i.e. Arg, Lys, His, etc.) groups present and determine the net charge on the peptide at a given pH than to actually determine its pI.

16. **Millar, R.P., Wormald, P.J., and Milton, R.C.,** Stimulation of gonadotropin release by a non-GnRH peptide sequence of the GnRH precursor, *Science,* 232, 68, 1986.

17. **Rivier, J., Rivier, C., and Vale, W.,** Synthetic competitive antagonists of corticotropin releasing factor: Effect on ACTH secretion in the rat, *Science,* 224, 889, 1984.

18. **Rivier, J., Rivier, C., Galyean, R., Yamamoto, G., and Vale, W.,** Potent long-acting growth hormone releasing factor (GRF) analogs, in *Annals of the New York Academy of Sciences, Conference on Vasoactive Intestinal Peptide and Related Peptides,* March 2—4, 1987, 44—50, 527, 1988.

PURIFICATION OF HISTIDINE-RICH HYDROPHILIC PEPTIDES

Anita L. Hong, Michael M. Brasseur, and Djohan Kesuma

I. INTRODUCTION

Histidine-rich peptides (HRPs) from human parotid saliva were reported by Pollock et al. to exert antifungal activity against *Candida albicans*.[1,2] Five of these peptides, HRP 2 through 6, were synthesized (Table 1). These peptides ranged in size from 20 to 38 amino acid residues, containing six or seven histidines. Synthesis was performed using the solid-phase strategy, whereby the t-butyloxycarbonyl (t-Boc) protected amino acids were activated as hydroxybenzotriazole (HOBT) esters. Coupling was performed in N-methylpyrrolidinone.[3] The HRPs obtained from the synthesis were contaminated with side products such as peptides of deleted sequences. These peptides are extremely hydrophilic because of the presence of multi-histidines and, therefore, exhibit short retention times in reversed-phase HPLC (RPC), making purification by this technique alone difficult. To improve the separation of the peptides, a multi-column approach was applied to their purification, involving the combined use of separation modes which utilize different selectivities. Thus, the peptides were first subjected to cation-exchange chromatography (CEC) which separates solutes according to their net positive charge, followed by RPC as a final purification and desalting step. This article describes the successful purification, in high yield, of HRP-6 (Table 1) using this multi-column approach to preparative peptide purification.

II. PROCEDURE

Following hydrogen fluoride (HF) cleavage of 2.14 g of the corresponding peptide-resin, crude HRP-6 was dissolved in dilute acetic acid (10% v/v). The pH of the peptide solution was adjusted to 3.0 using neat trifluoroacetic acid (TFA). This pH adjustment ensures binding of the desired peptide to a preparative PolySULFOETHYL Aspartamide (PolySULFOETHYL A) strong cation-exchange (SCX) column (Poly LC, Columbia, MD).[4] PolySULFOETHYL A is a silica-based material with a bonded coating of poly(2-sulfoethyl aspartamide) which is a hydrophilic, anionic polymer, containing sulfonate functionalities. The peptide solution was loaded onto the SCX column at a flow-rate of 15 ml/min and eluted with Buffer A (5 mM KH$_2$PO$_4$, pH 3.0, containing 20% [v/v] acetonitrile) until a steady baseline was obtained. The peptide was then subjected to linear gradient elution (45 mM NaCl/min for 10 min, followed by 5 mM NaCl/min for a further 90 min). Acetonitrile was included in the ion-exchange buffers since it reduces the viscosity of the mobile phase, thus lowering the back pressure. Figure 1 shows the elution profile of this strong cation-exchange preparative run.

The collected fractions from the preparative run were analyzed for the presence of the HRP product by linear gradient analysis (Figure 2). Fractions 1 through 16 (Figures 1 and 2) were pooled and then loaded directly onto a preparative reversed-phase column (Aquapore RP300 C$_8$; Brownlee Labs., Santa Clara, CA). The peptide was further purified, and desalted, using a linear water/ acetonitrile gradient (0.25% acetonitrile/min). The elution profile resulting from this reversed-phase separation is shown in Figure 3. Purity of the peptide fractions was determined by linear gradient analysis (Figure 4).

TABLE 1
Peptide Sequences

HRP-2: H-Asp-Ser-His-Glu-Lys-Arg-His-His-Gly-Tyr-Arg-Arg-Lys-Phe-His-Glu-Lys-His-His-Ser-His-Arg-Glu-
Phe-Pro-Phe-Tyr-Gly-Asp-Tyr-Gly-Ser-Asn-Tyr-Leu-Tyr-Asp-Asn-OH

HRP-3: H-Asp-Ser-His-Ala-Lys-Gly-His-His-Gly-Tyr-Lys-Arg-Lys-Phe-His-Glu-Lys-His-His-Ser-His-Arg-Gly-
Tyr-Arg-Ser-Asn-Tyr-Leu-Tyr-Asp-Asn-OH

HRP-4: H-Lys-Arg-His-His-Gly-Tyr-Lys-Arg-Lys-Phe-His-Glu-Lys-His-His-Ser-His-Arg-Gly-Tyr-Arg-Ser-Asn-
Tyr-Leu-Tyr-Asp-Asn-OH

HRP-5: H-Asp-Ser-His-Ala-Lys-Arg-His-His-Gly-Tyr-Lys-Arg-Lys-Phe-His-Glu-Lys-His-His-Ser-His-Arg-Gly-
Tyr-OH

HRP-6: H-Lys-Arg-His-His-Gly-Tyr-Lys-Arg-Lys-Phe-His-Glu-Lys-His-His-Ser-His-Arg-Gly-Tyr-OH

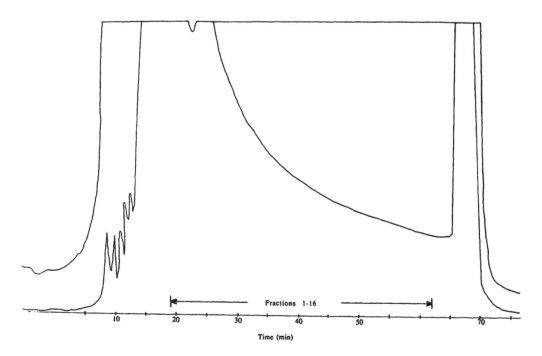

FIGURE 1. Preparative strong cation-exchange chromatography of HRP-6. The preparative HPLC consisted of Rainin analytical systems, a dual chamber mixer, a preparative flow cell, a Gilson Model 201 fraction collection, a Macintosh Plus computer and a control interface module. Column: PolySULFOETHYL A, 25 × 2.1 cm I.D., 12-μm particle size, 300-Å pore size. Conditions: linear AB gradient (0 to 30% B in 10 min [equivalent to 45 mM NaCl/min], followed by 30 to 60% B in 90 min [equivalent to 5 mM NaCl/min]), where Buffer A is 5 mM KH$_2$PO$_4$, pH 3.0, containing 20% (v/ v) acetonitrile and Buffer B is Buffer A with the addition of 1.5 M sodium chloride; flow-rate, 15 ml/min; absorbance at 220 nm. Fraction size: 45 ml.

From the preparative reversed-phase run, fractions 16 through 22 (Figures 3 and 4), containing peptide of the desired purity, were pooled. The acetonitrile was evaporated and the residual aqueous solution was lyophilized. Based on HPLC, HRP-6 with a purity greater than 99% was obtained (621 mg, equivalent to 70% yield; Table 2). Analytical reversed-phase chromatograms of the crude peptide and the purified peptide are shown in Figure 5.

III. DISCUSSION

Although IEC has become increasingly popular for the analysis of proteins in recent years, less attention has been paid to its application for peptide separations, due mainly to the

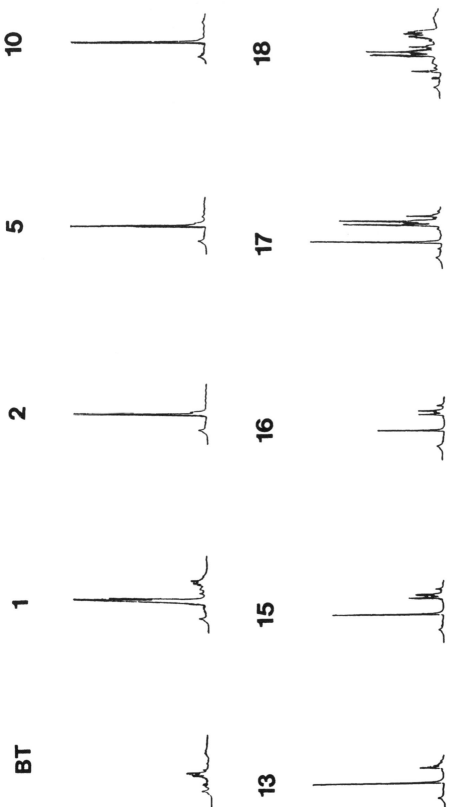

FIGURE 2. Fraction analysis of preparative cation-exchange chromatography run shown in Figure 1. Mobile phase buffers: see Figure 1. Fractions were analyzed used a 45-min gradient of 0 to 60% Buffer B (20 mM NaCl/min). Fractions 1 to 16 were subsequently pooled and applied to a preparative reversed-phase column for further purification (Figure 3). BT denotes breakthrough peak.

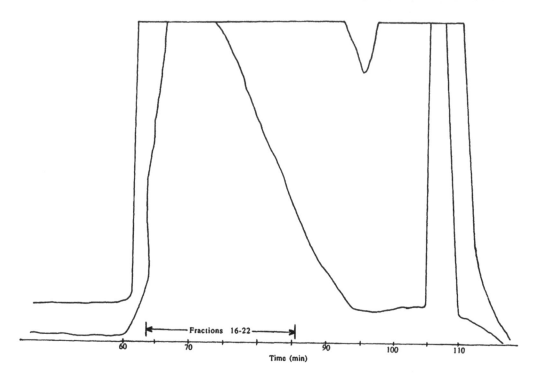

FIGURE 3. Preparative RPC of pooled HRP-6 fractions from Figure 1. Column: Aquapore RP300 C$_8$, 35 × 3.0 cm I.D., 20-μm particle size, 300-Å pore size. Conditions: linear AB gradient from 0% B (0.25% B/min), where Eluent A is 1% aq. trifluoroacetic acid (TFA) and Eluent B is 1% TFA in acetonitrile; flow-rate, 15 ml/min; absorbance, 220 nm. Fraction size: 45 ml.

availability of reversed-phase packings for such purposes and the necessity of sample desalting prior to further analysis. However, hydrophilic peptides which exhibit short retention times in RPC, such as those described in this article, are difficult to separate by this technique alone. Under these circumstances, IEC and RPC are often complementary, i.e., their combined use can provide optimal separation of a peptide mixture.[5,6]

Strong cation-exchange chromatography is the most useful mode of IEC for multistep peptide separations due to the ability of SCX packings to retain their negatively charged character in the acidic to neutral pH range. At pH 3.0 or below, basic peptide residues (His, Lys, Arg, amino-termini) are protonated and therefore positively charged, whereas acidic residues (Asp, Glu, carboxy-termini) remain unchanged. Thus, a strong cation-exchange column such as the PolySULFOETHYL A could be used in conjunction with RPC for separation of hydrophilic peptides containing basic residues (Table 1). Using a linear sodium chloride gradient, peptides were eluted in order of increasing net positive charge.

The final separation and desalting step was accomplished using RPC. For basic hydrophilic peptides, we recommend using 1% TFA in the eluting solvents instead of 0.1% TFA. The increased TFA concentration enhances the ion-pairing effect of the trifluoroacetate anionic counterion,[7] increasing the retention time of the peptides and facilitating the separation.

Using the methodology described in this article, we have successfully separated the histidine-rich peptides HRP-2 to HRP-6 (Table 2). HRP-6 is the shortest of all these peptides and the crude HRP-6 peptide was much better than the rest. Hence, the purification yield of this peptide was considerably higher than that of the other histidine-rich peptides.

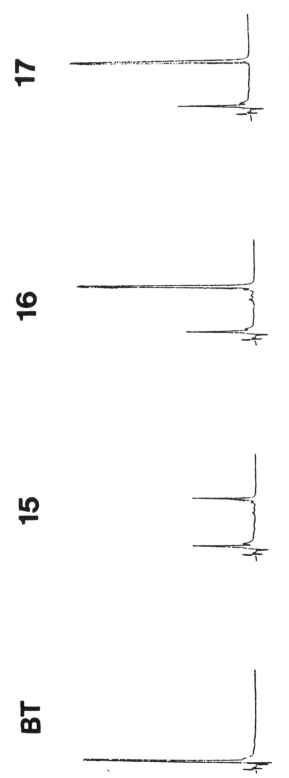

FIGURE 4. Fraction analysis of preparative reversed-phase chromatography run shown in Figure 3. Analyses were carried out on an Applied Biosystems Model 130A Separations System using an Aquapore RP300 C$_8$ column (220 × 2.1 mm I.D., 7-μm particle size, 300-Å pore size). Mobile phase solvents: see Figure 3. Fractions were analyzed using a 15-min gradient of 8 to 25% B (1.3% B/min). Fractions 16 to 22 were subsequently pooled. BT denotes breakthrough peak.

FIGURE 4 (continued)

TABLE 2
Purification Yield of the HRPS

Peptide name	No. of amino acids	Peptide-resin cleaved (g)	Peptide loaded (g)	Purified peptide (mg)
HRP-2	38	3.26	1.49	169 (11%)
HRP-3	30	3.00	1.53	282 (18%)
HRP-4	28	2.56	1.1	543 (50%)
HRP-5	24	2.55	1.05	441 (42%)
HRP-6	20	2.14	0.888	621 (70%)

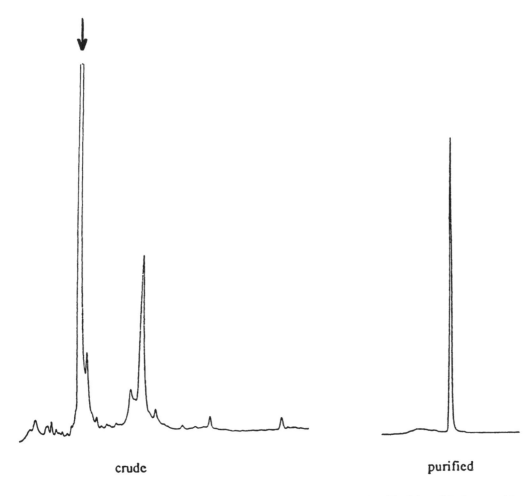

crude   purified

FIGURE 5. Comparison of crude and purified HRP-6. Column: Aquapore RP300 C_8, 220 × 2.1 mm I.D., 7-μm particle size, 300-Å pore size. Conditions: linear AB gradient from 0% B (1.3% B/min), where Eluent A is 0.1% aq. TFA and Eluent B is 0.08% TFA in acetonitrile; flow-rate, 0.23 ml/min; absorbance at 220 nm. The arrow denotes the peak of interest in the crude peptide.

REFERENCES

1. **Santarpia III, R. P., Brant, E. C., Lal, K., Brasseur, M. M., Hong, A. L., and Pollock, J. J.,** A comparison of the inhibition of blastospore viability and germ-tube development in *Candida albicans* by histidine peptides and ketoconazole, *Archs. Oral Biol.,* 33, 567, 1988.

2. **Lal, K., Colburn, J., Brasseur, M. M., Hong, A. L., and Pollock, J. J.,** Analytical separation of purified synthetic salivary HRPs, *J. Dental Res.,* Abstract 68, 405, 1989.

3. **Geiser, T.H., Bergot, J.B., and Otteson, K.M.,** Automation of solid-phase peptide synthesis, in *Macromolecular Sequencing and Synthesis, Selected Methods and Applications,* Alan R. Liss, 1988, 199.

4. **Alpert, A. J.,** Cation-exchange chromatography of peptides on poly(2-sulfoethyl aspartamide)-silica, *J. Chromatogr.,* 443, 85, 1988.

5. **Mant, C. T. and Hodges, R. S.,** Optimization of peptide separations in high-performance liquid chromatography, *J. Liq. Chromatogr.,* 12, 139, 1989.

6. **Alpert, A. J.,** Cation-exchange high-performance liquid chromatography of peptides, this publication.

7. **Guo, D., Mant, C. T., and Hodges, R. S.,** Effects of ion-pairing reagents on the prediction of peptide retention in reversed-phase high-performance liquid chromato-graphy, *J. Chromatogr.,* 386, 305, 1987.

PREPARATIVE REVERSED-PHASE GRADIENT ELUTION CHROMATOGRAPHY ON ANALYTICAL COLUMNS

Robert S. Hodges, T. W. Lorne Burke, Colin T. Mant, and
Sai M. Ngai

I. INTRODUCTION

There has been an exponential growth in recent years in the use of synthetic peptides and native protein fragments in biochemistry, immunology, and in the pharmaceutical and biotechnology industries. Thus, there has also been a concomitant requirement for rapid and efficient peptide purification procedures. The excellent resolving power and separation time of reversed-phase chromatography (RPC), coupled with the availability of volatile mobile phases, has made this mode of HPLC the favored method for both analytical and preparative separations of peptides. Most researchers would probably wish to carry out both analytical and preparative peptide separations on analytical equipment and columns, taking advantage of the powerful resolving capability of small particle size (5 to 10 μm) reversed-phase packings.[1-3] In addition, for purification of substantial levels of a desired peptide, researchers would ideally prefer to be able to optimize a separation based upon the retention time of the peptide(s) in an initial analytical elution profile to obtain maximum loading and resolution in the shortest time for preparative chromatography.[1-4]

The most common analytical method used in RPC of peptides involves linear gradient elution (gradient-rate of 1% eluent B/min at a flow-rate of 1 ml/min), where eluent A is 0.1% aq. trifluoroacetic acid (TFA) and eluent B is 0.1% TFA in acetonitrile. However, the required amount of purified peptide is generally in the range of 1 to 100 mg for most researchers, and the above analytical conditions will not suffice for the great majority of such preparative applications. This article examines the effect of varying gradient-rates and sample load, at a standard analytical flow-rate of 1 ml/min, on the yield of purified peptide from a peptide mixture. From these results, we describe procedures for rapid optimization of conditions for preparative RPC of peptides on analytical columns.

II. EXPERIMENTAL

A. INSTRUMENTATION

For the preparative work described in this article, the HPLC instrument consisted of a Varian Vista Series 5000 liquid chromatograph (Varian, Walnut Creek, CA) coupled to a Hewlett-Packard (Avondale, PA) HP 1040A detection system, HP 9000 Series 300 computer, HP 9133 disc drive and HP 7440A plotter. Samples were injected with a 5-ml injection loop (model 7125, Rheodyne, Cotati, CA). For the analysis of fractions and pooled fractions, analytical separations were carried out on a fully automated Hewlett-Packard HP 1090 liquid chromatograph, including auto sampler, HP 9000 Series 300 computer, HP 9133 disc drive and HP 7440A Plotter. This allowed for unattended overnight analysis of fractions.

B. COLUMNS

Separations were carried out on Aquapore RP 300 (C$_8$) analytical columns of various lengths as indicated in the figure legends (4.6 mm I.D., 7-μm particle size, 300-Å pore size) (Applied Biosystems, Foster City, CA).

III. RESULTS AND DISCUSSION

A major difficulty encountered in the preparative purification of peptides is the separation of impurities that are closely related structurally to the peptide of interest. During an analytical gradient run (1% acetonitrile/min) as shown in Figure 1, the impurities that are difficult to resolve in preparative gradient elution chromatography will be the impurities in the window of ± 2 min from the peptide peak of interest. Figure 1 (top panel) shows a peptide mixture where the product (P) (elution time of 21.4 min) is in the center of two impurities: a more hydrophilic impurity (I$_1$) (19.1 min) and a more hydrophobic impurity (I$_2$) (22.3 min). The product represents approximately 62% of the sample mixture and is representative of a separation problem of average degree of difficulty, since the impurities are completely resolved in this standard analytical run. The peptide product contains a tyrosine residue, hence the absorbance at 280 nm (Figure 1, bottom panel), while the two peptide impurities do not. This allows the peptide product to be easily located in the preparative run by monitoring the elution profile at 280 nm as well as 210 nm for the peptide bond absorption (Figure 2, panel B).

Comparison of Panels A and B of Figure 2 demonstrate the effect of sample load on peptide resolution. When 4 mg of crude peptide was applied to the column (100×4.6 mm I.D.), complete resolution of the impurities (I$_1$ and I$_2$) and product is no longer obtained when using the identical gradient conditions as in the analytical separation (1% acetonitrile/min; Panel A). The profile at 280 nm in Figure 2B, due to the presence of the Tyr residue in the desired peptide product, is below the 210 nm profile. The results shown in Figure 2B indicate a cross-contamination of I$_1$ on the front side of the product peak and of I$_2$ on the back side of the product peak. Thus, to obtain pure product, fraction analysis of the individual 1-ml fractions in the expected cross-over regions must be carried out. The portions containing homogeneous product were pooled (Pool 3) and contained 70% of the total product recovered. The cross-contaminated pools contained 15% product contaminated with impurity I$_1$ in Pool 2 and 15% product contaminated with impurity I$_2$ in Pool 4. Pools 1 and 5 contain homogeneous impurities I$_1$ and I$_2$, respectively. The question thus becomes, "can the researcher improve the yield of pure product and increase the sample load by changing the run conditions?" Elsewhere in this volume, we have shown that increasing flow-rate has little effect on the difference in retention time between peaks with analytical sample loads.[5] The main effect of increasing flow-rate is to decrease peak width and, thus, increase resolution. However, in the optimum flow rate range of 0.5 to 2.0 ml/min for an analytical column, there is very little effect on resolution. These results would suggest that flow-rate is really not a consideration for improving a preparative separation. On the other hand, gradient-rate has a dramatic effect on peptide separation.[5] Decreasing the gradient-rate increases the difference in retention time between peaks (Δt). At the same time, there is an increase in peak width with decreasing gradient-rate but the effect on peak width is less than on Δt; hence overall, resolution improves with decreasing gradient-rate.

Figure 3 shows the effect of decreasing gradient rate on the preparative separation of 4 mg of crude peptide mixture. The separation is shown as a histogram with area units/min plotted against elution time. As the gradient-rate is decreased from 1% acetonitrile/min to 0.5%/min and, finally, to 0.1%/min at the fixed sample load, the amount of homogeneous product recovered increases from 70 to 83 to 91%, respectively. With the increase in percent homogeneous product recovered, there is a corresponding sacrifice in run time. At the shallow gradient-rate of 0.1% B/min, the run is complete in 100 min compared to 25 min at a gradient-rate of 1%

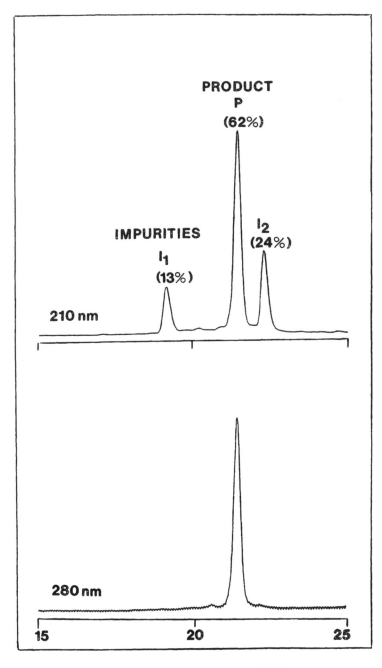

FIGURE 1. Analytical RPC elution profile of a synthetic peptide mixture. Column dimensions: 100×4.6 mm I.D. Conditions: linear AB gradient (1% B/min), where eluent A is 0.1% aq. trifluoroacetic acid (TFA) and eluent B is 0.1% TFA in acetonitrile; flow-rate, 1 ml/min. Top panel: elution profile at a detection wavelength of 210 nm; (P) denotes the peptide product of interest, and I_1 and I_2 denote more hydrophilic and hydrophobic peptides, respectively; the percentages (%) in brackets indicate the relative amount of each peptide in the mixture, as detected by integrated peak area. Bottom panel: elution profile at a detection wavelength of 280 nm, which detects tyrosine-containing peptide product (P). The peptide sequences are I_1 = (Ac-Arg-Gly-Ala-Gly-Gly-Leu-Gly-Leu-Gly-Lys-amide), I_2 = (Ac-Arg-Gly-Val-Ala-Gly-Leu-Gly-Leu-Gly-Lys-amide), and P = (Ac-Arg-Gly-Tyr-Gly-Gly-Ile-Gly-Leu-Gly-Lys-amide).

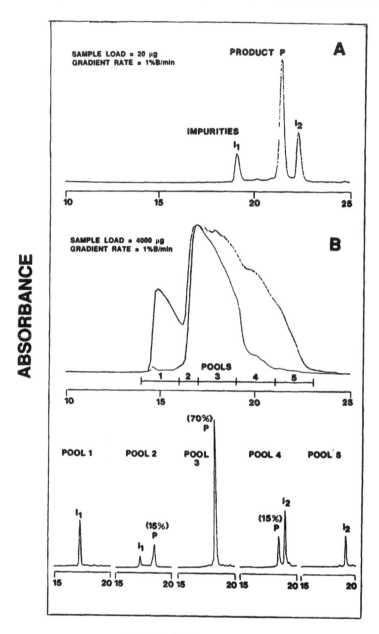

ELUTION TIME (min)

FIGURE 2. Linear gradient separation by RPC of synthetic peptide product, P, from more hydrophilic and more hydrophobic peptide impurities (I_1 and I_2, respectively). Column dimensions: same as Figure 1. Panel A: analytical elution profile of the peptide mixture; run conditions same as Figure 1. Panel B: preparative elution profiles; column as in panel A; conditions as in Figure 1. The elution profiles at 210 and 280 nm are shown, the latter beneath the former. Fractions (1 min, i.e., 1-ml volumes) were collected, analyzed by RPC and collected into 5 pools as shown. Bottom panel: analytical elution profiles of pools 1 to 5 (2% B/min gradient on a 220 × 4.6 min I.D. column). Peptide sequences are shown in Figure 1.

EFFECT OF GRADIENT RATE AT
CONSTANT SAMPLE LOAD

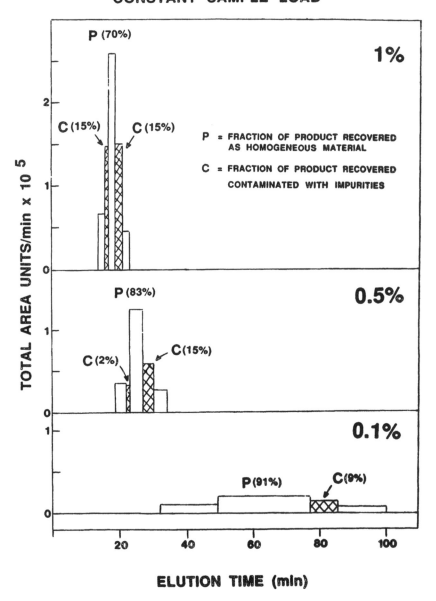

FIGURE 3. Histograms of three preparative RPC separations of a peptide mixture showing the effect of varying gradient rate at constant sample load (4 mg). Column dimensions and mobile phase: same as Figure 1. The profiles represent preparative runs carried out at gradient-rates of 1% B/min (top panel; same as run shown in Figure 2B), 0.5% B/min (middle panel) and 0.1% B/min (bottom panel); flow-rate, 1 ml/min. Each bar represents the pooled eluate collected over the corresponding time period. Area units were calculated from analytical aliquots of each pool run under the conditions of Figure 2, bottom panel, and represent the total area units recovered. Sequences of the peptide product (P) and more hydrophilic and hydrophobic impurities are shown in Figure 1.

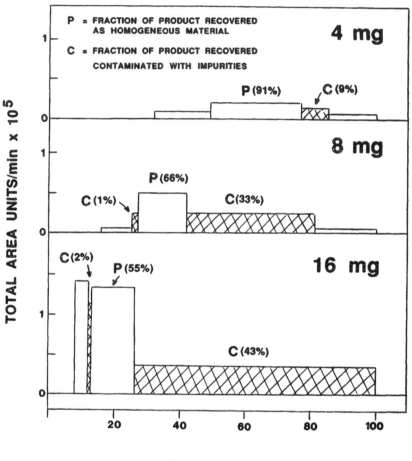

FIGURE 4. Histograms of three preparative RPC separations of a peptide mixture showing the effect of varying sample load at constant gradient-rate (0.1% B/min). Column dimensions and mobile phase: same as Figure 1. The profiles represent preparative runs carried out at sample loads of 4 mg (top panel), 8 mg (middle panel) and 16 mg (bottom panel); flow-rate, 1 ml/min. Each bar represents the pooled eluate collected over the corresponding time period. Area units were calculated from analytical aliquots of each pool run under the conditions of Figure 2, bottom panel, and represent the total area units recovered. Sequences of the peptide product (P) and more hydrophilic and hydrophobic impurities are shown in Figure 1.

B/min. From a practical consideration, it is our experience that a 100-min run time is acceptable for a preparative run in a research laboratory .

Figure 4 shows the effect of increasing the sample load from 4 to 8 to 16 mg of crude peptide at a fixed gradient rate of 0.1% B/min. As expected, the time taken for the front of the peptide to be eluted from the column decreases with sample load.[5,6] As sample load is increased, more and more of the stationary phase is utilized and eventually, at high loads, the excess peptide would be eluted in the void volume peak. In addition, as sample load is increased, the percent recovery of homogeneous product is decreased from 91 to 66 to 55% for the 4-, 8-, and 16-mg loads, respectively. The results of decreasing gradient-rate and increasing sample load on the

TABLE 1

Sample load (mg)	Gradient rate (%)	Homogeneous product recovered (%)	(mg)
4	1	70	1.7
4	0.5	83	2.1
4	0.1	91	2.3
8	0.1	66	3.3
16	0.1	55	5.5

percentage and weight (mg) of homogeneous product recovered are summarized in Table 1. As the gradient-rate is decreased, resolution is improved and, thus, the percent recovery and total milligram of homogeneous product recovered increases. As sample load is increased, the percent recovery of homogeneous product is decreased, but the recovery in terms of quantity (2.3, 3.3, and 5.5 mg of homogeneous peptide were recovered at sample loads of 4, 8, and 16 mg of crude peptide applied, respectively) is increased. Thus, at a load of 16 mg of crude peptide with 55% recovery of pure product, more than double the yield (mg) of homogeneous product is recovered compared to the 4 mg of crude peptide with 91% recovery of pure product (5.5 vs. 2.3 mg, respectively; Table 1).

The approach of using shallow gradients (0.1% acetonitrile/min starting from 0% acetonitrile) can substantially increase run times.[5,6] This is especially true if the components are reasonably hydrophobic. The maximum run time at any gradient-rate can be calculated from an initial standard analytical run (1% B/min) using the following equation:

Maximum run time (min) at a gradient-rate of X%/min

$$= \left(t_R^{1\%} - tg\right) \times \left(\frac{1\%}{X\%}\right) + tg$$

where $t_R^{1\%}$ is the retention time from an analytical run of the peptide of interest at a gradient-rate of 1% acetonitrile/min, starting from 0% acetonitrile; X% is the desired gradient-rate; and tg is the time for the gradient to reach the detector from the proportioning value, via pump, sample loop and column (tg consists of t_d plus t_o, where t_d is the gradient delay time, i.e., the time for the gradient to reach the top of the column, and t_o is the time for unretained compounds to be eluted from the column).

For example, if the retention time of the peptide of interest, $t_R^{1\%}$, is 21 min and the desired gradient rate of the preparative run is 0.1% acetonitrile/min and tg = 9 min, then, maximum run time at 0.1%/min

$$= (21-9) \times \left(\frac{1}{0.1}\right) + 9 = 120 + 9 = \underline{129 \text{ min}}$$

It should be remembered, that, in preparative chromatography at any given gradient-rate, the total sample elution time is independent of sample load. Sample load only affects the time at which the peptide begins to be eluted from the column.[5] As discussed by Burke et al. in the following article in this volume,[7] an excellent approach to ensuring that all runs are complete within 2 hr is to run a shallow gradient (0.1% B/min), starting 12% below the acetonitrile concentration required to elute the peptide in an analytical run. This arbitrary rule of thumb was shown to be successful in the preparative purification of 20 mg of crude 28-residue bovine parathyroid growth hormone fragment.[7]

Although shallow gradient elution for preparative purification of peptides is generally an advantage, Figure 5 clearly demonstrates the need for the use of two different gradient-rates in

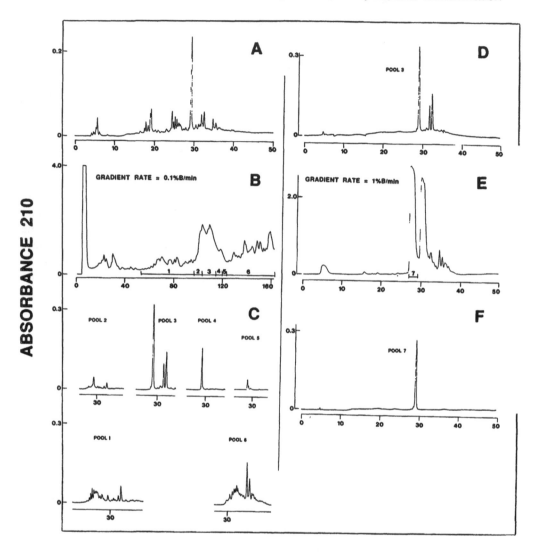

ELUTION TIME (min)

FIGURE 5. Isolation of peptide 'P' from a crude peptide mixture following solid-phase peptide synthesis. Column dimensions: 41 cm × 4.6 mm I.D., produced by linking several small columns in series; all runs, both analytical and preparative, were carried out on this multi-column arrangement. Mobile phase: same as Figure 1. Panel A: analytical run (1% B/min) of the crude peptide mixture. Panel B: preparative separation (80-mg sample) of the crude peptide mixture at a gradient-rate of 0.1% B/min. Fractions (3 min, i.e., 3-ml volumes) were collected, analyzed by RPC and collected into six pools as shown. Panel C: analytical elution profiles of pools 1 to 6 (2% B/min). Panel D: full elution profile of pool 3 (from panel C). Panel E: preparation separation of pool 3 (1% B/min). Fractions (1 min, i.e., 1-ml volumes) were collected and pure product was pooled as pool 7. Panel F: analytical elution profile (2% B/min) of pool 7. The sequence of the peptide product, P, is shown in Figure 1.

the purification of a synthetic peptide. Figure 5A shows the analytical separation at a gradient-rate of 1% acetonitrile/min. The synthesis indicates the presence of one major peak and a series of hydrophilic and hydrophobic impurities. Figure 5B displays the preparative elution profile of 80 mg of crude peptide injected onto the column and eluted at a gradient rate of 0.1% acetonitrile/min. Following fraction analysis, the resulting 6 pools were analyzed (Figure 5C). Homogeneous product "P" was found only in Pool 4, with the majority of product still contaminated with hydrophobic impurities. The hydrophobic impurities that are easily resolved at 1% acetonitrile/min on the analytical run are not resolved during a preparative run at a 0.1%/min gradient-rate. Figure 5D shows the analytical run of Pool 3 at a gradient-rate of 1% acetonitrile/min. A preparative run of this pool was separated at the same gradient-rate (Figure 5E). Fraction analysis of pool 7 indicates the presence of homogeneous peptide, P, (Figure 5F). This result can be explained by the difference in partitioning rates of the two hydrophobic impurities compared to the peptide product. The hydrophobic impurities are non-peptide (organic) in nature. These organic impurities arise from the synthesis and partition much more rapidly than the larger peptide product. Thus, at a shallow gradient-rate, the product and impurities have very similar retention times; whereas, at more rapid gradient-rates (1% acetonitrile/min), the greater hydrophobicity of the impurities results in resolution of the peptide product. This result clearly indicates and supports the results of Mant and Hodges[3] that peptides have much slower partitioning rates than smaller organic molecules. Thus, although shallow gradient-rates will, in general, provide better resolution of peptides, mixtures of peptides and molecules of non-similar structures may often be resolved best at quite different gradient-rates.

IV. CLOSING REMARKS

Preparative gradient elution chromatography is a compromise between run time, gradient-rate and sample load. If scale-up is required, the researcher can increase column length and/or increase column diameter. Thus, increasing the column length from 10 cm (used in this study) to 25 cm would allow an increased load of at least 2.5-fold. Doubling the column diameter enables a 4-fold increase in sample load. Thus, a simple switch from the analytical column (100 × 4.6 mm I.D.) used in this article to a semipreparative column (250 × 10 mm I.D), which can be used on an analytical instrument, allows for an approximate 10-fold increase in sample load.

A major disadvantage of preparative gradient elution chromatography is that the method does not make maximum use of the packing material and, thus, scale-up becomes extremely costly in terms of column packings and solvents, especially as one proceeds with scale-up to columns of internal diameters ≥ 2.5 cm. Also, columns of 2.5 cm. I.D. (or larger) require flow-rates in excess of 10 ml/min and, thus, require preparative instrumentation systems. The present article describes one approach to making more efficient use of column packings through the use of shallow gradients. In addition, other workers, notably Horváth and co-workers[8,9] and Hodges and co-workers,[10-12] have also been developing new methods for preparative chromatography on analytical columns.

ACKNOWLEDGMENTS

This work was supported by the Medical Research Council of Canada and by equipment grants from the Alberta Heritage Foundation for Medical Research.

REFERENCES

1. **Böhlen, P. and Kleeman, G.,** Analytical and preparative mapping of complex peptide mixtures by reversed-phase high-performance liquid chromatography, *J. Chromatogr.*, 205, 65, 1981.
2. **Hancock, W.S. and Sparrow, J.T.,** Use of mixed-mode, high-performance liquid chromatography for the separation of peptide and protein mixtures, *J. Chromatogr.*, 206, 71, 1981.

3. **Mant, C.T., Burke, T.W.L., and Hodges, R.S.,** Optimization of peptide separations in reversed-phase HPLC: isocratic versus gradient elution, *Chromatographia,* 24, 565, 1987.

4. **Sitrin, R., DePhillips, P., Dingerdissen, J., Erhard, K., and Filan, J.,** Preparative liquid chromatography. A strategic approach, *LC.GC.,* 4, 530, 1985.

5. **Burke, T.W.L., Mant, C.T., and Hodges, R.S.,** The effect of varying flow-rate, gradient-rate and detection wavelength on peptide elution profiles in reversed-phase chromatography, this publication.

6. **Parker, J.M.R., Mant, C.T., and Hodges, R.S.,** A practical approach to the preparative purification of peptides using analytical instrumentation with analytical and semipreparative columns, *Chromatographia,* 24, 832, 1987.

7. **Burke, T.W.L., Black, J.A., Mant, C.T., and Hodges, R.S.,** Preparative reversed-phase shallow gradient approach to the purification of closely-related peptide analogs on analytical instrumentation, this publication.

8. **Cramer, S. and Horváth, Cs.,** Displacement chromatography in peptide purification, *Prep. Chromatogr.,* 1, 29, 1988.

9. **Antia, F.D. and Horváth, Cs.,** Displacement chromatography of peptides and proteins, this publication.

10. **Burke, T.W.L., Mant, C.T., and Hodges, R.S.,** A novel approach to reversed-phase preparative high-performance liquid chromatography of peptides, *J. Liq. Chromatogr.,* 11, 1229, 1988.

11. **Hodges, R.S., Burke, T.W.L., and Mant, C.T.,** Preparative purification of peptides by reversed-phase chromatography. Sample displacement mode *versus* gradient elution mode, *J. Chromatogr.,* 444, 349, 1988.

12. **Mant, C.T. and Hodges, R.S.,** Preparative reversed-phase sample displacement chromatography of synthetic peptides, this publication.

PREPARATIVE REVERSED-PHASE SHALLOW GRADIENT APPROACH TO THE PURIFICATION OF CLOSELY-RELATED PEPTIDE ANALOGS ON ANALYTICAL INSTRUMENTATION

T. W. Lorne Burke, James A. Black, Colin T. Mant, and Robert S. Hodges

I. INTRODUCTION

The N-formyl group is the most commonly used protecting group used for tryptophan (Trp) during solid-phase peptide synthesis. In addition, N-formyl-Trp is often used as an analog of tryptophan in structure-function studies of peptide hormones. Two forms of a 28-residue bovine parathyroid growth hormone (bPTH) were prepared in the present study: Ac-bPTH (7-34)-amide (P2) and Ac(N-For-Trp^{23})-bPTH (7 to 34)-amide (P1) (Figure 1). Generally, by proper selection of the HF (hydrogen fluoride) cleavage conditions, all the side-chain protecting groups can be removed except for the N-formyl group on Trp. Using some of this peptide, the N-formyl group can then be selectively removed using an aqueous solution of 1 M (NH$_4$)HCO$_3$ to yield the native tryptophan-containing peptide (P2). Because of the presence of Trp and Met (methionine) residues, in addition to sequence-specific deprotection problems, it was determined that a significant yield of the desired N-formylated peptide (P1) could only be obtained by using a special combination of scavengers and temperature during HF cleavage. However, these special cleavage conditions resulted in the cleavage of the N-formyl group from ca. 50% of the crude peptide product, resulting in about a 1:1 molar ratio of P1 to P2. While both these peptides were desired, their presence in the same crude peptide mixture posed a difficult purification problem.

Neither ion-exchange HPLC (IEC) nor size-exclusion HPLC (SEC) would be effective separation modes for these peptide analogs, not even as an initial purification step, there being no difference in net charge or size between the peptides. Reversed-phase chromatography (RPC), with its powerful peptide resolving capabilities,[1,2] was the logical method of choice for attempting to separate the peptides quickly and with satisfactory yields of purified peptides. The preparative separation of the two peptide analogs appeared to represent an ideal application for the shallow gradient approach described in the previous article ("Preparative reversed-phase gradient elution chromatography on analytical columns" by Hodges et al.[3]) and by Mant et al.[4] In fact, this was a particularly stringent test — for both the resolving power of RPC as well as the efficiency of the preparative procedure — considering that the peptide analogs differed only by a formyl group on the single Trp residue (Figure 1).

This article describes the successful separation, in good yield, of the two bPTH analogs following application of the shallow gradient approach. In addition, the utility of multi-wavelength diode-array detection, coupled with gradient elution fraction analysis, during preparative peptide purification is highlighted.

```
PEPTIDE 1 (P1):  as P2 below except

       ....Glu-NForTrp-Leu....

PEPTIDE 2 (P2):  Ac-Phe-Met-His-Asn

     -Leu-Gly-Lys-His-Leu-Ser-Ser-Met-

     -Glu-Arg-Val-Glu-Trp-Leu-Arg-Lys-

     -Lys-Leu-Gln-Asp-Val-His-Asn-Phe-

     -amide
```

FIGURE 1. Sequences of peptide analogs. NForTrp in P1 denotes an N-formyl group on the single tryptophan residue.

II. EXPERIMENTAL

A. PEPTIDE SYNTHESIS

Peptides were synthesized on an Applied Biosystems (Foster City, CA) 430A synthesizer, using the general procedure for solid-phase synthesis outlined by Erickson and Merrifield[5] with modifications by Hodges et al.[6] The indole side-chain of the tryptophan residue was protected during peptide synthesis with an N-formyl group.

B. HF CLEAVAGE

The peptide resin was stirred at –5°C for 1 hr in HF: anisole: dimethylsulfide (DMS): parathiocresol: peptide-resin (10 ml: 1 ml: 0.5 ml: 0.2 ml: 1 g). Reagent grade anisole was distilled prior to use. 'Gold label DMS' was obtained from Aldrich Chemical (Milwaukee, WI). All other chemicals were reagent grade. These cleavage conditions resulted in a Trp_{native}: $Trp_{N-formyl}$ ratio of ca. 1:1 in the synthesized peptide.

III. RESULTS AND DISCUSSION

A. ANALYTICAL RPC OF CRUDE PEPTIDE MIXTURE

Figure 2A illustrates the analytical elution profile of the crude peptide mixture obtained on a C_8 column (220 × 4.6 mm I.D.). The peptide mixture was eluted from the column by a linear AB gradient (1% B/min) at a flow-rate of 1 ml/min, where Eluent A was 0.05% aq. trifluoroacetic acid (TFA) and Eluent B was 0.05% TFA in acetonitrile (pH 2). It can be seen that, despite the very similar hydrophobicities of the peptide analogs (P1 = N-formyl peptide; P2 = native peptide), the column was able to separate the two peptides even with a relatively steep gradient (1% B/min), although baseline resolution was not achieved. Apart from the two peptide analogs, various unknown impurities (e.g., I_1, I_2, I_3) were present with hydrophobicities similar to that of the two desired peptide products, thus complicating the purification problem. Increasing the sample load of the peptide mixture on the same analytical column and under the same chromatographic conditions would only lead to a worsening in peptide resolution. Hence, alternative run conditions were required for preparative RPC separation of the sample components using the same C_8 column.

B. PREPARATIVE RPC OF PEPTIDE ANALOGS

A decrease in the steepness of the gradient during gradient elution of peptide mixtures leads to an increase in peptide resolution,[2-4,7,8] thus allowing for greater sample loads to be applied. A

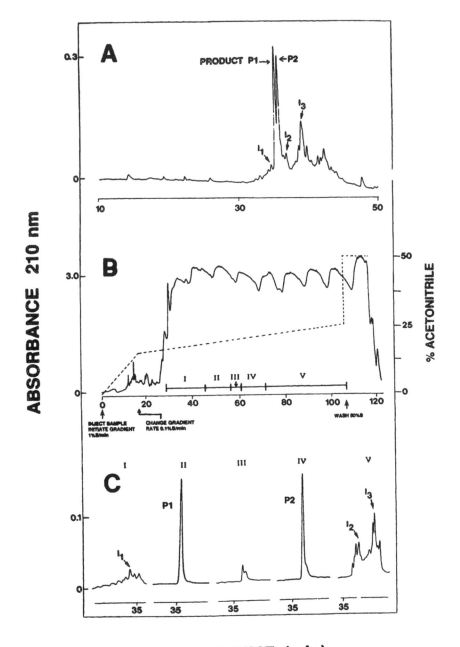

FIGURE 2. Preparative reversed-phase separation of peptide analogs, P1 and P2, from each other and from various hydrophilic (I$_1$) and hydrophobic (I$_2$,I$_3$) impurities. Column: Aquapore RP300 C$_8$, 220 × 4.6 mm I.D., 7-μm particle size, 300-Å pore size (Brownlee Labs., Santa Clara, CA). The HPLC instrument consisted of a Varian (Walnut Creek, CA) Vista Series 5000 liquid chromatograph, coupled to a Hewlett-Packard (Avondale, PA) HP1040A diode array detection system, HP9000 series 300 computer, HP9133 disc drive, HP2225A Thinkjet printer and HP7440A plotter. Samples were injected with a 500-μl (analytical runs) or 5-ml (preparative runs) injection loop (Rheodyne, Cotati, CA). Panel A: analytical chromatogram of crude peptide mixture. Conditions, linear AB gradient (1% B/min) at a flow-rate of 1 ml/min, where Eluent A is 0.05% aq. TFA and Eluent B is 0.05% TFA in acetonitrile. Panel B: preparative chromatogram of the same peptide mixture. Conditions, linear AB gradient (100% A/0% B to 84% A/16% B over 16 min [1% B/min], then to 75% A/25% B over a further 90 min [0.1% B/min]) at a flow-rate of 1 ml/min, followed by an isocratic wash with 50% A/50% B for 10 min prior to re-equilibration to 100% A/0% B. Fractions were collected every minute. Sample load, ca. 20 mg. Panel C: analytical chromatograms of Pools I to V from Panel B. Conditions, same as Panel A.

10-fold decrease in gradient steepness (1% B/min → 0.1% B/min) was subsequently chosen for the preparative run. A disadvantage of a shallow gradient-rate is the relatively long run time, although this is offset somewhat at high sample loads which result in a decrease in solute retention times. To shorten the run time in the preparative separation, we established an arbitrary rule of thumb that the shallow gradient (0.1% B/min) would start at an acetonitrile concentration of 12% below the acetonitrile concentration required to elute the peptide in the analytical run. For example, in our case, the peptide(s) were eluted at approximately 35 min in the analytical run using a gradient-rate of 1% B/min. The percent acetonitrile required for peptide elution in the analytical run (Figure 2A) is equal to the observed peptide(s) retention time (t_R = 35 min) minus the gradient delay time (the time, t_g, required for the gradient to travel from the proportioning value to the detector via pump, injector and column; t_g = 7 min) and multiplied by the gradient-rate (1% B/min):

$$(35 - 7) \times 1 = 28\%$$

Thus, the shallow gradient for the preparative run would start at an acetonitrile concentration of

$$28\% - 12\% = 16\%$$

Figure 2B shows the preparative RPC elution profile of ca. 20 mg of the crude peptide mixture on the analytical C_8 column. The dotted line represents the progress of the gradient. The column was equilibrated and the sample injected (5 ml sample volume) in 0.05% aq. TFA (100% A/0% B) to promote binding of the peptides to the column and to minimize partitioning of the peptide(s) during sample application. A linear gradient (1% B/min for 16 min) was then run up to 84% A/16% B, followed by a shallow gradient (0.1% B/min for 90 min) up to 75% A/25% B, during which the required peptide separation took place; a final 10-min wash with 50% A/ 50% B removed any remaining components from the column. Fractions (1 ml) were collected for analysis and subsequent pooling.

C. FRACTION ANALYSIS
1. Gradient Elution Analysis

The chromatogram of the preparative run at 210 nm was of little use in determining where the separation occurred, as the high load swamped the detection system (Figure 2B). Even at a less sensitive wavelength (i.e., 280 nm, to detect Trp and, to a lesser extent, Phe [phenylalanine] absorbance) the chromatogram yielded little useful information. To determine the location of P1 and P2 in the preparative elution profile, it was necessary to analyze the individual collected fractions. Isocratic runs are performed often in fraction analysis, since time is saved in not having to re-equilibrate the system.[9] However, it is difficult to establish the correct concentration of Eluent B that will ensure that all components in a fraction will be eluted from a column within a reasonable time and that these components are resolved without extensive peak broadening. Considerable precious peptide sample and time may be used up while establishing the optimum isocratic elution parameters.

The analytical elution profile shown in Figure 2A demonstrated that peptides P1 and P2 could be partially resolved within a reasonable run time (< 40 min) by a gradient-rate of 1% B/min, and these conditions were subsequently chosen for fraction analysis. The disadvantage of having to re-equilibrate the system after each run was compensated for by the certainty that all components in each fraction would be eluted as sharp peaks, with P1 and P2 resolved if present; in addition, eluted components could be tentatively identified by comparison with the chromatogram of the analytical crude run. Gradient elution has the advantage over isocratic elution in reproducibility of retention time, thereby making component identification easier.[10]

Figure 3 shows a selection of chromatograms obtained following analysis of the fractions

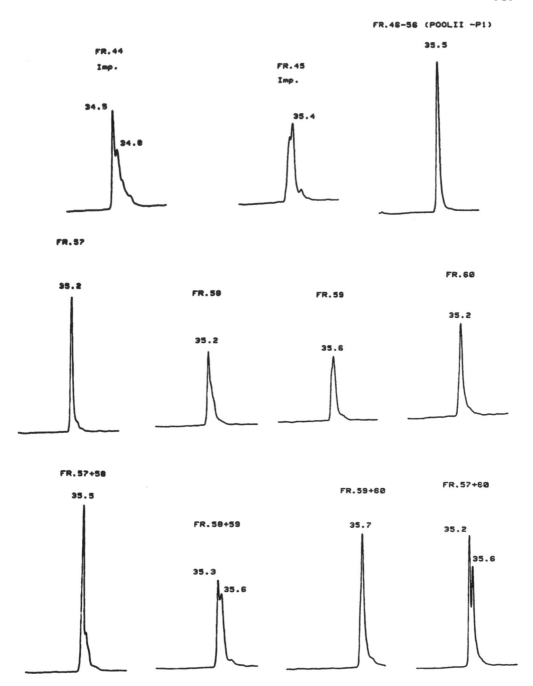

FIGURE 3. Analytical chromatograms of collected 1-ml fractions from preparative purification of peptide mixture shown in Figure 2B. Column and HPLC instrument: same as Figure 2. Conditions: same as Figure 2A. Sample volume, 1 µl. Fractions are labeled according to their elution time, e.g., the fraction collected between 57 and 58 min is labeled FR.58, etc. Some fractions were also co-injected (third row).

collected during preparative RPC of the crude peptide mixture (Figure 2B). The fractions are labeled according to their elution time (e.g., 44 to 45 min fraction = FR.45, 72 to 73 min fraction = FR.73, etc.).

Starting at the top left of Figure 3, FR.44 and FR.45 are the fractions immediately prior to the

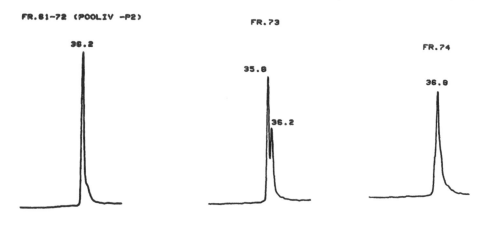

FIGURE 3 (continued)

first fraction (FR.46) containing pure P1. These two fractions formed part of Pool I (Figure 2B and C) and contained impurities denoted I_1 in Figure 2A and C. Following FR.44 and FR.45 is the pure P1 pool (FR.46 to 56; Pool II in Figure 2B and C).

The first fraction in the second row of Figure 3 shows FR.57 which also appeared to contain pure P1. However, this was not included in Pool II to ensure that no cross-contamination with P2 would occur. FR.58 and FR.59 (Pool III in figure 2B and C) are the crossover fractions between P1 and P2, followed by the first fraction (FR.60) which appeared to contain pure P2. It is important to note that the retention times of FR.57 and FR.60 were identical at 35.2 min. Without first analyzing the intervening fractions (FR.58 and FR.59), it would not be obvious that a separation had occurred and that FR.57 and FR.60 were P1 and P2, respectively. This result indicated that, even with gradient elution analysis, observed retention times are not always reproducible enough to make definite assignments of very closely-related peptides or impurities.

The difference between adjacent or other close fractions can usually be verified by co-injection of both fractions (Figure 3, third row). Using this technique, the difference between FR.57 and FR.60 is immediately apparent (third row, extreme right).

The bottom row in Figure 3 shows the P2 pool of FR.61 to 72 (Pool IV in Figure 2B and C). FR.60 was not included in this pool, although it appeared to contain pure P2, to ensure that no contamination with P1 would occur. The following impure fractions, FR.73 and FR.74, formed part of Pool V in Figure 2B and C. This pool contained impurities denoted I_2 and I_3 in Figure 2A and C. It should be noted that, although FR.59 was shown to contain more than one component (second row, third profile), its co-injection with FR.60 resulted in a single major peak (third row, third profile). Thus, even co-injection of different fractions has occasional limitations for assignment of peptide pools.

2. Wavelength-Scanning of Fractions

The wavelength-scanning capabilities of the diode array detector (DAD) can be extremely beneficial in aiding the researcher to interpret peptide elution profiles.[11,12] Figure 4 demonstrates how wavelength scans (210 to 310 nm) were employed to aid in fraction analysis of the preparative separation shown in Figure 2B. Figure 4A shows wavelength scans of peptides P1

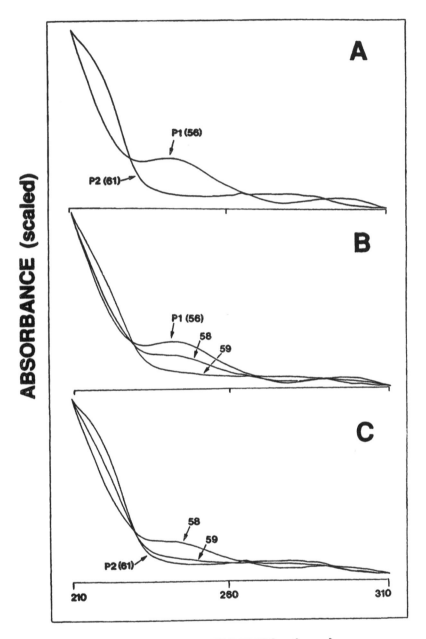

FIGURE 4. Normalized spectra from 210 to 310 nm of single main peak in each of four 1-ml fractions. The spectral scans are denoted by fraction number (see Figure 3). Panel A: comparison of spectra of pure peptide P1 ($Trp_{N-formyl}$; FR.56) and pure peptide P2 (Trp_{native}; FR.61). Panel B: comparison of spectra of pure peptide P1 (FR.56) with crossover fractions FR.58 and FR.59. Panel C: comparison of spectra of pure peptide P2 (FR.61) with crossover fractions FR.58 and FR.59.

(Trp$_{N-formyl}$) and P2 (Trp$_{native}$) from FR.56 and FR.61, respectively. The N-formyl group on the Trp residue produces the markedly different spectrum for P1 (with maxima at ca. 245 nm and 300 nm) compared to P2, enabling an easy distinction between the two desired products. This difference in spectral characteristics between the two peptide analogs quickly determined, for instance, that FR.57 and FR.60 contained different components, despite their observed identical retention times during gradient elution analysis (Figure 3, second row). Figure 4B and C show comparative wavelength scans of the crossover fractions, FR.58 and FR.59, with pure P1 (FR.56) and P2 (FR.61), respectively. These results demonstrate that spectral analysis is the easiest way to distinguish cross-contamination of P$_1$ and P$_2$ in the crossover fractions. The contaminated fraction has a spectrum different from that of either pure P1 or pure P2, since the spectrum of both peaks together is additive and the resulting net spectrum does not appear similar to either spectrum of the individual peptides (Figure 4B and C).

D. YIELD OF PURIFIED PEPTIDES

Pool II and Pool IV (Figure 2B and C) contained ca. 5 mg each of pure P1 and P2, respectively. These yields of pure peptide analogs, representing ca. 25% each of the original 20 mg of crude peptide mixture, were extremely gratifying considering the difficulties associated with preparative RPC of these very closely-related peptides. It is also interesting to note in Figure 2 how well pools I-V can be overlayed to form the preparative run (Figure 2B) with little loss of P1 or P2 to neighboring pools (Figure 2C).

IV. CONCLUSIONS

This article describes the successful preparative purification of ca. 20 mg of a crude peptide mixture of two closely-related 28-residue peptide analogs (differing only by an N-formyl group on the single Trp residue) on an analytical column and with analytical instrumentation. The purification, carried out in reversed-phase mode and using a shallow gradient (0.1% B/min) approach, yielded ca. 5 mg each of the two pure peptides. The presence of a Trp residue in one analog and an N-formyl-Trp residue in the other, with subsequently different spectral characteristics, enabled the use of multi-wavelength detection analysis for rapid, simple, and accurate identification of peptide products following gradient elution analysis of peptide fractions.

ACKNOWLEDGMENTS

This work was supported by the Medical Research Council of Canada and by equipment grants from the Alberta Heritage Foundation for Medical Research.

REFERENCES

1. **Mant, C.T. and Hodges, R.S.,** HPLC of peptides, in *High-Performance Liquid Chromatography of Biological Macromolecules: Methods and Applications,* Gooding, K. and Regnier, F., Eds., Marcel Dekker, New York, 1990, 301.
2. **Mant, C.T. and Hodges, R.S.,** Optimization of peptide separations by high-performance liquid chromatography, *J. Liq. Chromatogr.,* 12, 139, 1989.
3. **Hodges, R.S., Burke, T.W.L., Mant, C.T., and Ngai, S.M.,** Preparative reversed-phase gradient elution chromatography on analytical columns, this publication.

4. **Mant, C.T., Burke, T.W.L., and Hodges, R.S.,** Optimization of peptide separations in reversed-phase HPLC: isocratic versus gradient elution, *Chromatographia*, 24, 565, 1987.

5. **Erickson, B.W. and Merrifield, R.B.,** Solid-phase peptide synthesis, in *The Proteins*, Neurath, H., Hill, R.L., and Boeder, C.L., Eds., Academic Press, New York, 1976, 257.

6. **Hodges, R.S., Semchuk, P.D., Taneja, A.K., Kay, C.M., Parker, J.M.R., and Mant, C.T.,** Protein design using model synthetic peptides, *Peptide Res.*, 1, 19, 1988.

7. **Guo, D., Mant, C.T., Taneja, A.K., Parker, J.M.R., and Hodges, R.S.,** Prediction of peptide retention times in reversed-phase high-performance liquid chromatography. II. Correlation of observed and predicted peptide retention times and factors influencing the retention times of peptides, *J. Chromatogr.*, 359, 519, 1986.

8. **Burke, T.W.L., Mant, C.T., and Hodges, R.S.,** The effect of flow-rate and gradient-rate on peptide resolution in reversed-phase chromatography, this publication.

9. **Rivier, J., McClintock, R., Galyean, R., and Anderson, H.,** Reversed-phase high-performance liquid chromatography: preparative purification of synthetic peptides, *J. Chromatogr.*, 288, 303, 1984.

10. **Hodges, R.S., Burke, T.W.L., and Mant, C.T.,** Preparative purification of peptides by reversed-phase chromatography. Sample displacement mode *versus* gradient elution mode, *J. Chromatogr.*, 444, 349, 1988.

11. **Zavitsanos, P. and Goetz, H.,** The practical application of diode array UV-visible detectors to high-performance liquid chromatography analysis of peptides and proteins, this publication.

12. **Black, J.A. and Hodges, R.S.,** Peak identification in HPLC of peptides using spectral analysis and second order spectroscopy of aromatic amino acid residues, this publication.

PREPARATIVE REVERSED-PHASE SAMPLE DISPLACEMENT CHROMATOGRAPHY OF SYNTHETIC PEPTIDES

Colin T. Mant and Robert S. Hodges

I. INTRODUCTION

Synthetic peptides are a class of compounds with increasing therapeutic importance and this has led to a concomitant increase in the need for rapid and efficient peptide purification procedures. Although an efficient peptide synthesis should result in only a small number of synthetic impurities (deletion, terminated or chemically modified peptides), these impurities which are usually closely related structurally to the peptide of interest, often pose difficult purification problems. The excellent resolving power and separation time of reversed-phase chromatography (RPC), coupled with the availability of volatile mobile phases, has made this mode of HPLC the favored method for both analytical and preparative separations of synthetic peptides. The most common analytical method used in RPC of peptides involves linear AB gradient elution (gradient rate of 1% Eluent B/min at a flow-rate of 1 ml/min), where Eluent A is 0.1% aq. trifluoroacetic acid (TFA) and Eluent B is 0.1% TFA in acetonitrile. However, the gradient elution mode of RPC is handicapped by relatively poor utilization of the stationary and mobile phases.[1] Hence, large-scale gradient elution separations of closely related peptides generally necessitate the use of increasingly larger column volumes in order to maintain satisfactory levels of product and yield. This, in turn, leads to higher operating costs in terms of packings, equipment, and solvents. Most researchers would probably desire to carry out laboratory scale preparative separations on existing analytical equipment. This, in turn, tends to limit the researcher to the use of columns of analytical (50 to 250 × 4.0 to 4.6 mm I.D.) or semipreparative (250 × 7.5 to 10 mm I.D.) dimensions. However, under conditions of conventional linear gradient elution (0.5 to 2.0% organic modifier/min), a considerable number of reversed-phase runs with relatively low sample loads may have to be performed before the desired yield of pure product is obtained, due to the aforementioned inefficient utilization of column capacity.

In the past few years, Horváth and other researchers[1-11] have demonstrated significantly more efficient use of hydrophobic stationary phases by applying the displacement mode of chromatography to reversed-phase preparative-scale separations of several classes of compounds, including peptides and proteins.[4-7,9-11] This technique enables the separation of relatively large amounts of material on columns and instrumentation designed primarily for analytical work through significantly more efficient use of the resolving capabilities of column packings than can be achieved by gradient elution. Displacement chromatography involves sorption of a sample mixture near the inlet of the column, following its application in a carrier solvent that has low affinity for the stationary phase. A solution of a displacer, which has greater affinity for the stationary phase than any of the sample components, is then pumped slowly into the column. The sample components are thereby displaced from the surface of the stationary phase and move down the column preceding the displacer front, forming adjacent zones of purified solutes traveling at the same velocity (displacement train). Despite the advantages of displacement over elution chromatography in terms of sample load, the method is difficult to optimize in terms of choice of displacer, displacer concentration, and flow-rate.[1,3] In addition, the need for column regeneration, an operational step not contributing directly to solute separation, is an undesirable feature of the technique.[1] Finally, the mobile phase flow-rates, which are typically, and necessarily, low in displacement chromatography (0.1 to 0.2 ml/min), often lead to excessively long run times.

A perceived need for easier methods of preparative reversed-phase separations of closely related compounds, as well as expanding the researcher's options for solving a particular preparative separation problem, prompted the development in the authors' laboratory of RPC operated in sample displacement mode (SDM).[12-14] This novel approach to preparative-scale purification on analytical columns and instrumentation maximizes both the separation and loading potential of a reversed-phase packing, enabling rapid separation of a single peptide component, in high yield, from a complex multi-component mixture.

II. PRINCIPLES OF PREPARATIVE RPC IN SAMPLE DISPLACEMENT MODE

Preparative RPC, whether it is carried out in gradient elution or displacement mode, requires the optimization of several parameters in order to maximize separation and yield. Thus, gradient steepness, flow-rate, choice of organic modifier and, particularly, sample load are all critical factors in the efficient application of gradient elution to preparative separations. In displacement chromatography, column length, mobile phase flow-rate, sample concentration, the choice of displacer and displacer concentration are all important factors in producing the correct (isotachic) conditions required to form the solute displacement train.[1-11] The optimization procedure would be greatly simplified if the only variable was sample load, i.e., without the controlled increase of organic modifier typical of gradient elution or the addition of displacer in displacement chromatography.

Since peptides favor an adsorption stationary phase in RPC, an organic modifier is typically required for their elution under normal analytical load conditions. However, it appeared reasonable to suppose that, when a reversed-phase column is optimally loaded with a peptide sample mixture dissolved in a 100% aqueous mobile phase, there would be competition by the sample components for the adsorption sites on the reversed-phase sorbent. The more hydrophobic components would compete more successfully for these sites than less hydrophobic components, which should be displaced and quickly eluted from the column. Thus, operation in SDM is simply using the well-established general principles of displacement chromatography without using the displacer.

Separation of a major peptide component from various peptide impurities is a situation frequently encountered by researchers involved in peptide synthesis and purification. A crude peptide mixture typically produced by solid-phase peptide synthesis, may contain not only the desired product, but also hydrophilic and/or hydrophobic synthetic peptide impurities. Under conditions of optimal sample load, either the peptide of interest would be used to displace the impurities from the column or the impurities would be used to displace the peptide of interest. In the latter case, the component of interest is isolated in water, and the organic solvent is used only to wash the column free of impurities. In the former case, the organic solvent is used to elute the peptide of interest after the separation in water is complete. Situations where a desired peptide product is the most hydrophilic or most hydrophobic component of a crude peptide mixture represent the simplest applications for operation of a reversed-phase column in SDM and, hence, offer the most convenient model/system for demonstrating the potential of SDM as a preparative tool.

TABLE 1
Sequences of Model Synthetic Peptides

Peptide number	Peptide sequence
1	Arg-Gly|Ala-Gly|Gly-Leu-Gly-Leu-Gly-Lys-amide
2	Ac-Arg-Gly|Gly-Gly|Gly-Leu-Gly-Leu-Gly-Lys-amide
3	Ac-Arg-Gly|Ala-Gly|Gly-Leu-Gly-Leu-Gly-Lys-amide
4	Ac-Arg-Gly|Val-Gly|Gly-Leu-Gly-Leu-Gly-Lys-amide
5	Ac-Arg-Gly|Val-Val|Gly-Leu-Gly-Leu-Gly-Lys-amide

Note: Ac denotes Nα-acetyl; amide = Cα-amide. Peptide 1 has a free α-amino group. Variations in composition of the peptide analogs are shown between the two vertical lines. When referring to peptide product (P) or peptide impurity (I) in sample mixtures, the subscripts 1 to 5 denote peptides 1 to 5, respectively.

A. MODEL SYNTHETIC PEPTIDES

Table 1 shows the sequences of a series of model synthetic decapeptide analogs, peptides 1 to 5, mixtures of which were designed to simulate the crude peptide mixtures typically produced by solid-phase peptide synthesis. The peptides were synthesized on a Beckman (Berkeley, CA) Model 990 peptide synthesizer using the general procedure of solid-phase synthesis described by Parker and Hodges.[15] They were subsequently purified by RPC. The hydrophobicity of the peptides increases only slightly between peptide 2 and peptide 5 — between 2 and 3 there is a change from an α-H to a β-CH_3 group, between 3 and 4 there is a change from a β-CH_3 group to two methyl groups attached to the β-CH group, and between 4 and 5 there is a change from an α-H to an isopropyl group attached to the α-carbon — enabling a precise determination of the potency of preparative RPC in sample displacement mode. When referring to peptide product (P) or peptide impurity (I) in the sample mixtures, the subscripts 1 to 5 denote peptides 1 to 5 (Table 1), respectively.

B. DISPLACEMENT OF DESIRED HYDROPHILIC PEPTIDE COMPONENT BY HYDROPHOBIC IMPURITIES

Figure 1 demonstrates preparative separation in SDM of a four-peptide mixture on an analytical C_8 reversed-phase column. A short (30 mm in length) column was used to test the potential value of SDM in order to limit the amount of material required to saturate the hydrophobic stationary phase. Figure 1A shows an analytical separation profile of the four peptides, where peptide 2 is the desired component (or product) (P_2) and peptides 3 to 5 (I_3 to I_5, respectively) represent hydrophobic impurities. The ratio (w/w) of product to impurities was 3:1:1:1 (P_2:I_3:I_4:I_5). Total sample load was 6 mg, including 3 mg of P_2. This mixture, representing a somewhat poor synthesis yield of P_2 (50%), would be difficult to resolve by preparative gradient elution. The protocol for preparative separation in SDM of the sample mixture involved displacement in water (0.05% aqueous TFA) of P_2 by the more hydrophobic impurities (I_3, I_4, I_5), which should remain bound to the column. Figure 1B shows the SDM elution profile of the sample mixture. Following isocratic elution with 0.05% aqueous TFA at 1 ml/min, a gradient wash was initiated after 40 min (1% acetonitrile/min) to remove sample impurities from the column. The fractions pooled as Peaks I and II (Figure 1B) were subjected to gradient elution analysis and the results are shown in Figure 1C and D, respectively. Peak I (Figure 1C) was pure P_2 and contained 99% of recovered P_2, demonstrating excellent SDM separation. The remaining 1% was found in the 20 to 40 min region of the preparative run profile (Figure 1B), where it was

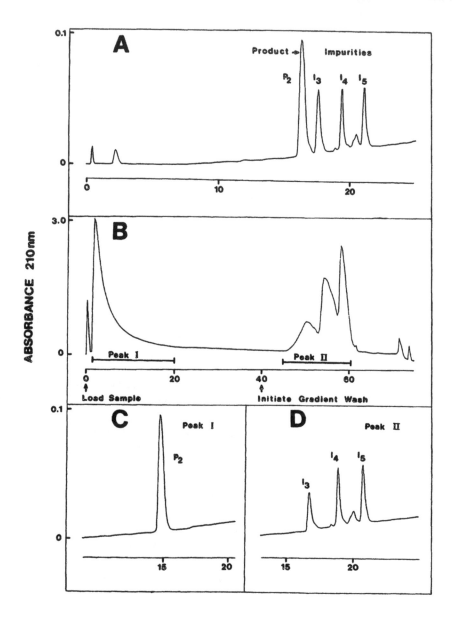

ELUTION TIME (min)

FIGURE 1. Separation by RPC operated in sample displacement mode of peptide product (P_2) from hydrophobic peptide impurities (I_3, I_4, I_5). The HPLC instrument consisted of a Varian Vista Series 5000 liquid chromatograph (Varian, Walnut Creek, CA) coupled to a Hewlett-Packard (Avondale, PA). HP1040A detection system, HP85B computer, HP9121 disc drive, HP2225A Thinkjet printer and HP7470 plotter. Column: Aquapore RP300 C_8, 30 × 4.6 mm I.D., 7-μm particle size, 300-Å pore size (Brownlee Labs., Santa Clara, CA). Panel A: analytical separation profile of peptide mixture; conditions, linear AB gradient (1% B/min) at a flow-rate of 1 ml/min, where Eluent A is 0.05% aqueous trifluoroacetic acid (TFA) and Eluent B is 0.05% TFA in acetonitrile. Panel B: preparative separation profile of peptide mixture; conditions, isocratic elution with 100% Eluent A for 40 min at a flow-rate of 1ml/min, followed by linear gradient elution at 1% B/min; sample load, 6.0 mg consisting of 3.0 mg of P_2 and 1.0 mg of each of I_3, I_4, and I_5 dissolved in 100 μl of Eluent A. Fractions were collected at 1-min intervals. Panels C and D show analytical elution profiles (same conditions as Panel A) of Peaks I and II (Panel B), respectively. The subscripts of P_2, I_3, I_4, and I_5 denote peptides 2 to 5, respectively (Table 1). (From Burke, T. W. L., Mant, C. T., and Hodges, R. S., *J. Liq. Chromatogr.*, 11, 1229, 1988. With permission.)

contaminated with I_3. Figure 1D demonstrates that the bulk of I_3, and all of I_4 and I_5, had, as expected, remained bound to the column during elution with 0.05% aqueous TFA, and were only removed by the addition of acetonitrile to the mobile phase.

C. DISPLACEMENT OF HYDROPHILIC IMPURITIES BY DESIRED HYDROPHOBIC PEPTIDE COMPONENT

Figure 2 demonstrates preparative separation in SDM of a five-peptide mixture, where peptide 5 is the desired product and peptides 1 to 4 (I_1 to I_4, respectively) represent hydrophilic impurities. Figure 2A shows an analytical separation profile of the five peptides. The ratio (w/w) of product to impurities was 1:1:1:1:2 (I_1:I_2:I_3:I_4:P_5), i.e., P_5 represented only 33% of the sample mixture. Total sample load was 21 mg, including 7 mg of P_5. Figure 2B shows the SDM elution profile of the sample mixture. Subsequent analysis of the fractions denoted Peaks I and II (Figure 2C and D, respectively) demonstrated excellent separation of P_5 in high yield from the four hydrophilic impurities. Peak II (Figure 2B), removed by gradient elution with acetonitrile following 40 min isocratic elution with water, was pure P_5 and accounted for 93% of total P_5 recovered. The remaining 7% was eluted prior to 47 min of the preparative run (Figure 2B) and was contaminated with I_4. The bulk of I_1 to I_4 was, as expected, eluted as an unretained fraction (Peak I) during isocratic elution with 0.05% aqueous TFA.

In addition to confirming the viability of the sample displacement approach to preparative purification of synthetic peptides, the results presented above also demonstrated that impressive yields of pure peptide products were obtainable even on small (30×4.6 mm I.D.) columns. Of course, the separations demonstrated in Figures 1 and 2, where the desired peptide product was the most hydrophilic or hydrophobic component, respectively, represented the simplest applications of a reversed-phase column in SDM. A stricter test of the effectiveness of SDM for preparative work would be its application to a crude peptide mixture containing impurities both more hydrophilic and hydrophobic than the desired peptide product. Thus, the next step in the furtherance of SDM as a preparative tool was to design a strategy for this more difficult, and more realistic, separation problem. In addition, scale-up to higher sample loads was also deemed a logical, and necessary, development.

III. PREPARATIVE PURIFICATION IN SDM OF A SYNTHETIC PEPTIDE FROM HYDROPHILIC AND HYDROPHOBIC IMPURITIES

A. STRATEGY

An outline of the strategy designed for more complex peptide separations is illustrated in Figure 3. In Step 1, the mixture to be separated has been applied at high load to the two-column system in 0.05% aq. TFA (pH 2.0) and is being eluted by the same mobile phase. Hydrophobic impurities have been retained by the precolumn; the desired peptide product has saturated the main column and displaced the hydrophilic impurities, resulting in their rapid elution from the system. In Step 2, the precolumn has been isolated at time t_1, trapping hydrophobic impurities. The valve enables isolation of the precolumn (C1) from the main column (C2) when only eluent flow through the main column is desired. At time t_2, all hydrophilic impurities have been washed off the main column by elution with 0.05% aq. TFA. The major separation of peptide components has now taken place and operation on SDM is now complete having only used the aqueous mobile phase. In Step 3, with the precolumn still isolated, a linear elution gradient has been initiated at time t_3 to remove the desired peptide product from the main column. In Step 4, the precolumn has been reconnected in series with the main column. Hydrophobic impurities have been removed from the precolumn by gradient elution and the whole two-column system is re-equilibrated to 100% Eluent A (0.05% [v/v] aq. TFA)

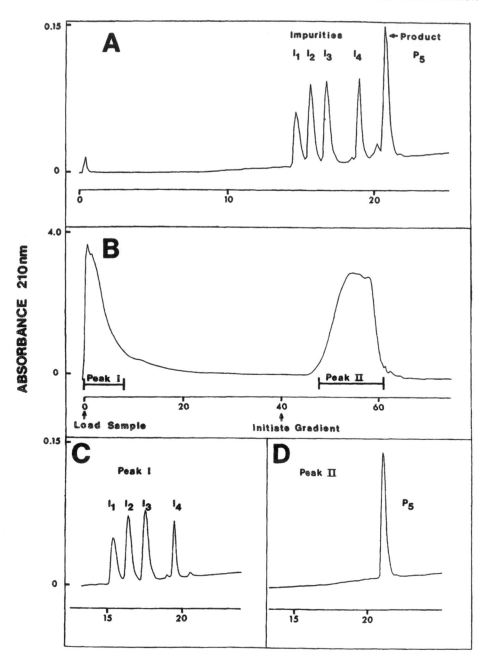

ELUTION TIME (min)

FIGURE 2. Separation by RPC operated in sample displacement mode of peptide product (P_5) from hydrophilic peptide impurities (I_1, I_2, I_3, I_4). HPLC instrument and column: as Figure 1. Panel A: analytical separation profile of peptide mixture; conditions, linear AB gradient (1% B/min) at a flow-rate of 1 ml/min, where Eluent A is 0.05% aqueous TFA and Eluent B is 0.05% TFA in acetonitrile. Panel B: preparative separation profile of peptide mixture; conditions, isocratic elution with 100% Eluent A for 40 min at a flow-rate of 1 ml/min, followed by linear gradient elution at 1% B/min; sample load, 21 mg consisting of 7.0 mg of P_5 and 3.5 mg of each of I_1, I_2, I_3 and I_4 dissolved in 500 µl of Eluent A. Fractions were collected at 1-min intervals. Panels C and D show analytical elution profiles (same conditions as Panel A) of Peaks I and II (Panel B), respectively. The subscripts of I_1, I_2, I_3, I_4, and P_5 denote peptides 1 to 5, respectively (Table 1). (From Burke, T. W. L., Mant, C. T., and Hodges, R. S., *J. Liq. Chromatogr.*, 11, 1229, 1988. With permission.)

STEP 1
Mode - Sample Displacement
Solvent - A = 0.05% TFA/H2O
Valve Position - Connect Precolumn C1
Valve
Sample Injector
Main Column C2
C1
Separation of hydrophobic impurities
from product
and hydrophilic impurities
STEP 2
Mode - Sample Displacement
Solvent - A = 0.05% TFA/H2O
Valve Position - Bypass Precolumn C1
C2
C1
Isolation of hydrophobic impurities on precolumn C1 at time t1.
Hydrophilic impurities are washed off column C2 at time t2.

FIGURE 3. Strategy for preparative RPC in sample displacement mode (SDM). In the present study, the precolumn (C1) consisted of two 30-mm (4.6 mm I.D.) Aquapore RP300 C_8 cartridges (7-µm particle size, 300-Å pore size; Brownlee Labs., Santa Clara, CA) connected in series; the main separating column (C2) was a 220 × 4.6 mm I.D. column, containing the same reversed-phase packing material as the precolumn. The valve was a Rheodyne sample injector (Model 7125; Rheodyne, Cotati, CA) arranged so that, when in load position, the precolumn was isolated; when in inject position, the mobile phase flow was directed through both the precolumn and main column. Other details are described in the text. (From Hodges, R. S., Burke, T. W. L., and Mant, C. T., *J. Chromatogr.*, 444, 349, 1988. With permission.)

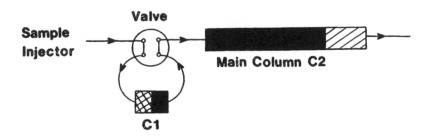

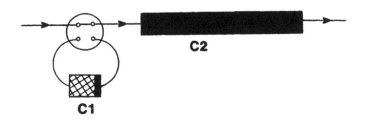

STEP 3

Mode   –   Linear AB Gradient – Isocratic Hold

Solvent –   A = 0.05% TFA/H_2O

B = 0.05% TFA/Acetonitrile

Valve Position   –   Bypass Precolumn C1

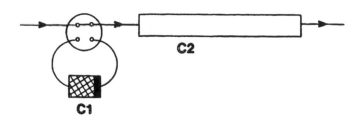

Initiate gradient at time t_3 followed by isocratic hold to elute product from column C2.

STEP 4

Mode   –   Gradient Wash and Equilibration

Solvent –   A = 0.05% TFA/H_2O

B = 0.05% TFA/Acetonitrile

Valve Position   –   Connect Precolumn C1

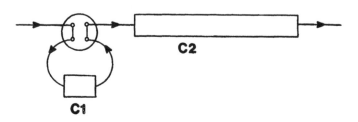

Removal of hydrophobic impurities from precolumn C1 and equilibration to 100% A.

FIGURE 3 (continued)

B. RESULTS OF SDM STRATEGY

Figure 4 shows the results obtained following application of the SDM strategy outlined above to the preparative separation of a single peptide component from a complex peptide mixture. Figure 4A shows the analytical separation profile of a mixture of the five synthetic peptides from Table 1. This mixture was again designed to represent crude peptide mixtures typically produced by solid-phase peptide synthesis. The desired product (P_3) is contaminated by hydrophilic (I_1, I_2) and hydrophobic (I_4, I_5) impurities. The ratio of product to impurities was 1:1:10:1:1 (I_1:I_2:P_3:I_4:I_5), i.e., P_3 represented ca. 71.4% of the crude peptide mixture. Figure 4B shows the

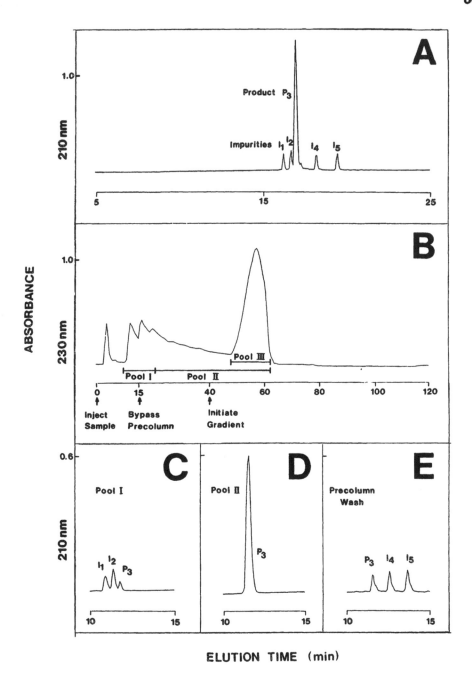

FIGURE 4. Preparative reversed-phase sample displacement chromatography of a peptide product (P_3) from hydrophilic (I_1, I_2) and hydrophobic (I_3, I_4) peptide impurities. HPLC instrument: same as Figure 1. Columns: see Figure 3. Panel A: analytical chromatogram of a synthetic peptide mixture on two-column system; conditions, linear AB gradient (2% B/min) at a flow-rate of 1 ml/min, where Eluent A is 0.05% aq. TFA and Eluent B is 0.05% TFA in acetonitrile. Panel B: preparative separation profile of peptide mixture; conditions, isocratic elution with 100% Eluent A for 40 min at a flow-rate of 1 ml/min, followed by linear AB gradient elution at 1% B/min; at 15 min, the precolumn was isolated from the main column. Sample load: 63 mg, consisting of 45 mg of P_3 and 4.5 mg of each of I_1, I_2, I_4, and I_5, dissolved in 1500 µl of Eluent A. Fractions were collected at 1-min intervals. Panels C and D: analytical chromatograms of pools I and II, respectively, on an Aquapore RP300 C_8 column (30 × 4.6 mm I.D.). Panel E: analytical chromatogram of peptide components retained by the precolumn (see Panel A for conditions). The subscripts of I_1, I_2, P_3, I_4, and I_5 denote peptides 1 to 5, respectively (Table 1). (From Hodges, R. S., Burke, T. W. L., and Mant, C. T., *J. Chromatogr.*, 444, 349, 1988. With permission.)

preparative chromatogram of 63 mg of the sample mixture. The sample contained 45 mg of P_3 and 4.5 mg of each of I_1, I_2, I_4, and I_5. Following injection, the sample mixture was eluted in SDM with 0.05% aq. TFA, the eluent flow being directed through both the precolumn and main column (Figure 3, Step 1). After 15 min (t_1 in Figure 3, Step 2), the precolumn was isolated to trap the hydrophobic impurities, I_4 and I_5. Continued elution with 0.05% aq. TFA ensured displacement of all hydrophilic impurities, I_1 and I_2, from the main column (t_2 in Figure 3, Step 2) and the operation in SDM was now complete. A linear elution gradient (1% B/min) was initiated at 40 min (t_3 in Figure 3, Step 3) to elute product P_3 from the main column. The final step involved removal of hydrophobic impurities, I_4 and I_5, from the precolumn by a linear gradient, followed by re-equilibration of the two-column system to 0.05% aq. TFA (Figure 3, Step 4). The fractions pooled as pools I, II, and III (Figure 4B) were subjected to gradient elution analysis. Pool I (Figure 4C) contained 100% of the hydrophilic impurities, I_1 and I_2, and 3% of recovered P_3 (Table 2). Interestingly, only one fraction (21 min) contained all three components. Pool II (Figure 4D) (incorporating Pool III, which contained 61% of the recovered P_3) contained pure P_3, representing 89% of recovered homogeneous product (Table 2). Analysis of components retained by the precolumn (Figure 4E) revealed the presence of 100% of hydrophobic impurities, I_4 and I_5, and 8% of recovered P_3.

C. EFFECT OF SAMPLE LOAD

The success of the SDM strategy outlined in Figure 3 was particularly encouraging, encompassing as it did not only a more difficult separation problem than those posed in Figures 1 and 2, but also a significantly higher sample load. The major parameter affecting the efficacy of SDM chromatography of peptides with water should be peptide load. Figure 5 shows preparative elution profiles of the crude peptide mixture (I_1, I_2, P_3, I_4, and I_5), applied to the two-column system outlined in Figure 3 at sample loads of 47 mg (Panel A), 63 mg (Panel B), and 84 mg (Panel C). The ratio of product to impurities remained constant (Figure 4). Following isolation of the precolumn at 15 min, elution in SDM with 0.05% aq. TFA was continued until the linear gradient (1% B/min) was started at 40 min. Recoveries of peptide product, P_3, in the various pooled fractions illustrated in Figure 5 are shown in Table 2. The recoveries of pure P_3 were impressive for all three sample loads. The major effect of increasing the sample load was to increase the rate at which the hydrophilic impurities, I_1 and I_2, were displaced from the system. Thus, for the 47-mg load, pure P_3 was obtained after 49 min; for the 63-mg load, pure P_3 was obtained after 24 min; for the 84-mg load, pure P_3 was obtained after only 14 min. In all three runs, the precolumn was sufficient to remove all hydrophobic impurities from the peptide mixture. As the load was increased, the amount of product, P_3, found on the precolumn decreased (13, 9, and 2% for the 47-, 63-, and 84-mg loads, respectively). This was expected, since increasing the sample load, increases the amount of hydrophobic impurities and more product is displaced from the precolumn trap.

IV. FEATURES OF SDM

A. CHROMATOGRAPHIC CONDITIONS
1. Flow-Rate

A flow-rate of 1 ml/min is possible on analytical columns in SDM, thereby allowing rapid separations, since the time required for the initial major separation in water is considerably less than that required to produce ideal (isotachic) conditions in traditional displacement chromatography. In addition, column regeneration to a 100% aqueous mobile phase is very rapid at a flow-rate of 1 ml/min.

TABLE 2
Recovery of Peptide Product (P₃) Following Preparative Reversed-Phase Sample Displacement Chromatography

Total load (mg)	Figure	Product (P_3) recovery in each pool (%)[a]					Total homogeneous product (P_3) recovered (%)
		Pool 1	Pool II	Pool III	Pool IV	Precolumn	
63	4B	3	89	61	—	8	89
47	5A	<<1	—	75	12	13	87
63	5B	2	80	49	9	9	90
89	5C	3	90	36	5	2	95

[a] Quantity of product (P_3) in each pool as a percentage of total product (P_3) recovered, using relative peak areas.

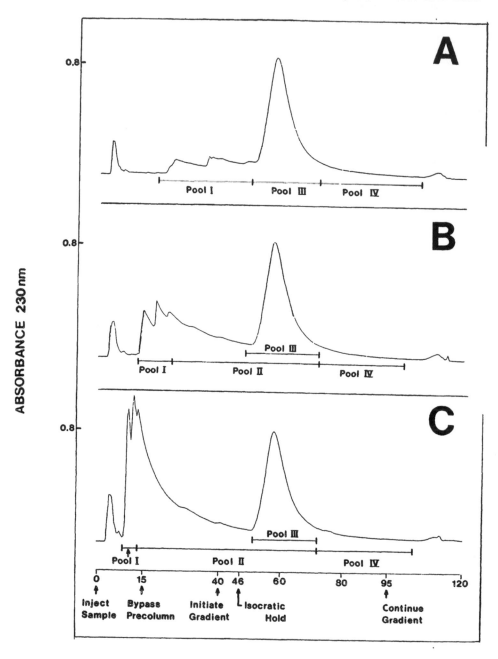

FIGURE 5. Effect of sample load on preparative reversed-phase sample displacement chromatography of a peptide mixture. HPLC instrument: same as Figure 1. Columns: see Figure 3. Conditions: isocratic elution with 100% Eluent A for 40 min at a flow-rate of 1 ml/min, followed by linear AB gradient elution at 1% B/min, isocratic hold (6% B) from 46 to 95 min, and continued linear gradient elution at 1% B/min (see text for discussion of optional use of an isocratic hold); at 15 min, the precolumn was isolated from the main column; Eluent A was 0.05% aq. TFA and Eluent B was 0.05% TFA in acetonitrile. Panel A: sample load, 47 mg, consisting of 34 mg of P_3 and 3.3 mg of each of I_1, I_2, I_4, and I_5, dissolved in 1700 μl of Eluent A. Panel B: sample load, 63 mg, consisting of 45 mg of P_3 and 4.5 mg of each of I_1, I_2, I_4, and I_5, dissolved in 2300 μl of Eluent A. Panel C: sample load, 84 mg, consisting of 60 mg of P_3 and 6 mg of each of I_1, I_2, I_4, and I_5, dissolved in 2300 μl of Eluent A. The subscripts of I_1, I_2, P_3, I_4, and I_5 denote peptides 1 to 5, respectively (Table 1). (From Hodges, R. S., Burke, T. W. L., and Mant, C. T., *J. Chromatogr.*, 444, 349, 1988. With permission.)

2. Organic Modifier

Treatment with an organic modifier, such as acetonitrile, is only required to wash retained components off the main separation column following preparative reversed-phase purification in SDM and takes no part in the major separation process. This treatment, usually in the form of a linear water/organic modifier gradient, requires minimal use of costly HPLC-grade solvents. In contrast, much larger volumes of these solvents would typically be required to generate yields of purified product similar to that achieved by SDM (Table 2) if traditional gradient elution techniques were applied to the kinds of sample loads shown in Figures 1, 2, 4, and 5.

3. Absence of Added Displacer

The difficulty of choosing the most suitable displacer for a particular separation in traditional displacement chromatography is well known. In addition, some of the more effective displacers tend to be somewhat toxic. In contrast, the sample components of a peptide mixture during preparative separation in SDM act as their own displacers.

B. DETECTION

The value of any preparative method should be assessed not only by its effectiveness in separating large loads of sample mixtures as rapidly as possible, but also by the ease with which desired homogeneous product(s) may be detected and pooled. An important characteristic of the preparative chromatograms shown in Figures 1, 2, 4, and 5 is the simple visualization of the major solute zones following operation in SDM and subsequent gradient elution. This is frequently not the case in traditional displacement chromatography, where the amount of material required for efficient development of the displacement train may overload the detector response, producing chromatographic profiles with no easily identifiable solute zones.[3,6] Similar problems in identifying solute zones also frequently occur when preparative amounts of a peptide mixture are separated by linear gradient elution.[14] Extensive fraction analysis is then necessary, unlike the relatively few fraction analyses required for complete assignment of components in the separation profiles presented in Figures 1, 2, 4, and 5. When analyzing fractions, isocratic elution is generally simpler if a large number of samples have to be analyzed, since there is no need for column regeneration. However, when only a small number of fractions have to be checked, such as is the case for SDM, gradient elution has the advantage of reproducibility of retention time, thereby making identification easier.

To maintain the UV profile at very high peptide loads, all that is required is to increase the wavelength typically utilized for detecting peptide bonds (205 to 210 nm) to 220 to 230 nm (Figures 4 and 5), or to decrease the pathlength of the flow-cell.

C. SAMPLE LOAD
1. Optimization

The ideal load for SDM operation for a given crude product on any given column is the load that will displace the hydrophilic impurities rapidly from the column and provide the maximum yield of homogeneous product that is easily visualized with a minimum number of analyses. In the case of the preparative separations illustrated in Figures 4 and 5, the maximum load of sample mixture was defined as the load in which the peptide product (P_3) was just eluted from the separating column (C2 in Figure 3) with 0.05% aq. TFA after 20 min. In the examples used in this article, a load of 47 to 60 mg of crude product produced a maximum yield of homogeneous product (in Pool III) and rapid elution of the hydrophilic impurities (within the first 20 min) (see Figure 5A and B and Table 2).

It is interesting to note that Pool III in the elution profiles shown in Figures 4 and 5 always contained large levels of homogeneous product, P_3, i.e., no analysis of this peak would be needed to guarantee the purity of the desired peptide product. The amount of P_3 in Pool III of these preparative chromatograms in SDM represented the maximum level of the peptide which could be applied to the main reversed-phase column (C2) without overloading the column. In this case, with the 22-cm analytical column, Pool III yielded ca. 25 mg of homogeneous P_3. Thus, a range of sample load between 47 and 84 mg quickly produced a significant amount of pure product, even though the highest load (84 mg) produced an overload situation according to the definition of maximum sample load applied to the SDM strategy (see above). In the case of the 84-mg sample load, peptide product was detected after just 14 min of isocratic elution with water instead of the preferred 20 min. Thus, there is considerable flexibility in the amount of sample that can be successfully separated in SDM, particularly in conjunction with the use of a precolumn to trap hydrophobic impurities (see below). Erring on the side of loading too much sample may not necessarily be a problem, particularly if the desired product is one of the more hydrophobic components. Any product eluted too quickly can be pooled and subjected to a further round of preparative RPC in SDM.

2. Value of Precolumn

Perhaps the most important consideration when attempting to optimize preparative SDM separations is the presence and size of the precolumn (Figure 3). When the preparative separation shown in Figure 4 was carried out without the use of a precolumn,[14] the level of recovered pure product (P_3) was still very high (98%), but the presence of hydrophobic impurities behind the product elution zone increased considerably the number of fractions that had to be analyzed to make the desired pool of pure product, compared to the small number required in the presence of the precolumn trap.

As illustrated in Figure 5, the use of a precolumn makes assessment of the proper sample load much less critical as long as the precolumn is long enough to trap the hydrophobic impurities. If the concentration of hydrophobic impurities in the crude peptide mixture is higher, the precolumn can be lengthened. Thus, the researcher can regulate the size of the precolumn (Figure 3, C1) depending on the amount of hydrophobic impurities in a particular sample, or the size of the main column (Figure 3, C2) depending on the amount of product desired.

3. Optional Isocratic Hold

An optional isocratic hold may be included at step 3 in the scheme shown in Figure 3, where the desired peptide product is removed from the main column, to ensure separation of product from any hydrophobic impurities which were not trapped on the precolumn because of an incorrect estimate of the size of the precolumn. A gradient-isocratic hold-gradient sequence required to isolate efficiently and rapidly the desired peptide product from closely related hydrophobic impurities is more difficult to optimize compared to simple gradient elution (as was utilized in Figure 4B), unless the researcher is sufficiently familiar with the elution properties of the peptide of interest. If gradient elution is employed, little knowledge is required about the percentage of organic modifier required to elute the product of interest, although the researcher must be confident that all hydrophobic impurities have been trapped on the precolumn. In fact, it is usually better to err on the long side with the precolumn if there is any uncertainty about the required length, since any desired product trapped on the precolumn may always be removed and subsequently pooled for a further round of SDM. However, if the researcher is familiar enough with a particular peptide mixture, the use of an isocratic hold (Figure 5), or even total isocratic elution of the desired peptide product, with subsequent reduction in consumption of expensive HPLC solvents, may contribute appreciably to cost savings. This could be particularly important for large-scale industrial applications.

V. CONCLUSIONS

The authors are currently investigating the effects on peptide resolution and yield of SDM run parameters, such as flow-rate, column length and diameter, sample concentration, and sample volume. In addition, the limits of SDM in terms of peptide chain length and hydrophobicity are being examined. A multi-column approach to sample displacement chromatography of peptides also appears extremely promising.[16] The potential of SDM as a preparative tool is considerable and should prove of great value to researchers involved in the purification of synthetic peptides, in the isolation of a single biologically active peptide from a crude peptide mixture or in the isolation of a single peptide component from a protein digest. SDM offers the potential of increased loading capacity, resolving power, and ease of localizing the desired pure component while yielding relatively rapid purification at standard flow-rates. Finally, a recent report[17] of ion-exchange sample displacement separations of proteins also suggests a future for this technique for separations of large polypeptides.

ACKNOWLEDGMENTS

This work was supported by the Medical Research Council of Canada and equipment grants from the Alberta Heritage Foundation for Medical Research.

REFERENCES

1. **Horváth, Cs., Nahum, A., and Frenz, J.H.,** High-performance displacement chromatography, *J. Chromatogr.,* 218, 365, 1981.
2. **Horváth, Cs. and Kalász, H.,** Preparative-scale separation of polymyxins with an analytical high-performance liquid chromatography system by using displacement chromatography, *J. Chromatogr.,* 215, 295, 1981.
3. **Verzele, M., Dewaele, C., van Dijck, J., and van Haver, D.,** Preparative-scale high-performance liquid chromatography on analytical columns, *J. Chromatogr.,* 249, 231, 1982.
4. **Horváth, Cs., Frenz, J., and El Rassi, Z.,** Operating parameters in high-performance displacement chromatography, *J. Chromatogr.,* 255, 273, 1983.
5. **Frenz, J., van der Schrieck, Ph., and Horváth, Cs.,** Investigation of operating parameters in high-performance displacement chromatography, *J. Chromatogr.,* 330, 1, 1985.
6. **Vigh, Gy., Varga-Puchony, Z., Szepesi, G., and Gazdag, M.,** Semi-preparative high-performance reversed-phase displacement chromatography of insulins, *J. Chromatogr.,* 386, 353, 1987.
7. **Cramer, S.M., El Rassi, Z., and Horváth, Cs.,** Tandem use of carboxypeptidase Y reactor and displacement chromatography for peptide synthesis, *J. Chromatogr.,* 394, 305, 1987.
8. **Valkó, K., Slégel, P., and Báti, J.,** Displacement chromatography of oligomycins, *J. Chromatogr.,* 386, 345, 1987.
9. **Subramanian, G., Phillips, M.W., and Cramer, S.M.,** Displacement chromatography of biomolecules, *J. Chromatogr.,* 439, 341, 1988.
10. **Viscomi, G., Lande, S., and Horváth, Cs.,** High-performance displacement chromatography of melanotropins and their derivatives, *J. Chromatogr.,* 440, 157, 1988.
11. **Cramer, S.M. and Horváth, Cs.,** Displacement chromatography in peptide purification, *Prep. Chromatogr.,* 1, 29, 1988.
12. **Burke, T.W.L., Mant, C.T., and Hodges, R.S.,** A novel approach to reversed-phase preparative high-performance liquid chromatography of peptides, *J. Liq. Chromatogr.,* 11, 1229, 1988.
13. **Hodges, R.S., Burke, T.W.L., and Mant, C.T.,** A novel approach to the purification of peptides by reversed-phase HPLC: sample displacement chromatography, in *Peptides: Chemistry and Biology—Proceedings of the Tenth American Peptide Symposium,* Marshall, G.R., Ed., Escom Science Publishers, Leiden, The Netherlands, 1988, 226.
14. **Hodges, R.S., Burke, T.W.L., and Mant, C.T.,** Preparative purification of peptides by reversed-phase chromatography. Sample displacement mode *versus* gradient elution mode, *J. Chromatogr.,* 444, 349, 1988.
15. **Parker, J.M.R. and Hodges, R.S.,** I. Photoaffinity probes provide a general method to prepare synthetic peptide-conjugates, *J. Protein Chem.,* 3, 465, 1985.
16. **Hodges, R.S., Burke, T.W.L., and Mant, C.T.,** Multi-column preparative reversed-phase sample displacement chromatography of peptides, *J. Chromatogr.,* in press.
17. **Veeraragavan, K., Bernier, A., and Braendli, E.,** Sample displacement mode chromatography: purification of proteins by use of a high-performance anion-exchange column, *J. Chromatogr.,* 541, 207, 1991.

DISPLACEMENT CHROMATOGRAPHY OF PEPTIDES AND PROTEINS

Firoz D. Antia and Csaba Horváth

I. LINEAR AND NONLINEAR CHROMATOGRAPHY

The term chromatography is usually associated with separations carried out in the elution mode when a relatively small amount of a mixture is applied to the column at the inlet and the components are separated into individual bands as they move down the column with different velocities upon the action of the eluent flow. Elution chromatography with samples so small that their equilibrium distribution between the stationary and mobile phases follows a linear law, i.e., it can be expressed by a constant distribution coefficient over the concentration range employed, has reached prominence in analytical applications in both gas chromatography and HPLC. A great advantage of such a linear chromatographic system is that retention times are independent of the amount injected, and the bands of quasi-Gaussian shape can be completely separated. Since they appear as distinct peaks, the chromatogram can be easily evaluated to extract analytical information about the sample components and the composition of the sample. Linear elution chromatography is well understood theoretically and has been employed in various forms of chromatography over the last 30 years. It is an admirably versatile analytical tool, and is mainly responsible for the wide popularity of chromatography.

In preparative separations, the employment of linear elution has the great advantage that the vast amount of experience accumulated in analytical chromatography can be used for the development of the separation process. On the other hand, in order to maintain linearity, the size of the sample has to be small with respect to the amount of stationary phase in the column so that only a small fraction of the chromatographic surface is occupied by the adsorbed eluite molecules. However, because of the low eluite concentrations in both the mobile and stationary phases, the column is poorly utilized. Furthermore, bandspreading in elution chromatography is tantamount to dilution which invariably occurs with isocratic elution and is associated with high solvent consumption. As a result, the product has to be recovered from a dilute solution and the economics of the process is adversely affected by the high solvent need. Increasing the amount of sample load can increase the throughput on a given column. However, the system becomes nonlinear, i.e., the equilibrium distribution no longer follows a linear law, and some of the above mentioned benefits of linear elution are lost. Nevertheless, it is often beneficial to conduct preparative separations under conditions of overload, in the nonlinear regime.

In nonlinear chromatography, the adsorption behavior of the sample components and their movement down the column is governed by the adsorption isotherm which relates the concentration in the stationary phase (q) to that in the mobile phase (c) at equilibrium and is shown schematically in Figure 1. Now the velocity of the component (u) will depend on its concentration and is determined by the slope of the straight line (the chord) drawn from the origin to the point on the isotherm which corresponds to the mobile phase concentration of the component, according to the relationship:

$$u = \frac{u_o}{1 + \phi \dfrac{\Delta q}{\Delta c}} \tag{1}$$

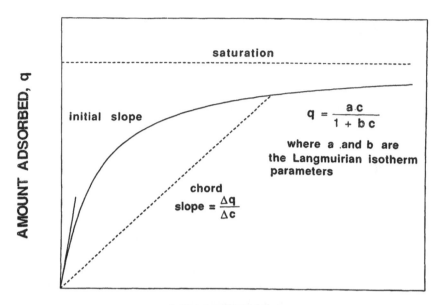

FIGURE 1.   Langmuirian adsorption isotherm of a single component. The initial slope representing the partition coefficient determines the velocity at infinite dilution (linear conditions), the saturation level represents the highest possible amount adsorbed on the stationary phase, and slope of the chord determines the velocity given in Equation 1 in the text.

where u_o is the velocity of the mobile phase, and ϕ is the phase ratio of the column and $\Delta q/\Delta c$ is the slope of the chord.[1]

In contradistinction, at sufficiently low concentrations the isotherm can be considered linear so that the ratio of concentrations in the stationary and mobile phases is constant, $q/c = K$, where K is the equilibrium constant for the component. Thus, under conditions of linear chromatography, the velocity of the molecules is constant in the concentration domain of interest. At higher concentrations, the nonlinear adsorption behavior has several consequences; the most important is that the velocities will be concentration dependent. It follows then that the substance band moves with velocities depending on the local concentrations that result in asymmetrical bands and the "retention time" will also be concentration dependent. With concave down isotherms such as the Langmuirian isotherm illustrated in Figure 1, self-sharpening front boundaries and diffuse rear boundaries, such as shown in Figure 2, will also be observed, and they are typical for column overload, the source of nonlinear behavior in chromatography.

So far, we have considered only a single component. In practice, the separation of two or more components is the goal. In linear chromatography of a multi-component mixture, the components do not interfere with each other's adsorption at the stationary phase and, as a result, they move down the column independently. In nonlinear chromatography, this is no longer the case. The interference manifests itself in a competitive adsorption behavior and the adsorption isotherm of each component depends on the concentration and isotherm parameters of all other components and this is expressed by an appropriate multi-component isotherm. From a simple model of adsorption, the multi-component Langmuir isotherm can be derived as:[2]

$$q_i = \frac{a_i c_i}{1 + \sum_{j=1}^{N} b_j c_j} \tag{2}$$

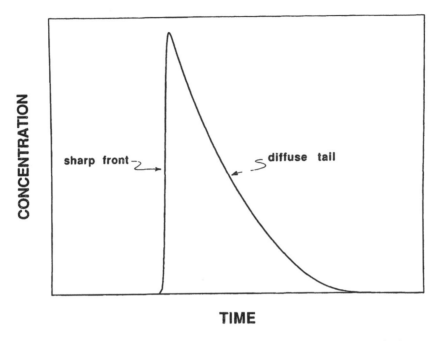

FIGURE 2. Elution profile of a single component having a Langmuirian adsorption isotherm. With such an isotherm, velocity increases with concentration causing the front to sharpen (self-sharpening front boundary) and the tail to spread (diffuse rear boundary) in the manner shown.

where q_i and c_i are the concentrations of component i in the stationary and mobile phases, respectively, N is the number of components under consideration, and a_i and b_j are the parameters of the single component isotherms for species i or j. In nonlinear elution chromatography, the separation is governed by the multicomponent isotherm relationship. It should be noted that although the adsorption behavior of a single component can often be fitted to the equation given in Figure 1, the competitive Langmuir adsorption isotherm in Equation 2 seldom reflects physical reality. Therefore more elaborate multicomponent isotherms have to be used in the treatment of nonlinear chromatography.

II. ELUTION AND DISPLACEMENT

In elution chromatography, the adsorption isotherms of the main eluent components underlie the isotherms of the sample components, i.e., the sample components bind more strongly to the chromatographic surface than does the eluent. For instance, in reversed-phase chromatography (RPC), water, the organic modifier, the inorganic and most organic buffer components bind to the stationary phase to a lesser extent than the substances to be separated. Only certain additives used under special circumstances in the eluent, e.g., amphiphilic substances in ion-pair chromatography, may exhibit stronger binding than the sample components to the chromatographic surface. In elution chromatography, the column is conditioned before sample injection so that the additive finds its equilibrium distribution between the two phases in the column. In isocratic elution, therefore, neither the mobile nor the stationary phase experience significant composition changes with respect to the mobile phase components barring that which is associated with the so-called system peaks. In gradient elution, there is a gradual increase in the concentration of the mobile phase modulator, e.g., the organic modifier in RPC or the salt in ion-exchange chromatography (IEC). However, since the isotherm of the modulator underlies those of the sample components, there is no qualitative difference in the two types of elution. In

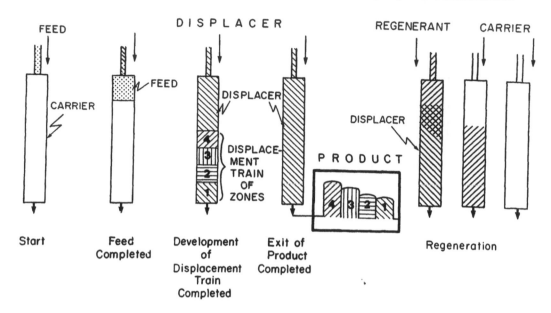

FIGURE 3. Operating steps in displacement chromatography. The column is first equilibrated with the carrier. The feed is then loaded followed by a continuous stream of displacer solution. The feed components separate into a train of adjacent bands, all moving at the velocity of the displacer front (hence the term 'isotachic'). After the purified components are collected, the column is regenerated and re-equilibrated with carrier. (From Horváth, Cs., Nahum, A., and Frenz, J., *J. Chromatogr.*, 218, 365, 1981. With permission.)

stepwise elution, the concentration of the mobile phase modulator is increased suddenly to a level such that all the substances or at least some of them are desorbed completely upon flushing the column with the mobile phase rich in the modifier. Because it binds more weakly to the stationary phase than the substance(s) it desorbs from the chromatographic surface as a result of the change in the thermodynamic conditions, this process is also elution although it has occasionally been called displacement in the literature.

In classical displacement chromatography, the feed is introduced into the column that is equilibrated with a mobile phase called the *carrier*, all the components of which bind more weakly to the stationary phase than any of the feed components. The introduction of a relatively large amount of sample (as compared with that customary in linear or quasi-linear elution chromatography) is followed by a continuous stream of the displacer, as shown in Figure 3. When the column is long enough, the sample components will form adjacent rectangular bands as they move down the column and develop an isotachic final pattern, as first elucidated by Tiselius[3] and illustrated in Figure 4. The displacer binds more strongly to the stationary phase surface than any of the feed components, i.e., its adsorption isotherm overlies theirs. The chord drawn from the origin to the point on the displacer isotherm corresponding to the displacer concentration is called the operating line, and its slope determines the velocity of the displacer front (*vide* Equation 1). The concentrations of the displaced substances in the final pattern are determined by the intersection of their respective isotherms with the operating line. The widths of the displaced bands depend on the amounts of the respective components in the feed, as mass conservation dictates. In contrast with isocratic elution, sample components may be concentrated during displacement. Departures from Tiselian, or ordinary, displacement may occur when the displacer is heterogeneous, with some constituents whose isotherms lie intercalated between those of the sample components so that groups of sample components are displaced by different displacer constituents.[4] Non-Tiselian displacement patterns may also occur if the feed component isotherms cross each other, if a multicomponent carrier is employed, or if there are

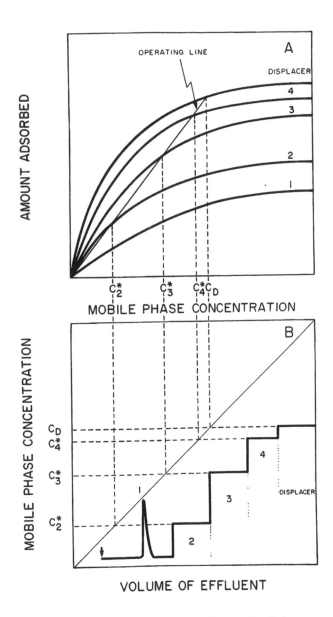

FIGURE 4. Relationship between the final pattern in displacement development and the isotherms of the displacer and feed components. The intersection of the operating line with the feed component isotherms determines their concentrations in the displacement train. The isotherm of component 1 lies beneath the operating line, hence it is eluted in the carrier. (From Horváth, Cs., Nahum, A., and Frenz, J., *J. Chromatogr.*, 218, 365, 1981. With permission.)

interactions between the components. Nevertheless, the key practical advantages of displacement can frequently be exploited even under such conditions.

During the feed step itself, in both elution and displacement chromatography, some separation takes place: the least retained component moves ahead of the others and, under overloaded conditions when the isotherms are concave downward, is concentrated to a level greater than in the feed. If the feed were to be introduced continuously, a pure fraction of this component could be recovered before the breakthrough of the other components, and this

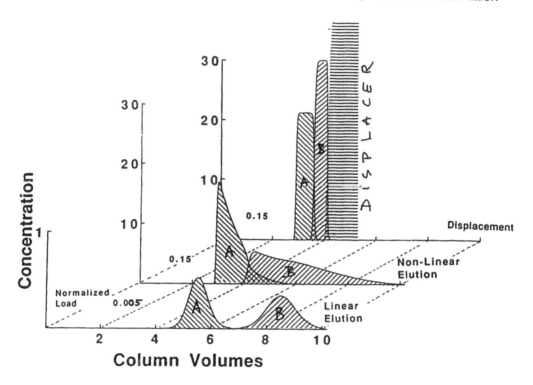

FIGURE 5. Comparison of isocratic linear elution, isocratic nonlinear elution and displacement chromatography of a 1:1 binary mixture. The profiles were generated on a computer by using a mathematical model described in References 10 and 14. Feed concentration of each component is 10 units on the ordinate. The normalized load represents the mass of each component normalized to the maximum saturation capacity of the column; one 'column volume' represents an effluent volume equal to the holdup volume of the column.

process has been called frontal chromatography by Tiselius.[5] A scheme employing frontal chromatography followed by stepwise or gradient elution has recently been termed 'sample displacement'.[6] Tandem methods utilizing frontal chromatography followed by either stepwise elution or displacement proper have also been used recently for protein separations.[7] All such schemes show the features of nonelution nonlinear chromatography that can be exploited profitably in preparative separations.

The displacement effect, i.e., the phenomenon of one component being driven ahead and concentrated due to the presence of a more strongly bound component, may be manifest in frontal chromatography or in overloaded isocratic or gradient elution[8-10] and this has sometimes been called 'sample self-displacement'.[11] The difference between linear elution, overloaded isocratic elution with sample self-displacement, and displacement proper is illustrated in Figure 5 for a binary sample. It is evident that in 'self-displacement' only the early eluted component is displaced, and the recovery of the trailing component is adversely affected by the tailing of the former, whereas in displacement proper both components are well separated and there is no excessive tailing.

Elution effects are likewise sometimes manifest in displacement chromatography and these usually occur when one or more of the feed component isotherms lie to the right of the operating line (*vide* Figure 4) or the process is terminated before the isotachic final pattern is reached. Experimental and theoretical illustrations of displacement chromatograms under such circumstances are available in the literature.[12-14] An exhaustive review of displacement chromatography, along with applications, can be found in Reference 15.

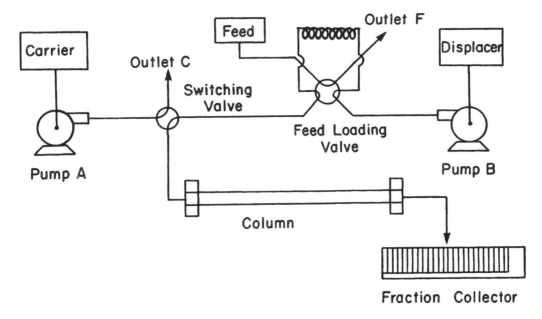

FIGURE 6. Flow sheet of a chromatograph for displacement. The switching valve controls whether the carrier or the displacer solution, delivered by pump A and B respectively, is fed to the column. The feed solution is stored in the feed loop; the loading valve permits pulseless introduction of feed into the column ahead of the displacer. Fractions of the effluent are collected and analyzed. (From Frenz, J. and Horváth, Cs., *AIChE J.*, 31, 400, 1985. With permission.)

III. EQUIPMENT AND OPERATING CONDITIONS

Save for the need for a large sample loop, and the means to switch from carrier to displacer to regenerant and back to carrier again, displacement chromatography on the 'preparative' scale is conveniently carried out in the laboratory with columns and equipment customary in analytical HPLC. A flow sheet is shown in Figure 6. Displacement exploits the binding capacity of the stationary phase more fully than do elution methods and consequently larger quantities can be purified with the same equipment.

The various operational steps in displacement chromatography are illustrated schematically in Figure 3. Following recovery of the products, the column must be stripped of the displacer in a regeneration step and re-equilibrated with the carrier before beginning the next run. A single regeneration step or multiple steps with suitable regenerants may be required.[16] Regeneration is facilitated by a change in conditions, e.g., in IEC the pH can be altered so that the displacer no longer binds strongly to the column.

The selection of a suitable displacer is a critical consideration in displacement chromatography. As discussed earlier, the displacer isotherm must overlie those of the sample components. In addition, the displacer must be sufficiently soluble in the carrier so that it can be used at relatively high concentrations to ensure that the operating line intersects the sample component isotherms at the appropriate concentrations. Furthermore, some displacers may form micelles and care must be taken to operate below the critical micelle concentration. Scouting for displacers can be tedious; thin layer chromatography (TLC) can be used fruitfully to screen several possible displacers simultaneously.[17] Most displacers employed so far are common chemicals; however, displacers can also be synthesized for a certain application.[18] A partial list of displacers that have been employed in RPC is given in Table 1.

TABLE 1
Displacers of Increasing
Affinity to Octadecyl-
Silica

Phenol
Butanol
Butoxyethanol
Propylene glycol
Pluracol o-285 polyether
Benzyldimethylammonium salts
Cetrimide
Palmitic Acid
Hexadecyltrimethylammonium salts

From Frenz, J. and Horváth, Cs., in
High Performance Liquid Chroma-
tography: Advances and Perspectives,
Vol. 5, Horváth, Cs., Ed., Academic
Press, New York, 1988, 211. With
permission.

The column length required for a given separation depends on the isotherms of the feed constituents, the quantity of feed, and the slope of the operating line, which is determined from the displacer concentration and its isotherm (*vide* Figure 4). As long as the displacer isotherm uniformly overlies those of the feed components, the particular initial slope and saturation level of the isotherm appears only to affect the sharpness of the boundary between the latest eluted feed component and the displacer and has little effect on the rest of the displacement train. The greater the displacer concentration, the lower is the slope of the operating line, and as a consequence the displacer front travels more rapidly through the column and the products are recovered at high concentrations. If product concentrations are too high, however, solubility limits might be exceeded and also the bands may be so narrow that their recovery is hindered by low column efficiency or by practical limitations in fraction collection.[13,19,20] The larger the part of the column occupied by the feed, the less is the available column length for displacement development. Clearly, operating conditions must be chosen such that a compromise is reached, and an optimum feed amount and displacer concentration can be selected to maximize production.[21]

At elevated temperatures, enhanced diffusivities, reduced fluid viscosity, and accelerated sorption kinetics tend to improve the efficiency of chromatographic separation.[22] For this reason, displacement chromatography has sometimes been carried out at temperatures up to 60°C.[23-26]

In preparative applications, monitoring the effluent by methods employed in analytical chromatography fails to yield the information required to recover the individual components. The practice most commonly followed is to collect fractions and analyze them individually. A more appropriate means is on-line analysis by rapid HPLC. This has already been described for displacement chromatography[21] and further advanced by progress in enhancement by speed of analysis with columns packed with micropellicular sorbents and by using elevated temperature as discussed in another chapter.

IV. SEPARATION OF PEPTIDES

Displacement chromatography was used as early as 1949 to fractionate protein hydroly-sates.[27] After a hiatus of nearly 30 years, high-performance displacement chromatography has been re-established as an important method for the preparative separation of peptides, and, as

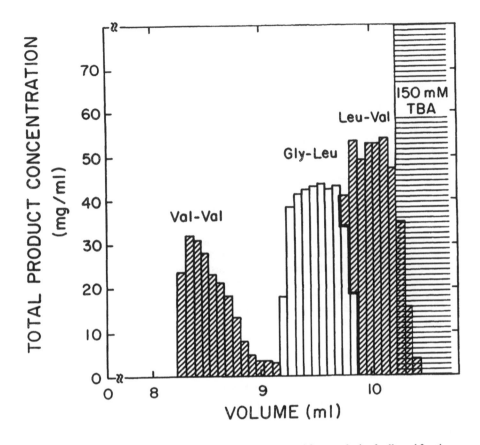

FIGURE 7. Displacement diagram of three dipeptides constructed from analysis of collected fractions. Column: 2×(250×4.6 mm); stationary phase: 5-μm Spherisorb-ODS; carrier: 50 m*M* phosphate buffer, pH 2.0; displacer: 150 m*M* tetrabutylammonium bromide in the carrier; flow rate: 0.72 ml/min; temperature: 30°C; fraction volume: 72 μl; feed: 20 mg valylvaline, 30 mg glycylleucine, and 30 mg leucylvaline in 1.0 ml of carrier. Under these conditions, valylvaline is eluted from the column, whereas glycylleucine and leucylvaline are displaced. (From Horváth, Cs., Frenz, J., and El Rassi, Z., *J. Chromatogr.*, 255, 273, 1983. With permission.)

in analytical HPLC, the most effective technique has been RPC with alkyl-silica stationary phases. Kalász and Horváth[28] used dodecyloctyldimethylammonium chloride as the displacer to separate more than 100 mg of commercial polymyxin B sulfate into its constituents on a 250 ×4.6 mm column in a single run. Horváth et al.,[23] while examining operating parameters in high-performance displacement chromatography, separated several dipeptides using a tetrabutylammonium bromide displacer. A chromatogram from this work is reproduced in Figure 7. Vigh et al.[29] purified bovine and porcine insulins on reversed-phase columns using hexadecyltrimethylammonium bromide (cetrimide) as the displacer. Cramer et al.[24,30,31] separated a number of di- and tri-peptides by displacement chromatography on an octadecyl-silica column using decyltrimethylammonium bromide, benzyltributylammonium bromide, butoxyethoxyethanol and cetrimide as displacers at 22 and 50°C. Viscomi et al.[32] used an aqueous solution of benzyldimethyldodecylammonium bromide to displace and purify 30 mg of a mixture containing the biologically active peptides α- and β- melanocyte stimulating hormone on a standard 250×4.6 mm octadecyl-silica column in a single run.

The introduction of microparticulate sorbents having a pellicular configuration, i.e., where the stationary phase proper is a thin spherical annulus around a fluid-impervious core, has made very rapid chromatographic analysis viable.[33-35] Such sorbents can also be used for small scale

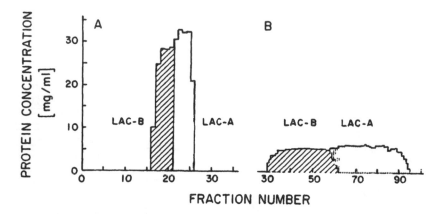

FIGURE 8. Displacement chromatograms of β-lactoglobulins (LAC) A and B with two different displacer concentrations. Column: TSK DEAE 5-PW, 75 × 7.5 mm; temperature: 22°C; carrier: 25 m*M* phosphate, pH 7.0; flow rate: 0.1 ml/min; fraction volume: 0.2 ml. (A) Feed: 62 mg of protein in 0.4 ml carrier; displacer: 20 mg/ml chondroitin sulfate dissolved in the carrier, and (B) feed: 70 mg protein in 0.4 ml carrier, displacer: 3 mg/ml chondroitin sulfate in the carrier. (From Liao, A. W., El Rassi, Z., LeMaster, D. M., and Horváth, Cs., *Chromatographia*, 24, 881, 1987. With permission.)

rapid preparative separations. With a 100 × 46 mm column packed with 2 μm pellicular octadecyl-silica, Fellegvári et al.[26] purified a mutant of the 26-residue polypeptide melittin, a constituent of bee venom, from a synthetic mixture using benzyldimethylhexadecylammonium chloride as the displacer. In 12 min, at 40°C and a flow rate of 0.6 ml/min, about 1.5 mg of pure product was obtained and the column was regenerated with formic acid, isopropanol, and acetonitrile-water (95:5 v/v) in less than 4 min.

Preparative separation of peptides has been accomplished by so-called sample displacement[6,36] and this is reviewed in the chapter by Mant and Hodges entitled "Preparative Reversed-Phase Sample Displacement Chromatography of Synthetic Peptides".

Displacement chromatography can be used to facilitate mass-spectrometric detection of trace sample constituents in LC-MS. By suitably adjusting the displacer concentration, the constituents can either be concentrated or made to be eluted in a relatively large eluent volume so that sufficient time is available for spectrometry. It promises to be a powerful tool for the characterization of complex mixtures and has been used recently to analyze tryptic digests of proteins produced by recombinant DNA technology.[37]

V. SEPARATION OF PROTEINS

Selection of a suitable displacer for the separation of proteinaceous compounds is challenging on several counts. Although in principle it is sufficient that the displacer isotherm overlies those of the feed, there is reason to believe that the particular adsorption behavior of large molecules may impede displacement. Thus, in addition to the properties mentioned previously, the displacer should preferably have molecular dimensions comparable with those of the sample components. Furthermore, particularly when the sample is a therapeutic protein, the displacer should be biologically inert.

The difficulty of finding macromolecular displacers with uniform molecular properties and narrow molecular weight distribution having the requisite attributes has limited somewhat the application of displacement methods in protein separations. Nevertheless, high-performance displacement separations of β-lactoglobulins A and B have been performed with a chondroitin sulfate displacer.[7,38] The well-resolved Tiselian displacement of the proteins by this method are shown in Figure 8. Other displacement separations of these proteins have been reported as well.[39-41] However, these used a multicomponent carboxymethyldextran displacer and, under the conditions employed, only β-lactoglobulin B was displaced, whereas A was eluted from the column. Anion-exchange columns and a chondroitin sulfate displacer have also been used to purify β-galactosidase.[42] A water-soluble coagulant, Nalcolyte 7105, and polyethyleneimine have been employed to displace mixtures of standard basic proteins on a cation-exchange column.[25,31]

The isotherms of some polypeptides and proteins on chromatographic surfaces are not uniformly concave downward, i.e., they are not Langmuirian, and this may hinder separation by displacement.[43] When isotherms cross, an adsorption azeotrope may form, and separation is impeded.[25,44-46] In either circumstance, the conditions may be altered, by adding or by changing the concentration of a carrier additive, to 'Langmuirize' or uncross the isotherms, respectively.

VI. DISPLACEMENT SEPARATIONS OF OTHER BIOMOLECULES

Several other important biomolecules have been separated by displacement chromatography with HPLC equipment. Corticosteroids,[17] nucleic acid fragments,[23,47] oligomycins,[48] and cephalosporin C[31] have been purified by displacement chromatography in reversed-phase systems with a variety of displacers. Chiral separations have also been performed in the displacement mode on reversed-phase columns with displacers expressly synthesized for the purpose.[18]

VII. OUTLOOK

Despite nearly three decades of dormancy while linear chromatography became the preeminent tool of chemical analysis, displacement, and other nonlinear methods are re-emerging as their advantages in preparative and other applications become apparent. Indeed, as the considerable body of work attests, displacement has proved itself as a powerful means for the purification of peptides and is likely to continue to be of increasing importance in this area. As research continues and appropriate displacers are identified or synthesized, displacement is likely to become an important technique in preparative protein purifications as well. It has already been demonstrated that displacement can be combined with other nonlinear chromatographic modes to achieve high throughout separation of proteins[6,7,49] and such hybrid schemes are expected to be used more frequently in the next decade, as biotechnology matures and more efficient and economic multicomponent purification methods become necessary.

Displacement chromatography has a sound theoretical foundation (cf. Reference 50) and efforts to expand the theory beyond ordinary displacement to include non-Tiselian effects and realistic multicomponent isotherms with a view towards optimization and scale-up are continuing. Because it is eminently suited to scale-up, displacement can be employed effectively at every stage in the development of a purification procedure. This is particularly useful when safety regulations require that the same process be used during product testing and production.

Besides its utility as a preparative procedure, recent work in coupling displacement with mass spectrometry[37] has demonstrated its usefulness as an analytical method. It appears that the vast potential of displacement chromatography as a preparative and analytical tool for the separation of biomolecules has only begun to be tapped.

ACKNOWLEDGMENTS

This work was supported by Grant No. GM20993 from the National Institutes of Health, U.S. Department of Health and Human Resources, as well as by the National Foundation for Cancer Research.

REFERENCES

1. **Helfferich, F. and Klein, G.**, *Multicomponent Chromatography*, Marcel Dekker, New York, 1970, 37.
2. **Schwab, G.M.**, *Ergebnisse der exacten Naturwisschaften*, Vol. 7, Julius Springer, Berlin, 1928, 276.
3. **Tiselius, A.**, Displacement development in adsorption analysis, *Ark. Kemi, Mineral Geol.*, 16A, No. 18 ,1, 1943.
4. **Torres, A. and Peterson, E.**, Displacement chromatography of simple protein mixtures, using carboxymethyldextrans, *J. Biochem. Biophys. Methods*, 1, 349, 1979.
5. **Tiselius, A.**, Studies on adsorption analysis, *Kolloid Z.*, 105, 101, 1943.
6. **Hodges, R.S., Burke, T.W.L., and Mant, C.T.**, Preparative purification of peptides by reversed-phase chromatography. Sample displacement mode versus gradient elution mode, *J. Chromatogr.*, 444, 349, 1988.
7. **Lee, A.L., Liao, A.W., and Horváth, Cs.**, Tandem separation schemes for preparative high-performance liquid chromatography of proteins, *J. Chromatogr.*, 443, 31, 1988.
8. **Guiochon, G. and Ghodbane, S.**, Computer simulation of the separation of a two-component mixture in preparative scale liquid chromatography, *J. Phys. Chem.*, 92, 3682, 1988.
9. **Snyder, L.R., Dolan, J.W., and Cox, G.B.**, Preparative purification of large biomolecules by reversed-phase gradient elution. Experimental and computer modeling studies, presented at "PREP-89" 6th Int. Symp. on Preparative Chromatography, Washington, D.C., May 8—10, 1989.
10. **Antia, F.D. and Horváth, Cs.**, Gradient elution in nonlinear preparative liquid chromatography, *J. Chromatogr.*, 484, 1, 1989.
11. **Newburger, J., Liebes, L., Colin, H., and Guiochon, G.**, Investigation of the influence of particle size on the productivity of preparative HPLC columns, *Sep. Sci. Technol.*, 22, 1933, 1987.
12. **Frenz, J. and Horváth, Cs.**, High performance displacement chromatography: calculation and experimental verification of zone development, *AIChE J.*, 31, 400, 1985.
13. **Katti, A.M. and Guiochon, G.A.**, Prediction of band profiles in displacement chromatography by numerical intergration of a semi-ideal model, *J. Chromatogr.*, 449, 25, 1988.
14. **Antia, F.D. and Horváth, Cs.**, Operational modes of chromatographic separation processes, *Ber. Bunsenges. Phys. Chem.*, 93, 961, 1989.
15. **Frenz, J. and Horváth, Cs.**, High-performance displacement chromatography, in *High Performance Liquid Chromatography: Advances and Perspectives*, Vol. 5., Horváth, Cs., Ed., Academic Press, New York, 1988, 211.
16. **Frenz, J. and Horváth, Cs.**, Movement of components in reversed-phase chromatography. III. Regeneration policies in liquid chromatography, *J. Chromatogr.*, 282, 249, 1983.
17. **Kalász, H. and Horváth, Cs.**, High-performance displacement chromatography of corticosteroids: scouting for displacer and analysis of the effluent by thin-layer chromatography, *J. Chromatogr.*, 239, 423, 1982.
18. **Camacho, P., Geiger, E., Farkas, G., Bartha, A., and Vigh, G.**, Efficient preparative-scale chiral separation using displacement chromatography, *J. Chromatogr.*, 1989, in press.
19. **Horváth, Cs., Nahum, A., and Frenz, J.**, High-performance displacement chromatography, *J. Chromatogr.*, 218, 365, 1981.
20. **Horváth, Cs.**, Displacement chromatography: yesterday, today and tomorrow, in *The Science of Chromatography*, Bruner, F., Ed., Elsevier, New York, 1985, 179.
21. **Frenz, J., van der Schrieck, P., and Horváth, Cs.**, Investigation of operating parameters in high-performance displacement chromatography, *J. Chromatogr.*, 330, 1, 1985.
22. **Antia, F.D. and Horváth, Cs.**, High-performance liquid chromatography at elevated temperatures: examination of conditions for the rapid separation of biopolymers, *J. Chromatogr.*, 435, 1, 1988.
23. **Horváth, Cs., Frenz, J., and El Rassi, Z.**, Operating parameters in high-performance displacement chromatography, *J. Chromatogr.*, 255, 273, 1983.
24. **Cramer, S. and Horváth, Cs.**, Displacement chromatography in peptide purification, *Prep. Chromatogr.*, 1, 29, 1988.
25. **Subramanian, G. and Cramer, S.M.**, Displacement chromatography of proteins under elevated flow rate and crossing isotherm conditions, *Biotechnol. Progress*, 5, 92, 1989.
26. **Fellegvári, I., Kalghatgi, K., and Horváth, Cs.**, Purification of melittin by fast displacement chromatography with micropellicular stationary phases, 1989, in preparation.
27. **Partridge, S.M.**, Displacement chromatography on synthetic ion-exchange resins, *Biochem. J.*, 45, 459, 1949.
28. **Kalász, H. and Horváth, Cs.**, Preparative-scale separation of polymyxins with an analytical high-performance liquid chromatography system by using displacement chromatography, *J. Chromatogr.*, 215, 295, 1981.
29. **Vigh, G., Varga-Puchony, Z., Szepesi, G., and Gazdag, M.**, Semi-preparative high-performance liquid chromatography of insulins, *J. Chromatogr.*, 386, 353, 1987.
30. **Cramer, S.M., El Rassi, Z., and Horváth, Cs.**, Tandem use of carboxypeptidase Y reactor and displacement chromatograph for peptide synthesis, *J. Chromatogr.*, 394, 305, 1987.

31. **Subramanian, G., Phillips, M.W., and Cramer, S.M.,** Displacement chromatography of biomolecules, *J. Chromatogr., 439,* 341, 1988.
32. **Viscomi, G., Lande, S., and Horváth, Cs.,** High-performance displacement chromatography of melanotropins and their derivatives, *J. Chromatogr., 440,* 157, 1988.
33. **Burke, D.J., Duncan, J.K., Dunn, L.C., Cummings, L., Siebert, J., and Ott, G.S.,** Rapid protein profiling with a novel anion-exchange material, *J. Chromatogr., 353,* 425, 1986.
34. **Unger, K.K., Jilge, G., Kinkel, J.N., and Hearn, M.T.W.,** Evaluation of advanced silica packings for the separation of biopolymers by high-performance liquid chromatography, *J. Chromatogr., 359,* 61, 1986.
35. **Kalghatgi, K. and Horváth, Cs.,** Rapid analysis of proteins and peptides by reversed-phase chromatography, *J. Chromatogr., 398,* 335, 1987.
36. **Burke, T.W.L., Mant, C.T., and Hodges, R.S.,** A novel approach to reversed-phase preparative high-performance liquid chromatography of peptides, *J. Liq. Chromatogr., 11,* 1229, 1988.
37. **Frenz, J., Bourell, J., and Hancock, W.S.,** High-performance displacement chromatography-mass spectrometry of tryptic fragments of biosynthetic proteins, presented at Ninth Int. Symp. on HPLC of Proteins, Peptides and Polynucleotides, Philadelphia, PA, November 6—8, 1989.
38. **Liao, A.W., El Rassi, Z., LeMaster, D.M., and Horváth, Cs.,** High-performance displacement chromatography of proteins: separations of β-lactoglobulins A and B, *Chromatographia, 24,* 881, 1987.
39. **Peterson, E.A.,** Ion-exchange displacement chromatography of serum proteins, using carboxymethyldextrans as displacers, *Anal. Biochem., 90,* 767, 1978.
40. **Peterson, E.A. and Torres, A.R.,** Ion-exchange displacement chromatography of proteins, using narrow-range carboxymethyldextrans and a new index of affinity, *Anal. Biochem., 130,* 271, 1983.
41. **Peterson, E.A. and Torres, A.,** Displacement chromatography of proteins, *Methods Enzymol., 104,* 113, 1984.
42. **Liao, A. and Horváth, Cs.,** Purification of β-galactosidase by combined frontal and displacement chromatography, *Anal. N.Y. Acad. Sci.,* 1989, in press.
43. **Huang, J.X. and Horváth, Cs.,** Adsorption isotherms on high-performance liquid chromatographic sorbents. I. Peptides and nucleic acid constituents on octadecyl-silica, *J. Chromatogr., 406,* 275, 1987.
44. **Horváth, Cs., Lee, A.L., Velayudhan, A., and Subramanian, G.,** Non-linear chromatography: elution and displacement development, paper #407 presented at Dal Nogare Symposium at the Pittsburgh Conference, Atlantic City, NJ, March 9—13, 1987.
45. **Lee, A.L. and Horváth, Cs.,** Adsorption behavior of biopolymers, poster #TH-P-488 presented at HPLC '88, Washington, D.C., June 19—24, 1988.
46. **Velayudhan, A.,** Ph.D. Thesis, Yale University, 1989.
47. **El Rassi, Z. and Horváth, Cs.,** High-performance displacement chromatography of nucleic acid fragments in a tandem enzyme reactor-liquid chromatograph system, *J. Chromatogr., 266,* 319, 1983.
48. **Valkó, K., Slégel, P., and Báti, J.,** Displacement chromatography of oligomycins, *J. Chromatogr., 386,* 345, 1987.
49. **Lee, A.L., Velayudhan, A., and Horváth, Cs.,** Preparative HPLC, in *8th International Biotechnology Symposium,* Durand, G., Bobichon, L. and Florent, J., Eds., Société Francaise de Microbiologie, Paris, 1988, 593.
50. **Helfferich, F.G.,** Theory of multicomponent chromatography. A state-of-the-art report, *J. Chromatogr., 373,* 45, 1986.

Section XIV
Analysis of Peptides and Proteins

PRACTICAL CONSIDERATIONS FOR ASSESSING PRODUCT QUALITY OF BIOSYNTHETIC PROTEINS BY HPLC

Rosanne C. Chloupek, John E. Battersby, and William S. Hancock

I. INTRODUCTION

HPLC is an important tool for the characterization and quality control of recombinant DNA-derived proteins in the biotechnology industry. Both reversed-phase HPLC (RPC) of the protein and HPLC peptide mapping of enzymatic digests by RPC have been used to confirm the identity of protein products and monitor for contaminants. Substances identifiable by these analyses range from protein variants to chemical modifications of the product, such as oxidation or deamidation. Ultimately, reproducibility is crucial to utilizing these methods for establishing identity and for monitoring product quality.

Many variables affect the reproducibility of HPLC separations, including instrument performance, separation conditions, column performance, sample stability and, in the case of peptide maps, enzyme quality. This discussion will focus on practical observations we have made in the application of an analytical RPC assay and an HPLC tryptic mapping assay of Protropin® (methionyl-rhGH)[1] or recombinant human growth hormone (rhGH).[2]

II. REVERSED-PHASE HPLC (RPC)

RPC is a highly selective chromatographic method that can be used to establish the identity and purity of proteins. The retention of proteins is related to the hydrophobic contact area upon binding to the non-polar chromatographic surface. Recombinant DNA-derived human growth hormone and pituitary-derived human growth hormone have the same chemical structure, and exhibit identical RPC chromatographic behavior (Figure 1, left). Shallow gradient (0.5% or less organic modifier/min) separations are useful because of good peak shapes, high recoveries, and the relative ease of reproducing simple gradients with good quality chromatographic equipment. Isocratic elution conditions can significantly improve resolution of closely related species, but can be more difficult to reproduce as a routine assay because of the limitations of solvent delivery systems.

A RPC method using a trifluoroacetic acid (TFA) and acetonitrile solvent system was developed to monitor growth hormone product quality, combining both isocratic and gradient elution parameters within a single run (Figure 1, left). The assay was validated by the protocol shown in Table 1. The reproducibility of the assay over several days and with two analysts (Figure 1, right) shows a small, but non-systematic drift of peak retention time. Under isocratic conditions, a change of 0.2% in organic solvent composition was found to affect significantly the retention of growth hormone (Figure 2, above), suggesting the sensitivity of a protein separation to the precise chromatographic conditions. Thus, equipment performance in delivering precise gradients as well as accurate solvent compositions for isocratic elution conditions can become crucial to the assay precision with respect to peak retention time. An assessment of the linearity of UV response for varying sample concentrations establishes the detection limits

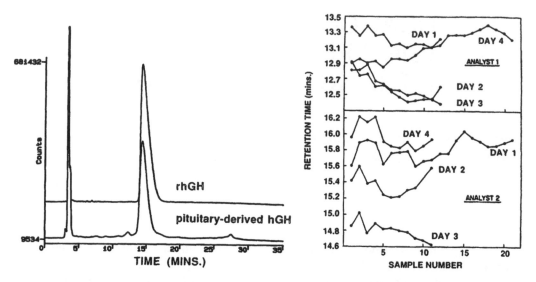

FIGURE 1. (Left) RPC of recombinant human growth hormone (rhGH) vs. pituitary-derived human growth hormone (pit-hGH). The separations were achieved using an automated Waters Associates gradient liquid chromatograph with Model 510 high pressure pumps. A Nelson Analytical data system was used for data acquisition. Column: Vydac C$_4$, 250 × 4.6 mm I.D., 5-μm particle size, 300-Å pore size. Mobile phase solvents: Eluent A, 0.1% aq. trifluoroacetic acid (TFA); Eluent B, 0.1% TFA in acetonitrile. Separation parameters: 49% B for 10 min followed by a 21-min linear gradient to 70% B (1% B/min). Equilibration parameters: 70% B for 10 min, returned to 49% B in 5 min and held at 49% B for 25 min before the next injection. Flow-rate: 1 ml/min. Wavelength: 214 nm. Temperature: 35°C. Sample loaded: 50 μg. (Right) The variation in retention time of Protropin® for analyses on different days and by different analysts. Conditions are described above, except a Hewlett-Packard 1090 liquid chromatograph was used by the second analyst.

TABLE 1
Validation of a Reversed-Phase HPLC Assay

Precision[a]
  Evaluation of peak retention time reproducibility
  Intra-assay retention time precision
  Inter-assay retention time precision
  Inter-assay and intra-assay peak area precision
  Interanalyst precision
Assay linearity[b]
Ruggedness[c]
  Column temperature
  System dependence
  Column lifetime
Separation from other proteins
Stability indicating properties[d]
  Hydrolytic (pH 4.5, 5.5, 10.5)
  Enzymatic (plasmin, carboxypeptidase, pronase, and trypsin)
Recovery

[a] Intra-assay = data within a single assay: inter-assay = data from assays performed on different days.
[b] Linearity of the recovery of sample within a specified concentration range.
[c] Investigation of factors that can contribute to assay variability.
[d] A test of the ability of the assay to detect degradation products.

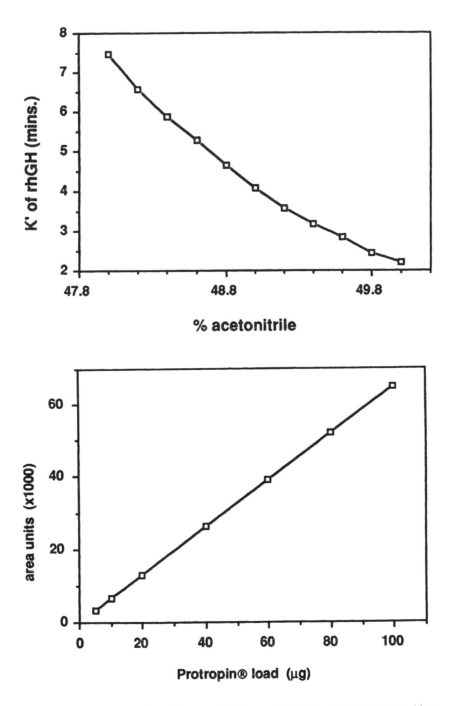

FIGURE 2. (Above) The effect of changes of 0.2% acetonitrile in the mobile phase composition on the retention time of rhGH. Separation conditions are described in Figure 1, except the separation parameters were isocratic for 30 min for each concentration of acetonitrile. The separations were achieved using a Hewlett-Packard 1090 liquid chromatograph. (Below) The linearity of the RPC assay, as measured by incremental amounts of Protropin® loaded on the column. Assay conditions are described in Figure 1, left .

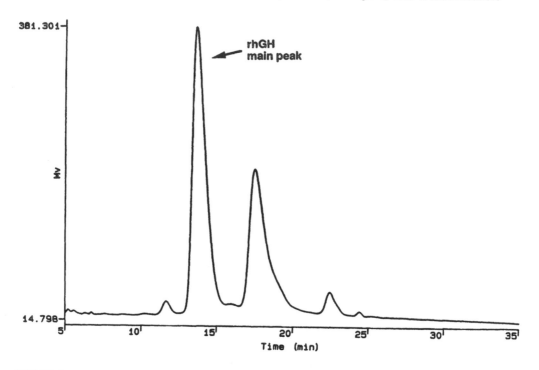

FIGURE 3. An example of a chemically-stressed sample of rhGH separated by RPC. Separation conditions are described in Figure 1, left. The sample was held for 4 weeks at pH 4.5 at 2 to 8°C. No change was detected in the primary structure by amino acid analysis, amino-terminal sequencing or in the tryptic map.

of the assay. This assay showed excellent linearity for 5 to 100 µg loads (Figure 2, below). The stability-indicating properties of this assay were investigated by chromatographing chemically and thermally stressed samples of Protropin®. These samples were analyzed by other techniques such as amino acid analysis, amino-terminal sequencing, and peptide mapping. Some degradation products were found to have an indistinguishable chemical structure by these techniques, yet were resolved in the RPC analysis (Figure 3). Recovery is another important parameter in evaluating protein stability, as aggregated forms may bind irreversibly to the column and not be detected. Metabolically labeled Protropin® (H, leucine) was loaded and gradient fractions were collected. The recovery was 99.9% of the injected radioactivity under these assay conditions.

III. HPLC PEPTIDE MAPPING

Peptide maps can be used to detect small changes in the primary structure of a protein. Methionine- and cysteine/cystine-containing tryptic peptides of human growth hormone change their retention time significantly when oxidized,[3] as do deamidated peptides.[4] The greatest challenge in peptide mapping is maximizing the detection limits of variants or other modified products, while minimizing non-specific cleavages and other artifacts that can complicate the analysis. In our experience, peptide peaks must be greater than 5% (mole:mole) to be readily detected as a new peak in a map, or greater than 15% (mole:mole) to be detected as a co-elution. Evaluating peaks based on retention time is less of a problem because the variation of retention time is typically less than 0.2 min. However, peptide mixtures run in different laboratories or at different times may show significant differences in chromatographic behavior.

Excellent reproducibility of the Protropin® tryptic map assay is routinely achievable, but after a great deal of experience with this assay several problems also became apparent. The initial

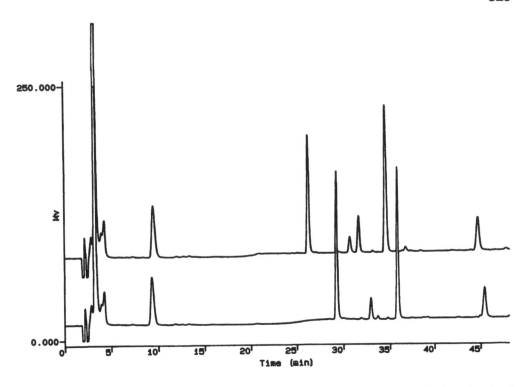

FIGURE 4. Variability of peptide retention times early in the Protropin® tryptic map between the first (above) and second (below) run. Pre-equilibration conditions were not consistent with the conditions used to re-equilibrate the column between sample injections. The separations were achieved using the same instrumentation described in Figure 1, left. Column: Nucleosil C$_{18}$, 150 × 4.6 mm I.D., 5-μm particle size, 300-Å pore size. Mobile phase solvents: as Figure 1. Separation parameters: linear gradient from 0 to 60% B in 120 min (0.5% B/min). Equilibration parameters: 60% B for 10 min, returned to 100% A in 5 min and held at 100% A for 20 min before the next injection. Flow-rate: 1 ml/min. Wavelength: 214 nm. Sample load: 100 μg. The digest buffer was 100 mM sodium acetate, 10 mM Tris, pH 8.3. The samples were incubated at 37°C and treated with two additions of Worthington TPCK-treated trypsin (1:100, w/w) at time zero and after 2 hr. The digest was terminated at 4 hr by the addition of acid a final pH of 2 to 3. Samples were then separated by RPC.

conditions used for the equilibration of the column affects the reproducibility of the first analysis. The degree of variability is greatest for the peptides that are eluted early in the gradient (Figure 4). This problem is avoided if the first analysis begins with exactly the same wash and re-equilibration parameters as is programmed for subsequent runs. The digestion of growth hormone with trypsin results in variable amounts of peptides formed by non-specific cleavages (Figure 5). Most of these are commonly found as very minor peaks, often hidden in the baseline noise. A comparison of digests of sample and a characterized reference material within each set of tryptic maps allows for the identification of any new peaks as product-related artifacts. Efforts must be made to minimize these additional cleavages, however, to avoid masking true contaminants. Various digest conditions involving different buffers, temperatures, and times of trypsin digestion were tested with no significant improvement in the amount of non-specific peptide cleavage. Since most of the non-specific cleavages were chymotrypsin-like,[2] bovine TPCK (L-1-tosylamide-2-phenylethylchloromethyl ketone)-treated trypsin was purified by RPC to remove any residual chymotrypsin contamination.[5] Two main peaks recovered were identified as the alpha and beta forms of trypsin.[6] The two forms were further purified by rechromatography of the isolated peaks. A comparison of the digest specificity of the two forms of trypsin for a sample of rhGH showed no difference in the number and retention time of the minor cleavage products, though amounts of each peak varied slightly.

In addition to minor cleavage products, variation in column selectivity introduces uncertain-

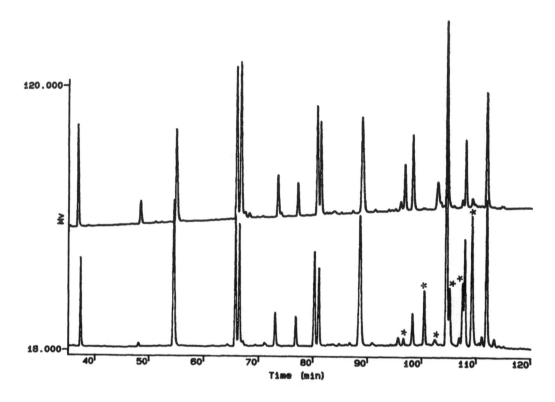

FIGURE 5. An example of the variability of non-specific cleavages in the Protropin® tryptic map. Peptides that are not indicated by an asterisk are predicted cleavages by trypsin. The digestion and separation conditions are described in Figure 4.

ties in the interpretation of the peptide map assay over time. A sample of Protropin® tryptic digest was separated under the same conditions comparing an "aged" HPLC column, exposed to greater than 100 hr to the acidic mobile phase, and an identical new column from the same manufacturer (Figure 6). Resolution of the peptides eluted between 60 to 70 min is lost with the old column, while resolution is gained for the peptide pair eluted between 75 to 85 min. The peak at 100 min on the old column is well resolved, but is eluted as a shoulder at 105 min on the new column. Another example of this problem shows the elution position of the des-Phe amino-terminal peptide of rhGH in chromatograms run 2 years apart (Figure 7). Such a marked change in the elution order was surprising and demonstrates the significance of column variability. Over this time period, it is possible that the manufacturing procedure changed sufficiently to alter the column properties, and the change in peptide elution position is not related to the time of exposure to the mobile phase. Given the unpredictable nature of such effects on the retention time observed for a given peptide, the presence of a new variant could be masked. In an effort to avoid such a disturbing possibility, detectors with increased sensitivity for analysis of peak composition are needed. This sensitivity could be accomplished by spectral matching of peaks utilizing a diode array detector[3] or with an on-line FAB-mass spectrometer.[7]

In conclusion, this chapter demonstrates that RPC is a valuable technique in the biotechnology industry for the analysis of both intact protein and the corresponding protease digest. However, the analytical protein chemist must validate the method before reliable monitoring can be performed on protein samples.

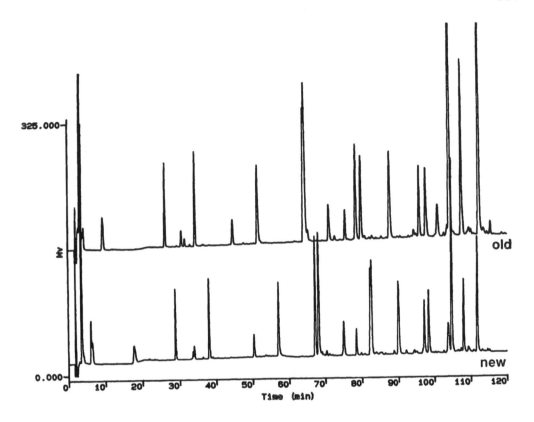

FIGURE 6. Tryptic maps of Protropin® comparing an extensively used Nucleosil C_{18}, reversed-phase column and a new Nucleosil C_{18} column. The HPLC tryptic mapping method is described in Figure 4.

ACKNOWLEDGMENTS

We wish to acknowledge Charles Du Mee, Brent Larson, and Kathleen Mulholland for their contributions to the investigation of trypsin purity and specificity.

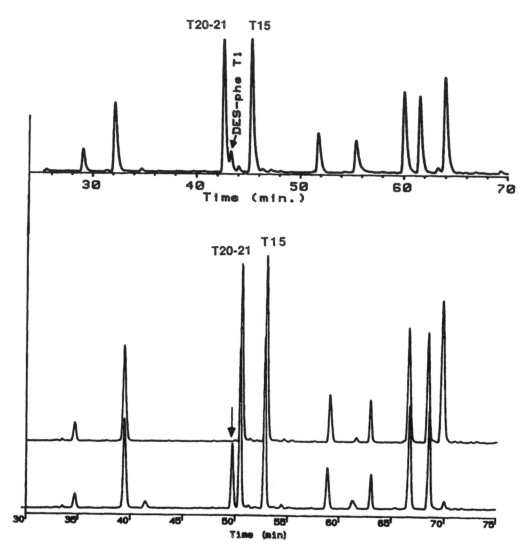

FIGURE 7. A comparison of the elution position of the des-Phe "T1" tryptic peptide (Pro-Thr-Ile-Pro-Leu-Ser-Arg) of rhGH for two chromatograms run approximately 2 years apart. The separations were achieved using a Hewlett-Packard 1090 liquid chromatograph (above) or an automated Waters Associates gradient liquid chromatograph with model 510 high pressure pumps (below). Column: same as Figure 4. Mobile phase solvents: Eluent A, 50 mM sodium phosphate, pH 2.85; Eluent B, acetonitrile. Separation parameters: linear gradient from 0 to 40% B in 120 min (0.33% B/min). Equilibration parameters: 60% B for 10 min, returned to 100% A in 5 min and held at 100% A for 20 min before the next injection. Flow-rate: 1 ml/min. Wavelength: 220 nm. Temperature: 40°C. Sample load: 100 μg. Tryptic digestion conditions are described in Figure 4.

REFERENCES

1. **Kohr, W.J., Keck, R., and Harkins, R.N.,** Characterization of intact and trypsin-digested human growth hormone by high-pressure liquid chromatography, *Anal. Biochem.*, 122, 348, 1982.

2. **Bennett, W.F., Chloupek, R., Harris, R., Canova-Davis, E., Keck, R., Chakel, J., Hancock, W.S., Gellefors, P., and Pavlu, B.,** Characterization of natural-sequence recombinant growth hormone, in *Advances in Growth Hormone and Growth Factor Research*, Muller, E.E., Cocchi, D., and Locatelli, V., Eds., Pythagora Press, Roma-Milano and Springer-Verlag, Berlin-Heidelberg, 1989, 29.

3. **Sievert, H.J.P., Wu, S.L., Chloupek, R., and Hancock, W.S.,** Automated evaluation of tryptic digest from recombinant human growth hormone (rhGH) using UV spectra and numeric peak information, *J. Chromatogr.*, 499, 221, 1990.

4. **Johnson, B.A., Shirokawa, J.M., Hancock, W.S., Spellman, M.W., Basa, L.J., and Aswad, D.W.,** Formation of isoaspartate at two distinct sites during *in vitro* aging of human growth hormone, *J. Biol. Chem.*, 264, 14262, 1989.

5. **Strickler, M.P., Gemski, M.J., and Dictor, B.P.,** Purification of commercially prepared bovine trypsin by reverse phase high performance liquid chromatography, *J. Liq. Chromatogr.*, 4, 1765, 1981.

6. **Schroeder, D.D. and Shaw, E.,** Chromatography of trypsin and its derivatives: characterization of a new active form of bovine trypsin, *J. Biol. Chem.*, 243, 2943, 1968.

7. **Canova-Davis, E., Chloupek, R.C., Baldonado, I.P., Battersby, J.E., Spellman, M.W., Basa, L.J., O'Connor, B., Pearlman, R., Quan, C., Chakel, J.A., Stultz, J.T., and Hancock, W.S.,** Analysis by FAB-MS and LC of proteins produced by either biosynthetic or chemical techniques, *Am. Biotech. Lab.*, May, 8, 1988.

PROTEIN AND PEPTIDE MAPPING BY TWO-DIMENSIONAL HIGH-PERFORMANCE LIQUID CHROMATOGRAPHY

Toshiaki Isobe, Nobuhiro Takahashi, and Frank W. Putnam

I. INTRODUCTION

Protein separation is a central technology in modern biological science. Although the efficiency of protein separation has been greatly improved by recent progress in electrophoresis and high-performance liquid chromatography (HPLC) techniques, more systematic and universal methods for total separation and analysis of proteins are required not only to reduce the time and labor for purifying proteins, but also to elucidate the molecular basis of various biological events or to determine the basis for the morphological and functional specificities of cells or tissues. Likewise, systematic peptide separation is important in order to survey and to isolate physiologically active peptides, and to separate peptide fragments of a protein in the strategy of protein sequence analysis.

Among the current technologies for protein and peptide separation, two-dimensional (2D) electrophoresis is probably the most powerful with respect to its universality and high resolution. Methods for sample recovery from the 2D gel by electroelution[1] or blotting techniques for protein sequencing[2,3] have also been developed. However, the procedures are relatively complicated and still have some limitations on sample preparation, including poor extraction efficiency and modification of some proteins during the electrophoresis, etc.

This chapter describes our continuing effort to develop an automated 2D-HPLC technique applicable to systematic separation of very complex protein or peptide mixtures. The method compares favorably with 2D-electrophoresis in high resolution and reproducibility, and has several advantages over the electrophoresis technique in ease of sample handling and recovery, and quantitation.

II. PRINCIPLE OF THE METHODS

The principle of 2D-HPLC is similar to that of 2D-electrophoresis which separates proteins according to their intrinsic charge and molecular weight by the combination of isoelectric focusing and polyacrylamide gel electrophoresis.[4-6] The chromatographic version of the 2D technique described here employs two columns having different separation specificities; one is an ion-exchange column which separates proteins and peptides according to charge,[7,8] and the other is a reversed-phase column on which the separation achieved depends on hydrophobicity,[9,10] the parameter that correlates strongly to the molecular weight.[11] Chromatography is performed sequentially by a time-dependent control of the flow system for each column. The resolution power of the chromatography is multiplied by the combination of two columns; e.g., if each column has a peak capacity of, for example, 50, then the theoretical separation capacity of the 2D system is $50 \times 50 = 2500$.

In principle, 2D-HPLC can be performed manually without any special equipment; e.g., the eluent from the first column is collected into numbers of fractions, and then each fraction is applied to the second column separately. This has actually been done by Takahashi et al. for the separation of a complex peptide mixture derived from a light chain of human immunoglobulin D,[12] as in the results presented in Section VII. In general, however, manual operation of 2D-HPLC may be susceptible to many problems such as precipitation during concentration of samples, sample loss by multihandling, increased time and labor, chance of human error, and decreased reproducibility through accumulation of experimental errors.

Several approaches seem to be possible to perform automated 2D-HPLC. One possible approach is to introduce automated sample injection equipment coupled with a fraction collector between the first and second columns, and the other is to connect many second columns in parallel after the first column through a column switching valve. In the latter method, the eluent from the first column is continuously applied little by little to all of the second columns by changing the valve automatically during the first chromatography, and then the chromatography of each of the second columns is performed sequentially. The third method is that only one second column is connected in tandem after the first column through a tee tube. In this case the elution of solutes from the first column is stopped in the midst of the chromatography in order to perform chromatography with the second column, to which the eluent from the first column is applied. After the completion of the chromatography of the second column, chromatography of the first column is started again, the eluent applied to the second column, and the chromatography of the second column is repeated as performed in the first step. These separation cycles are repeated over and over again.

Although each of the methods described above has its own advantages and disadvantages as discussed elsewhere,[13] we have chosen the last method because it does not need the excessive equipment that the other two require and can be carried out easily in many laboratories that have one or two ordinary HPLC assemblies with a computer-assisted controller.

III. APPARATUS

The system for our 2D-HPLC technique is composed of either two independent HPLC assemblies (System A; Figure 1), or one HPLC assembly with valve control equipment for buffer change (System B; Figure 2). In each system, programmed elution is carried out by a computer-assisted controller. A sample mixture is applied to a first dimensional ion-exchange column (C1) and eluted in a stepwise manner. The eluent from the first column is introduced directly into the second dimensional reversed-phase column (C2), which is connected in tandem through a tee tube (Figure 1) or an electrical column switching valve (Figure 2). After application of the eluent, the second dimensional reversed-phase separation is performed by linear gradient elution. Stepwise elution for ion-exchange chromatography and the gradient elution for reversed-phase chromatography are synchronized by a computer program. A more detailed description of each of the types of 2D-HPLC apparatus is given in the legends to Figure 1 and 2.

IV. PERFORMANCE OF 2D-HPLC

The performance of System A (Figure 1) is described in the legend to Figure 1, as well as in Reference 14. System B (Figure 2) processes the 2D-HPLC in essentially the same manner as System A. The process of 2D-HPLC, including the elution conditions for, e.g., protein mapping by System B is as follows. By starting the computer program, columns C1 and C2 are equilibrated with B1 (0.025 M Tris-HCl buffer, pH 7.5) and B3 (20% acetonitrile [CH_3CN] in 0.1% aq. trifluoroacetic acid [TFA]), respectively, at a flow-rate of 1.0 ml/min. After 40 min, pump 2 is stopped, and the column switching valve (3WV) moves to connect C1 and C2. A

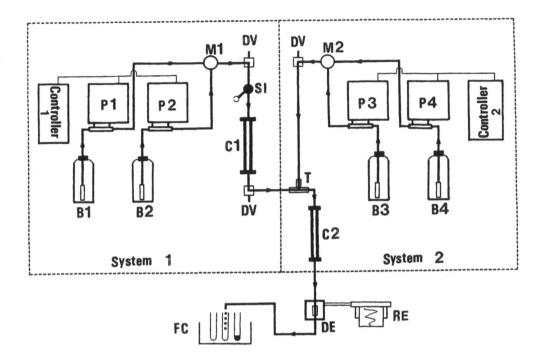

FIGURE 1. Schematic diagram of the automated 2D-HPLC system with two HPLC assemblies[14] (System A). For the first chromatography, a peptide mixture is applied in System 1 to the ion-exchange column (C1) and is eluted only with buffer B1. At this step, P2 does not pump buffer B2 (0% B2). After the eluate from C1 is applied directly into the reversed-phase column, C2, which is connected in tandem through a tee tube, System 1 is stopped. The second chromatography begins simultaneously as System 2 starts pumping to perform a linear gradient elution with acetonitrile. Column C2 is equilibrated again with B3 after the linear gradient elution is finished. Then the flow in System 2 is stopped. By this step, the first cycle of the 2D chromatography is completed. The next cycle is repeated exactly as described above except an increased concentration of B2 is used in the stepwise elution of the first chromatography. The pumping ratio of B2 is increased as follows: 0, 5, 10, 15, 20, 25, 30, 35, 40, 50, and 100% B2. The stepwise elution for ion-exchange chromatography and the linear gradient elution for reversed-phase chromatography are synchronized by computer programs of Controllers 1 and 2. The other symbols indicated are: DV = drain valve, FC = fraction collector, DE = wavelength-tunable UV detector, and RE = recorder.

sample mixture is applied to column C1 through a sample injector (I), and eluted with B1 for time t_1 (20 min in Figure 3) at a flow-rate of 1.0 ml/min. During this time, the eluent flows directly into the reversed-phase column C2. Pump 1 is stopped, and simultaneously the second chromatography begins as pump 2 starts pumping at a flow-rate of 1.0 ml/min with a linear gradient from B3 to B4 (20 to 60% of CH_3CN in 0.1% aq. TFA) during time t_2 (40 min in Figure 3, i.e., 1% CH_3CN/min). Column C2 is equilibrated again with B3 for 10 min after the linear gradient elution is finished; then pump 2 is stopped. By this step, the first cycle of the 2D-HPLC is completed. Then, pump 1 for the first chromatography starts again to elute proteins stepwise from the ion-exchange column, C1, by introducing and mixing a portion of buffer B2 (0.4 M NaCl in 0.025 M Tris-HCl buffer, pH 7.5) into B1. After applying the eluent to C2, the second chromatography is repeated exactly as described above. These procedures are repeated for a number of cycles (n), changing the mixture ratio of B1 and B2 with computer-assisted, time-dependent control of solenoid valves V1 and V2. For versatility, the computer program has been made open to changes in the elution times for C1 and C2 (t_1 and t_2), the cycle number (n) and the mixing ratio of B1/B2 in each cycle, so that an operator is able to input these parameters depending on the complexity and ionic distribution of a sample mixture.

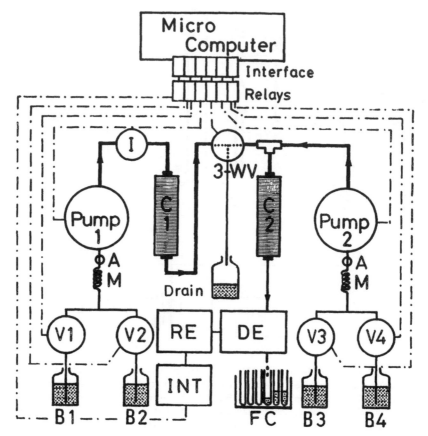

FIGURE 2. Schematic diagram of the 2D-HPLC system with one HPLC assembly (System B). The system is explained in the text. The symbols not indicated in the text are: A = air trap, M = coil solvent mixer, DE = wavelength-tunable UV detector, RE = Recorder, INT = integrator, and FC = fraction collector.

V. COLUMNS

The columns we selected for mapping proteins and peptides are of different types. For protein mapping, we used polymer-based columns for both dimensional separations, taking advantage of their durability and chemical stability in an alkaline solution. This is important for reproducible protein mapping because repeated application of crude protein mixtures such as tissue extracts tends to accumulate contaminants including lipids, nucleic acids, and very large proteins of low solubility, etc., on the column, causing a gradual decrease in peak resolution. In the case of a polymer-based column, however, these contaminants could easily be washed out by passing 0.5 M NaOH in 50% aqueous acetone through the column without detectable deterioration in peak resolution. Thus, the same set of columns could be used in more than 100 analytical runs. On the other hand, we selected silica-based columns for peptide mapping because the peptide mixture targeted for mapping is produced, in most cases, by digestion of a purified protein; thus it does not contain unknown contaminants to be cleaned up after running the chromatography. In addition, silica-based columns have been extensively used for separation of peptides, and the excellent resolution power as well as the retention behavior of peptides are well established for both ion-exchange and reversed-phase columns.

VI. APPLICATION TO PROTEIN MAPPING

Figure 3 shows a 2D protein map obtained by direct separation of soluble proteins in extracts of bovine cerebellum. The map is represented by a three-dimensional drawing of a series of elution profiles, in which each horizontal profile corresponds to one cycle in the chromatography. The NaCl concentration in the elution buffer for the first dimensional anion-exchange column is shown at the right side of each horizontal profile which, then, represents the result of the second-dimensional reversed-phase chromatography. Thus, the programmed number of cycles (n) was set at 10 in this separation, which required a total analysis time of 12 hr.

Under these conditions, the cerebellar extract, containing 6 mg of protein, is resolved into ca. 200 peaks. Quantitative analysis indicated that each of these peaks contained 0.5 to 60 μg of protein, suggesting that proteins present at a level of more than 0.01% of the total soluble proteins in bovine cerebellum are detectable with the detector sensitivity used (220 nm, 0.64 AUFS). For analytical purposes, this sensitivity could be increased ca. 50-fold simply by increasing the detector sensitivity (210 nm, 0.16 AUFS). This allows one to perform an analytical protein mapping of crude tissue or cell extracts with less than 10 mg of tissues, or with ca. 5×10^6 cells cultivated in a single well of a conventional 24-well plastic plate.

The efficiency of the 2D-HPLC technique as a separation method for proteins was examined by the analysis of 25 major peaks collected from the 2D map in Figure 3 by means of polyacrylamide gel electrophoresis: of these, twelve peaks contained proteins with sufficient purity for subsequent analysis. Although other peaks contained two or three proteins or contained some impurities giving rise to smeared bands in polyacrylamide gel, most of these proteins could be purified by a further reversed-phase chromatography step. The isolated proteins were then subjected to analysis including: 2D electrophoresis for the estimation of isoelectric point and molecular weight, dot-blot immunochemical analysis on a nitrocellulose membrane to establish a possible identity to a certain protein, analysis of amino acid composition, and analysis of amino-terminal sequence or internal amino acid sequence following chemical or enzymatic cleavage. The amino acid sequence data could be used for identification of the protein by computerized analysis of the sequence using NBRF, GenBank, and EMBL databases.

As a result of the analyses, 17 major peaks in the protein map of bovine cerebellum have been characterized so far (Figure 3). These include many known brain proteins such as the brain-type isozymes of glycolytic enzymes, e.g., neuron-specific enolase[15], etc., the "EF-hand" type calcium-binding proteins calmodulin and S100,[16] and the kinase II-dependent tyrosine and tryptophan hydroxylase activator protein 14-3-3.[17] Serum albumin and hemoglobin, which probably originated from the peripheral blood, and the non-histone protein HMG-1[18] were also identified. Not only were these known proteins isolated, but also several new proteins such as those designated as DEK, X-56, and so on, were purified by the chromatography.

The 2D-HPLC technique was useful not only for preparative protein separation, but also for detection of differences in the protein compositions of various tissues and cells. We have been applying this technique for the protein mapping of bovine cerebrum and some other tissues to search for proteins specific to the cerebellum, and also for the protein mapping of rat cerebellum at various developmental stages, in order to trace developmental changes in protein composition and to find proteins that may be related to the maturation and differentiation of this tissue.

VII. APPLICATION TO PEPTIDE MAPPING

Figure 4 shows a peptide map obtained by manual operation of the 2D-HPLC technique.[12]

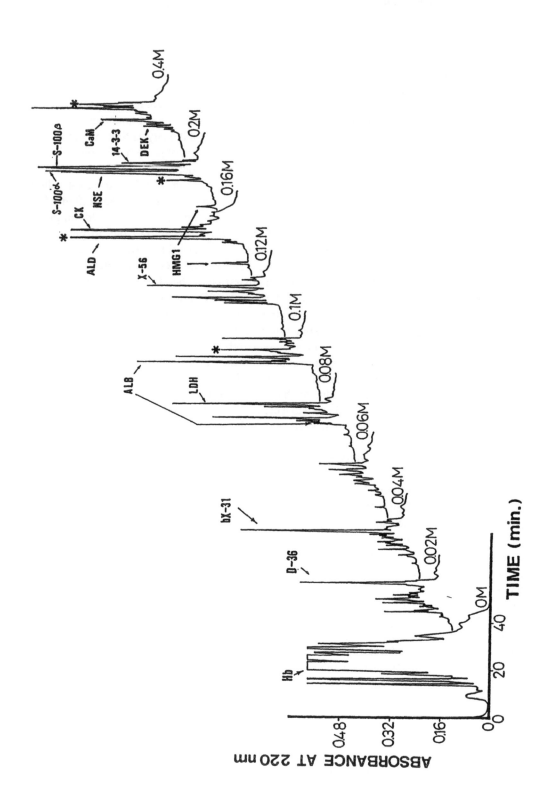

FIGURE 3. Three-dimensional representation of the protein map of bovine cerebellum. The 2D-HPLC was performed as described in the text, by the first dimensional anion-exchange chromatography on a TSK-Gel DEAE-5PW column (75 × 7.5 mm I.D., 10-μm particle size, 1000-Å pore size; Tosoh, Tokyo, Japan) followed by the second dimensional reversed-phase chromatography on a TSK-Gel Phenyl 5PW-RP column (75 × 4.6 mm I.D., 10-μm particle size, 1000-Å pore size; Tosoh, Tokyo, Japan). Peaks identified are designated as: Hb, hemoglobin; ALB, albumin; LDH, lactate dehydrogenase; ALD, aldolase; HMG-1, high mobility group I protein; CK, creatine kinase; NSE, neuron-specific enolase; S100α, alpha-subunit of S100 protein; S100β, beta-subunit of S100 protein; 14-3-3, 14-3-3 protein; and CaM, calmodulin. D-36, bX-31, X-56 and DEK are new proteins found by this method. The proteins detected in cerebellum, but not in cerebrum, are indicated by asterisks.

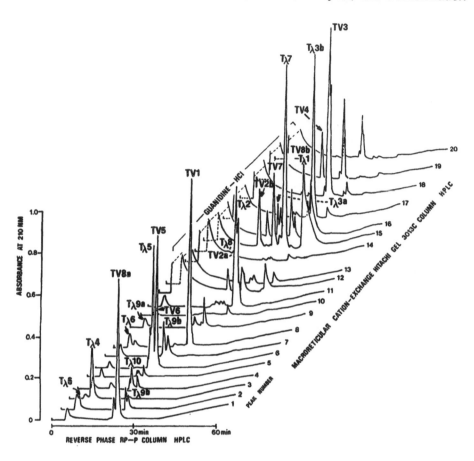

FIGURE 4. A three-dimensional visualization of the peptide map of a light chain of human immunoglobulin D obtained by a manual 2D-HPLC procedure.[12] The map was obtained by the combination of a cation-exchange Hitachi-Gel 3013C column (250 × 4.1 mm I.D., 5 to 7-μm particle size; Hitachi Ltd., Tokyo) and a reversed-phase SynChropak RP-P C$_{18}$ column (250 × 4.1 mm I.D., 6.5-μm particle size, 300-Å pore size; SynChrom, Linden, IN). For procedural details and buffer systems, see Reference 12.

In this case, the first separation was carried out on a cation-exchange carboxymethyl-column. Each of twenty peaks was collected manually into a fraction tube and then lyophilized. The second separation was then performed for each fraction on a reversed-phase column. The second chromatography was repeated a total of twenty times to obtain the complete separation and recovery of all peptides expected from tryptic digestion of a light chain of immunoglobulin D. The second chromatography may not be necessary to be performed for all of the peaks obtained from the first ion-exchange column in terms of the purity of the peptide; however, desalting must usually be carried out for the subsequent sequence analysis of the peptide because the elution of the peptides from the ion-exchange column is generally achieved with an increase of non-volatile salt concentration. Although an effective separation of a peptide mixture of a smaller protein may be achieved by only a single operation of reversed-phase column chromatography, this alone is usually not sufficient to separate all of the peptides obtained from an enzymatic digest of many proteins. When the manual 2D-HPLC is applied to peptide mapping of large proteins, many problems may be encountered, e.g., protein precipitation during lyophilization of some of the fractions from the ion-exchange column; thus, a denaturant such as guanidine chloride had to be used to dissolve some of the peptide fractions following reversed-phase chromatography. The main problem encountered was low reproducibility of the chromatogram

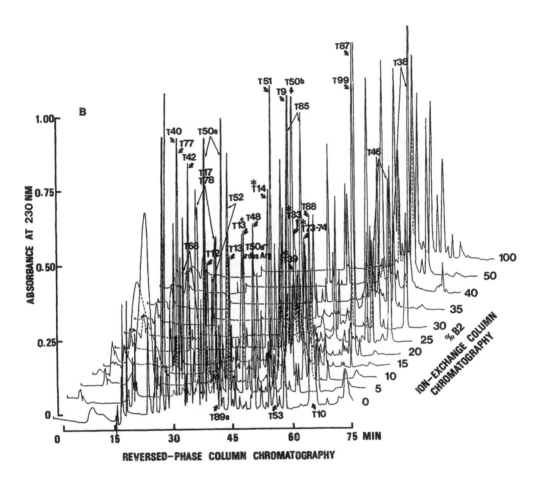

FIGURE 5. A three-dimensional visualization of the peptide map of human ceruloplasmin.[14] The 2D-HPLC was performed as described in the text and in the legend to Figure 1, by use of an anion-exchange Spherogel-TSK IEX-540 DEAE column (300 × 4.0 mm I.D. 5-μm particle size, 125-Å pore size; Altex, Berkeley, CA) and a reversed-phase Ultrasphere ODS column (150 × 4.0 mm I.D. 5-μm particle size, 120-Å pore size; Altex, Berkeley, CA). Some of the peptides identified in the chromatogram are indicated by the nomenclature used in Reference 19. Peptides identified as two peaks are shown by two fine arrows and peptides identified as a single peak by a thick arrow. Glycopeptides are indicated by an asterisk.

through accumulation of experimental errors because of the multistep handling of the method. Thus, the manual 2D method may be useful for comparative peptide mapping of only small proteins.

On the other hand, an automated 2D method we developed is designed to be applied to an extremely complex peptide mixture of a very large protein. Figure 5 shows a chromatogram obtained by an automated 2D-HPLC of a tryptic digest of human ceruloplasmin, the third largest plasma protein for which the complete amino acid sequence has been determined by direct protein sequencing.[19] The protein contains 1046 amino acids and has a potential for yielding 103 tryptic peptides. Although a long, continuous chromatogram is obtained by use of System A described in the Apparatus section, the chromatogram is shown in Figure 5 in three-dimensional visualization. By this method, the digest of ceruloplasmin was separated reproducibly into about 260 peaks within 16 hr. The number of peaks exceeds the expected 103 theoretical peptides because some peptides are duplicated in the adjacent stepwise elutions and also because of incomplete and/or non-specific cleavage. Although small peptides, di-, tri-, and tetrapeptides could not be separated by this method, two-thirds of the peptide peaks thus far analyzed for amino acid composition were pure and could be identified in the protein sequence after a single operation of the chromatography system. The limitations of the 2D method are described in detail in Reference 13. The automated method was successfully applied to purify glycoproteins and to identify the proteolytic microheterogeneity of ceruloplasmin. It was also used for comparative peptide mapping to identify the amino acid substitution in human albumin variants,[20] and we were able to identify the peptide with the amino acid substitution in six out of eleven different albumin variants which were collected in only small amounts in genetic studies of various populations.[21,22]

VIII. CONCLUSIONS

The automated 2D-HPLC technique described here is very powerful for systematic protein and peptide separations, as could be demonstrated in the protein mapping of bovine cerebellum and in the peptide mapping of ceruloplasmin. The high resolution of the chromatography is achieved by sequential ion-exchange and reversed-phase chromatography, while the reproducibility of the method is maintained by the automation. Although the technique still has some limitations in the resolution and recovery of very large proteins with molecular weights of more than 100,000 or in the resolution of small peptides with little hydrophobicity, we propose that it can be an alternative to 2D electrophoresis for systematic protein and peptide mapping, especially for preparative use. Such preparations are most suitable for use in the strategy of protein and gene sequence analysis, as well as in the strategy of immunochemical/immunohistochemical analysis for which the specific antibodies can be produced by, e.g., a cell fusion technique. As a protein mapping method, it can be used for detection of differences in the protein compositions of various tissues and cells without complicated pre-manipulation of samples and allows simultaneous analysis of a protein isolated from the map. As a peptide mapping method, the technique may be a complementary tool in looking for peptide probe sequences or for sequence confirmation in a strategy for DNA sequencing and, likewise, for the systematic purification of physiologically active peptides and of peptide fragments in the strategy for protein sequencing of very large proteins. Furthermore, it will be especially suitable for comparative peptide mapping of very closely related large proteins, such as genetic variants.

ACKNOWLEDGMENTS

One of the authors (T.I.) would like to thank Dr. T. Okuyama of Tokyo Metropolitan University for stimulating discussions and suggestions, and Dr. T. Manabe for the computer program for the control of 2D-HPLC system B. This work was supported in part by grants from the Protein Research Foundation and from the Iwaki Scholarship Foundation. This work was also supported in part by grants to F. W. Putnam from the National Institutes of Health (NIH grants DK19221 and CA08497).

REFERENCES

1. Hunkapiller, M., Lujan, W., Ostrander, F., and Hood, L.E., Isolation of microgram quantities of proteins from polyacrylamide gels for amino acid sequence analysis, in *Methods in Enzymology*, Vol. 91, Hirs, C.H.W. and Timasheff, S.N., Eds., Academic Press, New York, 1983, 227.

2. Matsudaira, P., Sequence from picomole quantities of proteins electroblotted onto polyvinylidene difluoride membranes, *J. Biol. Chem.*, 262, 10035, 1987.

3. Aebersold, R.H., Leavitt, J., Saavedra, R.A., Hood, L.E., and Kent, S.B.H., Internal amino acid sequence of proteins separated by one- or two-dimensional gel electrophoresis after in situ protease digestion on nitrocellulose, *Proc. Natl. Acad. Sci. U.S.A.*, 84, 6970, 1987.

4. O'Farrell, P.H., High resolution two-dimensional electrophoresis of proteins, *J. Biol. Chem.*, 250, 4007, 1975.

5. Manabe, T., Tachi, K., Kojima, K., and Okuyama, T., Two-dimensional electrophoresis of plasma proteins without denaturing agents, *J. Biochem. (Tokyo)*, 85, 649, 1979.

6. Anderson, N.L., Tracy, R.P., and Anderson, N.G., High-resolution two-dimensional electrophoretic mapping of plasma proteins, in *The Plasma Proteins*, Vol. IV, Putnam, F.W., Ed., Academic Press, Orlando, FL, 1984, 222.

7. Manabe, T., Utilization of microcomputer in protein chemistry, *Kagaku no Ryoiki*, 36, 470, 1982.

8. Kadoya, T., Isobe, T., Amano, Y., Kato, Y., Nakamura, K., and Okuyama, T., High performance anion-exchange chromatography of proteins, *J. Liq. Chromatogr.*, 8, 635, 1985.

9. Sasagawa, T., Okuyama, T., and Teller, D.C., Prediction of peptide retention times in reversed-phase high performance liquid chromatography during linear gradient elution, *J. Chromatogr.*, 240, 329, 1982.

10. Brown, C.A., Bennett, H.P.J., and Solomon, S., The isolation of peptides by high performance liquid chromatography using predicted elution positions, in *High Performance Liquid Chromatography of Proteins and Peptides*, Hearn, M.T.W., Regnier, F.E., and Wehr, C.T., Eds., Academic Press, New York, 1983, 65.

11. Ichimura, T., Amano, Y., Isobe, T., and Okuyama, T., High performance liquid chromatography of proteins and peptides on an octadecyl-bonded glass support, *Bunseki Kagaku*, 34, 653, 1985.

12. Takahashi, N., Takahashi, Y., and Putnam, F.W., Two-dimensional high-performance liquid chromatography and chemical modification in the strategy of sequence analysis: complete amino acid sequence of the lambda light chain of human immunoglobulin D, *J. Chromatogr.*, 266, 511, 1983.

13. Takahashi, N., Isobe, T., and Putnam, F.W., Multidimensional, microscale HPLC technique in protein sequencing, in *HPLC of Proteins, Peptides, and Polynucleotides*, Hearn, M.T.W., Ed., Verlag Chemie International, New York, in press.

14. Takahashi, N., Ishioka, N., Takahashi, Y., and Putnam, F.W., Automated tandem high-performance liquid chromatographic system for separation of extremely complex peptide mixtures, *J. Chromatogr.*, 326, 407, 1985.

15. Ishioka, N., Isobe, T., Kadoya, T., Okuyama, T., and Nakajima, T., Large scale preparation and crystallization of neuron-specific enolase, *J. Biochem. (Tokyo)*, 95, 611, 1984.

16. Isobe, T. and Okuyama, T., The amino acid sequence of the alpha subunit in bovine brain S100a protein, *Eur. J. Biochem.*, 116, 79, 1981.

17. Ichimura, T., Isobe, T., Okuyama, T., Takahashi, N., Araki, K., Kuwano, R., and Takahashi, Y., Molecular cloning of cDNA coding for brain-specific 14-3-3 protein, a protein kinase-dependent activator of tyrosine and trytophan hydroxylases, *Proc. Natl. Acad. Sci. U.S.A.*, 85, 7084, 1988.

18. Reeves, R., Transcriptionally active chromatin, *Biochim. Biophys. Acta*, 782, 343, 1984.

19. Takahashi, N., Ortel, T.L., and Putnam, F.W., Single-chain structure of human ceruloplasmin; the complete amino acid sequence of the whole molecule, *Proc. Natl. Acad. Sci. U.S.A.*, 81, 390, 1984.

20. Takahashi, N., Takahashi, Y., Ishioka, N., Blumberg, B.S., and Putnam, F.W., Application of an automated tandem high-performance liquid chromatographic system to peptide mapping of genetic variants of human serum albumin, *J. Chromatogr.*, 359, 181, 1986.

21. Takahashi, N., Takahashi, Y., Blumberg, B.S., and Putnam, F.W., Amino acid substitutions in genetic variants of human serum albumin and in sequences inferred from molecular cloning, *Proc. Natl. Acad. Sci. U.S.A.*, 84, 4413, 1987.

22. Takahashi, N., Takahashi, Y., Isobe, T., Putnam, F.W., Fujita, M., Satoh, C., and Neel, J.V., Amino acid substitutions in inherited albumin variants from Amerindian and Japanese populations, *Proc. Natl. Acad. Sci. U.S.A.*, 84, 8001, 1987.

AMINO ACID ANALYSES OF PROTEINS AND PEPTIDES: AN OVERVIEW

Lawrence B. Smillie and Michael Nattriss

A knowledge of the amino acid composition of a polypeptide or protein can be of critical importance for its identification, an understanding of its properties, an assessment of its purity and in establishing its concentration in solution. An automated system for amino acid analyses was first described by Spackman, Moore, and Stein[1] in 1958. The method, based on the ion-exchange chromatographic separation of the hydrolyzed amino acids followed by their on-line quantitation by reaction with ninhydrin, remains to this day as the most reliable methodology for amino acid analyses of peptides and proteins. While many improvements have been made in the instrumentation over the years to improve sensitivity, reproducibility, and analysis time, the chemistry of the system remains basically the same. Commercial instruments are capable of resolving and quantitating all of the amino acids normally found in protein hydrolysates as well as a number of their derivatives at a sensitivity of about 50 to 100 pmol for that amino acid present in least abundance.

In recent years and in response to the demand for ever more sensitivity and speed of analyses, various reagents, and instruments have been introduced for the precolumn derivatization of amino acids followed by their fractionation, detection, and quantitation on reversed-phase columns. In theory, these reagents should permit detection and quantitation of the amino acids at significantly greater sensitivity. However, in practice the increase in sensitivity to about 10 to 20 pmol that has been achieved with these methods has been disappointing, not because of an inability to detect the chromophoric or fluorogenic derivatives (in some cases in the low femtomole range) but rather to the variable and persistent background levels of contaminating amino acids arising from glassware, reagents, solvents, gases, and water. In spite of heroic efforts, it has not yet been possible to reduce this background to levels below a few picomoles.[2] Further, while in some cases these precolumn derivatization methods compare reasonably well with the postcolumn ninhydrin procedures at the > 100pmol level, the precision and accuracy achievable in the low picomole range are significantly less.

The purpose of the present article is to review the present status of development of the several precolumn methodologies for amino acid analyses in comparison with that of the long established and reliable postcolumn ninhydrin procedures. Hopefully this will enable researchers to assess the advantages and disadvantages of the various approaches and available instrumentation in light of their own requirements and financial resources.

I. HYDROLYSIS AND SPECIAL PROCEDURES FOR UNSTABLE AMINO ACIDS

Of critical importance to the reliability of any amino acid analysis of polypeptide or protein is the care taken in the preparation of its acid hydrolysate. In an early article, Moore and Stein[3] emphasized the importance of the exclusion of metals and O_2 from the hydrolysis mixture. Such precautions have taken on even more importance with the application of the methodology to very small amounts of protein. Corrections must also be applied for the partial destruction of certain amino acids (serine, threonine, tyrosine) and the incomplete release from peptide linkage of others (isoleucine, valine). Methionine may also be partially oxidized to its sulfoxide and the recovery of cysteine/cystine is not reproducible. Since tryptophan is largely degraded during a normal 6 *M* HCl hydrolysis, special hydrolytic conditions must be employed (see below). Because asparagine and glutamine are hydrolyzed quantitatively to aspartic and glutamic acid, these are reported as the totals of asparagine plus aspartate and of glutamine plus glutamate. No attempt is normally made to quantitate these separately.

The most commonly used hydrolytic procedure for amino acid analyses is the addition of reagent grade 6 *M* HCl (or constant boiling (5.7 *M* HCl)) containing 0.1% phenol to the dried protein or peptide sample in a pyrex tube. In many laboratories, tubes are previously pyrolyzed at 500 to 550°C for several hours to reduce contamination from this source. Following evacuation to 60 μm or less and sealing under a flame, hydrolysis is carried out at 110 ± 1°C for 20 hr. Since storage of dried samples can lead to loss of certain amino acids, this is best done with the unopened tubes at 4 or –20°C if analysis is not to be carried out immediately. Following opening and drying, the hydrolysate is taken up in starting buffer for application to the analyzer (postcolumn methodology) or in appropriate sample buffer for precolumn derivatization. Complete HCl removal is of particular importance if the hydrolysate is to be analyzed by precolumn derivatization using phenylisothiocyanate since this reaction appears to be particularly sensitive to pH.

Under these hydrolysis conditions a fraction of the serine (~ 10%) and threonine (~ 5%) will be degraded. These can be corrected approximately by applying the appropriate correction factor. Assuming that adequate sample is available however, it is better to extrapolate data from 24, 48, and 72 hr hydrolyses to zero time. Such an approach also permits more reliable estimates of valine and isoleucine since their release from peptide linkages and thus recovery normally only reaches a plateau value after 48 and 72 hr.

For total half-cysteine (cysteine plus cystine) determination, performic acid oxidation of the polypeptide followed by acid hydrolysis as described by Moore[4] has proven to be highly reliable. Methionine is converted to the sulfone which is more stable than methione itself. Alternatively the protein can be completely reduced with β-mercaptoethanol or dithiothreitol in the presence of a denaturant (6 *M* guanidine HCl or 8 *M* urea) and the resulting cysteines reacted with a low molar excess of appropriate alkylating agent. Iodoacetic acid, iodoacetamide, or 4-vinylpyridine are suitable for this purpose.[5-7] Following hydrolysis the derivatized cysteines can be detected and quantitated on most analytical systems.

Estimates of cystine content (disulfide bridges) are normally calculated from the difference in total half-cystine (as described above) and the cysteine content. Determination of the latter can be done on the intact protein by an appropriate colorimetric or spectrophotometric method or by appropriate alkylation in the presence of denaturant (without prior reduction) followed by acid hydrolysis. In the former case the most suitable approach is probably the use of a molar excess of DTNB reagent (5,5′-dithiobis(2-nitrobenzoic acid)).[8,9] At slightly alkaline pH, this disulfide interchange reaction results in the production of the intensely colored thiophenol in an amount equivalent to the thiol present in the protein. Since the reaction can be carried out in denaturing solvents, the reaction can be made quantitative in most cases if appropriate precautions are taken. However, because of the difficulties inherent in cysteine analyses, it is normally advisable to use at least two independent methods if reliable estimates of the cysteine and total half-cystine content are important to the conclusions of the investigation.

Analyses for tryptophan may also be carried out on the intact protein or by using hydrolytic conditions more suitable for tryptophan recovery in an undegraded form. In the former category a number of spectrophotometric methods have been described including the measurement of the absorbance spectrum of the protein in 0.1 *M* NaOH, the decrease in absorbance at 280 nm upon oxidation of the indole with N-bromosuccinimide, the reaction of the indole with *p*-dimethylaminobenzaldehyde and others. A critical description of these has been given by Scoffone and Fontana.[10]

A number of hydrolytic conditions have been described in which the tryptophan destruction is minimized but not completely prevented. These include the inclusion of a few percent of β-mercaptoacetic acid in the 6 *M* HCl (85 to 90% recovery)[11] and the use of 3 *M* p-toluenesulfonic acid[12] or 4 *M* methanesulfonic acid (plus 0.2% tryptamine as a scavenger)[13] instead of HCl. An additional hydrolytic method, described by Hugli and Moore,[14] employs sodium hydroxide with hydrolyzed starch added as a scavenging agent. Although consistent recoveries of better than 95% have been reported, the method has not been widely adopted, possibly because the procedure is somewhat more complicated than the organo sulfonic acid methods.

Variations on the classical hydrolysis procedure with 6 *M* HCl (20 hr at 110°C) have been described. These include the use of elevated temperature, usually 150 to 160°C to reduce hydrolysis time from 20 or more hours to less than 1 to 2 hr. While the advantages are obvious, several hydrolysis times are still required if corrections are to be applied for the destruction of serine and threonine and the incomplete release from peptide linkage of valine and isoleucine. As a partial remedy for the latter problem, especially for very nonpolar peptides, Tsugita and Scheffler[15] have recommended the use of a 1:2 mixture of trifluoroacetic and 12 *M* HCl at 166°C. The application of elevated temperature and reduced times has also been described for hydrolysis with the non-volatile 4 *M* methanesulfonic acid.[16] The application of microwave heating[17,18] represents a possible further extension of these trends but will probably require more adequate radiation control than is available with present household ovens.

For high sensitivity analyses using precolumn derivatization methodologies, vapor phase hydrolysis has been widely adopted.[19,20] The open tubes containing dried sample are sealed into a larger vessel containing the hydrolysis acid (usually constant boiling HCl). Since only the acid vapors come into contact with the sample, non-volatile contaminants arising from the acid are excluded. Before sealing, the larger vessel can be either evacuated or flushed with nitrogen or argon to provide an inert atmosphere. The corrections for serine, threonine, tyrosine, valine, and isoleucine based on different hydrolysis times can be applied as usual. Elevated temperatures and shorter hydrolysis times can also be used with vapor phase hydrolysis. However the method cannot be used with the non-volatile organo-sulfonic acids for tryptophan determinations.

II. POSTCOLUMN METHODOLOGY

As originally developed by Spackman, Stein, and Moore,[1] the mixture of amino acids present in the hydrolyzed protein/polypeptide is separated by cation exchange chromatography on columns of sulfonated polystyrene using buffers of increasing pH and ionic strength. Separation is dependent not only on the charge properties of the amino acids but also on nonpolar interactions of the side chains of the amino acids with the highly nonpolar polystyrene matrix. Thus neutral amino acids such as serine, threonine, glycine, and alanine are eluted earlier than more nonpolar neutrals (valine, leucine, isoleucine, and the aromatic amino acids). Following elution of the separated amino acids from the column the eluent stream is mixed on-line with ninhydrin reagent, passed through a heating coil (≥ 100°C) and the developed color monitored at appropriate wavelengths in colorimeters equipped with low volume flow through cells. Recording and data analyses in modern instrumentation are fully automated. Modern commercial adaptations of this basic separation methodology are described elsewhere in this volume.[21,22]

Successful application of the ninhydrin reagent by the Moore and Stein group for this purpose followed extensive investigations of the factors affecting overall yields of the colored product (Ruhemann's purple) (see Figure 1A). Exclusion of O_2 and maintenance of strongly reducing conditions were found to be mandatory both for storage of the reagent as well as for the reaction itself. For this reason reduced ninhydrin (hydrindantin) is included as a component of the reagent which is stored under an inert gas. Control of optimal pH at 5.0 is achieved by including a concentrated sodium acetate buffer in the reagent formulation. Since ninhydrin and hydrindantin are relatively insoluble in aqueous media, a high concentration of methyl cellosolve

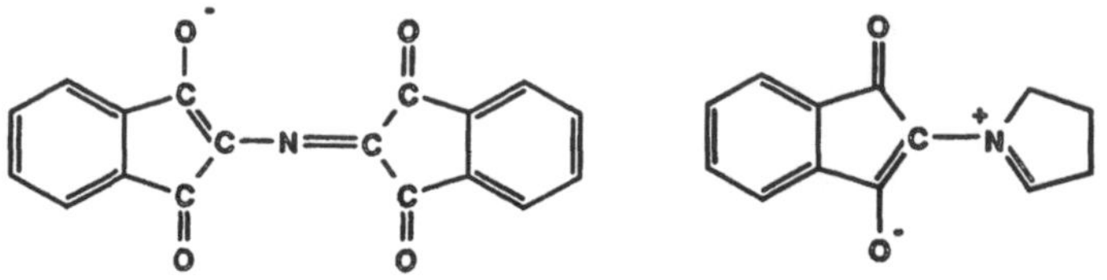

Ruhemann's purple   (λ_{max} = 570 nm)

A   B

FIGURE 1. The products of the ninhydrin reaction with (A) primary amino acids, (B) proline.

or dimethyl sulfoxide is included as a component of the reagent. Under appropriate conditions of temperature ($\geq$ 100°C) and time ($\leq$ 20 min) the yield of colored product is proportional to amino acid concentration over a wide range. However the "color yield" varies with different amino acids and individual correction factors must be applied for each. Proline and hydroxyproline yield a distinct product with a different absorption spectrum (see Figure 1B). For this reason, amino acid separation and quantitation are monitored at two wavelengths; normally 440 and 570 nm.

An important feature of the system is the relative tolerance of the methodology to the presence of salts and other impurities in the hydrolysate. This derives from the fact that the reaction of each amino acid with ninhydrin occurs only after such impurities have been separated from the amino acids by the ion-exchange process. The ninhydrin reaction thus occurs in a completely controlled environment of pH, temperature solvent, and reagent concentration. The presence of such contaminants can, however, affect the yields of various amino acids in the hydrolysis procedure especially when only small amounts of protein/polypeptide are being analyzed.

Modern instruments based on the postcolumn derivatization with ninhydrin are capable in experienced hands of providing highly reliable and reproducible analyses. By appropriate manipulation of column separation conditions (pH, salt concentration, temperature), all of the naturally occurring amino acids present in acid hydrolysates as well as many of their derivatives can be satisfactorily separated and quantitated. Analysis time is approximately 60 min for the analysis of a normal polypeptide hydrolysate with a reproducibility of ± 3% and a maximum sensitivity of approximately 50 to 100 pmol for the least abundant amino acid present in a polypeptide hydrolysate.

III. PRECOLUMN DERIVATIZATION METHODOLOGY

In response to the ever increasing need for high sensitivity amino acid analyses on proteins/ polypeptides available only in limited quantities, a number of methods have been introduced for the derivatization of amino acids present in acid hydrolysates, their separation by HPLC, and their quantitation based on their absorbance in the visible/ultraviolet range or their fluorescent properties. Inherently these offer the potential of higher sensitivity over the postcolumn ninhydrin procedure.

In assessing the usefulness of the various derivatization reagents presently available for this purpose a number of criteria should be considered. For example, do the reagent and its hydrolysis and/or side reaction products have similar absorbance or fluorescent properties as the derivatized amino acids and if so, do these interfere with the chromatographic fractionation, identification, and quantitation? Does the reagent react rapidly and quantitatively under mild conditions of pH and temperature with all of the naturally occurring amino acids including proline and hydroxyproline? Is the derivatization reaction unduly sensitive to pH and the

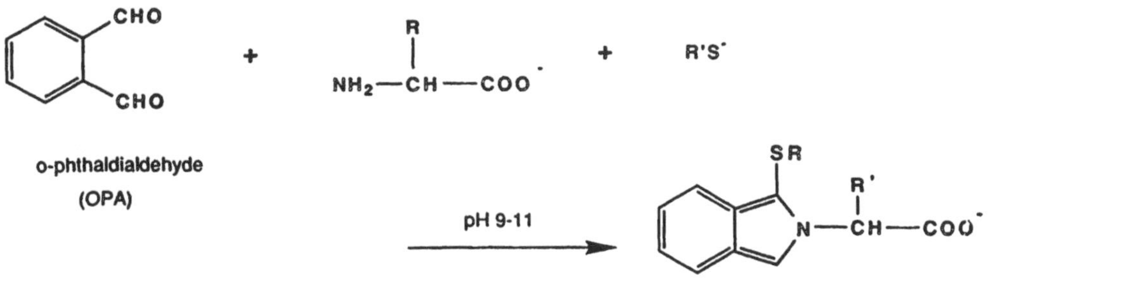

FIGURE 2. Reaction of an amino acid with *o*-phthaldialdehyde (OPA) in the presence of a thiol. The nature of the thiol (R') and R group determine the HPLC properties of the derivative.

presence of salts, metals, or other contaminants in the sample? Are the derivatized amino acids stable under the conditions of storage and/or HPLC separation? Does the reaction lead to a single mono derivative for each amino acid or are both mono and bis derivatives formed with some amino acids? In the latter case are there appropriate conditions for the selective formation of one form or the other? Do the various derivatized amino acids have similar molar extinction coefficients or fluorescent properties or are there wide differences between them? Is the absorbance or fluorescence linear with concentration over a wide range? Are there chromatographic systems available for the fractionation and quantitation of not only the amino acids normally occurring in proteins but also for the various derivatives of cysteine (e.g., cysteic acid, carboxymethyl cysteine, pyridylethyl cysteine, aminoethyl cysteine), and of methionine (methionine sulfone, methionine sulfoxide, homoserine, homoserine lactone), as well as the various post-translationally modified amino acids (e.g., phosphorylated, hydroxylated, methylated, etc.) and for glucosamine and galactosamine? An ideal precolumn derivatization system should satisfy all these criteria. Unfortunately perfection is not yet with us!

IV. ORTHO-PHTHALDIALDEHYDE

Fluorescamine and ortho-phthaldialdehyde (or ortho-phthalaldehyde) (OPA)[23,24] were originally introduced as postcolumn derivatizing reagents as alternatives to ninhydrin and provided a significant increase in sensitivity. Although OPA derivatives are relatively unstable, this is not a major problem in the postcolumn derivatization technique since the conditions of reaction, flow rates, and delay time between reaction and quantitation can be made reproducible. The major disadvantage of fluorescamine and OPA for this purpose is their failure to react with secondary amines (imino acids; proline and hydroxyproline). Although this can be circumvented by their oxidation on-line with chloramine-T or hypochlorite, this introduces additional complications to the plumbing and valving arrangements in the postcolumn system and reproducibility is a problem.

While fluorescamine proved unsuitable as a precolumn reagent since its fluorescent yields are low at pHs required for chromatography on silica-based columns, OPA has been used extensively for this purpose. Its major advantage is that while it itself has no inherent fluorescence, its reaction products with primary amines show very high fluorescent quantum yields with $\lambda_{ex} = 340$ nm and $\lambda_{em} = 455$ nm. At pH 9.5 and in the presence of a thiol, the reaction proceeds almost instantaneously at room temperature to form an isoindole product in which both the amino acid and thiol are incorporated (see Figure 2). Its chromatographic properties are thus dependent on both the nature of the amino acid as well as the nature of the thiol. Following quenching of the reaction at pH 7.2, the products must be immediately applied to the chromatographic system since some of these are quite unstable. This can be achieved with commercially available robotic or auto-sampler devices.[25] Various HPLC systems have been

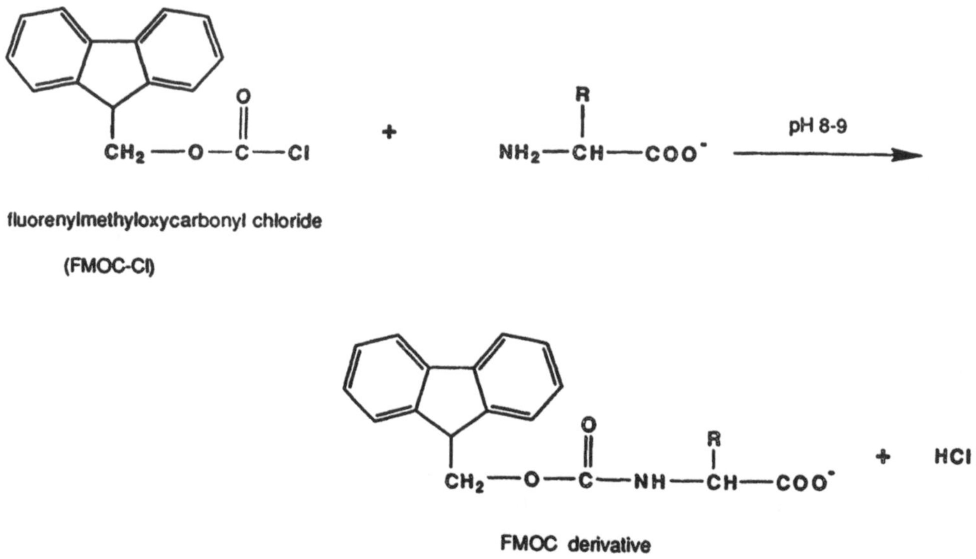

fluorenylmethyloxycarbonyl chloride

(FMOC-Cl)

FMOC derivative

FIGURE 3. Reaction of FMOC-Cl with amino acids to give highly fluorescent derivatives.

described for the separation of the amino acids and many of their derivatives. Mono derivatives are apparently obtained for all amino acids except cysteine and lysine whose two primary groups lead to the bis products. Even though these have significantly lower quantitative fluorescent yields because of intramolecular quenching, linear relationships between amino acid concentration and fluorescence have been reported over a concentration range of 1 to 100 pmol with a sensitivity of detection to < 100 fmol.[25] However, quantitation cannot be expected beyond the low picomole range for the practical reasons indicated earlier. In addition to the relative instability of the OPA reaction products the major disadvantage of the method is the failure of the reagent to react with the imino acid proline. For protein structural studies this is, of course, quite unacceptable. A strategy combining the use of OPA and FMOC-Cl (see below) has recently been introduced to circumvent the deficiency.

V. 9-FLUORENYLMETHYL CHLOROFORMATE (FMOC-Cl)

This highly reactive reagent, used as a protecting group in peptide synthesis, was first applied as a precolumn derivatizing agent for amino acid analyses by Einarsson et al.[26-29] Reaction (see Figure 3) occurs rapidly with all amino acids (including proline) at alkaline pH and ambient temperature, is complete in less than 1 min and is relatively tolerant to the presence of salts and other contaminants. The products are highly stable over many hours and have a fluorescent yield comparable to the OPA derivatives. However, since both mono and bis derivatives can be formed with histidine, tyrosine, and lysine, the relative proportions of these vary as a function of pH and the ratio of reagent to amino acids. Further, the reagent and its hydrolysis product(s) have very similar fluorescent properties to the derivatized amino acids. These interfere in the chromatographic separation and require their removal by organic solvent (e.g., pentane) extraction prior to application to the HPLC column. This can result in losses to the organic phase of some of the FMOC amino acids. An evaluation of a commercial instrument (Varian Amino Tag) based on this derivatization chemistry has recently been reported by Smith et al.[30] Under pH and reagent concentrations where only the bis derivatives were obtained, the FMOC method, at a sensitivity of 50 to 500 pmol protein, was found to compare favorably with the phenylthiocarbamyl (PTC) method (see below) and analyses based on postcolumn ninhydrin derivatization. At higher sensitivity (5 to 50 pmol protein) which was beyond the range where useful postcolumn ninhydrin analyses could be obtained, the two methods were comparable but showed considerable variability for some amino acids, including glycine (consistently overestimated; is eluted near FMOC-OH) and histidine. Methionine was not reported for reasons that are unclear but probably because of large variability arising from its oxidation to the sulfoxide under the condition of storage (pH 8.5) of hydrolysates prior to derivatization. Tyrosine recovery was also poor in some cases (bis derivative elutes near FMOC-Cl).

To circumvent the problem of interference by the elution of large peaks of FMOC-Cl and FMOC-OH in the same chromatographic region as several of the FMOC amino acids, Betner and Foldi[31] have used derivatization of the bulk of excess FMOC-Cl with a hydrophobic amine (1-aminoadamantane; ADAM) to produce a derivative which is eluted late in the chromatogram and after all of the FMOC amino acids. In addition to simplifying the chromatographic profile, this approach also eliminates the necessity for organic solvent extraction of the reaction mixture, a procedure that can lead to significant losses of some amino acids when high sensitivity analyses (< 50 pmol) are being attempted. Optimal reaction conditions including volumes, buffer composition, reagent volumes, and concentration have been reported by these authors[31] and should be consulted for further details. A brief description of a commercial adaptation (Pharmacia) of the FMOC-ADAM system, including a chromatographic profile for the HPLC separation of the FMOC amino acids has been included in the present volume.[22]

As indicated above, a combined OPA-FMOC approach has been described by Blankenship et al.[32] using a commercially available analyzer (Hewlett-Packard Amino Quant). Reaction of primary amino acids is carried out initially with OPA followed by reaction of proline (hydroxyproline) with FMOC. This is achieved with the robotic capability of the autosampler followed by immediate delivery of the reaction products to the HPLC chromatographic system. Fluorescence monitoring for the OPA primary amino acids and FMOC-proline derivatives is carried out at appropriate wavelengths. The FMOC-proline, which is eluted later than the OPA derivatives, appears as a peak against a rising background due to elution of the FMOC reagent and its hydrolysis product. In spite of this, successful quantitation can apparently be achieved down to 10 pmol. To date, however, a description of the satisfactory separation and quantitation of many of the derivatized amino acids has not been reported.

VI. PHENYLISOTHIOCYANATE (PITC)

This reagent, long used for the sequencing of polypeptides and proteins, was introduced for the analysis of amino acids in the early 1980s.[33-36] It is now the most extensively used reagent for precolumn derivatization in amino acid analysis (see Figure 4) and considerable information is available on the chemistry of the reaction, stability, fractionation and quantitation of the phenylthiocarbamyl (PTC) derivatives. The subject has been reviewed recently.[37-39]

Reaction of PITC with amino acids (including proline and hydroxyproline) is relatively rapid (< 20 min) and specific at alkaline pH and ambient temperature. Mono PTC derivatives are formed with the exception of lysine and cystine in which the bis PTC forms are recovered. The derivatives all have similar spectral properties and their separation and quantitation on HPLC columns can be monitored by absorbance in the 240 to 255 nm range. The products are relatively stable although some conversion to the phenylthiohydantoin can occur if the pH is not adequately controlled. While the PITC reagent is relatively volatile and is largely removed under vacuum, several of the UV-absorbing reagent side and degradation products must be dealt with in the chromatographic fractionation and identification of the PTC derivatives. These can include phenylthiourea, PTC-ethylamine, and aniline. Various reversed phase HPLC systems have been described to provide satisfactory separation of all amino acids found in protein/polypeptide hydrolysates as well as many of the amino acid derivatives of interest to the protein

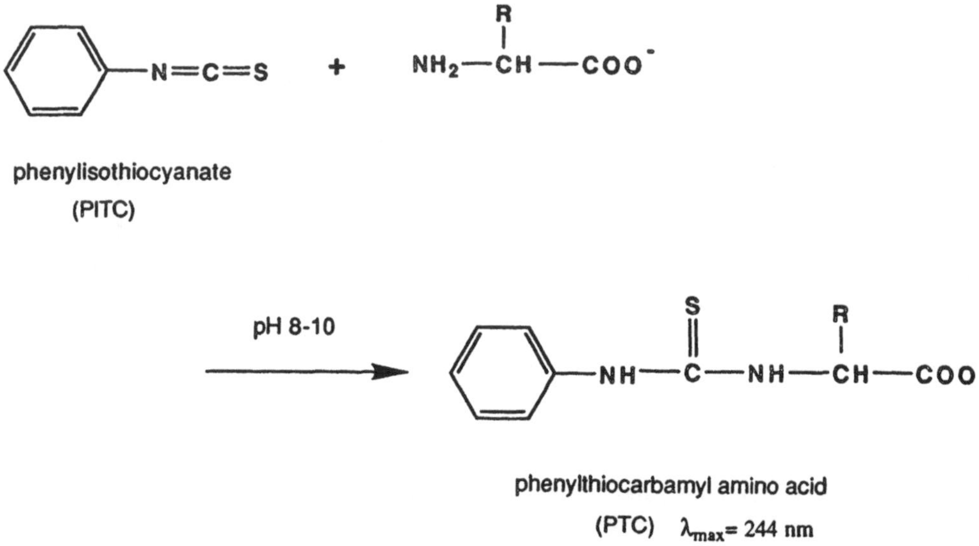

FIGURE 4. Reaction of phenylisothiocyanate (PITC) to form phenylthiocarbamyl (PTC) derivatives.

chemist. Response is linear over a wide range with a detection limit of ~ 1 pmol. However, because of other restrictions (described above) the practical quantitation limit is at the best ~ 10 pmol.

Although in a number of respects the PITC methodology can be considered superior to that of other derivatization techniques, it has a major disadvantage in that the yields of some PTC amino acids are markedly affected by the presence of some salts, divalent cations, metals, and buffer ions.[40] While ammonium acetate, sodium chloride, and sodium borate were found to have no effect on the recovery, others such as NH_4 acetate, sodium phosphate, and sodium bicarbonate were highly deleterious. Trace metal ion contamination has also been found to markedly reduce yields of PTC amino acids. The low and variable recoveries often seen for PTC-aspartate and -glutamate can sometimes be attributed to incomplete removal of acid after HCl hydrolysis but more seriously to the contamination of the hydrolyzed sample by the etching of the glass during the acid hydrolysis step. Mora et al.[41] have concluded that the extraction of cations (especially Ca^{2+}) during the acid etching of the glass led to the formation of metal salts of the acidic residues which were only slowly solubilized in the derivatization buffer. Yields were significantly improved in some situations by the inclusion of ethylenediamine tetraacetic acid (EDTA) or by increasing the aqueous content of the derivatization buffer. Improved yields by adding EDTA to the derivatization reaction have also been noted by Dupont et al.[40] who attribute low recoveries to the leaching of aluminum from the glass during hydrolysis.

Marketing of amino acid analyzers based on the PITC derivatization chemistry has had considerable commercial success (e.g., Waters Pico Tag and Applied Biosystems, Inc.) and these instruments are now being widely used for this purpose in North America and elsewhere. In attempts to reduce variability in the derivatization reaction and in the manual handling of multiple samples, the ABI instrument carries out the derivatization automatically followed by immediate on-line PTC analysis. The design is based on the principle of carrying out the derivatization reaction in a miniaturized reaction cell similar to that developed previously for the gas phase sequencer (see Reference 42 and review in this volume).[43] Hydrolysate is placed on one of three glass frits mounted in a glass slide. The latter is accommodated in a sample turntable having 24 slide positions. A moveable derivatization head with upper and lower jaws moves sequentially from sample to sample position to provide a controlled environment for the

dimethylaminoazobenzenesulfonyl chloride

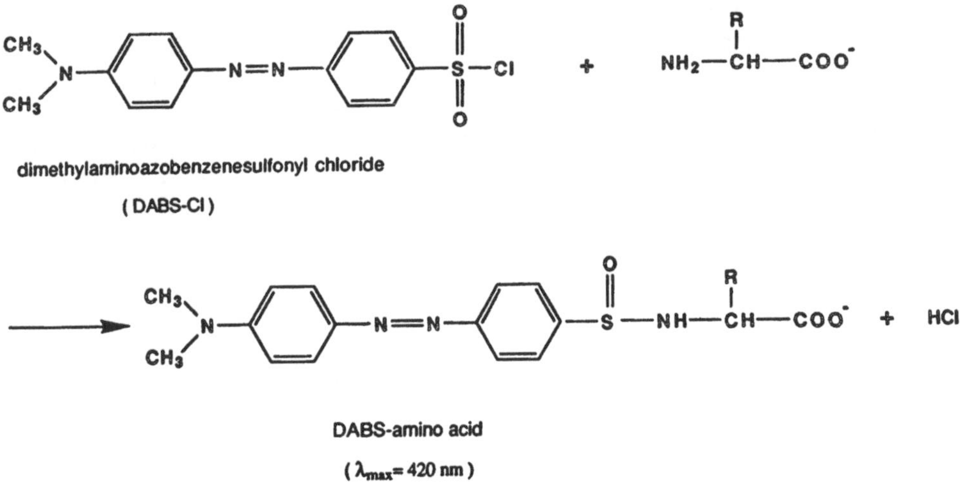

(DABS-Cl)

DABS-amino acid

(λ_{max}= 420 nm)

FIGURE 5. Reaction of DABS-Cl with amino acids to form colored derivatives.

derivatization reaction. Reagents and solvents are introduced through an inlet line in the upper jaw; following derivatization the PTC amino acids are extracted and delivered through an outlet line in the lower jaw to the analytical HPLC unit. HPLC separation is with a programmed acetonitrile gradient buffered with sodium acetate, pH 5.5. Identification and quantitation are by monitoring of absorbance at 254 nm. An on-line automated hydrolysis unit capable of carrying out gas-phase hydrolysis at elevated temperatures of protein/peptide samples applied to the glass frits has been under development for several years. A preliminary report[44] on its performance evaluation in an independent test laboratory is inconclusive. This report states, however, that the Applied Biosystems instrument (without automated hydrolysis) provides "reasonable quantitation and compositional accuracy (~ 10% error) ... around the 20 pmol amino acid level provided instrument performance is monitored and maintained". We may conclude that while this level of performance at this sensitivity may be adequate for providing compositional data on relatively small polypeptides (up to 20 to 30 residues) it is inadequate for meaningful compositional characterization of larger polypeptides and proteins.

VII. DIMETHYLAMINOAZONEBENZENESULFONYL CHLORIDE (DABS-Cl)

This reagent, which has been developed by Chang and his colleagues[45-51] and others[52-54] reacts with both primary and secondary amino acids to produce derivatives which have a high absorbance in the visible spectrum, its major advantage (see Figure 5). Reaction proceeds at ~ 70°C and pH 8.0 to 8.5 to produce highly stable derivatives in 10 to 12 min. Because of the solubility properties of the reagent, the reaction is carried out in 66% acetonitrile. Both mono and bis derivatives of lysine, tyrosine, and histidine are formed. Under optimal conditions of derivatization (dependent on reagent and amino acid concentrations) only the bis derivatives should be recovered. Fractionation of all the naturally occurring amino acids in acid hydrolysates as well as some derivatives has been described. The hydrolysis product of excess reagent (DABS-OH) appears in the chromatogram but is well enough separated not to interfere with quantitation of the amino acids. Response as monitored by absorbance at 436 nm appears to be linear over a wide range of amino acid concentration with detection limits extending into the femtomole range. However, to date the technique does not appear to have been successfully applied to the derivatization of amino acid standards or hydrolysates in the low picomole range. Typically derivatization has been carried out at the 0.5 to 1 nmol level and only a small fraction of the total derivatization mixture analyzed on the HPLC system. It is thus unclear as to how reproducible and quantitative the method can be made when applied to hydrolysates whose total content of any one amino acid may be in the 10 to 20 pmol range. Nor is it clear how the results would be affected by the presence of metals, salts, buffer ions, and other contaminants at this high level of sensitivity. A commercial adaptation of the method is available from Beckman Instruments, Inc.

VIII. SUMMARY AND CONCLUSIONS

The development of precolumn derivatization methodologies has been driven by the need for greater sensitivity and speed of analyses. The highly reliable and reproducible postcolumn ninhydrin methodology is limited to a sensitivity of ~ 100 pmol for that amino acid present in least abundance in a peptide or protein. For a molecule of Mr ~ 10,000 this corresponds to ~ 1 μg of polypeptide assuming only 1 residue/polypeptide for one or more amino acids. At the present time, the practical limits of sensitivity of precolumn derivatization methods are of the order of 10 to 20 pmol; i.e., an increase of 5- to 10-fold over the classical method even though in theory the limits of detection with the fluorescent reagents are much higher (low femtomole range). This limit is dictated by the variable background levels of contamination present in reagents, solvents, gases, and hardware which have been impossible to eliminate below a few picomole by even the most dedicated laboratories.

In assessing the advantages and disadvantages of the classical and newer precolumn derivatization methods, it is convenient to consider these at a sensitivity range of > 100 pmol and in the range 10 to 100 pmol. At the lower sensitivity range the postcolumn ninhydrin methodology has the clear advantage of being the most versatile in terms of its tolerance to the presence of impurities in the sample to be analyzed and in its proven accuracy and reproducibility based on many years of methodology development. In situations where the amount of sample is not a limiting factor, as for example in the quality assessment of the purity of synthetic polypeptides, this remains as the most reliable and satisfactory method.

In applying one of the precolumn derivatization methods at sensitivity levels < 100 pmol, it is clear that increased sensitivity comes at the expense of progressively decreasing reliability of the data. Thus while analyses at the 20-pmol level may be adequate for establishing the composition of a 20 to 30 amino acid residue peptide, the value and reliability of compositional data of such an analysis on a large protein are questionable.

ACKNOWLEDGMENTS

We thank Mr. M. Carpenter for help in the preparation of figures and the Medical Research Council of Canada for financial support.

REFERENCES

1. **Spackman, D.H., Stein, W.H., and Moore, S.,** Automatic recording apparatus for use in the chromatography of amino acids, *Anal. Chem.*, 30, 1190, 1958.
2. **Atherton, D.,** Successful PTC amino acid analysis at the picomole level, in *Techniques in Protein Chemistry*, Hugli, T.E., Ed., Academic Press, New York, 1989, 273.

3. **Moore, S. and Stein, W.H.,** Chromatographic determination of amino acids by the use of automatic recording equipment, in *Methods in Enzymology*, Vol. VI, Colowick, S.P. and Kaplan, N.O., Eds., Academic Press, New York, 1963, 819.

4. **Moore, S.,** Determination of cystine as cysteic acid, *J. Biol. Chem.,*, 238, 235, 1963.

5. **Crestfield, A.M., Moore, S., and Stein, W.H.,** The preparation and enzymatic hydrolysis of reduced and S-carboxymethylated proteins, *J. Biol. Chem.*, 238, 622, 1963.

6. **Hirs, C.H.W.,** Reduction and S-carboxymethylation of proteins, in *Methods in Enzymology*, Vol. XI, Hirs, C.H.W., Ed., Academic Press, New York, 1967, 199.

7. **Friedman, M., Krull, L.H., and Carins, J.F.,** The chromatographic determination of cystine and cysteine residues in proteins as S-β-(4-pyridylethyl) cysteine, *J. Biol. Chem.*, 245, 3868, 1970.

8. **Ellman, G.L.,** Sulfhydryl groups, *Arch. Biochem. Biophys.*, 82, 70, 1959.

9. **Habeeb, A.F.S.A.,** Reaction of protein sulfhydryl groups with Ellman's reagent, in *Methods in Ezymology*, Vol. 25, Hirs, C.H.W. and Timasheff, S.N., Eds., Academic Press, New York, 457, 1972.

10. **Scoffoni, E. and Fontana, A.,** Identification of specific amino acid residues, in *Protein Sequence Determination*, Needleman, S.B., Ed., Springer-Verlag, NY, 162, 1975.

11. **Matsubari, H. and Sasaki, R.M.,** High recovery of tryptophan from acid hydrolysates of proteins, *Biochem. Biophys. Res. Commun.*, 35, 175, 1969.

12. **Liu, T.Y. and Chang, Y.H.,** Hydrolysis of proteins with p-toluenesulfonic acid, *J. Biol. Chem.*, 216, 2842, 1971.

13. **Simpson, R.J., Neuberger, M.R., and Liu, T.Y.,** Complete amino acid analysis of proteins from a single hydrolysate, *J. Biol. Chem.*, 251, 1936, 1976.

14. **Hugli, T.E. and Moore, S.,** Determination of the tryptophan content of proteins by ion-exchange chromatography of alkaline hydrolysates, *J. Biol. Chem.*, 247, 2828, 1972.

15. **Tsugita, A. and Scheffler, J.-J.,** A rapid method for acid hydrolysis of protein with a mixture of trifluoroacetic acid and hydrochloric acid, *Eur. J. Biochem.*, 124, 585, 1982.

16. **Chiou, S.-H. and Wang, K.-T.,** Simplified protein hydrolysis with methanesulfonic acid at elevated temperature for the complete amino acid analysis of proteins, *J. Chromatogr.*, 448, 404, 1988.

17. **Chen, S.-T., Chiou, S-H., Chu, Y.-H., and Wang, K.-T.,** Rapid hydrolysis of proteins and peptides by means of microwave technology and its application to amino acid analysis, *Int. J. Peptide Protein Res.*, 30, 572, 1987.

18. **Chiou, S.-H. and Wang, K.-T.,** Peptide and protein hydrolysis by microwave irradiation, *J. Chromatogr.*, 491, 424, 1989.

19. **Bidlingmeyer, B.A., Cohen, S.A., and Tarvin, T.L.,** Rapid analysis of amino acids using precolumn derivatization, *J. Chromatogr.*, 336, 93, 1984.

20. **Meltzer, N.M., Tous, G.I., Gruber, S., and Stein, S.,** Gas-phase hydrolysis of proteins and peptides, *Anal. Biochem.*, 160, 336, 1987.

21. **Arrizon-Lopez, V., Biehler, R., Cummings, J., and Harbaugh, J.,** Beckman system 6300 high performance amino acid analyzer, this volume.

22. **Krapf, H.,** Amino acid analysis, this volume.

23. **Udenfriend, S., Stein, S., Böhler, P., Dairman, W., Leingruber, W., and Weigele, M.,** Fluorescamine: a reagent for assay of amino acids, peptides, proteins, and primary amines in the picomole range, *Science*, 178, 871, 1972.

24. **Roth, M.,** Fluorescence reaction for amino acids, *Anal. Chem.*, 43, 880, 1971.

25. **Dong, M.W.,** Analytical derivatization using robotics: amino acid analysis by precolumn o-phthaldehyde, *LC.GC.*, 5, 255, 1987.

26. **Einarsson, S., Josefsson, B., and Lagerkvist, S.,** Determination of amino acids with 9-fluorenylmethyl chloroformate and reversed-phase high-performance liquid chromatography, *Chromatography*, 282, 609, 1983.

27. **Einarsson, S.,** Selective determination of secondary amino acids using precolumn derivatization with 9-fluorenylmethyl chloroformate and reversed-phase high-performance liquid chromatography, *J. Chromatogr.*, 348, 213, 1985.

28. **Einarson, S., Folestad, S., Josefsson, B., and Lagerkvist, S.,** High-resolution reversed-phase liquid chromatography system for the analysis of complex solutions of primary and secondary amino acids, *Anal. Chem.*, 58, 1638, 1986.

29. **Betner, I. and Foldi, P.,** New automated amino acid analysis by HPLC: precolumn derivatization with fluorenylmethyloxycarbonylchloride, *Chromatography*, 22, 381, 1986.

30. **Smith, A.J., Presley, J.M., and McIntyre, W.,** An evaluation of an automated high sensitivity amino acid analyzer based on the 9-fluorenylmethylchloroformate (FMOC1) chemistry, in *Techniques in Protein Chemistry*, Hugli, T.E., Ed., Academic Press, New York, 255, 1989.

31. **Betner, I. and Foldi, P.,** The FMOC-ADAM approach to amino acid analysis, *LC.GC.*, 6, 832, 1988.

32. **Blackenship, T., Krivanek, M.A., Ackermann, B.L., and Cardin, A.D.,** High-sensitivity amino acid analysis by derivatization with o-phthalaldehyde and 9-fluorenyl chloroformate using fluorescence detection: applications to protein structure determinations, *Anal. Biochem.*, 178, 227, 1989.

33. **Koop, D.R., Morgan, E.T., Tarr, G.E., and Coon, M.J.,** Purification and characterization of an unique isozyme of cytochrome P-450 from liver microsomes of ethanol-treated rabbits, *J. Biol. Chem.,* 257, 8472, 1982.

34. **Tarr, G.E., Black, D.D., Fujita, V.S., and Coon, M.J.,** Complete amino acid sequence and predicted membrane topology of phenobarbital-induced cytochrome P-450 (isozyme 2) from rabbit liver microsomes, *Proc. Natl. Acad. Sci, U.S.A.,* 80, 6552, 1983.

35. **Bidlingmeyer, B.A., Cohen, S.A., and Tarvin, T.L.,** Rapid analysis of amino acids using precolumn derivatization, *J. Chromatogr.,* 336, 93, 1984.

36. **Heinrikson, R.L. and Meredith, S.C.,** Amino acid analysis by reverse-phase high-performance liquid chromatography: precolumn derivatization with phenylisothiocyanate, *Anal. Biochem.,* 136, 65, 1984.

37. **Cohen, S.A. and Strydom, D.J.,** Amino acid analysis utilizing phenylisothiocyanate derivatives, *Anal. Biochem.,* 174, 1, 1988.

38. **Tarr, G.E.,** Manual Edman sequencing system, in *Methods of Protein Micro-characterization,* Shively, J.E., Ed., Humana Press, Clifton, NJ, 1986, 155.

39. **Bidlingmeyer, B.A., Tarvin, T.L., and Cohen, S.A.,** Amino acid analysis of submicrogram hydrolysate samples, in *Methods in Protein Sequence Analysis,* Walsh, K., Ed., Humana Press, Clifton, NJ, 1986, 229.

40. **Dupont, D.R., Keim, P.A., Chui, A., Bello, R., Bozzini, M., and Wilson, K.J.,** A comprehensive approach to amino acid analysis, in *Techniques in Protein Chemistry,* Hugli, T.E., Ed., Academic Press, New York, 1989, 284.

41. **Mora, R., Berndt, K.D., Tsai, H., and Meredith, S.C.,** Quantitation of aspartate and glutamate in HPLC analysis of phenylthiocarbamyl derivatives, *Anal. Biochem.,* 172, 368, 1988.

42. **Hewick, R.M., Hunkapiller, M.W., Hood, L.E., and Dryer, W.J.,** A gas-liquid solid-phase peptide and protein sequenator, *J. Biol. Chem.,* 256, 7990, 1981.

43. **Smillie, L.B. and Carpenter, M.R.,** Protein and peptide sequencing by automated Edman degradation, this volume.

44. **West, K.A. and Crabb, J.W.,** Automatic hydrolysis and PTC amino acid analysis. a progress report, in *Techniques in Protein Chemistry,* Hugli, T.E., Ed., Academic Press, New York, 1989, 295.

45. **Lin, J.-K. and Chang, J.-Y.,** Chromophoric labeling of amino acids with 4-dimethylaminoazobenzene-4'-sulfonyl chloride, *Anal. Chem.,* 47, 1634, 1975.

46. **Chang, J.-Y., Knecht, R., and Braun, D.G.,** Amino acid analysis at the picomole level. Application to the C-terminal sequence analysis of polypeptides, *Biochem. J.,* 199, 547, 1981.

47. **Chang, J.-Y., Martin, P., Bernasconi, R., and Braun, D.G.,** High-sensitivity amino acid analysis: measurement of amino acid neurotransmitter in mouse brain, *FEBS Lett.,* 132, 117, 1981.

48. **Chang, J.-Y., Knecht, R., and Braun, D.G.,** A complete separation of dimethylaminoazo-benzenesulphonyl-amino acids. Amino acid analysis with low nanogram amounts of polypeptide with dimethylaminoazobenzene-sulfonyl chloride, *Biochem. J.,* 203, 803, 1982.

49. **Chang, J.-Y., Knecht, R., and Braun, D.G.,** Amino acid analysis in the picomole range by precolumn derivatization and high-performance liquid chromatography, in *Methods in Enzymol.,* Vol. 91, Hirs, C.H.W. and Timasheff, S.N., Eds., Academic Press, New York, 1983, 41.

50. **Knecht, R. and Chang, J.-Y.,** Liquid chromatographic determination of amino acids after gas-phase hydrolysis and derivatization with (dimethylamino)-azobenzenesulfonyl chloride, *Anal. Chem.,* 58, 2375, 1986.

51. **Chang, J.-Y., Knecht, R., Jenoe, P., and Vekemans, S.,** Amino acid analysis at the femtomole level using the dimethylaminoazobenzene sulfonyl chloride precolumn derivatization method: potential and limitation, in *Techniques in Protein Chemistry,* Hugli, T.H., Ed., Academic Press, New York, 1989, 305.

52. **Vendrell, J. and Avilés, F.,** Complete amino acid analysis of proteins by DABSYL derivatization and reversed-phase liquid chromatography, *J. Chromatogr.,* 358, 401, 1986.

53. **Hughes, G.J., Frutiger, S., and Fonck, C.,** Quantitative high-performance liquid chromatographic analysis of Dabsyl-amino acids within 14 min, *J. Chromatogr.,* 389, 327, 1987.

54. **Schneider, H.J.,** Amino acid analysis using DABS-Cl, *Chromatography,* 28, 45, 1989.

BECKMAN SYSTEM 6300 HIGH-PERFORMANCE AMINO ACID ANALYZER

Vivian Arrizon-Lopez, Ron Biehler, Judy Cummings, and Jon Harbaugh

The System 6300 series of amino acid analyzers are dedicated instruments using cation-exchange with postcolumn ninhydrin derivatization for separation and quantification of amino-containing components. This instrument represents a refinement of the Spackman, Stein, and Moore[1] technology that offers many advantages over alternative separation/quantification technologies. The accuracy, precision, and reproducibility of cation-exchange chromatography stem from the fact that derivatization proceeds continuously under the same conditions after chromatic resolution of each component. Therefore, salts, sugars, fats, and pH effects as well as mechanistic variations in derivatization time, temperature, drying, resuspension, extraction, and stability are not significant factors governing measurement precision.

The System 6300 console utilizes advanced technology to optimize the methodology that Spackman, Stein, and Moore developed almost 30 years ago. The System 6300 simultaneously monitors primary amines at 570 nm, secondary amines at 440 nm, and a reference channel at 690 nm. Many patented design features in the colorimeter and other components of the instrument ensure maximum sensitivity and resolution. Other features that contribute to the performance of the System 6300 are the narrow-bore column (3.0 mm I.D.), selection of up to six buffers, column temperature control, and refrigerated storage of buffers, reagent, and samples.

Columns, buffers, and reagents are formulated and manufactured specially for this instrument and manufactured to GMP (Good Manufacturing Practice) specifications, with the FDA verifying compliance. This results in high uniformity since each analytical column and batch of buffer or reagent is tested for conformity. The adherence to these standards makes the system suitable for diagnostic as well as bioresearch and biotechnology applications. The System 6300 is capable of separation and precise quantification of up to 52 known ninhydrin-positive components in 2.5 hr.

Another major advantage of the System 6300 is the "turnkey" concept of instrument use. Documented applications methods allow separation of all the amino acids required by the protein chemist with little operator intervention. See Figures 1 to 3 for typical separations.

The following points should be noted about the System 6300:

1. The cost of the instrument is $107,000. This includes start-up chemicals and column, installation, in-lab training, one-year warranty, and in-lab service.
2. Due to the microbore column design, solvent and reagent consumption are 20 and 10 ml/hr, respectively.
3. Low consumption of solvent and reagents results in a cost/analysis of less than $5.00 including buffers, reagent, column, and chart paper.
4. The System 6300 utilizes a split-beam photometer design with fixed optics. Visible colorimetry with ninhydrin offers several advantages over analysis methods which measure absorbance in the ultraviolet. Thousands of publications document the accuracy and validity of this design. In the System 6300, a single low-volume flow cell minimizes

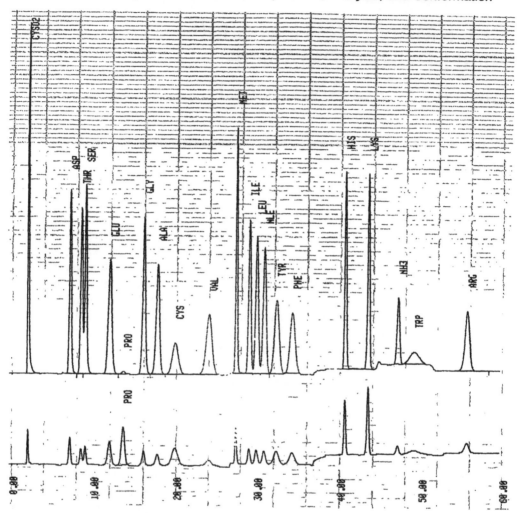

FIGURE 1. A typical chromatogram for the Beckman Calibration Standard Mixture with cysteic acid (CYSO2), norleucine (NLE), and tryptophan (TRP) added. Sodium citrate buffers Na-E™, Na-F™, Na-D™, and the Expanded Hydrolyzate Method II (available upon request from the Spinco Division of Beckman Instruments, Inc.) were used to generate the peak elution profile on a 25-cm column. Five nmol of each amino acid, except for cystine which is present in half that concentration, were delivered to the column for analysis at 1.0 absorbance units full scale (AUFS). The complete run cycle is 65 min.

sample diffusion while simultaneously monitoring at 440, 570, and 690 nm. This allows detection of both primary and secondary amines in addition to providing subtraction of a reference channel for reduction of baseline noise. The System 6300 can be used with ninhydrin detection to measure components present in concentrations as low as 50 pmol. Sensitivity can be increased fivefold with the use of an optional fluorometer.

5.   All reagents, including the sodium citrate buffers, sample dilution buffer, column regeneration buffer, ninhydrin, and calibration standards are manufactured by Beckman Instruments, Inc. These chemicals are packaged in the correct concentrations and are ready for immediate use by attaching directly to the instrument.

6.   Three sodium citrate buffers are used in sequence for separation of most multicomponent hydrolysates.

7.   Hydrolysis techniques are well established and have been published by a number of authors over the recent past. See References 2 to 4.

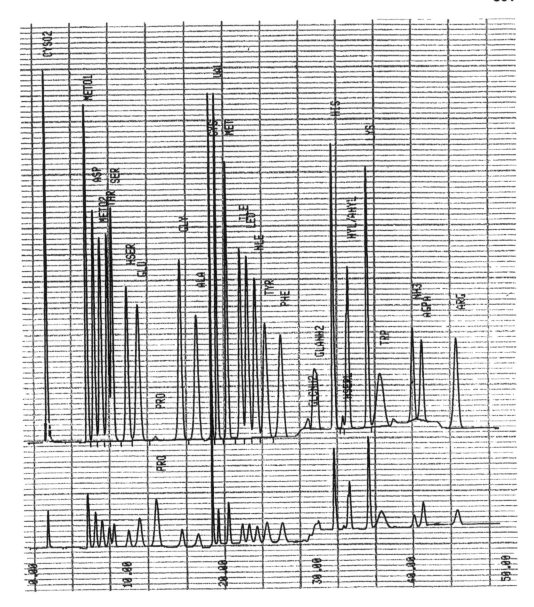

FIGURE 2. A typical chromatogram for the Beckman Complex Calibration Mixture, containing the standard 18 amino acids plus cysteic acid, methionine sulfoxide (METO1), methionine sulfone (METO2), homoserine (HSER), homoserine lactone (HSERL), glucosamine (GLCNH2), galactosamine (GLANH2), hydroxylysine (HYL), *allo*-hydroxylysine (AHYL) and internal standards norleucine and amino-β-guanidinopropionic acid (AGPA). Sodium citrate buffers Na-E™, Na-F™, Na-D™, and the Expanded Hydrolyzate Method I (available from the Spinco Division of Beckman Instruments, Inc.) were used to generate the peak elution profile on a 25-cm column. Five nanomoles of each amino acid, except for cystine which is present in half that concentration, were delivered to the column for analysis at 1.0 AUFS. The complete run cycle is 55 min.

8. Under-the-baseline artifacts are the following:

   a. **Buffer-change artifact.** These are the results of differences in pH and/or ionic strength of the eluents. These buffer-change artifacts are normal baseline shifts associated with step gradients and are positioned so they will not compromise the chromatography.

   b. **Ammonia plateau artifact.** Ammonia can be introduced to the buffers directly or

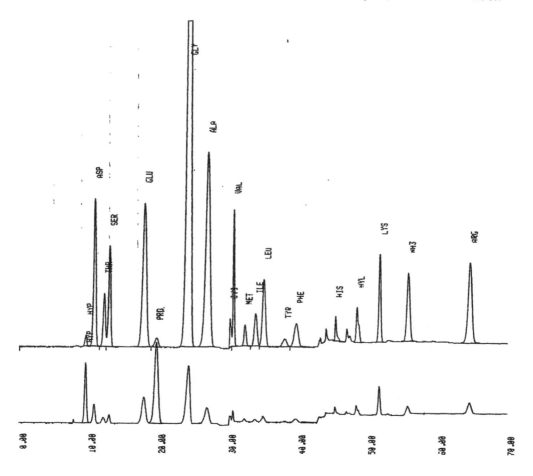

FIGURE 3. An achilles tendon hydrolysate sample profile using the Collagen Method (available from the Spinco Division of Beckman Instruments, Inc.). The pH-modified sodium citrate buffers Na-E™, Na-F™, and Na-D™ were used to generate the peak elution profile on a 25-cm column at 0.1 AUFS. The complete run cycle is 65 min. The tendon sample was hydrolyzed in 6 N HCl for 22 hr at 110°C. This chromatogram demonstrates the elution of hydroxyproline (HYP) and hydroxylysine (HYL) relative to the more common amino acids.

indirectly and results in a plateau in the third buffer region of the chromatogram. Consistent regeneration and equilibration cycles ensure reproducible baseline profiles which result in accurate amino acid quantitation. System 6300 methods have taken this phenomenon into consideration and peaks of interest are positioned away from artifacts.

9. Samples must be hydrolyzed, brought to dryness and reconstituted in pH 2.2 citrate buffer prior to injection. The 6 N HCl acid for 22 hr hydrolysis protocol is the most commonly used. Addition of mercaptoethanol helps preserve methionine and cysteine as well as minimizing tryptophan loss. Performic acid oxidation prior to hydrolysis is recommended for accurate quantitation of cystine and methionine.

    Samples with appreciable quantities of dissolved salts can be analyzed without difficulty on the ion-exchange columns of the System 6300 Amino Acid Analyzer. Approximately 0.25 mmol of dissolved salts can be tolerated per 50 µl of sample injected without causing distortion of the peaks in the resulting chromatogram.

    Introduction of ammonia contaminants during sample preparation may result in a higher ammonia content in the sample. However, ammonia will be eluted in the basic region of the chromatogram as a separate peak and can be present in high concentrations without affecting other peaks of interest.

10. There are two techniques for applying sample to the instrument: "overfill" and "sandwich". In both cases, samples are automatically transferred from the easily loaded sample coils to a fixed-volume metering loop on the sample injector valve. Metering loops are available in nominal volumes of 20, 50, and 100 µl. A recently introduced conductivity detection device surrounding the metering loop ensures maximum utilization of sample. Where sample utilization must be maximized, a technique of "sandwiching" sample between dilution buffer in the sample coils permits 100% of sample volumes to be applied to the column.

11. A sample is applied to the column by automatic injection from a metering loop into the high pressure separation stream. Repeatable transfer of the entire contents of the metering loop onto the separation column with every sample ensures maximum instrument precision. No sample derivatization is required prior to sample injection, therefore this complex and potentially destructive procedure is eliminated when using the System 6300.

12. The complete cycle time from one sample injection to the next sample injection is 55 to 65 min.

13. The concentration range that can be successfully derivatized and quantitated is from 5 nmol to 10 pmol per amino acid component per injection. The on-line ninhydrin postcolumn derivatization technique offers the advantage of area reproducibility. The instrument is linear within the above concentration range, and there is no need for multiple level calibrations. The System 6300 conservative specification for area reproducibility is ± 3% for samples of less than 100 pmol and ± 2% for larger samples.

14. The dynamic range of amino components that can be quantitatively handled in any one analysis is 1:500.

15. The System 6300 offers guaranteed in-lab performance with average precision of 3% C.V. (Coefficient of Variation) or better to as low as 50 pmol/component for the 18 calibration standard amino acids.

16. An IBM PC-based data system is available for the System 6300. System Gold Chromatography Software uses a window-based operator interface combining graphics capabilities and multitasking to provide simultaneous data collection and postrun data processing. System Gold Software prints peak name identification on the chromatogram, presents both channel outputs on a single page, and allows baseline subtraction and auto-zeroing. In addition to the advanced capabilities for postrun manipulation of data and calibration tables within this software environment, System Gold Software facilitates the downloading of data into other software environments for further data manipulation or database collection.

REFERENCES

1. **Spackman, D.H., Stein, W.H., and Moore, S.,** Automatic recording apparatus for use in the chromatography of amino acids, *Anal. Chem.* 30, 1190, 1958.
2. **Samata, T. and Matsuda, M.,** Contaminating peptides widely present in ion-exchanged water, reagents, experimental implements and natural sample, *Comp. Biochem. Physiol. B: Comp. Biochem.* 84B, 531, 1982.
3. **Schlesinger, D.H.,** High-sensitivity microsequencing of peptides and proteins, in *Macromolecular Sequencing and Synthesis*, Schlesinger, D.H., Ed., A.R. Liss, New York, 1988, 35.
4. **Tsugita, A. and Scheffler, J.J.,** A rapid method for acid hydrolysis of protein with a mixture of trifluoroacetic acid, *Eur. J. Biochem.* 124, 585, 1982.

PHARMACIA/LKB AMINO ACID ANALYZER SYSTEMS USING PRE-COLUMN OR POST-COLUMN DERIVATIZATION

Herman Krapf

Since the first automatic separation and quantitation of amino acids was performed in 1958,[1] several dedicated amino acid analyzers have become commercially available. They are based on ion-exchange separation and post-column derivatization with OPA or ninhydrin (the most widely used). The long experience of this method has made it very popular for use in routine analysis laboratories.

However, in the past 5 to 10 years, the analysis of amino acids has increased dramatically in different areas such as protein chemistry, food and beverage analysis, clinical analysis, and peptide synthesis. This increase has also been accompanied by requirements for high sensitivity, and faster and reproducible analysis. These demands have led to the use of pre-column derivatization in conjunction with reversed-phase HPLC (RPC).

The most common agents used for pre-column derivatization are phenylisothiocyanate (PITC)[2] or orthophthalaldehyde (OPA),[3] both of which have advantages and disadvantages.[4] For example, OPA does not react with secondary amines but offers high sensitivity and the possibility to automate fully the reaction and injection steps. PITC reacts with both primary and secondary amines but is less sensitive and cannot be fully automated.

The introduction over the past few years of 9-fluorenylmethoxycarbonyl chloride (FMOC)[5] as a new reagent for amino acid analysis has provided the possibility to increase sensitivity to the femtomole level.

I. POST-COLUMN DERIVATIZATION

The traditional approach to amino acid analysis, based on ion-exchange separation followed by post-column derivatization with ninhydrin or OPA, is not as sensitive as the RPC technique, but has proven reproducibility and reliability. The dedicated analyzer AlphaPlus[6-8] developed by Pharmacia LKB and utilizing the post-column derivatization technique, is a well-established instrument and is widely used in routine analysis for example in clinical laboratories and for the quality control of feedstuffs (Figures 1 and 2).

II. PRE-COLUMN DERIVATIZATION

The AminoSys™ system from Pharmacia LKB is a modular HPLC system for amino acid analysis with automated pre-column derivatization and computer control and evaluation.

A. DERIVATIZATION WITH PITC

Derivatization of amino acids with PITC, also known as Edman's reagent,[9] is widely used by protein chemists. The PTC amino acids, detected by their UV absorption at 245 nm, are stable but do not fluoresce. Consequently, this method is not as sensitive as the methods utilizing a fluorescence detector, and in practical terms is limited to samples of 5 to 10 pmol/amino acid.

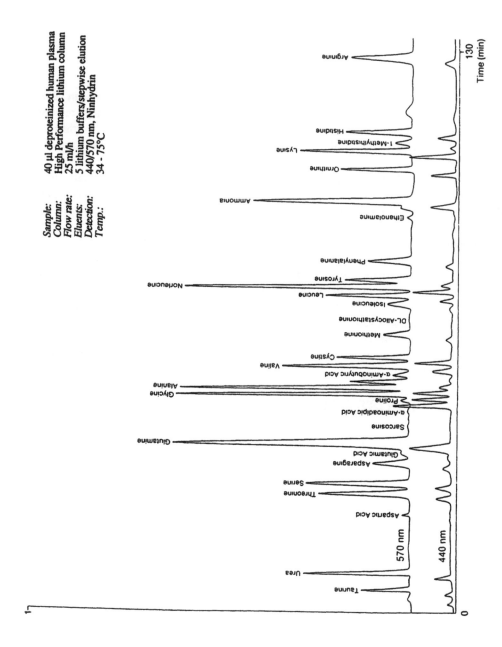

Sample: 40 μl deproteinized human plasma
Column: High Performance lithium column
Flow rate: 25 ml/h
Eluents: 5 lithium buffers/stepwise elution
Detection: 440/570 nm, Ninhydrin
Temp.: 34 - 75°C

FIGURE 1. A typical chromatogram of human plasma run on a 4151 AlphaPlus, with a lithium citrate buffer system.

Sample	10 nmol/amino acid, protein hydrolysate standard
Column	High Performance sodium column
Flow rate	25 ml/h
Eluents	3 sodium buffers/stepwise elution
Detection	440/570 nm, Ninhydrin
Temp	34 - 75° C

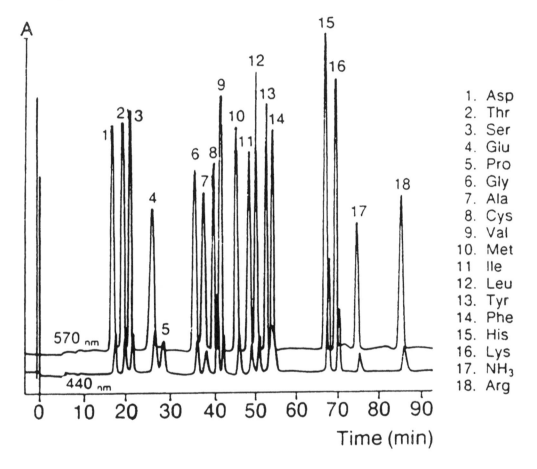

FIGURE 2. Analysis of protein hydrolysate standard on a 4151 AlphaPlus amino acid analyzer.

This means that the PITC method should be used when the amount of sample is not limited. PITC reacts with primary and secondary amines at alkaline pH and provides the possibility of using PITC in most application areas where the sensitivity is not the limiting factor.[10]

It is of utmost importance that the sample is free of hydrochloric acid prior to the addition of PITC and this is accomplished by a drying and re-drying step in a vacuum centrifuge. The excess of PITC is then removed in the vacuum centrifuge.

The sample is applied on a SuperPac C_{18} column and eluted with an acetonitrile gradient in sodium phosphate buffer at pH 6.4. Typical analysis times are between 10 and 30 min. An analysis of a protein hydrolysate standard is shown in Figure 3.

In conclusion, PITC has the advantage of being a well-known reagent which also reacts with secondary amines. Pre-column derivatization with PITC is a semi-automated procedure.

Sample: Protein hydrolysate standard, 62 pmol/amino acid, 5 μl
Column: SuperPac™ Spherisorb ODS II, 3 μm, 4×125 mm
Flow rate: 1.0 ml/min
Eluent A: 12.5 mM sodium phosphate, pH 6.4, 0.5 % CH$_3$CN
       B: 12.5 mM sodium phosphate, pH 6.4, 70 % CH$_3$CN
Gradient: 0 % B for 2 min, 0–32 % B in 12 min,
            32–100 % B in 2 min
Detection: UV 245 nm
Temp.:   38°C

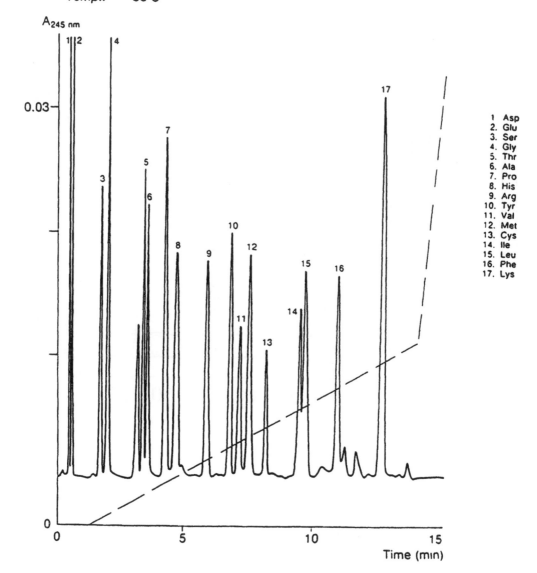

FIGURE 3. PTC amino acid standard run on AminoSys HPLC system. The 17 amino acids are effectively separated in 12 min.

B. DERIVATIZATION WITH OPA

The reaction of OPA with primary amines was described in 1971 by Roth.[11] The resulting fluorescent product is detected either by fluorescence or UV detection. Secondary amines do not react directly with OPA and must therefore be oxidized to primary amines prior to the derivatization.

Originally, this method was applied only on post-column derivatization, but is now widely used as a reagent for the pre-column derivatization of amino acids. the reaction occurs between pH 9.5 and 10.5, where OPA reacts with the primary amines to form an unstable fluorescent adduct. The reaction time is short (2 to 3 min), and requires the presence of a thiol. We recommend 3-mercaptopropionic acid because of its non-volatile features.

The OPA derivatives are rather unstable and, due to this, the derivatization and sample injection should preferably be automated using the Autosampler 2157. Chromatography should be carried out using a SuperPac C_{18} column in sodium phosphate buffer, pH 7.2, and a gradient of acetonitrile, since the lower pH in the buffer stabilizes the OPA derivatives somewhat.[12]

Detection of OPA amino acids can be easily performed using either fluorescence or UV detection, making it possible to have a broad working range from 100 fmol to 150 pmol. When using UV detection, the recommended wavelength is 330 nm. With the new Variable Wavelength Monitor 2141 from Pharmacia LKB, the detection limit is in the low picomole range. Fluorescence detection of OPA amino acids (excitation wavelength 330 nm and emission wavelength 450 nm) is 10- 100-fold more sensitive than UV detection, and this gives a limit of detection in the 100 fmol range.

OPA derivatization is recommended for routine amino acid analysis when high reproducibility is required and it is not necessary to detect the secondary amines such as proline and hydroxyproline. When high sensitivity is of importance, OPA combined with the Fluorescence Detector 2144 is a good choice. An analysis of a protein hydrolysate standard is shown in Figure 4.

C. DERIVATIZATION WITH FMOC

The demand has arisen for an alternative reagent to OPA and PITC for pre-column derivatization and which combines the advantages of both.[13,14] In 1983, FMOC was used for amino acid analysis for the first time and since then the method has been developed further.[5] FMOC derivatization of primary and secondary amino acids is fully automated using the Autosampler 2157. The derivatives are stable for more than 30 hr in an acid environment. As FMOC is a highly reactive reagent having an almost identical absorption spectra as its derivatives, it is essential that the excess is removed. The removal step is a simple derivatization of the excess FMOC using an amine (1-aminoadamantane[ADAM]), which is more hydrophobic than the most hydrophobic amino acid lysine to shift the interfering FMOC peak to the end of the chromatogram.[15,16] The derivatization step is completed in 45 sec. ADAM is added to capture the excess FMOC, and after another 45 sec the sample is ready for injection. This procedure should preferably be automated.

Elution of the FMOC amino acids from the column is executed with a sodium acetate buffer pH 4.2 and a gradient of acetonitrile (Figures 5 and 6). The column diameter has been reduced for the application shown in Figure 6 in order to increase sensitivity. In addition, the gradient has been optimized for this particular separation. The FMOC-ADAM peak in this separation is eluted very late in the chromatogram.

The derivatives are detected using the Fluorescence Detector 2144. FMOC has excitation maximum at 260 nm and emission maxima at 305 and 610 nm. The detection limit when using the Fluorescence Detector 2144 is lower than 100 fmol. The advantages of using FMOC are that it reacts with both primary and secondary amines, the resulting derivatives are stable at least 30

Sample: Protein hydrolysate standard, 25 fmol/amino acid, 1 μl
Column: Spherisorb ODS II, 3 μm, 1×250 mm
Flow rate: 80 μl/min
Eluent A: 12.5 mM sodium phosphate, pH 7.2, 0.5 % CH_3CN, 2.5 % MeOH
  B: 12.5 mM sodium phosphate, pH 7.2, 25 % CH_3CN, 5 % MeOH
Detection: Fluorescence, 330 nm Exc., 450 nm Em.
Temp.: 30°C

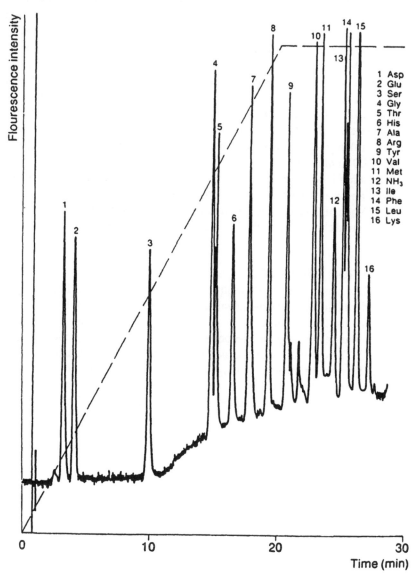

FIGURE 4. High sensitivity analysis of OPA derivatized amino acids with AminoSys using fluorescence detection.

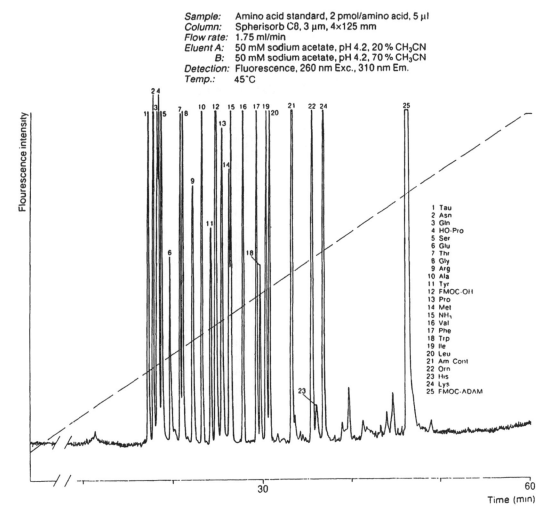

Sample: Amino acid standard, 2 pmol/amino acid, 5 µl
Column: Spherisorb C8, 3 µm, 4×125 mm
Flow rate: 1.75 ml/min
Eluent A: 50 mM sodium acetate, pH 4.2, 20 % CH$_3$CN
B: 50 mM sodium acetate, pH 4.2, 70 % CH$_3$CN
Detection: Fluorescence, 260 nm Exc., 310 nm Em.
Temp.: 45°C

1 Tau
2 Asn
3 Gln
4 HO-Pro
5 Ser
6 Glu
7 Thr
8 Gly
9 Arg
10 Ala
11 Tyr
12 FMOC-OH
13 Pro
14 Met
15 NH$_3$
16 Val
17 Phe
18 Trp
19 Ile
20 Leu
21 Am Cont
22 Orn
23 His
24 Lys
25 FMOC-ADAM

Time (min)

FIGURE 5. Elution profile of an artificial mixture of amino acids precolumn derivatized with FMOC. Observe that the excess FMOC has been removed to the end of the chromatogram by derivatization with ADAM.

hr, the reaction is easily automated and the sensitivity is very high. This method is still in need of further method development to achieve the level of reproducibility required in automated routine analysis.

In protein chemistry, where the demands on high sensitivity, small sample sizes and the need to detect secondary amines are of paramount importance, the FMOC derivatization technique is very useful.

III. DETAILS OF AMINOSYS™ SYSTEM

The amino acid analysis system AminoSys developed by Pharmacia LKB is a modular HPLC system that consists of HPLC manager software, two Pumps 2248, High Pressure Gradient Mixer, Solvent Conditioner 2156, Autosampler 2157, Column Oven 2155, and a Variable Wavelength Monitor 2141 or Fluorescence Detector 2144. The cartridge system recommended for the PITC and OPA techniques is a SuperPac cartridge (4.0 × 125 mm) prepacked with C$_{18}$-derivatized silica media and with a guard cartridge packed with the same material.

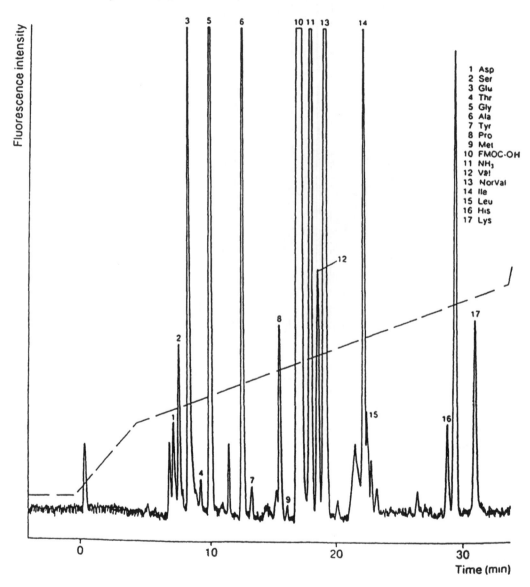

Sample: APC 1 peptide from brain tissue of Alzheimer's victims 200
           fmol/amino acid, 2 µl
Column: Supersphere CH8, 2×125 mm
Flow rate: 0.3 ml/min
Eluent A: 50 mM sodium acetate, pH 4.2
       B: 100% CH₃CN
Detection: Fluorescence, 263 nm Exc., 313 nm Em.
Temp.:   45°C

1 Asp
2 Ser
3 Glu
4 Thr
5 Gly
6 Ala
7 Tyr
8 Pro
9 Met
10 FMOC-OH
11 NH₃
12 Val
13 NorVal
14 Ile
15 Leu
16 His
17 Lys

Figure 6. High sensitivity analysis of FMOC amino acids from a hydrolysate of peptide APC I derived from brain tissue of Alzheimer's disease victims.

The AminoSys system is easily set up using the Net cables. Samples are loaded into the Autosampler prior to derivatization and analysis. Pre-column reaction immediately prior to the injection of the sample will then be automatically performed. A small, accurate and reproducible amount (as small as 1 µl) of sample is injected onto the column.

The eluents are delivered at a constant flow rate into the column by two dual piston pumps, while the gradient is reproducibly controlled by the HPLC manager. On increasing the concentration of eluent B (acetonitrile), the partitioning between the bonded phase and the mobile phase is decreased, causing the different amino acids to be eluted. To increase the sensitivity level and reproducibility of the analysis, a buffer de-gassing device (solvent conditioner) is employed.

The presence and concentration of eluted amino acids is measured by either a UV or fluorescence detector. Choose the detector suitable for your specific demands and depending on the sensitivity needed (femtomoles to nanomoles). The detector output is connected to the data handling system, where evaluations and calculations are made.

To ensure high sensitivity and reproducibility of the AminoSys system, it is essential to minimize dead volumes in the system as a whole. This is most important between the column and the detector, and also between the Autoinjector and the column. Generally, any diffusion of the sample, whether pre- or post-column, will cause peaks to broaden, flatten, and spread into each other, thereby losing sensitivity and resolution.

A major strength of HPLC for amino acid analysis is its inherent flexibility, with several options for optimizing the chromatography, e.g., choice of mobile and stationary phase, flow rate, and temperature. In addition, choice of derivatization agent is also a key factor in optimizing separation. Selection of stationary phase is influenced by the hydrophobicity and selectivity of the available materials. For PITC and OPA amino acids, a C_{18} phase, with its high hydrophobicity and selectivity, is always chosen. Usually, a C_8 phase is chosen for FMOC amino acids. For the ultimate performance, a 3 to 5-μm particle size is recommended for the amino acid analysis. The mobile phase is a binary gradient with sodium phosphate/acetonitrile for the PITC and OPA techniques, and sodium acetate/acetonitrile for the FMOC technique. Flow-rate is usually between 1 and 2 ml/min. The temperature chosen for all three techniques is between 25 and 45°C. Elevated temperatures are often used to decrease analysis time, but it also decreases the life time of the column.

In conclusion, the AminoSys system for amino acid analysis is suitable for any kind of pre-column derivatization technique with its reliability, flexibility and reproducibility. By adding the high sensitivity of the Variable Wavelength Monitor 2141 and Fluorescence Detector 2144, together with the control and evaluation software, you will get a powerful and convenient system for analysis of amino acids in most application areas.

REFERENCES

1. **Spackman, D.H., Stein, W.H., and Moore, S.,** Automatic recording apparatus for use in the chromatography of amino acids, *Anal. Chem.,* 30, 1190, 1958.
2. **Henriksson, R.L. and Meredith, S.C.,** Amino acid analysis by reversed-phase high-performance liquid chromatography: Precolumn derivatization with phenylisothiocyanate, *Anal. Biochem.,* 136, 65, 1983.
3. **Ogden, G. and Földi, P.,** Amino acid analysis: an overview of current methods, *LC.GC,* 1, 28, 1987.
4. **Betnér, I. and Földi, P.,** The FMOC-ADAM approach to amino acid analysis, *LC.GC,* 9, 832, 1988.
5. **Einarsson, S., Josefsson, B., and Lagerkvist, S.,** Determination of amino acids with 9-fluorenylmethylchloroformate and reversed-phase high-performance liquid chromatography, *J. Chromatogr.,* 282, 609, 1983.
6. **Blackburn, S.,** *Amino Acid Determination: Methods and Techniques,* Marcel Dekker, New York, 1978.
7. **Andrews, R.P. and Baldar, N.A.,** Amino acid analysis of feed constituents, *Science Tools,* 32, 44, 1985.
8. **Blom, W. and Huijmans, J.,** Differential diagnosis of (inherited) amino acid metabolism or transport disorders, *Science Tools,* 32, 10, 1985.
9. **Hill, W.D., Walters, F.H., Wilson, T.D., and Stuart, J.D.,** High-performance liquid chromatographic determination of amino acids in the picomole range, *Anal. Chem.,* 51, 1338, 1979.

10. **Einarsson, S.,** Selective determination of secondary amino acids using precolumn derivatization with 9-fluorenylmethylchloroformate and reversed-phase high-performance liquid chromatography, *J. Chromatogr.*, 348, 213, 1985.

11. **Edman, P. and Henschen, A.,** in *Determination in Protein Sequence*, 2nd ed. Needleman, S.B., Ed., Springer-Verlag, Berlin, 1975, p. 232.

12. **Graser, Th., Godel, H., Földi, P., and Fürst, P.,** An ultrarapid and sensitive HPLC method for determination of tissue and plasma free amino acids, *Anal. Biochem.*, 151, 142, 1985.

13. **Bidlingmeyer, B.A., Cohen, S.A., and Tarvin, T.L.,** Rapid analysis of amino acids using pre-column derivatization, *J. Chromatogr.*, 336, 93, 1984.

14. **Roth, M.,** Fluorescence reaction for amino acids, *Anal. Chem.*, 43, 880, 1971.

15. **Schneider, H.J., Hoepker, H.R., and Földi, P.,** *HPLC-Aminosäuren-Analyse-Pre-Column Derivatisierung*, LKB GmbH, Gräfelfing, Federal Republic of Germany, English version available, 1986.

16. **Betnér, I. and Földi, P.,** New automated amino acid analysis by HPLC precolumn derivatization with fluorenylmethyloxycarbonylchloride, *Chromatographia*, 22, 381, 1986.

PROTEIN AND PEPTIDE SEQUENCING BY AUTOMATED EDMAN DEGRADATION

Lawrence B. Smillie and Michael R. Carpenter

The amino acid sequence analysis of peptides and proteins has become one of the most widely used methodologies in modern biochemistry and molecular biology. Combined with DNA sequencing techniques, the generated information forms the data base on which many of our concepts of structure and biological function, molecular evolution, development, and regulation are founded. Progress in the development of ever more reliable and sensitive techniques for amino acid sequence analyses has been remarkable and continues unabated at the present time.

The most widely applied single technique for peptide and protein sequencing is that based on the chemistry developed by P. Edman over a period of decades.[1-3] The fundamentals of this chemistry, whether applied manually or in various versions of automated sequencing, involve three basic steps: (1) coupling; (2) cleavage; (3) identification. Coupling and cleavage constitute a single cycle of Edman degradation following which the truncated polypeptide chain (now shorter by the removal of one amino acid residue from its NH_2-terminus) may be recycled. Various stratagems are available for the identification process (see below).

Coupling involves the reaction of phenylisothiocyanate (PITC) with amino groups to form the phenylthiocarbamylated (PTC) derivatives (see Figure 1, coupling step). Careful pH control in the region of 9.0 to 9.5 is important since the reaction occurs with NH_2-groups in their nonprotonated form. At higher pH (≥ 10) the PITC becomes unstable. To avoid reaction with buffer components, various volatile or non-volatile tertiary amines have been used including pyridine, trimethylamine, dimethylallylamine, dimethylbenzylamine, or Quadrol (N,N,N′N′-tetrakis-(2-hydroxypropyl)-ethylene-diamine). Because of the limited solubility of PITC in water, it is usually introduced into the reaction as a solution in heptane. The reaction with terminal α-NH_2 groups under optimal conditions is normally close to quantitative at 55°C in 30 min.

Following coupling, excess reagent, buffer components, and reagent degradation products must be removed to facilitate the subsequent cleavage reaction and identification. The PTC derivative is also made anhydrous to minimize random polypeptide bond hydrolysis during the cleavage step. These objectives are normally met by taking the coupling reaction mixture to dryness (either under vacuum and/or under a stream of nitrogen or argon), extraction with organic solvents (benzene or heptane; ethyl acetate), and further drying. Specific cleavage at the NH_2-terminal peptide bond is achieved by exposure of the PTC-polypeptide to anhydrous acid (most commonly trifluoroacetic or heptafluorobutyric acid). The reaction, leading to the cyclic thiazolinone derivative and the truncated (by one residue) but otherwise intact polypeptide chain (Figure 1, cleavage step), proceeds rapidly and in most cases essentially quantitatively at moderate temperature (e.g., 3 min at 55°C). Following removal of the volatile acid by drying, the thiazolinone is extracted with 1-chlorobutane). The remaining peptide or protein may then be subjected to the next round of coupling and cleavage.

Various strategies have been used for identification of the amino acid derivative removed at each cycle of the degradation. In the subtractive Edman procedure, amino acid composition differences of the truncated peptide before and after each cycle permits such identification. However, this is generally restricted to the sequencing of relatively small peptides because of

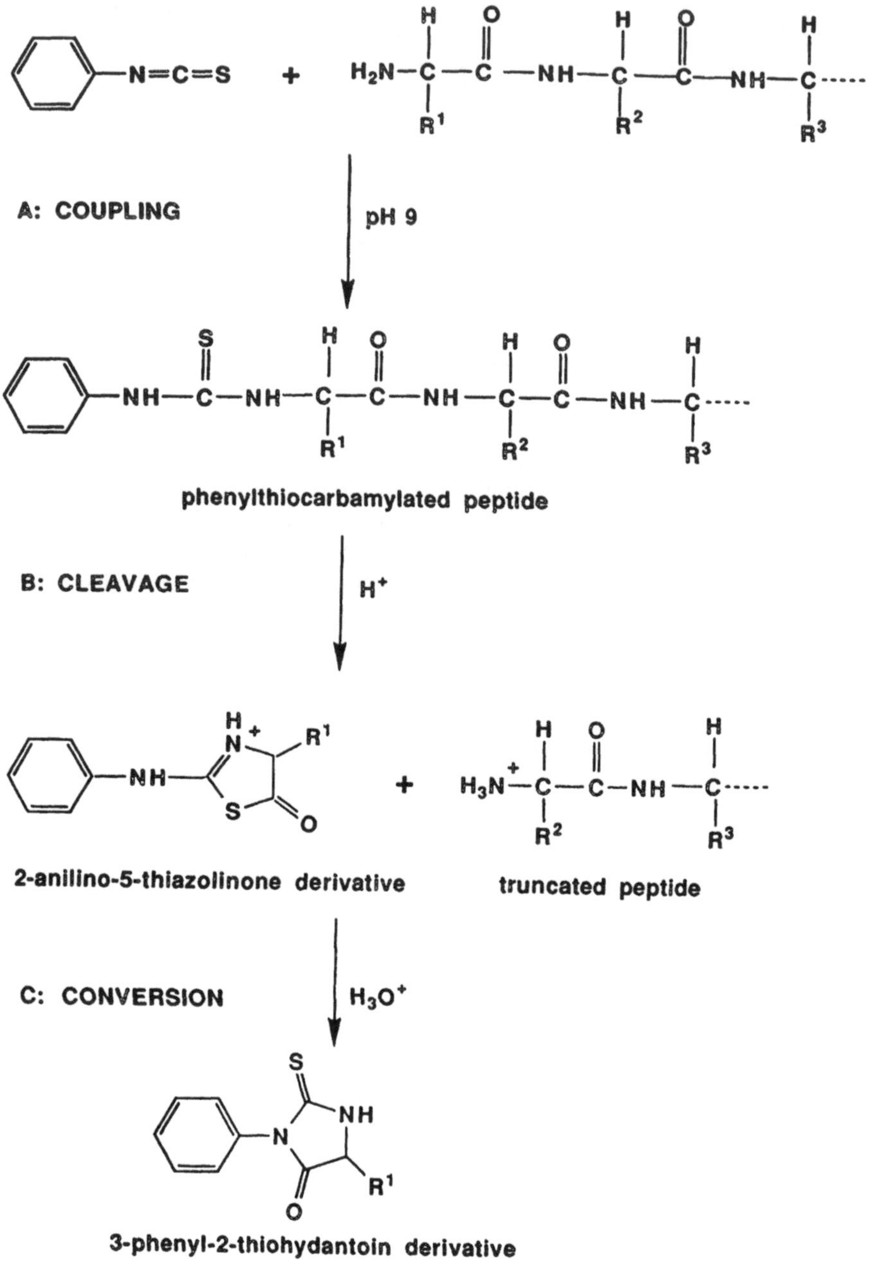

A: COUPLING

pH 9

phenylthiocarbamylated peptide

B: CLEAVAGE   H⁺

2-anilino-5-thiazolinone derivative

truncated peptide

C: CONVERSION   H₃O⁺

3-phenyl-2-thiohydantoin derivative

FIGURE 1. The Edman degradation chemistry.

the limitations in the precision of amino acid analyses. A more general and widely used approach has been to identify the newly formed NH_2-terminal residue of an aliquot of the truncated peptide after each cycle. Such reagents as DANSYL-Cl (dimethylaminonaphthylenesulfonyl chloride)[4] or DABSYL-Cl (dimethylaminoazobenzenesulfonyl chloride)[5] have been used for this purpose. Identification is also possible by the hydrolysis of the extracted thiazolinone to the free amino acid with acid or alkali followed by amino acid analysis.[1,6,7] However, with the development of modern HPLC methodology the identification of the cleaved product is now almost universally carried out by converting the thiazolinones to the more stable phenylthiohydantoin (PTH) derivatives (Figure 1, conversion step) and their separation on reversed-phase high-performance liquid chromatography (RPC) columns. Conversion (an isomerization reaction) can be achieved by treatment of the thiazolinone with 1 M HCl at 80° for 10 min or with 25% trifluoroacetic acid at 55°C for 30 min.

As a sequence analysis progresses through multiple cycles, there is an increase in the number and levels of spurious by-products. These are generally attributed to a low level of random peptide bond hydrolyses under the acid cleavage conditions leading to the generation of new NH$_2$-terminal residues. In addition, of course, the thiazolinone (or PTH) product recovered at each cycle will decrease progressively as the cycling reactions are continued. It is a combination of this decrease in the level of identifiable product (signal) coupled with increasing background (noise) that determines the number of cycles beyond which no further useful sequence information can be obtained. Small increases in the efficiency of the coupling and cleavage reactions combined with efforts to minimize impurities and by-products can lead to significantly increased sequence information. For example, it can be calculated that an increase in the average efficiency per cycle from 95 to 97% will lead after 40 cycles to an improvement in overall yield from 13 to 35%. Based on his investigations of the basic chemistry of the degradation procedures, Edman stressed the importance of eliminating O$_2$ and oxidizing agents from the reaction and of using high-purity reagents. The former can lead to desulfuration of the reagent or PTC polypeptide to yield the phenylcarbamylated derivative.[3] This will not readily cyclize in the cleavage reaction and thus essentially blocks the degradation. Contaminants in the buffers or reagents such as aldehydes, for example, will react with amino groups and significantly reduce the efficiency of coupling.

I. AUTOMATED SEQUENCING

Having achieved a high degree of efficiency in the coupling and cleavage steps of the manual degradation procedures, Edman turned his attention to the development of an automated system for the highly repetitive and tedious manipulations involved. The first such automated sequencer was described in a classic paper[8] published in 1967. The heart of the instrument is the concept of performing all of the reactions in a continuously spinning cylindrical cup under a nitrogen atmosphere and at controlled temperature (Figure 2). Maximal surface area contact between immiscible buffered aqueous protein and PITC in heptane is achieved through the distribution of these as a thin film over the lower surfaces of the cup from the centrifugal force. Drying is achieved by evacuation while all the reagent/solvent additions and extractions are through an inlet tube positioned near the bottom of the cup and an outlet scoop and line situated in a groove close to the cup's lip. The latter delivers wash solvents to waste, or in the case of the 1-chlorobutane extraction of the thiazolinones following the cleavage reaction, to a fraction collector. These are then manually converted to the PTH derivatives for identification. In more modern versions of the instrument,[9,10] an automated on-line conversion reaction cell is incorporated into the system. This increases recoveries and reproducibility because of the increased stability of the PTHs compared with the thiazolinones.

The spinning cup sequencer as first developed by Edman and as subsequently modified[9-12,31,34] was (and is) a highly successful instrument. It is capable of generating long NH$_2$-terminal sequences with up to 70 or 80 residues established in some cases. More routinely and in the hands of most experimenters, sequencing runs of 30 to 50 amino acid residues identified are obtained on amounts of protein of 1 to 10 nmol or better. It was the workhorse of most protein chemistry laboratories for well over a decade.

In spite of its success, the spinning cup sequencer has certain limitations. One of these is associated with the use of the non-volatile buffer Quadrol. This had been introduced by Edman because of its strong buffering capacity and its protein-solubilizing powers. The bulk of this

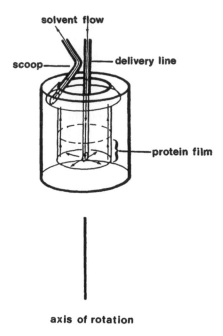

FIGURE 2. Spinning cup sequencer.

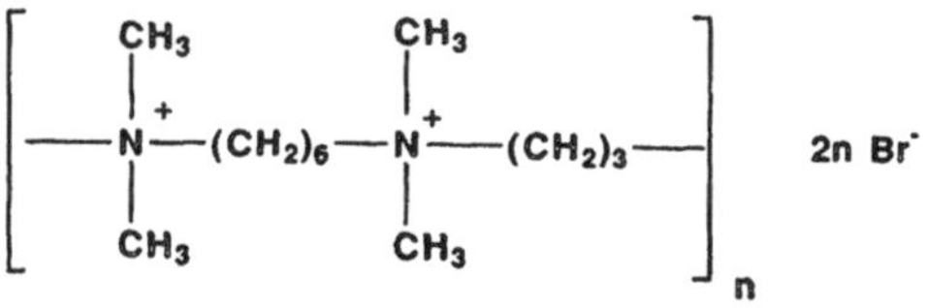

FIGURE 3. Polybrene.

reagent is normally removed during the organic solvent washes with benzene and ethyl acetate. However, any carry-over into the cleavage reaction and subsequent extraction into the 1-chlorobutane leads to interference in identification of the PTH derivatives by HPLC. The organic solvent extractions can also lead to the loss of polypeptide from the spinning cup. This is not normally a problem with long polypeptide chains and proteins because of their charged nature and relative insolubility in nonpolar solvents. However, for smaller polypeptide chains, particularly as the degradation approaches the COOH-terminus, losses can be substantial and can lead to the termination of useful sequence information before the analyses are complete. A notable exception is that for tryptic peptides terminating in Arg residues which, of course, retain their positive charge and remain in the aqueous phase of the spinning reaction vessel. It may be noted that Arg is the only amino acid whose side chain remains fully charged throughout the entire Edman degradation. Lysine side chains become at least partially derivatized during the coupling reaction, while aspartic and glutamic acid side chains will be protonated under the acid conditions of the cleavage reaction.

While various modifications such as the use of volatile buffers and sulfonated (charged) derivatives of PITC led to some improvement in this situation,[13] the most notable advance came with the introduction of Polybrene (Figure 3) into the reaction mixture.[14] This inert, highly positively charged (quaternary amine) polyelectrolyte also has considerable nonpolar character and presumably forms a tight complex with polypeptide chains both through ionic and hydrophobic interaction. It may also itself bind to the glass walls of the reaction cup. Whatever

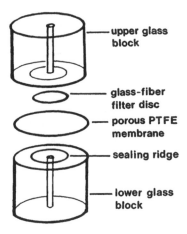

FIGURE 4A. Exploded view of reaction cartridge of gas-phase sequencer.

its mechanism of action, it markedly reduces the loss of peptides from the system as multiple cycles are carried out.

II. GAS-LIQUID-SOLID PHASE SEQUENCER

A further limitation of the spinning cup sequencer is the relatively large protein samples ($\geq$ 1 nmol) that are normally required for a successful degradation. This limitation has been overcome by the introduction of the gas-liquid-solid phase sequenator or sequencer in 1981.[15] Responding to the need for ever more sensitive sequence analyses of proteins available in only picomole quantities, Hewick et al.[15] developed an instrument in which, although using the same basic Edman chemistry, the spinning cup is replaced by a simple flow through cartridge type reaction cell in which the protein/peptide is immobilized on a small porous glass fiber disc (12 mm in diameter) (see Figure 4A). The reaction cell is mounted in an insulated oven thermostatted at the appropriate temperature. The glass fiber disc is supported by a porous Teflon sheet clamped between the top and bottom halves of the Pyrex glass reaction chamber. Gases and liquids are metered into the reaction cell through an inlet. The exit opening in the bottom half of the reaction cell allows for the escape of excess gases, reagents, and extraction solvents. These are transferred to waste or in the case of the thiazolinone to a conversion flask for transformation to the PTHs which, in turn, are dissolved and transferred to a fraction collector. In the more recent modification of the commercial instrumentation, the PTH derivatives are transferred directly on-line for analysis to an HPLC instrument (Figure 4B).

Procedures for applying the sample to the glass fiber disc are relatively simple. Initially a small volume of Polybrene solution (25 µl; 1.5 mg) containing 25 nmol glycyl-glycine is applied to the disc and dried down under vacuum. Following assembly into the cartridge, three to four cycles of Edman degradation are performed. This preliminary cycling is to clean up the Polybrene which contains contaminants which would otherwise interfere with the subsequent Edman chemistry. The cartridge is disassembled, the protein solution (5 pmol $\rightarrow$ 1 nmol in 30 µl) in an appropriate solvent is applied to the pretreated filter and air dried. The cartridge is reassembled and the cycling begun.

The basic chemistry and reaction steps are as in the spinning cup sequencer except as follows: only sufficient PITC in heptane (~ 20 µl of 5%) is metered into the reaction chamber to wet completely the glass fiber disc. Following removal of the heptane by flushing with argon,

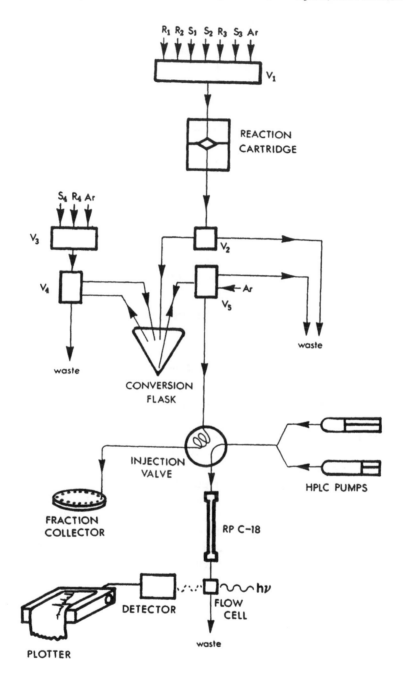

FIGURE 4B. A simplified schematic of a gas-phase sequencer with on-line HPLC for PTH identification. Arrows represent fluid flow. Fluid control valves are designated V_{1-5}. The sequencer reagents and solvents R_{1-4} and S_{1-4} are as follows: R_1 — 5% phenylisothiocyanate in heptane; R_2 — trimethylamine /water vapor; R_3 — trifluoroacetic acid vapor; R_4 — 25% trifluoroacetic acid in water; S_1 — heptane; S_2 — ethyl acetate; S_3 — 1-chlorobutane; and S_4 — 20% acetonitrile in water.

coupling is effected by bleeding trimethylamine/H_2O vapor through the cartridge to waste. This provides the necessary pH control for the coupling reaction to proceed. Since Quadrol is not used, many of the problems associated with its use in the spinning cup sequencer are avoided. In the more recent versions of the instrumentation, the benzene solvent wash is replaced by heptane (less toxic and more easily purified). The ethyl acetate wash remains but both extractions use much less volume. Drying is accomplished by flushing with argon or nitrogen (high purity); the vacuum pumps have been eliminated. Cleaving is achieved by slowly bleeding trifluoroacetic acid (TFA) vapor through the cartridge (as in the Applied Biosystems Inc. Model 470). Alternatively, anhydrous liquid TFA is metered onto the fritted disc and is said to provide higher repetitive yields (Applied Biosystems Model 477). Following drying, the final extraction with 1-chlorobutane is the same except that much less is required. Other organic solvent extractions of the thiazolinone derivatives are being investigated in different laboratories. Conversion to the PTH derivatives is automated and in the more modern commercial instruments the application and analyses of these by HPLC is automated and on-line.

The gas-liquid-solid phase sequencer was greeted enthusiastically by the scientific community and very widely applied. The original claims of its designers[15] have been largely corroborated by the experiences of others. Their original paper described the application of the method to varying amounts of polypeptides and proteins. Myoglobin, for example, was tested at loadings of 10 nmol to 5 pmol. With the former loading, an average repetitive yield of 98% was obtained with 90 residues identifiable. With 5 pmol, the average repetitive yield was 92% with partial sequence information for 22 residues. Serine and tryptophan were not identified. While most operators are not as successful as this with other polypeptides and proteins, the introduction of this instrumentation has clearly been a major improvement over previous methodologies.

While the efficiency of the Edman chemistry has not been improved dramatically, several features of the instrument design are clearly responsible for its success. The most important of these is undoubtedly miniaturization of the reaction chamber and the use of the porous glass fiber discs as a method of immobilizing the protein/polypeptide. The glass fiber provides a much larger surface area per unit volume to which the protein is adsorbed (either with or without Polybrene). This means that the polypeptide is immobilized in a more concentrated manner and yet as a thin film which, because of the porous nature of the glass fiber disc, is readily accessible to reagents and solvents. As a result, much less reagents and solvents are required to carry out the reactions and necessary extractions in an efficient manner. This reduces operating costs, increases the speed of analysis (< 1 hr/cycle) and reduces the build-up of impurities and side products. A further important finding was that the cleavage reaction could be carried out at a lower temperature (45°C) than that previously employed (55°C). This reduces the acid-catalyzed random cleavage of peptide bonds and this improves signal to noise ratios in the HPLC analysis of the PTH derivatives. The on-line analyses of these in the most modern commercial instruments has also meant an increase in reproducibility and sensitivity. Efficiency of instrument use is also enhanced since the operator is able to monitor the progress of the Edman degradations as they proceed. Thus, a sequencing run can be terminated as soon as it becomes obvious that useful information is no longer being obtained.

III. IDENTIFICATION OF PTH-DERIVATIVES BY HPLC

In the 24 years since the first automated protein sequencer design was published in 1967,[8] a variety of means has been employed to identify the sequencer products: one- and two-dimensional thin-layer chromatography,[8,16,17] gas-liquid chromatography,[18] mass spectrometry,[19] and hydrolysis back to the amino acid with subsequent identification by an amino acid analyzer.[6,7] However, the current method of choice is HPLC. First reported in 1973,[20,21] the analysis of PTH-derivatives by RPC on octadecylsilane,[22,23,24] octylsilane,[25,26] cyanopropylsilane,[27] or phenylalkylsilane[28] analytical columns (generally 250 × 4.6 mm, 5-μm particle size), with detection by ultra-violet absorption at 254 or 269 nm, has improved steadily, and now with the introduction of microbore (1 or 2 mm I.D.) columns[29] is capable of the following:

1. Keeping pace with the gas-phase sequencer cycle time of < 1 hr by separating all the common PTH-derivatives in a single run of < 30 min
2. Detecting quantitatively down to ~ 1 pmol
3. Being automated and made to operate in concert with the sequencer by having the sequencer deliver the newly formed PTH in solution into the sample loop of the HPLC apparatus (on-line HPLC operation)

Both isocratic[22,24,25,26,30] and gradient elution systems[23,27,28,29,30] have been successfully developed. Isocratic systems have the advantage of requiring less costly equipment, with lower operating costs, as the eluting solvent can be recycled, while still giving a flat baseline at the detector sensitivities (e.g., 0.005 AUFS, 10 mV chart recorder) required for the detection of < 10 pmol of PTH-derivatives. The major disadvantages of isocratic systems are the close bunching of the fast-eluted polar compounds (DTT [dithiothreitol, see below], D, E, N, Q, and T), and the peak broadening and hence loss in sensitivity seen with the slower-eluted derivatives (W, F, I, K, and L). The latter effect can be overcome to some extent by increasing the flow-rate for the second half of the analysis.[24] Whereas gradient systems are more complicated and expensive, they provide the means to obtain essentially the same sensitivity, in terms of peak width, across the whole chromatogram. They suffer from the disadvantage of a rising baseline as the percentage of organic solvent increases, but this defect can be minimized by the use of highly purified organic solvents and water, and by computer storage of the chromatograms with subtraction of the baseline obtained with a blank cycle.[31]

On-line HPLC analysis of PTH-derivatives has been accomplished by a number of self-assembled systems with solid-phase, liquid-phase, and gas-phase sequencers[32-37] as well as commercially available systems such as the Applied Biosystems Model 120A HPLC operated on-line with an Applied Biosystems Model 470A gas-phase sequencer. The latter combination is used in our laboratory.

This chromatographic system is a binary gradient system based upon two 10-ml syringe pumps driven by microprocessor-controlled stepping motors. This design eliminates detector noise due to flow pulsations produced by conventional reciprocating piston pumps. The gradient is generated by control of the syringe motors, instructions being stored in EEPROM, from a facile menu-presented program. The chromatographic separation is achieved by a 220×2.1 mm Brownlee Spheri-5 C_{18} column (5-μm particle size) kept in an air circulating oven at 55°C with a flow-rate of 200 μl/min. Solvent A is ~ 100 mM sodium acetate buffer, pH 3.95 to 4.00, with 3 to 5% tetrahydrofuran (THF). It is essential to use THF which is pure and contains neither ultraviolet adsorbing peroxide impurities nor stabilizer (e.g., Burdick and Jackson UV grade), and a suitable supply of HPLC-grade water (e.g., Milli-Q, Millipore Corp.), or the ready mixed 5% THF in water (Applied Biosystems). Solvent B is acetonitrile (HPLC grade). The PTH-derivatives are transferred by argon gas pressure from the conversion flask of the sequencer to the HPLC injection loop (50 μl) of a pneumatically activated injection valve (Rheodyne Model 7125) by dissolution in 120 μl 20% acetonitrile/water (sequencer solvent S4); the remaining ~ 70 μl being directed to the fraction collector where they are kept in case of failure during the initial HPLC analysis. The PTH-derivatives are detected by their absorption at 269 nm in a 12-μl flow-cell of pathlength 8-mm fitted in a Kratos Model 777 variable wavelength detector with a 1 V output for an integrator and 10 mV outputs for chart recorders. Using the 1 V output into a Spectra-Physics Model 4270 computing integrator, signal attenuations in the range 4 to 256 are generally used.

The precise operating conditions need to be tailored to each column and adjusted accordingly

883

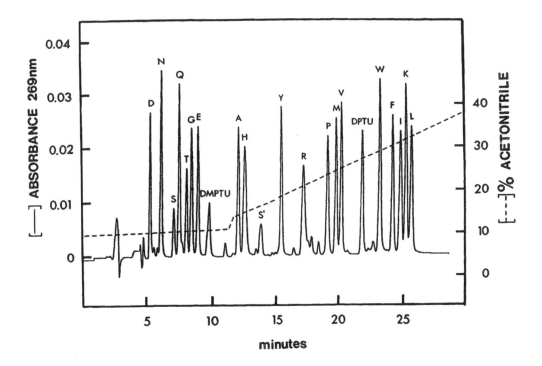

FIGURE 5A. HPLC separation of PTH-derivatives: 100 pmol of each component. Column: Brownlee Spheri-5 C_{18} reversed-phase column (220 × 2.1 mm I.D., 5-µm particle size). Mobile phase solvents: Solvent A, ~ 100 mM sodium acetate buffer, pH 3.95 to 4.00, with 3 to 5% tetrahydrofuran (THF); Solvent B, acetonitrile. Flow-rate: 200 µl/min. Temperature: 55°C. The single letter code is used to designate the amino acids. S' represents PTH-dehydroalanine; DMPTU denotes N,N-dimethyl-N'-phenylthiourea; DPTU denotes diphenylthiourea.

as the column ages (see below). Typical conditions for separation are given in Figure 5A. Three major impurities are present in all sequencing cycles. Oxidized DTT (DTT is added as a reducing agent to the sequencer chemicals R4 and S4) is eluted ~ 1 min before PTH-Asp. Diphenylthiourea (DPTU), the product of phenylisothiocyanate (PITC) reaction with aniline, the hydrolysis product of PITC

is eluted ~ 90s after PTH-Val. N,N-dimethyl-N'-phenylthiourea (DMPTU), the product of reaction of PITC with dimethylamine, a breakdown contaminant of the trimethylamine (R2)

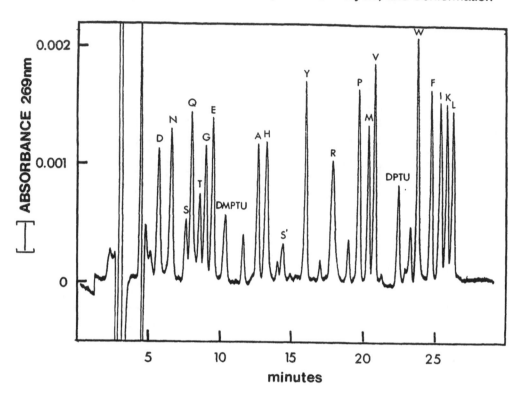

FIGURE 5B. HPLC separation of PTH-derivatives: 5 pmol of each component with baseline subtraction.

(DMPTU)

is eluted ~ 1 min after PTH-Glu.

Additional DMPTU (0.5 μM) is added to the acetonitrile to act as an oxidant scavenger in order to improve the yields of the PTH-derivatives when detection below 50 pmol is needed. None of these impurities interferes with the identification of the PTH-derivatives.

The most variable PTH-derivatives in terms of elution time are PTH-Arg and PTH-His which are eluted earlier with increased ionic strength buffer; the other derivatives being essentially insensitive to ionic strength. Column aging causes an increase in the elution times for these two derivatives which must be counteracted by a steady progression of increases in the sodium acetate concentration. It is convenient to record the conductivity of the current batch of buffer and to make small (e.g., 5%) increments in conductivity when necessary by the addition of more stock 3 M sodium acetate buffer. The other PTH-derivative whose position changes with column age (more slowly in this case) is PTH-Lys. It too is eluted later as the column ages, in time converging with PTH-Leu. It can be moved forward and positioned between PTH-Ile and PTH-Leu by a 1 or 2% increase in the final concentration of acetonitrile in the gradient. The derivatives of Asp and Glu are the most affected by pH: lowering the pH delays their elution and vice-versa. Columns have a lifetime of approximately 3000 injections, with peak broadening as the main reason for their replacement. All these subtleties in the setting up and operation of the system have been reviewed in detail by Hunkapiller.[31]

IV. SEQUENCE ASSIGNMENT

This is made by comparing the chromatogram for each cycle with that of all the standard PTH-derivatives. Cycle to cycle comparisons are made and towards the end of a protracted run (> 40 cycles) it is convenient to look for small changes by overlaying chromatograms on a light-box. Alternatively, if some means of computerized data storage and manipulation is available, subtraction of the chromatogram of cycle n from that of cycle (n + 1) identifies residue (n + 1) as a positive peak and residue n as a negative peak. As the run progresses, the size of the PTH-derivative peaks diminish, while the background pattern of all the derivatives, generally ascribed to the non-specific cleavage and exposure of new amino termini during TFA exposure,[8] very slowly increases. Thus, the task is one of picking out ever-decreasing signals above a slowly increasing background. The introduction of the on-line HPLC has aided this process considerably by reducing the cycle to cycle variability in sampling introduced by reconstituting dried-down derivatives when a stand-alone HPLC is used, as well as improving the yields of the derivatives—particularly the most labile ones (see below). An example of sequence assignment is presented in Figure 6.

Quantitation of the yield of a particular PTH-derivative can be made by comparing its peak height or area above that of the background with an external standard, if equipped with a computing integrator.

By obtaining the yields of the same PTH-derivative at two widely separated places in the sequence, an estimate of the efficiency of the sequencer in performing repeated degradations, usually expressed as a percentage and termed the repetitive yield (RY), can be obtained.

$$RY = \left(\frac{Yb}{Ya}\right)^{\frac{1}{b-a}} \cdot 100$$

where Ya = yield at cycle a
       Yb = yield at cycle b   a < b

Usually a residue such as alanine or leucine for which there are no complicating factors is chosen for such calculations.

Alternatively, the yields in all cycles can be calculated and the repetitive yield obtained as the slope of the best fit straight line when cycle number is plotted against ln (yield), or more easily by regression analysis with a suitable calculator, e.g., Hewlett Packard 11C or computer program.

These calculations are useful in monitoring the performance of the sequencer with standard proteins, e.g., myoglobin or β-lactoglobulin, but can be misleading with smaller peptides where wash-out of the peptide may cause severely reduced yields as the peptide is reduced in length.

A. SPECIAL CONSIDERATIONS
1. Serine

PTH-serine is always recovered in low yield (~ 25%) when compared with the yields of neighboring residues, together with PTH-dehydroalanine (25 to 50%), the product of β-elimination of the trifluoroacetylated thiazolinone during the conversion reaction.[38]

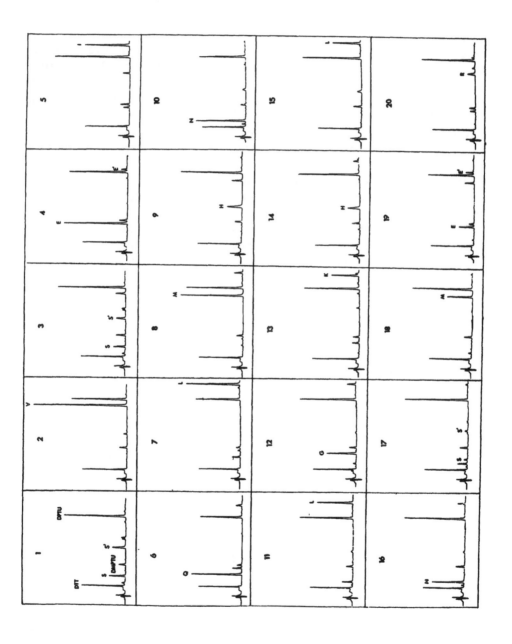

887

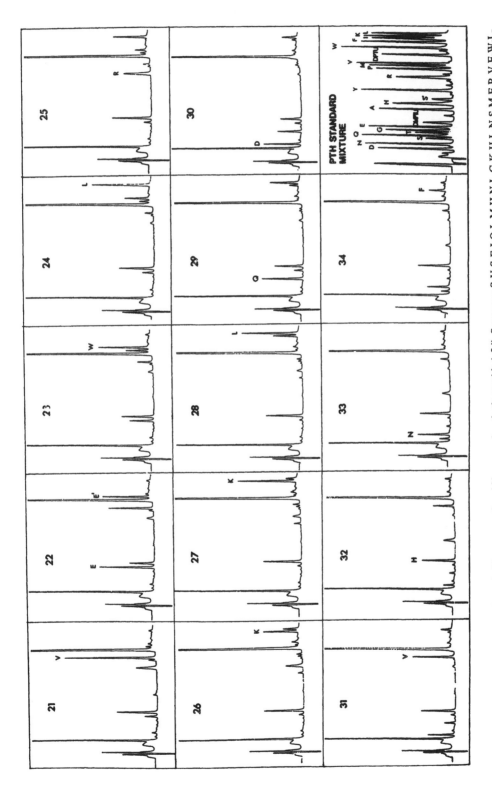

FIGURE 6. Sequence analysis of 250 pmol of human parathyroid hormone (synthetic peptide 1-34). Sequence: S-V-S-E-I-Q-L-M-H-N-L-G-K-H-L-N-S-M-E-R-V-E-W-L-R-K-K-L-Q-D-V-H-N-F

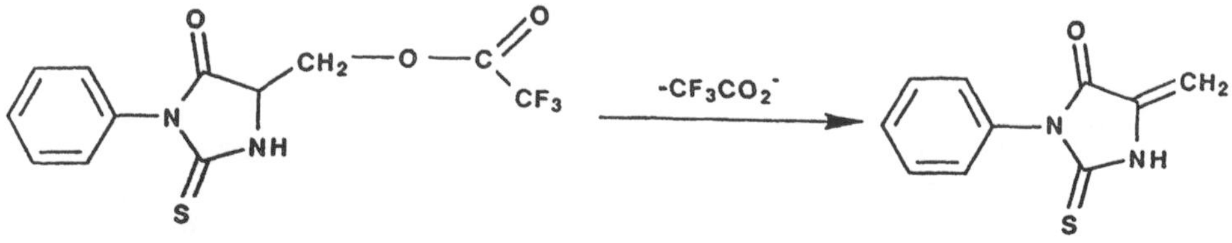

(PTH-dehydroalanine)

PTH-dehydroalanine forms an adduct with DTT, which is eluted midway between PTH-alanine and PTH-tyrosine.

2. Threonine

The β-elimination process takes place more slowly with threonine to give PTH-dehydro-α-aminoisobutyric acid, whose adducts with DTT are eluted as two sets of doublets between PTH-tyrosine and PTH-proline. The yield is very low (< 5%), whereas PTH-threonine is found in 30 to 50% yields.

3. Cysteine

The β-elimination of H_2S in cysteine appears to be complete as no PTH-cysteine is obtained, with PTH-dehydroalanine found in low yield (< 10%). This provides a possible means of distinguishing cysteine from serine (i.e., the presence of PTH-dehydroalanine without PTH-serine), but more reliable identification requires the reduction and alkylation of the cysteine sulfhydryls in the protein before sequencing is started. The preferred alkylating reagent is 4-vinylpyridine, which produces S-β-(4-pyridylethyl)cysteine,[39] the PTH derivative of which chromatographs just after PTH-valine.

It should be noted that when a stand-alone HPLC is used, the recoveries of these three particularly unstable derivatives may be even lower than stated above.

4. Asparagine and Glutamine

Up to 10% of the amide is deamidated during the conversion reaction and detected as the corresponding PTH-acid.

5. Aspartic Acid and Glutamic Acid

On extended gas-phase sequencing runs (> 20 cycles), an increasing proportion of the glutamic acid signal is seen as an unidentified peak which is eluted just after DPTU. As this peak is not seen in the spinning-cup sequencer where Quadrol-TFA buffer is used to maintain the coupling reaction at pH 9.0, it is presumably a salt of trimethylamine. Similarly, but to a much lesser extent with aspartic acid, an unknown peak is seen being eluted just in front of DPTU.

6. Tryptophan

The two major UV absorbing by-products seen on HPLC analysis are DPTU and DMPTU. These are well resolved from all the PTH-amino acids. However, a much smaller amount of N,N'-diphenylurea (DPU) is found in each cycle. The DPU is the result of desulfuration by atmospheric oxygen of either DPTU or the reagent PITC giving rise to phenylisocyanate which reacts with aniline produced by the hydrolysis of PITC during the aqueous coupling conditions.

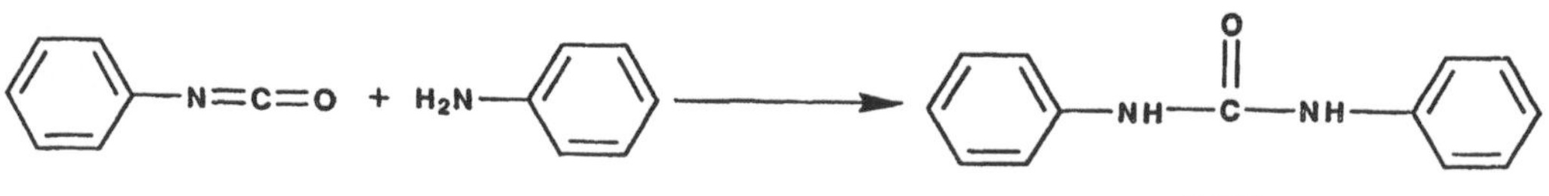

(DPU)

Unfortunately, DPU is eluted very close to tryptophan. This is not a problem during the early cycles when degrading relatively large samples (200 pmol to 2 nmol), where the PTH-tryptophan signal swamps that from DPU, but with low picomole sample size the identification of tryptophan above the somewhat variable DPU peak is much more difficult.

7. Proline

Because PTC-proline cleaves more slowly than the other amino acids,[38] overlap (lag or asynchrony) can be introduced into the sequencing pattern by failure to remove completely a proline residue. In degradations where some sequence is already known, e.g., a repeat analysis, longer cleavage times and a higher temperature (from 45 to 50°C) can be used at the known proline cycles. However, these conditions can be counter-productive by introducing more non-specific cleavage and consequently a higher background in the next cycle, and it may be better to stick to the standard conditions while being aware of the degree of asynchrony that has been introduced by the proline residue.

8. Arginine and Histidine

These two residues are generally detected in lower yields (< 50%) due to the poor solubility of their thiazolinone-derivatives in the extracting solvent (1-chlorobutane, S3). This problem is partially overcome by the inclusion of NaCl (30 µl 0.1 M NaCl) with the Polybrene (1 to 3 mg). The problems in identifying histidine and arginine are compounded by their poor chromatographic qualities. Both tend to give broader, tailing peaks. Thus, in longer sequencing runs (> 30 cycles), histidine, arginine, serine, and cysteine (if it is not alkylated) are likely to be the residues where identification cannot be made. Also, when degrading small amounts of protein (< 50 pmol) as is typically the case with an electroblotted sample (see further), gaps may occur after 10 cycles or less due to the poor recovery and/or poor chromatography of these residues.

As the column ages, the tailing of PTH-arginine and histidine increases more than with the other derivatives. Because the effect can be largely overcome by the addition of 0.5 to 1.0 ml of sequencer reagent R2 (12.5% trimethylamine in water) to 1 l of the eluting buffer, this is probably due to ion-pairing with exposed silanols. Trialkylamines have been documented as silanol ion-pairing reagents.[40]

B. FAILURE TO SEQUENCE DUE TO BLOCKED NH$_2$-TERMINUS

A significant percentage of proteins is blocked to Edman degradation. This may be due to inappropriate treatment during isolation or due to natural post-translational modification.

1. Pyroglutamate

If the NH$_2$-terminal amino acid is glutamine, this residue can cyclize to the pyrrolidone carboxylate (pyroglutamate) which is not accessible to reaction with PITC.

(pyroglutamyl peptide)

The condition may occur naturally or be due to the exposure of the polypeptide to acidic conditions during its isolation. In some cases, the pyroglutamate residues may be removed by the enzyme pyrrolidonyl peptidase.[41] Even though the efficiency of such cleavage may be low, the generation of the new NH$_2$-terminal residue in low yield at the penultimate position can be adequate to provide significant sequence information. The presence of glutamine at an internal position of the polypeptide chain is not a problem since little cyclization occurs under the reaction conditions for coupling and cleavage.

2. Acylation

Acetyl and formyl groups (e.g., N-formylmethionine in prokaryotes) are common blocking groups. The presence of acetylated NH_2-termini has been a particularly difficult problem, although recently a successful sequence was obtained in very low yield after treatment of the human plasma membrane Ca^{2+} pump with N-acylaminoacyl peptide hydrolase.[42,43] This approach is likely to be successful in only certain situations and cannot be considered a general method at this time.

3. Carbamylation

At alkaline pH, the amino groups of a protein are carbamylated and blocked by cyanate ions.[44] Such conditions exist in urea solutions at high pH. Therefore, proteins which are to be sequenced should only be exposed to urea at low pH, or in the presence of buffers such as Tris-HCl which act as scavengers for cyanate ions. Although this reaction is unlikely to lead to complete blockage of the NH_2-terminus, it could lead to significantly lower yields.

4. Methylation

A variety of methylated NH_2-terminal amino acids has been found in proteins from many sources;[45] but only a few, such as N-trimethylalanine,[46] block Edman degradation.

C. OTHER CONSEQUENCES OF POST-TRANSLATIONAL MODIFICATIONS

Blank Cycles — As well as the problems associated with poor chromatographic properties and/or low recoveries of some PTH-derivatives, there is also the possibility of obtaining a blank cycle in which just the background pattern of derivatives is seen. There are two well documented causes:

1. *Glycosylation site*: The thiazolinone of the glycosylated amino acid (usually Asn, Ser, or Thr) is insoluble in the extracting solvent (1-chlorobutane) and remains bound to the glass-fiber disc. When the blank is followed by a Ser or Thr in the second successive cycle, the presence of N-linked oligosaccharide can be assumed. However, this should be verified by amino-sugar analyses on an amino acid analyzer and/or sequencing of the polypeptide after treatment with N-glycanase. An Asp should now be recovered at the blank position.
2. *Phosphorylation or sulfation site*: Again the thiazolinone of the phosphorylated amino acid (e.g., Ser, Thr, Tyr) or the tyrosine sulfate is insoluble in the extracting solvent and remains on the disc. If a phosphorylated residue is suspected from other evidence, several possibilities are open. The polypeptide may be sequenced before and after treatment with alkaline phosphatase.[48] In the case of phosphoseryl peptides, β-elimination in the presence of NaOH and ethanethiol will lead to the stable derivative S-ethylcysteine.[49,50] Its PTH derivative is identifiable as a separate peak on the HPLC.

Unknown Peaks — As mentioned earlier, many N- and S-methylated amino acids have been found in proteins.[45] While these do not normally interfere with the Edman degradation, their presence would introduce an unknown peak into the HPLC analysis. While these may sometimes be identified by amino acid analysis of the peptide, mass spectrometry probably represents the most effective means towards solving the puzzle.

D. ELECTROBLOTTING

Due to its high resolution, speed, simplicity, and low cost, sodium dodecyl sulfate poly-acrylamide gel electrophoresis (PAGE) is probably the most widely used method for screening protein preparations. Several approaches have been employed to free the protein of interest from the polyacrylamide gel matrix for subsequent amino acid analysis and/or sequence determination, but the most effective has proven to be electroblotting in which the protein is driven electrophoretically from the gel onto a suitable support that can bind the protein non-covalently, and yet be unaffected by the sequencing reagents so that it can be placed in the reaction block of the sequencer. This was first achieved with Polybrene-coated glass fiber sheet[51] and covalently modified glass fiber sheets,[52] but is now most easily and successfully carried out with polyvinylidene difluoride (PVDF) membrane (Immobilon-P, Millipore).[53,54] Recently, the covalent bonding of proteins electroblotted to glass fiber activated with phenylisothiocyanate functional groups, with subsequent solid-phase microsequencing has been reported.[55]

In our experience, electroblotting onto PVDF membrane permits the routine gas-phase sequencing of 20 to 100 pmol of protein applied to the gel, with initial recoveries of ~ 25% (~ 5 to 25 pmol PTH-derivative). Low yields of PTH-lysine (< 50% compared with adjacent residues) suggest that the major cause of the poor recoveries is NH_2-group blockage. This appears to be borne out by comparisons between PAGE at pH 8.6 (Laemmli system) and pH 7.3,[56] in which electrophoresis at well below the pKa of protein NH_2-groups protects them from unwanted modification. Pre-electrophoresis with buffer containing thioglycolate as a scavenger for reactive species such as residual peroxides, free radicals, and acrylamide monomer, also improves the sequencing yields.[56] One of the difficulties encountered when sequencing PVDF electroblotted proteins is an increased variability in cycle to cycle yield. This is due to inadequate wetting and extractive flow over the PVDF membrane and can, to a large part, be overcome by modifications to the standard ABI sequencer program.[57]

V. SUMMARY AND CONCLUSIONS

Present methodologies for the automated sequencing of polypeptides and proteins now permit their analyses at the level of a few picomoles. This corresponds to the amounts that can be conveniently isolated by such techniques as HPLC, sodium dodecyl sulfate and two-dimensional polyacrylamide gel electrophoresis. While extended sequences (40 to 60 residues) cannot be expected from these amounts of polypeptide, sufficient information (10 to 30 residues) is often obtained to identify the protein, to establish homology with other proteins, or to provide sufficient data for the synthesis of oligonucleotides to be used in screening DNA libraries. In situations where blocked NH_2-termini are present, partial sequence information can often be obtained by isolating fragment(s) on an electrophoretic gel after suitable proteolytic or chemical (e.g., cyanogen bromide, hydroxylamine) cleavage. Blanks in the sequence can be expected for positions accompanied by the more labile PTH derivatives such as cysteine, serine, and threonine. The PTHs of tryptophan, arginine, and histidine are also difficult because of the coelution of the former with DPU and the poor solubility of the arginine and histidine thiazolinone derivatives in the extracting solvent and the poor chromatographic properties of their PTH derivatives on reversed-phase columns.

The identification of post-translational modifications can represent a special challenge and requires special procedures as outlined above. In this connection, the rapid advances in mass spectrophotometry presently underway offer a powerful new approach to such problems. The ability to measure molecular masses of polypeptides with a precision of less than one mass unit now permits the identification of disulfide bond positions, NH_2-terminal blocking groups, oligosaccharide and phosphorylation sites, and other post-translational changes. Sequencing from both the NH_2-terminal and COOH-terminal ends is also feasible. Since the principles of the methodology are completely different from the Edman degradation technique, the two are complementary. For further information, the reader is referred to recent reviews of the subject.[58-60]

REFERENCES

1. **Edman, P.,** Method for determination of the amino acid sequence in peptides, *Acta Chem. Scand.*, 4, 283, 1950.
2. **Edman, P.,** Note on stepwise degradation of peptides *via* phenyl thiohydantoins, *Acta Chem. Scand*, 7, 700, 1953.
3. **Ilse, D. and Edman, P.,** The formation of 3-phenyl-2-thiohydantoins from phenylthio-carbamyl amino acids, *Australian J. Chem.*, 16, 411, 1963.
4. **Gray, W.R. and Hartley, B.S.,** A fluorescent endgroup reagent for proteins and peptides, *Biochem. J.*, 89, 59, 1963.
5. **Lin, J-K. and Chang, J.Y.,** Chromophore labeling of amino acids with 4-dimethyl-aminoazobenzene-4'-sulfonyl chloride, *Anal. Chem.*, 47, 1634, 1975.
6. **Smithies, O., Gibson, D.M., Fanning, E.M., Goodfleish, R.M., Gillman, J.G., and Ballantyne, D.L.,** Quantitative procedures for use with the Edman-Begg sequenator. Partial sequences of two unusual immunoglobulin light chains, *Rzf. and Sac. Biochemistry*, 10, 4912, 1971.
7. **Mendez, E. and Lai, C.Y.,** Regeneration of amino acids from thiazolinones formed in the Edman degradation, *Anal. Biochem.*, 68, 47, 1975.
8. **Edman, P. and Begg, G.,** A protein sequenator, *Eur. J. Biochem.*, 1, 80, 1967.
9. **Wittmann-Liebold, B., Graffunder, H. and Kohls, H.,** Amino acid sequence studies on ten ribosomal proteins of *Escherichia coli* with an improved sequenator equipped with an automated conversion device, *Hoppe-Seyler's Z. Physiol. Chem.*, 354, 1415, 1973.
10. **Wittmann-Liebold, B., Graffunder, H., and Kohls, H.,** A device coupled to a modified sequenator for the automated conversion of anilinothiazolinones into PTH amino acids, *Anal. Biochem.*, 75, 621, 1976.
11. **Hunkapiller, M.W. and Hood, L.E.,** Direct microsequence analysis of polypeptides using an improved sequenator, a nonprotein carrier (Polybrene) and high pressure liquid chromatography, *Biochemistry*, 17, 2124, 1978.
12. **Bhown, A.S., Cornelius, T.W., Mole, J.E., Lynn, J.D., Tidwell, W.A., and Bennet, J.C.,** A simple modification on the vacuum system of the Beckman automated sequencer to improve the efficiency of Edman degradation, *Anal. Biochem.*, 102, 35, 1980.
13. **Braunitzer, G., Schwank, B., and Ruffus, A.,** Zur vollständigen automatischen Sequenzanalyse von Peptiden mit Quadrol, *Hoppe-Seyler's Z. Physiol. Chem.*, 352, 1730, 1971.
14. **Tarr, G.E., Beecher, J.F., Bell, M., and McKean, D.J.,** Polyquaternary amines prevent peptide loss from sequenators, *Anal. Biochem.*, 84, 622, 1978.
15. **Hewick, R.M., Hunkapiller, M.W., Hood, L.E., and Dryer, W.J.,** A gas-liquid solid-phase peptide and protein sequenator, *J. Biol. Chem.*, 256, 7990, 1981.
16. **Kulbe, K.D.,** Rapid separation of phenylthiohydantoin (PTH) amino acids by thin-layer chromatography on polyamide glass plates, *Anal. Biochem.*, 44, 548, 1971.
17. **Summers, M.R., Smythers, G.W., and Oroszlan, S.,** Thin-layer chromatography of sub-nanomole amounts of phenylthiohydantoin (PTH) amino acids on polyamide plates, *Anal. Biochem.*, 53, 624, 1973.
18. **Pisano, J.J. and Bronzert, T.J.,** Analysis of amino acid phenylthiohydantoins by gas chromatography, *J. Biol. Chem.*, 224, 5597, 1969.
19. **Fales, H.M., Nagai, Y., Milne, G.W.A., Brewer, H.B., Bronzert, T.J., and Pisano, J.J.,** Use of chemical ionization mass spectrometry in analysis of amino acid phenylthiohydantoin derivatives formed during Edman degradation of proteins, *Anal. Biochem.*, 43, 288, 1971.
20. **Graffeo, A.P., Haag, A., and Karger, B.L.,** High performance liquid chromato-graphic separation of phenylthiohydantoin derivatives of amino acids, *Anal. Lett.*, 6, 505, 1973.
21. **Zimmerman, C.L., Pisano, J.J., and Apella, E.,** Analysis of PTH amino acids by reversed-phase HPLC., *Biochem. Biophys. Res. Commun.*, 55, 1220, 1973.
22. **Zimmerman, C.L., Pisano, J.J., and Apella, E.,** Rapid analysis of amino acid phenylthiohydantoins by high-performance liquid chromatography, *Anal. Biochem.*, 77, 369, 1977.
23. **Somack, R.,** Complete phenylthiohydantoin amino acid analysis by high-performance liquid chromatography on Ultrasphere octadecyltrimethoxysilane, *Anal. Biochem.*, 104, 464, 1980.
24. **Tarr, G.E.,** Rapid separation of amino acid phenylthiohydantoins by isocratic high-performance liquid chromatography, *Anal. Biochem.*, 111, 27, 1981.
25. **Ashman, K. and Wittmann-Liebold, B.,** A new isocratic HPLC separation for PTH-amino acids based on 2-propanol, *FEBS Lett.*, 190, 129, 1985.
26. **Lottspeich, F.,** Identification of the phenylthiohydantoin derivatives of amino acids by high pressure liquid chromatography, using a ternary isocratic solvent system, *Hoppe-Seyler's Z. Physiol. Chem.*, 361, 1829, 1980.
27. **Johnson, N.D., Hunkapiller, M.W., and Hood, L.E.,** Analysis of phenylthiohydantoin amino acids by high-performance liquid chromatography on DuPont Zorbax cyanopropylsilane columns, *Anal. Biochem.*, 100, 335, 1979.

28. **Henderson, L.E., Copeland, T.D., and Oroszlan, S.,** Separation of amino acid phenylthiohydantoins by high-performance liquid chromatography on phenalkyl support, *Anal. Biochem.,* 102, 1, 1980.

29. **Cunico, R.L., Simpson, R., Correia, L., and Weber, C.T.,** High-sensitivity phenylthiohydantoin amino acid analysis using conventional and microbore chromatography, *J. Chromatogr.,* 336, 105, 1984.

30. **Lottspeich, F.,** Microscale isocratic separation of phenylthiohydantoin amino acid derivatives, *J. Chromatogr.,* 326, 321, 1985.

31. **Hunkapiller, M.W. and Hood, L.E.,** New protein sequenator with increased sensitivity, *Science,* 207, 523, 1980.

32. **Machleidt, W. and Hofner, H.,** Fully automated solid-phase sequencing with on-line identification of PTH's by high pressure liquid chromatography, in *Methods in Peptide and Protein Sequence Analysis,* Birr, C., Ed., Elsevier/North Holland, Amsterdam, 1980, 35.

33. **Wittmann-Liebold, B.,** An evaluation of the current status of protein sequencing in *Methods in Protein Sequence Analysis,* Elzinga, M., Ed., Humana Press, Clifton, NJ, , 1982, 27.

34. **Rodriguez, H., Kohr, W.J., and Harkins, R.M.,** Design and operation of a completely automated Beckman microsequencer, *Anal. Biochem.,* 140, 538, 1984.

35. **Ashman, K.,** The use of on-line high-performance liquid chromatography for phenylthiohydantoin amino acid identification, in *Advanced Methods in Protein Microsequence Analysis,* Wittmann-Liebold, B., Salnikow, J., and Erdmann, V.A., Eds., Springer Verlag, Berlin, 1986, 219.

36. **Gausepohl, H., Trosin, M., and Frank, R.,** An improved gas-phase sequenator including on-line identification of PTH amino acids, in *Advanced Methods in Protein Microsequence Analysis,* Wittman-Liebold, B., Salnikow, J., and Erdmann, V.A., Eds., Springer Verlag, Berlin, 1986, 149.

37. **Hunkapiller, M.W.,** Gas-phase sequence analysis of proteins/peptides, in *Protein/Peptide Sequence Analysis, Current Methodologies,* Bhown, A.S., Ed., CRC Press, Boca Raton, FL, 1988, 87.

38. **Tarr, G.E.,** Improved manual sequencing, *Methods Enzymol.,* 47, 335, 1977.

39. **Friedman, M., Krull, L.H., and Cavins, J.F.,** The chromatographic determination of cystine and cysteine residues in proteins as S-β-(4-pyridylethyl) cysteine, *J. Biol. Chem.,* 245, 3868, 1970.

40. **Wehr, C.T.,** Commercially available columns, in *CRC Handbook of HPLC for the Separation of Amino Acids, Peptides, and Proteins,* Vol. I, Hancock, W.S., Ed., CRC Press, Boca Raton, FL, 1984, 37.

41. **Padell, D.N. and Abraham, G.N.,** A technique for the removal of pyroglutamic acid from the amino terminus of proteins using calf liver pyroglutamic amino peptidase, *Biochem. Biophys. Res. Commun.,* 81, 176, 1978.

42. **Vermos, A.K., Filoteo, A.G., Stanford, D.R., Wieben, E.D., Penniton, J.T., Strehler, E.E., Fischer, R., Heim, R., Vogel, G., Mathews, S., Strehler-Page, M.-A., James, P., Vorherr, T., Krebs, J., and Carafoli, E.,** Complete primary structure of a human plasma membrane Ca^{2+} pump, *J. Biol. Chem.,* 263, 14152, 1988.

43. **Tsunasawa, S., Stewart, J.W., and Sherman, F.,** Amino-terminal processing of mutant forms of yeast iso-1-cytochrome c. The specificities of methionine amino peptidase and acetyltransferase, *J. Biol. Chem.,* 260, 5382, 1985.

44. **Stark, G.R., Stein, W.H., and Moore S.,** Reactions of the cyanate present in aqueous urea with amino acids and proteins, *J. Biol. Chem.,* 235, 3177, 1960.

45. **Paik, W.K. and Kim, S.,** *Protein Methylation,* John Wiley & Sons, New York, 1980.

46. **Lederer, F., Alix, J.H., and Hayes, D.,** N-trimethylalanine, a novel blocking group found in *E. coli* ribosomal proteins L11, *Biochem. Biophys. Res. Commun.,* 77, 470, 1977.

47. **Martensen, T.M.,** Phosphotyrosine in proteins. Stability and quantification. *J. Biol. Chem.,* 257, 9648, 1987.

48. **Myer, H.E., Hoffmann-Posorke, E., Korte, H., and Heilmeyer, L.M.G., Jr.,** Sequence analysis of phosphos-erine-containing peptides. Modification for picomolar sensitivity, *FEBS Lett.,* 204, 61, 1986.

49. **Holmes, C.F.B.,** A new method for the selective isolation of phosphoserine-containing peptides, *FEBS Lett.,* 215, 21, 1987.

50. **Vanderckhove, J., Bauw, G., Puype, M., Van Damme, J., and Van Montagne, M.,** Protein-blotting on polybrene-coated glass-fiber sheets. A basis for acid hydrolysis and gas-phase sequencing of picomole quantities of protein previously separated on sodium dodecyl sulfate/polyacrylamide gel, *Eur. J. Biochem.,* 152, 9, 1985.

51. **Aebersold, R.H., Teplow, D.B., Hood, L.E., and Kent, S.B.H.,** Electro-blotting onto activated glass: high efficiency preparation of proteins from analytical SDS-polyacrylamide gels for direct sequence analysis, *J. Biol. Chem.,* 261, 4229, 1986.

52. **Matsudaira, P.,** Sequence from picomole quantities of proteins electroblotted onto polyvinylidene difluoride membranes, *J. Biol. Chem.,* 261, 10035, 1987.

53. **LeGendre, N. and Matsudaira, P.,** Direct protein microsequencing from Immobilon-P transfer membrane, *BioTechniques,* 6, 154, 1988.

54. **Aebersold, R.H., Pipes, G.D., Nika, H., Hood, L.E., and Kent, S.B.H.,** Covalent immobilization of proteins for high-sensitivity sequence analysis: electroblotting onto chemically activated glass from sodium dodecyl sulfate-polyacrylamide gels, *Biochemistry,* 27, 6860, 1988.

55. **Moos, M., Nguyen, N.Y., and Liu, T-Y.,** Reproducible high yield sequencing of proteins electrophoretically separated and transferred to an inert support, *J. Biol. Chem.*, 263, 6005, 1988.
56. **Speicher, D.W.,** Microsequencing with PVDF membranes: efficient electroblotting, direct protein adsorption and sequencer modifications, in *Techniques in Protein Chemistry*, Hugli, T.E., Ed., Academic Press, Orlando, FL, 1988, 24.
57. **Biemann, K. and Martin, S.A.,** *Mass Spectrum Rev.*, 6, 1, 1987.
58. **Morris, H.R. and Greer, F.M.,** *Trends Biotechnol.*, 6, 140, 1988.
59. **Biemann, K.,** Sequencing by mass spectrometry, in *Protein sequencing. A Practical Approach.*, Findlay, J.B.C. and Geisow, M.J., Eds., IRL Press, Oxford, 1989, 99.

ANALYSIS OF PROTEINS AND PEPTIDES BY FREE SOLUTION CAPILLARY ELECTROPHORESIS (CE)

Joel C. Colburn, Paul D. Grossman, Stephen E. Moring, and Henk H. Lauer

Capillary electrophoresis (CE) is a relatively new separation method which has proven to be especially useful for the analysis of biomolecules.[1,2] This report is concerned mainly with the separation of peptides, although the literature reports many applications for protein,[2] amino acid,[1,3] and nucleic acid[4] research.

Free solution CE involves the migration of a charged solute in a narrow bore capillary under high electric field conditions. The method is primarily analytical because of the exceedingly small sample capacity. Only nanoliter volumes are applied, typically representing femtomoles (10^{-15} mol) of peptide analyte, thus rendering the method essentially non-destructive. Consequently, valuable biomolecules extracted from tissue sources can be analyzed without loss. CE is rapid, with run times in the order of minutes. In addition, turn-around times, i.e., the time required from the end of one run to the beginning of the next is approximately only 5 min. Because CE uses very high voltages in a < 100 μm internal diameter capillary, the efficiency is very high with theoretical plates up to 5×10^5 routinely observed, and a potential theoretical limit of 2.8×10^6 plates/m reported.[2] This efficiency is coupled to a high selectivity yielding excellent resolution characteristics.

A schematic for a CE instrument is given in Figure 1. The fused silica capillary is typically 50 to 100 cm in length and 50 to 100 μm in internal diameter; the ends of the capillary are mounted along with electrodes in buffer chambers. Electrodes are connected to a high voltage power supply capable of providing in excess of 30 kV. The capillary passes through a constant temperature compartment and then a detector. Various detectors used with CE include UV-VIS,[2] fluorescence,[5] electrochemical,[6] conductivity,[7] and mass spectrometry,[8] with UV being the most popular because of its flexibility. Output from the detector is fed to a recorder, integrator, or computerized data-acquisition system. Unlike slab-gel electrophoresis where materials are detected by time-consuming staining procedures and quantitated by densitometry, CE offers rapid on-line detection and quantitative data presentation (called an electropherogram) similar to an HPLC chromatogram. The CE system can be readily automated, and it is expected that several commercial automated instruments will soon become available.

The following is a discussion of some applications of CE that may be of interest to those in the area of biotechnology, protein and peptide chemistry, and analytical chemistry.

CE has been used to assess rapidly the purity of peptides and proteins from various sources. Although CE can be used routinely for analysis of synthetic peptides, because of the extremely small sample requirements it is ideally suited to the analysis of valuable materials derived from tissue. Since the mechanism of separation is different from HPLC, CE may provide information about sample homogeneity complementary to HPLC. Figure 2 shows the electropherograms obtained from analysis of different batches of a recombinant lymphokine. In all cases, the electrophoretic mobility of the major peak was $3.21 \pm 0.01 \times 10^{-4}$ cm²/volt.sec (see below). However, the electropherograms demonstrate a small degree of batch-to-batch variability and sample heterogeneity. It is interesting to note that reversed-phase HPLC (RPC) showed single, symmetrical peaks for each batch (data not shown). The importance of this data cannot be

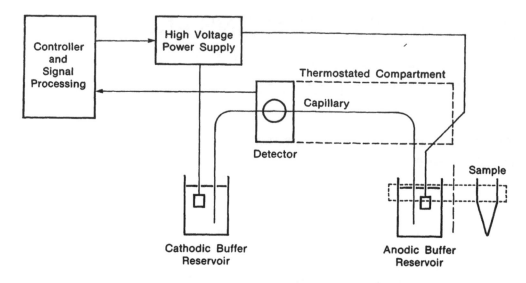

FIGURE 1. Schematic of a capillary electrophoresis system.

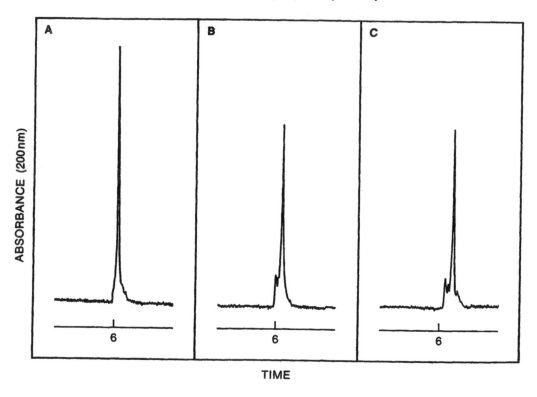

FIGURE 2. Electropherograms of three batches of a recombinant lymphokine. Conditions: field, 308 V/cm; current, 25 μA; buffer, monosodium citrate, 20 mM, pH 2.5; capillary length, 65 cm (45 cm to detector), 50 μm internal diameter; time is in minutes, and absorbance is measured at 200 nm at a range of 0.02 AUFS.

overstated considering that recombinant proteins are rapidly emerging into the therapeutic arena.

The HPLC analysis of protein digests as a means of identifying or fingerprinting the protein has been commonly used. CE can accomplish the same task, perhaps faster and with less sample.

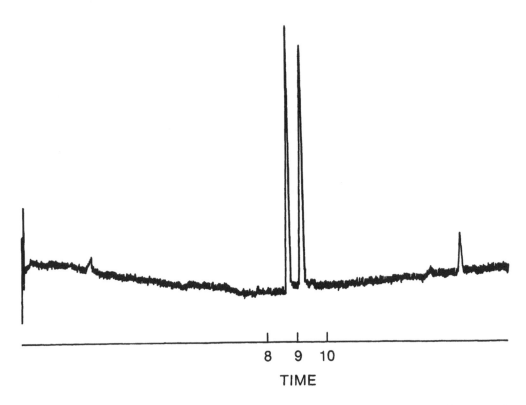

FIGURE 3. Electropherogram of a fraction obtained from the HPLC separation of a tryptic digest of β-lactoglobulin A. Conditions: same as Figure 2, except absorbance range was 0.005 AUFS.

More importantly, the use of both HPLC and CE methods for mapping provides different separation mechanisms, and therefore more reliable conclusions concerning sample integrity.

In certain instances, CE is a much easier and selective method than HPLC. The separation of β-endorphin, γ-endorphin, and their respective [Des-Tyr 1] derivatives are reported to be accomplished by HPLC only under very critical conditions, including a ternary gradient and specific reversed-phase column.[9] These conditions and the time required for the method development can be contrasted to the CE separation where all of the above endorphins and N-acetyl-β-endorphin were separated on the first attempt (data not shown).

One area we have investigated which demonstrates the use of both HPLC and CE is for the analysis of peptides prior to sequencing. Currently, protein sequence information is obtained by digesting the protein either enzymatically or chemically, then isolating the fragments, and finally sequencing the fragments. Isolation of the peptide fragments is most commonly performed by RPC, giving rise to fractions presumably containing a single peptide. However, sequencing of the peptide fragments is often confounded by contaminating peptides in the fraction; frequently one HPLC separation is insufficient for obtaining pure fractions. Since automated sequence analysis usually requires at least 3 hr before sequence homogeneity can be assessed, impure fractions lead to losses in time and productivity for the sequencer, as well as loss of valuable sample.

We have used CE as a quick screening tool for sample purity subsequent to HPLC fraction isolation. CE is ideal for this because (1) since it uses a separation mechanism different from HPLC, it should give additional information, (2) it uses only 5 nl of sample, allowing all material to be used for subsequent work, and (3) it is fast. A protein with a known sequence, β-lactoglobulin A, was digested with trypsin and the fractions were isolated by HPLC, followed by CE analysis. Figure 3 illustrates the importance of this procedure: fraction D from the digest appears as a single peak on the chromatogram, but as two peaks on the electropherogram. Sequence analysis of fraction D gave results consistent with the presence of two peptides. Results of the CE analysis of other fractions were also corroborated by sequencing. We can therefore conclude that the evaluation by CE will save sequencer time and sample; if fractions are found to be contaminated, a second HPLC separation method (different from the initial) can be utilized before sequencing.

Unlike HPLC, the retention time of a peak is not used for peak identification but rather an electrophoretic mobility (μ) is measured. The mobility is a function of the velocity of the solute as well as the electrical field strength. The velocity has an electrophoretic component and an electroosmotic component; the electroosmotic component is derived from the bulk flow of liquid in the capillary due to solvent ion-wall charge interactions. Electrophoretic mobility (μ) is calculated from the expression

$$\mu_{sample} = \frac{L_D \cdot L_T}{Volts. \, 60}\left[\frac{1}{t(sample)} - \frac{1}{t(std)}\right] + \mu_{std}$$

where L_D = length of the capillary from the injection point to the detector, L_T = total capillary length, t(sample) = observed time (min) for sample, t(std) = observed time (min) for internal standard, and μ_{std} = absolute electrophoretic mobility of standard. The absolute mobility of the standard used in the present work was 3.42×10^{-4} cm²/volt.sec.

The electrophoretic mobility of an analyte is a function of both charge and size. When comparing peptides of similar size but different charges, it is observed that the peptide with the greater positive charge moves quickly to the negative electrode and thus has a greater mobility. For example, at a buffer pH of 2.5, the unphosphorylated Kemptide, LRRASLG (a substrate for the enzyme cyclic-AMP dependent protein kinase), has both a greater charge and mobility (μ = 3.00×10^{-4} cm²/volt.sec) than does the phosphorylated form (μ = 2.01×10^{-4}). This peptide was first characterized and synthesized by Dr. Bruce Kemp, hence the name, and is commercially available from Sigma.

That mobility provides some real information about the nature of a peptide is a very important aspect of CE data. This can be illustrated by the following. Recently, a 13-amino-acid peptide, derived from leukocytes, was sequenced and the corresponding synthetic peptide prepared. Although both gave the same sequence, CE analysis of the mixture showed two peaks, with the native peptide demonstrating a higher mobility. Based on the mobilities of many other peptides of known charge and sequence, it was postulated that the mobility of the native peptide was consistent with its sequence containing a C-terminal amide, while the slower moving synthetic peptide was known to be the acid form. This was later confirmed by FAB-MS. Thus, CE can be used to test peptide heterogeneity, and also perhaps shed some light on peptide characteristics.

Since CE is a method based largely on charge differences between analyte species, one can manipulate the pH of the running buffer and by doing so manipulate the charges on the peptides. Peptides 1 and 2 have calculated charges of 1.41 and 1.37 respectively at pH 2.5, and are partially resolved as shown in Figure 4A.[10] Again, the peaks migrate in the order predicted: the more positive peptide 1 has a higher electrophoretic mobility. In Figure 4B, the effect of changing buffer pH is illustrated. At pH 4.0, peptides 1 and 2, with calculated charges of 1.02 and 0.46 respectively, are more fully resolved; a change of only 1.5 pH units dramatically changes the selectivity of the separation.

The selectivity of CE is again demonstrated by the resolution of peptides which differ by only one uncharged, non-polar amino acid. A mixture of three peptides each with seven amino acids and same overall charge were separated in free solution at pH 2.5 (Figure 4C). The only difference was the presence of either a Gly, Ala, or Trp positioned between adjacent lysines. It is postulated that the mechanism responsible for this unique selectivity involves a perturbation

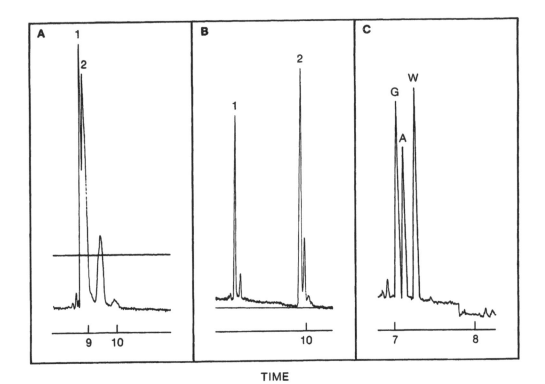

TIME

FIGURE 4. (A) Electropherogram of two peptides at pH 2.5. Peptide 1 has the sequence AFKAING and peptide 2 is AFKADNG. Conditions: field, 277 V/cm; current, 24 μA; buffer, monosodium citrate, 20 mM, pH 2.5; capillary length 65 cm (45 cm to detector), 50 μm internal diameter; time is in minutes and absorbance measured at 200 nm at a range of 0.02 AUFS. (B) Electropherogram of the same peptides at pH 4.0. Conditions: field 277 V/cm; current, 12 μA; buffer, monosodium citrate, 20 mM, pH 4.0. (C) Electropherogram of three peptides with the structure AFKXKNG, where X = G, A, or W. Conditions: same as Figure 4A.

of the more general idea of mobility based on charge. We believe that amino acids neighboring charged groups exert an influence such that the environment and solvation of the charged species is affected; therefore changes in the hydrophobic nature of uncharged amino acids can sufficiently affect the charge of a peptide to effect a separation.

The selectivity obtained with CE is further illustrated by the ability to separate peptide diastereomers. The diastereomers [L-Ala 2] enkephalin and [D-Ala 2] enkephalin are separated in free solution at pH 2.5; no additives, capillary modification, or special buffers are required. This technique was used to separate the diastereomers [L-Arg 6] dynorphin 1 to 13 from [D-Arg 6] dynorphin 1 to 13, (Figure 5A), as well as diastereomeric dipeptides and other peptides. These materials were obtained from different sources. Again, the idea of subtle influences on charge by neighboring amino acids is invoked in order to explain these findings. Bulky amino acid side chains should present some barrier to free rotation, and therefore the environment of a charged group should reflect the presence of an adjacent D- or L-amino acid. Very slightly differing environments should exert some influence on the charge of the peptide and ultimately the mobility of the peptide species. Although it is possible that D/L changes can affect the conformation of the peptide and that one is separating conformers as opposed to separating small local charge variations, this may be unlikely (at least some of the time) since the conformational forms of YAGF (D-Met) and YAGF (L-Met), with D/L changes on the C-terminus are expected to be identical, yet these materials are separable by CE.

The only peptide diastereomer pair tested that proved to be unresolved was [D-Trp 8]

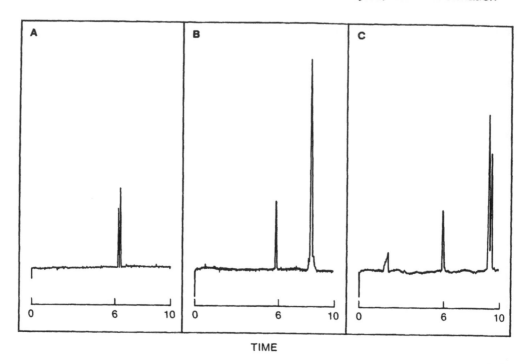

TIME

FIGURE 5. (A) Electropherogram of a mixture of [D-Arg 6] dynorphin 1 to 13 and [L-Arg 6] dynorphin 1 to 13. (B) Electropherogram of a mixture of [D-Trp 8] somatostatin and [L-Trp 8] somatostatin. (C) Electropherogram of the mixture in Figure 5B after treatment with 100 m*M* dithiothreitol. Conditions: same as Figure 2. Figures 5B and C contain an internal standard used for measuring electrophoretic mobility (μ) and which is eluted at ~ 6 min.

somatostatin from the [L-Trp 8] species (see Figure 5B). This molecule does differ from those already mentioned in that it contains a disulfide linkage which essentially cyclizes the molecule. We reasoned that this "conformational lock" somehow prevents differential interaction of the D or L tryptophan with neighboring charged groups. When the mixture of these two diastereomers was treated with a reducing agent, dithiothreitol, the disulfide link opened, and the diastereomers were then easily resolved (Figure 5C). Control experiments showed that separated materials were indeed diastereomers, and not oxidized and reduced forms of unresolved diastereomer species. The separation of amino acid enantiomers via additives to the running buffer has been reported.[11]

In addition to the free solution applications, the literature reports other uses for CE instrumentation. Gel-filled capillaries have enabled molecular weight determinations similar to that provided by slab-gel SDS-PAGE;[12] proteins and protein fragments smaller than 10 kDa gave linear plots of log molecular weight vs. mobility in a rapid quantitative fashion without the need for staining and densitometry. When an open capillary is filled with a buffer containing SDS in excess of the critical micelle concentration, micellar electrokinetic countercurrent chromatography can be performed.[13] Whereas free solution CE is unable to separate neutral molecules, the differential partitioning of neutrals into micelles coupled to a rapid electroosmotic flow for micelle mobility allows an efficient and selective separation. This method has also been extended to charged peptides[14] and provides yet another mechanism for peptide separation. The use of open tubes, gels, micelles, organic solvents in the running buffer,[15] and coated capillaries[16] have all been reported for the separation of proteins, peptides, nucleotides, and organic molecules, thus demonstrating the large variety of separation conditions, techniques, and applications possible with CE.

REFERENCES

1. Jorgenson, J.W. and Lukacs, K.D., Capillary zone electrophoresis, *Science*, 222, 266, 1983.
2. Lauer, H.H. and McManigill, D., Capillary zone electrophoresis of proteins in untreated fused silica tubing, *Anal. Chem.*, 58, 166, 1986.
3. Jorgenson, J.W., Zone electrophoresis in open tubular capillaries, *Trends in Anal. Chem.*, 3, 51, 1984.
4. Tsuda, T., Nakagawa, G., Sato, M., and Yagi, K., Separation of nucleotides by high-voltage capillary electrophoresis, *J. App. Biochem.*, 5, 330, 1983.
5. Green, J.S. and Jorgenson, J.W., Variable wavelength on-column fluorescence detector for open-tubular zone electrophoresis, *J. Chromatogr.*, 352, 337, 1986.
6. Wallingford, R.A. and Ewing, A.G., Capillary zone electrophoresis with electrochemical detection, *Anal. Chem.*, 59, 1762, 1987.
7. Huang, X., Pang, T.-K.J., Gordon, M.J., and Zare, R.N., On-column conductivity detector for capillary zone electrophoresis, *Anal. Chem.*, 59, 2747, 1987.
8. Smith, R.D., Olivares, J.A., Nguyen, N.T., and Udseth, H.R., Capillary zone electrophoresis-mass spectrometry using an electrospray ionization interface, *Anal. Chem.*, 60, 436, 1988.
9. van Nispen, J.W. and Janssen, P.S.L., Synthetic β-lipotropin fragments, in *CRC Handbook for the Separation of Amino Acids, Peptides, and Proteins*, Vol. II, Hancock, W.S. Ed., CRC Press, Boca Raton, FL, 1984, 229.
10. Grossman, P.D., Wilson, K.J., Petrie, G., and Lauer, H.H., Effect of buffer pH and peptide composition on the selectivity of peptide separations by capillary zone electrophoresis, *Anal. Biochem.*, 173, 265, 1988.
11. Gozel, P., Gassmann, E., Michelson, H., and Zare, R.N., Electrokinetic resolution of amino acid enantiomers with copper(II)-aspartame support electrolyte, *Anal. Chem.*, 59, 44, 1987.
12. Cohen, A.S. and Karger, B.L., High-performance sodium dodecyl sulfate polyacrylamide gel capillary electrophoresis of peptides and proteins, *J. Chromatogr.*, 397, 409, 1987.
13. Terabe, S., Otsuka, K., and Ando, T., Electrokinetic chromatography with micellar solution and open-tube capillary, *Anal. Chem.*, 57, 834, 1985.
14. Palmieri, R., Kalbag, S., and Osborne, J., Separation of Peptide Analogs Using Capillary Zone Electrophoresis in Micellar Solutions, presented at the Second Symp. of The Protein Society, San Diego, August 13—17, 1988.
15. Fujiwara, S. and Honda, S., Effect of addition of organic solvent on the separation of positional isomers in high-voltage capillary zone electrophoresis, *Anal. Chem.*, 59, 487, 1987.
16. Hjerten, S., High performance electrophoresis — elimination of electroendosmosis and solute absorption, *J. Chromatogr.*, 347, 191, 1985.

Indexes

STRUCTURE INDEX
Amino Acids Found in Proteins
(Including Three-Letter and One-Letter Codes)

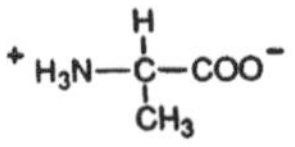

Alanine
(Ala, A)

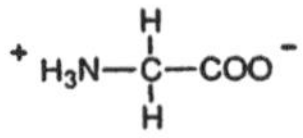

Glycine
(Gly, G)

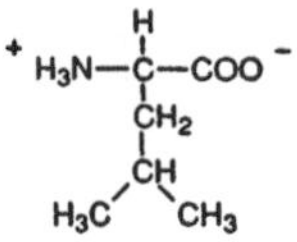

Leucine
(Leu, L)

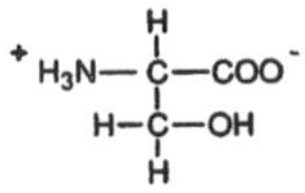

Serine
(Ser, S)

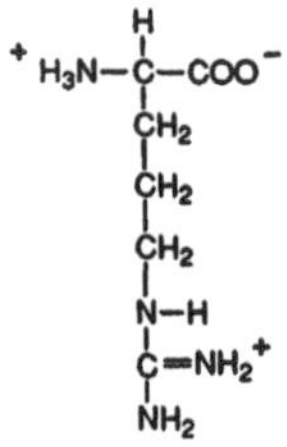

Arginine
(Arg, R)

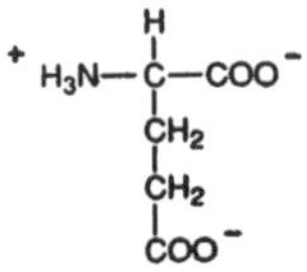

Glutamic acid
(Glu, E)

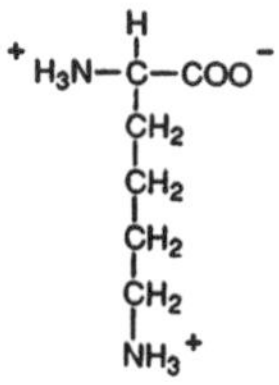

Lysine
(Lys, K)

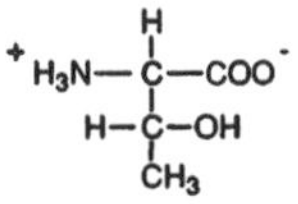

Threonine
(Thr, T)

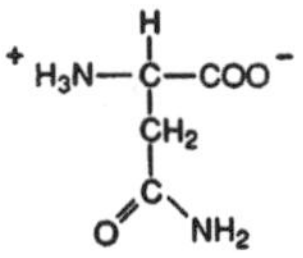

Asparagine
(Asn, N)

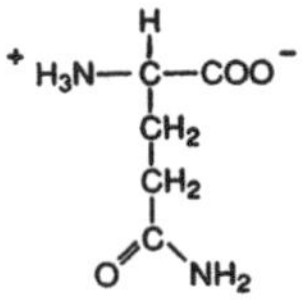

Glutamine
(Gln, Q)

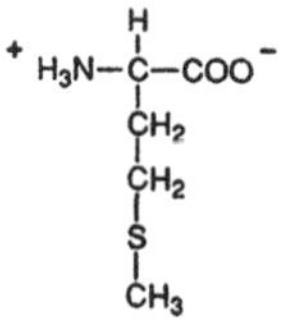

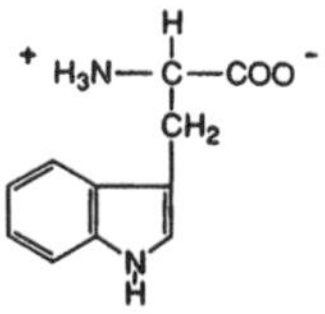

Tryptophan
(Trp, W)

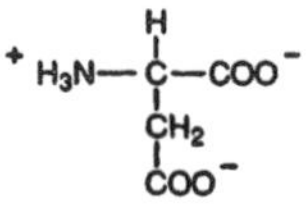

Aspartic acid
(Asp, D)

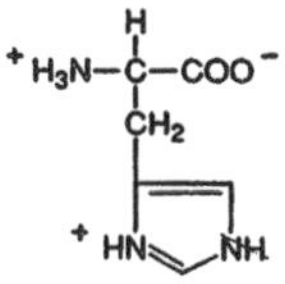

Histidine
(His, H)

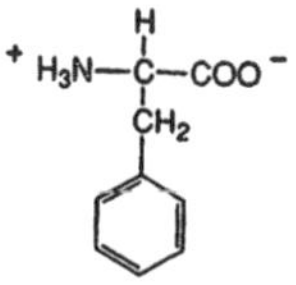

Methionine
(Met, M)

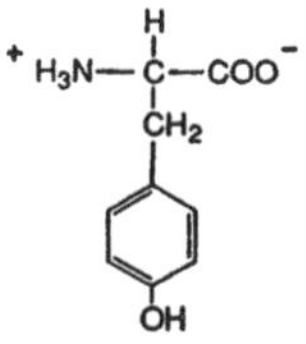

Tyrosine
(Tyr, Y)

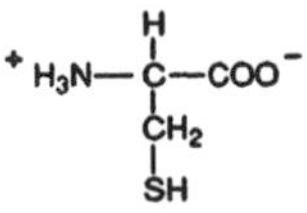

Cysteine
(Cys, C)

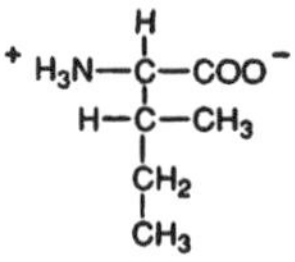

Isoleucine
(Ile, I)

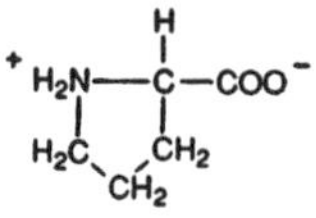

Phenylalanine
(Phe, F)

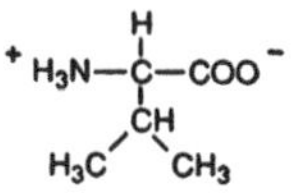

Valine
(Val, V)

Proline
(Pro, P)

Amino Acids not Found in Proteins

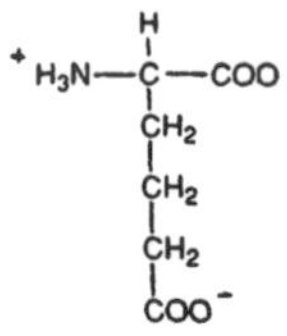

2-Aminoadipic acid

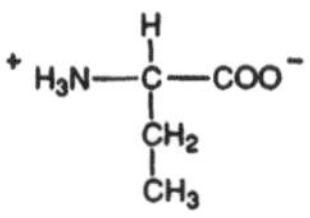

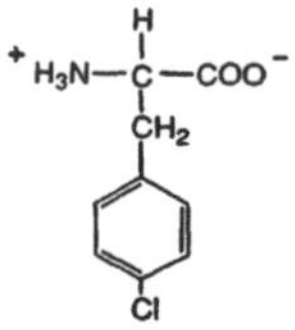

p-Chlorophenylalanine

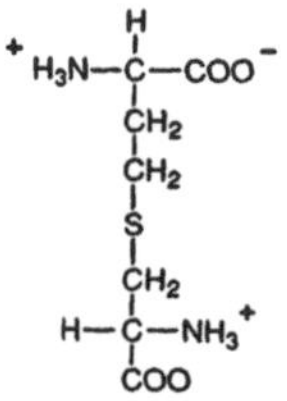

Homoserine lactone

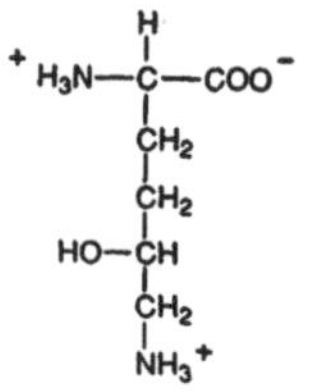

1-Methylhistidine

2-Aminobutyric acid

Cystathionine

5-Hydroxylysine

3-(2-Naphthyl)-alanine

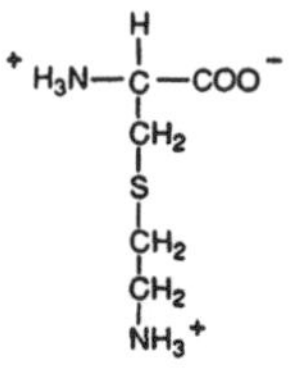

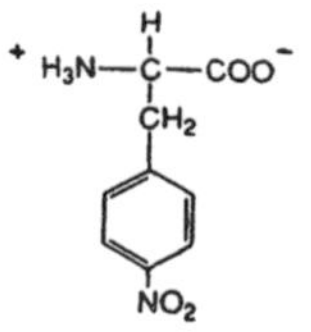

S-2-Aminoethyl-cysteine

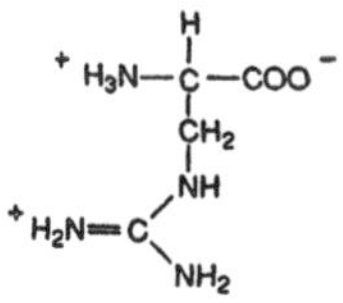

4-Hydroxyproline

p-Nitrophenylalanine

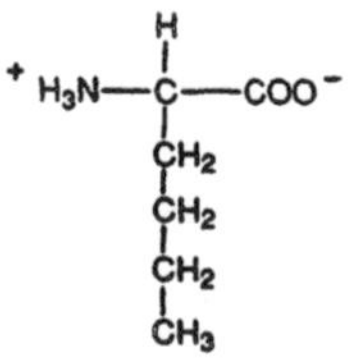

α-Amino-β-guanidinopropionic acid

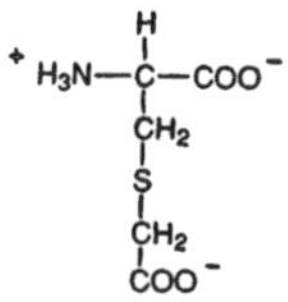

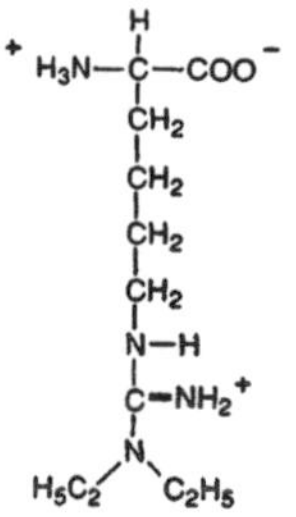

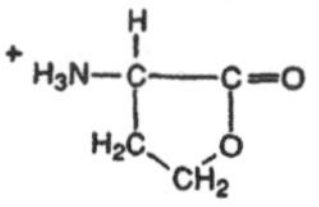

Diethylhomoarginine

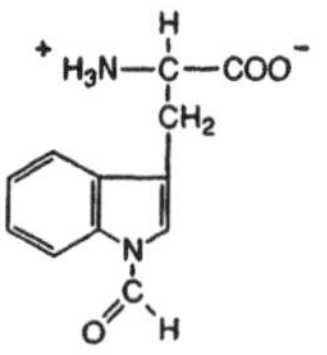

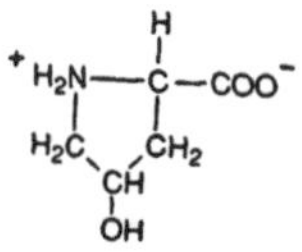

Methionine sulfone

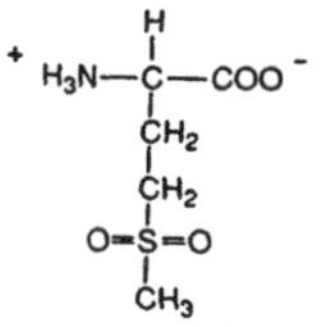

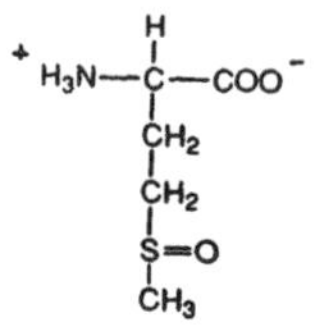

Norleucine

N-Formyl tryptophan

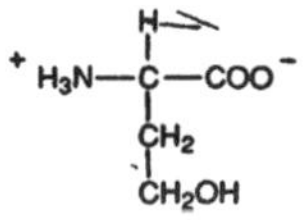

Methionine sulfoxide

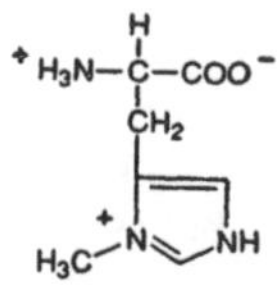

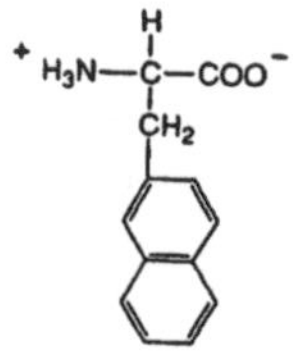

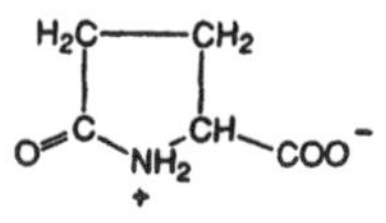

Pyroglutamic acid

S-Carboxymethyl cysteine

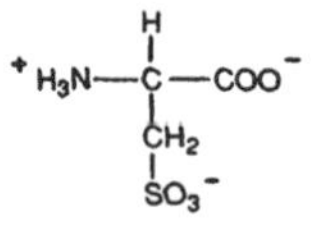

Homoserine

Cysteic acid

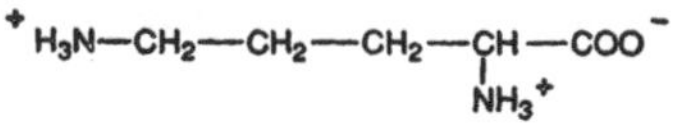

$^+$H$_3$N—CH$_2$—CH$_2$—CH$_2$—CH—COO$^-$
|
NH$_3$$^+$

Ornithine

$^+$H$_3$N—C—COO$^-$ (H above)
|
H$_3$C—C—CH$_3$
|
SH

Penicillamine

$^+$H$_3$N—C—COO$^-$ (H above)
|
CH$_2$
|
(pyridyl ring, N$^+$–H)

3-(3-Pyridyl)-alanine

$^+$H$_3$N—C—COO$^-$ (H above)
|
CH$_2$
|
S
|
CH$_2$
|
CH$_2$
|
(pyridyl ring, N$^+$–H)

S-(4-Pyridylethyl)-cysteine

H$_3$C—N—CH$_2$—COO$^-$ (H above N)

Sarcosine

$^+$H$_3$N—CH$_2$—CH$_2$—SO$_3$$^-$

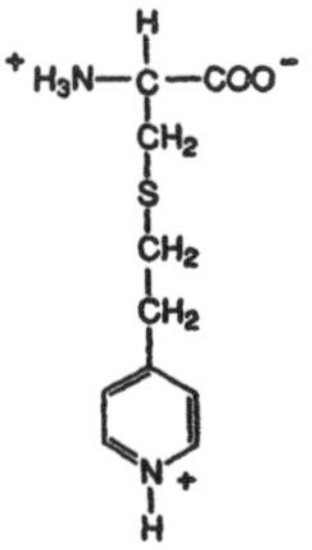

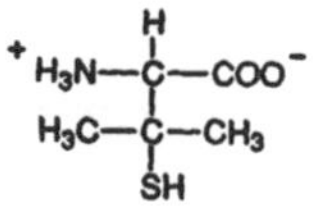

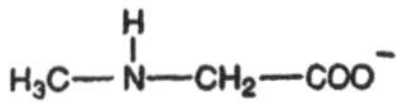

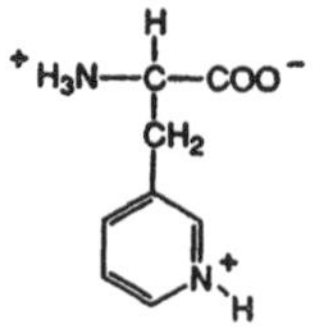

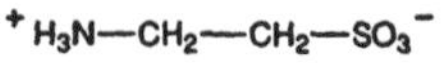

Taurine

H$_3$C—N$^+$—C—COO$^-$
(CH$_3$ above N, H above C; CH$_3$ below N, H below C)

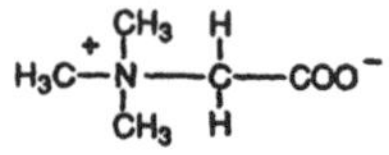

Trimethylglycine

General Structures

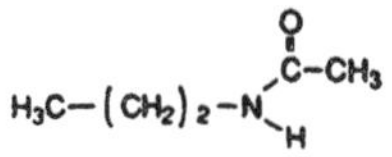

H$_3$C—(CH$_2$)$_2$—N—C—CH$_3$ (O above C, H below N)

N-Acetylpropylamine

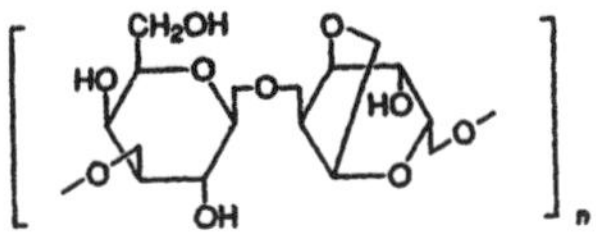

H$_2$C=C—C—NH$_2$ (H above, O above)

Acrylamide

(adamantane structure with NH$_2$)

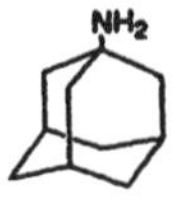

1-Adamantanamine
ADAM

(adenine–ribose structure)
NH$_2$
...CH$_2$—O—P—O—P—O—P—OH (O$^-$ on each P, O above)
OH OH

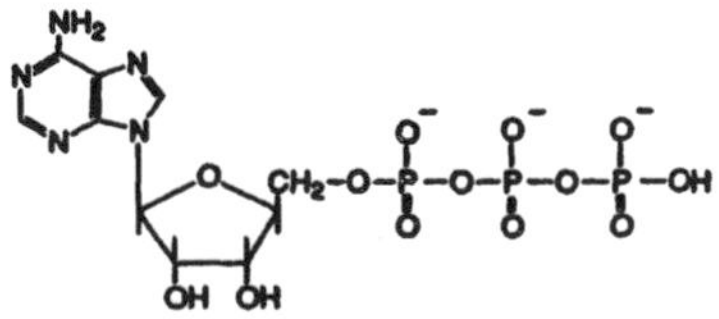

Adenosine triphosphate
ATP

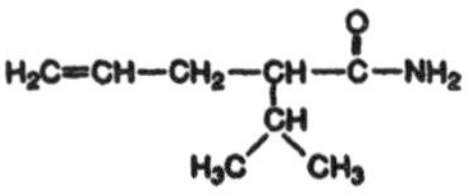

(agarose repeating unit)
CH$_2$OH
HO ... O ... HO ... O
OH O
[ ]$_n$

Agarose

H$_3$C—(CH$_2$)$_n$—SO$_3$H

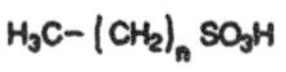

Alkylsulfonic acids
(homologous series of ion-pairing reagents)

n=0, Methanesulfonic acid
n=1, Ethanesulfonic acid
n=5, Hexanesulfonic acid
etc.

(generally available as the sodium
salt when n>0)

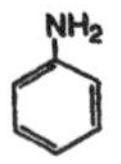

(benzene ring with C=O and R)

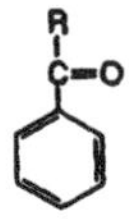

Alkylphenones
R = CH$_3$, C$_2$H$_5$, etc.

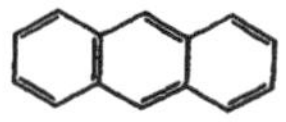

H$_2$C=CH—CH$_2$—CH—C—NH$_2$ (O above C)
|
CH
|
H$_3$C CH$_3$

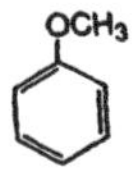

Allylisopropylacetamide

(coumarin ring with CH$_3$ and H$_2$N, two O)

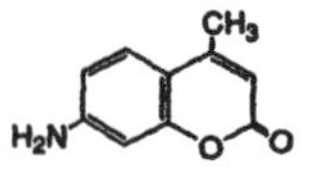

7-Amino-4-methylcoumarin
AMC

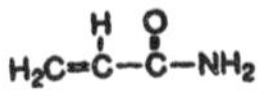

(benzene ring with NH$_2$)

Aniline

(benzene ring with OCH$_3$)

Anisole

(anthracene three-ring structure)

Anthracene

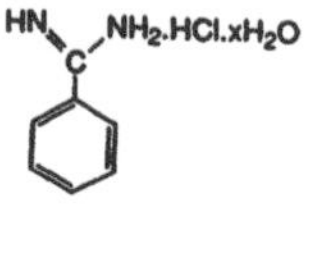

Benzamidine hydrochloride hydrate

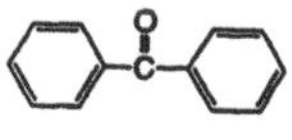

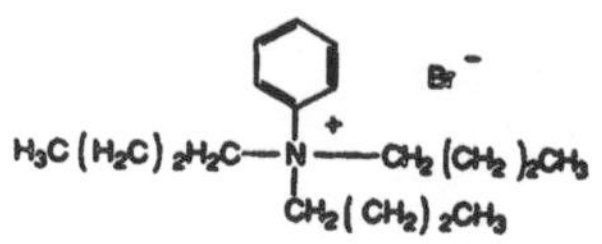

Benzyltributylammonium bromide

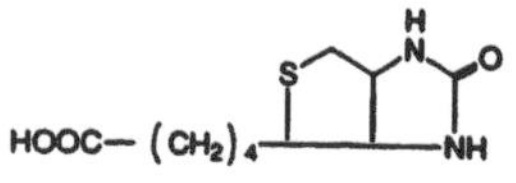

$CH_3(CH_2)_3OCH_2CH_2OCH_2CH_2OH$

2-(2-Butoxyethoxy)ethanol

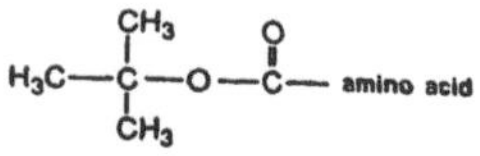

Benzophenone

Biotin

amino acid

t-Butyloxycarbonyl α-amino protecting group
t-Boc

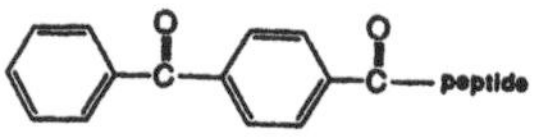

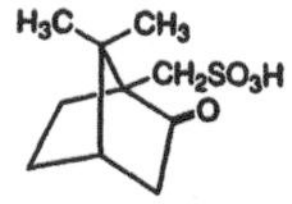

4-Benzoylbenzoyl photoprobe group

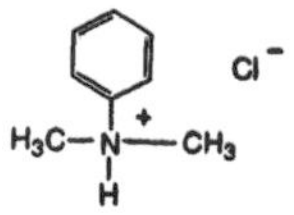

Biotin-hydrazine

Camphorsulfonic acid

Benzyldimethylammonium chloride

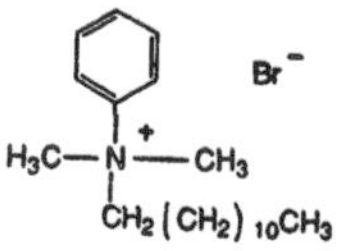

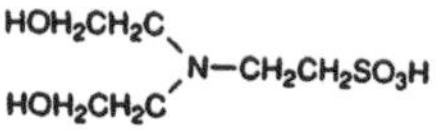

N,N-Bis(2-hydroxyethyl)-2-aminoethanesulfonic acid
BES

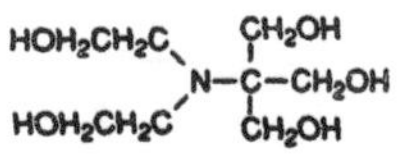

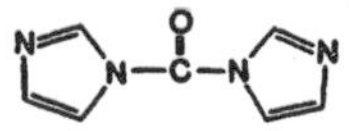

1,1'-Carbonyldiimidazole

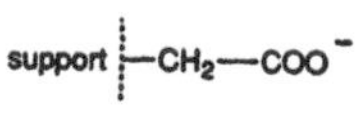

Benzyldimethyldodecylammonium bromide

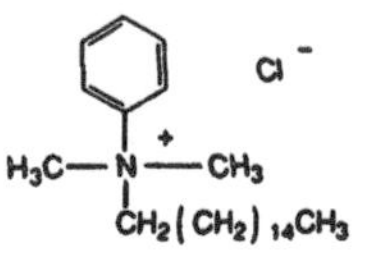

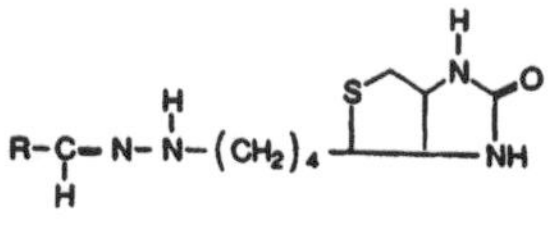

Bis(2-hydroxyethyl)iminotris(hydroxymethyl)methane
BIS-TRIS

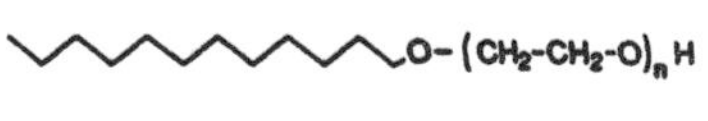

support—CH_2—COO^-

Carboxymethyl (CM)
cation-exchange group

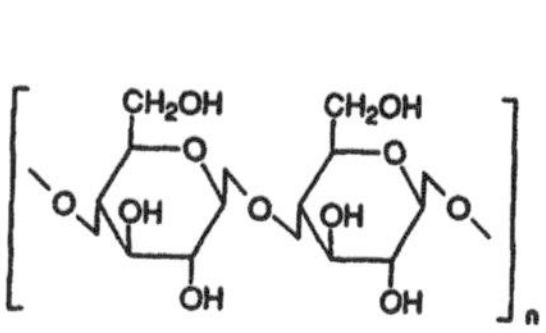

Benzyldimethylhexadecylammonium chloride

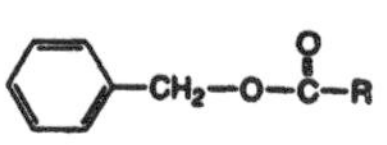

$O-(CH_2-CH_2-O)_n H$

Brij 35 (n=23), Dodecylpentaethyleneglycol-390,(n=5-9)
Polyoxyethylene alcohol

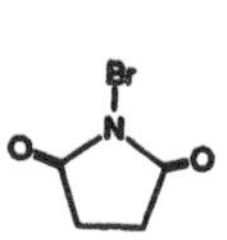

Cellulose

Benzyl ester: protection as benzyl ester
of carboxyl-containing side-chain

N-Bromosuccinimide

$CH_3(CH_2)_3OCH_2CH_2OH$

2-Butoxyethanol

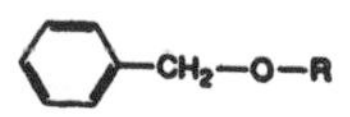

Benzyl ether: protection as benzyl ether
of carboxyl-containing side-chain

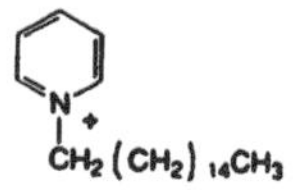

Cetylpyridinium cation

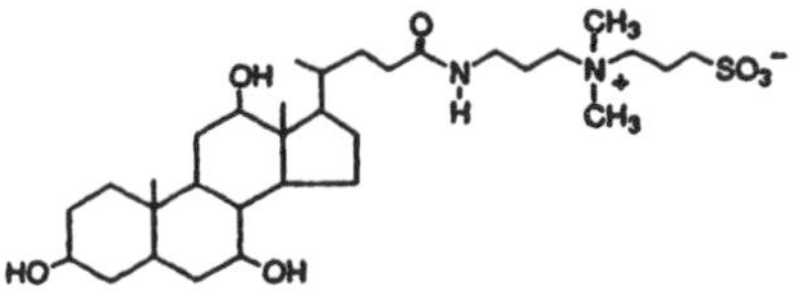

3-[(3-Cholamidopropyl)dimethylammonio]-1-propanesulfonate
CHAPS

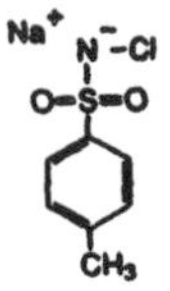

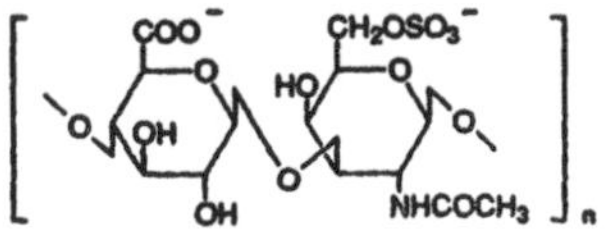

Chondroitin-6-sulfate

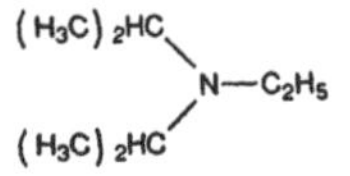

Diisopropylethylamine
DIEA

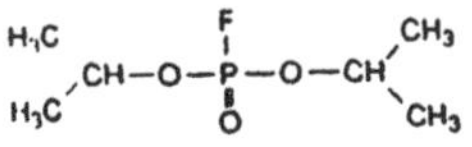

Chloramine-T

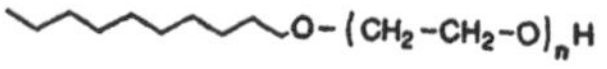

Decylpolyethyleneglycol-300 (n=5-9)

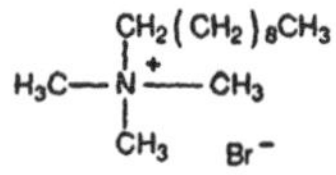

Diisopropyl fluorophosphate
DFP

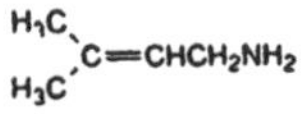

Decyltrimethylammonium bromide

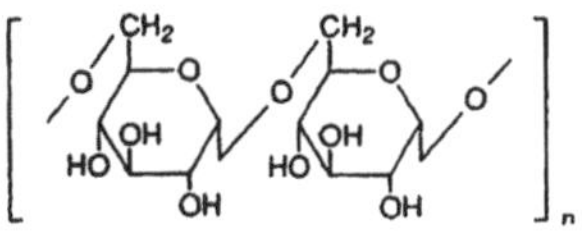

3,3-Dimethylallylamine

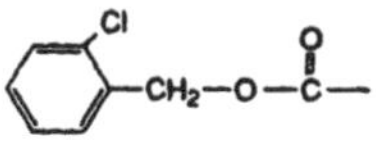

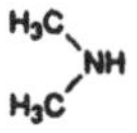

2-Chlorobenzyloxycarbonyl protecting
group of Lys side-chain

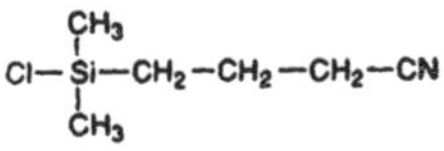

Dextran

Dimethylamine

Chloro(3-cyanopropyl)dimethylsilane

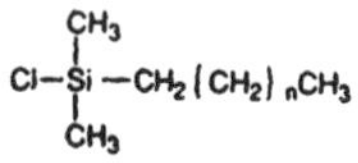

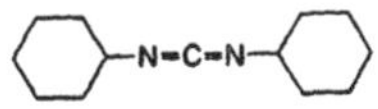

Dicyclohexylcarbodiimide
DCC

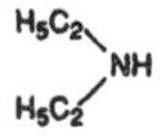

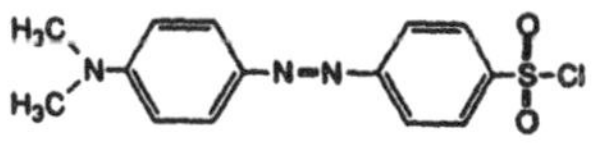

Dimethylaminoazobenzenesulfonyl chloride
Dabsyl chloride

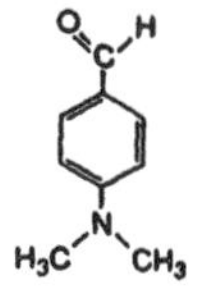

Diethylamine

n=15 , Chlorodimethyloctadecylsilane
n=6, Chlorodimethyloctylsilane
etc.

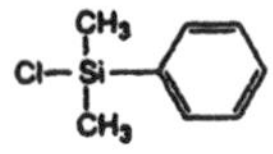

p-Dimethylaminobenzaldehyde

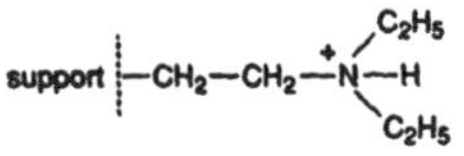

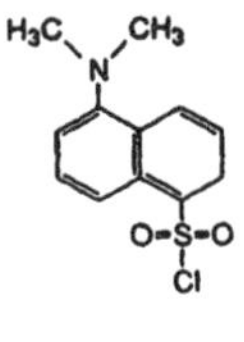

Diethylaminoethyl (DEAE)
anion-exchange group

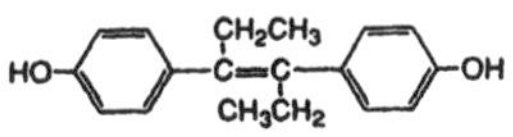

Chlorodimethylphenylsilane

5-Dimethylamino-1-naphthalenesulfonyl chloride
Dansyl chloride

Diethylstilbestrol
DES

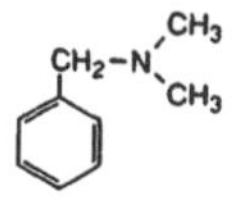

N-N '-Dimethylbenzylamine

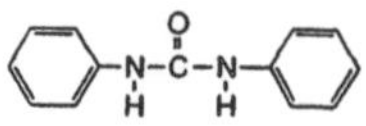

N,N' -Diphenylthiourea
DPTU

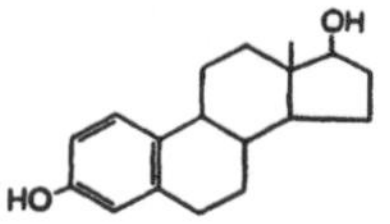

Estradiol

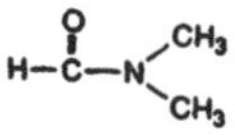

N,N-Dimethylformamide
DMF

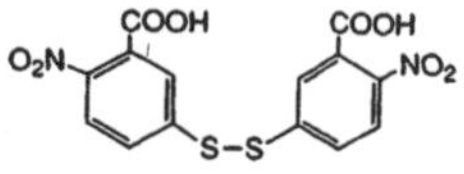

5,5'-Dithiobis(2-nitrobenzoic acid)
DTNB
Ellman's reagent

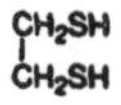

1,2-Ethanedithiol
EDT

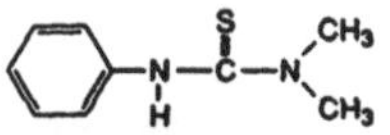

1-Dimethyl-3-phenyl-2-thiourea
DMPTU

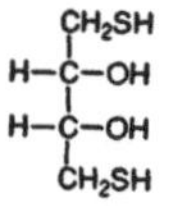

Dithioerythritol

Ethanolamine

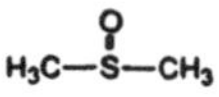

Dimethylsulfoxide
DMSO

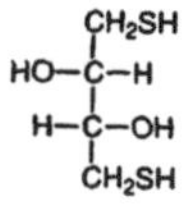

Dithiothreitol
DTT

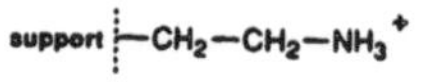

Ethylammonium (EA)
anion-exchange group

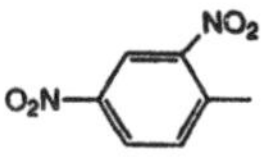

2,4-Dinitrophenyl protecting group
of His side-chain

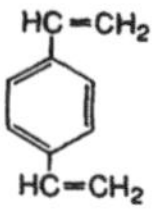

1,4-Divinylbenzene

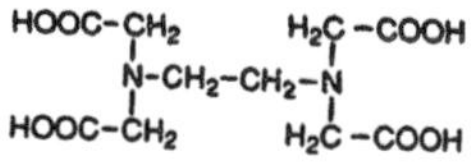

Ethylene diaminetetra-acetic acid
EDTA

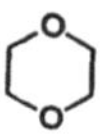

1,4-Dioxane

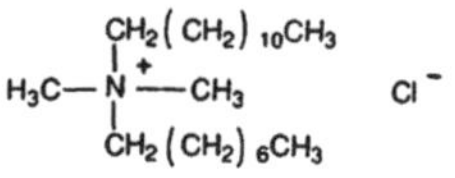

Dodecyloctyldimethylammonium chloride

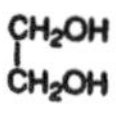

Ethylene glycol

Reduction with DTT:

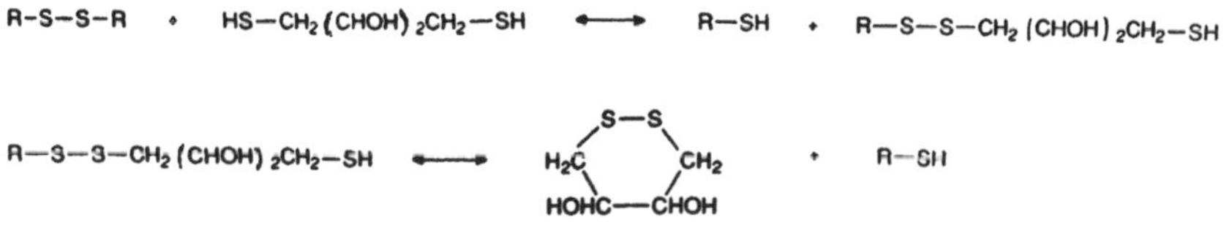

$$R-S-S-R \; + \; HS-CH_2(CHOH)_2CH_2-SH \longleftrightarrow R-SH \; + \; R-S-S-CH_2(CHOH)_2CH_2-SH \quad (1)$$

$$R-S-S-CH_2(CHOH)_2CH_2-SH \longrightarrow \; + \; R-SH \quad (2)$$

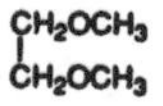

CH₂OCH₃
CH₂OCH₃

Ethylene glycol dimethyl ether

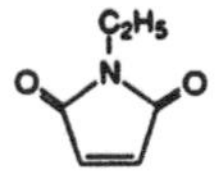

CH₂OH
CHOH
CH₂OH

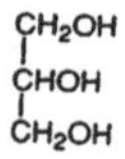

Glycerol

HOH₂CH₂C—N   N—CH₂CH₂SO₃H

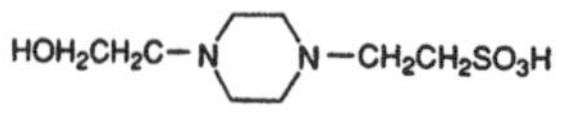

**N-(2-Hydroxyethyl)piperazine-
N'-(2-ethanesulfonic acid)
HEPES**

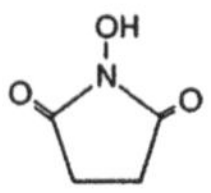

C₂H₅

N-Ethylmaleimide

H₂N   NH₂.HCl

NH

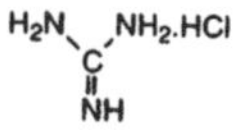

Guanidine hydrochloride

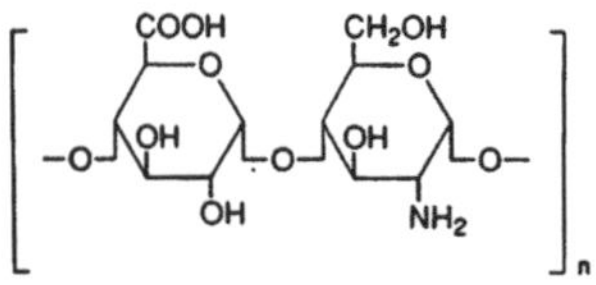

OH

N-Hydroxysuccinimide

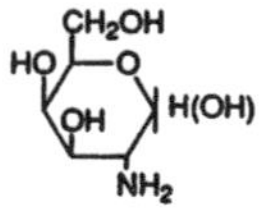

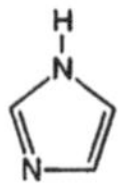

H₃C

O

Cl

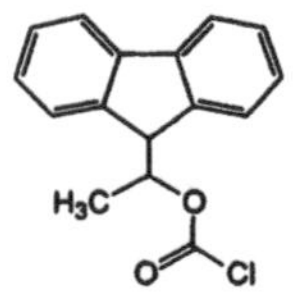

**1-(9-Fluorenyl)ethyl chloroformate
FLEC**

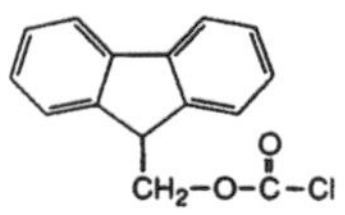

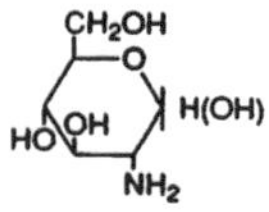

COOH   CH₂OH

OH   OH

OH   NH₂

Heparin

H

N

Imidazole

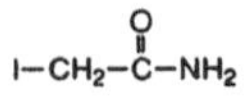

CH₂(CH₂)₁₄CH₃

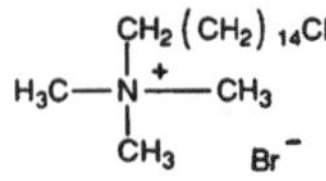

H₃C—N⁺—CH₃

CH₃   Br⁻

**Hexadecyltrimethylammonium bromide
Cetrimide**

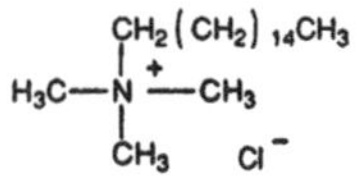

O

I—CH₂—C—NH₂

Iodoacetamide

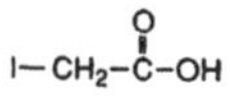

CH₂—O—C—Cl

**Fluorenylmethyloxycarbonyl chloride
FMOC-Cl**

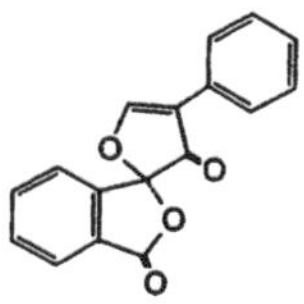

CH₂(CH₂)₁₄CH₃

H₃C—N⁺—CH₃

CH₃   Cl⁻

Hexadecyltrimethylammonium chloride

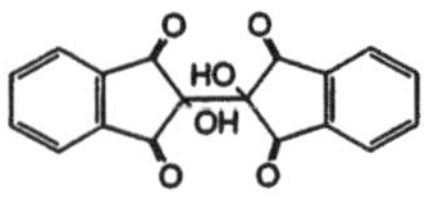

O

I—CH₂—C—OH

Iodoacetic acid

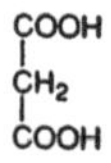

Fluorescamine

HO

Hydrindantin

COOH
CH₂
COOH

Malonic acid

CH₂OH

HO

H(OH)

NH₂

Galactosamine

N

N

.H₂O

N

OH

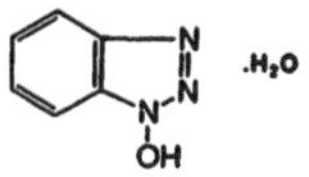

**1-Hydroxybenzotriazole hydrate
HOBT**

O

C—(CH₂)ₙ—CH₃

H₂C—N

CH₃

H—C—OH

HO—C—H

H—C—OH

H—C—OH

CH₂OH

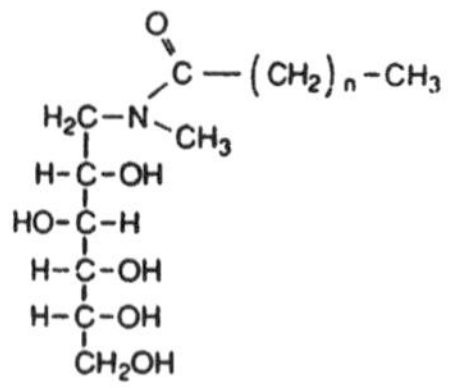

**MEGA-9 (n=7)
Nonanoyl-N-methylglucamide**

CH₂OH

HO

H(OH)

NH₂

Glucosamine

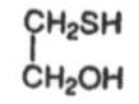

β-Mercaptoethanol
2-Mercaptoethanol

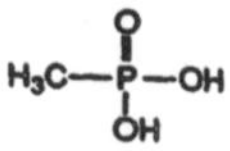

Methylphosphonic acid

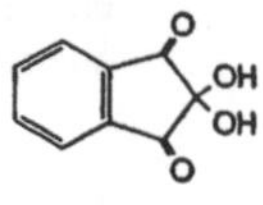

Ninhydrin

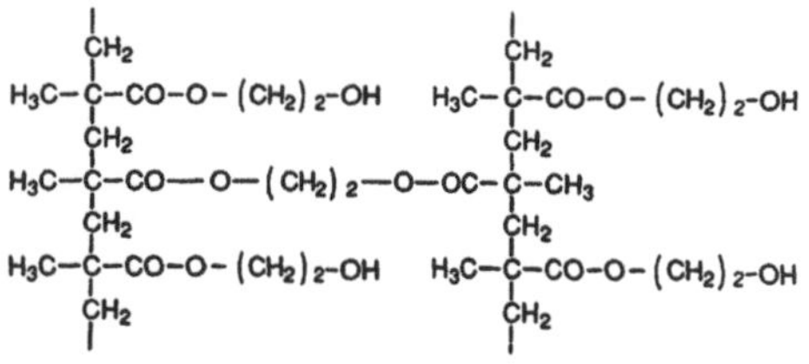

Methacrylate-based support
Example: co-polymer of 2-hydroxyethyl methacrylate
and ethylene dimethacrylate (Spheron,Separon Hema;
Tessek Ltd.,Czechoslovakia)

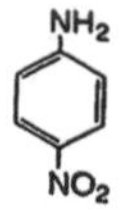

p-Nitroaniline

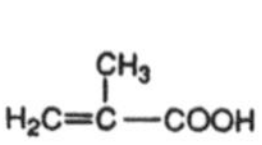

Methacrylic acid

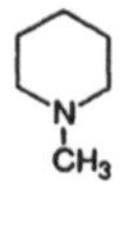

N-Methylpiperidine

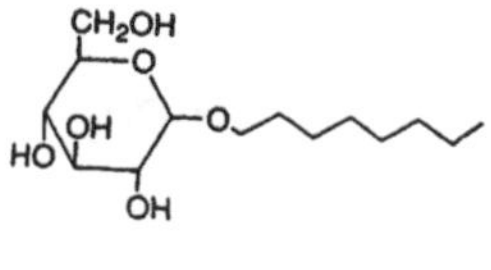

n-Octyl-β-D-glucopyranoside
Octylglucoside

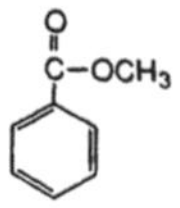

Methyl benzoate

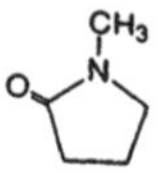

N-Methyl pyrrolidinone
1-Methyl-2-pyrrolidinone

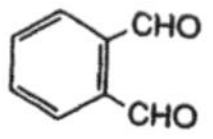

Orthophthalaldehyde
OPA

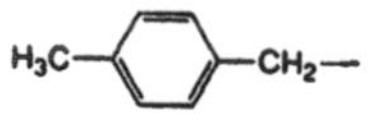

p-Methylbenzyl protecting group of
Cys side-chain

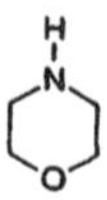

Morpholine

$CH_3(CH_2)_{14}COOH$

Palmitic acid

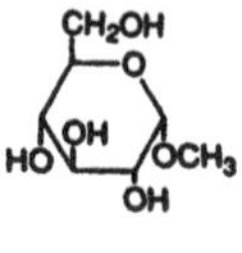

Methyl-α-D-glucopyranoside

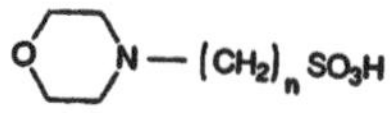

2-(N-Morpholino)ethanesulfonic acid
MES (n=2)

3-(N-Morpholino)propanesulfonic acid
MOPS (n=3)

$F_3C-(CF_2)_n COOH$

Perfluorinated carboxylic acids
(homologous series of Ion-pairing reagents)

n=0, Trifluoroacetic acid (TFA)
n=1, Pentafluoropropionic acid (PFPA)
n=2, Heptafluorobutyric acid (HFBA)

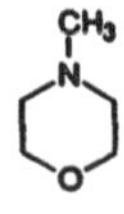

N-Methylmorpholinium

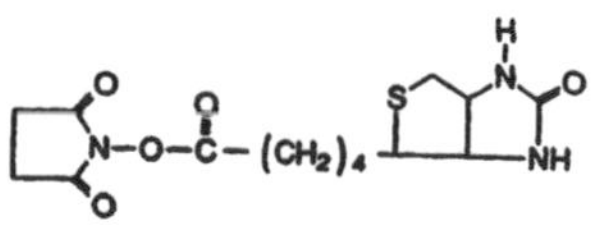

NHS-Biotin

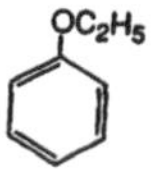

Phenetole

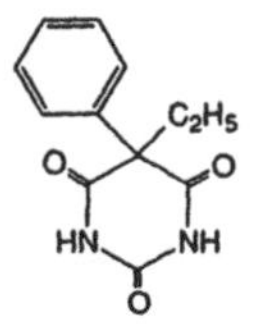

Phenobarbital

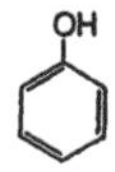

N-Phenylmaleimide

Polybutadiene

$\left[-CH_2-CH=CH-CH_2- \right]_n$

Polyethylene

$\left[-CH_2CH_2- \right]_n$

Phenol

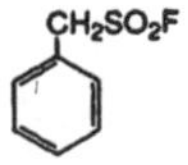

Phenylmethylsulfonyl fluoride
PMSF

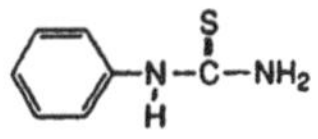

$H-\left(OCH_2CH_2 \right)_n -OH$

Poly(ethylene glycol) (PEG), where n>4, e.g., PEG 6000 (n≈158-204), PEG 400 (n=8.2-9.1). The numbers 6000 and 400 denote average MW

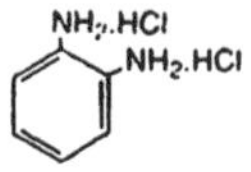

o-Phenylenediamine dihydrochloride
OPD

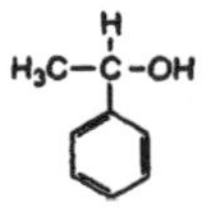

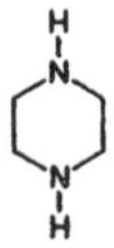

1-Phenyl-2-thiourea

$\left[-CH_2CH_2O- \right]_n$

Polyethylene oxide

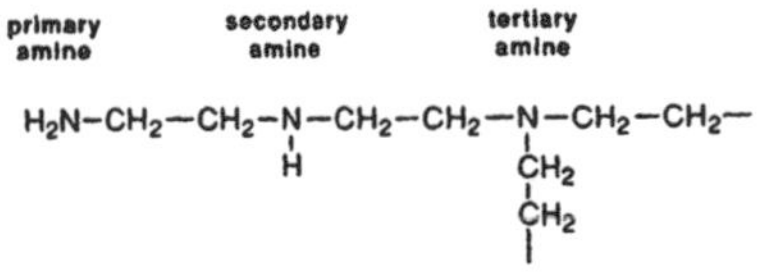

1-Phenylethanol

Piperazine

primary amine   secondary amine   tertiary amine

$H_2N-CH_2-CH_2-N-CH_2-CH_2-N-CH_2-CH_2-$

Polyethyleneimine (PEI)

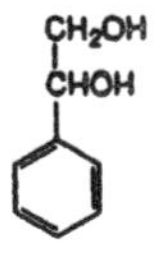

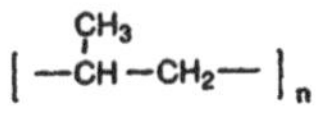

Phenylethylene glycol

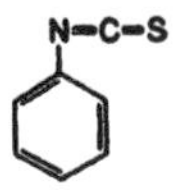

$HO_3SH_2CH_2C-N \underset{}{\bigcirc} N-CH_2CH_2SO_3H$

Piperazine-N,N'-bis(ethanesulfonic acid)
PIPES

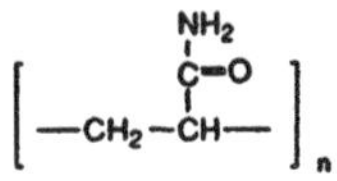

$\left[-CH-CH_2- \right]_n$ (with CH_3)

Polypropylene

Phenylisothiocyanate
PITC

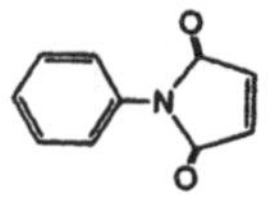

Polyacrylamide

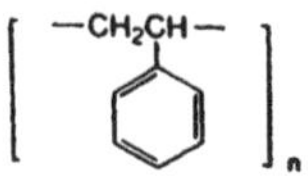

Polystyrene

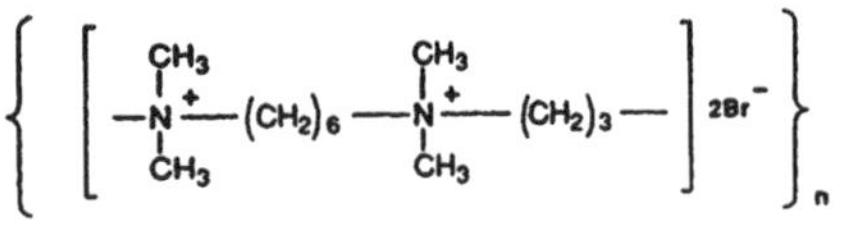

Polybrene

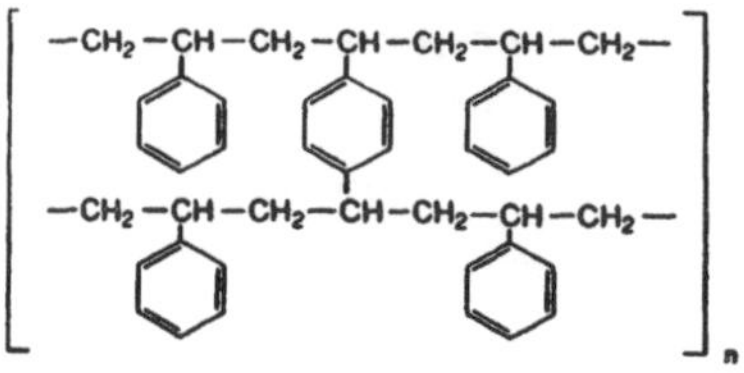

Polystyrene-divinylbenzene

Pyridine

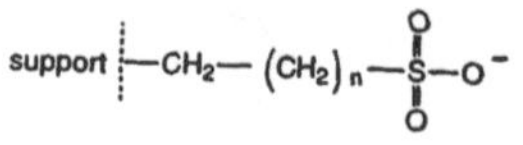

support $-CH_2-(CH_2)_n-\overset{\displaystyle O}{\underset{\displaystyle O}{S}}-O^-$

n=1 , Sulfoethyl (SE) cation-exchange group
n=2 , Sulfopropyl (SP) cation-exchange group

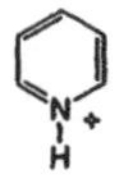

Pyridinium cation

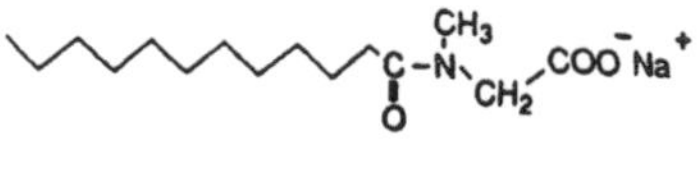

Sodium dodecyl-N-sarcosinate
Sarkosyl

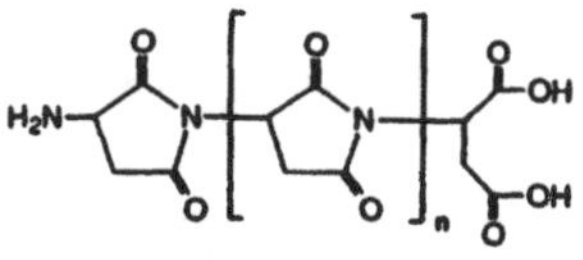

Poly(succinimide)

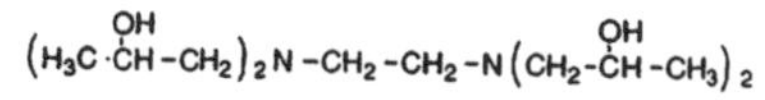

$\left| -CF_2CF_2- \right|_n$

Polytetrafluoroethylene
PTFE

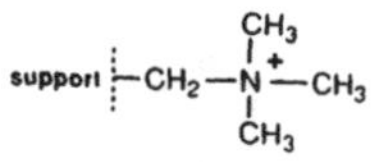

$(H_3C \cdot CH-CH_2)_2 \, N-CH_2-CH_2-N(CH_2-CH-CH_3)_2$
with OH on each

Quadrol
N,N,N',N'-Tetrakis(2-hydroxypropyl)ethylenediamine

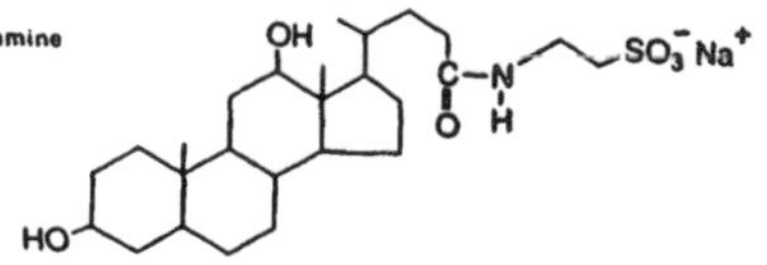

Sodium taurodeoxycholate

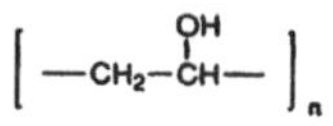

$\left[-CH_2-\overset{OH}{CH}- \right]_n$

Polyvinyl alcohol

support $-CH_2-\overset{CH_3}{\underset{CH_3}{\overset{+}{N}}}-CH_3$

Quaternarymethylammonium (QMA)
anion-exchange group

$\left| -CH_2CF_2- \right|_n$

Poly(vinylidene difluoride)
PVDF

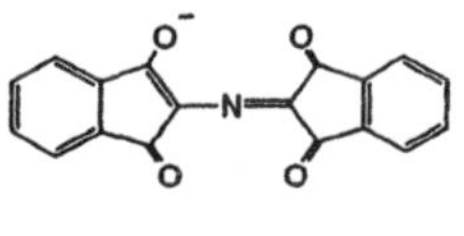

Ruhemann's purple

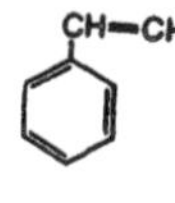

Styrene

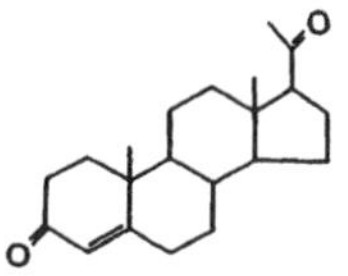

Progesterone (a progestin)

Sodium dodecyl sulfate
Sodium lauryl sulfate

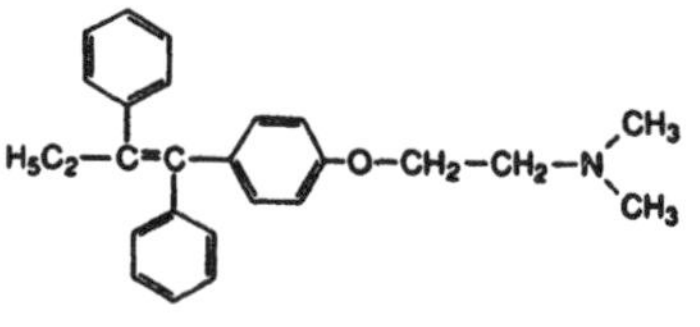

Tamoxifen

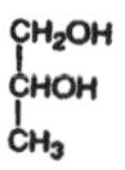

CH_2OH
$CHOH$
CH_3

Propylene glycol

CH_2OH
$H-C-OH$
$HO-C-H$
$H-C-OH$
$H-C-OH$
CH_2OH

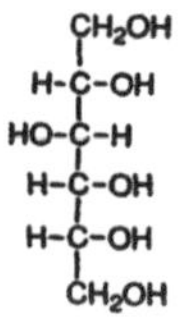

Sorbitol
D-Sorbitol
D-Glucitol

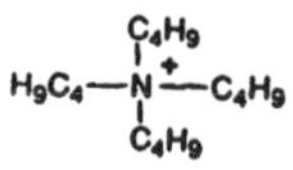

$H_9C_4-\overset{C_4H_9}{\underset{C_4H_9}{\overset{+}{N}}}-C_4H_9$

Tetrabutylammonium cation
($C_4H_9 = nC_4H_9$)

Tetrahydrofuran

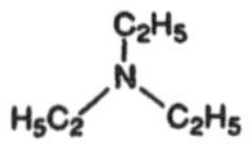

Triethylamine

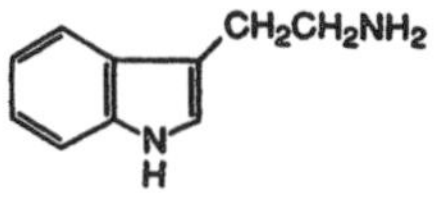

Tryptamine

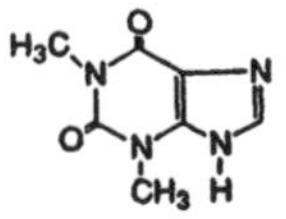

Theophylline

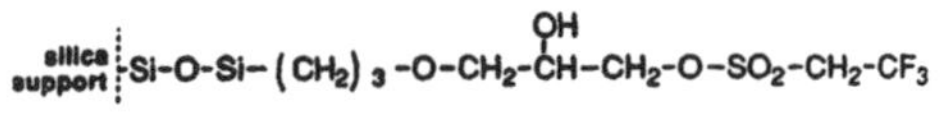

Activated Tresyl silica (Pierce Chemical Co.)

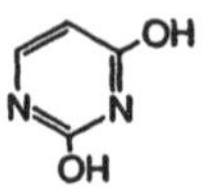

Uracil

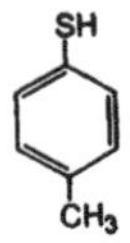

p-Thiocresol
p-Toluenethiol

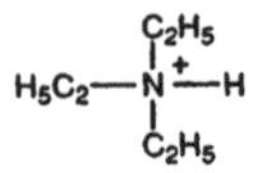

Triethylammonium cation

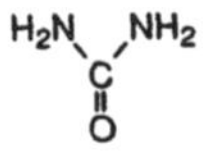

p-Toluenethiol

Urea

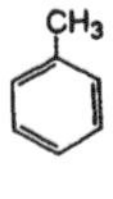

Toluene

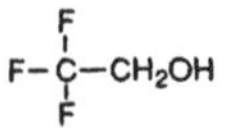

Trifluoroethanol
TFE

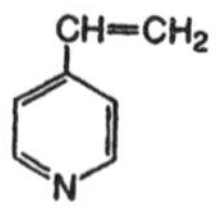

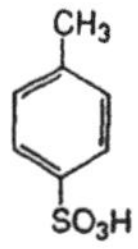

p-Toluenesulfonic acid

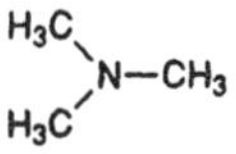

Trimethylamine

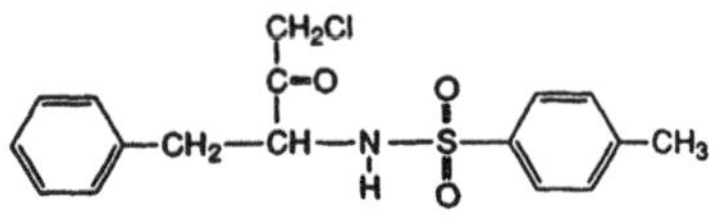

4-Vinylpyridine

L-1-p-Tosylamino-2-phenylethyl chloromethyl ketone
TPCK

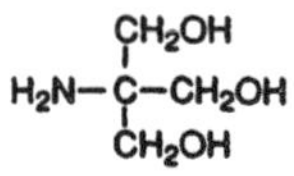

Tris(hydroxymethyl)aminomethane
TRIS

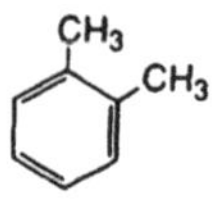

o-Xylene

INDEX

A

Elution effects, 814
Elution mode, 412
Elution time of unretained component, 86—88
Elution volume, 76, 97, 145
Embryonal carcinoma cells, 243
Emulgen 911, 211
Enantiomeric purity, 369
Enantiomers, 369
Endcapping, 76
Endfitting, 76
Endometrial carcinoma, 261
Endoproteinase Lys-C, 669
Endothelial cell growth factor, 501
Enrichment effect, 412
Enzymatic activity, 589
Enzyme-immunolinked assay (EIA), 265—268
Enzyme-linked immunosorbent assay (ELISA), 156—157, 160
Epitopes, 155, 267—268
Epoxide group, 482
Epoxide linkages, 482
Epoxide side chain, 508
Epoxy functional group, 482
Epoxy side chain, 508
Epoxy support group, 483
Epstein-Barr virus, 226, 228
Equilibration time, 546
Equilibrium distribution coefficient, 96, 98—99, 111
Equilibrium distribution constant, 106
Equilibrium profiles, 599—602, 604
[Erythro-D-β-Me-p-NO$_2$Phe^4]DPDPE, 384—386
[Erythro-D,L-β-Me-D-NO$_2$Phe^4]DPDPE, 383
[Erythro-L-β-Me-p-NO$_2$Phe^4]DPDPE, 384—386
Escherichia coli, 415, 730—733
Estradiol, 265—266, 268
Estradiol-17β, 470
Estrogen receptor, 255—259, 261—263, 266—268
  double isotope assay, 470—472
  high-performance hydrophobic interaction chromatography, 451—452
  microanalysis, 262—265
  N-ethylmaleimide (NEM), effect of, 466
  protein kinase activity with ligand binding form of, 467—470
  ribonuclease A, effect of, 467
  sodium molybdate, influence of, 463—466
  trypsin, effects of, 466—467
1,2-Ethanedithiol (EDT), 644
Ethanol, 442
Ethyl stationary phase, 443
Ethylene glycol, 514
Ethylenediamine tetraacetic acid (EDTA), 232, 854
Eukaryotic proteins, 669
Exclusion limit, 76
Exclusion volume, 76
Extra-column dispersion, 26
Extra column effects, 76, 104
Extra-column variables, 279 282—284
Extraction, see specific protein or peptide
Extrinsic membrane proteins, 231

F

Factor XII, 493—499
Fast LC, 76
Fast separations, 276
Faulty packing material, 535

Fc portion, 509
Ferritin, 151
Fibroblast growth factor, 243
Figure of merit (FOM), 572—573, 575, 577—578
Film diffusion, 100
Filters, 24
Filtration liquid chromatography, 24
Fittings, 26
Flow-rate, 76, 103—104, 115, 276, 628, see also Varying flow-rate and gradient-rate
  high-performance affinity chromatography, 488—489
  high-performance immunoaffinity chromatography, 513
  sample displacement mode, 802
Flow rate problems, 25
Flow-through monitoring, 258, 260
Flow-through radioisotope detector, 453
Flu analogs, 633—642
1-(9-Fluorenyl)ethyl chloroformate (FLEC) derivatization of amino acid enantiomers, 369—378
9-Fluorenylmethyl chloroformate (FMOC-C1), 369, 375
  amino acid analysis, 852—853
  precolumn derivatization, 869—872
Fluorescamine, 580
  amino acid analysis, 851—852
  post column reaction detection, 580—585,
Fluorescence, 406, 531
Fluorescence detection, 571—586
Fluorescence reagents, 369
Fluorescence response, 585
Fluorescent product, 580
Fluorimetric detector, 579
Fluorogenic reagents, 579
Flushing, 27
FMOC-ADAM, 869
Folding
  circular dichroism, 605—611
  HPLC analysis, 589—598
  pathway of, 589
  size-exclusion chromatography, 599—611
Folding profiles, 599—604
Folding/unfolding behavior, 118
Foot and mouth disease virus (FMDV), 225
Formic acid, 234, 236, 410
Formyl protecting group, 567
14-3-3, 14-3-3 protein, 841
Fraction analysis, 786—790
Fraction collection, 411
Fractionation of proteins, 404—406
Fractionation range, 76
Free energy changes, 99
Free programmable autoinjector, 410
Free sulfhydryl groups, 389
Freezing/thawing, 231
Frit, 76
Frontal analysis, 76—77
Frontal chromatography, 76—77, 814
Functional group coefficients, 115
Functional ligand
  hydrophobic interaction chromatography, 425—426
  hydrophobic matrices, 443—445
  protein confromation, 443—445

G

Gaps, 634, 638
Gas-liquid-solid phase sequencer, 879—881
Gas-phase hydrolysis, 371

H

S

Milton Keynes UK
Ingram Content Group UK Ltd.
UKHW051857071024
449327UK00025B/1998